Volume 2 (Cont'd)

OFDM: Technology 918
OFDM-Based Systems: Modem Architectures 934
Opportunistic Scheduling 942
Optical Broadband Services Networks 953
Optical Communication Networks: Trends.......................... 966
Optical Networks: Architectures....................................... 981
Optical Networks: Resource Management and
    Allocation 989
P2P: Overlay Multicast.. 1002
Packet Access: High-Speed Downlink............................... 1016
Rerouting Schemes: Performance Evaluation 1018
RF Systems Design 1023
RF Transceiver Architectures... 1035
Smart Antenna Systems: Architecture 1046
Smart Inter-Processor Communicator Protocol Stack 1055
Smartphones... 1065
Smartphones: Operating Systems 1073
Soft Handoff 1086

Volume 3

Spectral Efficiency 1095
Spectral Efficiency: Wireless Multicarrier
    Communications 1100
Spread Spectrum Communications 1114
TDMA Cellular Systems: Speech and Channel Coding...... 1122
TDMA/TDD Wireless Networks: Quality of Service
    (QoS) Mechanisms 1131
Telemetering: GSM Systems and Technologies................. 1141
Ultra-Wideband: Applications... 1150
Ultra-Wideband: Communications 1157
UMTS... 1165
Unlicensed Mobile Access: Architecture and Protocols...... 1168
Video Communications 1182
Video Streaming: Reliable Wireless Streaming with
    Digital Fountain Codes... 1193
Video Streaming: Wireless... 1200
Wi-Fi Walkman.. 1207
Wireless Ad Hoc Networks: Performance Analysis 1221
Wireless Ad Hoc Networks: Power Assignment 1234
Wireless Applications: Middleware Security 1242
Wireless Asynchronous Transfer Mode (ATM):
    Internetworking 1251
Wireless Asynchronous Transfer Mode (ATM): Mobility
    Management 1264

Volume 3 (Cont'd)

Wireless Asynchronous Transfer Mode (ATM): Quality
    of Service (QoS).. 1273
Wireless Channels: Link Adaptation Techniques............... 1284
Wireless Communications: Signal Processing Tools 1294
Wireless Communications: Spatial Diversity 1305
Wireless Data 1319
Wireless Home Entertainment Center: Protocol
    Communications and Architecture 1331
Wireless Internet: Applications 1339
Wireless Internet: Fundamentals..................................... 1346
Wireless LANs (WLANs) 1360
Wireless LANs (WLANs): Ad Hoc Routing Techniques 1369
Wireless LANs (WLANs): Real-Time Services with
    Quality of Service (QoS) Support................................. 1381
Wireless LANs (WLANs): Security and Privacy 1392
Wireless Mesh Networks: Architecture, Protocols, and
    Applications.. 1407
Wireless Multimedia Systems: Cross Layer
    Considerations 1417
Wireless Networks: Cross Layer MAC Design 1424
Wireless Networks: Error Resilient Video Transmission 1434
Wireless Networks: Routing with Intermittent
    Connectivity.. 1442
Wireless Networks: TCP 1452
Wireless Networks: Transmitter Power Control.................. 1464
Wireless Networks: VoIP Service.................................... 1478
Wireless Sensor Networks (WSNs).................................. 1487
Wireless Sensor Networks (WSNs): Central Node and
    Mobile Node Design and Development.......................... 1495
Wireless Sensor Networks (WSNs): Characteristics and
    Types of Sensors 1505
Wireless Sensor Networks (WSNs): Key Management
    Schemes 1509
Wireless Sensor Networks (WSNs): Optimization of the
    OSI Network Stack for Energy Efficiency...................... 1523
Wireless Sensor Networks (WSNs): Routing 1534
Wireless Sensor Networks (WSNs): Secure Localization ... 1545
Wireless Sensor Networks (WSNs): Security.................... 1554
Wireless Sensor Networks (WSNs): Self-Organization....... 1563
Wireless Sensor Networks (WSNs): Time
    Synchronization... 1580
Wireless Transceivers: Near-Instantaneously Adaptive...... 1588

Encyclopedias from Taylor & Francis Group

Agriculture Titles

Dekker Agropedia Collection (Eight Volume Set)
ISBN: 978-0-8247-2194-7   Cat. No.: DK803X

Encyclopedia of Agricultural, Food, and Biological Engineering
Edited by Dennis R. Heldman
ISBN: 978-0-8247-0938-9   Cat. No.: DK9381

Encyclopedia of Animal Science
Edited by Wilson G. Pond and Alan Bell
ISBN: 978-0-8247-5496-9   Cat. No.: DK2206

Encyclopedia of Pest Management
Edited by David Pimentel
ISBN: 978-0-8247-0632-6   Cat. No.: DK6323

Encyclopedia of Pest Management, Volume II
Edited by David Pimentel
ISBN: 978-1-4200-5361-6   Cat. No.: 53612

Encyclopedia of Plant and Crop Science
Edited by Robert M. Goodman
ISBN: 978-0-8247-0944-0   Cat. No.: DK1190

Encyclopedia of Soil Science, Second Edition (Two Volume Set)
Edited by Rattan Lal
ISBN: 978-0-8493-3830-4   Cat. No.: DK830X

Encyclopedia of Water Science
Edited by Stanley W. Trimble
ISBN: 978-0-8493-9627-4   Cat. No.: DK9627

Chemistry Titles

Encyclopedia of Chromatography, Second Edition (Two Volume Set)
Edited by Jack Cazes
ISBN: 978-0-8247-2785-7   Cat. No.: DK7851

Encyclopedia of Supramolecular Chemistry (Two Volume Set)
Edited by Jerry L. Atwood and Jonathan W. Steed
ISBN: 978-0-8247-5056-5   Cat. No.: DK056X

Encyclopedia of Surface and Colloid Science, Second Edition (Eight Volume Set)
Edited by P. Somasundaran
ISBN: 978-0-8493-9615-1   Cat. No.: DK9615

Engineering Titles

Encyclopedia of Chemical Processing (Five Volume Set)
Edited by Sunggyu Lee
ISBN: 978-0-8247-5563-8   Cat. No.: DK2243

Encyclopedia of Corrosion Technology, Second Edition
Edited by Philip A. Schweitzer, P.E.
ISBN: 978-0-8247-4878-4   Cat. No.: DK1295

Encyclopedia of Energy Engineering and Technology (Three Volume Set)
Edited by Barney L. Capehart
ISBN: 978-0-8493-3653-9   Cat. No.: DK653X

Dekker Encyclopedia of Nanoscience and Nanotechnology (Five Volume Set)
Edited by James A. Schwarz, Cristian I. Contescu, and Karol Putyera
ISBN: 978-0-8247-5055-8   Cat. No.: DK0551

Encyclopedia of Optical Engineering (Three Volume Set)
Edited by Ronald G. Driggers
ISBN: 978-0-8247-0940-2   Cat. No.: DK9403

Business Titles

Encyclopedia of Library and Information Science, Second Edition (Four Volume Set)
Edited by Miriam Drake
ISBN: 978-0-8247-2075-9   Cat. No.: DK075X

Encyclopedia of Library and Information Science, First Update Supplement
Edited by Miriam Drake
ISBN: 978-0-8493-3894-6   Cat. No.: DK8948

Encyclopedia of Public Administration and Public Policy, Second Edition (Three Volume Set)
Edited by Evan M. Berman
ISBN: 978-0-4200-5275-6   Cat. No.: AU5275

Encyclopedia of Wireless and Mobile Communications (Three Volume Set)
Edited by Borko Furht
ISBN: 978-0-4200-4326-6   Cat. No.: AU4326

These titles are available both in print and online. To order, visit:
www.crcpress.com
Telephone: 1-800-272-7737
Fax: 1-800-374-3401
E-Mail: orders@crcpress.com

Volume III

Encyclopedia of
Wireless *and* Mobile
Communications

Edited by

Borko Furht

Auerbach Publications
Taylor & Francis Group
Boca Raton   New York

Auerbach Publications is an imprint of the
Taylor & Francis Group, an **informa** business

Auerbach Publications
Taylor & Francis Group
6000 Broken Sound Parkway NW, Suite 300
Boca Raton, FL 33487-2742

© 2008 by Taylor & Francis Group, LLC
Auerbach is an imprint of Taylor & Francis Group, an Informa business

No claim to original U.S. Government works
Printed in the United States of America on acid-free paper
10 9 8 7 6 5 4 3 2 1

International Standard Book Number-13: 978-1-4200-4326-6 (Hardcover)

Library of Congress Cataloging-in-Publication Data

Encyclopedia of wireless and mobile communications / edited by Borko Furht.
    p. cm.
   Includes bibliographical references and index.
   ISBN 978-1-4200-4326-6 (entire set) -- ISBN 978-1-4200-5564-1 (v. 1) -- ISBN 978-1-4200-5565-8 (v. 2) -- ISBN 978-1-4200-5566-5 (v. 3)
   1. Wireless communication systems--Encyclopedias. 2. Mobile communication systems--Encyclopedias. I. Furht, Borivoje.

TK5103.2.E535 2008
621.38403--dc22
                                             2007035219

Visit the Taylor & Francis Web site at
http://www.taylorandfrancis.com

and the Auerbach Web site at
http://www.auerbach-publications.com

Encyclopedia of Wireless and Mobile Communications

Borko Furht, Editor-in-Chief

Editorial Advisory Board

Contributors

Karim Abed-Meraim / *Ecole Nationale Superieure des Telecommunications, Paris, France*

Raj Acharya / *Department of Computer Science and Engineering, Pennsylvania State University, University Park, Pennsylvania, U.S.A.*

George N. Aggélou / *Institute of Technology, Athens, Greece*

Dharma P. Agrawal / *OBR Research Center for Distributed and Mobile Computing, ECECS Department, University of Cincinnati, Cincinnati, Ohio, U.S.A.*

Shakeel Ahmad / *Department of Computer and Information Science, University of Konstanz, Konstanz, Germany*

Dewan Tanvir Ahmed / *School of Information Technology and Engineering (SITE), University of Ottawa, Ottawa, Ontario, Canada*

Syed Ahson / *iDEN Mobile Devices and Emerging Standards, Motorola Inc., Plantation, Florida, U.S.A.*

Marwan Al-Akaidi / *Faculty of Computing Sciences and Engineering, De Montfort University, Leicester, U.K.*

Fereshteh Amini / *University of Manitoba, Winnipeg, Manitoba, Canada*

Marios C. Angelides / *Information Systems and Computing, Brunel University, Uxbridge, U.K.*

Patroklos Argyroudis / *Centre for Telecommunications Value-Chain Research, Department of Computer Science, University of Dublin, Trinity College, Dublin, Ireland*

Hüseyin Arslan / *University of South Florida, Tampa, Florida, U.S.A.*

Chadi Assi / *Computer Security Laboratory, Concordia Institute for Information Systems Engineering, Concordia University, Montreal, Quebec, Canada*

Nadjib Badache / *Computer Systems Laboratory, Computer Science Department, University of Science and Technology, Algiers, Algeria*

Ahmed Badi / *Department of Computer Science and Engineering, Florida Atlantic University, Boca Raton, Florida, U.S.A.*

Shumeet Baluja / *Google Research, Google, Inc., Mountain View, California, U.S.A.*

Nicholas Bambos / *Stanford University, Stanford, California, U.S.A.*

Shudi Bao / *Department of Electronic Engineering, Chinese University of Hong Kong, Hong Kong, China*

Melbourne Barton / *Telcordia Technologies, Red Bank, New Jersey, U.S.A.*

Kalyan Basu / *Center for Research in Wireless Mobility and Networking, Department of Computer Science and Engineering, University of Texas at Arlington, Arlington, Texas, U.S.A.*

Jobaida Begum / *Department of Computer Science, University of Manitoba, Winnipeg, Manitoba, Canada*

A. Belouchrani / *Ecole Nationale Polytechnique, Algiers, Algeria*

David Benjamin / *Nortel Networks, St. Laurent, Quebec, Canada*

Watit Benjapolakul / *Department of Electrical Engineering, Chulalongkorn University, Bangkok, Thailand*

Janez Bester / *Faculty of Electrical Engineering, University of Ljubljana, Ljubljana, Slovenia*

Sumitha Bhandarkar / *Senior Staff Software Engineer, Mobile Devices Technology Office, Motorola Inc., Austin, Texas, U.S.A.*

Prabir Bhattacharya / *Computer Security Laboratory, Concordia Institute for Information Systems Engineering, Concordia University, Montreal, Quebec, Canada*

Naga Bhushan / *Corporate R & D, Qualcomm. Inc., San Diego, California, U.S.A.*

Ratnabali Biswas / *OBR Research Center for Distributed and Mobile Computing, ECECS Department, University of Cincinnati, Cincinnati, Ohio, U.S.A.*

Boualem Boashash / *Queensland University, Brisbane, Queensland, Australia*

Alan C. Bovik / *Department of Electrical and Computer Engineering, University of Texas at Austin, Austin, Texas, U.S.A.*

Madhukar Budagavi / *Texas Instruments, Dallas, Texas, U.S.A.*

John Buford / *Panasonic Princeton Laboratory, Princeton, New Jersey, U.S.A.*

Gruia Calinescu / *Department of Computer Science, Illinois Institute of Technology, Chicago, Illinois, U.S.A.*

Ionut Cardei / *Department of Computer Science and Engineering, Florida Atlantic University, Boca Raton, Florida, U.S.A.*

Mihaela Cardei / *Department of Computer Science and Engineering, Florida Atlantic University, Boca Raton, Florida, U.S.A.*

Jonathan Chan / *CSIRO Centre for Networking Technologies for the Information Economy, Collingswood, Victoria, Australia*

Li Fung Chang / *Mobilink Telecom, Middletown, New Jersey, U.S.A.*

M.C. Frank Chang / *SST Communications Corporation and University of California, Los Angeles, Los Angeles, California, U.S.A.*

Thomas M. Chen / *Department of Electrical Engineering, Southern Methodist University, Dallas, Texas, U.S.A.*

Xiao Chen / *Southwest Texas State University, San Marcos, Texas, U.S.A.*

Yu Chen / *Department of Electronic Engineering, Tsinghua University, Beijing, China*

Christine Cheng / *Department of Electrical Engineering and Computer Science, University of Wisconsin-Milwaukee, Milwaukee, Wisconsin, U.S.A.*

Liang Cheng / *Laboratory of Networking Group (LONGLAB), Department of Computer Science and Engineering, Lehigh University, Bethlehem, Pennsylvania, U.S.A.*

Matthew Cheng / *Mobilink Telecom, Middletown, New Jersey, U.S.A.*

Charles Chien / *Conexant Systems, Los Angeles, California, U.S.A.*

Peter Han Joo Chong / *Network Technology Research Centre, School of Electrical and Electronic Engineering, Nanyang Technological University, Singapore*

Konstantinos Chorianopoulos / *University of the Aegean, Syros, Cyclades, Greece*

Marco Conti / *Consiglio Nazionale delle Ricerche, Pisa, Italy*

Carlos de M. Cordeiro / *University of Cincinnati, Cincinnati, Ohio, U.S.A.*

Michele Covell / *Google Research, Google, Inc., Mountain View, California, U.S.A.*

Pedro Cuenca / *Albacete Research Institute of Informatics, University of Castilla-La Mancha, Albacete, Spain*

Antonio Cuevas Casado / *Departamento Ingeniería Telemática, Universidad Carlos III de Madrid, Madrid, Spain*

Igor D.D. Curcio / *Nokia Corporation, Tampere, Finland*

Sajal K. Das / *Center for Research in Wireless Mobility and Networking, Department of Computer Science and Engineering, University of Texas at Arlington, Arlington, Texas, U.S.A.*

A. P. de Vries / *Delft University of Technology, Amsterdam, The Netherlands*

Mourad Debbabi / *Computer Security Laboratory, Concordia Institute for Information Systems Engineering, Concordia UniversityMontreal, Quebec, Canada*

Francisco M. Delicado / *Albacete Research Institute of Informatics, University of Castilla-La Mancha, Albacete, Spain*

Jesús Delicado / *Albacete Research Institute of Informatics, University of Castilla-La Mancha, Albacete, Spain*

Jörg Diederich / *Distributed Systems Institute–Knowledge Based Systems, Leibniz Universität Hannover, Hannover, Germany*

Djamel Djenouri / *Basic Software Laboratory, CERIST Center of Research, Algiers, Algeria*

Christoph Dorn / *Distributed Systems Group (DSG), Institute of Information Systems, Vienna University of Technology, Vienna, Austria*

Schahram Dustdar / *Distributed Systems Group (DSG), Institute of Information Systems, Vienna University of Technology, Vienna, Austria*

Abdulmotaleb El Saddik / *Multimedia Communications Research Laboratory, University of Ottawa, Ottawa, Ontario, Canada*

Ahmed K. Elhakeem / *Concordia University, Montreal, Quebec, Canada*

Joseph B. Evans / *Distinguished Professor, Information and Telecommunication Technology Center, University of Kansas, Lawrence, Kansas, U.S.A.*

Zoheir Ezziane / *College of Information Technology, University of Dubai, Dubai, United Arab Emirates*

Ben Falchuk / *Telcordia Technologies, Inc., Piscataway, New Jersey, U.S.A.*

Dave Famolari / *Telcordia Technologies, Inc., Piscataway, New Jersey, U.S.A.*

Jian Feng / *Department of Computer Science, Hong Kong Baptist University, Hong Kong, China*

S. Ferretti / *Dipartimento di Scienze dell'Informazione, Università di Bologna, Bologna, Italy*

João Figueiras / *Department of Electronic Systems, Center for TeleInFrastruktur (CTIF), Aalborg University, Aalborg, Denmark*

Michael Fink / *Center for Neural Computation, Givat Ram Campus, Hebrew University of Jerusalem, Jerusalem, Israel*

Simone Frattasi / *Department of Electronic Systems, Center for TeleInFrastruktur (CTIF), Aalborg University, Aalborg, Denmark*

Ophir Frieder / *Department of Computer Science, Illinois Institute of Technology, Chicago, Illinois, U.S.A.*

Borko Furht / *College of Engineering and Computer Science, Florida Atlantic University, Boca Raton, Florida, U.S.A.*

Marco Furini / *Dipartimento di Informatica, Università Piemonte Orientale, Alessandria, Italy*

Qijie Gao / *Network Technology Research Centre, School of EEE, Nanyang Technological University, Singapore*

José Antonio Garcia-Macias / *CICESE Research Center, Esenada, Mexico*

Aysegül Gençata / *Department of Computer Engineering, Istanbul Technical University, Istanbul, Turkey*

Mladen Georgievski / *School of Computer Science and Mathematics, Victoria University, Melbourne, Victoria, Australia*

Savvas Gitzenis / *Stanford University, Stanford, California, U.S.A.*

Lal C. Godara / *Australian Defence Force Academy, School of Electrical Engineering University College, University of New South Wales, Canberra, Australian Capital Territory, Australia*

Steven Gorshe / *PMC-Sierra, Inc., Portland, Oregon, U.S.A.*

Steven D. Gray / *Nokia Research Center, Espoo, Finland*

Peter Green / *Nortel Networks, Ottawa, Ontario, Canada*

Stefanos Gritzalis / *University of the Aegean, Karlovassi, Samos, Greece*

Aditya Gupta / *OBR Research Center for Distributed and Mobile Computing, ECECS Department, University of Cincinnati, Cincinnati, Ohio, U.S.A.*

Sandeep K.S. Gupta / *Department of Computer Science and Engineering, Arizona State University, Tempe, Arizona, U.S.A.*

Amoakoh Gyasi-Agyei / *Faculty of Sciences, Engineering and Health, Central Queensland University, Australia*

Jyri Hämäläinen / *Nokia Networks, Oulu, Finland*

Mounir Hamdi / *Department of Computer Science, Hong Kong University of Science and Technology, Kowloon, Hong Kong*

Raouf Hamzaoui / *Faculty of Computing Sciences and Engineering, De Montfort University, Leicester, U.K.*

Lajos Hanzo / *University of Southampton, Southampton, U.K.*

Robert W. Heath, Jr. / *Department of Electrical and Computer Engineering, University of Texas at Austin, Austin, Texas, U.S.A.*

K. M. Ho / *Department of Electronic and Information Engineering, Hong Kong Polytechnic University, Hong Kong, China*

Bongkarn Homnan / *Department of Telecommunications Engineering, Dhurakit Pundit University, Bangkok, Thailand*

Masaaki Hoshino / *Information Technologies Laboratories, Sony Corporation, Tokyo, Japan*

M. Anwar Hossain / *Multimedia Communications Research Laboratory, University of Ottawa, Ottawa, Ontario, Canada*

Shamim Hossain / *Multimedia Communications Research Laboratory, University of Ottawa, Ottawa, Ontario, Canada*

Chih-Shun Hsu / *National Central University, Chung-Li, Taiwan*

Kevin Hung / *Department of Electronic Engineering, Chinese University of Hong Kong, Hong Kong, China*

Frederic Huve / *HP Technology Solutions, Hewlett-Packard Inc., Grenoble, France*

Mohamed Ibnkahla / *Queen's University, Kingston, Ontario, Canada*

Jamil Ibriq / *Department of Computer Science and Engineering, Florida Atlantic University, Boca Raton, Florida, U.S.A.*

Mohammad Ilyas / *College of Engineering and Computer Science, Florida Atlantic University, Boca Raton, Florida, U.S.A.*

Abul Faiz Muhad Ishaq / *Comsats Institute of Information Technology, Islamabad, Pakistan*

Bijan Jabbari / *George Mason University, Fairfax, Virginia, U.S.A.*

Neha Jain / *OBR Research Center for Distributed and Mobile Computing, ECECS Department, University of Cincinnati, Cincinnati, Ohio, U.S.A.*

Ravi Jain / *DoCoMo USA Labs, San Jose, California, U.S.A.*

Ramakrishna Janaswamy / *Naval Postgraduate School, Monterey, California, U.S.A.*

Tornaz Javornik / *Department of Communication Systems, Jozef Stefan Institute, Ljubljana, Slovenia*

Hai Jiang / *Department of Electrical and Computer Engineering, University of Alberta, Edmonton, Alberta, Canada*

Tao Jiang / *Department of Electrical and Computer Engineering, University of Michigan, Dearborn, Michigan, U.S.A.*

Alexandros Kaloxylos / *Department of Telecommunications Science and Technology, University of Peloponnese, Tripoli, Greece*

Hari Kalva / *Department of Computer Science and Engineering, Florida Atlantic University, Boca Raton, Florida, U.S.A.*

Ami Kanazawa / *Yokosuka Radio Communications Research Center, Communication Research Laboratory, Ministry of Posts and Telecommunications, Yokosuka, Japan*

Gorazd Kandus / *Department of Communication Systems, Jozef Stefan Institute, Ljubljana, Slovenia*

Aditya Karnik / *Indian Institute of Science, Bangalore, India*

Moazzam Khan / *University of Manitoba, Winnipeg, Manitoba, Canada*

Charbel Khawand / *Motorola Inc., Plantation, Florida, U.S.A.*

Lyes Khelladi / *Basic Software Laboratory, CERIST Center of Research, Algiers, Algeria*

Chaiyaporn Khemapatapan / *Computer and Telecommunication Engineering, Dhurakij Pundit University, Bangkok, Thailand*

Won-Ik Kim / *Electronics & Telecommunications Research Institute, Taejon, South Korea*

Manish Kochhal / *Department of Electrical and Computer Engineering, Wayne State University, Detroit, Michigan, U.S.A.*

Andrej Kos / *Faculty of Electrical Engineering, University of Ljubljana, Ljubljana, Slovenia*

Vassilios Koukoulidis / *Nokia Siemens Networks, Boca Raton, Florida, U.S.A.*

S.P.T. Krishnan / *Cryptography and Security Department, Institute for Infocomm Research (I2R), Singapore*

Anurag Kumar / *Indian Institute of Science, Bangalore, India*

Markku Kuusela / *Nokia Research Center, Nokia Group, Ruoholahti, Finland*

Dong-Hee Kwon / *Pohang University of Science and Technology, Pohang, South Korea*

Fabrice Labeau / *Electrical and Computer Engineering Department, McGill University, Montreal, Quebec, Canada*

Costos Lambrinoudakis / *University of the Aegean, Karlovassi, Samos, Greece*

George Lampropoulos / *Department of Informatics and Telecommunications, University of Athens, Athens, Greece*

Björn Landfeldt / *School of Information Technologies and School of Electrical and Information Engineering, University of Sydney, Sydney, New South Wales, Australia*

Marc-André Laverdière / *Computer Security Laboratory, Concordia Institute for Information Systems Engineering, Concordia University, Montreal, Quebec, Canada*

Gerry Leavey / *PMC-Sierra, Inc., Portland, Oregon, U.S.A.*

Ky Leng / *Department of Electrical Engineering, Chulalongkorn University, Bangkok, Thailand*

Jenq-Shiou Leu / *Department of Electronic Engineering, National Taiwan University of Science and Technology, Taipei, Taiwan*

Allen H. Levesque / *Worcester Polytechnic Institute, Worcester, Massachusetts, U.S.A.*

Xue Jun Li / *Network Technology Research Centre, School of EEE, Nanyang Technological University, Singapore*

Ting-Yu Lin / *National Chiao-Tung University, Hsinchu, Taiwan*

Nguyen Linh-Trung / *Aston University, Birmingham, U.K.*

Cong Liu / *Department of Computer Science and Engineering, Florida Atlantic University, Boca Raton, Florida, U.S.A.*

Kwok Tung Lo / *Department of Electronic and Information Engineering, Hong Kong Polytechnic University, Hong Kong, China*

Shoshana Loeb / *Telcordia Technologies, Inc., Piscataway, New Jersey, U.S.A.*

Angel Lozano / *Wireless Communication Research Department, Bell Laboratories (Lucent Technologies), Holmdel, New Jersey, U.S.A.*

Imad Mahgoub / *Department of Computer Science and Engineering, Florida Atlantic University, Boca Raton, Florida, U.S.A.*

Predrag Maksimovic / *Department of Computer Science and Engineering, Florida Atlantic University, Boca Raton, Florida, U.S.A.*

Mona Mehrandish / *Computer Security Laboratory, Concordia Institute for Information Systems Engineering, Concordia University, Montreal, Quebec, Canada*

Mehri Mehrjoo / *Center for Wireless Communications, Department of Electrical and Computer Engineering, University of Waterloo, Waterloo, Ontario, Canada*

Lazaros Merakos / *Department of Informatics and Telecommunications, University of Athens, Athens, Greece*

Paul Mermelstein / *INRS-Telecommunications, Montreal, Quebec, Canada*

Francisco Mico / *Faculty of Informatics, University of Valencia, Valencia, Spain*

Laurence B. Milstein / *University of California, La Jolla, California, U.S.A.*

Veljko Milutinović / *School of Electrical Engineering, University of Belgrade, Beograd, Serbia*

Gary J. Minden / *Information and Telecommunication Technology Center, University of Kansas, Lawrence, Kansas, U.S.A.*

Jelena Misic / *University of Manitoba, Winnipeg, Manitoba, Canada*

Vojislav Misic / *Department of Computer Science, University of Manitoba, Winnipeg, Manitoba, Canada*

Ryu Miura / *Yokosuka Radio Communications Research Center, Communications Research Laboratory, Ministry of Posts and Telecommunications, Yokosuka, Japan*

Ioannis Modeas / *Department of Informatics and Telecommunications, University of Athens, Athens, Greece*

Mohammad Mohsin / *Comsats Institute of Information and Technology, Islamabad, Pakistan*

Jose Ignacio Moreno / *Departamento Ingeniería Telemática, Universidad Carlos III de Madrid, Madrid, Spain and Deutsche Telekom Laboratories, Berlin, Germany*

Azzam Mourad / *Computer Security Laboratory, Concordia Institute for Information Systems Engineering, Concordia University, Montreal, Quebec, Canada*

Anindo Mukherjee / *OBR Research Center for Distributed and Mobile Computing, ECECS Department, University of Cincinnati, Cincinnati, Ohio, U.S.A.*

Biswanath Mukherjee / *Department of Computer Science, University of California, Davis, Davis, California, U.S.A.*

Timothy R. Newman / *Information and Telecommunication Technology Center, University of Kansas, Lawrence, Kansas, U.S.A.*

Nhut Nguyen / *Network Systems Lab, Samsung Telecommunications America, Richardson, Texas, U.S.A.*

Hiroyo Ogawa / *Communications Research Laboratory, Ministry of Posts and Telecommunications, Yokosuka, Japan*

Tero Ojanperä / *Nokia Research Center, Espoo, Finland*

Donal O'Mahony / *Centre for Telecommunications Value-Chain Research, Department of Computer Science, University of Dublin, Trinity College, Dublin, Ireland*

Luis Orozco-Barbosa / *Albacete Research Institute of Informatics, University of Castilla–La Mancha, Albacete, Spain*

Hadi Otrok / *Computer Security Laboratory, Concordia Institute for Information Systems Engineering, Concordia University, Montreal, Quebec, Canada*

Kaveh Pahlavan / *Worcester Polytechnic Institute, Worcester, Massachusetts, U.S.A.*

Kari Pajukoski / *Nokia Networks, Oulu, Finland*

C. E. Palazzi / *University of Bologna, Bologna, Italy, and Department of Computer Science, University of California Los Angeles, Los Angeles, California, U.S.A.*

Wasimon Panichpattanakul / *Faculty of Computer Engineering, Prince of Songkla University–Phuket Campus, Phuket, Thailand*

Symeon Papavassiliou / *New Jersey Institute of Technology, Newark, New Jersey, U.S.A.*

Bernd-Peter Paris / *George Mason University, Fairfax, Virginia, U.S.A.*

Nikos Passas / *Department of Informatics and Telecommunications, University of Athens, Athens, Greece*

G. Pau / *Department of Computer Science, University of California–Los Angeles, Los Angeles, California, U.S.A.*

George Pavlou / *Centre for Communication Systems Research (CCSR), University of Surrey, Surrey, U.K.*

Yong Pei / *Department of Computer Science and Engineering, Wright State University, Dayton, Ohio, U.S.A.*

Sreco Plevel / *Department of Communication Systems, Jozef Stefan Institute, Ljubljana, Slovenia*

Mohan Ponnada / *School of Computer Science and Mathematics, Victoria University, Melbourne, Victoria, Australia*

Min Qin / *Department of Computer Science, University of Southern California, Los Angeles, California, U.S.A.*

Gopal Racherla / *Photonics Division, Advanced Wireless Group, General Atomics, San Diego, California, U.S.A.*

Sridhar Radhakrishnan / *Department of Computer Science, University of Oklahoma, Norman, Oklahoma, U.S.A.*

Quazi Mehbubar Rahman / *Queen's University, Kingston, Ontario, Canada*

M. Yasin Akhtar Raja / *Physics and Optical Science Department, University of North Carolina, Charlotte, Charlotte, North Carolina, U.S.A.*

Bala Rajagopalan / *Tellium, Inc., Ocean Port, New Jersey, U.S.A.*

Rakesh Rajbanshi / *Information and Telecommunication Technology Center, University of Kansas, Lawrence, Kansas, U.S.A.*

Narasimha Reddy / *Department of Electrical and Computer Engineering, Texas A & M University, College Station, Texas, U.S.A.*

Marcel J.T. Reinders / *Delft University of Technology, Delft, The Netherlands*

Daniel Reininger / *Semandex Networks, Inc., Princeton, New Jersey, U.S.A.*

Marco Roccetti / *Dipartimento di Scienze dell'Informazione, University of Bologna, Bologna, Italy*

Andrew Roczniak / *Multimedia Communications Research Laboratory, University of Ottawa, Ottawa, Ontario, Canada*

Sumit Roy / *Engineering Department, Rhythm NewMedia, Mountain View, California, U.S.A.*

Muhammad Farooq Sabir / *Department of Electrical and Computer Engineering, University of Texas at Austin, Austin, Texas, U.S.A.*

Praveen Sanigepalli / *iDEN, Advanced Development Group, Motorola Inc., Plantation, Florida, U.S.A.*

Venkatesh Sarangan / *Department of Computer Science, Oklahoma State University, Stillwater, Oklahoma, U.S.A.*

Daniel Schall / *Distributed Systems Group (DSG), Institute of Information Systems, Vienna University of Technology, Vienna, Austria*

Loren Schwiebert / *Department of Computer Science, Wayne State University, Detroit, Michigan, U.S.A.*

Sandhya Sekhar / *OBR Research Center for Distributed and Mobile Computing, ECECS Department, University of Cincinnati, Cincinnati, Ohio, U.S.A.*

Aruna Seneviratne / *School of Electrical Engineering and Telecommunications, University of New South Wales, Kensington, New South Wales, Australia*

Mehul Shah / *T-Mobile USA Inc., Bellevue, Washington, U.S.A.*

Nalin Sharda / *School of Computer Science and Mathematics, Victoria University, Melbourne, Victoria, Australia*

Bo Shen / *HP Labs, Hewlett-Packard Inc., Palo Alto, California, U.S.A.*

Xuemin Sherman Shen / *Center for Wireless Communications, Department of Electrical and Computer Engineering, University of Waterloo, Waterloo, Ontario, Canada*

Shervin Shirmohammadi / *School of Information Technology and Engineering (SITE), University of Ottawa, Ottawa, Ontario, Canada*

Marvin K. Simon / *Jet Propulsion Laboratory, Pasadena, California, U.S.A.*

Suresh Singh / *Portland State University, Portland, Oregon, U.S.A.*

Narendra Singhal / *Department of Computer Science, University of California, Davis, Davis, California, U.S.A.*

Sivapathalingham Sivavakeesar / *Centre for Communication Systems Research (CCSR), University of Surrey, Surrey, U.K.*

Niels Snoeck / *INCA Group, Telematica Instituut, Enschede, The Netherlands*

Jaka Sodnik / *Faculty of Electrical Engineering, University of Ljubljana, Ljubljana, Slovenia*

Anastasis A. Sofokleous / *Information Systems and Computing, Brunel University, Uxbridge, U.K.*

Avinash Srinivasan / *Department of Computer Science and Engineering, Florida Atlantic University, Boca Raton, Florida, U.S.A.*

William Stallings / *Department of Computer Science and Engineering, Wright State University, Dayton, Ohio, U.S.A.*

Raymond Steele / *Multiple Access Communications, Southampton, U.K.*

Gordon L. Stüber / *Georgia Institute of Technology, Atlanta, Georgia, U.S.A.*

Mitja Stular / *Faculty of Electrical Engineering, University of Ljubljana, Ljubljana, Slovenia*

Mani Subramanian / *Georgia Institute of Technology, Atlanta, Georgia, U.S.A.*

Young-Joo Suh / *Pohang University of Science and Technology, Pohang, South Korea*

Raj Talluri / *Texas Instruments, Dallas, Texas, U.S.A.*

Masato Tanaka / *Kashima Space Research Center, Communications Research Laboratory, Ministry of Posts and Telecommunications, Kashima, Ibaraki, Japan*

Binh Thai / *School of Electrical Engineering and Telecommunications, University of New South Wales, Kensington, New South Wales, Australia*

Esa Tiirola / *Nokia Networks, Oulu, Finland*

Syrine Tlili / *Computer Security Laboratory, Concordia Institute for Information Systems Engineering, Concordia University, Montreal, Quebec, Canada*

Saso Tomazic / *Faculty of Electrical Engineering, University of Ljubljana, Ljubljana, Slovenia*

Leyla Toumi / *LSR-IMAG, CNSR/INPG, Grenoble, France*

Ha Duyen Trung / *Department of Electrical Engineering, Chulalongkorn University, Bangkok, Thailand*

Yu-Chee Tseng / *National Chiao-Tung University, Hsinchu, Taiwan*

Hiroyuki Tsuji / *Yokosuka Radio Communications Research Center, Communication Research Laboratory, Ministry of Posts and Telecommunications, Yokosuka, Japan*

Tomohiro Tsunoda / *Information Technologies Laboratories, Sony Corporation, Tokyo, Japan*

Anton Umek / *Electrical Engineering Department, University of Ljubljana, Ljubljana, Slovenia*

Eric van den Berg / *Applied Research, Telcordia Technologies, Morristown, New Jersey, U.S.A.*

Herma van Kranenburg / *INCA Group, Telematica Instituut, Enschede, The Netherlands*

Ramakrishna Vedantham / *Nokia Research Center, Palo Alto, California, U.S.A.*

Bharadwaj Veeravalli / *Department of Electrical and Computer Engineering, Computer Networks and Distributed Systems (CNDS) Lab, National University of Singapore, Singapore*

Aravindhan Venkateswaran / *Department of Computer Science and Engineering, Pennsylvania State University, University Park, Pennsylvania, U.S.A.*

José Villalón / *Albacete Research Institute of Informatics, University of Castilla-La Mancha, Albacete, Spain*

Jenjoab Virapanicharoen / *Department of Electrical Engineering, Chulalongkorn University, Bangkok, Thailand*

Mojca Volk / *Faculty of Electrical Engineering, University of Ljubljana, Ljubljana, Slovenia*

T. Bao Vu / *Department of Electronic Engineering, City University of Hong Kong, Kowloon, Hong Kong*

Peng-Jun Wan / *Department of Computer Science, Illinois Institute of Technology, Chicago, Illinois, U.S.A.*

Bin Wang / *Department of Computer Science and Engineering, Wright State University, Dayton, Ohio, U.S.A.*

Jun Wang / *Delft University of Technology, Delft, The Netherlands*

Ping Wang / *Department of Electrical and Computer Engineering, University of Waterloo, Waterloo, Ontario, Canada*

Martin Wibbels / *INCA Group, Telematica Instituut, Enschede, The Netherlands*

Risto Wichman / *Helsinki University of Technology, Helsinki, Finland*

Katharine Willis / *Mediacity Project, Bauhaus University of Weimar, Weimar, Germany*

Lawrence W.C. Wong / *Department of Electrical and Computer Engineering, Computer Networks and Distributed Systems (CNDS) Lab, National University of Singapore, Singapore*

Jie Wu / *Department of Computer Science and Engineering, Florida Atlantic University, Boca Raton, Florida, U.S.A.*

Alexander M. Wyglinski / *Department of Electrical and Computer Engineering, Worcester Polytechnic Institute, Worcester, Massachusetts, U.S.A.*

Zhiwei Xu / *SST Communications Corporation, Los Angeles, California, U.S.A.*

Lie-Liang Yang / *University of Southampton, Southampton, U.K.*

Yinying Yang / *Department of Computer Science and Engineering, Florida Atlantic University, Boca Raton, Florida, U.S.A.*

Qing Ye / *Laboratory of Networking Group (LONGLAB), Department of Computer Science and Engineering, Lehigh University, Bethlehem, Pennsylvania, U.S.A.*

Andrianus Yofy / *Department of Electrical Engineering, Chulalongkorn University, Bangkok, Thailand*

Heather Yu / *Panasonic Princeton Laboratory, Princeton, New Jersey, U.S.A.*

Jane Yang Yu / *Network Technology Research Centre, School of Electrical and Electronic Engineering, Nanyang Technological University, Singapore*

Hans Zandbelt / *Middleware Services, SURF net BV, Utrecht, The Netherlands*

Xudong Zhang / *Department of Electronic Engineering, Tsinghua University, Beijing, China*

Yuan-Ting Zhang / *Department of Electronic Engineering, Chinese University of Hong Kong, Hong Kong, China and Institute of Biomedical and Health Engineering, Shenzhen Institute of Advanced Technology, Chinese Academy of Sciences, Shenzhen, China*

Ding Zhemin / *Department of Computer Science, Hong Kong University of Science and Technology, Kowloon, Hong Kong*

Hong Zhou / *Faculty of Engineering and Surveying, University of Southern Queensland, Toowoomba, Queensland, Australia*

Jin Zhu / *New Jersey Institute of Technology, Newark, New Jersey, U.S.A.*

Weihua Zhuang / *Department of Electrical and Computer Engineering, University of Waterloo, Waterloo, Ontario, Canada*

Roger Zimmermann / *Department of Computer Science, University of Southern California, Los Angeles, California, U.S.A.*

Martina Zitterbart / *Institute of Telematics, Universität Karlsruhe (TH), Karlsruhe, Germany*

Contents

Contributors vii
Topical Table of Contents xxi
Preface xxxi
Acronyms xxxiii
About the Editor-in-Chief xlv

Volume 1

3G Systems: Accounting Principles and Billing Models / Zoheir Ezziane 1
3G Telephony Control Stack: Interactive Playback Control of Video / Michele Covell, Sumit Roy,
  Bo Shen and Frederic Huve 13
4G Mobile Systems: Multimedia Content Transmission / Antonio Cuevas Casado and
  Jose Ignacio Moreno 26
4G Mobile Systems: Radio Resource Management / Alexandros Kaloxylos, Ioannis Modeas,
  Nikos Passas and George Lampropoulos 40
Access Methods / Bernd-Peter Paris 50
Ad Hoc Networks: GPS-Based Routing Algorithm / Young-Joo Suh, Won-Ik Kim and
  Dong-Hee Kwon 58
Ad Hoc Networks: Power-Conservative Designs / Yu-Chee Tseng and Ting-Yu Lin 68
Ad Hoc Networks: Security / Patroklos Argyroudis and Donal O'Mahony 76
Ad Hoc Networks: Technologies / Marco Conti 89
Adaptation Techniques / Hüseyin Arslan 101
Ad-Coop Positioning System: Embedded Kalman Filter Data Fusion / Simone Frattasi and
  João Figueiras 120
Authentication and Privacy / Thomas M. Chen and Nhut Nguyen 132
Bluetooth Systems / William Stallings 141
Body Area Networks / Bin Wang and Yong Pei 148
Broadcasting Multimedia Transport: 3G and Future Mobile Networks / Ramakrishna Vedantham and
  Igor D.D. Curcio 157
CDMA Networks / Tero Ojanperä and Steven D. Gray 167
CDMA Networks: Microcells / Raymond Steele 180
Cellular Base Stations: Open Standards / Gerry Leavey and Steven Gorshe 186
Cellular Data Networks: Handoff and Rerouting / Gopal Racherla and Sridhar Radhakrishnan 196
Cellular Systems / Lal C. Godara 204
Cellular Systems: 1G / Lal C. Godara 214
Cellular Systems: 2G / Lal C. Godara 216
Cellular Systems: 3G / Lal C. Godara 220
Channel Assignment: Fixed and Dynamic / Bijan Jabbari 225
Cognitive Radio Implementation / Timothy R. Newman, Alexander M. Wyglinski and
  Joseph B. Evans 230
Compact Dual-Band Direct Conversion CMOS Transceiver / Zhiwei Xu, Charles Chien and
  M.C. Frank Chang 239
Content Repurposing: Multimedia / Shamim Hossain and Abdulmotaleb El Saddik 253
Content Repurposing: Ontology-Based Approach for Mobile Devices / Mohan Ponnada and
  Nalin Sharda 259
Content-Enriched Communication: Social Uses of Interactive TV / Konstantinos Chorianopoulos 267
Coordinated Multi-Device Presentations: Ambient-Audio Identification / Michael Fink,
  Michele Covell and Shumeet Baluja 274
Cordless Telephone Standard CT-2 / Lajos Hanzo 286
Diversity Techniques / Quazi Mehbubar Rahman and Mohamed Ibnkahla 296

Volume 1 (*cont'd.*)

Dynamic Wireless Sensor Networks (WSNs): Architecture and Modeling / Symeon Papavassiliou and Jin Zhu 300

EDGE / Syed Ahson. 312

Generic Context Management Framework: Contextual Reasoning / Herma van Kranenburg, Niels Snoeck, Hans Zandbelt and Martin Wibbels 313

GPRS / Syed Ahson. 324

GSM / George N. Aggélou 325

Handoff Prioritization: Early Blocking / Jörg Diederich and Martina Zitterbart 336

Heterogeneous Wireless Sensor Networks (WSNs) / Mihaela Cardei and Yinying Yang 345

High-Speed Wireless Internet Access: Multiantenna Technology / Angel Lozano 354

IEEE 802.11 Wireless LANs (WLANs) / José Antonio Garcia-Macias and Leyla Toumi. 362

IEEE 802.11 Wireless LANs (WLANs): Clustering and Roaming Techniques / Ahmed K. Elhakeem . .365

IEEE 802.11 Wireless LANs (WLANs): Quality of Service (QoS) Support / José Villalón, Francisco Mico, Pedro Cuenca and Luis Orozco-Barbosa 375

IEEE 802.11 Wireless LANs (WLANs): Security / Costos Lambrinoudakis and Stefanos Gritzalis 387

IEEE 802.15.4: MAC Layer Attacks in Sensor Clusters / Vojislav Misic and Jobaida Begum 396

IEEE 802.15.4: Signature-Based Intrusion Detection in Wireless Sensor Networks (WSNs) / Jelena Misic, Fereshteh Amini and Moazzam Khan 404

IEEE 802.16 Wireless Metropolitan Area Networks (WMANs) / Carlos de M. Cordeiro and Dharma P. Agrawal 412

IEEE 802.16 Wireless Metropolitan Area Networks (WMANs): Resource Management / Mehri Mehrjoo and Xuemin Sherman Shen. 417

IEEE 802.16d: Worldwide Interoperability for Microwave Access (WiMax) / Mani Subramanian 428

Image and Video Communication: Joint Source Channel Coding / Muhammad Farooq Sabir, Robert W. Heath, Jr. and Alan C. Bovik 429

Image and Video Communication: Power Optimized / Muhammad Farooq Sabir, Robert W. Heath, Jr. and Alan C. Bovik 436

Image and Video Communication: Source-Channel Distortion Modeling / Muhammad Farooq Sabir, Alan C. Bovik and Robert W. Heath, Jr. 443

Integrated Wireless LANs (WLANs) and Cellular Networks: Managing Handovers / George Lampropoulos, Nikos Passas, Alexandros Kaloxylos and Lazaros Merakos 452

IP Multimedia Subsystems (IMS) / Mojca Volk, Mitja Stular, Janez Bester, Andrej Kos and Saso Tomazic. 462

IP Networks: Personal Mobility Challenges / Björn Landfeldt, Jonathan Chan, Binh Thai and Aruna Seneviratne 474

IP Networks: Terminal Mobility Challenges / Björn Landfeldt, Jonathan Chan, Binh Thai and Aruna Seneviratne 476

IP Networks: User Mobility / Björn Landfeldt, Jonathan Chan, Binh Thai and Aruna Seneviratne 484

Land Mobile Communication Systems: Phased-Array Antennas / Hiroyo Ogawa, Hiroyuki Tsuji, Ami Kanazawa, Ryu Miura and Masato Tanaka. 491

Location-Based Services / Ben Falchuk, Dave Famolari and Shoshana Loeb 501

Volume 2

MANET / Marco Conti. 509

MANET: Clustering Techniques / Jane Yang Yu and Peter Han Joo Chong 513

MANET: Location-Aware Routing / Yu-Chee Tseng and Chih-Shun Hsu. .528

MANET: Multicast Protocols / Xiao Chen and Jie Wu. 537

MANET: Multicasting / Dewan Tanvir Ahmed and Shervin Shirmohammadi 546

MANET: Network Mobility as a Control Primitive / Aravindhan Venkateswaran, Venkatesh Sarangan, Sridhar Radhakrishnan and Raj Acharya. 556

MANET: Routing Protocols / Ha Duyen Trung and Watit Benjapolakul 563

MANET: Selfish Behavior on Packet Forwarding / Djamel Djenouri and Nadjib Badache. 576

Media Streaming: Transmission Strategies in P2P Networks / K. M. Ho, Kwok Tung Lo and Jian Feng. 588

Microcells / Raymond Steele 596
MIMO: Wireless Communications / Sreco Plevel, Saso Tomazic, Tornaz Javornik and
    Gorazd Kandus. 604
Mobile and Wireless Technologies: Spatial Presence / Katharine Willis 613
Mobile Commerce / Abul Faiz Muhad Ishaq and Mohammad Mohsin. 622
Mobile Communications: 4G / Jaka Sodnik, Mitja Stular, Veljko Milutinovic and Sašo Tomazic. 634
Mobile Communications: Call Admission Control / Jenjoab Virapanicharoen and
    Watit Benjapolakul. 643
Mobile Communications: Coding Techniques / Quazi Mehbubar Rahman and Mohamed Ibnkahla 652
Mobile Communications: Locating Position of Mobile Stations / Watit Benjapolakul. 659
Mobile Communications: Power Control / Wasimon Panichpattanakul and Watit Benjapolakul 666
Mobile Computing: Context-Aware / Daniel Schall, Christoph Dorn and Schahram Dustdar. 679
Mobile Devices: Quality of Service (QoS) Adaptation Using MPEG-21 / Marios C. Angelides and
    Anastasis A. Sofokleous 686
Mobile Games: Multiplayer / Predrag Maksimovic and Imad Mahgoub 694
Mobile Health: Wireless Body Sensor Network Integration / Kevin Hung, Shudi Bao and
    Yuan-Ting Zhang. 707
Mobile Internet Technologies / Christoph Dorn, Daniel Schall and Schahram Dustdar 718
Mobile IP / Andrianus Yofy and Watit Benjapolakul 726
Mobile Multimedia Transmission: Codecs / Nalin Sharda and Mladen Georgievski. 736
Mobile Music / Marco Furini 744
Mobile P2P: Computing / Dewan Tanvir Ahmed and Shervin Shirmohammadi. 751
Mobile P2P: Data Retrieval and Caching / Andrew Roczniak and Abdulmotaleb El Saddik. 759
Mobile Security: Game Theory / Hadi Otrok, Mona Mehrandish, Chadi Assi, Prabir Bhattacharya and
    Mourad Debbabi. 764
Mobile Sensor Networks / Dharma P. Agrawal, Ratnabali Biswas, Neha Jain, Anindo Mukherjee,
    Sandhya Sekhar and Aditya Gupta 771
Mobile Services / Jenq-Shiou Leu 774
Mobile Streaming: Performance Issues / Igor D.D. Curcio 782
Mobile Streaming: Standards / Igor D.D. Curcio 784
Mobile Systems: Migrating from 3G to 4G / Nalin Sharda and Mohan Ponnada 788
Mobile Systems: Quality of Service (QoS) / Nalin Sharda and Mladen Georgievski. 795
Mobile Terminals: Recommendation Services / Tomohiro Tsunoda and Masaaki Hoshino 805
Mobile Wireless Systems: Channel Characterization / Quazi Mehbubar Rahman and
    Mohamed Ibnkahla 817
Mobile Wireless Systems: Location Prediction Algorithms / Christine Cheng, Ravi Jain and
    Eric van den Berg. 825
Modulation Methods / Gordon L. Stüber 835
Multicarrier Transceivers: Peak-to-Average Power Radio Reduction / Rakesh Rajbanshi,
    Alexander M. Wyglinski and Gary J. Minden 843
Multihop Cellular Network Architectures / Xue Jun Li, Peter Han Joo Chong and Qijie Gao 856
**Multimedia Applications in Ad Hoc Networks: Quality of Service (QoS) Support Cross-Layered
    Approach** / Sivapathalingham Sivavakeesar and George Pavlou 870
Multimedia Broadcast Multicast Service (MBMS) / Praveen Sanigepalli, Hari Kalva and Borko Furht . .880
Multimedia Streaming / Abdulmotaleb El Saddik and Shamim Hossain 884
Multimedia Streaming: Mobile Networks / Igor D.D. Curcio 890
Multiple Access Techniques / Anton Umek, Mitja Stular and Saso Tomazic 900
Nomadic Users: Context-Aware Information Retrieval / M. Anwar Hossain and
    Abdulmotaleb El Saddik 911
OFDM: Technology / Tao Jiang 918
OFDM-Based Systems: Modem Architectures / Charles Chien 934
Opportunistic Scheduling / Amoakoh Gyasi-Agyei 942
Optical Broadband Services Networks / David Benjamin and Peter Green 953
Optical Communication Networks: Trends / Aysegül Gençata, Narendra Singhal and
    Biswanath Mukherjee 966
Optical Networks: Architectures / M. Yasin Akhtar Raja 981

Volume 2 (*cont'd.*)

Optical Networks: Resource Management and Allocation / Ding Zhemin and Mounir Hamdi 989

P2P: Overlay Multicast / Heather Yu and John Buford 1002

Packet Access: High-Speed Downlink / Jyri Hämäläinen, Risto Wichman, Markku Kuusela,
Esa Tiirola and Kari Pajukoski 1016

Rerouting Schemes: Performance Evaluation / Gopal Racherla and Sridhar Radhakrishnan 1018

RF Systems Design / Charles Chien 1023

RF Transceiver Architectures / Charles Chien 1035

Smart Antenna Systems: Architecture / T. Bao Vu 1046

Smart Inter-Processor Communicator Protocol Stack / Charbel Khawand .1055

Smartphones / Syed Ahson and Mohammad Ilyas 1065

Smartphones: Operating Systems / Syed Ahson and Mohammad Ilyas 1073

Soft Handoff / Bongkarn Homnan and Watit Benjapolakul 1086

Volume 3

Spectral Efficiency / Sašo Tomažič 1095

Spectral Efficiency: Wireless Multicarrier Communications / Alexander M. Wyglinski and
Fabrice Labeau 1100

Spread Spectrum Communications / Laurence B. Milstein and Marvin K. Simon 1114

TDMA Cellular Systems: Speech and Channel Coding / Paul Mermelstein .1122

TDMA/TDD Wireless Networks: Quality of Service (QoS) Mechanisms / Francisco M. Delicado,
Pedro Cuenca, Luis Orozco-Barbosa and Jesús Delicado 1131

Telemetering: GSM Systems and Technologies / Watit Benjapolakul and Ky Leng 1141

Ultra-Wideband: Applications / Gopal Racherla 1150

Ultra-Wideband: Communications / Chaiyaporn Khemapatapan and Watit Benjapolakul 1157

UMTS / Syed Ahson 1165

Unlicensed Mobile Access: Architecture and Protocols / Mehul Shah and Vassilios Koukoulidis 1168

Video Communications / Madhukar Budagavi and Raj Talluri 1182

Video Streaming: Reliable Wireless Streaming with Digital Fountain Codes / Raouf Hamzaoui,
Shakeel Ahmad and Marwan Al-Akaidi 1193

Video Streaming: Wireless / Min Qin and Roger Zimmermann 1200

Wi-Fi Walkman / Jun Wang, A. P. de Vries and Marcel J.T. Reinders 1207

Wireless Ad Hoc Networks: Performance Analysis / Anurag Kumar and Aditya Karnik 1221

Wireless Ad Hoc Networks: Power Assignment / Gruia Calinescu, Ophir Frieder and
Peng-Jun Wan 1234

Wireless Applications: Middleware Security / Marc-André Laverdière, Azzam Mourad, Syrine Tlili and
Mourad Debbabi 1242

Wireless Asynchronous Transfer Mode (ATM): Internetworking / Melbourne Barton,
Matthew Cheng and Li Fung Chang 1251

Wireless Asynchronous Transfer Mode (ATM): Mobility Management / Bala Rajagopalan and
Daniel Reininger 1264

Wireless Asynchronous Transfer Mode (ATM): Quality of Service (QoS) / Bala Rajagopalan and
Daniel Reininger 1273

Wireless Channels: Link Adaptation Techniques / Naga Bhushan 1284

Wireless Communications: Signal Processing Tools / Boualem Boashash, A. Belouchrani,
Karim Abed-Meraim and Nguyen Linh-Trung 1294

Wireless Communications: Spatial Diversity / Ramakrishna Janaswamy 1305

Wireless Data / Allen H. Levesque and Kaveh Pahlavan 1319

**Wireless Home Entertainment Center: Protocol Communications and
Architecture** / Marco Roccetti, C. E. Palazzi, S. Ferretti and G. Pau 1331

Wireless Internet: Applications / Daniel Schall, Christoph Dorn and Schahram Dustdar 1339

Wireless Internet: Fundamentals / Borko Furht and Mohammad Ilyas 1346

Wireless LANs (WLANs) / Suresh Singh 1360

Wireless LANs (WLANs): Ad Hoc Routing Techniques / Ahmed K. Elhakeem 1369

Wireless LANs (WLANs): Real-Time Services with Quality of Service (QoS) Support / Ping Wang, Hai Jiang and Weihua Zhuang 1381

Wireless LANs (WLANs): Security and Privacy / S.P.T. Krishnan, Bharadwaj Veeravalli and Lawrence W.C. Wong. 1392

Wireless Mesh Networks: Architecture, Protocols, and Applications / Vassilios Koukoulidis and Mehul Shah. 1407

Wireless Multimedia Systems: Cross Layer Considerations / Yu Chen, Jian Feng, Kwok Tung Lo and Xudong Zhang. 1417

Wireless Networks: Cross Layer MAC Design / Amoakoh Gyasi-Agyei and Hong Zhou 1424

Wireless Networks: Error Resilient Video Transmission / Yu Chen, Jian Feng, Kwok Tung Lo and Xudong Zhang. 1434

Wireless Networks: Routing with Intermittent Connectivity / Ionut Cardei, Cong Liu and Jie Wu 1442

Wireless Networks: TCP / Sumitha Bhandarkar and Narasimha Reddy 1452

Wireless Networks: Transmitter Power Control / Nicholas Bambos and Savvas Gitzenis. 1464

Wireless Networks: VoIP Service / Sajal K. Das and Kalyan Basu. 1478

Wireless Sensor Networks (WSNs) / Gopal Racherla 1487

Wireless Sensor Networks (WSNs): Central Node and Mobile Node Design and Development / Watit Benjapolakul and Ky Leng 1495

Wireless Sensor Networks (WSNs): Characteristics and Types of Sensors / Dharma P. Agrawal, Ratnabali Biswas, Neha Jain, Anindo Mukherjee, Sandhya Sekhar and Aditya Gupta 1505

Wireless Sensor Networks (WSNs): Key Management Schemes / Jamil Ibriq, Imad Mahgoub, Mohammad Ilyas and Mihaela Cardei. 1509

Wireless Sensor Networks (WSNs): Optimization of the OSI Network Stack for Energy Efficiency / Ahmed Badi and Imad Mahgoub 1523

Wireless Sensor Networks (WSNs): Routing / Dharma P. Agrawal, Ratnabali Biswas, Neha Jain, Anindo Mukherjee, Sandhya Sekhar and Aditya Gupta 1534

Wireless Sensor Networks (WSNs): Secure Localization / Avinash Srinivasan and Jie Wu 1545

Wireless Sensor Networks (WSNs): Security / Lyes Khelladi and Nadjib Badache 1554

Wireless Sensor Networks (WSNs): Self-Organization / Manish Kochhal, Loren Schwiebert and Sandeep K.S. Gupta 1563

Wireless Sensor Networks (WSNs): Time Synchronization / Qing Ye and Liang Cheng 1580

Wireless Transceivers: Near-Instantaneously Adaptive / Lie-Liang Yang and Lajos Hanzo. 1588

Index. I-1

Encyclopedia of Wireless and Mobile Communications
Topical Table of Contents

3G Systems

3G Systems: Accounting Principles and Billing Models / Zoheir Ezziane.................................1
3G Telephony Control Stack: Interactive Playback Control of Video / Michele Covell,
    Sumit Roy, Bo Shen and Frederic Huve..13
Authentication and Privacy / Thomas M. Chen and Nhut Nguyen..................................132
Broadcasting Multimedia Transport: 3G and Future Mobile Networks / Ramakrishna
    Vedantham and Igor D.D. Curcio...157
CDMA Networks: Microcells / Raymond Steele 180
Cellular Systems: 3G / Lal C. Godara 220
EDGE / Syed Ahson...312
Mobile Services / Jenq-Shiou Leu...774
Mobile Systems: Migrating from 3G to 4G / Nalin Sharda and Mohan Ponnada...........788
RF Systems Design / Charles Chien 1023
RF Transceiver Architectures / Charles Chien..1035
Smartphones / Syed Ahson and Mohammad Ilyas 1065
UMTS / Syed Ahson 1165
Unlicensed Mobile Access: Architecture and Protocols / Mehul Shah and
    Vassilios Koukoulidis...1168

4G Systems

4G Mobile Systems: Multimedia Content Transmission / Antonio Cuevas Casado and
    Jose Ignacio Moreno...26
4G Mobile Systems: Radio Resource Management / Alexandros Kaloxylos, Ioannis Modeas,
    Nikos Passas and George Lampropoulos...40
Mobile Communications: 4G / Jaka Sodnik, Mitja Stular, Veljko Milutinovic and Saso Tomazic.............634
Mobile Systems: Migrating from 3G to 4G / Nalin Sharda and Mohan Ponnada.......................788

Access Methods

Access Methods / Bernd-Peter Paris 50
Multiple Access Techniques / Anton Umek, Mitja Stular and Saso Tomazic..........................900
Wireless Communications: Signal Processing Tools / Boualem Boashash, A. Belouchrani,
    Karim Abed-Meraim and Nguyen Linh-Trung 1294

Ad Hoc Networks

Ad Hoc Networks: GPS-Based Routing Algorithm / Young-Joo Suh,
    Won-Ik Kim and Dong-Hee Kwon..58
Ad Hoc Networks: Power-Conservative Designs / Yu-Chee Tseng and Ting-Yu Lin 68
Ad Hoc Networks: Security / Patroklos Argyroudis and Donal O'Mahony............................76
Ad Hoc Networks: Technologies / Marco Conti 89

Ad Hoc Networks (*cont'd.*)

MANET: Clustering Techniques / Jane Yang Yu and Peter Han Joo Chong 513

Multihop Cellular Network Architectures / Xue Jun Li, Peter Han Joo Chong and Qijie Gao 856

Wireless Ad Hoc Networks: Performance Analysis / Anurag Kumar and Aditya Karnik........................ 1221

Wireless Ad Hoc Networks: Power Assignment / Gruia Calinescu, Ophir Frieder and
Peng-Jun Wan 1234

Wireless LANs (WLANs): Ad Hoc Routing Techniques / Ahmed K. Elhakeem 1369

Wireless Networks: TCP / Sumitha Bhandarkar and Narasimha Reddy.............................. 1452

Adaptation Techniques

Adaptation Techniques / Hüseyin Arslan 101

Wireless Channels: Link Adaptation Techniques / Naga Bhushan.............................. 1284

Antenna Systems

High-Speed Wireless Internet Access: Multiantenna Technology / Angel Lozano 354

Land Mobile Communication Systems: Phased-Array Antennas / Hiroyo Ogawa,
Hiroyuki Tsuji, Ami Kanazawa, Ryu Miura and Masato Tanaka.................................. 491

MIMO: Wireless Communications / Sreco Plevel, Saso Tomazic, Tornaz Javornik and
Gorazd Kandus... 604

Smart Antenna Systems: Architecture / T. Bao Vu 1046

Wireless Communications: Spatial Diversity / Ramakrishna Janaswamy 1305

Applications

Body Area Networks / Bin Wang and Yong Pei... 148

Cognitive Radio Implementation / Timothy R. Newman, Alexander M. Wyglinski and
Joseph B. Evans... 230

Compact Dual-Band Direct Conversion CMOS Transceiver / Zhiwei Xu, Charles Chien and
M.C. Frank Chang... 239

Content-Enriched Communication: Social Uses of Interactive TV /
Konstantinos Chorianopoulos.. 267

Coordinated Multi-Device Presentations: Ambient-Audio Identification / Michael Fink,
Michele Covell and Shumeet Baluja 274

Generic Context Management Framework: Contextual Reasoning / Herma van Kranenburg,
Niels Snoeck, Hans Zandbelt and Martin Wibbels 313

Location-Based Services / Ben Falchuk, Dave Famolari and Shoshana Loeb...................... 501

Mobile and Wireless Technologies: Spatial Presence / Katharine Willis 613

Mobile Commerce / Abul Faiz Muhad Ishaq and Mohammad Mohsin 622

Mobile Computing: Context-Aware / Daniel Schall, Christoph Dorn and Schahram Dustdar........ 679

Mobile Games: Multiplayer / Predrag Maksimovic and Imad Mahgoub 694

Mobile Health: Wireless Body Sensor Network Integration / Kevin Hung, Shudi Bao and
Yuan-Ting Zhang.. 707

Mobile Music / Marco Furini 744

Mobile Sensor Networks / Dharma P. Agrawal, Ratnabali Biswas, Neha Jain,
Anindo Mukherjee, Sandhya Sekhar and Aditya Gupta 771

Mobile Services / Jenq-Shiou Leu.. 774

Mobile Terminals: Recommendation Services / Tomohiro Tsunoda and Masaaki Hoshino......... 805

Mobile Wireless Systems: Location Prediction Algorithms / Christine Cheng,
Ravi Jain and Eric van den Berg 825

Nomadic Users: Context-Aware Information Retrieval / M. Anwar Hossain and
Abdulmotaleb El Saddik... 911

Smartphones / Syed Ahson and Mohammad Ilyas 1065

Ultra-Wideband: Applications / Gopal Racherla 1150

Wi-Fi Walkman / Jun Wang, A. P. de Vries and Marcel J.T. Reinders 1207

Wireless Applications: Middleware Security / Marc-André Laverdière, Azzam Mourad,
   Syrine Tlili and Mourad Debbabi 1242

Wireless Home Entertainment Center: Protocol Communications and Architecture /
   Marco Roccetti, C. E. Palazzi, S. Ferretti and G. Pau 1331

Wireless Internet: Applications / Daniel Schall, Christoph Dorn and Schahram Dustdar 1339

Wireless Mesh Networks: Architecture, Protocols, and Applications /
   Vassilios Koukoulidis and Mehul Shah 1407

Bluetooth

Bluetooth Systems / William Stallings 141

CDMA Networks

Authentication and Privacy / Thomas M. Chen and Nhut Nguyen 132

CDMA Networks / Tero Ojanperä and Steven D. Gray 167

CDMA Networks: Microcells / Raymond Steele 180

Cellular Systems: 2G / Lal C. Godara 216

Cellular Systems: 3G / Lal C. Godara 220

Channel Assignment: Fixed and Dynamic / Bijan Jabbari 225

Land Mobile Communication Systems: Phased-Array Antennas / Hiroyo Ogawa,
   Hiroyuki Tsuji, Ami Kanazawa, Ryu Miura and Masato Tanaka 491

Microcells / Raymond Steele 596

Mobile Wireless Systems: Channel Characterization / Quazi Mehbubar Rahman and
   Mohamed Ibnkahla 817

Packet Access: High-Speed Downlink / Jyri Hämäläinen, Risto Wichman,
   Markku Kuusela, Esa Tiirola and Kari Pajukoski 1016

Soft Handoff / Bongkarn Homnan and Watit Benjapolakul 1086

UMTS / Syed Ahson 1165

Cellular Systems

Cellular Base Stations: Open Standards / Gerry Leavey and Steven Gorshe 186

Cellular Data Networks: Handoff and Rerouting / Gopal Racherla and Sridhar Radhakrishnan 196

Cellular Systems / Lal C. Godara 204

Cellular Systems: 1G / Lal C. Godara 214

Cellular Systems: 2G / Lal C. Godara 216

Cellular Systems: 3G / Lal C. Godara 220

Channel Assignment: Fixed and Dynamic / Bijan Jabbari 225

GSM / George N. Aggélou 325

Integrated Wireless LANs (WLANs) and Cellular Networks: Managing Handovers /
   George Lampropoulos, Nikos Passas, Alexandros Kaloxylos and Lazaros Merakos 452

Microcells / Raymond Steele 596

Multihop Cellular Network Architectures / Xue Jun Li, Peter Han Joo Chong and Qijie Gao 856

TDMA Cellular Systems: Speech and Channel Coding / Paul Mermelstein 1122

Wireless Data / Allen H. Levesque and Kaveh Pahlavan 1319

Coding Techniques

Mobile Communications: Coding Techniques / Quazi Mehbubar Rahman and Mohamed Ibnkahla 652

Mobile Devices: Quality of Service (QoS) Adaptation Using MPEG-21 /
   Marios C. Angelides and Anastasis A. Sofokleous 686

Smartphones: Operating Systems / Syed Ahson and Mohammad Ilyas 1073

Coding Techniques (*cont'd.*)

Mobile Multimedia Transmission: Codecs / Nalin Sharda and Mladen Georgievski 736
Spectral Efficiency / Saso Tomazic 1095
Spectral Efficiency: Wireless Multicarrier Communications /
    Alexander M. Wyglinski and Fabrice Labeau 1100
Spread Spectrum Communications / Laurence B. Milstein and Marvin K. Simon 1114
TDMA Cellular Systems: Speech and Channel Coding / Paul Mermelstein 1122
Video Communications / Madhukar Budagavi and Raj Talluri 1182
Wireless Channels: Link Adaptation Techniques / Naga Bhushan 1284
Wireless Communications: Signal Processing Tools / Boualem Boashash,
    A. Belouchrani, Karim Abed-Meraim and Nguyen Linh-Trung 1294

Content Repurposing

Content Repurposing: Multimedia / Shamim Hossain and Abdulmotaleb El Saddik 253
Content Repurposing: Ontology-Based Approach for Mobile Devices /
    Mohan Ponnada and Nalin Sharda 259

Diversity Techniques

Diversity Techniques / Quazi Mehbubar Rahman and Mohamed Ibnkahla 296
Wireless Communications: Signal Processing Tools / Boualem Boashash,
    A. Belouchrani, Karim Abed-Meraim and Nguyen Linh-Trung 1294

IEEE 802.11 WLAN

Authentication and Privacy / Thomas M. Chen and Nhut Nguyen 132
Compact Dual-Band Direct Conversion CMOS Transceiver / Zhiwei Xu,
    Charles Chien and M.C. Frank Chang 239
IEEE 802.11 Wireless LANs (WLANs) / José Antonio Garcia-Macias and Leyla Toumi 362
IEEE 802.11 Wireless LANs (WLANs): Clustering and Roaming Techniques /
    Ahmed K. Elhakeem 365
IEEE 802.11 Wireless LANs (WLANs): Quality of Service (QoS) Support /
    José Villalón, Francisco Mico, Pedro Cuenca and Luis Orozco-Barbosa 375
IEEE 802.11 Wireless LANs (WLANs): Security / Costos Lambrinoudakis and Stefanos Gritzalis 387
IEEE 802.15.4: MAC Layer Attacks in Sensor Clusters / Vojislav Misic and Jobaida Begum 396
IEEE 802.15.4: Signature-Based Intrusion Detection in Wireless Sensor Networks
    (WSNs) / Jelena Misic, Fereshteh Amini and Moazzam Khan 404
IEEE 802.16 Wireless Metropolitan Area Networks (WMANs) /
    Carlos de M. Cordeiro and Dharma P. Agrawal 412
IEEE 802.16 Wireless Metropolitan Area Networks (WMANs): Resource Management /
    Mehri Mehrjoo and Xuemin Sherman Shen 417
IEEE 802.16d: Worldwide Interoperability for Microwave Access (WiMax) /
    Mani Subramanian 428
Wireless Data / Allen H. Levesque and Kaveh Pahlavan 1319
Wireless Networks: Cross Layer MAC Design / Amoakoh Gyasi-Agyei and Hong Zhou 1424

Location Based Systems and Services

Ad-Coop Positioning System: Embedded Kalman Filter Data Fusion /
    Simone Frattasi and João Figueiras 120
Location-Based Services / Ben Falchuk, Dave Famolari and Shoshana Loeb 501
Mobile Wireless Systems: Location Prediction Algorithms /
    Christine Cheng, Ravi Jain and Eric van den Berg 825

MANET

MANET / Marco Conti 509

MANET: Clustering Techniques / Jane Yang Yu and Peter Han Joo Chong 513

MANET: Location-Aware Routing / Yu-Chee Tseng and Chih-Shun Hsu 528

MANET: Multicast Protocols / Xiao Chen and Jie Wu...537

MANET: Multicasting / Dewan Tanvir Ahmed and Shervin Shirmohammadi 546

MANET: Network Mobility as a Control Primitive / Aravindhan Venkateswaran,
Venkatesh Sarangan, Sridhar Radhakrishnan and Raj Acharya 556

MANET: Routing Protocols / Ha Duyen Trung and Watit Benjapolakul 563

MANET: Selfish Behavior on Packet Forwarding / Djamel Djenouri and Nadjib Badache........576

Mobile Communications and Systems

EDGE / Syed Ahson...312

GPRS / Syed Ahson 324

GSM / George N. Aggélou 325

MANET / Marco Conti 509

Mobile Communications: 4G / Jaka Sodnik, Mitja Stular, Veljko Milutinovic and Saso Tomazic..............634

Mobile Communications: Call Admission Control / Jenjoab Virapanicharoen and
Watit Benjapolakul 643

Mobile Communications: Coding Techniques / Quazi Mehbubar Rahman and Mohamed Ibnkahla...........652

Mobile Communications: Locating Position of Mobile Stations / Watit Benjapolakul.................659

Mobile Communications: Power Control / Wasimon Panichpattanakul and Watit Benjapolakul 666

Mobile Systems: Migrating from 3G to 4G / Nalin Sharda and Mohan Ponnada.........................788

Smart Inter-Processor Communicator Protocol Stack / Charbel Khawand...................................1055

Mobility Management

IP Networks: Personal Mobility Challenges Björn Landfeldt,
Jonathan Chan, Binh Thai and Aruna Seneviratne 474

IP Networks: Terminal Mobility Challenges / Björn Landfeldt, Jonathan Chan,
Binh Thai and Aruna Seneviratne..476

IP Networks: User Mobility / Björn Landfeldt, Jonathan Chan, Binh Thai and Aruna Seneviratne 484

MANET: Network Mobility as a Control Primitive / Aravindhan Venkateswaran,
Venkatesh Sarangan, Sridhar Radhakrishnan and Raj Acharya 556

Wireless Asynchronous Transfer Mode (ATM): Mobility Management / Bala Rajagopalan and
Daniel Reininger 1264

Modulation Techniques

Modulation Methods / Gordon L. Stüber...835

Multicarrier Transceivers: Peak-to-Average Power Radio Reduction /
Rakesh Rajbanshi, Alexander M. Wyglinski and Gary J. Minden...843

OFDM: Technology / Tao Jiang..918

OFDM-Based Systems: Modem Architectures / Charles Chien 934

Spectral Efficiency / Saso Tomazic 1095

Spectral Efficiency: Wireless Multicarrier Communications /
Alexander M. Wyglinski and Fabrice Labeau 1100

Spread Spectrum Communications / Laurence B. Milstein and Marvin K. Simon 1114

Wireless Channels: Link Adaptation Techniques / Naga Bhushan...1284

Wireless Communications: Signal Processing Tools / Boualem Boashash,
A. Belouchrani, Karim Abed-Meraim and Nguyen Linh-Trung 1294

Multimedia

Multimedia Content Repurposing

Content Repurposing: Multimedia / Shamim Hossain and Abdulmotaleb El Saddik..................................253
Content Repurposing: Ontology-Based Approach for Mobile Devices /
Mohan Ponnada and Nalin Sharda..259

Multimedia Content Transmission

4G Mobile Systems: Multimedia Content Transmission / Antonio Cuevas Casado and
Jose Ignacio Moreno..26
Broadcasting Multimedia Transport: 3G and Future Mobile Networks /
Ramakrishna Vedantham and Igor D.D. Curcio 157
Image and Video Communication: Joint Source Channel Coding / Muhammad Farooq Sabir,
Robert W. Heath, Jr. and Alan C. Bovik 429
Image and Video Communication: Power Optimized / Muhammad Farooq Sabir,
Robert W. Heath, Jr. and Alan C. Bovik 436
Image and Video Communication: Source-Channel Distortion Modeling /
Muhammad Farooq Sabir, Alan C. Bovik and Robert W. Heath, Jr. 443
Mobile Communications: Coding Techniques / Quazi Mehbubar Rahman and Mohamed Ibnkahla 652
Mobile Devices: Quality of Service (QoS) Adaptation Using MPEG-21 /
Marios C. Angelides and Anastasis A. Sofokleous..686
Mobile Multimedia Transmission: Codecs / Nalin Sharda and Mladen Georgievski 736
Video Communications / Madhukar Budagavi and Raj Talluri..1182
Wireless Networks: Error Resilient Video Transmission /
Yu Chen, Jian Feng, Kwok Tung Lo and Xudong Zhang..1434

Multimedia Streaming

Media Streaming: Transmission Strategies in P2P Networks /
K. M. Ho, Kwok Tung Lo and Jian Feng 588
Mobile Streaming: Performance Issues / Igor D.D. Curcio 782
Mobile Streaming: Standards / Igor D.D. Curcio 784
Multimedia Streaming / Abdulmotaleb El Saddik and Shamim Hossain 884
Multimedia Streaming: Mobile Networks / Igor D.D. Curcio 890
Video Streaming: Reliable Wireless Streaming with Digital Fountain Codes /
Raouf Hamzaoui, Shakeel Ahmad and Marwan Al-Akaidi 1193
Video Streaming: Wireless / Min Qin and Roger Zimmermann..1200

Multimedia Systems and Services

3G Telephony Control Stack: Interactive Playback Control of Video /
Michele Covell, Sumit Roy, Bo Shen and Frederic Huve...13
IP Multimedia Subsystems (IMS) / Mojca Volk, Mitja Stular, Janez Bester,
Andrej Kos and Saso Tomazic...462
**Multimedia Applications in Ad Hoc Networks: Quality of Service (QoS) Support
Cross-Layered Approach** / Sivapathalingham Sivavakeesar and George Pavlou 870
Multimedia Broadcast Multicast Service (MBMS) / Praveen Sanigepalli, Hari Kalva and Borko Furht....880
TDMA Cellular Systems: Speech and Channel Coding / Paul Mermelstein 1122
Wi-Fi Walkman / Jun Wang, A. P. de Vries and Marcel J.T. Reinders 1207

Wireless Multimedia Systems: Cross Layer Considerations / Yu Chen, Jian Feng,
Kwok Tung Lo and Xudong Zhang...1417
Wireless Networks: VoIP Service / Sajal K. Das and Kalyan Basu...1478

OFDM Systems

OFDM: Technology / Tao Jiang...918
OFDM-Based Systems: Modem Architectures / Charles Chien...934

Optical Networks

Optical Broadband Services Networks / David Benjamin and Peter Green.............................953
Optical Communication Networks: Trends / Aysegül Gençata, Narendra Singhal and
Biswanath Mukherjee 966
Optical Networks: Architectures / M. Yasin Akhtar Raja..981
Optical Networks: Resource Management and Allocation / Ding Zhemin and Mounir Hamdi.................989

Peer-to-Peer Systems

Media Streaming: Transmission Strategies in P2P Networks /
K. M. Ho, Kwok Tung Lo and Jian Feng..588
Mobile P2P: Computing / Dewan Tanvir Ahmed and Shervin Shirmohammadi.....................751
Mobile P2P: Data Retrieval and Caching / Andrew Roczniak and Abdulmotaleb El Saddik....................759
P2P: Overlay Multicast / Heather Yu and John Buford 1002
Wi-Fi Walkman / Jun Wang, A. P. de Vries and Marcel J.T. Reinders 1207

Power Control

Ad Hoc Networks: Power-Conservative Designs / Yu-Chee Tseng and Ting-Yu Lin 68
Image and Video Communication: Power Optimized / Muhammad Farooq Sabir,
Robert W. Heath, Jr. and Alan C. Bovik...436
Mobile Communications: Power Control / Wasimon Panichpattanakul and Watit Benjapolakul 666
Wireless Ad Hoc Networks: Power Assignment / Gruia Calinescu, Ophir Frieder and
Peng-Jun Wan 1234
Wireless Networks: Transmitter Power Control / Nicholas Bambos and Savvas Gitzenis.....................1464

Quality of Services

Handoff Prioritization: Early Blocking / Jörg Diederich and Martina Zitterbart.........................336
IEEE 802.11 Wireless LANs (WLANs): Quality of Service (QoS) Support /
José. Villalón; Francisco Mico, Pedro Cuenca and Luis Orozco-Barbosa............................375
Mobile Communications: Call Admission Control / Jenjoab Virapanicharoen and
Watit Benjapolakul 643
Mobile Devices: Quality of Service (QoS) Adaptation Using MPEG-21 / Marios C. Angelides and
Anastasis A. Sofokleous 686
Mobile Systems: Quality of Service (QoS) / Nalin Sharda and Mladen Georgievski 795
**Multimedia Applications in Ad Hoc Networks: Quality of Service (QoS) Support
Cross-Layered Approach** / Sivapathalingham Sivavakeesar and George Pavlou 870
TDMA/TDD Wireless Networks: Quality of Service (QoS) Mechanisms /
Francisco M. Delicado, Pedro Cuenca, Luis Orozco-Barbosa and Jesús Delicado.............................1131
Wireless Asynchronous Transfer Mode (ATM): Quality of Service (QoS) / Bala Rajagopalan and
Daniel Reininger 1273

Quality of Services (*cont'd.*)

Wireless LANs (WLANs): Real-Time Services with Quality of Service (QoS) Support /
Ping Wang, Hai Jiang and Weihua Zhuang 1381

Radio Systems

4G Mobile Systems: Radio Resource Management / Alexandros Kaloxylos,
Ioannis Modeas, Nikos Passas and George Lampropoulos...40
Bluetooth Systems / William Stallings 141
Cognitive Radio Implementation / Timothy R. Newman,
Alexander M. Wyglinski and Joseph B. Evans..230
Compact Dual-Band Direct Conversion CMOS Transceiver /
Zhiwei Xu, Charles Chien and M.C. Frank Chang..239

Resource Management

4G Mobile Systems: Radio Resource Management / Alexandros Kaloxylos,
Ioannis Modeas, Nikos Passas and George Lampropoulos...40
Channel Assignment: Fixed and Dynamic / Bijan Jabbari...225
IEEE 802.16 Wireless Metropolitan Area Networks (WMANs): Resource Management /
Mehri Mehrjoo and Xuemin Sherman Shen...417
Mobile Communications: Call Admission Control / Jenjoab Virapanicharoen and
Watit Benjapolakul 643
Opportunistic Scheduling / Amoakoh Gyasi-Agyei 942
Optical Networks: Resource Management and Allocation / Ding Zhemin and Mounir Hamdi.................989

RF Systems

Cordless Telephone Standard CT-2 / Lajos Hanzo..286
RF Systems Design / Charles Chien 1023
RF Transceiver Architectures / Charles Chien...1035

Routing

Ad Hoc Networks: GPS-Based Routing Algorithm / Young-Joo Suh,
Won-Ik Kim and Dong-Hee Kwon..58
Cellular Data Networks: Handoff and Rerouting / Gopal Racherla and
Sridhar Radhakrishnan...196
MANET: Location-Aware Routing / Yu-Chee Tseng and Chih-Shun Hsu 528
MANET: Routing Protocols / Ha Duyen Trung and Watit Benjapolakul 563
Rerouting Schemes: Performance Evaluation / Gopal Racherla and
Sridhar Radhakrishnan..1018
Wireless Networks: Routing with Intermittent Connectivity /
Ionut Cardei, Cong Liu and Jie Wu 1442
Wireless Sensor Networks (WSNs): Routing / Dharma P. Agrawal, Ratnabali Biswas,
Neha Jain, Anindo Mukherjee, Sandhya Sekhar and Aditya Gupta.........................1534

Security

Ad Hoc Networks: Security / Patroklos Argyroudis and Donal O'Mahony.........................76
Authentication and Privacy / Thomas M. Chen and Nhut Nguyen....................................132
IEEE 802.11 Wireless LANs (WLANs): Security / Costos Lambrinoudakis and Stefanos Gritzalis............387

IEEE 802.15.4: Signature-Based Intrusion Detection in Wireless Sensor Networks (WSNs) /
Jelena Misic, Fereshteh Amini and Moazzam Khan...404
Mobile Security: Game Theory / Hadi Otrok, Mona Mehrandish, Chadi Assi,
Prabir Bhattacharya and Mourad Debbabi..764
Wireless Applications: Middleware Security / Marc-André Laverdière,
Azzam Mourad, Syrine Tlili and Mourad Debbabi 1242
Wireless LANs (WLANs): Security and Privacy / S.P.T. Krishnan,
Bharadwaj Veeravalli and Lawrence W.C. Wong 1392
Wireless Sensor Networks (WSNs): Key Management Schemes / Jamil Ibriq, Imad Mahgoub,
Mohammad Ilyas and Mihaela Cardei..1509
Wireless Sensor Networks (WSNs): Secure Localization / Avinash Srinivasan and Jie Wu...........1545
Wireless Sensor Networks (WSNs): Security / Lyes Khelladi and Nadjib Badache 1554

TDMA Systems

Cellular Systems: 2G / Lal C. Godara 216
Channel Assignment: Fixed and Dynamic / Bijan Jabbari...225
Mobile Communications: Call Admission Control / Jenjoab Virapanicharoen and
Watit Benjapolakul 643
Mobile Wireless Systems: Channel Characterization /
Quazi Mehbubar Rahman and Mohamed Ibnkahla..817
TDMA Cellular Systems: Speech and Channel Coding / Paul Mermelstein 1122
TDMA/TDD Wireless Networks: Quality of Service (QoS) Mechanisms /
Francisco M. Delicado, Pedro Cuenca, Luis Orozco-Barbosa and Jesús Delicado......................1131

Ultra-Wide Band Systems

Body Area Networks / Bin Wang and Yong Pei...148
Ultra-Wideband: Applications / Gopal Racherla...1150
Ultra-Wideband: Communications / Chaiyaporn Khemapatapan and Watit Benjapolakul 1157

Wireless Asynchronous Transfer Mode (ATM)

Wireless Asynchronous Transfer Mode (ATM): Internetworking / Melbourne Barton,
Matthew Cheng and Li Fung Chang..1251
Wireless Asynchronous Transfer Mode (ATM): Mobility Management /
Bala Rajagopalan and Daniel Reininger...1264
Wireless Asynchronous Transfer Mode (ATM): Quality of Service (QoS) /
Bala Rajagopalan and Daniel Reininger...1273

Wireless Internet

High-Speed Wireless Internet Access: Multiantenna Technology / Angel Lozano 354
Mobile Internet Technologies / Christoph Dorn, Daniel Schall and Schahram Dustdar.....................718
Mobile IP / Andrianus Yofy and Watit Benjapolakul...726
Wireless Internet: Applications / Daniel Schall, Christoph Dorn and Schahram Dustdar 1339
Wireless Internet: Fundamentals / Borko Furht and Mohammad Ilyas...1346
Wireless Networks: TCP / Sumitha Bhandarkar and Narasimha Reddy...1452
Wireless Networks: VoIP Service / Sajal K. Das and Kalyan Basu...1478

Wireless LAN

Authentication and Privacy / Thomas M. Chen and Nhut Nguyen...132
Integrated Wireless LANs (WLANs) and Cellular Networks: Managing Handovers /
George Lampropoulos, Nikos Passas, Alexandros Kaloxylos and Lazaros Merakos....................452

Wireless LAN (*cont'd.*)

Multihop Cellular Network Architectures / Xue Jun Li, Peter Han Joo Chong and
Qijie Gao 856
Unlicensed Mobile Access: Architecture and Protocols / Mehul Shah and Vassilios Koukoulidis 1168
Wireless LANs (WLANs): Ad Hoc Routing Techniques / Ahmed K. Elhakeem 1369
**Wireless LANs (WLANs): Real-Time Services with Quality of Service (QoS)
Support** / Ping Wang, Hai Jiang and Weihua Zhuang 1381
Wireless LANs (WLANs): Security and Privacy / S.P.T. Krishnan,
Bharadwaj Veeravalli and Lawrence W.C. Wong 1392

Wireless Mesh Networks

Wireless Mesh Networks: Architecture, Protocols, and Applications /
Vassilios Koukoulidis and Mehul Shah 1407
Wireless Networks: TCP / Sumitha Bhandarkar and Narasimha Reddy 1452

Wireless Sensor Networks (WSNs)

Dynamic Wireless Sensor Networks (WSNs): Architecture and Modeling /
Symeon Papavassiliou and Jin Zhu 300
Heterogeneous Wireless Sensor Networks (WSNs) / Mihaela Cardei and Yinying Yang 345
**IEEE 802.15.4: Signature-Based Intrusion Detection in Wireless Sensor
Networks (WSNs)** / Jelena Misic, Fereshteh Amini and Moazzam Khan 404
Mobile Sensor Networks / Dharma P. Agrawal, Ratnabali Biswas, Neha Jain,
Anindo Mukherjee, Sandhya Sekhar and Aditya Gupta 771
Telemetering: GSM Systems and Technologies / Watit Benjapolakul and Ky Leng 1141
Wireless Sensor Networks (WSNs) / Gopal Racherla 1487
**Wireless Sensor Networks (WSNs): Central Node and Mobile Node Design and
Development** / Watit Benjapolakul and Ky Leng 1495
Wireless Sensor Networks (WSNs): Characteristics and Types of Sensors / Dharma P. Agrawal,
Ratnabali Biswas, Neha Jain, Anindo Mukherjee, Sandhya Sekhar and Aditya Gupta 1505
Wireless Sensor Networks (WSNs): Key Management Schemes / Jamil Ibriq,
Imad Mahgoub, Mohammad Ilyas and Mihaela Cardei 1509
**Wireless Sensor Networks (WSNs): Optimization of the OSI Network Stack for Energy
Efficiency** / Ahmed Badi and Imad Mahgoub 1523
Wireless Sensor Networks (WSNs): Routing / Dharma P. Agrawal, Ratnabali Biswas,
Neha Jain, Anindo Mukherjee, Sandhya Sekhar and Aditya Gupta 1534
Wireless Sensor Networks (WSNs): Secure Localization / Avinash Srinivasan and Jie Wu 1545
Wireless Sensor Networks (WSNs): Security / Lyes Khelladi and Nadjib Badache 1554
Wireless Sensor Networks (WSNs): Self-Organization / Manish Kochhal,
Loren Schwiebert and Sandeep K.S. Gupta 1563
Wireless Sensor Networks (WSNs): Time Synchronization / Qing Ye and Liang Cheng 1580

Preface

Only a decade ago, wireless and mobile communication seemed like a new research field and an emerging new industry. Today, at the beginning of the new millennium, wireless and mobile communication has come of age, and the related industry has significantly grown.

The *Encyclopedia of Wireless and Mobile Communications* provides in-depth coverage of the important concepts, issues, and technology trends in the field of wireless and mobile technologies, systems, techniques, and applications. It is a comprehensive collection of articles that present current technologies and systems, perspectives, and future trends in the field from hundreds of leading researchers and experts in the field. The articles in the book describe a number of topics in wireless and mobile communication systems and applications—from 3G and 4G systems to OFDM systems, optical networks, wireless sensor networks, and to emerging applications.

The editor, working with the *Encyclopedia's* Editorial Board and a large number of contributors, surveyed and divided the field of wireless and mobile communications into specific topics that collectively encompass the foundations, technologies, applications, and emerging elements of this exciting field. The members of the Editorial Board and the contributors are world experts in the field of wireless and mobile communications from both academia and industry. The total number of contributors is more than 270, and they have written a total of 173 articles.

The *Encyclopedia's* intended audience is technically diverse and wide; it includes anyone concerned with wireless and mobile communication systems and their applications. Specifically, the *Encyclopedia* can serve as a valuable reference for researchers and scientists, system designers, engineers, programmers, and managers who are involved in wireless and mobile system design and their applications.

I thank the members of the Editorial Board for their help in creating this *Encyclopedia*, as well as the authors for their individual contributions. The members of the Editorial Board assisted in selecting the articles, in writing one or more long and short articles, and soliciting the other contributors. Without the expertise and effort of the contributors, this *Encyclopedia* would never have come to fruition. Special thanks are due to Taylor & Francis editors and staff, including Richard O'Hanley and Claire Miller. They deserve my sincere recognition for their support throughout the project.

Borko Furht
Florida Atlantic University
Boca Raton, Florida

Acronyms

Acronym	Definition
2G	second generation
3G	third generation
3GPP	third generation partnership project
4G	fourth generation
8-PSK	octagonal phase shift keying
A/D	analog to digital
AA	application agent
AAA	authentication, authorization and accounting
AAL	ATM adaptation layer
ABF	available bandwidth fraction
ABL	adaptive bit loading
ABR	available bit rate
AC	authentication center
ACELP	algebraic code-excited linear prediction
ACI	adjacent channel interference
ACK	acknowledgment
ACL	asynchronous connection-less
ACM	adaptive coded modulation
ACPR	adjacent channel power ratio
ACW	address code word
ADC	analog-to-digital converter
ADM	add/drop multiplexer(ing)
ADPCM	adaptive differential pulse code modulation
ADSL	asymmetric digital subscriber line
AES	advanced encryption standard
AFC	automatic frequency correction
AFH	adaptive frequency hopping
AGC	automatic gain control
AGEP	A-GSM encapsulation protocol
A-GSM	ad hoc GSM
AHP	analytic hierarchy process
AIC	Akaike information criteria
AIN	advanced intelligent network
AIR	adaptive intra refresh
AJ	antijam
AK	authentication key
AKA	authentication and key agreement
ALP	active link protection
AMC	adaptive modulation and coding; adaptive multi-hop clustering
AMI	ambient intelligence
AMPS	advanced mobile phone system (service)
AMPU	average margin per user
AMR	adaptive multi-rate

Acronym	Definition
AMRoute	adhoc multicast routing protocol
AMT	address mapping table
ANSI	American National Standards Institute
AoA	angle of arrival
AOB	asymptotically optimal backoff
AODV	ad hoc on-demand distance vector
AP	access point
API	application programming interface
APN	access point names
APON	ATM-based passive optical network
APS	automatic protection switching
APTEEN	adaptive periodic threshold sensitive energy efficient sensor network protocol
AR	adaptive routing
ARDIS	advanced radio data information service
ARIB	Association for Radio Industry and Business
ARP	address resolution protocol
ARPU	average revenue per unit
ARQ	automatic repeat request
ARS	ad hoc relay station
ART	adaptive rate transmission
ASE	amplified spontaneous emission
ASO	arbitrary slice sub-mode
ASP	application service provider
ASTN	automatic switched transport network
ATM	asynchronous transfer mode
AuC	authentication center
AUS	authentication server
AV	authentication vector
AWGN	additive white Gaussian noise
BAN	body area network
BBM	baseband module
BCCH	broadcast control channel
BCH	Bose–Chaudhuri–Hocquenghem
BCI	bandwidth change indication
BCM	block coded modulation
BCP	burst control packet
BCT	block cosine transform
BDA	blog delivering adapter
BER	bit-error rate
BF	basic frame
BFF	beamforming function
BHP	burst header packet
BI	blocking island
BIC	Bayesian information criteria
B-ICI	BISDN intercarrier interface
BIG	blocking island graph
BIH	blocking island hierarchy

BIP	broadcasting incremental power	CDV	cell delay variation
BISDN	broadband integrated services digital network	CDVCC	coded digital verification color code
		CDVT	cell delay variation tolerance
BISUP	BISDN user part	CEC	cluster-based energy conservation
BLAST	Bell Labs layered space–time	CELP	code excited linear prediction
BLER	block error rate	CEPT	Conference of European Posts and Telecommunications Administrations
BLSR	bidirectional line-switched rings		
BMC	block motion compensation	CES	circuit emulation service
BOBO	billing on behalf of	CFP	contention-free period; cordless fixed part
BoD	bandwidth-on-demand		
B-O-QAM	binary offset quadrature amplitude modulation	CH	cluster head; corresponding host
		CHM	channel marker
BP	beacon packet; burst period	CHMF	cordless fixed part channel marker
BPEL	business process execution language	CHMP	cordless portable part channel marker
BPSK	binary phase shift keying	CID	cluster identification
BR	blocking rate	CIP	cellular IP
BRAINS	blog rendering and accessing instantly	CIR	carrier-to-interference ratio; channel impulse response
BRAN	Broadband Radio Access Network	CITRIS	Center for Information Technology in the Interest of Society
BS	base station		
BSC	base station controller	CL-BI	connectionless bearer independent
BSHR	bidirectional self-healing ring	CLDC	connected limited device configuration
BSIC	base station identity code		
BSS	base station subsystem; base station system; basic service set; business support systems	CLR	cell loss ratio
		CM	configuration management
		CMC	combined-metrics-based clustering
BSSID	basic service set identification	cMCN	clustered multihop cellular network
BSSMAP	base station subsystem management application sub-part	CMEA	cellular message encryption algorithm
		CMOS	complementary metal oxide semiconductor
BTC	block turbo code		
BTS	base transceiver station	CMS	China MobileSoft
CA	collision avoidance; configuration agent	CN	core network; correspondent node
CAB	charging accounting and billing	CNR	carrier-to-noise ratio
CAC	call admission control; connection admission control	CO	central office; connection-oriented
		COA	care-of address
CAI	common air interface	CO-BI	connection oriented-bearer independent
CAMP	core-assisted mesh protocol	COFDM	coded orthogonal frequency division multiplexing
CAPEX	capital expense		
CAR	context-aware information retrieval	CoS	class of service; cross over switch
C-ART	channel quality-motivated adaptive rate transmission	CP	charge pump; contention period
		CPM	continuous phase modulation; cryptographic provider management
CAVE	cellular authentication and voice encryption		
		CPP	cordless portable part
CBR	constant bit rate	CPRI	common public radio interface
CC/PP	client-capability/preferences profile	CPU	central processing unit
CCI	co-channel interference; connection control interface	CQI	channel quality indicator
		CR	chip rate
CCK	complementary code keying	CRC	cyclic redundancy check
CCM	control and clock module	CRL	Communication Research Laboratory [ministry of posts and telecommunications, Japan]
CCS	common channel signaling		
CD	collision detection		
CDF	cumulative distribution function	CR-LDP	constraint-based routing–label distribution protocol
CDMA	code division multiple access		
CDPD	cellular digital packet data	CRM	customer relationship management
CDR	call data record	CRNSC	connected relay node single cover
CDS	connected dominating set	CRS	cell relay service

CS	circuit switching	DM	device management
C-SCDPD	circuit-switched cellular digital packet data	DMA	direct memory access
		DMTS	delay measurement time synchronization
CSCW	computer supported co-operative work	DNS	domain name service
CSMA	carrier sense multiple access	DOA	direction of arrival
CSMA/CA	carrier sense multiple access with collision avoidance	DoS	denial of service
		DP	distribution point; dynamic partition
CSMA/CD	carrier sense multiple access/collision detection	DPC	distributed power control
		DPCCH	dedicated physical control channel
CSP	context sharing platform	DPCM	differential pulse coded modulation
CT	cordless telephone	DPDCH	dedicated physical data channel
CT-2	Cordless Telephone-2	DPM	dynamic power management
CTIA	Cellular, Telecommunications and Internet Association	DPPM	differential pulse position modulation
		DQPSK	differential quadrature phase shift keying
CTS	clear to send		
CWDM	coarse wavelength division multiplexing	DREAM	distance routing effect algorithm for mobility
CWE	Collaborative Working Environment	DRM	Digital rights management
DA	data agent	drms	distance root mean square
DAC	device access code; digital-to-analog converter	DS	direct sequence; distribution system
		DSC	dominating-set-based clustering
DAM	data acquisition module	DSCP	DiffServ code point
DAMA	demand assigned multiple access	DSDV	destination sequence distance vector
DBPC-REQ	downlink burst profile change request	DSL	digital subscriber line
DBPC-RSP	downlink burst profile change response	DSMA	digital sense multiple access
DCA	dynamic channel allocation; dynamic channel assignment	DSP	digital signal processor(ing)
		DSR	dynamic source routing
DCC	digital control channel; distributed contention control	DSSS	direct-sequence spread spectrum
		DSVD	digital simultaneous voice and data
DCD	digital channel database	DTBR	dual threshold bandwidth reservation
DCF	distributed coordination function	DTC	digital traffic channel
DCS	digital cross-connects; distributed control system	DTMF	dual tone multi-frequency
		DTV	digital TV
DCT	discrete cosine transform	DVB	digital video broadcasting
DCW	data code word	DWDM	dense wavelength division multiplexing
DDCA	distributed dynamic clustering algorithm	E/O	electrical-to-optical
		EAP	extensive authentication protocol
DDM	differential destination multicast	EAPoL	extended authentication protocol over LAN
DECT	digital enhanced cordless telecommunication		
		ECMEA	Enhanced cellular message encryption algorithm
DES	data encryption standard		
DES-CBC	data encryption standard with cipher block chaining	ECS	efficient clustering scheme
		ECSBC	entropy constrained subband coding
DGPS	differential global positioning system	ECSD	Enhanced circuit-switched data
DHCP	dynamic host configuration protocol	EDC	efficient double-cycle
DHT	distributed hash table	EDFA	erbium-doped fiber amplifier
DIAC	dedicated inquiry access code	EDGE	enhanced data rates for GSM evolution
DiffServ	differentiated service	EDI	electronic data interchange
DIFS	distributed interframe space	EEC	Energy-efficient clustering
DIMIWU	dual-mode identity and internetworking unit	EEPROM	electrically erasable programmable read only memory
DIUC	downlink interval usage code	EGC	equal gain combiner(ing)
DL	downlink	E-GPRS	enhanced general packet radio service
DLBC	degree-load-balancing clustering	EIA/TIA	Electronic Industries Association and Telecommunication Industry Association
DLC	data link control		
DLE	dynamic lightpath establishment		

EIB	erasure indicator bit	FM	fault management; frequency modulation
EIFS	extended interframe space		
EIR	equipment identity register	FPLA	future probable location area
EIRP	effective isotropically radiated power	FPLMTS	Future Public Land Mobile Telecommunications Service [now IMT-2000]
E-OTD	enhanced observed time difference		
EP	extension point	FRS	frame relay service
EPON	ethernet passive optical network	FSAN	full-service access network
ERMES	enhanced radio message system	F-SCH	Forward supplemental channel
ERP	enterprise resource planning	FSK	frequency shift keying
ESA	enhanced subscriber authentication	FSMC	finite-state Markov channel
eSCO	extended synchronous connection oriented	FSR	fisheye routing
		FT	format type
ESCON	enterprise systems connection	FTP	file transfer protocol
ESN	equipment serial number	FTTB	fiber to the building
ESP	enhanced subscriber privacy	FTTC	fiber to the curb
ESPRIT	European Strategic Program on Research in Information Technology	FTTCab	fiber to the cabinet
		FTTH	fiber to the home
ESS	extended service set	FVC	forward voice channel
ETACS	European total access communication system	FWM	four-wave mixing
		GA	gathering agent; genetic algorithm
ETDM	electric time-domain multiplexing	GAF	geographic adaptive fidelity
ETSI	European Telecommunications Standards Institute	GAP	generic access protocol
		GCF	generic connection framework
EVM	error vector magnitude	GEC	Gilbert-Elliott channel
FA	foreign agent	GEDIR	geographic distance routing
FACCH	fast associated control channel	GEO	geostationary earth orbit
FastBP	fast beacon packet	GERAN	GSM/EDGE radio access network
FCA	fixed channel assignment	GFP	generic framing procedure
FCC	Federal Communications Commission; forward control channel	GFR	guaranteed frame rate
		GFSK	Gaussian frequency shift keying
F-CCCH	forward common control channel	GG	Gabriel graph
FCCH	frequency correction channel	GGSN	gateway GPRS support node
FCS	fast cell selection	GIAC	general inquiry access code
FDD	frequency division duplex	GIF	graphics interchange format
FDDI	fiber distributed data interface	GIS	Geographic Information Systems
FDMA	frequency division multiple access	GK	gatekeeper
FDO	forced dropout	GMSC	gateway mobile switching center
FDR	frame drop rate	GMSK	Gaussian minimum shift keying
FEC	forward error correction; forwarding equivalence class	GOB	group of blocks
		GP	guard period
FER	frame error rate	GPC	grant per connection
F-FCH	forward fundamental channel	GPM	general purpose module
FFH	fast frequency-hopped	GPRS	general packet radio service
FFT	fast Fourier transform	GPS	global positioning system
FGLP	fine grained loss protection	GPSR	greedy perimeter stateless routing
FGMP	forwarding group multicast protocol	GPSS	grant per subscriber station
FGS	fine granular scalability	GRA	geographical routing algorithm
FH	frequency hopping	GRAP	group RAP
FH/MC	DS-CDMA frequency hopping-based multicarrier direct spectrum code division multiple access	GS	group station
		GSM	Groupe Speciale Mobile
		GSTN	general switched telephone network
FHSS	frequency hopping spread spectrum	GTP	GPRS tunneling protocol
		GUI	graphical user interface
FICON	fiber connection	GW	gateway
FIFO	first-in, first-out	HA	home agent
FIR	finite-impulsive response	HAP	high altitude platform

HARQ	hybrid automatic repeat request	ILP	integer linear program
HCI	human-computer interaction	IM	instant messaging
HCS	hierarchical cellular structure	IMEI	International Mobile Equipment Identification
HDLC	high level data link control	IMP	inter-modulation product
HDMI	high-definition multimedia interface	IMS	IP multimedia subsystem
HDR	high data rate	IMSI	international mobile subscriber identity
HEC	header extension code	IMT2000	International Mobile Telephony 2000
HEMT	high electron mobility transistor	IMT-2000	International Mobile Telecommunications 2000
HF	hyperframe	IN	intelligent network
HFC	hybrid fiber coax	INT	interrupt line
HHO	horizontal handover	IntServ	integrated services
HIC	handset identification code	IOF	interoffice facilities
HIPERLAN	high performance radio local area network	IoP	importance-of-presence
HLP	hierarchical location prediction	IP	integer programming; internet protocol
HLR	home location register	IPIP	IP within IP
HMAC-MD5	hash-based message authentication code message digest 5	IPMOA	integrated personal mobility architecture
HMCN	hierarchical multihop cellular network	iPOP	ICEBERG point of presence
HO	handover	IPSec	IP security
HOL	head-of-the-line	IR	impulse radio; incremental redundancy; information retrieval; infrared
HOM	higher-order modulation	IRTF	Internet Research Task Force
HRA	heterogeneous range assignment problem	IRU	indefeasible right of use
HS	handshaking	ISD	independent segment decoding
HSCSD	high speed circuit switched data	ISDN	integrated services digital network
HSDPA	high-speed downlink packet access	ISI	inter-symbol interference
HS-DSCH	high-speed downlink shared channel	ISL	inter-satellite link
HSR	hierarchical state routing	ISM	industrial, scientific and medical
HSS	home subscriber server	ISMA	Internet Streaming Media Alliance
HTTP	hypertext transfer protocol	ISO	International Standards Organization
IAPP	interaccess point protocol	ISP	internet service provider
IBSS	independent basic service set	ITS	intelligent transportation system
IBT	in-band-terminator	ITU	International Telecommunications Union
IC	integrated circuit		
iCAR	integrated cellular and ad-hoc relaying	ITU-D	International Telecommunication Union —Telecommunication Development Sector
ICSP	in-circuit serial programming		
ICV	integrity check value		
IDE	integrated development environment	ITU-R	International Telecommunication Union —Radiocommunication Sector
IDFT	inverse discrete Fourier transform		
IDLBC	identification load balancing clustering	ITU-T	International Telecommunication Union —Telecommunications Standardization Sector
IDMP	intra-domain mobility management protocol		
IDS	intrusion detection systems	ITV	interactive TV
IE	information element	IV	initialization vector
IEC	International Electrotechnical Commission; International Engineering Consortium	IWF	wireless interworking function
		IWU	interworking unit
IEEE	Institute of Electrical and Electronic Engineers	JBD	joint block diagonalization
		JD	joint demodulation
IERP	interzone routing protocol	JOD	joint off-diagonalization
IETF	Internet Engineering Task Force	JPEG	joint photographic experts group
IF	information filtering; intermediate frequency	JSCC	joint source-channel coding
		JTAG	joint test action group
IFFT	inverse fast Fourier transform	JTWI	Java technology for the wireless industry
ILEC	incumbent local exchange carrier		

JVM	Java Virtual Machine	MC	movement circle; multicarrier; multicast capable; multiplex code
L2CAP	logical link control and adaptation protocol	MC-CDMA	multicarrier code division multiple-access
LAN	local area network	MCHO	mobile-controlled handover
LAR	location-aided routing	MCM	multilevel coded modulation
LBC	load-balancing clustering	MCN	multihop cellular network
LBS	location-based service	M-commerce	Mobile commerce
LCC	least cluster change	MCR	minimum cell rate
LCP	least congested path	MCS	modulation and coding schemes
LCR	level crossing rate	MC-TDMA	joint frequency–time
LDP	label distribution protocol	MCU	multipoint control unit
LEACH	low energy adaptive clustering hierarchy	MDBS	mobile data base station
LEO/MEO	low/mid-earth orbit	MDC	multiple description coding
LES	land earth station	MDF	media description file
LH	long-haul	MD-IS	mobile data intermediate system
LID	link identification	MDS	minimal dominating set
LLC	logical link control	ME-ATM	mobility-enabled ATM
LM	location management	MEMS	micro-electro-mechanical system
LMA	local mean algorithm	M-ES	mobile end system
LMC	low-maintenance clustering	MESFET	metal-semiconductor field effect transistor
LMDS	local multipoint distribution service		
LMM	local mobility model	MF	multiple field
LMN	local mean number of neighbors algorithm	MFS	managed fiber services
		MGCP	media gateway control protocol
LMP	link manager protocol	MH	mobile host
LMS	least mean square	MHN	multihop capable node
LO	local oscillator	MI	multicast incapable
LoD	level-of-detail	MIC	manufacturer identification code; message integrity code
LOS	line of sight		
LPC	linear prediction coding	MIDI	musical instrument digital interface
LPF	low pass filter		
LPI	low probability of intercept	MIDP	mobile information device profile
LPTSL	loss profile transport sublayer	MIM	metal-insulator-metal
LS	link state	MIMO	multiple-input multiple-output
LSE	least square error	MIN	mobile identification number
LSF	line-spectral frequency	MIS	management information systems
LSP	label switched path; link state packet	MISO	multiple-input single-output
LSR	label-switching router; low satisfaction rate	MK	master key
		ML	metering layer
LST	label switching table	MLAN	metropolitan local area network
M3I	Market Managed Multiservice Internet	MLSD	maximum likelihood sequence detection
MA	mobile agent		
MAC	media access control; mobility-aware clustering	MLSE	maximum likelihood sequence estimation
		ML-TCM	multilevel turbo coded modulation
MAHO	mobile-assisted handoff		
MAI	multiple-access interference	MM	motion marker
MAN	metropolitan area network	MMDS	multichannel multipoint distribution service
MANET	mobile ad hoc network		
MAODV	multicast ad hoc on-demand distance vector	MMP	mobile motion prediction
		MMS	multimedia message service
max CTD	maximum cell transfer delay	MMSE	minimum mean square error
MB	macroblock	MN	mobile node
MBA	macroblock address	MNO	mobile network operators
MBS	maximum burst size; mobile broadband system	MOS	mean opinion score

MPA	mobile people architecture; motion prediction algorithm	OAM&P	operations, administration, maintenance and provisioning
MPDU	MAC protocol data unit	OAN	optical access network
MPEG	Moving Picture Experts Group	OBEX	object exchange
MPEG-4 AAC	Moving Picture Experts Group-4 advanced audio coding	OBS	optical broadband services; optical burst switching
MPL	multiplex payload length	OBSAI	Open Base Station Architecture Initiative
MPLA	most-probable location area		
MPLS	multi-protocol label switching	OCC	optical connection controller
MRC	maximal ratio combiner(ing)	ODMA	opportunity driven multiple access
MRE	modify request	ODMRP	on-demand multicast routing protocol
MS	mobile station	ODWCA	on-demand weighted clustering algorithm
MSC	mobile switching center		
MSCHAP	Microsoft challenge handshake authentication protocol	OEM	original equipment manufacturers
		OEO	optical-to-electrical-to-optical; optoelectronic
MSDU	MAC service data unit		
MSE	mean squared error	OFDM	orthogonal frequency division multiplexing
MSS	mobile support station		
MST	minimal spanning tree	OFSGP	obstacle-free single-destination geocasting protocol
MT	mobile terminal; movement track		
MTMR	multiple-transmit multiple-receive	OGC	Open GIS Consortium
MTP	message transfer part	OIF	Optical Internetworking Forum
MTP3-B	message transfer part	OLAM	on-demand location aware multicast protocol
MTSO	mobile telephone switching office		
MUD	multi-user detection	OMA	Open Mobile Alliance
MUSIC	multiple signal classification	ONU	optical network unit
MUX	multiplex–demultiplex	OPEX	operational expense
MWI	Mobile Web Initiative	OPS	optical packet switching
MWS	managed wavelength service	OQAM	offset quadrature amplitude modulation
MZR	multicast zone routing protocol		
NA	network adaptation	OQPSK	offset quadrature phase-shift keying
NAI	network access identifier	OS	operating system
NAK	negative acknowledgment	OSA	open service access; open system authentication
N-AMPS	narrowband AMPS		
NAV	network allocation vector	OSGi	Open Services Gateway initiative
NBAP	node B application part	OSI	open system interconnection
NCC	network control center	OSP	optical service provider
NCNR	nearest common neighbor routing	OSPF	open-shortest-path-first
NE	network element	OSPF/ISIS	open-shortest-path-first/intermediate system to intermediate system
NG	next generation		
NIST	National Institute of Standards and Technology	OTA	over the air
		OTD	orthogonal transmit diversity
NMEA	National Maritime Electronics Association	OTDM	optical time domain multiplexing
		OTN	optical transport network
NMS	network management system	OVPN	optical virtual private network
NMT	Nordic mobile telephone	OXC	optical cross-connect
NNI	node-to-node interface	P2P	peer-to-peer
NOC	network operations center	PA	profile agent
NP	network planner	PACCH	packet associated control channel
nrt-VBR	non-real-time variable bit rate	PACS	personal access communication service
NTONC	National Transparent Optical Networking Consortium	PAGCH	packet access grant channel
		PAMAS	power-aware multi-access protocol with signaling
NTP	network time protocol		
NTT	Nippon Telephone and Telegraph	PAN	personal area network
O/E	optical-to-electrical	PANu	personal area network user
OADM	optical add/drop multiplexer(ing)	PBCCH	packet broadcast control channel

PBX	private branch exchange	PPS	precise positioning service
PC	passive clustering; personal computer	PRACH	packet random access channel, uplink
PCF	point coordination function	PRMA	packet reservation multiple access
PCM	pulse code modulation	PS	packet switching; portable station;
PCMA	power-controlled multiple access		power-saving
PCN	personal communications network;	PSD	power spectral density
	process change notification	PSK	phase shift keying
PCR	peak cell rate	PSMS	premium short messaging service
PCS	personal communication service	PSNR	peak signal to noise ratio
PDA	personal digital assistant	PSS	packet-switched streaming service
PDC	personal digital cellular [Japanese];	PSTD	phase sweep transmit diversity
	primary domain controller	PSTN	public switched telephony network
PDCH	packet data channel	PTK	pairwise transient key
PDCP	packet data convergence protocol	PUPA	portable device usage-based pervasive
pdf	probability density function		accounting
PDK	product development kit	PVR	personal video recorder
PDP	packet data protocol; policy decision	PVT	process-voltage-temperature
	point; power delay profile	QAM	quadrature amplitude modulation
PDTCH	packet data transfer channel	QCELP	Qualcom code excited linear prediction
PDU	protocol data unit; protocol description	QCIF	quarter common intermediate format
	unit	QDU	quantization distortion unit
PDU-SMS	protocol data unit-short message service	Q-O-QAM	quaternary offset quadrature amplitude
PEGASIS	power-efficient gathering in sensor		modulation
	information system	QoS	quality of service
PEP	performance enhancing proxy; policy	QPSK	quadrature phase shift keying
	enforcement point	R/W	read/write
PER	packet error rate; probability of error	RACH	random access channel
PFD	phase/frequency detector	R-ACH	reverse access channel
PHEMT	pseudomorphic high electron mobility	RAL	radio access layer
	transistor	RAM	random access memory
PHS	personal handy phone system	RAN	radio access network
PHY	physical	RANAP	radio access network application part
PIC	portable identity code; programmable	RAS	registration/admission/status
	integrated circuit	RBHSO	role-based hierarchical CDS-based self-
PIM	personal information management		organization
PKI	public key infrastructure	RC4	Rivest Cipher 4
PKM	privacy and key management	RCC	reverse control channel
PL	private line	R-CCCH	reverse common control channel
PLCM	private long code mask	RCPC	rate compatible punctured convolution
PLL	phase-lock loop	RCPSRC	rate compatible punctured systematic
PLMN	public land mobile network		recursive convolution
PLMR	public land mobile radio	RDA	regularity detection algorithm
PM	performance management	RD-LAP	radio data-link access protocol
PMK	pairwise master key	RECOVC	recognition-compatible voice coding
PN	pseudo-noise	RERR	route error
PNG	portable network graphics	RETRI	random, ephemeral transaction
PNNI	private network–network interface		identifier
PO	period overhead	RF	radio frequency
POCSAG	Post Office Code Advisory Group	R-FCH	reverse fundamental channel
POID	personal online identification	RFCOMM	radio frequency communications
PON	passive optical network	RFD	reserve a fixed duration
POS	packets over SONET; point-of-sale	RFID	radio frequency identification
PPCH	packet paging channel, uplink	RFM	RF module
PPM	prediction by partial match; pulse	RIAA	Recording Industry Association of
	position modulation		America
PPP	point to point protocol	RLC	radio link control

RLP	radio link protocol		SATATM	satellite ATM
RLS	recursive least squared		SC	selection combiner; single-carrier
rms	root mean square		SCCP	signaling connection control part
RNC	radio network controller		SCH	synchronization channel
RNG	relative neighborhood graph		SCL	serial clock line
RNG-REQ	ranging request		SCN	single-hop cellular network
RNG-RSP	ranging response		SCO	synchronous connection-oriented
RNL	radio network layer		SCR	sustainable cell rate
RNS	radio network subsystem		SCTP	stream control transmission protocol
RNSAP	RNS application part		SDA	serial data address
ROADM	reconfigurable optical add/drop multiplexer		SDCCH	stand alone dedicated control channel
ROF	radio over fiber		SDH	synchronous digital hierarchy
ROHC	robust header compression		SDK	software development kit
RP	radio port; reference point		SDMA	space division multiple access
RPE-LTP	residual excited linear predictive speech coding		SDP	service discovery protocol; session description protocol
RPG	role-playing games		SDR	software defined radio
RPS	reference picture selection		SDU	service data unit
RR	resource reservation		SER	symbol error rate
RREP	request response; route reply		SFH	slow frequency-hopped
RREQ	route request		SGSN	serving GPRS support node
RRH	remote radio head		SH	supervisor host
RRM	radio resource management		SHA–1	secure hashing algorithm 1
RS	Reed–Solomon; relaying station		SHR	self-healing ring
R-SCH	reverse supplemental channel		SIFS	short interframe space
RSN	robust security network		SIM	subscriber identity module
RSS	received signal strength; rectangular slice sub-mode		SINR	signal to interference plus noise ratio
RSSI	received signal strength indication		SIP	session initiation protocol
RSVP	resource reservation protocol		SIR	signal to interference ratio
RTC	real time control		SIS	strategic information systems
RTCP	real time control protocol		SISO	single-input single-output
R-TDMA	reservation-time division multiple access		SKA	shared key authentication
RTP	real time protocol		SLA	service level agreement
RTS	ready to transmit; request to send		SLE	static lightpath establishment
RTSP	real time streaming protocol		SLM	session layer mobility management
RTT	radio transmission technique		SLoP	spatial location protocol
rt-VBR	real-time variable bit rate		SM	software management
R–UIM	removable user identity module		SMDS	switched multimegabit data service
RVC	reverse voice channel		SMIL	synchronized multimedia integration language
RVCT	real view compilation tools		SMPTE	Society of Motion Picture and Television Engineers
RVLC	reversible variable length code		SMR	specialized mobile radio
RWA	routing and wavelength assignment		SMS	short message service
SA	service adaptation; simulated annealing algorithm		SMT	Steiner minimum tree
SAAL	signaling ATM adaptation layer		SNA	systems network architecture
SAAL-NNI	signaling ATM adaptation layer for network-to-network interface		SNDCP	subnetwork dependent convergence protocol
SACCH	slow associated control channel		SNI	subscriber network interface
SAN	storage area network		SNIR	signal to noise plus interference ratio
SAP	session announcement protocol		SNMP	simple network management protocol
S-ART	service-motivated adaptive rate transmission		SNR	signal-to-noise ratio
			SOA	service-oriented architecture
			SOAP	simple object access protocol [obsolete]
			SoC	system-on-a-chip
SAT	supervisory audio tone		SOLA	simple overlap add

SOMA	secure and open mobile agent	TEK	traffic encryption key
SONET	synchronous optical network	TELEREP	telecom research program
SP	shortest path	TETRA	terrestrial trunked radio
SPH	shortest path heuristic	TFSP	time–frequency signal processing
SPIHT	set partitioning into hierarchical trees	TFTP	trivial file transfer protocol
SP-MIDI	scalable polyphony MIDI	TH	time hopping
SPP	serial port profile	TIA	Telecommunications Industry Association
SPS	standard positioning service		
SQR	signal-to-quantization noise ratio	TIMIP	terminal independent mobility for IP
SR	signaling rate	TKIP	temporal key integrity protocol
SRBP	signaling radio burst protocol	TM	transport module
SRMC	source routing mobile circuit	TMSI	temporary mobile subscriber identity
SRNS	serving radio network subsystem	TNCP	transport network control plane
SS	signal strength; spread spectrum; subscriber station	TNL	transport network layer
		TOA	time of arrival
SS7	signaling system 7	TORA	temporally-ordered routing algorithm
SSCF	service-specific coordination function	TP	telepoint
SSCOP	service-specific connection oriented protocol	TPC	transmission power control
		TPSN	time-sync protocol for sensor networks
SSD	shared secret data	TS	time slot
SSID	service set identifier	TSL	transport layer security
SSMA	spread spectrum multiple access	TSM	time-scale modification
ST	signaling tone	TSN	transitional security network
STB	set-top box	T-TCM	turbo trellis coded modulation
STEM	sparse topology and energy management	TTDD	two tier data dissemination
		TTI	transport time interval
STFD	spatial time–frequency distribution	TTL	time-to-live
STP	signaling transfer point	TWME	two-way message exchanging
STS	space–time spreading; synchronous transport signal	UA	user agent
		UAProf	user-agent profile
S-UMTS	satellite-universal mobile telecommunications system	UBR	unspecified bit rate
		UCAN	unified cellular and ad-hoc network
SVD	singular value decomposition	UCD	uplink channel descriptor
SVG	scalable vector graphics	UDDI	universal description, discovery and integration
SYN	synchronization		
SYNC	synchronization	UDP	user datagram protocol
TACS	total access communication system	UDPCP	UDP communication protocol
TAG	tell-and-go	UE	user equipment
TBF	temporary buffer flow	UEA	UMTS encryption algorithm
TC	transmission convergence; turbo coding	UEP	unequal error protection
		UHF	ultra high frequency
TCAP	transaction capability application part	UIUC	uplink interval usage code
TCH	traffic channel	UL	uplink
TCM	trellis coded modulation	UMA	unlicensed mobile access
TCP	Transmission Control Protocol	UMP	user mobility pattern
TCP/IP	transmission control protocol/Internet protocol	UMTS	universal mobile telecommunications system
TDD	time division duplexing	UNI	user network interface
TDM	time division multiplex	UNII	unlicensed national information infrastructure
TDMA	time division multiple access		
TDOA	time difference of arrival	UPC	usage parameter control
TD-SCDMA	time-division synchronous code division multiple access	UPSR	unidirectional path-switched ring
		UPT	universal personal telecommunication
TDTD	time delay transmit diversity	URI	uniform resource identifiers
TEEN	threshold sensitive energy efficient sensor network	USA	user services assistant
		USHR	unidirectional self-healing ring

UTC	universal coordinated time	WAND	Magic Wireless ATM Network Demonstrator [European Union]
UTDoA	uplink time difference of arrival		
UTRA	UMTS terrestrial radio access	WANET	wireless ad hoc network
UTRAN	UMTS terrestrial radio access network	WAP	wireless application protocol
UVE	user virtual environment	WARM	wireless ad hoc real-time multicast
UWB	ultra-wideband	WATM	wireless asynchronous transfer mode
VBR	variable bit-rate	WBAN	wireless body area network
VC	virtual channel; virtual circuit; virtual container	WB-CDMA	wideband-code division multiple access
		WCDMA	wideband code division multiple access
VCAT	virtual concatenation	WCDS	weakly connected dominating set
VCC	virtual channel connection	WDM	wavelength division multiplexing
VCI	virtual channel identifier	WEP	wired equivalent privacy
VCO	voltage-control oscillator	Wi-Fi	wireless fidelity
VDO	voluntary dropout	WIM	WAP identity module
VDSL	very high data rate digital subscriber line	WiMAX	worldwide interoperability for microwave access
VGSN	virtual GPRS support node		
VHF	very high frequency	WLAN	wireless local area network
VHO	vertical handover	WLL	wireless local loop
VIP	virtual IP	WML	wireless markup language
VLAN	virtual LAN	WNIC	wireless network interface card
VLC	variable length coding	WPA	WiFi protected access
VLR	visitor location register	WPAN	wireless personal area network
VMAC	voice mobile attenuation code	WSDL	web service description language
VoD	video-on-demand	WSN	wireless sensor network
VoIP	voice over IP	WSP	wireless service provider
VPM	voice privacy mask	WUF	weight-updating function
VPN	virtual private network	WVS	Wigner–Ville spectrum
VS	virtual source	WWW	World Wide Web
VSELP	vector-sum excited linear prediction	xHTML	extensible hypertext markup language
WADM	wavelength add/drop multiplexer	XML	extensible mark up language
WAN	wide area network	ZRP	zone routing protocol

About the Editor-in-Chief

Borko Furht is a professor and chairman of the Department of Computer Science and Engineering at Florida Atlantic University (FAU) in Boca Raton, Florida. He also serves at FAU as Senior Assistant Vice President for Technology and Innovations. Before joining FAU, he was a vice president of research and a senior director of development at Modcomp (Ft. Lauderdale), a computer company of Daimler Benz, Germany, an associate professor at University of Miami in Coral Gables, Florida, and a senior researcher in the Institute Boris Kidric-Vinca, Yugoslavia. Professor Furht received his Ph.D. degree in electrical and computer engineering from the University of Belgrade. His current research is in multimedia systems and communications, video coding and compression, 3D video and image systems, video databases, wireless multimedia, and Internet computing. He is presently Principal Investigator and Co-Principal Investigator of several multiyear, multimillion dollar projects—on Coastline Security Technologies, funded by the Office on Naval Research, One Pass to Production, funded by Motorola, and High-Performance Computing, funded by NSF. He is the author of numerous books and articles in the areas of multimedia, communications, computer architecture, real-time computing, and operating systems. He is a founder and editor-in-chief of *the Journal of Multimedia Tools and Applications* (Springer). He has received several technical and publishing awards, and has consulted for many high-tech companies including IBM, Hewlett-Packard, Cisco, Xerox, General Electric, JPL, NASA, Honeywell, and RCA. He has also served as a consultant to various colleges and universities. He has given many invited talks, keynote lectures, seminars, and tutorials. He served on the Board of Directors of several high-tech companies.

Encyclopedia of

Wireless and Mobile Communications

Volume 3

Pages 1095 through 1598
Spectral–Wireless Trans

Spectral–
Telemet

Ultra–
Wi-Fi

Wireless
Ad Hoc–Asynch

Wireless
Channels–Internet

Wireless
LANs–Mesh

Wireless
Multi–Networks

WSNs–
WSNs: Optimiz

WSNs: Routing–
Wireless Trans

Spectral Efficiency

Saso Tomazic

Faculty of Electrical Engineering, University of Ljubljana, Ljubljana, Slovenia

Abstract

Spectral efficiency is a measure of the performance of channel coding methods. It refers to the ability of a given channel encoding method to utilize bandwidth efficiently. It is defined as the average number of bits per unit of time (bit rate) that can be transmitted per unit of bandwidth (bits per second per Hertz).

The term channel encoding refers to the procedure of mapping a bit stream to an analog signal, which can be transferred through a physical channel and later, after reception, decoded to yield the original bit stream, as shown in Fig. 1.

The received signal may be distorted in various ways and corrupted by noise and other interference on the channel. This can cause the bit stream at the output of the channel decoder to be different from the bit stream at the input of the channel encoder. The ratio of erroneously received bits to all transmitted bits is called the BER and is used as a measure of performance. A small BER is acceptable in practical systems. The value of acceptable BER varies from 10^{-10} to 10^{-1} depending on the system concerned.

The transmitted analog signal occupies a certain bandwidth (frequency range), which depends on the bit rate and on the way that the binary sequence is mapped to the continuous signal, i.e., on the channel encoding method. The ratio of bit rate to bandwidth is called spectral efficiency. Spectral efficiency cannot be made arbitrarily large. The maximal spectral efficiency depends on the power of the transmitted signal, on the acceptable BER, and on the characteristics of the channel (distortion, noise, interference). The coding method that achieves maximal spectral efficiency on a given channel is considered optimal for this channel.

THEORETICAL BOUNDS

The upper bound of the bit rate r_b over a noisy bandlimited channel for error-free transmission was first stated by Shannon.[1] One way to express this bound is:

$$r_b \leqslant \int_{f_l}^{f_u} C(f)\, df \tag{1}$$

where f_l and f_u are the lower and upper frequency bounds respectively, and $C(f)$ is the frequency-dependant capacity of the channel (bits per second per Hertz).

The performance of a channel coding method can be evaluated by comparing the bit rate obtained by that method to the Shannon bound. However, no method for determining $C(f)$ of an arbitrary channel is known. It can only be determined for a small number of special cases, for a channel with additive Gaussian noise, e.g, where it can be expressed as:

$$C(f) = \log_2(1 + \mathrm{SNR}(f)) \tag{2}$$

where $\mathrm{SNR}(f)$ is the frequency-dependant signal-to-noise ratio at the receiver input. Even in this case it can be hard or even impossible to determine $\mathrm{SNR}(f)$ at the receiver input when the channel is not linear.

Thus, in general one cannot tell how close to the Shannon bound different coding methods are. The performance of different channel coding methods is usually evaluated on an additive white Gaussian noise (AWGN) channel. On an AWGN channel, $\mathrm{SNR}(f)$ is constant and Eq. 1 can be rewritten as:

$$r_b \leqslant B \log_2\left(1 + \frac{S}{N}\right) \tag{3}$$

where B is the total bandwidth needed for transmission, S is the total signal power and N is the total noise power over bandwidth B. The above inequality is also known as the Shannon-Hartley theorem.

Signal power S is equal to the average bit energy E_b multiplied by the bit rate r_b:

$$S = E_b r_b \tag{4}$$

while noise power N over bandwidth B can be expressed as:

$$N = N_0 B \tag{5}$$

where N_0 is the noise level, i.e., the one-sided power spectral density of AWGN.

Encyclopedia of Wireless and Mobile Communications DOI: 10.1081/E-EWMC-120043448

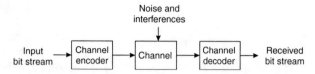

Fig. 1 Channel coding. The input bit stream is mapped to an analog signal (channel encoding) transferred through a channel with interference and decoded at the receiver (channel decoding). Ideally the received bit stream would be identical to the input bit stream.

Spectral efficiency u is given by the ratio:

$$u = \frac{r_b}{B} \tag{6}$$

Substituting Eqs. 4–6 into Eq. 3, after rearrangement, yields:

$$u \leqslant \log_2\left(1 + u\frac{E_b}{N_0}\right) \tag{7}$$

and

$$\frac{E_b}{N_0} \geqslant \frac{2^u - 1}{u} \tag{8}$$

The theoretically minimal ratio E_b/N_0 needed for error-free transmission over an AWGN channel is obtained when equality holds in the above expression. Denoting with BENR (bit energy to noise level ratio) the minimal E_b/N_0 in decibels, we can write:

$$BENR = 10\log_{10}\left(\frac{2^u - 1}{u}\right)B3u - 10\log_{10}(u) \tag{9}$$

BENR as a function of spectral efficiency u is shown in Fig. 2. Note that the BENR increases almost linearly with spectral efficiency u. By assigning more bandwidth to the transmission, the spectral efficiency drops and less power is needed for transmission. The minimal BENR i.e., the minimum energy required to transmit one bit of information over an AWGN channel, is obtained when the entire frequency range is assigned to the transmission. In this case, spectral efficiency approaches 0 and BENR approaches −1.592 dB, which is also known as the Shannon bound for an AWGN channel.

On the other hand, to increase spectral efficiency the power of the transmitter must be increased. To transmit one bit per second per Hertz ($u = 1$) the BENR is 0 dB, which means that the bit energy E_b must be at least equal to the noise level N_0. Further increasing the spectral efficiency can be very expensive in terms of transmitted power. To achieve a spectral efficiency of 57 Kbps for dial-up modems (u B 18.5), the BENR must already be more than

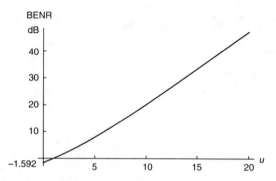

Fig. 2 The bit energy to noise level ratio (BENR) on an additive white Gaussian noise (AWGN) channel as a function of spectral efficiency u. To increase spectral efficiency, more energy is needed to transfer each bit, which also implies more transmitter power. The minimal BENR is −1.592 dB, which is also known as the Shannon bound.

43 dB, which means that the average bit energy must be approximately 20,000 times greater than the noise level.

The above results hold for an AWGN channel only; however, they are often used to estimate the bounds of spectral efficiency on other channels when the exact bounds and/or exact channel characteristics are not known.

CODING AND MODULATION

For practical purposes and also to simplify design and analysis of transmission systems, the process of channel coding (encoding/decoding) is usually divided into two parts: coding (in a narrow sense of the word) and modulation. It is important to note that within the scope of this article, the term encoding is used for mapping a bit stream to a real-valued symbol stream and the term modulation is used for mapping the symbol stream to an analog signal. Similarly, de-modulation maps the analog signal to a symbol stream and decoding maps the symbol stream to a bit stream. Other definitions of coding and modulation can be found in other sources. The divided system is shown in Fig. 3.

The input bit stream $b[n]$ at bit rate r_b is first encoded to a real valued symbol stream $s[m]$ at symbol rate r_s. The modulator maps the symbols $s[m]$ to an analog signal $s(t)$, which is then transmitted through the channel.

Speaking of spectral efficiency implicitly assumes band-limited modulation. If the modulation was not band-limited, spectral efficiency would be zero. When the channel is non-linear it produces out-of-band frequency components. These components can also carry information. Since we assume that the transmission is band-limited, the information transmitted out-of-band should be irrelevant for the receiver. To ensure that no information

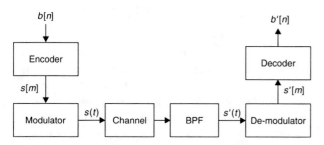

Fig. 3 Division of channel coding into coding and modulation. The input bit stream $b[n]$ is first encoded to a symbol stream $s[m]$ and then modulated to an analog signal $s(t)$. The received signal $s'(t)$ is de-modulated to a received symbol stream $s'[m]$ and then decoded to a received bit stream $b'[n]$. Ideally $b[n]$ and $b'[n]$ would be identical. BPF, band-pass filter.

is carried out of the transmission band, an ideal band pass filter (BPF) of bandwidth B is included at the output of the channel in front of the receiver. The band-limited signal $s'(t)$ at the receiver input is first de-modulated to the received symbol stream $s'[m]$ and then decoded to $b'[n]$. Ideally the received bit stream $b'[n]$ would be identical to the input bit stream $b[n]$. In practice a small number of errors is tolerated as long as BER does not exceed an acceptable level.

The purpose of encoding is twofold: to add redundancy (a dependency among the symbols at the encoder output) and to establish proper levels (values of the symbols) for modulation. The purpose of modulation is to map the values of the symbols to some properties of the analog signal in such a way that the symbol stream can be recovered from the received signal at the receiver.

Although Shannon determined the bounds for error-free transmission, he did not indicate how to achieve them. Different coding (encoding/decoding) techniques are in use to approach these bounds. Redundancy is added in the encoding process. It is used in the decoder to detect errors (error-detecting codes), to correct errors (error-correcting codes), or to minimize the probability of error (forward error correction codes). To approach the bound on a given channel, the code should also be adapted to the channel. Different adaptive techniques, such as adaptive equalization, adaptive inter-symbol interference canceling, adaptive filtering, and others are used for this purpose. There is no single optimal coding method for all channels, nor is any method currently known for determining the optimal code for an arbitrary channel.

The modulator encodes the values of the symbols at its input by varying different properties of the harmonic signal (the carrier) at its output: amplitude (AM, amplitude shift keying—ASK), frequency (FM, frequency shift keying—FSK), phase (PM, phase shift keying—PSK), or amplitude and phase at the same time (quadrature amplitude modulation—QAM). The spectral content of the modulated signal is centered around the frequency of the carrier. Its

bandwidth greatly depends on both encoding and modulation. In general, the bandwidth of frequency- and phase-modulated signals is greater than the bandwidth of amplitude-modulated signals. Sometimes, when the noise is not white and/or the channel characteristics are not flat, it is beneficial to modulate multiple carriers (multiple carrier modulation—MCM and orthogonal frequency division modulation—OFDM). Modulated signals with more bandwidth are usually more resistant to noise and other interference, thus robustness is obtained at the expense of lower spectral efficiency.

Spectral efficiency depends on coding and modulation, on one hand, and also depends on channel characteristics on the other. Thus, another way to improve spectral efficiency is to improve the channel characteristics. In wireless communications this can be done by using directional antennas at the transmitter and/or at the receiver. Another method is to use multiple antennas at both sides in so-called multiple input multiple output (MIMO) systems.

OPTIMAL BAND-LIMITED MODULATION

As stated in the previous section, no method is known for determining the optimal channel coding for an arbitrary channel. We also mentioned that spectral efficiency depends on coding and modulation. The latter has led to different attempts to discover a new modulation technique which would improve on all currently known methods in terms of spectral efficiency, e.g., FQPSK (Feher-patented quadrature phase shift keying),[2] VPSK (variable phase shift keying),[3] VMSK (very minimum shift keying)[4] and many other ultra-narrow band modulations. A simple proof that ultra-narrow band modulations may not be as spectrally efficient as claimed is given in Ref. 5.

Although there is no encoding method that would be optimal for every channel, this does not hold for band-limited modulation. If we accept the definition of modulation from the previous section, then QAM at symbol rate $r_s = 2B$ is optimal, at least from a theoretical point of view.

To be more specific: any band-limited modulation can be performed using a QAM modulator preceded by the appropriate encoder. In other words, the signal space at the output of the QAM modulator covers all possible band-limited signals. If it holds that any band-limited modulation can be performed using a QAM modulator, then it also holds that optimal band-limited modulation can be performed as well, which makes QAM itself optimal. An ideal QAM modulator is shown in Fig. 4.

The input symbol stream is split into two half symbol rate symbol streams that are used to modulate the amplitudes of two orthogonal carriers. The modulated carriers are summed up to form the output signal of the modulator. At the receiver the procedure is repeated in reverse order. The received analog signal is first multiplied by two

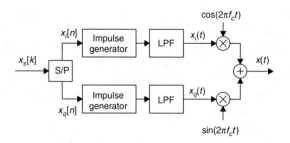

Fig. 4 Ideal QAM modulator. The input symbol stream $x_s[k]$ is split into two symbol streams $x_i[n]$ and $x_q[n]$ at half the symbol rate of the input symbol stream. The ideal Dirac impulses generated at the impulse generators are shaped at the ideal LPFs of bandwidth $B/2$. The in-phase signal $x_i(t)$ and $x_q(t)$ the quadrature signal are multiplied by two orthogonal carriers $\cos(2\pi f_c t)$ and $\sin(2\pi f_c t)$, respectively, and summed up to yield the output signal $s(t)$. S/P, serial-to-parallel converter.

coherent orthogonal carriers, one in each branch of the de-modulator. High frequencies are filtered out by two ideal LPFs and the signals at the output of the filters are sampled at half the symbol rate. The obtained symbol streams are combined into a single symbol stream at the output of the de-modulator. The QAM de-modulator is shown in Fig. 5.

To verify that QAM modulation is optimal, we should first recognize that the QAM de-modulator in Fig. 5 can be used as a band-limited signal sampler. The signals $x_i(t)$ and $x_q(t)$ are base-band signals with bandwidth $B/2$. If the sampling rate of both samplers is B samples per second, then the signals $x_i(t)$ and $x_q(t)$ can be perfectly re-constructed from the samples $x_i[n]$ and $x_q[n]$ respectively. This is known as the sampling theorem. A proof can be found in almost any elementary text on communications theory (see, e.g., Refs. 5, 6, or 7).

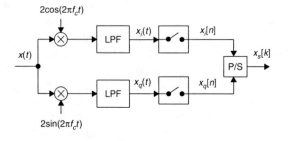

Fig. 5 QAM de-modulator. The input signal $x(t)$ is fed into two branches of the de-modulator. In one branch it is multiplied by $2\cos(2\pi f_c t)$ and by $2\sin(2\pi f_c t)$ in the other. The signals in both branches are then filtered with ideal LPFs and sampled at half the symbol rate. The symbol streams $x_i[n]$ and $x_q[n]$ at the output of the samplers are then combined in the P/S to yield the received symbol stream $x_s[k]$.

From the above it is then obvious that $x(t)$ can be perfectly re-constructed from $x_s[k]$ as well. The QAM modulator in Fig. 4 can be used for re-construction. The signals $x_i(t)$ and $x_q(t)$ are re-constructed from the samples $x_i[n]$ and $x_q[n]$ on ideal LPFs and then multiplied by orthogonal carriers and summed up to yield the original signal $x(t)$. The QAM de-modulator and QAM modulator at symbol rate $r_s = 2B$ can thus used for sampling and perfect re-construction of any band-limited signal with bandwidth less than or equal to B.

Without any change in performance, except for additional delay, we can add sampling with perfect re-construction at both sides of the transition system from Fig. 3, as shown in Fig. 6.

Suppose now that the original encoder and modulator in Fig. 6 were optimal for a given channel with regard to some criterion, e.g., maximal spectral efficiency. Since the performance of the system was not altered by the addition of QAM modulator/de-modulator pairs, the system is still optimal. In this new system the original encoder, original modulator, and QAM de-modulator form a new encoder (dotted box at the left-hand side of Fig. 6) which maps the input bit stream $b[n]$ to a symbol stream $s_s[k]$. The QAM modulator is then used for modulation. At the receiver the QAM de-modulator is used for de-modulation, and a new decoder, which includes the QAM modulator, original de-modulator and original decoder (dotted box at the

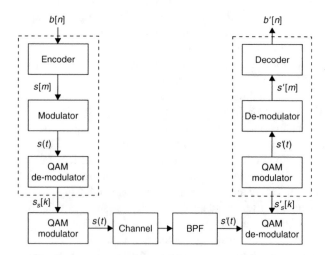

Fig. 6 Modified transmission system. The QAM de-modulator and QAM modulator are added at both sides. QAM de-modulators acts as samplers and QAM modulators perform perfect re-construction, thus the performance of the system is not altered by these additions, except for the delay of the filters. The original encoder, original modulator and QAM de-modulator form a new encoder, which encodes the bit stream $b[n]$ to the symbol stream $s_s[k]$. The QAM modulator, original de-modulator and original decoders form a new decoder, which decodes $s'_s[k]$ to $b'[n]$. BPF, band pass filter.

right-hand side of Fig. 6), is used to decode the received symbol stream $s'_s[k]$ to the received bit stream $b[n]$.

We suppose that the whole system is optimal. The modulation method used in this system is QAM. We can thus conclude that QAM at symbol rate $r_s = 2B$ is an optimal modulation method. It is important to note that the new encoder must be adapted to the channel characteristics; however, this does not hold for the QAM modulator. The same modulator can be used on any channel.

This result is important mainly from a theoretical point of view. We can only adapt to the characteristics of the given channel (e.g., mobile channel) through a suitable choice of coding, which indicates that research should be focused on finding optimal coding methods instead of being focused on discovering new modulation techniques.

Although, at least theoretically, any band-limited modulation can be implemented using QAM, this may not be the most practical solution. Other modulation methods may have practical advantages and thus will continue to be used in transmission systems.

SUMMARY

Spectral efficiency is an important measure of the performance of a digital transmission system, especially in wireless communications. The upper bound of spectral efficiency depends on the channel characteristics and is not known for all channels. To improve spectral efficiency, additional energy is needed to transmit one bit of information and/or channel coding must be adapted to the channel. If channel coding is split into coding and modulation, only the coding needs adaptation to the channel and QAM can be used for modulation. Other modulation methods can have practical advantages and will continue to be used in digital transmission systems.

REFERENCES

1. Shannon, C.E. Communication in presence of noise. IRE. **1949**, *37* (1), 10–21.
2. Feher, K. et al., Feher's quadrature phase shift keying (FQPSK). US Patents 5,784,402 and 5,491,457.
3. Walker, H.R. VPSK and VMSK modulation transmits audio and video at 15 b/s/Hz. IEEE Trans. Broadcast. **1997**, *43* (1), 96–103.
4. Walker, H.R.; Stryzak, B.; Walker, M.L. Attain high bandwidth efficiency with VMSK modulation. Microw. RF. **1997**, *36* (13), 173–186.
5. Tomazic, S. Comments on spectral efficiency of VMSK. IEEE Trans. Broadcast. **2002**, *48* (1), 61–62.
6. Papoulis, A. *The Fourier Integral and its Applications*; McGraw-Hill Book Co.: New York, 1962.
7. Proakis, G.; Salehi, M. *Communications Systems Engineering*; McGraw-Hill: MI, 2001.

Spectral Efficiency: Wireless Multicarrier Communications

Alexander M. Wyglinski
Department of Electrical and Computer Engineering, Worcester Polytechnic Institute, Worcester, Massachusetts, U.S.A.

Fabrice Labeau
Electrical and Computer Engineering Department, McGill University, Montreal, Quebec, Canada

Abstract

Adaptive allocation of multicarrier transceivers is a process by which the sub-carrier parameters are tailored to the operating conditions of the transmission environment.

INTRODUCTION

There currently exists a dichotomy between wireline and wireless communication networks. Whereas the data throughput supported by wireline networks is enormous, achieved using technologies such as fiber optic cables, the base station/mobile user links found in many wireless networks are constantly trying to keep up with the increasing demand for greater throughput. There are several significant restrictions when wireless modems transmit at high data rates, with the first being bandwidth usage. Since the spectrum that wireless systems use to transmit data is regulated by government agencies, such as the FCC in the United States, each licensed and unlicensed operator of a wireless data transmission infrastructure must abide by the established guidelines. This is done in order to avoid interference between different wireless operators. Therefore, the rate is constrained by the maximum bandwidth allotted to the operator. The second constraint is the channel environment that the data transmission system is operating through. At higher data rates, the amount of distortion introduced to the transmission becomes more pronounced, making it more difficult to compensate at the receiver.

To accommodate the growing demand for additional bandwidth from new and existing wireless services, several regulatory agencies have commenced work on redefining a number of frequency bands for use in a *dynamic spectrum access* (DSA) configuration.[1,2] With these DSA techniques, unoccupied portions of licensed spectrum can be temporarily "borrowed" by other unlicensed wireless users. Simultaneously, advances in *software-defined radio* (SDR) technology have resulted in agile and rapidly re-configurable wireless transceivers, which have enabled researchers to investigate advanced techniques for enhancing the performance of digital transmission systems operating in various channel conditions

(e.g., additive white Gaussian noise, multipath fading, impulse noise) that were not physically realizable several years ago.

One of these advanced techniques is *adaptive allocation*, which is employed by multicarrier transceivers in order to support robust, high-speed transmissions across dispersive communication channels. *Multicarrier modulation* (MCM) (currently at the core of most digital subscriber lines (DSLs) systems,[3] wireless LAN (WLAN) modems,[4–6] and wireless MAN (WMAN) transceivers.[7]) is a form of frequency division multiplexing (FDM), where data is transmitted in several narrowband streams at different carrier frequencies. However, unlike conventional FDM systems, where the narrowband sub-carrier signals are separated by frequency-domain guard bands,[8] MCM systems are spectrally efficient by allowing for the overlap of adjacent sub-carriers when a certain set of conditions are satisfied.[9–12]

Despite the advantages of MCM regarding high-speed transmissions, many conventional wireless systems do not fully exploit its potential. Specifically, most systems do not tailor the sub-carrier operating parameters to the channel environment, which may vary across the frequency domain. As a result, the overall error probability of the system is dominated by the error probabilities of the sub-carriers with the worst performance.[13] To solve this problem, adaptive allocation can be employed, which is a multicarrier transceiver optimization process designed to distribute resources so that the performance is enhanced while satisfying some constraints. Resources that are commonly allocated to the different sub-carriers are bits, i.e., through the choice of signal constellation, and power, i.e., transmit power levels. The techniques for adaptive allocation have originated from other areas, including financial analysis[14] and quantizer design.[15,16] However, a number of adaptive allocation algorithms have been developed and

Encyclopedia of Wireless and Mobile Communications DOI: 10.1081/E-EWMC-120043445

implemented for several data transmission systems, including DSL modems.[17–19]

Although the concept of adaptive allocation is relatively simple, devising an algorithm that can perform adaptive allocation which quickly provides a solution in a low complexity manner that is close to the optimal allocation is a challenge. There have been numerous proposed algorithms that attempt to solve for an allocation quickly and efficiently. Most of these solutions can be classified into several categories based on how they perform the search for the allocation. In this entry, several of these categories are presented within the context of wireless communication systems, with the algorithms belonging to each of these categories being evaluated using computer simulations and compared against each other in terms of throughput performance and error robustness. Simulations are performed when the algorithms possess either perfect or imperfect knowledge of the channel conditions.

The rest of this entry is organized as follows: the second section presents an overview of MCM transmission. A general description of adaptive allocation for MCM transceivers is covered in the third section, while three categories of algorithms are covered in the fourth, fifth, and sixth sections. Several models for imperfect channel estimates are discussed in the seventh section. Finally, simulation results for several algorithms are presented and compared in the eighth section, followed by some concluding remarks in the final section.

MCM OVERVIEW

The primary advantage of MCM is its ability to operate in a "divide-and-conquer" approach: by transmitting the data across the channel at a lower data rate in several frequency sub-carriers, the process of distortion compensation can be made simpler by treating each sub-carrier separately. From a time-domain perspective, this translates the wideband transmission system in a collection of parallel narrowband transmission systems each operating at a lower data rate.[8] From the frequency-domain perspective, MCM transforms the frequency-selective channel, i.e., non-flat spectrum across the frequency band of interest, into a collection of approximately flat sub-channels that the data gets transmitted over in parallel. Thus, MCM has become the technology of choice to combat frequency-selective fading channels.

A generic single input/single output MCM transceiver is shown in Fig. 1. A high-speed input data stream, $x(n)$, is parsed into N relatively slower streams and modulated using a prescribed signal constellation. The modulated streams, $d^{(k)}(n)$, $k = 0, \ldots, N - 1$, are then upsampled by a factor N, yielding the signals $y^{(k)}(n)$, $k = 0, \ldots, N - 1$. They are then filtered by a bank of synthesis filters, $g^{(k)}(n)$, $k = 0, \ldots, N - 1$, and the filtered signals are summed together to form the composite transmit

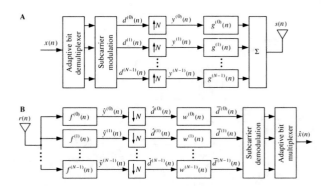

Fig. 1 Schematic of a generic single input/single output *multicarrier modulation* (MCM) system. (A) Single input/single output MCM transmitter. (B) Single input/single output MCM reciever.

signal, $s(n)$, namely:

$$s(n) = \sum_{k=0}^{N-1} \sum_{l=-\infty}^{\infty} g^{(k)}(l) y^{(k)}(n - l) \tag{1}$$

Between the transmitter and receiver lies a channel which introduces both noise (due to thermal excitation of the RF chain and antennas, atmospheric conditions, and interference from artificial and natural sources) and distortion (mainly due to multipath propagation) to the composite transmit signal. In general, the channel can be modeled as a FIR filter that possesses a frequency-selective fading characteristic. As a result, when $s(n)$ passes through the channel, the channel attenuates the spectrum of $s(n)$ non-uniformly in frequency.

At the receiver, a bank of analysis filters, $f^{(k)}(n)$, $k = 0, \ldots, N - 1$, are employed to separate the sub-carriers out of the received composite signal, $r(n)$, into N individual signals, $y^{(k)}$, $k = 0, \ldots, N - 1$. These signals are then downsampled by a factor of N, yielding $\hat{d}^{(k)}(n)$, $k = 0, \ldots, N - 1$. To remove the distortion introduced by the channel, equalizers $w^{(k)}(n)$, $k = 0, \ldots, N - 1$, are employed on a per-sub-carrier basis (although linear per-sub-carrier equalizers have been employed in Fig. 2, decision-feedback equalizers on each sub-carrier[20] or per-sub-carrier Tomlinson–Harashima precoding schemes[21–23] can also be used.). The outputs of the sub-carrier equalizers, $\bar{d}^{(k)}(n)$, $k = 0, \ldots, N - 1$, are then de-modulated and the resulting binary sequences combined using a multiplexer, yielding the reconstructed high-speed output $\hat{x}(n)$.[24,17,18]

Orthogonal FDM

An efficient form of MCM with respect to hardware implementation is *orthogonal FDM* (OFDM),[25–27] where DFT basis functions are employed to create the synthesis and analysis filterbanks (the DFT basis functions can be efficiently implemented using the FFT.). The filters in the filterbanks are *odd-stacked*, which means they are

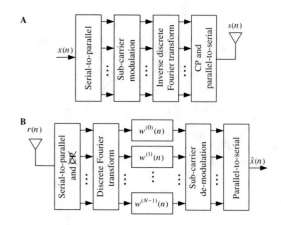

Fig. 2 Schematic of an OFDM system employing a CP. (A) OFDM transmitter with CP adder. (B) OFDM receiver with CP remover.

uniformly distributed throughout the frequency domain, with one filter centered at $\omega_0 = 0$ rad/s. Although OFDM systems could be implemented using a bank of sinusoid generators,[28] practical implementations employ the FFT and IFFT, which results in a significant complexity reduction.[25] As a result, OFDM has become a popular choice in many multicarrier applications, including digital audio broadcast (DAB), DSL, digital video broadcast (DVB), WLANs, e.g., IEEE 802.11a/g, European Telecommunications Standards Institute (ETSI) High performance radio local area network (HiperLAN/2), and MMAC high-speed wireless local area network (HiSWAN), and WMAN, e.g., IEEE 802.16.

A schematic of an OFDM transceiver is shown in Fig. 2. A high-speed input stream $x(n)$ is first de-multiplexed into N data streams, $x^{(k)}(n)$, $k = 0, \ldots, N-1$, using a serial-to-parallel converter, where $x^{(k)}(n)$ is the sub-carrier data for sub-carrier k. These streams are then individually modulated using multi-level quadrature amplitude modulation (M-QAM) constellations, to yield $y^{(k)}(n)$, $k = 0, \ldots, N-1$, where $y^{(k)}(n)$ is the M-QAM-modulated sub-carrier data for sub-carrier k. The IDFT is then applied to the sub-carriers, defined as:[29]

$$s^{(l)}(n) = \frac{1}{N} \sum_{k=0}^{N-1} y^{(k)}(n) e^{j2\pi kl/N} \tag{2}$$

where $l = 0, \ldots, N-1$, resulting in the sub-carriers being modulated to one of N evenly spaced center frequencies in the range $[0, 2\pi]$.

Since OFDM employs the DFT and its inverse, the filters applied to the sub-carriers have a low stopband attenuation since the frequency response of the filters are of the form $\text{sinc}(\omega x)$. Therefore, the performance of the OFDM system would significantly decrease if it were operating in a time-dispersive environment. To counteract the

time-dispersiveness of the channel, a cyclic extension is employed either before the symbol (i.e., cyclic prefix) or after it (i.e., cyclic suffix) to capture this effect. Without loss of generality, the system will add a cyclic prefix to the OFDM symbol prior to forming the composite transmit signal, $s(n)$.

At the receiver, the cyclic prefix is removed from the received composite signal, $r(n)$, and converted from a serial stream into a collection of parallel streams using a serial-to-parallel converter. The DFT is applied[29]

$$\hat{y}^{(k)}(n) = \sum_{l=0}^{N-1} r^{(l)}(n) e^{-j2\pi kl/N} \tag{3}$$

for $k = 0, \ldots, N-1$, where $r^{(k)}(n)$, $k = 0, \ldots, N-1$, are the parallel input streams to the DFT. The sub-carriers are then equalized with $w^{(k)}$, $k = 0, \ldots, N-1$, to compensate for the distortion introduced by the channel (any one-tap sub-carrier equalizer will be adequate when the cyclic prefix length is sufficient.). The equalized sub-carriers are then de-modulated before being multiplexed together using the parallel-to-serial converter, forming the output $\hat{x}(n)$.

ADAPTIVE ALLOCATION FOR MULTICARRIER TRANSCEIVERS

As mentioned earlier, multicarrier systems possess a "divide-and-conquer" quality which may be exploited under certain conditions in order to improve system performance. For instance, by sub-dividing a frequency-selective fading channel frequency response into a collection of relatively flat sub-channels, each sub-channel then has a different amount of distortion and a different instantaneous SNR value. Thus, by adapting the sub-carrier operating parameters to each sub-channel, such as the choice of signal constellation and/or power level, the transceiver can be optimized in one of three following ways:[30]

- **Throughput maximization**: Maximize the overall throughput of the system, subject to constraints on the overall transmit power level and error robustness.
- **Power minimization**: Minimize the overall transmit power level of the system, subject to constraints on the overall throughput and error robustness.
- **BER minimization**: Maximize the overall error robustness of the system, subject to constraints on the overall throughput and transmit power.

Specifically, the first category results in a spectrally efficient transmission by attempting to maximize the number of bits per symbol epoch while simultaneously maintaining an acceptable level of error robustness.

In this entry, we focus on *adaptive bit allocation*, also known as *adaptive bit loading* and *adaptive modulation*, where the sub-carrier signal constellations are individually varied across all the sub-carriers in order to achieve a performance objective while satisfying some prescribed constraint(s) (the sub-carrier transmit power levels are assumed to be constant and identical in this entry for the purpose of straightforward comparison. Nevertheless, combining adaptive bit allocation and adaptive power allocation together can achieve performance gains that are greater relative to those of the individual techniques.).

Adaptive Bit Allocation Example

To illustrate how adaptive bit allocation works, let us suppose the channel is sub-divided into N disjoint approximately flat sub-channels with complex gains H_i, $i = 0, \ldots, N - 1$. Furthermore, let the transmit power levels for the sub-carriers be specified as π_i, $i = 0, \ldots, N - 1$. Therefore, if the additive noise is white with variance σ_v^2 and the equalizer at the receiver is a single complex gain per sub-carrier, the SNR of sub-carrier i can be defined by

$$\gamma_i = \frac{\pi_i |H_i|^2}{\sigma_v^2} \tag{4}$$

where $|H_i|^2 \leqslant 1$ is always true due to path loss (For OFDM-type systems with a sufficiently long cyclic prefix, Eq. 4 becomes increasingly accurate as increases. However, for other multicarrier schemes, this approximation may be less accurate if other sources of distortion, such as intersymbol interference (ISI) and intercarrier interference (ICI), are not adequately suppressed.).

Given a set of different modulation schemes, let us suppose that the objective is to determine which scheme possesses the largest throughput for sub-carrier i, given the sub-carrier SNR in Eq. 4, while operating below the predefined error probability threshold P_T. An exhaustive search is performed by evaluating the closed form expressions of the probability of bit error, P_i, for all available signal constellations and sub-carriers. Furthermore, since the case of $|H_i|^2 \neq |H_k|^2$ for $i \neq k$ is very likely, the best choice of signal constellation for sub-carrier i may not be optimal for sub-carrier k and thus this exhaustive search procedure must be applied for each sub-carrier.

An example of bit allocation is shown in Fig. 3 for $N = 8$ sub-carriers. For $\Pi_i = 0.833$ mW, $i = 0, \ldots, N - 1$, and $\sigma_v^2 = 1 \times 10^{-13}$ W, the sub-carrier SNR values are computed using Eq. 4 and shown in Fig. 4. Note that there exists a deep spectral depression in the vicinity of sub-carrier 4, resulting in a relatively low sub-carrier SNR value. If 64-QAM modulation is applied to all sub-carriers, the resulting mean BER is $\bar{P} = 6.442 \times 10^{-4}$ and the overall throughput is 48 bits per symbol epoch. If the threshold P_T for acceptable performance is $P_T = 10^{-5}$,

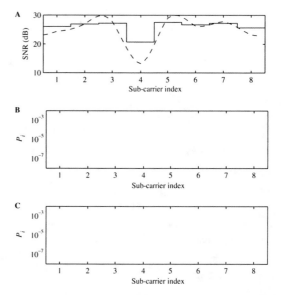

Fig. 3 Example of bit allocation performed on an eight sub-carrier system. (A) Sub-carrier SNR (in dB). Note the channel frequency response profile (dashed line). (B) 64-QAM applied across all sub-carriers. (C) 64-QAM for all sub-carriers except sub-carrier 4, which employs 16-QAM.

then this configuration is unacceptable since $\bar{P} > P_T$. As a result, all sub-carrier signal constellations would need to be reduced uniformly in a *conventional* (several wireless network standards[7] perform a form of adaptive bit allocation, where the sub-carrier signal constellations are all constrained to be equivalent.) MCM system in order to satisfy $\bar{P} \leqslant P_T$. On the other hand, if adaptive bit allocation is employed, the signal constellation of sub-carrier 4 would only be changed to 16-QAM, with 64-QAM still employed by all the other sub-carriers, in order to reduce the impact of sub-carrier 4 on the mean BER (refer to Fig. 4). Observing Fig. 4, all P_i are below P_T and, $\bar{P} = 1.767 \times 10^{-6}$, which satisfies the BER constraint. However, the overall throughput is reduced by 2 bits per symbol epoch.

CAPACITY APPROXIMATION-BASED BIT ALLOCATION

Although the principle of adaptive bit allocation is straightforward, obtaining the appropriate sub-carrier signal constellation configuration could potentially be a complex problem, especially for MCM transceivers employing a large number of sub-carriers. One approach for finding an acceptable sub-carrier configuration is to employ closed-form expressions that solve for the number of bits per sub-carrier per symbol epoch.[31,32] For instance, it has been shown that approximations to the Shannon channel capacity expression can provide solutions for sub-carrier bit allocations.[19,18,33]

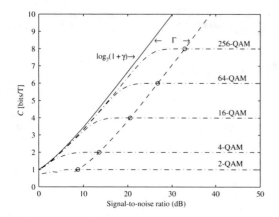

Fig. 4   Channel capacity C and the points for M-QAM given a P_T of 10^{-6} corresponding to an SNR gap Γ of 8.8 dB.

The maximum data rate for error-free transmission, or *capacity*, of a communication system transmitting in an additive white Gaussian noise channel is given by:

$$C = W\log_2(1 + \gamma) \tag{5}$$

where W is the signal bandwidth and γ is the SNR. Practical communication systems can be characterized by how close they are from achieving capacity. The distance between the SNR values for the maximum number of bits the system can sustain, given a target probability of error P_T, and the capacity normalized by the signal bandwidth is called the *SNR gap*, Γ. The maximum number of bits that can be sustained is (with error P_T):

$$b = \log_2\left(1 + \frac{\gamma}{\Gamma}\right) \tag{6}$$

The SNR gap can be expressed using the expression for the union bound on the error probability, yielding:[34]

$$\Gamma \approx \frac{1}{3}\left[Q^{-1}\left(\frac{P_T}{4}\right)\right]^2 \tag{7}$$

where $Q^{-1}(\cdot)$ is the inverse of the Q-function, defined as:

$$Q(x) = \frac{1}{\sqrt{2\pi}}\int_x^\infty e^{-t^2/2}\,\mathrm{d}t \tag{8}$$

An example of the SNR gap is shown in Fig. 4. The SNR gap between the normalized channel capacity and an uncoded M-QAM system operating at a P_T of 10^{-6} for $M \geq 2$, represented by the circles connected by the dashed-dotted line, is approximately 8.8 dB. Also plotted are several normalized capacity curves as computed by Ungerboeck[35] for M-QAM systems operating in bandlimited additive white Gaussian noise channels, with discrete-valued inputs and continuous-valued outputs, and assuming equiprobable occurrences of signal constellation points.

The allocation algorithm of Chow, Cioffi, and Bingham[18] makes use of the SNR gap to compute the number of bits for sub-carrier i, namely

$$b_i = \log_2\left(1 + \frac{\gamma_i}{\Gamma}\right) \tag{9}$$

where γ_i is the SNR of sub-carrier i. Assuming equal energy across all used sub-carriers, Γ is adjusted until the target bit rate is exceeded. For a geometric interpretation, we refer to the dashed line in Fig. 4. The dashed line represents the system operating at P_T. For a sub-carrier with SNR γ_i, this curve maps γ_i to a (non-integer) number of bits, which is rounded to the nearest integer value (the point on the curve is moved either vertically up or down). After the bit allocation, the transmission power levels are then adjusted in order to achieve the same sub-carrier BER, P_i, per non-nulled sub-carrier.

INCREMENTAL BIT ALLOCATION

Although the use of closed-form expressions can potentially achieve sub-carrier bit allocations in a short period of time, they also yield non-integer allocations, which require quantization. Thus, the quantization process may introduce rounding errors, resulting in a bit allocation that may be further from the optimal solution in terms of throughput. Discrete bit allocation is one possible solution to this problem. Although discrete bit allocation algorithms for communication systems have been around since 1987,[36] they have been influenced by discrete allocation algorithms developed in other areas, such as financial analysis[14] and quantization theory.[15]

Many discrete bit allocation algorithms are simply executed in an *incremental* fashion,[36,37] where the bits are distributed in an iterative fashion across the sub-carriers, 1 bit per iteration. Most incremental allocation algorithms can be considered as *greedy* algorithms, where the algorithm allocates 1 bit at a time to the sub-carrier that will do the most good for the allocation at that instant. The algorithm is called greedy since it only maximizes the quantity of interest for the current allocation without regard to the global effects of its choice.[16] Both Campello[38] and Fox[14] defined conditions which yield an optimal allocation (Campello showed for a bit rate maximization problem that the optimal bit allocation exists if the rate-distortion surface across all the sub-carriers is *efficient* and *E-tight*.[38] As for Fox, he showed that the optimal allocation can be determined if the objective function is concave and strictly increasing.[14]).

Although popular, incremental algorithms tend to be computationally expensive. As a result, there exists a need for practical and efficient loading algorithms. For instance, the discrete loading algorithm proposed in Ref. 39 used lookup tables and a fast Lagrange bisection search to determine the final bit and power allocation. Another incremental bit allocation algorithm is shown in

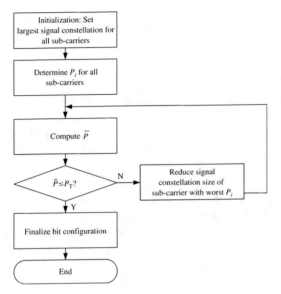

Fig. 5 Flow diagram of an incremental bit allocation algorithm based on a greedy approach.[40–42]

Fig. 5.[40–42] At the transmitter, the system first employs the modulation scheme with the largest signal constellation across all the sub-carriers. This is unlike most of the other incremental allocation algorithms in the literature, which usually initialize the allocation with the smallest modulation schemes. As a result, the algorithm in Fig. 5 has the advantage of fewer iterations in reaching the final allocation for medium to high SNR values. The sub-carrier BER values, P_i, $i = 0, \ldots, N - 1$, are then computed using a priori knowledge of the channel and the initial values of transmit power levels.

Averaging the sub-carrier BER values using:

$$\bar{P} = \frac{\sum_{i=1}^{N} b_i P_i}{\sum_{i=1}^{N} b_i} \qquad (10)$$

the mean BER $\bar{P}$ is compared against a BER threshold P_T. If the mean BER is below the threshold, the system configuration is kept. Otherwise, the signal constellation of the sub-carrier with the worst BER is reduced in size. The rationale behind choosing the worst-performing sub-carrier is that its BER dominates the overall average of the configuration, masking the performance of those sub-carriers with smaller BER values. Then, P_T is recomputed using Eq. 10 and compared against the threshold. This process is repeated until either the threshold is met or until all the sub-carriers are nulled.

To compute the probability of bit error for all sub-carriers, closed-form expressions are employed. For instance, the probability of bit error for binary phase shift keying (BPSK) is given by:[43]

$$P_{2,i}(\gamma_i) = Q(\sqrt{2\gamma_i}) \qquad (11)$$

while the probability of symbol error for quadrature phase-shift keying (QPSK) ($M_i = 4$), square 16-QAM ($M_i = 16$), and square 64-QAM ($M_i = 64$) is given by:[43]

$$P_{M_i,i}(\gamma_i) = 4\left(1 - \frac{1}{\sqrt{M_i}}\right)Q\left(\sqrt{\frac{3\gamma_i}{M_i - 1}}\right)$$
$$\cdot \left(1 - \left(1 - \frac{1}{\sqrt{M_i}}\right)Q\left(\sqrt{\frac{3\gamma_i}{M_i - 1}}\right)\right) \qquad (12)$$

where $\log_2(M_i)$ gives the number of bits to represent a signal constellation point (although Eqs. 11 and 12 depend on the interference being Gaussian, this is likely to be true in the case of OFDM-type systems with a sufficiently long cyclic extension.). To obtain the probability of bit error from the symbol error of Eq. 12, use the approximation $P_i \approx P_{M_i,i}/\log_2(M_i)$. Note that power allocation is not explicitly performed by this algorithm, although it can be easily modified to include it.[40,41]

Although the objective function of the algorithm shown in Fig. 5 is to maximize the overall throughput of the system, it is possible that the resulting bit allocation is not equal to the optimal allocation, i.e., the largest throughput satisfying the mean BER constraint. The following scenario illustrates how the algorithm could yield a sub-optimal bit allocation: For the final iteration of the algorithm, there exists two sub-carriers with the worst BER, i and j. Reducing the number of bits in either sub-carrier will result in $\bar{P} \leqslant P_T$. If sub-carrier i is chosen, 2 bits are removed. On the other hand, if sub-carrier j is chosen, only 1 bit is removed. This is possible since $\log_2(64) - \log_2(16) = \log_2(16) - \log_2(4) = 2$ bits per symbol epoch, while $\log_2(4) - \log_2(2) = \log_2(2) - 0 = 1$ bits per symbol epoch. Therefore, if $P_i < P_j$, then the algorithm will choose sub-carrier j to be decremented, resulting with an overall allocation that is optimal.

PEAK BER-CONSTRAINED BIT ALLOCATION ALGORITHM

Although incremental bit allocation algorithms could attain near-optimal solutions, their computational complexity is still rather high. For instance, the algorithm of Fig. 5 possesses a very large computational complexity at low SNR values. As a result, an algorithm is needed that accurately maps the sub-carrier SNR values to some final bit allocation in an iterative, low computational complexity fashion. This is the rationale behind the peak BER-constrained bit allocation algorithm.[42,44]

Instead of iteratively adding a bit to a sub-carrier and checking if the mean BER constraint is not violated, which is computationally expensive, the peak BER-constrained algorithm in Fig. 6 allocates bits to each sub-carrier so that all sub-carrier BER values, P_i, are below some peak BER constraint $\hat{p}$. The algorithm operates as follows: First, the

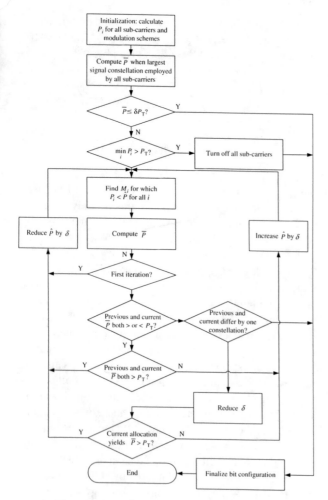

Fig. 6 Flow diagram of an adaptive bit allocation algorithm based on a peak BER constraint.[44,42]

value of P_i is evaluated for $i = 0, \ldots, N-1$ and all possible values sub-carrier signal constellations. Then, the largest sub-carrier signal constellation satisfying $P_i \leqslant \hat{P}$ is chosen for $i = 0, \ldots, N-1$. The initial value of the peak BER constraint $\hat{p}$ is chosen as a proxy to satisfy an average BER constraint $\bar{P}$. A first guess on $\hat{P}$ is taken, the bits b_i are allocated accordingly, and the resulting $\bar{P}$ is computed using Eq. 10. If is below (above) P_T, $\hat{P}$ is increased (decreased) by an amount δ in the logarithmic domain at every iteration. The value of $\hat{P}$ is adjusted in this way until $\bar{P}$ exceeds (goes below) P_T, in which case δ is reduced.

In the case that the previous and current $\bar{P}$ values straddle P_T, the allocations are compared in order to see if they differ by one signal constellation. If they do, it is obvious that the additional bit(s) is/are the cause of the violation of the mean BER constraint. Otherwise, δ is reduced until the case of one differing signal constellation is achieved. Note that the first two decision blocks provide a quick exit from the algorithm when the sub-carrier SNR values are either large enough to have the system operate

at maximum throughput or lower than the minimum SNR required to yield a non-zero throughput, respectively.

Initial Peak BER Threshold Calculation

The speed at which the peak BER-constrained algorithm reaches its final allocation depends on the choice of the initial $\hat{P}$ and the δ it uses. One approach to this problem is to determine how much any given sub-carrier can individually exceed P_T while $\bar{P}$ remains below it. An algorithm for finding the estimate of the initial peak BER $\hat{P}$ is shown in Fig. 7. Given that a sub-carrier can support B possible modulation schemes, resulting in B possible values for P_i, the largest P_i value that is below P_T is defined as β_i while the smallest value of P_i above P_T is defined as α_i. Therefore, knowing that the mean of β_i, $i = 0, \ldots, N-1$, is below P_T, we incrementally replace the smallest β_i with the corresponding α_i until $\bar{P} > P_T$. Thus, given β_i, $i \in S$, we need to solve for the amount of BER leeway, ΔP, namely:

$$\Delta P = \sum_{i \in S} b_i (P_T - \beta_i)$$

in order to determine by how much several sub-carriers can violate the condition $P_i \leqslant P_T$ while the system still satisfies $\bar{P} \leqslant P_T$. Once ∇P has been determined, the values of α_i are sorted in an increasing order and the largest value of I is computed for which:

$$\Delta P \geqslant \sum_{i=0}^{I} b_i (\alpha_i - P_T)$$

is true, where $0 \leqslant I < N$, and set α_i as the initial $\hat{P}$ for the algorithm.

The initial value of δ is proportional to the average SNR of the system, $\bar{\gamma}$. It has been observed in several simulations that for low $\bar{\gamma}$ values, small values for δ resulted in the algorithm converging quickly to a final solution, while for high $\bar{\gamma}$ values, large values of δ resulted in quickly obtaining the solution. Thus, choosing the values for δ between the two extremities, δ decreases linearly as a function of $\bar{\gamma}$.

Using these values of δ in conjunction with the initial $\hat{P}$ algorithm, the number of iterations required to find the final $\hat{P}$ can be reduced by as much as half when compared to a scheme using a random initialization.[44,42]

BIT ALLOCATION WITH IMPERFECT CHANNEL INFORMATION

When determining the performance improvements of adaptive bit allocation algorithms, most studies evaluated these algorithms when the channel conditions are perfectly known and the channel is time-invariant. However, the

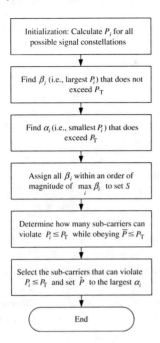

Fig. 7 Flow diagram of an algorithm used for determining the initial peak bit error rate.[44,42]

performance results may be overly optimistic and thus a more accurate performance analysis of bit allocation algorithms is required.

Most researchers have focused on developing a more realistic model of scenarios that these systems will experience in the real world. In particular, the impact of the following three effects are usually studied:

1. The effects of a time-varying channel and outdated channel state information.[45–50]
2. Channel estimation error and the propagation of that error.[49,51,45]
3. Imperfect feedback due to distortion and/or quantization.[52]

Since the effectiveness of the adaptive allocation algorithms are heavily dependent on the quality of the channel state information, any of these effects would have a serious impact on their performance.

Although many studies on adaptive bit allocation algorithms make the assumption that the sub-carrier SNR values are perfectly known, this is not the case in reality. As a result, the performance of the allocation algorithms can be significantly affected by imperfect channel information. There are a number of sources that can create imperfections in the channel information: channel estimation, time-varying channels, band-limited feedback channel, and inadequate training symbol design. Therefore, it is necessary to investigate the impact of the imperfect sub-carrier SNR information, which is derived from the channel information, on the throughput performance of systems employing adaptive bit allocation algorithms.

Gaussian Sub-carrier SNR Error Model

A derivation of the expression for the channel estimation error of an MCM system employing data-assisted channel estimation is presented in Ref. 53. It was observed that the channel estimation error consists of the term $\Delta^{(i)}(m)$, which refers to the estimation error on sub-carrier i. Therefore, the derivation for the SNR of sub-carrier i based on the channel estimates yields the final expression:

$$\hat{\gamma}^{(i)} = \gamma^{(i)} + \varepsilon^{(i)}(\gamma^{(i)}) \tag{13}$$

where the estimation error $\varepsilon^{(i)}(\gamma^{(i)})$ is a function of the sub-carrier SNR, $\gamma^{(i)}$.

Note that the sub-carrier channel estimation error $\varepsilon^{(i)}(\gamma^{(i)})$ can be approximated by a normal distribution with zero mean and variance σ^{2},[45] where the variance σ^{2} should be a function of the sub-carrier SNR, i.e., $\sigma^{2}(\gamma^{(i)})$. However, the negative values of $\varepsilon^{(i)}(\gamma^{(i)})$ must be constrained such that Eq. 13, being a ratio of powers, is never negative.

SNR Quantization Error Model

Adaptive bit allocation algorithms use the channel state information to determine the modulation configuration. However, the information is usually transformed into a metric that indicates the quality of the transmission across the different sub-carriers, such as the sub-carrier SNR values, which are then used to compute the sub-carrier BER values via closed form expressions for a given modulation scheme, e.g., Eq. 11 and 12. To reduce the implementation complexity, a look-up table can be used instead to translate sub-carrier SNR values into sub-carrier BER values, P_{i}, for each sub-carrier i. However, this implies that the sub-carrier SNR values must first be quantized before using the look-up table. Due to the quantization, additional errors are introduced to the sub-carrier SNR values that may cause the bit allocation to deviate further from the ideal. These errors can be classified as either *granular errors*, when the input signal is within the range of the quantizer but does not fall exactly on an output level, or *overload errors*, when the input signal falls outside of the range of the quantizer.[29]

To ensure that the impact of the quantization error is minimized when considering all the modulation schemes, the locations of the quantizer reproduction (i.e., output) levels, l_{k}, must be determined. If sub-carrier BER is used as a metric, an adequate resolution of the BER waterfall curves is desired around the target probability of bit error, P_{T}, so that the output levels are concentrated about that point. A sub-optimal l_{k} placement technique that tries to minimize the overall error while providing adequate resolution was proposed in Ref. 53 and is presented in the following section.

Quantization Reproduction Level Placement Technique

To obtain adequate resolution of the BER waterfall curves for the modulation schemes employed by the system, where the rate of decrease for each curve may vary drastically, the following technique is employed that tries to perform a sub-optimal placement of l_k:

1. Given q bits to represent a quantizer reproduction level, the number of levels is defined as 2^q, which corresponds to a 2^q-entry look-up table.
2. Determine the pair of SNR values to obtain the probability of bit error values, P_i, that are two orders of magnitude above and below P_T for each modulation scheme, thus forming regions Q_k, for $k = 1, \ldots, B$, where B is the number of possible modulation schemes.
3. For the B modulation schemes, put $2^q/B$ output levels uniformly in Q_k for all k. In the case of overlapping regions, combine them and their allocation of output levels, distributing the levels uniformly across the combined region.

An example of this procedure is shown in Fig. 8 for $P_T = 10^{-5}$ and $B = 4$. In this case, the P_i curves correspond to BPSK, QPSK, 16-QAM, and 64-QAM. If $P_i > 10^{-3}$, quantizing that part of the BER curve is not worthwhile since the P_i is so high that the sub-carrier would be nulled. On the other hand, if $P_i < 10^{-7}$, then P_i is so far below P_T that any quantization would not significantly affect the mean BER of the system, $\bar{P}$. Where Q_1 and Q_2 overlap, the output levels allocated to the two regions would be combined and distributed uniformly across the aggregate region. By distributing l_k, $k = 0, \ldots, 2^q - 1$ in this way, the BER waterfall curves can be quantized with sufficient resolution.

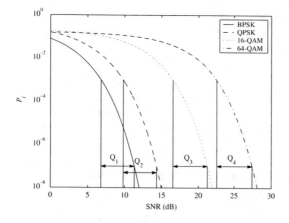

Fig. 8 Regions of uniform quantization of the P_i waterfall curves for $P_T = 10^{-5}$ and $B = 4$.

SIMULATION RESULTS

Experimental Setup

In this entry, we evaluated the performance of several adaptive bit allocation algorithm found in the literature when employed by indoor wireless MCM transceivers. The transceivers use some of the operating parameters specified in the WLAN standard, IEEE Standard 802.11a,[4] including the number of non-guard sub-carriers (52 sub-carriers), the frequency band of operation (5.15–5.25 GHz), the transmission bandwidth (16.6 MHz), and the available modulation schemes (BPSK, QPSK, rectangular 16-QAM, and rectangular 64-QAM). However, unlike the standard, where the same modulation scheme is employed across all the sub-carriers, the adaptive bit allocation algorithms can use a different modulation scheme for each sub-carrier. In addition, sub-carriers can be turned off. Results from the adaptive bit allocation algorithms were obtained for P_T values of 10^{-3} and 10^{-5}. Furthermore, an exhaustive search algorithm was also employed for a case with a reduced number of sub-carriers over a portion of the band, to keep the complexity manageable, in order to make it possible to determine the optimal solution to the bit allocation problem.

The transmitter/receiver separation was varied between 1 and 60 m, with the complex baseband representation of the channel having no line-of-sight component and assumed to be quasi-stationary and thus invariant during the bit allocation process. The channel impulse responses were generated using the method presented in Ref. 54. The channel model used a cluster arrival rate of 1.66×10^{-7} s^{-1}, a ray arrival rate of 1.66×10^{-8} s^{-1}, a cluster power-decay time constant of 40×10^{-8} s, a ray power-decay time constant of 8×10^{-8} s, and five clusters, each of which contained 100 rays. Only a single pair of antennas was employed. For each channel realization, the algorithms were operating at 70 different averaged SNR values ranging from −11 to 59 dB. The trials were repeated for 10,000 different channel realizations and the results averaged. Furthermore, the change in SNR corresponds to the change in transmitter/receiver separation distance.

The algorithms under study in this section are an incremental bit allocation algorithm via discrete optimization using marginal analysis,[14] a capacity approximation-based bit allocation algorithm,[19] an incremental bit allocation algorithm based on a greedy approach,[40–42] a peak BER-constrained bit allocation algorithm,[44,42] and an exhaustive search routine for the optimum bit allocation.

For a straightforward comparison, all the sub-carriers have a uniform power allocation. Therefore, several modifications are required for the capacity approximation-based algorithm. Specifically, the expression for the noise-to-signal ratio (NSR) used for determining which sub-carriers is to be turned off is modified such that constant uniform power allocation is employed. To achieve this, the NSR expression is modified to:

$$NSR = \frac{\pi_{\text{sc}}}{\Gamma} + \frac{1}{N_{\text{on}}} \sum_{n=1}^{N_{\text{on}}} \frac{1}{\gamma^{(n)}} \qquad (14)$$

where π_{sc} is the sub-carrier power (a constant value across all sub-carriers), N_{on} is the number of sub-carriers that are "on", Γ is the aforementioned SNR gap (in this work, $\Gamma = 6.06$ dB for $P_{\text{T}} = 1 \times 10^{-3}$, $\Gamma = 7.37$ dB for P_{T}, and $\Gamma = 8.42$ dB for $P_{\text{T}} = 1 \times 10^{-5}$), and $\gamma^{(n)}$ is the SNR of sub-channel n (i.e., $\gamma^{(n)} = |H_n|^2/\sigma_v^2$), where σ_v^2 is the noise variance, H_n is the channel frequency response across sub-carrier n). By defining the NSR this way, the sub-carrier power levels are fixed, instead of keeping the total power allocation fixed, as was done originally. Other than this modification, the algorithm is essentially the same as the original.

Perfect Sub-carrier SNR Information

In Fig. 9, the overall throughputs of the 5-bit allocation algorithms are presented for the case of eight sub-carriers. This reduced number of degrees of freedom allows for an exhaustive search of the optimal allocation. It is observed that the capacity approximation-based algorithm does not reach the same throughput as the other systems until high SNR values of 49 dB. This is due to the fact that the approximations in this algorithm are not very accurate, and is not an artifact of the modifications made to the algorithm to utilize uniform power allocation. As for the other algorithms, the difference in throughput between them is very small since they all perform discrete allocations rather than non-integer allocations followed by rounding. The largest throughput is produced by the exhaustive search algorithm, followed by both incremental allocation algorithms, and finally by the peak

BER-constrained algorithm. Since the objective function is not concave ("A function defined only on the integers is called concave if its first differences are decreasing." [14]) and the constraint function is not strictly convex, there is no guarantee that the incremental algorithm employing marginal analysis would reach the optimal allocation.[14]

The values of the mean BER $\bar{P}$ corresponding to the throughputs in Fig. 9 are shown in Fig. 10. It can be observed that all the algorithms, except for the capacity approximation-based algorithm, have approximately the same values as the exhaustive search. The error rates are low at low SNR values due to the nulling of poorly performing sub-carriers. This leaves the best-performing sub-carriers, combined with small signal constellations (e.g., BPSK), which results in low error rates. In the mid-range SNR region, all the sub-carriers are on, including the poorly performing ones. Thus, their BER values will dominate the mean BER of the system. At high SNR values, all sub-carriers are modulated with the largest constellation sizes and all have $P_i \geqslant P_{\text{T}}$.

When 52 sub-carriers are employed, as shown in Fig. 11, the algorithms, except for the capacity approximation-based algorithm, achieve nearly the same throughput with some small differences. The throughput of the capacity approximation-based algorithm is substantially less than that of the other methods, only reaching the performance of other algorithms at high SNR values. Note how at low SNR values, the throughput of this algorithm goes to zero. This is due to either the algorithm producing an allocation that exceeds P_{T} or the algorithm nulling all the sub-carriers. The corresponding $\bar{P}$ values are shown in Fig. 12.

As seen in Fig. 9, the throughput for most of the studied algorithms is very close to that of the optimal allocation produced by the exhaustive search algorithm. Furthermore, the peak BER-constrained algorithm executes more quickly than the incremental algorithm employing marginal analysis. Both the incremental algorithm based on a greedy approach and the peak BER-constrained algorithm perform similarly in terms of throughput and complexity at high SNR values. However, at low SNR values, the peak BER-constrained algorithm executes faster than the incremental algorithm (both mean and worst cases). This is due to the fact that the incremental algorithm is a bit removing algorithm, which means at low SNR values, it takes numerous iterations to reach the final allocation. Although the capacity approximation-based algorithm may execute at the same speed as the peak BER-constrained algorithm, the latter achieves far greater throughput. A summary of mean and worst-case computation times for a 52 sub-carrier system with a P_{T} of 10^{-5} is shown in Table 1 for several SNR values. Furthermore, the cumulative density functions of the computation times at SNR values of 10 and 40 dB are shown in Fig. 13. For a fair comparison, all algorithms were programed in the C programming language and executed on the same

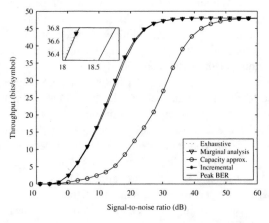

Fig. 9 Overall throughput of an $N = 8$ sub-carrier system satisfying a P_{T} of 10^{-3}. Except for the curve corresponding to capacity approximation-based algorithm, all the curves are superimposed.

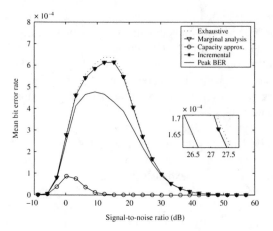

Fig. 10 Mean BER of an $N = 8$ sub-carrier system satisfying a P_T of 10^{-3}.

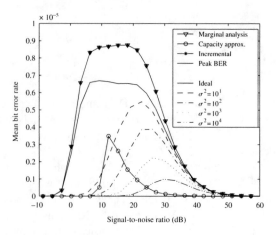

Fig. 12 Mean BER of an $N = 52$ sub-carrier system satisfying a P_T of 10^{-5} with and without Gaussian noise of variance σ^2 added to Γ_i, $i = 0, \ldots, N - 1$.

workstation (Intel Pentium IV 2 GHz processor). It should be noted that although the algorithms may vary in execution time, all the worst case execution times are of the same order of magnitude. This is due to the fact that the worst case computational complexity of all the algorithms under study are of, especially since the number of candidate signal constellations employed by most bit allocation algorithms is relatively small (e.g., in this work $B = 5$), its impact on the computational complexity of the algorithm is a multiplicative factor B at most.

Imperfect Sub-carrier SNR Information

The effect on throughput performance of adding Gaussian noise to the sub-carrier SNR values is studied first. Although results for both incremental algorithms, the peak BER-constrained algorithm, and the capacity

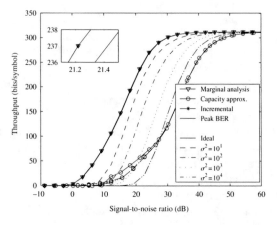

Fig. 11 Overall throughput of an $N = 52$ sub-carrier system satisfying a P_T of 10^{-5} using sub-carrier SNR values with Gaussian noise of variance σ^2 added. Note that the capacity approximation-based algorithm and the peak BER-constrained algorithm use the non-ideal sub-carrier SNR values.

approximation-based algorithm were obtained, the first three algorithms all had similar results. Therefore, only the throughput results for the peak BER-constrained algorithm as well as for the capacity approximation-based algorithm are presented in Fig. 11, while Figs. 12 and 14 show the $\bar{P}$ and outage probability results for the peak BER-constrained algorithm. The value of P_T was set to 10^{-5}. The results are obtained when the variance of the Gaussian noise is either $\sigma^2 = 10$, 10^2, 10^3, or 10^4 across all SNR values (although the error variance would be dependent on the nominal SNR, in this work the error variance was chosen to be constant for all SNR values in order to evaluate the robustness of these algorithms at different SNR values. Moreover, instead of employing a single error variance value across all SNR values, a range of error variance values were examined, with the smaller error variances realistically occurring at low nominal SNR values and the larger values occurring at higher SNR values.). Compared to the case where no Gaussian noise is added to the sub-carrier SNR values, the throughput of the system decreases as the variance increases. In particular, the throughput curves shift to the right as the noise variance σ^2 increases. Moreover, except at low SNR values for $\sigma^2 = 10^4$, the capacity approximation-based algorithm performs relatively poorly.

Since most of the adaptive bit allocation algorithms are close to the maximum achievable throughput given the maximum error constraint, the addition of Gaussian noise to the sub-carrier SNR values can either cause the system to violate the constraint (when $\hat{\gamma}_i > \gamma_i$) or decrease in throughput (when $\hat{\gamma}_i \leqslant \gamma_i$). Since $\bar{P} > P_T$ is not acceptable, when the former occurs, the throughput of the system is set to zero and a record of the number of times this occurs is kept. The fraction of realization violations is shown in Fig. 14. When the latter occurs, the throughput and are lower, as in Figs. 11 and 12. Note that at low SNR values, σ^2 is large enough that the algorithm experiences violations every time. As the SNR increases, the frequency of

Table 1 Mean (worst case) computation times in milliseconds at different SNR values, 52 sub-carriers, $P_T = 10^{-5}$.

Algorithm	10 dB	25 dB	40 dB	55 dB
Marginal analysis	1.13 (3.23)	1.48 (5.01)	1.41 (5.00)	1.37 (4.40)
Capacity approximation	0.94 (2.78)	0.96 (4.98)	0.93 (4.24)	0.90 (4.66)
Incremental	1.09 (2.86)	**0.91** (4.10)	**0.84** (2.09)	**0.80** (2.62)
Peak BER-constrained	**0.91** (2.96)	**0.91** (2.71)	0.86 (3.98)	0.82 (4.54)

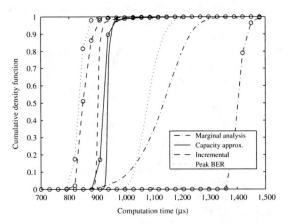

Fig. 13 Cumulative density function of computational time of an $N = 52$ sub-carrier system satisfying a P_T of 10^{-5} at SNR values of 10 dB (without circles) and 40 dB (with circles).

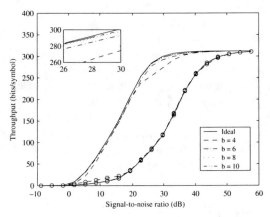

Fig. 15 Overall throughput of a multicarrier system employing the peak BER-constrained algorithm (no circles) and the capacity approximation-based algorithm (with circles) at P_T when the sub-carrier SNR values are quantized with 2^b levels. Note the latter uses another set of quantization reproduction levels.

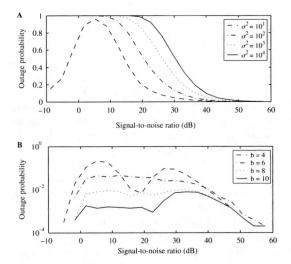

Fig. 14 Outage probability (fraction of realizations for which $\bar{P} < P_T$) of a multicarrier system employing the peak BER-constrained algorithm at $P_T = 10^{-5}$. (A) Gaussian noise of variance σ^2 added to the sub-carrier SNR values. (B) Sub-carrier SNR values quantized with 2^b levels.

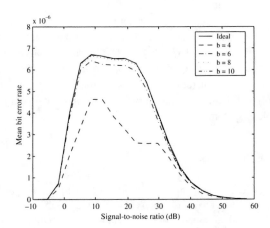

Fig. 16 Mean BER of a multicarrier system employing the peak BER-constrained algorithm at $P_T = 10^{-5}$ when the sub-carrier SNR values are quantized with 2^b levels.

violations decrease. Other than the case of $\sigma^2 = 10$, violations occurred 100% of the time due to the SNR values being the same order of magnitude as the Gaussian noise.

The outage probability, throughput, and $\bar{P}$ results of the peak BER-constrained algorithm using quantized sub-carrier SNR values are presented in Figs. 14–16, respectively. From Fig. 15 it can be observed that there is some degradation. However, the capacity approximation-based algorithm still has significantly lower throughput. For instance, at an SNR of 11 dB, the difference in throughput between the ideal case and the case where the sub-carrier SNR values are quantized to 2^3 output levels is 40 bps. However, as the number of output levels increases, the throughput approaches that of the ideal case. At an SNR of 19 dB, the difference in throughput between the ideal and a quantized sub-carrier SNR employing 2^{10} output levels is 0.2 bps. Equivalently, the $\bar{P}$ curves in Fig. 16 also approach the ideal case when the number of output levels increases. The difference in performance is dependent on granular error. More output levels results in a smaller granular error and correspondingly a result closer to the ideal case. The fraction of realization violations (when $\bar{P} > P_T$) are shown in Fig. 14.

CONCLUSION

In this entry, we presented on overview of spectrally efficient wireless communications for multicarrier transceivers employing adaptive bit allocation. Specifically, four adaptive bit allocation algorithms were studied with respect to overall throughput performance and error robustness. Simulation results were obtained for these algorithms and compared with each other in order to determine the design trade-offs for each algorithm. Overall, we observe that the use of adaptive bit allocation for wireless multicarrier transmission can substantially enhance the system performance, yielding greater throughput rates and error robustness relative to systems that do not employ these algorithms. Given the scarcity of wireless transmission spectrum, as well as the agility of current digital communications hardware, adaptive bit allocation is an obvious solution for future wireless systems.

REFERENCES

1. Federal Communications Commission. Spectrum policy task force report, ET Docket No. 02-135; 2002.
2. Federal Communications Commission. Unlicensed operation in the TV broadcast bands, ET Docket No. 04-186; 2004.
3. American National Standards Institute. Network to customer installation interfaces—asymmetric digital subscriber line (ADSL) metallic interface. ANSI Standard T1.413-1998; 1998.
4. Institute of Electronics and Electrical Engineers. Wireless LAN medium access control (MAC) and physical layer (PHY) specifications: High-speed physical layer in the 5 GHz band. IEEE Standard 802.11a; 1999.
5. European Telecommunications Standards Institute. Broadband radio access networks (BRAN): HIPERLAN type 2; physical (PHY) layer. ETSI TS 101 475; 2001.
6. Nee, R.V.; Awater, G.; Morikura, M.; Takanashi, H.; Webster, M.; Halford, K.W. New high-rate wireless LAN standard. IEEE Commun. Mag. **1999**, *37* (12), 82–88.
7. Institute of Electronics and Electrical Engineers. IEEE standard for local and metropolitan area networks part 16: Air interface for fixed and mobile broadband wireless access systems amendment for physical and medium access control layers for combined fixed and mobile operation in licensed bands. IEEE Standard 802.16e-2005; 2006.
8. Prasad, R. *Universal Wireless Personal Communications*; Artech House: Norwood, MA, 1998.
9. Ramachandran, R.P. *Bandwidth Efficient Filter Banks for Transmultiplexers*; Ph.D. thesis; McGill University: Montréal, QC, Canada, 1990.
10. Vaidyanathan, P.P. *Multirate Systems and Filter Banks*; Signal Processing Series. Prentice Hall: Upper Saddle River, NJ, 1993.
11. Fliege, N.J. *Multirate Digital Signal Processing: Multirate Systems, Filter Banks, Wavelets*; John Wiley and Sons: New York, 1994.
12. Sandberg, S.D.; Tzannes, M.A. Overlapped discrete multi-tone modulation for high speed copper wire communications. IEEE J. Sel. Areas Commun. **1995**, *13*, 1571–1585.
13. Keller, T.; Hanzo, L. A convenient framework for time-frequency processing in wireless communications. Proc. IEEE. **2000**, *88*, 611–640.
14. Fox, B. Discrete optimization via marginal analysis. Manage. Sci. **1996**, *13*, 210–216.
15. Segall, A. Bit allocation and encoding for vector sources. IEEE Trans. Inf. Theory. **1976**, *22*, 162–169.
16. Gersho, A.; Gray, R.M. *Vector Quantization and Signal Compression*; Kluwer Academic Publishers: Dordrecht, Netherlands, 1991.
17. Bingham, J.A.C. *ADSL, VDSL, and Multicarrier Modulation*; John Wiley and Sons: New York, 2000.
18. Chow, P.S.; Cioffi, J.M.; Bingham, J.A.C. A practical discrete multitone transceiver loading algorithm for data transmission over spectrally shaped channels. IEEE Trans. Commun. **1995**, *43*, 773–775.
19. Leke, A.; Cioffi, J.M. A maximum rate loading algorithm for discrete multitone modulation systems. IEEE Global Telecommunications Conference (GLOBECOM 1997), Phoenix, AZ, Nov 3–8, 1997; Vol. 3, 1514–1518.
20. Colieri, S.; Ergen, M.; Puri, A.; Bahai, A. A study of channel estimation in OFDM systems. 56th IEEE Vehicular Technology Conference (VTC 2002), British Columbia, Canada, Sept 24–28, 2002; Vol. 2, 894–898.
21. Benvenuto, N.; Tomasin, S.; Tomba, L. Equalization methods in OFDM and FMT systems for broadband wireless communications. IEEE Trans. Commun. **2002**, *50*, 1413–1418.
22. Cherubini, G.; Eleftheriou, E.; Olcer, S.; Cioffi, J.M. Filter bank modulation techniques for very high-speed digital subscriber lines. IEEE Commun. Mag. **2000**, *38*, 98–104.

23. Cherubini, G.; Eleftheriou, E.; Olcer, S. Filtered multitone modulation for very high-speed digital subscriber lines. IEEE J. Sel. Areas Commun. **2002**, *20*, 1016–1028.

24. Bingham, J.A.C. Multicarrier modulation for data transmission: An idea whose time has come. IEEE Commun. Mag. **1990**, *28*, 5–14.

25. Weinstein, S.B.; Ebert, P. Data transmission by frequency division multiplexing using the discrete fourier transform. IEEE Trans. Commun. **1971**, *19*, 628–634.

26. Hirosaki, B. An analysis of automatic equalizers for orthogonally multiplexed QAM systems. IEEE Trans. Commun. **1980**, *28*, 73–83.

27. Hirosaki, B. An orthogonally multiplexed QAM system using the discrete fourier transform. IEEE Trans. Commun. **1981**, *29*, 982–989.

28. Saltzberg, B.R. Performance of an efficient parallel data transmission system. IEEE Trans. Commun. **1967**, *15*, 805–811.

29. Proakis, J.G. *Digital Signal Processing: Principles, Algorithms, and Applications,* 3rd Ed.; Prentice Hall: Upper Saddle River, NJ, 1996.

30. Piazzo, L. Optimum adaptive OFDM systems. Eur. Trans. Telecommun. **2003**, *14*, 205–212.

31. Czylwik, A. Adaptive OFDM for wideband radio channels. IEEE Global Telecommunications Conference (GLOBE-COM 1996), London, UK, Nov 18–22, 1996; Vol. 1, 713–718.

32. Fischer, R.F.H.; Huber, J.B. A new loading algorithm for discrete multitone transmission. IEEE Global Telecommunications Conference (GLOBECOM 1996), London, UK, Nov 18–22, 1996; Vol. 1, 724–728.

33. Ding, Y.; Davidson, T.N.; Wong, K.M. On inproving the BER performance of rate-adaptive block-by-block transceivers, with applications to DMT. IEEE Global Telecommunications Conference (GLOBECOM 2003), San Francisco, CA, Dec 1–5, 2003; 1654–1658.

34. Garcia-Armada, A.; Cioffi, J.M. Multiuser constant-energy bit loading for M-PSK-modulated orthogonal frequency division multiplexing. IEEE Wireless Communications and Networking Conference, Orlando, FL, Mar 17–21, 2002; Vol. 2, 526–530.

35. Ungerboeck, G. Channel coding with multilevel/phase signals. IEEE Trans. Inform. Theory. **1982**, *28*, 55–67.

36. Hughes-Hartog, D.Ensemble Modem Structure for Imperfect Transmission Media. U.S. Patent 4,679,227, July 7, 1987; 4,731,816, Mar 15, 1988; and 4,833,706, May 30, 1989.

37. Keller, T.; Hanzo, L. Sub-band adaptive pre-equalized OFDM transmission. 50th IEEE Vehicular Technology Conference (VTC 1999), Amsterdam, Netherlands, Sept 19–22, 1999; Vol. 1, 334–338.

38. Campello, J. Optimal discrete bit loading for multicarrier modulation systems. International Symposium on Information Theory, Cambridge, MA, Aug 16–21, 1998; 19–3.

39. Krongold, B.S.; Ramchandran, K.; Jones, D.L. Computationally efficient optimal power allocation algorithms for multicarrier communication systems. IEEE Trans. Commun. **2000**, *48*, 23–27.

40. Wyglinski, A.M.; Kabal, P.; Labeau, F. Adaptive filterbank multicarrier wireless systems for indoor environments. 56th IEEE Vehicular Technology Conference, British Columbia, Canada, Sept 24–28, 2002; Vol. 1, 336–340.

41. Wyglinski, A.M.; Kabal, P.; Labeau, F. Adaptive bit and power allocation for indoor wireless multicarrier systems. International Conference on Wireless Communications, Alberta, Canada, July 9–11, 2003; 500–508.

42. Wyglinski, A.M.; Labeau, F.; Kabal, P. Bit loading with BER-constraint for multicarrier systems. IEEE Trans. Wirel. Commun. **2005**, *4*, 1383–1387.

43. Proakis, J.G. *Digital Communications,* 3rd Ed.; McGraw-Hill: New York, 1995.

44. Wyglinski, A.M.; Labeau, F.; Kabal, P. An efficient bit allocation algorithm for multicarrier modulation. IEEE Wireless Communications and Networking Conference, Atlanta, GA, Mar 21–25, 2004; B-13.

45. Leke, A.; Cioffi, J.M. Impact of imperfect channel knowledge on the performance of multicarrier systems. IEEE Global Telecommunications Conference, Sydney, Australia, Nov 8–12, 1998; Vol. 2, 951–955.

46. Su, Q.; Cimini, L.J.; Blum, R.S. On the problem of channel mismatch in constant-bit-rate adaptive modulation. 55th IEEE Vehicular Technology Conference, Birmingham, AL, May 6–9, 2002; Vol. 2, 585–589.

47. Goeckel, D.L. Strongly robust adaptive signaling for time-varying channels. IEEE International Conference on Communications, Atlanta, GA, June 7–11, 1998; Vol. 1, 454–458.

48. Goeckel, D.L. Adaptive coding for time-varying channels using outdated fading estimates. IEEE Trans. Commun. **1999**, *47*, 844–855.

49. Souryal, M.R.; Pickholtz, R.L. Adaptive modulation with imperfect channel information in OFDM. IEEE International Conference on Communications, Helsinki, Finland, June 11–15, 2001; Vol. 6, 1861–1865.

50. Falahati, S.; Svensson, A.; Ekman, T.; Sternad, M. Adaptive modulation systems for predicted wireless channels. IEEE Trans. Commun. **2004**, *52*, 307–316.

51. Ye, S.; Blum, R.S.; Cimini, L.J. Adaptive modulation for variable-rate OFDM systems with imperfect channel information. 55th IEEE Vehicular Technology Conference, Birmingham, AL, May 6–9, 2002; Vol. 2, 767–771.

52. Sen, S.; Pasupathy, S.; Kschischang, F.R. Quantized feedback information for the SVD filtered MIMO based FEC coded DS-CDMA multiuser detection receiver. International Conference on Wireless Communications, Alberta, Canada, July 7–9, 2003; 45–51.

53. Wyglinski, A.M.; Labeau, F.; Kabal, P. Effects of imperfect subcarrier SNR information on adaptive bit loading algorithms for multicarrier systems. IEEE Global Telecommunications Conference, Dallas, TX, Nov 29–Dec 3, 2004; Vol. 6, 3835–3839.

54. Saleh, A.A.M.; Valenzuela, R.A. A statistical model for indoor multipath propagation. IEEE J. Sel. Areas Commun. **1987**, *5*, 128–137.

Spread Spectrum Communications

Laurence B. Milstein
University of California, La Jolla, California, U.S.A.

Marvin K. Simon
Jet Propulsion Laboratory, Pasadena, California, U.S.A.

Abstract

Spread spectrum (SS) is a communication technique where the transmitted modulation is spread in bandwidth prior to transmission over the channel and then despread in bandwidth by the same amount at the receiver. Communication systems employ SS to reduce the communicator's detectability.

A BRIEF HISTORY

Spread spectrum (SS) has its origin in the military arena where the friendly communicator is 1) susceptible to detection/interception by the enemy, and 2) vulnerable to intentionally introduced unfriendly interference (jamming). Communication systems that employ SS to reduce the communicator's detectability and combat the enemy-introduced interference are respectively referred to as low probability of intercept (LPI) and antijam (AJ) communication systems. With the change in the current world political situation wherein the US Department of Defense (DOD) has reduced its emphasis on the development and acquisition of new communication systems for the original purposes, a host of new commercial applications for SS has evolved, particularly in the area of cellular mobile communications. This shift from military to commercial applications of SS has demonstrated that the basic concepts that make SS techniques so useful in the military can also be put to practical peacetime use. In the next section, we give a simple description of these basic concepts using the original military application as the basis of explanation. The extension of these concepts to the mentioned commercial applications will be treated later on in the entry.

WHY SS?

SS is a communication technique wherein the transmitted modulation is *spread* (increased) in bandwidth prior to transmission over the channel and then *despread* (decreased) in bandwidth by the same amount at the receiver. If it were not for the fact that the communication channel introduces some form of narrowband (relative to the spread bandwidth) interference, the receiver performance would be transparent to the spreading and despreading operations (assuming that they are identical inverses of each other). That is, after despreading the received signal would be identical to the transmitted signal prior to spreading. In the presence of narrowband interference, however, there is a significant advantage to employing the spreading/despreading procedure described. The reason for this is as follows.

Since the interference is introduced after the transmitted signal is spread, then, whereas the despreading operation at the receiver shrinks the desired signal back to its original bandwidth, at the same time it spreads the undesired signal (interference) in bandwidth by the same amount, thus reducing its power spectral density (PSD). This, in turn, serves to diminish the effect of the interference on the receiver performance, which depends on the amount of interference power in the despread bandwidth. It is indeed this very simple explanation which is at the heart of all SS techniques.

BASIC CONCEPTS AND TERMINOLOGY

To describe this process analytically and at the same time introduce some terminology that is common in SS parlance, we proceed as follows. Consider a communicator that desires to send a message using a transmitted power S Watt (W) at an information rate R_b bps. By introducing an SS modulation, the bandwidth of the transmitted signal is increased from R_b Hz to W_{ss} Hz, where $W_{ss} \gg R_b$ denotes the SS bandwidth. Assume that the channel introduces, in addition to the usual thermal noise (assumed to have a single-sided PSD equal to N_0 W/Hz), an additive interference (jamming) having power J distributed over some bandwidth W_J. After despreading, the desired signal bandwidth is once again now equal to R_b Hz and the interference PSD is now $N_J = J/W_{ss}$. Note that since the thermal noise is assumed to be white, i.e., it is uniformly distributed over all frequencies, its PSD is unchanged by the despreading operation and, thus, remains equal to N_0.

Encyclopedia of Wireless and Mobile Communications DOI: 10.1081/E-EWMC-120043935

Regardless of the signal and interferer waveforms, the equivalent bit energy-to-total noise spectral density ratio is, in terms of the given parameters,

$$\frac{E_b}{N_t} = \frac{E_b}{N_0 + N_J} = \frac{S/R_b}{N_0 + J/W_{SS}} \tag{1}$$

For most practical scenarios, the jammer limits performance, and, thus, the effects of receiver noise in the channel can be ignored. Thus, assuming $N_J \gg N_0$, we can rewrite Eq. 1 as

$$\frac{E_b}{N_t} \cong \frac{E_b}{N_J} = \frac{S/R_b}{J/W_{SS}} = \frac{S}{J} \frac{W_{SS}}{R_b} \tag{2}$$

where the ratio J/S is the *jammer-to-signal power ratio* and the ratio W_{ss}/R_b is the spreading ratio and is defined as the processing gain of the system. Since the ultimate error probability performance of the communication receiver depends on the ratio E_b/N_J, we see that from the communicator's viewpoint his goal should be to minimize J/S (by choice of S) and maximize the processing gain (by choice of W_{ss} for a given desired information rate). The possible strategies for the jammer will be discussed in the section on military applications dealing with AJ communications.

SS TECHNIQUES

By far the two most popular spreading techniques are direct sequence (DS) modulation and frequency hopping (FH) modulation. In the following subsections, we present a brief description of each.

DS Modulation

A DS modulation $c(t)$ is formed by linearly modulating the output sequence $\{c_n\}$ of a pseudorandom number generator onto a train of pulses, each having a duration T_c called the chip time. In mathematical form,

$$c(t) = \sum_{n=-\bullet} c_n p(t - nT_c) \tag{3}$$

where $p(t)$ is the basic pulse shape and is assumed to be of rectangular form. This type of modulation is usually used with binary phase-shift-keyed (BPSK) information signals, which have the complex form $d(t) \exp\{j(2\pi f_c t + \theta_c)\}$,

where $d(t)$ is a binary-valued data waveform of rate $1/T_b$ bps and f_c and θ_c are the frequency and phase of the data-modulated carrier, respectively. As such, a DS/BPSK signal is formed by multiplying the BPSK signal by $c(t)$ (see Fig. 1), resulting in the real transmitted signal

$$x(t) = \text{Re}\{c(t)d(t) \exp[j(2\pi f_c t + \theta_c)]\} \tag{4}$$

Since T_c is chosen so that $T_b \gg T_c$, then relative to the bandwidth of the BPSK information signal, the bandwidth of the DS/BPSK signal is effectively increased by the ratio $T_b/T_c = W_{ss}/2R_b$, which is one-half the spreading factor or processing gain of the system. For the usual case of a rectangular spreading pulse $p(t)$, the PSD of the DS/BPSK modulation will have $(\sin x/x)^2$ form with first zero crossing at $1/T_c$, which is nominally taken as one-half the SS bandwidth W_{ss}. At the receiver, the sum of the transmitted DS/BPSK signal and the channel interference $I(t)$ (as discussed before, we ignore the presence of the additive thermal noise) are ideally multiplied by the identical DS modulation (this operation is known as despreading), which returns the DS/BPSK signal to its original BPSK form whereas the real interference signal is now the real wideband signal $\text{Re}\{I(t)c(t)\}$. In the previous sentence, we used the word ideally, which implies that the PN waveform used for despreading at the receiver is identical to that used for spreading at the transmitter. This simple implication covers up a multitude of tasks that a practical DS receiver must perform. In particular, the receiver must first acquire the PN waveform. That is, the local PN random generator that generates the PN waveform at the receiver used for despreading must be aligned (synchronized) to within one chip of the PN waveform of the received DS/BPSK signal. This is accomplished by employing some sort of search algorithm which typically steps the local PN waveform sequentially in time by a fraction of a chip (e.g., half a chip) and at each position searches for a high degree of correlation between the received and local PN reference waveforms. The search terminates when the correlation exceeds a given threshold, which is an indication that the alignment has been achieved. After bringing the two PN waveforms into coarse alignment, a tracking algorithm is employed to maintain fine alignment. The most popular forms of tracking loops are the continuous time delay-locked loop and its time-multiplexed version the tau–dither loop. It is the difficulty in synchronizing the receiver PN generator to subnanosecond accuracy that limits PN chip rates to values on the order of hundreds of

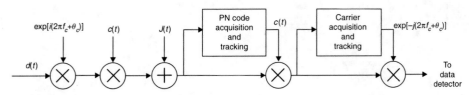

Fig. 1 A DS-BPSK system (complex form).

FH Modulation

A **FH modulation** $c(t)$ is formed by nonlinearly modulating a train of pulses with a sequence of pseudorandomly generated frequency shifts $\{f_n\}$. In mathematical terms, $c(t)$ has the complex form

$$c(t) = \sum_{n=-\bullet} \exp\{j(2\pi f_n + \phi_n)\}p(t - nT_h) \qquad (5)$$

where $p(t)$ is again the basic pulse shape having a duration T_h, called the hop time, and $\{\phi_n\}$ is a sequence of random phases associated with the generation of the hops. FH modulation is traditionally used with multiple-frequency-shift-keyed (MFSK) information signals, which have the complex form $\exp\{j2\pi(f_c + d(t))t\}$, where $d(t)$ is an M-level digital waveform (M denotes the symbol alphabet size) representing the information frequency modulation at a rate $1/T_s$ symbols/s (sps). As such, an FH/MFSK signal is formed by complex multiplying the MFSK signal by $c(t)$ resulting in the real transmitted signal

$$x(t) = \mathrm{Re}\{c(t) \exp\{j[2\pi(f_c + d(t))t]\}\} \qquad (6)$$

In reality, $c(t)$ is never generated in the transmitter. Rather, $x(t)$ is obtained by applying the sequence of pseudorandom frequency shifts $\{f_n\}$ directly to the frequency synthesizer that generates the carrier frequency f_c (see Fig. 2). In terms of the actual implementation, successive (not necessarily disjoint) k-chip segments of a PN sequence drive a frequency synthesizer, which hops the carrier over 2^k frequencies. In view of the large bandwidths over which the frequency synthesizer must operate, it is difficult to maintain phase coherence from hop to hop, which explains the inclusion of the sequence $\{\phi_n\}$ in the Eq. 5 model for $c(t)$. On a short term basis, e.g., within a given hop, the signal bandwidth is identical to that of the MFSK information modulation, which is typically much smaller than W_{ss}. On the other hand, when averaged over many hops, the signal bandwidth is equal to W_{ss}, which can be on the order of several gigaHertz, i.e., an order of magnitude larger than that of implementable DS bandwidths. The exact relation between W_{ss}, T_h, T_s and the number of frequency shifts in the set $\{f_n\}$ will be discussed shortly.

At the receiver, the sum of the transmitted FH/MFSK signal and the channel interference $I(t)$ is ideally complex multiplied by the identical FH modulation (this operation is known as dehopping), which returns the FH/MFSK signal to its original MFSK form, whereas the real interference signal is now the wideband (in the average sense) signal $\mathrm{Re}\{I(t)c(t)\}$. Analogous to the DS case, the receiver must acquire and track the FH signal so that the dehopping waveform is as close to the hopping waveform $c(t)$ as possible.

FH systems are traditionally classified in accordance with the relationship between T_h and T_s. Fast frequency-hopped (FFH) systems are ones in which there exist one or more hops per data symbol, i.e., $T_s = NT_h$ (N an integer), whereas slow frequency-hopped (SFH) systems are ones in which there exist more than one symbol per hop, that is, $T_h = NT_s$. It is customary in SS parlance to refer to the FH/MFSK tone of shortest duration as a "chip," despite the same usage for the PN chips associated with the code generator that drives the frequency synthesizer. Keeping this distinction in mind, in an FFH system where, as already stated, there are multiple hops per data symbol, a chip is equal to a hop. For SFH, where there are multiple data symbols per hop, a chip is equal to an MFSK symbol. Combining these two statements, the chip rate R_c in an FH system is given by the larger of $R_h = 1/T_h$ and $R_s = 1/T_s$ and, as such, is the highest system clock rate.

The frequency spacing between the FH/MFSK tones is governed by the chip rate R_c and is, thus, dependent on whether the FH modulation is FFH or SFH. In particular, for SFH where $R_c = R_s$, the spacing between FH/MFSK tones is equal to the spacing between the MFSK tones themselves. For noncoherent detection (the most commonly encountered in FH/MFSK systems), the separation of the MFSK symbols necessary to provide orthogonality is an integer multiple of R_s. Assuming the minimum spacing, i.e., R_s, the entire SS band is then

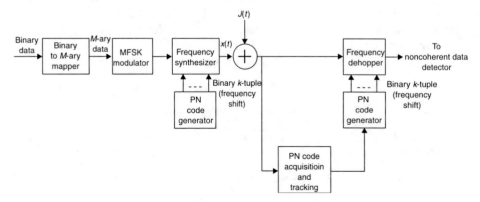

Fig. 2 An FH-MFSK system.

partitioned into a total of $N_t = W_{SS}/R_s = W_{SS}/R_c$ equally spaced FH tones. An optimum noncoherent MFSK detector consists of a bank of energy detectors each matched to one of the M frequencies in the MFSK set. In terms of this structure, the notion of orthogonality implies that for a given transmitted frequency there will be no crosstalk (energy spillover) in any of the other $M - 1$ energy detectors. One arrangement, which is by far the most common, is to group these N_t tones into $N_b = N_t/M$ contiguous, nonoverlapping bands, each with bandwidth $MR_s = MR_c$; see Fig. 3A. Assuming symmetric MFSK modulation around the carrier frequency, then the center frequencies of the $N_b = 2^k$ bands represent the set of hop carriers, each of which is assigned to a given k-tuple of the PN code generator. In this fixed arrangement, each of the N_t FH/MFSK tones corresponds to the combination of a unique hop carrier (PN code k-tuple) and a unique MFSK symbol. Another arrangement, which provides more protection against the sophisticated interferer (jammer), is to overlap adjacent M-ary bands by an amount equal to R_c; see Fig. 3B. Assuming again that the center frequency of each band corresponds to a possible hop carrier, then since all but $M - 1$ of he N_t tones are available as center frequencies,

the number of hop carriers has been increased from N_t/M to N_t–$(M - 1)$, which for $N_t \gg M$ is approximately an increase in randomness by a factor of M.

For FFH, where $R_c = R_h$, the spacing between FH/MFSK tones is equal to the hop rate. Thus, the entire SS band is partitioned into a total of $N_t = W_{SS}/R_h = W_{SS}/R_c$ equally spaced FH tones, each of which is assigned to a unique k-tuple of the PN code generator that drives the frequency synthesizer. Since for FFH there are R_h/R_s hops per symbol, then the metric used to make a noncoherent decision on a particular symbol is obtained by summing up R_h/R_s detected chip (hop) energies, resulting in a so-called *noncoherent combining loss*.

Time Hopping Modulation

Time hopping (TH) is to SS modulation what pulse position modulation (PPM) is to information modulation. In particular, consider segmenting time into intervals of T_f sec and further segment each T_f interval into M_T increments of width T_f/M_T. Assuming a pulse of maximum duration equal to T_f/M_T, then a **TH**-SS modulation would take the form

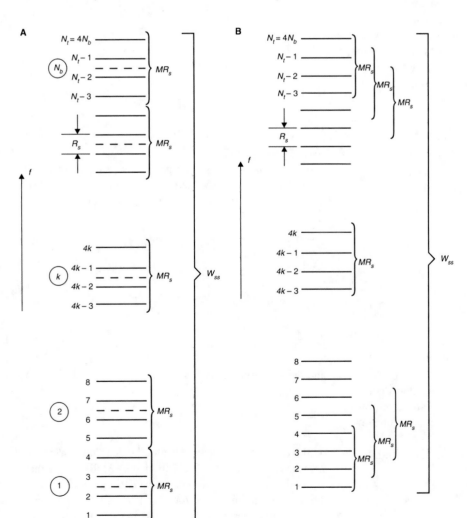

Fig. 3 Frequency distribution for FH-4FSK. (A) Nonoverlapping bands (dashed lines indicate location of hop frequencies). (B) overlapping bands.

$$c(t) = \sum_{n=-\bullet} p\left[t - \left(n + \frac{a_n}{M_T}\right)T_f\right] \qquad (7)$$

where a_n denotes the pseudorandom position (one of M_T uniformly spaced locations) of the pulse within the T_f second interval.

For DS and FH, we saw that *multiplicative* modulation, i.e., the transmitted signal is the product of the SS and information signals, was the natural choice. For TH, *delay* modulation is the natural choice. In particular, a TH-SS modulation takes the form

$$x(t) = \mathrm{Re}\{c(t - d(t)) \exp[j(2\pi f_c + \phi_\Gamma)]\} \qquad (8)$$

where $d(t)$ is a digital information modulation at a rate $1/T_s$ sps. Finally, the dehopping procedure at the receiver consists of removing the sequence of delays introduced by $c(t)$, which restores the information signal back to its original form and spreads the interferer.

Hybrid Modulations

By blending together several of the previous types of SS modulation, one can form hybrid modulations that, depending on the system design objectives, can achieve a better performance against the interferer than can any of the SS modulations acting alone. One possibility is to multiply several of the $c(t)$ wideband waveforms [now denoted by $c^{(i)}(t)$ to distinguish them from one another] resulting in a SS modulation of the form

$$c(t) = \prod_i c^{(i)}(t) \qquad (9)$$

Such a modulation may embrace the advantages of the various $c^{(i)}(t)$, while at the same time mitigating their individual disadvantages.

APPLICATIONS OF SS

Military

AJ communications

As already noted, one of the key applications of SS is for AJ communications in a hostile environment. The basic mechanism by which a **DS** SS receiver attenuates a noise jammer was illustrated in "Basic Concepts and Terminology" section. Therefore, in this section, we will concentrate on tone jamming.

Assume the received signal, denoted $r(t)$, is given by

$$r(t) = Ax(t) + I(t) + n_w(t) \qquad (10)$$

where $x(t)$ is given in Eq. 4, A is a constant amplitude,

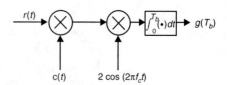

Fig. 4   Standard correlation receiver.

$$I(t) = \alpha \cos(2\pi f_c t + \theta) \qquad (11)$$

and $n_w(t)$ is additive white Gaussian noise (AWGN) having two-sided spectral density $N_0/2$. In Eq. 11, α is the amplitude of the tone jammer and θ is a random phase uniformly distributed in $[0, 2\pi]$.

If we employ the standard correlation receiver of Fig. 4, it is straightforward to show that the final test statistic out of the receiver is given by

$$g(T_b) = AT_b + \alpha \cos\theta \int_0^{T_b} c(t)\mathrm{d}t + N(T_b) \qquad (12)$$

where $N(T_b)$ is the contribution to the test statistic due to the AWGN. Noting that, for rectangular chips, we can express

$$\int_0^{T_b} c(t)\mathrm{d}t = T_c \sum_{i=1}^{M} c_i \qquad (13)$$

where

$$M \triangleq \frac{T_b}{T_c} \qquad (14)$$

is one-half of the processing gain, it is straightforward to show that, for a given value of θ, the signal-to-noise-plus-interference ratio, denoted S/N_{total}, is given by

$$\frac{S}{N_{\text{total}}} = \frac{1}{\frac{N_0}{2E_b} + \left(\frac{J}{MS}\right)\cos^2\theta} \qquad (15)$$

In Eq. 15, the jammer power is

$$J \triangleq \frac{\alpha^2}{2} \qquad (16)$$

and the signal power is

$$s \triangleq \frac{A^2}{2} \qquad (17)$$

If we look at the second term in the denominator of Eq. 15, we see that the ratio J/S is divided by M. Realizing that J/S is the ratio of the jammer power to the signal power before despreading, and J/MS is the ratio of the same quantity after despreading, we see that, as was the case for noise jamming, the benefit of employing DS SS signaling in the presence of tone jamming is to reduce the effect of the jammer by an amount on the order of the processing gain.

Finally, one can show that an estimate of the average probability of error of a system of this type is given by

$$P_e = \frac{1}{2\pi} \int_0^{2\pi} \phi\left(-\sqrt{\frac{S}{N_{\text{total}}}}\right) d\theta \qquad (18)$$

where

$$\phi(x) \overset{\Delta}{=} \frac{1}{\sqrt{2\pi} \int_{-\infty}^{x} e^{-y^2/2} dy} \qquad (19)$$

If Eq. 18 is evaluated numerically and plotted, the results are as shown in Fig. 5. It is clear from this figure that a large initial power advantage of the jammer can be overcome by a sufficiently large value of the processing gain.

LPI

The opposite side of the AJ problem is that of LPI, that is, the desire to hide your signal from detection by an intelligent adversary so that your transmissions will remain unnoticed and, thus, neither jammed nor exploited in any manner. This idea of designing an LPI system is achieved

in a variety of ways, including transmitting at the smallest possible power level, and limiting the transmission time to as short an interval in time as is possible. The choice of signal design is also important, however, and it is here that SS techniques become relevant.

The basic mechanism is reasonably straightforward; if we start with a conventional narrowband signal, say a BPSK waveform having a spectrum as shown in Fig. 6A, and then spread it so that its new spectrum is as shown in Fig. 6B, the peak amplitude of the spectrum after spreading has been reduced by an amount on the order of the processing gain relative to what it was before spreading. Indeed, a sufficiently large processing gain will result in the spectrum of the signal after spreading falling below the ambient thermal noise level. Thus, there is no easy way for an unintended listener to determine that a transmission is taking place.

That is not to say the spread signal cannot be detected, however, merely that it is more difficult for an adversary to learn of the transmission. Indeed, there are many forms of so-called intercept receivers that are specifically designed to accomplish this very task. By way of example, probably the best known and simplest to implement is a radiometer, which is just a device that measures the total power present in the received signal. In the case of our intercept problem, even though we have lowered the PSD of the transmitted signal so that it falls below the noise floor, we have not lowered its power (i.e., we have merely spread its power over a wider frequency range). Thus, if the radiometer integrates over a sufficiently long period of time, it will eventually determine the presence of the transmitted signal buried in the noise. The key point, of course, is that the use of the spreading makes the interceptor's task much more difficult, since he has no knowledge of the spreading code and, thus, cannot despread the signal.

Commercial

Multiple access communications

From the perspective of commercial applications, probably the most important use of SS communications is as a multiple accessing technique. When used in this manner, it becomes an alternative to either FDMA or TDMA and is typically referred to as either CDMA or SS multiple access (SSMA). When using CDMA, each signal in the set is given its own spreading sequence. As opposed to either FDMA, wherein all users occupy disjoint frequency bands but are transmitted simultaneously in time, or TDMA, whereby all users occupy the same bandwidth but transmit in disjoint intervals of time, in CDMA, all signals occupy the same bandwidth and are transmitted simultaneously in time; the different waveforms in CDMA are distinguished from one another at the receiver by the specific spreading codes they employ.

Since most CDMA detectors are correlation receivers, it is important when deploying such a system to have a set

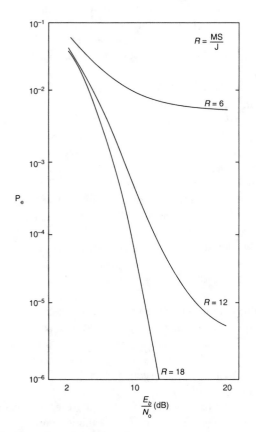

Fig. 5 Plotted results of Eq. 18.

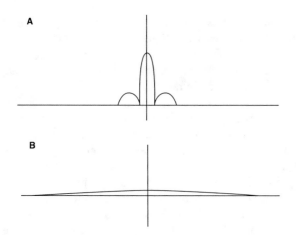

Fig. 6 LPI. (A) Conventional narrowband signal with BPSK waveform. (B) same signal with spread spectrum.

of spreading sequences that have relatively low-pairwise cross-correlation between any two sequences in the set. Further, there are two fundamental types of operation in CDMA, synchronous and asynchronous. In the former case, the symbol transition times of all of the users are aligned; this allows for orthogonal sequences to be used as the spreading sequences and, thus, eliminates interference from one user to another. Alternately, if no effort is made to align the sequences, the system operates asynchronously; in this latter mode, multiple access interference limits the ultimate channel capacity, but the system design exhibits much more flexibility.

CDMA has been of particular interest recently for applications in wireless communications. These applications include cellular communications, personal communications services (PCSs), and wireless local area networks. The reason for this popularity is primarily due to the performance that SS waveforms display when transmitted over a multipath fading channel.

To illustrate this idea, consider DS signaling. As long as the duration of a single chip of the spreading sequence is less than the multipath delay spread, the use of DS waveforms provides the system designer with one of two options. First, the multipath can be treated as a form of interference, which means the receiver should attempt to attenuate it as much as possible. Indeed, under this condition, all of the multipath returns that arrive at the receiver with a time delay greater than a chip duration from the multipath return to which the receiver is synchronized (usually the first return) will be attenuated because of the processing gain of the system.

Alternately, the multipath returns that are separated by more than a chip duration from the main path represent independent "looks" at the received signal and can be used constructively to enhance the overall performance of the receiver. That is, because all of the multipath returns contain information regarding the data that is being sent, that information can be extracted by an appropriately designed receiver. Such a receiver, typically referred to as a RAKE receiver, attempts to resolve as many

individual multipath returns as possible and then to sum them coherently. This results in an *implicit* diversity gain, comparable to the use of *explicit* diversity, such as receiving the signal with multiple antennas.

The condition under which the two options are available can be stated in an alternate manner. If one envisions what is taking place in the frequency domain, it is straightforward to show that the condition of the chip duration being smaller than the multipath delay spread is equivalent to requiring that the spread bandwidth of the transmitted waveform exceed what is called the coherence bandwidth of the channel. This latter quantity is simply the inverse of the multipath delay spread and is a measure of the range of frequencies that fade in a highly correlated manner. Indeed, anytime the coherence bandwidth of the channel is less than the spread bandwidth of the signal, the channel is said to be *frequency selective* with respect to the signal. Thus, we see that to take advantage of DS signaling when used over a multipath fading channel, that signal should be designed such that it makes the channel appear frequency selective.

In addition to the desirable properties that SS signals display over multipath channels, there are two other reasons why such signals are of interest in cellular-type applications. The first has to do with a concept known as the reuse factor. In conventional cellular systems, either analog or digital, in order to avoid excessive interference from one cell to its neighbor cells, the frequencies used by a given cell are not used by its immediate neighbors (i.e., the system is designed so that there is a certain spatial separation between cells that use the same carrier frequencies). For CDMA, however, such spatial isolation is typically not needed, so that so-called *universal reuse* is possible.

Further, because CDMA systems tend to be interference limited, for those applications involving voice transmission, an additional gain in the capacity of the system can be achieved by the use of *voice activity detection*. That is, in any given two-way telephone conversation, each user is typically talking only about 50% of the time. During the time when a user is quiet, he is not contributing to the instantaneous interference. Thus, if a sufficiently large number of users can be supported by the system, statistically only about one-half of them will be active simultaneously, and the effective capacity can be doubled.

Interference rejection

In addition to providing multiple accessing capability, SS techniques are of interest in the commercial sector for basically the same reasons they are used in the military community, namely their AJ and LPI characteristics. However, the motivations for such interest differ. For example, whereas the military is interested in ensuring that the systems they deploy are robust to interference generated by an intelligent adversary (i.e., exhibit jamming resistance), the interference of concern in commercial applications is unintentional. It is sometimes referred to as cochannel interference (CCI) and arises naturally as the result of many services using the same frequency band at the same time. And while such scenarios almost always allow for some type of spatial isolation between the interfering waveforms, such as the use of narrow-beam antenna patterns, at times the use of the inherent interference suppression property of a SS signal is also desired. Similarly, whereas the military is very much interested in the LPI property of a SS waveform, as indicated in "Basic Concepts and Terminology" section, there are applications in the commercial segment where the same characteristic can be used to advantage.

To illustrate these two ideas, consider a scenario whereby a given band of frequencies is somewhat sparsely occupied by a set of conventional (i.e., nonspread) signals. To increase the overall spectral efficiency of the band, a set of SS waveforms can be overlaid on the same frequency band, thus forcing the two sets of users to share a common spectrum. Clearly, this scheme is feasible only if the mutual interference that one set of users imposes on the other is within tolerable limits. Because of the interference suppression properties of SS waveforms, the despreading process at each SS receiver will attenuate the components of the final test statistic due to the overlaid narrowband signals. Similarly, because of the LPI characteristics of SS waveforms, the increase in the overall noise level as seen by any of the conventional signals, due to the overlay, can be kept relatively small.

Defining Terms

AJ communication system: A communication system designed to resist intentional jamming by the enemy.

Chip time (interval): The duration of a single pulse in a DS modulation; typically much smaller than the information symbol interval.

Coarse alignment: The process whereby the received signal and the despreading signal are aligned to within a single chip interval.

Dehopping: Despreading using a frequency-hopping modulation.

Delay-locked loop: A particular implementation of a closed-loop technique for maintaining fine alignment.

Despreading: The notion of decreasing the bandwidth of the received (spread) signal back to its information bandwidth.

DS modulation: A signal formed by linearly modulating the output sequence of a pseudorandom number generator onto a train of pulses.

DS SS: A spreading technique achieved by multiplying the information signal by a DS modulation.

FFH: A SS technique wherein the hop time is less than or equal to the information symbol interval, i.e., there exist one or more hops per data symbol.

Fine alignment: The state of the system wherein the received signal and the despreading signal are aligned to within a small fraction of a single chip interval.

Frequency-hopping modulation: A signal formed by nonlinearly modulating a train of pulses with a sequence of pseudorandomly generated frequency shifts.

Hop time (interval): The duration of a single pulse in a frequency-hopping modulation.

Hybrid SS: A spreading technique formed by blending together several SS techniques, e.g., DS, frequency-hopping, etc.

Low-probability-of-intercept communication system: A communication system designed to operate in a hostile environment wherein the enemy tries to detect the presence and perhaps characteristics of the friendly communicator's transmission.

Processing gain (spreading ratio): The ratio of the SS bandwidth to the information data rate.

Radiometer: A device used to measure the total energy in the received signal.

Search algorithm: A means for coarse aligning (synchronizing) the despreading signal with the received SS signal.

SFH: A SS technique wherein the hop time is greater than the information symbol interval, i.e., there exists more than one data symbol per hop.

SS bandwidth: The bandwidth of the transmitted signal after spreading.

Spreading: The notion of increasing the bandwidth of the transmitted signal by a factor far in excess of its information bandwidth.

Tau–dither loop: A particular implementation of a closed-loop technique for maintaining fine alignment.

Time-hopping SS: A spreading technique that is analogous to PPM.

Tracking algorithm: An algorithm (typically closed loop) for maintaining fine alignment.

BIBLIOGRAPHY

1. Cook, C.F.; Ellersick, F.W.; Milstein, L.B.; Schilling, D.L.; Spread Spectrum Communications, IEEE Press, 1983.
2. Dixon, R.C.; Spread Spectrum Systems, 3rd ed., John Wiley & Sons; New York, 1994.
3. Holmes, J.K.; Coherent Spread Spectrum Systems, John Wiley & Sons; New York, 1982.
4. Simon, M.K.; Omura, J.K.; Scholtz, R.A.; Levitt, B.K.; Spread Spectrum Communications Handbook, McGraw-Hill, 1994 (previously published as Spread Spectrum Communications, Computer Science Press, 1985).
5. Ziemer, R.E.; Peterson, R.L.; Digital Communications and Spread Spectrum Techniques, Macmillan; New York, 1985.

TDMA Cellular Systems: Speech and Channel Coding

Paul Mermelstein
INRS-Telecommunications, Montreal, Quebec, Canada

Abstract

This entry describes the techniques employed for speech transmission by the IS-54 digital cellular standard,
which are applied in TDMA cellular systems.

INTRODUCTION

The goals of this entry are to give the reader a tutorial introduction and high-level understanding of the techniques employed for speech transmission by the IS-54 digital cellular standard. It builds on the information provided in the standards document but is not meant to be a replacement for it. Separate standards cover the control channel used for the setup of calls and their handoff to neighboring cells, as well as the encoding of data signals for transmission. For detailed implementation information the reader should consult the most recent standards document.[1]

IS-54 provides for encoding bidirectional speech signals digitally and transmitting them over cellular and microcellular mobile radio systems. It retains the 30-kHz channel spacing of the earlier advanced mobile telephone service (AMPS), which uses analog frequency modulation for speech transmission and frequency shift keying for signaling. The two directions of transmission use frequencies some 45 MHz apart in the band between 824 and 894 MHz. AMPS employs one channel per conversation in each direction, a technique known as FDMA. IS-54 employs TDMA by allowing three, and in the future six, simultaneous transmissions to share each frequency band. Because the overall 30-kHz channelization of the allocated 25 MHz of spectrum in each direction is retained, it is also known as a FDMA–TDMA system. In contrast, the later IS-95 standard employs CDMA over bands of 1.23 MHz by combining several 30-kHz frequency channels.

Each frequency channel provides for transmission at a digital bit rate of 48.6 Kbps through use of differential quadrature-phase shift key (DQPSK) modulation at a 24.3-kBd channel rate. The channel is divided into six time slots every 40 ms. The full rate voice coder employs every third time slot and utilizes 13 Kbps for combined speech and channel coding. The six slots provide for an eventual half-rate channel occupying one slot per 40 ms frame and utilizing only about 6.5 Kbps for each call. Thus, the simultaneous call carrying capacity with IS-54 is increased by a factor 3 (factor 6 in the future) above that of AMPS.

All digital transmission is expected to result in a reduction in transmitted power. The resulting reduction in inter-cell interference may allow more frequent reuse of the same frequency channels than the reuse pattern of seven cells for AMPS. Additional increases in Erlang capacity (the total call carrying capacity at a given blocking rate) may be available from the increased trunking efficiency achieved by the larger number of simultaneously available channels. The first systems employing dual-mode AMPS and TDMA service were put into operation in 1993.

In 1996, the Telecommunications Industry Association TIA introduced the IS-641 enhanced full rate codec. This codec consists of 7.4 Kbps speech coding following the algebraic code-excited linear prediction (ACELP) technique,[2] and 5.6 Kbps channel coding. The 13 Kbps coded information replaces the combined 13 Kbps for speech and channel coding introduced by the IS-54 standard. The new codec provides significant enhancements in terms of speech quality and robustness to transmission errors. The quality enhancement for clear channels results from the improved modeling of the stochastic excitation by means of an algebraic **codebook** instead of the two trained vector-sum excited linear prediction (VSELP) codebooks. Improved robustness to transmission errors is achieved by employing predictive quantization techniques for the linear-prediction filter and gain parameters, and increasing the number of bits protected by forward error correction.

Modulation of Digital Voice and Data Signals

The modulation method used in IS-54 is $\pi/4$ shifted differentially encoded quadrature phase-shift keying (DPSK). Symbols are transmitted as changes in phase rather than their absolute values. The binary data stream is converted to two binary streams X_k and Y_k formed from the odd- and even-numbered bits, respectively. The quadrature streams I_k and Q_k are formed according to

$$I_k = I_{k-1} \cos[\Delta\phi(X_k, Y_k)] - Q_{k-1} \sin[\Delta\phi(X_k, Y_k)]$$
$$Q_k = I_{k-1} \sin[\Delta\phi(X_k, Y_k)] - Q_{k-1} \cos[\Delta\phi(X_k, Y_k)]$$

Encyclopedia of Wireless and Mobile Communications DOI: 10.1081/E-EWMC-120043945

where I_{k-1} and Q_{k-1} are the amplitudes at the previous pulse time. The phase change $\Delta\phi$ takes the values $\pi/4$, $3\pi/4$, $-\pi/4$, and $-3\pi/4$ for the dibit (X_k, Y_k) symbols $(0,0)$, $(0,1)$, $(1,0)$, and $(1,1)$, respectively. This results in a rotation by $\pi/4$ between the constellations for odd and even symbols. The differential encoding avoids the problem of $180°$ phase ambiguity that may otherwise result in estimation of the carrier phase.

The signals I_k and Q_k at the output of the differential phase encoder can take one of five values, $0, \pm 1, \pm 1/\sqrt{2}$ as indicated in the constellation of Fig. 1. The corresponding impulses are applied to the inputs of the I and Q baseband filters, which have linear phase and square root raised cosine frequency responses. The generic modulator circuit is shown in Fig. 2. The roll-off factor α determines the width of the transition band and its value is 0.35,

$$|H(f)|$$

$$= \begin{cases} 1 & 0 \leqslant f \leqslant (1-\alpha)/2T \\ \sqrt{1/2\{1-\sin[\pi(2fT-1)/2\alpha]\}} & (1-\alpha)/2T \leqslant f \leqslant (1+\alpha)/2T \\ 0 & f > (1+\alpha)/2T \end{cases}$$

Speech Coding Fundamentals

The IS-54 standard employs a VSELP coding technique. It represents a specific formulation of the much larger class of CELP coders [3] that have proved effective in recent years for the coding of speech at moderate rates in the range 4–16 Kbps. VSELP provides reconstructed speech with a quality that is comparable to that available with frequency modulation and analog transmission over the AMPS system. The coding rate employed is 7.95 Kbps. Each of the six slots per frame carry 260 bits of speech and channel coding information for a gross information rate of 13 Kbps. The 260 bits correspond to 20 ms of real time speech, transmitted as a single burst. (For a general

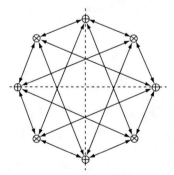

Fig. 1 Constellation for $\pi/4$ shifted QPSK modulation. *Source*: TIA, 1992. Cellular System Dual-mode Mobile Station–Base Station Compatibility Standard TIA/EIA IS-54. With permission.

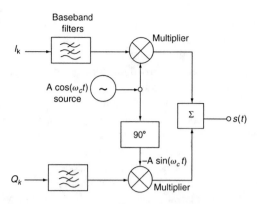

Fig. 2 Generic modulation circuit for digital voice and data signals. *Source*: TIA, 1992. Cellular System Dual-mode Mobile Station–Base Station Compatibility Standard TIA/EIA IS-54.

treatment of speech coding for telecommunications, see Ref. 4. For a more detailed treatment of linear prediction techniques, see Ref. 5)

For an excellent recent review of speech coding techniques for transmission, the reader is referred to Gersho, 1994.[6] Most modern speech coders use a form of analysis by synthesis coding where the encoder determines the coded signal one segment at a time by feeding candidate excitation segments into a replica of a synthesis filter and selecting the segment that minimizes the distortion between the original and reproduced signals. Linear prediction coding (LPC) techniques[7] encode the speech signal by first finding an optimum linear filter to remove the short-time correlation, passing the signal through that LPC filter to obtain a residual signal, and encoding this residual using much fewer bits than would have been required to code the original signal with the same fidelity. In most cases the coding of the residual is divided into two steps. First, the long-time correlation due to the periodic pitch excitation is removed by means of an optimum one-tap filter with adjustable gain and lag. Next, the remaining residual signal, which now closely resembles a white-noise signal, is encoded. Code-excited linear predictors use one or more codebooks from which they select replicas of the residual of the input signal by means of a closed-loop error-minimization technique. The index of the codebook entry as well as the parameters of all the filters are transmitted to allow the speech signal to be reconstructed at the receiver. Most code-excited coders use trained codebooks. Starting with a codebook containing Guassian signal segments, entries that are found to be used rarely in coding a large body of speech data are iteratively eliminated to result in a smaller codebook that is considered more effective.

The speech signal can be considered quasistationary or stationary for the duration of the speech frame, of the order of 20 ms. The parameters of the short-term filter, the LPC coefficients, are determined by analysis of the autocorrelation function of a suitably windowed segment of the input signal. To allow accurate determination of the time-varying pitch lag as well as simplify the computations, each speech frame is divided into four 5 ms subframes. Independent pitch filter computations and residual coding operations are carried out for each subframe.

The speech decoder attempts to reconstruct the speech signal from the received information as best possible. It employs a codebook identical to that of the encoder for excitation generation and, in the absence of transmission errors, would produce an exact replica of the signal that produced the minimized error at the encoder. Transmission errors do occur, however, due, to signal fading and excessive interference. Since any attempt at retransmission would incur unacceptable signal delays, sufficient error protection is provided to allow correction of most transmission errors.

CHANNEL CODING CONSIDERATIONS

The sharp limitations on available bandwidth for error protection argue for careful consideration of the sensitivity of the speech coding parameters to transmission errors. Pairwise interleaving of coded blocks and convolutional coding of a subset of the parameters permit correction of a limited number of transmission errors. In addition, a cyclic redundancy check (CRC) is used to determine whether the error correction was successful. The coded information is divided into three blocks of varying sensitivity to errors. Group 1 contains the most sensitive bits, mainly the parameters of the LPC filter and frame energy, and is protected by both error detection and correction bits. Group 2 is provided with error correction only. The third group, comprising mostly the fixed codebook indices, is not protected at all.

The speech signal contains significant temporal redundancy. Thus, speech frames within which errors have been detected may be reconstructed with the aid of previously correctly received information. A bad frame masking procedure attempts to hide the effects of short fades by extrapolating the previously received parameters. Of course, if the errors persist, the decoded signal must be muted

while an attempt is made to hand off the connection to a base station to/from which the mobile may experience better reception.

VSELP ENCODER

A block diagram of the VSELP speech encoder[8] is shown in Fig. 3. The excitation signal is generated from three components, the output of a long-term or pitch filter, as well as entries from two codebooks. A weighted synthesis filter generates a synthesized approximation to the frequency-weighted input signal. The weighted mean square error between these two signals is used to drive the error-minimization process. This weighted error is considered to be a better approximation to the perceptually important noise components than the unweighted mean square error. The total weighted square error is minimized by adjusting the pitch lag and the codebook indices as well as their gains. The decoder follows the encoder closely and generates the excitation signal identically to the encoder but uses an unweighted linear prediction synthesis filter to generate the decoded signal. A spectral postfilter is added after the synthesis filter to enhance the quality of the reconstructed speech.

The precise data rate of the speech coder is 7950 bps or 159 bits per time slot, each corresponding to 20 ms of signal in real time. These 159 bits are allocated as follows: 1) short-term filter coefficients, 38 bits; 2) frame energy, 5 bits; 3) pitch lag, 28 bits; 4) code words, 56 bits; and 5) gain values, 32 bits.

LINEAR PREDICTION ANALYSIS AND QUANTIZATION

The purpose of the LPC analysis filter is to whiten the spectrum of the input signal so that it can be better matched by the codebook outputs. The corresponding LPC

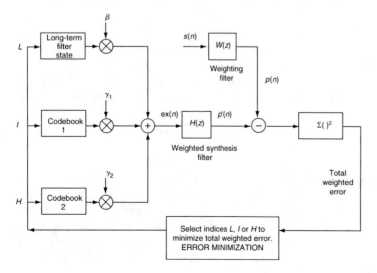

Fig. 3 Block diagram of the speech encoder in VSELP. *Source*: TIA, 1992. Cellular system Dual-mode Mobile Station–Base Station Compatibility Standard TIA/EIA IS-54.

synthesis filter $A(z)$ restores the short-time speech spectrum characteristics to the output signal. The transfer function of the tenth-order synthesis filter is given by

$$A(z) = \frac{1}{1 - \sum_{i=1}^{N_p} \alpha_i z^{-i}}$$

The filter predictor parameters $\alpha_1, \ldots, \alpha_{N_p}$ are not transmitted directly. Instead, a set of **reflection coefficients** $r_1, \ldots, r_{N_p}$ are computed and quantized. The predictor parameters are determined from the reflection coefficients using a well-known backward recursion algorithm.[9]

A variety of algorithms are known that determine a set of reflection coefficients from a windowed input signal. One such algorithm is the fixed point **covariance lattice,** FLAT, which builds an optimum inverse lattice stage by stage. At each stage j, the sum of the mean-squared forward and backward residuals is minimized by selection of the best reflection coefficient r_j. The analysis window used is 170 samples long, centered with respect to the middle of the fourth 5-ms subframe of the 20 ms frame. Since this center point is 20 samples from the end of the frame, 65 samples from the next frame to be coded are used in computing the reflection coefficient of the current frame. This introduces a lookahead delay of 8.125 ms.

The FLAT algorithm first computes the covariance matrix of the input speech for $N_A = 170$ and $N_p = 10$,

$$\phi(i,k) = \sum_{n=N_P}^{N_A^{-1}} s(n-i)s(n-k), \quad 0 \leqslant i, \, k \leqslant N_P$$

Define the forward residual out of stage j as $f_j(n)$ and the backward residual as $b_j(n)$. Then the autocorrelation of the initial forward residual $F_0(i,k)$ is given by $\phi(i,k)$. The autocorrelation of the initial backward residual $B_0(i,k)$ is given by $\phi(i+1, k+1)$ and the initial cross correlation of the two residuals is given by $C_0(i,k) = \phi(i, k+1)$ or $0 \leqslant i, k \leqslant N_{p-1}$. Initially j is set to 1. The reflection coefficient at each stage is determined as the ratio of the cross correlation to the mean of the autocorrelations. A block diagram of the computations is shown in Fig. 4. By quantizing the reflection coefficients within the computation loops, reflection coefficients at subsequent stages are computed taking into account the quantization errors of the previous stages. Specifically,

$$C'_{j-1} = C_{j-1}(0,0) + C_{j-1}(N_p - j, N_p - j)$$
$$F'_{j-1} = F_{j-1}(0,0) + F_{j-1}(N_p - j, N_p - j)$$
$$B'_{j-1} = B_{j-1}(0,0) + B_{j-1}(N_p - j, N_p - j)$$

and

$$r_j = \frac{-2 C'_{j-1}}{F'_{j-1} + B'_{j-1}}$$

Use of two sets of correlation values separated by $N_p - j$ samples provides additional stability to the computed reflection coefficients in case the input signal changes form rapidly.

Once a quantized reflection coefficient r_j has been determined, the resulting auto- and cross correlations can be determined iteratively as

$$F_j(i,k) = F_{j-1}(i,k) + r_j\left[C_{j-1}(i,k) + C_{j-1}(k,i)\right]$$
$$+ r_j^2 B_{j-1}(i,k)$$
$$B_j(i,k) = B_{j-1}(i+1, k+1) + r_j\left[C_{j-1}(i+1, k+1)\right.$$
$$\left. + C_{j-1}(k+1, i+1)\right] + r_j^2 F_{j-1}(i+1, k+1)$$

and

$$C_j(i,k) = C_{j-1}(i, k+1) + r_j\left[B_{j-1}(i, k+1)\right.$$
$$\left. + F_{j-1}(i, k+1)\right] + r_j^2 C_{j-1}(k+1, i)$$

These computations are carried out iteratively for $r_{j.} = 1, \ldots, N_p$.

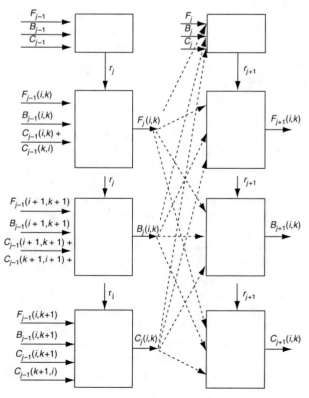

Fig. 4 Block diagram for lattice covariance computations.

BANDWIDTH EXPANSION

Poles with very narrow bandwidths may introduce undesirable distortions into the synthesized signal. Use of a binomial window with effective bandwidth of 80 Hz suffices to limit the ringing of the LPC filter and reduce the effect of the LPC filter selected for one frame on the signal reconstructed for subsequent frames. To achieve this, prior to searching for the reflection coefficients, the $\phi(i,k)$ is modified by use of a window function $w(j)$, $j = 1, \ldots, 10$, as follows:

$$\phi'(i, k) = \phi(i,k)w(|i - k|)$$

QUANTIZING AND ENCODING THE REFLECTION COEFFICIENTS

The distortion introduced into the overall spectrum by quantizing the reflection coefficients diminishes as we move to higher orders in the reflection coefficients. Accordingly, more bits are assigned to the lower order coefficients. Specifically, 6, 5, 5, 4, 4, 3, 3, 3, 3, and 2 bits are assigned to $r_1, \ldots, r_{10}$, respectively. Scalar quantization of the reflection coefficients is used in IS-54 because it is particularly simple. **Vector quantization** achieves additional quantizing efficiencies at the cost of significant added complexity.

It is important to preserve the smooth time evolution of the linear prediction filter. Both the encoder and decoder linearly interpolate the coefficients α_i for the first, second, and third subframes of each frame using the coefficients determined for the previous and current frames. The fourth subframe uses the values computed for that frame.

VSELP Codebook Search

The codebook search operation selects indices for the long-term filter (pitch lag L) and the two codebooks I and H so as to minimize the total weighted error. This closed-loop search is the most computationally complex part of the encoding operation, and significant effort has been invested to minimize the complexity of these operations without degrading performance. To reduce complexity, simultaneous optimization of the codebook selections is replaced by a sequential optimization procedure, which considers the long-term filter search as the most significant and therefore executes it first. The two vector-sum codebooks are considered to contribute less and less to the minimization of the error, and their search follows in sequence. Subdivision of the total codebook into two vector sums simplifies the processing and makes the result less

sensitive to errors in decoding the individual bits arising from transmission errors.

Entries from each of the two vector-sum codebooks can be expressed as the sum of basis vectors. By orthogonalizing these basis vectors to the previously selected codebook component(s), one ensures that the newly introduced components reduce the remaining errors. The subframes over which the codebook search is carried out are 5 ms or 40 samples long. An optimal search would need exploration of a 40-dimensional space. The vector-sum approximation limits the search to 14 dimensions after the optimal pitch lag has been selected. The search is further divided into two stages of 7 dimensions each. The two codebooks are specified in terms of the fourteen, 40-dimensional basis vectors stored at the encoder and decoder. The two 7-bits indices indicate the required weights on the basic vectors to arrive at the two optimum code words.

The codebook search can be viewed as selecting the three best directions in 40-dimensional space, which when summed result in the best approximation to the weighted input signal. The gains of the three components are determined through a separate error-minimization process.

LONG-TERM FILTER SEARCH

The long-term filter is optimized by selection of a lag value that minimizes the error between the weighted input signal $p(n)$ and the past excitation signal filtered by the current weighted synthesis filter $H(z)$. There are 127 possible coded lag values provided corresponding to lags of 20–146 samples. One value is reserved for the case when all correlations between the input and the lagged residuals are negative and use of no long-term filter output would be best. To simplify the convolution operation between the impulse response of the weighted synthesis filter and the past excitation, the impulse response is truncated to 21 samples or 2.5 ms. Once the lag is determined, the untruncated impulse response is used to compute the weighted long-term lag vector.

ORTHOGONALIZATION OF THE CODEBOOKS

Prior to the search of the first codebook, each filtered basis vector may be made orthogonal to the long-term filter output, the zero-state response of the weighted synthesis filter $H(z)$ to the long-term prediction vector. Each orthogonalized filtered basis vector is computed by subtracting its projection onto the long-term filter output from itself.

Similarly, the basis vectors of the second codebook can be orthogonalized with respect to both the long-term filter output and the first codebook output, the zero-state response of $H(z)$ to the previously selected summation of first codebook basis vectors. In each case the codebook excitation can be reconstituted as

$$u_{k,i}(n) = \sum_{m=1}^{M} q_{im} v_{k,m}(n)$$

where $k = 1, 2$ for the two codebooks, $i = I$ or H the 7-bits code vector received, $v_{k,m}$ are the two sets of basis vectors, and $\theta_{im} = +1$ if bit m of codeword $i = 1$ and -1 if bit m of codeword $i = 0$. Orthogonalization is not required at the decoder since the gains of the codebooks outputs are determined with respect to the weighted non-orthogonalized code vectors.

Quantizing the Excitation and Signal Gains

The three codebook gain values β, γ_1, and γ_2 are transformed to three new parameters GS, $P0$, and $P1$ for quantization purposes. GS is an energy offset parameter that equalizes the input and output signal energies. It adjusts the energy of the output of the LPC synthesis filter to equal the energy computed for the same subframe at the encoder input. $P0$ is the energy contribution of the long-term prediction vector as a fraction of the total excitation energy within the subframe. Similarly, $P1$ is the energy contribution of the code vector selected from the first codebook as a fraction of the total excitation energy of the subframe.

The transformation reduces the dynamic range of the parameters to be encoded. An 8-bits **vector quantizer** efficiently encodes the appropriate $(GS,P0,P1)$ vectors by selecting the vector which minimizes the weighted error. The received and decoded values β, γ_1, and γ_2 are computed from the received $(GS,P0,P1)$ vector and applied to reconstitute the decoded signal.

CHANNEL CODING AND INTERLEAVING

The goals of channel coding are to reduce the impairments in the reconstructed speech due to transmission errors. The 159 bits characterizing each 20 ms block of speech are divided into two classes, 77 in class 1 and 82 in class 2. Class 1 includes the bits in which errors result in a more significant impairment, whereas the speech quality is considered less sensitive to the class-2 bits. Class 1 generally includes the gain, pitch lag, and more significant reflection coefficient bits. In addition, a 7-bits CRC is applied to the 12 most perceptually significant bits of class 1 to indicate whether the error correction was successful. Failure of the CRC check at the receiver suggests that the received information is so erroneous that it would be better to discard it than use it. The error correction coding is illustrated in Fig. 5.

The error correction technique used is rate $1/2$ convolutional coding with a constraint length of 5.[10] A tail of 5 bits is appended to the 84 bits to be convolutionally encoded to result in a 178-bits output. Inclusion of the tail bits ensures independent decoding of

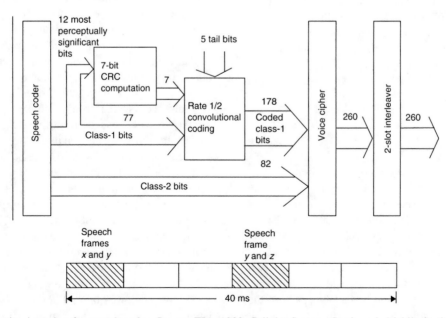

Fig. 5 Error correction insertion for speech coder. *Source*: TIA, 1992. Cellular Systems Dual-mode Mobile Station–Base Station Compatibility Standards TIA/EIA IS-54. With permission.

successive time slots and no propagation of errors between slots.

Interleaving the bits to be transmitted over two time slots is introduced to diminish the effects of short deep fades and to improve the error-correction capabilities of the channel coding technique. Two speech frames, the previous and the present, are interleaved so that the bits from each speech block span two transmission time slots separated by 20 ms. The interleaving attempts to separate the convolutionally coded class-1 bits from one frame as much as possible in time by inserting non-coded class-2 bits between them.

BAD FRAME MASKING

A CRC failure indicates that the received data is unusable, either due to transmission errors resulting from a fade, or from pre-emption of the time slot by a control message [fast associated control channel (FACCH)]. To mask the effects that may result from leaving a gap in the speech signal, a masking operation based on the temporal redundancy between adjacent speech blocks has been proposed. Such masking can at best bridge over short gaps but cannot recover loss of signal of longer duration. The bad frame masking operation may follow a finite state machine where each state indicates an operation appropriate to the elapsed duration of the fade to which it corresponds. The masking operation consists of copying the previous LPC information and attenuating the gain of the signal. State 6 corresponds to error sequences exceeding 100 ms, for which the output signal is muted. The result of such a masking operation is generation of an extrapolation in the gap to the previously received signal, significantly reducing the perceptual effects of short fades. No additional delay is introduced in the reconstructed signal. At the same time, the receiver will report a high frequency of bad frames leading the system to explore handoff possibilities immediately. A quick successful handoff will result in rapid signal recovery.

ACELP ENCODER

The ACELP encoder employs linear prediction analysis and quantization techniques similar to those used in VSELP and discussed in section "Linear Prediction Analysis and Quantization." The frame structure of 20 ms frames and 5 ms subframes is preserved. Linear prediction analysis is carried out for every frame. The ACELP encoder uses a long-term filter similar to the one discussed in section "Long-Term Filter Search" and represented as an adaptive codebook. The non-predictable part of the LPC residual is represented in terms of ACELP

codebooks, which replace the two VSELP codebooks shown in Fig. 3.

Instead of encoding the reflection coefficients as in VSELP, the information is transformed into line-spectral frequency pairs (LSPs).[11] The LSPs can be derived from linear prediction coefficients, a 10th order analysis generating 10 line-spectral frequencies (LSF), 5 poles, and 5 zeroes. The LSFs can be vector quantized and the LPC coefficients recalculated from the quantized LSFs. As long as the interleaved order of the poles and zeroes is preserved, quantization of the LSPs preserves the stability of the LPC synthesis filters. The LSPs of any frame can be better predicted from the values calculated and transmitted corresponding to previous frames, resulting in additional advantages. The long-term means of the LSPs are calculated for a large body of speech data and stored at both the encoder and decoder. First-order moving-average prediction is then used for the mean-removed LSPs. The time-prediction technique also permits use of predicted values for the LSPs in case uncorrectable transmissions errors are encountered, resulting in reduced speech degradation. To simplify the vector quantization operations, each LSP vector is split into 3 subvectors of dimensions 3, 3, and 4. The three subvectors are quantized with 8, 9, and 9 bits respectively, corresponding to a total bit assignment of 26 bpf for LPC information.

ALGEBRAIC CODEBOOK STRUCTURE AND SEARCH

Algebraic codebooks contain relatively few pulses having nonzero values leading to rapid search of the possible innovation vectors, the vectors which together with the Algebraic Code Book (ACB) output form the excitation of the LPC filter for the current subframe. In this implementation the 40-position innovation vector contains only four nonzero pulses and each can take on only values $+1$ and -1. The 40 positions are divided into four tracks and one pulse is selected from each track. The tracks are generally equally spaced but differ in their starting value, thus the first pulse can take on positions 0, 5, 10, 15, 20, 25, 30, or 35 and the second has possible positions 1, 6, 11, 16, 21, 26, 31, or 36. The first three pulse positions are coded with 3 bits and the fourth pulse position (starting positions 3 or 4) with 4 bits, resulting in a 17-bit sequence for the algebraic code of each subframe.

The algebraic codebook is searched by minimizing the mean square error between the weighted input speech and the weighted synthesized speech over the time span of each subframe. In each case the weighting is that produced by a perceptual weighting filter that has the effect of shaping the spectrum of the synthesis error signal so that it is better masked by spectrum of the current speech signal.

QUANTIZATION OF THE GAINS FOR ACELP ENCODING

The adaptive codebook gain and the fixed (algebraic) codebook gains are vector quantized using a 7-bit codebook. The gain codebook search is performed by minimizing the mean square of the weighted error between the original and the reconstructed speech, expressed as a function of the adaptive codebook gain and a fixed codebook correction factor. This correction factor represents the log energy difference between a predicted gain and an estimated gain. The predicted gain is computed using fourth-order moving-average prediction with fixed coefficients on the innovation energy of each subframe. The result is a smoothed energy profile even in the presence of modest quantization errors. As discussed above in case of the LSP quantization, the moving-average prediction serves to provide predicted values even when the current frame information is lost due to transmission errors. Degradations resulting from loss of one or two frames of information are thereby mitigated.

CHANNEL CODING FOR ACELP ENCODING

The channel coding and interleaving operations for ACELP speech coding are similar to those discussed in section "Channel Coding and Interleaving" for VSELP coding. The number of bits protected by both error-detection (parity) and error-correction convolutional coding is increased to 48 from 12. Rate 1/2 convolutional coding is used on the 108 more significant bits, 96 class-1 bits, 7 CRC bits and the 5 tail bits of the convolutional coder, resulting in 216 coded class-1 bits. Eight of the 216 bits are dropped by puncturing, yielding 208 coded class-1 bits which are then combined with 52 non-protected class-2 bits. As compared to the channel coding of the VSELP encoder, the numbers of protected bits is increased and the number of unprotected bits is reduced while keeping the overall coding structure unchanged.

CONCLUSIONS

The IS-54 digital cellular standard specifies modulation and speech coding techniques for mobile cellular systems that allow the interoperation of terminals built by a variety of manufacturers and systems operated across the country by a number of different service providers. It permits speech communication with good quality in a transmission environment characterized by frequent multi-path fading and significant inter-cell interference. Generally, the quality of the IS-54 decoded speech is better at the edges of a cell than the corresponding AMPS transmission due to the error mitigation resulting from channel coding. Near a base station or in the absence of significant fading and interference, the IS-54 speech quality is reported to be somewhat worse than AMPS due to the inherent limitations of the analysis–synthesis model in reconstructing arbitrary speech signals with limited bits. The IS-641 standard coder achieves higher speech quality, particularly at the edges of heavily occupied cells where transmission errors may be more numerous. At this time no new systems following the IS-54 standard are being introduced. Most base stations have been converted to transmit and receive on the IS-641 standard as well and use of IS-54 transmissions is dropping rapidly. At the time of its introduction in 1996 the IS-641 coder represented the state of the art in terms of toll quality speech coding near 8 Kbps, a significant improvement over the IS-54 coder introduced in 1990. These standards represent reasonable engineering compromises between high performance and complexity sufficiently low to permit single-chip implementations in mobile terminals.

Both IS-54 and IS-641 are considered second generation cellular standards. Third generation cellular systems promise higher call capacities through better exploitation of the time-varying transmission requirements of speech conversations, as well as improved modulation and coding in wider spectrum bandwidths that achieve similar bit-error ratios but reduce the required transmitted power. Until such systems are introduced, the second generation TDMA systems can be expected to provide many years of successful cellular and personal communications services.

DEFINING TERMS

Codebook: A set of signal vectors available to both the encoder and decoder

Covariance lattice algorithm: An algorithm for reduction of the covariance matrix of the signal consisting of several lattice stages, each stage implementing an optimal first-order filter with a single coefficient.

Reflection coefficient: A parameter of each stage of the lattice linear prediction filter that determines 1) a forward residual signal at the output of the filter stage by subtracting from the forward residual at the input a linear function of the backward residual; also 2) a backward residual at the output of the filter stage by subtracting a linear function of the forward residual from the backward residual at the input.

Vector quantizer: A quantizer that assigns quantized vectors to a vector of parameters based on their current values by minimizing some error criterion.

REFERENCES

1. Telecommunications Industry Association. EA/TIA Iterim Standard, Cellular System Dual-mode Mobile Station–Base Station Compatibility Standard IS-54B: TIA/EIA: Washington, D.C., 1992.

2. Salami, R.; Laflamme, C.; Adoul, J.P.; Massaloux, D. A toll quality 8 kb/s speech codec for the personal communication system (PCS). IEEE Trans. Vehic. Tech. **1994**, *43*, 808–816.

3. Jayant, N.S.; Noll, P. *Digital Coding of Waveforms*; Prentice Hall: Englewood, NJ, 1984.

4. Markel, J.; Gray, A. *Linear Prediction of Speech*; Springer–Verlag: New York, 1976.

5. Atal, B.S.; Schroeder, M. Stochastic coding of speech signals at very low bit rates. International Conference on Communication, Amsterdam, The Netherlands, May 14–17, 19841610–1613.

6. Gersho, A. Advances in speech and audio compression. IEEE. **1994**, *82*, 900–918.

7. Atal, B.S.; Hanauer, S.L. Speech analysis and synthesis by linear prediction of the speech wave. J. Acoust. Soc. Am. **1971**, *50*, 637–655.

8. Gerson, I.A.; Jasiuk, M.A. Vector sum excited linear prediction (VSELP) speech coding at 8 Kbps. IEEE International Conference on Acoustic, Speech and Signal Processing, ICASSP 1990, Albuquerque, NM, Apr 3–6, 1990; 461–464, 1990.

9. Makhoul, J. Linear prediction, a tutorial review. IEEE. **1975**, *63*, 561–580.

10. Lin, S.; Costello, D. *Error Control Coding: Fundamentals and Application*; Prentice Hall: Englewood Cliffs, NJ, 1983.

11. Soong, F.K.; Juang, B.H. Line spectrum pair (LSP) and speech data compression. IEEE International Conference on Acoustic, Speech and Signal Processing, ICASSP 1984, San Diego, CA, Mar 19–21, 19841.10.1–1.10.4.

TDMA/TDD Wireless Networks: Quality of Service (QoS) Mechanisms

Francisco M. Delicado
Pedro Cuenca
Luis Orozco-Barbosa
Jesús Delicado
Albacete Research Institute of Informatics, University of Castilla-La Mancha, Albacete, Spain

Abstract

Time division multiple access/time division duplex (TDMA/TDD) wireless networks all implement a centralized medium access control (MAC) protocol, based on time division multiplexing (TDM) [i.e., high-performance radio local area network (HiperLAN/2), IEEE 0802.16].

INTRODUCTION

Nowadays, wireless networks represent an alternative to wired networks. Current wireless networks operate at transmission rates that are able to support all types of applications: data, voice, video, etc. It is widely recognized that one of the main advantages of wireless networks is their great flexibility: the wire is done away with, allowing users to freely connect to the network. Standardization efforts have resulted in the definition of wireless networks standards, one of the main aims of which is to guarantee interoperability among equipments developed by different vendors. To date, various broadband wireless network standards have been defined.[1–4] We can classify wireless standards according to their medium access control (MAC) protocol, obtaining the following two types of MAC protocols: distributed and centralized. In the first case, the MAC is based on a contention process. This access control is used mainly by the family of IEEE 802.11 standards which implement a protocol based on a carrier sense multiple access with collision avoidance (CSMA/CA) algorithm with a "*backoff*" process. In this type of MAC protocol, a station has to determine the state of the channel before transmitting. A station may start to transmit after making sure that the channel is idle for an interval of time longer than the distributed interframe space (DIFS). On the other hand, if the channel is busy, once the transmission in course finishes and in order to avoid a potential collision with other active (waiting) stations, the station will wait a random interval of time (the backoff process) before

starting to transmit. The main problem with this type of access control is due to collisions, which increase rapidly with the number of mobile terminals (MTs). A station know that its transmission has suffered a collision if it does not receive an ACK (acknowledgment) message. In this case, the station re-transmits the information which has suffered the collision, until an ACK is received. To avoid further collisions with others, the station runs backoff periods. They run backoff period after the first collision.

Centralized MAC protocols have a "central controller" in the network, i.e., a device that decides which stations can obtain access to the medium and that controls the amount of data the stations can transmit. Usually, it is called BS (*base station*), and can also be a bridge between a wireless segment and a wired network. In this type of network, all the transmission is under the control of the BS, and the connections are multiplexed using a time division multiplexing (TDM) method.

Basically, the communications in a centralized MAC protocol are based on connections, so before transmitting a data packet, a connection has to be set up. The system defines two communication directions: the *uplink* sense, when data are transmitted from a terminal to the BS, and the *downlink* sense, in which data flow from the BS to the terminals. In both directions, the BS multiplexes connections using a TDM scheme, but there are some differences between them. The first is that in the downlink sense, all transmissions are broadcast, so all the terminals receive all the messages, and they must each select their data from

Encyclopedia of Wireless and Mobile Communications DOI: 10.1081/E-EWMC-120043586

these messages. To make this possible, the BS broadcasts information about which connections are sent, when the transmissions start, and the amount of data transmitted. In the uplink sense, the communication is in a unicast mode, from a terminal to the BS, and each terminal establishes when it must transmit and the amount of data to send by listening to a message broadcast by the BS. In this message, the BS gives details about the uplink connections: when each station can start and how much data each of them have been granted. All uplink transmissions are multiplexed using a TDMA scheme.

To multiplex both communication senses, the transmission is logically organized in frames with a fixed duration. Each frame is divided into two phases: the downlink phase, which includes the downlink data and the broadcast messages relative to the frame structure, and the uplink phase, which, basically, is composed of data in the uplink sense. Both phases could be multiplexed using a TDD (time division duplexing) or a FDD (frequency division duplexing) scheme. Independent of the physical implications about the use of TDD or FDD mechanisms, if an FDD scheme is used, a full-duplex communication could be done but, if a TDD scheme is used, only a half-duplex communication could be done.

These two types of MAC protocols have different characteristics. Distributed protocols are simple and easier to implement, but their performances depend on the number of terminals which have access to the network. On the other hand, a network is more robust against the number of terminals with a centralized MAC protocol, but it requires two types of terminal: a BS and user terminals, and, the access protocol is more complicated and difficult to implement. Furthermore, in a distributed protocol, direct communications between two terminals could exist; however, in a centralized protocol, all transmissions have to pass through the BS.

From the applications point of view, the fundamental difference between these two types of MAC is in how they could provide quality of service (QoS) support to the applications. In this way, the distributed MAC protocols are not enough appropriated to support QoS: it is very difficult to implement priority access with latency or jitter guarantees. The biggest inconvenience in providing QoS support in a distributed system is that the terminals have no information about the other applications running in the network, so each terminal can prioritize only its own applications in order to obtain access to the transmission channel, whereas the network cannot globally prioritize the entire set of connections.

On the other hand, the centralized wireless MAC protocols are good candidates to provide QoS for the connections,[5,6] because it is the central controller (BS) that decides which connections (applications) could transmit data during each period of time. In this way, if the BS makes decisions using QoS criteria then latency, jitter, packet-loss rate, or other QoS parameters of the connections

will be preserved. Obviously, the problem is how the BS identifies the active connections, and estimates, for these, the bandwidth needed at each instant. To solve this problem, the centralized protocols implement resource request (RR) mechanisms. Using these, the terminals can send to the BS the bandwidth resources that each one of its active connections needs. When the BS receives an RR, it enqueues the request, and, depending on the scheduler running, the resources may be granted totally or partially, or the concession may be postponed until the availability of bandwidth or until connections with higher priority have been served.

With centralized wireless MAC protocols, two main issues need be examined before defining a set of RR mechanisms for various service classes. The first issue has to do with the actual mechanisms allowing terminals to inform the BS of their bandwidth requirements on a timely basis. Two alternative mechanisms are made available to the terminals. First, a terminal may make use of an contention-based process. Depending on the network activity, in particular, on the number of RR instances making use of the contention-based method, the efficiency of this mechanism may dramatically degrade. The second alternative is to make use of a reserved channel, such that the access to this resource is based on a contention-free method. However, in order to be able to make use of such mechanism, the BS has to allocate the reserved channel to the requesting terminal. Clearly, therefore a compromise has to be made because the allocation of control slots to convey the RRs reduces the bandwidth available for the purpose of transmitting actual user data. The second issue has to do with the definition of the frequency at which the RRs are to be sent to the BS. There is a clear compromise to be respected. On one hand, the requests should be placed frequently enough to keep the error between the actual needs of the terminals and the requested resources to a minimum. On the other hand, in order to limit the overhead introduced by the signaling mechanism, the number of signaling messages should be kept within reasonable limits. It should therefore be clear that the proper setting of this system parameter requires a clear understanding not only of the needs of the service to be supported, but also of the traffic pattern generated by the application. It is also worth pointing out that the frequency at which the signaling messages are sent should respect the standards as well.

Currently, there exist some standards for wireless networks with centralized MAC. They could be classified in wireless LAN or wireless metropolitan area network (WMAN), depending on the coverage area of the networks. Two of the most popular of these standards are the following: high performance radio local area network (HiperLAN/2) in the case of WLAN and IEEE 802.16 for WMAN. In the following sections, we present a brief overview of these two standards.

This entry is structured as follows, we first present an overview of the general principles of the services and QoS mechanisms being incorporated into two of the most prominent broadband wireless network standards: Hiper-LAN/2 and IEEE 802.16. We then undertake the performance of a single-hop wireless supporting multiple services. We pay particular attention toward assessing the effectiveness of the overall QoS control structure by quantitatively evaluating the video quality as perceived by the end-user. Our results show that the QoS schemes are able to guarantee the QoS requirements of the various services as defined by most standards.

HIPERLAN/2

The HiperLAN/2 standard was developed by the *European Telecommunications Standards Institute* (ETSI) within the framework of its project *broadband radio access networks* (BRAN).[4] It operates in the band of 5 GHz with transmission rates from 6 up to 54 Mbps. HiperLAN/2 provides the means for the interconnection of the wireless network with a wide variety of network technologies, such as Ethernet. ATM, UMTS and IEEE 1394. HiperLAN/2 supports both of the following: infrastructure and ad hoc modes. When operating under the infrastructure mode, the standard distinguishes between two types of devices: the access point (AP) and the MTs. The AP is responsible for providing connectivity with the core network as well as for adapting the users' requirements by taking into account the characteristics of the core network and the services offered by HiperLAN/2. Furthermore, the AP takes care of the distribution of the resources and the coordination of all the MTs located within the cell. The MTs are the other type of device distinguished by the standard. They include end-user devices such as laptops and PDAs, among others.

Physical Layer

The transmission format on the physical layer is a burst, which consists of a preamble part and a data part. The standard chose *orthogonal frequency division multiplexing* (OFDM) because of its excellent performance on highly dispersive channel. Each channel has a bandwidth of 20 MHz, which allows high bit rates per channel with a reasonable number of channels in the allocated spectrum. Fifty-two sub-carriers are used per channel: of these, forty-eight are used to carry data and the other four are pilots.

A key feature of the physical layer is to provide several modulation and coding alternatives. This is both to adapt to current radio link quality and to meet the requirements for different physical layer properties as defined for the connections. The standard supports binary phase shift keying (BPSK), QPSK (quadrature phase shift keying), and 16QAM (16quadrature amplitude modulation) sub-carrier modulation schemes, and, optionally, 64QAM. A convolutional code with rate 1/2 and constraint-length seven performs forward error control. Three code rates 1/2, 9/16, and 3/4 could be obtained by puncturing; combining this three code rates with the four modulation schemes, seven physical layer modes are specified by the standard, which are listed in the Table 1.

MAC Protocol

The MAC protocol is the protocol used for access to the medium (the radio link) with the resulting transmission of data onto that medium. The control is centralized and given to the AP which informs the MTs at which point in time in the MAC frame they are allowed to transmit their data, which adapts according to the request for resources from each of the MTs.

The HiperLAN/2 MAC protocol[7] is based on a dynamic TDMA/TDD scheme with centralized control, using frames of 2 ms as logical transmission units. Given that the allocation of frame resources to each MT is done by the AP, the requirements of the application resources have to be known to these entities which are responsible for allocating the available resources according to user-needs. Toward this end, each MT has to request to the AP the required resources by issuing a RR message, while the AP informs the MT of the positive outcome by using a resource grant (RG) message.

Table 1 Physical layer modes for HiperLAN/2.

Mode	Modulation	Code rate	Physical bit rate (Mbps)
1	BPSK	1/2	6
2	BPSK	3/4	9
3	QPSK	1/2	12
4	QPSK	3/4	18
5	16QAM	9/16	27
6	16QAM	3/4	36
7 (optional)	64QAM	3/4	54

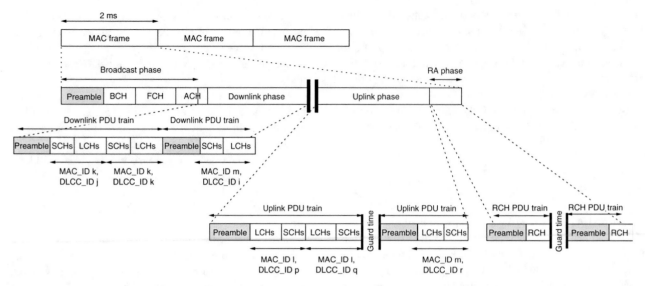

Fig. 1 MAC frame format of HiperLAN/2. BCH, broadcast channel; FCH, frame channel; ACH, access feedback channel; RA, random access phase; SCH, short transport channels; LCH, long transport channels; PDU, protocol data units.

Fig. 1 depicts the generic format of the time-frame used by HiperLAN/2 network, when it operates in infrastructure mode. As seen in the figure, the time-frame is divided into four phases:

1. ***Broadcast phase***: This phase is used to carry out the overall frame-control information. It is formed by a preamble to determine the beginning of the frame and two maps.
    a. ***Downlink (DL) map***: It contains the configuration parameters of the downlink phase of the frame such as the positions and number of slots allocated to each active connection. It is in this map that a *frame channel (FCH)* is typically included to convey the outcome of a successful RR, i.e., a RG message.
    b. ***Uplink (UL) map***: Similar to the DL map, the UL map contains the configuration parameters of the uplink phase. Furthermore, it also includes the configuration parameters of the contention phase (defined below).
2. ***Downlink phase***: This phase is formed by a group of downlink protocol data units (PDUs). Each PDU is made up of a preamble and a variable number of short transport channels (SCHs) and long transport channels (LCHs) dedicated to a given connection. The SCH channels are used to convey error-control codes as well as any other information affecting the connection parameters. The LCH channels are used to transport user data. Both types of channels are grouped into PDU trains (Fig. 1). Each PDU conveys the data pertaining to various connections associated to a given MT, properly identified by its ID.
3. ***Uplink phase***: Similar to the downlink phase, this phase is also formed by a set of uplink PDUs and is

composed of a preamble and a variable number of SCH and LCH channels dedicated to each one of the connections. The SCH channels convey RR messages, control codes, or other control messages. The LCH channels transport user data. Both types of channel are grouped into PDU trains (Fig. 1). Each PDU train conveys the data pertaining to various connections associated with a single MT identified by its MAC_ID.

4. ***Contention phase***: It consists of a number of random access channels (RCHs). Each RCH is used for the transmission of control information. Most standards require a minimum number of RCHs to be present in the frame at all times. To gain access to the RCH channels, a contention process, such as slotted-ALOHA scheme, is used.

The UL phase is handled by HiperLAN/2 MAC protocol as a demand–assignment protocol, which operates according to three phases: request, scheduling and data transmission. The MTs issue UL transmission requests to the AP using RR messages. The RR indicates the number of data pending in the MT queues for a particular connection. Based on these requests, the AP allocates resources for the MTs' UL transmissions according to the output of a scheduling algorithm, and advertises the corresponding UL phase structure in the FCH of the next frame. MTs then transmit their data in contention-free mode in accordance with the frame structure information given in the FCH. In a similar way, the same scheduling scheme is used to share the resource between the downlink connections, depending on the state of the connection queues in the AP.

The HiperLAN/2 standard does not define any particular scheduling mechanism. However, it provides a set of signaling messages that are sufficient to implement any scheduler algorithm. A particular implementation of the scheduler is out of the standard scope, and is thus left to the discretion of the manufacturers.

On the other hand, the standard specifies the RR mechanisms that must be used by MTs to communicate their transmission requests to the AP. HiperLAN/2 describes three mechanisms but neither says what kind of connections must use each of these mechanisms nor specifies the frequency at which the RR messages are sent. The three mechanisms are as follows:

- **Based on contract**: Strictly speaking this is not an RR scheme. It is based in a contract of resources, which are granted every frame or every given number of frames. The contract is established during setup time of connection.
- **Based on polling**: In this mechanism the AP polls periodically to each connection for its transmission requirements, granting to the connection resources to transmit a RR message on the present frame. The frequency at which the polling is done is not specified by the standard.
- **Based on contention**: In this scheme, when a connection needs resources to transmit the data, it sends an RR message across the random access phase on the present frame.

Obviously, aspects like the frequency of the requests, the resource requested per message, and which mechanisms are used by a connection are very important to support QoS; but they are out of the scope of the standard. So, many studies on these issues have been proposed in the literature [8,9] showing the importance of a good tuning of parameters and a good election of request mechanisms depending on the activity of the connections. The influence of the RR mechanisms on the QoS is probed by simulations below.

IEEE 802.16

The standard IEEE 802.16[10,2,11] defines the WMAN interface specification. It has been standardized by the IEEE 802.16 working group and the WiMAX (*worldwide interoperability for microwave access*) forum. The basic architecture consists of one BS and one or more SSs (*subscriber stations*). The BS is stationary and the SSs can be stationary or mobile. The fundamental architecture is a PMP (*point-to-multipoint*) architecture. On it, the BS is responsible for coordinating the whole communications process among SSs, so it is assumed that there is no direct communications between SSs.

The IEEE 802.16 can operate in 10–66 or 2–11 GHz band. The usage scenarios of IEEE 802.16 are from cellular backhaul to rural connectivity, passing by offshore communications and campus connectivity.

Physical Layer

The standard defines two bands: the 10–66 and 2–11 GHz. In the first one, line-of-sight (LOS) is required for communications and burst single-carrier modulation with adaptive burst profiling is used. The downlink and uplink directions are multiplexed using FDD or TDD schemes, using channel bandwidths of 20 or 25 MHz (USA) or 28 MHz (Europe). A Reed-Solomon GF(256) code is used like a forward error code. The standard specifies three modulations schemes, QPSK, 16QAM, and 64QAM, to form the burst profile with different robustness and efficiency. The scheme of modulation to be used by a connection can be chosen as a function of the conditions of the medium and robustness requirements of the applications.

Licensed and unlicensed bands could be used with the 2–11 GHz physical layer. In this case, the communication does not require LOS. There are three air specifications: WMAN–SC2, which uses a single-carrier modulation format; WMAN–OFDM, which uses OFDM with 256 sub-carriers; and WMAN–OFDMA, which uses OFDM with 2048 sub-carriers, multiplexing multiple transmissions addressing a sub-set of multiple carriers to individual receivers.

MAC Layer

The MAC access protocol in IEEE 802.16 is logically based in frames.[12] The duration of a frame is 0.5, 1, or 2 ms. The frame is divided into physical slots for the purpose of bandwidth allocation and identification of physical transitions.

Both TDD and FDD are supported by the IEEE 802.16 for allocating bandwidth on uplink and downlink channels into a frame. In TDD, uplink and downlink directions share the same frequency channel, so communication is half-duplex. For that, the MAC frame is divided in two sub-frames, uplink and downlink sub-frames. Physical slots allocated to different sub-frames may vary dynamically according to bandwidth need in each direction.

In FDD, uplink and downlink channels operate on different frequency bands, permitting full-duplex communications. So the duration of both sub-frames, uplink and downlink, are same as the frame duration. Independent of the method used to multiplex uplink and downlink channels, transmission in the downlink channel is according to a TDM mechanism, and in the uplink channel, a TDMA scheme is used to multiplex multiple connections.

The downlink sub-frame is shown in the Fig. 2. It starts with a preamble using for synchronization. After that, the DL-MAP (downlink map) and UL-MAP (uplink map) sections transmit. As the names indicate, these sections state a "*map*" of downlink and uplink sub-frames. The DL-MAP specifies frame duration and number, identifier

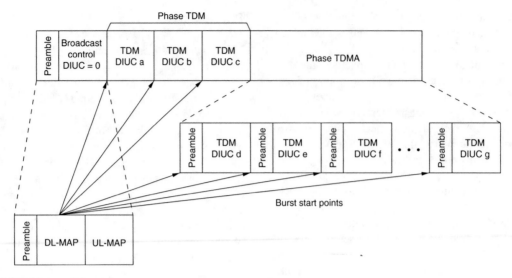

Fig. 2   IEEE 802.16 downlink subframe.

of connections and time when physical layer transitions, changes in the burst profile, occur within the downlink sub-frame. UL-MAP message specifies the start time of uplink sub-frame relative to the start of the frame and bandwidth grants to specific SSs. Specifically, it exits a control message (IE), in the UL-MAP, per grant, stating information about the physical slot where grant starts and the number of physical slots conceded.

Following maps, DCD and UCD messages could be introduced, which indicate the physical characteristics of burst profiles to be used in the present frame. So, both messages establish physical channel characteristics of the

transmission until news DCD and UCD messages appear in a future frame.

These four control messages are followed by a data-portion of the sub-frame. Transmission of data is according to TDM scheme, and it is organized in increasing order of burst-profile robustness. Each SS receives all the transmission in downlink direction, but it looks for MAC headers indicating data for it. After the TDM portion, in FDD systems, there may exist a TDMA segment to support half-duplex SSs, which may need to transmit before receiving data. As they transmit rather than receive, SSs lose synchronization with the frame. For that, it is necessary

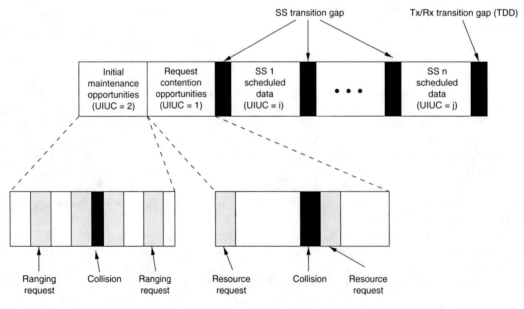

Fig. 3   IEEE 802.16 uplink sub-frame.

include extra preamble at the start of each new burst profile in order to allow SSs to regain synchronization.

In the same way, Fig. 3 shows the structure of the uplink sub-frame. This is divided into three parts: the first one is formed by slots reserved for initial maintenance. The SSs which want join the network send initial control messages in this part of the uplink sub-frame. The second part of the uplink sub-frame is reserved for request-contention slots. This part is used by SSs to respond to multicast and broadcast polls, or to send unicast RR messages. The access to these two uplink sub-frame parts is regulated by a slotted-ALOHA with a backoff algorithm. Finally, the third part in an uplink sub-frame is formed by the unicast transmissions from the SSs to the BS, which are described in the UL-MAP. Each transmission is separated by SS transition gaps. The gap allows ramping down of the previous burst, and it is following by the preamble of the next transmission, allowing the synchronization to the new SS transmission. The preamble and gap lengths are broadcast periodically in the upstream channel descriptor (UCD) message.

It is evident that the BS could grant resources in the downlink direction more easily than in the uplink direction because, in the downlink direction, the information about bandwidth requirements for each connection is directly got from the state of the connection queues in the BS but in the uplink direction, the knowledge about the queue's state is acquired from the RR messages sent by the SSs. So, it is very difficult for the BS to get updated information of the bandwidth requirements in the uplink direction. Also, if the request of bandwidth is not exact, probably the BS will not be able to support QoS for the connections.

Flow Types

As the IEEE 802.16 supports many traffic types, voice, rt-video, web traffic, etc., with different QoS requirements, the standard defines four types of service-flow which serves four different flow classes supported by the standard. An unique flow type is assigned to each connection in the network; this assignment only specifies which type of service will attend to connection but does not define any slot-allocation criteria or scheduling algorithm for this connection.

The four service types have distinct QoS requirements, these QoS specifications establish which connections are assigned to each services. These services are defined as follows:

- **UGS (*unsolicited grant service*):** tailored to carry connections that require fixed bandwidth. Here, the BS periodically grants the resources that the connection needs. These resources were negotiated at connection setup, and an explicit request for them is not required. This service is associated to *constant bit rate* (CBR) connections, such as voice over internet protocol (VoIP) without silence suppression. The parameters

used to specify the fixed bandwidth in the setup period are the following: minimum reserved traffic rate, maximum delay, and tolerated jitter.

- **rtPS (*real-time polling service*):** designed to support real-time connections with *variable bit rate* (VBR) traffic, such us MPEG video and VoIP with silence suppression. Due to the fact that the VBR connections are dynamic in nature, the BS must offer them periodic request opportunities to meet real-time requirements. To do that, the BS periodically polls rt-PS connections. The frequency of this polling is not specified by the standard, and its value will depend on variability of traffic. Parameters such as minimum reserved traffic rate, maximum burst size, and maximum delay are communicated to the BS in the connection setup by the SSs; they are used in traffic-shaping.

- **nrtPS (*non-real-time polling service*):** support non-real-time connections with VBR traffic and which tolerate longer delays and insensitive to jitter. In this case, the network reserves a minimum guaranteed rate to this connections and additional resources must to be requested using the contention phase at the beginning of the uplink sub-frame.

- **Best-effort service:** support best-effort traffic so, neither throughput nor delay guarantees are provided. All bandwidth for this service must be requested using the contention phase or polls done by the BS. The occurrence of polls for best-effort traffic is subject to network load, and the SS cannot rely on their presence.

Bandwidth Request Mechanisms

An SS could use different mechanisms to send requests for bandwidth to the BS. The standard defines three request mechanisms:

- **Base on contract:** It is used by UGS traffic. The request is made during connection setup, in that instant, the SS and BS reach an agreement about the bandwidth that the BS must grant to the connection. The resources contracted are granted while the connection is active.

- **Piggyback:** In this mechanism, bandwidth requests are piggybacked on the headers of data messages. This scheme is only used for incremental requests, i.e., the SS asks for greater bandwidth for a connection. However, this mechanism is effective only if the connection has some backlog for which bandwidth reservation has already been issued, i.e., rtPS or nrtPS.

- **Polling:** In this case the BS allocates the bandwidth needed to transmit bandwidth requests. There exist two types of poll depending on whether the bandwidth granted could be used by one or more SSs.

  — **Unicast polling**: Here the BS concedes resources to a specific connection for bandwidth request. The poll could be done only by this connection; so if the polled connection has no data to transmit, it will not reply to the poll, which is thus wasted.

— **Broadcast polling**: It is done by the BS to all uplink connections. A portion of the uplink subframe is reserved to the broadcast polling at the beginning of it, i.e., phase of contention slots. The drawback in this mechanism is that to have access to the contention phase the SSs must use a truncated binary exponential backoff algorithm, and collision between bandwidth requests from different connections could take place.

As it has been described above, the standard does not specify which request mechanism has to be used by each flow type. It only describes characteristics of the mechanism which could be use, i.e., if it is free of contention or not. So, an election of the optimum bandwidth request for each type of traffic is an important research topic, in order to provide QoS for IEEE 802.16 networks.

Grants

The IEEE 802.16 standard defines two ways for allocation of bandwidth grants:

- **GPC (*grant per connection*)**: The bandwidth is allocated to a connection, and this bandwidth must be used only by this connection.
- **GPSS (*grant per SS*)**: The bandwidth is allocated to a SS aggregating all the granted bandwidth for the connections of this SS. So it is the responsibility of the SS to distribute scheduler process to run.

In latter standard versions,[11] the GPC mechanisms have been eliminated because the GPSS is more robust, and allows adaptability to changes in state of connections. For instance, if the QoS situation at the SS has changed since the last request, the SS could send the higher QoS data along with a request to replace the bandwidth stolen from a lower QoS connection.

Obviously, the bandwidth granted by the BS and how this bandwidth is distributed by the SS depend on the scheduler that the BS and the SS are using. The specification about these schedulers is out of the scope of the standard.

SUPPORTING QoS IN TDMA/TDD WIRELESS NETWORKS

In the importance of schedulers and RR algorithms to provide QoS in a centralized MAC wireless network has been shown. This section shows how an intelligent election of the bandwidth request mechanism used by connections could increase the performance of the network and

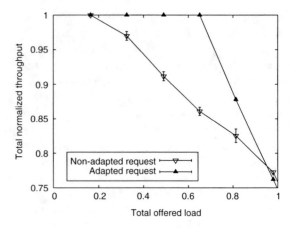

Fig. 4   Total normalized throughput vs. total offered load.

provide QoS support. The allocation of a request scheme to each one of the connections depends of the characteristics of the traffic associated with the connections, and the scheme must adapt itself to traffic pattern.

The following study has been carried out in order to prove the influence of the RR mechanisms in the performance of the network. It consists of a simulation of a centralized MAC wireless network, for instance, HIPERLAN/2, in two scenarios. In the first one, all the

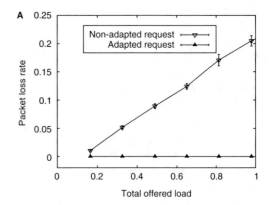

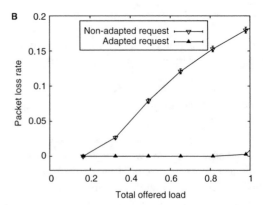

Fig. 5   PLR for (A) voice and; (B) video connections.

Table 2 Video quality scale.

Rating	Impairment	Quality
5	Imperceptible	Excellent
4	Perceptible, not annoying	Good
3	Slightly	Fair
2	Annoying	Poor
1	Very annoying	Bad

connections use a contention-based algorithm to request bandwidth. In the second scenario, each type of connection use a different bandwidth request mechanism.

The connections simulated in the two scenarios are: voice, video, http, and back-ground traffic. The voice traffic is considered constant, and the video traffic is generated by a VBR MPEG-4 video coder. In the scenario named **non-adapted request,** all the connections use the contention phase to sent their bandwidth requests. In the second scenario, named **adapted request**, the mechanisms of request used by the connections depend on the connection characteristics. In this way, voice bandwidth requests will be based on contract. The voice connections negotiate with the BS on the bandwidth that they need at the time of connection setup. The video bandwidth requests will be based on polling. The BS ensures polls at least with a frequency equal to the video frame frequency. In the case of http traffic, these connections use the polls and contention phase to send their requests. Finally, the background connections only use contention phase to request resources.

Fig. 4 represents the normalized (carried) throughput as a function of the offered load for both scenarios. As seen from the figure, as the load increases, the performance of the scenario with non-adapted request badly degrades. This situation can be simply explained as follows. Since the connections have to go through a contention

mechanism to place their requests, as the load increases, the number of collisions in the RA phase increases dramatically. So the request cannot reach the central controller. From this, the central controller infers that the connections have no need of bandwidth and does no grant any bandwidth to them, with the result that the throughput decreases rapidly.

Voice and video applications have hard delay restrictions. In this way, if a packet of video or voice is delayed more than a certain length of time, it is better, from the point of view of the application, to discard it than receive it with an excessive delay. So, in the study described here, all voice packets with an accumulated delay more than 10 ms will be discarded, and, in the case of video, this delay must be shorter than 100 ms.

An important metric to evaluate QoS when losses exist is the *packet-loss rate* (PLR), which is shown for voice and video in Fig. 5. The losses correspond to the packets dropped as soon as they exceed the maximum allowable delay. In the case of voice, Fig. 5A shows that the losses are completely avoided by statically allocating the contracted bandwidth. Then the quality of the voice connections in the receiver will be optima. In the video case, the losses are reduced considerably, avoiding them for loads up to 80%.

To measure the quality of video applications in the reception side, the moving pictures quality metric (MPQM) is used. The MPQM[13] video quality metric has been proved to behave consistently with human judgments according to the quality scale that is often used for subjective testing in the engineering community (see Table 2). The metric has been developed based on a spatio-temporal model of the human vision system. Therefore, the metric overcomes the lack of correlation of traditional metrics, such as PSNR among others, with human perception. The MPQM is based on the basic properties of human vision, mainly, that the human visual system is characterized by a collection of channels that mediate perception. Owing to the independent characteristic among the channels, the perception can be predicted channel by channel. In this way, the metric decomposes the original sequence and a distorted version of it into perceptual channels. It then computes a channel-based distortion measure for contrast sensitivity and masking. Throughout our experiments, we have confirmed that MPQM effectively assesses the spatio-temporal video quality degradation by rating the video sequence on a frame-by-frame basis. However, for the sake of clarity, we report the average MPQM for the overall video-clip.

Fig. 6 shows the quality of video in both scenarios using the MPQM metric. Clearly, the reader could see that the use of an resource request mechanism adapted to video traffic increase video quality. This quality is similar to the original quality ($\approx$4.4) for loads up to 80%, point where video begins to lose packets (Fig. 5).

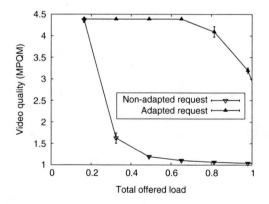

Fig. 6 Decoded video quality.

Spectral–
Telemet

CONCLUSIONS

In this entry, we have presented the general principles of the QoS mechanisms being incorporated into two of the most prominent TDMA/TDD broadband wireless network standards; HiperLAN/2 and IEEE 802.16. Then, we have shown the performance of a TDMA/TDD single-hop wireless network supporting multiple services. We have paid particular attention to assessing the effectiveness of the overall QoS control structure by quantitatively evaluating the video quality as perceived by the end-user. Our results have shown that the QoS schemes are able to guarantee the QoS requirements of the various services as defined by most of the standards.

REFERENCES

1. Broadband Radio Access Networks (BRAN); HIPERMAN; Data Link Control (DLC) Layer, ETSI Std. TS 102 178; 2003.
2. IEEE Standard for Local and Metropolitan Area Networks-Part 16: Air Interface for Fixed Broadband Wireless Access Systems—Amendment2: Medium Access Control Modifications and Additional Physical Layer Specifications for 2–11 GHz, IEEE Std. 802.16a-2003; 2003.
3. High-speed Physical Layer in the 5 GHz Band, IEEE Std. 802.11a; 1999.
4. Broadband Radio Access Networks (BRAN), HIgh PErformance Radio Local Area Networks (HIPERLAN) Type 2; System Overview, ETSI Std. TS 101 683; 2000.
5. Mingozzi, E. QoS support by the HiperLAN/2 MAC protocol: a performance evaluation. Cluster Comput. **2002**, *5*, 145–155.
6. Cicconetti, C.; Lenzini, L.; Mingozzi, E. Quality of service support in IEEE 802.16 networks. IEEE Netw. **2006**, *20* (2), 50–55.
7. Broadband Radio Access Networks (BRAN); HIPERLAN Type 2; Data Link Control (DLC) Layer; Part1: Basic Data Transport Functions, ETSI Std. TS 101 761-1; 2000.
8. Lenzini, L.; Mingozzi, E. Performance evaluation of capacity request and allocation mechanisms for HiperLAN2 wireless LANs. Comput. Netw. **2001**, *37*, 5–15.
9. Delicado, F.; Cuenca, P.; Orozco-Barbosa, L.; Garrido, A. Design and evaluation of a QoS-aware framework for HIPERLAN/2 networks. Wirel. Pers. Commun. J. **2005**, *34* (1–2), 67–90.
10. IEEE Standard for Local and Metropolitan Area Networks—Part 16: Air Interface for Fixed Broadband Wireless Access Systems, IEEE Std. 802.16-2001; 2001.
11. IEEE Standard for Local and Metropolitan Area Networks—Part 16: Air Interface for Fixed Broad band Wireless Access Systems, IEEE Std. 802.16-2004; 2004.
12. Eklund, C.; Marks, R.B.; Stanwood, K.L.; Wang, S. IEEE standard 802.16: a technical overview of the wireless MANTM air interface for broadband wireless access. IEEE Commun. Mag. **2002**, 98–107.
13. Van den Branden, C.J.; Verscheure, O. Perceptual measure using a spatio temporal model of human visual system. SPIE Conference on Electronic Imaging, Digital Video Compression: Algorithms and Technologies, San Jose, CA, Jan 28, 1996; Vol. 2668, 450–461.

Telemetering: GSM Systems and Technologies

Watit Benjapolakul
Ky-Leng
Department of Electrical Engineering, Chulalongkorn University, Bangkok, Thailand

Abstract
Telemetering is to collect data at a place and to relay the data to a point where the data may be evaluated.
Telemetering systems are special set of communication systems.

INTRODUCTION

An overview of a telemetering system is shown in Fig. 1. The overall system is composed of the following:[1]

1. Data collection system;
2. Multiplexing system;
3. Modulator, transmitter, and antenna;
4. Transmission channel;
5. Antenna, radio frequency receiver, intermediate frequency section, carrier de-modulator;
6. De-multiplexing system;
7. Data processing.

WIRELESS TRANSMISSION OF TELEMETERING DATA[2]

Wireless communication systems have been used for data transmission in distributed telemetering systems. This is the only possible transmission method where the object of telemetering is moving, or is a large distance away from the measurement system. Wireless measurement systems can also provide alternative when construction costs and/or operating costs of measurement line are high. Wireless data transmission is serial only. There are three types of measurement systems with wireless data transmission.

- Distributed telemetering systems with data transmission through a cellular communication network (mobile communications).
- Distributed telemetering systems with data transmission through dedicated radio channels (non-telephone).
- Telemetering systems with short distance wireless data transmission through infrared or radio frequency link. (infrared link, Bluetooth)

TELEMETERING WITH GSM-BASED DATA TRANSMISSION[2]

GSM Systems

The second mobile telephone systems, second generation (2G), were digital. One of these systems is called "global system for mobile communications (GSM)." GSM systems work in the 900 or 1800 MHz frequency bands. They used TDMA technique and their main function is voice transmission. The evolution of mobile communications is toward the implementation of GPRS in GSM. GSM networks with GPRS are referred to as 2.5 G mobile communication systems.

GSM systems can be used for data transmission whose data rate typically does not exceed 9.6 Kbps with one channel used. The system consists of mobile stations, base station controllers, a switching system, an operation and support system, and an interface to other communication systems.

Transmission in GSM system is performed in duplex mode. In GSM, separate frequency bands are allocated for uplink (mobile station–base station) and downlink channels (base station–mobile station). The uplink band ranges from 890 to 915 MHz and the downlink band ranges from 935 to 960 MHz. The number of uplink channels is 124, which is equal to the number of downlink channels. The width of each channel is 200 kHz.

GSM-Based Telemetering Data Transmission

The GSM system can also be used for digital data transmission. This second generation mobile telephone system was designed for both audio signal and digital data transmission. The block diagram of a typical mobile telephone is shown in Fig. 2.

Mobile telephones can be divided into three groups: MT0 (mobile terminal 0), MT1, and MT2, according to their capabilities of external digital data transmission (Fig. 3).

Encyclopedia of Wireless and Mobile Communications DOI: 10.1081/E-EWMC-120043443

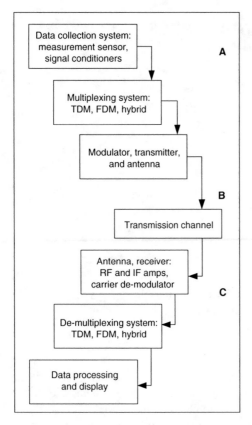

Fig. 1 Telemetering system overview: (A) Test vehicle package; (B) Transmission medium; (C) Ground equipment.[1] *Source*: Reproduced by permission from F. Carden, R. Jedlicka, and R. Henry, Telemetry Systems Engineering, Norwood, MA: Artech House, Inc., 2002. © 2002 by Artech House, Inc.

MT0 is a mobile telephone with no external data interface. In its simplest form, an MT0 telephone transmits voice and keyed SMS. It cannot transfer any data coming form another digital device.

MT1 telephones can transmit digital data. An MT1 communicates with an ISDN through an S interface, but requires a terminal adaptor (TA) for communication with a computer. A TA [e.g., PCMCIA (Personal Computer Memory Card International Association) card] is to adapt bi-directionally RS-232C Standard computer signals to the ISDN or mobile telephone standard.

MT2 telephones can transmit digital data from a computer via an RS-232C interface cable, via an IrDA (Infrared Data Association) link, or via a Bluetooth radio link.

GSM system provides a variety of digital data transmission services as follows:

SMS

SMS transmits alphanumeric messages of up to 160 characters in length, to GSM or e-mail users. SMS is a point-to-point service. The user can obtain the message directly into the mobile station without checking message box.

All SMS messages are transmitted via a SMS center (SMS-C), which is a part of the GSM switching system. For telemetering, SMS can be used in object monitoring with measurement data transmitted in the form of text. The maximum number of characters in an SMS message can be six times larger than the nominal number (equal $6 \times 160 = 960$) but messages longer than 160 characters will be divided into shorter messages.

Multimedia message service

MMS transmits multimedia files via the GSM network. Files can be transferred between users or between devices. MMS allows data transmission of text, graphic, sound, and video. The MMS standard uses a WAP as its transmission protocol. MMS is still not used in telemetering systems. However, it can be, e.g., in monitoring alarm systems in industrial production. At present, the size of MMS messages varies between 10 and 100 kilooctets.

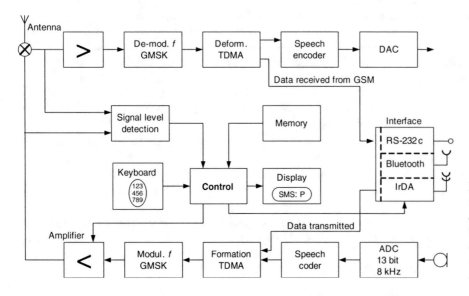

Fig. 2 The functional diagram of a mobile station (mobile telephone).[2] *Source*: Reproduced by permission from W. Nawrocki, Measurement Systems and Sensors, Norwood, MA: Artech House, Inc., 2005. © 2005 by Artech House, Inc.

Fig. 3 Different groups of mobile telephones, classified according to digital data transmission capability.[2] *Source*: Reproduced by permission from W. Nawrocki, Measurement Systems and Sensors, Norwood, MA: Artech House, Inc., 2005. © 2005 by Artech House, Inc.

Circuit switched data

Circuit switched data (CSD) transmission via a telephone radio channel has been in vogue since the creation of GSM. Its use in telemetering systems is limited to low data transfer speeds up to 9.6 Kbps. The cost efficiency of CSD is affected by the fact that this type of transmission occupies the entire traffic channel (see Fig. 4A).

High-speed CSD

High-speed CSD (HSCSD) is developed from CSD by allowing higher data rates. HSCSD are obtained by two methods: through data compression which enhances the capacity of a single transmission channel from 9.6 to 14.4 Kbps and through the combination of several (up to four) traffic channels within a single physical channel. The data rates can be up to 57.6 Kbps (see Fig. 4B).

In both CSD and HSCSD transmission modes, data are transmitted between two GSM network users or between a GSM network user and PSTN (public switched telephone network) user. Telemetering data can also be sent to an Internet address. The possible data rates in HSCSD are 14.4, 19.2, 28.8, 38.4, 43.2, or 56.0 Kbps.

GPRS

GPRS uses packet switching instead of circuit switching. A GPRS session can be activated in the "always connected" communication. Each packet can be transmitted

independently of the other packets, with the destination Internet address in the packet header (see Fig. 5).

A great advantage of GPRS is high data transfer speeds. It can use up to eight time slots allocated at transmission channel setup. The maximum GPRS data transfer speed is $8 \times 14.4 = 115.2$ Kbps. But it can be up to 170 Kbps with different methods of data coding.

EDGE

EDGE upgrades data transmission in GSM networks, providing data rate up to 384 Kbps. The high data rate is obtained by the combination of two modulation methods: 8-phase shift keying (8-PSK) and Gaussian minimum shift keying (GMSK).

The key features of EDGE transmission are:

- the ability to operate in 800, 900, 1800, 1900, or 2150 MHz frequency bands;
- standard maximum data transfer speed 384 Kbps;
- symmetric and asymmetric transmission channels;
- packet transmission function available.

AT commands

AT commands are used as a standard by international telecommunications organizations for communication between computers and modems. The command characters should be written in 8-bit words. The functional diagram

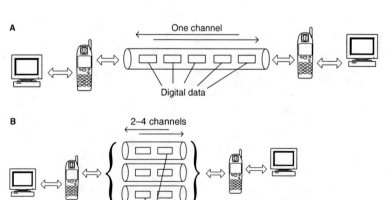

Fig. 4 Transmission in a GSM system: (A) CSD transmission; (B) HSCSD transmission. *Source*: Reproduced by permission from W. Nawrocki, Measurement Systems and Sensors, Norwood, MA: Artech House, Inc., 2005. © 2005 by Artech House, Inc.

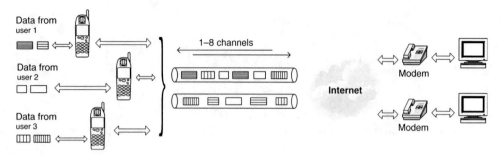

Fig. 5 GPRS transmission in a GSM system.[2] *Source*: Reproduced by permission from W. Nawrocki, Measurement Systems and Sensors, Norwood, MA: Artech House, Inc., 2005. © 2005 by Artech House, Inc.

for telemetering data transmission system in GSM-based systems is shown in Fig. 6.

GSM-based distributed telemetering systems

The block diagram of distributed telemetering system is shown in Fig. 7. An advantage in telemetering digital data transmission is the computer's role with the possibility of processing the telemetering data before forwarding it to the system center.

Telemetering system with SMS data transmission

There are two sections for the route of SMS from the computer to the SMS-C. The first section is from the terminal adaptor to the mobile phone. Here, SMS data do not use standard protocol to transfer. The other section is from the mobile phone to SMS-C. Here, SMS data are transferred in protocol data units (PDUs) frames. There are two types of PDU:

- SMS-SUBMIT type, conveying an SMS message from mobile phone to SMS-C.
- SMS-DELIVER type, conveying an SMS message from SMS-C to mobile phone.

AN EXAMPLE OF USING GSM SYSTEMS AND TECHNOLOGIES FOR TELEMETERING

In the following sections, the SMS in GSM system, as an example of using GSM systems and technologies for telemetering, will be discussed.

Introduction

Since the year 2000 a large number of researchers have been working on sensor network technologies primarily for the realization of large-scale environmental telemetering or monitoring systems for military use and/or scientific use. The sensor networks are designed with different technologies for different applications.[3–7] With the development of the GSM system and technology, there are many researchers working on the SMSs application. Ref. 8 has mentioned such a kind of application. However, their facilities and flexibilities still cannot satisfy our needs because they use code rather than script to control the output and though they have implemented something that resembles script, it is limited in length. For these reasons, as an example of using GSM systems and technologies for telemetering, our sensor node is implemented to display the most wanted telemetering parameter, the temperature on LCD screen, to send the temperature via

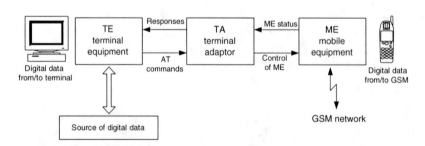

Fig. 6 Command flow between terminal equipment (e.g., a computer), terminal adaptor, and mobile equipment (e.g., mobile phone) in a GSM network.[2] *Source*: Reproduced by permission from W. Nawrocki, Measurement Systems and Sensors, Norwood, MA: Artech House, Inc., 2005. © 2005 by Artech House, Inc.

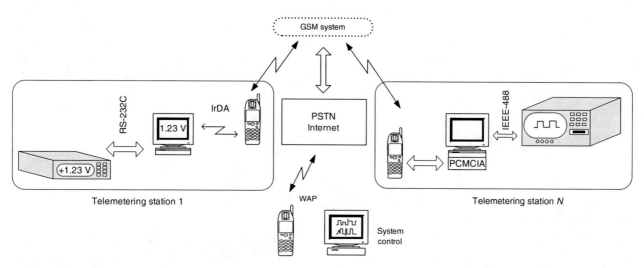

Fig. 7 Block diagram of distributed telemetering system with data transmission via a GSM network.[2] *Source*: Reproduced by permission from W. Nawrocki, Measurement Systems and Sensors, Norwood, MA: Artech House, Inc., 2005. © 2005 by Artech House, Inc.

SMS periodically or after receiving a command via SMS, to enable/disable the output port, to send an alarm while the temperature is higher and/or lower than some values (set by user before or after system installation), and update the setting (low threshold, high threshold, sending period, and alarm re-set) via SMS as the starting point of our development.[8] It is developed for the non-scientific user as well as the scientific user because it is easy to install and use even for those with limited working knowledge of computers or electronics, and for everyone who is familiar with the usage of SMS. The user can easily remember the text much better than the code or number. It is a very good starting point for our future application development. Thus, in this section, we discuss our application development of SMS in the GSM system for temperature telemetering.

Hardware and Software Design of Sensor Node

Hardware

This node is equipped with one 1-wire temperature sensor, one LCD display panel to display the information such as date-time from real time control (RTC) integrated circuit, temperature, and some information about CPU process (Read New SMS, Send SMS, Script Execution, and Delete SMS). RTC is also used in cooperation with sending period to send SMS in every sending period. The micro-controller (CPU) communicates with external EEPROM, calendar, and output module via I^2C bus and with the GSM mobile telephone via the RS-232 link. This I^2C bus is also used for writing the setting from a computer, therefore, an RS-232-to-I^2C interface adaptor is implemented. However, it is implemented in a separate board from the sensor node. A block diagram of this is shown in

Fig. 8 and the major components used by this node are listed in Table 1.

Software

In order to communicate with the computer and other components, the I^2C protocol and 1-wire protocol are implemented and in addition the script execution subroutine is implemented to support the facility and flexibility of the system. Moreover, as not every GSM mobile telephone supports the text-to-PDU (protocol data unit) mode, the text-to-PDU mode algorithm is implemented for then we can support the majority of GSM mobile telephone.

At start up, the kit (our sensor node) will check the presence of I^2C connection in order to decide whether we would like to change the setting of this kit or whether we would prefer this kit to do its job. If I^2C connection is

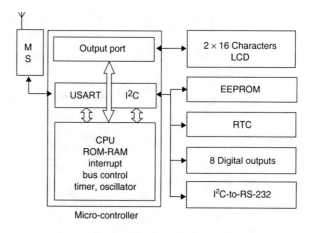

Fig. 8 Sensor node architecture.

Table 1 Major components.

Component	Function
PIC16F876	Micro-controller from microchip
DS1307	I²C real time control (calendar)
24C256	I²C EEPROM memory
PCF8574	I²C digital input/output
DS18S20	1-wire temperature sensor (-55 to $125°C$)

detected, the orange and green LED will turn on. To read/ set the configuration of the kit, a complete program written in Visual Basic language is implemented, so one can just run it and do what one wants. The appearance of this program and a short explanation is shown in Fig. 9. If no I²C connection is detected, the kit will continue to check the presence of the GSM mobile telephone. If the GSM mobile telephone works correctly and has already connected to the kit, our kit will read the SMS service center number for use during the SMS sending process and set all the necessary parameters to the GSM mobile telephone such as the selection of SMS format (PDU format is selected) and the selection of memory location (GSM mobile telephone's memory is selected).

After everything has been settled, the kit will start its main program in the endless loop beginning by reading the date/time from RTC, temperature value from 1-wire temperature sensor (DS1820 from Dallas company) and then display it on LCD. This process takes about 900 ms. After displaying the time and temperature, in the same loop, the kit will check the GSM mobile telephone whether a new message come in or not. If there is no new message in the GSM mobile telephone, the loop will re-start again. This process takes about 700 ms. orange LED indicates the time, temperature, and display process, whereas green LED indicates the process of checking, reading, and sending SMS. The red LED indicates the power supply and RTC status whether it is working correctly or not. The real picture of this kit is shown in Fig. 10.

To set the output port's name into the sensor node, another complete program is implemented. Right now, we use two programs to set each setting separately but in the near future, we will combine them to just only one program.

Now, let say the time of reporting the temperature has come, so the kit will start to send the SMS to the central GSM mobile telephone number or both (including the manager GSM mobile telephone number, depending on the user setting). The message format is:

Tempe = XXX.XX°C

(see Fig. 10)

In case this kit receives a script command " <READ> " from the user, it will send the temperature value to the central GSM mobile telephone number and/or manager GSM mobile telephone number and/or the user's number itself. All these options can be settled from the PC and completed at the same time; it will remain forever in the EEPROM external memory of the kit.

Operation

Before we discuss you the flow chart of this hardware, it is better to explain how this kit works and the methodology of using it by setting its options.

In order to facilitate a clear understanding for the reader, we create some scenarios to show how we can use "Read Temperature via SMS" program to change the configuration (option for the kit).

1. Let say, you want to get the report every 30 min with your central office's number 6612576490 and your number (manager's number) 6612576491. While the kit receives a command, you want it to reply to the GSM mobile telephone that you use to send this command. Therefore, you should fill data in our program as shown in Fig. 9.

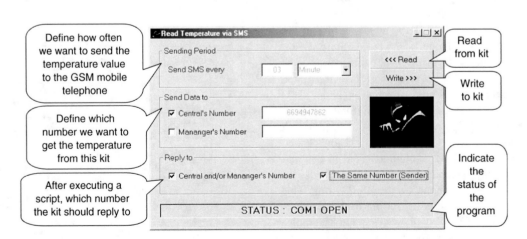

Fig. 9 "Read Temperature via SMS" setting program.

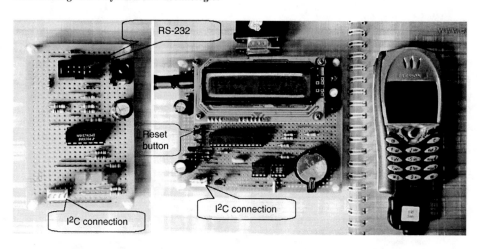

Fig. 10 RS-232-to-I²C interface adaptor, temperature kit and its GSM mobile telephone.

2. Let say, you want to get the report every 1 hr with your central office's number 6612576490 and when the kit receives a command, you want it to reply to the GSM mobile telephone that you use to send this command. Therefore, you should fill data in our program as shown in Fig. 9.

3. Let say, you want to get the report every 1 hr with your central office's number 6612576490 and your number (manager's number) 6612576491. When the kit receives a command, you want it reply to your central office's number and your number, so you should fill data in our program as shown in Fig. 9.

When you check both the central office's number and the manager's number, it means that you also use these numbers to get a reply when you send a command script <READ> to the kit. Just check only one "central and/or manager's number" in "Reply to" field, and you enable all numbers in "Send Data to" field. One more instruction—to send the script command you can use whatever number. There is no need to be restricted to just these two numbers that you set via this program.

After everything is completed (options that you want to set it in the kit), you just press "Write >>>" button to write all your options and your computer's time to the kit (options is written to external EEPROM, whereas the time is set to RTC IC).

To view your setting (options), you just press " <<< Read" button to get the data from the kit. With this feature, you can, not just only read but also review what you have done to check whether you have settled your options properly or not or whether the RS-232-to-I²C adaptor works correctly or not.

After everything has been done, you just unplug the I²C connector from RS-232-to-I²C adaptor; the kit will start its job as per your instructions.

All the keywords used by script execution sub-routine are listed in the Table 2. In one message, the user can combine many script commands together, however, they should keep in mind that the script is executed in order from the beginning to the end of the message.

In this work, we use PicBasic Pro Compiler and PIC16F876 as a core component. The flow chart in Fig. 11 shows the operation—an example of temperature sensor node.

Tests

The operation of this sensor node has been tested by setting the system configuration and running it in a real physical environment. The system setting is shown in Fig. 8. Every 3 min, the kit will send the temperature value to the central

Table 2 Script keywords.

Keyword	Descriptions
<READ>	Ask the sensor node to read the temperature then send this value back via SMS
<REPORT>	Ask the sensor node to read the output port then send this value back via SMS
<RESET>	Re-activate alarm for we can receive SMS alert while the temperature is critical
<RESET ALL>	Put all output port to 0 V and re-activate alarm
<TH=XXX>	Set threshold high for alarm (e.g., <TH=080>)
<TL=XXX>	Set threshold low for alarm (e.g., <TL=−05>)
<TS=XXU>	Set time scan (scanning period). "U" is time unit E.g., <TS = 03M>: set TS = 3 min or <TS = 03H>: set TS = 3 hr
<SSS=X>	Set output port. "SSS" represents the port's name. It can be any texts with the length not over 7 characters. e.g., <TV01 = 0>: turn off port named "TV1." <LAMPE04 = 1>: turn on port named "LAMPE04"

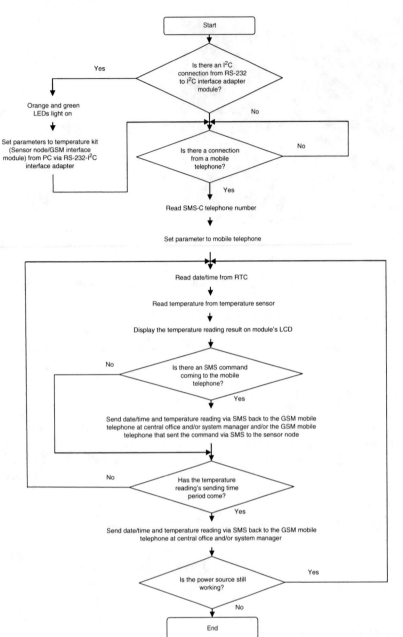

Fig. 11 Flow chart showing the operation of the example temperature sensor node.

GSM mobile telephone number 6694947862 and when the kit receives a command < READ >, it will reply via SMS to the GSM mobile telephone that we use to send this command and to the central GSM mobile telephone number.

The results of the test shown in Fig. 12 are the same as what we expected. Fig. 12A shows the SMS sending process and the temperature at that time. Fig. 12B shows the time and temperature received at GSM mobile telephone and Fig. 12C shows the time and temperature at GSM mobile telephone at 3 min later.

CONCLUSION

This application is the starting point for our future work. It looks simple, yet, it enables us to learn and understand more about how to use GSM and technologies for telemetering. We intend to design and develop a pilot platform of central and mobile nodes for telemetering sensor network, which can support up to 8 analog inputs, 8 digital inputs, 8 digital outputs, 2 analog outputs, and 1 RS-232 port for GPS.

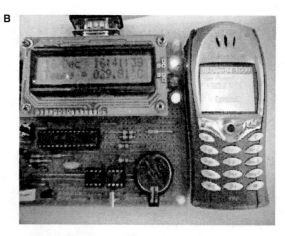

Fig. 12 The results of our evaluation. (A) The process of sending SMS with the present temperature. (B) The time and temperature received at GSM mobile telephone after the first sending. (C) The time and temperature received at 3 min later.

ACKNOWLEDGMENT

The authors wish to thank the Cooperative Project of Research between Department of Electrical Engineering, Chulalongkorn University and Private Sector under contract Koroau 7/2547 for supporting the work described above as an example of using GSM systems and technologies for telemetering.

REFERENCES

1. Carden, F.; Jedlicka, R.; Henry, R. *Telemetry Systems Engineering*; Artech House: Boston, London, 2002.

2. Nawrocki, W. *Measurement Systems and Sensors*; Artech House: Boston, London, 2005.

3. Akyildiz, I.F.; Su, W.; Sankarasubramaniam, Y.; Cayirci, E A survey on sensor networks. IEEE Commun. Mag. **2002**, *40* (8), 102–114.

4. Shepherd, D. Networked micro-sensors and the end of the world as we know it. IEEE Technol. Soc. Mag. **2003**, **22**(1) 17–21.

5. Kawahara, Y.; Minami, M.; Morikawa, H.; Aoyama, T. Design and implementation of a sensor network node for ubiquitous computing environment. 58th IEEE International Conference on Vehicular Technology, Orlando, Florida, Oct 4–9, 2003; Hyatt Orlando Hotel, Orlando, Florida, USA; Vol. 5, 3005–3009.

6. Mainwaring, A; Polastre, J; Szewczyk, R; Culler, D; Anderson, J: Design and implementation of wireless sensor networks for habitat monitoring, Intel Research Laboratory, Berkeley Intel Corporation, EECS Department, University of California at Berkeley, College of the Atlantic Bar Harbor: http://www.cs.berkeley.edu/~polastre/papers/masters.pdf, 2003.

7. Knaian, A.N. *A wireless sensor network for smart roadbeds and intelligent transportation systems*; Massachusetts Institute of Technology: Boston, London, 2000, http://www.media.mit.edu/ resenv/pubs/theses/AraKnai an-Thesis.pdf.

8. Benjapolakul, W. Final Report on Development of Application of Short Message Services in Mobile Telephone System for Telemetering (Year I), for the Cooperative Project of Research between Department of Electrical Engineering, Chulalongkorn University and Private Sector under Contract Koroau 7/2547, 2005.

Ultra-wideband: Applications

Gopal Racherla
Advanced Wireless Group, General Atomics, San Diego, California, U.S.A.

Abstract

Ultra-wideband (UWB) is a wireless communication scheme involving transmission of a series of very precisely timed, narrow, and low-power pulses over a wide band of radio spectrum without the use of the narrow-band modulation.

INTRODUCTION

Ultra-wideband (UWB) is a radio transmission scheme that uses extremely narrow and low-power pulses of radio energy spread across a wide spectrum of frequencies. UWB[1–16] has several advantages over conventional continuous wave radio communications including support for high data rates, robustness to multi-path interference and fading, and very high time resolution to support location and tracking applications. Additionally, as a spread spectrum technology, UWB offers a low probability of intercept (LPI) and a low probability of detection (LPD) which makes it well suited for covert military usage. Since UWB signals have extremely short bursts in time—with durations of 1ns or less—they are suited for applications requiring precision location. Application optimization, choice of the modulation scheme, and improvements on these UWB characteristics are design and implementation specific.

In the United States, according to the US Federal Communications Commission (FCC),[17] UWB communication systems are required to have a –10 dB fractional bandwidth of more than 20% or a –10 dB bandwidth of more than 500 MHz. The FCC issued a report and order in February 2002. This landmark decision, to permit UWB operation in the 3.1–10.6 GHz spectrum under Part 15 emission limits, with some additional restrictions, has catalyzed development and standardization processes as is evident by the large number of entities (companies, academic, and government institutions) associated with UWB. The FCC carefully chose the frequency band of operation to be above 3.1 GHz to avoid interfering with GPS and other life critical systems. Furthermore, the FCC ruled that emissions below Part 15 would provide for peaceful coexistence, the ability to have narrowband and UWB systems co-located on a non-interfering basis, because unintentional emissions from devices such as laptops are also limited to Part 15 rules. This ruling made it possible to have up to 15 UWB frequency bands in the 7.5 GHz allocated to unlicensed spectrum.[1–16]

UWB presents a great opportunity for data communications for today's media-rich consumer electronics and home entertainment systems that run on battery-powered handheld devices. It can form the basis of a low-cost, low-power and very high data rate solution as a wireless "cable replacement" technology for computer-to-peripherals, peripherals-to-peripherals, and digital home networking applications. A useful attribute of UWB technology is its ability to perform precision geolocation, which can aid in ad hoc or mesh networking, where the operations of the mobile hosts benefit by knowing the location of the other hosts. UWB technology aims to fill the void left by established standards like Bluetooth[18] and 802.11a/b/g.[19]

There are still several challenges to the wide adoption of UWB for wireless data communications including the relative infancy of the technology in the commercial arena, the lack of universal standards, and the lack of high-volume low-cost system-on-chip (SoC) implementations.

UWB technology has been around for many years in the form of impulse radio.[5] Dr. Gerald Ross's work in time-domain electromagnetics in 1962 focusing on the transient behavior of microwave networks through their characteristic impulse response marked the beginning of the UWB technology. In the late 1960s, Dr. Ross, at Sperry Rand Research Center, used UWB in various radar and communications applications. This technology was known as baseband, carrier-free or impulse radio. The US Department of Defense (DoD) coined the the term "ultra-wideband" in the late 1980s. Sperry Rand Corporation worked with UWB to expand the applications to collision avoidance, positioning systems, altimetry and other applications. UWB continued to thrive with active support from the US Government including the DoD and national research labs like Los Alamos National Labs (LANL) and Lawrence Livermore National Labs (LLNL). The Micropower Impulse Radar (MIR) invented by T.E. McEwan in 1994 at LLNL was a very low-power, small and inexpensive radar. The MIR heralded the incorporation of UWB into useful high-volume products.

In this entry, we look at UWB technology for data communications and location applications. We also present a brief synopsis of the regulatory and standardization efforts worldwide with special emphasis on the FCC. We

Encyclopedia of Wireless and Mobile Communications DOI: 10.1081/E-EWMC-120043606

present a detailed discussion on various UWB applications.

FCC's First Order and Report[17] establishes different technical standards and operating restrictions for three types of UWB devices based on their potential to cause interference. These three types of UWB devices are: 1) imaging systems including ground penetrating radars (GPRs), wall, through-wall, medical imaging, and surveillance devices; 2) vehicular radar systems (VRS); and 3) communications and measurement systems.

- **Imaging systems:** FCC allows operation of GPRs and other imaging devices subject to the following frequency and power restrictions. Imaging systems include:

  — **GPR systems:** GPRs can be operated either below 960 MHz or in the frequency band 3.1–10.6 GHz. FCC allows GPRs to operate only when in contact or in close proximity to the ground for the purpose of detecting or imaging buried objects.

  — **Wall-imaging systems:** Wall-imaging systems can be operated either below 960 MHz or in the frequency band 3.1–10.6 GHz. Wall-imaging systems are designed to detect the location of objects contained within a "wall," such as a concrete structure, the side of a bridge, or the wall of a mine.

  — **Through-wall imaging systems:** These systems must be operated below 960 MHz or in the frequency band 1.99–10.6 GHz. Through-wall imaging systems detect the location or movement of persons or objects that are located on the other side of a structure such as a wall.

  — **Medical systems:** These devices must be operated in the frequency band 3.1–10.6 GHz. A medical imaging system may be used for a variety of health applications to "see" inside the body of a person or animal.

  — **Surveillance systems:** FCC treats these in the same way as through-wall imaging and allows them to operate in the frequency band 1.99–10.6 GHz. Surveillance systems are allowed to operate as "security fences" by maintaining a fixed RF perimeter field and detecting the intrusions.

  — **VRS:** FCC allows for operation of VRS in the 24 GHz band using directional antennas on terrestrial transportation vehicles provided both the center frequency and the frequency at which the highest radiated emission occurs are greater than 24.075 GHz. VRS devices can be used to detect the location and movement of objects near a vehicle. These devices can be used for vehicular collision avoidance and better automobile control.

- **Communications and measurement systems:** FCC allows for use of a wide variety of other UWB devices, such as high-speed home and business networking devices in the frequency band 3.1–10.6 GHz. These devices are allowed for indoor use or they must consist of hand-held devices to be used in peer-to-peer operation.

Table 1 FCC equivalent isotropically radiated power (EIRP) limits for ultra-wideband (UWB).

Frequency (MHz)	EIRP (dBm)
960–1,610	−75.3
1,610–1,900	−63.3
1,900–3,100	−61.3
3,100–10,600	−41.3
>10,600	−61.3

A summary of the FCC emission limits can be found in Tables 1 and 2 below.[17] In both Tables 1 and 2, radiated emissions above 960 MHz must not exceed the limits as shown per Part 15, Section F using a resolution bandwidth of 1 MHz.

The radiated emissions at or below 960 MHz from a device operating under the provisions of this section shall not exceed the emission levels in Section 15.209 of this chapter. The radiated emissions above 960 MHz from a device operating under the provisions of this section shall not exceed the average limits shown above when measured using a resolution bandwidth of 1 MHz. Additionally, UWB transmitters shall not exceed the average limits shown below when measured using a resolution bandwidth of no less than 1 kHz.

UWB SYSTEM DESIGN CONSIDERATIONS

The basic waveform that employed in a UWB system is an approximation to an impulse, such as that shown in Fig. 1.

The short duration of the pulse is associated with large inherent bandwidth. Typical attributes of UWB waveforms are summarized in Table 3. The list of abbreviations is given in Table 4.

We consider the design considerations of an example UWB system namely a personal area sensor network.[20] Several considerations are needed when designing a personal area network (PAN). First, low power design is necessary because the portable devices within the network are battery-powered. Second, high data rate transmission is crucial for broadcasting multiple digital audio and video streams. Last, low cost is a prerequisite to broadening consumer adoption. In addition to these criteria, the

Table 2 Additional Federal Communications Commission (FCC) emission restrictions.

Frequency (MHz)	EIRP (dBm)
1164–1240	−85.3
1559–1610	−85.3

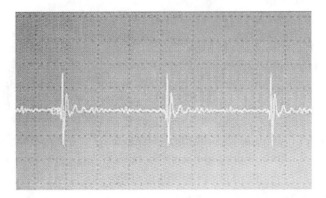

Fig. 1 An example ultra-wideband (UWB) pulse waveform.

UWB system designer must address synchronization and co-existence. Capturing and locking onto these short pulses make synchronization a non-trivial task. Co-existing peacefully with other wireless systems without interference is important; in particular, one needs to pay attention to the 802.11a wireless local area networks that operate in the 5 GHz ISM bands.

At the physical layer (PHY), additional challenges lie in the transceiver and antenna design. At the transmitter, pulse shaping is required to produce flat and wideband emission in the desired frequency bands. Although new integrated circuits provide less expensive forms of integration, the pulses can be affected by the parasitics from the component and packaging.[21] To accommodate the high data rates, tradeoffs between high and low pulse repetition frequency (PRF) and modulation schemes must be considered. The low PRF system with higher modulation (more bits per symbol) may require a more complex receiver, while the high PRF system with lower modulation may lead to performance degradation for delay spread in the channel. Finally, traditional antenna designs gear toward narrow band systems. To avoid dispersion at the receiver, the new wideband antennas need phase linearity and a fixed phase center.[21]

UWB STANDARDIZATION AND REGULATORY STATUS

There are several standards bodies incorporating, UWB for wireless solutions. The standards body[22] most advanced in terms of maturity and market deployment are WiMedia[23–26] and wireless USB (universal serial bus).[27]

Table 3 Typical parameters of a UWB system.

Parameter	Value
Fractional bandwidth	>20%
Pulse width	0.1–2 ns
Pulse repetition frequency	1 kHz–2 GHz
Average transmitted power	<1 mW

Table 4 Abbreviations.

Abbreviation	Expansion
ASIC	Application specific integrated circuit
AoA	Angle of arrival
DEVs	Devices
DoD	Department of Defense
DS-UWB	Direct sequence-UWB
EIRP	Equivalent isotropically radiated power
FCC	Federal Communications Commission
GPR	Ground penetrating radars
GHz	Gigahertz
HDTV	High-definition television
IEEE	Institute of Electrical and Electronic Engineers
IP	Internet protocol
ISM	Industrial, scientific, and medical
MHz	Megahertz
MIR	Micropower Impulse Radar
LANL	Los Alamos National Labs
LLNL	Lawrence Livermore National Labs
LPD	Low probability of detection
LPI	Low probability of interference
LS	Least square
MAC	Medium access control
MB-OFDM	Multi-band orthogonal frequency division multiplexing
MANET	Mobile ad hoc network
NLoS	Non line of sight
OS	Operating system
PAN	Personal area network
PHY	Physical layer
POS	Personal operating space
PNC	Piconet coordinator
PRF	Pulse repetition frequency
RF	Radio frequency
RFID	Radio frequency identity
RSS	Relative signal strength
SIG	Special Interest Group
SoC	System-on-chip
STB	Set top boxes
UAV	Unmanned aerial vehicle
ToA	Time of arrival
TDoA	Time difference of arrival
UWB	Ultra-wideband
USB	Universal serial bus
VLSI	Very large scale integration
VRS	Vehicular radar systems
WPAN	Wireless personal area network
WSN	Wireless sensor network

WiMedia/IEEE 802.15.3a

The original standard IEEE 802.15.3 defines medium access control (MAC) and PHY (2.4 GHz) specifications for a wireless PAN (WPAN). The standard is based on the concept of a piconet which is a network confined to a 10 m personal operating space (POS) around a person or object. A WPAN consists of one or more collocated piconets. Each piconet is controlled by a piconet co-ordinator (PNC) and may consist of devices. The PNC's functions include: the basic timing of the piconet using beacons, managing quality of service (QoS), managing the power save modes and security and authentication. The 802.15.3 PHY is defined for 2.4–2.4835 GHz band and has two defined channel plans. It supports five different data rates (11–55 Mbps). The base uncoded PHY rate is 22 Mbps. Institute of Electrical and Electronic Engineers (IEEE) established committees to initially explore and later specify an alternative PHY for short-range, high data rate applications. This PHY along with the original IEEE 802.15.3 MAC (with minor modifications as needed) was to form the IEEE 802.15.3a standard. The desired characteristics of this alternative PHY included:[23–26,28]

- Co-existence with all IEEE 802 PHY standards,
- data rate of more than 100 Mbps,
- robust performance with multi-path,
- location awareness,
- use of additional unlicensed spectrum for high data rate WPANs.

As a result, multiple proposal IEEE 802.15.3a divides PHYs into two proposals, namely 1) the multi-band orthogonal frequency division multiplexing (MB-OFDM) UWB, supported by the WiMedia Alliance; and 2) the direct sequence—UWB (DS-UWB), supported by the UWB Forum. However, in early 2006, because of technical gridlock, the IEEE standardization effort was stopped. The MB-OFDM approach, adopted by the WiMedia Alliance with over 100 companies supporting it, can support data rates up to 480 Mbps. The specification calls for bit rates of 110–200 Mbps at ranges up to 10 m, with the option to achieve 480 Mbps at 3 m. The power consumption requirement is presently set at 100–250 mW with $10e^{-5}$ bit error rate at the top of the PHY. Complexity/cost is comparable to Bluetooth and the PHY is required to support multiple co-located piconets. Co-existence is presently crucial (e.g., IEEE 802.11a) and the ability to scale the technology is key to a long-lasting and widely-adopted standard.

IEEE 802.15.4a

The IEEE 802.15.4a standard[29] was formed to provide a standard for ultra low-power sensor networks. These sensor nodes support high data rate communications, have the ability to perform location with at least 1m accuracy. The standard also is chartered to provide a solution that supports data rate scalability, longer-range, and lower power consumption and cost.

The 802.15.4a became an official Task Group in March 2004. In March 2005, the committee drafted an alternate PHY specification after selecting a baseline specification. The baseline specification allows two optional PHYs consisting of a UWB Impulse Radio and a 2.4 GHz Chirp Spread Spectrum radio.

Bluetooth 3.0

The Bluetooth Special Interest Group (SIG) has selected the WiMedia Alliance's UWB PHY[30] for integration with Bluetooth's future high-speed/high data rate version. This standard, dubbed "Bluetooth 3.0," will be used for high-speed, high-quality video and audio applications for portable devices, multimedia projectors, and television sets while continuing to cater to the needs of very low-power applications such as keyboards and mono headsets. The UWB option will allow devices to select the appropriate physical radio for the applications based on the requirements.

Wireless USB

This standard, formally called "Certified Wireless USB,"[27] is a wireless adaptation of USB that combines the speed and ease-of-use of USB 2.0. It is supported by many top consumer electronics and PC companies including Intel, Microsoft, NEC, Sony, Samsung, and Philips. Wireless USB is also based on the WiMedia Alliance's UWB radio. The Certified Wireless USB standard can support data rates up to 480 Mbps at 3 m and 110 Mbps up to 10 m.

Non-US Regulatory Effort

Apart from the US, extensive efforts have been going on throughout Europe [Centre for Environmental Planning and Technology (CEPT), European Telecommunications Standards Institute (ETSI), and the European Commission], Korea, and Japan (Association of Radio Industries and Businesses, and the Japanese Ministry of Telecommunications), and China (Ministry of Science and Technology). The Singapore UWB Program was launched in February 2003 as a two-year program to popularize UWB in Singapore. Singapore introduced a UWB-Friendly Zone and gave implementers and vendors experimental licenses.

In August 2006, the Japanese government allowed the use of UWB technology in the allocated frequency bands from 3.4 to 4.8 GHz and from 7.25 to 10.25 GHz. As for the 3.4–4.8 GHz band, there is a temporary waiver to permit the use of the 4.2–4.8 GHz band without an

Ultra–Wi-Fi

Ultra–
Wi-Fi

interference reduction technology until December 2008 after which interference caused by UWB to other radio technology in the permitted bands is not allowed. In Japan, the equivalent isotropically radiated power (EIRP) is limited to − 41.3 dBm/MHz on both bands. As for the devices without an interference technology that uses the 3.4–4.8 GHz band, however, the average transmission power is limited to − 70 dBm/MHz or lower.

APPLICATIONS

There is a plethora of UWB based applications[1–16,31] in commercial and military use. Based on the applicable market segments, UWB application can be classified as:

Commercial Applications

These applications, typically, are high-volume, low-cost solutions that can be used by the general population at large. Examples include:

- high data rate multimedia application,
- ad hoc networking,
- cable replacement,
- low-power sensors, and
- collision avoidance radar.

Defense/Military Applications

These applications, typically, are low-volume, relatively expensive solutions that are aimed at military, homeland defense, and security markets. Examples include:

- covert communications,
- low-power radar,
- sensor networks, and
- position location.

Another classification is UWB for 1)communication, and 2)location/position.[31–38] Based on the second classification based on functionality, UWB applications are discussed below.

Communication Applications

UWB is being used in many high data rate multimedia applications like connecting home audio video equipment like set top boxes (STB), high-definition televisions (HDTV), and wireless speakers. Major consumer electronics companies like Sony, Samsung, Philips, and Toshiba are looking at deploying UWB-based devices in the very near future. Another major application thrust for UWB is "cable replacement" like wireless USB and wireless IEEE 1394. All the major PC companies like Intel, Texas Instruments, Philips, Microsoft, Dell, Sony, and Samsung are

planning to provide wireless cable replacement solutions to consumers in 2007–2008.

Several companies are developing low-power sensor networks for industry and automotive and supply chain management using Zigbee which is based on the UWB IEEE 802.15.4a PHY. These companies include Texas Instruments, Motorola and several startups like Ember.

Pulse-link's CWave™ UWB whole-home connectivity is focused on in-home distribution of multiple HDTV multimedia streams using both coax and wireless connections. Possible applications include displaying content in any room in the home from any source in the home and wirelessly connecting HDTVs to PCs and video game consoles.

Tzero Technologies uses UWB technology to build IEEE 1394 based wireless extensions to provide wireless video to and access digital multimedia from anywhere in the home. Other companies like Wisair, Staccato Communications, WiQuest, Artimi, Alareon are developing WiMedia based solutions for certified wireless USB, and wireless multimedia.

Because of UWB's LPI/LPD, UWB has been used in covert high data rate military communications. Several companies like General Atomics, MSSI, and Time Domain have implemented UWB applications for defense and homeland security applications.

Location/Positioning

Since UWB has a high time resolution inherently, it allows for high-precision positioning/location information.[31–38] Traditional techniques that are used for location information include angle of arrival (AoA), comparing relative signal strength (RSS), and time-based techniques including time of arrival (ToA) or time difference of arrival (TDoA). Determining location and positioning of a node requires the installation of reference nodes whose position/location is known apriori. This known position/location information is used to calculate the position/location of a node in question. The Time based methods rely on measurements of flight times of signals between various UWB nodes. The advantages and disadvantages of each of these techniques has been studied in Refs. 3 and 4. Analyses in Refs. 3 and 4 show that time-based techniques are best suited for UWB based location and ranging. It should be noted that multi-path propagation, multiple access interference and Non Line-of-Sight (NLoS) propagation introduce errors that need to be mitigated.

ToA estimation technique uses a least square (LS) approach using a series of ToA measurements from N reference nodes to the node in question.[3,4] ToA can be used if all the nodes have the same reference clock. However, if there is clock synchronization between all the reference nodes and no synchronization between the reference nodes and the given node TDoA can be used.

It should be noted that ranging of a node, with respect to a reference node can be done using ToA measurements while positioning/location, with reference to the entire network requires triangulation using reference nodes along with ToA.

A two-way ranging protocol is presented in Ref. 32 to determine the round trip delay between two remote nodes which are not time-synchronized. IEEE 802.15.4a PHY provides for precision ranging capability which should auger well for a new generation of location-aware applications.

There are a variety of location-aware applications. These include:

- **Asset/inventory tracking**: Small radio frequency identity (RFID) tags can be used to locate inventory and assets in commercial and retailing settings like factories and department stores. The tracking can be used to track shipments and high-value assets allowing efficient supply chain management. Typical accuracy for these applications is 0.25–3 ft.[3]
- **Tracking people:** Tracking people indoors, emergency personnel and first responders in case of disaster recovery, medical personnel or high-risk patients are among the many possible applications for tracking people. Typical accuracy for these applications is about 5 ft.[3]
- **Security**: Security applications include 1) Home/office burglar alarms, and 2) intrusion detection with UWB based radars along national borders and around high-risk, high-value assets like oil pipelines and military/nuclear installations. Typical accuracy for these applications is less than 1–2 ft.[3]
- **Locating cars in a parking lot**: Locating cars in parking structures and parking lots can be implemented by installing a unique UWB based RFID when the car enters the parking area.
- **Wireless body area networking for medical and excercise/fitness**: These applications include monitoring motion and body movement, gathering people's vital statistics.
- **Real-time call forwarding**: By knowing people's location, a telephone call can be re-routed to the nearest pre-designated phone.
- **Sports**: Applications in the sports would allow athletes to be monitored with reference to the rest of the players, monitoring their health, and helping sports officials make the right calls.

SUMMARY

In this entry, the principles of UWB technology and its applications are explored. UWB technology has tremendous potential and applications involving low-power, high data rate communications and location applications. After FCC allowed commercial use of UWB in February 2002, there has been a proliferation of companies seeking to harness the capabilities of UWB for commercial and military applications. There has been a slow but steady work on standardization of UWB PHY and its incorporation in various consumer technologies. UWB is gaining world wide, regulatory approval which will enable UWB to be a globally pervasive wireless technology.

REFERENCES

1. Foerster, J.; Green, E.; Somayazulu, S.; Leeper, D. Ultra-wideband technology for short- or medium range wireless communications. Intel Technol. J. **2001**, *5* (3).
2. Foerster, J. Ultra-wideband Technology for Short-Range, High-Rate Wireless Communications, http://www.ieee.or.com/Archive/uwb.pdf (accessed Oct 2002).
3. Oppermann I. Hamalainen, M. Iinatti, J. (Eds), *UWB—Theory and Applications*; John Wiley & Sons: Hoboken, NJ, 2004.
4. Arslan H. Chen, Z.N., Di Benedetto, M.-G. (Eds), *Ultra Wideband Wireless Communication*; John Wiley & Sons: Hoboken, NJ, 2006.
5. Taylor J.D. (Eds), *Introduction to Ultra–Wideband Radar Technology*; CRC Press: Boca Raton, FL, 1995.
6. Win, M.Z.; Scholtz, R.A. Impulse radio: how it works. IEEE Commun. Lett. **1998**, *2* (1), 10–12.
7. Leeper, D.G. Wireless data blaster. Sci. Am. **2002**, *286* (5), 64–69, www.sciam.com/article.cfm?articleID=0002D51D-0A78-1CD4-B4A8809EC588EEDF (accessed oct 2002).
8. Hirt, W.; Moeller, D.L. UWB Radio Technology: The Global View of a Wireless System Integrator http:// www.zurich.ibm.com/pdf/ISART2002_IBM_UWB_Hirt.pdf (accessed Oct 2002).
9. Ray, S. An Introduction to Ultra-Wideband (Impulse) Radio, http://netlab1.bu.edu/~saikat/pdfs/ultra.pdf (accessed Oct 2002).
10. Jacobson, C. Ultra-Wideband Update and Review, Intel Developer's Forum, Sept 2002.
11. Paulo, G. The Promise of UWB: Early UWB Market Makers. Instat/MDR, May **2002**, http://www.instat.com (accessed Oct 2002).
12. Aiello, G.R.; Ho, M.; Lovette, J. Ultra-wideband: an emerging technology for wireless communications, http://www.uwbgroup.ru/eng/articles/communic.htm (accessed Jan 2001).
13. Reed, J.; Buehrer, R.M.; McKinstry, D. Introduction to UWB: Impulse Radio for Radar and Wireless Communications, www.mprg.org/people/buehrer/ultra/UWB%20tutorial.pdf.
14. Barrett, T.W. History of Ultra-WideBand (UWB) Radar & Communications: Pioneers and Innovators, http://www.ntia.doc.gov/osmhome/uwbtestplan/barret_history_(piersw-figs).pdf.
15. Ultra Wideband (UWB) Frequently Asked Questions (FAQ), www.multispectral.com/UWBFAQ.html.
16. Aiello, G.R.; Rogerson, G.D. Ultra-wideband wireless systems. IEEE Microw. Mag. **2003**, *4* (2), 36–47.
17. FCC First Order and Report on Ultra-Wideband Technology, http://www.fcc.gov/Bureaus/Engineering_Technology/Orders/2002/fcc02048.pdf (accessed Oct 2002).
18. Official Bluetooth Web Page, (accessed Oct 2002). http://www.bluetooth.com.

19. IEEE 802.11 Working Group, http://grouper.ieee.org/groups/802/11/ (accessed Oct 2002).

20. Akylidiz, I.F.; Su, W.; Sankarasubramaniam, Y.; Cayirci, E. A survey on sensor networks. IEEE Commun. Mag. **2002**, *40* (8), 102–114.

21. Radhakrishnan, S.; Racherla, G.; Furuno, D. Mobile ad hoc networks: principles and practices. *Wireless Internet Handbook: Technologies, Standards, and Application*; CRC Press, Inc.: Boca Raton, FL, 2003, 381–405.

22. Wood, S. UWB Standards, WiMedia Alliance White Paper, June 2006.

23. WiMedia Alliance, http://www.wimedia.org/en/index.asp.

24. IEEE 802.15.3 WPAN Task Group, http://www.ieee802.org/15/pub/TG3.html (accessed Oct 2002).

25. IEEE 802.15.3 Study Group 'a' Technical requirements document (02104r13P802-15_SG3a-Technical-Requirements), http://ieee802.org/15/pub/2002/ (accessed Oct 2002).

26. IEEE 802.15.3 Study Group 'a' Schedule (02022r0P802-15_SG3a-AltPHY-Study-Group-Schedule), http://ieee802.org/15/pub/2002/ (accessed Oct 2002).

27. Wireless USB, www.usb.org/wusb.

28. Akahane, M.; Huang, B.; Sugaya, S.; Takamura, K. , CE Requirements for Alternative PHY CFA, http://www.ieee802.org/15/pub/SG3a.html (accessed Oct 2002).

29. www.ieee802.org/15/pub/TG4.html.

30. Bluetooth and UWB, http://bluetooth.com/Bluetooth/Press/SIG/BLUETOOTH_SIG_SELECTS_WIMEDIA_ ALLIANCE_ULTRAWIDEBAND_TECHNOLOGY_FOR_HIGH_SPEED_BLUETOOTH_APPLICATION.htm.

31. Fontana, R.J. Recent Applications of Ultra Wideband Radar and Communications Systems, http://www.multispectral.com/pdf/AppsVGs.pdf (accessed Oct 2002).

32. Lee, J.-Y.; Scholtz, R.A. Ranging in a dense multipath using an UWB radio link. IEEE Trans. Sel. Areas Commun. **2002**, *20* (9), 1677–1683.

33. Aetherwire Inc., Low-Power, Miniature, Distributed Position Location and Communication Devices Using Ultra-Wideband, Nonsinusoidal Communication Technology, Semi-Annual Technical Report, ARPA Contract J-FBI-94-058, July 1995, http://www.aetherwire.com/PI_Report_95/awl_pi95.pdf (accessed Oct 2002).

34. Sherf, K.; Robison, N. Networks in the Home: Analysis and Forecast: 3rd Edition, http://www.parksassociates.com (accessed Oct 2002).

35. Tewfik, A.H.; Saberinia, E. Multicarrier-UWB, document number 03147r1P802-15_TG3a; IEEE 802.15 March 2003 Plenary, Dallas, TX, Mar 2003.

36. Tekinay, S. Wireless geolocation systems and services. IEEE Commun. Mag. **1998**, *36* (4), 2–8.

37. Lee, J.-Y.; Scholtz, R.A. Ranging in a dense multipath environment using an UWB radio. IEEE J. Sel. Areas Commun. **2002**, *20* (9).

38. Fontana, R.J. Experimental Results from an Ultra Wideband Precision Geolocation System, Ultra-Wideband, Short-Pulse Electromagnetics. Jan 2000.

BIBLIOGRAPHY

1. A Brief History of UWB Communications, http://www.multispectral.com/history.html (accessed Jan 2007).

2. Alereon, http://www.alereon.com. (accessed Jan 2007).

3. An introduction to Ultra Wideband (UWB) wireless,http://www.deviceforge.com/articles/AT8171287040.html (accessed Jan 2007).

4. An introduction to Wireless USB (WUSB), http://www.deviceforge.com/articles/AT9015145687.html (accessed Jan 2007).

5. Fontana, J.; Gunderson, S.J. Ultra-wideband precision asset location system. IEEE Conference on Ultra Wideband Systems and Technologies, Baltimore, MD, May, 20–23, 2002.

6. General Atomics, http://photonics.ga.com/uwb/ (accessed Jan 2007).

7. Pulse-Link, http://www.pulse-link.net/. (accessed Jan 2007).

8. Staccato Communications, http://staccatocommunications.com/ (accessed Jan 2007).

9. Tzero Inc., http://www.tzti.com (accessed Jan 2007).

10. Ultra-Wideband (UWB) Technology, http://www.intel.com/technology/comms/uwb/ (accessed Jan 2007).

11. Ultrawideband Planet, http://www.ultrawidebandplanet.com/ (accessed Jan 2007).

12. Ultra-Wideband, http://en.wikipedia.org/wiki/Ultra-wideband (accessed Jan 2007).

13. Ultrawideband: A Better Bluetooth, http://www.computerworld.com/mobiletopics/mobile/technology/story/0,10801,-94896,00.html (accessed Jan 2007).

14. UWB Applications www.multispectral.com/pdf/AppsVGs.pdf (accessed Jan 2007).

15. UWB-Intel in Standards, http://www.intel.com/standards/case/case_uwb.htm #momentum (accessed Jan 2007).

16. UWB Forum, www.uwbforum.org (accessed Jan 2007).

17. WiMedia Alliance, http://www.wimedia.org/en/index.asp (accessed Jan 2007).

18. Wi-Quest Inc., http://www.wiquest.com/ (accessed Jan 2007).

19. Wisair Inc., http://www.wisair.com/ (accessed Jan 2007).

Ultra-Wideband: Communications

Chaiyaporn Khemapatapan
Computer and Telecommunication Engineering, Dhurakij Pundit University, Bangkok, Thailand

Watit Benjapolakul
Department of Electrical Engineering, Chulalongkorn University, Bangkok, Thailand

Abstract

Ultra-wideband communication is a new technique for transmitting data over short distance by using impulse-like, wideband, and noise-level signals. It is a potential candidate for use in wireless personal area networks.

INTRODUCTION

Earlier, ultra-wideband (UWB) was used for military purpose such as short-distance radio detection and ranging (radar) applications. Since the approval of UWB for communication by the FCC in March 2002 for unlicensed operation in the 3.1–10.6 GHz band subject to modified Part 15 rules, many UWB communication systems have been studied and proposed. The subject of UWB communication systems is to transfer massive data in short distance using low-power consumption.

Contemporary personal communication is a key for new era of human life. UWB in communication systems becomes the most important candidate since the FCC has allocated a spectrum mask for UWB communications in February 2002.[1] With the characteristics of UWB signal, massive data can be transferred in a short duration and its energy consumption is much less than other communication systems. Therefore, the use of UWB communications is merely suitable for modern applications of short-distance communications such as new consumer devices that require multimedia streaming and bulky data transmission.

Research and development of UWB systems have been done firstly for military purposes. Sight and specifying of the hidden objects under the ground or in the forest or behind walls can be certainly possible due to the penetrating behavior of UWB signals. In modern human life, bulky data transmission for small areas such as high-definition television, massive data transfer between computers, are unavoidable. Wireless personal area network (WPAN) and wireless USB (WUSB) are examples of data networking in small areas. Of course, UWB communication systems are standardized as parts of WPAN and WUSB. However, both WPAN and WUSB have various versions. Thus, various UWB communications are studied. Variety of UWB communications tradeoff high data rate transmission with long-life battery of device or complexity of transmission technique.

UWB CHARACTERISTICS

FCC has allocated UWB spectrum masks for using in communication systems as shown in Fig. 1. FCC defines the bandwidth of UWB signal for communication systems from 3.1 to 10.6 GHz and its spectrum power density is less than –41.3 dBm/MHz. According to the FCC definition, UWB signal system is defined as a system with a –10 dB fractional bandwidth greater than 0.2 MHz, or with a –10 dB bandwidth greater than 500 MHz. Through the late 1980s, UWB technology was referred to as baseband, carrier-free or impulse technology. It was not until 1989 that the Department of Defense applied the term "ultra-wideband." By that time, UWB techniques had been under development for more than 30 years. Recently, UWB applications were permissible only under special license, but on February 14, 2002, the FCC issued a Report and Order authorizing commercial deployment of UWB technology. Time Domain Corporation, in October 2002, became the first company to win federal approval to sell and deploy products based on UWB technology.

From Fig. 1, the spectrum masks have been skipped in the range of 1–3 GHz in order to avoid the interference to/from the existing radio systems especially for GPS. Moreover, the spectrum masks are classified into two masks: indoor and outdoor or handheld. Thus, different masking is for different devices depended on different environment, e.g., indoor or outdoor usage. From the definition, the UWB signal will be similar to noise signal because its energy is much smaller than other communication signals' energy. Thus, without transmitter being in the know of the discrimination of UWB signals from noise signal is almost impossible. UWB communication systems are very

Encyclopedia of Wireless and Mobile Communications DOI: 10.1081/E-EWMC-120043440

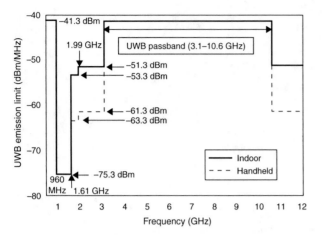

Fig. 1 The FCC's spectrum masks for UWB communication systems.

The transmitted signal in Fig. 2 is of impulse form or Gaussian monocycle waveform. Normally, impulse signal is used in the UWB communication systems. Instead of traditional sine waves, UWB communication sends several coded pulses per second across its spectrum. Gaussian pulse and its derivatives as shown in Fig. 3A are the pulse signals mostly used in UWB communication systems. Gaussian pulse signal is certainly one type of natural signal and it is easily generated by basic electronic devices such as the spark gap switching system or MOSFET switched capacitor or fast recovery diode.[2,3] Normally, the generated signal from these devices is filtered by RC devices in order to derive the Gaussian signal. The spectra of Gaussian pulse and its derivatives are shown in Fig. 3B. It can be noted that the spectrum of each signal may not be compiled with the FCC's spectrum mask. Thus, in order to make the transmission signal compile with the FCC's spectrum mask, the combination of these signals, or the transmitted signal shaped by Gaussian pulse or its derivative may be applied to practical UWB communication systems. In addition, a new technique such as orthogonal

difficult to eavesdrop by undefined users, i.e., the security of UWB communication systems has been already improved by the characteristics of UWB signals. An example of UWB signal is shown in Fig. 2. Without the highlight of the signal, it cannot discriminate the received signal from the noise signal.

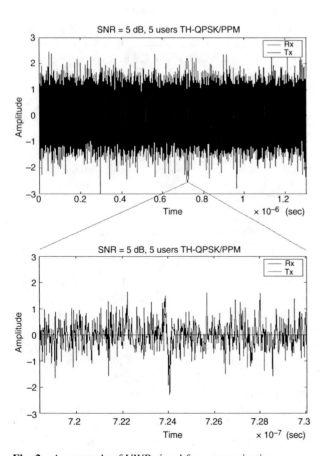

Fig. 2 An example of UWB signal for communication.

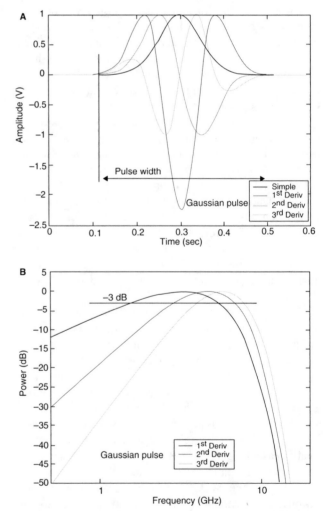

Fig. 3 (A) Gaussian pulse and its derivatives. (B) Their spectra.

frequency division multiplexing (OFDM) is used to generate UWB signal in some practical UWB systems where it is easier to design the signal to fit the spectrum into the FCC's spectrum mask.

Because of impulse transmission, UWB signal causes less interference than narrowband radio designs while yielding excellent multipath immunity. As an illustration, monocycle travels from a transmitter to a receiver along two paths. Since the paths are of two different lengths, the second pulse will arrive after the first pulse, i.e., there is no intra-pulse interference caused by the multipath channel.

TRANSMISSION TECHNOLOGIES

In the earlier stages of development, transmitter emits very short Gaussian monocycle pulse with precise pulse-to-pulse intervals. This technique can be found in impulse radio UWB (IRUWB) systems. With some modulation schemes, IRUWB will apply spread spectrum (SS) techniques such as direct-sequence SS (DS-SS) or time-hopping-SS (TH-SS) to transmit data. However, to ease the implementation, a new technique using OFDM technology has been proposed to UWB communication systems by a group of companies proposing the use of OFDM in UWB system, called multi-band OFDM alliance (MBOA).

Earlier, the most popular technique for multiple-access (MA) in IRUWB system was the time hopping spread spectrum (THSS) scheme. TH-SSMA with pulse-position modulation (PPM)[4] is an original idea for UWB communication system. TH-SSMA PPM UWB system can provide outstanding performance in additive white Gaussian noise (AWGN) channel, i.e., the system capacity is approximately 27,488 users with BER of 10^{-3} and a user data rate of 19.2 Kbps.

Fig. 4 shows the receiver diagram of TH-SSMA PPM UWB system. At the output of transmitter's antenna, the transmitted data will be modulated by the position of Gaussian monocycle pulse, e.g., pulse will be shifted for bit "1" but will not be for bit "0". Moreover, each Gaussian monocycle pulse will be hopped in the time slot of transmission frame. The hopping pattern is controlled by pseudo-noise (PN) code for each user. An example of UWB signals for TH-SSMA UWB system is shown in Fig. 5.

At receiver, the received signal is assumed to be an idealized Gaussian monocycle doublet pulse or the second derivative of Gaussian pulse. Demodulation process of TH-SSMA PPM UWB system is not similar to the traditional narrowband communication system. Template signal derived from the received Gaussian monocycle doublet pulse is used to find output data. Pulse correlator is used to find correlation between the template signal and the received signal. Different position of the transmitted Gaussian monocycle pulse provides different sign of the correlated output. However, high precision-time circuit is needed for TH-SSMA UWB system.

The other modulation techniques are pulse-amplitude modulation (PAM), on–off keying (OOK), and binary-phase shift-keying (BPSK). Typically, all mentioned modulation techniques can be applied with TH-SSMA UWB system. However, BPSK modulation can also be used with the DS-SSMA UWB system. The modulation process of DS-SSMA BPSK UWB system operates by spreading input data symbol into sequence of monocycle pulses using PN or other orthogonal codes in order to obtain

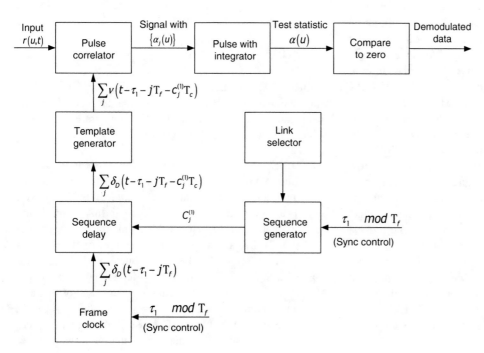

Fig. 4 Receiver diagram of TH-SSMA PPM UWB system.[4]

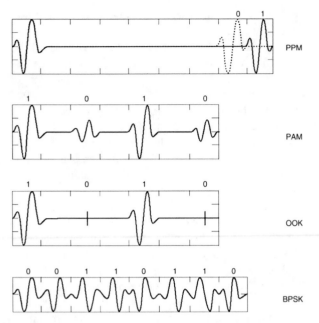

Fig. 5 Example modulated signals used in IRUWB systems.

higher processing gains. Demodulation process of DS-SSMA UWB system can be performed in the same manner as in the traditional communication systems. DS-SSMA BPSK UWB system is one of the candidates for WPAN. Moreover, the DS-SSMA UWB system may apply the additional modulation called M-ary binary-orthogonal keying (M-BOK).[5] M-BOK modulation is a technique used in SS communications where one basic spreading function, such as a PN sequence, is phase modulated on a carrier. The spreading function is modified by another certain function, such as Walsh or Hadamard function to cause a modulation to be orthogonal to the basic function and to every other modulating function used. The output signal from a matched filter with respect to input orthogonal signals is zero, while the response of a matched filter to non-orthogonal signals is maximum. Actually, M-BOK modulation has already been used in existing communication systems such as IEEE802.11b. However, M-BOK technique needs more complex circuits and is used extensively in secure communications. In Fig. 6, DS-SSMA UWB communication system splits spectrum into low and high bands. Low band is designed for low rate transmission but long distance service, while high band is designed for higher data rate but short distance service. However, the use of two bands will provide the highest data rate of 1.35 Gbps. Note that in Fig. 6, U-NII is "Unlicensed National Information Infrastructure". The FCC has made available 300 MHz of spectrum for U-NII band to support short-range, high speed wireless digital communications such as IEEE802.11a. The U-NII spectrum is located at 5.15–5.35 GHz and 5.725–5.825 GHz.

In addition, many techniques for IRUWB have been studied and proposed, such as transmitted reference UWB system,[6] modulating signal UWB systems using pulse or spectral shaping.[7–9] UWB systems using pulse shaping are mostly studied by researchers in university's laboratories. Actually, pulse or spectral shaping can also be applied to other UWB systems because this method alleviates the coexistence of UWB communication systems and other traditional communication systems. Normally, pulse shaping function may be implemented by filter circuit. Root raised cosine filter[9,10] is an example of a filter to shape IRUWB signals.

The other high potential technique is of course MBOA UWB systems.[11] MBOA UWB is a major candidate for high data rate UWB communications. By support from many corporations, such as Microsoft, Intel, MBOA UWB may be the first UWB communication system in commercial markets. OFDM modulation is widely used in contemporary communication systems, such as IEEE802.11a and IEE802.11g wireless LANs (WLANs), digital-video broadcasting (DVB), digital-audio broadcasting (DAB). Typically, OFDM modulation can be implemented by IFFT and FFT functions programmed in a DSP chip. At transmitter, BPSK or quadrature phase-shift keying (QPSK) symbol is applied to the input of IFFT function in order to convert this symbol into frequency domain signal before modulation with carrier. However, MBOA UWB systems applies multi-band spectrum having bandwidth over 500 MHz.

The MBOA UWB communication system consists of 13 band spectra as shown in Fig. 7. Each spectrum has a bandwidth of 528 MHz. Thus, MBOA UWB system can transfer a total data rate of 480 Mbps. At receiver, FFT function is applied to reverse the transmission process.

STANDARDS

Since 2002, the IEEE establishes task groups for drafting the standards of WPANs and these standards will be in the standard group of IEEE802.15. The standard composes of four sub standards as follows:

- IEEE802.15.1. To standardize the existent WPAN or Bluetooth. Detail of the standard is focused on air interfacing and media access control (MAC) layer. The lower transport layer [logical link control and adaptation protocol (L2CAP), link manager protocol (LMP), Baseband and radio interface] of Bluetooth is newly improved by this task group.
- IEEE802.15.2: To study and define the coexistences between UWB communication systems and other traditional communication systems such as GPS, WLANs, cellular phone networks. Because UWB signal has very wideband spectrum and can interfere with other

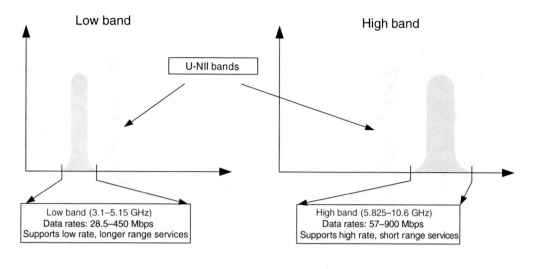

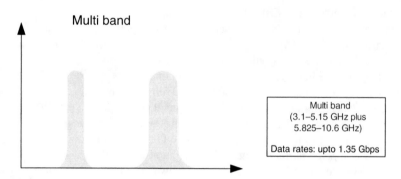

Fig. 6   DS-SSMA UWB spectrum design.

traditional communication signals, the study of coexistence between UWB communication systems and other traditional communication systems is necessary. The task group also developed a coexistence model to quantify the mutual interference of a WLAN and a WPAN.

- IEEE802.15.3: To propose new topologies of the future WPANs with the maximum data rate of 55 Mbps and the coverage radius of 20 m. However, the next sub standard, IEEE802.15.3a, has been proposed in order to overcome very high data rate of 1 Gbps with lower

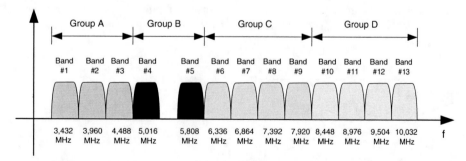

Fig. 7   Spectrum band of MBOA UWB communications.

coverage radius of 10 m. IEEE802.15.3 task group also studied channel characteristics. Then, four channel models[12] based on indoor and multipath channels have been suggested by this task group.

- IEEE802.15.4: To propose new topologies of the future WPANs with multi-month or multi-year battery life and low complexity devices. It is operating in an unlicensed, international frequency band. Potential applications are wireless sensors, interactive toys, smart badges, remote controls, and home automation.

Of course, IEEE802.15.1 is currently standardized and already in use in our lives because the standard came from industrial product, Bluetooth. In the future, IEEE802.15.3 or IEEE802.15.3a and IEEE802.15.4 will become high potential technologies in life. Many consumer devices, such as digital camera, home TV, mobile phone, merge these technologies in the basic functions. IEEE802.15.3 and IEEE802.15.3a are designed for a high data rate WPAN. Bulky data transmission is possible by using these standards. Many techniques are proposed to the task groups of IEEE802.15.3 and IEEE802.15.3a, such as DS-SSMA UWB, MBOA UWB. Some proposals are already in commercial trademark. Similar to Wi-Fi® for WLANs, WiMedia®[13] is the commercial trademark of MBOA UWB system[14] designed for future WPANs. It is established by very strong company partners, such as Microsoft, Intel, Nokia, Sony. However, IEEE802.15.3 task group also accepted other techniques in the draft of standards. IEEE802.15.4 is designed for multi-month or multi-year battery life with data rate no more than 1 or 2 Mbps. With the definition, IEEE802.15.4 will become indispensable not only to many human-life devices with very low-power consumption, such as wireless earphone, smart mobile phone, watch, interactive toys, and smart badges, but also in industrial and business applications, such as wireless sensors, remote controls, and home automation. Wireless sensor is a new technology that contributes the solution to the question—how to feedback environment's parameters without wire.

However, WPANs have a small coverage area. Thus, routing data throughout the network with mobility is somewhat complex. Typically, WPANs provide the connectivity between devices called piconet.[14] Piconet is a general, low powered, ad hoc radio network. It provides a base level of connectivity to even the simplest of sensing and computing objects. Topology of network devices is indefinite as shown in Fig. 8. Note that in Fig. 8, PNC is piconet coordinator which acts like a network manager or access point and DEV is a device. Thus, the present connectivity between these devices may be changed from previous time. Ad hoc routing is unavoidable. IEEE802.15.3 and IEEE802.15.3a task groups have already expressed concern about piconet and ad hoc routing in the proposals.

APPLICATIONS

UWB communication system is a new technology that uses signal with very low-power level and very wide bandwidth. UWB signal can be considered as noise signal. Thus, signal capture by unwanted users is not easy. The comparisons of battery life and data rate among IEEE802.15.3a and IEEE802.15.4, and other existing communication systems are shown in Fig. 9. However, UWB communication systems can only be used in a small coverage area like WPANs. Thus, most applications of UWB communication systems are for personal use. In Fig. 10, wireless home network is an example of application using UWB communication systems. Data relaying between home or consumer devices, such as music and video files, and massive data from/to computer, will be inevitable in the future of human life.

WiMedia is a high potential trademark of WPAN that is fast approaching. As mentioned earlier, MBOA consists of many industrial companies promoting WiMedia in commercial markets. It can provide a maximum data rate of 480 Mbps with low-power consumption. WUSB[15] is an application stacked on WiMedia. The protocol stack of WUSB is shown in Fig. 11. WiMedia consists of MBOA UWB physical layer, MBOA UWB MAC layer, and convergence layer. The functions of MBOA UWB physical layer are based on air-interfacing, channel and data rate selection, and power control. MBOA UWB MAC concerns itself with addressing, data flow control that is similar to other wireless network protocols. Convergence layer

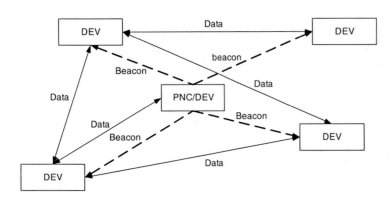

Fig. 8 An example of piconet and ad hoc routing.

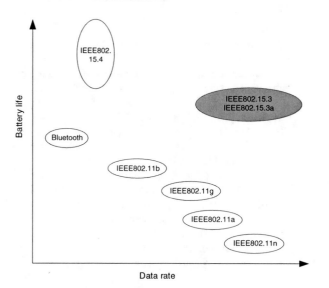

Fig. 9 Comparisons between UWB and other communication systems.

converts and formats data from/to a higher layer because different applications have different format. WUSB framework is designed to be compatible with USB2.0, i.e., MA technology is token-based control between host and device, and maximum number of devices connected to host is 127 devices. Maximum data rates are 480 and 110 Mbps for coverage areas of 3 and 10 m, respectively. However, if coverage area is less than 1 m, maximum data rate can reach 1 Gbps. Moreover, it can be noted that WiMedia supports other application layers, such as Internet protocol and W-1394 protocol.

CONCLUSION

UWB communication system is a key for WPAN in the future of human life. It is different from traditional communication systems, i.e., continuous, narrowband, and

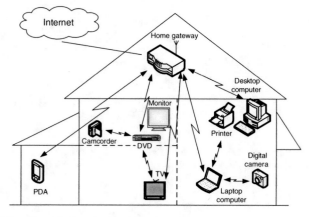

Fig. 10 Wireless home network.

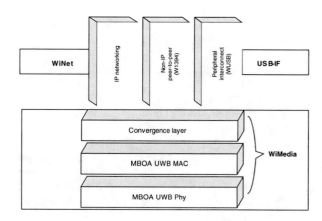

Fig. 11 Protocol stack of WUSB.

much higher-power signals than noise-level signals are used in the traditional communication systems, while impulse-like, wideband, and noise-level signals are used in UWB communication systems. The current WPAN system is Bluetooth. However, there are many techniques to be used for WPAN in UWB communication systems as described in this entry. UWB communication system can provide higher data rate and performance, i.e., long-life battery and high data rate devices is possible by using UWB communication systems.

ACKNOWLEDGMENTS

The authors wish to thank the Thailand Research Fund (TRF) for supporting research works on Ultra-wideband Signal Transmission[8–10] under the Royal Golden Jubilee (RGJ) program, contract no. PHD/0206/2543.

REFERENCES

1. Spectrum Policy Task Force Report; FCC: Washington, DC, ET Docket 02-135, 2002.
2. Marsden, K.; Lee, J.; Lee, H. Low power CMOS re-programmable pulse generator for UWB system. IEEE Conference on Ultra Wideband Systems and Technologies, Reston, VA, Nov 6–19, 2003; Hyatt Regency Reston, Reston, Virginia, USA; 443–447.
3. Han, J.; Nguyen, C. A new ultra-wideband, ultra-short monocycle pulse generator with reduced ringing. IEEE Microw. Wirel. Compon. Lett. **2002**, *12* (6), 206–208.
4. Win, M.Z.; Scholtz, R.A. Ultra-wide bandwidth time-hopping spread-spectrum impulse radio for wireless multiple-access communications. IEEE Trans. Commun. **2000**, *48* (4), 679–691.
5. DS-UWB Physical Layer Submission to 802.15 Task Group 3a, Institute of Electrical and Electronics Engineers, Inc.802.15-04/0137r3, 2004.

Ultra–
Wi-Fi

6. Hoctor, R.; Tomlinson, H. Delay-hopped transmitted reference RF communications. IEEE Conference on UWB Systems and Technologies, Digest of Papers, 2002, Baltimore, MD, May 20–23, 2002; Wyndham Baltimore Inner Harbor, Baltimor, Maryland, USA; 265–270.

7. Nakache, Y.P.; Molisch, A.F. Spectral shaping of UWB signals for time-hopping impulse radio. IEEE J. Sel. Areas Commun. **2006**, *24*, 738–744.

8. Khemapatapan, C.; Benjapolakul, W.; Araki, K. Time hopping QPSK impulse signal transmission for ultra wideband communication system in the presence of multipath channel. IEEE/ACES Conference on Wireless Communications and Applied Computational Electromagnetics (IEEE/ACES 2005), Honolulu, HI, Apr 3–7, 2005; Hilton Hawaiian Village, Honolulu, Hawaii; 5–8.

9. Khemapatapan, C.; Benjapolakul, W.; Araki, K. QPSK impulse signal transmission for ultra wide band communication systems in multipath channel environments. IEICE Trans. Fundamentals. **2005**, *E88-A* (11), 3100–3109.

10. Khemapatapan, C. Multi-band ultra-wide bandwidth data transmission using raised cosine pulse shaping. International Symposium on Information Theory and its Applications (ISITA 2006), Seoul, Korea, Oct 29–Nov 1, 2006; 444–448.

11. TG4a UWB-H-IR Modulation Schemes and preliminary Simulation Results, Institute of Electrical and Electronics Engineers, Inc.802.15-04/0429, 2005.

12. Foerster, J. et al. *Channel Modeling Sub-committee Report Final*, Institute of Electrical and Electronics Engineers, Inc.802.15-02/490, 2002.

13. http://www.wimedia.org (accessed August 17, 2007).

14. *Multi-band OFDM Physical Layer for IEEE802.15 Task Group 3a*, IEEE802.15-03/268r3, 2004.

15. http://www.usb.org/wusb/ (accessed August 17, 2007).

UMTS

Syed Ahson
iDEN Subscriber Division of Motorola Inc., Plantation, Florida, U.S.A.

Abstract

UMTS is a network protocol for 3G wireless systems.

INTRODUCTION

A 3G UMTS network consists of three interacting domains: core network (CN), UMTS terrestrial radio access network (UTRAN), and user equipment (UE). A UMTS system is divided into a set of domains and reference points that interconnect them. The 3G UMTS protocol structure is based on the principle that the layers and planes are logically independent of each other. We describe the 3G UCN architecture, reference points, UTRAN, and protocol structure.

UCN ARCHITECTURE

UCN is composed of a circuit-switched (CS) domain and a packet-switched (PS) domain. The CS domain consists of a mobile switching center (MSC) and a gateway MSC (GMSC). Fig. 1 illustrates the entities present in a UCN. The PS domain contains the serving GPRS support node (SGSN), the gateway GPRS support node (GGSN), the domain name service (DNS), the dynamic host configuration protocol (DHCP) server, packet charging gateway, and firewalls. The home location register (HLR) interfaces with both domains over signaling system 7 (SS7) links. Other components required for operation of the UCN include billing systems, provisioning systems, and service/element management systems. The 3G-MSC is responsible for mobility management. It handles international mobile subscriber identity (IMSI) attach, authentication, HLR updates, serving radio network subsystem (SRNS) relocation, and intersystem handovers. The 3G-MSC handles call setup messages from and to mobile users and provides supplementary services such as call waiting. The 3G-MSC also provides CS data services for services such as fax.

The 3G-SGSN provides functionality similar to the 3G-MSC for the PS domain. The 3G-SGSN handles GPRS attach, authentication, visitor location register (VLR) updates, SRNS relocation, and intersystem handover for user packet data session. It accepts session setup messages and enforces admission control. The 3G-SGSN is also responsible for tunneling user TCP/IP data to the 3G-GGSN using GTP. It collects statistics relating to mobile users' internal data usage, which may be used for charging. The 3G-GGSN provides internetworking with the external PS network, and also offers packet filtering services. The 3G-GGSN may allocate dynamic IP addresses to mobile users on PDP context activation, and is also responsible for tunneling user TCP/IP data to the 3G-SGSN using GPRS tunneling protocol (GTP). The 3G-GGSN collects statistics relating to mobile users' external data usage, which may be used for charging. The DNS server translates access point names (APN) to the 3G-GGSN IP addresses. A DHCP server may be present to automatically allocate IP addresses for mobile users at packet data protocol (PDP) context activation. A packet-data firewall is present to protect the GPRS PS domain.

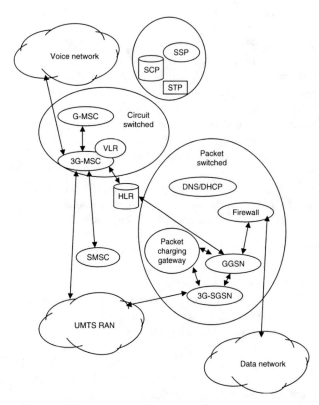

Fig. 1   UMTS core network architecture.

Encyclopedia of Wireless and Mobile Communications DOI: 10.1081/E-EWMC-120043959

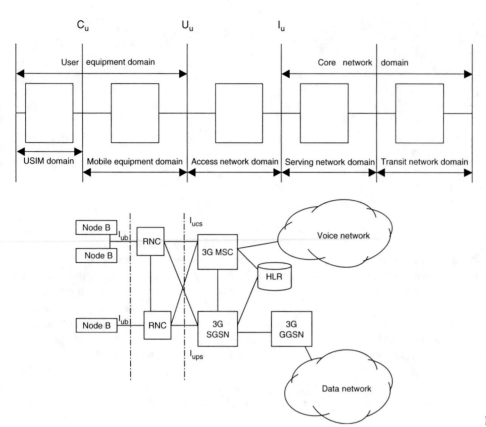

Fig. 2 UMTS reference points.

UMTS NETWORK REFERENCE ARCHITECTURE

A set of reference points and domains have been defined for the UMTS network. C_u is defined as the reference point between the UMTS subscriber identity module (USIM) domain and the mobile equipment domain. U_u is defined as the reference point between the mobile equipment domain and the UMTS radio interface domain. I_u is defined as the reference point between the UMTS radio interface domain and serving network domain. The I_u reference point is split into I_{ucs} and I_{ups}. I_{ucs} connects the UMTS radio interface domain to the CS domain. I_{ups} connects the UMTS radio interface domain to the PS domain. Fig. 2 illustrates UMTS domains and reference points.

UTRAN

UTRAN consists of a set of radio network subsystems (RNSs) (Fig. 3). An RNS is responsible for radio resources and coverage in a set of cells. It has two main elements: Node B and radio network controller (RNC).

RNC enables autonomous radio resource management (RRM) by UTRAN. It assists in soft handover of the UE when a mobile user moves from one cell to another. It combines and splits I_{ub} data streams received from multiple Node Bs, and is also responsible for frame synchronization, outer loop power control, and SRNS relocation.

Node B is the physical unit of radio transmission and reception with cells. It can support both time division duplex (TDD) and frequency division duplex (FDD) modes and can be collocated with a GSM base transceiver system (BTS) to reduce implementation costs. It connects to the UE via the U_u interface, and the RNC via the I_{ub} interface. The main task of Node B is the conversion to and from the U_u radio interface, including forward error correction (FEC), rate adaptation, WCDMA spreading and de-spreading, and quadrature phase shift keying (QPSK) modulation on the air interface. It measures the quality and strength of the connection and determines the frame error rate (FER), transmitting these data to the RNC as a measurement report. Node B also participates in power control,

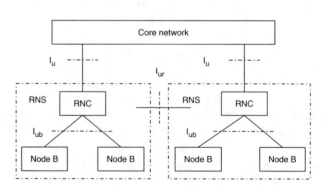

Fig. 3 UMTS architecture.

because it enables the UE to adjust its power using down link (DL) transmission power control (TPC) commands via the inner-loop power control on the basis of up link (UL) TPC information. The predefined values for the inner-loop power control are derived from the RNC via the outer-loop power control.

UTRAN LOGICAL INTERFACES

The general protocol model for UTRAN interfaces is shown in Fig. 4. The structure is based on the principle that the layers and planes are logically independent of each other.[1-3] The protocol structure consists of two main layers: the radio network layer (RNL), and the transport network layer (TNL). All UTRAN-related issues are visible only in the RNL, and the TNL represents standard transport technology that is selected for use by UTRAN but without any UTRAN-specific requirements. The control plane includes radio access network application part (RANAP) at I_u, RNS application part (RNSAP) at I_{ur}, or Node B application part (NBAP) at I_{ub}, and the signaling bearer for transporting the application protocol messages. Among other things, the application protocol is used for setting up bearers (i.e., radio access bearer or radio link) in the RNL. The user plane includes the data streams and the data bearers for the data streams. The data streams are characterized by one or more frame protocols specified for that interface.

The transport network control plane (TNCP) does not include any RNL information, and is completely in the transport layer. It includes the access link control application part (ALCAP) protocols that are needed to set up the transport bearers (data bearer) for the user plane. It also

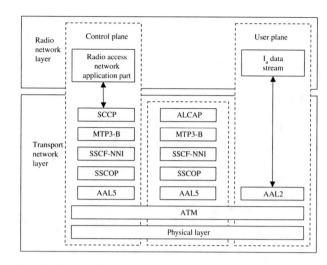

Fig. 5 I_{ucs} interface protocols.

includes the appropriate signaling bearers needed for the ALCAP protocols. The TNCP is a plane that acts between the control plane and the user plane. The introduction of TNCP is performed in such a way that the application protocol in the radio network control plane is kept completely independent of the technology selected for data bearer in the user plane.

The UMTS I_{ucs} logical interface interconnects the UTRAN to the UMTS CS core network. The circuit-switched protocol architecture on the I_{ucs} interface is illustrated in Fig. 5. The RNL control plane consists of RANAP. The transport network user plane consists of SS7 protocols. Signaling connection control part (SCCP), message transfer part (MTP3-B), and signaling ATM adaptation layer for network-to-network interface (SAAL-NNI) is present in the transport network user plane. SAAL-NNI is divided into service-specific coordination function (SSCF), service specific connection oriented protocol (SSCOP), and ATM adaptation layer 5 (AAL5). SSCF and SSCOP are designed for signaling transport in ATM networks. AAL5 is responsible for segmenting data into ATM cells.

REFERENCES

1. 3GPP Technical Specification 25.410 UTRAN I_u Interface: General Aspects and Principles.
2. 3GPP Technical Specification 25.420 UTRAN I_{ur} Interface: General Aspects and Principles.
3. 3GPP Technical Specification 25.430 UTRAN I_{ub} Interface: General Aspects and Principles.

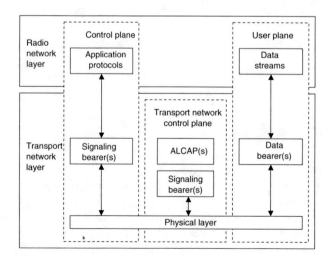

Fig. 4 UMTS protocol model.

Unlicensed Mobile Access: Architecture and Protocols

Mehul Shah
T-Mobile USA Inc., Bellevue, Washington, U.S.A.

Vassilios Koukoulidis
Nokia Siemens Networks, Boca Raton, Florida, U.S.A.

Abstract

Unlicensed mobile access (UMA) is the third generation partnership project (3GPP) standard for cellular/ Wi-Fi convergence which allows for the seamless handover of voice and data between mobile networks and WLANs with a dual-mode handset.

INTRODUCTION

After years of enjoying a rapidly expanding customer base, wireless service providers are now entering a phase where they need to find new avenues of growth, reduce churn and unlock potential new customer bases. Subscribers often cite lack of sufficient radio coverage at home as the main reason for switching service providers. The next logical step then is to tap into the "indoor voice" market to achieve the goals mentioned above. The traditional way to do this is to build new cell sites and improve radio penetration in buildings. However, the prohibitive cost of rolling out new cell sites has limited providers in aggressively pursuing this path.

On the other hand the broadband penetration rates in the US and across the world have been rapidly gaining momentum. Home and commercial Wi-Fi zones or hotspots have mushroomed and are gaining in popularity. The availability of cheap broadband equipment such as cable and digital subscriber line (DSL) modems and access points (APs) and a growing base of equipment manufacturers have only increased the uptake of broadband services at home. These developments did not escape the attention of mobile operators and equipment vendors and the concept of providing existing GSM services over unlicensed spectrum was born.

The unlicensed mobile access (UMA) technology enables the user access to voice and data services provided by the cellular network over unlicensed spectrum technologies such as 802.11 [WLAN] and Bluetooth. UMA provides for the seamless handover between GSM/GPRS and WLAN so that users can receive a consistent level of service as they roam between these networks.

We shall explain the brief history of UMA and also explain how this technology came to be adopted by the third generation partnership project (3GPP) standardization body, followed by a discussion of the protocols and architecture used in the UMA network. Finally we conclude the chapter by presenting call flow explanations for some basic UMA procedures.

HISTORY

As explained earlier, the UMA standard was born from the need of mobile operators to deliver voice and data services to subscribers at home or in the office. In late 2003, with this goal in mind, a group of operators and vendors founded the UMA consortium and mutually agreed to develop a set of specifications to deliver GSM services over an IP access network. In September 2004, the initial UMA specifications were published and were introduced to the 3GPP standardization body. On April 8, 2005, just 8 mo after formal introduction, the UMA specifications were approved as 3GPP standards for generic access to A/Gb interfaces for 3GPP Release 6.

After the acceptance of the UMA standard by 3GPP, the consortium was disbanded and its members now continue to develop and refine the specifications under the aegis of the 3GPP.

UMA OVERVIEW

UMA introduces a new network element called the UMA network controller (UNC) which anchors the WLAN access network, provides secure access to the GSM core network over standardized interfaces and facilitates roaming between licensed and unlicensed networks (Fig. 1).

The working of UMA can be briefly summarized as follows:

1. A user with a dual-mode handset moves into the coverage area of an AP to which it can connect.
2. Once the user has access to the Wi-Fi network, it then contacts the UNC to be authenticated for UMA service.

Encyclopedia of Wireless and Mobile Communications DOI: 10.1081/E-EWMC-120043504

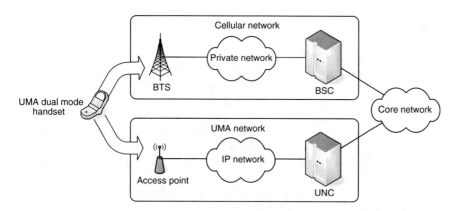

Fig. 1 Unlicensed mobile access (UMA) overview.

3. After the UNC has authenticated and successfully registered the user, its location information is updated in the network and all further traffic will be routed through the UNC over the broadband IP access network. The UNC communicates to the core network over standardized interfaces.

4. As the handset moves out of range (or moves into range) of Wi-Fi coverage, the UNC facilitates the handover between the two networks in a seamless manner. This can happen during an active voice or data call. The user is completely unaware of the underlying process and can maintain session continuity transparently.

UMA NETWORK ARCHITECTURE

The UMA network (Fig. 2) comprises of:

1. One or more APs. These may be either Bluetooth or 802.11.
2. One or more UNCs.

The handset used to access the UMA network is dual-mode (GSM and unlicensed) and is able to switch between these two technologies. It supports either Bluetooth [personal area networking (PAN) profile] or 802.11.

The AP provides the local wireless link over the unlicensed spectrum for the dual-mode handset and provides the Bluetooth (PAN profile) or 802.11 functionality. The AP does not have any UMA specific functionality and any off-the-shelf unit can be used to interconnect the handset to the UNC via a broadband network. The IP connectivity extends all the way to the handset through the AP and is defined as the Up-interface.

The UNC appears as a GSM EDGE (enhanced data rates for global evolution) radio access network (GERAN) base station controller (BSC) to the core network and includes a functional unit called the security gateway (SGW). This unit terminates the IPSec tunnel from the handset providing authentication, encryption, and data integrity for voice and data traffic.[1] All user and control plane traffic is transported to the UMA network over a broadband IP network secured by the IPSec association between the handset and the SGW.

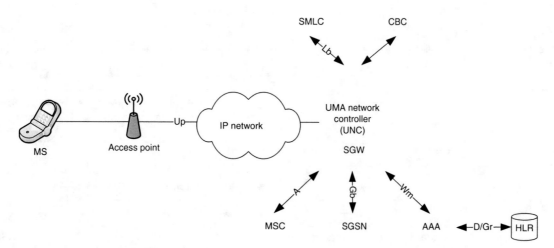

Fig. 2 Unlicensed mobile access (UMA) architecture.

The UNC interacts with the core network over the following 3GPP standardized interfaces:

- A-interface with the mobile switching centre (MSC)/ visitor location register (VLR) for voice services;[2]
- Gb-interface with the serving GPRS support node (SGSN) for data services;[3]
- Wm-interface with the authentication, authorization and accounting (AAA) server for the authentication of the user when setting up a secure tunnel;[4]
- Lb-interface for location based services;[5]
- Cell broadcast center (CBC)–BSC interface for cell broadcast services.[6]

SIGNALING AND USER PLANE ARCHITECTURE (Up-INTERFACE)

The Up-interface operates over an IP transport network and is responsible for the tunneling of GSM/GPRS signaling and user plane traffic from the mobile subscribers (MS) to the core network.

The Up-interface circuit switched (CS) domain signaling plane architecture is as shown below in Fig. 3.[1]

Notable Points

- GSM mobility management (MM) layers and above are carried transparently between the MS and the core MSC.
- The GSM radio resource (RR) protocol is replaced by the UMA radio resource (UMA-RR) protocol, which has been specifically created for the unlicensed radio link. The UNC terminates the UMA-RR protocol and interworks it to the A-interface using the base station system application part (BSSAP) protocol as a BSC.

The potential implementation of the CS signaling plane architecture in the MS is available in Ref. 1.

The Up-interface CS domain user plane architecture is as shown below in Fig. 4.[1]

Notable Points

- Real-time transport protocol (RTP)[7–9] is used to transport audio from the MS to the unlicensed mobile access network (UMAN). Adaptive multi-rate (AMR) is mandatory when the MS operates in UMA mode.

The Up-interface GPRS signaling plane architecture is as shown below in Fig. 5.[1]

Notable Points

- Signaling protocol data units (PDUs) [logical link control (LLC) layer and higher] are relayed transparently between the MS and the SGSN. All GPRS services are available to the MS, only the underlying radio access method has changed.
- A new protocol UMA radio link control protocol (UMA-RLC) is used to carry the LLC signaling PDUs from the MS to the UNC. The UNC interworks it to the standard Gb protocol to communicate with the SGSN.
- TCP is the transport protocol for UMA-RLC signaling between the MS and the UNC.

The Up-interface GPRS user plane architecture is as shown below in the Fig. 6.[1]

Notable Points

- GPRS data PDUs (LLC layer and above) are carried transparently between the MS and the UNC.
- The UMA-RLC protocol is used to carry the LLC data PDUs between the MS and the UNC, which then interworks it to the base station system (BSS) GPRS

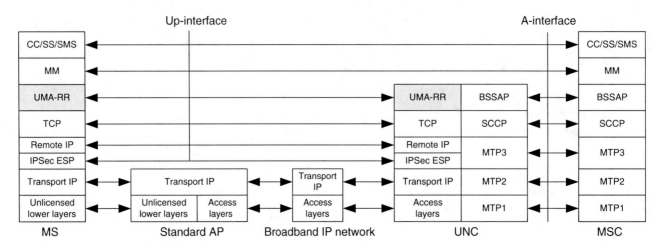

Fig. 3 CS domain signaling plane protocol architecture.

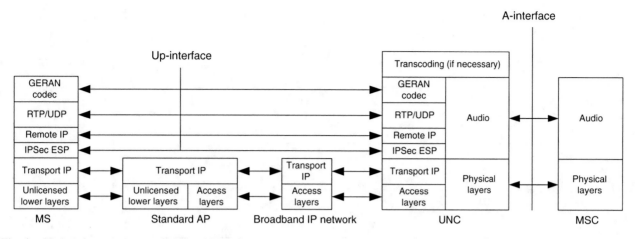

Fig. 4 CS domain user plane protocol architecture.

(BSSGP) protocol over the standard Gb-interface for communication with the SGSN.

- User datagram protocol (UDP) is the transport protocol for UMA-RLC user plane traffic between the MS and the UNC.

PROTOCOLS

The protocols used in the UMA network can be broadly classified in the following four categories:

1. Standard 3GPP protocols:
   a. All GSM MM and connection management protocols are used without any modification between the MS and the MSC;
   b. All GPRS protocols above the LLC layer are used without modification between the MS and the SGSN;

   c. The A-interface protocols are used without any changes between the UNC and the MSC;
   d. The Gb-interface protocols are used without any changes between the UNC and the SGSN;
   e. The UNC uses a subset of the protocols offered by the Wm-interface when communicating with the AAA server without any changes.

2. UMA specific protocols: there are two protocols in this category:
   a. **UMA-RR:** The UMA-RR is the equivalent of the GSM radio resource (GSM-RR) protocol in the handset and the UNC. This protocol has been customized to take advantages of the special characteristics of the unlicensed spectrum over which it operates. It provides the mechanisms to support functions such as registration with the UNC, setting up bearer paths for CS traffic, support of handover between GERAN and UMAN, GPRS suspension, paging etc.

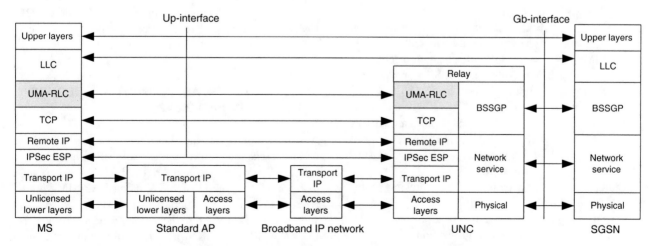

Fig. 5 Up-interface GPRS signaling plane protocol architecture.

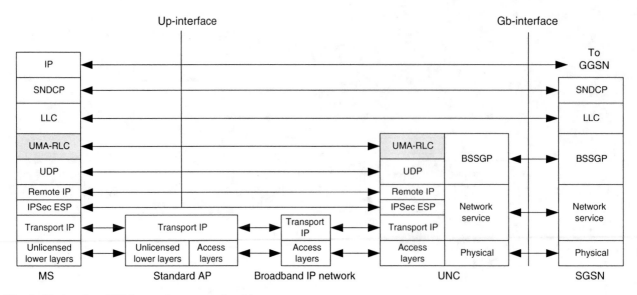

Fig. 6   Up-interface GPRS user plane protocol architecture.

b. **UMA-RLC:** The UMA-RLC protocol provides the delivery of GPRS signaling and user plane traffic. It also provides traffic channel setup and termination procedures that are required for data transfer.

3. Standard IP: UMA uses several IP-based protocols developed by the IETF:

   a. TCP[10] is used for CS and GPRS signaling.

   b. UDP[11] is used for GPRS user plane traffic and for CS voice in conjunction with RTP. UDP is also used for IPSec network address translation (NAT) traversal.

   c. A specific profile[12] of IPSec encapsulating security payload (ESP)[13] is used to secure traffic between the MS and the UNC. This profile is aligned with 3GPP WLAN interworking scenario.[14] For NAT traversal, the UDP encapsulation for ESP tunnel mode is supported as specified in Ref 15.

   d. A UMA specific profile of the Internet key exchange (IKEv2) and extensible authentication protocol–subscriber identity modules (EAP-SIM)[16] is used as the authentication mechanism for those handsets that have a SIM or universal SIM (USIM), but no support for UMTS authentication and key agreement (AKA). For the handsets that have a SIM or USIM and can support UMTS AKA, EAP-AKA[17] and IKEv2 are used for mutual authentication. NAT traversal is supported as specified in Ref 18.

4. Unlicensed spectrum protocols: the following protocols are used without any modification:

   a. 802.11 protocols for authentication, encryption, data transfer and traffic prioritization.[19]

b. Bluetooth protocols for discovery, paging, pairing and encryption. The Bluetooth network encapsulation protocol (BNEP) is used to provide ethernet emulation over Bluetooth asynchronous communication links (ACL) as per the PAN profile as defined in Ref 20.

BASIC UMA PROCEDURES

When an MS wants to start using UMA services, it has to establish a secure connection with the UMAN and then identify the serving UNC (S-UNC) to which it can register. In order to do so, it first has to connect to the provisioning UNC (P-UNC) and then discover the default UNC (D-UNC). After this the MS may be redirected to a S-UNC. The D-UNC may also be the S-UNC and is dependant on operator preference. We shall explain the IPSec tunnel establishment (with EAP-SIM and IKEv2), UMAN discovery and registration and handover procedures in detail in the following sections.

IPSec TUNNEL ESTABLISHMENT PROCEDURE WITH EAP-SIM AND IKEv2

Authentication over the Up-interface uses GSM or UMTS credentials of the MS. An MS which has a SIM or USIM, but no support for UMTS AKA is authenticated with the UNC-SGW using EAP-SIM within IKEv2. For an MS which has a SIM or USIM and support for UMTS AKA the authentication protocol is IKEv2, while EAP-AKA provides mutual authentication and key generation. We shall discuss the former case in this section.

The MS is provisioned with the IP addresses or the fully qualified domain name (FQDN) of the P-UNC and the provisioning SGW (P-SGW) and sets up a secure connection with the P-SGW as shown in Fig. 7 below:

1. The MS obtains connectivity with the local Bluetooth or Wi-Fi AP. If it has been provisioned with the FQDN of the P-SGW, it then interacts with a domain name system (DNS) to resolve the FQDN to an IP address. Using the derived IP address, it

initializes the IKEv2 authentication procedure with the SGW.

2. The authentication procedure is initiated by the exchange of IKE_SA_INIT messages by the MS and the SGW. The encryption algorithm to be used is mutually agreed upon by the exchange of these messages.

3. The MS indicates that EAP is to be used for mutual authentication by leaving out the AUTH payload from the IKE_AUTH message parameters. It sends the International mobile subscriber identity (IMSI) as the initiator identity in the network access identifier (NAI) format as specified in Ref. 21.

Ultra–
Wi-Fi

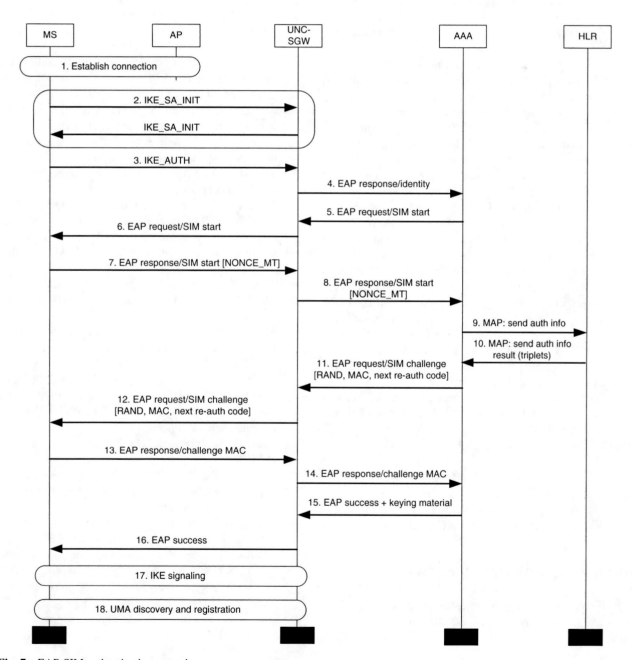

Fig. 7   EAP-SIM authentication procedure.

4. The SGW contacts the AAA server with the MS identity (IMSI) received in 3 above. This is the start of the EAP-SIM procedure.

5. The AAA responds with the EAP request /SIM start message which signals that the subscriber has been identified as a candidate for authentication with EAP-SIM.

6. The SGW relays this message to the MS.

7. The MS generates a random number (NONCE_MT) and sends it to the SGW. The random number is used in network authentication.

8. The SGW relays this message to the AAA server.

9. The AAA server requests the authentication triplets from the home location register (HLR) for this subscriber.

10. The HLR responds with multiple authentication triplets.

11. The AAA server computes the message authentication code (MAC) based on the random number (NONCE_MT) and Cipher Key (Kc). A new re-authentication code may be generated using EAP-SIM generated keying material. The AAA then composes the EAP Request/SIM challenge message which includes multiple RAND challenges, MAC and the re-authentication code and sends it to the SGW.

12. The SGW forwards this message to the MS.

13. The MS runs the GSM A3/A8 algorithm in the SIM once for each received RAND from the SGW. This gives rise to a number of SRES and Kc values. The MS generates MAC codes based on the generated Kc values and compares it with the received MAC from the AAA. If the MAC does not match, the MS cancels the authentication. It will only proceed with the exchange if the MAC is correct. The re-authentication code, if received, is stored for future use.

14. The SGW forwards this message to the AAA.

15. The AAA verifies the received MAC from the MS and if correct, sends the EAP Success message to the SGW. The AAA server also includes derived keying material for confidentiality and integrity protection between the MS and the SGW.

16. The SGW forwards the EAP success message to the MS.

17. After successful EAP-SIM authentication, the remaining IKE signaling is completed.

18. The MS is now ready to initiate the registration and discovery process with the UNC.

UMAN DISCOVERY AND REGISTRATION

The UMAN discovery and registration procedures are the ones initiated by the MS in order to obtain UMA services from the network. The discovery procedure refers to the process of identifying the S-UNC. In order to do this, the MS has to first establish a connection to the P-UNC and then discover the D-UNC, which in turn can redirect the MS to the S-UNC. The registration procedure is performed between the MS and the UNC and fulfils the following:

1. Informs the UNC that an MS has joined the UMAN through a particular AP and has been assigned an IP address;

2. Informs the MS of certain UMAN specific parameters that may be used during the course of the UMA session.

The discovery and registration procedures can be summarized in the following Fig. 8:

1. The MS is either provisioned with the IP address of the P-SGW or the FQDN of the P-UNC and the associated SGW. In case the SIM is provisioned with the FQDN, the MS interacts with a DNS server to obtain the IP address of the P-SGW. Alternatively, the MS can also derive the FQDN of the P-UNC using rules specified in Ref. 1.

2. The MS establishes a secure connection with the P-SGW as explained in the earlier section.

3. If the MS has a provisioned or derived FQDN of the P-UNC, it will interact with a DNS server (within the secure tunnel) to resolve it to an IP address. This step is omitted, if the MS already has the IP address of the P-UNC provisioned.

4. The DNS returns the IP address of the P-UNC.

5. The MS sends the UMA radio resource (URR) DISCOVERY REQUEST to the P-UNC via a TCP connection established to a well known port. This message is used to elicit the IP address of the D-UNC.

6. The P-UNC responds with the IP address of the D-UNC and the associated D-SGW in the URR DISCOVERY ACCEPT message. The location of the MS is used as criteria to direct it to a local UNC to optimize network operations.

7. The MS establishes a secure connection with the D-SGW.

8. The URR REGISTER REQUEST message is sent out to the D-UNC by the MS in an attempt to register itself to the UMA network.

9. If the D-UNC wants to redirect the MS to another S-UNC, it responds with the URR REGISTER REDIRECT message containing the IP address of the S-UNC and the associated serving SGW (S-SGW). Alternatively, the D-UNC may respond with the URR REGISTER ACCEPT, in which case it is also the S-UNC. The determination of whether the D-UNC also acts as the S-UNC is

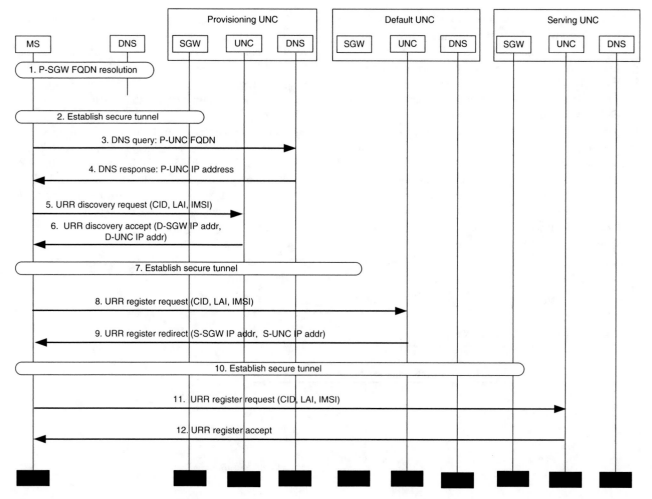

Fig. 8 UMAN discovery and registration.

dependant on criteria such as operator policy, load balancing, and roaming agreements in case of a roaming MS.

10. Using the received IP address of the S-SGW, the MS establishes a secure connection with it.

11. The URR REGISTER REQUEST message is sent out to the S-UNC.

12. If the S-UNC decides to accept the registration, it will respond with the URR REGISTER ACCEPT message containing UMAN specific system information. The location area and routing area updates (RAUs) will then be triggered by the MS. In case the register message is rejected, the MS will again establish a connection with the D-UNC and try to get the IP address of another S-UNC in the network. If the MS does not receive a response to the registration request sent to the D-UNC, it will attempt the discovery procedure with the P-UNC in order to obtain a new D-UNC.

HANDOVER TO UMAN

The following sequence (Fig. 9) explains the procedure for the handover of an active voice call from GERAN to UMAN.[1]

1. The MS is on an active call on GERAN and has detected UMA coverage and successfully registered on the UMAN. As a result of this, it has all the UMAN specific system information which includes UMAN cell information. Also, the GERAN provides the MS with neighboring cell information, such that one of the entries [absolute radio frequency channel number (ARFCN) and base station identity code (BSIC)] in the list matches that of the UMAN cell.

2. The measurement report to GERAN includes UMA cell information. The MS reports the highest signal level for that UMAN cell.

3. Based on the received information, the GERAN decides to handover to UMAN, using an internal mapping of the ARFCN and BSIC to common gateway interface (CGI). The handover procedure is initiated by the GERAN by sending the *Handover Required* command to the core network including the target UMAN cell information.

4. The core network sends the *Handover Request* message to the target UNC requesting allocation of resources.

5. The target UNC responds with a *Handover Request Acknowledge* message indicating that it can support the requested handover. It also sends a *Handover Command* message to the core network which includes the radio channel information to which the MS should be directed to.

6. The core network forwards this message to the GERAN.

7. The GERAN sends the *Handover Command* message to the MS to initiate the handover to UMAN. The message contains target UMAN cell information such as broadcast control channel (BCCH), ARFC, public land mobile network (PLMN) color code and BSIC. Upon reception of this information the MS does not immediately switch the audio path from GERAN to UMAN, but waits till it sends the URR *Handover Complete* message to keep the audio interruption as short as possible.

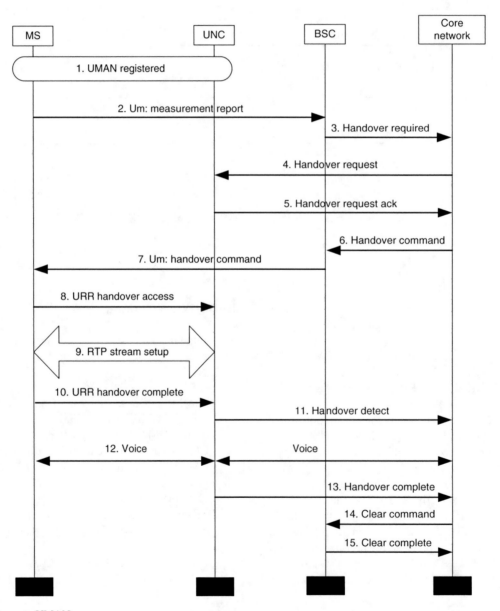

Fig. 9 Handover to UMAN.

8. The MS communicates with the S-UNC using the URR *Handover Access* message which includes the entire *Handover Command* message received from the GERAN. The handover reference in the command message helps the UNC to correlate the earlier *Handover Request Acknowledge* message sent to the core network and aids in identifying the completion of the handover process.

9. The S-UNC sets up the RTP bearer path with the MS.

10. The MS sends the URR *Handover Complete* message indicating that handover process has been completed. It then switches the bearer from the GERAN to UMAN user plane.

11. The UNC sends the *Handover Detect* message to the core network indicating that the MS has been detected. This allows the core network to switch the voice path from GERAN to the S-UNC.

12. Voice traffic now flows between the MS, UNC and the core network.

13. The S-UNC sends the *Handover Complete* message indicating that the handover is complete.

14. The core network sends the *Clear Command* and tears down the GERAN connection.

15. The GERAN responds with the *Clear Complete* message and releases its resources.

HANDOVER TO GERAN

The following sequence (Fig. 10) explains the procedure for the handover of an active voice call from UMAN to GERAN.[1]

1. The MS has an ongoing call on the UMAN network and is continuously monitoring the signal strength of the local unlicensed radio coverage. The handover from GERAN to UMAN is always triggered by the MS and may be dependant on the outcome of this measurement.

2. The S-UNC also monitors the quality of the radio signal and may send the URR *Uplink Quality Indication* message to the MS if it believes that there is a problem with the uplink quality for the ongoing call. The criterion used for such a decision is explained in Ref. 1.

3. The MS sends the URR *Handover Required* message to the S-UNC, which includes a list of GERAN cells, identified by the CGI, sorted in descending order of signal strength (ranked by path-loss parameter).

4. The UNC indicates the initiation of the handover process by sending the *Handover Required* message to the core network, including the GERAN cell list given to it by the MS.

5. The core network selects a particular target GERAN cell and requests that resources be allocated by sending the *Handover Request* message.

6. The target GERAN constructs a *Handover Command* message and sends it to the core network within the *Handover Request Acknowledge* message. The command message contains information on the channel allocated.

7. The core network sends the received *Handover Command* message to the S-UNC signaling the MS be handed over to GERAN.

8. The S-UNC sends the URR *Handover Command* message to the MS including the GERAN resource allocation details.

9. The MS sends the *Um: Handover Access* message to the GERAN which contains the handover reference element which allows the GERAN to correlate the handover access with the *Handover Command* message transmitted earlier to the core network.

10. The GERAN confirms that the *Handover Request* has been detected by sending the *Handover Detect* message to the core network.

11. The core network may switch the voice path to the target BSS.

12. The GERAN provides some physical information such as timing advance to allow the MS to synchronize with the GERAN.

13. The MS sends the *Um: Handover Complete* message to the GERAN indicating that the handover is completed.

14. The GERAN in turn confirms to the core network that the handover has completed.

15. Voice traffic now flows between the MS, GERAN and the core network.

16. The core network notifies the UNC to release resources via the *Clear Command*.

17. The UNC notifies the MS to clean up its resources by sending the URR *Release* message.

18. The UNC acknowledges that all the resources have been cleared.

19. The MS confirms that all its resources have been released by sending the URR *Release Complete* message.

20. Finally, the MS sends the URR *Deregister* message to the S-UNC.

GPRS SIGNALING AND DATA TRANSFER

As explained earlier, the UMAN uses the UMA radio link control (URLC) protocol for the transport of GPRS related signaling and user data between the MS and the UNC. GPRS signaling messages are transferred by the URLC protocol over TCP and user plane traffic over UDP. The URLC protocol offers certain transport channel

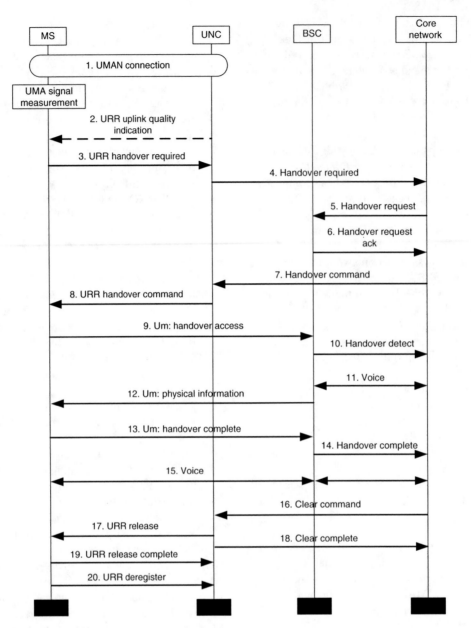

Fig. 10 Handover from UMAN to GERAN.

management procedures that are invoked by the MS before it can initiate data transfer. A transport channel can be said to be setup between the MS and the S-UNC when:

1. The MS knows the destination IP address and the UDP ports for GPRS data transfer;
2. The MS learns of the transport channel timer (URLC-CHANNEL-TIMER);
3. The UNC knows the IP address and the UDP ports of the MS for GPRS data transfer.

The transport channel management procedures facilitate the above. The following Fig. 11 explains the

sequence of events when the MS wants to initiate data transfer.

1. The MS indicates its desire to attach to the GPRS network by transmitting the *Attach Request* as part of the payload of the URLC-DATA message. This message is a signaling message and TCP is used as the transport protocol.
2. The UNC forwards the *Attach Request* to the SGSN over the Gb-interface using standard BSSGP messages.
3. The *Attach Accept* message is received by the UNC on the Gb-interface.

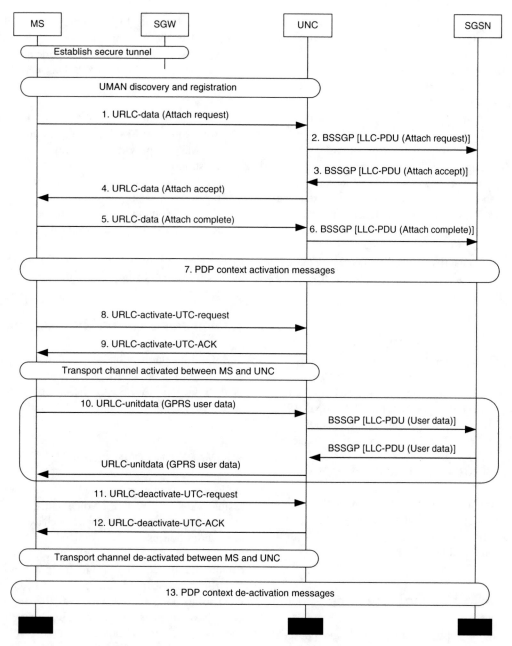

Fig. 11 GPRS signaling and data transfer in UMA.

4. The UNC forwards the *Attach Accept* message to the MS encapsulated in the URLC-DATA message.

5. The MS responds with the *Attach Complete* message.

6. The UNC forwards the *Attach Complete* message to the SGSN.

7. The MS activates the PDP context using URLC messages to the UNC, which in turn relays them to the SGSN over the Gb-interface. This is similar to steps 1–6 above.

8. Prior to data transfer, the MS will activate a transport channel by transmitting the URLC-ACTIVATE-UTC-REQ message to the UNC. This message

contains the UDP port that the MS wants to use for data transfer.

9. The UNC responds with the URLC-ACTIVATE-UTC-ACK message indicating its IP address and UDP port that will be used for transmitting and receiving data packets from the MS. After this exchange, a transport channel has been activated and both peers have IP address and port information to begin initiating data exchange.

10. The MS transmits user data encapsulated in the URLC-UNITDATA message to the UNC. The UNC in turn strips off the UMA-RLC protocol header and forwards the received data to the SGSN over

the Gb-interface using standard BSSGP messages. All data transfer is done over UDP as the transport protocol.

11. Once active data transfer is completed, the MS de-activates the transport channel by sending the URLC-DEACTIVATE-UTC-REQ message to the UNC. The cause parameter indicates the reason for the de-activation.

12. The UNC responds with the URLC-DEACTIVATE-UTC-ACK message indicating the transport channel was successfully de-activated. It is important to note that even though the TC has been deactivated, the PDP context may still be active between the MS and the GPRS core network [SGSN, gateway GPRS support node (GGSN)].

13. The MS may deactivate the PDP context. This exchange is done in a similar fashion to the attach and PDP context activation procedures.

GPRS SHORT MESSAGE SERVICE AND OTHER PROCEDURES

The transfer of GPRS SMS messages is based on the same mechanism as that for the transfer of MM/SM signaling messages. The MS transfers SMS messages using LLC SAPI = 7. The SM-CP protocol is tunneled between the MS and the SGSN using URLC-DATA messages. The UNC then interworks these messages to the SGSN.

UMA also provides dedicated URR and URLC messages for supporting GPRS functions such as Suspend, packet switched (PS)-paging, CS-paging and flow control in both the uplink and downlink directions.

SUPPORT FOR EMERGENCY SERVICES

UMA specifications support both phase 1 and phase 2 emergency call requirements as mandated by the FCC. For phase 1, the supported mechanism is as follows:

1. The UNC indicates whether GERAN/UTRAN or UMAN mode is preferred for emergency calls.

2. If UMAN is the preferred mode, then the call is placed over UMAN by the MS. Operator policy will decide whether the call shall be accepted or rejected.

3. If GERAN/UTRAN is the preferred mode, the emergency call is placed over GERAN/UTRAN, if coverage is available. The call is placed over UMAN if the GERAN/UTRAN network is not available.

The phase 2 mechanism is as follows:

The requirements for the emergency call remain as mentioned in the previous section. If the call is placed over UMAN, the UNC needs to provide accurate location information to the E911 PSAP. This may be done in a number of ways:[1]

1. The UNC maintains a database of AP locations. This is provided by the MS during registration.

2. The MS can provide its location information during registration.

3. The UNC can lookup the location of the AP based on the public IP address it obtains during registration.

4. The UNC communicates with the serving mobile location center (SMLC) to determine the location of the MS.

The UNC communicates the location of the subscriber to the core network when required.

3GPP SPECIFICATIONS INFORMATION

The 3GPP standardization body is now responsible for the evolution of the original UMA standard. It is now known as the "Generic Access to the A/Gb-Interface." More details can be obtained from Refs. 22 and 23.

SUMMARY AND CONCLUSION

Unlicensed mobile access technology provides for the seamless delivery of GSM voice and data services over unlicensed spectrum technologies such as Wi-Fi and Bluetooth. It introduces a new broadband IP radio access method to the mobile core network for service providers to exploit and monetize by providing new data services and meet the subscribers' communications needs by enabling them to use their mobile phone as their primary phone. Users have long desired to maintain a single number, one address book, one mailbox for receiving voice mails etc. UMA makes this a reality.

In this chapter we have discussed the UMA technology in detail including the architecture and protocols used on the various interfaces and provided information on the basic procedures such as IPSec tunnel establishment, UMAN Discovery and Registration, handover procedures from GERAN to UMAN and vice versa and GPRS sessions over UMA.

REFERENCES

1. UMA Architecture (Stage 2), R1.0.4.
2. 3GPP TS 48.008: Mobile Switching Centre—Base Station system (MSC-BSS) interface; Layer 3 specification.

3. 3GPP TS 48.018: General Packet Radio Service (GPRS); Base Station System (BSS)—Serving GPRS Support Node (SGSN); BSS GPRS protocol (BSSGP).

4. 3GPP TS 29.234: 3GPP system to Wireless Local Area Network (WLAN) interworking; Stage 3.

5. 3GPP TS 43.059: Functional Stage 2 description of Location Services (LCS) in GERAN.

6. 3GPP TS 23.041: Technical Realization of Cell Broadcast Service (CBS).

7. RFC 3550: RTP: A Transport Protocol for Real-time Applications, July 2003.

8. RFC3551: RTP Profile for Audio and Video Conferences with Minimal Control, July 2003.

9. RFC 3267: Real-time Transport Protocol (RTP) Payload Format and File Storage Format for the Adaptive Multirate (AMR) and Adaptive Multi-rate Wideband (AMR-WB) Audio Codecs, June 2002.

10. RFC793: Transmission Control Protocol, Sept 1981.

11. RFC 768: User Datagram Protocol, Aug 1980.

12. RFC4308: Cryptographic Suites for IPSec, Dec 2005.

13. RFC4303: IP Encapsulating Security Payload (ESP), Dec 2005.

14. 3GPP 33.234: Wireless Local Area Network (WLAN) Interworking Security.

15. RFC3948: UDP Encapsulation of IPSec ESP Packets, Jan 2005.

16. RFC4186: Extensible Authentication Protocol Method for Global System for Mobile Communications (GSM) Subscriber Identity Modules (EAP-SIM), Jan 2006.

17. RFC4187: Extensible Authentication Protocol Method for 3rd Generation Authentication and Key Agreement (EAP-AKA), Jan 2006.

18. RFC4306: Internet Key Exchange (IKEv2) Protocol, Dec 2005.

19. 802.11-1999: Standard for Information Technology—Telecommunications and information exchange between systems—Local and Metropolitan Area Networks—Specific requirements—Part 11: Wireless LAN Medium Access Control (MAC) and Physical Layer (PHY) specifications, 1999.

20. Bluetooth SIG: Personal Area Network Profile, http://www.bluetooth.com/Bluetooth/Learn/Technology/Specifications.

21. RFC4282: The Network Access Identifier, Dec 2005.

22. 3GPP TS 43.318: Generic Access to the A/Gb interface; Stage 2.

23. 3GPP TS 44.318: Generic Access (GA) to the A/Gb interface; Mobile GA interface layer 3 specification.

BIBLIOGRAPHY

1. The official Bluetooth membership site, http://www.bluetooth.org.

2. The present state of UMA, http://www.umatoday.com.

3. The UMA technology homepage, http://www.umatechnology.org/.

Ultra–Wi-Fi

Video Communications

Madhukar Budagavi
Raj Talluri
Texas Instruments, Dallas, Texas, U.S.A.

Abstract

This entry discusses international standards, such as MPEG-4 and H.263, which are making wireless video communications possible.

INTRODUCTION

Recent advances in technology have resulted in a rapid growth in mobile communications. With this explosive growth, the need for reliable transmission of mixed media information—audio, video, text, graphics, and speech data—over wireless links is becoming an increasingly important application requirement. The bandwidth requirements of raw video data are very high (a 176 × 144 pixels, color video sequence requires over 8 Mbps). Since the amount of bandwidth available on current wireless channels is limited, the video data has to be compressed before it can be transmitted on the wireless channel. The techniques used for video compression typically utilize predictive coding schemes to remove redundancy in the video signal. They also employ variable length coding (VLC) schemes, such as Huffman codes, to achieve further compression.

The wireless channel is a noisy fading channel characterized by long bursts of errors.[8] When compressed video data is transmitted over wireless channels, the effect of channel errors on the video can be severe. The VLC schemes make the compressed bitstream sensitive to channel errors. As a result, the video decoder that is decoding the corrupted video bitstream can easily lose synchronization with the encoder. Predictive coding techniques, such as **block motion compensation (BMC),** which are used in current video compression standards, make the matter worse by quickly propagating the effects of channel errors across the video sequence and rapidly degrading the video quality. This may render the video sequence totally unusable.

Error control coding,[5] in the form of **forward error correction (FEC)** and/or **automatic repeat request (ARQ),** is usually employed on wireless channels to improve the channel conditions.

FEC techniques prove to be quite effective against random bit errors, but their performance is usually not adequate against longer duration burst errors. FEC techniques also come with an increased overhead in terms of the overall bitstream size; hence, some of the coding efficiency gains achieved by video compression are lost. ARQ techniques typically increase the delay and, therefore, might not be suitable for real-time videoconferencing. Thus, in practical video communication schemes, error control coding is typically used only to provide a certain level of error protection to the compressed video bitstream, and it becomes necessary for the video coder to accept some level of errors in the video bitstream. Error resilience tools are introduced in the video codec to handle these residual errors that remain after error correction.

The emphasis in this entry is on discussing relevant international standards that are making wireless video communications possible. We will concentrate on both the error control and source coding aspects of the problem. In the next section, we give an overview of a wireless video communication system that is a part of a complete wireless multimedia communication system. The International ITU-T H.223[11] standard that describes a method of providing error protection to the video data before it is transmitted is also described. It should be noted that the main function of H.223 is to multiplex/demultiplex the audio, video, text, graphics, etc., which are typically communicated together in a videoconferencing application—error protection of the transmitted data becomes a requirement to support this functionality on error-prone channels. In section "Error Resilient Video Coding," an overview of error resilient video coding is given. The specific tools adopted into the ISO/IEC MPEG version 4 (i.e., MPEG-4)[7] and the ITU-T H.263[3] video coding standards to improve the error robustness of the video coder are described in sections "MPEG-4 ErrorResilience Tools" and "H.263 Error Resilience Tools," respectively.

Table 1 provides a listing of some of the standards that are described or referred to in this entry.

Encyclopedia of Wireless and Mobile Communications DOI: 10.1081/E-EWMC-120043947

WIRELESS VIDEO COMMUNICATIONS

Fig. 1 shows the basic block diagram of a wireless video communication system.[10] Input video is compressed by the video encoder to generate a compressed bitstream. The transport coder converts the compressed video bitstream into data units suitable for transmission over wireless channels. Typical operations carried out in the transport coder include channel coding, framing of data, modulation, and control operations required for accessing the wireless channel. At the receiver side, the inverse operations are performed to reconstruct the video signal for display.

In practice, the video communication system is part of a complete multimedia communication system and needs to interact with other system components to achieve the desired functionality. Hence, it becomes necessary to understand the other components of a multimedia communication system in order to design a good video communication system. Fig. 2 shows the block diagram of a wireless multimedia terminal based on the ITU-T H.324 set of standards.[4] We use the H.324 standard as an example because it is the first videoconferencing standard for which mobile extensions were added to facilitate use on wireless channels. The system components of a multimedia terminal can be grouped into three processing blocks: 1) audio, video, and data (the word *data* is used here to mean still images/slides, shared files, documents, etc.), 2) control, and 3) multiplex-demultiplex blocks.

1. **Audio, video, and data processing blocks:** These blocks basically produce/consume the multimedia information that is communicated. The aggregate bit rate generated by these blocks is restricted due to limitations of the wireless channel and, therefore, the total rate allowed has to be judiciously allocated among these blocks. Typically, the video blocks use

Fig. 1 A wireless video communication system.

up the highest percentage of the aggregate rate, followed by audio and then data. H.324 specifies the use of H.261/H.263 for video coding and G.723.1 for audio coding.

2. **Control block:** This block has a wide variety of responsibilities all aimed at setting up and maintaining a multimedia call. The control block facilitates the set-up of compression methods and preferred bit rates for audio, video, and data to be used in the multimedia call. It is also responsible for end-to-network signaling for accessing the network and end-to-end signaling for reliable operation of the multimedia call. H.245 is the control protocol in the H.324 suite of standards that specifies the control messages to achieve the above functionality.

3. **Multiplex–Demultiplex (MUX) block:** This block multiplexes the resulting audio, video, data, and control signals into a single stream before transmission on the network. Similarly, the received bitstream is demultiplexed to obtain the audio, video, data, and control signals, which are then passed to their respective

Table 1 List of relevant standards.

ISO/IEC 14496-2 (MPEG-4)	Information technology—coding of audio-visual objects: visual
H.263 (version 1 and version 2)	Video coding for low bit rate communication
H.261	Video codec for audio-visual services at p × 64 Kbps
H.223	Multiplexing protocol for low bit rate multimedia communication
H.324	Terminal for low bit rate multimedia communication
H.245	Control protocol for multimedia communication
G.723.1	Dual rate speech coder for multimedia communication transmitting at 5.3 and 6.3 Kbps

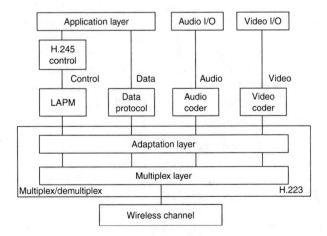

Fig. 2 Configuration of a wireless multimedia terminal.

processing blocks. The MUX block accesses the network through a suitable network interface. The H.223 standard is the multiplexing scheme used in H.324.

Proper functioning of the MUX is crucial to the operation of the video communication system, as all the multimedia data/signals flow through it. On wireless channels, transmission errors can lead to a breakdown of the MUX resulting in, e.g., non-video data being channeled to the video decoder or corrupted video data being passed on to the video decoder. Three annexes were specifically added to H.223 to enable its operation in error-prone environments. Below, we give a more detailed overview of H.223 and point out the levels of error protection provided by H.223 and its three annexes. It should also be noted that MPEG-4 does not specify a lower-level MUX like H.223, and thus H.223 can also be used to transmit MPEG-4 video data.

Recommendation H.223

Video, audio, data, and control information is transmitted in H.324 on distinct logical channels. H.223 determines the way in which the logical channels are mixed into a single bitstream before transmission over the physical channel (e.g., the wireless channel). The H.223 multiplex consists of two layers—the multiplex layer and the adaptation layer, as shown in Fig. 2. The multiplex layer is responsible for multiplexing the various logical channels. It transmits the multiplexed stream in the form of packets. The adaptation layer adapts the information stream provided by the applications above it to the multiplex layer below it by adding, where appropriate, additional octets for the purposes of error control and sequence numbering. The type of error control used depends on the type of information (audio/video/data/control) being conveyed in the stream. The adaptation layer provides error control support in the form of both FEC and ARQ.

H.223 was initially targeted for use on the benign general switched telephone network (GSTN). Later on, to enable its use on wireless channels, three annexes (referred to as Levels 1–3, respectively), were defined to provide improved levels of error protection. The initial specification of H.223 is referred to as Level 0. Together, Levels 0–3 provide for a trade-off of error robustness against the overhead required, with Level 0 being the least robust and using the least amount of overhead and Level 3 being the most robust and also using the most amount of overhead.

1. **H.223 Level 0—default mode:** In this mode the transmitted packet sizes are of variable length and are delimited by an 8-bit HDLC (high-level data link control) flag (**01111110**). Each packet consists of a 1-octet header followed by the payload, which consists of a variable number of information octets. The header octet includes a multiplex code (MC) which specifies, by indexing to a multiplex table, the logical channels to which each octet in the information field belongs. To prevent emulation of the HDLC flag in the payload, bitstuffing is adopted.

2. **H.223 Level 1 (Annex A)—communication over low error-prone channels:** The use of bitstuffing leads to poor performance in the presence of errors; therefore in Level 1, bitstuffing is not performed. The other improvement incorporated in Level 1 is the use of a longer 16-bit pseudo-noise synchronization flag to allow for more reliable detection of packet boundaries. The input bitstream is correlated with the synchronization flag and the output of the correlator is compared with a correlation threshold. Whenever the correlator output is equal to or greater than the threshold, a flag is detected. Since bitstuffing is not performed, it is possible to have this flag emulated in the payload. However, the probability of such an emulation is low and is outweighed by the improvement gained by not using bitstuffing over error-prone channels.

3. **H.223 Level 2 (Annex B)—communication over moderately error-prone channels:** When compared to the Level 1 operation, Level 2 increases the protection on the packet header. A multiplex payload length (MPL) field, which gives the length of the payload in bytes, is introduced into the header to provide additional redundancy for detecting the length of the video packet. A (24,12,8) extended Golay code is used to protect the MC and the MPL fields. Use of error protection in the header enables robust delineation of packet boundaries. Note that the payload data is not protected in Level 2.

4. **H.223 Level 3 (Annex C)—communication over highly error-prone channels:** Level 3 goes one step above Level 2 and provides for protection of the payload data. Rate compatible punctured convolutional (RCPC) codes, various cyclic redundancy check (CRC) polynomials, and ARQ techniques are used for protection of the payload data. Level 3 allows for the payload error protection overhead to vary depending on the channel conditions. RCPC codes are used for achieving this adaptive level of error protection because RCPC codes use the same channel decoder architecture for all the allowed levels of error protection, thereby reducing the complexity of the MUX.

ERROR RESILIENT VIDEO CODING

Even after error control and correction, some amount of residual errors still exist in the compressed bitstream fed to the video decoder in the receiver. Therefore, the video decoder should be robust to these errors and should provide acceptable video quality even in the presence of some residual errors. In this section, we first describe a standard video coder configuration that is the basis of many international standards and also highlight the potential problems that are encountered when compressed video from these systems is transmitted over wireless channels. We then give an overview of the strategies that can be adopted to overcome these problems. Most of these strategies are incorporated in the MPEG-4 video coding standard and the H.263 (version 2) video coding standard.[3] The original H.263 standard[2] which was standardized in 1996 for use in H.324 terminals connected to GSTN is referred to as version 1. Version 2 of the H.263 standard provides additional improvements and functionalities (which include error resilience tools) over the version 1 standard. We will use H.263 to refer to both version 1 and version 2 standards and a distinction will be made only when required.

A Standard Video Coder

Redundancy exists in video signals in both spatial and temporal dimensions. Video coding techniques exploit this redundancy to achieve compression. A plethora of video compression techniques have been proposed in the literature, but a hybrid coding technique consisting of BMC and discrete cosine transforms (DCT) has been found to be very effective in practice. In fact, most of the current video coding standards such as H.263 and MPEG-4, which provide state-of-the-art compression performance, are all based on this hybrid coding technique. In this hybrid BMC/DCT coding technique, BMC is used to exploit temporal redundancy and the DCT is used to reduce spatial redundancy.

Fig. 3 illustrates a standard hybrid BMC/DCT video coder configuration. Pictures are coded in either of two modes—inter-frame (INTER) or intra-frame (INTRA) mode. In intra-frame coding, the video image is encoded without any relation to the previous image, whereas in inter-frame coding, the current image is predicted from the previous image using BMC, and the difference between the current image and the predicted image, called the residual image, is encoded. The basic unit of data which is operated on is called a macroblock (MB) and is the data (both **luminance and chrominance** components)

corresponding to a block of 16×16 pixels. The input image is split into disjoint MBs and the processing is done on a MB basis. Motion information, in the form of **motion vectors,** is calculated for each MB. The motion compensated prediction residual error is then obtained by subtracting each pixel in the MB with its motion shifted counterpart in the previous frame. Depending on the mode of coding used for the MB, either the image MB or the corresponding residual image MB is split into blocks of size 8×8 and an 8×8 DCT is applied to each of these 8×8 blocks. The resulting DCT coefficients are then quantized. Depending on the quantization step-size, this will result in a significant number of zero-valued coefficients. To efficiently encode the DCT coefficients that remain nonzero after quantization, the DCT coefficients are zig-zag scanned, and run-length encoded and the run-lengths are variable length encoded before transmission. Since a significant amount of correlation exists between the neighboring MBs' motion vectors, the motion vectors are themselves predicted from already transmitted motion vectors and the motion vector prediction error is encoded. The motion vector prediction error and the mode information are also variable length coded before transmission to achieve efficient compression.

The decoder uses a reverse process to reconstruct the MB at the receiver. The variable length code words present in the received video bitstream are decoded first. For INTER MBs, the pixel values of the prediction error are reconstructed by inverse quantization and inverse DCT and are then added to the motion compensated pixels from the previous frame to reconstruct the transmitted MB. For INTRA MBs, inverse quantization and inverse DCT directly result in the transmitted MB. All MBs of a given picture are decoded to reconstruct the whole picture.

Error Resilient Video Decoding

The use of predictive coding and VLC, though very effective from a compression point of view, makes the video

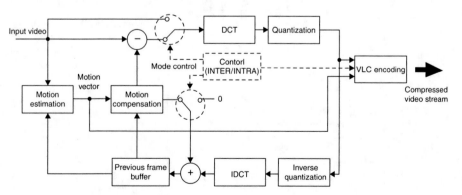

Fig. 3 A standard video coder.

decoding process susceptible to transmission errors. In VLC, the boundary between code words is implicit. The compressed bitstream has to be read until a full code word is encountered; the code word is then decoded to obtain the information encoded in the code word. When there are transmission errors, the implicit nature of the boundary between code words typically leads to an incorrect number of bits being used in VLC decoding and, thus, subsequently results in a loss of synchronization with the encoder. In addition, the use of predictive coding leads to the propagation of these transmission errors to neighboring spatial blocks and to subsequently decoded frames, which leads to a rapid degradation in the reconstructed video quality.

To minimize the disastrous impact that transmission errors can have on the video decoding process, the following stages are incorporated in the video decoder to make it more robust:

- error detection and localization,
- resynchronization,
- data recovery,
- error concealment.

Fig. 4 shows an error resilient video decoder configuration. The first step involved in robust video coding is the detection of errors in the bitstream. The presence of errors in the bitstream can be signaled by the FEC used in the multiplex layer. The video coder can also detect errors whenever illegal VLC code words are encountered in the bitstream or when the decoding of VLC code words leads to an illegal value of the decoded information (e.g., occurrence of more than 64 DCT coefficients for an 8 × 8 DCT block). Accurate detection of errors in the bitstream is a very important step, since most of the other

error resilience techniques can only be invoked if an error is detected.

Due to the use of VLC, the location in the bitstream where the decoder detects an error is not the same location where the error has actually occurred but some undetermined distance away from it. This is shown in Fig. 5. Once an error is detected, it also implies that the decoder is not in synchronization with the encoder. Resynchronization schemes are then employed for the decoder to fall back into lock step with the encoder. While constructing the bitstream, the encoder inserts unique resynchronization words into the bitstream at approximately equally spaced intervals. These resynchronization words are chosen such that they are unique from the valid video bitstream. That is, no valid combination of the video algorithm's VLC tables can produce these words. The decoder, upon detection of an error, seeks forward in the bitstream looking for this known resynchronization word. Once this word is found, the decoder then falls back in synchronization with the encoder. At this point, the decoder has detected an error, regained synchronization with the encoder, and isolated the error to be between the two resynchronization points. Since the decoder can only isolate the error to be somewhere between the resynchronization points but not pinpoint its exact location, all of the data that corresponds to the MBs between these two resynchronization points needs to be discarded. Otherwise, the effects of displaying an image reconstructed from erroneous data can cause highly annoying visual artifacts.

Some data recovery techniques, such as "reversible decoding," enable the decoder to salvage some of the data between the two resynchronization points. These techniques advocate the use of a special kind of VLC table at the encoder in coding the DCTs and motion vector

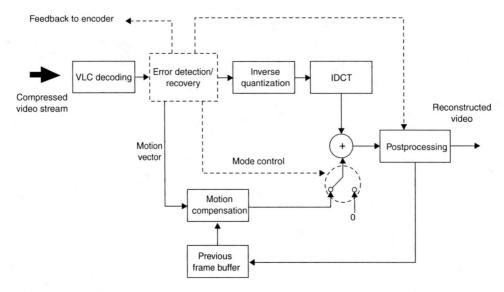

Fig. 4   Error resilient video decoder.

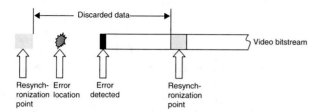

Fig. 5 At the decoder, it is usually not possible to detect the error at the actual error occurrence location; hence, all the data between the two resynchronization points may need to be discarded.

information. These special VLCs have the property that they can be decoded both in the forward and reverse directions. By comparing the forward and reverse decoded data, the exact location of the error in the bit stream can be localized more precisely and some of the data between the two resynchronization points can be salvaged. The use of these reversible VLCs (RVLCs) is part of the MPEG-4 standard and will be described in greater detail in the following sections.

After data recovery, the impact of the data that is deemed to be in error needs to be minimized. This is the error concealment stage. One simple error concealment strategy is to simply replace the luminance and chrominance components of the erroneous MBs with the luminance and chrominance of the corresponding MBs in the previous frame of the video sequence. While this technique works fairly well and is simple to implement, more complex techniques use some type of estimation strategies to exploit the local correlation that exists within a frame of video data to come up with a better estimate of the missing or erroneous data. These error concealment strategies are essentially postprocessing algorithms and are not mandated by the video coding standards. Different implementations of the wireless video systems utilize different kinds of error concealment strategies based on the available computational power and the quality of the channel.

If there is support for a decoder feedback path to the encoder as shown in Fig. 3, this path can be used to signal detected errors. The feedback information from the decoder can be used to retransmit data or to influence future encoder action so as to stop the propagation of detected errors in the decoder. Note that for the feedback to take place, the network must support a back channel.

Classification of Error Resilience Techniques

In general, techniques to improve the robustness of the *video* coder can be classified into three categories based on whether the encoder or the decoder plays a primary part in improving the error robustness.[10] *Forward error resilience* techniques refer to those techniques where the encoder plays the primary part in improving the error robustness, typically by introducing redundancy in the transmitted information. In *postprocessing* techniques,

the decoder plays the primary part and does concealment of errors by estimation and interpolation (e.g., spatial-temporal filtering) using information it has already received. In *interactive error* resilience techniques, the decoder and the encoder interact to improve the error resilience of the video coder. Techniques that use decoder feedback come under this category.

MPEG-4 ERROR RESILIENCE TOOLS

MPEG-4 is an ISO/IEC standard being developed by the motion pictures expert group. Initially MPEG was aimed primarily at low bit rate communications; however, its scope was later expanded to be much more of a multimedia coding standard.[7] The MPEG-4 video coding standard is the first video coding standard to address the problem of efficient representation of visual objects of arbitrary shape. MPEG-4 was also designed to provide "universal accessibility," i.e., the ability to access audio-visual information over a wide range of storage and transmission media. In particular, because of the proliferation of wireless communications, this implied development of specific tools to enable error resilient transmission of compressed data over noisy communication channels.

A number of tools have been incorporated into the MPEG-4 video coder to make it more error resilient. All these tools are basically forward error resilience tools. We describe below each of these tools and its advantages.

Resynchronization

As mentioned earlier, a video decoder that is decoding a corrupted bitstream may lose synchronization with the encoder (i.e., it is unable to identify the precise location in the image where the current data belongs). If remedial measures are not taken, the quality of the decoded video rapidly degrades and becomes unusable. One approach is for the encoder to introduce resynchronization markers in the bitstream at various locations. When the decoder detects an error, it can then look for this resynchronization marker and regain synchronization.

Previous video coding standards such as H.261 and H.263 (version 1) logically partition each of the images to be encoded into rows of MBs called group of blocks (GOBs). These GOBs correspond to a horizontal row of MBs for quarter common intermediate format (**QCIF**) images. Fig. 6 shows the GOB numbering scheme for H.263 (version 1) for QCIF resolution. For error resilience purposes, H.263 (version 1) provides the encoder an option of inserting resynchronization markers at the beginning of each of the GOBs. Hence, for QCIF images these resynchronization markers are allowed to occur only at the left edge of the images. The smallest region that the error can be isolated to and concealed in this case is thus one row of MB's.

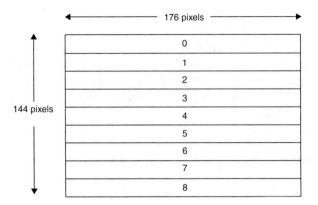

Fig. 6 H.263 GOB numbering for a QCIF image.

In contrast, the MPEG-4 encoder is not restricted to inserting the resynchronization markers only at the beginning of each row of MBs. The encoder has the option of dividing the image into video packets. Each video packet is made up of an integer number of consecutive Ms in raster scan order. These MBss can span several rows of MBs in the image and can even include partial rows of MBs. One suggested mode of operation for the MPEG-4 encoder is for it to insert a resynchronization marker periodically at approximately every K bits. Note that resynchronization markers can only be placed at a MB boundary and, hence, the video packet length cannot be constrained to be exactly equal to K bits. When there is a significant activity in one part of the image, the MBs corresponding to these areas generate more bits than other parts of the image. If the MPEG-4 encoder inserts the resynchronization markers at uniformly spaced bit intervals, the MB interval between the resynchronization markers is a lot closer in the high activity areas and a lot farther apart in the low activity areas. Thus, in the presence of a short burst of errors, the decoder can quickly localize the error to within a few MBss in the important high activity areas of the image and preserve the image quality in these important areas. In the case of H.263 (version 1), where the resynchronization markers are restricted to be at the beginning of the GOBs, it is only possible for the decoder to isolate the errors to a fixed GOB independent of the image content. Hence, effective coverage of the resynchronization marker is reduced when compared to the MPEG-4

scheme. The recommended spacing of the resynchronization markers in MPEG-4 is based on the bit rates. For 24 Kbps, it is recommended to insert them at intervals of 480 bits and for bit rates between 25 and 48 Kbps, it is recommended to place them at every 736 bits. Figs. 7A and B illustrate the placement of resynchronization markers for H.263 (version 1) and MPEG-4.

Note that in addition to inserting the resynchronization markers at the beginning of each video packet, the encoder also needs to remove all data dependencies that exist between the data belonging to two different video packets within the same image. This is required so that even if one of the video packets in the current image is corrupted due to errors, the other packets can be decoded and utilized by the decoder. In order to remove these data dependencies, the encoder inserts two additional fields in addition to the resynchronization marker at the beginning of each video packet, as shown in Fig. 8. These are 1) the absolute MB number of the first MB in the video packet, MB No. (which indicates the spatial location of the MB in the current image), and 2) the quantization parameter, QP, which denotes the initial quantization parameter used to quantize the DCT coefficients in the video packet. The encoder also modifies the predictive encoding method used for coding the motion vectors such that there are no predictions across the video packet boundaries. Also shown in Fig. 8 is a third field, labeled header extension code (HEC). Its use is discussed in a later section.

Data Partitioning

Data partitioning in MPEG-4 provides enhanced error localization and error concealment capabilities. The data partitioning mode partitions the data within a video packet into a motion part and a texture part (DCT coefficients) separated by a unique motion marker (MM), as shown in Fig. 9. All the syntactic elements of the video packet that have motion-related information are placed in the motion partition and all the remaining syntactic elements that relate to the DCT data are placed in the texture partition. If the texture information is lost, data partitioning enables the salvation of motion information, which can then be used to conceal the errors in a more effective manner.

The MM is computed from the motion VLC tables using a search program such that it is Hamming distance 1 from any possible valid combination of the motion VLC

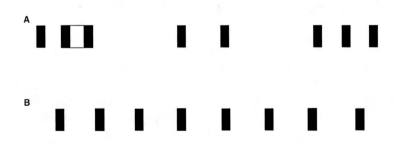

Fig. 7 Position of resynchronization markers in the bitstream for (A) H.263 (version 1) encoder with GOB headers and for; (B) an MPEG-4 encoder with video packets.

Resync. marker	MB No.	QP	HEC	Macroblack data

Fig. 8 An MPEG-4 video packet.

tables.[9] The MM is uniquely decodable from the motion VLC tables, and it indicates to the decoder the end of the motion information and the beginning of the DCT information. The number of MBs in the video packet is implicitly known after encountering the MM. Note that the MM is only computed once based on the VLC tables and is fixed in the standard. Based on the VLC tables in MEPG-4, the MM is a 17-bit word whose value is **1 1111 0000 0000 0001.**

RVLCs

As was shown in Fig. 5, if the decoder detects an error during the decoding of VLC code words, it loses synchronization and hence typically has to discard all the data up to the next resynchronization point. RVLCs are designed such that they can be instantaneously decoded both in the forward and the backward direction. When the decoder detects an error while decoding the bitstream in the forward direction, it jumps to the next resynchronization marker and decodes the bitstream in the backward direction until it encounters an error. Based on the two error locations, the decoder can recover some of the data that would have otherwise been discarded. This is shown in Fig. 10, which shows only the texture part of the video packet—only data in the shaded area is discarded. Note that if RVLCs were not used, all the data in the texture part of the video packet would have to be discarded. RVLCs thus enable the decoder to better isolate the error location in the bitstream.

Fig. 11 shows the comparison of performance of resynchronization, data partitioning, and RVLC techniques for 24 Kbps QCIF video data. The experiments involved transmission of three video sequences, each of duration 10 sec, over a bursty channel simulated by a 2-state Gilbert model.[6] The burst duration on the channel is 1 ms and the burst occurrence probability is 10^{-2}. Fig. 11, which plots the average peak signal-to-noise ratios of the received video frames, shows that data partitioning and RVLC provide improved performance when compared to using only resynchronization markers.

HEC

Some of the most important information that the decoder needs in order to decode the video bitstream is in the video

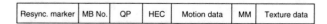

Resync. marker	MB No.	QP	HEC	Motion data	MM	Texture data

Fig. 9 A data partitioned MPEG-4 video packet.

frame header data. This data includes information about the spatial dimensions of the video data, the time stamps associated with the decoding and the presentation of this video data, and the type of the current frame (INTER/INTRA). If some of this information is corrupted due to channel errors, the decoder has no other recourse but to discard all the information belonging to the current video frame. In order to reduce the sensitivity of this data, a technique called HEC was introduced into the MPEG-4 standard. In each video packet, a 1-bit field called HEC is present. The location of HEC in the video packet is shown in Fig. 8. For each video packet, when HEC is set, the important header information that describes the video frame is repeated in the bits following the HEC. This information can be used to verify and correct the header information of the video frame. The use of HEC significantly reduces the number of discarded video frames and helps achieve a higher overall decoded video quality.

AIR

Whenever an INTRA MB is received, it basically stops the temporal propagation of errors at its corresponding spatial location. The procedure of forcefully encoding some MBs in a frame in INTRA mode to flush out possible errors is called INTRA refreshing. INTRA refresh is very effective in stopping the propagation of errors, but it comes at the cost of a large overhead. Coding a MB in INTRA mode typically requires many more bits when compared to coding the MB in INTER mode. Hence, the INTRA refresh technique has to be used judiciously.

For areas with low motion, simple error concealment by just copying the previous frame's MBs works quite effectively. For MBs with high motion, error concealment becomes very difficult. Since the high motion areas are perceptually the most significant, any persistent error in the high motion area becomes very noticeable. The adaptive intra refresh (AIR) technique of MPEG-4 makes use of the above facts and INTRA refreshes the motion areas more frequently, thereby allowing the corrupted high motion areas to recover quickly from errors.

Depending on the bit rate, the AIR approach only encodes a fixed and predetermined number of MBs in a frame in INTRA mode (the exact number is not standardized by MPEG-4). This fixed number might not be enough to cover all the MBs in the motion area; hence, the AIR technique keeps track of the MBs that have been refreshed (using a "refresh map") and in subsequent frames refreshes any MBs in the motion areas that might have been left out.

H.263 ERROR RESILIENCE TOOLS

In this section, we discuss four error resilience techniques which are part of the H.263 standard—*slice structure mode*

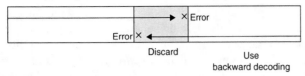

Fig. 10 Use of reversible variable length codes.

and *independent segment decoding* (ISD), which are forward error resilience features, and *error tracking* and *reference picture selection* (RPS), which are interactive error resilience techniques. Error tracking was introduced in H.263 (version 1) as an appendix, whereas the remaining three techniques were introduced in H.263 (version 2) as annexes.

Slice Structure Mode (Annex K)

The slice structured mode of H.263 is similar to the video packet approach of MPEG-4 with a slice denoting a video packet. The basic functionality of a slice is the same as that of a video packet—providing periodic resynchronization points throughout the bitstream. The structure of a slice is shown in Fig. 12. Like an MPEG-4 video packet, the slice consists of a header followed by the MV data. The SSC is the slice start code and is identical to the resynchronization marker of MPEG-4. The MB address (MBA) field, which denotes the starting MB number in the slice, and the SQUANT field, which is the quantizer scale coded non-predictively, allow for the slice to be coded independently.

The slice structured mode also contains two sub-modes which can be used to provide additional functionality. The sub-modes are

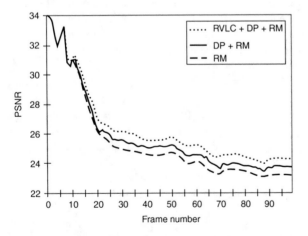

Fig. 11 Performance comparison of resynchronization, data partitioning, and RVLC over a bursty channel simulated by a 2-state Gilbert model. Burst durations are 1 ms long and the probability of occurrence of a burst is 10^{-2}.

- **Rectangular slice sub-mode (RSS):** This allows for rectangular shaped slices. The rectangular region contained in the slice is specified by slice width indication SWI + 1 (see Fig. 12 for the location of the SWI field in the slice header), which gives the width of the rectangular region, and MBA, which specifies the upper left MB of the slice. Note that the height of the rectangular region gets specified by the number of MBs contained in the slice. This mode can be used, for example, to subdivide images into rectangular regions of interest for region-based coding.

- **Arbitrary slice sub-mode (ASO):** The default order of transmission of slices is such that the MBA field is strictly increasing from one slice to the next transmitted slice. When ASO is used, the slices may appear in any order within the bitstream. This mode is useful when the wireless network supports prioritization of slices which might result in out-of-order arrival of video slices at the decoder.

ISD (Annex R)

Even though the slice structured mode eliminates decoding dependency between neighboring slices, errors in slices can spatially propagate to neighboring slices in subsequent frames due to motion compensation. This happens because motion vectors in a slice can point to MBs of neighboring slices in the reference picture. Independent segment decoding eliminates this from happening by restricting the motion vectors within a predefined segment of the picture from pointing to other segments in the picture, thereby helping to contain the error to be within the erroneous segment. This improvement in the localization of errors, however, comes at a cost of a loss of coding efficiency. Because of this restriction on the motion vectors, the motion compensation is not as effective, and the residual error images use more bits.

For ease of implementation, the ISD mode puts restrictions on segment shapes and on the changes of segment shapes from picture to picture. The ISD mode cannot be used with the slice structured mode (Annex K) unless the RSS of Annex K is active. This prevents the need for treating awkward shapes of slices that can otherwise arise when Annex K is not used with RSS. The segment shapes are not allowed to change from picture to picture unless an INTRA frame is being coded.

Error Tracking (Appendix I)

The error tracking approach is an INTRA refresh technique but uses decoder feedback of errors to decide

SSC	...	MBA	...	SQUANT	SWI	Macroblock data

Fig. 12 Structure of a slice in H.263/Annex K.

which MBs in the current image to code in INTRA mode to prevent the propagation of these errors. When there are no errors on the channel, normal coding (which usually results in the bit-efficient INTER mode being selected most of the time) is used. The use of decoder feedback allows the system to adapt to varying channel conditions and minimizes the use of forced INTRA updates to situations when there are channel errors.

Because of the time delay involved in the decoder feedback, the encoder has to track the propagation of an error from its original occurrence to the current frame to decide which MBs should be INTRA coded in the current frame. A low complexity algorithm was proposed in Appendix I of H.263 to track the propagation of errors. However, it should be noted that the use of this technique is not mandated by H.263. Also, H.263 itself does not standardize the mechanism by which the decoder feedback of error can be sent. Typically, H.245 control messages are used to signal the decoder feedback for error tracking purposes.

RPS (Annex N)

The RPS mode of H.263 also relies on decoder feedback to efficiently stop the propagation of errors. The back channel used in RPS mode can be a separate logical channel (e.g., by using H.245), or if two-way communication is taking place, the back channel messages can be sent multiplexed with the encoded video data. In the presence of errors, the RPS mode allows the encoder to be instructed to select one of the several previously correctly received and decoded frames as the reference picture for motion compensation of the current frame being encoded. This effectively stops the propagation of error. Note that the use of RPS requires the use of multiple frame buffers at both the encoder and the decoder to store previously decoded frames. Hence, the improvement in performance in the RPS mode has come at the cost of increased memory requirements.

DISCUSSION

In this entry, we presented a broad overview of the various techniques that enable wireless video transmission. Due to the enormous amount of bandwidth required, video data is typically compressed before being transmitted, but the errors introduced by the wireless channels have a severe impact on the compressed video information. Hence, special techniques need to be employed to enable robust video transmission. International standards play a very important role in communications applications. The two current standards that are most relevant to video applications are ISO MPEG-4 and ITU H.263. In this entry, we detailed these two standards and explained the error resilient tools that are part of these standards to enable robust video

communication over wireless channels. A tutorial overview of these tools has been presented and the performance of these tools has been described.

There are, however, a number of other methods that further improve the performance of a wireless video codec that the standards do not specify. If the encoder and decoder are aware of the limitations imposed by the communication channel, they can further improve the video quality by using these methods. These methods include encoding techniques such as rate control to optimize the allocation of the effective channel bit rate between various parts of video to be transmitted and intelligent decisions on when and where to place INTRA refresh MBs to limit the error propagation. Decoding methods such as superior error concealment strategies that further conceal the effects of erroneous MBs by estimating them from correctly decoded MBs in the spatiotemporal neighborhood can also significantly improve the effective video quality.

This entry has mainly focused on the error resilience aspects of the video layer. There are a number of error detection and correction strategies, such as FEC, that can further improve the reliability of the transmitted video data. These FEC codes are typically provided in the systems layer and the underlying network layer. If the video transmission system has the ability to monitor the dynamic error characteristics of the communication channel, joint source-channel coding techniques can also be effectively employed. These techniques enable the wireless communication system to perform optimal trade-offs in allocating the available bits between the source coder (video) and the channel coder (FEC) to achieve superior performance.

Current video compression standards also support *layered* coding methods. In this approach, the compressed video information can be separated into multiple layers. The *base* layer, when decoded, provides a certain degree of video quality and the *enhancement* layer, when received and decoded, then adds to the base layer to further improve the video quality. In wireless channels, these base and enhancement layers give a natural method of partitioning the video data into more important and less important layers. The base layer can be protected by a stronger level of error protection (higher overhead channel coder) and the enhancement layer by a lesser strength coder. Using this unequal error protection (UEP) scheme, the communication system is assured of a certain degree of performance most of the time through the base layer, and when the channel is not as error prone and the decoder receives the enhancement layer, this scheme provides improved quality.

Given all these advances in video coding technology, coupled with the technological advances in processor technology, memory devices, and communication systems, wireless video communications is fast becoming a very compelling application. With the advent of higher bandwidth third generation wireless communication systems, it will be possible to transmit compressed video in many

Ultra–
Wi-Fi

wireless applications, including mobile videophones, videoconferencing systems, personal digital assistants (PDAs), security and surveillance applications, mobile Internet terminals, and other multimedia devices.

DEFINING TERMS

Automatic repeat request (ARQ): An error control system in which notification of erroneously received messages is sent to the transmitter which then simply retransmits the message. The use of ARQ requires a feedback channel and the receiver must perform error detection on received messages. Redundancy is added to the message before transmission to enable error detection at the receiver.

Block motion compensation (BMC): Motion compensated prediction that is done on a block basis; that is, blocks of pixels are assumed to be displaced spatially in a uniform manner from one frame to another.

Forward error correction (FEC): Introduction of redundancy in data to allow for correction of errors without retransmission.

Luminance and chrominance: Luminance is the brightness information in a video image, whereas chrominance is the corresponding color information.

Motion vectors: Specifies the spatial displacement of a block of pixels from one frame to another.

QCIF: Quarter common intermediate format (QCIF) is a standard picture format that defines the image dimensions to be 176×144 (pixels per line $\times$ lines per picture) for luminance and 88×72 for chrominance.

REFERENCES

1. International Telecommunications Union—Telecommunications Standardization Sector. Recommendation H.223: Multiplexing protocol for low bitrate multimedia communications, Geneva, 1996.
2. International Telecommunications Union—Telecommunications Standardization Sector. Recommendation H.263: Video coding for low bit rate communication, Geneva, 1996.
3. International Telecommunications Union—Telecommunications Standardization Sector. Draft Recommendation H.263 (version 2): video coding for low bit rate communication, Geneva, 1998.
4. International Telecommunications Union—Telecommunications Standardization Sector. Recommendation H.324: terminal for low bit rate multimedia communications, Geneva, 1996.
5. Lin, S.; Costello, D.J., Jr. *Error Control Coding: Fundamentals and Applications*; Prentice-Hall: Englewood Cliffs, NJ, 1983.
6. Miki, T.; Kawahara, T.; Ohya, T. Revised error pattern generation programs for core experiments on error resilience, *ISO/IEC JTC1/SC29/WG11 MPEG96/1492*, Maceio, Brazil, 1996.
7. International Organization for Standardization. Committee draft of Tokyo (N2202): information technology-coding of audio-visual objects: visual, ISO/IEC 14496-2, 1998.
8. Sklar, B. Rayleigh fading channels in mobile digital communication systems, Pt. I: Characterization, IEEE Commun. Mag. **1997**, *35*, 90–100.
9. Talluri, R.; Moccagatta, I.; Nag, T.; Cheung, G. Error concealment by data partitioning. Signal Process., Image Commun. **1998**, *14*.
10. Wang, Y.; Zhu, Q. Error control and concealment for video communication: a review. IEEE Trans. Circuits Syst. Video Technol. **1998**, *86* (5), 974–997.

FURTHER INFORMATION

A broader overview of wireless video can be found in the special issue of *IEEE Communications Magazine*, June 1998. Wang and Zhu[10] provide an exhaustive review of error concealment techniques for video communications. More details on MPEG-4 and ongoing version 2 activities in MPEG-4 can be found on the web page http://drogo.cselt.it/mpeg/standards/mpeg-4/mpeg-4.htm. H.263 (version 2) activities are tracked on the web page http://www.ece.ubc.ca/spmg/research/motion/h263plus/. Most of the ITU-T recommendations can be obtained from the web site http://www.itu.org. The special issue of *IEEE Communications Magazine*, December 1996, includes articles on H.324 and H.263.

Current research relevant to wireless video communications is reported in a number of journals including *IEEE Transactions on Circuits and Systems for Video Technology, IEEE Transactions on Image Processing, IEEE Transactions on Vehicular Technology, Signal Processing: Image Communication. The IEEE Communications Magazine* regularly reports review articles relevant to wireless video communications. Conferences of interest include the IEEE International Conference on Image Processing (ICIP), IEEE Vehicular Technology Conference (VTC), and IEEE International Conference on Communications (ICC).

Video Streaming: Reliable Wireless Streaming with Digital Fountain Codes

Raouf Hamzaoui
Faculty of Computing Sciences and Engineering, De Montfort University, Leicester, U.K.

Shakeel Ahmad
Department of Computer and Information Science, University of Konstanz, Konstanz, Germany

Marwan Al-Akaidi
Faculty of Computing Sciences and Engineering, De Montfort University, Leicester, U.K.

Ultra–
Wi-Fi

Definition

Video streaming refers to a video transmission method that allows the receiver to view the video continuously after only a short delay.

INTRODUCTION

3G cellular systems currently offer video streaming services to mobile users through unicast transmission in which a separate point-to-point connection with each recipient is established and maintained. However, due to limited server capacity and reduced available spectrum, the point-to-point approach does not scale well when the number of subscribers increases. Recently, the 3G partnership project (3GPP) introduced the multimedia broadcast/multicast service (MBMS),[1] which allows efficient point-to-multipoint transmission of streaming video over the existing GPRS/EDGE and wideband code division multiple access (WCDMA) 3G networks. MBMS is an IP-based technology that uses forward error correction (FEC) with Raptor codes[2] at the application layer to protect the video bitstream against packet loss.

Raptor codes are a new class of erasure codes (i.e., codes used against symbol erasure[3]) with many desirable features. Whereas traditional erasure codes have a fixed code rate that must be chosen before the encoding begins, Raptor codes are rateless as the encoder can generate on-the-fly as many encoded symbols as needed. This is an advantage when the channel conditions are unknown or varying because the use of a fixed channel code rate would lead to either bandwidth waste if the packet loss rate is overestimated or poor performance if the packet loss rate is underestimated. Similarly, in broadcast and multicast systems where the same data is sent to many users over heterogeneous links, the choice of an appropriate fixed channel code rate is not obvious. Note that strategies based on re-transmission of the lost packets would also be not appropriate for such applications because if too many receivers request re-transmission of the data, the server will be overwhelmed.

Another (and probably the most) attractive property of Raptor codes is their low encoding and decoding complexity as an encoded symbol is generated from k source symbols independently of the other encoded symbols in only $O(\log(1/\varepsilon))$ time, and the k source symbols are recovered from $k(1 + \varepsilon)$ encoded symbols with high probability in $O(k \log(1/\varepsilon))$ time for any positive number ε. This is a tremendous speed up over Reed–Solomon codes (the standard erasure codes), which typically have $O(k(n - k)q)$ encoding and decoding complexity if k source symbols are encoded into n codeword symbols for a symbol alphabet of size q, where q should be larger than n.

Raptor codes are an extension and improvement to Luby transform (LT) codes.[4–6] Both codes are known in the literature as digital fountain codes.[7] The goal of this entry is to describe digital fountain codes and to explain how they are used in MBMS. The entry is organized as follows. The following three sections describe respectively, notations and terminology, LT codes, and Raptor codes. Section "Video Streaming with MBMS" focuses on the MBMS video streaming framework. The last section overviews recent research on wireless video transmission with digital fountain codes.

NOTATIONS

In this section, we introduce the notations and terminology used in the entry. Digital fountain codes are based on bipartite graphs, i.e., graphs with two disjoint sets of vertices such that two vertices in the same set are not connected by an edge. The first set of vertices contains the source symbols, whereas the second set contains the encoded

Encyclopedia of Wireless and Mobile Communications DOI: 10.1081/E-EWMC-120043593

symbols. The symbols are binary vectors, and arithmetic on symbols is defined modulo 2, in particular, $\oplus$ denotes modulo-2 addition. If the number of source symbols is k, the degree of an encoded symbol is given by a degree distribution $\Omega(x) = \sum_{i=0}^{k} \Omega_i x^i$ on $\{0, \ldots, k\}$, where Ω_i is the probability that degree i is chosen. For example, suppose that

$$\Omega_i = \begin{cases} 0 & \text{if } i = 0 \\ \frac{1}{k} & \text{if } i = 1 \\ \frac{1}{i(i-1)} & \text{otherwise} \end{cases}$$

Then $\Omega(x)$ is called the ideal soliton distribution.[6] A more practical distribution is the robust soliton distribution $\Delta(x)$[6] given by $\Delta_i = (\Omega_i + \Gamma_i)/d$, where $\Omega(x)$ is an ideal soliton distribution, $\Gamma(x)$ is given by

$$\Gamma_i = \begin{cases} \frac{s}{ki} & \text{if } i = 1, \ldots, \lfloor \frac{k}{s} \rfloor - 1 \\ \frac{s}{k} \ln\left(\frac{s}{\delta}\right) & \text{if } i = \lfloor \frac{k}{s} \rfloor \\ 0 & \text{otherwise} \end{cases}$$

$d = \sum_{i=1}^{k} \Omega_i + \Gamma_i$, and $s = c \ln(k/\delta)\sqrt{k}$. Here c and δ are parameters (see[6] for an interpretation of these parameters).

Any distribution $\Omega(x)$ on $\{0, \ldots, k\}$ induces a distribution on F_2^k, the set of binary vectors of length k, by $\text{Prob}(v) = \Omega_{w(v)} / \binom{k}{w(v)}$, where $v \in F_2^k$ and $w(v)$ is the weight of v (i.e., the number of nonzero components of v).

A reliable decoding algorithm for a digital fountain code is an algorithm that ensures that all k source symbols can be recovered with probability larger than or equal to $1 - (1/k^c)$ for a positive constant c.

LT-CODES

The LT encoder takes a set of k source symbols and generates a potentially infinite sequence of encoded symbols of the same alphabet. The symbol alphabet may consist of any set of l-bit symbols, e.g., bits or bytes. Each encoded symbol is computed independently of the other encoded symbols. The LT decoder takes a little more than k encoded symbols and recovers all source symbols with probability $1 - \varepsilon$. Here ε is a positive number, which gives the tradeoff between the loss recovery property of the code and the encoding and decoding complexity. Luby[6] proves that for the robust soliton distribution, an encoded symbol can be computed in $O(\log(k/\varepsilon))$ time and all source symbols can be recovered from $k + O(\sqrt{k}\log^2(k/\varepsilon))$ encoded symbols with probability $1 - \varepsilon$ in $O(k\log(k/\varepsilon))$ time. The following subsection gives the details of the encoding.

Encoding

Given k source symbols $s_1, \ldots, s_k$, and a suitable degree distribution $\Omega(x)$ on $\{0, \ldots, k\}$, a sequence of encoded symbols c_i, $i \geqslant 1, \ldots$, is generated as follows. For each $i \geqslant 1$

1. Select randomly a degree $d_i \in \{1, \ldots, k\}$ according to the degree distribution $\Omega(x)$.
2. Select uniformly at random d_i distinct source symbols and set c_i equal to their modulo-2 bitwise sum.

Fig. 1 illustrates the encoding procedure.

In the following subsection, we describe the decoding algorithm.

Decoding

We assume that a transmitted encoded symbol is either lost or received correctly. We also assume that the decoder can determine both the degree of a received encoded symbol and the source symbols connected to this encoded symbol. This can be done, e.g., by using a pseudo-random generator with the same seed as the one used in the encoding. Suppose now that n encoded symbols have been received. Then the decoder proceeds as follows.

1. Find an encoded symbol c_i which is connected to only one source symbol s_j. If there is no such encoded symbol (i.e., an encoded symbol with degree one),

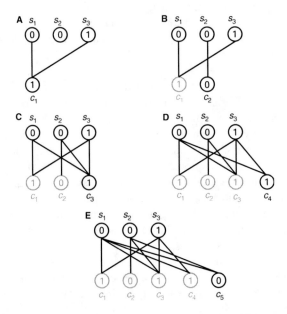

Fig. 1 LT encoding of $k = 3$ source symbols (here bits) $s_1 = 0$, $s_2 = 0$, and $s_3 = 1$ to $n = 5$ encoded symbols $c_1, \ldots, c_5$. Each step shows a new encoded symbol. (A) $d_1 = 2$, $c_1 = s_1 \oplus s_3$; (B) $d_2 = 1$, $c_2 = s_2$; (C) $d_3 = 3$, $c_3 = s_1 \oplus s_2 \oplus s_3$; (D) $d_4 = 2$, $c_4 = s_1 \oplus s_3$; (E) $d_5 = 2$, $c_5 = s_1 \oplus s_2$.

stop the decoding and wait until more encoded symbols are received before proceeding with the decoding.

a. Set $s_j = c_i$,

b. set $c_x := c_x \oplus s_j$ for all indices $x \neq i$ such that c_x is connected to s_j,

c. remove all edges connected to s_j.

2. Repeat step 1 until all k source symbols are recovered.

Fig. 2 illustrates the decoding procedure. Fig. 3 shows the performance of an LT code for the robust soliton distribution.

RAPTOR CODES

Raptor codes[2] are an extension of LT-codes that allow faster encoding and decoding. The complexity improvement is obtained by reducing the number of edges in the code graph. Since the number of edges in an LT code is determined by the underlying distribution $\Omega(x)$, the idea is to choose a distribution that generates a small number of edges. However, by using such a distribution, the recovery performance of the code is worsened. In particular, it is shown in Ref. 2 that if an LT code with distribution $\Omega(x)$ on $\{0, \ldots, k\}$ has a reliable decoding algorithm, then its code graph must have at least $ck \log k$ edges, where c is a positive number. To tackle this problem a traditional block

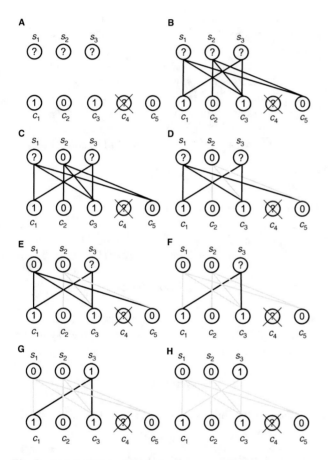

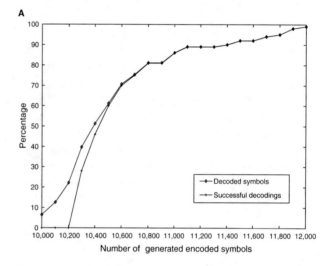

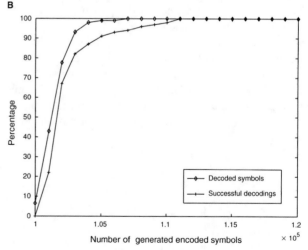

Fig. 2 (A) An LT decoder trying to recover three source symbols (here bits): s_1, s_2, and s_3 from four received encoded symbols c_1, c_2, c_3, and c_5, while c_4 is lost. (B) Determine the degree and the source symbols associated to each received encoded symbol. (C) Find an encoded symbol with degree one. The only one here is c_2. Set $s_2 = c_2 = 0$. Since s_2 is also connected to c_3 and c_5, set $c_3 := c_3 \oplus s_2$ and $c_5 := c_5 \oplus s_2$. (D) Remove all edges connected to s_2. (E) The next encoded symbol with degree one is c_5, which is connected to s_1, so set $s_1 = c_5$. Since s_1 is also connected to c_1 and c_3, set $c_1 := c_1 \oplus s_1$ and $c_3 := c_3 \oplus s_1$. (F) Remove all edges connected to s_1. (G) There are two symbols with degree one: c_1 and c_3. Both are connected to s_3. Hence s_3 can be decoded with either of them. Set s_3 equal to either c_1 or c_3. (H) All source symbols have been recovered. The decoding stops.

Fig. 3 Performance of an LT code for a robust soliton distribution with parameters $c = 0.1$ and $\delta = 0.5$. The number of source symbols is 10^4 in (A) and 10^5 in (B). The curves titled "decoded symbols" show the average percentage of recovered source symbols as a function of the number of encoded symbols for 100 simulations. The curves titled "successful decodings" show the percentage of simulations where all source symbols were recovered.

Ultra–
Wi-Fi

code called precode is introduced before the LT code. In this way, a Raptor code consists of a concatenation of two codes: a precode and an LT code (Fig. 4). More precisely, a Raptor code with parameters $(k,n,C,\Omega(x))$ is the concatenation of a traditional erasure code C of dimension k and length n and an LT code with distribution $\Omega(x)$ on n symbols. The precode considers all codeword symbols of the precode that were not successfully decoded by the LT code as erasures.

Examples

1. An LT code with distribution $\Omega(x)$ on k source symbols can be seen as a $(k,k,F_2^k,\Omega(x))$ Raptor code.
2. Precode only Raptor codes. These Raptor codes have parameters (k,n,C,x).
3. Fast Raptor codes. These Raptor codes have parameters $(k,n,C,\Omega_D(x))$ where $D = \lceil 4(1+\varepsilon)/\varepsilon \rceil$ and $\Omega_D(x) = 1/(\mu+1)\left(\mu x + \sum_{i=2}^{D} x^i/((i-1)i) + x^{D+1}/D\right)$ with $\mu = \varepsilon/2 + (\varepsilon/2)^2$. One of the main results of Ref. 2 is that if C has code rate $R = (1+\varepsilon/2)/(1+\varepsilon)$, $n = \lceil k/(1-R) \rceil$ and C can be decoded on a binary erasure channel with erasure probability $(1-R)/2$ with $O(n\log(1/\varepsilon))$ arithmetic operations, then the fast Raptor code can decode all k source symbols from $(1+\varepsilon)k$ encoded symbols with high probability in $O(k\log(1/\varepsilon))$ time. Examples of such precodes C are given in Ref. 2.

Systematic Raptor Codes

The fountain codes described in the previous section do not provide systematic encoding, i.e., if $s_1,\ldots,s_k$ are the source symbols and $c_1,\ldots,c_n$ are the encoded symbols, then there do not necessarily exist indices $i_1,\ldots,i_k$ such that $s_j = c_{i_j}, j = 1,\ldots,k$ (the source symbols do not appear in the sequence of encoded symbols). This is a limitation because systematic encoding allows the decoder to immediately exploit any received symbol that corresponds to a source symbol. In the following subsection, we describe an algorithm[2] that provides systematic Raptor encoding.

Systematic encoding of Raptor codes

Let $(k,n,C,\Omega(x))$ be a Raptor code. Let G be an $n \times k$ generator matrix for C. The encoding algorithm takes source symbols $x_1, x_2,\ldots,x_k$ and computes a set of k indices $i_1,\ldots,i_k$ between 1 and $k(1+\varepsilon)$ and a sequence of encoded symbols $z_1,z_2,\ldots$, satisfying $z_{i_1} = x_1, z_{i_2} = x_2,\ldots,z_{i_k} = x_k$. This is done as follows:

1. **Preprocessing step:** Compute a matrix R and indices $i_1,\ldots,i_k$ in $\{1,\ldots,k(1+\varepsilon)\}$ as follows:
   a. Get $k(1+\varepsilon)$ vectors $v_1,\ldots,v_{k(1+\varepsilon)}$ in F_2^n by sampling $k(1+\varepsilon)$ times independently from the distribution $\Omega(x)$ on F_2^n;
   b. Compute a $k(1+\varepsilon) \times n$ matrix S whose rows are the vectors $v_1,\ldots,v_{k(1+\varepsilon)}$;
   c. Compute the $k(1+\varepsilon) \times k$ matrix $T = SG$. Use Gaussian elimination to find the rank of T. If the rank of T is less than k, output an error message. Otherwise, find a submatrix R of T consisting of k rows $i_1,\ldots,i_k$.
2. Compute $y = (y_1,\ldots,y_k)$ with $y^T = R^{-1}x^T$, where $x = (x_1,\ldots,x_k)$;
3. Compute $u = (u_1,\ldots,u_n)$ with $u^T = Gy^T$;
4. Compute $z_i = v_i u^T, 1 \le i \le k(1+\varepsilon)$;
5. Apply the LT code with distribution $\Omega(x)$ on u to generate the output symbols $z_i, i > k(1+\varepsilon)$.

Decoding of systematic Raptor codes

Given the received encoded symbols $u_1,\ldots,u_{k(1+\varepsilon)}$, the algorithm recovers the original source symbols $x_1,x_2,\ldots,x_k$ as follows:

1. Use the decoding algorithm of the Raptor code to obtain $y_1,\ldots,y_k$;
2. Recover the input symbols $x_1,\ldots,x_k$ from $x^T = Ry^T$.

VIDEO STREAMING WITH MBMS

3GPP defines three functional layers for the delivery of MBMS-based services. The first layer, called bearers, provides a mechanism to transport data over IP. Bearers is based on point-to-multipoint data transmission (MBMS bearers), which can also be used in conjunction with point-to-point transmission. The second layer is called delivery method and offers two modes of content delivery: download and streaming. Delivery also provides reliability with FEC. The third layer (user service/application) enables applications to the end-user and allows him to activate or deactivate the service. An MBMS session consists of the following three phases:

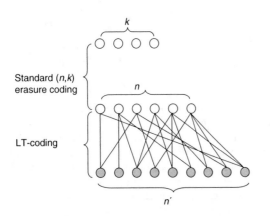

Fig. 4 Raptor code.

1. **User service discovery phase:** MBMS services are announced to the end-user using either 2-way point-to-point TCP-IP-based communication or 1-way point-to-multipoint user datagram protocol (UDP)-IP-based transmission.

2. **Delivery phase:** Multimedia contents are delivered (in either the streaming or the download mode) using 1-way point-to-multipoint UDP-IP-based transmission.

3. **Post-delivery phase:** A user may report on the quality of the received contents or request a file repair service (if in the download delivery mode) using 2-way point-to-point TCP-IP-based communication.

During the delivery phase, a UDP packet may be discarded by the physical layer if the bit errors cannot be corrected, and it can be lost due to, e.g., network congestion or hardware failure. Because there is no feedback channel in the delivery phase, ARQ-based protocols cannot be used. Instead, the media data is protected at the application layer using FEC with systematic Raptor codes.

The MBMS technical specifications recommend the H.264/AVC baseline profile as a video coder. The primary unit generated by the H.264/AVC codec is called the network abstraction layer (NAL) unit. At the transport level, real-time transport protocol (RTP), as specified by RFC1889 of the Internet Engineering Task Force is used. In general, one NAL unit is encapsulated in a single RTP packet according to the RTP payload specification.[8] However, one NAL unit may also be fragmented into a number of RTP packets or one RTP packet may contain more than one NAL unit. FEC is applied to the incoming stream of RTP packets.

A copy of each RTP packet is forwarded to the FEC encoder to construct a source block. A source block is a two-dimensional array of size $T \times k$, where the source block length (SBL) k is the number of symbols in the block, and T is the symbol size in bytes. To each incoming RTP packet, a 3-byte identifier is prepended, and the resulting block is inserted in the source block, starting from the first available empty row. The prepended identifier contains the RTP flow ID f and the length l of the RTP packet. The RTP flow ID f allows multiplexing several streams and protecting them together. If for an RTP packet, $l + 3$ is not a multiple of T, then the block must be padded with additional zeros. The padded zeros are not transmitted and can be inserted by the receiver to duplicate the original 2-D array. The source block is filled with incoming RTP packets until the number of source symbols reaches k. The value of k for a source block is flexible and computed dynamically during the source block construction. However, for any source block, the constraints $k \geqslant k_{min} = 1024$ (as the performance of Raptor codes is low for smaller k) and $k \leqslant k_{max} = 8192$ must be satisfied. Fig. 5 shows an example of source block construction.

0	11		$B_{0,0}$	$B_{0,1}$	$B_{0,2}$	$B_{0,3}$	$B_{0,4}$
$B_{0,5}$	$B_{0,6}$	$B_{0,7}$	$B_{0,8}$	$B_{0,9}$	$B_{0,10}$	0	0
0	9		$B_{0,0}$	$B_{0,1}$	$B_{0,2}$	$B_{0,3}$	$B_{0,4}$
$B_{0,5}$	$B_{0,6}$	$B_{0,7}$	$B_{0,8}$	0	0	0	0
1	20		$B_{1,0}$	$B_{1,1}$	$B_{1,2}$	$B_{1,3}$	$B_{1,4}$
$B_{1,5}$	$B_{1,6}$	$B_{1,7}$	$B_{1,8}$	$B_{1,9}$	$B_{1,10}$	$B_{1,11}$	$B_{1,12}$
$B_{1,13}$	$B_{1,14}$	$B_{1,15}$	$B_{1,16}$	$B_{1,17}$	$B_{1,18}$	$B_{1,19}$	0

$k = 7$

$T = 8$

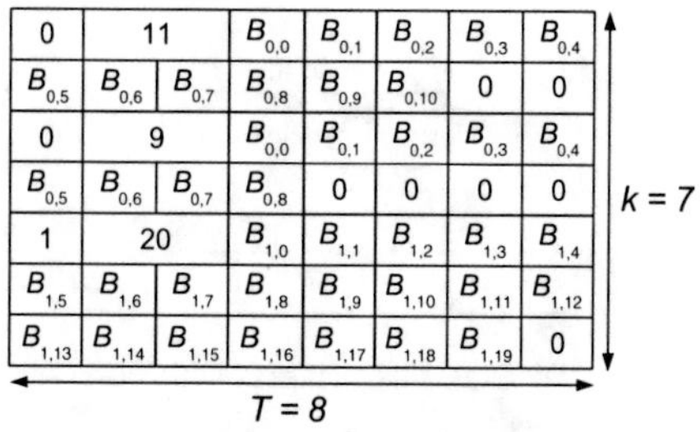

Fig. 5 MBMS source block example. Three payloads of lengths 11, 9, and 20 bytes are placed in a source block of SBL $k = 7$ with symbol size $T = 8$ bytes. The first two payloads are from RTP flow $f = 0$, and the third one is from RTP flow $f = 1$. Each cell in the block is a byte. $B_{i,j}$ denotes the $(j + 1)$-th byte of the $(i + 1)$-th RTP flow.

Once a source block is completed, the FEC encoder generates $N - k$ repair (redundant) symbols, each of size T, by applying a systematic Raptor code on the k symbols of the source block. A pseudo-random number generator is used to generate the graph of the Raptor code. The pseudo-random number generator is based on a fixed set of 512 random numbers[1] that must be available to both sender and receivers. The value of N is not fixed and may vary with the source block.

Each symbol (source and repair) has two associated fields called source block number (SBN) and encoding symbol ID (ESI). The fields SBN and ESI, each of size two bytes, indicate the associated source block number and the position of the symbol within the block, respectively. ESI values of source symbols are in $\{0, \ldots, k - 1\}$, while ESI values of repair symbols are larger than or equal to k. A source FEC payload ID is appended at the end of each RTP packet to create an FEC source packet, which is then encapsulated by UDP and sent to the receiver by the MBMS bearers (Fig. 6). A number G of consecutive repair symbols are concatenated, and a repair FEC payload ID is prepended to the resulting block, yielding an FEC repair packet (Fig. 7). Each FEC repair packet is encapsulated by UDP and sent to the recipients by the MBMS bearers.

At the receiver side, the received stream of source and repair packets is processed in blocks. If some of the source packets in a block are lost but sufficient repair packets

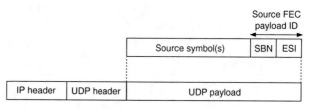

Fig. 6 FEC source packet. SBN is a 16-bit integer that identifies the source block related to the RTP packet. ESI is a 16-bit integer that gives the index of the first source symbol in this packet.

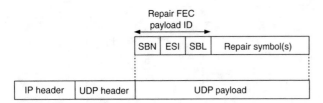

Fig. 7 FEC repair packet. The packet contains G repair symbols of size T each. SBN is a 16-bit integer that identifies the source block related to the repair symbols. ESI is a 16-bit integer that gives the index of the first repair symbol in this packet. SBL is a 16-bit integer that gives the number of source symbols k in the source block.

from the same block are received, then the original source block can be reconstructed by the Raptor decoder. The original RTP packets in individual streams can be recovered using f and l. These RTP packets are passed to the RTP layer, which extracts the NAL units and forwards them to the H.264 decoder.

Fig. 8 illustrates the general framework of MBMS video streaming.

The MBMS specifications[1] provide recommendations for the values of T and G. These recommendations are based on the input parameters B (maximum source block size in bytes), A (symbol alignment factor in bytes), P (the maximum repair packet payload size which is a multiple

of A), k_{min}, k_{max}, and G_{max} (maximum number of repair symbols per repair packet), which should satisfy $\lceil (B/P) \rceil \leqslant k_{max}$. The number of repair symbols per repair packet is estimated as $G = \min(\lceil (P \times k_{min}/B \rceil, P/A, G_{max})$. The symbol size is estimated as $T = \lfloor (P/(A \times G)) \rfloor \times A$. Table 1 shows some recommended settings for the input parameters $A = 4$, $k_{min} = 1024$, $k_{max} = 8192$, $G_{max} = 10$, and $P = 512$.

FURTHER READING

Apart from the 3GPP MBMS technical specifications[1] presented in this entry, only a few works have considered the application of digital fountain codes to wireless video transmission. Afzal et al.[9] studied the effect of the MBMS video streaming parameters on the system performance. In particular, they suggest to determine the value of the SBL k by inserting as many RTP packets into the source block as possible subject to the constraint that the maximum end-to-end delay is not exceeded. Wagner et al.[10] applied Raptor codes to efficiently stream scalable video data from multiple servers to a client. One limitation of the MBMS framework is that the information contained in packets that contain unrecoverable bit errors is ignored since these packets are discarded at the physical layer. Gasiba et al.[11] show that by modifying the receiver and the protocol stack, one can forward this information to the

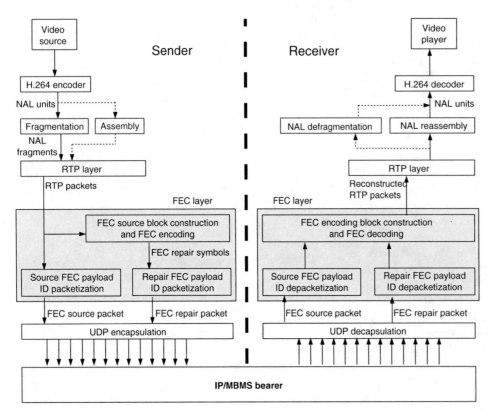

Fig. 8 MBMS video streaming framework.

Table 1 MBMS recommendations for a maximum source block size B.

B (KB)	G	T
40	10	48
160	4	128
640	1	512

application layer and significantly improve the performance of MBMS video streaming at the cost of higher decoder complexity.

REFERENCES

1. 3GPP TS 26.346 V6.6.0; Technical Specification Group Services and System Aspects; Multimedia Broadcast/Multicast Service (MBMS); Protocols and codecs, 2006.
2. Shokrollahi, A. Raptor codes. IEEE Trans. Inf. Theory. **2006**, *52* (6), 2551–2567.
3. Rizzo, L. Effective erasure codes for reliable computer communication protocols. ACM Comput. Commun. Rev. **1997**, *27* (2), 24–36.
4. Luby, M. *Information Additive Code Generator and Decoder for Communication Systems*. US Patent No. 6,307,487, Oct 23, 2001.
5. Luby, M. *Information Additive Code Generator and Decoder for Communication Systems*. US Patent No. 6,373,406, Apr 16, 2002.
6. Luby, M. LT-Codes. 43rd Annual IEEE Symposium on Foundations of Computer Science, Vancouver, BC, Canada, Nov 16–19, 2002; 271–280.
7. Byers, J.W.; Luby, M.; Mitzenmacher, M. A digital fountain approach to asynchronous reliable multicast. IEEE J. Sel. Areas Commun. **2002**, *20* (8), 1528–1540.
8. Wenger, S.; Stockhammer, T.; Hannuksela, M.M.; Westerlund, M.; Singer, D. RTP payload format for H.264 video, RFC3984; IETF, 2005.
9. Afzal, J.; Stockhammer, T.; Gasiba, T.; Xu, W. Video streaming over MBMS: a system design approach. J. Multimed. **2006**, *1* (5), 25–35.
10. Wagner, J.; Chakareski, J.; Frossard, P. Streaming of scalable video from multiple servers using rateless codes. IEEE international Conference on Multimedia and Expo 2006, Ontario, Canada, July 9–11, 2006.
11. Gasiba, T.; Stockhammer, T.; Afzal, J.; Xu, W. System design and advanced receiver techniques for MBMS broadcast services. IEEE International Conference on Communications 2006, Istanbul, Turkey, June 11–15, 2006.

BIBLIOGRAPHY

1. Digital Fountain, Inc., http://www.digitalfountain.com/.
2. 3GPP Multimedia Broadcast/Multicast Service (MBMS); Protocols and codecs, http://www.3gpp.org/ftp/Specs/html-info/26346.htm.

Ultra–Wi-Fi

Video Streaming: Wireless

Min Qin
Roger Zimmermann
Department of Computer Science, University of Southern California, Los Angeles, California, U.S.A.

Abstract

Many new challenges appear when streaming video content over wireless links is concerned. These challenges include high loss rate, node mobility, limited bandwidth, etc.

Online video streaming services have become increasingly popular in recent years. For example, more than 100 million video clips are watched everyday on YouTube. As a result, streaming video data is becoming crucial to modern communication systems. Recent development of 3G and 802.11 wireless networks has provided a new platform for video streaming applications along with many new challenges. Limited bandwidth, unreliable data delivery, and wireless interferences all make traditional streaming protocols susceptible to quality degradations. In this chapter, we identify several main challenges in wireless video streaming applications. Additionally, recent technologies for improving wireless streaming performances are surveyed.

AVAILABLE BANDWIDTH

Unlike wired environments, wireless networks are often constrained by their limited bandwidth. Wireless bandwidth can be affected by many factors, including distance, interference, obstacles, and mobility. Collision avoidance also has a huge impact on the bandwidth that each wireless device can utilize, especially for 802.11-based wireless networks. For each packet to be sent, adjacent wireless senders have to compete with each other to access the shared channel and avoid collisions. This leads to a reduced channel utilization.

Fig. 1 illustrates how collision avoidance works in 802.11 protocols. To avoid collisions, 802.11 protocols use the CSMA/CA (carrier sense multiple access/collision avoidance) with RTS/CTS (request to send/clear to send) technique. Before sending a data packet, a node needs to send an RTS frame to the destination and the destination will reply with a CTS frame. Any other node receiving the CTS frame should refrain itself for a certain amount of time before trying to access the medium. As shown in Fig. 1, the actual amount of time used to send a packet is t_d, which is determined by the packet size and the maximum raw data rate. Therefore, the channel utilization in Fig. 1 is only t_d/t_s. As a result, a large portion of the wireless bandwidth is wasted.

Table 1 lists various characteristics of existing 802.11 technologies. Due to collision avoidance, most 802.11 protocols have a typical bandwidth that is about half of the maximum bandwidth. Also, higher-level transport protocols have an additional impact on the achievable bandwidth. For example, the extremely popular 802.11b protocol has a maximum raw data rate of 11 Mbps. However, the maximum data rate it can achieve is 7.1 Mbps over user datagram protocol (UDP) and 5.9 Mbps over TCP.

As more participants in a wireless network result in more contentions, bandwidth drops significantly as multiple streaming applications co-exist in the same region. In Ref. 1, the authors demonstrated that if k participants are in the same radio range, each of the links get about $1/k$ of the available bandwidth. Therefore, the streaming data rate will be significantly affected if multiple participants in the same wireless network try to access the service at the same time.

Except collision avoidance, distance may also affect the actual bandwidth of a wireless link. As the distance between two wireless endpoints increases, signal strength is reduced continuously. Wireless protocols often use different modulation schemes as signal strength changes. For example, the 802.11b protocol uses BPSK (binary phase-shift keying) for 1 Mbps, QPSK (quadrature phase-shift keying) for 2 Mbps, and CCK (complementary code keying) for 5.5 and 11 Mbps. As the modulation scheme changes only when signal strength drops below a certain threshold, wireless bandwidth is not a continuous function of distance. It drops abruptly as distance increases. Additionally, for indoor environments, signals

Encyclopedia of Wireless and Mobile Communications DOI: 10.1081/E-EWMC-120043507

Ultra–
Wi-Fi

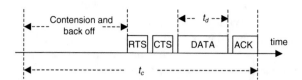

Fig. 1 802.11 CSMA/CA with RTS/CTS.

can bounce off from walls and thus increase the distance they can travel. However, the loss rate is higher when used indoors due to multi-path fading and various obstacles.

Another important factor that can seriously affect wireless bandwidth is interference. Wireless channels can often interfere with each other by sharing the same frequency band. For example, both the 802.11b and 802.11g share the 2.4–2.48 GHz frequency band. The Federal Communications Commission (FCC) divides the band into 11 channels in North America whereas Europe and Japan have 13 and 14 channels, respectively. Table 2 lists the channel frequencies for the 802.11 b/g protocols in North America. As shown, channels 1, 6, and 11 do not overlap with each other. Therefore, there can be a maximum of three 802.11 b/g transmissions that co-exist in the same region without interfering with each other. To maintain the required streaming rate, some video streaming applications use multiple non-overlapping channels to increase the achievable bandwidth. For networks with overlapping frequencies, the throughput of each network is largely affected by others. Because of the popularity of 802.11b network, it is very hard for an 802.11g link in an office environment to achieve 54 Mbps due to the interferences from the surrounding 802.11b networks. On the other hand, 802.11a links are normally free from interference as most commercial access points only support 802.11b/g. This can have a significant impact on many streaming applications. For example, the HYDRA live streaming project, as shown in Fig. 2, transmits high definition (HD) video streams with a constant data rate of 19.2 Mbps. Although both 802.11a and 802.11g can support a data rate of up to 23 Mbps, users prefer to choose 802.11a as 802.11g often cannot reach 20 Mbps due to interferences from adjacent 802.11b networks.

PACKET LOSSES

Packet losses are very common in wireless networks. In wired environments, most packet losses are caused by network congestions. However, in wireless links, packet losses can be caused by a number of factors, including interferences, hidden terminals, link breaks, or delays. Packet losses have a considerable impact on video streaming applications. To reduce the bandwidth requirement, many video streaming applications use block-based compression algorithms to reduce temporal redundancies in video sequences. However, one lost packet can cause picture quality degradations in multiple frames. Therefore, packet loss recovery is very important for wireless streaming applications. There are three ways to ensure packet delivery: using re-transmission-based error control, forward error correction (FEC), and error concealment.

To guarantee packet delivery in wireless environments, one solution is to use reliable transmission protocols such as TCP. TCP has many desirable features such as flow control and congestion control. However, TCP is not well suited for wireless streaming applications. This is because many streaming protocols such as real-time transport protocol (RTP) are mainly based on UDP for media streaming. Additionally, TCP behaves differently in wired and wireless environments. In wired environments, packet losses are often caused by network congestions. As a result, TCP will use its congestion control protocol to slow down the sender when it detects packet losses. However, in wireless environments, packet losses can be caused by various reasons. Therefore, the underlying assumption made by TCP is no longer true. Slowing down the sender may only reduce the network throughput. Thus, the traditional TCP protocol may not be efficient in wireless environments.

To improve TCP performance in wireless streaming applications, many improvements have been proposed. For example, the end-to-end WMSTFP (wireless multimedia streaming TCP-friendly protocol)[2] can differentiate erroneous packet losses from congestive losses and thus, improve the throughput of wireless streaming applications. In Ref. 3, the author introduced another method by modifying the lower layers instead of changing TCP. The technique incorporates simultaneous MAC packet

Table 1 Characteristics of existing 802.11 technologies.

Protocol	Frequency (GHz)	Raw bandwidth (Mbps)	Typical bandwidth (Mbps)	Range (indoors) (m)	Modulation techniques
802.11b	2.4	11	6	30	DSSS
802.11g	2.4	54	23	30	OFDM
802.11a	5	54	23	10	OFDM
802.11n	2.4	200	100+	30	OFDM with MIMO

Table 2 Channel frequencies of 802.11 b/g in North America.

Channel	Center frequency (GHz)	Minimum frequency (GHz)	Maximum frequency (GHz)
1	2.412	2.401	2.423
2	2.417	2.405	2.428
3	2.422	2.411	2.433
4	2.427	2.416	2.438
5	2.432	2.421	2.443
6	2.437	2.426	2.448
7	2.442	2.431	2.453
8	2.447	2.436	2.458
9	2.452	2.441	2.463
10	2.457	2.446	2.468
11	2.462	2.451	2.473

transmission (SMPT) as the link layer protocol. When bursty errors happen, SMPT uses additional channels (CDMA codes) to transmit packets until all the MAC (medium access control) packets have been successfully transmitted. Therefore, it can stabilize the throughput of wireless links. As a result, TCP performance will not be seriously harmed when used in a wireless environment.

In UDP-based streaming protocols, packet delivery is not guaranteed by the protocol itself. To recover packet losses in such situations, Papadopoulos et al.[4] Introduced a semi-reliable re-transmission-based error control for media streaming applications. Packet losses are recovered by re-transmitting the packet from the source to the destination. However, this requires one round-trip delay to recover a lost packet. It is possible that the re-transmitted packet arrives later than the time when the corresponding frame should be displayed. One

Fig. 2 The HYDRA live streaming project can stream HD video between two ad hoc laptops through 802.11a networks.

way to reduce this problem is to increase the buffer size, so that the frames stay in the buffer for a longer time before being rendered. However, this may significantly affect the performance of delay-sensitive applications. Further improvements for re-transmission-based error control were introduced in Ref. 5. Instead of recovering end-to-end packet losses, this approach tries to localize link errors. Lost packets are re-transmitted on the link where the error is identified rather than requesting re-transmission from the source. This can largely save network bandwidth and reduce error recovery time. Fig. 3 shows the video quality of the HYDRA HD live streaming system with and without re-transmissions. The transmitted video format is 720p encoded in MPEG-2 (moving picture experts group-2) format.

Another technique to recover packet loss is to use advanced encoding mechanisms. For example, the FEC techniques are widely accepted for video streaming applications in wired environments. For wireless links, several new FEC encoding[6,7] techniques have been proposed in recent years. In addition to the original video stream, redundant data for recovering partial packet losses are also transmitted to the receiver. As a result, no re-transmission is required if the packet loss rate is low. This is very useful to delay-sensitive applications such as video conferencing systems. However, FEC adds additional overhead to the bandwidth requirements of streaming applications. To recover bursty packet losses, it may require up to 30% additional bandwidth. As wireless links are constrained by their bandwidth, increasing the bandwidth requirement may introduce extra problems such as collision and interference.

Error concealment does not recover lost data. It tries to preserve the video quality by re-constructing a video frame from earlier frames through interpolation or approximation. This is based on the assumption that two adjacent video frames are not significantly different to human eyes. As video streams are often compressed, performing video interpolation requires the receiver to have enough computational power. In Ref. 8, the author introduced a
bit-stream-level error concealment algorithm. Instead of recovering video frames through interpolation, this algorithm tries to guess the error bit from the video stream and flip it. One benefit of error concealment is that it does not add additional overhead to the bandwidth requirements. However, it is often hard to correct bursty errors.

ENCODING

Video encoding is a very important way to reduce the bandwidth requirement of wireless streaming applications. Recent encoding algorithms such as MPEG-4 can achieve

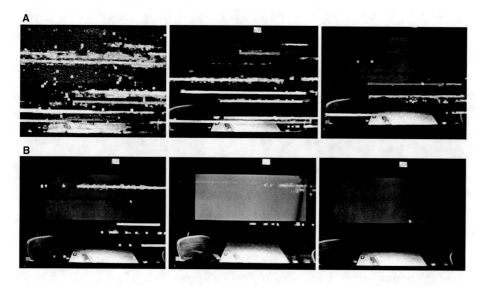

Ultra–Wi-Fi

Fig. 3 HD video quality with or without re-transmission algorithm. (A) Video quality without re-transmissions (10, 5, 1% loss rate, from left to right). (B) Video quality with re-transmissions (10, 5, 1% loss rate, from left to right).

a good compression ratio while keeping the video quality reasonably good. To reduce the bandwidth requirements, most video streaming applications adopt differential encoding schemes. For example, the MPEG-4 format is one of the commonly used techniques. MPEG video encoding algorithms group multiple frames together into a structure called group of pictures (GOP). A GOP consists of one I (intra-coded) picture and multiple P (predictive-coded) and B (bidirectionally coded) pictures. Fig. 4 shows a typical GOP structure. An I picture contains detailed information as it captures an entire picture. For P and B pictures, they only describe the temporal difference to frames that precede or follow them. As a result, losing one packet may degrade the picture quality in the subsequent frames.

To solve the problem of one packet loss causing future picture quality degradation, different encoding schemes can be adopted. For example, M-JPEG compresses each frame into an independent JPEG image. As a result, losing several packets in one frame will not affect the picture quality in future frames. However, as video frames have a lot of redundancies, encoding them separately increases the bandwidth requirement of the streaming application. As discussed in Ref. 9, replacing differential encoding schemes with I mechanisms does not introduce severe

video quality changes at the receiver side. This is because I pictures are usually larger in size and are thus susceptible to transmission errors. However, B and P pictures are usually small and unlikely to cause transmission errors. Another approach to minimize picture quality degradation is to localize errors to part of the pictures. For example, Qureshi et al.[10] introduced an encoding scheme that divides each video frame into a number of sub-frames. Each sub-frame is independently encoded and transmitted through the wireless network. To reduce the possibility of losing the I pictures of all sub-frames at the same time and to reduce bandwidth usage, sub-frame encoders are out of phase with each other. For each frame to be sent, some of the sub-frames are I while the rests are either P or B pictures, as shown in Fig. 5. Therefore, losing one packet only affects a small part of the video for a certain amount of time.

For heterogeneous networks with both wired and wireless participants, using a single encoding algorithm may not suit the needs of all clients. This is because wireless links often have a lower bandwidth than wired links. Also, wireless devices may be computationally incapable of decoding high-quality video streams. As a result, video transcoding becomes very important. As shown in Fig. 6, a server streams the high-quality video to Receiver 1 through a wired network. In addition to displaying the video, Receiver 1 can relay the stream to Receiver 2 through a wireless access point. However, due to the wireless bandwidth limitation, Receiver 1 has to transcode the video stream before relaying it. The conventional transcoding process includes: decoding the video stream first, transcoding the video format in the pixel domain, and then re-encoding the video stream. This often requires a huge amount of computing power. In Ref. 11, the author introduced a secure scalable streaming (SSS) coder

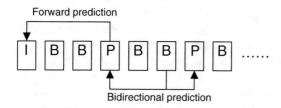

Fig. 4 A typical GOP structure of MPEG video streams.

Fig. 5 Dividing a video frame into sub-frames and then using encoders that are out of phase with each other.

for such environments. Unlike the traditional transcoding procedure, the SSS transcoder only needs to truncate or remove data packets without decoding and re-encoding the stream. Therefore, transcoding can be carried out on a low-power node. Also, security is enforced by the SSS technique.

NODE MOBILITY

Most wireless technologies are omni-directional. They allow users to move freely and still be connected to the network. However, mobility can have a significant impact on wireless performance. It can cause wireless bandwidth variations along with a high packet loss rate. In Refs. 11 and 12, the authors tested the bandwidth variations of streaming applications as users move around. Wireless bandwidth tends to change abruptly rather than gradually as users move. A major reason for such sudden change is that a user can easily move out of the coverage area of a base station or an access point. To provide a better streaming performance for mobile users, link duration prediction is very important, so that handoffs[7] can be smoothly achieved.

With the widespread use of handheld devices, streaming video content between two wireless ad hoc peers is becoming a popular application. For example, the Sony PlayStation Portable (PSP) and most PDAs can both store and play MPEG videos. Therefore, users can share the video content they have with others by streaming it to others. When operating in ad hoc mode, wireless devices can communicate with each other without the support of an access point. As a result, it is easier to set up the connection and to distribute the content. For example, the MStream streaming architecture[13] combines ad hoc streaming with a centralized server to distribute multimedia content in a large wireless network. It is designed to provide a location-sensitive media streaming service to multiple mobile devices. It can serve as guides for a university campus, a park, a zoo, or a botanical garden, where users can get relevant media information based on their physical location and time. As shown in Fig. 7, MStream divides the geographical space into multiple regions and each region is associated with a dedicated audio/video stream. The server can stream media data to clients within its communication range in a client-server fashion. On the other hand, clients within the same streaming area may cluster themselves and construct a peer-to-peer network. Therefore, clients can stream the data from each other even if they are outside the server's communication range.

One of the most important challenges in streaming multimedia content among mobile ad hoc peers is to deliver the content, usually large in size, over a dynamic peer-to-peer-based network where link availability is constantly changing. As handheld devices have a limited communication range, it is possible that the link between two ad hoc peers breaks before the streaming process completes, if both peers are moving. As shown in Fig. 8, the link between peer A and C and that between B and C are constrained by C's communication range. If both peer A and B have the multimedia content that C has requested, it is better to stream the content from A as the link between A and C is likely to last longer. Therefore, to achieve

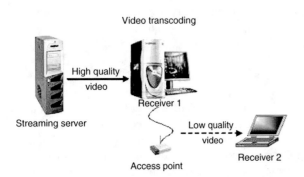

Fig. 6 Video stream transcoding to support a mix of wired and wireless environments.

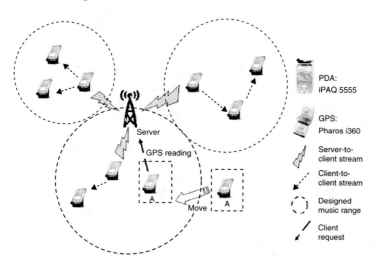

Fig. 7 The MStream streaming architecture.

a smooth media performance, link availability prediction is of crucial importance to mobile streaming applications. It can also help mobile ad hoc peers determine handoffs or to select the best choice among a number of sources. An iterative algorithm for predicting continuous link availability between two mobile ad hoc peers is presented in Ref. 13.

As a single streaming server cannot cover a large area, the question of how to replicate the streaming service to different locations to guarantee service coverage in a mobile ad hoc network is very important. There are two strategies: one is to replicate the streaming service to some fixed nodes or fixed locations. However, fixed service replication often cannot adapt to changing network topologies. The other scheme is to use dynamic service replications as introduced in Refs. 14 and 15. Instead of relying on fixed servers, service is replicated when a node is about to leave the service coverage area. However, this approach assumes that all nodes have the same capability of becoming a streaming server. As shown in Ref. 14 and 15, dynamic service replication can adapt to the changing topology of a network and still achieve a good service coverage ratio.

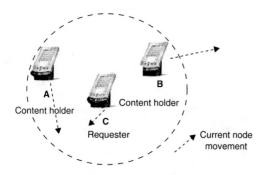

Fig. 8 Multimedia content sharing among PDAs.

ACKNOWLEDGMENTS

The HYDRA research has been funded in part by NSF grants EEC-9529152 (IMSC ERC) and IIS-0534761.

REFERENCES

1. Bararia, S.; Ghandeharizadeh, S.; Kapadia, S. Evaluation of 802.11a for streaming data in ad-hoc networks. 4th Workshop on Applications and Services in Wireless Networks, Boston, MA, Aug 9–11, IEEE, Piscataway (New Jersey): 2004.

2. Yang, F.; Zhang, Q.; Zhu, W.; Zhang, Y. End-to-end TCP-friendly streaming protocol and bit allocation for scalable video over wireless Internet. IEEE J. Sel. Areas Commun. **2004**, *22* (4), 777–790.

3. Fitzek, F.H.P.; Reisslein, M. Wireless video streaming with TCP and simultaneous MAC packet transmission (SMPT). Int. J. Commun. Syst. **2004**, *17* (5), 421–435.

4. Papadopoulos, C.; Parulkar, G.M. Retransmission-based error control for continuous media applications. 6th International Workshop on Network and Operating System Support for Digital Audio and Video (NOSSDAV), Zushi, Japan, Apr 23–26, ACM Press, New York: 1996.

5. Eckhardt, D.A.; Steenkiste, P. Improving wireless LAN performance via adaptive local error control. 6th IEEE International Conference on Network Protocols, Austin, Texas, Oct 13–16, IEEE, Piscataway (New Jersey): 1998.

6. Chen, L.-J.; Sun, T.; Sanadidi, M.Y.; Gerla, M. Improving wireless link throughput via interleaved FEC. The 9th IEEE Symposium on Computers and Communications (ISCC 2004), Alexandria, Egypt, June 28–July 1, IEEE, Piscataway (New Jersey): 2004.

7. Huang, C.; Xu, L. Study of a practical FEC Scheme for wireless data streaming. IASTED Internet and Multimedia Systems and Applications (EuroIMSA 2005), Grindelwald,

Switzerland, Feb 21–23, ACTA Press; Calgary, Anaheim, Zurich: 2005.

8. Ding, X.; Roy, K. A novel bitstream level joint channel error concealment scheme for realtime video over wireless networks. IEEE INFOCOM 2004, Hong Kong, China, Mar 7–11, IEEE, Piscataway (New Jersey): 2004.

9. Fitzek, F.H.P.; Seeling, P.; Reisslein, M. Video streaming in wireless Internet. In *Wireless Internet: Technologies and Applications Series: Electrical Engineering & Applied Signal Processing Series,* 2004.

10. Qureshi, A.; Carlisle, J.; Guttag, J. Tavarua: video streaming with WWAN striping. ACM Multimedia 2006, Santa Barbara, Oct 23–27, 2006; ACM Press, New York: 2006.

11. Wee, S.; Apostolopoulos, J. Secure scalable video streaming for wireless networks. IEEE International Conference on Acoustics, Speech, and Signal Processing, Salt Lake City, UT, May 7–11, 2001; IEEE, Piscataway (New Jersey): 2001.

12. Bai, G.; Williamson, C. The effects of mobility on wireless media streaming performance. Wireless Networks and Emerging Technologies (WNET), Alberta, Canada, July 8–10, 2004; ACTA Press, Calgary, Anaheim, Zurich: 2004; 596–601.

13. Qin, M.; Zimmermann, R.; Liu, L.S. Supporting multimedia streaming between mobile peers with link availability prediction. ACM Multimedia 2005, Singapore, Nov 6–12, ACM Press, New York: 2005; 956–965.

14. Li, B.; Wang, K.H. Nonstop: continuous multimedia streaming in wireless ad hoc networks with node mobility. IEEE J. Sel. Areas Commun. **2003**, *21* (10), 1627–1641.

15. Qin, M.; Zimmermann, R.; Liu, L.S. Supporting guaranteed continuous media streaming in mobile ad-hoc networks with link availability prediction. ACM Multimedia 2006, Santa Barbara, CA, Oct 23–27, ACM Press, New York: 2006.

BIBLIOGRAPHY

1. www.youtube.com
2. Pan, Y.; Lee, M.; Kim, J.B.; Suda, T. An end-to-end multipath smooth handoff scheme for stream media. 1st ACM International Workshop on Wireless Mobile Applications and Services on WLAN Hotspots, San Diego, CA, Sept 19, 2003; IEEE Journal on Selected Areas in Communications. IEEE, Piscataway (New Jersey): 2003.

Ultra–
Wi-Fi

Wi-Fi Walkman

Jun Wang
Delft University of Technology, Delft, The Netherlands

A.P. de Vries
Delft University of Technology, Amsterdam, The Netherlands

Marcel J.T. Reinders
Delft University of Technology, Delft, The Netherlands

Abstract

Wireless home entertainment center refers to a device able to handle heterogeneous media and to connect client devices located within the house and the outside world (i.e., the Internet).

INTRODUCTION

In recent years, with the rapid progress in information processing, communications, and storage technologies, the amount of information that we deal with in our daily lives has rapidly increased. Although we enjoy the entertainment and convenience brought to us by a variety of sources, the volume of information is increasing far more quickly than our ability to digest it. For instance, the World Wide Web is the most significant media source for most Internet users and is growing at an exponential speed. However, the ability of obtaining useful information grows slowly. Most importantly, the retrieval of information relevant to the user's interest remains an unsolved problem. Above all, the gap between the amount of information that is available and the information that people are able to extract is increasing.[1] Unfortunately, today's computers merely act as information providers. One of the solutions to close this information gap is to increase the ability of computers to steer the user's interests and select/represent relevant information on the user's behalf.

To this regard, the research on information filtering is aroused to filter out, refine and systematically represent the relevant information and intuitively ignore superfluous computations on redundant data. One of the solutions for overcoming the information overload is to provide personalized suggestions based on the history of a user's likes and dislikes. In the domain of human–computer interactions (HCIs), especially for the interface of e-commerce, the information overload has created increasing interests in recommender systems that recommend products such as books, CDs, movies, TV programs, and music.

Personalized Services

In human society, people are extremely experienced in person-to-person communication. People have a common understanding of each other both conceptually and perceptually, i.e., it is easy for them to obtain an understanding of each other in terms of the interests, tastes and expectations. Therefore, it is easy between people to provide different services. This, however, comes at a price: it costs a lot in terms of time and labor which severely hampers to cover required demands from large masses of people.

In recent times, more and more services are available in the form of HCIs, especially due to the increasing interest in Internet. For example, business-to-consumer (B2C) services in the e-commerce domain broadly extend the range of the traditional services and provide a convenient and low-cost way to deliver services to a large group of consumers. For instance, people can purchase books, CDs, electronics and other items from the Web anytime, anywhere, without the need to go to the different shops accordingly.

When compared with the person-to-person service, the current computer-based service is not operating in a friendly manner. The computer just acts as the information provider and provides the necessary transactions. The computer does not know the interest and intent of the individual user and therefore is only able to supply services of a general nature, i.e., not adapted to the specific customer that it is dealing with. As a result it may damage the quality of the service (in contrast with the person-to-person communication) when there is a large amount of options and the user is overloaded with this information and is not able to make an instant decision. To this end, it is necessary to develop methods that allow the computer to

Encyclopedia of Wireless and Mobile Communications DOI: 10.1081/E-EWMC-120043607

infer the user's interest or intent such that it can provide personalized services. This will eventually support companies to realize a shift from offering mass products and services to offering customizing goods and services that efficiently fulfill the desires and needs of individual customers.[2] Obviously, recommender systems can be one of the solutions to such a shift.

Folk Computing

Personalization is sited in the folk computing environment. Original computing environments were designed for scientists and intensive training was needed before users were able to use them. The recent progress in the information and communication technology (ICT) has supplied a pervasive or ubiquitous computing environment. In this environment, the computer user need not necessarily be the "mouse-clicker" in front of the desktop anymore. The target user in the folk computing environment is the common user. The interactions between a human and a computer in this environment are required to be natural: "users apply their senses to observe data and information of interest related to an event (conceptual and perceptual analysis) and interact with the data based on what they find interesting."[1]

In the folk computing environment the central issue is to realize a natural interaction between user and computer. For example, one of the important design strategies is "what you see is what you get." Another strategy is that the interface should be more compelling and natural and less intimidating to people than a keyboard and mouse. This could be achieved by making use of multimedia (audio, images, graphics, video, and touch) and multiple sensors (camera, motion detector, voice capture, GPS, etc.). Therefore, in this environment, personalization should be based on this natural interaction instead of the traditional "mouse-click" styled interaction.

Moreover, for personalization, one of the benefits in this environment is that having the possibility of multimodal input devices will make it possible to infer the intent of a user through such sources, e.g., the emotional state of a user can be interpreted from recognizing his/her expressions from video recordings. Having information about the intent of the user also opens up the possibility to react on this intent, e.g., by recommending the desired services to the user.

Peer-to-Peer Networks

Not only the availability of the sufficient types of the information, but also the way people access information is changing. Peer-to-peer and ad hoc networks, as new network topology, become a new way for people to distribute, exchange, and consume resources from their local storage devices in many different locations, such as the future home, office, or university campuses. There are two significant advantages of peer-to-peer and ad hoc networks: 1) the replicas of the content among peers increases the content availability; 2) for the exchange of information, no centralized storage and management from third parties is necessary, which makes these networks very low-cost. In recent years, these attributes have attracted a large body of people in the Internet domain. For instance, Internet based peer-to-peer networks have increased rapidly and they have given a large number of people the possibility of sharing resources in their local storage devices.[3,4] Recently, sharing resources in wireless networks has received some attention. The TunA system[5] allows users to "tune in" to other nearby TunA music players and listen to what someone else is listening to. Another system, SoundPryer[6] allows drivers to jointly listen to music shared between cars on the road. Interestingly, these two applications show that the upcoming technologies have started to care about their social impact on everyday life, i.e., they bring people together that have been socially separated by the technologies for the last decades (such as TV, Internet, portable music player, etc.) Clearly, these technologies[3–6] are different from the traditional technologies in that they encourage people to make social interactions such as sharing and exchanging information. However, those applications are implemented on devices that are far away from so-called intelligent devices which aims to provide personalized services on user's behalf. We present here a different system that has the ability to react to the user's interests and select relevant information on the user's behalf accordingly.

In ad hoc network environments, the volume of information is increasing far more quickly than our ability to digest it. The traditional textual keywords-based information retrieval approaches[7–10] can no longer be used as filter mechanisms since they suffer from three major problems. First, the transition from textual data to heterogeneous data requires large amount of textual metadata on the one hand. It is practically intractable to ask people to provide content as well as associated metadata at the same time. On the other hand, automatic content analysis on the non-textual data is far from being efficient to get the metadata that we need. Second, keywords are not semantically expressive enough to enable a seamless search, i.e., people hardly issue a textual query when they cannot exactly express what they are looking for. Thirdly, in mobile environments, the user interface is constrained and consequently does not permit complex interactions between users and their handheld devices.

WI-FI WALKMAN

The Wi-Fi walkman that we developed is a case study that investigates the technological and usability aspects of

HCI with personalized, intelligent and context-aware wearable devices in ad hoc wireless environments such as the future home, office, or university campuses. It is a small handheld device with a wireless link that contains music content in the environment or from the user. Users carry their own Wi-Fi walkman around and listen to the music content. All this music content can be shared using mobile ad hoc networking. The Wi-Fi walkman is situated in a peer-to-peer environment and naturally interacts with the users. Without annoying interactions with users, it can learn the users' music taste and consequently provide personalized music resources to fit the user's interest according to the user's current situated context.

Music Recommendation

In the Wi-Fi walkman scenario, the multimedia content data the user intends to access are music files (MP3 formatted). Those music files are possible stored in the hard disk of each Wi-Fi walkman and can be accessed through the Wi-Fi mobile network. Users are able to share music content through the network. However, as the network size increases, the music content available is increasing as well. This consequently causes an information overload problem. To address this, in this scenario, music recommendation is implemented as a user oriented music file filter to help user to find relevant or desired music files according to current situated context and learned user interest.

Scenarios

We now discuss detail descriptions of some possible concise scenarios for Wi-Fi walkman:

A business man called Frank is a music fan. He has just bought a Wi-Fi walkman attached with a personalized music recommender system (MRS). This personalized MRS can recommend music files (such as play-lists) to Frank based on his interests (profiles) and the context anytime anywhere.

Scenario 1 (*During jogging in the morning*). As usual, Frank, bringing along with his favorite Wi-Fi walkman, is jogging in a nearby park. Due to the fact that the mobile network in the park area does not have good quality, Frank's Wi-Fi Walkman may not download music in this area. However, since Wi-Fi walkman knows Frank usually enjoy sport music during this time. The MRS knows Frank's long-term interest (profile) and the current situation (that Frank is engaged in sports and the network quality is poor). So the system has already pre-cached and recommends a bunch of music (Frank's favorite sport music) which best fits Frank's interest and current situation.

Scenario 2 (*During a trip*). Frank with his friends joins a tourist group to a church. He switches on his Wi-Fi walkman and asks the MRS to recommend some music to fit this environment. The MRS knows he is in the church by communicating with both the situated network and his friends' Wi-Fi walkmans and consequently recommend some church music fitting Frank's favorites. By using his Wi-Fi walkman, Frank enjoys a complete church experience.

Scenario 3 (*Before sleeping*). Frank usually sleeps at 12 midnight. The Wi-Fi walkman knows his schedule. At 12 midnight, the system finds and recommends desired music from the Internet. Since Frank been having trouble recently getting to sleep, the system recommends some light music to help Frank go to sleep.

Problem Definition and Formalization

From above mentioned scenarios, we can simply articulate the problem as:

According to user's interest or taste and current situated context, the system recommends the appropriate music service to the user.

In this definition, some factors need to be clarified.

Interest

Many aspects affect the interest of a user. If we treat interest as a whole, it is difficult to grasp the latent patterns behind. Hence, we classify the user's interest into long-term interest and short-term interest.

Long-term interest is the user's preference or taste. It evolves slowly and smoothly as the user experiences and socially interacts with other people and the outside world. We assume it is comparatively static and diverse. On the contrary, short-term interest is the current intent or task at hand. It evolves sharply based on the context and the user's willingness. It is not stable but focused. If we model the two types of interest differently, it could help us to accurately understand the user's interest.

Context

In the definition, there are two important factors that need to be considered: user's interest or preference and context. Context plays an important role to understand the user's current short-term interest or task.

Services

Here we would like to state that, rather than the music/song itself, the service of the music should be the target item. This is because we are in a pervasive computing environment (in particular owing to the foreseen ad hoc network). The deliverability and quality of the service are also important factors to be taken into account. For instance, as shown in the first scenario, if the recommended music is hard to reach, the system may pre-cache it at a suitable time. This extends the system to consider the service rather than to provide only the content.

Ultra–
Wi-Fi

Resources to Build up User Preferences

To provide personalized content services, the starting point, clearly, is to understand the user's interests and/or preferences. The more information we have about the user and the content itself, the better we know what the user wants in a certain context. There are three channels of information to acquire such information: the people-to-people correlation, the music-to-music correlation, and the demographic data of the users.

The people-to-people correlation information

The people-to-people correlation information reflects the correlations among people's "tastes" for the music. A collaborative filtering-based recommender system is able to recommend music to a user based on the other users in the system who have similar "tastes." The music play-list of a user reflects the taste of that user. The information in the play-list could be useful including song's name, playing times, playing frequency etc. Initially, we will utilize a dataset in the AudioScrobbler community.[3] Currently this dataset has 857.020 tracks and 4.175.146 playback actions.

Since collaborative filtering has been widely investigated, there are other datasets available but in other (than music) domains. For instance, the movie lens research group at the University of Minnesota provides two ratings datasets (Each movie and Movie lens). Numerous collaborative filtering publications have employed those datasets for evaluations. If necessary, we will use these "standard" datasets for evaluation purposes.

Music-to-music correlation information

Correlation between different pieces of music can be obtained by analyzing the content description of those music pieces. Automatic content analysis is, however, still an unsolved problem. On the other hand, manual annotation is a time-consuming and annoying task.

Fortunately, a community, called MusicBrainz, provides a music meta-database of the content description of music songs purpose. Some learning methods can be applied on this content description to find the correlations among music.

In addition, for the new MP3 files which are not annotated in the database, we can use acoustic fingerprinting (FFT transformation compressed to a few bytes), unique to each piece of music to find the correct metadata for them as well.

Demographic data

Demographic data about users (such as age, gender, social position etc.) is useful for categorizing users. This data could especially be useful to solve the start-cold problem, i.e., providing recommendations when the system does not yet have any information about the user ratings or play-lists. The demographic data allows that users to be compared. However, user's demographic data need to be collected from other applications.

RECOMMENDER SYSTEM FOR WI-FI WALKMAN

There are two approaches for implementing recommendation, namely data-driven, and rule-based or knowledge-driven approaches. Both of them have to understand user beforehand either explicitly given or implicitly learned. Data-driven approach achieves recommendation by observing the behavior of the user and learning patterns in this behavior while the rule-based/knowledge-driven approach implements recommendation where a user defines his/her likes/dislikes through questionnaires.

Recommender system is a popular form to generate personalization. It is employed in the e-commerce domain (Web-based shops) to create personalized environment for selling items such as books, CDs. In our research here, we also focus on the recommender system as a form of personalization.

Definition

A recommender system can be defined as:

A system which "has the effect of guiding the user in a personalized way to interesting or useful objects in a large space of possible options."[11]

This definition makes it clear that user oriented guidance is critical in a recommender system. This means that, during the interactions between a human and the computer, the computer needs to provide not only the information but also the guidance toward that information on user's behalf.

Characteristics of User and Product Information

To be able to perform a personalized recommendation one needs to understand the user's interest or preference. Hereto we need to acquire and analyze information about the user. On the other hand, we also need information about the products to be customized.

Generally, the information about the user and the products has the following characteristics:

1. It comes in tremendous volumes;
2. It is dynamic, i.e., it varies over time;
3. It may be continuous in time, i.e., streaming information;
4. It does not exist in isolation, i.e., it exists in its ambient context with other data;
5. It is inherently heterogeneous, i.e., it is collected or sensed from a set of distinct sources.

For example, multimedia (audio, images, graphics, video, and touch) and multiple sensors (camera, motion detector, voice capture, GPS, etc.) are usually employed to furnish the computing environment for more compelling and natural and less intimidating to people.

1. The quality varies (greatly).

When designing a recommender system that obtains relevant information from the user and product information, we need to take these characteristics info full consideration.

Research Issues

Clearly, there are numerous issues related to recommender systems. Here, we would like to mention some of them which do exist in our domain and are of particular interest to us. In addition, we also clarify some research issues which have not been stated clearly in literature.

Context awareness (task focus)

Human perception is greatly aided by the ability to probe the environment through various sensors along with the use of the situated context. In return, the context has a large influence on the interest and intent of one particular user. This causes the interest and intent of a user to vary dynamically over time. Thus, knowing the current context of the users is critical to correctly understand the interest and intent of a user. Context awareness is thus a major factor when dealing with personalization and recommender systems.

The user's preference is determined by both the general taste of the user (long-term taste) and the current task of the user (his/her context or short-term taste). Ignoring one of them reduces the quality of the recommendation considerably. When exploiting this context in the recommendation, there are two major problems: 1) the determination of the current context; and 2) the integration of this context with the general taste.

It is certainly not trivial to acquire information about the current task of the user. Clearly, this information can be revealed either implicitly, i.e., derived from other services contacted by the user, or explicitly, i.e., when the user indicates his/her current task (e.g., defining the task type using a menu-driven mechanism).

When information about the current context is available it still should be combined with the general taste of the user. One simple approach could be that the general taste recommendation is filtered based on the information about the context, or the other way around. We will address the question of how both types of information should be combined such that an optimal and efficient recommendation can be provided to the user.

Proactive resource caching

Within an ad hoc mobile network the availability of resources is not guaranteed. The recommendation system should take this into account, either 1) by incorporating the availability of the data into the recommendation engine; or 2) by predicting near-future recommendation so that the necessary data can be pre-cached. Both forms of adaptation change fundamentally the way the recommendations are done.

Adaptability

The interest or taste of a user may change over time. A recommender system should be aware of this change and consequently adapt the recommendations accordingly. When the change in the ratings of a user are known, collaborative filtering is quite capable of adapting to these changes since the position of the user in the rating space changes and consequently the recommendations change. However, this requires that we know the change in the ratings of the user. Thus we need some kind of feedback mechanism to let the system know the changes in the user's tastes. One way is to ask the user feedback on the recommendations made. Clearly, this is not a desirable way. Therefore, suitable and user friendly feedback mechanisms should be developed.

Sparsity

Since the amount of items is extremely large and most users do not rate most items, the matrix for measuring the people-to-people correlation is typically very sparse. Therefore, there is no guarantee of finding a set of neighbors who have similar taste. This typically happens when the ratio between the number of items to the number of users is very high or when the system is in the initial stage of use.

Some potential solutions include making use of the content descriptions or the, demographic data. These, and possibly others, should be investigated with respect to the music recommendation scenario.

Scalability

The standard collaborative filtering approach needs to know the user-to-user correlations based on the ratings. However, correlations need to change when new users are added. Therefore the computation of these correlations has to be on-line. To this end, using collaborative filtering to generate recommendations is a computationally expensive task. The nearest neighbor algorithm that is used in traditional collaborative filtering requires an amount of computations that grows with both the number of users and items. The algorithms which achieve fast results do not guarantee computationally efficient results when they are applied on large practical datasets.

To deal with this scalability problem, some solutions are proposed. For instance, one can reduce the data size by exploiting dimensionality reduction techniques such as principal component analysis (PCA)[12] or singular value decomposition (SVD).[13] As these methods approximate the data, they have the side-effect of reducing the recommendation quality.

As an alternative solution, instead of computing the user-to-user correlations on-line, item-based collaborative filtering, which computes item-to-item correlations (based on ratings as well), has recently been proposed to improve the scalability.[14,15] Due to the fact that the correlations between items are relatively static, they can be computed off-line. Therefore, the item-based collaborative filtering approach could make most of the computation off-line. This intuitively improves the scalability in large datasets.

Cold-start problem

The cold-start problem[16] is one of the common difficulties in a collaborative filtering-based recommender system. This problem can be divided into user cold-start problem and item cold-start problem. Since the correlations are obtained by the ratings, the algorithm fails where there are no/less correlations available. User cold-start problem happens when there is a new user in the system on whom no or few rating information is available. The collaborative filtering method then does not have enough information to reliably estimate a similarity (correlation) between users so that only poor or even no recommendations can be made. Similarly, the item cold-start problem occurs when there is no (or few) rating information on a new item. Then the similarity estimates between items is very inaccurate.

As already mentioned, correlations coming from the content information (for instance, the item-to-time correlation regarding to the content) link the old items and new items and intuitively provides some solutions to the item cold-start problem. For instance, when a new item is added, there is no rating information available about this item. Collaborative filtering cannot recommend this item to the users. However, by knowing the content information of the new item and the old items, the correlations between them can be built and we can actually recommend the new item to the users who like the highly correlated old items.

Meanwhile, demographics of a new user can categorize the user into some classes and as well correlate the existing users to the new user. Those findings could help us for the further research on this issue.

A Basic System

The Wi-Fi walkman is implemented on the Sharp Zaurus PDA (personal digital assistant), see Fig. 1, by using C++. It runs on an ad hoc wireless network. It features audio playback, audio storage, audio recommendation, and ad hoc wireless connectivity for audio exchange.

The Wi-Fi walkman itself contains an audio agent, a transport agent, and a wireless interface shown in Fig. 2. The audio agent is responsible for the communication with the recommendation services, manages the MP3 files on the storage devices (e.g., a fresh card), and selects which MP3 to play. The transport agent uses the wireless ad hoc network to communicate with other transport agents and enables the sharing of the music files. Due to the dynamic nature of an ad hoc network, the transport agents must keep track of the other walkmans around them. The enhanced ad hoc wireless interface also informs the transport agent of new walkmans and walkmans that can no longer be reached.

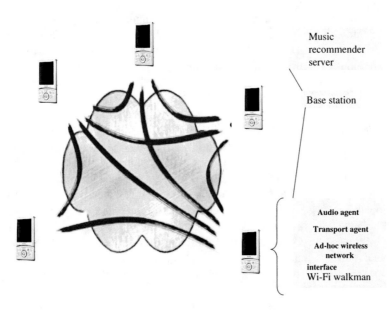

Music
recommender
server

Base station

Audio agent

Transport agent

Ad-hoc wireless
network
interface
Wi-Fi walkman

Fig. 1 Illustration of the Wi-Fi walkman in client/server model.

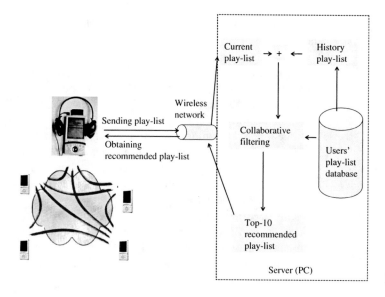

Fig. 2 Recommendation in the client/server model.

Peers and play-lists

Each peer represents a Wi-Fi walkman used by a particular user. Let's define the set of peers as:

$$P_i, i = \{1, \ldots, M\} \tag{1}$$

where M is the number of the peers currently on-line in the peer-to-peer network. That means they can be located and accessed with the available bandwidth. Since the peers (Wi-Fi walkman) and users exist in pairs, we will use the term peer and user interchangeably.

The music content in the network is defined as a set of items, denoted by the set I. Each item has a specific physical location, i.e.,

$$I = \{I^{i,j} \mid i = \{1, \ldots, M\}; j = \{1, \ldots, N_i\}\} \tag{2}$$

where N_i is the number of items physically located in the local storage device by the peer P_i. $I^{i,j}$ denotes the j-th item owned by user P_i. The set of items owned by peer P_q is denoted as:

$$I_q = \{I^{i,j} \mid i = q; j = \{1, \ldots, N_q\}\} \tag{3}$$

Users will retrieve music content according to their own interests. At a particular time, a user, however, will have a particular interest. The interest can be obtained either explicitly or implicitly. For instance, it could be explicitly obtained by asking users to rate items. Alternatively, this can also be implicitly indicated by the music items that the user is playing. In our Wi-Fi walkman, we assume the user's music play-list to be indicative of the user's music interest. Formally, we use a vector $V_q = \{v_q^{i,j}\}$, $i = \{1, \ldots, M\}; j = \{1, \ldots, N_i\}$ to represent the

play-list of the user P_q, where the element $v_q^{i,j} = 1$, if user P_q played the item $I^{i,j}$; otherwise $v_q^{i,j} = 0$.

It may be noted that generally the interest of the user will change over time. It in fact depends on the current context. Therefore, the play-list (representing the current users' interest) should ideally be dependent on the time also, i.e., $V_q \to V_q(t)$.

We utilize a sliding time window to ignore the old music items users have played, as shown in Fig. 2. By doing so, the system focuses on the user's current interest.

The current recommender system is implemented by using the collaborative filtering technique. Collaborative filtering utilizes the correlations (commonalities) between users on the basis of their ratings (in this case, the play-lists of the users) to predict and recommend music items which have the highest correlations to the user's preference.

The accuracy of the collaborative filtering directly relies on the number of users, who provide their ratings. In mobile networks, the density of peers may vary strongly depending on the local situation. For instance, on a bus, there are only a dozen people while at an airport there are thousands. Depending on the current density of peers, we perform recommendation by two different approaches, namely the flooding model and the client/server model.

Flooding model

When the density of peers is large (i.e., thousands of users) and the play-lists from those users are enough to obtain a good recommendation, we use the flooding approach to find the correlations between users.

By using the correlation,[17,18] the similarity between the play-lists V_q and V_p of two users is calculated as follows:

$$\text{Sim}(V_q, V_p) = \frac{\sum_{i,j}^{M,N_i} (v_q^{i,j} - \bar{v}_q)(v_p^{i,j} - \bar{v}_p)}{\sqrt{\sum_{i,j}^{M,N_i} (v_q^{i,j} - \bar{v}_q) \sum_{i,j}^{M,N_i} (v_p^{i,j} - \bar{v}_p)}} \quad (4)$$

where $\bar{v}_q$ and $\bar{v}_p$ are the mean rating of the user P_q and P_p respectively, that are used for removing the bias.

$$\bar{v}_q = \frac{1}{\sum_i^M N_i} \sum_i^M \sum_j^{N_i} v_q^{i,j}, \bar{v}_p = \frac{1}{\sum_i^M N_i} \sum_i^M \sum_j^{N_i} v_p^{i,j} \quad (5)$$

The distance measurement between a music item $I^{i,j}$, not known to user P_q, and the play-list from user P_q can be calculated as the weighted average rating,[17,18] as follows:

$$d(I^{i,j}, V_q) = \bar{v}_q + k \cdot \sum_{\{V_p | V_p \in N_q, \text{sim}(V_q,V_p) > T\}}$$
$$\cdot \text{sim}(V_q, V_p)(v_p^{i,j} - \bar{v}_p) \quad (6)$$

where k is a normalization constant. In the flooding model, the play-list V_q of the user P_q is broadcast to all its neighbors P_p to determine the recommendation for that user. The neighboring peers check the similarity (using in Eq. 5) between the received play-list and their own play-list. They decrease the TTL (time to live) field of the broadcast play-list and then pass it to their neighboring peers until the TTL count reaches 0. We use set N_q to denote all the neighboring peers that the querying play-list V_q can reach. If one of the neighboring peers has a play-list that has a similarity to the broadcasted play-list that is higher than T, then the items in the play-list of the neighbor P_p (including the locations) are sent back to the peer P_q that posed the query V_q. We use I_q^* to denote the set of these returned items. Finally all items I_q^* received by the querying peer are ranked according to the distance measurement (Eq. 6) and consequently the top-N ranked items are recommended to the user (Eq. 7).

$$\text{Rec}_q^N = \text{Top}N\{\text{rank}\{d(I^{i,j}, V_q)|I^{i,j} \in I_q^*, I^{i,j} \notin I_q\}\} \quad (7)$$

Client/server model

When the density of the peers is small and consequently the play-lists (rating) from those users are not enough to obtain a good recommendation, we have to access a predefined rating database and use that to calculate the recommendation. In this model, we assume the peer has a chance to access a server which has a rating database.

The rating database stores the play-lists of all the users in the networks.

Fig. 3 illustrates the procedure of obtaining the recommended play-list. In order to reduce the computational complexity, we apply the item-based recommendation algorithm proposed in Refs. 14, and 15 to calculate the recommendations.

In item-based recommendation, each music item can be represented by who has played it. More formally, each item $I^{i,j}$ can be represented by a vector $U^{i,j}$, where its element $u_q^{i,j} = 1$, if the item $I^{i,j}$ has been played by the peer P_q and zero otherwise.

Item-based recommendation is then performed by exploring the correlations between the items rather than the correlations between users. Recommendations are created by finding items that are similar to other items that the user prefers according to:

$$\text{sim}(I^{i,j}, I^{i',j'}) = \frac{\text{Freq}(I^{i,j}, I^{i',j'})}{\text{Freq}(I^{i,j}) \times \text{Freq}(I^{i',j'})} \quad (8)$$

where $\text{Freq}(I^{i,j})$ is the number of times that item $I^{i,j}$ is in any of the play-lists. $\text{Freq}(I^{i,j}, I^{i',j'})$ is the number of times that item $I^{i,j}$ and $I^{i',j'}$ are in the same play-list.

Due to the fact that the item-to-item matrix is relatively static, it is possible to compute this matrix off-line, which extremely reduces the computational demands. That is, by applying Eq. 8, for each item $I^{i,j}$, its top N similar items can be obtained off-line and it is denoted as $I_q^{\text{Top}N}$.

When the play-list V_q of user P_q is sent to the server, the recommendation then is calculated according to the following equation:

$$\text{Rec}_q^N = \text{Top}N\{\text{rank}\{\text{sim}(I^{i,j}, I^{i',j'}) | I^{i',j'}$$
$$\in I_{\text{Top}N}^{i,j}, I^{i,j} \notin I_q; I^{i,j} = 1 \cap I^{i,j}$$
$$\in V_q\}\} \quad (9)$$

Implementation

The recommendation is implemented in the server part. We utilize a dataset of the AudioScrobbler community[3] as our play-list dataset. Currently this dataset has 857,020 tracks and 4,175,146 playback actions. The interaction between each peer and the server is illustrated in Fig. 4.

Snap-shots of the Wi-Fi walkman application are shown in Fig. 5. The procedure to obtain the music files

$$\dots I_i^{10}(t-T-1), I_i^{11}(t-T),\dots, I_i^{12}(t-2), I_i^{13}(t-1),$$
$$I_i^{14}(t) \qquad \underbrace{\qquad\qquad} \qquad \text{Play sequence}$$
$$\text{Time window}$$

Fig. 3 Time window for forgetting.

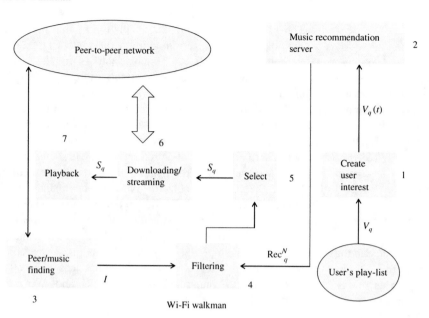

Fig. 4 System diagram of the Wi-Fi walkman application.

that fit the user's interest is illustrated in Fig. 4 and each step is described in the following flowchart:

Wi-fi_walkman()
Begin

1. Create $V_q(t)$ to represent the user's current interest from the play-list by utilizing a time window
2. Get recommendation Rec_q from server
3. Find on-line peers and obtain the music item list I (resources) from those peers

4. Filter the music list I to get the recommended list Rec_q by the top N recommended items Rec_q^N.
$$Rec_q = I \cap Rec_q^N$$

1. Select the downloading/streaming items by users through GUIs $S_q \subset Rec_q$

1. Locate the recommended items S_q and download/stream them
2. Playback the obtained items S_q

End.

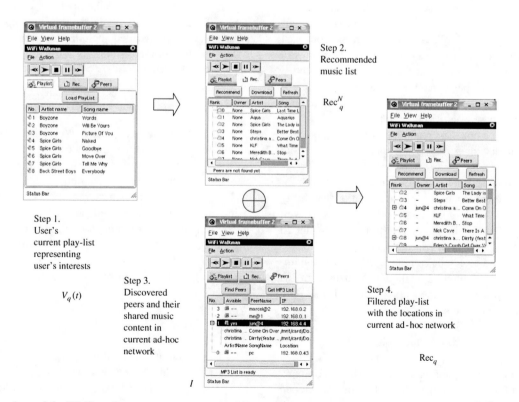

Fig. 5 Snap-shots of the Wi-Fi walkman prototype.

STATE-OF-THE-ART IN RECOMMENDER SYSTEMS

In this section, we will review the state-of-the-art of commercially available recommender systems as well as existing research solutions. Since most of the methods are not domain-constrained and can be further generalized, this review is not constrained to music recommendation. Other recommender systems for movies, books, CDs, TV programs, etc. as well as the general recommender system in the e-commerce field are included as well.

Commercially Available Recommender Systems

Recommender systems are best known for their use in the e-commerce domain. Here, it is employed to furnish personalized environments for selling items such as books, CDs, etc. Many on-line retailers utilize recommender systems, like, e.g., Amazon, CDnow, BarnesAndNoble, IMDB, etc. It is greatly successful as the appeal of personalized content created by recommender system exceeds those of untargeted content such as banner advertisements and top-seller lists which are usually used on the Web. This success has also boosted a number of successful start-up companies like Firefly Net Perceptions, LikeMinds and ChoiceStream to provide recommending solutions.

Amazon.com

Based on item-to-item collaborative filtering, Amazon.com developed a practical book recommender system.[19] Rather than the traditional user-based collaborative filtering (matching current user to similar customers), the item-to-item (item-based) collaborative filtering approach matches each of the items a user purchased and ratings to other similar items in the database and then constructs a recommendation list based on those similar items.

In the Amazon.com homepage, users can obtain recommendation lists based on the items in their shopping carts and can also filter out their recommendations by product line and subject area based on previous rates and purchases.

For large on-line retails, the two important factors for designing a recommender system are: 1) the scalability with respect to the (large) number of customers as well as items; and 2) the ability for real-time processing. In contrast to the similarity matrix of users-to-users, the matrix of items-to-items can be computed off-line. The item-to-item collaborative filtering method in Amazon.com has proven to produce recommendations in real-time and scales to massive datasets.

The MyBestBets personalization platform

ChoiceStream, Inc., a software development company headquartered in Cambridge, Massachusetts, developed The MyBestBet personalization platform. This platform provides a personalization solution for content providers to deliver personalized content such as movies, TV, music, commerce, and community throughout their applications.

The MyBestBets personalization engine combines content-based classification and users' preferences to match each individual with the contents which is best suited to his or her particular tastes and preferences. Users should complete a survey about their preferences (rating the items) before the system is able to give the recommendations. However, when filling such survey one needs some knowledge about the item classification, which common users usually lack. Moreover, in order to make the recommendations more accurate, the user should keep providing feedback, constantly re-valuating their preferences. Currently this engine is integrated on the AOL and Winamp websites.

MyBest TV

MyBest TV is a category-specific recommendation service currently embedded on AOL. This magazine-like service provides users with TV program recommendation that are delivered on-line or via e-mail. This service narrows options for consumers, helping them choosing a TV program they will really enjoy.

MyBestBets for music

MyBestBets for music personalization engine is currently utilized as a tab in the Winamp browser to provide a personalized music experience for the Winamp community. The services provided include: recommendations for CDs, short lists of radio station or music on TV, discovery of music buddies, etc.

Smart radio

The smart radio system[20] is a Web-based client-server application. It ties together the concept of a music program with a personalized recommendation to allow users to have personalized stream music programs.

User ratings in the smart radio system are gathered using either explicit user feedback (explicitly rating track items or individual programs) or an implicit way (scratching an initial play-list). After the system collects user ratings, it performs a users-to-users top-N collaborative filtering algorithm to construct and recommend the music programs to the user.

PTV

PTV is an Internet system offering personalized TV guides for each individual user. Its embedded ClixSmart personalization engine is a hybrid recommendation system combining content-based and collaborative filtering approaches. The benefits of this hybrid system are the ability to make diverse program recommendations, to cope with new or one-of-a-kind programs, and to cope with new or unusual users.

In this system, there are two databases used: a program database and a schedule database. The program database provides the content information regarding the programs such as program name, genre, country of origin, cast, studio, director, writer, and so on while schedule database stores the current channel schedules such as program name, its channel and time information, and a textual episode description. The contents in both databases are vital for obtaining the recommendations.

TiVo

TiVo is an automatic personal video recorder (PVR) that also adapts to user's interest. TiVo allows user to rate what he/she enjoys by using "Thumbs Up" and "Thumbs Down" buttons on the remote. With the user's preferences stored on the local receiver, TiVo matches those preferences with the program data it receives from the TiVo service and meanwhile searches through thousands of programs to create the own personalized suggestions and record programs into the hard disks accordingly.

Basic Solutions

To close the increasing gap between the amount of information that is available and the amount of useful information that one is able to extract, recommender systems have aroused more and more attention in the fields of electronic commerce as well as information retrieval. There are two prevalent approaches: content-based filtering and collaborative filtering. Recently, more and more research aims to combine the two approaches in order to gain better (more accurate) performance with fewer drawbacks than any of the individual approaches.

Collaborative filtering

One of the most promising, widely implemented and familiar technologies is collaborative filtering.[14,15,17–19] Collaborative filtering-based approaches utilize the correlations (commonalities) between customers on the basis of their ratings, to predict and recommend items which have the highest correlations to the user's rated /purchased items (user's preference).

In RINGO,[21] a personalized music recommendation system, similarities between the tastes of different users are utilized to recommend music items. This user-based collaborative filtering approach works as follows: A new user is matched against the database to discover neighbors, who are other customer who, in the past, have had a similar taste as the user, i.e., who have bought similar items as the new user. Items (unknown to the new user) that these neighbors like are then recommended to the new user.

Since the relationships between users are relatively dynamic (they continuously buy new products), it is hard to calculate the user-to-user matrix on-line. This causes the user-to-user (user-based) collaborative filtering approach to be relatively computationally expensive.

To address this, item-based algorithms[14,15] are introduced that explore the correlations between the items rather than the correlations between users. Recommendations are created by finding items that are similar to other items that the user likes (has already bought). Due to the fact that the item-to-item matrix is relatively static, it is possible to compute this matrix off-line. This extremely reduces the computational demands. This method has been successfully applied in the on-line retails such as Amazom.com.[19]

One of the drawbacks that influence the performance of this technique is that the recommendation is solely based on the historical rating data. The current task at hand or current context is ignored, even though it greatly affects the current interest or intent of the user. Authors in Ref. 22 proposed a pure collaborative filtering task-focused recommendation method to tackle this problem. In their approach, besides the long-term user's interest profile, a task profile is established by either explicitly providing some items associated with the current task or implicitly observing the user's behavior (intent). By utilizing the item-to-item correlation matrix, items that resemble the items in the task profile are obtained for recommendation. As they match the task profile, these items fit the current task of the user. These items will be re-ranked to fit the user's interests based on the interest prediction before recommending them to the user.

Content-based filtering

Content-based approaches recognize the correlation between contents of different items to predict and recommend items that have the most correlations to the user's rated/purchased items (user's preference).[23]

One of the requirements for the content-based method is the content description. Usually, contents of items are represented by metadata in the form of textual information. Lang in Ref. 24 uses words as features to filter newsgroups in a newsgroup filtering system: NewsWeeder. Similarly, a machine learning method for text-categorization is applied to content-based recommending of Web pages in Ref. 25 and to book recommending in Ref. 26.

Nevertheless, other researchers apply content-based analysis methods directly to the raw media data where the metadata is absent. In Ref. 27, music is categorized based on features extracted from the raw music file such as the pitch, tempo and loudness. These features are then used to build up a recommendation. This is achieved by recommending the music that has similar features to the music the user has recently listened to. Alternatively, in Ref. 28, authors try to learn the user's preferences by mining the melody patterns from the music access behavior. Music recommendation is achieved through a melody preference classifier. However, due to the difficulty of feature extractions, their experiments are all based on midi files and not real music.

In contrast to collaborative filtering, the item-to-item correlation is learned based on the representations (as features) of the items' content rather than based on the users' ratings.

Other approaches

There are many other recommendation approaches available. Here we just mention demographic recommender systems and knowledge-based recommender systems since they have unique attributes to help solving some problems (such as user cold-start etc.) that content-based and collaborative approaches cannot tackle.

Demographic recommender systems are aimed to categorize the user regarding to personal demographic data (e.g., age and gender) and classify items into the user classes. Approaches falling into this group can be found in Ref. 29 for book recommendation, and in Ref. 30 for marketing recommendation. Like collaborative filtering, demographic techniques also employ user-to-user correlations but differ in the fact that they do not require a history of user ratings.

Knowledge-based recommender systems attempt to reason about the relationship between a need and a possible recommendation. The user profile should encompass some knowledge structure that supports this inference. The system proposed by Ref. 31 tried to employ case-based reasoning to achieve the knowledge-based recommendation.

Hybrid approach

Although the collaborative filtering approach has significant advantages, often it is combined with other techniques to improve the recommendations. The other techniques include weighting, switching, mixing, feature combination, cascade, feature augmentation, meta-level, etc. A complete review about these techniques can be found in Ref. 11.

In Ref. 16, authors employed aspect modeling to model both the user rating based item-to-item correlation and content-to-user correlation. In this general probabilistic framework, content information and user rating information are systematically integrated in an attempt to solve the cold-start problem.

In Ref. 23, authors tried to boost the pure collaborative filtering by utilizing a content-based predictor. In order to provide content-based predictions, they treat the prediction task as a text-categorization problem. According to this, a bag-of-words naïve Bayesian text classifier is employed on the textual metadata to construct the content-based predictor. Consequently, the predictions from the predictor are treated as pseudo user-ratings vector and added to the user-based item-to-item correlation matrix to fill up the sparse spaces. This intuitively provides some pseudo ratings to the new items which are not rated by the customers.

Comparison and Analysis

All recommendation techniques have their advantages and drawbacks. Table 1 shows a comparison of the different recommendation techniques.

Collaborative filtering utilizes the correlations of users to recommend items liked by other similar users. The user profile is only based on the user ratings about the items and there is no content knowledge needed. This makes the technique extremely simple and general. In addition, it has the ability to provide cross-genre recommendation

Table 1 The advantages and drawbacks of the recommending technologies.

	Advantages	**Drawbacks**
User-based collaborative filtering	Can cross-genre recommending; non domain constrained	User/item cold-start problem; sparse problem; rating and history interaction data required; non task focus/context aware-less; expensive computation; less scalability
Item-based collaborative filtering	Can cross-genre recommending; no domain-constrained; off-line computations; scalability	User/item cold-start problem; sparse problem; rating and history interaction data required; non task focus/less context-aware
Content-based filtering	No item cold-start problem; no sparse problem; task-oriented	User cold-start problem; computation expensive; content description/training required; domain constrained
Demographic	Can cross-genre recommending; non domain constrained	Item cold-start problem; demographic data required; less accurate
Knowledge based	Non item/user cold-start problem; adaptive; including non-product features	Knowledge discovery required

(recommending items which are significantly different from previously obtained items according to the contents). The scalability problem of collaborative filtering can be solved by item-based collaborative filtering. The off-line computation of the item correlation matrix allows on-line processing of a large amount of items and users.

Collaborative filtering, however, also has some significant drawbacks. First, it suffers from the user and item cold-start problem. Lack of rating information on new items and new users cause that new items and new users cannot be categorized. The item cold-start problem can be tackled by utilizing the content information of new items while the user cold-start problem can be dealt with by extracting demographic data of new users. Second, collaborative filtering suffers from the so-called sparsity problem since the recommendation depends on the neighbors of the user. If the user rating space is sparse or when the target user is "an unusual user" the algorithm will fail since in both cases there are no relating neighbors.

Content-based recommendation approaches ignore the user correlation and only consider the correlations between the contents. They overcome the item cold-start problem by matching the content descriptions between the new item and the existing items. In addition, it is easy to be task-oriented by matching the user task and the content descriptions (metadata).

Nevertheless, content-based recommendation approaches suffer from some drawbacks as well. First, given the fact that most of the media data (audio, video) is opaque to the system, obtaining content descriptions is a problem. Second, the content-based user profile constrains (prunes) the region of the item space to a particular content. It effectively hampers the recommendation from dealing with the diverse taste of a user. For instance, a user could possible like both jazz music and dance music but a content-based approach will not be able to find the correlations between jazz music and dance music since their content descriptions are far apart.

CONCLUSIONS

In this entry, we presented the state-of-the-art in and our view on recommender systems. We then introduced a new wireless application called Wi-Fi walkman. Without bothering users with any annoying keywords input, the Wi-Fi walkman can steer user's music interest and recommend appropriate music in the peer-to-peer networks. We described scenarios for an MRS in the Wi-Fi walkman, and gave our scope and basic MRS by using the collaborative filtering technique.

REFERENCES

1. Jain, R. Folk Computing. IEEE Multimedia. **2002**.
2. Pine, B.J., II. *Mass Customization*, Harvard Business School Press: Boston, MA, 1993.
3. Audioscrobbler, http://last.fm.
4. FreeNet, http://freenet.sourceforge.net.
5. Bassoli, A.; Cullinan, C.; Moore, J.; Agamanolis, S. TunA : a mobile music experience to foster local interactions (poster). Fifth International Conference on Ubiquitous Computing (UbiComp 2003), Seattle, October 12–15, 2003.
6. Östergren, M. Sound pryer field trials: learning about adding value to driving. Workshop Designing for ubicomp in the wild: Methods for exploring the design of mobile and ubiquitous services (MUM'2003), Norrköping, Sweden, Dec 10–12, 2003.
7. Gnawali, O.D. *A keyword set search system for peer-to-peer networks*; Master's thesis, Massachusetts Institute of Technology, June 2002.
8. Li, J.; Loo, B.; Hellerstein, J.; Kaashoek, F.; Karger, D.; Morris, R. On the feasibility of peer-to-peer web indexing and search. 2nd International Workshop on Peer-to-Peer Systems, 2003, Berkeley, CA, Feb 20–21, 2003.
9. Cooper, B.; Garcia-Molina, H. Studying search networks with SIL. 2nd International Workshop on Peer-to-Peer Systems, 2003, Berkeley, CA, Feb 20–21, 2003.
10. Bhattacharjee, B.; Chawathe, S.; Gopalakrishnan, V.; Keleher, P.; Silaghi, B. Efficient peer-to-peer searches using result-catching. 2nd International Workshop on Peer-to-Peer Systems, 2003, Berkeley, CA, Feb 20–21, 2003.
11. Burke, R. Hybrid recommender systems: survey and experiments. User Model. User-Adapt. Interact. **2002**, *12* (4), 331–370.
12. Goldberg, K.; Roeder, T.; Gupta, D.; Perkins, C. Eigentast: a constant time collaborative filtering algorithm. Information Retrieval J. **2001**, *4* (2), 133–151.
13. Sarwar, B.M.; Karypis, G.; Konstan, J.A.; Riedl, J. Application of dimensionality reduction in recommender system—a case study. Web Mining for E-Commerce Workshop, 2000 (WebKDD 2000), Boston, MA, Aug 20, 2000.
14. Sarwar, B.; Karypis, G.; Konstan, J.; Riedl, J. Item-based collaborative filtering recommendation algorithms. WWW10 Conference, Hong Kong, May 1–5, 2001; 285–295.
15. George, K. Evaluation of item-based top-N recommendation algorithms, Technical Report #00-046. Department of Computer Science, University of Minnesota, 1999.
16. Schein, A.I.; Popescul, A.; Ungar, L.H.; Pennock, D.M. Generative models for cold-start recommendations. 2001 SIGIR Workshop on Recommender Systems, 2001, New Orleans, LA, Sept 13, 2001.
17. Konstan, J.; Miller, Bo; Maltz, D.; Herlocker, J.; Gordon, L.; Riedl, J. GroupLens: applying collaborative filtering to Usenet news. Commun. ACM. **1997**, *40* (3), 77–87.
18. Breese, J.S.; Heckerman, D.; Kadie, C. Empirical analysis of predictive algorithms for collaborative filtering. 14th Conference on Uncertainty in Artificial Intelligence (UAI-98), Madison, WI, July 24–26, 1998; Cooper, G.F., Moral, S., Eds.; Morgan-Kaufmann: San Francisco, CA, 1998; 43–52.
19. Linden, R.; Smith, B.; York, J. Amazon.com recommendations. IEEE Internet Comput. **2003**, *7* (1), 76–80.
20. Hayes, C.; Cunningham, P. Smart Radio—Building Music Radio on the Fly. Expert Systems 2000, Cambridge, UK, 2000.
21. Shardanand, U.; Maes, P. Social information filtering: algorithms for automating 'word of mouth'. Conference on

Human Factors in Computing Systems (CHI95), Denver, CO, May 7–11, 1995; ACM Press: 1995; 210–217.

22. Herlocker, J.L.; Konstan, J.A. Content-independent task-focused recommendation. IEEE Internet Comput. **2001**, *5* (6), 40–47.

23. Melville, P.; Mooney, R.; Nagarajan, R. Content-boosted collaborative filtering for improved recommendations. AAAI 2002, Mar 25–27, 2002.

24. Lang, K. NewsWeeder: learning to filter Netnews. 12th International Conference on Machine Learning, Tahoe City, CA, 1995.

25. Pazzani, M.; Muramatsu, J.; Billsus, D. Syskill & Webert: identifying interesting web sites. 13th National Conference on Artificial Intelligence, Portland, OR, 1996; 54–61.

26. Mooney, R.J.; Roy, L. Content-based book recommending using learning for text categorization. 5th ACM Conference on Digital Libraries, San Antonio, TX, June 2–7, 2000; 195–204.

27. Chen, H.C.; Chen, A.L.P. A music recommendation system based on music data grouping and user interests. ACM International Conference on Information and Knowledge Management (CIKM 2001), Atlanta, Georgia, Nov 5–10, 2001.

28. Kuo, F.F.; Shan, M.K. A personalized music filtering system based on Melody style classification. IEEE International Conference on Data Mining, (ICDM 2002), Japan, Dec 9–12, 2002.

29. Rich, E. User modelling via stereotypes. Cogn. Sci. **1979**, *3* (4), 329–354.

30. Krulwich, B. Lifestyle finder: intelligent user profiling using large-scale demographic data. Artif. Intell. Mag. **1997**, *18* (2), 37–45.

31. Schmitt, S.; Bergmann, R. Applying case-base reasoning technology for product selection and customization in electronic commerce environments. 12th Bled Electronic Commerce Conference, Bled, Slovenia, June 7–9, 1999.

BIBLIOGRAPHY

1. Gnutella. http://www.gnutella.com.

Wireless Ad Hoc Networks: Performance Analysis

Anurag Kumar
Aditya Karnik
Indian Institute of Science, Bangalore, India

Abstract

This entry discusses the major issues in the performance of wireless ad hoc networks (WANETs) including stochastic capacity and capacity scaling, Bluetooth performance, and the performance of the transmission control protocol.

INTRODUCTION

For the purpose of performance analysis, one needs to abstract out the essential (performance governing) aspects of a system so as to build a mathematical model. Wireless ad hoc networks (WANETs) are communication networks in which even the network topology depends on the way the network is operated. Evidently, unlike the situation with wired and fixed topology networks, understanding and optimizing the performance of WANETs is a difficult undertaking, owing to the complex interaction between the various "layers" of the network. Nevertheless, in order to talk about the issues systematically, we can begin by developing a layered view of a WANET, akin to the one utilized so effectively by the OSI (open system interconnection) model for communication network protocols. This will also set down our view of a WANET, for our subsequent discussions.

Key features of WANETs include the following:

Wireless physical communication. At the physical level, a WANET comprises several nodes (e.g., handheld, laptop, or personal computers), each equipped with a digital radio unit (see Fig. 1 for a schematic depiction). The antennas are (typically) omnidirectional, and the radios use a portion of the spectrum that does not require complicated spectrum licensing, coordination, and planning. Thus, e.g., the popular IEEE 802.11b standard and the emerging Bluetooth system use the unlicensed 2.4 GHz ISM (industrial, scientific and medical) band. To combat multi-path and interference, spread-spectrum modulation is usually employed. Beginning with 1 or 2 Mbps, with modulation schemes such as OFDM and better coding techniques, data rates greater than 50 Mbps are now achievable. We note that the digital communication system employed, the transmit power used, and the radio propagation characteristics of the environment determine the "links" in the network, and hence the topology of a WANET (see section "WANET Topology, Spatial Reuse, and Connectivity"). Thus the most basic performance aspects of a WANET (which nodes are

neighbors, which nodes have a path between them, and at what speeds neighbor nodes can communicate) are determined by the physical communication layer.

Multiple access mechanism. Since all the radios share the same band and the antennas are omnidirectional, clearly nodes cannot arbitrarily communicate with each other. Hence a WANET employs a multiple access mechanism which permits the nodes to coordinate their transmissions in a decentralized manner. A simple random access mechanism can be used, in which the nodes just attempt, and then resend if there is a collision. Such extreme lack of coordination obviously leads to poor performance. Such a simple mechanism, however, is amenable to analysis, at least to yield performance bounds (such an analysis is presented in section "Stochastic Capacity"). The IEEE 802.11 standard employs the CSMA/CA (carrier sense multiple access/collision avoidance) multiple access algorithm. A large class of multiple access schemes is based on the "RTS-CTS" dialogue. A node wanting to transmit first requests the destination node [request to send (RTS)] and transmits only when it receives a grant [clear to send (CTS)] from it. This ensures that the destination node is in the receiving mode during transmission. Other nodes that hear this dialogue refrain from transmitting for some duration, thus reducing interference. The Bluetooth system, essentially designed for ad hoc wireless interconnection of devices (laptop, cordless phone, printer, computer) in an office environment, uses a polling-based multiple access mechanism. In each so-called *piconet* there is a master device that polls the other devices; all data transfer takes place through the piconet master. We will present an analysis of the Bluetooth system in section "Performance of Bluetooth."

Multihop packet communication. Nodes in an ad hoc network are usually battery-powered handheld devices. This constrains their transmission power. Even though a node may not be able to directly send a packet to its destination, it can forward that packet to a node to which

Encyclopedia of Wireless and Mobile Communications DOI: 10.1081/E-EWMC-120043891

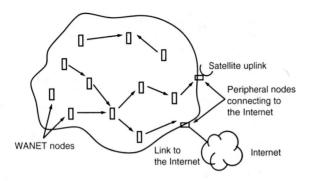

Fig. 1 A schematic of a wireless ad hoc network, showing packet flow between the nodes and attachments to the wide-area communication infrastructure via satellite or terrestrial links.

it can, hoping that the latter node has a path to the final destination; hence we have a multihop packet radio network. It is not necessarily good for a node to send as far as it possibly can, as this may increase the interference it causes to other transmissions. For effective communication (low probability of error) and also to conserve battery power, nodes need to choose appropriate transmission ranges that determine their neighboring nodes.

Adaptive routing protocols. A routing protocol, typically derived from one of the Internet routing protocols, permits nodes to forward packets in such a way that if there is a path between a pair of nodes, then packets can be sent between those nodes. Since such a routing protocol is adaptive, the nodes can actually move, and new routes are learned as the topology changes. If nodes fail (owing to damage or loss of power), the network will automatically learn new paths between nodes that can still communicate. Routing protocols vary in their handling of topology updates. Some protocols are proactive; they constantly keep learning new routes and updating routing tables. Other protocols are "on-demand"; they search for routes when such a request is made. We will not touch on this topic any further in this entry. A recent book[11] is devoted mainly to ad hoc network routing protocols. Several simulation studies comparing various ad hoc routing protocols are available (see Refs. 2 and 3).

Ad hoc network applications. We can broadly classify WANETs into wireless ad hoc internets and special purpose ad hoc networks. Nodes in an ad hoc Internet would like to run the same applications as any other node attached to the Internet. Hence the TCP/IP suite has to be extended across such a WANET. Ad hoc Internet nodes encapsulate their data into IP packets. The TCP, in the applications in the end-systems, takes care of lost packets and prevents network congestion. By adopting the TCP/IP networking protocols, applications on ad hoc network nodes can communicate seamlessly with applications in the Internet. The performance of TCP's end-to-end adaptive window based packet transmission mechanism has been a subject of much concern over point-to-point wireless links (we will summarize an analytical approach to such problems in section "Performance of TCP Controlled Transfers over a WANET"). The problems with TCP over multihop packet radio links are only recently beginning to be understood (for simulation based performance analysis, see Refs. 4 and 5, and for some scaling results, see Ref. 6). We will also touch upon the performance of TCP over a multihop radio path in section "Performance of TCP Controlled Transfers over a WANET."

An important emerging class of applications of WANETs arises from the availability of inexpensive miniature multifunction devices. Such a device would have embedded sensors (for light, chemicals, temperature, for example) and would have embedded computing, a digital radio, and a long idle-life battery. These devices could be independent units to be randomly strewn into an area or could be deliberately embedded in common systems such as watches, appliances, building walls, etc. Functions of such ad hoc sensor networks could be:

- Monitoring contamination levels after radioactive or chemical leaks;
- Forest fire detection, timber management, and wildlife tracking;
- Materials and people tracking and security management in factories, airports, and hospitals;
- Distributed instrumentation in large machines or vehicles (such as ships).

Ad hoc sensor networks would not need to use TCP. In fact, the challenge would be for them to achieve their function by only explicitly communicating with their neighbors. See Ref. 7 for a survey of this emerging area.

Ad hoc networks are hard to analyze, and most analytical models are intractable because of the various interdependencies that need to be accounted for. The earliest papers aimed at determining the network capacity for a given topology and a combination of channel access and capture schemes[8–10] (see Ref. 11 for a survey). These were essentially extensions of the analyses for single-hop networks; hence the effort was made to formulate Markovian models. The general models assumed exogenous traffic as independent Poisson processes, exponentially distributed packet lengths, and transmission scheduling process for packets from any node i to any node j as independent Poisson processes which included new and rescheduled packets. The network topology was specified as a graph where two nodes were connected by a link if they could hear each other's transmissions. Throughputs of ALOHA, CSMA, and CDMA schemes with various collision management and capture assumptions were analyzed. Results, however, could be obtained only for simple network topologies since the state description became formidable for large networks.

Since the devices will, in general, be randomly located, the graph of a WANET is actually a *random graph*. Hence, the "specification" of the topology in the above analyses turned out to be a major limitation of these approaches. Recent papers, therefore, have focused on obtaining bounds on the network capacity for random topologies.[12-15] These analyses have led to the so-called "scaling laws", i.e., scaling of the network capacity with the number of nodes in the network. These results will be discussed in sections "WANET Topology, Spatial Reuse, and Connectivity" and "The Capacity of a WANET".

The following is a summary of the results surveyed in this entry. In section "WANET Topology, Spatial Reuse, and Connectivity", we discuss the basic tradeoff between spatial reuse and WANET connectivity. We discuss how rapidly the transmission range can decrease when the number of network nodes increases in a given area so that the network stays connected. In section "The Capacity of a WANET", we first review how this leads to a scaling law that shows that the per node capacity of a WANET scales poorly with the number of nodes. Then, turning to stochastic capacity, we provide a packet flow model of WANET and show how the performance of the network depends on the physical layer and the way nodes organize themselves and operate. In section "Performance of Bluetooth", we analyze a specific WANET, namely Bluetooth, and seek its scaling law, i.e., how the performance of a cluster of piconets degrades as the number of piconets increases in a given area. We answer this question in terms of the stationary outage probability and the temporal correlation in the outage process. In section "Performance of TCP Controlled Transfers over a WANET," we first review the performance analysis of TCP over a single-hop wireless link. Then we discuss the performance of multihop TCP connections over WANETs. Using a simple example, we illustrate the complexity involved in this task. We conclude in section "Conclusion".

WANET TOPOLOGY, SPATIAL REUSE, AND CONNECTIVITY

One of the most basic questions about a network is whether any two nodes in the network are able to communicate, perhaps by routing their data through other nodes. Such a question is formally answered by representing the topology of the network by a graph on the set of nodes of the network. In the context of a WANET, where a radio spectrum is shared among several contending nodes, there is the additional, and conflicting, issue of spatial reuse. To promote spatial reuse, the transmission ranges of the nodes need to be kept small, while the opposite is required for ensuring that the network is connected. We will begin by examining this issue.

The WANET Graph

Let $N (= \{1, 2, \ldots, n\})$ denote the WANET nodes. In the WANET topology graph, there is a directed link (i,j) (from node i to node j) if node i can send data to node j with the desired level of reliability (which may, e.g., be specified as a bit error rate of 10^{-4}, or perhaps a packet error rate of 0.01). This definition of a link is not as straightforward as it may at first seem. If several nodes are simultaneously transmitting, and if node i is attempting to transmit to node j, then this transmission may not succeed owing to excessive radio interference at j. On the other hand, if i is the only transmitting node, then the signal power to receiver noise ratio at j may be sufficient to provide reliable communication. For the purpose of defining the WANET topology, we will define a link in this more "optimistic" sense. The point is that if two nodes are not connected in this graph, then they cannot be connected when arbitrary sets of nodes are allowed to transmit. Further, we will assume in our discussions that if link (i,j) exists then so does link (j,i), and hence we can take the WANET topology graph to be undirected. (This assumption basically requires that the propagation channel between the two nodes is reciprocal.) Let L denote these links, and let $G = (N,L)$ denote the WANET graph.

Thus, we say that the WANET is connected if the graph G is connected. (There is a path between every pair of nodes of the graph.)

Spatial Reuse and Connectivity

Given a particular transmitter and receiver design and a transmitter power, there is a range (say, r) over which two nodes can communicate. Now let us consider n nodes distributed uniformly in a given area, and the effect of increasing n. If r is kept fixed as we increase n, then transmissions from any node interfere with a large number of nodes, and the number of simultaneous transmissions possible in the network (or the spatial reuse) is just $O(1/r^2)$, which does not increase with increasing n. Evidently, in order to increase the spatial reuse, the transmitter powers must be reduced so that the range [say, $r(n)$] is a decreasing function of n. Fig. 2 shows a WANET in which

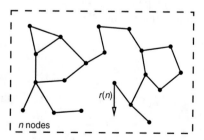

Fig. 2 The graph of a wireless ad hoc network. There are n nodes in a given area. The radio communication range is $r(n)$.

the nodes have transmission range $r(n)$; any two nodes that are within a distance $r(n)$ of each other are shown as being neighbors in the WANET graph. Observe that, if $r(n)$ is too small, then the WANET may not be connected. An important question that follows from this discussion is, "How fast can $r(n)$ decrease so that the network remains connected?"

Consider n nodes uniformly distributed over a square field of unit area, and a graph g on these nodes obtained by putting a link between any two nodes that are separated by a distance no more than $r(n)$. What we obtain is a probabilistic model, called a *random graph*. Let I_i denote the event that node i is isolated (i.e., has no links to any other node), and consider the probability that at least one node is isolated, i.e., $p(\cup_{i=1}^{n} I_i)$. We proceed in the spirit of the arguments in Ref. 16. Since the area within the range of a node, and also within the square field, is at least $1/4\,\pi\,r^2\,(n)$, we have for all i, $1 \leqslant i \leqslant n$,

$$P(I_i) \leqslant \left(1 - \frac{1}{4}\pi r^2(n)\right)^{(n-1)}$$

Using the union bound we then have

$$
\begin{aligned}
P\!\left(\cup_{i=1}^{n} I_i\right) &\leqslant \sum_{i=1}^{n} P(I_i) \\
&\leqslant n\left(1 - \frac{1}{4}\pi r^2(n)\right)^{(n-1)} \\
&= e^{\left(\ln n - (n-1)\ln\left(\frac{1}{4}\pi r^2(n)\right)\right)} \\
&= e^{\left(\ln n - (n-1)O\left(\frac{1}{4}\pi r^2(n)\right)\right)} \\
&= e^{\ln n\left(1 - \frac{(n-1)}{\ln n}O\left(\frac{1}{4}\pi r^2(n)\right)\right)}
\end{aligned}
$$

where we have used the fact that $r(n)$ decreases to zero as $n \to \infty$. It follows that if $r(n)$ shrinks more slowly than

$$\sqrt{\frac{\ln n}{n}}$$

then the probability of some node being isolated goes to zero as $n \to \infty$. In fact, using some results from random graph theory, it has been shown in Ref. 16 that in order for the WANET graph to stay *connected* with probability 1, $r(n)$ must shrink more slowly than

$$\sqrt{\frac{\ln n}{n}}$$

More precisely, it has been shown in Ref. 17 that if the range $r(n)$ scales as

$$O\!\left(\sqrt{\frac{\ln n + c(n)}{n}}\right)$$

then the graph stays connected, as $n \to \infty$, if and only if $c(n) \to \infty$. Note that $c(n)$ can go to ∞ arbitrarily slowly.

THE CAPACITY OF A WANET

Consider again n nodes constituting a WANET in a given area Fig. 3 is a depiction of such a network. Each node is running some applications (e.g., packet telephony, web browsing, or e-mail) that generate packets that need to be transported to other nodes in the network. We can view these application-generated packets as arrivals into the network. These arrivals are shown by the straight arrows pointing into the nodes in Fig. 3. Packets are transmitted to neighboring nodes, from which they may need to be relayed to other nodes; i.e., some packets are multihopped. Finally, each packet reaches its destination node, and this can be viewed as a departure, shown by the straight arrows pointing away from some of the nodes in Fig. 3.

With the above view of the traffic flow, it is natural to think of the WANET as a service system into which customers (i.e., packets) arrive, get serviced (i.e., transported over the wireless medium), and finally depart. Fig. 4 shows a queuing model for the entire WANET. The model shows one queue at each node, and this queue holds all the packets (new or transit) waiting to be transmitted out of the node. The digital radio communication and multiple access mechanism of the network can be viewed as a complex service mechanism that serves packets from these queues. Multihopped packets are fed back to other queues.

For this queuing model, the first question that can be asked is: "How fast can packets arrive into the nodes so that the nodal queues remain stable?" Stability could be defined in the usual sense of the joint queue length random process converging to a proper distribution.

A Scaling Law from Spatial Reuse

Let the rate of arrival of new packets at each node be $\lambda(n)$. The transmission range $r(n)$ scales with n, as discussed above. Let $h(n)$ be the average number of hops over which a packet is relayed. Hence, the total rate of packets that need to be transmitted by the WANET is $n\lambda(n)h(n)$.

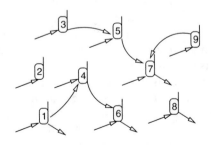

Fig. 3   Packet flow in a multihop wireless ad hoc network.

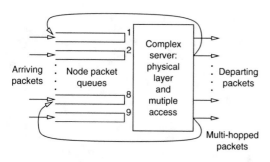

Fig. 4 A queuing model representing the entire wireless ad hoc network.

Assuming that the nodes are uniformly distributed, and all destinations are equally likely, it follows that $h(n) = O(r(n)^{-1})$.

Let us denote the service capacity of the network by $S(n)$; i.e., $S(n)$ is the total rate at which the network can serve packets from the queues. In general, $S(n)$ will depend on the amount of spatial reuse, on the physical layer and the multiple access, on the routing protocol, and on the prevailing propagation conditions. In any case, the rate of packets to be transported can be no more than $S(n)$; i.e.,

$$n\lambda(n)h(n) \leqslant S(n)$$

If we consider only spatial reuse, then $S(n) \leqslant O(r(n)^{-2})$. Hence if $r(n)$ scales down no faster than

$$\sqrt{\frac{\ln n}{n}}$$

(in order to maintain connectivity), we can put together the above expressions to yield

$$\lambda(n) \leqslant O\left(\frac{1}{\sqrt{n \ln n}}\right)$$

Thus the per node capacity of a WANET scales very poorly with n, and this can be seen to be primarily because of multihopping.

It follows that dense WANETs would serve the best for applications in which there is primarily local communication between nodes. The basic wireless ad hoc networking paradigm is extremely well suited to ad hoc sensor networks. When strewn over an area, these devices sense their immediate environment and can be viewed as constituting a *data space*. The sensor network can then be queried ("What is the maximum level of chemical contamination?"), and by using a distributed algorithm, which only requires the devices to communicate with their neighbors, the query can be answered and the results returned.

Stochastic Capacity

After discussing the above fairly general scaling results, we now turn to a packet flow model of a WANET. Packet flow models are of interest because various performance measures can be directly calculated from them. The question of particular interest is the following: If the network consists of n randomly distributed nodes in a given area A, then for given input traffic rates $\{\gamma_1, \gamma_2, \ldots \gamma n\}$, does there exist a channel access, routing, and packet scheduling scheme so that the network is stable? In this section, however, we will instead discuss the following simpler questions: For a given multiple access, routing, and packet scheduling scheme, what is the throughput capacity of the network, and what is the maximum input traffic rate for which the network is stable?

We consider the following network model. Nodes are randomly distributed in the plane according to a Poisson point process of intensity λ per m^2. In the simulation results presented, we consider a Poisson field of 5000 network nodes. Each radio node transmits on the same frequency f with power P using an omnidirectional antenna. Also, it cannot transmit and receive simultaneously. For the propagation model, we assume only path loss with exponent $\eta = 4$. We say that a transmission can be decoded if its signal to interference ratio (SIR) exceeds a given threshold β [Given a modulation and coding scheme, β actually specifies the maximum BER, assuming interference to be Gaussian (which is a fair assumption with a large number of interferers). In this way, details of the actual modulation and coding scheme can be abstracted.] Typical values of β are in the range of 10–20 dB. Channel access is random, i.e., nodes make decisions to transmit or receive independently, and time is slotted. In each slot, a node decides to transmit with probability α and decides to receive with probability $(1 - \alpha)$ independently; α is called the attempt probability. If the transmission is successful, the packet is removed from the queue (i.e., this is an assumption of instantaneous acknowledgments). A successfully received packet at a node joins the queue for next hop transmission with probability v, otherwise the packet departs the network. A node addresses (By actually inserting a physical or IP address in the packet header.) a transmission only to one of its neighbors; the neighbors are the nodes within a given distance R from a node. Thus, a transmission is successful when the node to which it is addressed is in receive mode, and the SIR of the transmission exceeds β at the addressed node. Note that R need not be the radio range of a node, in fact, we show later that a smaller R improves the performance. In each slot, the one-hop destination for the head-of-the-line (HOL) packet is chosen randomly from among the neighbors, i.e., given the number of neighbors K, a neighbor is chosen as the destination of the HOL packet with probability $1/K$ (for a typical node, the number of its neighbors is Poisson distributed with mean $\lambda \pi R^2$). Then the random selection of nodes to which to transmit is basically the assumption that the routing and traffic pattern are uniform and homogeneous. Thus, each packet will traverse a geometrically distributed number of hops before departing the network.

Interference analysis

With the channel attempt model described above, in each slot, the nodes that are in the transmit mode form an independent Poisson process (In general, the spatial point processes of transmitters and receivers depend upon the channel access scheme.) (see Ref. 18 for some early work in this framework). At a typical node, the average power received in a slot, from transmitters further than a distance d_0, is proportional to

$$\lambda \alpha \int\limits_{d_0}^{\infty} \left(\frac{r}{d_0}\right)^{-\eta} 2\pi r \, \mathrm{d}r$$

where d_0 denotes the near field cut-off distance. The integral converges for path loss exponents $\eta > 2$, and hence even with infinite number of transmitters the average power received is finite. We now assume that $\eta = 4$. It can be further shown that the characteristic function of the distribution of total received power at a node in a slot is given by

$$\Psi(\varpi) = \exp^{\lambda \pi d_0^2 j \varpi \int_0^1 t^{-1/2} \exp^{j\varpi t} \, \mathrm{d}t}$$

It can be shown from this that the mean interference is $2\lambda \alpha \pi d_0^2$, and the variance is $4/3\lambda \alpha \pi d_0^2$

We approximate the interference as Gaussian

$$N\left(2\lambda \alpha \pi d_0^2, \frac{4}{3}\lambda \alpha \pi d_0^2\right)$$

where $N(\mu,\sigma^2)$ denotes a Gaussian distribution with mean μ and variance σ^2. Note that the mean is linear in both λ and α; this is shown in Fig. 5, where simulation results are plotted along with the analytical values. The value of d_0 is taken to be 0.1 m, and $\lambda = 1$ nodes per square meter. For the simulation, 5000 nodes were generated, and the simulation was run for 20,000 slots. An average was taken over several realizations of the node positions and for several target nodes in the middle of the network. We observe that the analysis and simulation results are qualitatively the same, and the analysis is also a good approximation.

Consider a transmitter–receiver pair at a distance r in this Poisson field of points. Their communication is

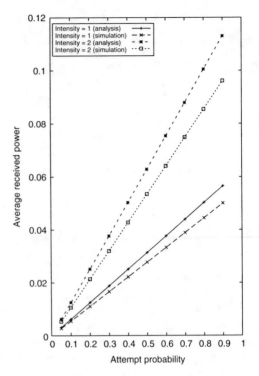

Fig. 5 Variation of mean interference at a node with attempt probability, for $\lambda = 1$ and 2 per square meter.

successful with probability

$$p_s(r,\alpha) := p\left(\frac{p_s}{N+1} \geq \beta\right)$$

where $P_s = (r/d_0)^{-\eta}$ denotes the intended signal power (normalized to the power at the reference distance d_0), I the interference [distributed as $N(2\lambda \alpha \pi d_0^2, 4/3\lambda \alpha \pi d_0^2)$], and N the thermal noise at the receiver.

Fig. 6 shows the variation of $p_s(r,\alpha)$ with r for various values of α, for $\lambda = 1$ per m². For these calculations, we

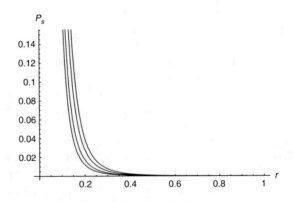

Fig. 6 Variation of probability of successful transmission with distance for $\alpha = 0.1, 0.2, 0.5, 1$, and $\lambda = 1$ per m².

take the transmit power to be 1 mW, and the thermal noise $N = 5 \times 10^{-7}$ mW. The curves are decreasing with α; here $\alpha = 1$ is just indicative of the performance if all the nodes in the network were transmitting simultaneously and thus interfering with reception at our reference node. Note the sharp decay of $p_s(r, \alpha)$ with r. The probability of success is very poor beyond a few centimeters from a node; clearly random access is not a good way to operate a WANET.

Saturation throughput

The saturation throughput of a network is defined as the rate of transmission in the network when a transmitted packet from the HOL of a queue at a node is immediately replaced by another packet. Saturation throughput is akin to a *service rate* provided by the complex channel *server* (see Fig. 4). In some models, the saturation throughput is shown to yield a sufficient condition for network stability. Let γ_i denote the saturation throughput of the i-th node (The network is assumed to be homogenous; therefore, the subscript i can be omitted. If the network consists of N nodes, then the network throughput is $\sum_{j=1}^{N} \gamma_j$). Then,

$$\gamma_{i:} = \lim_{n \to \infty} \frac{1}{n} \sum_{k=1}^{n} Z_k = p_t$$

where $Z_k = 1$ if the transmission is successful in slot k, and p_t denotes the probability of successful transmission. Fig. 7 shows the probability of successful transmission as α

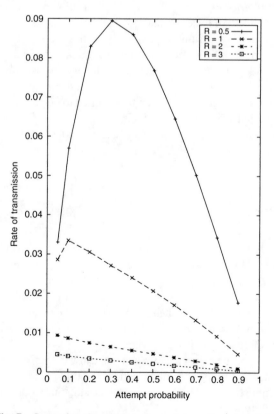

Fig. 7 Saturation throughput of a node vs. attempt probability; simulation results. $R = 0.5, 1, 2, 3$ m.

varies for various values of R. Recall that we use a Poisson field of 5000 nodes. Here $\lambda = 1$ per m^2. We have again used $\eta = 4$. R specifies the "neighborhood" of a node; the mean number of neighbors is $\lambda \pi R^2$. Observe that, as R increases, the probability decreases. This is to be expected. A node attempts transmissions uniformly randomly to its neighbors. With a smaller R, a node's neighbors are closer, and the probability of successful transmission to these nodes is higher. We also see from Fig. 7 that, for a given R, there is an *optimum* attempt probability. Thus, for a range of 0.5 m, an attempt probability $\alpha = 0.3$ provides the best throughput. This analysis shows that the performance of the network can be optimized based on how nodes organize themselves.

The maximum rate of transmission is obtained when, instead of addressing its transmission to a particular neighbor (Recall that the success of a transmission depends on the addressed node being in receive mode.), a node broadcasts it. Its transmission is, thus, successful *if at least one node in its radio range* decodes it. Let C denote the maximum rate of transmission of a node in saturation (normalized to average number of nodes in its radio range). Then, it can be shown that

$$C < \int_{0}^{\frac{1}{\beta}N} \frac{\alpha(1 - \alpha)}{\pi \sqrt{2.6 \lambda \alpha d_0^e}} \, e^{-\left(y - 2\lambda \alpha \pi d_0^2\right)} \mathrm{d}y \qquad (1)$$

This bound, in a sense, is a measure of network capacity of our model. Fig. 8 shows its variation with α for $\lambda = 1$.

It appears from Figs. 7 and 8 that when $R = 0.5$, the throughput exceeds the capacity bound. The answer to this apparent paradox lies in the fact that when R is small, the network is not connected and consists of many isolated nodes. Since isolated nodes do not transmit at all (since these nodes do not have neighbors to address their transmissions to), the overall interference to transmitting nodes is reduced, thereby greatly increasing the probability of success. The capacity bound, on the other hand, is calculated assuming every node is able to transmit. This leads to important questions regarding the performance of the network and the organization of nodes, which are beyond the scope of this entry.

The above analysis was all for WANETs in which nodes use random access. There appears to be little literature available on the analytical performance evaluation of RTS-CTS type protocols in dense WANETs with random node placement.

PERFORMANCE OF BLUETOOTH

The Model and Its Analysis

We consider a single circular room (radius R) office environment [line-of-sight (LOS) propagation], with a number

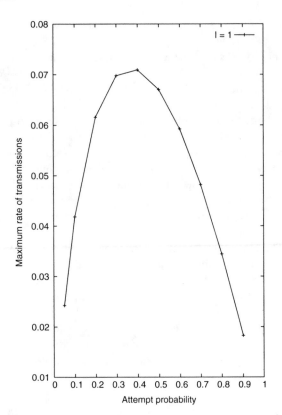

Fig. 8 Capacity bound vs. attempt probability (Eq. 1). $\lambda = 1 \text{ m}^{-2}$.

of Bluetooth devices distributed randomly in it to form M piconets. Because of the ad hoc nature of the formation of piconets, the devices in a piconet are spatially randomly distributed in the room leading to overlap among the piconets. Note that in our results M will be a parameter, i.e., we are interested in analyzing the performance of the Bluetooth system as the number of piconets increases in a given area.

The propagation of radio waves inside a building is a complicated process and depends significantly on the indoor environment (e.g., office, factory), and topography (LOS, obstructed). The statistics of the indoor channel vary with time due to motion of people and equipment. The indoor channel is characterized by high path losses (path loss exponent η is 1.5–1.8 for LOS; see Ref. 19) and large variations in losses. Motion causes signal fading which compares well with the Rician distribution with K equal to 6.8 dB for the office environment. If the degree of motion is small (as in an office environment), the fading is extremely slow and the channel is said to be quasi-static. For practical data rates, fading is quasi–wide-sense stationary. In our analysis, since we consider LOS propagation in an open office environment, we take $\eta = 2$ and a Rician signal distribution with $K = 6$ dB.

Since Bluetooth operates in the ISM 2.4 GHz band, there are many sources of interference other than Bluetooth devices, e.g., microwave ovens and IEEE 802.11 Wavelans. We, however, focus on the interference generated within the Bluetooth system only. A piconet experiences interference when a frequency hit occurs, i.e., the transmission frequency of a packet in a piconet matches with that being used in one or more other piconets for some overlapping duration. We assume an ideal frequency hopping pattern (with N_f frequency hops), and model it as a discrete time Markov chain; a state denotes a frequency. Hopping patterns of piconets are independent and identically distributed. T denotes the hop time. Note that the piconets are not synchronized in time. We assume that the time offset of the i-th piconet, denoted by t_i, with respect to the reference piconet are uniformly distributed in $[0,T]$.

Since the devices in the piconets are spatially distributed, the *interfering piconet* is not a fixed location or device but a random device in that piconet transmitting at a given time. We assume a physically stationary environment (little or infrequent movement) so that the channel is quasi-static and quasi–wide-sense stationary. Hence, for the *duration of a frequency hit* on the single frequency, constant interference power can be assumed to be received from the individual interferers.

We consider a reference piconet and denote by $\{\Gamma(t), t \geqslant 0\}$ the SIR process:

$$\Gamma(t) = \frac{p_{s(t)}}{\sum_{j=1}^{N(t)} p_I^j(t) + \frac{N_0 W}{2}}$$

where, at time t, $P_s(t)$ is the desired signal power, $P^j_1(t)$ is the interference power due to the j-th interfering piconet, and $N(t)$ is the number of interferers. Recall that in Bluetooth a device can transmit in alternate slots. In addition, a slave can transmit only when polled. Hence, the receiver is not a fixed device, and the SIR process we characterize is of the reference piconet as a whole.

Outage occurs when the bit SIR falls below the resistance ratio (denoted by β) specified by the standard (11–14 dB). We denote by Γ_b the bit SIR and seek the stationary outage probability P_{out}.

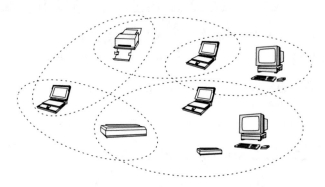

Fig. 9 A system of overlapping Bluetooth piconets.

$$P_{out}: P = \left[\frac{p_s}{\sum_{j=1}^{N_b} P_I^j + \frac{N_0 W}{2}} < \beta \right]$$

P_s is the desired signal power and P^j_I denotes the power received from the j-th interferer. N_b is a random variable denoting the number of interferers in a bit duration. We assume that the devices are distributed uniformly in the room with respect to the center of the room (A Bluetooth receiver has -70 dBm sensitivity level to meet BER <0.001. With 1 Mbps bit rate, the noise power is approximately 5×10^{-7} mW.).

In order to characterize the temporal correlation in the outage process, we consider a bit b and obtain the one step joint distribution of bit b and $b + 1$ being in outage. This outage process is then approximated by a Markov process. This gives us a two-state Markov model as follows. We say that the channel is in *good* state, denoted by "g", if $\Gamma_b \geq \beta$ or BER on the channel is below the specified value, and that it is in *bad* state, denoted by "b", when $\Gamma_b < \beta$, i.e., when the outage occurs. Let P_{ij} denote the transition probability of going from state i to state j. Then, the outage duration (in number of bits) is geometrically distributed with mean $1/(1 - P_{bb})$. Also the outage probability calculated from the Markov model, denoted by P^M_{out}, is given by $P_{gb}/(P_{gb} + P_{bg})$.

Details of the modeling assumptions and the analysis can be found in Ref. 20.

Results and Discussion

We take $R = 5$ m. Since the calculation of exact outage probability is complicated, even numerically, we resort to finding bounds. Table 1 shows lower bounds on P_{out} with signal fading.

Using the two-state Markov chain approximation and and lower bounds on the outage probability, we calculate approximate values of P_{gb} and P_{bg} for $\beta = 14$ dB and 400 bits packet size. We also calculate the outage probability from this Markov model (P^M_{out}). Results are shown in Table 2. Table 2 also shows in bits the mean outage duration (EB) and the mean duration in good state (EG). Note that for Bluetooth, $N_f = 79$.

It is very complicated to calculate even the approximate values of P_{out} as the number of piconets increases. We,

therefore, have results only for $M \leq 3$ piconets. However, the insight gained with these results allows us to predict the performance for larger values of M. The results indicate that when a frequency hit occurs it is difficult to maintain the resistance ratio of 14 dB as specified by the standard. Hence, co-channel interference is avoided mainly through the use of a large number of frequencies. For larger values of M, we expect that P_{out} will increase almost linearly, in the form of $(M - 1)/N_f$. Though the Markov model is only an approximation (as seen from Table 2), we expect that when the number of piconets is small, outages are infrequent (once in 20,000 bits). This is because piconets are not time-synchronized, and frequency hits occur with probability $1/N_f$. However, outages persist for approximately 200 bits, which is a considerable duration considering $1/3$ and $2/3$ FEC used in Bluetooth. Also in certain packet types FEC is not mandatory. As M increases, the duration of the good state will reduce and there will be longer outage durations.

PERFORMANCE OF TCP CONTROLLED TRANSFERS OVER A WANET

One of the applications of WANETs is to create ad hoc internets. Each WANET node in such a network is an endpoint, as well as a packet forwarding device (even a fully fledged router, participating in a distributed routing protocol). In an Internet, most store-and-forward applications (e-mail, file transfers, web browsing) operate over the end-to-end TCP. The TCP protocol serves three functions in the Internet:

1. TCP converts the non-sequential and unreliable packet transport service provided at the IP layer into a sequential and reliable service.
2. TCP implements end-to-end flow control, thereby preventing a fast sender (e.g., a high performance computer) from swamping out a slow receiver (e.g., a slow mechanical device such as a printer).
3. The applications that use TCP's services are *elastic*, in the sense that they can work satisfactorily even if the network offers them a time-varying packet transfer rate. TCP determines the way the network shares bandwidth among the competing flows.

Since TCP sessions comprise over 90% of the Internet traffic, it can be said that to a large extent TCP governs the dynamics and the performance of the Internet.

Since elastic applications will continue to be the most popular ones to be deployed on ad hoc Internets, it is important to understand the impact of TCP on the performance of these networks.

We divide our discussion into two parts. First we consider the situation in which the TCP session is over a

Table 1 Lower bounds on outage probability for $M = 2,3$ piconets.

M − 1	Lower bound on P_{out}	
	$\beta = 14$ dB	$\beta = 11$ dB
1	0.0114	0.0103
2	0.0227	0.0204

Table 2 Parameters of the two-state Markov model.

$M-1$	P_{gb}	P_{bg}	P^{M}_{out}	EB	EG
1	0.00005	0.005	0.010	200	20,000
2	0.000055	0.0035	0.016	285	18,018

single-hop between neighboring nodes. This will be by far the most common situation within offices and homes, as one can expect that the WANET node will be near a wired access point. On the other hand, ad hoc internets can be expected to be deployed in situations where there is no wired infrastructure (e.g., in disaster zones). The performance of multihop TCP sessions over such a network will, therefore, also be of interest.

Single-Hop Performance

The motivating scenario is shown in Fig. 10. A WANET node is downloading a large file from a server on a high-speed LAN. The propagation delay over this wireless link is much smaller than the packet transmission times. We identify random epochs $(T_1, T_2, \ldots, T_k, \ldots)$ as follows. Let $T_0 = 0$, and the connection starts off with the congestion window $W = 1$. As packets are transmitted, the window grows (slow-start, if necessary, and then congestion avoidance). Eventually a loss occurs on the lossy link at the epoch l_1; we denote the window achieved at this epoch by X_1. Owing to the local area and high-speed LAN assumptions, this window includes all increments due to any successful transmissions before the lost packet. Some random time later (depending on the version of the protocol and the recovery method used), at the epoch T_1, the normal congestion window algorithm resumes with a slow-start threshold of $[X_1/2]$ In this way we identify the sequence of random vectors $[(X_0,T_0), (X_1,T_1), \ldots, (X_k,T_k), \ldots]$; see Fig. 11. With Bernoulli packet loss,[21] or Markov modulated packet loss[22] and appropriate assumptions (the window increase during congestion avoidance is taken as a probabilistic increase), we can show that:

1. $\{X_k\}$ is a Markov chain.
2. $\{(X_k,T_k)\}$ is a Markov renewal process.

Denoting the number of successful packets in the k-th cycle by a "reward" V_k, we find that the V_k depends only on $X_{(k-1)}$, and we have a Markov renewal reward process. Letting U denote the random variable for the cycle length, and taking expectations under the stationary distribution $\pi(\cdot)$ of $\{X_k\}$, the throughput in packets per second is given by

$$\gamma = \frac{E_\pi V}{E_\pi U}$$

The various versions of TCP are distinguished by the structure of the Markov chain $\{X_k\}$, and the cycle times $\{T_k\}$; e.g., for the original Van Jacobson algorithm (TCP "Old" Tahoe) $T_k = l_k +$ timeout.

In spite of many approximations, the analysis is quite accurate as demonstrated in Fig. 12 (taken from Ref. 21), where analytically obtained results are compared with those obtained from actual TCP code running over an emulated lossy link.

Performance of Ad Hoc Internets with Multihop TCP Connections

A detailed model, such as the one above, for multihop TCP connections in a WANET is not yet available. In fact, several basic issues that have been well studied for TCP over wired networks are not yet clear in the context of TCP over multihop wireless links. For example, if there is a single TCP connection over an h hop wired connection, with each link's propagation delay being very small compared to the packet transmission time, then the TCP window should optimally be h [ignoring the acknowledgment (ack) transmission times]. In a multihop wireless connection, however, the links are half-duplex, and furthermore arbitrary links cannot be simultaneously used owing to interference.

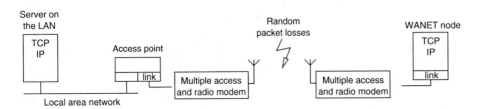

Fig. 10 A WANET node transferring data over a TCP connection from a server on a LAN.

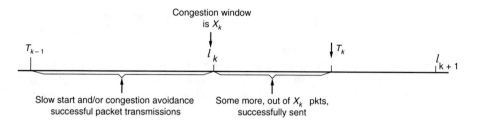

Fig. 11 Analysis of TCP performance over a lossy link. The epochs $\{T_k\}$ and the Markov chain $\{X_k\}$.

In Figs. 13 and 14, we show the evolution of end-to-end window controlled packet transmission, over a three-hop connection, with the nodes labeled a,b,c,d. This would be the case if TCP with a fixed window is used. It is assumed that each transmission takes one slot, whether it is a TCP packet or a TCP ack, and there is perfect centralized scheduling of transmissions (i.e., which nodes should transmit and to whom). We further assume that nodes more than two hops away (e.g., a and c, or b and d) are out of range; this implies that if a transmits to b, and d to c, the interference at b, caused by the transmission from d to c, is acceptable. The numbers to the left of Fig. 14 are the slot numbers. The open triangles are packets, and the shaded ones are acknowledgments. A triangle with an arrow indicates that the packet or ack is being transmitted. A triangle without an arrow means that the packet or ack is queued. The triangles "point" in the direction in which the packet is traveling.

In Fig. 13, the window is 1, and it takes six slots to transmit one packet and get back its ack, thus yielding a connection throughput of 1/6 packets per slot. Can we use a window of 2, and thus improve the throughput? We note

that not all hops can be active at the same time. For example, node a cannot send to node b while node c sends to node d, as the transmission from node c will interfere at node b with the transmission from node a. Fig. 14 shows a transmission schedule with a window of 2. The schedule is assumed to have "started" some time back and will continue even after slot 4. At the end of schedule, the ack reaching the TCP transmitter and the packet reaching the TCP receiver would generate a new packet and an ack, thus bringing the schedule back to the state that we had in slot 1. Notice that two "half" packets were transmitted in four slots, thus yielding a throughput of $1/4$.

The above simple example illustrates the complexity of the question "What is the optimal TCP window, and what is the corresponding throughput?" The answer will depend on the topology of the path, the way the links interfere, and the multiple access protocol (i.e., the link scheduling mechanism).

We notice from the above discussion that even if a link can be scheduled, and if TCP is being used, this link may not be used; see the example in Fig. 13. Consider a WANET confined to a fixed (say, unit) area. There are n nodes. Suppose that TCP is being used, and each node, when given a turn to transmit, is only able to use a fraction of turns $\Phi(n)$. It follows, by continuing the arguments in section "A Scaling Law from Spatial Reuse," that the per node throughput will scale as

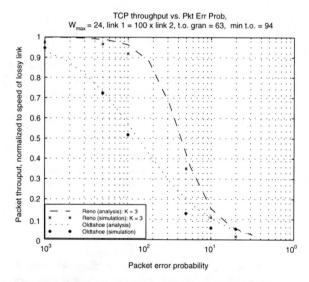

Fig. 12 Throughput of TCP Old Tahoe and TCP Reno vs. packet loss probability; comparison of simulation and analysis. (Maximum congestion window is 24 packets; the LAN is 100 times faster than the lossy link; K is the duplicate ack threshold for fast retransmit.)

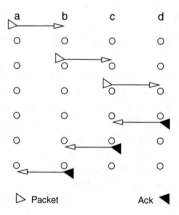

Fig. 13 Window controlled transmission over a three-hop wireless route; window $= 1$.

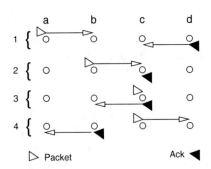

Fig. 14 Window controlled transmission over a three-hop wireless route; window = 2.

$$\lambda(n) \leq O\left(\frac{1}{\sqrt{n\ln n}}\right)\Phi(n)$$

Some directions to determining $\Phi(n)$ have been provided in Ref. 6. In Ref. 14, the authors have reported empirical results, where it has been found that the per node throughput in an ad hoc Internet (of UDP controlled sessions, i.e., no error recovery) scales as

$$O\left(\frac{1}{n^{1.68}}\right)$$

Since the transfers do not take place over TCP, this scaling probably captures the inefficiencies of the multiaccess protocol. One would expect that the performance would be even worse with TCP.

Recent research (see Refs. 4 and 5) has shown that TCP over multihop wireless connections has serious performance limitations. A single TCP connection provides good throughput only if the window is properly tuned; several TCP connections sharing a WANET tend to lock each other out (almost completely) over long periods of time, and these lockouts occur at random times. We may wish to conclude that dense WANETs should not be used for building traditional Internet-like packet networks.

CONCLUSION

In the performance evaluation of ad hoc networks, we need to work with models that incorporate the properties of all the layers. Models that are general and address these interdependencies are hard to analyze. We have surveyed some of the recent progress in the area. Much recent work has emphasized network scaling laws, which show how the network performance changes with increasing node density. We have also provided a model for stochastic capacity and a detailed analysis of a Bluetooth scatternet, taking into account propagation phenomena. The stochastic model for a general WANET assumed network synchronization along slots; although this is impractical, the

model provides useful insights. This analysis shows that in order for the network to operate efficiently, optimization of performance needs to be done simultaneously at various levels. The performance of the network depends on the physical layer and the way nodes organize themselves and operate. Similar insights were obtained from our discussion of TCP over multihop wireless routes. The most challenging task is to build upon such results and insights and develop distributed algorithms for efficient resource management of WANETs.

REFERENCES

1. Perkins, C.E., Ed. *Ad Hoc Networking*; Addison Wesley: Reading, MA, 2001.

2. Broch, J.; Maltz, D.A.; Johnson, D.B.; Hu, Y.-C.; Jetcheva, J. A performance comparison of multi-hop ad hoc network routing protocols. 4th Annual ACM/IEEE International Conference on Mobile Computing and Networking, Dallas, TX, Oct 25–30, 1998.

3. Perkins, C.E.; Royer, E.M.; Das, S.R.; Marina, M.K. Performance comparison of two on-demand routing protocols for ad hoc networks. IEEE Commun. Mag. **2001**, *8* (1), 16–28.

4. Gerla, M.; Bagrodia, R.; Zhang, L.; Tang, K.; Wang, L. TCP over wireless multi-hop protocols: simulations and experiments. IEEE International Conference on Communications (ICC'99), Vancouver, Canada, June 6–10, 1999.

5. Xu, S.; Saadawi, T. Does the IEEE 802.11 MAC protocol work well in multihop wireless ad hoc networks?. IEEE Commun. Mag. **2001**, *39* (6), 130–137.

6. Bansal, S.; Gupta, R.; Shorey, R.; Ali, I.; Razdan, A.; Misra, A. Energy efficiency and throughput for TCP traffic in multihop wireless networks. Conference of the IEEE Computer and Communications Societies (IEEE INFOCOM'02), New York, June 23–27, 2002.

7. Akyildiz, I.F.; Su, W.; Sankarasubramaniam, Y.; Cayirci, E. Wireless sensor networks: a survey. Comput. Netw. **2002**, *38*, 393–422.

8. Chen, M.-S.; Boorstyn, R. Throughput analysis of code division multiple access (CDMA) multihop packet radio networks in the presence of noise. Conference of the IEEE Computer and Communications Societies (IEEE Infocom'85), Washinton, DC, 1985.

9. Boorstyn, R.R.; Kershenbaum, A. Throughput analysis of multihop packet radio. IEEE International Conference on Communications (IEEE ICC'80), Seattle, WA, June 8–12, 1980.

10. deSouza, O.; Sen, P.; Boorstyn, R.R. Performance analysis of spread spectrum packet radio networks. IEEE Military Communications Conference (IEEE MILCOM'88), San Diego, CA, Oct 23–26, 1988.

11. Tobagi, F.A. Modeling and performance analysis of multihop packet radio networks. IEEE. **1987**, *75* (1), 135–155.

12. Gupta, P.; Kumar, P.R. The capacity of wireless networks. IEEE Trans. Inf. Theory. **2000**, *46* (2), 388–404.

13. Grossglauser, M.; Tse, D. Mobility increases the capacity of ad-hoc wireless networks. Conference of the IEEE Computer and Communications Societies (IEEE INFOCOM'00), Tel Aviv, Israel, Mar, 26–30, 2000.

14. Gupta, P.; Gray, R.; Kumar, P.R. An Experimental Scaling Law for Ad Hoc Networks. Technical report, Technical Report, ECE Department, University of Illinois at Urbana-Champaign. May 2001.

15. Li, J.; Blake, C.; De Couto, D.S.J.; Lee, H.I.; Morris, R. Capacity of ad hoc wireless networks. 7th Annual International Conference on Mobile Computing and Networking (ACM MobiCom'01), Rome, Italy, July 16–21, 2001.

16. Panchapakesan, P.; Manjunath, D. On the transmission range in dense ad hoc radio networks. Conference on Signal Processing and Communications SPCOM, Bangalore, India, July 15–18, 2001.

17. Gupta, P.; Kumar, P.R. Critical power for asymptotic connectivity in wireless networks. In *Stochastic Analysis, Control, Optimisation and Applications: A Volume in Honour of W.H. Fleming*; McEneany, W.M., Yin, G., Zhang, Q., Eds.; Birkhauser: Boston, MA, 1998; 547–566.

18. Sousa, E.S.; Silvester, J.A. Optimum transmission ranges in a direct-sequence spread-spectrum multihop packet radio network. IEEE J. Sel. Areas Commun. **1990**, *8*, 762–771.

19. Hashemi, H. The indoor radio propagation channel. IEEE. **1993**, *81* (7), 943–968.

20. Karnik, A.; Kumar, A. Performance of the Bluetooth physical layer. IEEE International Conference on Wireless Personal Communications (ICPWC'00), Hyderabad, India, Dec 17–20, 2000.

21. Kumar, A. Comparative performance analysis of versions of TCP in a local area network with a lossy link. IEEE/ACM Trans. Netw. **1998**, *6*, 485–498.

22. Kumar, A.; Holtzman, J.M. Performance analysis of versions of TCP in a local network with a mobile radio link. SADHANA: Indian Acad. Sci. Proc. Eng. Sci. **1998**, *23*, 113–129 (also a WINLAB Technical Report, Rutgers University, New Brunswick, NJ, 1996).

**Wireless
Ad Hoc—Asynch**

Wireless Ad Hoc Networks: Power Assignment

Gruia Calinescu
Ophir Frieder
Peng-Jun Wan
Department of Computer Science, Illinois Institute of Technology, Chicago, Illinois, U.S.A.

Abstract

This entry presents approximation algorithms for minimizing the total power consumption in wireless ad hoc networks.

INTRODUCTION

One of the major concerns in ad hoc wireless networks is reducing node power consumption. In fact, nodes are usually powered by batteries of limited capacity. Once the nodes are deployed, it is very difficult or even impossible to recharge or replace their batteries in many application scenarios. Hence, reducing power consumption is often the only way to extend network lifetime. For the purpose of energy conservation, each node can (possibly dynamically) adjust its transmitting power, based on the distance to the receiving node and the background noise.

In the most general model, a weighted directed graph $H = (V,E)$ with power requirements $c: E \to R^+$ is given by the positioning of the n wireless nodes, where $c(u,v)$ represents the power requirement for the node u to establish a unidirectional link to node v. But, reflecting the broadcast nature of ad hoc wireless networks, once a node u transmits with power $p(u)$, all nodes v with $c(u,v) \leqslant p(u)$ receive the signal. A function $p: V \to R^+$ is called a power assignment, and it induces a directed graph, always denoted by $G = (V,F)$, with links uv whenever $p(u) \geqslant c(u,v)$. The *symmetric restriction* of a directed graph $G = (V,F)$ is the undirected graph, always denoted by $\bar{G}$, with vertex set V and having an edge uv if and only if G has both uv and vu. An example is depicted in Fig. 1.

Typically, we require that the graph induced by the power assignment respects certain connectivity constraints; assigning the power level to ensure that G or $\bar{G}$ have certain graph properties is called topology control. Many types of connectivity constraints have been studied in the literature and we will later address several. Here we mention just two:

1. Strong connectivity: G must be strongly connected, thus guaranteeing that a packet can be relayed from any source to any destination with the current power assignment.
2. Symmetric connectivity: $\bar{G}$ must be connected. This connectivity requirement is more stringent than strong connectivity, and is motivated by the advantage of having one-link acknowledgment packets.

Given a connectivity constraint, the objective of Power Assignment is minimizing the total power, given by $\sum_{u \in V} p(u)$. In addition to reducing energy consumption, having reduced transmission power creates less interference.

The following special cases of the power requirement have been considered in the literature, presented starting with the most general.

1. Arbitrary symmetric: an undirected graph $H = (V,E)$ with power requirements $c: E \to R^+$.
2. Euclidean: the nodes of the graph are embedded in the two-dimensional plane and $c(uv) = d(u,v)^\ell$, where d is the Euclidean distance.
3. Line: same power requirements as Euclidean, but the nodes lie on a line.

The Euclidean case is motivated by the fact that in the most common power-attenuation model, the signal power falls as $d^{-\ell}$, where d is the Euclidean distance from the transmitter antenna and ℓ is the pass-loss exponent of the wireless environment, a real constant typically between 2 and 5. The line case is a special case of the Euclidean case motivated by ad hoc networks with the nodes following a highway or other "linear" pattern.

The arbitrary symmetric cases handle the cases when the nodes are in three-dimensional space or obstacles completely block the communication between certain pairs of nodes, or if the signal attenuation is not uniform,

Encyclopedia of Wireless and Mobile Communications DOI: 10.1081/E-EWMC-120043926

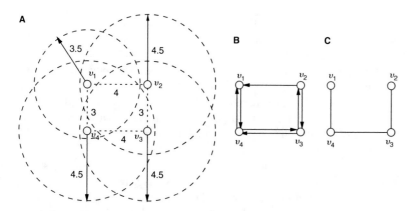

Fig. 1 The network topology: (A) the nodes and their transmission power; (B) The directed graph G induced by the power assignment, and (C) The undirected graph $\bar{G}$ induced by the power assignment.

Wireless
Ad Hoc—Asynch

or if there is a maximum transmission power, or if there is a discrete number of possible transmission power levels.

In addition, wireless nodes can have nonuniform power thresholds for signal detection, and we call *sensitivity* $s(v)$ the threshold of node v. Also, wireless nodes can have nonuniform transmission efficiency, and we use $e(v)$ to denote the efficiency of node v. Efficiency can also be used to simulate the fact that certain nodes have at some moment higher battery reserves.[1] With these definitions, node u reaches node v with power level $p(u)$ if and only if

$$p(u) \geq \frac{c(u,v)}{e(u)s(v)}$$

Adding sensitivity or efficiency to any symmetric power requirement case creates asymmetric power requirements, while asymmetric power requirements can be adjusted to handle sensitivity and efficiency by simply redefining $c_{new}(u,v) = c(u,v)/e(u)s(v)$.

In this survey, we concentrate on centralized algorithms for static power assignment in static networks. We do point out when certain algorithms are suitable for distributed implementation or for mobility of the nodes, but space limitations do not allow a more general treatment. Also, we do not consider adjusting the transmission power for each packet transmitted, an approach suggested in the literature for further reducing power consumption.

Minimizing total power under most connectivity constraints generates NP-hard problems. Thus, efficient algorithms computing the exact optimum are unlikely to exist. Moreover, to our knowledge, Althaus et al.[2] is the only work that has computed optimum solutions, working with what we believe is the easiest of these NP-hard problems: symmetric connectivity in the Euclidean case. Customized state-of-the-art integer linear programming software solves randomly generated instances with up to 40 nodes in (at most) 1 h. Our experience is that guaranteeing optimum solutions for 100 nodes or more is very unlikely.

This motivates the design of efficient heuristics and approximation algorithms. An approximation algorithm is a polynomial-time algorithm whose output is guaranteed to be (at most) α the optimum value, for some value α (which could depend on n, the number of vertices, but does not depend otherwise on the input). This value α is called the *approximation ratio*; and the smaller it is, the better the approximation algorithm. We call heuristic a proposed algorithm without a proven approximation ratio; such algorithms could have good practical performance.

At the end of the survey, we present a table with the best-known approximation ratios for the variety of existing problems. Due to the tight limitation on the number of references, only the best-known ratios will be quoted.

Before we proceed, we mention that in order to simplify the mathematical proofs, many articles use an alternative description of the optimization problem, described below: given (directed) spanning subgraph Q, define $p_Q(v) = \max_{uv \in Q} c(uv)$ and $p(Q) = \sum_{v \in V} p_Q(v)$; we call $p(Q)$ the *power* of Q. Since assigning $p(v) \geq p_Q(v)$ is necessary to produce the (directed) spanning subgraph Q, and $p(v) > p_Q(v)$ is just wasting power, the power assignment problem is equivalent to finding the (directed) graph Q satisfying the connectivity constraint with minimum $p(Q)$.

As an example, Min-Power Symmetric Connectivity asks for the spanning tree T with $p(T)$ minimum. The broadcast nature of the wireless communication explains the difference from classical graph optimization problems, such as the polynomial-time solvable minimum spanning tree (MST). However, methods inspired by classical Steiner trees were useful in devising approximation algorithms.

STRONG CONNECTIVITY

The study of the min-power power assignment for strong connectivity problems was initiated by Chen and Huang.[3] Assuming symmetric power requirements, they prove that a minimum (cost) spanning tree (*MST*) of the input graph H has power at most twice the optimum, and therefore the MST algorithm has approximation ratio at most 2. The fact that the power of the output is at most twice the optimum is simple enough to present in this survey, and we use the reversed model as follows.

Consider G, the digraph induced by an optimal power assignment. Let s be an arbitrary node and B be an inward branch rooted at s (a tree with edges oriented toward the root) contained in G. As every node $v \neq s$ with parent v' has $p_B(v) = c(vv')$, we have $c(B) = \sum_{v \in V \setminus \{s\}} c(vv') = \sum_{v \in V \setminus \{s\}} p_B(v) = p(B)$. On the other side,

$$p(MST) = \sum_{v \in V} P_{MST}(v) = \sum_{v \in V} \max_{vu \in MST} c(vu)$$
$$\leqslant \sum_{v \in V} \sum_{vu \in MST} c(vu) = 2c(MST)$$

since every edge of MST is counted twice in $\sum_{v \in V} \sum_{vu \in MST} c(vu)$. Therefore, $p(MST) \leqslant 2c(MST) \leqslant 2c(B) = 2p(B) \leqslant 2p(G)$, where $c(MST) \leqslant c(B)$ follows from the fact that MST is a minimum spanning tree.

The following example shows that the ratio of 2 for the MST algorithm is tight. Consider $2n$ points located on a single line such that the distance between consecutive points alternates between 1 and $\epsilon < 1$ (see Fig. 2) and let $\ell = 2$. Then the MST connects consecutive neighbors and has power $p(MST) = 2n$. On the other hand, the tree T with edges connecting each other node (see Fig. 2B) has power equal $p(T) = n(1 + \epsilon)^2 + (n - 1)\epsilon^2 + 1$. When $n \to \infty$ and $\epsilon \to 0$, we obtain $p(MST)/p(T) \to 2$.

In the line case, Kirousis et al.[4] present a dynamic programming algorithm for min-power strong connectivity, and show the NP-hardness of the three-dimensional Euclidean case.

Min-power strong connectivity with symmetric power requirements is approximable (APX)-hard. This means that there is an $\epsilon > 0$ such that the existence of an approximation algorithm with ratio $1 + \epsilon$ implies that $P = NP$. Clementi et al.[5] showed that min-power strong connectivity in the Euclidean case is NP-hard. With asymmetric power requirements (or even with symmetric power requirements modified by nonuniform efficiency), a standard reduction shows that strong connectivity is as hard as set cover, which implies that there is no polynomial-time

algorithm with approximation ratio $(1 - \epsilon) \ln n$ for any $\epsilon > 0$ unless $P = NP$.

Calinescu et al.[1] present a greedy approximation algorithm with ratio $2 \ln n + 3$ for strong connectivity with asymmetric power requirements. This algorithm picks an arbitrary vertex s, and uses an approximation algorithm for Broadcast from s, which we discuss later, and Edmunds' algorithm for minimum cost incoming branch rooted at s.

Improving the approximation ratio under 2 in the symmetric power requirements case appears to be a very difficult problem. For the Euclidean power requirements case, we have a candidate for an algorithm with approximation ratio under 2: run the two algorithms below and output the better solution. The two algorithms are the best approximation algorithm from the next section, and Christofidies' algorithm for the traveling salesman problem, followed by orienting all the edges of the Hamiltonian cycle to obtain a directed circuit.

SYMMETRIC CONNECTIVITY

The connectivity of $\bar{G}$, the symmetric restriction of G, implies the connectivity of G. The reverse is not true, and in general it is harder for a power assignment to ensure symmetric connectivity. In fact, the power for the min-power strong connectivity can be half the power for min-power symmetric connectivity, as illustrated by the following example, in the Euclidean case with $\ell = 2$. The terminal set (see Fig. 3) consists of n groups of $n + 1$ points each, located on the sides of a regular $2n$-gon. Each group has two terminals in distance 1 of each other (represented as thick circles in Fig. 3) and $n - 1$ equally spaced points (dashes in Fig. 3) on the line segment between them. It is easy to see that the minimum power assignment ensuring strong connectivity assigns power of 1 to the one thick terminal in each group and power of $\epsilon^2 = (1/n)^2$ to all other points in the group. The total power then equals $n + 1$. For symmetric connectivity it is

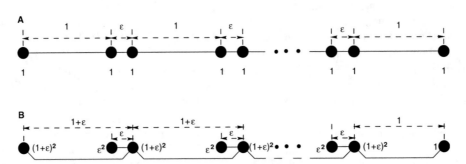

Fig. 2 Tight example for the performance ratio of the minimum spanning tree (MST) algorithm ($\ell = 2$). (A) The MST-based power assignment needs total power $2n$. (B) Optimum power assignment has total power $n(1 + \epsilon)^2 + (n - 1)\epsilon^2 + 1 \to n + 1$.

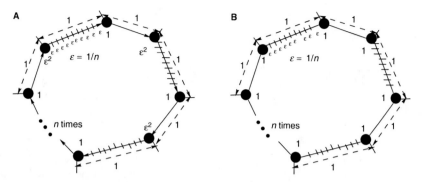

Fig. 3 Total power for min-power strong connectivity can be almost as small as half the total power for min-power symmetric connectivity ($\ell = 2$) for the same input. (A) Minimum power assignment ensuring strong connectivity has total power $n + n^2\epsilon^2 = n + n^2\frac{1}{n^2} = n + 1$. (B) Minimum power assignment ensuring symmetric connectivity has the total power $(2n-2) + (n^2 - n + 2)\epsilon^2 = 2n - 1 - \frac{1}{n} + \frac{2}{n^2}$.

necessary to assign power of 1 to all but two thick points, and power of ϵ^2 to the remaining points, which results in total power of $2n - 1 - 1/n + 2/n^2$.

Note that the *MST* algorithm of the previous section produces a symmetric output and therefore *MST* has an approximation ratio 2 for symmetric connectivity in the symmetric power requirements case, and this is tight, as shown in the example in Fig. 2.

Better approximation algorithms were presented by Althaus et al.,[2] with the best achieving a ratio of $5/3 + \epsilon$ for any $\epsilon > 0$. This algorithm is based on an existing $1 + \epsilon$ approximation algorithm for the following "3-hypertree" problem (a particular case of matroid parity): given a weighted hypergraph $Q = (V, J)$ with the hyperedges of J being subsets of V of size 2 or 3, find a minimum weight set of edges K such that the hypergraph (V, K) is connected. A hypergraph is connected if one can reach any vertex from any other vertex by a path in which any two consecutive vertices are included in a hyperedge.

The "3-hypertree" algorithms are not practical. On random instances, both uniformly and with skewed distribution, the following heuristic, adapted from one of the approximation algorithms in Ref. 2 (but without having a proof of approximation ratio better than 2), produced the best results:

1. Maintain a spanning tree T, initially the MST.
2. Find three vertices uvw and two edges e_1 and e_2 of T such that the $T' = T\backslash\{e_1, e_2\} \cup \{uv, uw\}$ has $p(T')$ mininmum.
3. If $p(T') < p(T)$, replace T with T' and go to Step 2.

A simpler heuristic that only introduces one new edge uv in T with u and v at bounded distance in T, removes one edge e from T, as long as there is some improvement in the power, and also produces reasonable improvement over the MST and is suitable for distributed implementation.

With asymmetric power requirements (or even with symmetric power requirements modified by nonuniform

efficiency), min-power symmetric connectivity is again as hard as set cover,[1] and the algorithm described below has approximation ratio at most $2 \ln n + 2$.

The algorithm starts iteration i with a graph G_i, seen as a set of edges with vertex set V. Unless G_i is connected, a star S (details below) is computed such that it achieves the biggest reduction in the number of components divided by the power of the star. The algorithm then adds the star (seen as a set of edges) to G_i to obtain G_{i+1}.

A *star* is a tree consisting of one *center* and several *leaves* adjacent to the center. Note that the power of the star is the maximum power requirement of the arcs from the center to the leaves plus the sum of power requirements of the arcs from the leaves to the center. With respect to G_i, let $d(S)$ be the number of different components of G_i to which the vertices of the star belong.

See Fig. 4 of a star and its power. Our algorithm appears in Fig. 5.

Next we describe how to find the star S minimizing $p(S)/(d(S) - 1)$. We search all vertices v and all power levels $p(v) \in \{c(vu)|u \in V\}$. For every connected component C_j of G_i, we find the vertex u_j (in case such a vertex

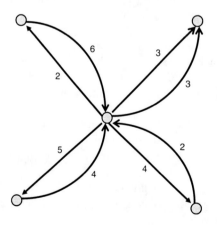

Fig. 4 A star with four leaves, of power max.$\{2,3,4,5\} + 6 + 3 + 2 + 4 = 20$.

> **Input:** A complete directed graph $H = (V,E)$ with power requirements $c: E \rightarrow R^+$
> **Output:** An undirected connected spanning graph G (seen as a set of edges, with $V(G) = V$)
>
> **0.** Initialize $G = \emptyset$
> **1.** While G has at least two connected components
> **1.1.** Find the star S which minimizes $p(S) / (d(S) - 1)$ with respect to G
> **1.2** Set $G \leftarrow G \cup S$
> **2.** For all vertices v, assign $p(v) = \max_{vu \in G} c(vu)$

Fig. 5 The greedy algorithm for symmetric connectivity with asymmetric power requirements.

exists; note that u_j might be v) such that $c(vu_j) \leqslant p(v)$ and $c(u_j v)$ is minimum. We sort the satisfied components C_j for which u_j exists in nondecreasing order of $c(u_j v)$. Then, for every $d \geqslant 2$, we try the star with center v and leaves $u_1, u_2, \ldots, u_d$. Thus, we search a total of at most n^3 stars and pick the one minimizing $p(S)/(d(S) - 1)$.

We conclude the section with a constant-ratio approximation algorithm for min-power symmetric connectivity in the Euclidean with efficiency model.[1] Define $w(u,v) = d(u,v)^\ell/e(u) + d(u,v)^\ell/e(v)$ and construct a MST T of the node set V with respect to weight w. Assign power to vertices according to T; that is, set $p(v) = \max_{vu \in E(T)} d(v,u)^\ell/e(v)$.

BICONNECTIVITY

In this section we discuss two connectivity constraints:

1. Biconnectivity: for any two vertices u and v, G must contain two paths from u to v that do not share nodes except for u and v. The biconnectivity requirement is motivated by reliability issues.
2. Symmetric Biconnectivity: $\bar{G}$ must be two-connected (i.e., removing any vertex from G' results in a connected graph).

All the algorithms presented in this section produce outputs that satisfy the more stringent symmetric biconnectivity constraint. However, their approximation ratio is computed with respect to the optimum to min-power biconnectivity.

Ramanathan and Rosales-Hain[6] proposed the first heuristic, which we describe in a slightly improved version. It uses the idea of Kruskal's algorithm for MSTs, and assumes symmetric power requirements:

- Sort the edges of H in non-decreasing cost order.
- Use binary search to find minimum i such that Q, the spanning subgraph of H with edge set $\{e_1, e_2, \ldots, e_i\}$, is biconnected.
- Starting with e_i and going downward, remove from Q any edge e for which $E(Q)\backslash\{e\}$ induces a biconnected spanning subgraph.
- Assign power to vertices according to Q; that is, set $p(v) = \max_{vu \in E(Q)} c(vu)$.

- For every vertex, reduce its power as long as the induced graph $\bar{G}$ is biconnected.

In the symmetric power requirements case, we have examples showing that the Ramanathan and Rosales-Hain heuristic has an approximation ratio of at least $n/2$. Its approximation ratio in the Euclidean power requirements case is not known, and our experiments show it is quite good on uniform random instances in the unit square.

Lloyd et al.[7] proposed using the approximation algorithm of Khuller and Vishkin,[8] which we refer to as Algorithm KV and which was designed for minimum-weight biconnected spanning subgraph, and proved an approximation ratio of $2(2 - 2/n)(2 + 1/n)$ for min-power symmetric biconnectivity. Calinescu and Wan[9] showed that the power of the output of Algorithm KV is within 4 of the power of the (possibly nonsymmetric) best solution of the Min-Power Biconnectivity problem.

The Algorithm KV is complicated. In the Euclidean case, Calinescu and Wan[9] proposed MSTAugmentation, another algorithm with a constant approximation ratio, much faster, simpler, and better suited for distributed implementation. This $O(n \log n)$ algorithm first constructs a MST T over V. Then at any non-leaf node v of T, a local Euclidean MST T_v over all the neighbors of v in T is constructed. The output is Q, the union of T and T_v's for all non-leaf nodes v of the T; power is assigned to vertices according to Q. Another advantage of this algorithm is its independence of the path-loss exponent.

k-EDGE-CONNECTIVITY

In this section we discuss two connectivity constraints:

1. k-Edge-connectivity: for any two vertices u and v, G must contain k edge-disjoint paths from u to v.
2. Symmetric k-edge-connectivity: $\bar{G}$ must be k-edge-connected (i.e., removing any $k - 1$ edges from $\bar{G}$ results in a connected graph).

The results of this section are similar to those in the previous section. All the algorithms we present produce outputs that satisfy the more stringent symmetric k-edge-connectivity constraint. However, their approximation ratio is computed with respect to the optimum to k-edge-connectivity.

The Ramanathan and Rosales-Hain heuristic can be applied (with approximation ratio in the Euclidean case unknown) to k-edge-connectivity by checking for k-edge-connectivity instead of biconnectivity when removing the edges.

Lloyd et al.[7] proposed using the approximation algorithm of Khuller and Raghavachari,[8] which was designed for Minimum-Weight k-Edge-Connected Spanning Subgraph, and proved an approximation ratio of $8(1 - 1/n)$ for min-power symmetric k-Edge-Connectivity. Calinescu and Wan[9] showed that the power of the output of the Khuller and Raghavachari algorithm is within $2k$ of the power of the (possibly non-symmetric) best solution of the min-power k-edge-connectivity problem.

MST-augmentation can also be used for 2-edge-connectivity, with a constant approximation ratio in the Euclidean case.

SYMMETRIC UNICAST

The min-power unicast problem requires that G contains a directed path from s to t, where s and t are given nodes. It is easily solvable by Dijkstra's shortest paths algorithm in the graph H.

In min-power symmetric unicast, we are given s and t, and $\bar{G}$ must contain a path from s to t. The following example in the Euclidean case shows that a straightforward application of Dijkstra's algorithm does not work; i.e., a minimum cost (with cost function c given by the power requirements) $s - t$ path does not always have minimum power. Consider a network consisting of three nodes, $s = (0,3)$, $t = (4,0)$, and $x = (0,0)$ (see Fig. 6), and assume $\ell = 2$. Then, the two $s - t$ paths, namely, (s,t) and (s, v, t), have the same cost of 25 but different powers; the power; of (s,t) is $25 + 25 = 50$ while the power of (s, v, t) is $9 + 16 + 16 = 41$.

Althaus et al.[2] present a solution of MIN-POWER SYMMETRICUNICAST that first modifies the given graph $H = (V,E,c)$ and then applies Dijkstra's algorithm to the resultant directed graph H'. We now describe the construction of the directed graph $H' = (V',E',c')$, and note that it does not assume that the cost function is symmetric.

For any $u \in V$, we sort all adjacent vertices $\{v_1, \ldots, v_k\}$ in ascending order of costs of edges connecting them to u, i.e., $c(u, v_i) \leqslant c(u, v_i + 1)$. The vertex v is replaced by a *gadget* (see Fig. 7A) as follows:

1. Each edge (u,v) is replaced by two vertices: $[u,v]$ and $[v,u]$.
2. For each u, we connect all vertices $[u,v_i]$'s by two directed paths:

   $$P_1 = (u, [u, v_1], \ldots, [u, v_{k-1}], [u, v_{k-1}]) \quad \text{and}$$
   $$P_2 = ([u, v_{k-1}], [u, v_{k-1}], \ldots, [u, v_1], u)$$

3. The costs of the arcs on path P_1 are $c(u, v_2), c(u, v_1) - c(u, v_1), \ldots, c(u, v_k) - c(u, v_{k-1})$, respectively; and the cost of all arcs on the path P_2 is zero.

Finally, each edge (u,v) of H is replaced in H' by one arc $([u,v], [v,u])$ of cost $c(v,u)$.

Fig. 7B shows the graph H' for the example in Fig. 6. It is easy to see that a shortest $s - t$ path in H' corresponds to a minimum power $s - t$ undirected path.

BROADCAST AND MULTICAST

In this section we discuss two related connectivity constraints:

1. Broadcast: G must contain a directed path from a given node called the root to every other node.
2. Multicast: G must contain a directed path from the root to a given set of nodes called terminals.

Min-Power Broadcast was first studied by Wieselthier et al.[10] They proposed three heuristics but did not prove approximation ratios. The first heuristic, SPT, uses a shortest-path tree from the root. The second heuristic, MST, uses a MST. In both cases, the resulting undirected graph is oriented away from the root, and power is assigned accordingly. The third heuristic, called broadcasting incremental power (BIP), is a Prim-like heuristic that starts with an outgoing branch consisting of the root, and iteratively adds an arc connecting the set of nodes currently reached to an outside node with minimum increase in power.

For the Euclidean case, Wan et al.[11] studied the approximation ratios of the above three heuristics. An instance was constructed to show that the approximation ratio of SPT is as large as $(n/2) - o(1)$. On the other hand, both MST and BIP have constant approximation ratios. The following geometric constant plays an important role in their analysis. Let

$$\sigma = \sup_{o \in P \subset \mathcal{D}, |P| < \infty} \sum_{e \in mst(P)} \|e\|^2$$

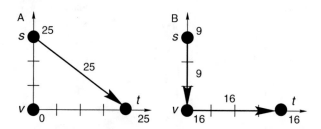

Fig. 6 An example of two paths with the same cost and different powers: (A) The path (s, t) assigns powers 25 to s and to t. (B) The path (s, v, t) assigns powers 9 to v and 16 to v and t.

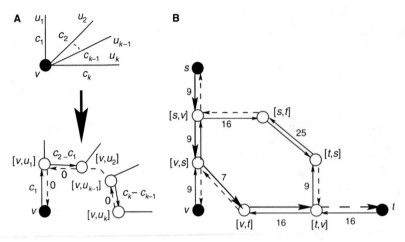

Fig. 7 **A** A vertex v adjacent to k vertices $u_1, \ldots, u_k$ via edges of cost $c_1, c_2, \ldots, c_k$ and a gadget replacing v with a bidirectional path. The solid edges of the path $(v, [v, u_2]), ([v, u_2]), [v, u_3]), \ldots ([v, uk - 1], [v, uk])$ have cost $c1, c2, cl, \cdots, ck - ck - 1$ respectively. The dashed edges have zero cost. **B** The graph H' for the example in Fig. 6. As the example in Fig. 6 has symmetric power requirements, the pair of opposite directed edges of the type $([u, v], [v, u])$ and $([v, u], [u, v])$ have the same cost and the figure only has one number next to such a pair. Thick edges belong to the shortest path corresponding to the path (s, v, t) in H.

where $\mathcal{D}$ is the closed disk of radius one centered at the origin **o** and $mst(P)$ is a Euclidean minimum spanning tree of P. It is proved that $6 \leqslant \sigma \leqslant 12$. In addition, any unidirectional broadcast routing has power at least $c(MST)/\sigma$. This immediately implies that the approximation ratio of MST is at most σ. It is also proved[11] that $p(BIP) \leqslant c(MST)$ and therefore BIP also has an approximation ratio at most σ. Wan et al.[11] also constructed two instances that lead to a lower bound of 6 on the approximation ratio of MST and a lower bound of $13/4$ on the approximation ratio of BIP, respectively. We conjecture that $\sigma = 6$ and therefore MST has an approximation ratio of 6.

For the Euclidean case, it follows immediately from the analysis in Ref. 11 that an α-approximate Steiner tree gives an approximation ratio $\alpha \cdot \sigma$ for min-power multicast. With current best-known $\alpha = 1 + (\ln 3/2) + \epsilon$, we obtain an 18.59 approximation algorithm. Faster algorithms have been proposed by Wan (unpublished manuscript) for Multicast, based on faster Minimum Steiner Tree algorithms.

Now we move from the Euclidean case to symmetric power requirements. A standard reduction from Set Cover shows that no approximation ratio better than $O(\log n)$ is possible. It turns out that the best approximation algorithm[11] also works for min-power broadcast with

Table 1 Upper bounds (UBs) and lower bounds (LBs) of the power assignment complexity.

	Complexity of the min-power assignment problems					
	Asymmetric power reqs.		**Symmetric power reqs.**		**Euclidean ($\alpha \geqslant 2$)**	
Connectivity constraint	UB	LB	UB	LB	UB	LB
Strong connectivity	$3 + 2\ln(n - 1)^2$	SCH[1]	$2^{[3,4]}$	APX-hard (APXH)*	2	NPH[5]
Broadcast	$2 + 2\ln(n - 1)^2$	SCH	$2 + 2\ln(n - 1)$	SCH[11]	$12^{[11]}$	NPH*
Multicast	DST*	DSTH[1]	$O(\ln n)$**[12]	SCH**[11]	18.59**[11]	NPH*
Symmetric conn.	$2 + 2\ln(n - 1)^2$	SCH[1]	$5/3 + \epsilon^1$	APXH*	$5/3 + \epsilon$	NPH*
By connectivity		NPH*	$4^{[9]}$	NPH*	4	
Symm. biconn.		APXH*	$4^{[9]}$	APXH*	4	NPH[9]
k-Edge-conn.		NPH*	$2k^{[9]}$	NPH*	2k	
Symm. k-edge-conn.		APXH*	$2k^{[9]}$	APXH*	2k	NPH[9]

*Folklore results.
**Indicate the result is implicit in the respective articles.

Wireless
Ad Hoc—Asynch

asymmetric power requirements. The algorithm is an extension of the algorithm for min-power symmetric connectivity presented in Fig. 5, except that it uses a more complicated structure called *spider* and it aims to pick a spider with minimum ratio of its weight to the number of strongly connected components with no incoming edge it "hits." For min-power multicast in the symmetric power requirements case, the construction devised for broadcast by Caragiannis and Kaklamanis[12] together with approximation algorithms for node-weighted Steiner tree can be used to obtain an $O(\log n)$ approximation ratio. This result cannot be easily extended to min-power multicast in the asymmetric power requirements case, as in fact this most general power assignment problem is equivalent to directed Steiner trees,[1] a problem seemingly harder than node-weighted Steiner tree.

SUMMARY OF APPROXIMABILITY RESULTS

We summarize the known results on power assignment in Table 1. Each cell of the table describes the best-known approximation ratio and lower bounds for each combination given by the connectivity constraint type and the power requirement case. NPH means NP-hard, and APXH means APX-hard (there is an $\epsilon > 0$ such that the existence of an approximation algorithm with ratio $1 + \epsilon$ implies that $P = NP$). SCH means the problem is as hard as SetCover, which implies that there is no polynomial-time algorithm with approximation ratio $(1 - \epsilon) \ln n$ for any $\epsilon > 0$, unless $P = NP$. DST means that the problem reduces (approximation-preserving) to directed Steiner tree and DSTH means directed Steiner tree reduces (approximation-preserving) to the problem given by the cell. The best-known approximation ratio for directed Steiner tree is $O(n)$ for any $\epsilon > 0$ and finding a poly-logarithmic approximation ratio remains a major open problem in the field of approximation algorithms.

We omit the line case, where all problems can be solved in polynomial time. In line with the efficiency case, all the problems have dynamic programming polynomial-time algorithms,[1] while with sensitivity, surprisingly, even for the line nothing better than general asymmetric power requirements is known. We also omit unicast, where shortest paths algorithms directly solve the problem, and symmetric unicast, where shortest paths algorithms in a specially constructed graph solve the problem (see "Symmetric Unicast" section).

In the Euclidean case with efficiency, constant-ratio algorithms exist for Min-power symmetric connectivity,[1]

min-power symmetric biconnectivity, and min-power symmetric 2-edge-connectivity; while for the "directed" connectivity types, nothing better than the results for asymmetric power requirements is known.

REFERENCES

1. Calinescu, G.; Kapoor, S.; Olshevsky, A.; Zelikovsky, A. Network lifetime and power assignment in ad-hoc wireless networks. 11th European Symposium on Algorithms, Budapest, Hungary, Sept 16–19, 2003.

2. Althaus, E.; Calinescu, G.; Mandoiu, I.; Prasad, S.; Tchervenski, N.; Zelikovsky, A. Power efficient range assignment in ad-hoc wireless networks. IEEE Wireless Communications and Networking Conference, New Orleans, LA, Mar 16–20, 2003.

3. Chen, W.T.; Huang, N.F. The strongly connecting problem on multihop packet radio networks. IEEE Trans. Commun. **1989**, *37* (3), 293–295.

4. Kirousis, L.M.; Kranakis, E.; Krizanc, D.; Pelc, A. Power consumption in packet radio networks. Theor. Comput. Sci. **2000**, *243*, 289–305.

5. Clementi, A.E.F.; Penna, P.; Silvestri, R. On the Power Assignment Problem in Radio Networks. Electronic Colloquium on Computational Complexity (ECCC), (054), 2000.

6. Ramanathan, R.; Rosales-Hain, R. Topology control of multihop wireless networks using transmit power adjustment. IEEE INFOCOM 2000, Tel Aviv, Israel, Mar 26–30, 2000; 404–413.

7. Lloyd, E.; Liu, R.; Marathe, M.; Ramanathan, R.; Ravi, S.S. Algorithmic aspects of topology control problems for ad hoc networks. 3rd ACM International Symposium on Mobile Ad Hoc Networking and Computing (MobiHoc 2002), Lausanne, Switzerland, June 9–11, 2002.

8. Khuller, S. Approximation algorithms for finding highly connected subgraphs. In *Approximation Algorithms for NP-Hard Problems*; Hochbaum, D.S., Ed.; 1996; 236–265.

9. Calinescu, G.; Wan, P.-J. Range Assignment for High Wireless Ad Hoc Networks, connectivity 2nd International Conf., ADHOC-NOW. Montreal, Canada, Oct 8–10, 2003.

10. Wieselthier, J.E.; Nguyen, G.D.; Ephremides, A. On the construction of energy-efficient broadcast and multicast trees in wireless networks. IEEE INFOCOM 2000, Tel Aviv, Israel, Mar 26–30, 2000; 585–594.

11. Wan, P.-J.; Calinescu, G.; Li, X.-Y.; Frieder, O. Minimum energy broadcast routing in static ad hoc wireless networks. Wirel. Netw. **2002**, *8* (6), 607–617.

12. Caragiannis, P.K.I.; Kaklamanis, C. New results for energy-efficient broadcasting in wireless networks. ISAAC 2002, Vancouver, Canada, Nov 20–23, 2002.

Wireless Applications: Middleware Security

Marc-André Laverdière
Azzam Mourad
Syrine Tlili
Mourad Debbabi
Computer Security Laboratory, Concordia Institute for Information Systems Engineering,
Concordia University, Montréal, Quebec, Canada

Abstract

In this entry, we survey the security of wireless applications and middleware. Nowadays, we are more and more relying on the use of mobile and wireless systems such as PDAs, cell-phones, pagers, etc., for communication, work, and entertainment. The popularity of these devices is increasing day by day. These devices now offer increased bandwidth and processing power, so that a great variety of applications can be executed on them. Alongside, the security issues typically associated with wired computing have now migrated to the mobile systems' world. Many middleware frameworks [e.g., Java 2 Micro Edition (J2ME)] that have been adapted to respect the resource limitations of such type of devices offer high-level security features. In this context, we examined security objectives of applications and middleware, vulnerabilities they face, detection techniques used to uncover them and hardening practices performed to remedy them.

INTRODUCTION

As traditional software met the world of wireless systems, and as they started offering networking capabilities that enabled features previously available only on wired systems, they brought along security vulnerabilities that threaten both devices and networks. These systems have proven to be vulnerable to software-related attacks common to PCs.[1,2] BlueSnarfing, for instance, is a method using buffer overflows (BoFs) on certain phones allowing to steal information remotely. Furthermore, high-level security, such as, the principle of least privilege is lacking on handheld computers and cell-phones giving full capabilities to any piece of code executing on it, malicious or not.

The security of wireless communication itself has been the focus of a lot of study, which resulted in strong medium security technologies[3] that are now used in the industry; however, some security concerns remain in the integration of mobile systems into traditional network infrastructure such as the Internet.[4,5] Middleware frameworks are of great help, as they offer interoperable features that often

have been built with security in mind. Other useful features offered by such technologies include fault tolerance, standardized communication mechanisms, adaptability, scalability and resource sharing. They can allow developers to create richer applications and systems in a shorter time, while not sacrificing security. Although middleware is mostly known in the field of distributed systems in classical environments with technologies such as CORBA, .NET and J2EE, it is nevertheless present in mobile devices via lightweight variants, such as, increasingly popular, Java 2 Micro Edition (J2ME).

Overall, mobile and wireless systems of all kinds are now facing the same security challenges as in traditional computing environments, and must solve them by taking in to consideration limited resources. Traditional protection methods, such as, firewalls and execution monitors, are not always realistic options owing to such resource constraints or unfitting system architecture. Furthermore, the traditional code-exploit-fix cycle is unrealistic for devices and systems that rarely offer automatic updating. Economic pressures force shorter time-to-market constraints on product development and force the reuse of

Encyclopedia of Wireless and Mobile Communications DOI: 10.1081/E-EWMC-120043598

existing software. In short, the realities of mobile systems require security improvements of the operating system, middleware, and applications both before and after deployment. The process used must be simple, systematic, rapid, and effective, in order to be used in-depth during development and maintenance.

We refer to such improvements as security hardening and will show how vulnerabilities can be automatically detected both statically and dynamically, and how security engineers are able to specify security hardening plans to remedy both low-level and high-level security weaknesses.

In this entry, we will examine how software vulnerabilities can be found and corrected. We will look at the related work in next section, and then briefly survey software vulnerabilities in section "Security Requirements and Vulnerabilities." Afterward, we will study automatic detection of low-level security vulnerabilities in section "Security Evaluations." We then move on to section "Security Hardening" to determine how existing mobile applications and middleware can be hardened in a rigorous and systematic manner in response to the discovered vulnerability. We finally bring concluding remarks.

RELATED WORKS

Traditional middleware have been shown to be imperfect solutions when directly ported to wireless systems, and significant research has been invested in adapting those frameworks for the mobile world, as well as into building new solutions that were more adapted to platform requirements for both nomadic and ad hoc systems.[6] With such improvements, middleware has been shown to be useful in enabling new ways to interact with enterprise information systems.[5] Adaptation middleware (e.g., Odyssey, Mobiware, and Puppeteer), mobile agent systems (e.g., Telescript), service discovery (e.g., UPnP, Jini, SLP, Salutation, and Ninja), tuple spaces (e.g., Lime and L2imbo), data replication (e.g., Xmiddle), Object-request brokers (e.g., LW-IOP) and transaction support (e.g., Kangaroo) are some of the middleware services for mobile systems in the literature.[6,7]

The study of middleware security has typically been related to the enforcement of access control in distributed systems,[8,9] although smart card support via middleware has been demonstrated.[10] Some systems offer more security services, such as .NET and J2EE that both offer a five-prong approach to security: code-based access control, role-based access control, secure code verification and execution, secure communication mechanisms and secure code and data protection.[9]

The field of both low and high-level security vulnerabilities has been extensively studied for classical applications, and taxonomies have emerged.[11–16] Explanations of low-level vulnerabilities abound,[16–19] and a few authors detail how to exploit them (Refs. 20–23 to name a few).

Secure programming has also been studied especially for C and C++ programs, a technology popular for both traditional and embedded systems. The authors[16–19,24–28] usually explain what kind of code creates security vulnerabilities and how such vulnerabilities should be prevented by better coding.

Many techniques external to flawed programs have been developed. The reader will find many libraries for C/C++, some specifically compensating C memory manipulation vulnerabilities,[29,30] compiler extensions,[31,32] as well as operating system extensions that complicate the exploitation vulnerabilities.[33,34,35]

To complement the previous contributions, authors have developed automatic tools to statically detect software vulnerabilities, mostly for C/C++ programs.[36–45] Others have developed tools that perform code instrumentation to detect security vulnerabilities at runtime.[46,47]

SECURITY REQUIREMENTS AND VULNERABILITIES

High-level security features are of tremendous importance for a sustainable mobile environment. Sadly, the implementation of software is never free of errors, and some lower-level errors can cause vulnerabilities that defeat the existence of security mechanisms. Each middleware has its own models, paradigms and techniques that provide its security capabilities and services. In this context, the authentication, authorization, confidentiality, integrity, and non-repudiation are mainly the high-level security requirements that are enforced by the security components of the major wireless middleware such as J2ME, CORBA, .NET etc. In the following list, we present the definitions of these requirements provided by the ISO standard 7498-2.[48]

1. **Authentication:** Corroboration of the identity of an entity or source of information;
2. **Authorization:** Restricting access to resources to privileged entities;
3. **Confidentiality:** Keeping information secret from all but those who are authorized to see it;
4. **Integrity:** Ensuring information has not been altered by unauthorized or unknown means;
5. **Non-Repudiation:** Preventing the denial of previous commitments or actions.

Other requirements, such as, availability, anonymity, auditing, certification, privacy, revocation, timestamping, etc. exist in the security literature.

Many vulnerabilities could be created during the development phase of the software allowing security threats that can compromise the security capabilities of the applications and middleware. Flaws in the security models or in any of their components could be the cause of such vulnerabilities. For instance, if an application is using a particular application programming interface (API) or protocol implementation in order to provide one of the aforementioned requirements and this API or protocol contains some flaws, then the application as a whole will be considered as non-secure. Implementation flaws may lead to dangerous security vulnerabilities that are called low-level security or safety vulnerabilities. Their exploitation is considered as dangerous threats because they can affect high-level security capabilities provided by the applications and middleware.

The low-level security vulnerabilities are extremely dependent on the programming language and platform. In this context, the C/C++ programming language has a bad reputation in the security world because it was designed for maximal performance, at the expense of some safety-enhancing techniques. Its memory management, which is left to the programmer, and the lack of type safety are the major causes of security flaws. Such flaws do not exist in many other technologies, such as Java applications and middleware because they offer better built-in security options: type verification and garbage collection in the virtual machine. In the sequel, we list the major safety vulnerabilities that are introduced in the source code during the implementation and discuss their impacts on the applications and middleware security.

BoF

BoF results from common programming errors that arise mostly from weak or nonexistent bounds checking of input being stored in memory buffers. Attacks that exploit these vulnerabilities are considered as one of the most dangerous security threats since it can compromise the integrity, confidentiality and availability of the target system often via code injection. Buffers on both the stack and the heap can get corrupted.[16,17] The following are the common causes of BoFs: boundary conditions errors, input validation errors, assumption of Null-termination and improper format string.

Integer Overflow

Integer security issues arise on the conversion (either implicit or explicit) of integers from one type to another, and because of their inherently limited range.[18] C compilers distinguish between signed and unsigned integer types and silently perform operations such as implicit casting, integer promotion, integer truncation, overflows and underflows. Such silent operations are typically overlooked, which can cause various security vulnerabilities. Integer vulnerabilities may be used to write to an unauthorized area of memory. A first instance is the allocation of less memory than sought, allowing writing to unwanted parts of the heap. Another instance is to access invalid memory

areas with a negative index or memory copying operation. In some cases, if the access is to an invalid page, the result will be a denial of service via an application crash. They can also cause other security problems by bypassing preconditions and expected protocol values that are specific to the program exploited. The following are the common causes of these vulnerabilities: converting between signed and unsigned, signedness errors, truncation errors, overflow and underflow.

Memory Management

The C programming language allows programmers to dynamically allocate memory for objects during program execution. C memory management is an enormous source of safety and security problems. The programmer is in charge of pointer management, buffer dimensions, and allocation and de-allocation of dynamic memory space. Thus, memory management functions must be used with precaution in order to prevent memory corruption, unauthorized access to memory space, BoF, etc. The following are the major errors caused by improper memory management in C: using un-initialized memory, accessing freed memory, freeing unallocated memory, and memory leaks.

File Management

File management problems occur when access to or modification of a restricted file happens. Some problems are closely related to race conditions. File Management errors can lead to many security vulnerabilities such as data disclosure, data corruption, code injection and denial of service. The following are two major sources of vulnerabilities in file management: unsafe temporary file and improper file creation access control flags.[19]

SECURITY EVALUATIONS

The main security mechanisms to consider for applications and middleware security are access control, authentication, message protection and audit. So far, the focus was on security policies and security protocols. However, there are many issues related to the implementation of these mechanisms. Growing applications and middleware security requirements have raised the stakes on software security. Building a secure software focuses on techniques and methodologies of designing and implementing a software in order to avoid exploitable security vulnerabilities. The C language is considered the de facto standard for system programming. Most of the existing middleware are written in C. These include Core ORB, J2ME, and HPCM. C fulfills performance, flexibility, strong support, and portability requirements for these middleware. However, security features are either absent or badly supported in C programming. Lack of type safety and leaving memory management to the programmer's discretion are sources of many critical security vulnerabilities such as BoFs and format strings. These vulnerabilities enable an intruder to take a complete control over the target machine and to circumvent all deployed security mechanisms for middleware. Despite the huge availability of books and documents that guide programmers in writing secure code, implementation errors still exist in C source code. Therefore, automated tools for vulnerability detection are very helpful for programmers in detecting and fixing errors in source code. There are different techniques and approaches for verification and validation of software security requirements.

Security Code Inspection

Code inspection as introduced by Michael Fagan,[49] is widely recognized as an effective technique for finding software defects and bugs. It is carried out by a technically competent team of four persons: moderator, designer, coder and tester. The moderator, an experienced software engineer but preferably not involved in the project, is the key person and plays the role of a coach in the inspection process. The Fagan inspection process is based on the following five steps: 1) Overview: the designer describes what he has designed to all the participants; 2) Preparation: each member, using the work documentation, individually tries to understand the design, its intent and logic; 3) Inspection: all the participants get together as a group and attempt to find as many defects as possible; 4) Re-work: the designer or the coder resolves all the errors or problems noted in the inspection report; 5) Follow-up: the moderator checks the quality of the work and determines whether the component needs to be re-inspected.

Static Analysis

The source code is examined statically without any execution of the source code.[50] The security properties that hold during static analysis are supposed to hold true for all executions of the analyzed source code. The objective of static analysis is to detect ahead of time the vulnerabilities in the code. There are four main approaches for static analysis: abstract interpretation, type systems, flow analysis and model checking. Each of these approaches has its advantages [11] and drawbacks. The abstract interpretation has a strong formal semantics foundation which is more than 30 yr old. PolySpace[41] and AbsInt[51] are the two prominent commercial tools based on abstract interpretation. These tools can detect run-time errors that others static analysis tools fail to. PolySpace has served many clients in telecommunications industries such as France Telecom, Siemens, LG Electronics and Samsung. Coverity[42] is another commercial tool that uses flow based static analysis to detect security vulnerabilities in source code. Flow analysis does not have strong formal

foundation as abstract interpretation but it does scale to large programs. Companies in telecommunication industry such as NOKIA, NOMADIX and ShoreTel have used Coverity tools to verify the security of their software. There are many other academic tools that can be used to assess the security of open source middleware. Some of these tools focus on C pointers and type management for being the main causes of security violations. We cite a representative set of these tools: Ccured,[52] Cyclone,[53] and Fail-Safe.[54] Their approach is based on type based static analysis. In other words, the standard C type is extended with additional type annotation and type check to tackle statically at compile time execution errors. Most of these tools resort to dynamic analysis by inserting runtime checks when it is not possible to decide statically about the safety of an operation in the program. Type based analysis has also been used to check authorization and authentication properties which are the main security requirements of middleware. In fact, Cqual[55] has been used to check the placement of security hooks in Linux Security Modules that are used to enforce access control security policies. The intent is to make sure that each security critical operation is performed after being authorized by the security policy. Hence, Cqual can also be used for the implementation of access control policies in middleware. Model checking is also an appealing approach for checking security properties of middleware. There are many success stories of verifying protocols by model checking. Therefore, the approach can easily be adopted for checking the safe behavior of authentication, access control and cryptographic protocols of middleware. Microsoft uses SLAM model checker to check temporal properties of its drivers and interfaces. Since middleware are acting as an interface between application and protocol stack, it is very useful to consider using a tool such as SLAM[56] to ensure a safe behavior of the middleware. There are also many open source model checkers that can be used to check temporal properties for C source code. The most stable model checker for that purpose is MOPS.[57]

Dynamic Analysis

The source code is examined during its execution. Dynamic analysis is done through code instrumentation to collect information on the program as it runs. In contrast to static analysis, the properties derived from dynamic analysis hold for the current execution of the analyzed program and may not be generalized for all other possible executions. Software testing is the most common approach for dynamic analysis. Basically, the underlying principle of software testing is to exercise the system/software under test using a subset of its input domain called test cases, in order to unveil the defects and bugs. An appealing approach is to use aspect oriented program to insert checks into the source code. The complexity and efficiency varies with respect to the nature (centralized, distributed) of the application under test. Similar to static analysis, there are many tools that use dynamic analysis to check security properties of software. Parasoft offers a wide range of testing framework that can automatically detect errors in source code. Parasoft SOAtest is intended for Service Oriented Architectures which are the essence of middleware. Parasoft[58] has been used by IBM to check security and interoperability features of their web services. Aspect oriented programming (AOP) can be considered as a promising approach for middleware. The company near-Infinity uses AOP for auditing J2EE middleware and many API such as Servlet, JNDI and EJB.

SECURITY HARDENING

Security hardening is a relatively unknown term in the current literature and, as such, we first provide a definition for it. We also propose taxonomy of security hardening methods that refer to areas to which the solution is applied. We established our taxonomy by studying the solutions of software security problems in the literature. Even though our reading included a significant bias towards C,[16,18,19,27] we believe that our taxonomy is language independent. We also investigated the security engineering of applications at different levels, including specification and design issues.[16,59,60] From this information on how to correctly build new programs, and some hardening advice existing in the literature, we were able to draw out a classification for software hardening methodologies.

We define software security hardening as any process, methodology, product or combination thereof that is used to increase the security of existing software. In this context, the following constitutes the detailed classification of security hardening methods:

1. Code-level hardening changes in the source code in a way that prevents vulnerabilities without altering the design;
2. Software process hardening replacement of the development tools, the use of stronger implementation of libraries and the use of code weaving tools in a way that does not change the original code;
3. Design-level hardening re-engineering of the application in order to integrate security features that were absent or insufficient;
4. Operating environment hardening improvements to the security of the execution context (network, operating systems, libraries, utilities, etc.) that is relied upon by the software.

Code-level hardening. Code-level hardening is the topic closest to what has been covered extensively in the literature. Code-level hardening constitutes of removing the programming-related vulnerabilities in a systematic way by implementing the proper coding standards that were not enforced originally.

Software process hardening. Software process hardening refers to the inclusion of hardening practices within the software development process, notably on the matter of choosing appropriate platforms, statically-linked libraries, compilers, etc. that result in increased security. It is possible that a more secure implementation of the same library is used, instead of modifying the underlying source code. This will allow externalizing the security issues, keeping the solution independent of the using software, maximizing code reuse and facilitating the hardening of multiple applications relying on this library. It is also possible to use compilers and aspects that add some protections in the object code, which were not specified in the source code, and that prevent or complicates the exploitation of vulnerabilities existing in the program. In all cases, the original source code is not modified.

Design-level hardening. Design-Level Hardening refers to changes in the application design and specification. Some security vulnerabilities cannot be resolved by a simple change in the code or by a better environment, as they are due to a fundamentally flawed design or specification. Changes in the design are thus necessary to solve the vulnerability or to ensure that a given security policy is enforceable. Moreover, some security features need to be added for new versions of existing products. This category of hardening practices naturally target more high-level security issues. In this context best practices, known as security design patterns,[60] can be used to guide the re-design effort. Although such patterns are targeting the security engineering of new systems, such approach can also be re-directed and mapped to cover deploying security into existing software.

Operating environment hardening. The Operating Environment Hardening refers to practices that impact the security of the software in a way that is unrelated to the program itself. This addresses the hardening of the operating system, the protection of the network layer, the configuration of middleware, the use of security-related operating system extensions, the normal system patching, etc.[19,61] Many security appliances can be deployed and integrated into the operating environment in a way that provides some high-level security services. These hardening practices fall within the scope of proper management of an IT department, and as much as they can prevent exploitation of vulnerabilities, they do not remedy them.

Hardening of High-Level Security

Many mobile wireless devices lack high-level security properties that make them ideal backdoors that compromise an organization's network. Consumers reacted to this situation by increasingly demanding security features for new and existing systems. In this context, many wireless middleware provide the security capabilities needed to prevent some of these attacks. Other security features such as privilege isolation can only be implemented at the operating system level, making middleware an incomplete although solution. Thus, applying low- and high-level security becomes a synergic approach for overall security in the wireless computing universe.

When we talk about high-level security, we usually refer to authentication, authorization, confidentiality, availability, non-repudiation and integrity. These are the major security objectives and properties that need to be achieved. For instance, we can choose between password-based or certificate-based authentication, role-based access control (RBAC) or multilevel authorization, Kerberos or SSL for confidentiality, checksum or atomic transaction for integrity, etc. Typically, the lack of security at that level is mainly caused by either the absence of these features or the improper implementation of the mechanisms enforcing them. We focus in this subsection on the first case since the later one belongs to low-level security hardening.

Hardening of high-level security consists of configuration changes at the relevant levels, changes in the application source up to a complete re-design, or integrating new components or libraries to provide the required features. The aforementioned options can be applied either manually or automatically by using AOP. In order for this approach to be truly relevant, the threats should be determined first, so that only the needed counter-measures are put in place. Once the threat is well identified and categorized, it is possible to determine the appropriate technique(s) to mitigate it. In the literature, it is possible to find mapping between categories of threats and known counter-measures addressing them. Choosing the best techniques will be mostly based on the state of the art of weaknesses and mitigation methods as well as security patterns. For instance, in Ref. 16, the authors provide a list of mitigation techniques for each category of threats of their classification. Table 1 provides an excerpt of this mapping.

Table 1 Mapping between threats and mitigation.

Threat type	Mitigation techniques
Spoofing identity	Appropriate authentication, protect secret data
Tampering with data	Appropriate authorization, hashes, message authentication codes, digital signatures
Repudiation	Digital signatures, timestamps, audit trails
Information disclosure	Authorization, encryption, protect secrets
Denial of service	Appropriate authentication and authorization, filtering, throttling, quality of service
Elevation of privilege	Run with least privilege

The hardening of counter-measures should be specified by the security architects into security hardening plans. These plans determine what the security improvements are, where they should be applied and how they should be applied. These plans offer a separation of concerns between the specification of the hardening and its implementation. The detail steps needed to perform the hardening are the responsibility of security experts. In many cases, they are encapsulated in patterns, libraries, and middleware frameworks. For instance, security patterns encapsulate expert knowledge in the form of proven solutions to common problems. Catalog of patterns offer a structured repository of solutions that are inherently formulated in a form similar to the weakness-mitigation mapping we detailed previously. Many works have been published in this domain, which we previously surveyed in Ref. 62.

Hardening of Low-Level Security

Based on our classification, deploying security at that level is mainly categorized as code-level and software process hardening. As such, this type of hardening will be extremely dependent on the programming language and the platform. In this context, we discuss in this subsection the hardening techniques used to remedy the low-level security vulnerabilities in C programs.

Hardening for BoF vulnerabilities

Many APIs and tools have been deployed to solve the problem of BoF or to make its exploitation harder.[17,33,35] More methods for secure coding can be found in Ref. 16. In this context, the following are some design and programming tips and assumptions that can help to solve the BoF problem:[14]

1. Always assume that input may overflow a buffer and design the program in a way that provides proper input validation conditions;
2. Use functions that respect buffer bounds such as fgets, strncpy, strncat;
3. Ensure NULL-termination of strings, even when using those functions;
4. Invalidate stack execution, since stack-based BoFs are the easiest to exploit;
5. The number of arguments of printing functions must be checked to make sure that the format string argument is explicitly specified.

Table 2 summarizes the security hardening solutions for BoFs.

Hardening for integer vulnerabilities

Integer vulnerabilities can be solved using sound coding practices. The generalized use of unsigned integers can

Table 2   Hardening for BoFs.

Hardening level	Product/method
Code	Bound-checking, memory manipulation functions with length parameter, null-termination, ensuring proper loop bounds
Software process	Compile with canary words, inject bound-checking aspects
Design	Input validation, input sanitization
Operating environment	Disable stack execution, use libsafe, enable stack randomization

simplify things for the programmer, and the addition of range checking before sensitive operations can avoid unexpected results. Some compilers provide built-in support for detection of integer issues, and it is possible to replace integer operations with safer calls.[18] The security hardening solutions that we described above are summarized in Table 3.

Hardening for memory management vulnerabilities

There are no known API or library solutions solving memory management problems as a whole. However, hardened memory managers can prevent multiple freeing vulnerabilities. Other than that, only improvements in programming practices can be useful in hardening against such problems. The following are some hints and best practices: initialize each declared pointer and make it point to a valid memory location, do not allow a process to de-reference or operate on a freed pointer, and apply error checking on memory allocation calls. The security hardening solutions that we described above are summarized in Table 4.

Hardening for file management vulnerabilities

Vulnerabilities related to unsafe temporary file management and creation can be minimized by using secure

Table 3   Hardening for integer vulnerabilities.

Hardening level	Product/method
Code	Use of functions detecting integer overflow/underflow, migration to unsigned integers, ensuring integer data size in assignments/casts
Software process	Compiler option to convert arithmetic operation to error condition-detecting functions
Design	
Operating environment	

Table 4   Hardening for memory management vulnerabilities.

Hardening level	Product/method
Code	NULL assignment on freeing and initialization, error handling on allocation
Software process	Using aspects to inject error handling and assignments, compiler option to force detection of multiple-free errors
Design	
Operating environment	Use a hardened memory manager (e.g., dmalloc, phkmalloc)

library calls.[19] In some cases, we can re-design the application to use inter-process communication instead of temporary files. File creation mask vulnerabilities, in UNIX-like systems, can be resolved using proper file creation-related system calls and specifying appropriate access rights. The security hardening solutions that we described above are summarized in Table 5.

CONCLUSIONS

In response to the growth of mobile system use and the need to improve their security, we provided in this entry a survey on the security of wireless applications and middleware. We first presented the related work in section "Related Works", and then briefly surveyed software vulnerabilities in Section "Security Requirements and Vulnerabilities." Afterwards, we studied automatic detection of low-level security vulnerabilities in Section "Security Evaluations." In Section "Security Hardening," we determined how existing mobile applications and middleware can be hardened in a rigorous and systematic manner in response to the discovered vulnerability.

Table 5   Hardening for file management vulnerabilities.

Hardening level	Product/method
Code	Use of proper temporary file functions, default use of restrictive file permissions, setting a restrictive file creation mask, use of ISO/IEC TR 24731 functions
Software process	Set a wrapper changing file creation mask
Design	Refactor to avoid temporary files
Operating environment	Restricting access rights to relevant directories

REFERENCES

1. Jamaluddin, J.; Zotou, N.; Edwards, R.; Coulton, P. Mobile phone vulnerabilities: a new generation of malware. 2004 IEEE International Symposium on Consumer Electronics, Reading, UK, Sept 1–3, 2004; 199–202.

2. Leavitt, N. Malicious code moves to mobile devices. Computer. 2000, 33 (12), 16–19.

3. Das, S.; Anjum, F.; Ohba, Y.; Salkintzis, A.K. Security issues in wireless IP networks. In *Mobile Internet: Enabling Technologies and Services*; Salkintzis, A.K., Ed.; CRC Press: Boca Raton, FL, 2004; Vol. Chapter 9, 9-1–9-31.

4. Nikander, P.; Arko, J. Secure mobility in wireless IP networks. In *Mobile Internet: Enabling Technologies and Services*; Salkintzis, A.K., Ed.; CRC Press: Boca Raton, FL, 2004; Vol. Chapter 8.

5. Wang, G.; Chen, A.; Sripada, S.; Wang, C. Application of middleware technologies to mobile enterprise information services. In *Middleware for Communications*; Mahmoud, Q.H., Ed.; John Wiley & Sons Ltd.: West Sussex, UK, 2004; Vol. Chapter 12, 281–304.

6. Mascolo, C.; Capra, L.; Emmerich, W. Principles of mobile computing middleware. In *Middleware for Communications*; Mahmoud, Q.H., Ed.; John Wiley & Sons Ltd.: West Sussex, UK, 2004; Vol. Chapter 11, 261–280.

7. Adelstein, F.; Gupta, S.K.S.; Richard III, G.G.; Schwiebert, L. *Fundamentals of Mobile and Pervasive Computing*; McGraw-Hill: New York, 2005; 40–2.

8. Bacon, J.; Moody, K.; Yao, W. Access control and trust in the use of widely distributed services. Middleware 2001: IFIP/ACM International Conference on Distributed Systems Platforms, Heidelberg, Germany, Nov 12–16, 2001; Springer: 2001.

9. Demurjian, S.; Bessette, K.; Doan, T.; Phillips, C. Concepts and capabilities of middleware security. In *Middleware for Communications*; Mahmoud, Q.H., Ed.; John Wiley & Sons Ltd.: West Sussex, UK, 2004; Vol. Chapter 9, 211–236.

10. Vogt, H.; Rohs, M.; Kilian-Kehr, R. Middleware for smart cards. In *Middleware for Communications*; Mahmoud, Q. H., Ed.; John Wiley & Sons Ltd.: West Sussex, UK, 2004; Vol. Chapter 15, 359–392.

11. Aslam, T. A taxonomy of security faults in the unix operating system, M.S. thesis; Purdue University: West Lafayette, IN, Aug 1995.

12. Aslam, T.; Krsul, I.; Spafford, E.H. Use of a taxonomy of security faults. 19th NIST-NCSC National Information Systems Security Conference, Baltimore, MD, Oct 22–25, 1996; 551–560.

13. Bishop, M. A Taxonomy of Unix System and Network Vulnerabilities. Technical Report CSE-9510, Department of Computer Science, University of California at Davis, May 1995.

14. Bishop, M.; Bailey, D. A Critical Analysis of Vulnerability Taxonomies. Technical Report CSE-96-11, Department of Computer Science, University of California at Davis, 1996.

15. McDermott, J.P.; Landwehr, C.E.; Bull, A.R.; Choi, W.S. A taxonomy of computer program security flaws. ACM Comput. Surv. 1994, 26 (3), 211–254, http://doi.acm.org/10.1145/185403.185412.

16. Howard, M.; LeBlanc, D.E. *Writing Secure Code*; Microsoft Press: Redmond, WA, 2002.

17. Younan, Y.; Joosen, W.; Piessens, F. Code Injection in C and C++: a Survey of Vulnerabilities and Countermeasures. Technical Report CW386, Department of Computer Science, Katholieke Universiteit Leuven, July 2004.

18. Seacord, R. *Secure Coding in C and C++*; Addison-Wesley: Boston, MA, 2005.

19. Wheeler, D. Secure Programming for Linux and Unix HOWTO—Creating Secure Software v3.010. 2003, http://www.dwheeler.com/secure-programs/.

20. Blexim. Basic integer overflows. Phrack 2002, 0x0b (0x3c), http://www.phrack.org/phrack/60/p60-0x0a.txt.

21. Aleph, One Smashing the stack for fun and profit. Phrack. 1996, 7 (49), http://www.phrack.org/phrack/49/P49-14.

22. Friedl's, S. SQL Injection Attacks by Example, 2005, http://www.unixwiz.net/techtips/sql-injection.html.

23. Twitch. Taking advantage of non-terminated adjacent memory spaces. Phrack 2000, 0xA (0x38), http://www.phrack.com.

24. Bishop, M. How to Write a Setuid Program. Technical Report Technical Report 85.6, Research Institute for Advanced Computer Science, Moffett Field, May 1985.

25. Shinagawa, T. Implementing A Secure Setuid Program; Dept. of Computer, Information and Communication Sciences. Tokyo University of Agriculture and Technology: Japan.

26. Wheeler, D.A. Secure Programmer: Prevent Race Conditions; 2004, http://www-128.ibm.com/developerworks/library-combined/l-sprace.html.

27. Bishop, M. How Attackers Break Programs, and How to Write More Secure Programs, http://nob.cs.ucdavis.edu/Ÿbishop/secprog/sans2002/index.html.

28. Viega, J.; Messier, M. *Secure Programming Cookbook For C and C++*; O'Reilly Media Inc.: Sebastopal, CA, 2003.

29. Watson, G. Debug Malloc Library; 2004, http://dmalloc.com/docs/5.4.2/online/dmalloc_toc.html.

30. Avaya Labs Research. Libsafe, http://www.research.avaya-labs.com/project/libsafe.

31. IMMUNIX. Stackguard: Protecting Systems From Stack Smashing Attacks, http://www.cse.ogi.edu/DISC/projects/immunix/StackGuard/.

32. Ashcraft, K.; Engler, D. Using programmer-written compiler extensions to catch security holes. 2002 IEEE Symposium on Security and Privacy, Oakland, CA, May 12–15, 2002.

33. Bhatkar, S.; DuVarne, D.C.; Sekar, R. Address obfuscation: an efficient approach to combat a broad range of memory error exploits. 12th USENIX Security Symposium, Washington, DC, Aug 4–8, 2003.

34. Xu, J.; Kalbarczyk, Z.; Iyer, R.K. Transparent runtime randomization for security. 22nd International Symposium on Reliable Distributed Systems, 2003, Florence, Italy, Oct 6–8, 2003.

35. McCormick, J. Openbsd declares war on buffer overruns. TechRepublic. 2003, http://techrepublic.com.com/5100-1035_11-5034831.html.

36. Shankar, U.; Talwar, K.; Foster, J.S.; Wagner, D. Detecting format string vulnerabilities with type qualifiers. 10th USENIX Security Symposium, Washington, DC, Aug 13–17, 2001.

37. Viega, J.; Bloch, J.T.; Kohno, Y.; McGraw, G. ITS4: a static vulnerability scanner for C and C++ code. 16th Annual Computer Security Applications Conference (ACSAC '00), New Orleans, LA, Dec 11–15, 2000; IEEE Computer Society: 2000; 25–7.

38. Evans, D.; Guttag, J.V.; Horning, J.J.; Tan, Y.M. LCLint: a tool for using specifications to check code. 2nd ACM SIG-SOFT Symposium on the Foundations of Software Engineering, New Orleans, LA, Dec 6–9, 1994.

39. Larochelle, D.; Evans, D. Statically detecting likely buffer overflow vulnerabilities. 10th USENIX Security Symposium, Washington, DC, Aug 13–17, 2001; 177–190.

40. Chen, H.; Wagner, D. MOPS: an infrastructure for examining security properties of software. 9th ACM Conference on Computer and Communications Security (CCS'02), Washington, DC, Nov 18–22, 2002; 235–244.

41. PolySpace. Automatic Detection of Run-Time Errors at Compile Time, http://www.polyspace.com/.

42. Coverity. Coverity Prevent for C and C++, http://www.coverity.com/main.html.

43. Zhang, X.; Edwards, A.; Jaeger, T. Using CQUAL for static analysis of authorization hook placement. 11th USENIX Security Symposium, San Francisco, CA, Aug 5–9, 2002; 33–48.

44. Engler, D.; Ashcraft, K. Racerx: effective, static detection of race conditions and deadlocks. 19th ACM symposium on Operating systems principles (SOSP '03), Sagamore, NY, Oct 19–22, 2003; ACM Press: 2003; 237–252, http://doi.acm.org/10.1145/945445.945468.

45. Evans, D. Static detection of dynamic memory errors. ACM SIGPLAN 1996 conference on Programming language design and implementation (PLDI '96), Philadelphia, PA, May 22–24, 1996; ACM Press: New York, 1996; 44–53.

46. Hasting, R.; Joyce, B. Purify: fast detection of memory leaks and access errors. Winter USENIX Conference, San Francisco, CA, Jan 1992; 125–136.

47. Seward, J.; Nethercote, N. Using valgrind to detect undefined value errors with bit-precision. USENIX'05 Annual Technical Conference, Anaheim, CA, Apr 10–15, 2005.

48. ISO/IEC 7498-2:1989. Information processing systems—open systems interconnection—basic reference model—part 2: Security architecture, 1989.

49. Fagan, M.E. Design and code inspections to reduce errors in program development. IBM Syst. J. **1999**, *38* (2–3), 258–287.

50. Nielson, F.; Nielson, H.R.; Hankin, C. *Principles of Program Analysis*; Springer-Verlag, New York Inc.: Secaucus, NJ, 1999.

51. AbsInt. Advanced Compiler Technology for Embedded Systems, http://www.absint.com/.

52. Necula, G.C.; McPeak, S.; Weimer, W. CCured: type-safe retrofitting of legacy code. 29th SIGPLAN-SIGACT Symposium on Principles of Programming Languages, Portland, OR, Jan 16–18, 2002; 128–139.

53. Grossman, D.; Morrisett, G.; Jim, T.; Hicks, M.; Wang, Y.; Cheney, J. Region-based memory management in cyclone. ACM SIGPLAN 2002 Conference on Programming language design and implementation (PLDI '02), Berlin, Germany, June 17–19, 2002; ACM Press: New York, 2002; 282–293.

54. Yutaka, O.; Sekiguchi, T.; Sumii, E.; Yonezawa, A. Fail-safe ANSI-C compiler: an approach to making C programs secure: progress report. Software security: theories and systems: Mext-NSF-JSPS international symposium, ISSS 2002, Tokyo, Japan, Nov 8–10, 2002; Springer-Verlag: Berlin/Heidelberg, 2002; 133–153.

55. Shankar, U.; Talwar, K.; Foster, J.; Wagner, D. Detecting format string vulnerabilities with type qualifiers. 10th USENIX Security Symposium, Washington, DC, Aug 13–17, 2001; 201–220.

56. SLAM. A Symbolic Model Checker for Boolean Programs, http://research.microsoft.com/slam/.

57. Chen, H.; Wagner, D.A. MOPS: an Infrastructure for Examining Security Properties of Software. Technical Report UCB/CSD-02-1197, EECS Department, University of California, Berkeley, 2002.

58. Parasoft. Parasoft AEP: Automated Error Prevention Solutions for Business and Information Technology Initiatives, http://www.parasoft.com/.

59. Bishop, M. *Computer Security: Art and Science*; Addison-Wesley Professional: Boston, MA, 2003; pp. 1084.

60. Blakley, B.; Heath, C. Security Design Patterns. Technical Report G031, Open Group, 2004.

61. Bastille linux, 2006, http://www.bastille-linux.org/.

62. Laverdière, M.-A.; Mourad, A.; Hanna, A.; Debbabi, M. Security design patterns: Survey and evaluation. IEEE Canadian Conference on Electrical and Computer Engineering, Ottawa, Canada, May 7–10, 2006.

Wireless Asynchronous Transfer Mode (ATM): Internetworking

Melbourne Barton
Telcordia Technologies, Red Bank, New Jersey, U.S.A.

Matthew Cheng
Li Fung Chang
Mobilink Telecom, Middletown, New Jersey, U.S.A.

Abstract

This entry describes the ATM networking in wireless systems. A generic personal communications services to ATM internetworking and its architectural features and models are presented.

INTRODUCTION

The ATM Forum's wireless asynchronous transfer mode (WATM) Working Group (WG) is developing specifications intended to facilitate the use of ATM technology for a broad range of wireless network access and interworking scenarios, both public and private. These specifications are intended to cover the following two broad WATM application scenarios:

- *End-to-End WATM*—This provides seamless extension of ATM capabilities to mobile terminals, thus providing ATM virtual channel connections (VCCs) to wireless hosts. For this application, high data rates are envisaged, with limited coverage, and transmission of one or more ATM cells over the air.
- *WATM interworking*—Here the fixed ATM network is used primarily for high-speed transport by adding mobility control in the ATM infrastructure network, without changing the non-ATM air interface protocol. This application will facilitate the use of ATM as an efficient and cost-effective infrastructure network for next generation non-ATM wireless access systems, while providing a smooth migration path to seamless end-to-end WATM.

This entry focuses on the ATM interworking application scenario. It describes various interworking and non-ATM wireless access options and their requirements. A generic PCS-to-ATM interworking scenario is described which enumerates the architectural features, protocol reference models, and signaling issues that are being addressed for mobility support in the ATM infrastructure network. The term PCS is being used in a generic sense to mean emerging digital wireless systems, which support mobility in microcellular and other environments. It is currently defined in ANSI T1.702-1995 as "a set of capabilities that allows some combination of terminal mobility, personal mobility, and service profile management." Evolution strategies intended to eventually provide end-to-end WATM capabilities and a methodology to consistently support a range of quality of service (QoS) levels on the radio link are also described.

BACKGROUND AND ISSUES

ATM is the switching and multiplexing standard for broadband integrated services digital network (BISDN), which will ultimately be capable of supporting a broad range of applications over a set of high capacity multiservice interfaces. ATM holds out the promise of a single network platform that can simultaneously support multiple bandwidths and latency requirements for fixed access and wireless access services without being dedicated to any one of them. In today's wireline ATM network environment, the user network interface (UNI) is fixed and remains stationary throughout the connection lifetime of a call. The technology to provide fixed access to ATM networks has matured. Integration of fixed and wireless access to ATM will present a cost-effective and efficient way to provide future tetherless multimedia services, with common features and capabilities across both wireline and wireless network environments. Early technical results[1–3] have shown that standard ATM protocols can be used to support such integration and extend mobility control to the subscriber terminal by incorporating wireless specific layers into the ATM user and control planes (C-Planes).

Integration of wireless access features into wireline ATM networks will place additional demands on the

Encyclopedia of Wireless and Mobile Communications DOI: 10.1081/E-EWMC-120043950

fixed network infrastructure due primarily to the additional user data and signaling traffic that will be generated to meet future demands for wireless multimedia services. This additional traffic will allow for new signaling features including registration, call delivery, and handoff during the connection lifetime of a call. Registration keeps track of a wireless user's location, even though the user's communication link might not be active. Call delivery, establishes a connection link to/from a wireless user with the help of location information obtained from registration. The registration and call delivery functions are referred to as location management. Handoff is the process of switching (rerouting) the communication link from the old coverage area to the new coverage area when a wireless user moves during active communication. This function is also referred to as mobility management.

In June 1996 the ATM Forum established a WATM WG to develop requirements and specifications for WATM. The WATM standards are to be compatible with ATM equipment adhering to the (then) current ATM Forum specifications. The technical scope of the WATM WG includes development of: 1) radio access layer (RAL) protocols for the PHY, medium access control (MAC), and data link control (DLC) layers; 2) wireless control protocols for radio resource management; 3) mobile protocol extensions for ATM (mobile ATM) including handoff control, routing considerations, location management, traffic and QoS control, and wireless network management; and 4) wireless interworking functions (IWFs) for mapping between non-ATM wireless access and ATM signaling and control entities. Phase-1 WATM specifications are being developed for short-range, high-speed, end-to-end WATM devices using wireless terminals that operate in the 5 GHz frequency band. Operating speeds will be up to 25 Mbps, with a range of 30–50 m indoor and 200–300 m outdoor. The ETSI broadband radio access networks (BRANs) project is developing the RAL for Phase-1 WATM specifications using the high performance radio LAN (HIPERLAN) functional requirements. The ATM Forum plans to release the Phase-1 WATM specifications by the second quarter of 1999.

There are a number of emerging wireless access systems (including digital cellular, PCS, legacy LANs based on the IEEE 802.11 standards, satellite, and IMT-2000 systems), that could benefit from access to, and interworking with, the fixed ATM network. These wireless systems are based on different access technologies and require development of different IWFs at the wireless access network and fixed ATM network boundaries to support WATM interworking. For example, a set of network interfaces have already been identified to support PCS access to the fixed ATM network infrastructure, without necessarily modifying the PCS air interface protocol to provide end-to-end ATM capabilities.[4–6] The WATM WG might consider forming sub-WGs which could work in parallel to identify other network interfaces and develop IWF

specifications for each of (or each subset of) the wireless access options that are identified. These WATM interworking specifications would be available in the Phase-2 and later releases of the WATM standards.

Some service providers and network operators see end-to-end WATM as a somewhat limited service option at this time because it is being targeted to small enterprise networks requiring high-speed data applications, with limited coverage and low mobility. On the other hand, WATM interworking can potentially support a wider range of services and applications, including low-speed voice and data access, without mandatory requirements to provide over-the-air transmission of ATM cells. It will allow for wider coverage, possibly extending to macrocells, while supporting higher mobility. WATM interworking will provide potential business opportunities, especially for public switched telephone network (PSTN) operators and service providers, who are deploying emerging digital wireless technologies such as PCS. Existing wireless service providers (WSPs) with core network infrastructures in place can continue to use them while upgrading specific network elements to provide ATM transport. On the other hand, a new WSP entrant without such a network infrastructure can utilize the public (or private) ATM transport network to quickly deploy the WATM interworking service, and not be burdened with the cost of developing an overlay network. If the final goal is to provide end-to-end WATM services and applications, then WATM interworking can provide an incremental development path.

WIRELESS INTERWORKING WITH TRANSIT ATM NETWORKS

Fig. 1 shows one view of the architectural interworking that will be required between public/private wireless access networks and the fixed ATM network infrastructure. It identifies the network interfaces where modifications will be required to allow interworking between both systems. A desirable objective in formulating WATM specifications for this type of wireless access scenario should be to minimize modifications to the transit ATM network and

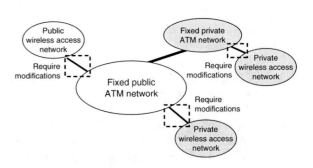

Fig. 1 Wireless ATM interworking architecture.

existing/emerging wireless access system specifications. This objective can be largely met by limiting major modifications to the network interfaces between the boundaries of the transit ATM network and public/private wireless networks, and where possible, adopting existing network standard processes (i.e., SS7, IS-41, MAP, AIN.3, Q.931, Q.932, Q.2931, etc.) to minimize development costs and upgrades to existing service providers' network infrastructure. Development of standard network interfaces that allow interworking of a reasonable subset of non-ATM digital wireless access systems with the fixed ATM network infrastructure ensure that:

- Large-scale revisions and modifications are not necessary to comply with later versions of the WATM specifications to accommodate other emerging digital wireless access systems that do not require end-to-end ATM connectivity.
- WATM systems are supported by open interfaces with a rich set of functionality to provide access to both ATM and non-ATM wireless access terminal devices.
- WATM services can reach a much larger potential market including those markets providing traditional large-scale support for existing voice services and vertical voice features.

Integrated Wireless–Wireline ATM Network Architecture

Fig. 2 shows one example of a mature, multifunctional ATM transport network platform, which provides access to fixed and mobile terminals for wide-area coverage. Four distinct network interfaces are shown supporting: 1) fixed access with non-ATM terminal, 2) fixed access with ATM terminal, 3) wireless access with non-ATM terminal (WATM interworking), and 4) wireless access with ATM terminal (end-to-end WATM).

ITU and ATM Forum standard interfaces either exist today or are being developed to support fixed access to ATM networks through various network interfaces. These include frame relay service (FRS) UNI, cell relay service (CRS) UNI, circuit emulation service (CES) UNI, and switched multimegabit data service (SMDS) subscriber NI (SNI). The BISDN intercarrier interface (B-ICI) specification[7] provides examples of wired IWFs that have been developed for implementation above the ATM layer to support intercarrier service-specific functions developed at the network nodes, and distributed in the public ATM/BISDN. These distributed, service-specific functions are defined by B-ICI for FRS, CRS, CES, and SMDS. Examples of such functions include ATM cell conversion, clock recovery, loss of signal and alarm indication detection, virtual channel identifier (VCI) mapping, access class selection, encapsulation/mapping, and QoS selection. In addition to the BICI, the private network–network interface (PNNI) specification[8] defines the basic call control signaling procedures (e.g., connection setup and release) in private ATM networks. It also has capabilities for autoconfiguration, scalable network hierarchy formation, topology information exchange, and dynamic routing.

On the wireless access side, existing ITU specifications provide for the transport of wireless services on public ATM networks (see, e.g., Refs. 9 and 10). For example, if the ATM UNI is at the mobile switching center (MSC), then message transfer part (MTP) 1 and 2 would be replaced by the PHY and ATM layers, respectively, the BISDN user part (BISUP) replaced by MTP 3, and a common channel signaling (CCS) interface deployed in the ATM node. BISUP is used for ATM connection setup and any required feature control. If the ATM UNI is at the base station controller (BSC), then significant modifications are likely to be required. Equipment manufacturers have not implemented, to any large degree, the features that are available with the ITU specifications. In any case, these features are not sufficient to support the WATM scenarios postulated in this entry.

The two sets of WATM scenarios postulated in this entry are shown logically interfaced to the ATM network through a mobility-enabled ATM (ME-ATM) switch. This enhanced ATM access switch will have capabilities to support mobility management and location management. In addition to supporting handoff and rerouting of ATM connections, it will be capable of locating a mobile user anywhere in the network.

It might be desirable that functions related to mobility not be implemented in standard ATM switches so that network operators and service providers are not required to modify their ATM switches in order to accommodate WATM and related services. A feasible strategy that has been proposed is to implement mobility functions in servers (i.e., service control modules or SCMs) that are logically separated from the ATM switch. In these servers, all mobility features, service creation logic, and service

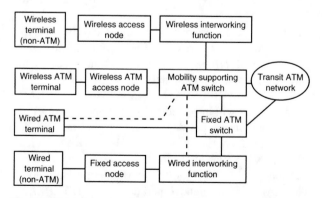

Fig. 2 Wireless and wireline system integration with transit ATM network.

management functions will be implemented to allow logical separation of ATM switching and service control from mobility support, service creation, and management. The open network interface between the ATM access switch and SCM will be standardized to enable multivendor operation. It would be left up to the switch manufacturer to physically integrate the SCM into a new ATM switching fabric, or implement the SCM as a separate entity.

Wireless Access Technology Options

Fig. 3 identifies various digital wireless systems that could be deployed to connect mobile terminals to the transit ATM network through IWFs that have to be specified. The main emphasis is on developing mobility support in the transit ATM infrastructure network to support a range of wireless access technologies. Significant interests have been expressed, through technical contributions and related activities in the WATM WG, in developing specifications for the IWFs shown in Fig. 3 to allow access to the fixed ATM network through standard network interfaces. The main standardization issues that need to be addressed for each wireless access system are described below.

PCS

This class of digital wireless access technologies include the low-tier PCS systems as described in Cox.[11] Digital cellular systems which are becoming known in the US as high-tier PCS, especially when implemented in the 1.8–1.9 GHz PCS frequency bands, are addressed in the "Digital Cellular" section below. The PCS market has been projected to capture a significant share of the huge potential revenues to be generated by business and residential customers. In order to provide more flexible, widespread, tetherless portable communications than can be provided by today's limited portable communications approaches, low-power exchange access radio needs to be integrated

with network intelligence provided by the wireline network.

Today, the network for supporting PCS is narrowband ISDN, along with network intelligence based on advanced intelligent network (AIN) concepts, and a signaling system for mobility/location management and call control based on the signaling system 7 (SS7) network architecture. It is expected that the core network will evolve to BISDN/ATM over time with the capability to potentially integrate a wide range of network services, both wired and wireless, onto a single network platform. Furthermore, there will be no need for an overlay network for PCS. In anticipation of these developments, the WATM WG included the development of specifications and requirements for PCS access to, and interworking with, the ATM transit network in its charter and work plan. To date, several contributions have been presented at WATM WG meetings that have identified some of the key technical issues relating to PCS-to-ATM interworking. These include 1) architectures, 2) mobility and location management signaling, 3) network evolution strategies, and 4) PCS scenarios.

Wireless LAN

Today, wireless LAN (WLAN) is a mature technology. WLAN products are frequently used as LAN extensions to access areas of buildings with wiring difficulties and for cross-building interconnect and nomadic access. Coverage ranges from tens to a few hundreds of meters, with data rates ranging from hundreds of Kbps to more than 10 Mbps. Several products provide 1 or 2 Mbps. ATM LAN products provide LAN emulation services over the connection-oriented (CO) ATM network using various architectural alternatives.[12] In this case, the ATM network provides services that permit reuse of existing LAN applications by stations attached directly to an ATM switch, and allow interworking with legacy LANs. Furthermore, the increasing importance of mobility in data access networks and the availability of more usable spectrum are expected to speed up the evolution and adoption of mobile access to WLANs. Hence, it is of interest to develop WLAN products that have LAN emulation capabilities similar to wireline ATM LANs.

The ETSI BRAN project is developing the HIPERLAN RAL technology for wireless ATM access and interconnection. It will provide short-range wireless access to ATM networks at approximately 25 Mbps in the 5 GHz frequency band. HIPERLAN is an ATM-based WLAN technology that will have end-to-end ATM capabilities. It does not require the development of an IWF to provide access to ATM. A previous version of HIPERLAN (called HIPERLAN I), which has been standardized, supports data rates from 1 to 23 Mbps in the 5 GHz band using a non-ATM RAL.[13] This (and other non-ATM HIPERLAN standards being developed) could benefit from

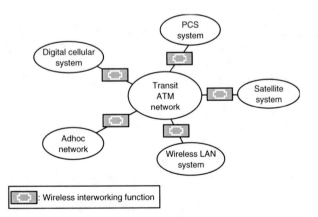

: Wireless interworking function

Fig. 3 Various wireless access technologies supported by wireless ATM interworking.

interworking with the backbone ATM as a means of extending the marketability of these products in the public domain such as areas of mass transit and commuter terminals.

Several proposals have been submitted to IEEE 802.11 to provide higher speed extensions of current IEEE 802.11 systems operating in the 2.4 GHz region and the development of specifications for new systems operating in the 5 GHz frequency band. The proposed 2.4 GHz extensions support different modulation schemes, but are interoperable with the current IEEE 802.11 low rate PHY and are fully compliant with the IEEE 802.11 defined MAC. The 5 GHz proposals are not interoperable with the current 2.4 GHz IEEE 802.11 systems. One of the final three 5 GHz proposals being considered is based on orthogonal frequency division multiplexing (OFDM), or multicarrier modulation. The other two are single carrier proposals using offset quadrature phase-shift keying (OQPSK)/offset quadrature amplitude modulation (QAM) (OQAM), and differential pulse position modulation (DPPM). The OFDM proposal has been selected. With 16-QAM on each sub-carrier and rate-3/4 convolutional coding, the OFDM system has a peak data rate capability of 30 Mbps.

It is clear that a whole range of WLAN systems either exist today, or are emerging, that are not based on ATM technology, and therefore cannot provide seamless access to the ATM infrastructure network. The development of IWF specifications that allow these WLANs to provide such access through standard network interfaces will extend the range of applications and service features provided by WLANs. In Pahlavan,[14] a number of architectural alternatives to interconnect WLAN and WATM to the ATM and/or LAN backbone are discussed, along with service scenarios and market and product issues.

Digital Cellular

Digital cellular mobile radio systems include the 1.8–1.9 GHz (high-tier PCS) and the 800–900 MHz systems that provide high-mobility, wide-area coverage over macrocells. Cellular radio systems at 800–900 MHz have evolved to digital in the form of GSM in Europe, personal digital cellular (PDC) in Japan, and IS-54 time division multiple access (TDMA) and IS-95 CDMA in the US. The capabilities in place today for roaming between cellular networks provide for even wider coverage. Cellular networks have become widespread, with coverage extending beyond some national boundaries. These systems integrate wireless access with large-scale mobile networks having sophisticated intelligence to manage mobility of users.

Cellular networks (e.g., GSM) and ATM networks are evolving somewhat independently. The development of IWFs that allow digital cellular systems to utilize ATM transport will help to bridge the gap. Cellular networks have sophisticated mechanisms for authentication and handoff, and support for rerouting through the home network. In order to facilitate the migration of cellular systems to ATM transport, one of the first issues that should be addressed is that of finding ways to enhance the basic mobility functions already performed by them for implementation in WATM. Contributions presented at WATM WG meetings have identified some basic mobility functions of cellular mobile networks (and cordless terminal mobility) which might be adopted and enhanced for WATM interworking. These basic mobility functions include rerouting scenarios (including path extension), location update, call control, and authentication.

Satellite

Satellite systems are among the primary means of establishing connectivity to untethered nodes for long-haul radio links. This class of applications has been recognized as an essential component of the National Information Infrastructure (NII).[15] Several compelling reasons have been presented in the ATM Forum's WATM WG for developing standard network interfaces for satellite ATM (SATATM) networks. These include 1) ubiquitous wide area coverage, 2) topology flexibility, 3) inherent point-to-multipoint and broadcast capability, and 4) heavy reliance by the military on this mode of communications. Although the geostationary satellite link represents only a fraction of satellite systems today, WATM WG contributions that have addressed this interworking option have focused primarily on geostationary satellites. Some of these contributions have also proposed the development of WATM specifications for SATATM systems having end-to-end ATM capabilities.

Interoperability problems between satellite systems and ATM networks could manifest themselves in at least four ways.

1. Satellite links operate at much higher BERs with variable error rates and bursty errors.
2. The approximately 540 ms round trip delay for geosynchronous satellite communications can potentially have adverse impacts on ATM traffic and congestion control procedures.
3. Satellite communications bandwidth is a limited resource, and might be incompatible with less bandwidth efficient ATM protocols.
4. The high availability rates (at required BERs) for delivery of ATM (e.g., 99.95%) is costly, hence the need to compromise between performance relating to availability levels and cost.

A number of experiments have been performed to gain insights into these challenges.[16] The results can be used to guide the development of WATM specifications for the satellite interworking scenario. Among the work items that have been proposed for SATATM access using geostationary satellites links are 1) identification of requirements for

RAL and mobile ATM functions; 2) study of the impact of satellite delay on traffic management and congestion control procedures; 3) development of requirements and specifications for bandwidth efficient operation of ATM speech over satellite links; 4) investigation of various WATM access scenarios; and 5) investigation of frequency spectrum availability issues.

One interesting SATATM application scenario has been proposed to provide ATM services to multiuser airborne platforms via satellite links for military and commercial applications. In this scenario, a satellite constellation is assumed to provide contiguous overlapping coverage regions along the flight path of the airborne platforms. A set of interworked ground stations form the mobile enhanced ATM network that provides connectivity between airborne platforms and the fixed terrestrial ATM network via bent-pipe satellite links. Key WATM requirements for this scenario have been proposed, which envisage among other things modifications to existing PNNI signaling and routing mechanisms to allow for mobility of (ATM) switches.

In related work, the Telecommunications Industry Association (TIA) TR34.1 WG has also proposed to develop technical specifications for SATATM networks. Three ATM network architectures are proposed for bent-pipe satellites and three others for satellites with onboard ATM switches.[17] Among the technical issues that are likely to be addressed by TR34.1 are protocol reference models and architecture specifications, RAL specifications for SATATM, and support for routing, rerouting, and handoff of active connections. A liaison has been established between the ATM Forum and TR34.1, which is likely to lead to the WATM WG working closely with TR34.1 to develop certain aspects of the TR34.1 SATATM specifications.

Ad Hoc Networks

The term ad hoc network is used to characterize wireless networks that do not have dedicated terminals to perform traditional BS and/or wireless resource control functions. Instead, any mobile terminal (or a subset of mobile terminals) can be configured to perform these functions at any time. Ad hoc networking topologies have been investigated by WLAN designers, and are part of the HIPERLAN and IEEE 802.11 wireless LAN specifications.[18] As far as its application to WATM is concerned, low-cost, plug-and-play, and flexibility of system architecture are essential requirements. Potential application service categories include rapidly deployable networks for government use (e.g., military tactical networks, rescue missions in times of natural disasters, law enforcement operations, etc.), ad hoc business conferencing devoid of any dedicated coordinating device, and ad hoc residential network for transfer of information between compatible home appliances.

There are some unique interworking features inherent in ad hoc networks. For example, there is a need for location management functions not only to identify the location of terminals but also to identify the current mode of operation of such terminals. Hence, the WATM WG is considering proposals to develop separate requirements for ad hoc networks independent of the underlying wireless access technology. It is likely that requirements will be developed for an ad hoc RAL, mobility management signaling functions, and location management functions for supporting interworking of ad hoc networks that provide access to ATM infrastructure networks.

The range of potential wireless access service features and wireless interworking scenarios presented in the five wireless access technologies discussed above is quite large. For example, unlicensed satellite systems could provide 32 Kbps voice, and perhaps up to 10 Mbps for wireless data services. The main problem centers around the feasibility of developing specifications for IWFs to accommodate the range of applications and service features associated with the wireless access options shown in Fig. 3. The WATM WG might consider forming sub-WGs which would work in parallel to develop the network interface specifications for each (or a subset of each) of the above non-ATM wireless access options.

THE PCS-TO-ATM INTERWORKING SCENARIO

This section presents a more detailed view of potential near-term and longer-term architectures and reference models for PCS-to-ATM interworking. A signaling link evolution strategy to support mobility is also described. Mobility and location management signaling starts with the current CCS network, which is based on the SS7 protocol and 56 Kbps signaling links, and eventually migrates to ATM signaling. The system level architectures, and the mobility and location management signaling issues addressed in this section serve to illustrate the range of technical issues that need to be addressed for the other WATM interworking options.

Architecture and Reference Model

The near-term approach for the PCS-to-ATM interworking scenario is shown in Fig. 4. The near-term approach targets existing PCS providers with network infrastructures in place, who wish to continue to use them while upgrading specific network elements (e.g., MSCs) to provide ATM transport for user data. The existing MSCs in the PCS network are upgraded to include fixed ATM interfaces. ATM is used for transport and switching of user data, while mobility/location management and call control signaling is carried by the SS7 network. No mobility support is required in the ATM network. Synchronization

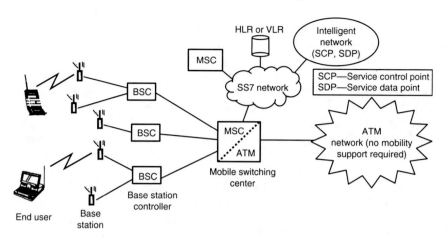

Fig. 4 A near-term architecture for PCS-to-ATM interworking.

problems might develop because different traffic types relating to the same call may traverse the narrow-band SS7 network and the broadband ATM network and arrive at the destination in an uncoordinated manner. Upgrading the SS7 network to broadband SS7 (e.g., T1 speeds or higher) should partially alleviate this potential problem.

A longer-term approach for PCS-to-ATM interworking has been proposed in some technical contributions to the WATM WG. This is illustrated in Fig. 5, together with the protocol stacks for both data and signaling. The ATM UNI is placed at the BSC, which acts as the PCS-to-ATM gateway. The ATM network carries both data and signaling. ATM cells are not transmitted over the PCS link. Communications between the BS and BSC are specific to

the PCS access network, and could be a proprietary interface. Compared with existing/emerging BSCs, additional protocol layer functionality is required in the BSC to provide 1) transfer/translation and/or encapsulation of PCS protocol data units (PDUs), 2) ATM to wireless PDU conversion, and 3) a limited amount of ATM multiplexing/demultiplexing capabilities.

The BSC is connected to the ATM network through a ME-ATM access switch instead of a MSC. The ME-ATM switch provides switching and signaling protocol functions to support ATM connections together with mobility and location management. These functions could be implemented in servers that are physically separate from, but logically connected to, the ME-ATM switch. The WATM

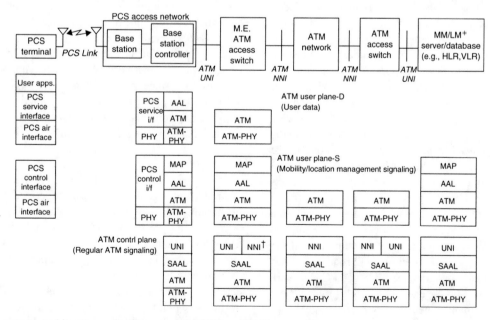

Fig. 5 A longer-term architecture and reference model PCS-to-ATM interworking.
† NNI here is used in a generic sense, to refer to both public and private networks. In public networks, there will be an additional MTP3 layer between the NNI and SAAL layers. ‡ MM—mobility management.

WG is expected to formulate requirements and specifications for these mobile-specific functions. On the other hand, an IWF can be introduced between the BSC and the ME-ATM switch shown in Fig. 5. In this case, the UNI is between the IWF and the ME-ATM switch, and another standard interface (not necessarily ATM) can be used to connect the BSC to the IWF. The BSC then requires no modification, but a new entity (i.e., the IWF) is required. The IWF will perform protocol conversion and it may serve multiple BSCs.

A unique feature of this architecture is that modifications to network entities to allow for interworking of PCS with the transit ATM network are only required at the edges of the ATM and wireless access networks, i.e., to the BSC and the ME-ATM access switch. In order to minimize the technical impact of mobility on existing/emerging transit ATM networks and PCS specifications, an initial interworking scenario is envisaged in which there are no (or only very limited) interactions between PCS and ATM signaling entities. PCS signaling would be maintained over the air interface, traditional ATM signaling would be carried in the C-Plane and PCS signaling would be carried over the ATM network as user traffic in the user plane (U-Plane). In the long term, this architecture is expected to eventually evolve to an end-to-end ATM capability.

On the ATM network side, mobility and location management signaling is implemented in the U-Plane as the mobile application part (MAP) layer above the ATM adaptation layer (AAL), e.g., AAL5. The MAP can be based on the MAP defined in the existing PCS standards (e.g., IS41-MAP or GSM-MAP), or based on a new set of mobile ATM protocols. The U-plane is logically divided into two parts, one for handling mobility and location management signaling messages and the other for handling traditional user data. This obviates the need to make modifications to the ATM UNI and NNI signaling protocols, currently being standardized. The MAP layer also provides the necessary end-to-end reliability management for the signaling because it is implemented in the U-Plane, where reliable communication is not provided in the lower layers of the ATM protocol stack as is done in the signaling AAL (SAAL). The MAP functionality can be distributed at individual BSCs or ME-ATM access switches or centralized in a separate server to further reduce modifications to existing implementations. The setup and release of ATM connections are still handled by the existing ATM signaling layer. This approach allows easy evolution to future wireless ATM architectures, which will eventually integrate the ATM and PCS network segments to form a homogeneous network to carry both mobility/location management and traditional ATM signaling in the C-Plane. issues relating to mobility support for PCS interworking with ATM networks are addressed in more detail in Cheng.[19]

Signaling Link Evolution

Signaling will play a crucial role in supporting end-to-end connections (without location restrictions) in an integrated WATM network infrastructure. The signaling and control message exchanges required to support mobility will occur more frequently than in wireline ATM networks. Today's CCS network, which is largely based on the SS7 protocol and 56 Kbps signaling links, will not be able to support the long-term stringent service and reliability requirements of WATM. Two broad deployment alternatives have been proposed for migrating the CCS/SS7 network (for wireline services) to using the broadband signaling platform.[20]

- Migration to high-speed signaling links using the narrowband signaling platform supported by current digital transmission (e.g., DS1) facilities. The intent of this approach is to support migration to high-speed signaling links using the existing CCS infrastructure with modified signaling network elements. This would allow the introduction of high-speed (e.g., 1.5 Mbps) links with possible minimal changes in the CCS network. One option calls for modifications of existing MTP2 procedures while maintaining the current protocol layer structure. Another option replaces MTP2 with some functionality of the ATM/SAAL link layer, while continuing to use the same transport infrastructure as DS1 and transport messages over the signaling links in variable length signal units delimited by flags.
- Alternative two supports migration of signaling links to a broadband/ATM signaling network architecture. Signaling links use both ATM cells and the ATM SAAL link layer, with signaling message transported over synchronous optical network (SONET) or existing DS1/DS3 facilities at rates of 1.5 Mbps or higher. This alternative is intended primarily to upgrade the current CCS network elements to support an ATM-based interface, but could also allow for the inclusion of signaling transfer point (STP) functions in the ATM network elements to allow for internetworking between ATM and existing CCS networks.

The second alternative provides a good vehicle for the long-term goal of providing high-speed signaling links on a broadband signaling platform supported by the ATM technology. One signaling network configuration and protocol option is the extended Q.93B signaling protocol over ATM in associated mode.[20] The PSTN's CCS networks are currently quasi-associated signaling networks. Q.93B is primarily intended for point-to-point bearer control in user-to-network access, but can also be extended for (link-to-link) switch-to-switch and some network-to-network applications. Standards activities to define high-speed signaling link characteristics in the SS7 protocol have been largely finalized. Standards to support SS7 over ATM are at various stages of completion. These activities provide a good basis for further evolving the SS7 protocol to provide the mobility and location management features that will be required to support PCS (and other wireless systems) access to the ATM network.

The functionality required to support mobility in cellular networks is currently defined as part of the MAP. Both IS-41 and GSM MAPs are being evolved to support PCSs with SS7 as the signaling transport protocol.[21] Quite similar to the two alternatives described above, three architectural alternatives have been proposed for evolving today's IS-41 MAP on SS7 to a future modified (or new) IS-41 on ATM signaling transport platform.[6] They are illustrated in Fig. 6. In the first (or near-term) approach, user data is carried over the ATM network, while signaling is carried over existing SS7 links. The existing SS7 network can also be upgraded to broadband SS7 network (e. g., using T1 links) to alleviate some of the capacity and delay constraints in the narrowband SS7 network. This signaling approach can support the near-term PCS-to-ATM interworking described in the previous subsection. In the second (or midterm) approach, a hybrid-mode

operation is envisaged, with the introduction of broadband SS7 network elements into the ATM network. This results in an SS7-over-ATM signaling transport platform. Taking advantage of the ATM's switching and routing capabilities, the MTP3 layer could also be modified to utilize these capabilities and eliminate the B-STP functionality from the network. In the third phase (or long-term approach), ATM replaces the SS7 network with a unified network for both signaling and user data. No SS7 functionality exists in this approach. Here, routing of signaling messages is completely determined by the ATM layer, and the MAP may be implemented in a format other than the transaction capability application part (TCAP), so that the unique features of ATM are best utilized. The longer-term PCS-to-ATM interworking approach is best supported for this signaling approach. The performance of several signaling protocols for PCS mobility support for this long-term architecture is presented in Cheng.[5,19] The above discussion mainly focuses on public networks. In private networks, there are no existing standard signaling networks. Therefore, the third approach can be deployed immediately in private networks.

**Wireless
Ad Hoc—Asynch**

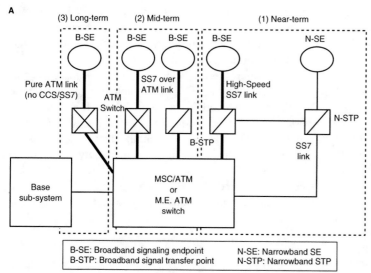

A

B-SE: Broadband signaling endpoint   **N-SE:** Narrowband SE
B-STP: Broadband signal transfer point   **N-STP:** Narrowband STP

Note: A signaling endpoint is the mobility control at a mobile switch, server or database.

Architecture

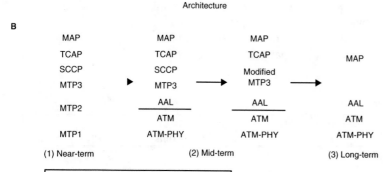

B

TCAP—Transaction capabilities application part
SCCP—Signaling connection control part

Protocol stack

Fig. 6 Signaling link evolution for mobility/location management over a public ATM network.

There are several ways to achieve the third signaling approach in ATM networks. The first approach is to overlay another "network layer" protocol (e.g., IP) over the ATM network, but this requires the management of an extra network. The second approach is to enhance the current ATM network–network interface (e.g., B-ICI for public networks and PNNI for private networks) to handle the new mobility/location management signaling messages and information elements, but this requires modifications in the existing ATM network. The third approach is to use dedicated channels (PVCs or SVCs) between the mobility control signaling points. This does not require any modifications in existing ATM specifications. However, signaling latency may be high in the case of SVCs, and a full mesh of PVCs between all mobility control signaling points is difficult to manage. The fourth approach is to use the generic functional protocols [e.g., CO-bearer independent (CO-BI) or connectionless-BI (CL-BI) transport mechanism] defined in ITUT's Q.2931.2.[22] This cannot be done in existing ATM networks without modifications, but these functions are being included in the next version of PNNI (PNNI 2.0) to provide a generic support for supplementary services. There are also technical contributions to the ATM Forum proposing "connectionless ATM",[23] which attempt to route ATM cells in a connectionless manner by using the routing information obtained through the PNNI protocol. However, the "connectionless ATM" concept is still being debated.

QoS SUPPORT

One immediate impact of adding mobility to an otherwise fixed ATM infrastructure network is the need to manage the changing QoS levels that are inherent in a mobile environment due to the vagaries of the wireless link, the need for rerouting of traffic due to handoff, and available bandwidth, etc. Dynamic QoS negotiation and flow control will be required to flexibly support QoS guarantees for multimedia service applications that are likely to be encountered in this environment. QoS provisioning is based on the notion that the wireless channel is likely to demand more stringent measures than the fixed ATM network to support end-to-end QoS. QoS metrics include 1) throughput, 2) delay sensitivity, 3) loss sensitivity, and 4) BER performance. issues relating to end-to-end QoS

provisioning in multimedia wireless networks are discussed in Naghshineh[24] and articles therein. Here, the focus is on BER maintenance in the context of PCS-to-ATM interworking using forward error correction (FEC) at the radio PHY layer. This is the first step toward developing a hybrid automatic repeat request (ARQ)/FEC protocol for error control of the wireless link, with FEC at the PHY supplemented by ARQ at the DLC layer. A comparison of commonly used FEC and ARQ techniques and their potential application to WATM is presented in Ayanoglu.[25]

One adaptive FEC coding scheme that has been proposed for PCS-to-ATM interworking is based on the use of rate-compatible punctured convolution (RCPC), punctured Bose-Chaudhuri-Hocquenghem (BCH) or Reed–Solomon (RS) coding at the wireless PHY layer to provide unequal error protection of the wireless PDU.[26] These coding schemes can support a broad range of QoS levels consistent with the requirements of multimedia services, minimize the loss of information on the wireless access segment, and prevent misrouting of cells on the fixed ATM network. Code rate puncturing is a procedure used to periodically discard a set of predetermined coded bits from the sequence generated by an encoder for the purposes of constructing a higher rate code. With the rate-compatibility restriction, higher rate codes are embedded in the lower rate codes, allowing for continuous code rate variation within a data frame.

An example set of three wireless PDU formats that might be appropriate for a PCS system that provides access to the ATM network is shown in Table 1. It is desirable to establish a tight relationship between the wireless PDU and wireline ATM cell to minimize incompatibilities between them. This will reduce the complexity of the IWF at the PCS-to-ATM gateway by limiting the amount of processing required for protocol conversion. The wireless PDU can be tightly coupled to the wireline ATM cell in two ways.

1. Encapsulate the equivalent of a full 48-byte ATM information payload, along with wireless-specific overhead (and a full/compressed ATM header, if required by the mobile terminal), in the wireless PDU (e. g., PDU-3).
2. Transmit a submultiple of 48-byte ATM information payload, along with wireless-specific overhead (and a

Table 1 Examples of wireless PDU formats.

Protocol data unit (PDU) type	PDU header (bits)	Information payload (bits)	PDU trailer (bits)	PDU size (bits)
PDU-1	24	128	—	152
PDU-2	40	256	8	304
PDU-3	56	384	16	456

Note: Information payloads are limited to submultiples of a 48-byte ATM cell information payload.

compressed ATM header, if required by the mobile terminal), in the wireless PDU (e.g., PDU-1 and PDU-2).

If the second option is used, then the network has to decide whether to send partially filled ATM cells over the ATM infrastructure network, or wait until enough wireless PDUs arrive to fill an ATM cell.

The performance of RCPC and punctured BCH (and RS) codes have been evaluated in terms of the decoded bit BER[26] on a Rayleigh fading channel. For RCPC coding, the Viterbi upper bound on the decoded BER is given by:

$$P_b \leqslant \frac{1}{P} \sum_{d=d_f}^{\infty} b_d P_d \qquad (1)$$

where P is the puncturing period, β_d is the weight coefficient of paths having distance d, P_d is the probability of selecting the wrong path, and d_f is the free distance of the RCPC code. Closed-form expressions for P_d are presented in Hagenauer[27] for a flat Rayleigh fading channel, along with tables for determining β_d. Since the code weight structure of BCH codes is known only for a small subset of these codes, an upper BER performance bound is derived in the literature independent of the structure. Assume that for a t-error correcting BCH(n,k) code, a pattern of i channel errors ($i > t$) will cause the decoded word to differ from the correct word in $i + t$ bit positions. For flat Rayleigh fading channel conditions and differential QPSK (DQPSK) modulation, the decoded BER for the BCH(n, k) code is Ref. 28:

$$P_b = \frac{2}{3} \sum_{i=t+1}^{n} \frac{i+t}{n}(1-P_s)^{n-1} P_s^i \qquad (2)$$

where P_s is the raw symbol error rate (SER) on the channel. An upper bound on the decoded BER for a t-error correcting RS(n,k) code with DQPSK signaling on a flat

Rayleigh fading channel is similar in form to Eq. 2. However, P_s should be replaced with the RS-coded digit error rate P_e, and the unit of measure for the data changed from symbols do digits. A transmission system that uses the RS code [from GF(2^m)] with M-ary signalling generates $r = m/\log_2 M$ symbols per m-bit digit. For statistically independent symbol errors, $P_e = 1 - (1 - P_s)^r$. For the binary BCH(n,k) code, $n = 2^m - 1$. GF(2^m) denotes the Galois Field of real numbers from which the RS code is constructed. Multiplication and addition of elements in this field are based on modulo-2 arithmetic, and each RS code word consists of m bits.

Table 2 shows performance results for RCPC and punctured BCH codes that are used to provide adaptive FEC on the PHY layer of a simulated TDMA-based PCS system that accesses the fixed ATM network. The FEC codes provide different levels of error protection for the header, information payload, and trailer of the wireless PDU. This is particularly useful when the wireless PDU header contains information required to route cells in the fixed ATM network, for example. Using a higher level of protection for the wireless PDU header increases the likelihood that the PDU will reach its destination, and not be misrouted in the ATM network.

The numerical results show achievable average code rates for the three example PDU formats in Table 1. The PCS system operates in a microcellular environment at 2 GHz, with a transmission bit rate of 384 Kbps, using DQPSK modulation. The channel is modeled as a time-correlated Rayleigh fading channel. The PCS transmission model assumes perfect symbol and frame synchronization, as well as perfect frequency tracking. Computed code rates are shown with and without the use of diversity combining. All overhead blocks in the wireless PDUs are assumed to require a target BER of 10^{-9}. On the other hand, the information payload has target BERs of 10^{-3} and 10^{-6}, which might be typical for voice and data, respectively. Associated with this target BERs is a design goal of 20 dB for the SNR. The mother code rate for the RCPC code is

Table 2 Average code rates for RCPC coding[a] and punctured BCH coding[b] for a 2-GHz TDMA-based PCS system with access to an ATM infrastructure network.

Coding scheme	Average code rate (10^{-3} BER for information payload)			Average code rate (10^{-6} BER for information payload)		
	PDU-1	PDU-2	PDU-3	PDU-1	PDU-2	PDU-3
RCPC: no diversity	0.76	0.77	0.78	0.61	0.62	0.63
RCPC: 2-branch diversity	0.91	0.93	0.93	0.83	0.84	0.85
Punctured BCH: no diversity	0.42	0.44	0.51	0.34	0.39	0.46
Punctured BCH: 2-branch diversity	0.83	0.81	0.87	0.75	0.76	0.82

[a]With soft decision decoding and no channel state information at the receiver.
[b]With and without 2-branch frequency diversity.

$R = 1/3$, the puncturing period $P = 8$, and the memory length $M = 6$. For BCH coding, the parameter $m \geqslant 8$.

The numerical results in Table 2 show the utility of using code rate puncturing to improve the QoS performance of the wireless access segment. The results for punctured RS coding are quite similar to those for punctured BCH coding. Adaptive PHY layer FEC coding can be further enhanced by implementing an ARQ scheme at the DLC sublayer, which is combined with FEC to form a hybrid ARQ/FEC protocol[27] to supplement FEC at the PHY layer. This approach allows adaptive FEC to be distributed between the wireless PHY and DLC layers.

CONCLUSIONS

The ATM Forum is developing specifications intended to facilitate the use of ATM technology for a broad range of wireless network access and interworking scenarios, both public and private. These specifications are intended to cover requirements for seamless extension of ATM to mobile devices and mobility control in ATM infrastructure networks to allow interworking of non-ATM wireless terminals with the fixed ATM network. A mobility-enhanced ATM network that is developed from specifications for WATM interworking may be used in near-term cellular/PCS/satellite/wireless LAN deployments, while providing a smooth migration path to the longer-term end-to-end WATM application scenario. It is likely to be cost-competitive with other approaches that adopt non-ATM overlay transport network topologies.

This entry describes various WATM interworking scenarios where the ATM infrastructure might be a public (or private) transit ATM network, designed primarily to support broadband wireline services. A detailed description of a generic PCS-to-ATM architectural interworking scenario is presented, along with an evolution strategy to eventually provide end-to-end WATM capabilities in the long term. One approach is described for providing QoS support using code rate puncturing at the wireless PHY layer, along with numerical results. The network architectures, protocol reference models, signaling protocols, and QoS management strategies described for PCS-to-ATM interworking can be applied, in varying degrees, to the other WATM interworking scenarios described in this entry.

DEFINING TERMS

BISDN B-ICI: A carrier-to-carrier public interface that supports multiplexing of different services such as SMDS, frame relay, circuit emulation, and CRSs.

BISDN: A cell-relay-based information transfer technology upon which the next-generation telecommunications infrastructure is to be based.

HIPERLAN: Family of standards being developed by the ETSI for high-speed wireless LANs, to provide short-range and remote wireless access to ATM networks and for wireless ATM interconnection.

IWFs: A set of network functional entities that provide interaction between dissimilar subnetworks, end systems, or parts thereof, to support end-to-end communications.

Location management: A set of registration and call delivery functions.

MAP: Application layer protocols and processes that are defined to support mobility services such as intersystem roaming and handoffs.

ME-ATM: An ATM switch with additional capabilities and features to support location and mobility management.

Mobility management: The handoff process associated with switching (rerouting) of the communication link from the old coverage area to the new coverage area when a wireless user moves during active communication.

PCSs: Emerging digital wireless systems which support mobility in microcellular and other environments, which have a set of capabilities that allow some combination of terminal mobility, personal mobility, and service profile management.

PDUs: The PHY layer message structure used to carry information across the communications link.

PNNI: The interface between two private networks.

RAL: A reference to the PHY, MAC, and DLC layers of the radio link.

RCPC: Periodic discarding of predetermined coded bits from the sequence generated by a convolutional encoder for the purposes of constructing a higher rate code. The rate-compatibility restriction ensures that the higher rate codes are embedded in the lower rate codes, allowing for continuous code rate variation to change from low to high error protection within a data frame.

SATATM: A satellite network that provides ATM network access to fixed or mobile terminals, high-speed links to interconnect fixed or mobile ATM networks, or form an ATM network in the sky to provide user access and network interconnection services.

UNI: A standardized interface providing basic call control functions for subscriber access to the telecommunications network.

WATM: An emerging wireless networking technology that extends ATM over the wireless access segment, and/or uses the ATM infrastructure as a transport network for a broad range of wireless network access scenarios, both public and private.

REFERENCES

1. Raychaudhuri, D.; Wilson, N.D. ATM-based transport architecture for multiservices wireless personal communication networks. IEEE J. Sel. Areas Commun. **1994**, *12* (8), 1401–1414.

2. Raychaudhuri, D. Wireless ATM networks: architecture, system design, and prototyping. IEEE Pers. Commun. **1996**, *3* (4), 42–49.

3. Umehira, M.; Hashimoto, A.; Matsue, H. An ATM wireless system for tetherless multimedia services. ICUPC 1995, Tokyo, Japan, Nov 6–10, 1995.

4. Barton, M. Architecture for wireless ATM networks. PIMRC 1995, Toronto, Canada, Sept 27–29, 1995778–782.

5. Cheng, M. Performance comparison of mobile assisted network controlled, and mobile-controlled hand-off signalling for TDMA PCS with an ATM backbone. ICC 1997, Québec, Canada, Jun 8–12, 1997.

6. Wu, T.H.; Chang, L.F. Architecture for PCS mobility management on ATM transport networks. ICUPC 1995, Tokyo, Japan, Nov 6–10, 1995; 763–768.

7. ATM. B-ICI Specification, v.2.0, ATM Forum, 1995.

8. ATM. Private Network-Network Interface (PNNI) Specification, v.1.0., ATM Forum, 1996.

9. ITU. Message Transfer Part Level 3 Functions and Messages Using the Service of ITU-T Recommendations, Q.2140; International Telecommunications Union-Telecommunications Standardization Sector, Geneva, Switzerland. TD PL/11–97. 1995.

10. ITU. B-ISDN Adaptation Layer—Service Specific Coordination Function for Signaling at the Network Node Interface (SCCF) at NNI, ITU-T Q.2140. International Telecommunications Union, Telecommunications Standardization Sector, Geneva, Switzerland. 1995.

11. Cox, D.C. Wireless personal communications: what is it?. IEEE Pers. Commun. **1995**, *2* (2), 20–35.

12. Truong, H.L.; Ellington, W.W., Jr.; Le Boudec, J-K.; Meier, A.X. LAN emulation on an ATM network. IEEE Commun. Mag. **1995**, *33*, 70–85.

13. Wilkinson, T.; Phipps, T.; Barton, S.K. A report on HIPER-LAN standardization. Int. J. Wirel. Inf. Netw. **1995**, *2* (2), 99–120.

14. Pahlavan, K.; Zahedi, A.; Krishnamurthy, P. Wideband local wireless: wireless LAN and wireless ATM. IEEE Commun. Mag. **1997**, *35*, 34–40.

15. NSTC. Strategic Planning Document—Information and Communications, National Science and Technology Council. 10, 1995.

16. Schmidt, W.R. Optimization of ATM and legacy LAN for high speed satellite communications. Transport Protocols for High-Speed Broadband Networks Workshop. GLOBE-COM 1996, London, UK, Nov 18–22, 1996.

17. TIA. TIA/EIA Telecommunications Systems Bulletin (TSB)—91. Satellite ATM Networks: Architectures and Guidelines, TIA/EIA/TSB-91; Telecommunications Industry Association, 1998.

18. LaMaire, R.O.; Krishna, A.; Bhagwat, P.; Panian, J. Wireless LANs and mobile networking: standards and future directions. IEEE Commun. Mag. **1996**, *34* (8), 86–94.

19. Cheng, M.; Rajagopalan, S.; Chang, L.F.; Pollini, G.P.; Barton, M. PCS mobility support over fixed ATM networks. IEEE Commun. Mag. **1997**, *35* (11), 82–92.

20. SR. Alternatives for Signaling Link Evolution, Bellcore Special Report. Bellcore SR-NWT-002897 (1), 1994.

21. Lin, Y.B.; Devries, S.K. PCS network signalling using SS7. IEEE Commun. Mag. **1995**, *2* (3), 44–55.

22. ITU. Digital Subscriber Signaling System No. 2—Generic Functional Protocol: Core functions, ITU-T Recommendation Q.2931.2; International Telecommunications Union, Telecommunications Standardization Sector, Geneva, Switzerland. 1996.

23. Veeraraghavan, M.; Pancha, P.; Eng, K.Y. Connectionless Transport in ATM Networks. ATM Forum Contribution. ATMF/97-0141. 9–14, 1997.

24. Naghshineh, M.; Willebeek-LeMair, M. End-to-end QoS provisioning in multimedia wireless/mobile networks using an adaptive framework. IEEE Commun. Mag. **1997**, *35*, 72–81.

25. Ayanoglu, E.; Eng, K.Y.; Karol, M.J. Wireless ATM: limits, challenges, and protocols. IEEE Pers. Commun. **1996**, *3* (4), 18–34.

26. Barton, M. Unequal error protection for wireless ATM applications. GLOBECOM 1996, London, UK, Nov 18–22, 1996; 1911–1915.

27. Hagenauer, J. Rate-compatible punctured convolution codes (RCPC codes) and their applications. IEEE Trans. Commun. **1988**, *COM-36* (4), 389–400.

28. Proakis, J.G. *Digital Communications*; McGraw-Hill: New York, USA, 1989.

Wireless
Ad Hoc—Asynch

FURTHER INFORMATION

Information supplementing the wireless ATM standards work may be found in the ATM Forum documents relating to the Wireless ATM Working Group's activities (web page http://www.atmforum.com). Special issues on wireless ATM have appeared in the August 1996 issue of *IEEE Personal Communications*, the January 1997 issue of the *IEEE Journal on Selected Areas in Communications*, and the November 1997 issue of *IEEE Communications Magazine*.

Reports on proposals for higher speed wireless LAN extensions in the 2.4 GHz and 5 GHz bands can be found at the IEEE 802.11 web site (http://grouper.ieee.org/groups/802/11/Reports). Additional information on HIPERLAN and related activities in ETSI BRAN can be obtained from their web site (http://www.etsi.fr/bran).

Wireless Asynchronous Transfer Mode (ATM): Mobility Management

Bala Rajagopalan
Tellium Inc., Ocean Port, New Jersey, U.S.A.

Daniel Reininger
Semandex Networks Inc., Princeton, New Jersey, U.S.A.

Abstract

This entry describes mobility management protocols in the wireless asynchronous transfer mode.

INTRODUCTION

Allowing end system mobility in ATM networks gives rise to the problem of mobility management, i.e., maintaining service to end systems regardless of their location or movement. A fundamental design choice here is whether mobility management deals with user mobility or terminal mobility. When a network supports user mobility, it recognizes the user as the subscriber with an associated service profile.

The user can then utilize any MT for access to the subscribed services. This results in flexibility in service provisioning and usage, but some extra complexity is introduced in the system implementation, as exemplified by the GSM system.[1] Support for user mobility implies support for terminal mobility. A network may support only terminal mobility and not recognize the user of the terminal, resulting in a simpler implementation. In either case, the mobility management tasks include:

- *Location management (LM)*: keeping track of the current location of an MT in order to permit correspondent systems to set up connections to it. A key requirement here is that the correspondent systems need not be aware of the mobility or the current location of the MT.
- *Connection handover*: maintaining active connections to an MT as it moves between different points of attachment in the network. The handover function requires protocols at both the radio layer and at the network layer. The issue of preserving quality of service (QoS) during handovers was described earlier and this introduces some complexity in handover implementations.
- *Security management*: authentication of mobile users (or terminals) and establishing cryptographic procedures for secure communications based on the user (or terminal) profile.[2]

- *Service management*: maintaining service features as a user (or terminal) roams among networks managed by different administrative entities. Security and service management can be incorporated as part of LM procedures.

Early wireless ATM (WATM) implementations have considered only terminal mobility.[3,4] This has been to focus the initial effort on addressing the core technical problems of mobility management, i.e., LM and handover.[5] Flexible service management in wide-area settings, in the flavor of GSM, has not been an initial concern in these systems. The mobility management protocol standards being developed by the ATM Forum WATM working group may include support for user mobility. In the following, therefore, we concentrate on the LM and handover functions required to support terminal mobility in WATM.

LM IN WATM

LM in WATM networks is based on the notions of permanent and temporary ATM addresses. A permanent ATM address is a location-invariant, unique address assigned to each MT. As the MT attaches to different points in a WATM network, it may be assigned different temporary ATM addresses. As all ATM end system addresses, both permanent and temporary addresses are derived from the addresses of switches in the network, in this case end system mobility-supporting ATM switches (EMASs). This allows connection set-up messages to be routed toward the MT, as described later. The EMAS whose address is used to derive the permanent address of an MT is referred to as the home EMAS of that MT. The LM function in essence keeps track of the current temporary address corresponding to the permanent address of each MT. Using this function, it becomes possible for

Encyclopedia of Wireless and Mobile Communications DOI: 10.1081/E-EWMC-120043952

correspondent systems to establish connections to an MT using only its permanent address and without knowledge of its location.

NETWORK ENTITIES INVOLVED IN LM

The LM functions are distributed across four entities:

- *The location server (LS)*: This is a logical entity maintaining the database of associations between the permanent and temporary addresses of MTs. The LS responds to query and update requests from EMASs to retrieve and modify database entries. The LS may also keep track of service-specific information for each MT.
- *The authentication server (AUS)*: This is a logical entity maintaining a secure database of authentication and privacy related information for each MT. The authentication protocol may be implemented between EMASs and the AUS, or directly between MTs and the AUS.
- *The MT*: The MT is required to execute certain functions to initiate location updates and participate in authentication and privacy protocols.
- *The EMAS*: Certain EMASs are required to identify connection set-up messages destined to MTs and invoke location resolution functions. These can be home EMASs or certain intermediate EMASs in the connection path. All EMASs in direct contact with MTs [via their radio ports (RPs)] may be required to execute location update functions. Home EMASs require the ability to redirect a connection set-up message. In addition, all EMASs may be required to participate in the redirection of a connection set-up message to the current location of an MT.

There could be multiple LSs and AUSs in a WATM network. Specifically, an LS and an AUS may be incorporated with each home EMAS, containing information on all the MTs that the EMAS is home to. On the other hand, an LS or an AUS may be shared between several EMASs, by virtue of being separate entities. These choices are illustrated in Fig. 1, where the terms "integrated" and "modular" are used to indicate built-in and separated LS

and AUS. In either case, protocols must be implemented for reliably querying and updating the LS, and mechanisms to maintain the integrity and security of the AUS. NEC's WATMnet[3] and BAHAMA[4] are examples of systems implementing integrated servers. The modular approach is illustrated by GSM[1] and next-generation wireless network proposals.[6]

LM FUNCTIONS AND CONTROL FLOW

At a high level, the LM functions are registration, location update, connection routing to home or gateway EMAS, location query, and connection redirect.

Registration and Location Update

When an MT connects to a WATM network, a number of resources must be instantiated for that mobile. This instantiation is handled by two radio layer functions: association, which establishes a channel for the MT to communicate with the edge EMAS, and registration, which binds the permanent address of the MT to a temporary address. In addition, the routing information pertaining to the mobile at one or more LSs must be updated whenever a new temporary address is assigned. This is done using location updates.

The authentication of a MT and the establishment of encryption parameters for further communication can be done during the location updating procedure. This is illustrated in Fig. 2 which shows one possible control flow when an MT changes location from one EMAS to another. Here, the Broadcast_ID indicates the identity of the network, the location area, and the current RP. Based on this information, e.g., by comparing its access rights and the network ID, the MT can decide to access the network. After an association phase, which includes the setting up of the signaling channel to the EMAS, the MT sends a registration message to the switch. This message includes the MT's home address and authentication information. The location update is initiated by the visited EMAS and the further progression is as shown. The LS/AUS are shown logically separate from the home EMAS for generality.

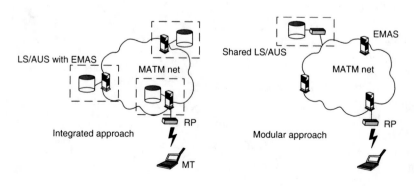

Fig. 1 Server organizations.

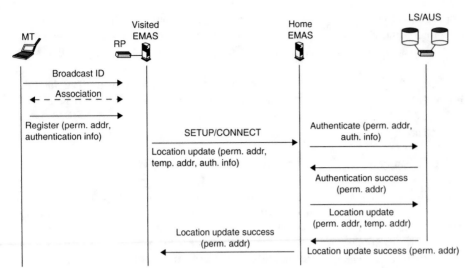

Fig. 2   Location update control flow.

They can be integrated with the home EMAS. Details on the implementation of a similar location update scheme can be found in Ref. 7.

Now, there are other possible configurations of LSs that give rise to different control message flow. For example, a two-level hierarchical LS arrangement can be used. Under this organization, the LS in the visiting network is updated as long as the MT remains in this network, and the home LS is updated only when the MT moves to a different WATM network. The information kept in the home LS must, therefore, point to a gateway EMAS in the visited network, since precise location in the visited network will not be available at the home LS. GSM LM is an example of this scheme.[1]

Connection Forwarding, Location Query, and Connection Redirect

After a location update, a LS handling the MT has the correct association between its permanent and temporary ATM addresses. When a new connection to the MT is established, the set-up message must be routed to some EMAS that can query the LS to determine the current address of the MT. This is the connection forwarding function. Depending on how MT addresses are assigned, the location query can occur very close to the origination of the connection or it must progress to the home EMAS of the MT. For instance, if MT addresses are assigned from a separately reserved ATM address space within a network, a gateway EMAS in the network can invoke location query when it processes a set-up message with a destination address known to be an MT address. To reach some EMAS that can interpret an MT address, it is sufficient to always forward connection set-up messages toward the home EMAS. This ensures that at least the home EMAS can invoke the query if no other EMAS enroute can do this. The location query is simply a reliable control message exchange between an EMAS and an LS. If the

LS is integrated with the EMAS, this is a trivial operation. Otherwise, it requires a protocol to execute this transaction.

The control flow for connection establishment when the MT is visiting a foreign network is shown in Fig. 3. The addresses of various entities shown have been simplified for illustration purposes. Here, a fixed ATM terminal (A.1.1.0) issues a SETUP toward the MT whose permanent address is C.2.1.1. The SETUP message is routed toward the home EMAS whose address is C.2.1. It is assumed that no other EMAS in the path to the home EMAS can detect MT addresses. Thus, the message reaches the home EMAS which determines that the end system whose address is C.2.1.1 is an MT. It then invokes a location query to the LS which returns the temporary address for the MT (B.3.3). The home EMAS issues a redirected SETUP toward the temporary address. In this message, the MT is identified by its permanent address thereby enabling the visited EMAS to identify the MT and proceed with the connection SETUP signaling.

It should be noted that in the topology shown in Fig. 3, the redirection of the connection set-up does not result in a

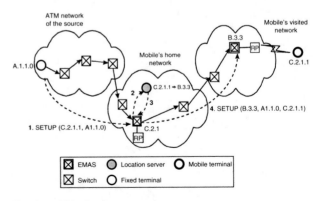

Fig. 3   MT in foreign network.

nonoptimal path. But, in general, this may not be the case. To improve the overall end-to-end path, redirection can be done with partial teardown in which case a part of the established path is released and the connection set-up is redirected from an EMAS that occurs further upstream of the home EMAS. This is shown in Fig. 4. Here, the EMAS labeled cross over switch (COS) occurs in the original connection path upstream of the home EMAS. To redirect the set-up to B.3.3, the connection already established to the home EMAS is torn down up to COS, and new segment is established from the COS to B.3.3. This requires additional signaling procedures.

Finally, in Fig. 5, the case of hierarchical LSs is illustrated. Here, as long as the MT is in the visited network, the address of the gateway EMAS (B.1.1) is registered in its home LS. The connection

set-up is sent via the home EMAS to the gateway EMAS. The gateway then queries its local LS to obtain the exact location (B.3.3) of the MT. It is assumed that the gateway can distinguish the MT address (C.2.1.1) in a SETUP message from the fact that this address has a different network prefix (C) than the one used in the visited network (B).

Signaling and Control Messages for LM

The signaling and control messages required can be derived from the scenarios above. Specifically, interactions between the EMAS and the LS require control message exchange over a virtual channel (VC) established for this purpose. The ATM signaling support needed for connection set-up and redirection is described in.[7]

Connection Handover in WATM

WATM implementations rely on mobile-initiated handovers whereby the MT is responsible for monitoring the radio link quality and decides when to initiate a

handover.[5] A handover process typically involves the following steps:

1. *Link quality monitoring*: When there are active connections, the MT constantly monitors the strength of the signal it receives from each RP within range.
2. *Handover trigger*: At a given instance, all the connections from/to the MT are routed through the same RP, but deterioration in the quality of the link to this RP triggers the handover procedure.
3. *Handover initiation*: Once a handover is triggered, the MT initiates the procedure by sending a signal to the edge EMAS with which it is in direct contact. This signal indicates to the EMAS the list of candidate RPs to which active connections can be handed over.
4. *Target RP selection*: The edge EMAS selects one RP as the handover target from the list of candidates sent by the MT. This step may make use of network-specific criteria for spreading the traffic load among various RPs and interaction between the edge EMAS and other EMASs housing the candidate RPs.
5. *Connection rerouting*: Once the target RP is selected, the edge EMAS initiates the rerouting of all connections from/to the MT within the MATM network to the target RP. The complexity of this step depends on the specific procedures chosen for rerouting connections, as described next. Due to constraints on the network or radio resources, it is possible that not all connections are successfully rerouted at the end of this step.
6. *Handover completion*: The MT is notified of the completion of handover for one or more active connections. The MT may then associate with the new RP and begin sending/receiving data over the connections successfully handed over.

Specific implementations may differ in the precise sequence of events during handover. Furthermore, the handover complexity and capabilities may be different.

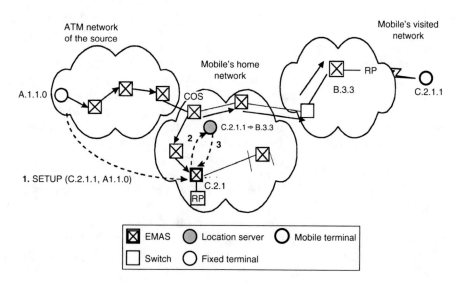

Fig. 4 Connection redirect.

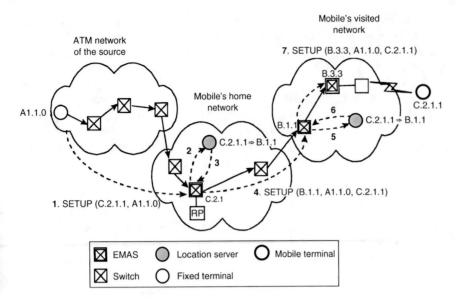

Fig. 5   Hierarchical LS configuration.

For instance, some systems may implement lossless handover whereby cell loss and missequencing of cells are avoided during handover by buffering cells inside the network.[8] The handover control flow is described in detail below for two types of handovers:

- *Backward handover*: The MT initiates handover through the current RP it is connected to. This is the normal scenario.
- *Forward handover*: The MT loses connectivity to the current RP due to a sudden degeneration of the radio link. It then chooses a new RP and initiates the handover of active connections.

The following description allows only hard handovers, i.e., active connections are routed via exactly one RP at a given instance, as opposed to soft handovers in which the MT can receive data for active connections simultaneously from more than one RP during handover.

Backward Handover Control Flow

Fig. 6 depicts the control sequence for backward handover when the handover involves two different EMASs. Here, "old" and "new" EMAS refer to the current EMAS and the target EMAS, respectively. The figure does not show handover Steps (1) and (2), which are radio layer functions, but starts with Step (3). The following actions take place:

1. The MT initiates handover by sending an *HO_RE-QUEST* message to the old EMAS. With this message, the MT identifies a set of candidate RPs. Upon receiving the message, the old EMAS identifies a set of candidate EMASs that house the indicated RPs. It then sends an *HO_REQUEST_QUERY* to each

candidate EMAS, identifying the candidate RP as well as the set of connections (including the traffic and QoS parameters) to be handed over. The connection identifiers are assumed to be unique within the network.[7]

2. After receiving the *HO_REQUEST_QUERY* message, a candidate EMAS checks the radio resources available on all the candidate RPs it houses and selects the one that can accommodate the most number of connections listed. It then sends an *HO_REQUEST_RESPONSE* message to the old EMAS identifying the target RP chosen and a set of connections that can be accommodated (this may be a subset of connections indicated in the *QUERY* message).

3. After receiving an *HO_REQUEST_RESPONSE* message from all candidate EMASs, the old EMAS selects one target RP, based on some local criteria (e.g., traffic load spreading). It then sends an *HO_RESPONSE* message to the MT, indicating the target RP. At the same time, it also sends an *HO_COMMAND* to the new EMAS. This message identifies the target RP and all the connections to be handed over along with their ATM traffic and QoS parameters. This message may also indicate the connection rerouting method. Rerouting involves first the selection of a COS which is an EMAS in the existing connection path. A new connection segment is created from the new EMAS to the COS and the existing segment from the COS to the old EMAS is deleted. Some COS selection options are:
   VC extension: The old EMAS-E itself serves as the COS.
   Anchor-based rerouting: The COS is determined a priori (e.g., a designated EMAS in the network) or during connection set-up (e.g., the EMAS that first

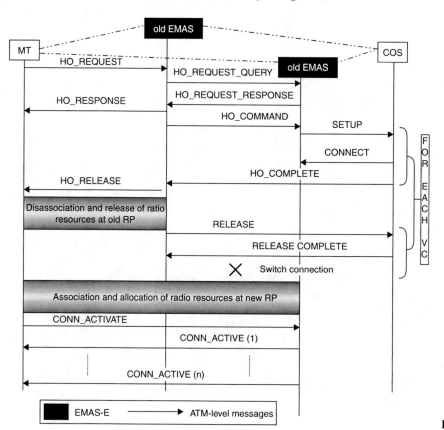

Fig. 6   Backward handover control flow.

served the MT when the connection was set up). The selected COS is used for all handovers during the lifetime of the connection.

Dynamic COS discovery: The COS is dynamically determined during each handover.

These procedures are illustrated in Fig. 7. While anchor-based rerouting and VC extension result in the same COS being used for all the connections being handed over, dynamic COS selection may result in different COSs for different connections. The *HO_COMMAND* message indicates which COS selection method is used, and if VC extension or anchor-based rerouting is used, it also includes the identity of the COS. If dynamic COS selection is used, the message includes the identity of the first EMAS in the connection path from the source (this information is collected during connection set-up[7]). For illustrative purposes, we assume that the dynamic COS selection procedure is used.

4. Upon receiving the *HO_COMMAND* message, the new EMAS allocates radio resources in the target RP for as many connections in the list as possible. It then sends a *SETUP* message toward the COS of each such connection. An EMAS in the existing connection path that first processes this message becomes the actual COS. This is illustrated in Fig. 8. This action, if successful, establishes a new

segment for each connection from the new EMAS to the COS.

5. If the set-up attempt is not successful, new EMAS sends an *HO_FAILURE* message to the old EMAS, after releasing all the local resources reserved in Step (4). This message identifies the connection in question and it is forwarded by the old EMAS to the MT. What the MT does in response to an *HO_FAILURE* message is not part of the backward handover specification.

6. If the COS successfully receives the *SETUP* message, it sends a *CONNECT* message in reply, thereby completing the partial connection establishment procedure. It then sends an *HO_COMPLETE* message to old EMAS-E. The *HO_COMPLETE* message is necessary to deal with the situation when handover is simultaneously initiated by both ends of the connection when two MTs communicate (for the sake of simplicity, we omit further description of this situation, but the reader may refer to Ref. 9 for further details).

7. The old EMAS-E waits to receive *HO_COMPLETE* messages for all the connections being handed over. However, the waiting period is limited by the expiry of a timer. Upon receiving the *HO_COMPLETE* message for the last connection, or if the timer expires, the old EMAS sends an *HO_RELEASE*

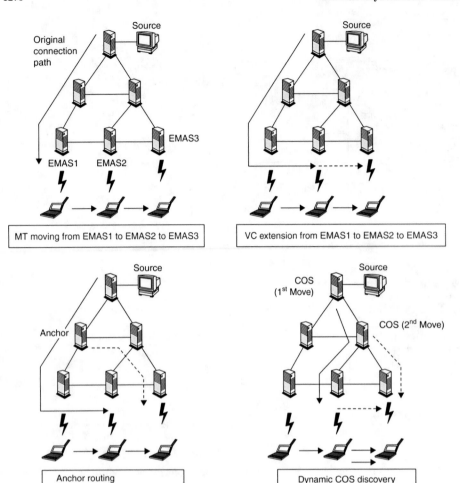

Fig. 7 COS discovery.

message to MT. Waiting for the *HO_RELEASE* message allows the MT to utilize the existing connection segment as long as possible thereby minimizing data loss. However, if the radio link deteriorates rapidly, the MT can switch over to the new RP without receiving the *HO_RELEASE* message.

8. The old EMAS initiates the release of each connection for which an *HO_COMPLETE* was received by sending a *RELEASE* message to the corresponding COS.

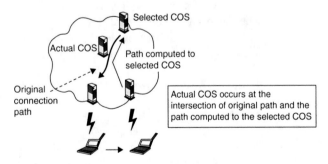

Fig. 8 COS selection when MT moves.

9. Upon receiving the *RELEASE* message, the COS sends a *RELEASE COMPLETE* to the previous switch in the path (as per regular ATM signaling) and switches the data flow from the old to the new connection segment.

10. Meanwhile, after receiving the *HO_RELEASE* message from the old EMAS or after link deterioration, the MT dissociates from the old RP and associates with the new RP. This action triggers the assignment of radio resources for the signaling channel and user data connections for which resources were reserved in Step (4).

11. Finally, the MT communicates to the new EMAS its readiness to send and receive data on all connections that have been handed over by sending a *CONN_ACTIVATE* message.

12. Upon receiving the *CONN_ACTIVATE* message from the MT, new EMAS responds with a *CONN_ACTIVE* message. This message contains the identity of the connections that have been handed over, including their new ATM VC identifiers. Multiple *CONN_ACTIVE* messages may be generated, if all

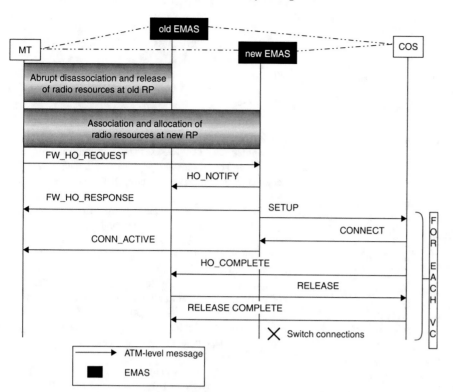

Fig. 9 Forward handover control flow.

the connections have not been handed over when the *CONN_ACTIVATE* message was received. However, handover of remaining connections and the subsequent generation of *CONN_ACTIVE* signals are timer-bound: if the MT does not receive information about a connection in a *CONN_ACTIVE* message before the corresponding timer expires, it assumes that the connection was not successfully handed over. The recovery in this case is left up to the MT.

The description above has left open some questions. Among them: what mechanisms are used to reliably exchange control messages between the various entities that take part in handover? What actions are taken when network or radio link failures occur during handover? How can lossless handover be included in the control flow? What effect do transient disruptions in service during handover have on application behavior?, and what are the performance impacts of signaling for handover? The short answers to these questions are: reliability can be incorporated by implementing a reliable transfer protocol for those control messages that do not already use such a transport (*SETUP, RELEASE,* etc., do, but *HO_REQUEST_QUERY* e.g., requires attention). Actions taken during network failures require further analysis, but forward handover can be used to recover from radio link failures during handover. Lossless handover requires inband signaling within each connection and buffering in the network. Details on this

can be found in Ref. 8. The effect of transient disruptions on applications can be minimal, depending on how rerouting is implemented during handover. This is described in detail in Ref. 10. Finally, some of the performance issues related to mobility management are investigated in Ref. 11.

Forward Handover

Forward handover is considered as the measure of last resort, to be invoked when the current radio link deteriorates suddenly. The forward handover procedure is simpler than backward handover, as illustrated in Fig. 9. Here,

1. The radio link with the old EMAS degrades abruptly. This results in the MT being dissociated with the old RP. The MT then chooses a new RP and associates with it. In the example shown, the new RP is housed by a different EMAS (the new EMAS).
2. The MT initiates forward handover by sending a *FW_HO_REQUEST* message to the new EMAS. This message indicates the active connections by their network-wide unique identifiers, along with their ATM traffic and QoS parameters, the identity of the previous EMAS (the "old" EMAS), and the COS information (this information may be obtained when the connection is initially set up).
3. The new EMAS sends an *HO_NOTIFY* message to the old EMAS indicating the initiation of handover.

This serves to keep the old EMAS from prematurely releasing the existing connections.

4. The new EMAS reserves radio resources for as many listed connections as possible on the RP to which the MT is associated. It then sends a *FW_HO_RESPONSE* message to the MT identifying the connections that can be handed over.

5. For each such connection, the new EMAS generates a SETUP message toward the COS to establish the new connection segment. This message includes the identity of the new EMAS. An EMAS in the existing connection path that first processes this message becomes the COS (Fig. 8).

6. Upon receiving the *SETUP* message, the COS completes the establishment of the new connection segment by sending a *CONNECT* message to the new EMAS.

7. After receiving the *CONNECT* message, new EMAS sends a *CONN_ACTIVE* message to the MT, indicating the connection has become active. Reception of *CONN_ACTIVE* by the MT is subject to a timer expiry: if it does not receive information about a connection in any *CONN_ACTIVE* message before the corresponding timer expires, it may initiate any locally defined recovery procedure.

8. If the new connection segment cannot be setup, the new EMAS sends an *HO_FAILURE* message to the old EMAS and the MT, after releasing all the local resources reserved for the connection. Recovery in this case is left up to the MT.

9. If the COS did send a *CONNECT* in Step 7, it switches the connection data to the new segment and sends an *HO_COMPLETE* message to old EMAS. As in the case of backward handover, the *HO_COMPLETE* message is necessary to resolve conflicts in COS selection when handover is simultaneously initiated by both ends of the connection when two MTs communicate.

10. Upon receiving the *HO_COMPLETE* message, the old EMAS releases the existing connection segment by sending a *RELEASE* message to the COS. In response, the COS sends a *RELEASE COMPLETE* to the previous switch.

REFERENCES

1. Mouly, M.; Pautet, M.-B. *The GSM System for Mobile Communications*; Cell & Sys: Palaiseau, France, 1992.
2. Brown, D. Techniques for privacy and authentication in personal communication systems. IEEE Pers. Commun. **1995**, *2*, 6–10.
3. Raychaudhuri, D.; French, L.J.; Siracusa, R.J.; Biswas, S.K.; Yuan, R.; Narasimhan, P.; Johnston, C.A. WATMnet: a prototype wireless ATM system for multimedia personal communication. IEEE J. Sel. Areas Commun. **1997**, *15*, 83–95.
4. Veeraraghavan, M.; Karol, M.J.; Eng, K.Y. Mobility and connection management in a wireless ATM LAN. IEEE J. Sel.Areas Commun. **1997**, *15*, 50–68.
5. Acharya, A.; Li, J.; Rajagopalan, B.; Raychaudhuri, D. Mobility management in wireless ATM networks. IEEE Commun. Mag. **1997**, *37*, 100–109.
6. Tabbane, S. Location management methods for third-generation mobile systems. IEEE Commun. Mag. **1997**, *35*, 72–84.
7. Acharya, A.; Li, J.; Bakre, A.; Raychaudhuri, D. Design and prototyping of location management and handoff protocols for wireless ATM networks. ICUPC 1997, San Diego, CA, Oct 12–16, 1997.
8. Mitts, H.; Hansen, H.; Immonen, J.; Veikkolainen, S. Lossless handover in wireless ATM. Mobile Netw. Appl. **1996**, *1*, 299–312.
9. Rajagopalan, B. Mobility management in integrated wireless ATM networks. Mobile Netw. Appl. **1996**, *1* (3), 273–286.
10. Mishra, P.; Srivastava, M. Effect of connection rerouting on application performance in mobile networks. IEEE Conference on Distributed Computing Systems, Baltimore, MD, May 27–30, 1997.
11. Pollini, G.P.; Meier-Hellstern, K.S.; Goodman, D.J. Signalling traffic volume generated by mobile and personal communications. IEEE Commun. Mag. **1995**, *33*, 60–65.

Wireless Asynchronous Transfer Mode (ATM): Quality of Service (QoS)

Bala Rajagopalan
Tellium Inc., Ocean Port, New Jersey, U.S.A.

Daniel Reininger
Semandex Networks Inc., Princeton, New Jersey, U.S.A.

Abstract

This entry focuses on asynchronous transfer mode quality of service for connections that terminate on mobile end systems over a radio link.

Wireless
Ad Hoc—Asynch

INTRODUCTION

The type of quality of service (QoS) guarantees to be provided in wireless ATM (WATM) systems is debatable.[1] On the one hand, the QoS model for traditional ATM networks is based on fixed terminals and high quality links. Terminal mobility and error-prone wireless links introduce numerous problems.[2] On the other hand, maintaining the existing QoS model allows the transparent extension of fixed ATM applications into the domain of mobile networking. Existing prototype implementations have chosen the latter approach.[3–5] This is also the decision of the ATM Forum WATM working group.[6] Our discussion, therefore, is oriented in the same direction, and to this end we first briefly summarize the existing ATM QoS model (Fig. 1).

ATM QoS MODEL

Five service categories have been defined under ATM.[7] These categories are differentiated according to whether they support constant or variable rate traffic, and real-time or non-real-time constraints. The service parameters include a characterization of the traffic and a reservation specification in the form of QoS parameters. Also, traffic is policed to ensure that it conforms to the traffic characterization, and rules are specified for how to treat nonconforming traffic. ATM provides the ability to tag nonconforming cells and specify whether tagged cells are policed (and dropped) or provided with best-effort service.

Under UNI 4.0, the service categories are constant bit rate (CBR), real-time variable bit rate (rt-VBR), non-real-time VBR (nrt-VBR), unspecified bit rate (UBR) and available bit rate (ABR). The definition of these services can be found in Ref. 7. Table 1 summarizes the traffic descriptor parameters and QoS parameters relevant

to each service category in ATM traffic management specifications version 4.0.[8] Here, the traffic parameters are peak cell rate (PCR), cell delay variation tolerance (CDVT), sustainable cell rate (SCR), maximum burst size (MBS) and minimum cell rate (MCR). The QoS parameters are cell loss ratio (CLR), maximum cell transfer delay (max CTD) and delay variation (CDV). The explanation of these parameters can be found in Ref. 8.

Functions related to the implementation of QoS in ATM networks are usage parameter control (UPC) and connection admission control (CAC). In essence, the UPC function (implemented at the network edge) ensures that the traffic generated over a connection conforms to the declared traffic parameters. Excess traffic may be dropped or carried on a best-effort basis (i.e., QoS guarantees do not apply). The CAC function is implemented by each switch in an ATM network to determine whether the QoS requirements of a connection can be satisfied with the available resources. Finally, ATM connections can be either point-to-point or point-to-multipoint. In the former case, the connection is bidirectional, with separate traffic and QoS parameters for each direction, while in the latter case it is unidirectional. In this entry, we consider only point-to-point connections for the sake of simplicity.

QOS APPROACH IN WATM

QoS in WATM requires a combination of several mechanisms acting in concert. Fig. 2 illustrates the various points in the system where QoS mechanisms are needed:

- *At the radio interface*: A QoS-capable medium access control (MAC) layer is required. The mechanisms here are resource reservation and allocation for ATM virtual circuits under various service categories, and

Encyclopedia of Wireless and Mobile Communications DOI: 10.1081/E-EWMC-120043951

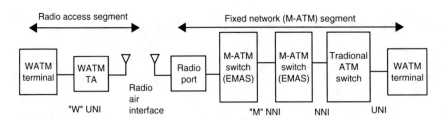

Fig. 1 WATM reference model.

scheduling to meet delay requirements. Furthermore, an error control function is needed to cope with radio link errors that can otherwise degrade the link quality. Finally, a CAC mechanism is required to limit access to the multiple access radio link in order to maintain QoS for existing connections.

- *In the network*: ATM QoS mechanisms are assumed in the network. In addition, a capability for QoS renegoti-ation will be useful. This allows the network or the mobile terminal (MT) to renegotiate the connection QoS when the existing connection QoS cannot be maintained during handover. Renegotiation may also be combined with *soft* QoS mechanisms, as described later. Finally, mobility management protocols must include mechanisms to maintain QoS of connections rerouted within the network during handover.

- *At the MT*: The MT implements the complementary functions related to QoS provisioning in the MAC and network layers. In addition, application layer functions may be implemented to deal with variations in the available QoS due to radio link degradation and/or terminal mobility. Similar functions may be implemen-ted in fixed terminals communicating with MTs.

In the following, we consider the QoS mechanisms in some detail. We focus on the MAC layer and the new network layer functions such as QoS renegotiation and soft QoS. The implementation of existing ATM QoS mechanisms have been described in much detail by others (e.g., see Ref. 9).

MAC LAYER FUNCTIONS

The radio link in a WATM system is typically a broadcast multiple access channel shared by a number of MTs. Different multiple access technologies are possible, for instance frequency, time, or code division multiple access (FDMA, TDMA, and CDMA, respectively). A combina-tion of FDMA and *dynamic* TDMA is popular in WATM implementations. That is, each radio port (RP) operates on a certain frequency band and this bandwidth is shared dynamically among ATM connections terminating on multiple MTs using a TDMA scheme.[10] ATM QoS is achieved under dynamic TDMA using a combination of a resource reservation/allocation scheme and a scheduling mechanism. This is further explained using the example of two WATM implementations: NEC's WATMnet 2.0 pro-totype[10] and the European Union's Magic Wireless ATM Network Demonstrator (WAND) project.[5]

Resource Reservation and Allocation Mechanisms (WATMnet 2.0)

WATMnet utilizes a TDMA/TDD scheme for medium access. The logical transmission frame structure under this scheme is shown in Fig. 3. As shown, this scheme allows the flexibility to partition the frame dynamically for downlink (from EMAS to MTs) and uplink (from MTs to EMAS) traffic, depending on the traffic load in each direc-tion. Other notable features of this scheme are:

Table 1 ATM traffic and QoS parameters.

| Attribute | ATM service category | | | | |
	CBR	rt-VBR	nrt-VBR	UBR	ABR
Traffic parameters					
PCR and CDVT	Yes	Yes	Yes	Yes	Yes
SCR and MBS	N/A	Yes	Yes	N/A	N/A
MCR	N/A	N/A	N/A	N/A	Yes
QoS parameters					
CDV	Yes	Yes	No	No	No
Maximum CTD	Yes	Yes	No	No	No
CLR	Yes	Yes	Yes	No	No

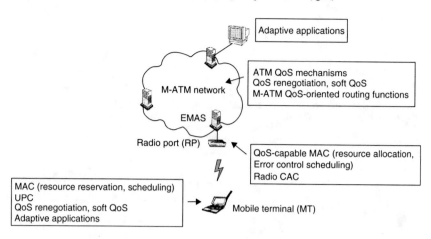

Fig. 2 QoS mechanisms in WATM.

Wireless
Ad Hoc—Asynch

- A significant portion of each slot is used for forward error control (FEC)
- A separate contention region in the frame is used for MTs to communicate with the EMAS
- 8-byte control packets are used for bandwidth request and allocation announcements. An MT can tag request packets along with the WATM cells it sends or in the contention slots
- WATM cells are modified ATM cells with data link control (DLC) and cyclic redundancy check (CRC) information added

In the downlink direction, the WATM cells transported belong to various ATM connections terminating on different MTs. After such cells arrive at the EMAS from the fixed network, the allocation of TDMA slots for specific connections is done at the EMAS based on the connections' traffic and QoS parameters. This procedure is described in the next section. In the uplink direction, the allocation is based on requests from MTs. For bursty traffic, an MT makes a request only after each burst is generated. Once uplink slots are allocated to specific MTs,

the transmission of cells from multiple active connections at an MT is again subject to the scheduling scheme. Both the request for uplink slot allocation from the MTs and the results from the EMAS are carried in control packets whose format is shown in Fig. 4. Here, the numbers indicate the bits allocated for various fields. The sequence number is used to recover from transmission losses. Request and allocation types indicate one of four types: CBR, VBR, ABR or UBR. The allocation packet has a start slot field which indicates where in the frame the MT should start transmission. The number of allocated slots is also indicated.

The DLC layer implementation in WATMnet is used to reduce the impact of errors that cannot be corrected using the FEC information. The DLC layer is responsible for selective retransmission of cells with uncorrectable errors or lost cells. Furthermore, the DLC layer provides request/reply control interface to the ATM layer to manage the access to the wireless bandwidth, based on the instantaneous amount of traffic to be transmitted. The WATM cell sent over the air interface is a modified version of the standard ATM cell with DLC and CRC information, as shown in Fig. 5. The same figure also shows the

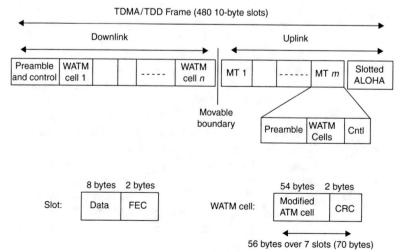

Fig. 3 TDMA logical frame format.

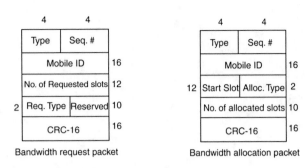

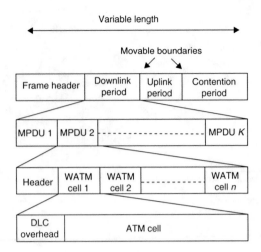

Fig. 4 Bandwidth control packet formats.

Fig. 6 Magic WAND TDMA frame structure.

acknowledgment packet format for implementing selective retransmission. In this packet, the VCI field specifies the ATM connection for which the acknowledgment is being sent. The sequence number field indicates the beginning sequence number from which the 16-bit acknowledgment bitmap indicates the cells correctly received (a "1" in the bit map indicates correct reception).

The TDMA/TDD scheme thus provides a mechanism for dynamic bandwidth allocation to multiple ATM connections. How active connections are serviced at the EMAS and the MTs to maintain their QoS needs is another matter. This is described next using Magic WAND as the example.

Scheduling (Magic WAND)

The Magic WAND system utilizes a TDMA scheme similar to that used by the WATM-net. This is shown in Fig. 6. Here, each MAC protocol data unit (MPDU) consists of a header and a sequence of WATM cells from the same MT (or the EMAS) referred to as a *cell train*. In the Magic WAND system, the scheduling of both uplink and downlink transmissions is done at the EMAS. Furthermore, the scheduling is based on the fact that the frame length is variable. A simplified description of the scheduling scheme is presented below. More details can be found in Refs. 11 and 12.

At the beginning of each TDMA frame, the scheduling function at the EMAS considers pending transmission

requests, uplink and downlink, from active connections. The scheduler addresses two issues:

1. The determination of the number of cells to be transmitted from each connection in the frame, and;
2. The transmission sequence of the selected cells within the frame.

The objective of the scheduler is to regulate the traffic over the radio interface as per the declared ATM traffic parameters of various connections and to ensure that the delay constraints (if any) are met for these connections over this interface.

The selection of cells for transmission is done based on the service categories of active connections as well as their traffic characteristics. First, for each connection, a priority based on its service category is assigned. CBR connections are assigned the highest priority, followed by rt-VBR, nrt-VBR, ABR. In addition, for each active connection that is not of type UBR, a token pool is implemented. Tokens for a connection are generated at the declared SCR of the connection and tokens may be accumulated in the pool as long as their number does not exceed the declared MBS for the connection. The scheduler services active connections in two passes: in the first pass, connections are considered in priority order from CBR to ABR (UBR is omitted) and within each priority class only connections with a positive number of tokens in their pools are considered. Such connections are serviced in the decreasing order of the number of tokens in their pools. Whenever a cell belonging to a connection is selected for transmission, a token is removed from its pool. At the end of the first pass, either all the slots in the downlink portion of the frame are used up or there are still some slots available. In the latter case, the second pass is started. In this pass, the scheduler services remaining excess traffic in each of CBR, rt and nrt-VBR and ABR classes, and UBR traffic in the priority order. It is clear that in order to adequately service active connections, the mean

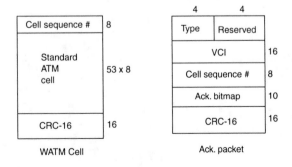

Fig. 5 WATM formats at the DLC layer.

bandwidth requirement of the connections cannot exceed the number of downlink slots available in each frame. The CAC function is used to block the setting up of new connections over a radio interface when there is a danger of overloading. Another factor that can result in overloading is the handover of connections. The CAC must have some knowledge of the expected load due to handovers so that it can limit new connection admissions. Preserving the QoS for handed over connections while not degrading existing connections at a radio interface requires good network engineering. In addition, mechanisms such as QoS renegotiation and soft QoS (described later) may be helpful.

Now, at the end of the selection phase, the scheduler has determined the number of cells to be transmitted from each active connection. Some of these cells are to be transmitted uplink while the others are to be transmitted downlink. The scheduler attempts to place a cell for transmission within the frame such that the cell falls in the appropriate portion of the frame (uplink or downlink, Fig. 6) and the delay constraint (CDT) of the corresponding connection is met. To do this, first the delay allowed over the radio segment is determined for each connection with a delay constraint (it is assumed that this value can be obtained during the connection routing phase by decomposing the path delay into delays for each hop). Then, for downlink cells, the arrival time for the cell from the fixed network is marked. For uplink cells, the arrival time is estimated from the time at which the request was received from the MT.

The deadline for the transmission of a cell (uplink or downlink) is computed as the arrival time plus the delay allowed over the radio link.

The final placement of the cells in the frame is based on a three-step process, as illustrated by an example with six connections (Fig. 7). Here, D_n and U_n indicate downlink and uplink cells with deadline = slot n, respectively, and D^i_j and U^i_j indicate downlink and uplink cells of the ith connection with deadline = slot j, respectively. In the first step, the cells are ordered based on their deadlines (Fig. 7A). Several cells belonging to the same connection may have been selected for transmission in a frame. When assigning a slot for the first cell of such a "cell train" the scheduler positions the cell such that its transmission will be before and as close to its deadline as possible. Some cells may have to be shifted from their previously allocated slots to make room for the new allocation. This is done only if the action does not violate the deadline of any cell. When assigning a slot for another cell in the train, the scheduler attempts to place it in the slot next to the one allocated to the previous cell. This may require shifting of existing allocations as before. This is shown in Fig. 7B–D. Here, the transmission frame is assumed to begin at slot 5.

At the end of the first step, the cell sequencing may be such that uplink and downlink cells may be interleaved. The second step builds the downlink portion of the frame by first shifting all downlink cells occurring before the first

uplink cell as close to the beginning of the frame as possible. In the space between the last such downlink cell and the first uplink cell, as many downlink cells as possible are packed. This is illustrated in Fig. 7E. A slot for *period overhead* (PO) is added between the downlink and uplink portions. Finally, in the last step, the uplink cells are packed, by moving all uplink cells occurring before the next downlink cell as shown in Fig. 7F. The contention slots are added after the last uplink cell and the remaining cells are left for the next frame.

Thus, scheduling can be a rather complicated function. The specific scheduling scheme used in the Magic WAND system relies on the fact that the frame length is variable. Scheduling schemes for other frame structures could be different.

NETWORK AND APPLICATION LAYER FUNCTIONS

Wireless broadband access is subject to sudden variations in bandwidth availability due to the dynamic nature of the service demand (e.g., terminals moving in and out of RPs coverage area, VBR interactive multimedia connections) and the natural constraints of the physical channel (e.g., fading and other propagation conditions). QoS control mechanisms should be able to handle efficiently both the mobility and the heterogeneous and dynamic bandwidth needs of multimedia applications. In addition, multimedia applications themselves should be able to adapt to terminal heterogeneity, computing limitations, and varying availability of network resources.[13]

In this section, the network and application layer QoS control functions are examined in the context of NECs WATMnet system.[3] In this system, a concept called *soft-QoS* is used to effectively support terminal mobility, maintain acceptable application performance and high network capacity utilization. Soft-QoS relies on a QoS control framework that permits the allocation of network resources to dynamically match the varying demands of mobile multimedia applications. Network mechanisms under this framework for connection admission, QoS renegotiation, and handoff control are described. The soft-QoS concept and the realization of the soft-QoS controller based on this concept are described next. Finally, experimental results on the impact of network utilization and soft-QoS provisioning for video applications are discussed.

A Dynamic Framework for QoS Control

The bit-rates of multimedia applications vary significantly among sessions and within a session due to user interactivity and traffic characteristics. Contributing factors include the presence of heterogeneous media (e.g., video, audio, and images) compression schemes (e.g., MPEG, JPEG), presentation quality requirements

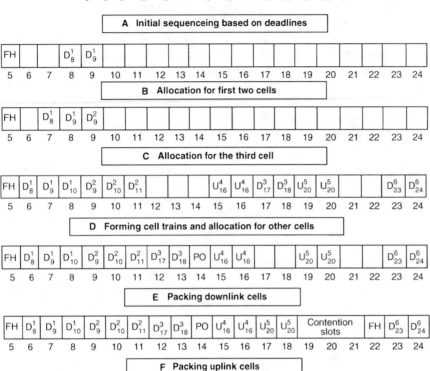

Fig. 7   Scheduling example.

(e.g., quantization, display size), and session interactivity (e.g., image scaling, video cassette recorder (VCR)-like control). Consider, e.g., a multimedia application using several media components or media objects (such as a multiwindow multimedia user interface or future MPEG-4 encoded video) which allows users to vary the relative importance-of-presence (IoP) of a given media object to match the current viewing priorities. In this case there would be a strong dependency of user/application interaction on the bandwidth requirements of individual media components. Fig. 8 shows the bit-rate when the user changes the video level-of-detail (LoD) during a session. A suitable network service for these applications should support bandwidth renegotiation to simultaneously achieve high network utilization and maintain acceptable performance. For this purpose, an efficient network service model should support traffic contract renegotiation during a session. It has been experimentally verified that bandwidth renegotiation is key for efficient QoS support of network-aware adaptive multimedia applications, and that the added implementation complexity is reasonable.[14,15]

In the mobile multimedia communication scenario, bandwidth renegotiation is particularly important. Conventional network services use static bandwidth allocation models that lack the flexibility needed to cope with multimedia interactivity and session mobility. These session

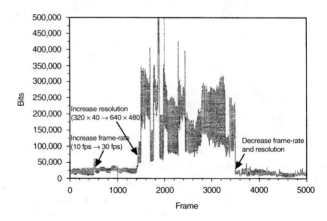

Fig. 8   Video bit-rate changes on an interactive multimedia session.

properties enlarge the dynamic range of bandwidth requirements and make dynamic bandwidth management protocols a requirement for effective end-to-end QoS support. Renegotiation may be required during handover, as well as when resource allocation changes are warranted due to instantaneous application needs and sudden changes in network resource availability.

Fig. 9 shows the system and application programming interface (API) model for QoS control with bandwidth renegotiation. The API between the adaptive application and the QoS control module is dynamic, i.e., its parameters can be modified during the session. For example, the Winsock 2 API under Microsoft Windows[16] allows the dynamic specification of QoS parameters suitable for the application. In addition, the API between the QoS controller and the network allows the traffic descriptor to be varied to track the bit-rate requirements of the bitstream. A new network service, called VBR$^+$ allows renegotiation of traffic descriptors between the network elements and the terminals.[17] VBR$^+$ allows multimedia applications to request "bandwidth-on-demand" suitable for their needs.

Soft-QoS Model

Although multimedia applications have a wide range of bandwidth requirements, most can gracefully adapt to sporadic network congestion while still providing acceptable performance. This graceful adaptation can be quantified by a softness profile.[14] Fig. 10 shows the characteristics of a softness profile. The softness profile is a function defined on the scales of two parameters: satisfaction index and bandwidth ratio. The satisfaction index is based on the subjective mean-opinion-score

(MOS), graded from 1 to 5; a minimum satisfaction divides the scale in two operational regions: the acceptable satisfaction region and the low satisfaction region. The bandwidth ratio is defined by dividing the current bandwidth allocated by the network to the bandwidth requested to maintain the desired application performance. Thus, the bandwidth ratio is graded from 0 to 1; a value of 1 means that the allocated bandwidth is sufficient to achieve the desired application performance. The point indicated as B is called the critical bandwidth ratio since it is the value that results in minimum acceptable satisfaction. As shown in Fig. 10, the softness profile is approximated by piecewise linear "S-shaped" function consisting of three linear segments. The slope of each linear segment represents the rate at which applications performance degrades (satisfaction index decreases) when the network allocates only a portion of the requested bandwidth: the steeper the slope is, the "harder" the corresponding profile is.

The softness profile allows efficient match of application requirements to network resource availability. With the knowledge of the softness profile, network elements can perform soft-QoS control—QoS-fair allocation of resources among contending applications when congestion arises. Applications can define a softness profile that best represents their needs. For example, the softness profile for digital compressed video is based on the nonlinear relationship between coding bit-rate and quality, and the satisfaction index is correlated to the user perception of quality.[18,19] While video-on-demand (VoD) applications may, in general, tolerate bit-rate regulations within a small dynamic range, applications such as surveillance or teleconference may have a larger dynamic range for bit-rate control. Other multimedia applications may allow a larger range of bit-rate control by resolution scaling.[15] In these

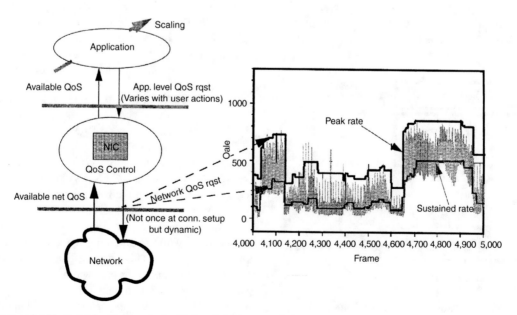

Fig. 9　System and API model for QoS control with bandwidth renegotiation.

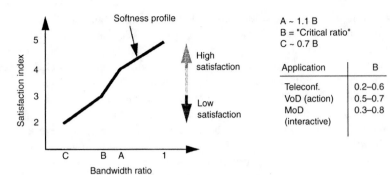

A ~ 1.1 B
B = "Critical ratio"
C ~ 0.7 B

Application	B
Teleconf.	0.2–0.6
VoD (action)	0.5–0.7
MoD (interactive)	0.3–0.8

Fig. 10   Example softness profile.

examples, VoD applications are matched to a "harder" profile than the other, more adaptive multimedia applications. Users on wireless MTs may select a "softer" profile for an application in order to reduce the connection's cost, while a "harder" profile may be selected when the application is used on wired desktop terminal. Thus, adaptive multimedia applications able to scale their video quality could specify their soft-QoS requirements dynamically to control the session's cost.

Fig. 11 conceptually illustrates the role of application QoS/bandwidth renegotiation, service contract, and session cost in the service model. The soft-QoS service model is suitable for adaptive multimedia applications capable of gracefully adjusting their performance to variable network conditions. The service definition is needed to match the requirements of the application with the capabilities of the network. The service definition consists of two parts: a usage profile that specifies the target regime of operation and the service contract that statistically quantifies the soft-QoS to be provided by the network. The usage profile, e.g., can describe the media type (e.g., MPEG video), interactivity model (e.g., multimedia browsing, video conference), mobility model (indoors, urban semi-mobile, metropolitan coverage area), traffic, and softness profiles. The service contract quantifies soft-QoS in terms of the probability that the satisfaction

of a connection will fall outside the acceptable range (given in its softness profile), the expected duration of "satisfaction outage," and the new connection blocking probability.

Network resource allocation is done in two phases. First, a CAC procedure, called soft-CAC, checks the availability of resources on the terminals coverage area at connection set-up time. The necessary resources are estimated based on the service definition. The new connection is accepted if sufficient resources are estimated to be available for the connection to operate within the service contract without affecting the service of other ongoing connections. Otherwise the connection is blocked. Second, while the connection is in progress, dynamic bandwidth allocation is performed to match the requirements of interactive VBR traffic. When congestion occurs, the soft-QoS control mechanism (re)-allocates bandwidth among connections to maintain the service of all ongoing connections within their service contracts. The resulting allocation improves the satisfaction of undersatisfied connections while maintaining the overall satisfaction of other connections as high as possible.[14] Under this model, connections compete for bandwidth in a "socially responsible" manner based on their softness profiles. Clearly, if a cost model is not in place, users would request the maximum QoS possible. The cost model provides feedback on session cost to the applications; the user can adjust the long-term QoS requirements to maintain the session cost within budget.

Soft-QoS Control in the WATMnet System

In the WATMnet system, soft-QoS control allows effective support of mobile multimedia applications with high network capacity utilization. When congestion occurs, the soft-QoS controller at the EMASs allocates bandwidth to connections based on their relative robustness to congestion given by the applications softness profiles. This allocation improves the satisfaction of undersatisfied connections while maintaining the overall satisfaction of other connections as high as possible. Within each EMAS, connections compete for bandwidth in a "socially responsible" manner based on their softness profiles.

ATM UNI signaling extensions are used in the WATMnet system to support dynamic bandwidth management.

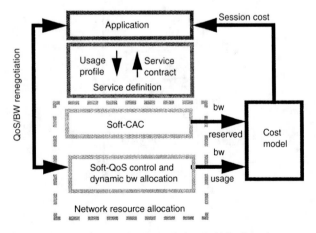

Fig. 11   Service model for dynamic bandwidth allocation with soft-QoS.

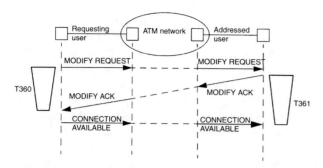

Fig. 12 Successful Q2963 modification of ATM traffic parameters with (optional) confirmation.

These extensions follow ITU-T recommendations for ATM traffic parameter modification while the connection is active.[20] Although these procedures are not finalized at the time of this writing, an overview of the current state of the recommendation is given next with an emphasis on its use to support soft-QoS in the mobile WATM scenario.

ITU-T Q.2963 allows all three ATM traffic parameters, PCR, SCR, and MBS, to be modified during a call. All traffic parameters must be increased or decreased; it is not possible to increase a subset of the parameters while decreasing the others. The user who initiates the modification request expects to receive from the network a new set of traffic parameters that are greater than or equal to (or less than) the existing traffic parameters if the modification request is an increase (or decrease). Traffic parameter modification is applicable only to point-to-point connections and may be requested only by the terminal that initiated the connection while in the active state.

The following messages are added to the UNI:

- MODIFY REQUEST (MRE) message is sent by the connection owner to request modification of the traffic descriptor; its information element (IE) is the ATM traffic descriptor.
- MODIFY ACKNOWLEDGE message is sent by the called user or network to indicate that the MRE is accepted. The broadband report type IE is included in the message when the called user requires confirmation of the success of modification.
- CONNECTION AVAILABLE is an optional message issued by the connection owner to confirm the connection modification performed in the addressed user to requesting user direction. The need for explicit confirmation of modification is indicated by the "modification confirmation" field in the MODIFY AC-KNOWLEDGE broadband report IE.
- MODIFY REJECT message is sent by the called user or network to indicate that the modify connection request is rejected. The cause of the rejection is informed through the cause IE.

Figs. 12–14 show the use of these messages for successful, addressed user rejection and network rejection of modification request, respectively.

Additionally, the soft-QoS control framework of the WATMnet system uses the following modifications to the Q2963 signaling mechanisms:

- Bandwidth change indication (BCI) message supports network-initiated and called user-initiated modification. The message is issued by the network or called user to initiate a modification procedure. The traffic descriptor to be used by the connection owner when issuing the corresponding MRE message is specified in the BCIs ATM traffic descriptor IE. Fig. 15 illustrates the use of BCI for called user-initiated modification. Timer T362 is set when issuing the BCI message and cleared when the corresponding MRE message is received; if T362 expires, the terminal and/or network element can modify the traffic policers to use the ATM traffic descriptor issued in the BCI message.
- Specification of softness profile and associated minimum acceptable satisfaction level (sat_{min}) in the MRE message. The softness profile and sat_{min} are used for QoS-fair allocation within the soft-QoS control algorithm.
- Specification of available bandwidth fraction (ABF) for each ATM traffic descriptor parameter. ABF is defined as the ratio of the available to requested traffic descriptor parameter. This results in ABF-PCR, ABF-SCR, and ABF-MBS for the peak, sustained, and MBS, respectively. These parameters are included in the MOD-IFY REJECT message. Using the ABF information, the connection owner may recompute the requested ATM traffic descriptor and reissue an appropriate MRE message.

Two additional call states are defined to support modification. An entity enters the MRE state when it issues a MRE of BCI message to the other side of the interface. An entity enters the modify received state when it receives a MRE of BCI message from the other side of the interface.

Soft-QoS control is particularly useful during the handover procedure as a new MT moves into a cell and places

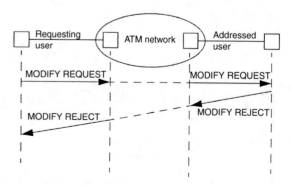

Fig. 13 Addressed user rejection of modification.

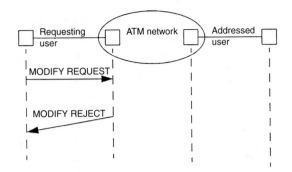

Fig. 14 Network rejection of modification request.

demands on resources presently allocated to connections from other MTs. A flexible way of prioritizing the bandwidth allocation to various session VCs is through their softness profiles. If a MT faces significant drop in bandwidth availability as it moves from one cell to another, rather than dropping the handover connections, the EMAS might be able to reallocate bandwidth among selected active connections in the new cell.

Within the soft-QoS framework, the soft-QoS controller selects a set of connections, called donors, and changes their bandwidth reservation so as to ensure satisfactory service for all.[14] This process is called network-initiated renegotiation. Network-initiated renegotiation improves the session handover success probability since multiple connections within and among sessions can share the available resources at the new EMAS, maintaining the satisfaction of individual connections above the minimum required. This mechanism allows multimedia sessions to transparently migrate the relative priority of connections as the MT moves across cells without a need to further specify details of the media session's content.

Fig. 16 shows an MT moving into the coverage area of a new RP under a new EMAS and issuing a MRE message. As a result, the EMAS might have to reallocate bandwidth of other connections under the RP to successfully complete the handover. This is accomplished by issuing BCI

messages to a selected set of connections, called donors. At the time of receiving the first MRE message for a connection being handed over, no state exists for that connection within the new EMAS. This event differentiates the MRE messages from ongoing connections and connections being handed over. Different algorithmic provisions can be made to expedite bandwidth allocation to MRE messages of connections being handed over, reducing the probability of handover drop. For example, upon identifying a MRE from such connection, the soft QoS controller can use cached bandwidth reserves to maintain the satisfaction of the connection above the minimum. The size of the bandwidth cache is made adaptive to the ratio of handed-over to local bandwidth demand. The bandwidth cache for each RP can be replenished offline using the network-initiated modification procedure. In this way, a handed-over connection need not wait for the network-initiated modification procedure to end before being able to use the bandwidth. The outcome of the reallocation enables most connections to sustain a better than minimum application performance while resources become available. Short-term congestion may occur due to statistical multiplexing. If long-term congestion arises due to the creation of a hot spot, dynamic channel allocation (DCA) may be used to provide additional resources. It is also possible that if connection rerouting is required inside the network for handover, the required resources to support the original QoS request may not be available within the network along the new path. Renegotiation is a useful feature in this case also.

An important performance metric for the soft-QoS service model is the low satisfaction rate (LSR). LSR measures the probability of failing to obtain link capacity necessary to maintain acceptable application performance. Fig. 17 compares the LSR with and without network-initiated modification over a wide range of link utilization for an MPEG-based interactive video application. The softness profiles for MPEG video were derived from empirical results reported in ref. 18 and 19. The figure shows that network-initiated modification has an important contribution to soft-QoS control performance: robust operation (LSR $< 10^{-3}$) is achievable while maintaining

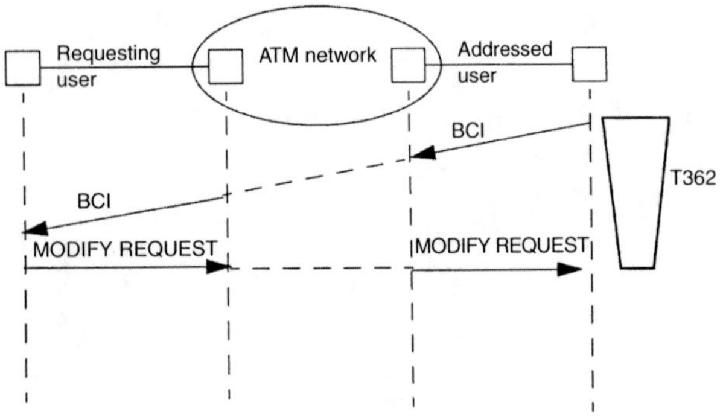

Fig. 15 Use of BCI message for addressed user–initiated modification.

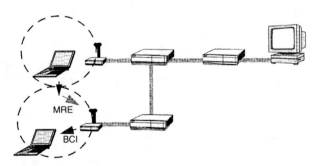

Fig. 16 Handover procedure with soft-QoS in the WATMnet system.

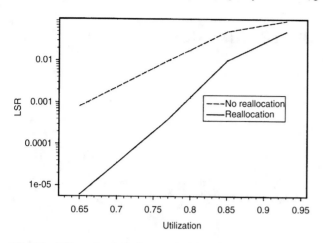

Fig. 17 Effect of soft-QoS control with and without network-initiated modification.

70–80% utilization. In the WATM scenario, the handoff success probability with soft-QoS control is related to the LSR by Prob(handoff success) = $P(\text{sat}^+ > \text{sat}_{min}) > 1 - \text{LSR} \gg 1 - P_b$, where sat^+ represents the satisfaction after handover to the new AP completes. Since it is better to block a new connection than to drop an existing connection for lack of capacity, the condition $\text{LSR} \ll P_b$ is used, where P_b is the connection blocking probability. The operating goal for the system is to maintain utilization > 70%, $\text{LSR} \sim 10^{-5}$, $P_b \sim 10^{-3}$.

Although the results presented are based on softness profiles for video, the definition of soft-QoS is appropriate for adaptive multimedia applications in general. Various representative softness profiles can be defined and refined as users' experience with distributed multimedia applications grows. New profiles can easily be incorporated within the framework as they become available.

REFERENCES

1. Singh, S. Quality of service guarantees in mobile computing. Comput. Commun, **1996**, *19*, 359–371.
2. Naghshineh, M.; Schwartz, M.; Acampora, A.S. issues in wireless access broadband networks. In *Wireless Information Networks, Architecture, Resource Management, and Mobile Data*; Holtzman, J.M., Ed.; Kluwer: 1996.
3. Raychaudhuri, D.; French, L.J.; Siracusa, R.J.; Biswas, S. K.; Yuan, R.; Narasimhan, P.; Johnston, C.A. WATMnet: a prototype wireless ATM system for multimedia personal communication. IEEE J. Sel. Areas Commun, **1997**, *15*, 83–95.
4. Veeraraghavan, M.; Karol, M.J.; Eng, K.Y. Mobility and connection management in a wireless ATM LAN. IEEE J. Sel.Areas Commun, **1997**, *15*, 50–68.
5. Ala-Laurila, J.; Awater, G. The magic WAND: wireless ATM network demonstrator. ACTS Mobile Summit 1997, Aalborg, Denmark, Oct 7–10, 1997.
6. The ATM Forum. *Wireless ATM Capability Set O Spec*, 2002.
7. The ATM Forum Technical Committee. *ATM User-Network Signalling Specification*, Version 4.0, AF-95-1434R9; 1996.
8. The ATM Forum Technical Committee. *Traffic Management Specification*, Version 4.0, AF-95-0013R11; 1996.
9. Liu, K.; Petr, D.W.; Frost, V.S.; Zhu, H.; Braun, C.; Edwards, W.L. A bandwidth management framework for ATM-based broadband ISDN. IEEE Commun. Mag, **1997**, *35*, 138–145.
10. Johnston, C.A.; Narasimhan, P.; Kokudo, J. Architecture and implementation of radio access protocols in wireless ATM networks. IEEE ICC 1998, Atlanta, GA, June 7–11, 1998.
11. Passas, N.; Paskalis, S.; Vali, D.; Merakos, L. Quality-of-service-oriented medium access control for wireless ATM networks. IEEE Commun. Mag, **1997**, *35*, 42–50.
12. Passas, N.; Merakos, L.; Skyrianoglou, D. Traffic scheduling in wireless ATM networks. IEEE ATM 1997 Workshop, Lisbon, Portugal, May 25–28, 1997.
13. Raychaudhuri, D.; Reininger, D.; Ott, M.; Welling, G. Multimedia processing and transport for the wireless personal terminal scenario. SPIE Visual Communications and Image Processing Conference (VCIP 1995), Taiwan, China, May 23–26, 1995.
14. Reininger, D.; Izmailov, R. Soft Quality-of-Service with VBR+ video. 8th International Workshop on Packet Video (AVSPN 1997), Aberdeen, Scotland, Sept 15–16, 1997.
15. Ott, M.; Michelitsch, G.; Reininger, D.; Welling, G. An architecture for adaptive QoS and its application to multimedia systems design. Comput. Commun, **1998**, *21*, 334–349.
16. Microsoft Corporation, Windows Quality of Service Technology; White paper, http://www.microsoft.com/ntserver/.
17. Reininger, D.; Raychaudhuri, D.; Hui, J. Dynamic bandwidth allocation for VBR video over ATM networks. IEEE J. Sel. Areas Commun, **1996**, *14* (6), 1076–1086.
18. Lourens, J.G.; Malleson, H.H.; Theron, C.C. Optimization of bit-rates, for digitally compressed television services as a function of acceptable picture quality and picture complexity. IEE Colloquium on Digitally Compressed TV by Satellite, 1995.
19. Nakasu, E.; Aoi, K.; Yajima, R.; Kanatsugu, Y.; Kubota, K. A statistical analysis of MPEG-2 picture quality for television broadcasting. SMPTE J, **1996**, *105*, 702–711.
20. ITU-T. ATM Traffic Descriptor Modification by the connection owner, ITU-T Q.2963.2; International Telecommunication Union-Telecom, 1997.

**Wireless
Ad Hoc—Asynch**

Wireless Channels: Link Adaptation Techniques

Naga Bhushan
Corporate R & D, Qualcomm Inc., San Diego, California, U.S.A.

Abstract

Link adaptation is a process of changing the code rate, modulation, and other parameters on a packet-to-packet basis or even during the transmission of a single packet, in response to channel conditions.

The mobile wireless channel is characterized by very large variations in link quality over time, induced by shadowing and small-scale fading of the radio signals. A mobile wireless system that uses a fixed transmission format to communicate with a user would have to be designed conservatively, so as to ensure reliable performance at the worst-case channel conditions; this conservative approach leads to significant inefficiency in the power and bandwidth utilization needed to communicate at the desired throughput and latency requirements. Hence, a mobile wireless communication needs to be able to modify its transmission parameters in response to fluctuations in the channel quality, so as to achieve reliable communication in a bandwidth- and power-efficient manner. Various physical and media access control (MAC) layer procedures designed to achieve this channel-adaptive transmission are broadly referred to as link adaptation techniques.

DIGITAL COMMUNICATION BASICS

In order to gain an understanding of link adaptation techniques in wireless communications, it is useful to consider the conceptual architecture of a digital communication system. The main functional elements of a digital communication system are shown in Fig. 1. The source generates information to be transmitted by the communication system, such as voice, video, graphics, text, and such. The source encoder extracts the relevant information from the source, which is transformed into a stream of binary symbols (bits). The output of the source encoder feeds into a channel encoder, which converts the binary sequence into a sequence of real or complex-valued modulation symbols, through a process or error control coding and modulation. The signal modulator accepts the sequence of modulation symbols, and transforms the symbols into a signal/waveform that is suitable for transmission over the communication channel (medium). The communication channel subjects its input signal to various impairments (attenuation, distortion, noise/interference etc.) and delivers the

resulting signal/waveform to the receiver. At the receiver, the signal demodulator uses the channel output along with the knowledge of channel statistics, to estimate the modulation symbols that may be sent by the transmitter. This estimate is represented by a sequence of demodulation symbols. The demodulation symbols are used by the channel decoder, which generates the decoded bit stream based on the demodulation sequence, exploiting the characteristics of the error control code and modulation. Finally, the source decoder uses the decoded bit stream to reconstruct the source data, which is delivered to the recipient.

In the context of link adaptation, we focus on the three functional blocks—channel encoder, channel decoder, and the channel. Link adaptation is essentially a technique of dynamically reconfiguring the channel encoder and decoder, in response to changes in the channel statistics. The benefit of this technique is that the channel encoding/decoding processes are optimized for the prevalent channel statistics, as it changes over time. In order for the communication system to exploit link adaptation, we need a mechanism to determine and track the changes in channel statistics/characteristics, and convey the relevant information to the transmitter and receiver, so that the channel encoder and decoder may be reconfigured in a mutually consistent manner, and in a manner that optimizes system performance for the given channel statistics. In this article, we explore some of the details of the channel coding and channel models, in order to describe explicit techniques that may be used for link adaptation, and performance gains that may be achieved from these techniques.

In Fig. 2, we see a more detailed structure of the channel encoder. As indicated before, the encoding process may usually be realized as a two- or three-step process. First, the sequence of information bits (generated by the source encoder) is segmented into fixed-length blocks called packets. Each packet is fed to an error control code, which generates a potentially longer block of code bits, by a strategic insertion of redundancy in its output sequence. The error control code may be based on classical

Encyclopedia of Wireless and Mobile Communications DOI: 10.1081/E-EWMC-120043479

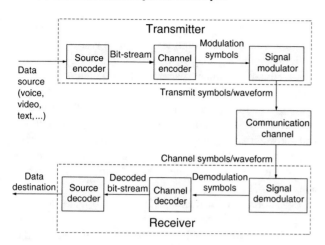

Fig. 1 Functional blocks of a digital communication system.

(algebraic) block codes over Galois fields, (terminated) convolutional codes based on linear state machines over the binary field, turbo codes based on parallel/serial concatenation of convolutional code fragments connected by interleavers, regular/irregular low-density parity-check (LDPC) codes based on Tanner graphs or protograph realizations, or any other class of error control codes. In some cases, the code symbols are passed through an interleaver before they are applied to the modulation/signal-mapping block. The interleaver, if used, permutes the (finite-length) binary sequence at its input, so that consecutive bits at the input to the interleaver are spread out over the span of the output sequence. The interleaved code bits are applied to a discrete modulator or signal-mapping block, which segments the coded bits into small groups of say 2, 3, or 4 bits, and maps each group of bits to a real or complex-valued modulation symbol. In this manner, the sequence of potentially random information bits from the source encoder are transformed into a sequence or modulation symbols with some built-in redundancy or statistical correlation, intended to improve the error resilience of the communication system. The above framework may be interpreted in a general sense, so as to include pure modulation codes (e.g., orthogonal/bi-orthogonal signaling) on one extreme, and also trellis codes and other schemes where discrete modulation is tightly integrated into the channel coding process.

Redundancy that is deliberately injected into channel encoder output sequence is exploited by the channel decoder, as it combats the effects of impairments introduced by the channel. The introduction of redundancy

leads to transmission overhead, which refers to increased demand of the signal resources needed to transmit a certain amount of information on the channel. The amount of transmission overhead introduced by the channel encoder may be measured by its effective rate, expressed in units of information bits per modulation symbol. Suppose the binary error control code accepts a packet of length k_e bits, and generates an output sequence of length n_e bits. Suppose further that the discrete modulator maps each group of k_m code bits into a sequence of n_m modulation symbols. Then, the effective rate of the channel encoder is given by $R_{eff} \equiv (k_e/n_e) \times k_m/n_m$ information bits per modulation symbol. The individual ratios $R_c \equiv k_e/n_e \leqslant 1$ and $R_m \equiv k_m/n_m$ are referred to as the code rate and modulation order of the coding and modulation schemes respectively. Code rate provides a direct measure of the redundancy introduced by the binary code, while modulation order indicates how many binary symbols are packed into each real or complex-valued modulation symbol that is transmitted over the channel. A well-designed channel coding scheme tries to minimize the redundancy or transmission overhead, while maximizing the error resilience of the signal, taking into account the impairments introduced by the channel on a statistical basis. Error resilience of the channel code may be expressed in terms of the packet error rate (PER), which represents the chance (probability) that the receive is unable to determine the actual block (packet) of information symbols that was indendently coded and transmitted over the channel, based on the signal observed at the output of the channel.

In order to understand how well a channel coding scheme performs in the presence of a communication channel, let us study some of the most rudimentary channel models. A simple channel model, such as the one described in Fig. 3, consists of signal attenuation, followed by signal distortion and additive noise/interference.

Signal attenuation accounts for the difference between the average power of the signal at the transmitter and the average power of the signal at the receiver. The signal from the transmitter is attenuated by cabling/connector losses at the transmitter and receiver, directivity of the transmit/receive antennas, propagation loss (inverse square law over free space, or other power laws over terrestrial medium), as well as shadowing induced by obstacles between the transmitter and receiver.

The attenuated signal from the transmitter may be subject to linear distortion, arising from inter-symbol interference due to finite bandwidth of the channel, as well as

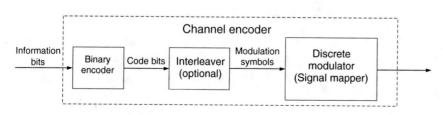

Fig. 2 Structure of a typical channel encoder.

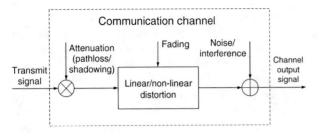

Fig. 3 A simple model of a communication channel.

multi-path fading distortion induced by wireless signal scatters. In particular, the linear distortion may simply consist of short-term signal attenuation accompanied by a phase-shift, caused by single-path (flat) fading arising from a small cluster of scatters near the transmitter/receiver. Non-linear distortion arises mostly due to the dynamic range limitations or the transmitter and receivers (conveniently lumped with the channel for the ease of modeling), and may generally be ignored.

The attenuated, distorted signal is further subjected to additive impairments, induced by thermal noise at the antenna, as well as receiver imperfections such as device noise and quantization (also lumped with the channel for modeling purposes). Furthermore, signals transmitted by other users in the system may also appear as an additive interference term at the channel output, and many cases behave essentially in the same manner as the noise term. Sometimes, the interference term may have a slightly different correlation structure from noise, which may be exploited by the receiver to suppress the effect of interference during signal demodulation.

COMMUNICATION OVER AN ADDITIVE WHITE GAUSSIAN NOISE CHANNEL

The additive white Gaussian noise (AWGN) channel is a simple and very useful, special case of the above channel model, in which the channel attenuation is fixed (time-invariant) and known to the receiver, the signal distortion is absent, and the additive noise is given by a zero-mean (circularly symmetric), Gaussian random process, that is stochastically independent from symbol to symbol. The extent of impairment induced by an AWGN channel may be quantified by its signal-to-noise ratio (SNR), defined as follows:

$$\text{SNR} \equiv \frac{\text{Transmit signal power}}{\text{Additive noise power}} \times \text{signal power attenuation}$$

$$= \left(\frac{P_S}{P_N}\right) \times G \qquad (1)$$

For a given channel coding scheme, the PER may be expressed as a monotonically decreasing function of the channel SNR. Moreover, for well-designed channel coding

schemes, the PER at a given SNR may be reduced by communicating at a lower effective rate, i.e., by increasing the redundancy induced by the binary code or by reducing the modulation order (packing fewer binary symbols in one modulation symbol). In 1948, Claude Shannon explored the fundamental trade-offs between the effective rate, channel SNR, and communication reliability (PER), and showed that it is possible to communicate at arbitrarily low PER on an AWGN channel with a given SNR, (only) as long as the effective rate of the channel code satisfies $R_{\text{eff}} < \log_2(1 + \text{SNR})$.

For more general channels, it is possible to define a number called channel capacity C, such that communication at arbitrarily low PER is possible if and only if the effective rate R_{eff} of the channel code is less than the channel capacity C. In particular, it is possible to define the channel capacity of an AWGN channel with the additional restriction that the channel code can employ a binary code, followed by a given modulation scheme [QPSK (quadrature phase-shift keying), 16-QAM (quadrature amplitude modulation) etc.]. In this case, the channel capacity is a function of SNR, represented by $C_m(\text{SNR})$, where m refers to a given modulation scheme (QPSK, 16-QAM, 64-QAM etc.). Fig. 4 shows the constrained capacity $C_m(\text{SNR})$ of the AWGN channel, for different modulation constraints on the channel code. As expected, the unconstrained AWGN capacity increases with the modulation order, and approaches the unconstrained AWGN capacity shown in Eq. 1. Fig. 4 also shows that the constrained AWGN capacity approximately equals the unconstrained AWGN capacity at low SNR; more precisely, at SNRs for which the unconstrained capacity is significantly lower than the modulation order (2 bits/symbol for QPSK, 4 bits/symbol for 16-QAM, and 6 bits/symbol for 64-QAM).

The operational significance of Shannon's result on channel capacity lies in the fact that practical channel coding schemes achieve low PERs operating at effective rates equal to the channel capacity, offset by a gap parameter. More precisely, binary error correcting codes followed by standard modulation schemes achieve low

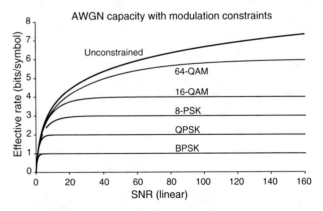

Fig. 4 AWGN channel capacity under different modulation constraints.

PERs (on the order of 1% or 0.1%), while operating at an effective rate given by $R_{eff} = C_m(\text{SNR}/\Gamma_m)$, where Γ_m represents the gap from channel capacity. In fact, well-designed turbo or LDPC codes with moderately large block lengths achieve low PERs at a channel capacity gap Γ_m of about 1 dB or less.

In any case, the main implication is that a well-designed communication system for an AWGN channel with a given modulation constraint uses a channel code whose effectiveness is completely determined by the constrained capacity associated with the channel SNR and the modulation scheme. The choice of modulation is in turn dictated by the following considerations: In a practical system, realistic channel estimation overheads and low complexity decoding requirements may lead to a somewhat larger gap parameter, somewhere between 1 and 2 dB. Furthermore, for a given code rate, higher order modulations impose additional decoding complexity (or additional performance loss with simplified decoding procedures), not to mention increased sensitivity to channel estimation errors. As a result, for a given demodulation/decoding scheme, the gap parameter Γ_m tends to be an increasing function of the modulation order m. The optimal modulation scheme and effective rate for an AWGN channel with a given SNR is obtained by solving the maximization problem $R_{opt}(\text{SNR}) = \max_m C_m \text{SNR}/\Gamma_m \equiv C_{opt}(\text{SNR})$, over the set of available modulation schemes. The solution depends not only on constrained capacity functions $C_m(\cdot)$, but also on the gap parameters Γ_m, which in turn depend on the transmission scheme and receiver design. A well-designed channel code for an AWGN channel employs the simplest (smallest) modulation scheme that is consistent with the channel SNR; in other words, the modulation order is chosen to be just high enough to ensure negligible difference between constrained capacity and unconstrained AWGN capacity (offset by the appropriate gap parameter).

If the communication channel is time-invariant and AWGN, then a fixed coded-modulation scheme consistent with the channel SNR is sufficient to operate efficiently on the communication channel. In this case, no link adaptation techniques are necessary. We now take a closer look at time-varying channels, and how link adaptation techniques may be designed to extract optimal performance out of those channels.

MODELING OF TIME-VARYING CHANNELS

In order to describe a time-varying channel model, we refer back to Fig. 3. Suppose the transmit signal is scaled by a time-invariant attenuation factor (induced by path-loss/shadowing) and then subjected to linear distortion, described by the tap-delay line $h(t) = \sum_{l=1}^{L} \alpha_l(t)\delta(t - \tau_l)$. This describes a multi-path channel with L propagation paths, where the l-th multi-path component is associated with a complex amplitude $\alpha_l(t)$ and delay τ_l. The complex amplitude $\alpha_l(t)$ captures the amplitude scaling of the signal on the l-th multi-path component, as well as the phase-shift on the component. The attenuated signal is convolved with the impulse response of the tapped delay line, and the result is further corrupted with additive noise/interference, which may be modeled as a Gaussian random process, as in the case of the AWGN channel.

Note that the time-varying nature of the above channel model comes from the time-dependence of the complex amplitude $\alpha_l(t)$. A time-varying channel is considered a slowly time-varying channel if the complex amplitude $\alpha_l(t)$ remains essentially constant over several packet durations. If the channel is slowly time varying, then it can be regarded as a static channel with respect to individual packets. This forms the basis of extending Shannon's theory of static channels to time-varying channels we have described above.

A NON-ADAPTIVE COMMUNICATION SYSTEM OVER A TIME-VARYING CHANNEL

To simplify our analysis, let us first consider a single-path, slowly time-varying channel. This channel is characterized by the impulse response $h(t) = \alpha_1(t)\delta(t - \tau_1)$ where $\alpha_1(t)$ does not vary noticeably over the duration of several packets. We assume that $\alpha_1(t)$ is a random process with unit average power; i.e., $E\lfloor|\alpha_1(t)|^2\rfloor = 1$. In this case, a packet transmitted at time t essentially sees an AWGN channel, with an overall channel attenuation $G|\alpha_1(t)|^2$. The channel SNR for this packet is given by

$$\text{SNR}(t) = \left(\frac{P_S}{P_N}\right) \times G|\alpha_1(t)|^2 = \text{SNR}_{mean}|\alpha_1(t)|^2 \qquad (2)$$

where P_S denotes the signal power at the transmitter, P_N denotes the noise power at the receiver, channel amplitude of the first (and only) multi-path component of the channel, at time t. The main difference from the ideal AWGN channel is the time-dependence of the channel SNR. In this case, the probability of error (PER) for the packet transmitted at time t is equal to the PER of the same packet on an AWGN channel with the SNR given in Eq. 2. It follows that error-free transmission of the packet is possible at time t requires that the effective rate of the channel code be less than the channel capacity at time t, given by $\log_2(1 + \text{SNR}(t)) = \log_2(1 + \text{SNR}_{mean}|\alpha_1(t)|^2)$ in the absence of any modulation constraints, or

$C_m(\text{SNR}(t)) = C_m(\text{SNR}_{\text{mean}}|\alpha_1(t)|^2)$ in the presence of modulation constraints. In practice, essentially error-free communication is achieved at time t if $R_{\text{eff}} \leqslant C_m(\text{SNR}(t)/\Gamma_m) = C_m((\text{SNR}_{\text{mean}}/\Gamma_m)|\alpha_1(t)|^2)$, where the gap parameter Γ_m accounts for the non-ideality of the channel encoding scheme at the transmitter, as well as imperfections of channel estimation, demodulation, and decoding techniques employed at the receiver. Suppose the channel varies over time due to random variation of the multi-path coefficient $\alpha_1(t)$ over a slow-time scaling (spanning many packet durations), and the coding scheme is fixed (time-invariant), with an effective rate R_{eff}. In order to guarantee essentially error-free communication with probability exceeding $1-\varepsilon$, the mean SNR of the channel must be so high that the instantaneous (gap-adjusted) capacity of the channel remains above R_{eff}, with probability exceeding $1-\varepsilon$. In other words, we need a fading margin on the mean channel SNR, sufficient to combat fading of the multi-path channel coefficient $\alpha_1(t)$. More precisely, we need the mean channel SNR to satisfy the condition $\Pr\{C_m(\text{SNR}_{\text{mean}}/\Gamma_m)|\alpha_1(t)|^2 < R_{\text{eff}}\} \leqslant \varepsilon$. Equivalently, we must have $\Pr\{|\alpha_1(t)|^2 < \Gamma_m C_m^{-1}(R_{\text{eff}})/\text{SNR}_{\text{mean}}\} \leqslant \varepsilon$, or $\Gamma_m C_m^{-1}(R_{\text{eff}})/\text{SNR}_{\text{mean}} \leqslant F^{-1}(\varepsilon)$ where $F(\cdot)$ denotes the cumulative density function (CDF) of the squared multi-path coefficient $|\alpha_1(t)|^2$. The above equation implies,

$$R_{\text{eff}} \leqslant C_m\left(F^{-1}(\varepsilon)\frac{\text{SNR}_{\text{mean}}}{\Gamma_m}\right) \leqslant \max_m C_m\left(F^{-1}(\varepsilon)\frac{\text{SNR}_{\text{mean}}}{\Gamma_m}\right)$$
$$= C_{\text{opt}}\left(F^{-1}(\varepsilon)\text{SNR}_{\text{mean}}\right) \qquad (3)$$

By contrast, on a time-invariant AWGN channel, the same channel coding scheme would have required $R_{\text{eff}} \leqslant C_{\text{opt}}(\text{SNR}_{\text{mean}})$, in order to achieve the same communication reliability.

It follows that the fixed (non-adaptive) channel coding scheme incurs a penalty of $1/F^{-1}(\varepsilon) \geqslant 1$ on the mean SNR, with respect to AWGN channel, in order to combat channel variations. This mean SNR penalty could indeed be quite significant for the fading channels commonly encountered in mobile wireless communications. For example, if $\alpha_1(t)$ is a Rayleigh fading process, then $1/F^{-1}(\varepsilon) = 1/(-\ln(1-\varepsilon)) \approx 1/\varepsilon$ for small ε. Thus, on a single-path Rayleigh fading channel, in order to guarantee reliable communication with 99% probability, the required mean SNR is 100 times (i.e., 20 dB) higher than the SNR needed on the AWGN channel (with the same channel coding scheme). This is indeed a huge price to pay for the luxury of working with a fixed (time-varying) channel coding scheme. It will be seen shortly that much of this loss can be recouped with a link-adaptive channel coding scheme, that varies the code rate and modulation scheme in concert with the variations in channel SNR.

A LINK-ADAPTIVE COMMUNICATION SYSTEM FOR A SLOWLY TIME-VARYING CHANNEL

Now consider a communication system in which the transmitter sends a reference signal (pilot signal) with a known transmit power and signature. The receiver may use this reference signal to measure the instantaneous SNR of the pilot signal, and report this value as a channel quality metric to the transmitter (using a separate feedback channel). If the channel is slowly time varying (as was assumed in the previous discussion), the SNR of the pilot signal would be close to the reported value over the next few packet transmission periods. In this case, the transmitter may control the SNR of the next data packet by transmitting the data packet at a suitable power relative to the pilot signal. In addition, the transmitter may select the code rate and modulation scheme of the data packet so as to optimize the effective rate of the packet, on an AWGN channel with the anticipated SNR. The conceptual structure of such a link-adaptive communication scheme is depicted conceptually in Fig. 5. In this communication system, the transmitter uses a larger gain for the data packet relative to the pilot, and/or picks a low code rate and a low modulation order [e.g., BPSK (binary phase shift key)/QPSK] for the next data transmission, whenever the multi-path channel coefficient $\alpha_1(t)$ has a small amplitude (compared with 1). On the other hand, when the multi-path channel coefficient $\alpha_1(t)$ has a large amplitude (compared with 1), the transmitter uses a smaller gain for the data packet relative to the pilot, and/or picks a relatively high code rate and a high modulation order (e.g., 16-QAM/64-QAM) for the next data transmission.

Note that the choice of code rate and modulation scheme determines the effective rate $R_{\text{eff}}(t)$ at time t (associated with the given packet transmission), while the relative power $P_r(t)$ of the data packet with respect to

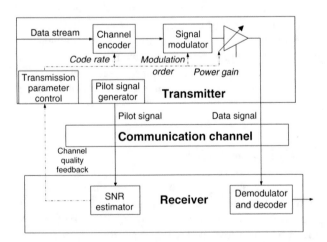

Fig. 5 Conceptual view of a link-adaptive system for a slow-fading channel.

the pilot signal determines the short-term SNR of the data packet. These two parameters are chosen so as to satisfy $C_{opt}(\text{SNR}_{mean,pilot} P_r(t)|\alpha_1(t)|^2) \geqslant R_{eff}(t)$, which ensures reliable (virtually error-free) reception of the data packet.

The link adaptation formalism above is quite general, as it includes both power control and rate control. In many cases, the power of the data packet relative to the pilot signal is held fixed, and only the effective rate (combination of code rate and modulation order) is changed in response to the SNR report from the receiver. This corresponds to the case of pure rate control. In this case, the instantaneous data rate is determined by the channel, and the objective is to maximize the average data rate (over time), for a given average SNR of the channel. On the other hand, there are situations where the transmitter does not have the luxury of changing the data rate as a function of channel conditions. This would be the case if the data is delay-sensitive and the data rate is fixed, irrespective of the channel. In this case, the transmitter needs to work with the power allocated to the data packet (as compared to the pilot signal), in order to ensure that the SNR of the data packet is fixed, although the channel gain varies over time.

Let us first focus on the case of pure rate control. In a pure rate-controlled system, we must satisfy $C_{opt}(\text{SNR}_{mean,pilot} P_r(0)|\alpha_1(t)|^2) = C_{opt}(\text{SNR}_{mean}|\alpha_1(t)|^2) \geqslant R_{eff}(t)$ where $\text{SNR}_{mean,pilot} P_r(0) \equiv \text{SNR}_{mean}$ represents the average SNR of the data packets. In this case, the long-term average data rate achieved by the communication system is given by

$$R_{avg} = R_{eff} \equiv E[R_{eff}(t)] \leqslant E\left\lfloor C_{opt}\left(\text{SNR}_{mean}|\alpha_1(t)|^2\right)\right\rfloor \quad (4)$$

Ideally, the average rate achieved by the system is equal to the upper bound on the right hand side, referred to as the *ergodic capacity* of the channel. This is to be contrasted with Eq. 3, where the upper bound on right hand side represents the *outage capacity* of the channel. In Fig. 6, we compare the ergodic capacity and outage capacity of a single-path Rayleigh fading channel, assuming for simplicity, that $C_{opt}(\text{SNR}) \approx \log_2(1 + (\text{SNR}/\Gamma_{nom}))$ for some nominal gap parameter Γ_{nom}, and for a target error (outage) probability $\varepsilon = 1\%$.

We now consider the case of fixed data rate and pure power control. In this case, we must satisfy $C_{opt}(\text{SNR}_{mean,pilot} P_r(t)|\alpha_1(t)|^2) \geqslant R_{eff} \equiv R_{target}$. Hence, the long-term average SNR of the data packets is given by $\text{SNR}_{avg} \equiv E\left[\text{SNR}_{mean,pilot} P_r(t)\right] \geqslant \Gamma_{opt} C_{opt}^{-1}(R_{target}) \cdot E[1/|\alpha_1(t)|^2]$, or equivalently,

$$R_{target} \leqslant C_{opt}\left(E\left[\frac{1}{|\alpha_1(t)|^2}\right]^{-1} \text{SNR}_{avg}\right) \quad (5)$$

This is to be compared with Eqs. 3 and 4. If the mean (average) SNR is low ($\text{SNR} \ll 1$), then the capacity function is nearly linear and hence Eqs. 4 and 5 may be approximated by

$$R_{avg} \leqslant E\left\lfloor C_{opt}\left(\text{SNR}_{mean}|\alpha_1(t)|^2\right)\right\rfloor$$
$$\approx C_{opt}(\text{SNR}_{mean}) E\left\lfloor|\alpha_1(t)|^2\right\rfloor \quad (4')$$

and

$$R_{target} \leqslant C_{opt}\left(E\left[\frac{1}{|\alpha_1(t)|^2}\right]^{-1} \text{SNR}_{avg}\right)$$
$$\approx C_{opt}(\text{SNR}_{avg}) E\left[\frac{1}{|\alpha_1(t)|^2}\right]^{-1} \quad (5')$$

As the arithmetic mean $E\lfloor|\alpha_1(t)|^2\rfloor$ is always greater than (or equal to) the harmonic mean $E[1/|\alpha_1(t)|^2]^{-1}$, it follows that pure rate control provides a better SNR-data rate tradeoff than pure power control, at least when the SNR is low. The above argument can be extended for high SNR whenever $C_{opt}(\text{SNR})$ is a convex function of $(1/\text{SNR})$, which is generally satisfied for most communication system models, which implies pure rate control is more efficient than pure power control. Hence, pure control is recommended only if latency variations induced by rate control is unacceptable for end-user experience.

For the single-path Rayleigh fading channel, $E[1/|\alpha_1(t)|^2]$, implying that it is impossible to maintain a constant data rate (data SNR) on this channel at all times, with any finite average power constraint. In this case, the pure power-controlled system is modified so as to ensure

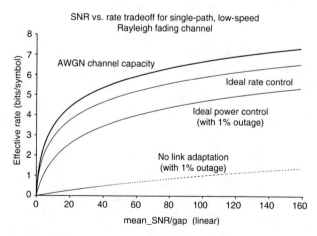

Fig. 6 SNR vs. rate for single-path, low-speed Rayleigh fading channel.

that the target data rate is maintained with high probability $1 - \varepsilon$, where the parameter ε denotes communication outage probability, analogous to the error probability criterion in the original, non-adaptive communication system example. In this case, the performance of the power-controlled system may be expressed as

$$R_{\text{target}} \leqslant C_{\text{opt}} \left(E\left[\frac{1_{\left\{ |\alpha_1(t)|^2 \geqslant F^{-1}(\varepsilon) \right\}}}{|\alpha_1(t)|^2} \right]^{-1} \text{SNR}_{\text{avg}} \right) \tag{5''}$$

where $F(\cdot)$ denotes the CDF of the squared channel coefficient $|\alpha_1(t)|^2$. In other words, the harmonic mean of the squared channel coefficient in Eq. 5 is replaced by a truncated harmonic mean. In Fig. 6, we compare the SNR-data rate tradeoff provided by pure power control with outage criterion ($\varepsilon = 1\%$) with the corresponding tradeoff provided by the pure rate-controlled system (ergodic capacity), as well as a non-adaptive communication with the same outage criterion, operating over a single-path Rayleigh fading channel.

EXTENSION TO SLOW RAYLEIGH FADING CHANNELS WITH MULIPLE PATHS AND RECEIVE ANTENNAS

The above analysis can be extended to the case where the linear distortion induced by the channel involves multiple channel coefficients with different delays. If the multiple coefficients are time-invariant, with a channel impulse response $h(t) = \sum_{l=1}^{L} \alpha_l \delta(t - \tau_l)$, the capacity of the resulting static, frequency-selective channel is given by $R_{\text{eff}} = (1/W) \int C_m(\text{SNR}(f)|H(f)|^2) \, df$, where $H(f) \equiv \alpha_l \exp(-j2\pi f \tau_l)$ represents the frequency response of the multi-path channel, $\text{SNR}(f)$ is the ratio of signal and noise power spectral densities at frequency f, and W is the (nominal) bandwidth over which the signal is transmitted. This channel capacity may be realized (offset by a reasonable gap parameter) which transmitting the modulation symbols in frequency domain, using orthogonal frequency-division multiplexing (OFDM). A note on the choice of optimal modulation scheme for a multi-path channel. The optimal choice is governed by the maximization $C_{\text{opt}}(\text{SNR}(\cdot), H(\cdot)) \equiv \max_m (1/W) \int C_m((\text{SNR}(f)/\Gamma_m) \cdot |H(f)|^2) \, df$. As in the case of AWGN channel, the optimal choice of modulation m is influenced by the lower saturation value of the constrained capacity function $C_m(\cdot)$ for small values of m, and the larger value of gap parameter Γ_m for large values of m. Unlike the case of AWGN channel, we need to choose the modulation order m so that the saturation of $C_m(\cdot)$ is minimized not only at the

nominal SNR, $\text{SNR}_{\text{nom}} \equiv (1/W) \int \text{SNR}(f)|H(f)|^2 \, df$, but also when $\text{SNR}(f)|H(f)|^2$ exceeds the nominal SNR in a statistically significant manner. As a result, the optimal modulation order for a frequency selective channel is generally higher than that for an AWGN with the same (nominal) SNR.

If the multi-path channel is slowly time-varying, individual data packets see an instantaneous (static) snapshot of the multi-path channel, with an achievable rate $R_{\text{eff}}(t) = (1/W) \int C_m(\text{SNR}_{\text{mean}}(f)/\Gamma_m|H(f,t)|^2) \, df$ at time t, where $H(f,t) \equiv \alpha_l(t) \exp(-j2\pi f \tau_l)$. Suppose the receiver employs multiple antennas, each of which sees a different fading process on the multi-path components, we have a vector multi-path channel $h_i(t) = \sum_{l=1}^{L} \alpha_{l,i}(t) \delta(t - \tau_l)$, with an instantaneous achievable rate of

$$R_{\text{eff}}(t) = \frac{1}{W} \int C_m \left(\frac{\text{SNR}_{\text{mean}}(f)}{\Gamma_m} \sum_{i=1}^{N_R} |H_i(f,t)|^2 \right) df$$

$$\equiv C_m \left(\frac{\text{SNR}_{\text{equiv}}(t)}{\Gamma_m} \right) \tag{6}$$

where $H_i(f,t) \equiv \alpha_{l,i}(t) \exp(-j2\pi f \tau_l)$, and $\text{SNR}_{\text{equiv}}(t)$ is defined as the short-term equivalent SNR of the time-varying, frequency-selective channel.

The analysis and link adaptation techniques of the previous sections carry over to the time-varying, multi-path, multi-antenna fading channels, with the capacity expression in Eq. 6 replacing the corresponding capacity (achievable-rate) functions for the single-path channels. The achievable rate as a function of mean SNR for non-adaptive and link-adaptive systems (with pure rate control) are shown in Fig. 7, for a single-path low-speed Rayleigh channel with 1, 2, and 4 receive antennas. In these results, the fading coefficients $\alpha_{1,i}(t)$ on different antennas are assumed to be statistically uncorrelated (independent). Note that as the number of (uncorrelated) receive antennas is increased, the mean SNR penalty $1/F^{-1}(\varepsilon)$ incurred by the non-adaptive system reduces considerably. Nevertheless, a link-adaptive system achieves significant gains over the non-adaptive system even with multiple receiver antennas, approaching the capacity of an AWGN channel with increasing number of receive antennas.

LINK ADAPTATION TECHNIQUES FOR RAPIDLY TIME-VARYING CHANNELS

The link adaptation techniques described in the previous sections assume that the channel quality measured on the pilot signal is essentially the same as the channel experienced by the next data packet. If the channel conditions

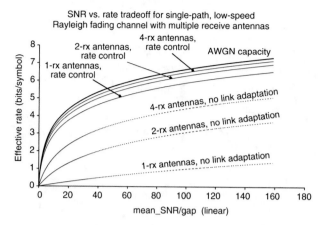

Fig. 7 Outage and ergodic capacities of multi-path low-speed Rayleigh fading channels.

change significantly between the channel quality feedback from the receiver and the next opportunity to transmit the packet on the channel, we need to employ a different link adaptation approach in order to communicate efficiently over the channel. In fact, if the channel changes significantly between successive transmission opportunities, then we may transmit each data packet in multiple installments, sufficiently spaced in time, so that different installments of the packet experience nearly independent (stochastic) realizations of the channel. The decodability of the packet is governed by a certain average of multiple channel realizations, thereby the SNR penalty is reduced in much the same way as in a slow-fading channel with multiple, uncorrelated receive antennas. Thus, significant performance improvements can be achieved on a rapidly varying channel without any (receiver-aided) link adaptation technique.

However, further performance improvements can be achieved if the packet encoding at the transmitter includes an error-detection code (e.g., CRC), the receiver tries to decode the packet after each new installment (packet fragment) is received, and informs the transmitter once the packet is successfully decoded. The transmitter sends successive fragments of a packet only until it receives an acknowledgement from the receiver (indicating that the receiver was able to decode the packet based on previously transmitted fragments of that packet). Whenever a packet is acknowledged before all its fragments are transmitted, the transmission of that packet is early terminated, so that the effective rate of that packet is increased. For example, if a packet is acknowledged after 2 out a maximum of 4 packets, the data rate of the packet is doubled, with respect to its nominal data rate corresponding to 4 transmissions. This is known as the automatic repeat request (ARQ) procedure. This is a very effective way of exploiting statistical variations of the channel, while maintaining a low error rate in the presence of severe channel variations. Fig. 8 shows the effective rate achieved on a rapidly

time-varying channel (with independent 1-path Rayleigh realizations during successive transmit slots), when packets are transmitted in multiple installments, with and without early termination from ARQ. It may be seen that a close approximation to the ergodic capacity of the Rayleigh fading channel may be realized with about 16 rounds of ARQ.

We now describe various forms in which ARQ technique may be employed in a communication system. In the basic ARQ system (i.e., type 0 ARQ), the receiver tries to decode the most recent packet installment on its own, without making use of the previously received installments. This technique provides time-diversity advantages against channel fading, in the sense that the packet is successfully decoded as soon a fragment is received at a time when the short-term channel SNR is sufficiently large. In the soft ARQ system (i.e., type 1 ARQ), the transmitter sends the same modulation symbols in each packet fragment, and the receiver combines the modulation symbols from each installment, in order to decode the packet—a decoding technique known as code combining or chase combining. A soft ARQ system exploits channel diversity over time, as well as energy combining gains. In other words, the packet is successfully decoded if the sum of the short-term channel SNR accumulated the successive packet fragment transmissions is sufficiently high. In a more advanced, hybrid ARQ (H-ARQ) system (i.e., type 2/3 ARQ), the transmitter sends different fragments of a long sequence of modulation symbols at each transmit opportunity. The data packet is encoded and modulated so that for each packet installment n, the first n fragments of the modulation symbol sequence correspond to an efficient channel code for the associated code rate and modulation order—a procedure known as incremental redundancy. At the receiver, the data packet is decoded based on all the modulation symbols received up to the given point in time. A H-ARQ scheme exploits channel diversity over time, energy combining gains, as well as coding gains realized by (nearly) optimizing the code rate and modulation order for each stage of the installment-based packet

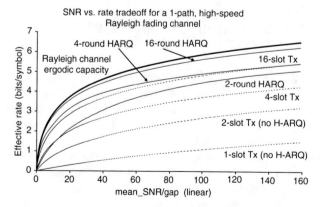

Fig. 8 Rate vs. SNR on a fast-fading Rayleigh channel with and without.

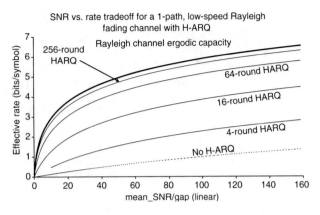

Fig. 9 SNR vs. rate for a slow-fading channel with pure H-ARQ.

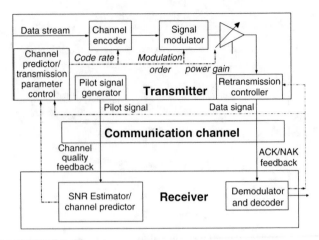

Fig. 10 Conceptual architecture of different link adaptation components.

transmission. Fig. 9 was generated assuming this H-ARQ design, with incremental redundancy in successive transmissions of the packet. The channel metrics governing the success of packet reception in the different ARQ schemes after the n-th packet installment is given by

$$R_{\text{eff}} \leqslant \frac{1}{n} \max_{1 \leqslant i \leqslant n} C_{\text{opt}}(\text{SNR}_i) \leqslant \frac{1}{n} C_{\text{opt}}\left(\sum_{i=1}^{n} \text{SNR}_i\right)$$

$$\leqslant \frac{1}{n} \sum_{i=1}^{n} C_{\text{opt}}(\text{SNR}_i) \leqslant C_{\text{opt}}\left(\frac{1}{n} \sum_{i=1}^{n} \text{SNR}_i\right) \qquad (7)$$

where SNR_i represents equivalent SNR of the channel during the i-th transmission attempt (see Eq. 6). The inequalities follow from the monotonicity and concavity of the capacity function $C_{\text{opt}}(\cdot)$. The second inequality (from the left) establishes the superiority of soft ARQ (chase combining, energy gain) over basic ARQ; the third inequality establishes the further advantages of H-ARQ (incremental redundancy, coding gain) over soft ARQ based on repetition combining. Incidentally, the last inequality represents the gain from channel hardening, as can be obtained from multi-antenna receiver diversity.

LINK ADAPTATION FOR CHANNELS WITH INTERMEDIATE TIME-VARIABILITY

At first glance, it appears as though H-ARQ is a common link adaptation technique that works with both fast and slowly time-varying channels. On the other hand, achieving sufficient H-ARQ gains on a slowly varying channel requires an unacceptably large number of H-ARQ transmit installments, leading to unacceptably large delays in most cases. This is illustrated in Fig. 9, where it is assumed that the 1-path Rayleigh fading channel is constant over multiple transmissions of the packet.

Hence, a combination of channel quality feedback and H-ARQ is needed for efficient operation on a wide class of time-varying channels, including channels with intermediate time-variability, where the SNR changes somewhat between channel quality feedback and the next packet transmission, and between successive transmission of packet fragments. In general, link adaptation includes an open-loop link adaptation component based on input from the receiver (e.g., channel quality feedback) prior to packet reception, and a closed loop link adaptation component based on packet decoding status [acknowledged/not acknowledged (ACK/NAK) feedback from the receiver] during the course of packet retransmission. Open-loop link adaptation is used to predict the channel quality during the subsequent packet transmission, while the close-loop adaptation is used to compensate for errors in the open-loop prediction through early termination. Packet decoding feedback (ACK/NAK) from the receiver may also be used to fine-tune the open-loop prediction process for the prevalent channel statistics. Fig. 10 illustrates the conceptual architecture and interactions between different link adaptation components of a wireless communication system, as described in this article.

REFERENCES

1. Cover, T.M.; Thomas, J.A. *Elements of Information Theory*; John Wiley and Sons: 1991.

2. Tse, D.; Vishwanath, P. *Fundamentals of Wireless Communications*; Cambridge University Press: 2005.

3. Goldsmith, A.J.; Chua, S.-G. Adaptive coded modulation for fading channels. IEEE Trans. Commun. **1998**, *46* (5), 595–602.

4. Chase, D. A maximum likelihood decoding approach for combining an arbitrary number of packets. IEEE Trans. Commun. **1985**, *33* (5), 385–393.

5. Cheng, J.-F. On the coding gain of incremental redundancy over chase combining. IEEE Global Telecommunications Conference, Research Triangle Park, NC, Dec, 1–5, 2003; Ericsson Res: 2003; Vol. 1, 107–112.

6. Sindhushayana, N.T.; Black, P.J. Forward link coding and modulation for CDMA2000 1xEV-DO (IS-856). The 13th IEEE International symposium on Personal, Indoor and Mobile Radio Communications, San Diego, CA, Sep 15–18, 2002; QUALCOMM Inc.: 2002; Vol. 4, 1839–1846.

7. Das, A.; Khan, F.; Sampath, A.; Su, H.-J. Adaptive, asynchronous incremental redundancy (A2IR) with fixed transmission time intervals (TTI) for HSDPA. IEEE International Symposium on Personal, Indoor and Mobile Radio Communications, 2002; Vol. 3, 1083–1087.

8. Qiu, X.; Chuang, J.; Chawla, K.; Whitehead, J. Performance comparison of link adaptation and incremental redundancy in wireless data networks. IEEE Wireless Communications and Networking Conference, New Orleans, LA, Sept 21–24, 1999.

9. Lau, V.K.N. Channel capacity and error exponents of variable rate adaptive channel coding for Rayleigh fading channels. IEEE Trans. Commun. **1999**, *47* (9), 1345–1356.

10. Zhou, Z.; Vucetic, B.; Dohler, M.; Li, Y. MIMO systems with adaptive modulation. IEEE Trans. Veh. Technol. **2005**, *54* (5), 1828–1842.

11. Zhou, Q.; Dai, H. Joint antenna selection and link adaptation for MIMO systems. IEEE Trans. Veh. Technol. **2006**, *55* (1), 243–255.

Wireless Communications: Signal Processing Tools

Boualem Boashash
Queensland University, Brisbane, Queensland, Australia

A. Belouchrani
Ecole Nationale Polytechnique, Algiers, Algeria

Karim Abed-Meraim
Ecole Nationale Superieure des Telecommunications, Paris, France

Nguyen Linh-Trung
Aston University, Birmingham, U.K.

Abstract

This entry describes the time–frequency signal processing (TFSP) techniques and tools used in wireless communications, with special emphasis on time-frequency array processing.

TIME–FREQUENCY ARRAY SIGNAL PROCESSING

Conventional array signal processing algorithms assume stationary signals and mainly employ the covariance matrix of the data array. When the frequency content of the measured signals is time varying (i.e., nonstationary signals), this class of approaches can still be applied. However, the achievable performances in this case are reduced with respect to those that would be achieved in a stationary environment. In the last decades, the stationarity hypothesis was motivated by the crucial need in practice of estimating sample statistics by resorting to temporal averaging under the additional assumption of ergodic signals. Instead of considering the nonstationarity as a shortcoming and trying to design algorithms robust with respect to nonstationarity, it would be better to take advantage of the nonstationarity by considering it as a source of information. The latter can then be exploited in the design of efficient algorithms in such nonstationary environments.

The question now is, How can we exploit the nonstationarity in array processing? This can be done by resorting to the spatial time–frequency distributions (STFDs), which are a generalization of the TFDs to a vector of multisensor signals (see Fig. 1). Under a linear model, the STFDs and the commonly known covariance matrix exhibit the same eigen structure. In wireless communications involving multiantennae, the aforementioned structure is often exploited to estimate some signal parameters through subspace-based techniques.

Algorithms based on STFDs properly use the time–frequency information to significantly improve performance. This improvement comes essentially from the fact that the effects of spreading the noise power while localizing the source energy in the time–frequency domain increase the SNR. STFD-based algorithms exploit the time–frequency representation of the signals together with the spatial diversity provided by the multiantennae.

The concept of the STFD was introduced for the first time in 1996.[1] It was used successfully in solving the problem of the blind separation of nonstationary signals.[1-4] This concept was then applied to solve the problem of direction-of-arrival (DOA) estimation.[5] Since then, several works were conducted in this area using the new concept of STFD.[6-15]

The following notations are used throughout the rest of this chapter. For a given matrix $\mathbf{A}$, the symbols $\mathbf{A}^T, \mathbf{A}^*, \mathbf{A}^H$, $\mathbf{A}^\#$, trace($\mathbf{A}$), and norm($\mathbf{A}$) respectively denote the transpose, conjugate, conjugate transpose, Moore–Penrose pseudoinverse, trace, and (Euclidean) norm of $\mathbf{A}$.

STFD

Given an analytic vector signal $\mathbf{z}(t)$, the spatial instantaneous autocorrelation function is defined as

$$\mathbf{K}_{\mathbf{zz}}(t, \tau) = \mathbf{z}\left(t + \frac{\tau}{2}\right)\mathbf{z}^H\left(t - \frac{\tau}{2}\right) \tag{1}$$

Define also the smoothed spatial instantaneous autocorrelation function as

$$\mathbf{S}_{\mathbf{zz}}(t, \tau) = G(t, \tau) \underset{t}{\star} \mathbf{K}_{\mathbf{zz}}(t, \tau) \tag{2}$$

Encyclopedia of Wireless and Mobile Communications DOI: 10.1081/E-EWMC-120043924

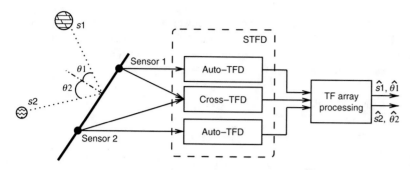

Wireless
Channels—Internet

Fig. 1 Time–frequency array signal processing.

where $G(t,\tau)$ is some time-lag kernel. The time convolution operator $\underset{t}{\star}$ is applied to each entry of the matrix $\mathbf{K}_{zz}(t,\tau)$. The class of quadratic STFDs then defined as

$$\mathbf{D_{zz}}(t,f) = \mathcal{F}_{\tau \to f}\{\mathbf{S_{zz}}(t,\tau)\} \tag{3}$$

where the Fourier transform $\mathcal{F}$ is applied to each entry of the matrix $\mathbf{S}_{zz}(t,\tau)$.

The discrete-time definition equivalent to Eqs. 2 and 3 leads to the simple implementation of STFD and is expressed as

$$\mathbf{D_{zz}}(n,k) = \mathcal{DF}_{m \to k}\{G(n,m) \underset{n}{\star} \mathbf{K_{zz}}(n,m)\} \tag{4}$$

which can also be expressed as

$$\mathbf{D_{zz}}(n,k) = \sum_{m=-M}^{M} \sum_{p=-M}^{M} G(p-n,m)$$
$$\cdot \mathbf{z}(p+m)\mathbf{z}^H(p-m)e^{-j4\pi\frac{mk}{N}} \tag{5}$$

where the discrete Fourier transform $\mathcal{DF}$ and the discrete-time convolution operator $\underset{n}{\star}$ are applied to each entry of the matrix $G(n,m) \underset{n}{\star} \mathbf{K_{zz}}(n,m)$ and matrix $\mathbf{K_{zz}}(n,m)$, respectively.

Note that the STFD of a vector signal is a matrix whose diagonal entries are the classical auto-TFDs of the vector components, and the off-diagonal entries are the cross-TFDs.

A more general definition of the STFD can be given as

$$\mathbf{D_{zz}}(n,k) = \sum_{m=-M}^{M} \sum_{p=-M}^{M} \mathbf{G}(p-n,m)$$
$$\odot \left(\mathbf{z}(p+m)\mathbf{z}^H(p-m)e^{-j4\pi\frac{mk}{N}}\right) \tag{6}$$

where $\odot$ designates the Hadamard product and $[\mathbf{G}(n,m)]_{ij} = G_{ij}(n,m)$ is the time-lag kernel associated with the pair of sensor signals $z_i(n)$ and $z_j(n)$ [kernels G_{ij} might be chosen according to the nature of considered signals $z_i(n)$ and $z_j(n)$ if such a priori information is available].

Structure under linear model

Consider the following linear model of the vector signal $\mathbf{z}(n)$:

$$\mathbf{z}(n) = \mathbf{As}(n) \tag{7}$$

where $\mathbf{A}$ is a $K \times L$ matrix ($K \geqslant L$) and $\mathbf{s}(n)$ is a $L \times 1$ vector, which is referred to as the source signal vector.

Under this linear model the STFDs take the following structure:

$$\mathbf{D_{zz}}(n,k) = \mathbf{A}\mathbf{D_{ss}}(n,k)\mathbf{A}^H \tag{8}$$

where $\mathbf{D_{ss}}(n,k)$ is the source STFD of vector $\mathbf{s}(n)$ whose entries are the auto- and cross-TFDs of the source signals.

The auto-STFD denoted by $\mathbf{D}_{zz}^a(n,k)$ is the STFD, $\mathbf{D}_{zz}(n,k)$, evaluated at autoterm points only. Correspondingly, the cross STFD $\mathbf{D}_{zz}^c(n,k)$ is the STFD, $\mathbf{D}_{zz}(n,k)$, evaluated at cross-term points.

Note that the diagonal (off-diagonal) elements of $\mathbf{D_{ss}}(n,k)$ are autoterms (cross-terms). Thus, the auto- (cross-) STFD $\mathbf{D}_{ss}^a(n,k)\mathbf{D}_{ss}^c(n,k)$ is diagonal [off-diagonal (A matrix is off-diagonal if its diagonal entries are zeros.)] for each time–frequency point that corresponds to a source autoterm (cross-term), provided the window effect is neglected.

Structure under Unitary Model

Denote by $\mathbf{W}$ a $L \times K$ whitening matrix such that

$$(\mathbf{WA})(\mathbf{WA})^H = \mathbf{UU}^H = \mathbf{I} \tag{9}$$

Pre- and postmultiplying the STFD $\mathbf{D_{zz}}(n,k)$ by $\mathbf{W}$ leads to the whitened STFD, defined as

$$\underline{\mathbf{D_{zz}}}(n,k) = \mathbf{W}\mathbf{D_{zz}}(n,k)\mathbf{W}^H = \mathbf{U}\mathbf{D_{ss}}(n,k)\mathbf{U}^H \tag{10}$$

where the second equality stems from the definition of $\mathbf{W}$ and Eq. 8. This above whitening leads to a linear model with a unitary mixing matrix.

Note that the whitening matrix can be computed in different ways. It can be obtained, for example, as an inverse square root of the observation covariance matrix[4] or computed from the STFD matrices as shown in Ref. 16.

At autoterm points, the whitened auto-STFD has the following structure:

$$\underline{\mathbf{D}}_{\mathbf{zz}}(n, k) = \mathbf{U}\mathbf{D}_{\mathbf{ss}}^a(n, k)\mathbf{U}^H \tag{11}$$

where $\mathbf{D}_{\mathbf{ss}}^a(n, k)$ is diagonal. However, at cross-term points, the whitened cross STFD exhibits the following structure:

$$\underline{\mathbf{D}}_{\mathbf{zz}}(n, k) = \mathbf{U}\mathbf{D}_{\mathbf{ss}}^c(n, k)\mathbf{U}^H \tag{12}$$

where $\mathbf{D}_{\mathbf{ss}}^c(n, k)$ is off-diagonal.

The above-defined STFDs permit the application of subspace techniques to solve a large class of channel estimation and equalization, blind source separation, and high-resolution DOA estimation problems. For the blind source separation problem, the STFDs allow the separation of Gaussian sources with identical spectral shape but with different time–frequency signatures.[4] In the area of DOA finding, the estimation of the signal and noise subspaces from the STFDs highly improves the angular resolution performance.

STFD Structure in Wireless Communications

In wireless communications, when L user signals arrive at a K-element antenna, the linear data model

$$\mathbf{z}(n) = \mathbf{A}\mathbf{s}(n) + \mathbf{n}(n) \tag{13}$$

is commonly assumed, where we recall that $\mathbf{z}(n)$ is the $K \times 1$ data vector received at the antennae and $\mathbf{s}(n)$ is the $L \times 1$ user data vector, the spatial matrix $\mathbf{A} = [\mathbf{a}_1, \ldots, \mathbf{a}_n]$ represents the propagation matrix, (This matrix is also known as the mixing matrix.) $\mathbf{a}_i$ is the steering vector corresponding to the ith user, and $\mathbf{n}(n)$ is an additive noise vector whose entries are modeled as stationary, temporally and spatially white, zero-mean random processes, and independent of the user-emitted signals.

Under the above assumptions, the expectation of the TFD matrix between the user signal and the noise vectors vanishes, i.e.,

$$E\{\mathbf{D}_{\mathbf{sn}}(n, k)\} = 0 \tag{14}$$

and it follows that

$$\tilde{\mathbf{D}}_{\mathbf{zz}}(n, k) = \mathbf{A}\tilde{\mathbf{D}}_{\mathbf{ss}}(n, k)\mathbf{A}^H + \sigma^2\mathbf{I} \tag{15}$$

with

$$\tilde{\mathbf{D}}_{\mathbf{zz}}(n, k) = E\{\mathbf{D}_{\mathbf{zz}}(n, k)\} \tag{16}$$

$$\tilde{\mathbf{D}}_{\mathbf{ss}}(n, k) = E\{\mathbf{D}_{\mathbf{ss}}(n, k)\} \tag{17}$$

where σ^2 is the noise power and $\mathbf{I}$ is the identity matrix. Under the same assumptions, the data covariance matrix, which is commonly used in array signal processing, is given by

$$\mathbf{R}_{\mathbf{zz}} = \mathbf{A}\mathbf{R}_{\mathbf{ss}}\mathbf{A}^H + \sigma^2\mathbf{I} \tag{18}$$

where

$$\mathbf{R}_{\mathbf{zz}} = E\{\mathbf{z}(n)\mathbf{z}(n)^H\} \tag{19}$$

$$\mathbf{R}_{\mathbf{ss}} = E\{\mathbf{s}(n)\mathbf{s}(n)^H\} \tag{20}$$

From Eqs. 15 and 18, it becomes clear that the STFDs and the covariance matrix exhibit the same eigen-structure. This structure is often exploited to estimate some signal parameters through subspace-based techniques.

Advantages of STFDs over Covariance Matrix

The STFDs allow the processing of the received data in both the spatial domain and the two-dimensional time–frequency domain simultaneously. In time–frequency array signal processing, the STFDs are eigen-decomposed, instead of the traditional covariance matrix $\mathbf{R}_{\mathbf{zz}}$, to separate the signal subspace and noise subspace. Thanks to the availability of time-varying filtering in the time–frequency domain, the STFD-based approaches can handle signals corrupted by interference occupying the same frequency band or the same time slot, but with different time–frequency signatures; thus, signal selectivity is increased with respect to covariance matrix-based methods. In addition, the effect of spreading noise power while localizing the user energy in the time–frequency plane amounts to increased robustness of the STFD-based approaches with respect to noise. In other words, the eigenvectors of the signal subspace obtained from an STFD matrix that is made up of signal autoterms are more robust to noise than those obtained from the covariance matrix. Hence, the performance of the STFD-based approaches can be significantly improved, particularly when the input SNR is low (Subspace analysis of the STFDs vs. the covariance matrix is provided in Ref. 8) (typically, an SNR of 0 dB or lower). Moreover, in Ref. 12 it is proved that the traditional covariance-based subspace methods are low-dimensional cases of the STFD subspace methods.

If one selects the kernel $G(n,m)$ in Eq. 5 so that the corresponding TFD satisfies the marginal condition

$$\sum_k \tilde{\mathbf{D}}_{z_i z_j}(n, k) = E\{z_i(n)z_j(n)^*\} \tag{21}$$

then

$$\sum_k \tilde{\mathbf{D}}_{\mathbf{zz}}(n,k) = E\{\mathbf{z}(n)\mathbf{z}(n)^H\} = \mathbf{R}_{\mathbf{zz}} \qquad (22)$$

The above equation shows that the projection of the STFD over the time domain is nothing more than the traditional covariance matrix. Hence, the space spanned by $\mathbf{R}_{\mathbf{zz}}$ is the projection of the space spanned by $\tilde{\mathbf{D}}_{\mathbf{zz}}(n,k)$ over the space that is orthogonal to the frequency dimension. This means that the space spanned by $\tilde{\mathbf{D}}_{\mathbf{zz}}(n,k)$ is the extension of the space spanned by $\mathbf{R}_{\mathbf{zz}}$ toward a higher-dimension space. Therefore, the $\mathbf{R}_{\mathbf{zz}}$-based techniques can be seen as a low-dimension special case of the $\tilde{\mathbf{D}}_{\mathbf{zz}}(n,k)$-based ones. This is quite straightforward since the STFD-based methods are multidimensional (spatial-time–frequency) processing methods. Obviously, the details and signatures of the signal will be described more accurately and finely in higher-dimension space. In fact, this is the reason that the STFD-based methods have better performance, such as signal selectivity, interference suppression, and high resolution, over the conventional covariance matrix-based approaches.

Selection of Autoterms and Cross-terms in the Time–Frequency Domain

STFD-based methods require computation of STFDs at different time–frequency points. At autoterm points, where the diagonal structure of the source STFD is enforced, the data STFDs are either incorporated into a joint diagonalization (JD) technique or eigen decomposed after simple averaging over the source signatures of interest to estimate the mixing, or the array manifold matrix. At cross-term points, where this time the off-diagonal structure of the source STFD is enforced, the data STFDs are incorporated in an off-diagonalization technique to achieve the task of the mixing/propagation matrix identification.

An intuitive procedure to select the autoterms is to consider the time–frequency points corresponding to the maximum energy in the time–frequency plane.[1] The above intuitive procedure has shown some limitations in practical situations. A projection-based selection procedure of cross-terms and autoterms has been proposed in Ref. 9. The latter exploits the off-diagonal structure of the cross-source STFDs and proceeds on whitened data STFDs. More precisely, for a cross-source STFD, we have

$$\mathrm{Trace}(\mathbf{D}_{\mathbf{zz}}^c(n,k)) = \mathrm{Trace}(\mathbf{U}\mathbf{D}_{\mathbf{ss}}^c(n,k)\mathbf{U}^H)$$
$$= \mathrm{Trace}(\mathbf{D}_{\mathbf{ss}}^c(n,k)) \approx 0 \qquad (23)$$

Based on this observation, the following testing procedure applies:

$$\mathrm{if}\ \frac{\mathrm{Trace}\{\hat{\mathbf{D}}_{\mathbf{zz}}(n,k)\}}{\mathrm{norm}\{\hat{\mathbf{D}}_{\mathbf{zz}}(n,k)\}} < \epsilon \rightarrow \mathrm{decide\ that}\ (n,k)$$

is a cross-term point

where the threshold ϵ is a positive scalar (typically $\epsilon = 0.9$). In the underdetermined case (i.e., $K < L$), the matrix $\mathbf{U}$ (see "Structure under Unitary Model" section) is non-square (with more columns than rows), and consequently, $\mathbf{U}^H\mathbf{U} \neq \mathbf{I}$ represents the projection matrix onto the row space of $\mathbf{U}$. Therefore, Eq. 23 becomes only an approximation. An alternative solution consists of exploiting the existence of only one source at some autoterm points. At such points, each autoterm STFD matrix is of rank one, or at least has one large eigenvalue compared to its other eigenvalues. Therefore, one can use rank selection criteria, such as MDL (minimum description length) or AIC (Akaike information criterion),[17] to select autoterm points as those corresponding to STFD matrices of selected rank equal to one. For simplicity, the following criterion can be used:

$$\mathrm{if}\ \left|\frac{\lambda_{\max}\{\hat{\mathbf{D}}_{\mathbf{zz}}(n,k)\}}{\mathrm{norm}\{\hat{\mathbf{D}}_{\mathbf{zz}}(n,k)\}} - 1\right| > \epsilon \rightarrow \mathrm{decide\ that}\ (n,k)$$

is a cross-term point

where ϵ is a small positive scalar (typically, $\epsilon = 1\mathrm{E}{-}4$) and $\lambda_{\max}\{\hat{\mathbf{D}}_{\mathbf{zz}}(n,k)\}$ represents the largest eigenvalue of $\hat{\mathbf{D}}_{\mathbf{zz}}(n,k)$.

A statistical test to decide whether a time–frequency point is dominated by auto- or cross-terms is proposed in Ref. 18. The latter consider the following test statistic:

$$\frac{\mathrm{Trace}\{\hat{\mathbf{D}}_{\mathbf{zz}}(n,k)\}}{\mathrm{norm}\{\hat{\mathbf{D}}_{\mathbf{zz}}(n,k)\}} \qquad (24)$$

To discriminate between noise and either auto- or cross-term, the variance of the test statistic is used. Because only a single value of the test statistic is known at the time–frequency point under test, the variance is estimated using a bootstrap resampling technique.[14,18] Once the noise regions in the time–frequency domain are identified, a threshold is set to distinguish the autoterms from the cross-terms. In Ref. 19, array averaging of the STFDs is used to reduce the cross-terms without smearing the autoterms, allowing the autoterms to be more pronounced and easier to detect in the time–frequency plane.

In Ref. 20, it is shown that for real-valued signals, the imaginary parts of the STFDs, when not equal to zero, only correspond to cross-terms whatever the considered point in the time–frequency plane. This result was exploited to derive a criterion for the auto- and cross-term selection. In the case of noisy signals, Ref. 11 describes a detection criteria of cross- and auto-terms in the time–frequency plane by introducing two thresholds based on the Bayesian and Neyman–Pearson approaches.

The selection of autoterms in the time–frequency domain is still an open problem. And the success of any STFD-based technique depends highly on the performance of the employed autoterm selection procedure.

Time–Frequency Direction-of-arrival Estimation

In order to obtain the mobile users' spatial information and achieve the space-division multiple access (SDMA), the DOA estimation of far field sources from the multiantenna outputs is one of the important issues in next-generation wireless communications. Thanks to their super resolution and robustness, the subspace-based methods, such as MUSIC[21] and ESPRIT,[22] are considered the most popular techniques in traditional array processing. However, all these subspace methods assume the signals impinged on the antennae stationary, while typical signals in wireless communications, such as frequency-hopping signals or frequency-modulated signals, are nonstationary with some a priori known information on their time-varying frequency content. In addition, several nonspatial features such as time and frequency signatures of the signals are ignored in conventional methods. These defects may result in unaffordable estimation error.

In most wireless communications systems, the signals are man-made and hence much information contained in these signals is known or can be obtained a priori. One can exploit this information not only in the spatial domain but also in the time–frequency domain in order to improve the performance. One of these techniques is the STFD-based DOA estimation method. Recently, several traditional DOA estimation techniques have been extended to nonstationary signals thanks to the use of STFD instead of the covariance matrix. Hence, time–frequency MUSIC (TF-MUSIC) was first introduced in Ref. 5; then time–frequency maximum likelihood (TF-ML), time–frequency signal subspace fitting (TF-SSF), and time–frequency ESPRIT (TF-ESPRIT) were introduced in Refs. 12, 15, and 23, respectively. Below, we describe only TF-MUSIC as an illustrative example on how the STFDs can be exploited for DOA estimation.

Data Model

Consider again Eq. 13, which is often encountered in wireless communications:

$$\mathbf{z}(n) = \mathbf{A}(\theta)\mathbf{s}(n) + \mathbf{n}(n) \qquad (25)$$

Herein, the propagation matrix $\mathbf{A}(\theta) = [\mathbf{a}(\theta_1), \dots, \mathbf{a}(\theta_L)]^T$, also known as steering matrix, is parameterized by the parameter vector $\theta = [\theta_1, \dots, \theta_L]^T$, where $\mathbf{a}(\theta_k)$ and θ_k define the steering vector and the DOA of the kth user, respectively.

Assuming a noise-free environment, the structure of the STFD associated with the above model is given by

$$\mathbf{D}_{\mathbf{zz}}(n, k) = \mathbf{A}(\theta)\mathbf{D}_{\mathbf{ss}}(n, k)\mathbf{A}(\theta)^H \qquad (26)$$

TF-MUSIC

By performing the singular value decomposition (SVD) of the steering matrix,

$$\mathbf{A}(\theta) = [\mathbf{E}_s \mathbf{E}_n][\mathbf{D} \quad \mathbf{0}]^T \mathbf{V}^H \qquad (27)$$

and incorporating the results in Eq. 26, it is easily shown that

$$\mathbf{D}_{\mathbf{zz}}(n, k) = [\mathbf{E}_s \mathbf{E}_n]\mathbf{D}(n, k)[\mathbf{E}_s \mathbf{E}_n]^H \qquad (28)$$

where $\mathbf{D}(n,k)$ is a block-diagonal matrix given by

$$\mathbf{D}(n, k) = \mathrm{diag}[\mathbf{D}\mathbf{V}^H \mathbf{D}_{\mathbf{ss}}(n, k)\mathbf{V}\mathbf{D} \quad \mathbf{0}] \qquad (29)$$

Since $\mathbf{E}_s$ and $\mathbf{E}_n$, which span the signal subspace and noise subspace, respectively, are fixed and independent of the time–frequency point (n,k), Eq. 28 reveals that any matrix $\mathbf{D}_{\mathbf{zz}}(n,k)$ is block-diagonalized by the unitary transform $\mathbf{E} = [\mathbf{E}_s \ \mathbf{E}_n]$.

A simple way to estimate $\mathbf{E}_s$ and $\mathbf{E}_n$ is to perform the SVD on a single matrix $\mathbf{D}_{\mathbf{zz}}(n,k)$ or an averaged version of $\mathbf{D}_{\mathbf{zz}}(n,k)$ over the source signatures of interest. But indeterminacy arises in the case where $\mathbf{D}_{\mathbf{ss}}(n,k)$ is singular. To avoid this problem, a joint block diagonalization (JBD) of the combined set of $\{\mathbf{D}_{\mathbf{zz}}(n_l, k_l) | l = 1, \dots, P\}$ can be performed by exploiting the joint structure (Eq. 28) of the STFDs. This JBD is achieved by the maximization under unitary transform of the following criterion:

$$C(\mathbf{U}) \overset{\Delta}{=} \sum_{l=1}^{P} \sum_{i,j=1}^{L} \left| \mathbf{u}_i^H \mathbf{D}_{\mathbf{zz}}(n_l, k_l)\mathbf{u}_j \right|^2 \qquad (30)$$

over the set of unitary matrices $\mathbf{U} = [\mathbf{u}_1, \dots, \mathbf{u}_k]$. Note that in Ref. 24, an efficient algorithm for solving Eq. 30 exists. Once the signal and noise subspaces are estimated, one can use any subspace-based technique to estimate the DOAs. The MUSIC algorithm[21] is then applied to the noise subspace matrix $\hat{\mathbf{E}}_n$ estimated from Eq. 30. Hence, the TF-MUSIC algorithm estimates the DOAs by finding the L largest peaks of the localization function

$$f(\theta) = \left| \hat{\mathbf{E}}_n^H \mathbf{a}(\theta) \right|^{-2} \qquad (31)$$

Time–Frequency Source Separation

Currently, blind source separation is considered one of most promising techniques in wireless communications and more specifically in multiuser detection. The underlying problem consists of recovering the original waveforms of the user-emitted signals without any knowledge on their linear mixture. This mixture can be either instantaneous or convolutive. The problem of blind source separation has two inherent indeterminacies such that source signals can only be identified up to a fixed permutation and some complex factors.[25]

So far, the problem of blind source separation has been solved using statistical information available on the source signals. The first solution was based on the cancellation of higher-order moments assuming non-Gaussian and independent and identically distributed (i.i.d.) signals.[26] Other solutions based on minimization of cost functions, such as contrast functions[27] or likelihood function,[28] have been proposed. In the case of non-i.i.d. signals and even Gaussian sources, solutions based on second-order statistics were also proposed.[25]

When the frequency content of the source signals is time varying, one can exploit the powerful tool of the STFDs to separate and recover the incoming signals. In this context, the underlying problem can be regarded as signal synthesis from the time–frequency plane with the incorporation of the spatial diversity provided by the antennae.

In contrast to conventional blind source separation approaches, the STFD-based signal separation techniques allow separation of Gaussian sources with identical spectral shape provided that the sources have different time–frequency signatures. Below, we describe applications of the STFDs for the separation of both instantaneous and convolutive mixtures.

Separation of instantaneous mixture

The multiantenna signal $\mathbf{z}(n)$ is assumed to be nonstationary and to obey the linear model in Eq. 7. The problem under consideration consists of identifying the matrix $\mathbf{A}$ and recovering the source signals s(n) up to a fixed permutation and some complex factors.

By selecting autoterm points, the whitened auto-STFDs have the structure in Eq. 11 that we recall herein:

$$\underline{\mathbf{D}}_{\mathbf{zz}}(n,k)= \mathbf{U}\mathbf{D}^a_{\mathbf{ss}}(n,k)\mathbf{U}^H \qquad (32)$$

with $\mathbf{D}^a_{\mathbf{ss}}(n,k)$ a diagonal matrix. The missing unitary matrix $\mathbf{U}$ is retrieved up to permutation and phase shifts by JD of a combined set $\{\underline{\mathbf{D}}_{\mathbf{zz}}(n_a,k_a) \mid a=1,\ldots,P\}$ of P auto-STFDs. The incorporation of several autoterm points in the JD reduces the likelihood of having degenerate eigenvalues and increases robustness to a possible additive noise. The above JD is defined as the maximization of the following criterion:

$$C_{JD}(\mathbf{V}) \overset{\Delta}{=} \sum_{a=1}^{P} \sum_{i=1}^{L} \left| \mathbf{v}_i^H \underline{\mathbf{D}}_{\mathbf{zz}}(n_a,k_a)\mathbf{v}_i \right|^2 \qquad (33)$$

over the set of unitary matrices $\mathbf{V} = [\mathbf{v}_1,\ldots,\mathbf{v}_L]$.

The selection of cross-term points leads to the whitened cross STFD (Eq. 12),

$$\underline{\mathbf{D}}_{\mathbf{zz}}(n,k)= \mathbf{U}\mathbf{D}^c_{\mathbf{ss}}(n,k)\mathbf{U}^H \qquad (34)$$

with $\mathbf{D}^c_{\mathbf{ss}}(n,k)$ an off-diagonal matrix. The unitary matrix $\mathbf{U}$ is found up to permutation and phase shifts by joint off-diagonalization (JOD) of a combined set $\{\underline{\mathbf{D}}_{\mathbf{zz}}(n_c,k_c)|c=1,\ldots,Q\}$ of Q cross-STFDs. This JOD is defined as the maximization of the following criterion:

$$C_{JOD}(\mathbf{V}) \overset{\Delta}{=} - \sum_{c=1}^{Q} \sum_{i=1}^{L} \left| \mathbf{v}_i^H \underline{\mathbf{D}}_{\mathbf{zz}}(n_c,k_c)\mathbf{v}_i \right|^2 \qquad (35)$$

over the set of unitary matrices $\mathbf{V} = [\mathbf{v}_1,\ldots,\mathbf{v}_L]$.

The unitary matrix $\mathbf{U}$ can also be found up to permutation and phase shifts by a combined JD/JOD of the two sets $\{\underline{\mathbf{D}}_{\mathbf{zz}}(n_a,k_a)|a=1,\ldots,P\}$ and $\{\underline{\mathbf{D}}^c_{\mathbf{zz}}(n_c,k_c)|c=1,\ldots,Q\}$.

Once the unitary matrix $\mathbf{U}$ is obtained from either the JD, JOD, or combined JD/JOD, an estimate of the mixing matrix $\mathbf{A}$ can be computed by the product $\mathbf{W}^{\#}\mathbf{U}$, where $\mathbf{W}$ is the whitening matrix (see "Structure under Unitary model" section). An estimate of the source signals s(n) can then be obtained by the product $\mathbf{A}^{\#}\mathbf{z}(n)$.

Separating more sources than sensors

A challenging problem consists of the blind separation of more sources than sensors (i.e., $L > K$); this problem, also known as the underdetermined blind source separation problem, was pointed out for the first time in Ref. 28 while separating discrete sources. Since then, several other works based on a priori knowledge of the probability density functions of the sources[29,30] were conducted. In Ref. 31, an approach for the resolution of the aforementioned problem exploits the concept of disjoint orthogonality of short Fourier transforms. Herein, for the resolution of the underdetermined problem, we review a STFD-based blind source separation method.[10]

We start by selecting autoterm points where only one source exists, as described in "Selection of Autoterms and Cross-terms in the Time–Frequency Domain" section. The corresponding STFD then has the following form:

$$\mathbf{D}_{\mathbf{zz}}(n,k) = D_{s_i s_i}(n,k)\mathbf{a}_i\mathbf{a}_i^H, \quad \text{where } (n,k) \in \Omega_i \qquad (36)$$

where Ω_i denotes the time–frequency support of the ith source. The idea of the algorithm consists of clustering together the autoterm points associated with the same principal eigenvector of $\mathbf{D}_{\mathbf{zz}}(n,k)$ representing a particular source signal. Once the clustering and classification of the autoterms are done, the estimates of the source signals are obtained from the selected autoterms using a time–frequency synthesis algorithm.[32] Note that the missing autoterms in the classification, often due to intersection points, are automatically interpolated in the synthesis process. An advanced clustering technique of the above autoterms based on gap statistics is proposed in Ref. 33.

Separation of Convolutive Mixtures

Consider a convolutive multiple-input multiple-output linear time-invariant model given by

$$z_i(n) = \sum_{j=1}^{L} \sum_{c=0}^{C} a_{ij}(c) s_j(n-c) \quad \text{for } i = 1, \ldots, K \quad (37)$$

where $s_j(n)$, $j = 1, \ldots, L$, are the L source signals; $z_i(n)$, $i = 1, \ldots, K$, are the $K > L$ sensor signals; and $a_{ij}(c)$ is the transfer function between the jth source and the ith sensor with an overall extent of $(C + 1)$ taps. The sources are assumed to have different time–frequency signatures, and the channel matrix $\mathbf{A}$ defined below in Eq. 39 is full column rank.

In matrix form, Eq. 37 becomes

$$\mathbf{z}(n) = \mathbf{A}\mathbf{s}(n) \quad (38)$$

where

$$\mathbf{s}(n) = [s_1(n), \ldots, s_1(n - (C + C') + 1), \ldots,$$
$$s_L(n - (C + C') + 1)]^T$$

$$\mathbf{z}(n) = [z_1(n), \ldots, z_1(n - C' + 1), \ldots, z_K(n - C' + 1)]^T$$

$$\mathbf{A} = \begin{bmatrix} \mathbf{A}_{11} & \cdots & \mathbf{A}_{1L} \\ \vdots & \ddots & \vdots \\ \mathbf{A}_{K1} & \cdots & \mathbf{A}_{KL} \end{bmatrix} \quad (39)$$

with

$$\mathbf{A}_{ij} = \begin{bmatrix} a_{ij}(0) & \cdots & a_{ij}(C) & \cdots & 0 \\ & \ddots & \ddots & \ddots & \\ 0 & \cdots & a_{ij}(0) & \cdots & a_{ij}(C) \end{bmatrix} \quad (40)$$

Note that $\mathbf{A}$ is a $[KC' \times L(C + C')]$ matrix and $\mathbf{A}_{ij}$ are $[C' \times (C + C')]$ matrices. C' is chosen such that $KC' \geqslant L(C + C')$.

Herein, the same formalism as in the instantaneous mixture case is retrieved and the data STFDs still have the same expression as in Eq. 8. However, the source auto-STFDs, $\mathbf{D}_{ss}^a(n, k)$, are not diagonal but block diagonal with diagonal blocks of size $(C + C') \times (C + C')$. Note that the block diagonal structure comes from the fact that the cross-terms between $s_i(n)$ and $s_i(n - d)$, where d is some delay, are not zero and depend on the local

correlation structure of the signal. This block diagonal structure is exploited to achieve the separation of the convolutive mixture.

STFD-based Separation

First the data vector $\mathbf{z}(n)$ is whitened. The whitening matrix $\mathbf{W}$ is of size $[L(C' + C) \times KC']$ and verifies

$$\mathbf{W}E\{\mathbf{z}(n)\mathbf{z}(n)^H\}\mathbf{W}^H = \mathbf{W}\mathbf{R}_{zz}\mathbf{W}^H$$
$$= (\mathbf{W}\mathbf{A}\mathbf{R}_{ss}^{\frac{1}{2}})(\mathbf{W}\mathbf{A}\mathbf{R}_{ss}^{\frac{1}{2}})^H = \mathbf{I} \quad (41)$$

where $\mathbf{R}_{zz}$ and $\mathbf{R}_{ss}$ denote the covariance matrices of $\mathbf{z}(n)$ and $\mathbf{s}(n)$, respectively. Eq. 41 shows that if $\mathbf{W}$ is a whitening matrix, then

$$\mathbf{U} = \mathbf{W}\mathbf{A}\mathbf{R}_{ss}^{\frac{1}{2}} \quad (42)$$

is a $L(C' + C) \times L(C' + C)$ unitary matrix where $\mathbf{R}_{ss}^{\frac{1}{2}}$ (Hermitian square root matrix of $\mathbf{R}_{ss}$) is block diagonal. The whitening matrix $\mathbf{W}$ can be determined from the eigendecomposition of the data covariance matrix $\mathbf{R}_{zz}$ as in Ref. 4.

Now by considering the whitened STFD matrices $\underline{\mathbf{D}}_{zz}(n, k)$ and the above relations, we obtain the key relation

$$\underline{\mathbf{D}}_{zz}(n, k) = \mathbf{U}\mathbf{R}_{ss}^{-\frac{1}{2}}\mathbf{D}_{ss}(n, k)\mathbf{R}_{ss}^{-\frac{1}{2}}\mathbf{U}^H = \mathbf{U}\mathbf{D}(n, k)\mathbf{U}^H \quad (43)$$

where $\mathbf{D}(n, k) = \mathbf{R}_{ss}^{-\frac{1}{2}}\mathbf{D}_{ss}(n, k)\mathbf{R}_{ss}^{-\frac{1}{2}}$. Since the matrix $\mathbf{U}$ is unitary and $\mathbf{D}(n, k)$ is block diagonal, the latter just means that any whitened STFD matrix is block diagonal in the basis of the column vectors of matrix $\mathbf{U}$. The unitary matrix can be retrieved by computing the block diagonalization of some matrix $\underline{\mathbf{D}}_{zz}(n, k)$. But to reduce the likelihood of indeterminacy and increase the robustness of determining $\mathbf{U}$, we consider the JBD of a set $\{\underline{\mathbf{D}}_{zz}(n_l, k_l); l = 1, \ldots, P\}$ of P whitened STFD matrices. This JBD is achieved by the maximization under unitary transform of the following criterion:

$$\mathbf{C}(\mathbf{U}) \overset{\Delta}{=} \sum_{l=1}^{P} \sum_{m=1}^{L} \sum_{i,j=(C'+C)(m-1)+1}^{(C'+C)m} \left| \mathbf{u}_i^H \underline{\mathbf{D}}_{zz}(n_l, k_l) \mathbf{u}_j \right|^2 \quad (44)$$

over the set of unitary matrices $\mathbf{U} = [\mathbf{u}_1, \ldots, \mathbf{u}_{L(C'+C)}]$. Note that an efficient Jacobi-like algorithm for the minimization of Eq. 44 exists in Refs. 24 and 34.

Once the unitary matrix $\mathbf{U}$ is determined up to a block diagonal unitary matrix $\mathbf{D}$ coming from the inherent indeterminacy of the JBD problem,[35] the recovered signals are obtained up to a filter by

$$\hat{s}(n) = \mathbf{U}^H \mathbf{W} \mathbf{z}(n) \qquad (45)$$

According to Eqs. 38 and 42, the recovered signals verify

$$\hat{\mathbf{s}}(n) = \tilde{\mathbf{D}} s(n) \qquad (46)$$

with

$$\tilde{\mathbf{D}} = \mathbf{D} \mathbf{R}_{ss}^{-\frac{1}{2}} \qquad (47)$$

where we recall that the matrix $\mathbf{R}_{ss}^{-\frac{1}{2}}$ and $\mathbf{D}$ are the block diagonal matrix and unitary block diagonal matrix, respectively. Consequently, $\tilde{\mathbf{D}}$ is also a block diagonal matrix, and the above STFD-based technique leads to the separation of the convolutive mixture up to a filter instead of a full MIMO deconvolution procedure.

Note that if needed, a SIMO (single-input multi-output) deconvolution/equalization[36] can be applied to the estimated sources of Eq. 46.

OTHER TFSP APPLICATIONS IN WIRELESS COMMUNICATIONS

Precoding for LTV Channels

Linear precoding is a useful signal processing tool for coping with frequency-selective propagation channels encountered with high-rate wireless transmission.

Precoding consists of mapping each incoming block of M symbols onto a P-long vector through a $P \times M (P > M)$ matrix referred to as the precoding matrix. Each received block is then multiplied by a $M \times P$ decoding matrix to retrieve the original symbols under the condition $M > L$ and $P = M + L$, where L is the overall channel length. To avoid interblock interference, guard intervals can be used, as in OFDM, for example. This can be done by forcing either the last L rows of the precoding matrix or the first L columns of the decoding matrix to zero.[37]

Precoding of LTV channels can be optimized by a priori knowledge of the channel temporal evolution. This knowledge can be provided by a feedback channel such that the receiver estimates periodically the channel parameters, also called channel status information (CSI),[37] and sends them back to the transmitter. The latter uses this CSI to predict the channel evolution within a finite time interval and commutes the optimal precoder. The optimality herein should be understood in the sense of maximizing the information rate over the linear channel affected by additive Gaussian noise. Under the constraint of a fixed average transmit power, the optimal precoder, i.e., the optimal precoding matrix, is obtained from the SVD of the channel matrix (The channel matrix is the transfer matrix from the transmitted block vector to the received blocjk vector).[38,39]

In Ref. 37, a physical interpretation of the optimal precoding for time- and frequency-dispersive channels is provided thanks to an approximate analytic model for the eigenfunctions of LTV channels. ($v(t)$ is said to be an eigenfunction of $f(t)$ if and only if $\lambda v(t) = \int_{-\infty}^{+\infty} f(t - t')v(t')dt'$, where λ is known as the associated eigenvalue.). The approximate model is valid for multipath channels with finite Doppler and delay spread. Under the above model and using a time–frequency representation of the eigenfunctions, the latter are shown to be characterized by an energy distribution along curves, in the time–frequency plane, given by contour lines of the time–frequency representation of the LTV channel. In the same reference, it is also shown under mild conditions often met in practice that the TFDs of the right singular vectors of the LTV channel are mainly concentrated along the curves where the energy in the time–frequency domain of the channel equals the square of the associated singular value. The TFDs of the left singular vectors are simply time- and frequency-shift versions of the TFDs of the right singular vectors. This interpretation clearly establishes the optimal power allocation in the time–frequency domain as a generalization of the well-known water-filling principle.[40,41] The above interpretation also allows an approximate computation of the channel singular vectors and values directly from the time–frequency representation of the LTV channel, without computing the SVD.

Signaling Using Chirp Modulation

TFSP tools can be used for the receiver design and for optimizing the design parameters of a spreading system using a chirp modulation scheme.

Indeed, chirp signals or, equivalently, linear FM signals have been widely used in sonar applications for range and Doppler estimation, as well as in radar systems for pulse compression. Thanks to their particular time–frequency signatures, these signals provide high interference rejection and inherent immunity against

Doppler shift and multipath fading[42] in wireless communications. In addition, they are bandwidth efficient.[43] It have been shown[44] that for a same SNR and Doppler shift, the chirp signaling outperforms the frequency-shift keying (FSK) signaling, thanks to their better cross-coherence properties, compared to FSK.

Signaling using chirp modulation is also seen as a spread-spectrum technique, which is defined as "a mean of transmission in which the signal occupies a bandwidth in excess of the minimum necessary to send the information."[45] Chirp modulation was first suggested in Ref. 43 by using a pair of linear chirps with opposite chirp rates for binary signaling. A system for multiple access within a common frequency band was proposed in Ref. 46 by assigning pairs of linear chirps with different chirp rates to several users. In this system, the number of users simultaneously accessing the shared resources is limited by the MAI for a given time–bandwidth product. To reduce this shortcoming, the chirp signals are selected in Ref. 47 such that they all have the same power as well as the same bandwidth, offering inherent protection against frequency-selective fading. Further, the combination of chirp modulation signaling with frequency-hopping (FH) multiple access was proposed in Ref. 42. The obtained hybrid system (Fig. 2) improves communications system performance, especially in multipath fading-dispersive channels. Note that this system was extended to FH-CDMA in Ref. 48 and compared to the FSK-FH-CDMA system, leading to the same conclusion as in Ref. 42. In Ref. 49, the chirp parameters, to be used in chirp modulation signaling, are selected under the actual time–bandwidth requirements so as to reduce significantly multiple-access interference and bit error rates.

Detection of FM Signals in Rayleigh Fading

Diversity reception is currently one of the most effective techniques for coping with the multipath Rayleigh fading effect in mobile environments.[50] It requires a number of signal transmission paths that carry the same information but have uncorrelated multipath fadings. A circuit to combine the received signals or select one of the paths is necessary. Diversity techniques take advantage of the fact that signals exhibit fades at different places in time, frequency, or space, depending on different situations. However, a diversity scheme normally requires a number of antennae at the transmitter or receiver, resulting in high cost and redundancy of information. Herein, we review a method that can overcome this problem.

Given a transmitted signal $s(t)$ under such an environment, the received signal $\underline{x}(t)$ (Random terms are hereafter underlined to distinguish them from deterministic terms.)

is then considered random and may be modeled as[50]

$$\underline{x}(t) = \sum_{i=1}^{N} \underline{a}_i s(t - \underline{\tau}_i) \exp\{j\underline{\theta}_i\} \tag{48}$$

where N is the number of received waves and $\underline{a}_i$, $\underline{\tau}_i$, and $\underline{\theta}_i$ are the random attenuation, multipath delay, and phase shift associated with the ith path, respectively. When considering narrowband communications with frequency or phase modulation, the transmitted signals have the following form:

$$s(t) = \exp\{j[2\pi f_o t + \Psi_s(t)]\} \tag{49}$$

where $\Psi_s(t)$ represents the baseband signal and f_o is the carrier frequency. The received signal $\underline{x}(t)$ will then be expressed as

$$\underline{x}(t) = \underline{r}(t) \exp\{j[2\pi f_o t + \Psi_s(t) + \underline{\Psi}_r(t)]\} \tag{50}$$

For non-Rician channels (in general, channels with no line-of-sight path), envelope $\underline{r}(t)$ can be assumed to be Rayleigh distributed and hence has an autocorrelation function approximated by[51]

$$R_r(\tau) \approx \alpha(1 + 1/4\eta(\tau)) \tag{51}$$

where α is some constant and $\eta(\tau)$ is some particular function. The Fourier transform of the latter, referred to as S_η, exhibits a peak at the zero frequency.[50] The spectrum of the signal envelope is then given by

$$S_R(f) = \alpha(\delta(f) + \frac{1}{4}S_\eta(f)) \tag{52}$$

In Refs. 52 and 53, it is shown that for various types of frequency modulation, the second-order Wigner–Ville spectrum (WVS)(The second-order time-varying Wigner–Ville spectrum is defined as[54]) (or spectra of other TFDs) of the received FM signal $\underline{x}(t)$ has a delta concentration along the IF of the transmitted signal. The special structure of the envelope spectrum (Eq. 52) makes the delta concentration possible in the TF plane of the WVS (or spectra of other TFDs). Consequently, the detection of FM signals through the Rayleigh flat fading environment can be achieved in the above time–frequency plane without the use of the conventional diversity techniques or higher-order spectra approach.[55]

Mobile Velocity/Doppler Estimation Using Time–Frequency Processing

Many wireless communications systems require prior knowledge of the mobile velocity. This knowledge allows compensation of distortions introduced by the communications channel. In addition, reliable estimates of the mobile velocity are useful for effective dynamic channel

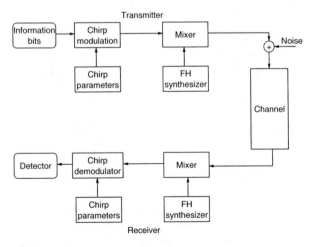

Fig. 2 Chirp modulation signaling and FH communications system.

assignment and for the optimization of adaptive multiple access. In another chapter of Ref. 56, an overview of existing velocity estimators is given, particularly an estimator based on the estimation of the IF of the received signal. In contrast to approaches based on the envelope of the received signal, the IF-based velocity estimators are robust to shadow fading, which is produced by variations of the average of the received signal envelope over few wavelengths. Interested readers can refer to Ref. 56 for more details.

$$w_{\underline{x}}^{(2)}(t,f) \overset{\Delta}{=} \int_{-\infty}^{\infty} E\{\underline{x}(t+\tau/2)\underline{x}^*(t-\tau/2)\}e^{-j2\pi f\tau}d\tau$$

REFERENCES

1. Belouchrani, A.; Amin, M.G. A new approach for blind source separation using time frequency distributions. SPIE Conference on Advanced Algorithms and Architectures for Signal Processing, Denver, CO, Aug 4–9, 1996.

2. Belouchrani, A.; Amin, M.G. Blind source separation using time-frequency distributions: algorithm and asymptotic performance. IEEE ICASSP 1997, Munich, Germany, Apr 21–24, 1997.

3. Belouchrani, A.; Amin, M.G. On the use of spatial time frequency distributions for signal extraction. Multidimens. Syst. Signal Process.(Special Issue of the Journal). **1998**, *9* (4), 349–354.

4. Belouchrani, A.; Amin, M.G. Blind source separation based on time-frequency signal representation. IEEE Trans. Signal Process, **1998**, *6* (11), 2888–2897.

5. Belouchrani, A.; Amin, M.G. Time-frequency MUSIC: a new array signal processing method based on time-frequency signal representation. IEEE Signal Process. Lett, **1999**, *6*, 109–110.

6. Kayhan, A.S.; Amin, M.G. Spatial evolutionary spectrum for DOA estimation and blind signal separation. IEEE Trans. Signal Process, **2000**, *48* (3), 791–798.

7. Leyman, A.R.; Kamran, Z.M.; Abed-Meraim, K. Higher order time frequency based blind source separation technique. IEEE Signal Process. Lett, **2000**, *7* (7), 193–196.

8. Zhang, Y.; Ma, W.; Amin, M.G. Subspace analysis of spatial time frequency distribution matrices. IEEE Trans. Signal Process, **2001**, *49* (4), 747–759.

9. Belouchrani, A.; Abed-Meraim, K.; Amin, M.G.; Zoubir, A. Joint anti-diagonalization for blind source separation. ICASSP 2001, Salt, Lake City, UT, May 7–11, 2001; Vol. 5, 2789–2792.

10. Nguyen, L.-T.; Belouchrani, A.; Abed-Meraim, K.; Boashash, B. Separating more sources than sensors using time frequency distributions. ISSPA 2001, Kuala-Lampur, Malaysia, Aug 13–16, 2001; Vol. 2, 583–586.

11. Giulieri, L.; Thirion-Moreau, N.; Arques, P.-Y. Blind sources separation based on bilinear time-frequency representations: a performance analysis. ICASSP 2002, Orlando, FL, May 13–17, 2002; Vol. 2, 1649–1652.

12. Jin, L.; In, Q.-Y.; Wang, W.-J. Time-frequency signal subspace fitting method for direction of-arrival estimation.

13. Wang, G.; Xia, X.-G. Iterative algorithm for direction of arrival estimation with wideband chirp signals. IEE Proc., Radar Sonar Navig, **2000**, *147*, 233–238.

14. Cirillo, L.; Zoubir, A.; Ma, N.; Amin, M.G. Automatic classification of auto- and cross-terms of time-frequency distributions in antenna arrays. ICASSP 2002, Orlando, FL, May 13–17, 2002.

15. Hassanien, A.; Gershman, A.B.; Amin, M.G. Time-frequency ESPRIT for direction-of-arrival estimation of chirp signals. Sensor Array and Multichannel Signal Processing Workshop, Rosslyn, VA, Aug 4–6, 2002; 337–341.

16. Zhang, Y.; Amin, M.G. Blind separation of sources based on their time-frequency signatures. ICASSP 2000, Istanbul, Turkey, June 5–9, 2000.

17. Wax, M.; Kailath, T. Detection of signals by information theoretic criteria. IEEE Trans. Acoust. Speech Signal Process. **1985**, *ASSP-33*, 387–392.

18. Cirillo, L.; Zoubir, A.; Amin, M.G. Selection of auto- and cross-terms for blind non-stationary source separation. IEEE International Symposium on Signal Processing and Information Technology (ISSPIT 2001), Cairo, Egypt, Dec 28–30, 2001.

19. Mu, W.; Amin, M.G.; Zhang, Y. Bilinear signal synthesis in array processing. IEEE Trans. Signal Process, **2003**, *51*, 90–100.

20. Giulieri, L.; Thirion-Moreau, N.; Arques, P.-Y. Blind sources separation using bilinear and quadratic time-frequency representations. ICA 2001, San Diego, CA, Dec 9–13, 2001.

21. Schmidt, R. Multiple emitter location and signal parameter estimation. IEEE Trans. Antennas Propag, **1986**, *34*, 276–280.

22. Roy, R.; Kailath, T. ESPRIT: estimation of signal parameters via rotational invariance techniques. IEEE Trans. Acoust. Speech Signal Process, **1989**, *37*, 984–995.

23. Zhang, Y.; Mu, W.; Amin, M.G. Time-frequency maximum likelihood methods for direction finding. J. Franklin Institue. **2000**, *337*, 483–497.

24. Belouchrani, A.; Amin, M.G.; Abed-Meraim, K. Direction finding in correlated noise fields based on joint block-diagonalization of spatio-temporal correlation matrices. IEEE Signal Process. Lett, **1997**, *4*, 266–268.

25. Belouchrani, A.; Abed-Meraim, K.; Cardoso, J.F.; Moulines, E. A blind source separation technique using second order statistics. IEEE Trans. Signal Process, **1997**, *45* (2), 434–444.

26. Jutten, C.; Hérault, J. Détection de grandeurs primitives dans un message composite par une architecture de calcul neuromimétrique en apprentissage non supervisé. GRETSI 1985, Nice, France, May 20–24, 1985.

27. Comon, P. Independent component analysis, a new concept?. Signal Process, **1994**, *36*, 287–314.

28. Belouchrani, A.; Cardoso, J.-F. Maximum likelihood source separation for discrete sources. EUSIPCO 1994, Edinburgh, UK, Sept 13–16, 1994; 768–771.

29. Comon, P.; Grellier, O. Nonlinear inversion of underdetermined mixtures. ICA 1999, Aussois, France, Jan 11–15, 1999; 461–465.

IEEE International Symposium on Circuits and Systems (ISCAS 2000), Geneva, Switzerland, May 28–31, 2000; Vol. 3, 375–378.

30. Diamantaras, K.I. Blind separation of multiple binary sources using a single linear mixture. International Conference on Acoustics, Speech, and Signal Processing (ICASSP 2000), Istanbul, Turkey, June 5–9, 2000; Vol. 5, 2657–2660.

31. Jourjine, A.; Rickard, S.; Yilmaz, O. Blind separation of disjoint orthogonal signals: demixing n sources from 2 mixtures. International Conference on Acoustics, Speech, and Signal Processing (ICASSP 2000), Istanbul, Turkey, June 5–9, 2000; Vol. 5, 2657–2660.

32. Boudreaux-Bartels, G.F.; Marks, T.W. Time-varying filtering and signal estimation using Wigner distributions. IEEE Trans. Acoust. Speech Signal Process. **1986**, *34*, 422–430.

33. Luo, Y.; Chambers, J. Active source selection using gap statistics for underdetermined blind source separation. 7th International Symposium on Signal Processing and its Applications (ISSPA 2003), Paris, France, July 1–4, 2003; Vol. 1, 137–140.

34. Belouchrani, A.; Abed-Meraim, K.; Hua, Y. Jacobi-like algorithms for joint block diagonalization: application to source localization. ISPACS 1998, Melbourne, Australia, Nov 5–6, 1998.

35. Bousbia-Saleh, H.; Belouchrani, A.; Abed-Meraim, K. Jacobi-like algorithm for blind signal separation of convolutive mixtures. Electron. Lett, **2001**, *37*, 1049–1050.

36. Abed-Meraim, K.; Qiu, W.; Hua, Y. Blind system identification. IEEE. **1997**, *85*, 1310–1322.

37. Barbarossa, S.; Scaglione, A. Optimal precoding for transmissions over linear time-varying channels. GLOBECOM 1999, Rio De Janeiro, Brazil, Dec 5–9, 1999; Vol. 5, 2545–2549.

38. Scaglione, A.; Barbarossa, S.; Giannakis, G.B. Filterbank transceivers optimizing information rate in block transmissions over dispersive channels. IEEE Trans. Inf. Theory. **1999**, *45*, 1019–1032.

39. Medard, M. *The Capacity of Time-Varying Multiple User Channels in Wireless Communications*; Ph.D. Thesis, MIT: Cambridge, MA, 1995.

40. Gallager, R.G. *Information Theory and Reliable Communication*; John Wiley & Sons: New York, 1968.

41. Goldsmith, A. *Design and Performance of High-Speed Communication Systems over Time-Varying Radio Channels*; Ph.D. Thesis: University of California, Berkeley, 1994.

42. El-Khamy, S.E.; Shahban, S.E.; Thabet, E.A. Frequency hopped multi-user chirp modulation (FH/MCM) for multi-path fading channels. IEEE Symposium Antennas and Propagation, Orlando, FL, July 11–16, 1999; Vol. 1, 996–999.

43. Winkler, M.R. Chirp Signals for Communications. In *WESCON convention record*; 1962 Paper 14.2.

44. Berni, A.J.; Gregg, W.D. On the utility of chirp modulation for digital signaling. IEEE Trans. Commun, **1973**, *21* (6), 748–751.

45. Pickholtz, R.L.; Milstein, L.B.; Schilling, D.L. Spread spectrum for mobile communications. IEEE Trans. Veh. Technol, **1991**, *40*, 313–322.

46. Cook, C.E. Linear FM signal formats for beacon and communication systems. IEEE Trans. Aerosp. Electron. Syst, **1974**, *10*, 471–478.

47. El-Khamy, S.E.; Shahban, S.E.; Thabet, E.A. Efficient multiple access communications using multi-user chirp modulation signals. IEEE 4th International Symposium on Spread Spectrum Techniques and Applications, Mainz, Germany, Sept 22–25, 1996; Vol. 3, 1209–1213.

48. Gupta, C.; Papandreou-Suppappola, A. Wireless CDMA communications using time-varying signals. 6th International Symposium on Signal Processing and Its Applications, Kuala-Lampur, Malaysia, Aug 13–16, 2001; Vol. 1, 242–245.

49. Hengstler, S.; Kasilingam, D.P.; Costa, A.H. A novel chirp modulation spread spectrum technique for multiple access. IEEE 7th International Symposium on Spread Spectrum Techniques and Applications, Prague, Czech Republic, Sept 2–5, 2002; Vol. 1, 73–77.

50. *Microwave Mobile Communications*; Jakes, W., (Ed).; IEEE Press: Washington, DC, 1998 (reprint).

51. Davenport, W.B.; Root, W.L. *An Introduction to the Theory of Random Signals and Noise*; IEEE Press: Washington, DC, 1987.

52. Nguyen, L.-T.; Senadji, B. Analysis of nonlinear signals in the presence of Rayleigh fading. 5th International Symposium on Signal Processing and Its Applications (ISSPA 1999), Queensland, Australia, Aug 22–25, 1999; Vol. 1, 411–414.

53. Nguyen, L.-T.; Senadji, B. Detection of frequency modulated signals in Rayleigh fading channels based on time-frequency distributions. International Conference on Acoustics, Speech, and Signal Processing (ICASSP 2000), Istanbul, Turkey, June 9, 2000; Vol. II, 729–732.

54. Boashash, B.; Ristic, B. Polynomial time-frequency distributions and time-varying higher order spectra: application to the analysis of multicomponent FM signals and to the treatment of multiplicative noise. Signal Process, **1998**, *67*, 1–23.

55. Senadji, B.; Boashash, B. A mobile communications application of time-varying higher order spectra to FM signals affected by multiplicative noise. International Conference on Information, Communications, and Signal Processing (ICICS 1997), Singapore, Sep 9–12, 1997; Vol. 3 1489–1492.

56. Senadji, B.; Azemi, G.; Boashash, B. Mobile velocity estimation for wireless communications (Ch.5). In *Signal Processing for Mobile Communications Handbook*; Ibnkahla, M., Ed.; CRC Press: Boca Raton, FL, 2003.

Wireless Communications: Spatial Diversity

Ramakrishna Janaswamy
Naval Postgraduate School, Monterey, California, U.S.A.

Abstract

Spatial diversity reduces the ill-effect of multiple fading in wireless systems.

INTRODUCTION

The ill-effects of multipath fading in wireless systems can be reduced by employing spatial diversity, where a multi-element antenna at the receiver is employed. When fading is present, a higher than otherwise average carrier power is needed to perform at a given bit-error rate (BER). In an array antenna, the signals received by the various elements can be weighted appropriately to result in a combined signal that fluctuates less rapidly than the individual signals. This array then requires less power to achieve a given BER than the case where only one element is used. For spatial diversity to work effectively, the signals, received by the various antenna branches must be sufficiently decorrelated so that if one of the elements is in deep fade, there is still hope of recovering the signal by receiving it at other antenna terminals. This can always be achieved in practice by choosing the element spacing appropriately. The spacing required between antenna elements to maintain certain decorrelation depends on mutual coupling and the disposition of scatterers causing the multipath transmission. For instance, in the absence of mutual coupling, spacings of about $\lambda/2$ should be sufficient at a mobile terminal that is usually surrounded uniformly by scatterers. On the other hand, spacings of the order of 10λ or more may be necessary to maintain the same decorrelation value at an elevated base station. A second condition that is required for diversity benefits is that the mean signal strengths of the diversity paths should approximately be the same. An in-depth treatment of linear combining techniques is given in Ref. 1, and to a lesser extent, in Ref. 2. Some developments of antenna diversity with applications to mobile communications are treated in Ref. 3.

In this entry, the basic principles of spatial diversity combining are treated. By assuming that the elements are spaced appropriately so that perfect decorrelation among various branches exists—a branch is assumed to mean an antenna—the improvement that can be accomplished by using diversity arrays with various combing techniques is first demonstrated. The effect of branch correlation caused by either mutual coupling between elements or angular despread of the incoming ways is shown later. The BER performance of basic modulation schemes with a diversity array is also shown. An attractive alternative to spatial diversity where space is a limiting factor is polarization diversity where two antennas with orthogonal polarization are almost colocated. In a highly scattering environment, the signal is received in almost all polarizations irrespective of the transmitter polarization. Although the decorrelation between the signals received in two orthogonally polarized states is not as good as in well-separated spatially diverse antennas, the loss in diversity improvement at the 1% probability level was shown to be less than 1 dB at 842 MHz.[4] Polarization diversity is not treated in this entry.

GENERAL RECEIVER ARRAY THEORY

Let us consider an array comprised of N equispaced elements placed along the x-axis as shown in Fig. 1. The position of the n-th element is $x_n = (n-1)d$. Although the elements of the array could be arbitrary, we consider the case where the elements are identical dipole antennas, with the dipole axes along the z-axis. A plane wave of polarization $\hat{\theta}$ and radian frequency ω is assumed incident from angles (θ, ϕ):

$$\mathbf{E}^{\text{inc}} = \hat{\theta} E_{\text{O}} e^{jk_{\text{O}}(x\sin\theta\cos\phi + y\sin\theta\sin\phi + z\cos\theta)} \tag{1}$$

where $k_{\text{o}} = \omega\sqrt{\mu_0\varepsilon_0}$ is the wave number in free space. The n-th element is assumed to be terminated in a center-fed load impedance Z_{Ln}. In the absence of mutual coupling, the open-circuited voltage induced in the isolated n-th dipole can be written as[5]

$$V_n^{\text{oc}} = \mathbf{h} \cdot \mathbf{E}^{\text{inc}}(x_n, 0, 0) \tag{2}$$

where $\mathbf{h}$ is the vector effective length of the dipole given by

$$\mathbf{h} = \frac{\mathbf{E}^{\text{rad}}}{jk_{\text{o}}\eta_{\text{o}}I_{\text{in}}} 4\pi r e^{jk_{\text{o}}r} \tag{3}$$

Encyclopedia of Wireless and Mobile Communications DOI: 10.1081/E-EWMC-120043855

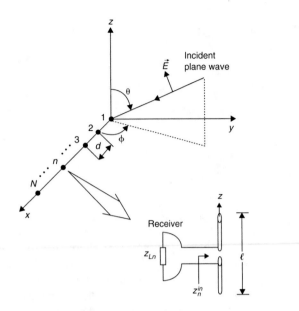

Fig. 1 An array comprised of N elements along x-axis.

$\eta_o \approx 120\pi$ is the free-space intrinsic impedance, I_{in} is the input current to the antenna and $\mathbf{E}^{\text{rad}}$ is the field radiated by the antenna, both in the transmit mode, and r is the distance from the center of the dipole to the observation point. For a dipole of length $\mathbb{L}$ carrying a sinusoidal current distribution, $I(z) = I_o \sin[k_o(\mathbb{L}/2 - |z|)]$, the vector effective length can be obtained as

$$\mathbf{h} = \hat{\theta}\frac{\lambda}{\pi}\frac{F(\theta,\phi)}{\pi \sin k_o \ell} \tag{4}$$

where $F(\theta,\phi)$ is the radiation pattern of the dipole antenna:

$$F(\theta, \phi) = \frac{\cos((k_o\ell/2)\cos\theta) - \cos(k_o\ell/2)}{\sin\theta} \tag{5}$$

The azimuth plane pattern ($\theta = \pi/2$) of a z-directed dipole is uniform. Inserting the preceding and Eq. 1 into Eq. 2, we get

$$V_n^{\text{oc}} = \frac{\lambda}{\pi}E_o\frac{F(\theta, \phi)}{\sin(k_o\ell/2)}e^{j(n-1)k_o d \sin\theta\cos\phi}$$

$$V_n^{\text{oc}} = \frac{\lambda}{\pi}E_o\frac{F(\theta = \pi/2, \phi)}{\sin(k_o\ell/2)}e^{j(n-1)k_o d \cos\phi} \tag{6}$$

for incidence in the xy-plane.

For simplicity, let us assume that the plane wave is incident from the xy-plane so that Eq. 6 applies. The open-circuited voltage drives a series equivalent network system composed of the load impedance Z_{Ln} and the antenna input impedance Z_n^{in}, Fig. 2. The voltage signal presented to the load is

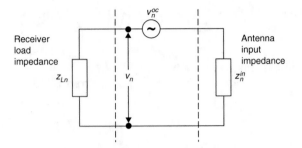

Fig. 2 Antenna open-circuited voltage driving a series system composed of input impedance and receiver.

$$V_n = \frac{Z_{Ln}}{Z_{Ln} + Z_n^{\text{in}}}V_n^{\text{oc}} \tag{7}$$

Let us now assume that the input impedance of the dipoles are all the same and equal to Z^{in}, the load impedances Z_{Ln} are all equal to Z_L, and the corresponding V_{on}'s are all equal to V_o. Denoting the vector $[V_1, V_2, \ldots, V_N]^T$ by $\mathbf{v}$, and the vector $\left[1, e^{jk_o d\cos\phi}, \ldots, e^{j(N-1)k_o d\cos\phi}\right]^T$ by ψ, we can express the received signals compactly as

$$\mathbf{v} = V_o F(\pi/2, \phi)\psi \tag{8}$$

where the superscript T denoting the matrix transpose operator. The vector ψ is known as the *steering vector* of the array and is seen to depend only on the properties of the incident wave and the geometry of the array. The other factor $V_o F(\pi/2,\phi)$ contains information about the nature of the element and is known as the element factor. If the elements are changed from dipoles to loops, e.g., the element factor gets changed; the steering vector of the array remains the same. The voltage vector presented to the load is different for different angles ϕ of the incident wave. Hence, the array has a directional response to the incoming plane waves. In the study of array antennas, one often assumes the elements to be isotropic [i.e., $F(\theta,\phi) = 1$] and focuses on the study of the array factor. In the case of isotropic elements,

$$\mathbf{v} = V_0\psi \tag{9}$$

The more general case is described by Eq. 7.

COMBINING TECHNIQUES

Basically, the three ways of combining the signals are

1. selection combining,
2. maximal ratio combining,
3. equal gain combining.

We briefly look at each of these combining techniques and show the diversity improvement by considering the statistics of the combined signal. Although the treatment is valid for both uplink (mobile-to-base station) or downlink (base station-to-mobile), we specifically consider the case of uplink only where the mobile transmits and the base station receives.

The signals received by the elements of the receive antenna are combined linearly as shown in Fig. 3. The weights for combining are chosen to be w_1^*, w_2^*, ..., w_N^*, where * denotes complex conjugation. A complex conjugate is included in the definition of the weights only for mathematical convenience so that the combined output appears in a convenient form as $\mathbf{w}^\dagger \mathbf{s}$.

It is assumed that the spacing is chosen so that signals received are sufficiently decorrelated. The signal received at one antenna element constitutes one diversity branch. Because of multipath and movement of the mobile, the signal received at each element is not a constant, but fluctuates at the fading rate. The fading rate depends on the speed of the mobile and frequency of the radio signal and is approximately equal to the maximum Doppler shift.[6] The maximum Doppler shift f_{dM} corresponding to a radio frequency of f_G GHz for a mobile moving at a speed v mi/h is

$$f_{dM} = 1.4815 f_G^v \tag{10}$$

For a mobile moving at 100 mi/h, the maximum Doppler shift is 133.33 Hz at 900 MHz and 266.67 Hz at 1800 MHz. Coherence time of the radio channel is approximately equal to the reciprocal of the fading rate. In the preceding example, the coherence time is 7.5 ms at 900 MHz and 3.75 ms at 1800 MHz. The multipath medium between the transmitting antenna and the receiving antenna can be thought of as a linear time-varying filter and each of the N branches is characterized by an equivalent low-pass transfer function $T_n(f;t)$, $n = 1, ..., N$, with the t argument signifying the time variations of the radio channel responses and the f argument signifying frequency selective nature of the channel. Assuming frequency non-selective or flat fading on each diversity branch allows us to express this transfer function as $T_n(f;t) = g_n(t)$, where each $g_n(t)$ is a zero-mean complex Gaussian process. Thus, the signals received in the diversity branches may be

represented in the form

$$s_n(t) \triangleq \Re\left[r_n(t)e^{j2\pi f_c t}\right] = \Re\left[g_n(t)u(t)e^{j2\pi f_c t}\right] \tag{11}$$

where f_c is the nominal carrier frequency, $u(t)$ is the complex envelope of the transmitted signal, and $r_n(t)$ is the complex envelope of the received signal. The assumption of flat fading is valid for narrowband transmission where the delays experienced by the various multipath components are all much less than the symbol duration. It is seen that flat fading results in a multiplicative propagation factor for the transmitted signal. This is characteristic of Rayleigh faded narrowband signals. Furthermore, it is assumed that the symbol period T_s is much less than the reciprocal of the fading rate so that the fading pattern does not change over the symbol duration. For convenience we normalize the transmitted signal such that mean power is constant, namely,

$$E_T\left(|u(t)|^2\right) \triangleq \frac{1}{T_s} \int_{-T_s/2}^{T_s/2} |u(t)|^2 dt = 1 \tag{12}$$

where E_T is the time-expectation or time-averaging operator in the sense of the preceding integral. The complex envelope of additive noise in the n-th receiver branch is assumed to be $v_n(t)$ with the mean intensity over each symbol period or over any longer interval given by

$$\frac{1}{2}E_T\left(|v_n(t)|^2\right) = \frac{1}{2}\left\langle|v_n(t)|^2\right\rangle = P_n = P_{no} \tag{13}$$

where the angle brackets denote statistical means and it has been assumed that all P_n are equal to P_{no}. Implicit in the preceding equation is the assumption of ergodicity of the noise process. We define the instantaneous, γ_n, and the average, Γ_n, carrier-to-noise ratios (CNRs) for the n-th branch as

$$\gamma_n \triangleq \frac{local\ mean\ \text{carrier power per branch}}{\text{mean noise power per branch}} \tag{14}$$

$$\gamma_n = \frac{E_T\left(\frac{1}{2}|r_n(t)|^2\right)}{P_{no}} \tag{15}$$

$$\gamma_n \approx \frac{|g_n(t)|^2}{2P_{no}} \quad [\text{since } g_n(t) \text{ is approximately constant over } T_s] \tag{16}$$

$$\Gamma_n \triangleq \frac{statistical\ mean\ \text{carrier power per branch}}{\text{mean noise power per branch}} \tag{17}$$

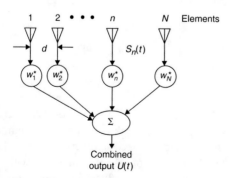

Fig. 3 Diversity array consisting of N elements.

$$\langle \gamma_n \rangle \overset{\Delta}{=} \frac{P_o}{P_{no}} \qquad (18)$$

where P_o is the statistical average of $|g_n|^2/2$ over the fading interval (coherence time). Note that as a result of the normalization Eq. 12, we have used the fact that the local average of the received signal envelope across each branch is

$$E_T\left(|r_n(t)|^2\right) = E_T\left(|g_n(t)|^2 |u(t)^2|\right) \approx |g_n(t)|^2 \qquad (19)$$

Thus, $|g_n(t)|$ may be regarded as the local mean envelope of the received signal. By assuming Rayleigh fading, the probability density function (**pdf**) for the received signal envelope is

$$P(|g_n|) = \frac{|g_n|}{P_o} e^{-|g_n|^2/2P_o} \qquad (20)$$

and that of γ_n is

$$P(\gamma_n) = \frac{1}{\Gamma_n} e^{-\frac{\gamma_n}{\Gamma_n}} \qquad (21)$$

SELECTION COMBINING

The ideal selection combiner is defined as one where the branch with the highest CNR is chosen for the system output. As far as the statistics of the system output are concerned, it is immaterial where the selection is done at the intermediate frequency (IF) (predetection) or at post-detection. Because it is assumed that all the receivers have the same noise level, the highest value of γ_n also corresponds to the highest value of $|g_n|^2/2 + P_n$. Accordingly, the receiver selected is that which also has the highest total *instantaneous* power. Mathematically, we state the weight condition as

$$w_n = \begin{cases} 1, & n = M, \text{ where } M|\gamma_M(t) = \max[\gamma_n(t)] \\ 0, & \text{otherwise} \end{cases} \qquad (22)$$

In selection combining the unused branches at any instant do not contribute to the output signal. These unused terminals may then be left open to minimize the effects of antenna mutual coupling. In terms of implementation, selection combiner would need no more than a comparator and a fast switch. Most modern cellular systems make use of dual diversity ($N = 2$) with selection combining. A practical alternative to picking the branch with the highest CNR is to switch to the first branch that remains above a certain threshold relative to Γ_n. This is known as *switched combining*. Of course the branch picked is not necessarily the one with the highest CNR.

The probability that the CNR in one branch is less than or equal to specified value γ_s is

$$\text{Prob}\,(\gamma_n \leqslant \gamma_s) = \int_0^{\gamma_s} P(\gamma_n)\,d\gamma_n$$

$$\text{Prob}\,(\gamma_n \leqslant \gamma_s) = 1 - e^{\frac{\gamma_s}{\Gamma_n}} \qquad (23)$$

Over the short-term fading, one can now describe all those events in which the selector output CNR γ is less than or equal to γ_s as those events in which each of the branch CNRs is simultaneously below γ_s. Because fading is assumed independent in each of the N branches, this would be equal to the probability that the CNR in *all* branches is simultaneously less than or equal to γ_s, i.e.,

$$\begin{aligned} \text{Prob}\,(\gamma_n \leqslant \gamma_s) &= \text{Prob}(\gamma_1, \ldots, \gamma_N \leqslant \gamma_s) \\ &= \prod_{n=1}^N \text{Prob}\,(\gamma_n \leqslant \gamma_s) = \prod_{n=1}^N \left[1 - e^{-\frac{\gamma_s}{\Gamma_N}}\right] \\ &= \left[1 - e^{-\frac{\gamma_s}{\Gamma_N}}\right] \overset{\Delta}{=} P_N^{sc}(\gamma_s) \end{aligned} \qquad (24)$$

where it has been assumed that all the diversity branches have equal mean CNR over the short-term fading (i.e., $\Gamma_n = \Gamma$). Eq. 24 is the cumulative distribution function (**cdf**) of the variation in γ. The subscript N in Eq. 24 stands for the number of elements and the superscript sc stands for selection combining. The assumption of equal mean CNRs at the branches is very reasonable in view of the fact that the array length, in practice, is very small compared with the distance from the transmitter so that all elements experience approximately the same path loss and shadowing loss.

The probability density function **pdf** for γ is obtained by differentiating the **cdf** with respect to γ_s. It is

$$P_N^{sc}(\gamma) = \frac{N}{\Gamma} e^{-\frac{\gamma}{\Gamma}} \left[1 - e^{-\frac{\gamma}{\Gamma}}\right]^{N-1} \qquad (25)$$

The mean CNR of the selected signal is

$$\begin{aligned} \langle \gamma \rangle &= \int_0^\infty \gamma p_N^{sc}(\gamma)\,d\gamma \\ &= \Gamma \sum_{n=1}^N \frac{1}{n} \sim \left(C + \ln N + \frac{1}{2N}\right)\Gamma \end{aligned}$$

$$\text{for large } N(N \geqslant 3) \qquad (26)$$

where $C = 0.577215\ldots$ is the Euler's constant. Hence the mean CNR improves logarithmically with the number of branches in the case of selection diversity. Fig. 4 shows a plot of the **cdf**, $P_N^{sc}(\gamma_s)$, of selection diversity with N as a parameter. To interpret the results, let us consider the point

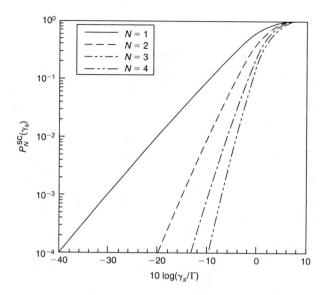

Fig. 4 Cumulative distribution function of γ_s vs. γ_s/Γ for selection diversity with N as a parameter.

$10 \log (\gamma_s/\Gamma) = -10$ dB on the horizontal axis. For $N = 1$ there is a 10% probability that the instantaneous output CNR falls 10 dB below the mean branch CNR Γ. For $N = 4$ with selection combining, the probability is reduced to 0.01%. Thus, combining has significantly reduced the depth of fades. It is also seen that at 99% reliability [i.e., $1 - P_N^{sc}(\gamma_s) = 0.99$], selection diversity provides a 10 dB $[-10 - (-20)]$ savings in power with two branches and 16-dB savings with four branches.

On knowing $p(\gamma)$, it is straightforward to derive the error rate performance for digital signaling. For instance, BER of ideal coherent binary phase shift keying (BPSK) assuming constant envelope γ is[7]

$$P_{BPSK} = \frac{1}{2} \text{erfc}(\sqrt{\gamma}) \tag{27}$$

where γ is the CNR per information bit. In the presence of fading, the envelope fluctuates and we view the preceding formula as a conditional probability, where the condition is that γ is fixed. To obtain error probabilities when γ is random, we must average P_{BPSK} over the **pdf** of γ as given in Eq. 25. One then obtains the following result:

$$P_{BPSK}^{sc} = \int_0^\infty p_{BPSK}(\gamma) p_N^{sc}(\gamma) d\gamma$$

$$= \frac{N}{2} \int_0^\infty e^{-x} [1 - e^{-x}]^{N-1} \text{erfc}\left(\sqrt{\Gamma x}\right) dx$$

$$= \frac{1}{2\sqrt{\pi}} \int_0^\infty \frac{e^{-\gamma} [1 - e^{-\frac{\gamma}{\Gamma}}]^N}{\sqrt{\gamma}} d\gamma \tag{28}$$

The preceding equation may be evaluated in a closed form for $N = 1$ and expressed as a series for $N > 1$. Experimental verification of the performance improvement of selection diversity at 920 MHz with two vehicle mounted monopole antennas is demonstrated in Ref. 8.

As an aside, for $N = 1$, the substitution $\sqrt{\gamma} = t$ converts the integral into a Gaussian type of integral with the result:

$$P_{BPSK} = \frac{1}{2}\left(1 - \sqrt{\frac{\Gamma}{1 + \Gamma}}\right) \tag{29}$$

Eq. 29 gives the probability of bit error for BPSK in the presence of Rayleigh fading with one branch. This is in contrast to the expression in Eq. 27, which is valid without fading. Fig. 5 shows a plot of the BER for BPSK with and without fading. From the plot it is seen that to maintain a BER of 10^{-3}, the required CNR at the receiver is approximately 6 and 24 dB without and with Rayleigh fading, respectively. In other words, to maintain the same BER with and without fading, approximately 18 dB more carrier power is required under Rayleigh faded conditions. Conversely, for a CNR of 10 dB, the BER without and with fading is 5×10^{-6} and 2×10^{-2}, respectively. Clearly, fading worsens BER by four orders of magnitude. It is thus seen that fading causes severe degradation in the error performance of modulation schemes.

MAXIMUM RATIO COMBINING

In maximum ratio combining, the branch signals are weighted and combined so as to yield in the highest

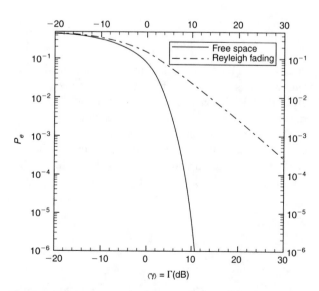

Fig. 5 BER for coherent BPSK with and without Rayleigh fading.

Wireless Channels—Internet

instantaneous CNR possible with any linear combining technique. By using Eq. 11 the total complex envelope at the n-th branch in the presence of additive noise $\mathbf{v}_n(t)$ can be written as

$$z_n(t) = g_n(t)u(t) + \mathbf{v}_n(t) \tag{30}$$

If the received signals are weighted by weights w_n^* the combined output $U(t)$ of the array is

$$U(t) = \mathbf{u}(t)\mathbf{w}^\dagger\mathbf{g} + \mathbf{w}^\dagger\mathbf{v} \tag{31}$$

where † denotes Hermitian conjugate (transpose and complex conjugate), $\mathbf{w} = [w_1, \ldots, w_N]^T$, $\mathbf{v} = [\mathbf{v}_1, \ldots, \mathbf{v}_N]^T$, and $\mathbf{g} = [g_1, \ldots, g_N]^T$. Assuming that the individual noise components are mutually independent, the total output noise power $P_{no}(o/p)$ is

$$P_{no}(o/p) = \frac{1}{2}\left\langle\left|\boldsymbol{\omega}^\dagger\mathbf{v}\right|^2\right\rangle = \sum_{n=1}^{N}\left|\mathbf{w}_n^*\right|^2 P_n \tag{32}$$

Therefore, the instantaneous output CNR is

$$\gamma = \frac{(1/2)E_T\left(\left|u(t)\boldsymbol{\omega}^\dagger\mathbf{g}\right|^2\right)}{\sum_{n=1}^{N}\left|\mathbf{w}_n^*\right|^2 P_n} = \frac{(1/2)\left|\sum\limits_{n=1}^{N}\mathbf{w}_n^* g_n\right|^2}{\sum_{n=1}^{N}\left|\mathbf{w}_n^*\right|^2 P_n} \tag{33}$$

The optimum weights are determined by the condition that the variation in γ with respect to the real and imaginary parts of w_n is zero. Alternately, the weights may be obtained by applying the Schwarz inequality to Eq. 33. By writing $w_n = \xi + j\eta$ and differentiating γ successively with respect to the two parameters ξ and η setting the result to zero, we get

$$w_n^* = \frac{g_n^*}{P_n} \tag{34}$$

which implies that the signals must be combined with weights made directly proportional to the complex conjugate of the branch signals and inversely proportional to the branch noise power. Thus, the branches with high signal strength are weighted more than the branches with weak signal strength. It is also noted that the weighted signals are all in phase and thus add coherently. The output CNR with the preceding weights is then

$$\gamma = \frac{(1/2)\left|\sum\limits_{n=1}^{N}|g_n|^2\Big/P_n\right|^2}{\sum_{n=1}^{N}\left|g_n^*\right|^2\Big/P_n} = \frac{1}{2}\sum_{n=1}^{N}\frac{|g_n|^2}{P_n} = \sum_{n=1}^{N}\gamma_n \tag{35}$$

Thus, the output CNR is the sum of the CNR of the individual branches. Implementation of the maximum

ratio combiner is expensive because the weights need both amplitude and phase tracking of the channel response [i.e., $g_n(t)s$]. Furthermore, linear amplifiers and phase shifters over a large dynamic range of input signals are needed. In this regard, maximum ratio combining is primarily of theoretical interest. However, it serves as a benchmark against which the performance of other practical linear combining techniques can be assessed.

Because of the sum relation expressed in Eq. 35, the statistical distributions of the output CNR in this case can be easily derived from its characteristic function. By considering, once again, the case where all branches have equal mean CNR $\Gamma_n = \Gamma$, the **pdf** for an N-branch maximum ratio combiner can be shown to be

$$P_N^{mrc}(\gamma) = \frac{1}{(N-1)!}\frac{\gamma^{N-1}}{\Gamma^N}e^{-\gamma/\Gamma}$$
$$\Gamma_n = \Gamma \tag{36}$$

The **pdf** is the well-known Erlang distribution,[9] which is obtained by the addition of N independent and identical exponential distributions. The mean CNR at the output of the combiner is

$$\langle\gamma\rangle = \sum_{n=1}^{N}\langle\gamma_n\rangle = \sum_{n=1}^{N}\Gamma = NT \tag{37}$$

The corresponding **cdf** is

$$P_N^{mrc}(\gamma_s) = \text{Prob}\,(\gamma < \gamma_s) = \int_0^{\gamma_s} P_N^{mrc}(\gamma)\mathrm{d}\gamma$$

$$= \frac{1}{(N-1)!}\int_0^{\gamma_s/\Gamma} x^{N-1}e^{-x\,dx}$$

$$= 1 - e^{-\gamma_s/\Gamma}\sum_{n=0}^{N-1}\left(\frac{\gamma_s}{\Gamma}\right)^n\frac{1}{n!} \tag{38}$$

$$= e^{-\gamma_s/\Gamma}\sum_{n=N}^{\infty}\left(\frac{\gamma_s}{\Gamma}\right)^n\frac{1}{n!} \tag{39}$$

$$= P_{N-1}^{mrc}(\gamma_s) - \frac{e^{-\gamma_s/\Gamma}}{(N-1)!}\left(\frac{\gamma_s}{\Gamma}\right)^{N-1} \tag{40}$$

Fig. 6 shows a plot of the **cdf** $P_N^{mrc}(\gamma_s)$, of maximum ratio combining with N as a parameter. At 99% reliability maximum ratio diversity provides a 12-dB savings in power with two branches and 19-dB savings with four branches. This is in contrast to the savings of 10 and

16 dB achievable with selection diversity. Clearly, maximum ratio combining is more efficient than selection combining.

It is interesting also to look at the **pdf** in the limit as $N \to \infty$. It is very easy to show from Eqs. 36 and 37 that the **pdf** approaches a delta function for large N

$$P_{\infty}^{\text{mrc}}(\gamma) \overset{\Delta}{=} \lim_{N \to \infty} p_N^{\text{mrc}}(\gamma) = \delta(\gamma - \langle \gamma \rangle) \tag{41}$$

i.e., the **pdf** reduces to that of a signal received in a free-space situation with no fading.

As in the selection diversity case, the performance of any digital scheme under diversity can be evaluated by looking at the BER for a constant γ and averaging the result over the **pdf** of γ. For instance, for coherent BPSK, the BER with maximum ratio combining assuming identical noise in each branch is

$$
\begin{aligned}
P_{\text{BPSK}}^{\text{mrc}} &= \int_0^\infty P_{\text{BPSK}}(\gamma)\, p_N^{\text{mrc}}(\gamma)\, d\gamma \\
&= \int_0^\infty \frac{1}{2}\text{erfc}(\sqrt{\gamma})\, \frac{1}{(N-1)!}\, \frac{\gamma^{N-1}}{\Gamma^N}\, e^{-\gamma/\Gamma} d\gamma \\
&= \frac{1}{2(N-1)!} \int_0^\infty erfc(\sqrt{\Gamma x}) x^{N-1} e^{-x}\, dx
\end{aligned} \tag{42}
$$

$$
\begin{aligned}
&= \frac{1}{(N-1)!}\left(\frac{1-\mu}{2}\right)^N \sum_{n=0}^N \frac{(N-1+n)!}{n!}\left(\frac{1+\mu}{2}\right)^2 \\
&\sim \left(\frac{1}{4\Gamma}\right)^N \frac{(2N-1)!}{N!(N-1)!} \quad \text{for } \Gamma \gg 1
\end{aligned} \tag{43}
$$

where

$$\mu = \sqrt{\frac{\Gamma}{1+\Gamma}} = \sqrt{\frac{\langle \gamma \rangle}{N + \langle \gamma \rangle}} \tag{44}$$

It is seen that the BER decreases with γ as $1/\gamma^N$ for large enough N and γ. In the limit as $N \to \infty$ the BER becomes

$$P_{\text{BPSK}}^{\text{mrc}} = \frac{1}{2}\text{erfc}(\sqrt{\gamma}) = P_{\text{BPSK}}^{\text{nofade}} \tag{45}$$

This result is to be expected because the output of the combiner would approach a steady value as N becomes large. Fig. 7 shows a plot of the BER vs. $\langle \gamma \rangle = N\gamma$ for BPSK with N as a parameter.

EQUAL GAIN COMBINING

In equal gain combining all of the weights have the same magnitude but a phase opposite to that of the signal in the respective branch, i.e., $w_n = \exp(-j \angle g_n)$. The combined output CNR with equal gain combining is

$$\gamma = \frac{1}{2}\frac{\left[\sum_{n=1}^N |g_n|\right]^2}{\sum_{n=1}^N P_n} = \frac{r^2}{2NP_{\text{no}}} \tag{46}$$

assuming equal noise in the branches, and where

$$r = \sum_{n=1}^N |g_n| \tag{47}$$

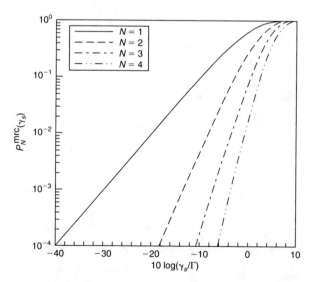

Fig. 6 cdf of γ_s vs. γ_s/γ for maximum ratio combining with N as a parameter.

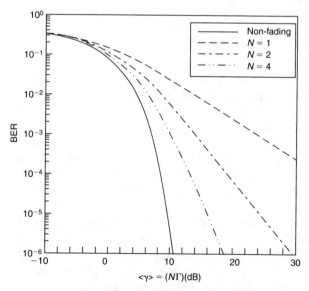

Fig. 7 BER vs. average CNR per bit $\langle \gamma \rangle = N\gamma$ for coherent BPSK under Rayleigh fading with maximum ratio diversity.

is the envelope (local mean) of the combined signal. It is seen equal to the sum of N Rayleigh variables $|g_n|$. Unfortunately, no closed form expression exits for the **pdf** or **cdf** of the combiner output in the case of equal gain combiner. However, a closed form expression for the *mean CNR* at the output of the combiner can be obtained easily from

$$\langle \gamma \rangle = \frac{1}{2NP_{no}} \left\langle \left[\sum_{n=1}^{N} |g_n| \right]^2 \right\rangle$$

$$= \frac{1}{2NP_{no}} \sum_{n=1}^{N} \sum_{m=1}^{N} \langle |g_n||g_m| \rangle$$

$$= \frac{1}{2NP_{no}} \left[2NP_{o} + N(N-1)\frac{\pi P_{o}}{2} \right]$$

$$= \left[1 + (N-1)\frac{\pi}{4} \right] \Gamma \qquad (48)$$

where we have used the fact that $\langle |g_n|^2 \rangle = 2P_o \langle |g_n| \rangle = \sqrt{\pi P_o / 2}$, and $\langle |g_n||g_m| \rangle = \langle |g_n| \rangle \langle |g_m| \rangle$ since the g_n are assumed uncorrelated. Similar to maximum ratio combining, equal gain diversity improves the output mean CNR in proportion to N. Reiterating once again, for large N, mean CNR improvement in selection diversity is proportional to $\ln(N)$, whereas in both maximum ratio and equal gain diversities, it increases linearly as N. Fig. 8 shows a plot of the improvement of average CNR of a diversity combiner for selection diversity, Eq. 26, maximum ratio diversity, Eq. 37, and equal gain diversity Eq. 48. Both the maximum gain diversity and equal gain diversity provide superior improvement compared with the selection diversity and that the results for equal gain combining are within 1 dB of those of maximum gain combining for up to 10 branches.

DIVERSITY GAIN

Diversity gain of an N-element array is defined as the improvement in link margin for certain performance criterion. Normally, the performance criterion is taken as the BER. For example, with reference to Fig. 7, it is seen that to provide a 10^{-2} BER with coherent phase shift keying (PSK), an average CNR per bit of 4.3 and 13.8 dB is needed without fading and with Rayleigh fading, respectively. Clearly, an additional 9.5 dB higher average output power is needed with Rayleigh fading. Using two antennas (i.e., two-branch diversity) reduces the required power to 8.4 dB and we say that the two-element array provides a diversity gain of 5.4 dB (= 13.8 − 8.4). Obviously, the maximum diversity gain that one can achieve with multiple antennas under maximum ratio combining is 9.5 dB at this BER and this value will be reached asymptotically for large N as evident from Eq. 45.

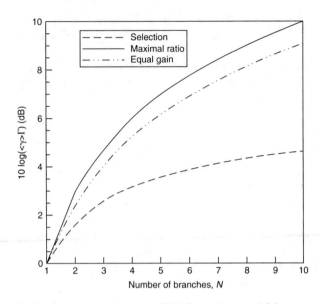

Fig. 8 Improvement in mean CNR for various combining techniques.

ANTENNA GAIN

The diversity gain must be distinguished from the antenna gain that is defined as the ratio of the output CNR of the array to the output CNR ratio of a single element for highly correlated input signals (single-incident plane wave). Recall that for a single-input plane wave, the response at various elements only differs by a phase factor $\exp(j\alpha)$, where $\alpha = k_o d\cos\phi$ depends on element spacing, the radio frequency (RF), and the angle of arrival of the plane wave with respect to the array axis. The input signal for the purpose of antenna gain then is assumed to be of the form $u(t)$

$$\sqrt{2P_o}[1,\exp(j\alpha),\exp(j2\alpha),...,\exp(j[N-1]\alpha)]^{T} \overset{\Delta}{=} u(t), \sqrt{2P_o}\Psi,$$

where P_o is the mean power at each b ranch. In the cases of maximum ratio combining and equal gain combining, the weights will be equal and proportional to $w=\sqrt{2P_o}\Psi/P_{no}$, respectively, where P_{no} is the input noise power at each branch. The combined signal plus noise voltage for a single plane wave incident is

$$U(t) = \frac{\sqrt{2P_o}}{P_{no}} \left[\sqrt{2P_o}u(t)\Psi^{\dagger}\Psi + \Psi^{\dagger}v \right]$$

$$= \frac{2NP_o}{P_{no}}u(t) + \frac{\sqrt{2P_o}}{P_{no}}\Psi^{\dagger}v \qquad (49)$$

The average carrier power at the output is $E_T(|[2NP_o u(t)/P_{no}]^2|)/2 = 2N^2(P_o/P_{no})^2$, whereas the noise power at the output $P_o \langle \Psi^{\dagger}vv^{\dagger}\Psi \rangle /P_{no}^2 = 2NP_o/P_{no}$ is assuming noise to be uncorrelated at various branches and using $\langle v_n^2(t) \rangle = 2P_{no}$ The output CNR is then equal to

$$\gamma = \frac{NP_o}{P_{no}} \qquad (50)$$

from which it is clear that the antenna array gain to equal to N. Note that the *mean* CNR improvement of a maximum ratio combiner is the same whether or not the branches are correlated and the same result stated earlier would have also been obtained for the mean CNR in the absence of branch correlation. This is actually shown in the next section.

The result could also be established from standard antenna theory, which predicts that the directivity of an antenna array is directly proportional to its length. Fig. 9 shows the directivity of an N-element, uniformly excited, broadside array as a function of d/λ with N as a parameter.

For $d = m\lambda/2$ with m an integer, the directivity is equal to N. For other spacings, the directivity varies around N, reaching N asymptotically for large d/λ. For $d = m\lambda$ with m and integer, the directivity takes a sharp dip because of the onset of grating lobes.

EFFECT OF BRANCH CORRELATION

The performance with diversity is somewhat degraded when the various branches are not perfectly decorrelated. A general account of this is given in Ref. 3 or 10. The branch correlation may be caused by a number of factors such as angular despreading of incoming waves or mutual coupling between antenna elements. For example, with a single plane wave incident, the two branches are always correlated no matter how large the spacing is.

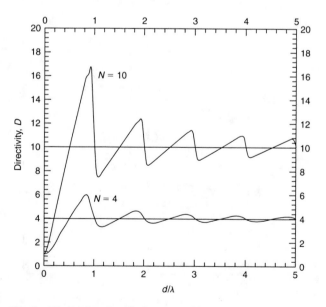

Fig. 9 Directivity of an N-element, uniformly excited broadside array.

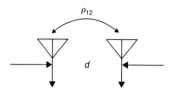

Fig. 10 Two elements with correlated signals.

To illustrate the effect, consider the two-branch case with maximum ratio combining where the complex Gaussian signals at the branches are correlated with a complex correlation coefficient ρ_{12}. The correlation coefficient measures the degree of correlation between signals received at two spatial points, separated by some distance d. For waves incident from the xy-plane, it is defined in Eq. 67 as

$$\rho_{12} = \left\langle e^{-jk_o d \cos\phi} \right\rangle \qquad (51)$$

where the angle brackets denote statistical average with respect to the angle variable ϕ. The distance d in our case corresponds to the interelement spacing d. Fig. 10 illustrates this case. The *envelope* correlation between the two branches is $\rho_e \approx |\rho_{12}|^2$.[6] In the presence of branch correlations, the **cdf** of the combined signal presented in Eq. 39 will get modified to[5]

$$p(\gamma_s) = 1 - \frac{1}{2|\rho_{12}|} \left[(1 + |\rho_{12}|)e^{-\gamma_s/(1+|\rho_{12}|)\Gamma} - 1 \right.$$
$$\left. - |\rho_{12}|e^{-\gamma_s/(1-|\rho_{12}|)\Gamma} \right]$$

and the corresponding **pdf** changes to

$$p(\gamma) = \frac{1}{2|\rho_{12}|\Gamma} \left[e^{-\gamma/(1+|\rho_{12}|)\Gamma} e^{-\gamma/(1+|\rho_{12}|)\Gamma} \right] \qquad (53)$$

The mean CNR of the combined signal remains at $\langle\gamma\rangle = 2\Gamma$—independent of $|\rho_{12}|$—as can be easily verified from Eq. 53. However, the distribution of γ depends on $|\rho_{12}|$ as seen from Fig. 11. The BER for a basic modulation scheme such as coherent BPSK can be carried out as in the previous section and the result is

$$P_e = \int_0^\infty \frac{1}{2}\text{erfc}(\sqrt{\gamma})P(\gamma)d\gamma$$

$$= \frac{1}{4|\rho_{12}|} \left[(1 + |\rho_{12}|)\left(1 - \frac{1}{\sqrt{1 + \frac{1}{(1+|\rho_{12}|)\Gamma}}}\right) \right.$$

$$\left. + (1 - |\rho_{12}|)\left(1 - \frac{1}{\sqrt{1 + \frac{1}{(1+|\rho_{12}|)\Gamma}}}\right) \right] \qquad (54)$$

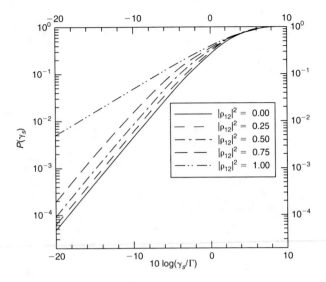

Fig. 11 Effect of branch correlation on the distribution of output power in a dual-diversity, maximum ratio combiner.

For the special case of perfect correlation (i.e., $\rho_{12} = 1$) the preceding expression reduces to Eq. 29 with Γ replaced by 2Γ.

The correlation coefficient depends on the distribution of incoming waves in the azimuth plane, which in turn depends on the disposition of scatterers about the transmitting and the receiving antennas. If waves arrive uniformly in the azimuth plane from all angles, as in the case of a mobile receiver situation in a highly cluttered environment, one gets the classical Clarke's two-dimensional (2-D) model.[2] On the other hand, waves received at a tall base station antenna arrive from a relatively narrow angle and one of the models used to describe such a situation is the circular scattering model.[10] In the circular scattering model, scatterers are assumed to be distributed uniformly within a radius R about the mobile. Propagation from the mobile to be base station is assumed to take place via single scattering off the scatterers. Because of the finite area of scattering, waves transmitted from the mobile arrive from a small angular region about the mean mobile direction. If the distance between the mobile and base station is D, the spatial correlation between the two elements for the circular scattering model can be derived as[6]

$$\rho_{12} = \frac{2J_1(k_o dR/D)}{k_o dR/D} \qquad (55)$$

where $J_n(\cdot)$ is the Bessel function of the first kind of order n. The angular spread of waves depends on the relative sizes of R and D. Using $d/\lambda = 5$, the BER performance as computed from Eq. 54 for various values of angular spread is shown in Fig. 12 as a function of (γ), expressed in decibels. For the model chosen, a rms angular spread of $1°$, as seen from the base station, gives rise to an envelope correlation of 0.74. Clearly, as the angular spread

increases, the branch signals become more and more decorrelated and the BER curve approaches the ideal ($\rho_{12} = 0$) 2-branch diversity curve. Comparing with Fig. 7 it is seen that an r.m.s. spread of $2°$ almost completely decorrelates the two signals. At a BER value of 10^{-2}, the diversity gain for a two-branch diversity reduces by about 5 dB ($= 14 - 9$ dB) when the signals change from being uncorrelated to completely correlated. In the next section, we look at another cause of correlation, namely, mutual coupling between elements, which becomes particularly important for small spacings.

MUTUAL COUPLING

Because of electromagnetic coupling the signals received by the elements of an array will no longer be independent, but become dependent on each other. Mutual coupling will influence the cross-correlation between the received signals and will be particularly important when spacings are small. Such would be the case for antenna diversity employed at the mobile station. One way to incorporate the presence of element mutual coupling is by means of an impedance matrix. The use of impedance matrix is most convenient for the wire type of antennas. As before, we consider an array of vertical dipoles for illustration. The terminal voltages V_n and the open-circuited voltages V_n^{oc} are related through the mutual impedance Z_{nm} as

$$V_1 = Z_{11} I_1 + Z_{12} I_2 + \cdots + Z_{1N} I_N + V_1^{oc}$$
$$V_2 = Z_{21} I_1 + Z_{22} I_2 + \cdots + Z_{2N} I_N + V_2^{oc}$$
$$\vdots$$
$$V_N = Z_{N1} I_1 + Z_{N2} I_2 + \cdots + Z_{NN} I_N + V_N^{oc} \qquad (56)$$

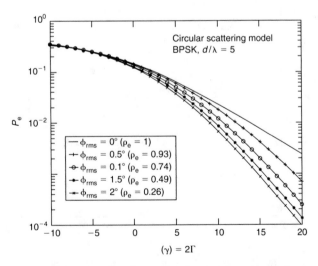

Fig. 12 Bit-error rate BER vs. (dB) for dual-branch, maximum ratio combiner with correlated fadings.

The preceding equation could be written in a matrix form as

$$\mathbf{v} = \mathbf{Z}\mathbf{i} + \mathbf{v}^{oc} \tag{57}$$

where $\mathbf{v}$, $\mathbf{v}^{oc}$, and $\mathbf{i}$ are the vectors of terminal voltages, open-circuited voltages, and terminal currents, respectively, and $\mathbf{Z}$ is the impedance matrix. All these quantities are in an *array environment*. Fig. 13 illustrates the T-equivalent representation of the impedance matrix for a two-element array. The concept of impedance matrix, devised originally for lumped circuit elements, presents some difficulties when applied directly to wave propagation problems. Although we continue to use it later, the reader must at least be made aware of the difficulties. We elaborate on these by considering center-fed broadside elements in the transmit mode. According to Eq. 56,

$$Z_{11} = \frac{V_1}{I_1}\Big|I_2, \ldots, I_N = 0 \tag{58}$$

That is, Z_{11} is the driving point impedance of element 1 when the *terminal* currents at all the other elements are made zero. However, unlike in circuits, making the terminal current of an antenna to go to zero by open-circuiting its terminals does not force the current on its entire structure to go to zero. Hence, Z_{11} is not the isolated self impedance of element 1, but includes the effects of the induced currents in any and all the other elements that can flow with the terminals open. Thus, the driving point impedance Z_{nn}, $n = 1, \ldots, N$, depends on the number of elements, the interelement spacing, the element orientations, and the relative positions of the element. Even in the case of identical elements, it has a different value for different elements of the array. Similarly, Z_{mn} is slightly different because elements m and n take up different positions in the same array. Although there are means of avoiding these difficulties by resorting to full numerical approaches such as the method of moments,[11] antenna engineers have continued to adopt the ideas of impedance matrix in design, with success, because the intricacies mentioned earlier tend to have a second-order effect. In a like manner, the open circuited voltage V_n^{oc} is the voltage induced by the incident plane wave across the n-th terminal when all the other terminals are open circuited. It, too, is not the voltage induced in an isolated dipole, but depends on the currents flowing on the other antenna

structures under open-circuit conditions. In the subsequent analysis we, however, ignore these second-order effects.

For a load impedance Z_{Ln}, the terminal voltage V_n and the current I_n (defined to flow *into* the terminals) are related by

$$V_n = -Z_{L_n} I_n \tag{59}$$

By using this in Eq. 57, we get

$$\mathbf{v} = \mathbf{Z}_L (\mathbf{Z} + \mathbf{Z}_L)^{-1} \mathbf{v}^{oc} \tag{60}$$

where $\mathbf{Z}_L$ is the diagonal matrix containing the load impedances. For convenience, we define $\mathbf{Z}^c$ as

$$\mathbf{Z}^c = \mathbf{Z}_L (\mathbf{Z} + \mathbf{Z}_L)^{-1} \tag{61}$$

and rewrite Eq. 60 as

$$\mathbf{v} = \mathbf{Z}^c \, \mathbf{v}^{oc} \tag{62}$$

It is clear that the presence of mutual coupling can be simply accounted for by replacing the vector $\mathbf{v}^{oc}$ with $\mathbf{Z}^c \mathbf{v}^{oc}$. For example, in the case of isotropic elements, the received signal vector (terminal voltages) in terms of the steering vector becomes

$$\mathbf{v} = V_o \mathbf{Z}^c \, \psi \tag{63}$$

Eq. 63 is a generalization of Eq. 9. For a two-element array composed of vertical dipoles oriented as in Fig. 1, the received signals for a plane wave incident from the xy-plane are

$$V_1 = Z_{11}^c V_1^{oc} + Z_{12}^c V_2^{oc} \tag{64}$$

$$V_2 = Z_{21}^c V_1^{oc} + Z_{22}^c V_2^{oc} \tag{65}$$

where

$$V_n^{oc} = V_o e^{j(n-1)k_o d \cos \phi} \tag{66}$$

$n = 1.2$

Define the spatial cross-correlation between the two received signals without and with mutual coupling as

$$\rho_{12}^{oc} = \frac{\langle V_1^{oc} V_2^{*oc} \rangle}{\sqrt{\langle |V_1^{oc}|^2 \rangle \langle |V_2^{oc}|^2 \rangle}}$$

$$= \langle e^{-jk_o d \cos \phi} \rangle \quad \text{(without mutual coupling)} \tag{67}$$

$$\rho_{12} = \frac{\langle V_1 V_2^* \rangle}{\sqrt{\langle |V_1|^2 \rangle \langle |V_2|^2 \rangle}} \quad \text{(with mutual coupling)} \tag{68}$$

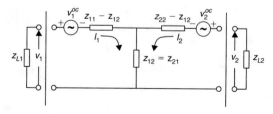

Fig. 13 T-equivalent representation of antenna mutual coupling for a two-element array.

Wireless
Channels—Internet

where the angle brackets denote statistical average with respect to the angle variable Φ. In the following it is assumed that the **pdf** of the arriving waves in the azimuthal plane is symmetrical about the broadside to the array axis $\Phi = \pi/2$ so that ρ^{oc}_{12} is a real quantity. This is true both for the Clarke's model and the circular scattering model with the scatter region directly ahead of the array. By assuming identical elements so that $Z_{11} = Z_{22}$ and using $Z_{12} = Z_{21}$, $Z_{L1} = Z_{L2} = Z_L$, the spatial correlation expression can be simplified as

$$\rho_{12} = \frac{\rho^{oc}_{12} + \zeta}{1 + \zeta \rho^{oc}_{12}} \tag{69}$$

where

$$\zeta = \frac{2\Re e(\zeta_1)}{1 + |\zeta_1|^2} \tag{70}$$

and

$$\zeta_1 = \frac{-Z_{12}}{Z_{11} + Z_L} \tag{71}$$

It is clear that in the absence of mutual coupling, $Z_{12} = 0 \Rightarrow \zeta = 0$ so that $\rho_{12} = \rho^{oc}_{12}$ as expected. For center-fed, half-wavelength dipoles, $Z_{11} \approx 73 + j42.5\ \Omega$. Mutual impedance $Z_{12} = R_{12} + jX_{12}$ between identical center-fed dipoles arranged side by side is given by Ref. 12

$$R_{12} = 30\left[\text{Cin}\left[k_o\left(\sqrt{e^2 + d^2} + e\right)\right]\right.$$
$$\left. + \text{Cin}\left[k_o\left(\sqrt{e^2 + d^2} - e\right)\right] - 2\text{Cin}(k_o d)\right] \tag{72}$$

$$X_{12} = 30\left[\text{Si}\left[k_o\left(\sqrt{e^2 + d^2} + e\right)\right]\right.$$
$$\left. + \text{Si}\left[k_o\left(\sqrt{e^2 + d^2} - e\right)\right] - 2\text{Si}(k_o d)\right] \tag{73}$$

where $\text{Cin}(x)$ and $\text{Si}(x)$ are cosine and sine integrals defined as[13]

$$\text{Cin}(x) = \int_0^x \frac{1 - \cos t}{t}\, dt \tag{74}$$

$$\text{Si}(x) = \int_0^x \frac{\sin t}{t}\, dt \tag{75}$$

The real and imaginary parts of Z_{12} are plotted in Fig. 14 as a function of element spacing. Note that for

$d = 0$, $Z_{12} = Z_{11}$. Figs. 15 and 16 show the envelope correlation $\rho_e = |\rho_{12}|^2$ for Clarke's 2-D model[2] [$\rho^{oc}_{12} = J_0(k_o d)$] and the circular scattering model (azimuth angular spread of about 15°) with and without mutual coupling for load impedance $Z_L = Z^*_{11}$. As mentioned previously, the former model would correspond to array antennas employed at the mobile station, whereas the latter would correspond to array antennas employed at the base station. From the figures, the envelope cross-correlation calculated with mutual coupling differs substantially from that calculated assuming to mutual coupling. The exact effect depends on the interplay between mutual coupling and angular spread of arrival of the incoming waves and could lead to either increased or decreased cross-correlation between the two antennas. From Fig. 15, mutual coupling actually decreases the correlation between the elements when the waves arrive uniformly from all directions (Clarke's model). However, with the circular scattering model, it t ends to increase the correlation for $\rho_e < 0.6$. To provide an envelope correlation value of 0.7 with Clarke's 2-D model, element separation of 0.14 and 0.05 λ is needed without and with mutual coupling. For the circular scattering model, the spacing required with and without mutual coupling is 0.18 and 0.4 λ, respectively.

Fig. 17 shows the envelope correlation calculated using Eq. 69 for the Clarke's model and compared with measurements reported in Refs. 8 and 14. Vaughan and Scott[14] conducted experiments in the city of Wellington, New Zealand, with quarter-wave monopoles terminated in $Z_{L1} = Z_{L2} = 50\ \Omega$. Error bars shown are for the 95% confidence interval. Miki and Hata[8] conducted experiments in the city of Tokyo and provide data for envelope

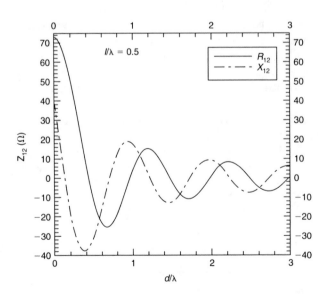

Fig. 14 Mutual impedance between two side-by-side $\lambda/2$ dipoles.

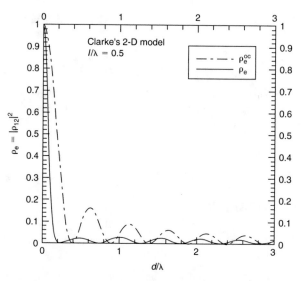

Fig. 15 Spatial correlation between two elements with and without mutual coupling, Clarke's model.

correlation for and element spacing of 0.2 λ. Once again, their results are for quarter-wave monopoles. Eq. 69, derived for center-fed dipoles, can also be used for monopoles by noting that the values of Z_{11} and Z_{12} for monopoles are half of those dipoles. The values computed with mutual coupling agree quite well with the measured one.

SUMMARY

The use of spatial diversity, whereby antenna arrays are employed at the receiver, to combat the ill-effects

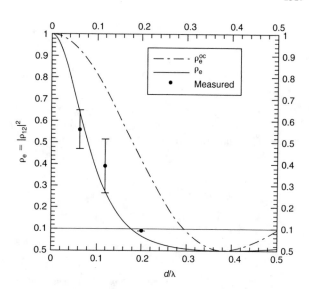

Fig. 17 Spatial correlation between two elements with and without mutual coupling, Clarke's model. Measured values are from Ref. 8 and 14.

of fading on the reception of narrowband signals was discussed. Three techniques of combining, namely, selection combining, maximum ratio combining, and equal gain combining, together with their effects on the bit error performance of uncoded modulation schemes such as BPSK was discussed. Inclusion of angular spreading and element mutual coupling in analysis was demonstrated for arrays composed of vertical dipole elements.

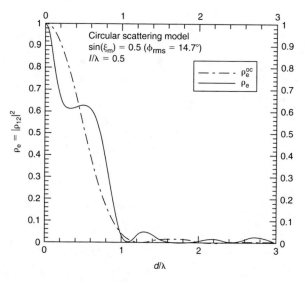

Fig. 16 Spatial correlation between two elements with and without mutual coupling, circular scattering model.

REFERENCES

1. Schwartz, M.; Bennett, W.R.; Stein, S. *Communication Systems and Techniques*; IEEE Press (Classic Reissue): Piscataway, NJ, 1996.

2. Jakes, W.C. *Microwave Mobile Communications*; IEEE Press (Classic Reissue): Piscataway, NJ, 1994.

3. Vaughan, R.G.; Andersen, J.B. Antenna diversity in mobile communications. IEEE Trans. Veh. Technol. **1987**, *36* (4), 149–171.

4. Dietrich, C.B., Jr.; Stutzman, W.L.; Byung-Ki, K.; Dietze, K. Smart antennas in wireless communications: base station diversity and handset beamforming. IEEE Antennas Propag. Mag. **2000**, *42* (5), 142–151.

5. Collin, R.E.; Zucker, F.J. *Antenna Theory*; McGraw-Hill: New York, 1969 Pt. 1, Ch. 5.

6. Janaswamy, R. *Radiowave Propagation and Smart Antennas for Wireless Communications*; Kluwer Academic Publishers: Boston, 2000.

7. Proakis, J.G. *Digital Communications, *, 2nd Ed.; McGraw-Hill: New York, 1989.

8. Miki, T.; Hata, M. Performance of 16 kbit/s GMSK transmission with post detection selection diversity in land mobile radio. IEEE Trans. Veh. Technol. **1984**, *VT-33* (3), 128–133.

9. Papoulis, A. *Probability, Random Variables, and Stochastic Processes*; McGraw-Hill: New York, 1984.

10. Petrus, P.; Reed, J.H.; Rappaport, T.S. Geometrically based statistical channel model for macro cellular mobile environments. IEEE Globecommun, London, England, Nov, 18–22, 1996; 1197–1201.

11. Harrington, R.F. *Field Computation by Moment Methods*; Krieger: Malabar, FL, 1968.

12. Kraus, J.D. *Antennas, *, 2nd Ed.; McGraw-Hill: New York, 1988.

13. Abramowitz, M.; Stegun, I. *Handbook of Mathematical Functions*; Dover: New York, 1964.

14. Vaughan, R.G.; Scott, N.L. Closely spaced monopoles for mobile communications. Radio Sci. **1993**, *28* (6), 1259–1266.

BIBLIOGRAPHY

1. Collin, R.E. *Antennas and Radiowave Propagation*; McGraw-Hill: New York, 1985.

2. Weeks, W.L. *Antenna Engineering*; McGraw-Hill: New York, 1968.

Wireless Data

Allen H. Levesque
Kaveh Pahlavan
Worcester Polytechnic Institute, Worcester, Massachusetts, U.S.A.

Abstract

This entry describes the existing and evolving wireless data networks and the related standards and services.

INTRODUCTION

Wireless data services and systems represent a steadily growing and increasingly important segment of the communications industry. While the wireless data industry is becoming increasingly diverse, one can identify two mainstreams that relate directly to users' requirement for data services. On one hand, there are requirements for relatively low-speed data services provided to mobile users over wide geographical areas, as provided by private mobile data networks and by data services implemented on common-carrier cellular telephone networks. On the other hand, there are requirements for high-speed data services in local areas, as provided by cordless private branch exchange (PBX) systems and wireless LANs, as well as by the emerging personal communications services (PCSs). In this entry we mainly address wide-area wireless data systems, commonly called *mobile data systems,* and touch upon data services to be incorporated into the emerging digital cellular systems. We also briefly address paging and messaging services.

Mobile data systems provide a wide variety of services for both business users and public safety organizations. Basic services supporting most businesses include e-mail, enhanced paging, modem and facsimile transmission, remote access to host computers and office LANs, information broadcast services and, increasingly, Internet access. Public safety organizations, particularly law-enforcement agencies, are making increasing use of wireless data communications over traditional very high frequency (VHF) and ultra high frequency (UHF) radio dispatch networks, over commercial mobile data networks, and over public cellular telephone networks. In addition, there are wireless services supporting vertical applications that are more or less tailored to the needs of specific companies or industries, such as transaction processing, computer-aided delivery dispatch, customer service, fleet management, and emergency medical services. Work currently in progress to develop the national ITS includes the definition of a wide array of new traveler services, many of which will be supported by standardized mobile data networks.

Much of the growth in use of wireless data services has been spurred by the rapid growth of the paging service industry and increasing customer demand for more advanced paging services, as well as the desire to increase work productivity by extending to the mobile environment the suite of digital communications services readily available in the office environment. There is also a desire to make more cost-efficient use of the mobile radio and cellular networks already in common use for mobile voice communications by incorporating efficient data transmission services into these networks. The services and networks that have evolved to date represent a variety of specialized solutions and, in general, they are not interoperable with each other. As the wireless data industry expands, there is an increasing demand for an array of attractively priced standardized services and equipment accessible to mobile users over wide geographic areas. Thus, we see the growth of nationwide privately operated service networks as well as new data services built upon the first and second generation cellular telephone networks. The implementation of PCS networks in the 2-GHz bands as well as the eventual implementation of 3G wireless networks will further extend this evolution.

In this entry we describe the principal existing and evolving wireless data networks and the related standards activities now in progress. We begin with a discussion of the technical characteristics of wireless data networks.

CHARACTERISTICS OF WIRELESS DATA NETWORKS

From the perspective of the data user, the basic requirement for wireless data service is convenient, reliable, low-speed access to data services over a geographical

Encyclopedia of Wireless and Mobile Communications DOI: 10.1081/E-EWMC-120043949

area appropriate to the user's pattern of daily business operation. By low speed we mean data rates comparable to those provided by standard data modems operating over the public switched telephone network (PSTN). This form of service will support a wide variety of short-message applications, such as notice of e-mail or voice mail, as well as short file transfers or even facsimile transmissions that are not overly lengthy. The user's requirements and expectations for these types of services are different in several ways from the requirements placed on voice communication over wireless networks. In a wireless voice service, the user usually understands the general characteristics and limitations of radio transmission and is tolerant of occasional *signal fades* and brief dropouts. An overall level of acceptable voice quality is what the user expects. In a data service, the user is instead concerned with the accuracy of delivered messages and data, the time-delay characteristics of the service network, the ability to maintain service while traveling about, and, of course, the cost of the service. All of these factors are dependent on the technical characteristics of wireless data networks, which we discuss next.

Radio Propagation Characteristics

The chief factor affecting the design and performance of wireless data networks is the nature of radio propagation over wide geographic areas. The most important mobile data systems operate in various land-mobile radio bands from roughly 100 to 200 MHz, the specialized mobile radio (SMR) band around 800 MHz, and the cellular telephone bands at 824–894 MHz. In these frequency bands, radio transmission is characterized by distance-dependent field strength, as well as the well-known effects of *multipath fading,* signal shadowing, and signal blockage. The signal coverage provided by a radio transmitter, which in turn determines the area over which a mobile data receiving terminal can receive a usable signal, is governed primarily by the *power–distance relationship,* which gives signal power as a function of distance between transmitter and receiver. For the ideal case of single-path transmission in free space, the relationship between transmitted power P_t and received power P_r is given by

$$\frac{P_r}{P_t} = G_t G_r \left(\frac{\lambda}{4\pi d}\right)^2 \qquad (1)$$

where G_t and G_r are the transmitter and receiver antenna gains, respectively, d is the distance between the transmitter and the receiver, and λ is the wavelength of the transmitted signal. In the mobile radio environment, the power–distance relationship is in general different from the free-space case just given. For propagation over an Earth plane at distances much greater than either the signal wavelength or the antenna heights, the relationship

between P_t and P_r is given by

$$\frac{P_r}{P_t} = G_t G_r \left(\frac{h_1^2 h_2^2}{d^4}\right) \qquad (2)$$

where h_1 and h_2 are the transmitting and receiving antenna heights. Note here that the received power decreases as the fourth power of the distance rather than the square of distance seen in the ideal free-space case. This relationship comes from a propagation model in which there is a single signal reflection with phase reversal at the Earth's surface, and the resulting received signal is the vector sum of the direct line-of-sight signal and the reflected signal. When user terminals are deployed in mobile situations, the received signal is generally characterized by rapid fading of the signal strength, caused by the vector summation of reflected signal components, the vector summation changing constantly as the mobile terminal moves from one place to another in the service area. Measurements made by many researchers show that when the fast fading is averaged out, the signal strength is described by a Rayleigh distribution having a log-normal mean. In general, the power–distance relationship for mobile radio systems is a more complicated relationship that depends on the nature of the terrain between transmitter and receiver.

Various propagation models are used in the mobile radio industry for network planning purposes, and a number of these models are described in Ref. 1. Propagation models for mobile communications networks must take account of the terrain irregularities existing over the intended service area. Most of the models used in the industry have been developed from measurement data collected over various geographic areas. A very popular model is the *Longley–Rice model*.[2,3] Many wireless networks are concentrated in urban areas. A widely used model for propagation prediction in urban areas is one usually referred to as the *Okumura–Hata model*.[4,5]

By using appropriate propagation prediction models, one can determine the range of signal coverage for a base station of given transmitted power. In a wireless data system, if one knows the level of received signal needed for satisfactory performance, the area of acceptable performance can, in turn, be determined. Cellular telephone networks utilize base stations that are typically spaced 1–5 mi apart, though in some mid-town areas, spacings of 1/2 mi or less are now being used. In packet-switched data networks, higher power transmitters are used, spaced about 5–15 mi apart.

An important additional factor that must be considered in planning a wireless data system is the in-building penetration of signals. Many applications for wireless data services involve the use of mobile data terminals inside buildings, e.g., for troubleshooting and servicing computers on customers' premises. Another example is wireless communications into hospital buildings in support of emergency medical services. It is usually estimated that in-building signal penetration losses will be in the range of 15–30 dB. Thus, received signal strengths can be satisfactory in the outside areas around a building but totally unusable inside the building. This becomes an important issue when a service provider intends to support customers using mobile terminals inside buildings.

One important consequence of the rapid fading experienced on mobile channels is that errors tend to occur in bursts, causing the transmission to be very unreliable for short intervals of time. Another problem is signal dropouts that occur, e.g., when a data call is handed over from one base station to another, or when the mobile user moves into a location where the signal is severely attenuated. Because of this, mobile data systems employ various error-correction and error-recovery techniques to ensure accurate and reliable delivery of data messages.

MARKET ISSUES

There are two important trends that are tending to propel growth in the use of wireless data services. The first is the rapidly increasing use of portable devices such as laptop computers, pen-pads, notebook computers, and other similar devices. Increasingly, the laptop or notebook computer is becoming a standard item of equipment for traveling professional or business person, along with the cellular telephone and pager. This trend has been aided by the steady decrease in prices, increases in reliability, and improvements in capability and design for such devices. The second important trend tending to drive growth in wireless data services is the explosive growth in the use of the Internet. As organizations become increasingly reliant upon the Internet for their everyday operations, they will correspondingly want their employees to have convenient access to the Internet while traveling, just as they do in the office environment. Wireless data services can provide the traveler with the required network access in many situations where wired access to the public network is impractical or inconvenient. Mobile data communication services and newer messaging services discussed here provide a solution for wireless access over wide areas. Recent estimates of traffic composition indicate that data traffic now accounts for less than 1% of the traffic on

wireless networks, compared to 50% for wireline networks. Therefore, the potential for growth in the wireless data market continues to be strong.

MODEM SERVICES OVER CELLULAR NETWORKS

A simple form of wireless data communication now in common use is data transmission using modems or facsimile terminals over analog cellular telephone links. In this form of communication, the mobile user simply accesses a cellular channel just as he would in making a standard voice call over the cellular network. The user then operates the modem or facsimile terminal just as would be done from office to office over the PSTN. A typical connection is shown in Fig. 1, where the mobile user has a laptop computer and portable modem in the vehicle, communicating with another modem and computer in the office. Typical users of this mode of communication include service technicians, real estate agents, and traveling sales people. In this form of communication, the network is not actually providing a data service but simply a voice link over which the data modem or fax terminal can interoperate with a corresponding data modem or fax terminal in the office or service center. The connection from the mobile telephone switching office (MTSO) is a standard landline connection, exactly the same as is provided for an ordinary cellular telephone call. Many portable modems and fax devices are now available in the market and are sold as elements of the so-called "mobile office" for the traveling business person. Law enforcement personnel are also making increasing use of data communication over cellular telephone and dispatch radio networks to gain rapid access to databases for verification of automobile registrations and drivers' licenses. Portable devices are currently available that operate at transmission rates up to 9.6 or 14.4 Kbps. Error-correction modem protocols such as MNP-10, V.34, and V.42 are used to provide reliable delivery of data in the error-prone wireless transmission environment.

In another form of mobile data service, the mobile subscriber uses a portable modem or fax terminal as already described but now accesses a modem provided by the cellular service operator as part of a *modem pool*, which is connected to the MTSO. This form of service is

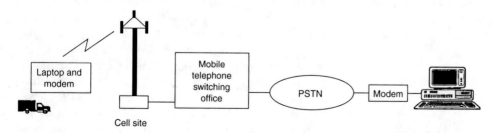

Fig. 1 Modem operation over an analog cellular voice connection.

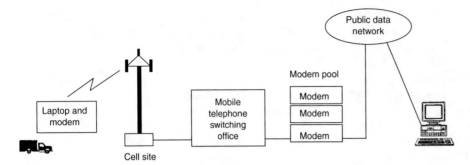

Fig. 2 Cellular data service supported by modem pools in the network.

shown in Fig. 2. The modem pool might provide the user with a choice of several standard modem types. The call connection from the modem pool to the office is a digital data connection, which might be supported by any of a number of public packet data networks, such as those providing X.25 service. Here, the cellular operator is providing a special service in the form of modem pool access, and this service in general carries a higher tariff than does standard cellular telephone service, due to the operator's added investment in the modem pools.

In this form of service, however, the user in the office or service center does not require a modem but instead has a direct digital data connection to the desktop or host computer.

Each of the types of wireless data transmission just described is in effect an appliqué onto an underlying cellular telephone service and, therefore, has limitations imposed by the characteristics of the underlying voice-circuit connection. That is, the cellular segment of the call connection is a circuit-mode service, which might be cost effective if the user needs to send long file transfers or fax transmissions but might be relatively costly if only short messages are to be transmitted and received. This is because the subscriber is being charged for a circuit-mode connection, which stays in place throughout the duration of the communication session, even if only intermittent short message exchanges are needed. The need for systems capable of providing cost-effective communication of relatively short message exchanges led to the development of wireless packet data networks, which we describe next.

PACKET DATA AND PAGING/ MESSAGING NETWORKS

Here we describe two packet data networks that provide mobile data services to users in major metropolitan areas of the United States and briefly discuss paging/messaging networks.

Advanced Radio Data Information Service (Motient Data Network)

Advanced radio data information service (ARDIS) is a two-way radio service developed as a joint venture between IBM and Motorola and first implemented in 1983. In mid-1994, IBM sold its interest in ARDIS to Motorola and early in 1998 ARDIS was acquired by the American Mobile Satellite (now Motient Corporation). The ARDIS network consists of four network control centers (NCCs) with 32 network controllers distributed through 1250 base stations in 400 cities in the US The service is suitable for two-way transfers of data files of size less than 10 KB, and much of its use is in support of computer-aided dispatching, such as is used by field service personnel, often while they are on customers' premises. Remote users access the system from laptop radio terminals, which communicate with the base stations. Each of the ARDIS base stations is tied to one of the 32 radio network controllers, as shown in Fig. 3. The backbone of the network is implemented with leased telephone lines. The four ARDIS hosts, located in Chicago, New York, Los Angeles, and Kentucky, serve as access points for a customer's mainframe computer, which can be linked to an ARDIS host using async, bisync, SNA, or X.25 dedicated circuits.

The operating frequency band is 800 MHz, and the RF links use separate transmit and receive frequencies spaced by 45 MHz. The system was initially implemented with an RF channel data rate 4800 bps per 25-kHz channel, using the MDC-4800 protocol. In 1993 the access data rate was upgraded to 19.2 Kbps, using the radio data-link access protocol (RD-LAP), which provides a user data rate of about 8000 bps. In the same year, ARDIS implemented a nationwide roaming capability, allowing users to travel between widely separated regions without having to preregister their portable terminals in each new region. The ARDIS system architecture is cellular, with cells overlapped to increase the probability that the signal transmission from a portable transmitter will reach at least one base station. The base station power is 40 W, which provides line-of-sight coverage up to a radius of 10–15 miles. The portable units operate with 4 W of radiated power. The overlapping coverage, combined with designed power levels, and error-correction coding in the transmission format, ensures that the ARDIS can support portable communications from inside buildings, as well as on the street. This capability for in-building coverage is an important characteristic of the ARDIS. The modulation technique is frequency-shift keying (FSK), the access

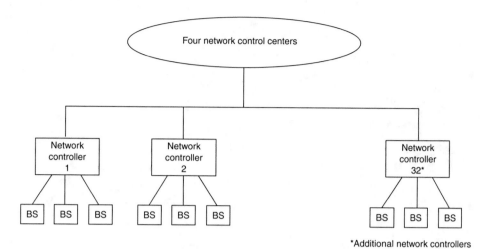

*Additional network controllers
can be installed

Fig. 3 ARDIS network
architecture.

method is FDMA, and the transmission packet length is 256 bytes.

Although the use of overlapping coverage, almost always on the same frequency, provides reliable radio connectivity, it poses the problem of interference when signals are transmitted simultaneously from two adjacent base stations. The ARDIS network deals with this by turning off neighboring transmitters, for 0.5–1 sec, when an outbound transmission occurs. This scheme has the effect of constraining overall network capacity.

The laptop portable terminals access the network using a random access method called digital sense multiple access (DSMA).[6] A remote terminal listens to the base station transmitter to determine if a "busy bit" is on or off. When the busy bit is off, the remote terminal is allowed to transmit. If two remote terminals begin to transmit at the same time, however, the signal packets may collide, and retransmission will be attempted, as in other contention-based multiple access protocols. The busy bit lets a remote user know when other terminals are transmitting and, thus, reduces the probability of packet collision.

MOBITEX (Cingular Wireless)

The MOBITEX system is a nationwide, interconnected trunked radio network developed by Ericsson and Swedish Telecom. The first MOBITEX network went into operation in Sweden in 1986, and networks have either been implemented or are being deployed in 22 countries. A MOBITEX operations association oversees the open technical specifications and coordinates software and hardware developments.[7] In the US, MOBITEX service was introduced by RAM Mobile Data in 1991. In 1992 Bell South Enterprises became a partner with RAM. Currently, MOBITEX data service in the US is operated by Cingular Wireless, formerly known as Bell South Wireless Data, LP. The Cingular MOBITEX service now covers over 90% of the US urban business population with about 2000 base

stations, and it provides automatic "roaming" across all service areas. By locating its base stations close to major business centers, the system provides a degree of in-building signal coverage. Although the MOBITEX system was designed to carry both voice and data service, the US and Canadian networks are used to provide data service only.

MOBITEX is an intelligent network with an open architecture that allows establishing virtual networks. This feature facilitates the mobility and expandability of the network.[8,9]

The MOBITEX network architecture is hierarchical, as shown in Fig. 4. At the top of the hierarchy is the NCC, from which the entire network is managed. The top level of switching is a national switch (MHX1) that routes traffic between service regions. The next level comprises regional switches (MHX2s), and below that are local switches (MOXs), each of which handles traffic within a given service area. At the lowest level in the network, multichannel trunked-radio base stations communicate with the mobile and portable data sets. MOBITEX uses packet-switching techniques, as does ARDIS, to allow multiple users to access the same channel at the same time. Message packets are switched at the lowest possible network level. If two mobile users in the same service area need to communicate with each other, their messages are relayed through the local base station, and only billing information is sent up to the NCC.

The base stations are laid out in a grid pattern using the same frequency reuse rules as are used for cellular telephone networks. In fact, the MOBITEX system operates in much the same way as a cellular telephone system, except that handoffs are not managed by the network. That is, when a radio connection is to be changed from one base station to another, the decision is made by the mobile terminal, not by a network computer as in cellular telephone systems.

To access the network, a mobile terminal finds the base station with the strongest signal and then registers with that

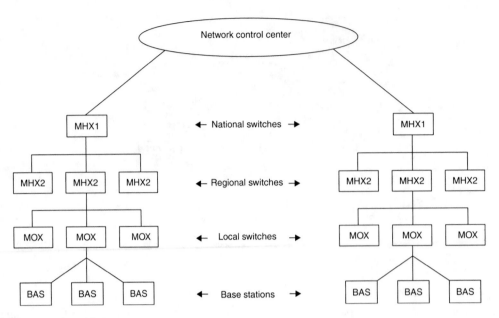

Fig. 4 MOBITEX network architecture.

base station. When the mobile terminal enters an adjacent service area, it automatically re-registers with a new base station, and the user's whereabouts are relayed to the higher level network nodes. This provides automatic routing of messages bound for the mobile user, a capability known as *roaming*. The MOBITEX network also has a store-and-forward capability.

The mobile units transmit at 896–901 MHz and the base stations at 935–940 MHz. The base stations use a trunked radio design employing 2–30 radio channels in each service area. The system uses dynamic power setting, in the range of 100 mW–10 W for mobile units and 100 mW–4 W for portable units. The Gaussian minimum shift keying (GMSK) modulation technique is used, with $BT = 0.3$ and noncoherent demodulation. The transmission rate is 8000 bps half-duplex in 12.5-kHz channels, and the service is suitable for file transfers up to 20 KB. The MOBITEX system uses a proprietary network-layer protocol called MOBITEX packet (MPAK), which provides a maximum packet size of 512 bytes and a 24-b address field. Forward-error correction, as well as retransmissions, are used to ensure the bit-error-rate quality of delivered data packets. Fig. 5 shows the packet structure at various layers of the MOBITEX protocol stack. The system uses the reservation-slotted ALOHA (R-S-ALOHA) random access method.

Paging and Messaging Networks

The earliest paging networks, first launched in the early 1960s, supported simple one-way tone pagers. Tone paging was soon replaced by special paging codes, beginning with numeric codes introduced in North American networks in the 1970s.[10] In 1976, European and US manufacturers and paging industry representatives formed the Post Office Code Advisory Group (POCSAG), and they subsequently defined the POCSAG code. The POC-SAG code supported tone, numeric, and alpha-text services and was widely adopted for paging networks. In 1985 the organization of European Post, Telephone and Telegraph administrations, known as CEPT, began work to develop a unified European paging code standard. In 1993 their recommendation for an enhanced protocol was approved by the ETSI as the enhanced radio message system (ERMES). The ERMES code was based upon the POC-SAG code, and incorporated enhancements for multicasting and multifrequency roaming as well as for transmission of longer text messages. However, ERMES protocol did not receive widespread support, and the market demands for simpler short numeric message paging tended to prevail.

At present, a suite of alphanumeric paging protocols called FLEX™, developed by Motorola in 1993, has become the de facto world standard for paging services. While the FLEX protocols bear some similarity to the POCSAG and ERMES protocols, they overcome many of the weaknesses of those protocols. FLEX carries less transmission overhead than does ERMES, providing improved alphanumeric messaging capacity. The FLEX protocol is a synchronous time-slotted protocol referenced to a global positioning system (GPS) time base. Each pager is assigned to a base frame in a set of 128 frames transmitted during a 4-min cycle (32 frames per min, 1.875 sec per frame). Fifteen FLEX cycles occur each hour and are synchronized to the start of the GPS hour. Each FLEX frame consists of a synchronization portion and a data portion. The synchronization portion consists of a sync signal and an 11-bit frame information word allowing the

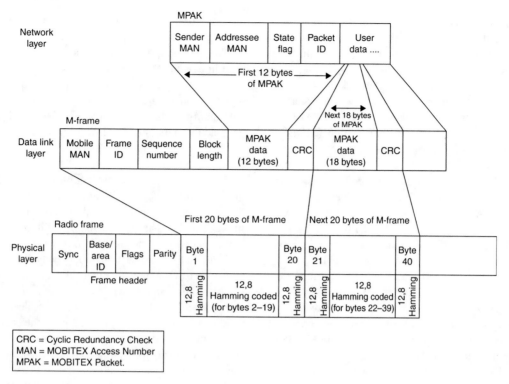

Fig. 5 MPAK and frame structure at three layers of the protocol stack.

pager to identify the frame and cycle in which it resides. A second sync signal indicates the rate at which the data portion is transmitted.

Three data rates are provided, 1600, 3200, or 6400 bps, and the modulation used is either two-level or four-level FSK. For two-way paging service, which is becoming increasingly popular in the US, the FLEX forward channel (in the 931 MHz band) protocol is combined with a return channel (901 MHz band) protocol called ReFLEX™.

CELLULAR DATA NETWORKS AND SERVICES

Cellular Digital Packet Data

The cellular digital packet data (CDPD) system was designed to provide packet data services as an overlay onto the existing analog cellular telephone network, which is called advanced mobile phone service (AMPS). CDPD was developed by IBM in collaboration with the major cellular carriers. Any cellular carrier owning a license for AMPS is free to offer its customers CDPD service without any need for further licensing. A basic concept of the CDPD system is to provide data services on a noninterfering basis with the existing analog cellular telephone services using the same 30-kHz channels. This is accomplished in either of two ways. First, one or a few AMPS channels in each cell site can be devoted to CDPD service. Second, CDPD is designed to make use of a

cellular channel that is temporarily not being used for voice traffic and to move to another channel when the current channel is assigned to voice service. In most of the CDPD networks deployed to date, the fixed-channel implementation is being used.

The compatibility of CDPD with the existing cellular telephone system allows it to be installed in any AMPS cellular system in North America, providing data services that are not dependent on support of a digital cellular standard in the service area. The participating companies issued release 1.0 of the CDPD specification in July 1993, and release 1.1 was issued in late 1994.[11] At this writing (early 2002), CDPD service is implemented in many of the major market areas in the US typical applications for CDPD service include e-mail, field support servicing, package delivery tracking, inventory control, credit card verification, security reporting, vehicle theft recovery, traffic and weather advisory services, and a wide range of information retrieval services. Some CDPD networks support palm handheld devices.

Although CDPD cannot increase the number of channels usable in a cell, it can provide an overall increase in user capacity if data customers use CDPD instead of voice channels. This capacity increase results from the inherently greater efficiency of a connectionless packet data service relative to a connection-oriented service, given bursty data traffic. That is, a packet data service does not require the overhead associated with setup of a voice traffic channel in order to send one or a few data packets. In the following paragraphs we briefly describe the CDPD network architecture and the principles of operation of the system. Our discussion follows Ref. 12, closely.

The basic structure of a CDPD network (Fig. 6) is similar to that of the cellular network with which it shares transmission channels. Each mobile end system (M-ES) communicates with a mobile data base station (MDBS) using the protocols defined by the air-interface specification, to be described subsequently. The MDBSs are typically collocated with the cell equipment providing cellular telephone service to facilitate the channel-sharing procedures. All of the MDBSs in a service area are linked to a mobile data intermediate system (MD-IS) by microwave or wireline links. The MD-IS provides a function analogous to that of the mobile switching center (MSC) in a cellular telephone system. The MD-IS may be linked to other MD-ISs and to various services provided by ESs outside the CDPD network. The MD-IS also provides a connection to a network management system and supports protocols for network management access to the MDBSs and M-ESs in the network.

Service endpoints can be local to the MD-IS or remote, connected through external networks. A MD-IS can be connected to any external network supporting standard routing and data exchange protocols. A MD-IS can also provide connections to standard modems in the PSTN by way of appropriate modem interworking functions (modem emulators). Connections between MD-ISs allow routing of data to and from M-ESs that are roaming, i.e., operating in areas outside their home service areas. These connections also allow MD-ISs to exchange information

required for mobile terminal authentication, service authorization, and billing.

CDPD employs the same 30-kHz channelization as used in existing AMPS cellular systems throughout North America. Each 30-kHz CDPD channel supports channel transmission rates up to 19.2 Kbps. Degraded radio channel conditions, however, will limit the actual information payload throughput rate to lower levels, and will introduce additional time delay due to the error-detection and retransmission protocols.

The CDPD radio link physical layer uses GMSK modulation at the standard cellular carrier frequencies, on both forward (base-to-mobile) and reverse (mobile-to-base) links. The Gaussian pulse-shaping filter is specified to have bandwidth-time product $B_b T = 0.5$. The specified $B_b T$ product ensures a transmitted waveform with bandwidth narrow enough to meet adjacent-channel interference requirements, while keeping the intersymbol interference small enough to allow simple demodulation techniques. The choice of 19.2 Kbps as the channel bit rate yields an average power spectrum that satisfies the emission requirements for analog cellular systems and for dual-mode digital cellular systems.

The forward channel carries data packets transmitted by the MDBS, whereas the reverse channel carries packets transmitted by the M-ESs. In the forward channel, the MDBS forms data frames by adding standard high level data link control (HDLC) terminating flags and inserted 0 bits, and then segments each frame into blocks of 274 b. These 274 b, together with an 8-b *color code* for MDBS and MD-IS identification, are encoded into a 378-b coded block using a (63,47) Reed–Solomon code over a 64-ary

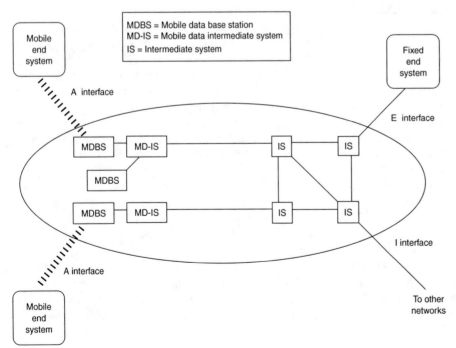

Fig. 6 Cellular digital packet data network architecture.

MDBS = Mobile data base station
MD-IS = Mobile data intermediate system
IS = Intermediate system

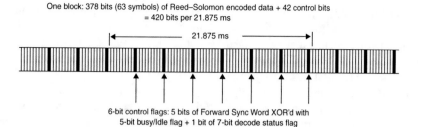

One block: 378 bits (63 symbols) of Reed–Solomon encoded data + 42 control bits
= 420 bits per 21.875 ms

21.875 ms

6-bit control flags: 5 bits of Forward Sync Word XOR'd with
5-bit busy/Idle flag + 1 bit of 7-bit decode status flag

Fig. 7 Cellular digital packet data forward link block structure.

alphabet. A 6-b synchronization and flag word is inserted after every nine code symbols. The flag words are used for reverse link access control. The forward link block structure is shown in Fig. 7.

In the reverse channel, when an M-ES has data frames to send, it formats the data with flags and inserted zeros in the same manner as in the forward link. That is, the reverse link frames are segmented and encoded into 378-b blocks using the same Reed–Solomon code as in the forward channel. The M-ES may form up to 64 encoded blocks for transmission in a single reverse channel transmission burst. During the transmission, a 7-b transmit continuity indicator is interleaved into each coded block and is set to all ones to indicate that more blocks follow, or all zeros to indicate that this is the last block of the burst. The reverse channel block structure is shown in Fig. 8.

The media access control (MAC) layer in the forward channel is relatively simple. The receiving M-ES removes the inserted zeros and HDLC flags and reassembles data frames that were segmented into multiple blocks. Frames are discarded if any of their constituent blocks are received with uncorrectable errors.

On the reverse channel (M-ES to MDBS), access control is more complex, since several M-ESs must share the channel. CDPD uses a multiple access technique called DSMA, which is closely related to the carrier sense multiple access/collision detection (CSMA/CD) access technique.

The network layer and higher layers of the CDPD protocol stack are based on standard ISO and Internet protocols. The CDPD specification stipulates that there be no changes to protocols above the network layer of the seven-layer ISO model, thus ensuring the compatibility of applications software used by CDPD subscribers.

The selection of a channel for CDPD service is accomplished by the radio resource management entity in the MDBS. Through the network management system, the MDBS is informed of the channels in its cell or sector that are available either as dedicated data channels or as potential CDPD channels when they are not being used for analog cellular service, depending on which channel allocation method is implemented. For the implementation in which CDPD service is to use "channels of opportunity," there are two ways in which the MDBS can determine whether the channels are in use. If a communication link is provided between the analog system and the CDPD system, the analog system can inform the CDPD system directly about channel usage. If such a link is not available, the CDPD system can use a forward power monitor ("sniffer" antenna) to detect channel usage on the analog system. Circuitry to implement this function can be built into the cell sector interface.

Another version of CDPD called circuit-switched CDPD (C-SCDPD) is designed to provide service to subscribers traveling in areas where the local cellular service provider has not implemented the CDPD service. With C-SCDPD, the subscriber establishes a standard analog cellular circuit connection to a prescribed number, and then transmits and receives CDPD packets over that circuit connection. The called number is a gateway that provides connection to the CDPD backbone packet network.

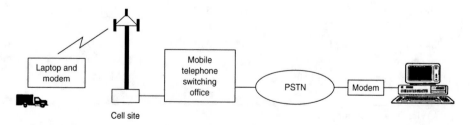

Fig. 8 Cellular digital packet data reverse link block structure.

Digital Cellular Data Services

In response to the rapid growth in demand for cellular telephone service throughout the US and Canada, the CTIA and the TIA have developed standards for new digital cellular systems to replace the existing analog cellular system (AMPS). Two air-interface standards have now been published. The IS-54 standard specifies a three-slot time division multiple access (TDMA) system, and the I-95 standard specifies a CDMA spread spectrum system. In both systems, a variety of data services are provided or planned.

Following the development of the IS-95 standard for CDMA voice service, the cellular industry has worked on defining various data services to operate in the same networks. The general approach taken in the definition of IS-95 data services has been to base the services on standard data protocols, to the greatest extent possible.[13] The previously-specified physical layer of the IS-95 protocol stack was adopted for the physical layer of the data services, with an appropriate radio link protocol (RLP) overlaid. The first CDMA data service to be defined was asynchronous ("start-stop" interface) data and Group-3 facsimile. This service provides for interoperability with many standard PSTN data modems as well as standard office fax machines. This service is in the category of *circuit-mode service*, since a circuit connection is first established, just as with a voice call, and the circuit connection is maintained until the user disconnects.

Following the standardization of the asynchronous data service, the industry defined a service that carries packet-formatted data over a CDMA circuit connection. It is important to note that this is not a true packet-data service over the radio link, since the full circuit connection is maintained regardless of how little packet data is transmitted. One potential application for this type of service is to provide subscribers with CDPD access from a CDMA network.

It is recognized that in order to make use more efficient, it will be highly desirable to provide a contention-based packet data service in CDMA cellular networks. This is currently a subject of study in CDMA data services standardization groups.

In parallel with the CDMA data services efforts, another TIA task group, TR45.3.2.5, has defined standards for digital data services for the TDMA digital cellular standard IS-54.[14,15] As with the IS-95 data services effort, initial priority was given to standardizing circuit-mode asynchronous data and Group-3 facsimile services.[16]

Table 1 Characteristics and parameters of five mobile data services.

System	ARDIS	MOBITEX	CDPD	IS-95[a]	TETRA[a]
Frequency band					
Base to mobile, (MHZ).	(800 band,45-kHz sep.)	935–940[b]	869–894	869–894	(400 and 900 bands)
Mobile to base, (MHZ).		896–901	824–849	824–894	
RF channel spacing	25 kHz (US)	12.5 kHz	30 kHz	1.25 MHz	25 kHz
Channel access/ multiuser access	FDMA/DSMA	FDMA/dynamic-R-S-ALOHA	FDMA/DSMA	FDMA/CDMA-SS	FDMA/DSMA and SAPR[c]
Modulation method	FSK, 4-FSK	GMSK	GMSK	4-PSK/DSSS	Π/4-QDPSK
Channel bit rate, (Kbps)	19.2	8.0	19.2	9.6	36
Packet length	Up to 256 bytes (HDLC)	Up to 512 bytes	24–928 b	(Packet service-TBD)	192 b (short) 384 b (long)
Open architecture	No	Yes	Yes	Yes	Yes
Private or public carrier	Private	Private	Public	Public	Public
Service coverage	Major metro. areas in US	Major metro. areas in US	All AMPS areas	All CDMA cellular areas	European trunked radio
Type of coverage	In-building and mobile	In-building and mobile	Mobile	Mobile	Mobile

[a]IS-95 and TETRA data services standardization in progress.
[b]Frequency allocation in the US In the U.K., 380–450 MHz band is used.
[c]Slotted-ALOHA packet reservation.

OTHER DEVELOPING STANDARDS

Terrestrial Trunked Radio

As has been the case in North America, there is interest in Europe in establishing fixed wide-area standards for mobile data communications. Whereas the Pan-European standard for digital cellular, termed GSM, will provide an array of data services, data will be handled as a circuit-switched service, consistent with the primary purpose of GSM as a voice service system. Therefore, the ETSI began developing a public standard in 1988 for trunked radio and mobile data systems, and this standardization process continues today. The standards, which are now known generically as terrestrial trunked radio (TETRA) (formerly Trans-European trunked radio), were made the responsibility of the ETSI RES-6 subtechnical committee.[17] In 1996, the TETRA standardization activity was elevated within RES-6 with the creation of project TETRA.

TETRA is being developed as a family of standards. One branch of the family is a set of radio and network interface standards for trunked voice (and data) services. The other branch is an air-interface standard optimized for wide-area packet data services for both fixed and mobile subscribers and supporting standard network access protocols. Both versions of the standard will use a common physical layer, based on $\pi/4$ differential quadrature phase shift keying ($\pi/4$-DQPSK) modulation operating at a channel rate of 36 Kbps in each 25-kHz channel. The composite data rate of 36 Kbps comprises four 9 Kbps user channels multiplexed in a TDMA format. The TETRA standard provides both connection-oriented and connectionless data services, as well as mixed voice and data services.

TETRA has been designed to operate in the frequency range from VHF (150 MHz) to UHF (900 MHz). The RF carrier spacing in TETRA is 25 kHz. In Europe, harmonized bands have been designated in the frequency range 380–400 MHz for public safety users. It is expected that commercial users will adopt the 410–430 MHz band. The Conference of European Posts and Telecommunications Administrations (CEPT) has made additional recommendations for use in the 450–470 MHz and 870–876/915–921 MHz frequency bands.

Table 1 compares the chief characteristics and parameters of the wireless data services described.

CONCLUSIONS

Mobile data radio systems have grown out of the success of the paging-service industry and the increasing customer demand for more advanced services. The growing use of portable, laptop, and palmtop computers and other data services will propel a steadily increasing demand for wireless data services. Today, mobile data services provide length-limited wireless connections with in-building penetration to portable users in metropolitan areas. The future direction is toward wider coverage, higher data rates, and capability for wireless Internet access.

REFERENCES

1. Bodson, D., McClure, G.F., McConoughey, S.R., Eds.; *Land-Mobile Communications Engineering*; (Selected Reprint Series): IEEE Press: New York, 1984.

2. Longley, A.G.; Rice, P.L. Prediction of tropospheric radio transmission over irregular terrain. A computer method— 1968, Environmental Sciences and Services Administration Tech. Rep. ERL 79-ITS 67. **1968**, U.S. Government Printing Office, Washington, DC.

3. Rice, P.L.; Longley, A.G.; Norton, K.A.; Barsis, A.P. Transmission loss predictions for tropospheric communication circuits. National Bureau of Standards, Tech. Note 101. **1967**, Boulder, CO.

4. Hata, M. Empirical formula for propagation loss in land-mobile radio services. IEEE Trans. Veh. Technol. **1980**, *29* (3), 317–325.

5. Okumura, Y.; Ohmori, E.; Kawano, T.; Fukuda, K. Field strength and its variability in VHF and UHF land-mobile service. Rev. Electron. Commun. Lab. **1968**, *16*, 825–873.

6. Pahlavan, K.; Levesque, A.H. *Wireless Information Networks*; J. Wiley & Sons: New York, 1995.

7. Khan, M.; Kilpatrick, J. MOBITEX and mobile data standards. IEEE Commun. Mag. **1995**, *33* (3), 96–101.

8. Kilpatrick, J.A. Update of RAM Mobile Data's packet data radio service. 42nd IEEE Vehicular Technology Conference (VTC 1992), Denver, CO, May 11–13, 1992; IEEE: New York, NY, 1992; 898–901.

9. Parsa, K. The MOBITEX packet-switched radio data system. Personal, Indoor and Mobile Radio Conference (PIMRC 1992), Boston, MA, Oct 19–21, 1992; IEEE: New York, NY, 1992; 534–538.

10. Beaulieu, M. *Wireless Internet Applications and Architecture*; Addison-Wesley: Boston, MA, 2002.

11. Cellular Digital Packet Data Industry Coordinator. Cellular Digital Packet Data Specification, Release 1.1, November 1994, Pub. Cellular Digital Packet Data; Kirkland, WA, 1994.

12. Quick, R.R., Jr.; Balachandran, K. Overview of the cellular packet data (CDPD) system. Personal, Indoor and Mobile Radio Conference (PIMRC 1993), Yokohama, Japan, Sept 9–11, 1993; IEEE: New York, NY, 1993; 338–343.

13. Tiedemann, E. Data services for the IS-95 CDMA standard. Personal, Indoor and Mobile Radio Conference (PIMRC 1993), Yokohama, Japan, Sept 9–11, 1993.

14. Sacuta, A. Data standards for cellular telecommunications—a service-based approach. 42nd IEEE Vehicular Technology Conference (VTC 1992), Denver, CO, May 11–13, 1992; IEEE: New York, NY, 1992; 263–266.

15. Weissman, D.; Levesque, A.H.; Dean, R.A. Interoperable wireless data. IEEE Commun. Mag. **1993**, *31* (2), 68–77.

16. Telecommunications Industry Association. Async data and fax, Project No. PN-3123, and Radio link protocol 1. Project No. PN-3306; Nov. 14. Issued by Telecommunication Industry Association, Washington, DC, 1994.

17. Haine, J.L.; Martin, P.M.; Goodings, R.L.A. A European standard for packet-mode mobile data. Personal, Indoor, and Mobile Radio Conference (PIMRC 1992), Boston, MA, Oct 19–21, 1992; IEEE: New York, NY, 1992.

BIBLIOGRAPHY

1. International Standards Organization (ISO). Protocol for providing the connectionless-mode network service, Pub. International Standards Organization 8473; 1987.

2. Pahlavan, K.; Levesque, A.H. Wireless data communications. IEEE. **1994**, *82* (9), 1398–1430.

FURTHER INFORMATION

Ref. 6 provides a comprehensive survey of the wireless data field as of mid-1994. Ref. 10 covers the wireless data field up to 2001, with emphasis on current and emerging applications for wireless data networking technology. The monthly journals *IEEE Communications Magazine and IEEE Personal Communications Magazine,* and the bimonthly journal *IEEE Transactions on Vehicular Technology* report advances in many areas of mobile communications, including wireless data. For subscription information contact: IEEE Service Center, 445 Hoes Lane, P.O. Box 1331, Piscataway, NJ 08855-1131. Phone (800) 678-IEEE.

Wireless Home Entertainment Center: Protocol Communications and Architecture

Marco Roccetti
Dipartimento di Scienze dell'Informazione, Università di Bologna, Bologna, Italy

C.E. Palazzi
Dipartimento di Scienze dell'Informazione, Università di Bologna, Bologna, Italy, and Department of Computer Science, University of California Los Angeles, Los Angeles, California, U.S.A.

S. Ferretti
Dipartimento di Scienze dell'Informazione, Università di Bologna, Bologna, Italy

G. Pau
Department of Computer Science, University of California Los Angeles, Los Angeles California, U.S.A.

Abstract

Wireless home entertainment center refers to a device able to handle heterogeneous media and connect client devices located within the house and the outside world (i.e., the Internet).

Technologies such as TiVo and Media Center have introduced the concept of DVR in millions of homes.[1,2] Consumers are now able to pause live-TV programs and watch them at their own convenience without the need of obsolete VCRs. These devices and their future evolutions can be grouped into the class of HECs. Specifically, a HEC can be defined as a hub for all in-home entertainment experiences, able to handle heterogeneous media and perform as a gateway between client devices located within the house and the outside world (i.e., the Internet).

Besides the DVR functionalities, a HEC can be considered as a *magic box* that provides several other services such as IPTV, web radio, game console, picture viewer, electronic program guide, CD/DVD/video player, music jukebox, web browser, e-mail handler, instant messenger etc. Contents related to these services can be locally available or distributed over the Internet to be dynamically retrieved based on the consumer's need with just one click of a button. Several streams are thus produced by these active services, all distributed by the HEC throughout the house.

At the same time, wireless home connectivity is becoming more and more popular, offering mobility, flexibility, and high transmission rates (e.g., 54 Mbps for the IEEE 802.11g and 100 Mbps for the IEEE 802.11n). We can hence assume that every HEC will be endowed with an access point (AP) in order to guarantee wireless connectivity to the various devices (e.g., screens, speakers, joypads).

Unfortunately, protocols run on APs have been designed based on the belief that the main application run by users on the wireless channel was browsing the Internet and, in general, downloading files. Therefore, buffers and local re-transmissions are extensively used, with the aim of providing reliability and high throughput to this kind of applications. On the other hand, these solutions have been demonstrated to be harmful toward real-time applications as they increase the per-packet delivery latency.[3] Simply stated, the current state of the art for in-home wireless connectivity systems does not allow to play on-line games or watch real-time stream videos when another application is simultaneously downloading a big file through the same AP.

We demonstrate here that these applications' co-existence can be achieved with a simple and smart modification to the AP that has a minimal impact on existing architecture and protocols. From this point of view, this technique resembles the journeys taken by hippies in the 1960s and 1970s, as key characteristics of hippie trails were those of traveling toward a destination as cheaply and respectfully with the environment as possible. Analogously, the proposed *smart AP* aims at reaching an efficient utilization and co-existence of the various entertainment applications run in a wireless home with a solution that costs a little and has a minimal impact on surrounding architecture and protocols.

Specifically, the AP and the associated HEC are in a strategic position that allows them to gather information

Encyclopedia of Wireless and Mobile Communications DOI: 10.1081/E-EWMC-120043481

Wireless
Channels—Internet

about the channel condition and the ongoing traffic. This information can be put to good use to regulate the transmission flow of downloading applications in order to produce a smooth traffic that utilize the channel efficiently but, at the same time, does not create queues at the AP. Furthermore, to ensure an easy deployability, this scheme exploits only existing features of standard protocols and a modified AP.

DIGITAL ENTERTAINMENT IN A HOME NETWORKING SCENARIO

The current market is heading toward a wireless house where all the devices (e.g., computers, televisions, intelligent fridges, etc.) are wirelessly connected to the home network and possibly controlled by the HEC.

In this context, take, e.g., a mid-class American household where a family of four people lives: two teenage kids and the hardworking parents. Each family member presumably owns several networked personal portable devices such as PDAs, MP3 players, game consoles and digital cameras, all these being also connected to the home network.

Based on the market trends, we also consider that all those devices are wirelessly connected through a Wi-Fi IEEE 802.11g link to a HEC that controls the in-house media distribution and provides access to the Internet as well as to the cable television and companies providing external services (e.g., the alarm company). We also assume that several family members will be accessing the household network at the same time according to their work or leisure needs. In particular, we can consider family scenario summarized by Fig. 1: (a) one teenager is watching the movie "Star Wars," streaming it from the close-by HEC; (b) the other one is playing with his latest on-line game against a crowd of buddies across the Internet; (c) the dad is having a conversation through an IP based video-chat; and (d) the mum is downloading the last U2 greatest hit compilation from the Apple iTunes music store. In the above everyday-life picture, it is worth noticing that each of the aforementioned employed applications features different requirements in terms of network performance, as well as suffers from very specific problems all due to the best effort nature of the Internet transport protocols.

These are as follows:

- **Video streaming:** Streaming applications are affected by the jitter phenomenon but are resilient to some packet loss; a network designed mainly for video streaming should minimize the jitter.
- **Video-chat and massive multiplayer on-line games:** Both these applications require a high degree of interactivity; they greatly suffer from delays and packet jitter although may tolerate some packet loss.
- **iTunes music download:** A music download activity is typically resilient to jitter and delays but decreases the sending rate in presence of packet losses: it hence does not tolerate any error losses (losses that are not generated because of congestion).

DOWNLOADING AND REAL-TIME FLOWS ON A WIRELESS HEC: A DIFFICULT CO-EXISTENCE

Applications can be grouped into two main classes depending on their performance metrics: *downloading* and *real-time*. The first one is concerned with transferring data in a reliable way. Even if data is generally chunked into packets before being transmitted, the performance of these applications are generally measured in terms of how much time is required to have the *whole* file transferred. Examples of this class of applications are file transfer (e.g., FTP, HTTP, SMTP) and remote control (e.g., Telnet). On the other hand, a second class of applications is concerned with ensuring a quick delivery of *every single* packet transmitted by the application. For these applications, performances are measured in terms of the percentage of packets that reach the destination within a certain time threshold. Interactive on-line games, real-time IPTV, video/audio chatting etc. represent typical instances of this class of applications.

Downloading and real-time applications can be distinguished also by the employed transport protocol: TCP or UDP. The former is a protocol that guarantees the reliable

Fig. 1   Digital entertainment delivery in a wireless home.

and ordered delivery of every packet sent; to this aim it establishes a session and performs re-transmissions of lost packets. Because of these features, TCP is utilized by downloading applications. Where not differently stated, with the term TCP we refer to the two most common versions, i.e., TCP New Reno and TCP SACK.

A very important component of TCP is represented by its congestion control functionality. Through it, every TCP flow probes the link with higher and higher data rates eventually filling up the channel. At that point, packets will be queued at the buffer associated with the bottleneck of the link until it overflows causing packet losses. TCP re-transmits the lost packets, and halves its sending rate to reduce the congestion level. Finally, the regular increase of the sending rate is re-established and so forth.

UDP is simpler: packets are immediately sent toward the receiver with a data rate decided by the sender. UDP does not guarantee reliable and ordered delivery of packets but, at the same time, its small overhead and lack of re-transmissions make it less prone to generate delays in the packets' delivery. For this reason, UDP is usually employed by real-time applications.

The lack of congestion control functionalities of UDP had lead the scientific community to wisely consider UDP as unfair toward TCP. Indeed, citing from Ref. 4: *"Although commonly done today, running multimedia applications over UDP is controversial to say the least. [...] the lack of congestion control in UDP can result in high loss rates between a UDP sender and receiver, and the crowding out of TCP sessions—a potentially serious problem."*

However, even if this is true when the available bandwidth is very scarce, the larger and larger bandwidth offered today to home consumers overturns this situation. Indeed, with this broadband connectivity, the traffic generated by UDP-based applications can be accommodated; yet, a problem emerges when real-time applications (UDP-based) co-exist with downloading ones (TCP-based) on a wireless channel, causing the former to experience a scattered flow progression.[5]

Major causes for this problem can be found in the TCP's congestion control functionality. TCP continuously probes for higher transfer rates, eventually queuing packets on the buffer associated with the bottleneck of the connection. Since the same wireless connection might be shared by several devices and applications, it is even more evident how the congestion level and queue lengths can increase, thus delaying the delivery of packets stuck in queue and jeopardizing requirements of real-time applications.

This negative situation is further worsened by the following three factors due to the wireless nature of the link. First, the wireless medium allows the transmission of only one packet at a time and is not full-duplex as wired links. Packets have hence to wait their turns to be transmitted. Second, as interference, errors, fading, and mobility may cause packet loss, the IEEE 802.11 medium access control

(MAC) layer reacts through local re-transmissions (4 at most,[6]) which, in turn, cause subsequent packets to wait in queue until the preceding ones or their re-transmissions eventually reach the receiver. Last but not the least, the back-off mechanism of the IEEE 802.11 introduces an increasing amount of time before attempting a transmission again.[6]

As an example of problems caused by this mixture of causes, we show in Fig. 2 the jitter experienced by on-line game packets in a realistic wireless in-home environment, with an AP configured in an off-the-shelf fashion and a constant inter-departure time of 50 ms. In the considered scenario, the on-line game application (UDP-based) is started at 45 sec, when a video-stream application (UDP-based) was already active, and both last for the whole experiment. At 90 sec, a video-chat conversation (UDP-based) is added but, still, the traffic generated by the combination of these three applications is far from consuming all the available bandwidth. Instead, at 135 sec an FTP application (TCP-based) starts to download a file and quickly saturates the channel.

In Fig. 2 it is shown that the jitter experienced by on-line game packets remains regular until the FTP/TCP flow takes action. At that point, a multitude of packets start to experience high delays that cause a scattered progression of the on-line game flow.

What technically happens is even more evident in Fig. 3 that reports the *congestion/sending window* (swnd) (We use the term "congestion/sending window" to indicate when the sending window exactly corresponds to the congestion window.) and *slow start threshold* (ssth) of the TCP flow, as well as the pipe size, i.e., the Bandwidth (BW)-round-trip time (RTT) product (RTT × BW) of the channel. By overlapping Figs. 2 and 3 we notice that the irregularity in the inter-arrival time of on-line game packets is directly proportional to the size of the congestion/sending window (i.e., the transfer rate). Every time the

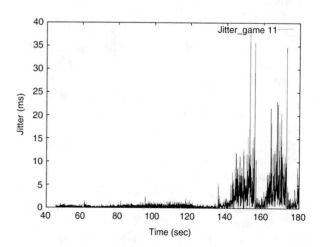

Fig. 2 Measured on-line game jitter; from 135 sec, a regular TCP flow is competing for the channel.

Wireless Channels—Internet

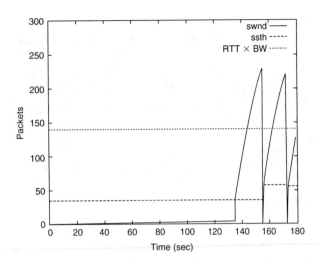

Fig. 3 Measured TCP congestion/sending window with a regular TCP flow.

congestion/sending window exceeds the pipe size, packets are queued at the buffer of the bottleneck link. Delay increments of on-line game packets can hit also tens of milliseconds thus representing a huge waste of time when trying to deliver real-time information for entertainment services. For instance, transmission delays of interactive on-line games should be less than 100 ms, with a maximum endurable value of 150 ms.[7]

RELATED WORK

In the attempt to protect real-time applications from queuing delays caused by the co-existence with downloading ones, the scientific community has designed the IEEE 802.11e protocol,[3] which is a particular version of the well known IEEE 802.11b/g. The IEEE 802.11e is able to assign different priorities to the various flows traversing the AP in order to privilege real-time applications with respect to downloading ones. Unfortunately, this protocol requires the source to mark each packet with a priority level value that can then be used to discriminate among flows. Obviously, this poses serious obstacles to its effective deployment since current applications do not have this functionality. Even if in future all the new applications would be developed with the ability to mark packets, still problems will arise if some designer chooses the wrong priority level for her/his novel application or whenever a user will utilize an "old" application.

Another possible solution is represented by the utilization of TCP Vegas in place of the legacy TCP to download files.[8] Indeed, this protocol embodies one of the most cited alternatives to regular TCP in scientific papers and its applicability to the considered problem lies in the fact that TCP Vegas tries to avoid congestion before it happens. In particular, TCP Vegas augments its congestion window

until buffers along the path between sender and receiver have a low utilization, whereas it reduces its congestion window when queuing is sensed. Therefore, TCP Vegas perfectly fits real-time applications' need for low buffer utilization. Nonetheless, even if TCP Vegas has been proven to fairly share the channel with other TCP Vegas flows, it behaves too conservatively in presence of simultaneous regular TCP flows. If a legacy TCP and a TCP Vegas flows are sharing the same bottleneck, the former fully exploits the available buffer while the latter interprets the consequent RTT trend as an indicator of excessive congestion, thus progressively reducing its sending rate to very low values.[9] In essence, the dramatic efficiency decrease experienced when competing with regular TCP traffic impedes TCP Vegas' factual deployment.

Finally, curtailing the large TCP advertised windows to maximize performance (CLAMP) is a protocol that achieves a fair and efficient share of a wireless channel through a distributed algorithm.[10] In particular, the AP and the various wireless nodes cooperate to have transmission flows generated/received by the various nodes limited by the value of their respective fair share. As a result, the total traffic never exceeds the wireless channel capacity thus avoiding queuing and relative delays in packet delivery. Unfortunately, fundamental condition for CLAMP to work is that the AP and *all* the connected nodes have to be modified to incorporate the algorithm. As it is clear, this strict requirement makes CLAMP not factually deployable on a large scale as we cannot assume that in a network there will be no device implementing standard protocols.

LIMITING THE ADVERTISED WINDOW TO AVOID QUEUING DELAYS

Aiming at finding the best solution to the tradeoff relationship existing between TCP throughput and real-time application delays, the two types of traffic should be able to co-exist without affecting each other and the employed solution should be easily and factually deployable.

Starting from the last point, i.e., deployability, it is evident how a technique that would exploit existing features of the already utilized protocols could be easily implemented in a real scenario. A possible solution could hence be that of utilizing the advertised window to limit the TCP flow's sending rate and hence the consumed bandwidth.

As it is well known, in fact, the actual sending rate of a TCP flow depends on its current sending window; this value is determined as the minimum between the congestion window (continuously recomputed by the sender) and the advertised window (provided by the receiver via returning acknowledgment code (ACK) packets).[11] It is hence evident how the advertised window perfectly embodies a natural upper bound for the sending rate of TCP flows.

Limiting the maximum sending rate of a TCP connection may greatly improve the performance of the HEC.[12,13] In practice, an optimal tradeoff between the needs for high throughput and low delays could be achieved by maintaining the sending rate of the TCP flows high enough to efficiently utilize the available bandwidth but, at the same time, limited in its growth so as not to utilize buffers. In this way, the throughput is maximized by the absence of packet losses which would halve the congestion window, while the delay is minimized by the absence of queues.

To better understand how limiting the sending window could guarantee the same or even a higher throughput with respect to utilizing regular TCP, we show in Fig. 4 a typical saw-tooth shaped sending window of a regular TCP and overlap it with one limited by the advertised window; these two lines also correspond to the data sending rate. The limited window is more stable than the regular one and, if appropriately chosen, can guarantee the same final throughput. However, as with the limited window the total traffic never exceeds the pipe size, packets will not be queued at the bottleneck.

To put this into practice, we need to address two important issues: how to determine the appropriate upper bound for the sending window and how to factually apply it.

Regarding the first point, the most appropriate formula can be derived from the two main goals we want to achieve: full utilization of the available bandwidth and no queuing delays. Real-time traffic generally exploits UDP and this transport protocol has no congestion control mechanism. Some smart UDP-based applications, however, implement congestion control at the application layer.[14] In any case, to avoid queuing delays, the aggregate bandwidth utilized by TCP flows cannot exceed the total capacity of the bottleneck link diminished by the portion of the channel occupied by the concurrent (UDP-based) real-time traffic.

In essence, the maximum sending rate for each TCP flow at time t, namely maxTCPrate(t), is represented by:

$$\text{maxTCPrate}(t) = \frac{(C - \text{UDPtraffic}(t))}{\#\text{TCPflows}(t)} \tag{1}$$

where UDPtraffic(t) corresponds to the amount of bandwidth occupied by UDP-based traffic at time t, #TCPflows (t) is the concurrent number of TCP flows, and C represents the capacity of the bottleneck link.

The second issue that we need to address is how to factually employ this formula in order to have it working in a real scenario. This means 1) identifying the location for its implementation, and 2) proposing a method to compute the value of the various variables.

Regarding (1), the advertised window is generally imposed by the receiver; however, this could not represent the most suitable place to set it. Determining the most appropriate value for the advertised window requires a comprehensive knowledge about all the flows that are transiting through the bottleneck. Since all flows have to pass through the AP, this represents the most appropriate node on which to implement our scheme. Indeed, the AP is integrated with the HEC and the mechanism can take advantage of this, to retrieve all the necessary information.

Focusing on (2), in any commercial operating system it is possible to know which kind of connection is in use and which its nominal speed is just by looking at the status of the network interface. Through snooping the channel or exploiting information known at the HEC we can also infer the number of active TCP connections and the aggregate amount of current UDP traffic. The AP can hence easily compute the best maxTCPrate(t) utilizing Eq. 1 and accordingly modify the advertised window included in the transiting ACKs.

We refer to this scheme as Smart Access Point with Limited Advertised Window (SAP-LAW).

SIMULATION ASSESSMENT

To analyze the wireless home, the NS-2 network simulator (version ns-2.28) was utilized with MAC layer parameters set so as to simulate the IEEE 802.11g. Moreover, the wireless environment was simulated through the Shadowing Model as it represents the most realistic wireless model available in NS-2. Following directions provided by the official NS-2 manual, a home environment partitioned into several rooms was represented through the *path loss exponent* and the *shadowing deviation* parameters equal to 4 and 9, respectively. The signal attenuation grows with the increase of these two parameters; therefore, by setting the shadowing deviation value at the highest suggested by the NS-2 manual for an indoor environment (i.e., 9), transmission will suffer from more packet losses.

The HEC is positioned at the center of the house and endowed with an AP to deliver entertainment to four mobile nodes. Devices inside the house are connected with servers, reachable through the Internet and the bottleneck is situated corresponding to the last hop (i.e., the wireless connection at home).

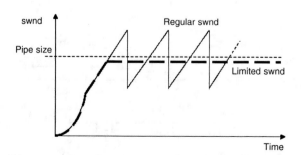

Fig. 4 Comparison between regular and limited swnd.

As for the simulated traffic, a video-stream delivery immediately starts from the HEC to a wirelessly connected screen. Then, at subsequent intervals of 45 sec other three applications are run, specifically: an on-line game session, a video-chat conversation, and a persistent download.

Real trace files were used for the movie stream and for the video-chat; these trace files determine packet size and rate for high quality MPEG-4 Star Wars IV (for the movie stream) and for two VBR H.263 Lecture Room-Cam (for the video-chat).[15] Moreover, parameters characterizing the game-generated traffic were inspired by directions provided in scientific literature. Assuming that the player in the house is engaged in the popular first person shooter game Quake Counter Strike with other 25 players through the Internet, game-traffic parameters were set as inspired by real measurements reported in Ref. 16.

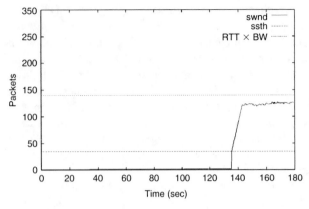

Fig. 6 Measured congestion window and advertised window of a TCP flow; the AP employs SAP-LAW with $C = 18$ Mbps.

EXPERIMENTAL RESULTS

In order to implement SAP-LAW, the simulated scenario was enhanced by enabling the AP to modify the advertised window (included in returning ACKs) according to Eq. 1. In particular, the average UDP-based aggregate traffic was computed through a simple low-pass filter and the new advertised window was determined every 200 ms.

When employing SAP-LAW, the AP is able to keep track of the concurrent real-time traffic and determine the most appropriate advertised window; as a result, the jitter experienced by on-line game packets is kept low. See, for instance, Fig. 5 where C was set equal to 18 Mbps, i.e., 90% of the maximum achievable bandwidth and the maximum jitter value was, in this way, bounded under 10 ms.

At the same time, SAP-LAW guarantees a high TCP throughput by continuously maintaining an appropriately high value of the sending window. To this aim, Fig. 6 shows how the advertised window, exploited by the TCP,

limits the sending window, thus eliminating the deleterious saw-tooth shape and guaranteeing a high but also smooth traffic that does not create queues at the bottleneck.

Obviously, results obtainable by employing SAP-LAW depends on the chosen C value in Eq. 1. For a deeper comprehension, various values for the parameter C have been tested and results are reported in Figs. 7 and 8. In particular. Fig. 7 shows the average, the standard deviation, and the maximum value for the jitter experienced by the game flow directed from the server to the client, while Fig. 8 presents the throughput trend of the concurrent TCP-based flow.

As clearly shown, both the average and the standard deviation of the on-line game flow increase with C; this is coherent with the fact that higher C values decrease the resilience of the scheme to TCP bursts thus leading to some queuing at the AP. Moreover, while the average results are very low for all C values, the standard deviation sensibly increases with higher values of C thus indicating the presence of many peaks of very high delay in the

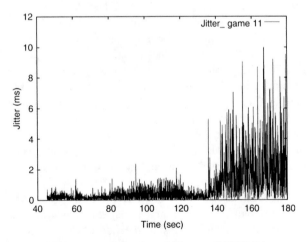

Fig. 5 On-line game jitter; from 135 sec, a TCP flow is competing for the channel; the AP employs SAP-LAW with $C = 18$ Mbps.

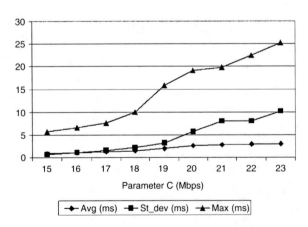

Fig. 7 Jitter statistics of the game flow when employing SAP-LAW.

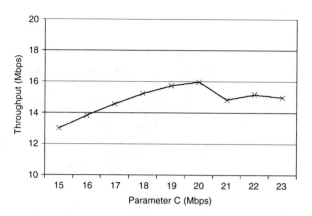

Fig. 8 Throughput achieved by the TCP flow when employing SAP-LAW.

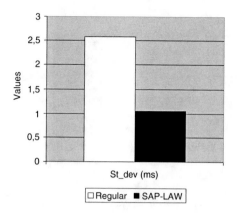

Fig. 10 Standard deviation of the jitter experienced by on-line game packets: comparison between regular scheme and SAP-LAW (with $C = 18$ Mbps).

packet delivery. This is confirmed also by the maximum delay values experienced by packets.

Fig. 8 demonstrates how the throughput decreases when C is set too low. Instead, if C is set higher than the maximum achievable throughput on the channel (in this case, 20 Mbps effectively available), then the sender will be allowed to send more packets than those bearable by the bottleneck link causing queuing delays. Thus, it happens that some packets may overflow the buffer and consequent losses cause the reduction of the sending window and average throughput.

Figs. 9–11 summarize statistical results obtained by comparing regular protocols (Regular) and SAP-LAW. In particular, considered statistical parameters are the average (Fig. 9), the standard deviation (Fig. 10), and the maximum value of the jitter experienced by on-line game packets (Fig. 11). Needless to say, the lower these values, the better the performance. Results obtained by the other real-time applications running in the simulated scenario (i.e., video-stream and video-chat) are coherent with the showed ones and need no further explanation; we hence skip presenting their

outcomes. Rather, we show in Fig. 12 the average throughput achieved by the concurrent (TCP-based) FTP connection. In this case, higher value gives better performance.

As it is evident, employing SAP-LAW conspicuously improves performance both in terms of lowest per-packet delay and achieved throughput. Moreover, SAP-LAW can be easily implemented as it only requires the presence of slightly "smarter" APs. The modifications to the AP are very limited, thus minimally impacting on their cost and, at the same time, SAP-LAW can perfectly co-exist with the current Internet and its employed protocols. Considering this and the remarkable results achieved, SAP-LAW is definitely a candidate for enhancing computer-centered home entertainment in a wireless scenario.

SUMMARY

Considering a scenario where in-home entertainment is delivered to wireless devices through a HEC, even a single

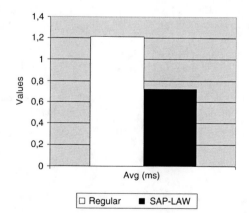

Fig. 9 Average of the jitter experienced by on-line game packets: comparison between regular scheme and SAP-LAW (with $C = 18$ Mbps).

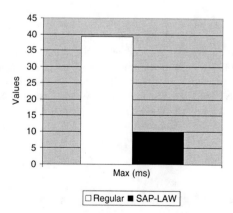

Fig. 11 Maximum jitter experienced by on-line game packets: comparison between regular scheme and SAP-LAW (with $C = 18$ Mbps).

Wireless Channels—Internet

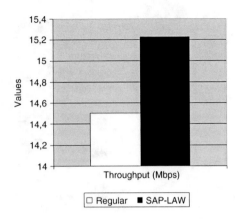

Fig. 12 Total throughput achieved by the TCP flow run simultaneously with the real-time applications: comparison between regular scheme and SAP-LAW (with $C = 18$ Mbps).

downloading flow can conspicuously increase the queuing delay suffered by concurrent real-time applications. This constitutes the reverse of the well known argument by which UDP's lack of congestion control would harm TCP; whereas even the TCP's lack of buffering control is harmful toward UDP-based applications.

To solve this problem, SAP-LAW utilizes an enhanced AP that does not need to modify existing Internet's protocols. Results showed that SAP-LAW is able to consistently ameliorate the global performance of computer-centered home entertainment services. Moreover, different from other possible solutions (i.e., IEEE 802.11e, TCP Vegas, CLAMP), SAP-LAW is fully compatible with the Internet and requires only the plugging-in of an enhanced AP with no protocol modifications at the Internet side. It hence emerges as the optimal candidate for enhancing computer-centered home entertainment in a wireless scenario.

REFERENCES

1. The TiVo Homepage, http://www.tivo.com/.
2. Windows XP Media Center Edition 2005 Home Page, http://www.microsoft.com/windowsxp/mediacenter/.
3. IEEE Standard for Information Technology. Telecommunications and Information Exchange between Systems—Local and Metropolitan Area Networks—Specific Requirements Part 11: Wireless LAN Medium Access Control (MAC) and Physical Layer (PHY) Specifications: Amendment: Medium Access Control (MAC) Quality of Service Enhancements, P802.11e/D13.0, Jan 2005.
4. Kurose, J.F.; Ross, K.W. *Computer Networking: A Top-Down Approach Featuring the Internet*; Addison Wesley Longman: Boston, MA, 2001.
5. Palazzi, C.E.; Pau, G.; Roccetti, M.; Gerla, M. In-home on-line entertainment: analyzing the impact of the wireless MAC-transport protocols interference. IEEE 2005 International Conference on Wireless Networks, Communications and Mobile Computing (WIRELESSCOM 2005), Maui, HI, June 13–16, 2005; IEEE Press: 2005.
6. IEEE. Standard for Wireless LAN Medium Access Control (MAC) and Physical Layer (PHY) Specifications, Specifications, ISO/IEC 8802-11:1999(E), 1999.
7. Pantel, L.; Wolf, L.C. On the impact of delay on real-time multiplayer games. 12th International Workshop on Network and Operating Systems Support for Digital Audio and Video, Miami, FL, May 12–14, 2002; ACM Press: 2002, 23–29.
8. Low, S.; Peterson, L.; Wang, L. Understanding vegas: a duality model. J. ACM. **2002**, *49* (2), 207–235.
9. Marfia, G.; Palazzi, C.E.; Pau, G.; Gerla, M.; Sanadidi, M.Y.; Roccetti, Roccetti, M. TCP Libra: Exploring RTT-Fairness for TCP. UCLA CSD Technical Report #TR050037, 2005.
10. Andrew, L.L.H.; Hanly, S.V.; Mukhtar, R.G. CLAMP: active queue management at wireless access points. 11th European Wireless Conference 2005, Nicosia, Cyprus, Apr 10–13, 2005 VDE Verlag: 2005.
11. Stevens, W.R. The Protocols. In *TCP/IP Illustrated*; Addison Wesley, 1994; Vol. 1.
12. Palazzi, C.E.; Pau, G.; Roccetti, M.; Ferretti, S.; Gerla, M. Wireless home entertainment center: reducing last hop delays for real-time applications. ACM SIGCHI International Conference on Advances in Computer Entertainment Technology (ACE 2006), Hollywood, CA, June 14–16, 2006; ACM Press: 2006.
13. Palazzi, C.E.; Ferretti, S.; Roccetti, M.; Pau, G.; Gerla, M. What's in that magic box? The home entertainment center's special protocol potion, revealed. IEEE Trans. Consum. Electron, IEEE Consumer Electronics Society. **2006**, *52* (4), 1280–1288.
14. Balk, A.; Gerla, M.; Sanadidi, M.; Maggiorini, D. Adaptive video streaming: pre-encoded MPEG-4 with bandwidth scaling. Comput. Netw. Int. J. Comput. Telecommun. Netw. **2004**, *44* (4), 415–439.
15. Movie trace files, http://www-tkn.ee.tu-berlin.de/research/trace/ltvt.html.
16. Färber, J. Traffic modelling for fast action network games. Multimedia Tools Appl. **2004**, *23* (1), 31–46.

Wireless Internet: Applications

Daniel Schall
Christoph Dorn
Schahram Dustdar
Distributed Systems Group, Institute of Information Systems, Vienna University of Technology, Vienna, Austria

Abstract

Web services (WS)-based wireless Internet applications enable mobile users to provide or access services, which may be hosted on a fixed or mobile network infrastructure, seamlessly using a standard XML based protocol.

Web services (WS) provide the ability to design new exciting Internet applications and innovative mobile applications. WS are a family of XML-based standards designed for the communication of loosely coupled, dynamically bound applications. They can be seen as the evolution of the web where not only humans interact with applications via HTML forms, but also applications interact directly with one another. Alternatively, WS can be seen as a generalization of distributed object middleware such as common object request broker architecture (CORBA).[1] In whichever way one considers WS, there is a strong trend toward the wide adoption of the technology as means for addressing inter-operability issues.

WS are not only a recent technological hype, but are also defining a new way for designing applications and information systems. They are the most known incarnation of the programming paradigm known as service-oriented computing,[2] which has the goal of enabling service-oriented architectures (SOA, e.g., see Ref. 3). However, inter-operability comes at a price. Most notably, the use of XML-based protocols for the communication is computationally expensive and bandwidth consuming. Furthermore, using XML does not solve per se the ontological problem of having independent applications inter-operate. Nonetheless, WS are ready to be used on mobile devices to design new kind of applications, and to address inter-operability issues.[4]

Considering a complex setting of mobile devices and embedded systems which communicate in peer-to-peer fashion, one has to consider a number of issues such as heterogeneity and inter-operability, determined by the device's functional properties (e.g., hardware capabilities made available through a descriptive interface), and on the other hand, continuously changing operational parameters such as CPU utilization, memory usage, etc. which in turn limit the ability to contribute to a task at hand. A task may require composition of a set of services that are provided by mobile devices to aggregate functionality and to meet task-specific requirements. The challenge is to find a feasible device configuration that satisfies task-specific needs, and the execution of tasks in a satisfactory fashion considering task deadlines and quality of service (QoS) agreements.[5,6]

WS provide an ideal framework to address issues such as heterogeneity and inter-operability. However, the prime concern is to understand performance limits and constraints, in terms of resource requirements, imposed by various WS toolkits, to estimate expected performance at run-time.

We discuss WS as means for enabling innovative wireless Internet applications such as sharing of content on mobile devices and mobile collaboration scenarios. We propose to use WS as an enabling infrastructure for the inter-operation of heterogeneous mobile devices.

WIRELESS APPLICATIONS AND SERVICES

There are many scenarios in which heterogeneous mobile devices need to inter-operate. For instance, application scenarios in the mobility domain and in the home domain, where we find an increasing number of embedded devices with network capabilities.

Wireless networks with varying coverage, including 3G, WiFi, and Bluetooth provide data access on the move. In essence, the type of mobile application can be categorized by the data traffic. According to UMTS QoS traffic classes, we distinguish between 1) conversational, 2) streaming, 3) interactive, and 4) background classes.[7]

Encyclopedia of Wireless and Mobile Communications DOI: 10.1081/E-EWMC-120043505

In the following sections, we highlight typical applications and services that can be found in aforementioned traffic classes. The distinct characteristic is real-time versus non-real-time data traffic.

Real-time Multimedia

Multimedia content such as real-time audio and video fall into the conversational or streaming class. The session initiation protocol (SIP) (see Additional Reading 4) and the real-time transmission protocol (RTP) (see Additional Reading 3) provide the negotiation/transport framework to deliver delay sensitive A/V multimedia content. Various architectures, for instance situated devices and resources,[8] can be used to overcome limits of mobile devices (e.g., display size, computational power, etc.) and deliver media to mobile users in an optimized fashion.

Interactive mobile applications

In contrast to real-time multimedia applications, WS-based applications follow the request/response pattern and thus, are set up in the interactive or best effort class. In general, we categorize mobile applications, based on the interactive traffic class, in information **consumer** and **provider**. The former includes information access through search queries, document download, etc. performed by mobile users; whereas the latter refers to content or data that is provided or shared on mobile devices, for instance, captured multimedia content such as still images or audio files.

Location-based services are among the first mobile applications that have been developed by the ubiquitous computing community.[9] Application scenarios include localized queries,[10] as an example, finding restaurants in proximity, or providing information pro-actively to the mobile user through a location-based push mechanism.[11] This information can be customized based on user's position determined by GPS coordinates or GSM cell Id information. For example, the Google WS application program interface (API) (detailed information can be found under Additional Reading 2) provides access to the Google search engine from mobile devices and thus, can be used to obtain localized information, if combined with a form of localized search.

Location information is only a sub-set of the user's context. Various architectures and systems have been proposed to collect and aggregate sensory data and infer high-level contextual information (a survey on context-aware systems is given in Ref. 12).

NETWORKED DEVICES AND APPLIANCES AT HOME

Domotics is the field where housing meets technology in its various forms (informatics, but also robotics,

mechanics, ergonomics, and communication) to provide better homes from the safety and comfort point-of-view. Traditionally, domotic solutions were provided by a single vendor, using proprietary solutions for communication, mostly closed and expensive. This is no longer true. Domotic elements are heterogeneous in all aspects. Devices come from various vendors, have different hardware capabilities, network interfaces, and operating systems, and yet need to have the ability to inter-operate. Users need to have a unique view on all hardware elements and devices located at their homes. In addition, various tasks such as self-configuration should be done automatically, through internal communication and coordination, without requiring human intervention or manual configuration.

To solve the complexity of the home-distributed system and to allow cooperation among the set of independent devices, it is necessary to have standard protocols and ways of communicating among them. WS are an enabling technology for this task, as proposed in Ref. 4. The solution to the inter-operability comes at the cost of computational resources for handling XML messages, though, the growth in computational power, communication abilities, battery life of the devices, together with the lowering of the prices makes the approach evermore feasible.

CONTEXT-AWARE MOBILE COLLABORATION

A team is a group of people working collaboratively on tasks and activities. In mobile scenarios, individual team members are geographically distributed, collaborating with one another using mobile devices. In real life, a person may be a member of multiple teams at the same time. In fact, a person may work on more than one task or activity simultaneously. This means that users need to handle a number of activities, switching back and forth between activity contexts and gather all information relevant for a task at hand. Context-aware collaboration enables mobile teams to collaborate effectively by taking mobility aspects and hardware limitations into account. Context information is used to establish team awareness and to filter relevant information considering the user's or team's performed activities. WS on mobile devices provide means to discover and interact with peers seamlessly and an effective way to share and exchange context information.[13] Fig. 1 shows such a case where distributed team members utilize WS to setup a meeting.

The overall goal is to schedule a meeting among participants. Data that helps the meeting scheduling algorithm to determine suitable time and location may be saved in calendars (e.g., calendar WS 1 and 2) as well as on end user devices such as laptops or smartphones. Let us walk through the activity steps as depicted in Fig. 1. Suppose the **initiator** uses a smartphone to invoke the meeting

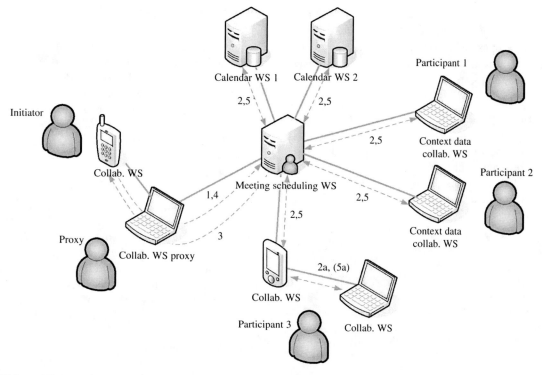

Fig. 1 WS in mobile collaboration.

scheduling WS. In Fig. 1, we show the concept of a **proxy**. A proxy is an entity that interacts with the meeting scheduling WS *on behalf* of the initiator. The initial request 1 is sent through the proxy to the meeting scheduling WS, which in turn interacts with calendar WS 1, calendar WS 2, collaboration WS on participant 1's and participant 2's laptop, and the collaboration WS on participant 3's smartphone—illustrated by request 2. Participant 1 and participant 2 share context information or data that is saved on their laptops through collaboration WS. In this step, we assume that the meeting scheduling WS knows how to find these WS and which calendar WS and which collaboration WS (located on potential participants devices) to consume. Step 2a shows that participant 3 has two devices at hand, smartphone and laptop, which communicate in peer-to-peer style to synchronize information such as local calendar items. This peer-to-peer link can be established on demand (dynamic "composition") or upon request (consuming the WS on participant 3's PDA). The scheduling algorithm retrieves all data and computes an optimal meeting schedule which will then be returned to the initiator (step 3 via the proxy). Step 4 is the conformation that the meeting should be scheduled and in step 5 notifications are sent to the calendar WS and to all participants (participants 1–3). Step 5a (optional) denotes peer-to-peer notification or synchronization between participant 3's PDA and laptop.

In the following section, we will further elaborate on concepts such as WS (mobile) proxies and WS that can be deployed on mobile devices suitable for collaboration scenarios such as meeting scheduling.

SERVICE INTERACTIONS

In this section, we discuss various service interaction scenarios that can be found in mobile Internet applications. Fig. 2 illustrates on the left side the mobile device that can be either a WS client or a WS provider and, what we define as "Stationary Node," on the right side. The arrows call/return depict a request/response pattern.

Mobile Client

Fig. 2A depicts the case where the WS client is located on the mobile device. The web service (WS provider) is

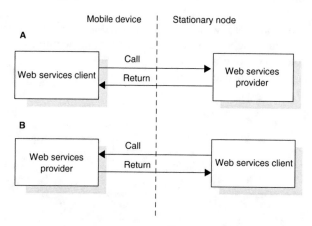

Fig. 2 Direct interaction between web services client and provider. (A) Web services client located on the mobile device. (B) Web service hosted on a mobile device.

located on a fixed network infrastructure (e.g., Internet or Intranet). Requests or WS calls can be made in synchronous (blocking) or asynchronous fashion. Asynchronous service invocation is achieved by spawning a new thread for each blocking request.

Mobile Server

In Fig. 2B, a web service is hosted on a mobile device. The web service consumer or client may be a stationary node (as shown in Fig. 2B) or may also be located on mobile device. To illustrate an application example, a set of WS hosted by mobile entities or embedded devices could be aggregated or federated to provide a more complex service. A sensor network platform could be equipped with WS to deliver real-time data from the field to service consumers. Moreover, a web service deployed on mobile devices can be used to register for and receive asynchronous notifications (e.g., through a publish/subscribe mechanism). Context-dependent information (e.g., by utilizing location and other context information) can be pushed to mobile entities.

Mobile Proxy

Fig. 3 shows a middleware approach where a WS mobile proxy acts as an intermediate entity between a WS client and a set of WS providers. A considerable number of WS exists offering only a complex interface [WS description language application program interface (WSDL API)], thus preventing entities such as mobile devices with limited hardware capabilities from consuming these services.

A mobile proxy provides a "view" on such services by offering a simplified API or WSDL interface which is specifically tailored to the requirements of mobile devices. In addition to providing a sub-set of a single service provider's features, the proxy is able to aggregate (and abstract) a number of services depending on the mobile users' needs. A mobile proxy may be deployed on a "Stationary Node" or on powerful mobile computers such as laptops or PDAs.

A mobile proxy can also be employed in an interaction scenario as depicted in Fig. 2B, thereby functioning as a protection proxy and preventing a mobile WS provider from being flooded with a large number of requests.

WS ON MOBILE DEVICES

Implementing the web service stack on mobile devices is reality today. A number of WS toolkits exist for mobile devices in C++ as well as Java/Java 2 Platform Micro Edition (J2ME) (e.g., Symbian-based devices), and furthermore .NET implementations on Microsoft platforms. As mobile devices have constraint resources, the choice of the toolkit is crucial for the application's performance.

C++ Toolkits

gSOAP is a platform-independent toolkit for WS,[14] which includes a WSDL parser *wsdl2h* (creates header files) and a stub/skeleton compiler *soapcpp2*. Depending on the SOAP client/server requirements, C/C++ files can be generated. The run-time library, *stdsoap2*, serializes and de-serializes calls and is the only dependency needed on the target platform. Fig. 4 shows the gSOAP (client) run-time and shows the development and deployment cycle.

The development process starts with C/C++ header file creation based on the service's WSDL file. Next, the gSOAP compiler is used to create the code files. At run-time (i.e., the deployment), remote procedure calls (RPCs) are made on client side proxies.

The gSOAP toolkit is suitable for development of WS clients and providers on various platforms including platforms such as Symbian. In addition, the Symbian platform provides a WS framework (WSF) API that allows the developer to implement WS clients (see Additional Reading 7 for details). C++ code can be automatically generated from WSDL files (WSDL-to-C++). A wizard takes a WSDL file as input and generates matching C++ proxy code. The code provides a C++ method call for each web service described in the WSDL file.

Toolkits for J2ME

The J2ME platform is a set of standard Java APIs defined through the Java community process (JCP). The J2ME specifications define the connected device configuration (CDC) (a sub-set of J2SE) and the connected limited device configuration (CLDC). In contrast to CDC, CLDC provides libraries such as the *Connection Framework*,

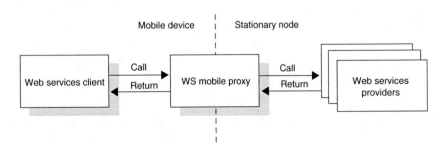

Fig. 3 Interactions through WS middleware.

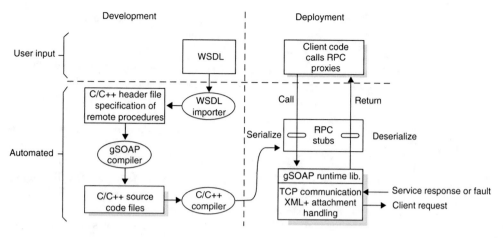

Fig. 4 gSOAP web services run-time.

which are suitable for devices with a small memory foot-print (not part of J2SE). CLDC targets hardware platforms with 128–512 Kb memory and 16-bit or 32-bit CPUs. The mobile information device profile (MIDP) is specifically designed for cell phones and provides the user interface, network connectivity, local data storage, and application management needed by these devices. Following SOAP APIs and WS toolkits are suitable for J2ME/MIDP-based devices:

kSOAP is an open source SOAP API for J2ME devices (Additional Reading 8). It provides a lightweight way to access SOAP-based WS. However, kSOAP cannot gener-ate client side stubs from the web service's WSDL file.

JSR-172 is a set of WS APIs (WSA) for J2ME[15] available in Sun's wireless tool kit (WTK) 2.2. In contrast to kSOAP, client side stubs can be generated automatically using WSDL files, which accelerate the development pro-cess. WSA for J2ME has thus similar capabilities (e.g., stub generator) when compared with gSOAP's client run-time or the WSDL-to-C++ wizard for Symbian plat-forms. However, it is important to note that WSA or WSDL-to-C++ are only suitable for consumption of WS on mobile devices and cannot provide a service on a mobile host (service interaction depicted in Fig. 2B).

WS for .NET Compact Framework

The .NET Compact Framework (CF) is a sub-set of Micro-soft's .NET framework. The .NET CF is supported on various devices/platforms that are based on pocket PC and Smartphone architectures. WS on .NET CF support the use of synchronous or asynchronous invocation [Microsoft developer network (MSDN) Additional Read-ing 6]. Development of embedded WS is analog to imple-menting WS clients in .NET. A *web reference* (a reference to the actual service) has to be added to a project and code is automatically generated.

MOBILE WS PERFORMANCE LIMITS

A number of parameters can be selected upon which per-formance metrics can be established.[16] For example:

- Average time needed to execute a given request;
- Latency, given a number of requests to be executed;
- Maximum number of concurrent requests that can be executed;
- Overhead of using Java (e.g., multi-threaded Java ap-plication) in terms of start-up overhead, CPU usage, and memory consumption (allocation, given a number of requests to be executed).

We conducted an empirical (*online*) performance study to compare kSOAP and the gSOAP toolkit (see more details in Ref. 16) where the mobile devices interact with a web service provider located on the Internet (as shown in Fig. 2A). However, discussions on performance results have analog significance for service providers implemen-ted on mobile devices (see Fig. 2B). We used the Google WS API to invoke search queries from a Symbian device. By using a Symbian-based cell phone, we limit the choice of available WS toolkits to Java and C++-based implementations.

A straight forward way to obtain packet statistics is to use the **ping** utility [using the Internet control message protocol (ICMP)]. For a given packet size, we measure roundtrip delay [roundtrip time (RTT)] for each packet and some statistics such minimum, maximum, and average RTT in milliseconds. For our experiments, we take a similar approach. We inject a *performance context* into the SOAP stack for each SOAP request. We evaluate performance by obtaining latency through roundtrip delay measurements of SOAP messages per second. To trace a request, we add a time stamp at method invocation on the SOAP client, upon sending the SOAP message

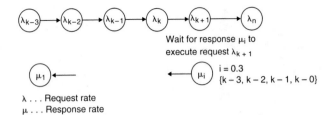

λ ... Request rate
μ ... Response rate

Fig. 5 Concurrent execution of SOAP.

through the socket interface, upon receiving a response on the socket interface, and finally when getting the actual result of the SOAP call. We can calculate timer intervals such as time needed to create a SOAP message (time spent in the WS stack), time to receive a response on the socket interface (delta t Network), and the time needed to process the response message in the stack.

We execute multiple SOAP requests concurrently. Therefore, we use multiple threads to achieve non-blocking operation. The native C++ Symbian API supports multi-threading through the *RThread* class. On J2ME, a standard thread pool can be used to accomplish concurrent execution of requests. Fig. 5 shows the multi-threading behavior of our WS applications.

A new thread is spawned (RThread) or assigned to an idle thread in the thread pool (Java) for each request. If the maximum number of concurrent requests in execution is reached (limited to four concurrent requests, otherwise, a Symbian exception is thrown), we wait until a response μ_i is received and continue execution of waiting requests. In our performance study, we executed up to sixteen requests in a batch manner. Fig. 5 shows some performance results

by comparing C++ and Java-based toolkits for a set of request rates.

Fig. 6 shows best, worst, and average processing time, given a number of requests to be executed. In this diagram, the WS stack processing is shown (neglecting random network delays and processing time on the server). Java has in general larger deviations in MIN/MAX values (time needed to process a particular request). Reasons are non-deterministic garbage collection on the Java kilobyte virtual machine (KVM) and code optimization at run-time. Repeatedly, parts of the code, called hotspots, are optimized.[17] This gives up to 50% faster execution of subsequent requests, compared with the first ones (e.g., first 4 executions if thread pool size is set to 4). The overall average performance of kSOAP, however, is comparable with gSOAP.

SUMMARY

As high speed 3G networks such as UMTS and metropolitan WiFi are increasingly deployed, ubiquitous data access is becoming a reality. A large number of embedded systems (e.g., sensors and actuators) at home are equipped with wireless interfaces and may organize themselves autonomously. WS on embedded devices address heterogeneity and enable inter-operability among disparate systems. We discussed various application scenarios in both domains as well as a collaboration scenario, utilizing WS, in detail. We provided an overview of different WS toolkits available for the Symbian platform. Our empirical performance study compares the performance of Java/J2ME-based WS with toolkits available in C++. In addition, we highlighted the concept of a mobile proxy

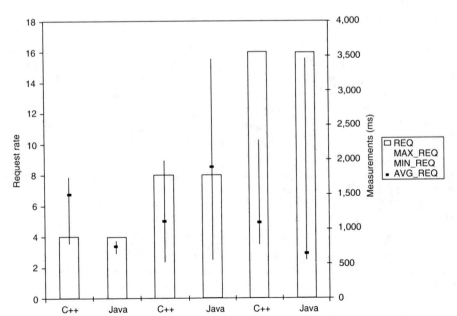

Fig. 6 Comparison of web services toolkits.

which can help to consume complex services by providing a simplified API that can be accessed by mobile devices. The mobile proxy can furthermore adapt the service behavior of "legacy services" to make service consumption and provisioning context-aware, e.g., by considering users' location information or by personalizing the service content. We believe that WS on mobile devices provide great flexibility and inter-operability. WS play an important role in providing new kind of mobile services.

ACKNOWLEDGMENTS

Part of this work was supported by the EU STREP Project inContext (FP6-034718) and Austrian Science Fund (FWF) grant P18368-N04 Project OMNIS.

REFERENCES

1. Vinoski, S. CORBA: integrating diverse applications within distributed heterogeneous environments. IEEE Commun. Mag. **1997**, *35* (2), 46–55.

2. Papazoglou, M.P.; Georgakopoulos, D. Service-oriented computing. Commun. ACM. **2003**, *46* (10), 24–28.

3. Chappell, D. *Enterprise Service Bus*; O'Reilly: , 2004.

4. Aiello, M. The role of web services at home. IEEE Web Service Based Systems and Applications (WEBSA) 2006, Guadeloupe, French Caribbean, Feb 23–25, 2006; IEEE Computer Society: 1997, .

5. Lazovik, A.; Aiello, M.; Papazoglou, M. Planning and monitoring the execution of web service requests. In *ICSOC-03, Lecture Notes in Computer Sciences 2910*; Springer: , 2003; 335–350.

6. Rosenberg, F.; Platzer, C.; Dustdar, S. Bootstrapping performance and dependability attributes of web services. IEEE International Conference on Web Services (ICWS'06), Chicago, IL, Sept 18–22, 2006; IEEE Computer Society: 2006, .

7. Dixit, S.; Prasad, R. *Wireless IP and Building the Mobile Internet*; Artech House: , 2003.

8. Pham, T.L.; Schneider, G.; Goose, S.; Pizano, A. Composite device computing environment: a framework for augmenting the PDA using surrounding resources. Workshop on Situated Interaction in Ubiquitous Computing at CHI 2000, The Hague, Netherlands, Apr 1–6, 2000; ACM Press: , 2000.

9. Hightower, J.; Borriello, G. A survey and taxonomy of location systems for ubiquitous computing. Computer. *34* (8), 57–66.

10. Hariharan, R.; Krumm, J.; Horvitz, E. Web-enhanced GPS. International Workshop on Location- and Context-Awareness (LoCA 2005), Oberpfaffenhofen, Germany, May 12–13, 2005; Springer: , 2005.

11. Miladinovic, I.; Pospischil, G.; Stadler, J. A location-based push architecture using SIP. 4th International Symposium on Wireless Personal Multimedia Communications (WPMC 2001), Aalborg, Denmark, Sept 9–12, 2001; WPMC 2001, .

12. Baldauf, M.; Dustdar, S.; Rosenberg, F. A survey on context aware systems. Int. J. Ad Hoc Ubiquitous Comput. Inderscience Publishers, January 2006.

13. Dorn, C.; Schall, D.; Dustdar, S. Granular context in collaborative mobile environments. International Workshop on Context-Aware Mobile Systems CAMS'06, Montpellier, France, Oct 31–Nov 1, 2006; Springer: , 2006.

14. Engelen, R.A. gSOAP: C/C++ Web Services Toolkit, 2004, http://gsoap2.sourceforge.net/ (accessed on May 24, 2007).

15. Sun Microsystems. J2ME Web Services Technical White Paper, July 2004, http://java.sun.com/j2me/reference/whitepapers/Web_Svcs_wp072904.pdf.

16. Schall, D.; Aiello, M.; Dustdar, S. Web services on embedded devices. Int. J. Web Inf. Syst. **2006**, *2* (1), 1–6.

17. Tierno, J.; Campo, C. Smart camera phones: limits and applications. IEEE Pervasive Comput. **2005**, *4* (2), 84–87.

BIBLIOGRAPHY

1. Engelen, R.A. SOAP/XML Web Service Performance, http://www.cs.fsu.edu/engelen/soapperformance.html.

2. Google API specifications, http://www.google.com/apis/.

3. IETF. RTP: a transport protocol for real-time applications, January 1996, http://www.ietf.org/rfc/rfc1889.txt.

4. IETF. SIP, http://www.ietf.org/html.charters/sip-charter.html.

5. Jammes, F.; Mensch, A.; Smit, H. Service-oriented device communications using the devices profile for web services. 3rd International Workshop on Middleware for Pervasive and Ad-hoc Computing (MPAC '05), Grenoble, France, Nov 28–Dec 2, 2005; ACM Press, 2005; 1-8.

6. MSDN. Consuming WS with the Microsoft .NET CF, March 2003, http://msdn.microsoft.com/library/default.asp?url=/library/en-us/dnnetcomp/html/netcfwebservices.asp.

7. Nokia WSDL-to-C++ Wizard for S60, http://forum.nokia.com.

8. SOAP implementation for J2ME, http://kobjects.sourceforge.net/.

Wireless Internet Fundamentals

Borko Furht
Mohammad Ilyas
College of Engineering and Computer Science, Florida Atlantic University,
Boca Raton, Florida, U.S.A.

Abstract

This entry presents a comprehensive introduction to the field of wireless systems and their applications. We begin with fundamental principles of wireless communications including modulation techniques, wireless system topologies, and performance elements of wireless communication. Then, we present three generations of wireless systems based on access techniques, and we introduce the basic principles of FDMA, TDMA, and CDMA techniques. We discuss various wireless Internet networks and architectures, including wireless personal area networks (W-PANs), LANs, and WANs. We present common wireless devices and their features as well as wireless standards, such as wireless application protocol (WAP). A survey of present and future wireless applications is given, from messaging applications to mobile commerce (M-commerce), entertainment, and mobile web services. We also briefly discuss the future trends in wireless technologies and systems.

INTRODUCTION

The wireless Internet is coming of age! Millions of people worldwide are already using web phones and wireless handheld devices to access the Internet. Nations and corporations are making enormous efforts to establish a wireless infrastructure, including declaring new wireless spectrum, building new towers, and inventing new handset devices, high-speed chips, and protocols.

The adoption of the wireless Internet strictly depends on the mobile bandwidth—the bandwidth of access technologies. The current 2G wireless access technologies transmit at 9.6–19.2 Kbps. These speeds are much slower than the dial-up rates of desktop PCs connecting to the Internet. However, 2.5G wireless technologies, that are already in use and provide speeds of 100 Kbps, and 3G technologies with speeds of 2–4 Mbps, will allow wireless connections to run much faster than the wired cable and digital subscriber line (DSL) services today. Fig. 1 illustrates the transmission speeds of wired networks and their applications. This figure also includes wireless access networks that show that 2G networks are basically used for voice and text messaging, but 2.5G networks, and particularly 3G networks will open doors for many new wireless applications that use streaming video and multimedia.

Today, the number of subscribers with fixed Internet access is much higher than those with mobile Internet access. However, according to the forecast from Ericsson, in several years the number of mobile subscribers to the Internet will reach 1 billion and will be higher than those having fixed access (see Fig. 2).

In this entry, we introduce fundamental concepts of wireless Internet. In the second section, we describe the basic principles of wireless communications including wireless network technologies. In the third section, the modulation techniques and basic access technologies are described. Wireless Internet networks are described in the fourth section, whereas wireless devices and their functionality are presented in the fifth section. The sixth section gives an overview of current and potential wireless Internet applications, whereas some future trends in wireless technologies are discussed in the seventh section. Concluding remarks are given in the last section.

PRINCIPLES OF WIRELESS COMMUNICATIONS

In this section, we describe fundamental principles of wireless communications and related wireless technologies including wireless radio and satellite communications. We introduce basic modulation techniques used in radio communications and two fundamental wireless system topologies—point-to-point and networked topologies. We also discuss performance elements of wireless communications.

Encyclopedia of Wireless and Mobile Communications DOI: 10.1081/E-EWMC-120043608

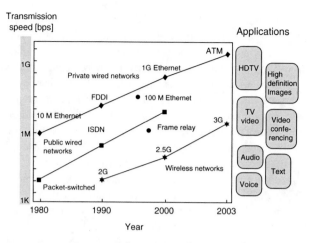

Fig. 1 Wired and wireless networks and their applications.

Wireless Technologies

Today, there are many wireless technologies that are used for a variety of applications. *Wireless radio communications* are based on transmission of radio waves through the air. The radio waves between 30 MHz and 20 GHz are used for data communications. The range lesser than 30 MHz could also support data communication; however, it is typically used for FM and AM radio broadcasting, because these waves reflect on the earth's ionosphere to extend the communications. Radio waves over 20 GHz may be absorbed by water vapor, and therefore they are not suitable for long distance communications. Table 1 shows RFs used for wireless radio applications from AM and FM radio, to TV, and GPS and cell phones.[1]

Microwave transmission is based on the same principles as radio transmission. The microwave networks require a direct transmission path, high transmission towers, and antennae. Microwave equipment in the US operates at 18–23 GHz. There are 23,000 microwave networks in the US alone.

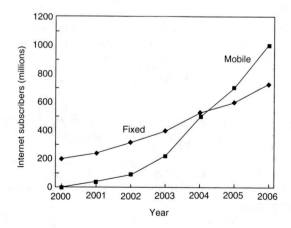

Fig. 2 Mobile Internet access (source: Ericsson).

Table 1 Radio spectrum and applications.

Applications	Frequency spectrum
AM	535–1700 kHz
FM	88–108 MHz
TV	54–88, 174–220 MHz
GPS	1200–1600 MHz
Cell phones	800–1000 MHz
	1800–2000 MHz

Satellite communications are used for a variety of broadcasting applications. Two most popular frequency bands for satellite communications are C-band (frequency range 5.9–6.4 GHz for uplink and 3.7–4.2 GHz for downlink) and Ku-band (frequency range 14–14.5 GHz for uplink and 11.7–12.2 GHz for downlink). Recently, the Ku-band spectrum has been opened up to the US satellite communications, which receives at 30 GHz and sends at 20 GHz.

The radio transmission system consists of a radio transmitter and a radio receiver. The main components of a radio transmitter are transducer, oscillator, modulator, and antenna. A transducer converts the information to be transmitted to an electrical signal. Example is a microphone, or a camera. An oscillator generates a reliable frequency that is used to carry the signal. A modulator embeds the desired signal (voice or data) into the carrier frequency. An antenna is used to radiate an electrical signal into space in the form of electromagnetic waves.

A radio receiver consists of an antenna, oscillator, demodulator, and amplifier. An antenna captures radio waves and converts them into an electrical signal. An oscillator generates electrical waves at the carrier frequency that is used as a reference wave to extract the signal. A demodulator detects and restores modulated signals. An amplifier amplifies the received signal that is typically very weak.

Modulation Techniques

Modulation techniques embed a signal into the carrier frequency. They can be classified into 1) analog modulations, and 2) digital modulations. Traditional analog modulations include AM and FM. In digital modulations, binary "1"s and "0"s are embedded in the carrier frequency by changing their amplitude, frequency, or phase. Subsequently, digital modulations, called keying techniques, can be amplitude shift keying (ASK), frequency shift keying (FSK), and phase shift keying (PSK).

Some new popular keying techniques include Gaussian minimum shift keying (GMSK) and differential quadrature PSK (DQPSK). GMSK is a type of FSK modulation that uses continuous phase modulation, so it can avoid abrupt changes. It is used in GSM, and digital enhanced cordless telecommunications (DECTs). Differential PSK (DPSK) is a type of phase modulation, which defines four rather than two phases. It is used in TDMA systems in the US.

A significant drawback of traditional RF systems is that they are quite vulnerable to sources of interference. Spread spectrum modulation techniques resolve the problem by spreading the information over a broad frequency range. These techniques are very resistant to interference. Spread spectrum techniques are used in CDMA systems, and are described in more detail in "Wireless Internet Architectures" section.

Wireless System Topologies

Two basic wireless system topologies are 1) point-to-point (or ad hoc), and 2) networked topology. In the point-to-point topology, two or more mobile devices are connected using the same air interface protocol. Fig. 3A illustrates the full mesh point-to-point configuration, where all devices are inter-connected. Limitations of this topology are that the wireless devices cannot access the web, send e-mail, or run remote applications.

In the networked topology, there is a link between wireless devices connected in the wireless network and the fixed public or private network. A typical configuration, shown in Fig. 3B, includes wireless devices (or terminals), at least one bridge between the wireless and the physical network, and the number of servers hosting applications used by wireless devices. The bridge between the wireless and the physical network is called base station, or access point.

Performance Elements of Wireless Communications

Wireless communication is characterized with several critical performance elements:

- Range,
- Power used to generate the signal,
- Mobility,
- Bandwidth, and
- Actual data rate.

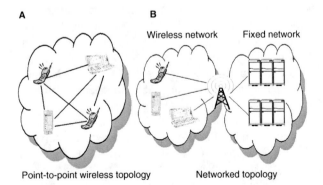

Fig. 3 Wireless topologies: (A) Point-to-point topology; (B) Networked topology.

The range is a critical factor that refers to the coverage area between the wireless transmitter and the receiver. The range is strongly correlated with the *power* of the signal. Simplified approximation is that for 1 mW of power, the range is 1 m in radius. For example, 1 W of power will allow the range of 1 km in radius. As the distance from the base station increases, the signal will degrade, and data may incur a high error rate. Using part of the spectrum for error correction can extend the range. Also, the use of multiple base stations can extend the range.

Mobility of the user refers to the size of the wireless device. Miniaturization of the wireless device will provide better mobility. This can be achieved by reducing the battery size and consequently by minimizing power consumption. However, this will cause that the generated signal is weaker that will give reduced range. In summary, there should be a trade-off between the range and the mobility—the extended range will reduce the mobility, and vice versa the better mobility will reduce the range of wireless devices.

Bandwidth refers to the amount of frequency spectrum available per user. Using wider channels gives more bandwidth. Transmission errors could reduce the available bandwidth, because the part of the spectrum will be used for error correction.

Actual data rate mostly depends on the bandwidth available to the user; however, there are some other factors that also influence it, such as the movement of the transceiver, position of the cell, and density of users. The actual data rate is typically higher for stationary users than for walking users. The users traveling at high speed (such as in cars or trains) have the lowest actual data rate. The reason for this is that the part of the available bandwidth must be used for error correction due to greater interference that traveling users may experience.

Similarly, the interference depends on the position of the cell, and with higher interference the actual data rate will be reduced. Optimal location is when there is direct line of sight between the user and the base station and the user is not far from the base station. Then there is no interference and the transmission requires minimum bandwidth for error correction.

Finally, if the density of the users is high, there will be more users transmitting within a given cell, and consequently there will be less aggregate bandwidth per user. This also reduces the actual data rate.

GENERATIONS OF WIRELESS SYSTEMS BASED ON WIRELESS ACCESS TECHNOLOGIES

From the late 1970s till today, there were three generations of wireless systems that are based on different access technologies:

- 1G wireless systems—based on FDMA;

- 2G wireless systems—based on TDMA and CDMA;
- 3G wireless systems—mostly based on wide CDMA (W-CDMA).

In "Future of Wireless Technology" section, we introduce the future efforts in building the 4G wireless systems.

The 1G Wireless Systems

The 1G of wireless systems was introduced in late 1970s and early 1980s and was built for voice transmission only. It was an analog, circuit-switched network that was based on *FDMA* air interface technology. In FDMA, each caller has a dedicated frequency channel and related circuits. For example, three callers use three frequency channels (see Fig. 4A). An example of a wireless system that employs FDMA is advanced mobile phone service (AMPS).

The 2G Wireless Systems

The 2G of wireless systems was introduced in the late 1980s and the early 1990s with the objectives to improve transmission quality, system capacity, and the range. Major multiple access technologies used in 2G systems are TDMA and CDMA. These systems are digital, and they use circuit-switched networks.

TDMA technology

In TDMA systems, several callers timeshare a frequency channel. A call is sliced into a series of time slots, and each caller gets one time slot at regular intervals. Typically, a 39 kHz channel is divided into three time slots, which allows three callers to use the same channel. In this case, nine callers use three channels. Fig. 4 illustrates the operation of FDMA and CDMA technologies.

The main advantage of the TDMA systems is the increased efficiency of transmission, and there are some

additional benefits compared with the CDMA-based systems. First, TDMA systems can be used for transmission of both voice and data. They offer data rates from 64 Kbps to 120 Mbps, which enable operators to offer personal communication services such as fax, voice-band data, and SMSs. TDMA technology separates users in time, thus ensuring that they will not have interference from other simultaneous transmissions. TDMA also provides extended battery life, as the transmission occurs only a portion of time. One of the disadvantages of TDMA is caused from the fact that each caller has a pre-defined time slot. This can cause that when callers are roaming from one to another cell, all time slots in the next cell are already occupied, and the call might be disconnected.

GSM

GSM is the best known, European implementation of services that uses TDMA air interface technology. It operates at 900 and 1800 MHz in Europe, whereas there is the US GSM operating at 1900 MHz. European GSM has been also exported to the rest of the world. GSM has also applied the frequency hopping (FH) technique, which provides switching the call frequency many times per second for security. The other systems that deploy TDMA are digital enhanced cordless telephony (DECT), IS-136 standard, and integrated digital enhanced network (iDEN).

CDMA technology

CDMA is radically different air interface technology that uses FH spread spectrum technique. The signal is randomly spread across the entire allocated 1.35 MHz bandwidth, as illustrated in Fig. 5. The randomly spread sequences are transmitted all at once, which gives higher data rate and improved capacity of the channels compared with TDMA and FDMA. It gives eight to ten times more callers per a channel than FDMA/TDMA air interface. The CDMA also provides better signal quality and secure communications. The transmitted signal is dynamic bursty, and ideal for data communication. Many mobile phone standards currently being developed are based on CDMA.

The 3G Wireless Systems

The 3G wireless systems are digital systems based on packet-switched network technology intended for wireless transmission of voice, data, images, audio, and video. These systems typically employ W-CDMA and CDMA2000 air interface technologies.

Packet switching vs. circuit switching

In circuit switching networks, resources needed along a path for providing communication between the end

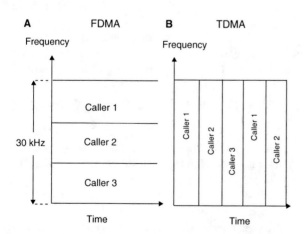

Fig. 4 FDMA vs. TDMA. (A) In FDMA a 30-kHz channel is dedicated to each caller. (B) In TDMA a 30-kHz channel is timeshared by three callers.

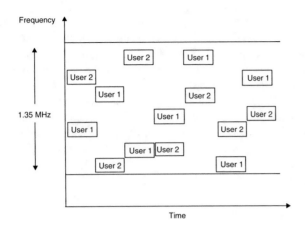

Fig. 5 Frequency-hopped spread spectrum applied in CDMA air interface.

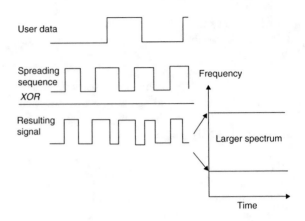

Fig. 6 Direct sequence spread spectrum applied in W-CDMA air interface.

systems are reserved for the entire duration of the session. These resources are typically buffers and bandwidth. In packet switching networks, several users share these resources, and various messages use the resources on demand. Therefore, packet switching offers better sharing of bandwidth, it is simpler, more efficient, and less costly to implement. On the other hand, packet switching is not suitable for real-time services, because of its variable and unpredictable delays.

W-CDMA technology

W-CDMA uses direct sequence (DS) spread spectrum technique. DS spread spectrum uses a binary sequence to spread the original data over a larger frequency range as illustrated in Fig. 6. The original data is multiplied by a second signal called spreading sequence or spreading code, which is a pseudo random code (PRC) of much wider frequency. The resulting signal is as wide as the spreading sequence, but carries the data of the original signal.

2.5G wireless systems

An intermediate step in employing fully packet switching 3G systems is the 2.5G wireless systems. They use separate air interfaces—circuit switching for voice and packet switching for data—designed to operate in 2G network spectrum. The 2.5G provides an increased bandwidth to about 100 Kbps, much larger than 2G systems, however much lower than the expected bandwidth of 3G systems. GPRS is the 2.5G implementation of IP packet switching on European GSM networks.[2] It is also an upgrade for the IS-136 TDMA standard, used in North and South America. GRPS combines neighboring 19.2 Kbps time slots, typically one uplink and two or more downlink slots per GRPS tower. The rate can potentially reach 115

Kbps. Enhanced data for GSM enhancement (EDGE) is another packet-switched technology that is a GRPS upgrade based on TDMA. The theoretical pick of this technology is 384 Kbps, but the tests show that the practical rates are in the range 64–100 Kbps. EDGE is a standard of AT&T Wireless in the US.

Universal mobile telecommunications system

Universal mobile telecommunications system (UMTS) is 3G wireless standard that supports two different air interfaces—W-CDMA and time-division CDMA (TD-CDMA). W-CDMA will be used for the cellular wide area coverage and high-mobility service, whereas TD-CDMA will be used for low mobility, local in-building services, asymmetrical data transmission, and typical office applications. GSM, IS-136, and personal digital cellular (PDC) operators have all adopted the UMTS standard, but Qualcomm has developed a similar standard CDMA2000, which could attract existing IS-95 carriers. Basic concepts of W-CDMA radio access network are described in Ref. 3.

Table 2 presents the characteristics of three generations of wireless systems, while Fig. 7 shows most possible migration path from 2G to 3G wireless systems.

In summary, the target features of 3G wireless systems include:

- High data rates, which are expected to be 2–4 Mbps for indoor use, 384 Kbps for pedestrians, and 144 Kbps for vehicles;
- Packet-switched networks will provide that the users will be always connected;
- Voice and data network will be dynamically allocated;
- The system will offer enhanced roaming;
- The system will include common billing and will have user profiles;

Table 2 Basic characteristics of generations of wireless systems.

Features	First generation (1G)	Second generation (2G)	2.5G	Third generation (3G)
Bandwidth		~10 Kbps	~100 Kbps	~2–4 Mbps
Voice traffic	Circuit switched	Circuit switched	Circuit switched	Packet switched (VoIP)
Data traffic	No data	Circuit switched	Packet switched	Packet switched
Modulation	Analog	Digital	Digital	Digital
Air interfaces	FDMA	TDMA CDMA	TDMA	W-CDMA TD-CDMA CDMA 200
Examples of services	AMPS	GSM IS-136 PDC IS-95	GPRS EDGE	UMTS CDMA2000

- The system will be capable to determine the geographic position of the users via mobile terminals and networks;
- The system will be well suited for transmission of multimedia and will offer various services such as bandwidth on demand, variable data rates, and quality sound.

WIRELESS INTERNET ARCHITECTURES

The general wireless system architecture, which includes connections to the Internet, is shown in Fig. 8, adapted from Ref. 4. A wireless device is connected to a base station through one of the wireless Internet networks (see "Wireless Internet Networks" section). The base station is wired to a telecommunication switch. In 2.5G systems, the telecommunication switch is used to send voice calls through the circuit-switched telephone network, and data through the packet-switched Internet. However, 3G systems use packet-switched Internet for both voice and data.

Wireless Internet Networks

The wireless part of the wireless Internet architecture, shown in Fig. 8, is referred as wireless Internet network. Wireless Internet networks can be classified as:

- Wireless personal area networks (W-PANs),
- Wireless LANs (WLANs), and
- Wireless WANs (wireless WANs).

The main difference among these networks is in the range they cover. W-PANs and W-LANs operate on unlicensed spectrum, whereas wireless WANs are licensed, well regulated public networks. They can all be used as access networks to the Internet, as it is discussed in "Wireless Internet Topologies" section.

W-PANs

W-PANs have a very short range of up to 10 m. They are used to connect mobile devices to send voice and data to

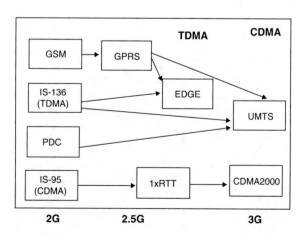

Fig. 7 Migration path from 2G to 3G wireless systems.

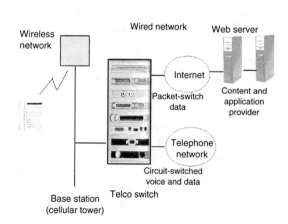

Fig. 8 Wireless system architecture.

perform transactions, data transfer, or voice relay functions. They are also used in PCs to replace cables and connectors, e.g., for keyboards and printers. Two popular technologies for W-PANs are infrared and Bluetooth technologies. Infrared devices use infrared data association (irDA) standard and are used to transmit data among a variety of devices including cell phones, notebooks, PDAs, digital cameras, and others.

The Bluetooth network, called a *piconet*, is used to connect up to eight devices. It uses FH spread spectrum technique implemented with Gaussian FSK (GFSK). The Bluetooth network is intended for wireless connection between mobile devices, fixed computers, and cellular phones.

Wireless LANs

Wireless LANs are used to substitute fixed LANs in the range of about 100 m. They are used in office buildings and homes to connect devices using a WLAN protocol. Typically, WLANs have a fixed transceiver, which is a base station that connects the WLAN to a fixed network. Popular WLANs include DECT, home RF, and 802.11 networks.

DECT is a standard for cordless phones that operate in the frequency range from 1880 to 1900 MHz in a range of 50 m. It is based on TDMA technology. Home RF network is used to connect home appliances. It uses shared wireless access protocol (SWAP), which is similar to DECT but carries both data and voice. It supports up to 127 devices in the range of about 40 m. The 802.11 is a standard developed for WLANs that covers an office building or a group of adjacent buildings. Standard 802.11b (that is a revision of an original 802.11 standard) sub-divides its frequency band of 2.4–2.483 GHz into several channels. Its specification supports DS spread spectrum technique.

Wireless WANs

Wireless WANs are licensed public wireless networks that are used by web cell phones and digital modems in handheld devices. With a single transceiver (also called base station or cellular tower), the range is about 2500 m, however WLANs usually have multiple receivers that make their range to be practically unlimited. The most popular wireless WANs are cellular networks that consist of multiple base stations positioned in a hexagon (see Fig. 9). Cellular networks can be classified as mobile phone networks that carry primarily voice, and they are typically using circuit switching technology, and packet data networks, that carry primarily data and use packet switching technology.

Table 3, adapted from Ref. 1, summarizes basic features of these three wireless networks.

Wireless Internet Topologies

A typical wireless device that has one radio and one antenna, can either connect to a public, cellular phone network, or to a private WLAN, or to a PAN. However, all these devices, referred to as WAN, LAN, or PAN, can connect to the wireless Internet. One of the recent trends is that some wireless devices have multiple antennae and, thus, multiple air interfaces (MAIs). This approach allows the devices to connect to various wireless networks to optimize coverage.

Fig. 8, presented earlier in this section, is a typical wireless Internet topology that consists of a wireless and fixed network. This architecture can be further expended into a star topology, shown in Fig. 10 (Ref. 5). In this topology, a centralized radio network controller (RNC) is connected by point-to-point links with the base stations that handle connectivity for a particular geographic area or cell. RNCs are interconnected to allow mobile users to roam between geographical areas controlled by different RNCs. RNCs are further connected to a circuit switching network for voice calls (in 2G and 2.5G systems), and to packet switching network for data and access to the Internet. One of the drawbacks of this architecture is that the RNC present a single point of failure, so if an RNC fails, the entire geographical region will loose the service. This problem is addressed in Ref. 5, and some new architectures for future 4G of wireless systems are proposed.

Fig. 11 illustrates a network topology that includes a combination of wireless PANs, LANs, and WANs, all connected to the Internet, through base stations and fixed networks. Some devices, such as one denoted in Fig. 10 as the MAIs device, can be connected to several wireless networks including a satellite network. MAIs in this case can complement each other to provide optimized coverage of a particular area.

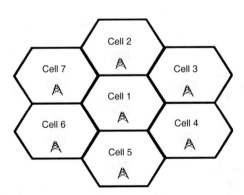

Fig. 9 Cellular network is a (WLAN) that has multiple base stations positioned in a hexagon.

Table 3 Wireless Internet networks

Wireless networks	Range	Frequency spectrum	Examples of networks
PAN	~10 m	Unlicensed	IrDA Bluetooth
LAN	~100 m	Unlicensed	DECT HomeRF 802.11b
WAN	~2,500 m—One transceiver Unlimited—multiple transceivers	Licensed	Cellular networks GSM IS-95 IS-136 PDC

WIRELESS DEVICES AND STANDARDS

In this section, we introduce the most common wireless devices and their applications. We also discuss the wireless application protocol (WAP), which is a common standard for presenting and delivering services on wireless devices. We describe Java-enabled wireless devices, which use Java technology to run applications on wireless devices.

Wireless Devices

Wireless (or mobile) devices can be classified in six groups: web phones, wireless handheld devices, two-way pagers, voice portals, communication appliances, and web PCs.[4]

A web phone is the most common device that is a cellular phone with an Internet connection. There are three major web phones: HDML & WAP phone in the USA, WAP phone in Europe, and I-mode phone in Japan. Web phones can exchange short messages, access websites with a minibrowser, and run personal service applications such as locating nearby places of interest. Web phones can operate only when they have a network connection, however advanced web phones can run their own applications.

A wireless handheld device is another common device (such as Palm) that can exchange messages and use a minibrowser to access the Internet. Industrial handheld devices, such as Symbol and Psion, can perform complex operations such as completing orders.

A two-way pager allows users to send and receive messages and the use of a minibrowser. They are typically used in business applications.

Voice portals allow users to have a conversation with an information service using kind of telephone or mobile phone.

Communication appliances are electronic devices that use wireless technology to access the Internet. Examples include wireless cameras, watches, radios, pens, and others.

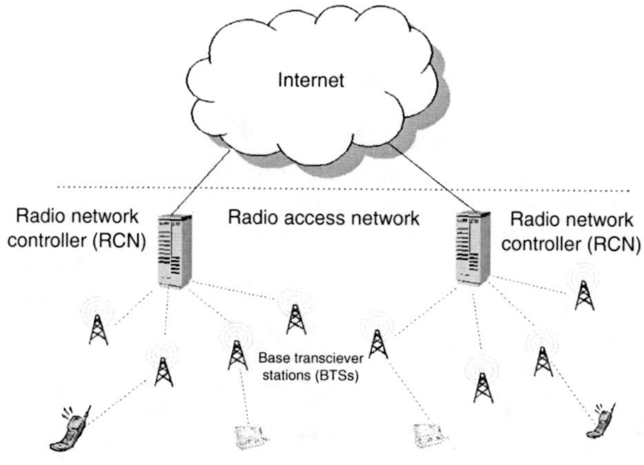

Fig. 10 Wireless Internet architecture using star topology.

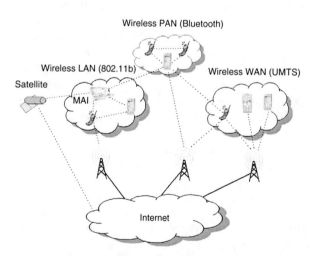

Fig. 11 A network topology with various wireless networks connected to the Internet. A wireless device (MAI) with multiple air interfaces can be connected to the Internet through W-LAN, W-PAN, or even through a satellite network.

Web PCs are standard PCs connected to the Internet that can access mobile services wirelessly.

Wireless devices typically use an embedded real-time OS. The most common OSs for wireless devices include PalmOS (used in Palm handheld devices), Windows CE and Windows NT Embedded by Microsoft (used in a variety of devices such as handheld PCs, pocket PCs, webTV, and Smart Phone), and Symbian OS. We present a brief description of Symbian OS that was re-named from Epoc OS and it was used for many years in Psion handheld devices. It is currently used in many wireless devices including Nokia 9200 Communicator Series. The architecture of the Symbian OS, shown in Fig. 12, consists of four layers. The Symbian core is common for all devices and consists of a kernel, file server, memory management, and device drivers. The system layer consists of data service enablers that provide communications and computing services such as TCP/IP, IMP4, SMS, and database management. User interface software is made and licensed by manufacturers (e.g., for the Nokia 9200 Platform). Application engines enable software developers to create user interfaces. Various applications are at the last layer.

Fig. 13 shows several representative wireless devices: Nokia 9210 and 9290, Sony Ericsson R520, and Palm. The Nokia 9290 Communicator is a wireless device that combines wireless phone and handheld device. The user can send and receive e-mails with attachments, and has an access to the Internet. It also has many applications built-in, such as MS Word, PowerPoint, and Excel. An interesting feature is that the user can take notes using the keyboard while holding conference call on built-in, hands-free speakerphone.

WAP

Wireless Internet protocol (WAP) is a de facto standard for presenting and delivering wireless services on mobile devices. It is developed by mobile and wireless communication companies (Nokia, Motorola, Ericsson, and Unwired Planet) and includes a minibrowser, scripting language, access function, and layered communication specification. Most wireless device manufacturers as well as service and infrastructure providers have adopted the WAP standard.

There are three main reasons why wireless Internet needs a different protocol: 1) transfer rates, 2) size and readability, and 3) navigation.

The 2G wireless systems have data transfer rates of 14.4 Kbps or less which is much less than 56 Kbps modems, or DSL connections and cable modems. Therefore, loading existing web pages at these speeds will take very long time.

Another challenge is a small size of the screens of wireless devices (phones or handheld devices). Web pages are designed for desktops and laptops that have

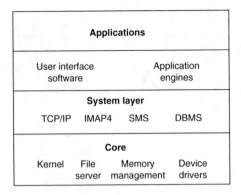

Fig. 12 The architecture of Symbian OS.

resolution of 640×480 pixels. Wireless devices may have a resolution of 150×150 pixels, and the page could not fit on the display.

Navigation is also quite different on wireless devices. On desktops and laptops, the navigation is performed using point and click action of a mouse, whereas typical wireless devices (specifically phones) use the scroll keys.

Therefore, WAP is created to provide web pages to typical wireless devices having in mind these limitations. Instead of using HTML, WAP uses wireless markup language (WML), which is a small sub-set of XML. WML is used to create and deliver content that can be deployed on small wireless devices. WML is scalable and extensible, because, like XML, it lets users to add new markup tags.

WAP stack

WAP stack consists of six layers as illustrated in Fig. 14. The wireless application environment (WAE) consists of the tools for wireless Internet developers. These tools include WML and WMLScript. WMLScript is a scripting language (similar to JavaScript or VBScript) that provides interactivity of web pages presented to the user.

Web phone—Pal VII   Nokia 9210 for GSM

Sony Ericsson   Nokia 9290 communicator

Fig. 13 Contemporary wireless phones and handheld devices.

The wireless session protocol (WSP) specifies a type of session between the wireless device and the network, which can be either connection-oriented or connectionless. Typically, a connected-oriented session is used in two-way communications between the device and the network. Connectionless session is commonly used for broadcasting or streaming data to the device.

The wireless transaction protocol (WTP) is used to provide data flow through the network. WTP also determines each transaction request as reliable two ways, reliable one way, or unreliable one way.

The wireless transport layer security (WTLS) provides some security features similar to the TLS in TCP/IP. It checks data integrity, provides data encryption, and performs client and server authentication.

The wireless datagram protocol (WDP) works in conjunction with the network carrier layer and provides WAP to adapt to a variety of bearers. Network carriers or bearers depend on current technologies used by the wireless providers.

WAP topology

Fig. 15 shows a typical WAP topology. The wireless device, which is a WAP client, through its minibrowser sends a radio signal searching for service. A connection is established with the service provider, and the user selects a website to be viewed. The universal resource locator (URL) request from the WAP client is sent to the WAP gateway server, which is located between the carrier's network and the Internet. The WAP gateway server retrieves the information from the web server. It consists of the WAP encoder, script compiler, and protocol adapters to convert the HTML data into WML. There are two possible scenarios how the WAP gateway server operates:

1. If the web server provides content in WML, the WAP gateway server transmits this data directly to the WAP client.
2. If the web server delivers content in HTML, the WAP gateway server first encodes the HTTP data into WML and then transmits to the client device.

In both cases, the WAP gateway server also encodes the data from the web server into a compact binary form for transmission over low-bandwidth wireless channels.

With the development of 3G wireless systems, there is a question whether WAP will be still needed. WAP was primarily developed for 2G systems that provide limited data rates of 9.6–14.4 Kbps. UMTS network, a 3G wireless system, with expected data rates of 2–4 Mbps will resolve the problem of the limited bandwidth.

On the other hand, WAP form arguments that even in 3G systems the bandwidth will play a crucial role and that WAP will be beneficial for the UMTS network as well. The WAP features that could be useful for the UMTS network include screen size, low power consumption, carrier independence, multi-device support, and intermittent coverage. Another argument is that new applications will require higher bandwidth and data rates, so WAP will still play a crucial role.

Java-enabled Wireless Devices

New wireless devices that have recently emerged are referred to as Java-enabled wireless devices. While WAP wireless devices run new applications remotely using WAP, Java-enabled wireless devices allow users to download applications directly from the Internet. In addition, these devices allow users to download Java applets that can customize their devices. Another benefit of Java-enabled wireless devices is that they run applications and services from different platforms.

Java-enabled wireless devices use Java 2 Platform Mobile Edition (J2ME) that allows Java to work on small devices. J2ME includes some core Java instructions and APIs; however, its graphics and database access are less sophisticated than in J2SE and J2EE.

Wireless application environment (WAE)
WML   WMLScript
Wireless session protocol (WSP)
Wireless transaction protocol (WTP)
Wireless transport layer security (WTLS)
Wireless datagram protocol (WDP)
Network carrier method
SMS USSD CSD IS-136 CDMA CDPD PDC-P

Fig. 14 WAP stack consisting of six layers.

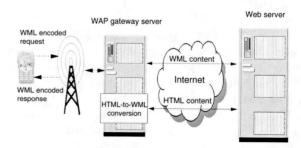

Fig. 15 The WAP topology.

Java technology can be implemented either in software or in hardware. In a software implementation, the CPU of the wireless device runs the Java code, whereas the hardware implementation is based on either a specialized Java acceleration chip, or a core within the main processor. Hardware approach typically increases the performance of Java applications by running it more efficiently and thus reduces power demands. Several companies are currently developing hardware chips that run Java or can be used as Java co-processors, including ARC Cores, ARM Ltd., Aurora VLSI, and Zucotto Wireless.

Korea's LG Telecom developed the first Java-enabled phone in 2000. Java phones are presently produced by Nextel in US, NTT BoCoMo in Japan, and British Telecom. Nokia plans to ship 50 million Java phones in 2002 and 100 million in 2003 (Ref. 6).

WIRELESS INTERNET APPLICATIONS

The wireless Internet will keep large number of people in motion. Four wireless applications drive the wireless Internet: messaging, browsing, interacting, and conversing.[4] In messaging applications, a wireless device is used to send and receive messages. The device uses SMS and other e-mail protocols. In browsing applications, a wireless device uses a minibrowser to access various web sites and receives web services. In interacting applications, the applications run on wireless devices and include business and personal applications, and standalone games. In conversing applications, a wireless device calls voice portals (such as Tellme or Wildfire) to get voice information from web services.

However, there are still a number of challenges in the development of wireless applications. The desktop computer will continue to be a dominant platform for generating content, however professionals and consumers will increasingly use wireless devices to access and manage information. The great challenge for developers is to tailor content to the unique characteristics of wireless devices. The main objective is to provide quick and easy access to the required information, rather than to provide complex directory tree, where the user will easily get lost. Another challenge for developers is the design of user interfaces, which should be simple, because of the limited size of the wireless devices.

The 2.5G and 3G wireless systems will allow new applications to include rich graphical content. Software vendors have been developing authoring tools for creating WAP-compatible WAP sites that include rich graphical content and animations. Examples include Macromedia and Adobe that are offering WAP and i-mode versions of their products. Macromedia Spectra product for creating dynamic, interactive, and content-rich websites has been

extended, so a developer can easily add wireless Internet by creating WML code instead HTML.

Firepad developed a vector-based graphics application for mobile devices. This application uses high-speed vector rendering engine for complex applications such as geographic information systems and CAD drawings, as illustrated in Fig. 16.

In this section, we present several wireless applications that, in our opinion, are a major force in further driving the development of wireless Internet.

Messaging Applications

Messaging in mobile networks today mainly involves short text using SMS protocol. The GSM Association has estimated that 24 billion SMS are sent each month.[7] However, it is expected that soon wireless devices will support pictures, audio, and video messages. At the same time, the popular messaging services on the Internet, such as e-mail, chat, and instant messaging, are extending to wireless environments.

Mobile Commerce

Mobile commerce (M-commerce) applications refer to conducting business and services using wireless devices. These applications can be grouped into: 1) transaction management applications, 2) digital content delivery, and 3) telemetry services.

Transaction management applications include online shopping tailored to wireless devices with online catalogs, shopping carts, and back office functions. Other transaction applications include micro transactions and low-cost purchases for subway or road toll, parking tickets, digital cash, and others.

Digital content delivery includes a variety of applications such as:

- Information browsing on weather, travel, schedules, sport scores, stock prices, etc.;
- Downloading educational and entertainment products;
- Transferring software, images, and videos;
- Innovative multimedia applications.

According to the recent study by HPI Research Group,[7] the following is the top ten mobile entertainment features:

- Sending SMS,
- Local traffic and weather information,
- Use of a still camera,
- Getting latest news headlines,
- Sending photos to a friend,
- Use of a video camera,

Fig. 16 Firepad software comprises of a high-speed vector rendering engine that can be used in CAD drawings.

- Book and buy tickets for movies,
- Getting information on movies,
- Listening to radio, and
- Requesting specific songs.

Entertainment on mobile devices is attractive because it is almost always with the user, whether commuting, traveling, or waiting.

Telemetry services include a wide range of new applications such as:

- Transmission of status, sensing, and measurement information;
- Communications with various devices from homes, offices, or in the field;
- Activation of remote recording devices or service systems.

Corporate Applications

Banks and transport companies were among the first businesses to deploy wireless applications, based on WAP, for their customers and employees. In banks, the goal was to reduce consumer banking transaction costs, whereas transport companies wanted to track transportation and delivery status online.

Gartner research group expects most corporations to implement wireless applications in four overlapping phases:[7]

- The first group of applications are readily justifiable and include high-value, vertical niche solutions such as field force automation.
- The second phase includes horizontal applications such as e-mail and personal information management applications.

- The third wave of applications consist of vertical applications such as mobile extensions to customer relationship management (CRM), sales force automation, and enterprise resource planning systems.
- In the long term, Gartner expects that 40–60% of all corporate systems to involve mobile elements.

Wireless Application Service Providers

Wireless application service providers (WASPs) allow wireless access to various software products and services. Business WASP applications are targeted to mobile business people, field personnel, and sales staff. Other WASP applications include[8]

- Mobile entertainment services,
- Wireless gaming,
- Wireless stock trading, and
- In-vehicle services such as traffic control and car management.

Mobile Web Services

Web services include well-defined protocol interfaces through which businesses can provide services to their customers and business partners over the Internet. Web services specify a common and inter-operable way for defining, publishing, invoking, and using application services over networks. They are built on emerging technologies such as XML, simple access object protocol (SOAP), web service description language (WSDL), universal description, discovery and integration (UDDI), and HTTP.

Mobile web services provide content delivery, location discovery, user authentication, presence awareness, user profile management, data synchronization, terminal profile management, and event notification services. Initially, wireless terminals are likely to access mobile web services indirectly, through application servers. The application server will manage the interactions with the required web services.

Wireless Teaching and Learning

Web-based distance learning could be extended to wireless systems. For example, the project Numina at University of North Carolina at Wilmington is intended to explore how the wireless technology can be used to facilitate learning of abstract scientific and mathematical concepts.[9] Students use handheld computers with appropriate software, which are connected to the wireless Internet. The system provides interactive exercises, and integrates various media and hypertext materials.

FUTURE OF WIRELESS TECHNOLOGY

The major trend that is already emerging is the migration of mobile networks to fully IP-based networks. The next generation of wireless systems, 4G systems, will use new spectrum and emerging wireless air interfaces that will provide a very high bandwidth of 10 + Mbps. It will be entirely IP-based and use packet switching technology. It is also expected that 4G systems will increase usage of wireless spectrum. According to Cooper's law, on average, the number of channels has doubled every 30 mo since 1985.

Fig. 17, adapted from Ref. 10, shows the user mobility and data rates for different generations of wireless systems, and for wireless PANs and LANs. The 3G and later 4G systems will provide multimedia services to users everywhere, whereas WLANs provide broadband services in hot spots, and W-PANs connect personal devices together at very short distances.

Spread spectrum technology is presently used in 3G systems; however there are already research experiments with multi-carrier modulation (MCM), which is a step further from spread spectrum. MCM simultaneously transmits at many frequencies.

New types of smart antennae are currently under investigation. Most current antennae are omni-directional, which means that they transmit in all directions with similar intensity. New directional antennae transmit primarily in one direction, whereas adaptive antennae vary directions to maximize performance.

New generation of software radios will dynamically adapt to wireless technology. They apply digital signal processors, so they can update the software with new versions of transmission techniques.

The transition from circuit-switched to packet-switched networks provides increased efficiency of the network and higher overall throughput. However, packet-switched networks operate on a best-effort basis and therefore cannot guarantee the service (specifically when the load is high). This will require the development of new quality of service (QoS) approaches to handle various network scenarios.

New wireless multimedia applications will require new solutions related to error resilience, network access, adaptive decoding, and negotiable QoS.

Error resilience solutions should enable delivering rich digital media over wireless networks that have high error rates and low and varying transmission speeds. Network access techniques should provide the delivery of rich media without adversely affecting the delivery of voice and data services. Innovative adaptive decoding techniques should optimize rich media for wireless devices with limited processing power, limited battery life, and varying display sizes. New negotiable QoS algorithms should be developed for IP multimedia sessions as well as for individual media components.

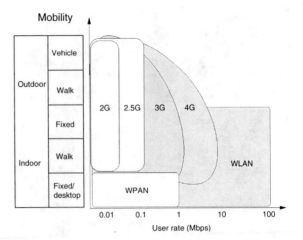

Fig. 17 User mobility and data rates for different generations of wireless systems.

CONCLUSIONS

In this entry, we presented fundamental concepts and technologies for wireless communications, and introduced various architectures and three generations of wireless systems. We are currently at the transition between 2G and 3G systems (2.5G systems). The 3G systems will soon offer higher data rates suitable for a variety of applications dealing with multimedia. Services and applications are driving 3G systems. With 3G systems, users will be able to send graphics, play games, locate a restaurant, book a ticket, read news updates, check a bank statement, watch their favorite soap operas, and many other exciting applications.

In July 2002, Ericsson delivered 15 real-life 3G applications, including real-time sport applications, face-to-face video calling, and exciting team games, to 40 operators so they can demonstrate to their customers what the wireless Internet is all about.

In the mean time, researchers are already working on 4G systems that will provide even higher data rates, will be entirely IP-based, and will include many other new features.

REFERENCES

1. Rhoton, J. *The Wireless Internet Explained*; Digital Press: 2002.
2. Park, J.-H. Wireless Internet access for mobile subscribers based on GPRS network. IEEE Commun. Mag. **2002**, *40* (4), 38–49.
3. Basic Concepts of WCDMA Radio Access Network; Ericsson, White paper: 2002, www.ericsson.com.
4. Beaulieu, M. *Wireless Internet Applications and Architecture*; Addison-Wesley, 2002.

5. Kempf, J.; Yegani, P. OpenRAN: a new architecture for mobile wireless Internet radio access network. IEEE Commun. Mag. **2002**, *40* (5), 118–123.

6. Lawton, G. Moving Java into mobile phones. IEEE Comput. **2002**, *35* (6), 17–20.

7. Mobile Terminal Software—Markets and Technologies for the Future; Nokia, White paper: 2002, www.nokia.com.

8. Steemers, P. Critical Success Factors for Wireless Application Service Providers; Cap Gemini Ernst & Young, White paper, 2002.

9. Shotsberger, P.G.; Vetter, R. Teaching and learning in the wireless classroom. IEEE Comput. **2001**, *34* (3), 110–111.

10. Pahlavan, K.; Krishnamurthy, P. *Principles of Wireless Networks: A Unified Approach*; Prentice Hall: 2002.

BIBLIOGRAPHY

1. Buracchini E. The software radio concept. IEEE Commun. Mag. **2000,** *38* (9), 138–143.

2. Hanzo L. Cherriman, P.J.; Streit, J. *Wireless Video Communications*; IEEE Press: New York, 2001.

3. Krikke J. Graphics applications over the wireless web: Japan sets the pace. IEEE Comput. Graph. Appl. **2001**, *21* (3), 9–15.

Wireless Channels—Internet

Wireless LANs (WLANs)

Suresh Singh
Portland State University, Portland, Oregon, U.S.A.

Abstract

Wireless local area networks along with cellular networks, connected to high-speed networks, allow users with portable devices and computers to be connected with communication networks and service providers.

INTRODUCTION

A proliferation of high-performance portable computers combined with end-user need for communication is fueling a dramatic growth in wireless LAN technology. Users expect to have the ability to operate their portable computer globally while remaining connected to communications networks and service providers. Wireless LANs and cellular networks, connected to high-speed networks, are being developed to provide this functionality.

Before delving deeper into issues relating to the design of wireless LANs, it is instructive to consider some scenarios of user mobility.

1. A simple model of user mobility is one where a computer is physically moved while retaining network connectivity at either end. For example, a move from one room to another as in a hospital where the computer is a hand-held device displaying patient charts and the nurse using the computer moves between wards or floors while accessing patient information.
2. Another model situation is where a group of people (at a conference, for instance) set up an ad hoc LAN to share information as in Fig. 1.
3. A more complex model is one where several computers in constant communication are in motion and continue to be networked. For example, consider the problem of having robots in space collaborating to retrieve a satellite.

A great deal of research has focused on dealing with physical layer and **medium access control (MAC)** layer protocols. In this entry we first summarize standardization efforts in these areas. The remainder of the chapter is then devoted to a discussion of networking issues involved in wireless LAN design. Some of the issues discussed include routing in wireless LANs (i.e., how does data find its destination when the destination is mobile?) and the problem of providing service guarantees to end users (e.g., error-free data transmission or bounded delay and bounded bandwidth service).

PHYSICAL LAYER DESIGN

Two media are used for transmission over wireless LANs, IR and RF. RF LANs are typically implemented in the industrial, scientific, and medical (ISM) frequency bands 902–928 MHz, 2400–2483.5 MHz, and 5725–5850 MHz. These frequencies do not require a license allowing the LAN product to be portable, i.e., a LAN can be moved without having to worry about licensing.

IR and RF technologies have different design constraints. IR receiver design is simple (and thus inexpensive) in comparison to RF receiver design because IR receivers only detect the amplitude of the signal not the frequency or phase. Thus, a minimal of filtering is required to reject interference. Unfortunately, however, IR shares the electromagnetic spectrum with the sun and incandescent or fluorescent light. These sources of modulated IR energy reduce the signal-to-noise ratio of IR signals and, if present in extreme intensity, can make the IR LANs inoperable. There are two approaches to building IR LANs.

1. The transmitted signal can be focused and aimed. In this case the IR system can be used outdoors and has an area of coverage of a few kilometer.
2. The transmitted signal can be bounced off the ceiling or radiated omnidirectionally. In either case, the range of the IR source is 10–20 m (i.e., the size of one medium-sized room).

RF systems face harsher design constraints in comparison to IR systems for several reasons. The increased demand for RF products has resulted in tight regulatory constraints on the allocation and use of allocated bands. In the US, e.g., it is necessary to implement spectrum spreading for operation in the ISM bands. Another design constraint is the requirement to confine the emitted

Encyclopedia of Wireless and Mobile Communications DOI: 10.1081/E-EWMC-120043948

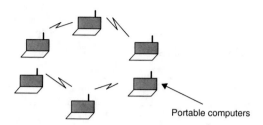

Fig. 1 Ad hoc wireless LAN.

spectrum to a band, necessitating amplification at higher carrier frequencies, frequency conversion using precision local oscillators, and selective components. RF systems must also cope with environmental noise that is either naturally occurring, e.g., atmospheric noise or man-made, e.g., microwave ovens, copiers, laser printers, or other heavy electrical machinery. RF LANs operating in the ISM frequency ranges also suffer interference from amateur radio operators.

Operating LANs indoors introduces additional problems caused by multipath propagation, Rayleigh fading, and absorption. Many materials used in building construction are opaque to IR radiation resulting in incomplete coverage within rooms (the coverage depends on obstacles within the room that block IR) and almost no coverage outside closed rooms. Some materials, such as white plasterboard, can also cause reflection of IR signals. RF is relatively immune to absorption and reflection problems. Multipath propagation affects both IR and RF signals. The technique to alleviate the effects of multipath propagation in both types of systems is the same use of aimed (directional) systems for transmission enabling the receiver to reject signals based on their angle of incidence. Another technique that may be used in RF systems is to use multiple antennas. The phase difference between different paths can be used to discriminate between them.

Rayleigh fading is a problem in RF systems. Recall that Rayleigh fading occurs when the difference in path length of the same signal arriving along different paths is a multiple of half a wavelength. This causes the signal to be almost completely canceled out at the receiver. Because the wavelengths used in IR are so small, the effect of

Rayleigh fading is not noticeable in those systems. RF systems, on the other hand, use wavelengths of the order of the dimension of a laptop. Thus, moving the computer a small distance could increase/decrease the fade significantly.

Spread spectrum transmission technology is used for RF-based LANs and it comes in two varieties: direct-sequence spread spectrum (DSSS) and frequency-hopping spread spectrum (FHSS). In a FHSS system, the available band is split into several channels. The transmitter transmits on one channel for a fixed time and then hops to another channel. The receiver is synchronized with the transmitter and hops in the same sequence; see Fig. 2A. In DSSS systems, a random binary string is used to modulate the transmitted signal. The relative rate between this sequence and user data is typically between 10 and 100; see Fig. 2B.

The key requirements of any transmission technology is its robustness to noise. In this respect DSSS and FHSS show some differences. There are two possible sources of interference for wireless LANs: the presence of other wireless LANs in the same geographical area (i.e., in the same building, etc.) and interference due to other users of the ISM frequencies. In the latter case, FHSS systems have a greater ability to avoid interference because the hopping sequence could be designed to prevent potential interference. DSSS systems, on the other hand, do exhibit an ability to recover from interference because of the use of the spreading factor (Fig. 2B).

It is likely that in many situations several wireless LANs may be collocated. Since all wireless LANs use the same ISM frequencies, there is a potential for a great deal of interference. To avoid interference in FHSS systems, it is necessary to ensure that the hopping sequences are orthogonal. To avoid interference in DSSS systems, on the other hand, it is necessary to allocate different channels to each wireless LAN. The ability to avoid interference in DSSS systems is, thus, more limited in comparison to FHSS systems because FHSS systems use very narrow subchannels (1 MHz) in comparison to DSSS systems that use wider subchannels (e.g., 25 MHz), thus, limiting the number of wireless LANs that can be collocated. A summary of design issues can be found in Ref. 1.

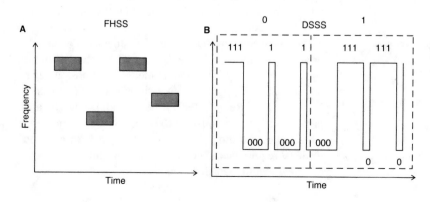

Fig. 2 Spread spectrum.

MAC LAYER PROTOCOLS

MAC protocol design for wireless LANs poses new challenges because of the in-building operating environment for these systems. Unlike wired LANs (such as the Ethernet or token ring), wireless LANs operate in strong multipath fading channels where channel characteristics can change in very short distances resulting in unreliable communication and unfair channel access due to capture. Another feature of the wireless LAN environment is that carrier sensing takes a long time in comparison to wired LANs; it typically takes between 30 and 50 μs (see Ref. 2), which is a significant portion of the packet transmission time. This results in inefficiencies if the carrier sense multiple access (CSMA) family of protocols is used without any modifications.

Other differences arise because of the mobility of users in wireless LAN environments. To provide a building (or any other region) with wireless LAN coverage, the region to be covered is divided into cells as shown in Fig. 3. Each cell is one wireless LAN, and adjacent cells use different frequencies to minimize interference. Within each cell there is an access point called a **mobile support station** (MSS) or base station that is connected to some wired network. The mobile users are called **mobile hosts** (MHs). The MSS performs the functions of channel allocation and providing connectivity to existing wired networks; see Fig. 4. Two problems arise in this type of an architecture that are not present in wired LANs.

1. The number of nodes within a cell changes dynamically as users move between cells. How can the channel access protocol dynamically adapt to such changes efficiently?
2. When a user moves between cells, the user has to make its presence known to the other nodes in the cell. How can this be done without using up too much bandwidth? The protocol used to solve this problem is called a handoff protocol and works along the following lines: a switching station (or the MSS nodes working together, in concert) collects signal strength information for each MH within each cell.

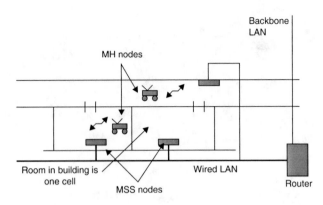

Fig. 4 In-building LAN (made up of several wireless LANs). MH, mobile host; MSS, mobile support station.

Note that if a MH is near a cell boundary, the MSS node in its current cell as well as in the neighboring cell can hear its transmissions and determine signal strengths. If the MH is currently under the coverage of MSS M1 but its signal strength at MSS M2 becomes larger, the switching station initiates a handoff whereby the MH is considered as part of M2's cell (or network).

The mode of communication in wireless LANs can be broken in two: communication from the mobile to the MSS (called *uplink* communication) and communication in the reverse direction (called *downlink* communication). It is estimated that downlink communication accounts for about 70–80% of the total consumed bandwidth. This is easy to see because most of the time users request files or data in other forms (image data, etc.) that consume much more transmission bandwidth than the requests themselves. In order to make efficient use of bandwidth (and, in addition, guarantee service requirements for real-time data), most researchers have proposed that the downlink channel be controlled entirely by the MSS nodes. These nodes allocate the channel to different mobile users based on their current requirements using a protocol such as TDMA. What about uplink traffic? This is a more complicated problem because the set of users within a cell is dynamic, thus making it infeasible to have a static channel allocation for the uplink. This problem is the main focus of MAC protocol design.

What are some of the design requirements of an appropriate MAC protocol? The IEEE 802.11 recommended standard for wireless LANs has identified almost 20 such requirements, some of which are discussed here (the reader is referred to Ref. 3, for further details). Clearly any protocol must maximize throughput while minimizing delays and providing fair access to all users. In addition to these requirements, however, mobility introduces several new requirements.

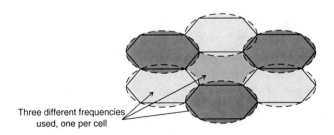

Three different frequencies used, one per cell

Fig. 3 Cellular structure for wireless LANs (note frequency reuse).

1. The MAC protocol must be independent of the underlying physical layer transmission technology adopted (be it DSSS, FHSS or IR).
2. The maximum number of users can be as high as a few hundred in a wireless LAN. The MAC protocol must be able to handle many users without exhibiting catastrophic degradation of service.
3. The MAC protocols must provide secure transmissions because the wireless medium is easy to tap.
4. The MAC protocol needs to work correctly in the presence of collocated networks.
5. It must have the ability to support ad hoc networking (as in Fig. 1).
6. Other requirements include the need to support priority traffic, preservation of packet order, and an ability to support multicast.

Several contention-based protocols currently exist that could be adapted for use in wireless LANs. The protocols currently being looked at by IEEE 802.11 include protocols based on **CSMA**, polling, and TDMA. Protocols based on CDMA and FDMA are not considered because the processing gains obtained using these protocols are minimal while, simultaneously, resulting in a loss of flexibility for wireless LANs.

It is important to highlight an important difference between networking requirements of ad hoc networks (as in Fig. 1) and networks based on cellular structure. In cellular networks, all communication occurs between the MHs and the MSS (or base station) within that cell. Thus, the MSS can allocate channel bandwidth according to requirements of different nodes, i.e., we can use centralized channel scheduling for efficient use of bandwidth. In ad hoc networks there is no such central scheduler available. Thus, any multiaccess protocol will be contention based with little explicit scheduling. In the remainder of this section we focus on protocols for cell-based wireless LANs only.

All multiaccess protocols for cell-based wireless LANs have a similar structure; see Ref. 3.

1. The MSS announces (explicitly or implicitly) that nodes with data to send may contend for the channel.
2. Nodes interested in sending data contend for the channel using protocols such as CSMA.
3. The MSS allocates the channel to successful nodes.
4. Nodes transmit packets (contention-free transmission).
5. MSS sends an explicit ACK for packets received.

Based on this model we present three MAC protocols.

Reservation-TDMA

This approach is a combination of TDMA and some contention protocol [see packet reservation multiple access (PRMA) in Ref. 4]. The MSS divides the channel into slots (as in TDMA), which are grouped into frames. When a node wants to transmit it needs to reserve a slot that it can use in every consecutive frame as long as it has data to transmit. When it has completed transmission, other nodes with data to transmit may contend for that free slot. There are four steps to the functioning of this protocol.

a. At the end of each frame the MSS transmits a feedback packet that informs nodes of the current reservation of slots (and also which slots are free). This corresponds to Steps 1 and 3 from the preceding list.
b. During a frame, all nodes wishing to acquire a slot transmit with a probability ρ during a free slot. If a node is successful it is so informed by the next feedback packet. If more than one node transmits during a free slot, there is a collision and the nodes try again during the next frame. This corresponds to Step 2.
c. A node with a reserved slot transmits data during its slot. This is the contention-free transmission (Step 4).
d. The MSS sends ACKs for all data packets received correctly. This is Step 5.

The reservation-TDMA (R-TDMA) protocol exhibits several nice properties. First and foremost, it makes very efficient use of the bandwidth, and average latency is half the frame size. Another big benefit is the ability to implement power conserving measures in the portable computer. Since each node knows when to transmit (nodes transmit during their reserved slot only) it can move into a power-saving mode for a fixed amount of time, thus increasing battery life. This feature is generally not available in CSMA-based protocols. Furthermore, it is easy to implement priorities because of the centralized control of scheduling. One significant drawback of this protocol is that it is expensive to implement (see Ref. 5).

Distributed Foundation Wireless MAC

The CSMA/CD protocol has been used with great success in the Ethernet. Unfortunately, the same protocol is not very efficient in a wireless domain because of the problems associated with cell interference (i.e., interference from neighboring cells), the relatively large amount of time taken to sense the channel (see Ref. 6) and the hidden terminal problem (see Refs. 7 and 8). The current proposal is based on a CSMA/collision avoidance (CA) protocol with a four-way handshake; see Fig. 5.

The basic operation of the protocol is simple. All MH nodes that have packets to transmit compete for the channel by sending ready to transmit (RTS) messages using nonpersistent CSMA. After a station succeeds in transmitting a RTS, the MSS sends a clear to transmit (CTS) to the MH. The MH transmits its data and then receives an ACK. The only possibility of collision that exists is in the RTS phase of the protocol and inefficiencies occur in the

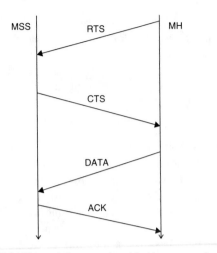

Fig. 5 CSMA/CA and four-way handshaking protocol.

protocol, because of the RTS and CTS stages. Note that unlike R-TDMA it is harder to implement power saving functions. Furthermore, latency is dependent on system load making it harder to implement real-time guarantees. Priorities are also not implemented. On the positive side, the hardware for this protocol is very inexpensive.

Randomly Addressed Polling

In this scheme, when a MSS is ready to collect uplink packets it transmits a READY message. At this point all nodes with packets to send attempt to grab the channel as follows.

a. Each MH with a packet to transmit generates a random number between 0 and P.
b. All active MH nodes simultaneously and orthogonally transmit their random numbers (using CDMA or FDMA). We assume that all of these numbers are received correctly by the MSS. Remember that more than one MH node may have selected the same random number.
c. Steps a and b are repeated L times.
d. At the end of L stages, the MSS determines a stage (say, k) where the total number of distinct random numbers was the largest. The MSS polls each distinct each random number in this stage in increasing order. All nodes that had generated the polled random number transmit packets to the MSS.
e. Since more than one node may have generated the same random number, collisions are possible. The MSS sends a ACK or NACK after each such transmission. Unsuccessful nodes try again during the next iteration of the protocol.

The protocol is discussed in detail in Ref. 2 and a modified protocol called group RAP (GRAP) is discussed

in Ref. 3. The authors propose that GRAP can also be used in the contention stage (Step 2) for TDMA- and CSMA-based protocols.

NETWORK LAYER ISSUES

An important goal of wireless LANs is to allow users to move about freely while still maintaining all of their connections (network resources permitting). This means that the network must route all packets destined for the mobile user to the MSS of its current cell in a transparent manner. Two issues need to be addressed in this context.

- How can users be addressed?
- How can active connections for these mobile users be maintained?

Ioanidis et al.[9] propose a solution called the IPIP (IP-within-IP) protocol. Here each MH has a unique IP address called its home address. To deliver a packet to a remote MH, the source MSS first broadcasts an address resolution protocol (ARP) request to all other MSS nodes to locate the MH. Eventually some MSS responds. The source MSS then encapsulates each packet from the source MH within another packet containing the IP address of the MSS in whose cell the MH is located. The destination MSS extracts the packet and delivers it to the MH. If the MH has moved away in the interim, the new MSS locates the new location of the MH and performs the same operation. This approach suffers from several problems as discussed in Ref. 10. Specifically, the method is not scaleable to a network spanning areas larger than a campus for the following reasons.

1. IP addresses have a prefix identifying the campus subnetwork where the node lives; when the MH moves out of the campus, its IP address no longer represents this information.
2. The MSS nodes serve the function of routers in the mobile network and, therefore, have the responsibility of tracking all of the MH nodes globally causing a lot of overhead in terms of message passing and packet forwarding; see Ref. 11.

Teraoka and Tokoro,[10] have proposed a much more flexible solution to the problem called virtual IP (VIP). Here every MH has a VIP address that is unchanging regardless of the location of the MH. In addition, hosts have physical network addresses (traditional IP addresses) that may change as the host moves about. At the transport layer, the target node is always specified by its VIP address only. The address resolution from the VIP address to the current IP address takes place either at the network layer of the same machine or at a gateway. Both the host machines and the gateways maintain a cache of VIP to IP mappings with associated timestamps. This information is in the form of a table called *address mapping table* (AMT). Every MH has an associated *home gateway*. When a MH moves into a new subnetwork, it is assigned a new IP address. It sends this new IP address and its VIP address to its home gateway via a *VipConn* control message. All intermediate gateways that relay this message update their AMTs as well. During this process of updating the AMTs, all packets destined to the MH continue to be sent to the old location. These packets are returned to the sender, who then sends them to the home gateway of the MH. It is easy to see that this approach is easily scaleable to large networks, unlike the IPIP approach.

Alternative View of Mobile Networks

The approaches just described are based on the belief that mobile networks are merely an extension of wired networks. Other authors[12] disagree with this assumption because there are fundamental differences between the mobile domain and the fixed wired network domain. Two examples follow.

1. The available bandwidth at the wireless link is small; thus, end-to-end packet retransmission for TCP-like protocols (implemented over datagram networks) is a bad idea. This leads to the conclusion that transmission within the mobile network must be connection oriented. Such a solution, using virtual circuits (VCs), is proposed in Ref. 11.

2. The bandwidth available for a MH with open connections changes dynamically since the number of other users present in each cell varies randomly. This is a feature not present in fixed high-speed networks where, once a connection is set up, its bandwidth does not vary much. Since bandwidth changes are an artifact of mobility and are dynamic, it is necessary to deal with the consequences (e.g., buffer overflow, large delays, etc.) locally to both, i.e., shield fixed network hosts from the idiosyncrasies of mobility as well as to respond to changing bandwidth quickly (without having to rely on end-to-end control). Some other differences are discussed in Ref. 12.

A Proposed Architecture

Keeping these issues in mind, a more appropriate architecture has been proposed in Ghai and Singh,[11] and Singh.[12] Mobile networks are considered to be different and separate from wired networks. Within a mobile network is a three-layer hierarchy; see Fig. 6. At the bottom layer are the MHs. At the next level are the MSS nodes (one per cell). Finally, several MSS nodes are controlled by a **supervisor host (SH)** node (there may be one SH

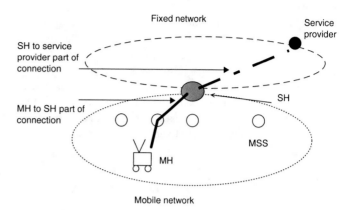

Fig. 6 Proposed architecture for wireless networks.

node per small building). The SH nodes are responsible for flow control for all MH connections within their domain; they are also responsible for tracking MH nodes and forwarding packets as MH nodes roam. In addition, the SH nodes serve as a *gateway* to the wired networks. Thus, any connection setup from a MH to a fixed host is broken in two, one from the MH to the SH and another from the SH to the fixed host. The MSS nodes in this design are simply connection endpoints for MH nodes. Thus, they are simple devices that implement the MAC protocols and little else. Some of the benefits of this design are as follows.

1. Because of the large coverage of the SH (i.e., a SH controls many cells) the MH remains in the domain of one SH much longer. This makes it easy to handle the consequences of dynamic bandwidth changes locally. For instance, when a MH moves into a crowded cell, the bandwidth available to it is reduced. If it had an open FTP connection, the SH simply buffers undelivered packets until they can be delivered. There is no need to inform the other endpoint of this connection of the reduced bandwidth.

2. When a MH node sets up a connection with a service provider in the fixed network, it negotiates some quality of service (QoS) parameters such as bandwidth, delay bounds, etc. When the MH roams into a crowded cell, these QoS parameters can no longer be met because the available bandwidth is smaller. If the traditional view is adopted (i.e., the mobile networks are extensions of fixed networks) then these QoS parameters will have to be renegotiated each time the bandwidth changes (due to roaming). This is a very expensive proposition because of the large number of control messages that will have to be exchanged. In the approach of Singh,[12] the service provider will never know about the bandwidth changes since it deals only with the SH that is accessed via the wired network. The SH bears the responsibility of handling bandwidth changes by either buffering packets until

the bandwidth available to the MH increases (as in the case of the FTP example) or it could discard a fraction of real-time packets (e.g., a voice connection) to ensure delivery of most of the packets within their deadlines. The SH could also instruct the MSS to allocate a larger amount of bandwidth to the MH when the number of buffered packets becomes large. Thus, the service provider in the fixed network is shielded from the mobility of the user.

Networking Issues

It is important for the network to provide connection-oriented service in the mobile environment (as opposed to connectionless service as in the Internet) because bandwidth is at a premium in wireless networks, and it is, therefore, inadvisable to have end-to-end retransmission of packets (as in TCP). The proposed architecture is well suited to providing connection-oriented service by using VCs.

In the remainder of this section we look at how VCs are used within the mobile network and how routing is performed for connections to MHs. Every connection set up with one or more MH nodes as a connection endpoint is routed through the SH nodes and each connection is given a unique VC number. The SH node keeps track of all MH nodes that lie within its domain. When a packet needs to be delivered to a MH node, the SH first buffers the packet and then sends it to the MSS at the current location of the MH or to the predicted location if the MH is currently between cells. The MSS buffers all of these packets for the MH and transmits them to the MH if it is in its cell. The MSS discards packets after transmission or if the SH asks it to discard the packets. Packets are delivered in the correct order to the MH (without duplicates) by having the MH transmit the expected sequence number (for each VC) during the initial handshake (i.e., when the MH first enters the cell). The MH sends ACKs to the SH for packets received. The SH discards all packets that have been acknowledged. When a MH moves from the domain of SH1 into the domain of SH2 while having open connections, SH1 continues to forward packets to SH2 until either the connections are closed or until SH2 sets up its own connections with the other endpoints for each of MH's open connections (it also gives new identifiers to all these open connections). The detailed protocol is presented in Ref. 11.

The SH nodes are all connected over the fixed (wired) network. Therefore, it is necessary to route packets between SH nodes using the protocol provided over the fixed networks. The VIP appears to be best suited to this purpose. Let us assume that every MH has a globally unique VIP address. The SHs have both a VIP as well as a fixed IP address. When a MH moves into the domain of a SH, the IP address affixed to this MH is the IP address of the SH. This ensures that all packets sent to the MH are routed through the correct SH node. The SH keeps a list of all VIP addresses of MH nodes within its domain and a list of open VCs for each MH. It uses this information to route the arriving packets along the appropriate VC to the MH.

TRANSPORT LAYER DESIGN

The transport layer provides services to higher layers (including the application layer), which include connectionless services like user datagram protocol (UDP) or connection-oriented services like TCP. A wide variety of new services will be made available in the high-speed networks, such as continuous media service for real-time data applications such as voice and video. These services will provide bounds on delay and loss while guaranteeing some minimum bandwidth.

Recently variations of the TCP have been proposed that work well in the wireless domain. These proposals are based on the traditional view that wireless networks are merely extensions of fixed networks. One such proposal is called I-TCP[5] for indirect TCP. The motivation behind this work stems from the following observation. In TCP the sender times out and begins retransmission after a timeout period of several hundred milliseconds. If the other endpoint of the connection is a MH, it is possible that the MH is disconnected for a period of several seconds (while it moves between cells and performs the initial greeting). This results in the TCP sender timing out and transmitting the same data several times over, causing the effective throughput of the connection to degrade rapidly. To alleviate this problem, the implementation of I-TCP separates a TCP connection into two pieces—one from the fixed host to another fixed host that is near the MH and another from this host to the MH (note the similarity of this approach with the approach in Fig. 6). The host closer to the MH is aware of mobility and has a larger timeout period. It serves as a type of gateway for the TCP connection because it sends ACKs back to the sender before receiving ACKs from the MH. The performance of I-TCP is far superior to traditional TCP for the mobile networks studied.

In the architecture proposed in Fig. 6, a TCP connection from a fixed host to a MH would terminate at the SH. The SH would set up another connection to the MH and would have the responsibility of transmitting all packets correctly. In a sense this is a similar idea to I-TCP except that in the wireless network VCs are used rather than datagrams. Therefore, the implementation of TCP service is made much easier.

A problem that is unique to the mobile domain occurs because of the unpredictable movement of MH nodes (i.e., a MH may roam between cells resulting in a large variation of available bandwidth in each cell). Consider the following example. Say nine MH nodes have opened 11-Kbps connections in a cell where the available bandwidth is

100 Kbps. Let us say that a tenth MH M10, also with an open 11-Kbps connection, wanders in. The total requested bandwidth is now 110 Kbps while the available bandwidth is only 100 Kbps. What is to be done? One approach would be to deny service to M10. However, this seems an unfair policy. A different approach is to penalize all connections equally so that each connection has 10-Kbps bandwidth allocated.

To reduce the bandwidth for each connection from 11 to 10 Kbps, two approaches may be adopted:

1. Throttle back the sender for each connection by sending control messages.
2. Discard 1-Kbps data for each connection at the SH. This approach is only feasible for applications that are tolerant of data loss (e.g., real-time video or audio).

The first approach encounters a high overhead in terms of control messages and requires the sender to be capable of changing the data rate dynamically. This may not always be possible; for instance, consider a teleconference consisting of several participants where each mobile participant is subject to dynamically changing bandwidth. In order to implement this approach, the data (video or audio or both) will have to be compressed at different ratios for each participant, and this compression ratio may have to be changed dynamically as each participant roams. This is clearly an unreasonable solution to the problem. The second approach requires the SH to discard 1-Kbps of data for each connection. The question is, how should this data be discarded? That is, should the 1 Kb of discarded data be consecutive (or clustered) or uniformly spread out over the data stream every 1 sec? The way in which the data is discarded has an effect on the final perception of the service by the mobile user. If the service is audio, e.g., a random uniform loss is preferred to a clustered loss (where several consecutive words are lost). If the data is compressed video, the problem is even more serious because most random losses will cause the encoded stream to become unreadable resulting in almost a 100% loss of video at the user.

A solution to this problem is proposed in Seal and Singh,[13] where a new sublayer is added to the transport layer called the *loss profile transport sublayer* (*LPTSL*). This layer determines how data is to be discarded based on special transport layer markers put by application calls at the sender and based on negotiated loss functions that are part of the QoS negotiations between the SH and service provider. Fig. 7 illustrates the functioning of this layer at the service provider, the SH, and the MH. The original data stream is broken into *logical segments* that are separated by markers (or flags). When this stream arrives at the SH, the SH discards entire logical segments (in the case of compressed video, one logical segment may represent one frame) depending on the bandwidth available to the MH. The purpose of discarding entire logical segments is that discarding a part of such a segment of data makes the rest of the data within that segment useless—so we might as well discard the entire segment. Observe also that the flags (to identify logical segments) are inserted by the LPTSL via calls made by the application layer. Thus, the transport layer or the LPTSL does not need to know encoding details of the data stream. This scheme is currently being implemented at the University of South Carolina by the author and his research group.

CONCLUSIONS

The need for wireless LANs is driving rapid development in this area. The IEEE has proposed standards (802.11) for the physical layer and MAC layer protocols. A great deal of work, however, remains to be done at the network and transport layers. There does not appear to be a consensus regarding subnet design for wireless LANs. Our work has indicated a need for treating wireless LAN subnetworks as being fundamentally different from fixed networks, thus resulting in a different subnetwork and transport layer designs. Current efforts are under way to validate these claims.

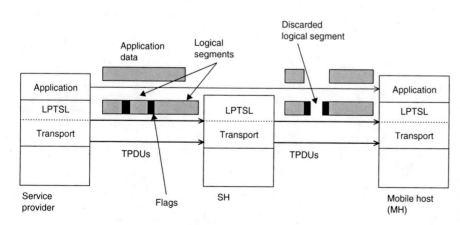

Fig. 7 LPTSL, an approach to handle dynamic bandwidth variations.

DEFINING TERMS

CSMA: Protocols such as those used over the Ethernet.

MAC: Protocols arbitrate channel access between all nodes on a wireless LAN.

MH nodes: The nodes of wireless LAN.

SH: The node that takes care of flow-control and other protocol processing for all connections.

REFERENCES

1. Bantz, D.F.; Bauchot, F.J. Wireless LAN design alternatives. IEEE Netw. **1994**, *8* (2), 43–53.
2. Chen, K.-C.; Lee, C.H. RAP: a novel medium access control protocol for wireless data networks. IEEE GLOBECOM 1993, Houston, TX, Nov 29–Dec 2, 1993; IEEE Press: Piscataway, NJ, 1993; 1713–1717.
3. Chen, K.-C. Medium access control of wireless LANs for mobile computing. IEEE Netw. **1994**, *8* (5), 50–63.
4. Goodman, D.J. Cellular packet communications. IEEE Trans. Commun. **1990**, *38* (8), 1272–1280.
5. Barke, A.; Badrinath, B.R. I-TCP: indirect TCP for mobile hosts. Tech. Rept. **1994**, DCS-TR-314, Dept. Computer Science, Rutgers University, Piscataway, NJ, 1994.
6. Glisic, S.G. 1-Persistent carrier sense multiple access in radio channel with imperfect carrier sensing. IEEE Trans. Commun. **1991**, *39* (3), 458–464.
7. Tobagi, F.; Kleinrock, L. Packet switching in radio channels: part I carrier sense multiple access models and their throughput delay characteristic. IEEE Trans. Commun. **1975**, *23* (12), 1400–1416.
8. Tobagi, F.; Kleinrock, L. Packet switching in radio channels: part II the hidden terminal problem in CSMA and busy-one solution. IEEE Trans. Commun. **1975**, *23* (12), 1417–1433.
9. Ioanidis, J.; Duchamp, D.; Maguire, G.Q. IP-based protocols for mobile internetworking. ACM SIGCOMM 1991, Zürich, Switzerland, Sept 3–6, 1991; ACM Press: New York, NY, 235–245.
10. Teraoka, F.; Tokoro, M. Host migration transparency in IP networks: the VIP approach. ACM SIGCOMM 1993, San Francisco, CA, Sept 13–17, 1993; ACM Press: New York, NY, 1993; 45–65.
11. Ghai, R.; Singh, S. An architecture and communication protocol for picocellular networks. IEEE Pers. Commun. Mag. **1994**, *1* (3), 36–46.
12. Singh, S. Quality of service guarantees in mobile computing. J. Comput. Commun. **1996**, *19*, 359–371.
13. Seal, K.; Singh, S. Loss profiles: a quality of service measure in mobile computing. J. Wirel. Netw. **1996**, *2*, 45–61.

FURTHER INFORMATION

A good introduction to physical layer issues is presented in Bantz and Bauchot[1] and MAC layer issues are discussed in Chen.[3] For a discussion of network and transport layer issues, see Singh[12] and Ghai and Singh.[11]

Wireless LANs (WLANs): Ad Hoc Routing Techniques

Ahmed K. Elhakeem
Concordia University, Montreal, Quebec, Canada

Abstract

This entry provides an overview of ad hoc routing techniques for wireless local area networks.

INTRODUCTION

There has been a growing interest in providing mobile users traversing nested wireless LANs (LANs) with free roaming and minimum call blocking. Existing IEEE 802.11 based wireless LANs[1] have minimal capabilities when it comes to providing wireless connectivity among neighboring wireless LAN segments. These networks have always found applications in the military and disasterous relief fields. However, the future may carry more surprises given the need to deploy the cellular based macrosystems in remote areas, e.g., underground stations of public transportation. The high bandwidth these wireless LANs provide is another attribute to their utilization and interconnection to the generally narrower band macrosystems. This paves the way to wider deployment of the bandwidth-thirsty multimedia applications such as full IP web access and many other services of the intelligent transportation systems.[2] Many such LAN segments may be interconnected by means of a terrestrial Ethernet or other networks utilizing the standard access point (AP) mode of operation. However, such connectivity is lacking within the IEEE 802.11 ad hoc mode. In the latter, users establish peer-to-peer communications without seeking the help of a central station AP. One can envision that the distributed ad hoc mode will be adopted by the multivendor multidesign based deployment of such nested LANs. This will make the need for efficient routing of users' packets by wireless means even greater. For mobile users, faster roaming between the microcells of the nested wireless LANs necessitates the devising of efficient routing techniques so as to mitigate the combined effects of logical and physical link failures.

Ad hoc networks are easier to deploy compared to cellular counterparts which necessitate expensive planning, support ground networks, and expensive base stations (BSs). Failure of a few APs (wireless LAN BSs) may not lead to overall network failure. On the other hand, the limited communication range wireless LANs may have is more than offset by the high rate and cheaper operations in the unlicensed 900 MHz, 1900 MHz, 2.4 GHz, and 6 GHz bands. Ad hoc routing techniques for wireless LANs can be categorized as follows:[3]

- **Global precomputed routing (sometimes called flat or proactive):** All routes to all nodes are computed a priori and updated periodically at each node. Distance vector routing distance-vector routing (destination-sequenced distance-vector routing, or DSDV)[4] and links state (LS)[5] routing fit into this category.
- **Hierarchical routing:**[6] Here, two kinds of nodes exist: end points and switches. End points select the switch (similar to the BS of the cellular system) and form a cell around it. In turn, the switches form clusters among themselves. Cluster heads are appointed and form a higher level cluster and so on.
- **Flooding and limited flooding techniques:** A node hearing a data racket rebroadcasts to all neighbors and so on. To limit the amount of overhead traffic, limited flooding techniques are proposed.[2]
- **On-demand routing techniques (sometimes called reactive):** Here, nodes issue route requests (R-requests) and receive route replies as the need arises with no build-up of routing tables nor a forwarding data basis.

AD HOC ON-DEMAND DISTANCE VECTOR ROUTING

The Routing Algorithm

This technique is a mix of proactive/reactive ad hoc routing approaches. It establishes a balance between full proactivity where the wireless node stores all information about every node in the wireless Internet (complete routing tables) and full reactivity where there is no information about any node. In the latter case, excessive routing query-response will have to be carried out leading to long latency (call waiting delay).[7]

Encyclopedia of Wireless and Mobile Communications DOI: 10.1081/E-EWMC-120043956

Wireless
LANs—Mesh

Source	Dest	Source Seq_no	Dest Seq_no	# of Hops	Prev hop

Fig. 1 A typical route request packet of AODV.

Reply to	Source	Dest	Dest Seq_no	# of Hops	Prev hop

Fig. 3 A typical route reply message in AODV.

The routing algorithm can be summarized as follows:

1. The source node consults its routing table for the intended call destination. If it finds this destination, the call will commence, otherwise a R-request packet is sent. Fig. 1 shows the routing related fields in this packet. Other typical fields such as SYNC, length, cyclical redundancy check (CRC), etc. are not shown.

2. Every intermediate node hearing one or more R-request packets for the lifetime of the same source-destination will flood only the first heard request (and if and only if it does not know the route to destination). Each intermediate node will also update its routing table to include the source of the R-request in preparation for the reverse-path setup (Fig. 2). By flooding, multicast, or broadcast we mean that the packet is physically heard by all neighbors and processed by all neighbors.

3. If the node hearing the R-request finds a route to the specified destination in its routing tables or finds itself as the requested destination, it will unicast (unicast means that the packet is physically heard by all neighbors but processed only by the intended node) back an route reply (R-reply) packet to the immediate source through which the R-request was heard; it will not multicast the R-request. Fig. 3 shows a typical R-reply packet, again not showing the SYNC, CRC, and other fields not related to routing. The sequence number in Fig. 2 is the same as that of the corresponding R-request packet.

4. An intermediate node hearing an R-reply packet will repeat it only unicast to the neighboring node through which it has heard the corresponding R-request packet in the first place. It will also insert the destination of the R-reply and the corresponding intermediate node that delivered this R-reply in its routing table. The later information forms the forward path, Fig. 4, which is subsequently used for forwarding regular data packets from source to destination.

5. Intermediate nodes hearing an R-request but no corresponding R-reply will time out after a certain time and discard the bending R-request and the corresponding reverse path.

6. Intermediate nodes receiving an R-reply for which no corresponding R-request was received will discard it. However, they will update their routing tables to include the source and destination of the heard R-reply.

7. The source may finally receive many R-reply and, so, many routes to the same destination. The route with smallest number of hops is then selected. However, discovered routes with a more recent sequence number are preferred even if they have more hops as compared to an R-reply with older sequence numbers.

8. Nodes store route information for certain source-destination pairs for a certain lifetime beyond which these routes are eliminated.

The Build-up of the Routing Table of Ad Hoc on-demand Distance Vector Routing

1. Every node periodically sends a hello packet to its neighbors (Fig. 5). Ad hoc on-demand distance vector routing (AODV) also has R-request error and other types of packets.[7]

2. All neighbors update their lists of neighbors reacheability set and their routing tables for each hello packet received.

3. Each entry in the reacheability set formed at each node has a lifetime. Upon exceeding this, the entry and all involved routes are erased unless a fresh hello packet is received before expiry of this lifetime. The expiry of these entries is an indication that certain nodes have moved out of reach, in which case link failure may occur and involved source–destination pairs have to be notified.

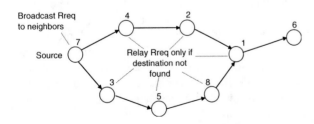

Fig. 2 Route request propagation: reverse path setup.

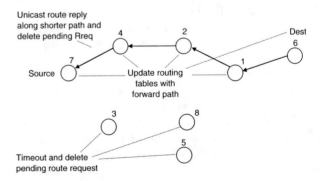

Fig. 4 Route reply propagation: forward path setup.

Source	Seq_no	List of neighbors

Fig. 5 A typical "hello" message in AODV.

Notes

AODV[7] has other types of packets such as route error (RERR) packets, R-reply ACK, and R-reply-ACK packets. AODV offers features such as local repair of a route (due to intermediate or end node mobilities). A gratitutous mode, where an intermediate node who knows the route to a certain destination would unicast this back immediately to the requesting source without further broadcasting the R-request to the destination, also exists in AODV.

FISHEYE ROUTING

The Routing Algorithm

Though the subject node stores routing entries in its table for all possible nodes, it communicates its routing table contents less frequently to more distant nodes.[3,8] As in Fig. 6, the subject node #11 is surrounded by the nodes in the first oval (first scope). Routing information (table of contents) between node #11 and nodes of first scope is exchanged at the highest rate. Routing information is exchanged at a lesser rate between node #11 and member nodes of second scope (second oval or second hop) and so on. This way, large latencies (waiting for routes to be exchanged) will occur if the applicable node #11 wishes to communicate with nodes in second or third scopes (hops), but the flooding overhead of link state (LS) routing[5] for all nodes (near and far) is eliminated in fisheye routing (FSR). This LS flooding is caused by the fact that each node has to transmit its routing table contents at the same rate irrespective of distance from the applicable node meaning more routing overhead.

Performance of FSR

The computer simulations in Ref. 3 are conducted for five kinds of routing techniques including FSR. Two scopes (hops) were taken, for a network of 200 nodes. Data packets were generated from a Possion kind of traffic with an inter-arrival period of 2 sec amounting to a data rate of 4 Kbps per node source. Arriving data packets consist of 10 Kb, nodes moved at a speed of 60 kmph, maximum radio range is 100 m, and the data rate on channel is 2 Mbps. The routing tables are refreshed at a rate of once every 2 sec for first scope and once every 6 sec for second scope nodes.

Figs. 9–13 show the routing performance results obtained. Figs. 7 and 8 show that the routing overhead (control) per cluster does not change much with active nodes' numbers or their mobilities (speeds). However, FSR control overhead is relatively higher than the other routing techniques (to be discussed).

Fig. 9 shows that the average data packet delay does not rise much with nodes' mobilities. FSR performance is the best in this regard compared to other techniques. Similar results are drawn from Fig. 10 which shows the average number of hops traveled by a typical data packet.

Fig. 11 shows that the FSR overhead ratio does not increase as much as other techniques as the network size increases (while keeping the same user density). In comparison, Fig. 7 uses the same geographical area while increasing the number of users.

Figs. 7–11 also show the performance of simulated on-demand routing techniques such as AODV, dynamic source routing (DSR), and temporally-ordered routing algorithm (TORA) (to be presented). In types A and B, the on-demand routing tables are updated every 3 and 6 sec, respectively.

HIERARCHICAL STATE ROUTING

The Routing Algorithm

Hierarchical state routing (HSR) combines dynamic, distributed multilevel hierarchical grouping (clustering) and efficient location management. Clustering (dividing nodes into groups and different kinds of nodes) at the media access control (MAC) and network layers helps the routing of data as the number of nodes grow and, subsequently, the cost of processing (nodes memory and processing required).[3]

HSR keeps a hierarchical topology, where selected cluster heads at the lowest level become members of the next higher level and so on. While this clustering is based on network topology (i.e., physical), further logical partitioning of nodes eases the location management problem

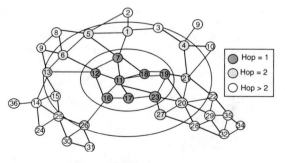

Fig. 6 Scope of fisheye.[3] *Source*: ©IEEE. Iwata, A.; Chiang, C.C.; Pei, G.; Gerla, M.; Chan, T.W. Scalable routing strategies for ad hoc wireless networks. IEEE J. Sel. Areas Commun. 1999, 17 (8), 1369.

Wireless
LANs—Mesh

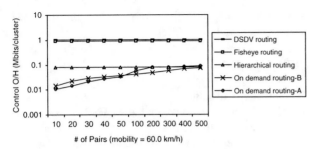

Fig. 7 Control O/H vs. traffic pairs fixed area.[3]
Source: ©IEEE. Iwata, A.; Chiang, C.C.; Pei, G.; Gerla, M.;
Chan, T.W. Scalable routing strategies for ad hoc wireless
networks. IEEE J. Sel. Areas Commun. 1999, 17 (8), 1369.

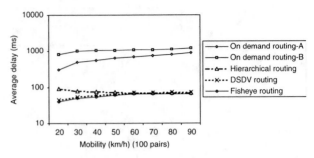

Fig. 9 Average delay vs. mobility (100 pairs).[3]
Source: ©IEEE. Iwata, A.; Chiang, C.C.; Pei, G.; Gerla, M.;
Chan, T.W. Scalable routing strategies for ad hoc wireless
networks. IEEE J. Sel. Areas Commun. 1999, 17 (8), 1369.

in HSR. Fig. 14 shows four physical level clusters, namely, CO-1, CO-2, CO-3, and CO-4. Cluster heads are selected either manually (by the network administrator) or via the appropriate real time distributed voting mechanism, e.g., Refs. 9 and 10. In Fig. 12, nodes 1, 2, 3, and 4 are assumed to be cluster heads, nodes 5, 9, and 10 are internal nodes, and nodes 6, 7, 8, and 11 are gateway nodes.

Gateway nodes are those belonging to more than one physical cluster. At the physical cluster level, all cluster heads, gateways, and internal nodes use the MAC address, but each also has a hierarchical address as will be discussed.

Cluster heads exchange the LS information with other cluster heads via the gateway nodes. For example, in Fig. 12, cluster heads 2 and 3 exchange LS information via gateway 8. Cluster heads 4 and 3 via gateway 11 and so on. In the sequel, logical level C1-1 is formed of cluster heads 1, 2, 3, and 4. However, in level C1-1, only 1 and 3 are cluster heads.

Similarly, at level C2-1, only 1 is a cluster head while node 3 is an ordinary node at the C1-2 level. Routing table storage at each node is reduced by the aforementioned hierarchical topology. For example, for node 5 to deliver a packet to node 10, it will forward it to cluster head 1

which has a tunnel (route) to cluster head 3, which finally delivers the packet to node 10. But how would cluster head 1 know that node 10 is a member of a cluster headed by 3? The answer is that nodes in each cluster exchange virtual LS information about the cluster (who is the head, who are the members) and lower cluster (with less details), and the process repeats at lower clusters. Each virtual node floods this LS information down to nodes within its lower level cluster. Consequently, each physical node would have hierarchical topology information (actually, summary of topology including cluster heads and member nodes) rather than the full topology of flat routing where all individual routes to each node in the networks are stored at all nodes and are exchanged at the same rate to all nodes.

The hierarchical address (a sequence of MAC addresses) is sufficient for packet delivery to any node in the network. For example, in Fig. 11, node HID (5) = (1, 1, 5) going from the top to the lowest cluster, 1 is the cluster head of clusters C1-1 and CO-1, and node 5 is an interior node of CO-1. Similarly, HID (6) = (3, 2, 6) and HID (10) = (3, 3, 10). Returning back to the example above where node 5 seeks a route to node 10, it will ask 1 (its cluster head). Node 1 has a virtual link or tunnel, i.e., the

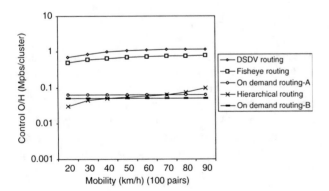

Fig. 8 Control O/H vs. mobility (100 pairs).[3]
Source: ©IEEE. Iwata, A.; Chiang, C.C.; Pei, G.; Gerla, M.;
Chan, T.W. Scalable routing strategies for ad hoc wireless
networks. IEEE J. Sel. Areas Commun. 1999, 17 (8), 1369.

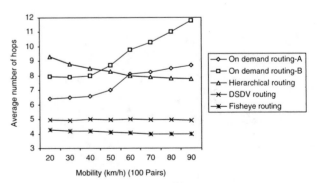

Fig. 10 Average hops vs. mobility (100 pairs).[3]
Source: ©IEEE. Iwata, A.; Chiang, C.C.; Pei, G.; Gerla, M.;
Chan, T.W. Scalable routing strategies for ad hoc wireless
networks. IEEE J. Sel. Areas Commun. 1999, 17 (8), 1369.

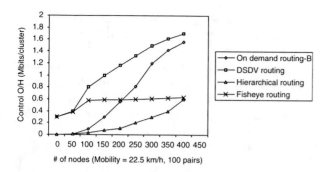

Fig. 11 Control O/H vs. number of nodes (area increase with number of hops).[3] *Source*: ©IEEE. Iwata, A.; Chiang, C.C.; Pei, G.; Gerla, M.; Chan, T.W. Scalable routing strategies for ad hoc wireless networks. IEEE J. Sel. Areas Commun. 1999, 17 (8), 1369.

succession of nodes (1, 6, 2, 8, 3) to node 3 which is the cluster head of the final destination. This tunnel is computed from the LS information flooded down from higher cluster heads, as above. Finally, node 3 delivers the packet to node 10 along the downward hierarchical path which is a mere single hop. Gateways have multiple HID since they belong to more than one physical cluster.

Performance of HSR

Figs. 7–11 show the performance of HSR. The refresh rate of routing tables is 2 sec. Fig. 7 reveals that the overheard ratio in HSR is a relatively constant (w.r.t) number of communicating pairs. Also, it is better than that of FSR or digital simultaneous voice and data (DSVD). Fig. 7

shows that the HSR overhead ratio is better than FSR and DSDV as the nodes increase their speed. Fig. 9 also shows better delay for HSR, but the average number of hops in Fig. 10 does not reveal marked improvement compared to other routing techniques. Finally, Fig. 11 shows a superior overhead ratio performance of HSR, as the number of communicating nodes rises (while keeping the user density the same). Figs. 7–11 also show the performance of simulated on-demand routing techniques such as AODV, DSR, and TORA. In types A and B, the on-demand routing tables are updated every 3 and 6 sec, respectively.

Drawbacks

Utilization of long hierarchical addresses and the cost of continuously updating the cluster hierarchical and hierarchical addresses as nodes move (memory and processing of nodes) are the shortcomings of HSR. Also, clustering and voting for cluster heads may consume more overhead not to mention creating processing, security, and reliability hazards at cluster heads at different hierarchical levels.

DESTINATION-SEQUENCED DSDV

Description

DSDV is based on the routing information protocol (RIP)[11] used within the Internet.[3,4,11] Each node keeps a routing table containing routes (next hop nodes) to all possible destination nodes in the wireless network. Data packets are routed by consulting this routing table. Each entry in the table (Fig. 13) contains a destination node address, the address of the next hop node enroute to this destination, the number of hops (metric) between destination and next hop node, and the lifetime of this entry (called the sequence number). Each node must periodically transmit its entire routing table to its neighbors using update packets.

Build-up and update of routing tables is done via the update packets to be described shortly. Fig. 13 shows how a typical data packet is forwarded from node 1 to final destination node 3. Node 1 consults its routing table and finds node 4 as the immediate node to reach destination node 3. Node 1 then forward the data packet (Fig. 14A) with the destination field equal to 3 and the next hop field set to 4. All nodes in the neighborhood of node 1 hear this data packet (recall the physical broadcast nature of wireless channels), but only node 4 processes this packet. Node 4 consults its routing table (Fig. 14B) and finds the next hop node 5 corresponding to destination field set to 3. Node 4 builds a new packet with the destination field set to 3 and the next hop field set to 5 and transmits this packet. This packet is similarly processed upon its reception at node 5 and so on until the packet reaches the final destination.

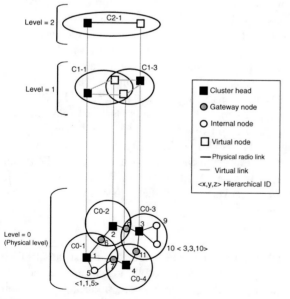

Fig. 12 An example of physical/virtual clustering.[3] *Source*: ©IEEE. Iwata, A.; Chiang, C.C.; Pei, G.; Gerla, M.; Chan, T.W. Scalable routing strategies for ad hoc wireless networks. IEEE J. Sel. Areas Commun. 1999, 17 (8), 1369.

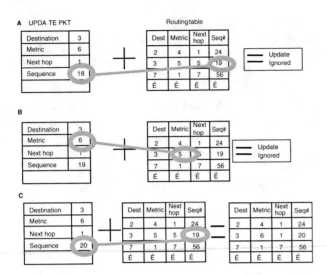

Fig. 13 A node receiving three update packets.

Routing tables, as such, are updated at a slower rate within the terrestrial based Internet. However, in the case of the mobile wireless LANs and due to nodes mobility, the routing table's update has to occur at faster rate. A great deal of DSDV processing and bandwidth (overhead) is consumed by table management (generation and maintenance). Routing tables have to react faster to topology changes or routing loops may result. The route update packets (Fig. 13) contain a destination field which has either the address of the first node (called destination) originating the route update or the address of one of its neighbors. The metric field is set to 1 by the first node originating the route update. This field denotes the number of wireless hops to the declared destination. This field is incremented by 1 by each node receiving this update and repeating it to its neighbors. The next hop field has the address of the node which has just retransmitted the route

update packet. The sequence number is set by the originating node. As the node moves, it sends fresh route update packets with higher sequence numbers. Nodes hearing these will update their tables according to the information in the newer packets from the same originator (destination), i.e., from packets having higher sequence numbers (for same destination). Fig. 13 shows three scenarios corresponding to a certain intermediate node receiving three different update packets from neighboring nodes. In Fig. 13A, the received sequence (18) is less than what the subject node has already in routing table (19) to the same destination (3), i.e., it is older and so is discarded.

In Fig. 13B, the sequence number is the same in both of the route update packets and the routing table, but the metric is higher in the route update packet. In this case, the route update packet is discarded (meaning that this update packet looped too much and so the smaller metric already existing in the table is preferred).

In Fig. 13C, the received route update packet has a sequence number (20) which is higher than the existing value in the routing table (for the same destination). In this case, the received route update packet replaces the existing entry in the routing table even if the metric in the table is smaller than that in the received route update packet. Higher sequence numbers mean fresh information and are preferred even if this new routing information has a shorter path to its destination, i.e., a smaller metric.

Performance of DSDV

This is shown in Figs. 7–11. Fig. 7 shows that the DSDV overhead performance is worse than FSR, HSR, etc. Fig. 8 shows that the DSDV overhead performance vs. speed of nodes is prohibitive. Figs. 7 and 8 also show that DSDV performance is somewhat independent of the nodes traffic conditions; this is due to the fact that all nodes are busy with the transmission of their routing tables. DSDV exhibits good delay and an average number of hops (Figs. 9 and 10). However, for highly deployed networks where the number of nodes increases as the area enlarges, the overhead ratio is much worse than FSR, HSR, etc. (Fig. 11). Figs. 7–11 also show the performance of simulated on-demand routing techniques such as AODV, DSR, and TORA. In types A and B, the on-demand routing tables are updated every 3 and 6 sec, respectively.

Comments

DSDV is a flat routing technique that propagates 100% of the routing table contents of all nodes and with the same frequency for all table entries. The frequency of transmitting the routing tables becomes higher as node speeds increase. Ref. 4 contains procedures for handling different network layer addresses, damping routing oscillations due to topology changes, etc. For example, broken links (a

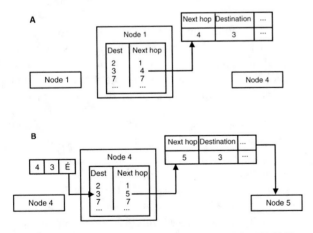

Fig. 14 Routing example in distance-vector routing (DSDV): (A) Node 1 transmits packet to node 4 for forwarding; (B) Node 4 looks up the destination in its routing table.

node does not hear from a neighbor for some time) are repaired by transmitting a new route update packet with a larger sequence number and a metric of ∞.

ZONE ROUTING PROTOCOL

Description

This is a hybrid reactive/proactive routing technique. Routing is flat and not hierarchical (as in HSR) thus leading to reducing the processing and overhead. Nodes keep routing tables of a node's neighborhood and not of the whole network.[12]

Query/reply messaging is needed if the destination is not found in the routing table. The query/reply process is handled only by certain selected nodes (called border nodes of the zone). On the other hand, all interior nodes repeat (physical broadcast) the query/reply packet but do not process it. The border nodes are not gateways nor cluster heads such as the case in HSR. They are plain nodes located at the borders of the routing zone of the applicable node. The routing zone of a certain node of radius $r > 1$ is defined as the collection of these nodes which can be reached via 1, 2, or 4, ..., r hops from the applicable node. Nodes within one hop from the center node (e.g., node S in Fig. 15) are those that can directly hear the radio transmission of node S. These are nodes A, B, C, D, and E (one-hop nodes). The higher the power of a typical node S, the larger the number of one hop nodes. This may lead to increasing the control traffic necessary for these nodes to exchange their routing tables. Also, this increases the level of contention of the IEEE 802.11 carrier sense multiple access (CSMA) based access.[1] The routing zone in Fig. 15 corresponds to a routing zone of two hops ($r = 2$). Each node in this routing zone exchanges routing packets only with members in its routing zone (nodes A–K in Fig. 15). This takes place according to the intrazone protocol (IARP) which forms the proactive part of the ZRP.

Nodes G–K in Fig. 15 are called the peripheral nodes of node S. These again are ordinary nodes but just happened to be at 2 hops, i.e., at the border of the routing zone of node S ($r = 2$, zone radius of routing zone of Fig. 15). For a different node, the peripheral nodes will change. The IARP provides nodes with routing tables with a limited number of entries corresponding to nodes in the same routing zone (nodes A–K in Fig. 15). If the destination lies outside the routing zone such as node L in Fig. 15, the calling node (say node S) will issue a query to search for a route to node L. This query packet will be retransmitted by nodes A–E to the peripheral nodes. These are the only nodes that will process and respond to the route query packet.

This process is called bordercasting and the underlying protocol is called interzone routing protocol (IERP). The identities, distances, and number of hops of the peripheral

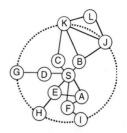

Fig. 15 A routing zone of two hops radius.[13] *Source*: ©IEEE. Pearlman, M.R.; Haas, Z.J. Determining the optimal configuration for the zone routing protocol. IEEE J. Sel. Areas Commun. 1999, 17 (8), 1395.

nodes of the applicable node as well as all other nodes (interior nodes) in its routing zone are assumed to be known to each node in the wireless Internet. Query packets are sent unicast or broadcast to all peripheral nodes if the applicable node does not find the destination in its routing tables. Further, if these peripheral nodes do not find the required destination in their zone, they would forward the query to their own peripheral nodes and so on. If one of the peripheral nodes know the route to destination, it will return a route-reply packet to the node that sent the query. In Fig. 16 node S does not find the destination node D in its routing zone.

S bordercasts a query packet to nodes C, G, and H. Each of these retransmits this query packet to its peripheral nodes after finding that D is not in their routing zones. The process repeats until node B finds the destination node D in its routing zone. Node B returns a reply containing the ID sequence S–H–B–D.

In this route accumulation routine process, each node adds its ID to the query packets then transmits it to its peripheral nodes. The destination node uses the reversed accumulated sequence of node IDs to send a R-reply back to the source node.

The accumulated ID sequence is likened to source routing techniques in terrestrial networks. To alleviate longer R-reply packets due to this accumulation, intermediate peripheral nodes may store temporary short routing

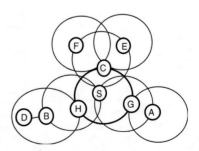

Fig. 16 An example of IERP operation.[12] *Source*: ©IEEE. Pearlman, M.R.; Haas, Z.J. Determining the optimal configuration for the zone routing protocol. IEEE J. Sel. Areas Commun. 1999, 17 (8), 1395.

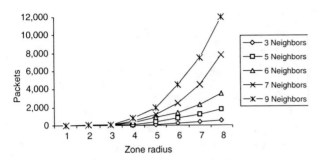

Fig. 17 IARP traffic generated per neighbor.[12]
Source: ©IEEE. Pearlman, M.R.; Haas, Z.J. Determining the optimal configuration for the zone routing protocol. IEEE J. Sel. Areas Commun. 1999, 17 (8), 1395.

tables. These contain the IDs of the previous peripheral nodes who have sent the query. The intermediate peripheral nodes will also overwrite the ID of the previous peripheral node from which they have received the query before retransmitting the query packet rather than appending its ID to the query packet. Once this intermediate node receives a R-reply, it will send it to the previous peripheral node whose ID is stored in the short temporary table. Needless to say, even with bordercasting, there is still some flooding of the route query (but less than pure flooding) and the source node of the query may select the route with minimum number of hops.

Performance of ZRP

Figs. 17–21 show some of the simulation results in Ref. 12. In all of these, *v* denotes the speed of the node expressed in neighbor acquisition per second, *d* is the distance from transmitter to call destination, and Rrout_failure is the rate of route failure per second. Fig. 17 shows the volume of traffic in packets generated by a neighbor node vs. the zone radius *r* due to the proactive IARP traffic which increases with the number of neighboring nodes. Fig. 18 shows a similar result for the IERP traffic volumes (due to route query/reply traffic). However, the IERP traffic decreases with the zone radius as opposed to Fig. 17, thus reflecting one of the trade-offs in ZRP and other ad hoc routing

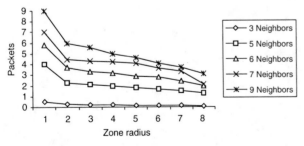

Fig. 18 IERP traffic received by each node per query.[12]
Source: ©IEEE. Pearlman, M.R.; Haas, Z.J. Determining the optimal configuration for the zone routing protocol. IEEE J. Sel. Areas Commun. 1999, 17 (8), 1395.

techniques (i.e., proactivity vs. reactivity). Fig. 19 shows the route failure rate, which decreases with *r*. The route failure is caused by node mobilities and topology changes. Failure rate increases slightly for higher numbers of nodes. Fig. 20 shows the ZRP traffic per node vs. *r*. Optimal values for *r* are possible, as shown (yielding minimum traffic in packets/sec). The dependence of optimal values of zone radius r on nodes speeds is shown in Fig. 21.

Comments

ZRP is a good example of an ad hoc routing technique that tries to strike a balance between proactivity and reactivity. The MAC layer neighbor discovery control overhead (which may not be belittled) is called association traffic in IEEE 802.11.[1] This overhead, being related to the MAC layer, was not accounted for in all of the results above. Bordercasting is seen to decrease the volumes of IARP traffic. Pearlman [1999] also presents excellent ways of adapting/adjusting the zone radius depending on the ratio of reactive IARP traffic to proactive IARP traffic volumes.[12] However, it was noted[12] that accuracy of the optimal zone radius is a very complex function of underlying wireless Internet and its too many parameters (mobilities, user density, traffic volumes and types, etc.).

DSR

Description

In source routing techniques (wireless and wire based), each data packet contains the complete route identification (list of nodes addresses, IDs, or route paths) from source to destination node (Fig. 22) DSR does not require periodic route advertisement or link status packets, implying less routing overhead.[13,14] An intermediate node hearing a data packet, and finding its ID within, preceded by the ID of the node that transmitted the packet will retransmit this data packet. Otherwise this intermediate node discards the data packet. If the intermediate node is listed as the final destination of the data packet, the packet will not be retransmitted. This is seen as an example of physical multicast (broadcast), due to the nature of the wireless channel, but logical unicast since the lonely node meant by this data packet repeats it. Each node has to find and store an end-to-end route to each destination it wishes to communicate with.

Route discovery is the second axiom of DSR, where the source node broadcasts a R-request whenever it places a call to this destination. Each node hearing this R-request and not finding itself the destination of this request nor finding its address within the route list of the packet (series of intermediate node addresses) will append its address (ID) to the route list then multicasts this packet (physically and logically). The R-request hops this way until the

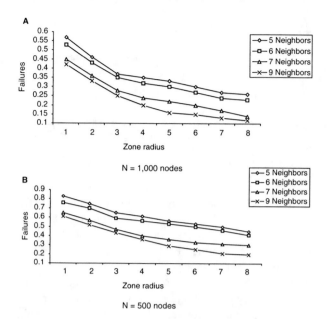

Fig. 19 Route failure rate traffic (normalized to node velocity).[12] *Source*: ©IEEE. Pearlman, M.R.; Haas, Z.J. Determining the optimal configuration for the zone routing protocol. IEEE J. Sel. Areas Commun. 1999, 17 (8), 1395.

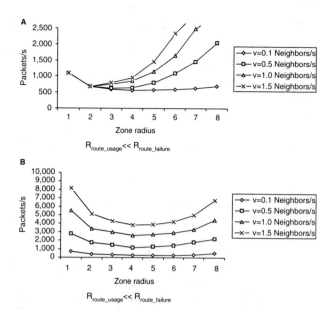

Fig. 21 ZRP traffic per node ($N = 1,000$ nodes, $d = 6.0$ neighbors).[12] *Source*: ©IEEE. Pearlman, M.R.; Haas, Z.J. Determining the optimal configuration for the zone routing protocol. IEEE J. Sel. Areas Commun. 1999, 17 (8), 1395.

destination is reached or one intermediate node has the information regarding the route to destination. In both cases, the R-request is not retransmitted but a R-reply packet is sent back to the source node along the same (but reversed) path of the route list of the route-request

packet which the subject destination node has received. Each intermediate node multicasts a certain R-request from a certain source to a certain destination only once and discards similar R-requests.

All nodes maintain complete routing information (route path) to the destination they recently communicated with as well as routes for which they have handled the R-requests. Similar to wire based networks, the processes above may generate multiple route paths, but the destination node has the option of returning only the first discovered route. Other options include the returning of all possible routes or the best route (which has the minimum number of hops) to the source node.

The third axiom of DSR is route maintenance. Here, a node receiving a data or R-request type packet and detecting a RERR (e.g., an intermediate node moved and can no longer handle certain routes) will send a RERR packet to

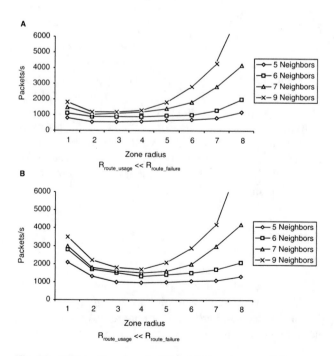

Fig. 20 ZRP traffic per node ($N = 1,000$ nodes, $v = 0.5$ neighbor/s).[12]

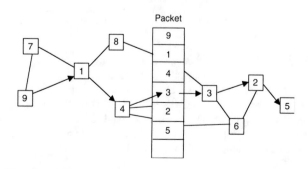

Fig. 22 Path based routing from node 9 to node 5 in DSR.

Wireless LANs—Mesh

Table 1   Latency of first route reply by type [all optimal DSR, rectangular site, constant motion].[14]

	Latency				
	#replies	Mean (ms)	Minimum (ms)	99 Percentile (ms)	Maximum (s)
Neighbor replies	3,136	7.1	1.3	457	1.78
Cache replies	1,524	45.9	1.3	752	2.4
Target replies	12	87.6	23.6	458	458.3

Source: © IEEE. Maltz, D.A.; Broch, J.; Jetcheva, J.; Johnson, D. The efforts of on-demand behavior in routing protocols for multitop wireless ad hoc networks. IEEE J. Sel. Areas Commun. 1999, 17(8), 1439.

the source. All intermediate nodes hearing a RERR packet will remove from their routing tables the broken link for all involved routes (of other source–destination pairs) which use this broken link. All such routes will then become truncated at this point.

In networks with full duplex connections, RERR packets follow the reverse path back to the source. If only half duplex connections are available,[13,14] provide other mechanisms for returning RERR packets back to the source node.

Route maintenance packets are acknowledged either end-to-end [via the transmission control protocol (TCP) transport layer software, for example] or on a hop-by-hop basis (via the MAC layer).

Performance of DSR

A network with 50 nodes was simulated.[14] The nodes moved at variable speeds from 0 to 20 m/s. Fig. 23 shows the packet delivery ratio and routing overhead vs. node mobility (900 in Fig. 23 means a stationary node and 0 means a continuously moving node).

Routing overhead include all kinds of route-request, route-reply, etc. Packet delivery ratio is the total number of packets delivered to all destinations during simulations divided by total number of packets actually generated.

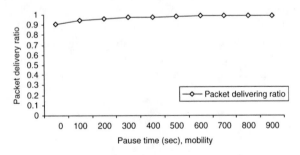

Fig. 23   (A) Baseline dynamic source routing (DSR) packet delivery ratio (all-optimal DSR, rectangular site).[14] (B) Baseline DSR routing packet overhead (all-optimal DSR, rectangular site).[14] *Source*: © IEEE. Maltz, D.A.; Broch, J.; Jetcheva, J.; Johnson, D. The efforts of on-demand behavior in routing protocols for multitop wireless ad hoc networks. IEEE J. Sel. Areas Commun. 1999, 17(8), 1439.

Table 1 shows the latency of first route-reply (time for route discovery). "Neighbor" refers to all nodes which are at one hop from the subject node. "Cache replies" refers to all route replies returned back by an intermediate rather than the destination node. "Target reply" is the route returned by the destination node. Table 2 summarizes the route discovery costs. Non-propagated R-requests are defined as those requests forwarded only to neighbors and not transmitted further. Propagated R-requests are forwarded to all nodes and are transmitted hop-by-hop until the destination.

"Request Og" is the number of R-request originators. "Fw Rep" is the number of replies forwarded and so on. Containment is defined as the percentage of nodes that are not neighbors of the source node. The cost is defined as:

$$\frac{\left(\sum \text{Og Req} + \sum \text{Fw Req} + \sum \text{Og Rep} + \sum \text{Fw Rep}\right)}{\text{Og Req}}$$

Comments

Intelligent caching techniques[13,14] can be used to reduce the route discovery and route maintenance overheads at the expense of memory and CPU times of nodes. Yet the greatest overhead loss of DSR is attributed to the long route path in each packet. Compression techniques may be used to cut such overhead, with the penalty of more frequent occurrence of long bursts of bit errors. The security of DSR is also inferior compared to other non-source coding routing techniques.

SUMMARY

A variety of ad hoc routing techniques were presented. The ideal or standard ad hoc routing technique does not exist yet. Extensive research activities are still going on in this important area. Pure proactive or pure reactive routing did not provide the efficiency nor the dynamicity required in

Table 2 Summary of route discovery costs (all optimal DSR, rectangular site, constant motion).[14]

	Nonpropagating	Propagating	Total
Request Og	957	316	1,273
Request Fw	0	6,115	6,115
Reply Og	3,912	3,215	7,127
Reply Fw	0	7,002	7,002
Containment	77.0%	41.0%	68.0%
Cost	5.06	52.69	16.90

Source: © IEEE. Maltz, D.A.; Broch, J.; Jetcheva, J.; Johnson, D. The efforts of on-demand behavior in routing protocols for multitop wireless ad hoc networks. IEEE J. Sel. Areas Commun. 1999, 17(8), 1439.

interconnected wireless LANs, so most of the recent works are concentrated around hybrid techniques (partly proactive/partly reactive). Establishing the best mix of proactive and reactive techniques is a complex function of the traffic amounts and types, the channel, required quality-of-service (QoS), etc.

The difficulty of standardizing the routing performance criteria over the various routing techniques is one of the main obstacles standing in the way of adapting a standard for ad hoc wireless routing. The processing speeds and memory requirement of the real-time operation of routing candidate remains to be evaluated. It is not uncommon to discard good routing techniques that are too complex for the small battery power, small processors, and memories these handheld devices have. As a matter of fact, for these devices, routing is just one of the many functions the portable node is supposed to handle.

DEFINING TERMS

AP mode: Centralized mode of operation of wireless LANs where the AP resembles the functionalities of the cellular system BS.

Ad hoc routing: Routing techniques amenable to wireless LANs with no predefined borders, no base stations, and with all nodes performing the same distributed routing algorithms based on peer-to-peer communications.

Hierarchical routing: Nodes are sorted into logical groupings; certain nodes assume the functionality of cluster heads and gateways so as to reduce the routing overhead.

MAC: Set of techniques and standards for controlling the right to transmit on a certain shared channel such as TDMA, FDMA, CSMA, etc.

On-demand routing: Similar to reactive routing, e.g., AODV[7] and DSR.[13]

Proactive (flat) routing: Where modes build and update complete routing tables with entries corresponding to all nodes in the wireless Internet.

Reactive routing: Where nodes issue R-requests and receive route replies as the need arises, with no buildup of routing tables nor forwarding data basis.

Unicast: The packet is physically heard by all neighbors but processed by only one intended node and discarded by the other nodes.

REFERENCES

1. Gier, J. *Wireless LANs, Implementing Interoperable Networks*; Macmillan Technical Publishing: New York, USA, 1999.
2. Elhakeem, A.K.; Ali, S.M.; Aquil, F.; Li, Z.; Zeidi, S.R.A. New forwarding data basis and adhoc routing techniques for nested clusters of wireless LANs. Wirel. Commun. J.
3. Iwata, A.; Chiang, C.C.; Pei, G.; Gerla, M.; Chan, T.W. Scalable routing strategies for ad hoc wireless networks. IEEE J. Sel. Areas Commun. **1999**, *17* (8), 1369.
4. Perkins, C.E.; Bhagwat, P. Highly dynamic destination-sequenced distance-vector routing (DSDV) for mobile computers. SIGCOM 1994 Conference on Communication Architectures, Protocols and Applications, London, UK, Aug 31–Sept 2, 1994; 234.
5. Jaquet, P.; Muhlethaler, M.; Qayyum, A. Optimized link state routing protocol; 2000 ITEF, Internet Draft, draft-itef-manet-olsr-01.txt.
6. Kasera, K.K.; Ramanathan, R. A location management protocol for hierarchically organized multihop mobile wireless networks. IEEE ICUPC 1997, San Diego, CA, Oct 12–16, 1997; 158.
7. Perkins, C.E. Ad hoc demand distance vector routing; 1997 ITEF, Internet Draft, draft-itef-manet-adov-00.txt.
8. Kleinrock, L.; Stevens, K. Fisheye: a lenslike computer display transformation. Tech. Report, University of California, Los Angeles, CA, 1971.
9. Chiang, C.C.; Wu, H.K.; Liu, W.; Gerla, M. Routing in clustered multihop, mobile, wireless networks. IEEE

Singapore International Conference on Networks, Singapore, Apr 14–17, 1997; 197.

10. Gerla, M.; Tsai, J. Multiuser, mobile, multimedia radio network. ACM Baltzer J. Wirel. Netw. **1995**, *1* (3), 255.

11. Hedrick, C. Routing information protocol, RFC1058; 1988.

12. Pearlman, M.R.; Haas, Z.J. Determining the optimal configuration for the zone routing protocol. IEEE J. Sel. Areas Commun. **1999**, *17* (8), 1395.

13. Broch, J.; Johnson, D.B.; Maltz, D.A. The dynamic source routing protocol for mobile ad hoc networks; 1998 Internet Draft, draft-itef-manet-dsr-01.txt.

14. Maltz, D.A.; Broch, J.; Jetcheva, J.; Johnson, D. The efforts of on-demand behavior in routing protocols for multihop wireless ad hoc networks. IEEE J. Sel. Areas Commun. **1999**, *17* (8), 1439.

BIBLIOGRAPHY

1. Lin, C.R.; Lui, J.S. QoS routing in ad hoc wireless networks. *IEEE J. Sel. Areas Comm* **1999**, *17* (8), 1426.

2. McDonald, A.B.; Znati, T.F. A mobility-based framework for adaptive clustering in wireless ad hoc networks. *IEEE J. Sel. Areas Comm* **1999**, *17* (8), 1466.

3. Ng, M.J.; Lu, I.T. A peer-to-peer zone-based two-level link state routing for mobile ad hoc networks. *IEEE J. Sel. Areas Comm* **1999**, *17* (8), 1415.

4. Park, V.D.; Corson, M.S. 1997, A highly adaptive distributed routing algorithm for mobile wireless networks INFOCOM '97, Kobe, Japan. 1405.

5. Sivakumar, R.; Sinha, P.; Bharaghavan, V. CEDAR: a core-extraction distributed ad hoc routing algorithm. *IEEE J. Sel. Areas Comm* **1999**, *17* (8), 1454.

Wireless LANs (WLANs): Real-Time Services with Quality of Service (QoS) Support

Ping Wang
Department of Electrical and Computer Engineering, University of Waterloo, Waterloo, Ontario, Canada

Hai Jiang
Department of Electrical and Computer Engineering, University of Alberta, Edmonton, Alberta, Canada

Weihua Zhuang
Department of Electrical and Computer Engineering, University of Waterloo, Waterloo, Ontario, Canada

Abstract

Voice over wireless local area network (VoWLAN) is an emerging application taking advantage of the promising voice over Internet Protocol (VoIP) technology and the wide deployment of WLANs all over the world. The real-time nature of voice traffic determines that controlled access rather than random access should be adopted. Further, to fully exploit the capacity of WLAN supporting voice traffic, it is essential to explore statistical multiplexing and to suppress the large overhead. In this article, we propose mechanisms to enhance the WLAN with voice quality of service (QoS) provisioning capability when supporting hybrid voice/data traffic. Voice multiplexing is achieved by a polling mechanism in the contention-free period and a deterministic priority access for voice traffic in the contention period. Header overhead for voice traffic is also reduced significantly. Delay-tolerant data traffic is guaranteed an average portion of service time in the long run. A session admission control algorithm is presented to admit voice traffic into the system with QoS guarantee. Analytical and simulation results demonstrate the effectiveness and efficiency of our proposed solutions.

Wireless LANs—Mesh

INTRODUCTION

Although originally designed for data services, the Internet can also support real-time traffic such as voice and video. The technology of voice over Internet protocol (VoIP), also known as Internet telephony, IP telephony, or packet voice, enables real-time voice conversations over the Internet. It has attracted much interest from the academia and industry because of the following facts:[1] 1) VoIP has much lower cost than the traditional telephone services; 2) The universal presence of IP makes it convenient to launch VoIP applications; 3) There is an increasing demand for the networks to interact with end users having real-time data, voice, and video images, leading to the requirement of integrated voice, data, and video services; and 4) The emerging digital signal processing (DSP) and voice coding/decoding techniques make VoIP more and more mature and feasible. Therefore, VoIP is anticipated to offer a viable alternative to traditional public switched telephone networks (PSTN).

To provide person-to-person (instead of place-to-place) connections anywhere and anytime, the Internet is expected to penetrate into the wireless domain. One most promising wireless network is the wireless LANs, which have shown the potential to provide high-rate data services at low cost over local area coverage. Working at the license-exempt 2.4-GHz ISM frequency band, the IEEE 802.11b WLAN offers a data rate up to 11 Mbps, while IEEE 802.11a WLAN and the ETSI HIPERLAN/2 can support a data rate up to 54 Mbps at the 5-GHz frequency band. As a wireless extension to the wired Ethernet, WLANs typically cover small geographic areas, in hotspot local areas where the traffic intensity is usually much higher than that in other areas.

The promising VoIP technology and the wide deployment of WLANs are driving the application of voice over WLAN (VoWLAN), which is expected to experience a dramatic increase in the near future.[2] Fig. 1 shows a typical VoWLAN system where voice conversations are supported through the access point (AP). At the sender, the analog voice signal is compressed and encoded by a codec. After the inclusion of the RTP (real-time transport protocol)/UDP (user datagram protocol)/IP headers during the packetization procedure at the transport and network layers, the voice packets are transmitted over the networks and finally to the receiver end. At the receiver, a playout buffer is usually used to alleviate the effect of delay jitter. Then the receiver applies de-packetization and decoding to recover the original voice signal.

Encyclopedia of Wireless and Mobile Communications DOI: 10.1081/E-EWMC-120043590

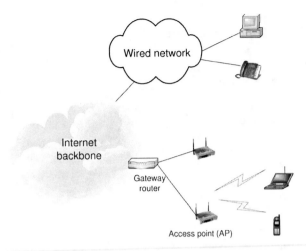

Fig. 1 The architecture for WLANs supporting voice and data services.

There exist two major challenges for VoWLAN. One challenge is how to increase the system capacity for voice users. It has been found out (Refs. 3–5) that the system capacity for voice users is quite low in current WLANs, far from what is needed. Originally designed for data traffic, the WLANs experience bandwidth inefficiency when supporting voice traffic due to the large overhead. Hence, it is essential to enlarge the VoIP capacity supported by WLANs. The other challenge is quality of service (QoS) provisioning for voice users. As a real-time application, VoWLAN is delay-sensitive but can tolerate a certain level of packet loss. Hence, delay and delay jitter are the main QoS measures. Each voice packet should be transmitted within a delay bound. Also, the delay jitter (i.e., variation of voice packet delay) should be carefully controlled as it may degrade voice quality more severely than delay. However, the current WLANs cannot provide guaranteed priority access to voice traffic. When data traffic load is high, the voice traffic may suffer delay bound violation or large delay variance.

The focus of this research is to address the two challenges. We base our work on the IEEE 802.11e,[6] since it is the most promising technology for QoS provisioning in WLANs. With minor modifications to the IEEE 802.11e standard, we can increase the system capacity significantly for voice traffic, provide guaranteed QoS to voice users, and provide data traffic a certain level of service share. Specifically, the contributions of this work are as follows:

- With minor modifications to the IEEE 802.11e, we propose an efficient resource allocation scheme that can provide guaranteed QoS to voice traffic and, at the same time, guarantee an average portion of service time to data traffic.
- The proposed scheme can increase the system capacity significantly in terms of the number of voice sessions by achieving voice traffic multiplexing and effectively reducing the system overhead.

- We propose an analytical model to study the performance of the proposed scheme. The accuracy of the analytical model is verified by extensive simulations. Voice capacity is also derived to facilitate session admission control on voice traffic.

BACKGROUND

Limitations of the Current WLAN Standard (IEEE 802.11) in Supporting Voice

As the most popular WLAN standard, the IEEE 802.11 defines a mandatory distributed coordination function (DCF) and an optional centralized point coordination function (PCF). DCF is based on the carrier sense multiple access with collision avoidance (CSMA/CA), where the collision is resolved by binary exponential backoff. The optional RTS/CTS dialogue can also be applied to further deal with the hidden terminal problem. Mainly designed for data transmission, DCF does not take into account the delay-sensitive nature of real-time services. On the other hand, with PCF, a contention-free period (CFP) and a contention period (CP) alternate periodically. During the CFP, when polled, a station gets the permission to transmit its DATA frames (the DATA frame means the information frame of voice or data traffic). The main drawbacks of PCF include uncontrolled transmission time of a polled station and unpredictable CFP start time.[7]

To enhance the legacy IEEE 802.11 medium access control (MAC), the IEEE 802.11e proposes new features with QoS provisioning to real-time applications. As an extension of DCF, the enhanced distributed channel access (EDCA) provides a priority scheme to distinguish different traffic categories by classifying the arbitration inter-frame space (AIFS), and the initial and maximum window sizes in the backoff procedures. In the IEEE 802.11e, the hybrid coordination function (HCF) can assign specific transmission durations by a polling mechanism. A station can be polled in either CFP or CP.

DCF or EDCA are neither effective nor efficient in supporting the delay-sensitive voice traffic. The contention-based nature and exponential backoff mechanism cannot guarantee that a voice packet is successfully delivered within the delay bound. In addition, the time to transmit the payload of a voice packet is only a very small portion of the total time to transmit the packet due to the overhead such as the RTP/UDP/IP headers, the MAC and physical (PHY) layer headers, and the IFSs. Subsequently, the capacity to accommodate voice traffic in the DCF or EDCA is very limited. For example, the IEEE 802.11b can support approximately 10 simultaneous two-way voice calls if a G.711 codec is used.[8]

In order to guarantee the delay requirement of voice service, controlled access is preferred in WLANs in which the AP polls each voice station periodically. To utilize the radio resources efficiently, two challenging issues need to be addressed:

- *Voice multiplexing.* Generally, voice traffic can be represented by an on/off model: active voice users (at the on state) transmit at a constant rate and inactive users (at the off state) do not transmit, and the durations of the states are independent and exponentially distributed. It is desired to achieve the statistical multiplexing based on this property in VoWLAN.
- *Overhead reduction.* The overhead due to RTP/UDP/IP headers and the polling procedure may significantly degrade the system efficiency and should be reduced as much as possible.

Related Work

In recent years, VoWLAN has drawn a lot of attention from the R&D community. To provide service priority to real-time traffic, EDCA is defined in the IEEE 802.11e. It applies different initial and maximum contention window (CW) sizes and different IFS values to provide differentiation to different types of traffic. However, it provides only statistical rather than guaranteed priority access to high-priority traffic such as real-time voice. In other words, the priority for high-priority traffic is only guaranteed in a long term, but not for every contention. Since each station continues to count down its backoff timer once the channel becomes idle for an IFS, a low-priority packet with a probably large initial backoff timer will eventually count down its backoff timer to a small value, most likely smaller than the backoff timer of a newly backlogged high-priority packet. Then the low-priority packet wins the channel, resulting in the high-priority packet waiting for a long time for the next competition.[9] Such statistical priority access is difficult to satisfy the delay requirement of each voice packet. Furthermore, when applying the EDCA, with an increase of low-priority traffic load, the collision probability seen by the high-priority traffic increases. High-priority traffic can suffer performance degradation when the traffic load of low priority is heavy.[10]

Many efforts have been made on voice traffic capacity analysis. Both experimental results[5] and analytical results[4] have demonstrated that system capacity for voice traffic is very limited in WLANs due to the large header overhead and the inefficiency of IEEE 802.11 MAC protocol. Most of previous work assumes that the voice traffic is constant-rate traffic, which is not the case in reality. For more accurate capacity estimate, the on/off model should be used and the voice traffic multiplexing should be considered. Further, all the above work focuses on DCF based (or contention based) WLANs. However, as mentioned, DCF is neither effective nor efficient to support the delay-sensitive voice traffic. Controlled access is more suitable for voice traffic delivery because of its less overhead and guaranteed delay performance.

Unfortunately, only very limited work in the literature focuses on controlled access.

In order to improve the capacity of voice traffic over WLANs, various solutions have been proposed.[3,11–14] A cyclic shift and station removal polling scheme is proposed (Ref. 11) to take advantage of the multiplexing of voice packets. Without changing the IEEE 802.11 MAC protocol, VoIP capacity is increased by reducing the header overhead of voice packets (Refs. 3, 12). A voice multiplex-multicast (M-M) scheme is proposed (Ref. 3) in which the AP multiplexes packets from several VoIP streams into one multicast packet for transmission. However, an additional delay is expected in composing such a composite packet. Compressed RTP is used to reduce the VoIP header (Ref. 12); however, the overhead incurred by IP, MAC, and PHY remains high. On the other hand, some research[13,14] increases the capacity by introducing new MAC protocols. By reducing the number of collisions or reducing the idle time caused by backoff, these new MAC protocols achieve a larger throughput than the IEEE 802.11 MAC protocol, resulting in an increased capacity. These MAC protocols are all contention based, and none of the studies focus on controlled access.

Since the system capacity is very limited in WLANs, session admission control is important and necessary to maintain the QoS of existing sessions. As revealed in Ref. 5, an additional session that exceeds the system capacity will cause unacceptable quality for all ongoing sessions. Previous research on session admission control in WLANs can be classified into two categories. One is analysis based and the other is measurement based. Analysis based admission control algorithms, including those in Refs. 15 and 16 and the reference model provided in the IEEE 802.11e, make admission decision based on the knowledge of the system capacity derived from analysis. The measurement based algorithms[12,17,18] make admission decisions based on the measurement or estimate of channel utilization. Based on the measurements of the fraction of time in each time unit needed to transmit the flow over the network,[17] collision statistics of each flow,[18] or the transmission time of each traffic type,[12] available/residual capacity budgets are calculated for admission control.

THE PROPOSED MAC SCHEME

Consider a WLAN supporting both voice and data services. We use the same structure of the IEEE 802.11e HCF service interval, and the beacon interval is equal to the service interval. In each service interval, there are two periods: CFP and CP, as shown in Fig. 2. The CFP is used to accommodate voice stations in the downlink (from the AP to the mobile stations) and uplink (from mobile stations to the AP). For the uplink transmission, the AP sends a CF-Poll frame that grants each polled station a transmission opportunity (TXOP). Since voice traffic is delay-sensitive

Wireless LANs—Mesh

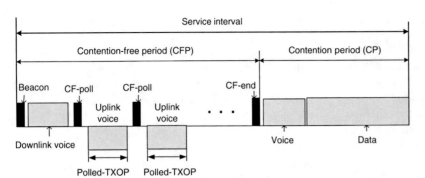

Fig. 2 The structure of a service interval.

and can tolerate a certain level of packet loss, acknowledgment (ACK)/retransmission is not required for voice transmission in order to avoid the retransmission delay. In the CP, the AP and all the stations can contend for the channel. The CP is mainly used to serve data stations and to transmit the first few packets of each voice station's uplink talk spurts. To guarantee the priority of voice over data in the CP, voice packets are always transmitted ahead of data packets; this will be discussed in the "Guaranteed Priority Access to Voice over Data in the CP" section. The length of a service interval is fixed and depends on the delay bound of voice traffic. The lengths of the CFP and CP vary with time, depending on the voice and data traffic loads.

The proposed MAC scheme consists of three components: voice traffic multiplexing, guaranteed priority access to voice traffic, and overhead reduction, as elaborated in the following sections.

Voice Traffic Multiplexing

In order to achieve high resource utilization, the network designers should consider the on/off characteristic of voice traffic, so that resources are allocated to stations only when they are in a talk spurt. However, the IEEE 802.11e does not describe a polling method in HCF to achieve voice traffic multiplexing. Generally, it is easy for the AP to recognize the ending moment of a talk spurt, but it is difficult to know the exact starting moment of a talk spurt. The AP may still need to poll a voice station even during its silent period in order not to miss the beginning of a talk spurt, which is not efficient considering the polling overhead. Here we propose a more efficient polling scheme to achieve the voice traffic multiplexing.

Consider the case when a station initiates a voice session. If the session can be admitted, the AP will add the station to its polling list. Since the duration of each service interval (T_{SI}) is fixed and the voice packet inter-arrival time is a constant in a talk spurt, each station (in the on state) will be granted a fixed TXOP just enough to accommodate the generated voice packets during a service interval. If a polled station has no packet to send or cannot use up all the time of TXOP, the AP considers the station being in the silent period and removes it from its polling list, except the newly added (to the polling list) stations. When

a previously off station has voice packets to send, the station will contend for the channel during the next CP. Once it gets the channel, it will send out all the voice packets in the buffer (as long as the transmission time does not exceed the TXOP). The AP monitors all the packets transmitted in the CP. For every voice packet, the AP records the sender address (or ID) and adds it to the polling list. If the station is newly added to the list during the last service interval, the AP will retain it in the list, even though it may not use up all the TXOP or has no packet to send in the current CFP, since a few voice packets at the beginning of a talk spurt were sent during the last CP.

Once a voice station is added to the polling list, all the subsequent voice packets in the same talk spurt will be transmitted in the CFP. Hence, the voice station does not need to contend for the channel anymore for the current talk spurt.

Guaranteed Priority Access to Voice over Data in the CP

Another challenging issue is raised from the voice multiplexing: To meet the strict delay requirement of uplink voice traffic, it should be guaranteed that a voice station could access the channel successfully during the CP when needed.

To provide QoS guarantee for voice traffic regardless of the data traffic load in WLANs, data stations should not transmit in the CP until no voice station contends for the channel. As discussed in the "Related Work" section, statistical priority access (e.g., the EDCA) cannot meet this requirement. As a result, a guaranteed priority access is more appropriate. Only a few voice packets at the beginning of each talk spurt need to contend in the CP, which should not significantly degrade the QoS of data traffic if a guaranteed priority access is provided to voice.

A simple way to provide guaranteed priority access is to modify the EDCA so that the AIFS of data access category (AC) (AIFS[AC_data]) is equal to (AIFS[AC_voice] + CW_{max}[AC_voice]), the summation of the AIFS of voice AC, and the maximum CW size of voice AC. However, it is not efficient in terms of channel utilization. The number of contending voice packets is expected to be small in the CP, and all the data packets have to wait for a long time before getting the channel, resulting in a waste of resources.

Inspired by the idea of black-burst contention,[19] we propose here an efficient scheme to provide guaranteed priority access to voice, by minor modifications to the IEEE 802.11e EDCA. In our scheme, the AIFSs for voice traffic and data traffic remain the same as those in the EDCA. In addition, the contention behaviors for data stations remain the same as in the EDCA. The contention behaviors of voice stations are modified as follows. For a contending voice station, after waiting for the channel to be idle for AIFS[AC_voice], instead of further waiting for the channel to be idle for a duration of backoff time, the voice station will send a busy tone, and the length of the busy tone (in the unit of slot time) is equal to its backoff timer. After the completion of its own busy tone, the station monitors the channel. If the channel is still busy (which means that at least one other voice station is sending busy tone), the station will quit the current contention and wait for the channel to be idle for AIFS[AC_voice] again. Otherwise, the station (which sends the longest busy tone) will send its voice packets. It is possible that two or more voice stations happen to send the same longest busy tone, resulting in a collision. CWs of collided stations evolve in the same way as that in the EDCA, and each collided voice station chooses a backoff timer randomly from its CW for the next contention. Since there is no ACK frame sent back to acknowledge the successful voice transmission, it is difficult for the sender to recognize the collision. To address the problem in our scheme, for the first packet from a voice station received in the CP, the receiver should send an ACK frame to the sender. The voice sender continues to contend in the CP if no ACK is received.

For data stations, if there exists at least one voice contender, they will sense the busy tone during the AIFS[AC_data] (> AIFS[AC_voice]) and defer their transmissions. When a collision happens between voice stations, the data stations will wait for the channel to be idle for the duration of ACK timeout plus AIFS[AC_data] before they attempt to acquire the channel, which ensures that voice stations will not lose the channel access priority to the data stations even when a collision happens. Furthermore, when all the active voice stations are included in the polling list, the data stations can make full use of the CP resources.

By using the above scheme, it seems that the waiting time (before getting the channel) for a voice station is larger than that in the EDCA, since the voice station with the largest backoff timer, instead of the smallest backoff timer (as in the EDCA), gets the channel. However, as the number of voice stations simultaneously contending for the channel is very likely to be small, the initial and maximum window sizes for voice AC can be set to small values, so the negative effect of our scheme should be negligible.

Voice Overhead Reduction

To support VoWLANs, it is important to reduce the overhead and improve the transmission efficiency over the radio link. The large packet header overhead can significantly affect the capacity of WLANs in supporting voice service. For example, if a GSM 6.10 codec is used, voice packet payload is 33 bytes while the RTP/UDP/IP overhead is 40 bytes. In addition, the PHY preamble, MAC header, and control packets all consume bandwidth. As a result, the overall efficiency is less than 3%.[3] Actions need to be taken to alleviate the effect of the overhead.

Recently, various header compression techniques for VoIP have been proposed. The RTP/UDP/IP header can be compressed to as little as 2 bytes.[20] The compression technique is adopted in our research.

In our proposed scheme, the PHY and MAC layer header overheads are further reduced by aggregating the buffered voice packets from or to a voice station together and transmitting them by one MAC frame. For uplink transmission as an example, the AP polls each voice station periodically after every service interval, within which several voice packets may be generated and buffered at each voice station. In order to increase the efficiency, we combine the payload of these packets together and add a common MAC layer header instead of sending them one by one. It reduces the overall MAC layer header and PHY preamble overhead.

PERFORMANCE ANALYSIS

As discussed, in each service interval, the CFP is used to transmit voice traffic only; and in the CP, voice traffic has guaranteed priority over data traffic. To provide data traffic certain level of QoS, it is required that the average service time in each CP for data traffic is at least a pre-specified fraction (ϕ) of the whole service interval. Hence, we need to determine the maximum number of voice sessions (a "voice session" means a two-way voice session) that can be supported by the average fraction ($1 - \phi$) of time in each service interval.

Average Time Required to Serve Contending Voice Sessions in the CP

In a CP, consider n voice sessions contending for the channel. For simplicity of presentation, the CW of each voice session takes values from the set {CW_1, CW_2} where $CW_2 = 2 \times (CW_1 + 1) - 1$, and at the beginning of each CP, all contending voice sessions are with CW_1. Our analysis can be easily extended to cases with three or more choices for the CW size. Define state (n_1, n_2), where n_1 and n_2 are the numbers of voice sessions with CW_1 and CW_2, respectively. Hence, in each CP, the beginning state is (n,0). When a voice session contends successfully (i.e., it is the only one with the maximum backoff timer), it will leave the contention. After each successful transmission or collision, the state will evolve, remaining in the current state, or moving to the next one. The state transition is

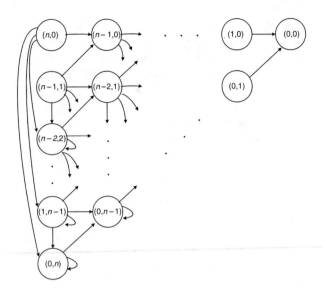

Fig. 3 The state transition diagram for (n_1, n_2) in a contention period.

shown in Fig. 3, where the state $(0,0)$ is the absorbing state when all voice sessions are served. To understand the state transition diagram, we use state $(1, n-1)$ as an example. Its next state is $(0, n-1)$ if the voice session with CW_1 transmits successfully, $(1, n-2)$ if one voice session with CW_2 transmits successfully, $(0,n)$ if the session with CW_1 collides with one or more other sessions, or remains in $(1, n-1)$ if two or more sessions with CW_2 collide. From the diagram, the transition probability and the average time needed for transition can be derived. Obviously, the average time needed for transitions from state $(n,0)$ to the absorbing state $(0,0)$ is the average time to serve all the n contending voice sessions in a CP, denoted by $T(n, 0)$.

Packet Loss Rate Bound in CFP

First, we estimate the number (X) of voice packets that can be supported in each CFP. We call the packets of the uplink or downlink of a voice session ready for transmission in the CFP *a burst*. Let M denote the maximum number of downlink or uplink voice packets generated in a service interval from a voice session. The average burst size is given by:

$$B = \frac{\sum_{i=1}^{M} P_d(i) \cdot i + \sum_{i=1}^{M} P_u(i) \cdot i}{\sum_{i=1}^{M} P_d(i) + \sum_{i=1}^{M} P_u(i)} \tag{1}$$

where $P_d(i)$ and $P_u(i)(0 \leqslant i \leqslant M)$ denote the probability of generating i downlink and i uplink voice packets respectively for transmission in a CFP. They can be determined based on the on/off model. Here, $P_d(i)$ and $P_u(i)$ are different because, in the uplink, the first several packets in each talk spurt are transmitted in a CP.

The average number of bursts in each CFP is X/B. A burst is in uplink transmission (thus requiring CF-Poll) with

a probability, $\sum_{i=1}^{M} P_u(i) / [\sum_{i=1}^{M} P_d(i) + \sum_{i=1}^{M} P_u(i)]$. The duration of the CFP, T_{CFP}, is then,

$$T_{\text{CFP}} = (X/B) \cdot (T_o + \frac{L \cdot B}{R})$$
$$+ (X/B) \frac{\sum_{i=1}^{M} P_u(i)}{\sum_{i=1}^{M} P_d(i) + \sum_{i=1}^{M} P_u(i)} \cdot T_{\text{poll}} \tag{2}$$

where T_o is the overhead due to IFS, PHY preamble, and MAC overhead, L is the voice payload size, R is the transmission rate of voice payload, and T_{poll} is the polling overhead. Then,

$$X = B \cdot \frac{T_{\text{CFP}}}{T_o + \frac{L \cdot B}{R} + \frac{\sum_{i=1}^{M} P_u(i)}{\sum_{i=1}^{M} P_d(i) + \sum_{i=1}^{M} P_u(i)} \cdot T_{\text{poll}}} \tag{3}$$

Let X_i denote the number of up- and downlink voice packets from the ith voice session ready for transmission in the CFP of a service interval and $Y = \sum_{i=1}^{N} X_i$, where N is the total number of voice sessions. The expectation $E[X_i]$ and variance $\text{Var}[X_i]$ of X_i can be determined based on the on/off model. If the packet loss rate in the CFP is required to be bounded by P_L, the following inequality should hold.

$$\frac{\sum_{Y > X} (Y - X) P(Y)}{E[Y]} \leqslant P_L \tag{4}$$

According to the central limit theorem, the random variable Y can be approximated as a Gaussian random variable with mean $N \cdot E[X_i]$ and variance $N \cdot \text{Var}[X_i]$ when N is large. The maximum N satisfying the above inequality is the maximum voice session number supported by the CFP.

Session Admission Control for Voice

In order to guarantee QoS of voice traffic, it is critical to have an appropriate session admission control algorithm. The AP is responsible for admitting or rejecting a new voice session based on the available resources to ensure that the QoS requirements (such as delay and packet loss rate) of all the admitted voice sessions are satisfied. For session admission control, it is essential to have the capacity region. The IEEE 802.11e has given a reference design. When there are k existing voice sessions in a WLAN, a new voice session indexed by $k+1$ can be admitted if the following inequality holds.

$$\frac{\text{TXOP}_{k+1}}{T_{\text{SI}}} + \sum_{i=1}^{k} \frac{\text{TXOP}_i}{T_{\text{SI}}} \leqslant 1 - \rho_{\text{CP}} \tag{5}$$

where ρ_{CP} is the minimum percentage of time used for the EDCA during each beacon interval and TXOP_i the minimum time that needs to be allocated for session i to ensure its QoS requirements. The value of TXOP_i depends on the voice packet size and the packet arrival rate. This algorithm is suitable only for constant-rate voice traffic without statistical multiplexing. Based on the algorithm, variable-rate voice traffic (represented by the on/off model) requires much more resources than what is actually needed. Here, we propose another algorithm to determine the capacity region that takes into account statistical multiplexing and at the same time guarantees the delay and packet loss rate requirements of voice traffic.

For a WLAN supporting voice/data traffic, the QoS requirement of the low-priority data traffic should also be guaranteed. In our system, data traffic is guaranteed the average service time in each service interval, i.e., an average fraction ϕ of time in a service interval is used by data traffic in the long run. Hence, we need to determine how many voice sessions can be admitted with average service time $(1 - \phi)T_{SI}$ (in both CFP and CP) in each service interval and with the required packet loss rate guaranteed. Let $\overline{T^v_{CP}}$ denote the average time in a CP used to serve contending voice sessions. Thus,

$$T_{CFP} + \overline{T^v_{CP}} \leqslant (1 - \phi)T_{SI}. \tag{6}$$

In the "Average Time Required to Serve Contending Voice Sessions in the CP" section, we derive the average time required in a CP to serve a fixed number, n, of voice sessions contending in the CP. However, with N voice sessions in service, the number of contending voice sessions in a CP varies (due to the voice on/off nature), so does the required service time in the CP. The average service time for contending voice sessions in a CP is given by

$$\overline{T^v_{CP}}(N) = \sum_{n=1}^{N} \binom{N}{n}(1 - P_c)^n (P_c)^{N-n} \cdot T(n,0) \tag{7}$$

where $T(n,0)$ is the average time to serve n contending voice sessions in a CP (as defined in the "Average Time Required to Serve Contending Voice Sessions in the CP" section), and P_c is the probability that a voice station contends for the channel in a CP, given by,

$$P_c = \frac{\alpha}{\alpha + \beta}[1 - \exp(-\beta T_{SI})], \tag{8}$$

where α^{-1} and β^{-1} are the mean durations of voice on state and off state, respectively. In this work, the duration of the CP in a service interval, T_{CP}, is larger than $\overline{T^v_{CP}}(N)$, as the difference between them is the average service time for data traffic in a CP. The performance of contending voice sessions in a CP can be evaluated by the outage probability that T_{CP} is not sufficient to serve all the contending voice sessions, as shown in the "Numerical Results and Discussion" section.

From the analysis in the "Packet Loss Rate Bound in CFP" section, if N voice sessions are admitted, we can

Table 1　Simulation Parameters.

Parameter	Value
Slot time τ	20 μs
T_{SI}	100 ms
SIFS	10 μs
AIFS[AC_voice]	40 μs
AIFS[AC_data]	60 μs
PHY preamble	192 μs
MAC header	34 bytes
ACK	14 bytes
CF-Poll	36 bytes
R	11 Mbps
Basic rate	2 Mbps
$1/\alpha$	352 ms
$1/\beta$	650 ms
Data packet payload	1000 bytes
P_L	1%

determine the minimum value of T_{CFP}, denoted by $T^m_{CFP}(N)$, in order to guarantee the voice packet loss rate bound. From the constraint (Eq. 6), the capacity region for voice is the maximum integer N (denoted by N^*) satisfying

$$T^m_{CFP}(N) + \overline{T^v_{CP}}(N) \leqslant (1 - \phi)T_{SI}. \tag{9}$$

and the service interval should be configured with a CFP having duration $T^m_{CFP}(N^*)$ and a CP having duration $T_{SI} - T^m_{CFP}(N^*)$.

NUMERICAL RESULTS AND DISCUSSION

To validate the analysis and evaluate the performance of our proposed scheme, computer simulations are carried out using Matlab. The simulation for each run consists of 1000 service intervals. We choose the GSM 6.10 codec as the voice source as an example. The voice payload size is 33 bytes and the packet inter-arrival period is 20 ms. Compressed RTP/UDP/IP headers with size 4 bytes are used in all the simulations. Other simulation parameter values are listed in Table 1. We first vary the contending voice session number (i.e., n) in the CP (where each voice session has one MAC frame to send) and analyze and/or simulate the time to serve all the MAC frames (i.e., the time to serve the contending voice sessions in a CP). Then we evaluate how the packet loss rate in the CFP changes with the number of voice sessions N. In the evaluation, the first several packets of an uplink talk spurt are not transmitted in a CFP (but are transmitted in a CP by contention). Finally, we evaluate the capacity of the whole system, and compare it with that of IEEE 802.11e. We

Wireless LANs—Mesh

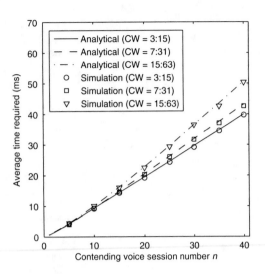

Fig. 4 Average time required to serve contending voice sessions in a CP in our scheme.

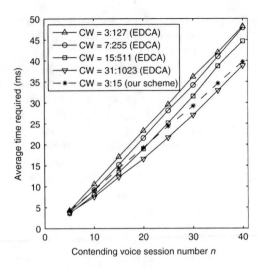

Fig. 6 Average time required to serve voice sessions in the CP in the EDCA compared with our scheme.

obtain the portion of time required to serve different numbers of admitted voice sessions and obtain the system capacity.

Time to Serve Contending Voice Sessions in the CP

For uplink voice transmission in our scheme, the first several packets in each talk spurt are transmitted in the CP. With the system parameters, the probability of a voice session contending in the CP is around 9% according to the analysis. Hence, if the total voice session number is 200, there are on average 18 voice sessions contending in each CP. Fig. 4 and Fig. 5 show the average time required to serve contending voice sessions [i.e., $T(n,0)$] and average

voice collision number in a CP vs. the contending voice session number n with different settings for initial and maximum CWs (CW_{min} and CW_{max}), respectively. It is clear that our analysis matches well with the simulations. CW settings are critical for contention-based channel access. In our scheme, when the voice sessions have smaller CW_{min} and CW_{max}, the time to transmit busy tone is smaller, at the cost of more collisions. Via the analysis and simulations, we find out that $CW_{min} = 3$ and $CW_{max} = 15$ can lead to a minimal average time to serve all the voice sessions. Via simulations, we also compare this best case with the cases if the EDCA of IEEE 802.11e is applied in the CP and demonstrate the results in Fig. 6. It can be seen

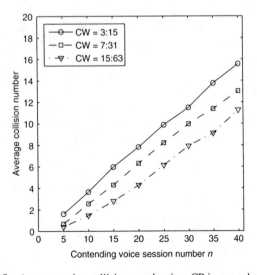

Fig. 5 Average voice collision number in a CP in our scheme.

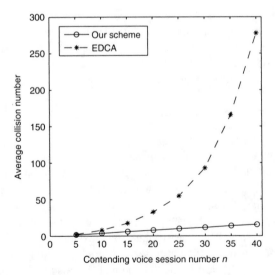

Fig. 7 Average collision number in a CP when our guaranteed priority scheme or the EDCA is applied with $CW_{min} = 3$ and $CW_{max} = 15$.

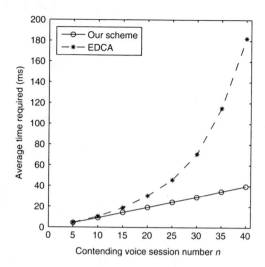

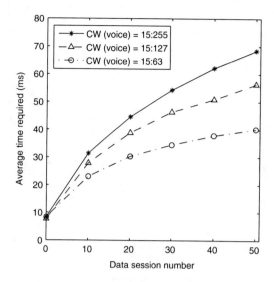

Fig. 8 Average time required to serve contending voice sessions in a CP when our guaranteed priority scheme or the EDCA is applied with $CW_{min} = 3$ and $CW_{max} = 15$.

Fig. 9 Average time required so that all contending voice sessions can be served in a CP when the EDCA is used to support ten voice sessions and variable number of long-lived data sessions with CW [data] = 31:1023.

that there is not much difference between our scheme and the EDCA. Only the EDCA with $CW_{min} = 31$ and $CW_{max} = 1023$ has a non-trivial gain over our proposed scheme. However, to obtain priority in the EDCA, voice AC is very likely to have a smaller CW_{min} (<31) and CW_{max} (<1023). Although a voice station with the largest backoff timer instead of the smallest one (as in the EDCA) gets the channel in our scheme, voice performance is not degraded much. The reason is that our proposed scheme can use very small CW_{min} and CW_{max} but the EDCA cannot. If our proposed scheme and the EDCA use the same CW_{min} and CW_{max}, the backoff waiting time in our scheme is larger. However, our scheme has a smaller collision probability. If multiple nodes choose the same backoff timer, a collision will occur in the EDCA, but a collision happens in our scheme only when the multiple nodes are with the largest backoff timer (among all the nodes). Fig. 7 and Fig. 8 show the average collision number and required time to serve contending voice sessions, respectively, in the EDCA and our scheme with $CW_{min} = 3$ and $CW_{max} = 15$ via simulations. It can be seen that, as the contending voice session number increases, the collision number increases rapidly in the EDCA, but relatively slowly in our scheme. Hence, the time required to serve contending voice sessions in our proposed scheme is much smaller than that in the EDCA when the contending session number is large in the example.

In addition, in the above comparison, the EDCA is applied with no contending data sessions. As discussed in the "Related Work" section, when data sessions are added, the voice performance of the EDCA will be degraded, but our scheme will not be affected, because it can provide guaranteed priority access to voice sessions over data sessions. Fig. 9 shows the effect of data traffic on the average

time required in a CP so that all contending voice sessions can be served in the EDCA. Long-lived data sessions use the initial and maximum CW pair (31, 1023), while voice sessions choose initial CW size 15 and maximum CW size 63, 127, or 255. We can see that when the number of data sessions increases, the average time required increases accordingly, and the negative effect is more significant if voice sessions choose a larger contention window pair [(15, 255) in the example].

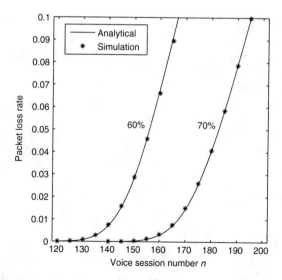

Fig. 10 Packet loss rate in the CFP for $T_{CFP} = 60\% T_{SI}$ and $T_{CFP} = 70\% T_{SI}$ in our scheme.

Wireless LANs—Mesh

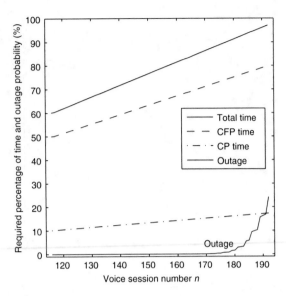

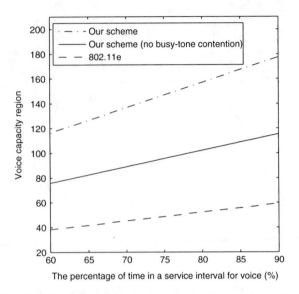

Fig. 11 The percentage of time [in the CP, CFP, and totally] in a service interval needed to serve the voice sessions with QoS guarantee and the outage probability that not all the contending voice sessions in the CP can be served.

Fig. 12 The analytical results for voice traffic capacity region of our proposed scheme, our scheme without busy-tone contention mechanism, and IEEE 802.11e.

Packet Loss Rate in the CFP

Fig. 10 shows the analytical results of the packet loss rate versus voice session number N with T_{CFP} equal to 60 and 70% of T_{SI} in our scheme. Simulations are also carried out for selected values of N. It can be seen that our analysis matches well with the simulations. From Fig. 10, when the voice session number is equal to or less than 141 when $T_{CFP} = 60\%T_{SI}$ or 167 when $T_{CFP} = 70\%T_{SI}$, the packet loss rate in the CFP is bounded by 1%.

Capacity Region of Voice

To determine the capacity region of voice in our scheme, we vary the number N of voice sessions in the system and calculate the average time in the CP $\overline{T_{CP}^v}(N)$ and the duration of CFP $T_{CFP}^m(N)$ in order to guarantee that voice packet loss rate in CFP is bounded by 1%. We further obtain the total average time in a service interval needed to serve the N voice sessions with QoS guarantee. The analytical results are shown in Fig. 11, which also gives the outage probability that the CP with duration $[T_{SI} - T_{CFP}^m(N)]$ cannot serve all contending voice sessions. It is shown that, when data traffic requires average 30% service time (thus 70% time for voice) in a service interval, we should configure $T_{CFP} \approx 57\%T_{SI}$ and $T_{CP} \approx 43\%T_{SI}$ with a maximum admitted voice session number of 136. The outage probability is negligible (<1%) if the total average time for voice is less than 90%.

We further obtain the voice capacity when the percentage of time (in both CFP and CP) used by voice in each service interval varies from 60 to 90% and get the

analytical results as shown in Fig. 12. The analytical results for voice capacity of the IEEE 802.11e polling scheme with the same percentage of time for voice are also included in Fig. 12. For a comparison, Fig. 12 also shows the voice capacity when our scheme is applied without the busy-tone contention mechanism (i.e., with only the overhead reduction mechanism). For uplink transmissions, all voice stations are polled. If a polled voice station has no packets to transmit, it will respond with a NULL frame. From Fig. 12, it can be seen that both of our proposed overhead reduction and busy-tone contention mechanisms can significantly improve system capacity as compared with IEEE 802.11e.

CONCLUSION

To meet the delay requirement of VoWLAN applications, the controlled channel access is preferred to the contention-based access. In this entry, we address QoS provisioning and capacity enhancement issues for WLANs supporting voice services. At the same time, we guarantee a certain level of channel resources to data traffic. Our proposed solution provides guaranteed priority access to voice stations, avoids unnecessary polling of silent voice stations, and reduces the overheads significantly. This research can be smoothly incorporated in the implementation of IEEE 802.11e, as only minor modifications are needed. We provide an analytical model for the performance evaluation, which is validated by extensive simulations. A session admission control scheme is presented to admit voice sessions into the system. Our solutions are shown to significantly improve the capacity of IEEE 802.11e WLANs supporting voice and data services. This research should provide helpful insights to the development and deployment of VoIP technologies over WLANs (which were originally designed for data services).

ACKNOWLEDGMENTS

This work was supported by a research grant from the Natural Science and Engineering Research Council (NSERC) of Canada.

REFERENCES

1. Black, U. *Voice Over IP*; Prentice Hall: Upper Saddle River, New Jersey, 2000.
2. Voice over wireless LAN: 802.11x hears the call for wireless VoIP. Market Research Report. In-Stat, 2002.
3. Wang, W.; Liew, S.C.; Li, V.O.K. Solutions to performance problems in VoIP over a 802.11 wireless LAN. IEEE Trans. Veh. Technol. **2005**, *54* (1), 366–384.
4. Hole, D.P.; Tobagi, F.A. Capacity of an IEEE 802.11b wireless LAN supporting VoIP. IEEE ICC 2004, Paris, France, June 20–24, 2004; 196–201.
5. Garg, S.; Kappes, M. An experimental study of throughput for UDP and VoIP traffic in IEEE 802.11b networks. IEEE WCNC 2003, New orleans, LA, Mar 16–20, 2003; 1748–1753.
6. IEEE Standard for Information technology-Telecommunications and information exchange between systems-Local and metropolitan area networks-Specific requirements-Part 11: Wireless Medium Access Control (MAC) and Physical Layer (PHY) specifications: Amendment 7: Medium Access Control (MAC) Quality of Service (QoS) Enhancements, IEEE 802.11 WG, IEEE 802.11e/D11; 2004.
7. Ni, Q.; Romdhani, L.; Turletti, T. A survey of QoS enhancements for IEEE 802.11 wireless LAN. Wirel. Commun. Mob. Comput. **2004**, *4* (5), 547–566.
8. Cai, L.; Shen, X.; Mark, J.W.; Cai, L.; Xiao, Y. Voice capacity analysis of WLAN with unbalanced traffic. 2nd International Conference on QoS in Heterogeneous Wired/Wireless Networks (QShine 2005), Orlando, FL, Aug 22–24, 2005.
9. Yang, X.; Vaidya, N.H. Priority scheduling in wireless ad hoc networks. 3rd ACM International Symposium Mobile ad hoc networking and computing, Lausanne, Switzerland, June 09–11, 2002; 71–79.
10. Robinson, J.W.; Randhawa, T.S. Saturation throughput analysis of IEEE 802.11e enhanced distributed coordination function. IEEE J. Sel. Areas Commun. **2004**, *22* (5), 917–928.
11. Ziouva, E.; Antonakopoulos, T. A dynamically adaptable polling scheme for voice support in IEEE802.11 networks. Comput. Commun. **2003**, *26* (2), 129–142.
12. Xiao, Y.; Li, H.; Choi, S. Protection and guarantee for voice and video traffic in IEEE 802.11e wireless LANs. IEEE INFOCOM 2004, Hong Kong, China, Mar 7–11, 2004; 2152–2162.
13. Kim, H.; Hou, J.C. Improving protocol capacity for UDP/TCP traffic with model-based frame scheduling in IEEE 802.11-operated WLANs. IEEE J. Sel. Areas Commun. **2004**, *22* (10), 1987–2003.
14. Bladwin, R.O.; Davis, N.J., IV; Midkiff, S.F.; Raines, R.A. Packetized voice transmission using RT-MAC, a wireless real-time medium access control protocol. Mob. Comput. Commun. Rev. **2001**, *5* (3), 11–25.
15. Fan, W.F.; Tsang, D.H.K.; Bensaou, B. Admission control for variable bit rate traffic using variable service interval in IEEE 802.11e WLANs. IEEE ICCCN 2004, Chicago, IL, Oct 11–13, 2004; 447–453.
16. Kuo, Y.L.; Lu, C.H.; Wu, E.H.K.; Chen, G.H. An admission control strategy for differentiated services in IEEE 802.11. IEEE GLOBECOM 2003, San Francisco, CA, Dec 1–5, 2003; 707–712.
17. Garg, S.; Kappes, M. Admission control for VoIP traffic in IEEE 802.11 networks. IEEE GLOBECOM 2003, San Francisco, CA, Dec 1–5, 2003; 3514–3518.
18. Pong, D.; Moors, T. Call admission control for IEEE 802.11 contention access mechanism. IEEE GLOBECOM 2003, San Francisco, CA, Dec 1–5, 2003; 174–178.
19. Sobrinho, J.L.; Krishnakumar, A.S. Quality-of-service in ad hoc carrier sense multiple access wireless networks. IEEE J. Sel. Areas Commun. **1999**, *17* (8), 1353–1368.
20. Casner, S.; Jacobson, V. Compressing IP/UDP/RTP headers for low-Speed serial links. IETF RFC 2508. **1999**.

Wireless LANs—Mesh

Wireless LANs (WLANs): Security and Privacy

S.P.T. Krishnan
Cryptography and Security Department, Institute for Infocomm Research (I2R), Singapore

Bharadwaj Veeravalli
Lawrence W.C. Wong
Department of Electrical and Computer Engineering, Computer Networks and Distributed Systems (CNDS) Lab, National University of Singapore, Singapore

Abstract

Wireless LAN (WLAN) security refers to providing LAN services over a wireless medium in a secured way and ensuring privacy during communications.

INTRODUCTION

Keeping a wireless LAN (WLAN) in a safe and secured way is an imperative issue in administering wireless communications. WLANs being a counterpart of their wired infrastructure poses several challenging issues in seeking a secured wireless connection between any two devices communicating on a WLAN. The underlying medium being "free air" raises issues that have seldom been issues in wired networks. To understand this, consider two devices seeking connection to the Internet via a wireless access point (WAP) [access point (AP) is a device stationed to effect connections between devices and serves as an interconnection point.], and let us suppose these two devices share the same frequency band. Since each of these devices requests connection to the Internet, one of the primary issues is that the AP must schedule these devices by some means to provide contention free access and also must make sure that either of the devices does not occupy the channel forever. Also, since the data (packets) are accessible to every device, content protection must be ensured and must also eliminate possibility of hijacking the channels by jammers and attackers. Also an associated problem is identifying an intruder and cleverly avoiding him or eliminating him. Security threats pose serious problems in assuring a fail-safe infrastructure under wireless mode. The technology being new, currently standards are being introduced in place to safeguard and to provide a fail-safe connection.

In this entry, we will attempt to make a systematic categorization of the types of WLAN standards and their security related standards and tools, and also see the vulnerabilities associated with each of them. In a way, the material in this entry attempts to provide a comprehensive overview of the state-of-the-art in WLAN security and privacy. A complete list of abbreviations and their definitions that are used in this entry is listed in Appendix for the ease of reference.

HISTORY OF IEEE 802.11

In 1997, the IEEE approved a standard for WLAN called 802.11 which specified the characteristics of radio devices with a signaling rate of 1 and 2 Mbps. The standard specifies the media access control (MAC) and PHY layers for transmissions in the 2.4 GHz band. The standard defines the MAC procedures for accessing the physical medium, which can be infrared or radio frequency. Mobility is handled at the MAC layer, so handoff between adjacent cells is transparent to layers built on top of an IEEE 802.11 device. The spectrum used ranges from 2.4 to 2.4835 GHz in the United States and Europe, while in Japan it ranges from 2.471 to 2.497 GHz. The available bandwidth is divided into 14 partially overlapping channels, each 22 MHz wide. Only 11 of these channels are available in the United States, 13 in Europe, and just one in Japan. All the devices in the same basic service set (BSS) use the same channel.

In 1999, IEEE published the specifications of a new amendment of the 802.11 family, 802.11a. The specifications still refer to the MAC and PHY layers, and the band used is 5 GHz, which is unlicensed in the United States but not so in most other countries. The signaling rates for these specifications are 6, 9, 12, 18, 24, 36, 48, and 54 Mbps. Due to frequency band licensing restrictions, the take-up of 802.11a was slower on a global scale. Again in 1999, IEEE ratified a new amendment, with better performance,

Encyclopedia of Wireless and Mobile Communications DOI: 10.1081/E-EWMC-120043609

referred to as IEEE 802.11b, which works at additional signal rates of 5.5 and 11 Mbps; most devices currently deployed are based on this technology. 802.11b specifies some coding modifications, leaving the lower-layer radio characteristics unmodified, and making very small changes to the upper MAC layers, thus facilitating compatibility with 802.11 devices. This specification allows for the theoretical transmission of approximately 11 Mbps of raw data at indoor distances of several dozen to several hundred feet, and outdoor distances of several to tens of miles as an unlicensed use of the 2.4 GHz wireless band.

Due to the overlapping nature of the channel frequencies which extend up to two adjacent channels and to obtain maximum throughput and minimize interference, deployments typically do not use all the channels. Also, it is well-known that Bluetooth devices, cordless phones and microwave ovens share spectrum with 802.11b/g, and hence can interfere with an 802.11b/g network. Therefore, the number of orthogonal channels provided by 802.11b/g is 3 (namely channels 1, 6, and 11), while 802.11a supports 12 orthogonal channels.

In 2003, the IEEE approved 802.11g as a further evolution of the 802.11 standard. 802.11g provides the same performance as 802.11a, while working in the 2.4 GHz band, which makes it deployable in Europe. Compatibility with 802.11b devices is guaranteed.

WLAN OVERVIEW

An IEEE 802.11 WLAN is a group of stations (wireless network nodes) located within a limited physical area. There are two WLAN design structures: Ad hoc and Infrastructure networks. The foundation of an 802.11 network is the BSS. Any cluster of stations inter-communicating between themselves form an ad hoc network or **independent BSS** (IBSS). In the ad hoc mode, each client communicates directly with the other clients within the network and it is designed to facilitate communication between clients within transmission range of each other. If a client in an ad hoc network wishes to communicate outside the cell, a member of the cell must operate as a gateway and perform routing. Creating a multi-hop network requires higher-level protocols and is not captured as an 802.11 standard. A cluster of one or more stations connected with a **WAP** forms an infrastructure WLAN. In this mode, all stations communicate with each other through one AP at a time—no direct station-to-station communication is permitted. The AP acts as an Ethernet bridge and is responsible for routing packets between stations. By extension, a cluster of BSSs where interconnected APs serve as bridges between service areas form an **extended service set** (ESS). The APs also relay traffic from wireless nodes to the infrastructure, the **distribution system** (DS), which allows access to external wired

networks. The vast majority of installations use infrastructure-based WLANs. Fig. 1 shows a typical enterprise wireless network comprising a secured internal WLAN and an open WLAN for visitors. It captures a basic infrastructure and describes how different mobile entities interact/communicate with one another. In order to access the Internet via a WLAN, the routing established will be via a router, preferably through a firewall. Within the WLAN, there could be several overlapping frequency ranges among the interacting or sharing devices and hence multi-hop routing is expected to happen. However, from an infrastructure WLAN (ESS), an attacker first seeks connectivity to a wireless node (active node) which facilitates connection to an AP and to the LAN. It is also possible that these two wireless domains may interact or establish connectivity. Of course, the overall delay expected would have a cumulative effect and may encounter large delays at intermediate nodes.

Prior to communicating, in infrastructure mode, wireless clients and APs must establish a relationship, or an association. As defined by the standard, association is *"the service used to establish access point/station (AP/STA) mapping and enable STA invocation of the distribution system services."* An association is basically a connection at layer 2 (link layer) of the OSI model. A node finds a network using a very simple procedure. All APs transmit a beacon management frame at fixed intervals. To associate with an AP and join a BSS, a client listens for beacon messages to identify the APs within range. The client then selects the BSS to join in a vendor-independent manner. A client may also send a probe request management frame to find an AP affiliated with a desired **service set Id** (SSID). The SSID is a 32 byte or less network name of a service set. APs may be configured as "open" or "closed." In the open mode, the AP broadcasts its SSID; in closed mode, it does not and only those clients with knowledge of the network name, or SSID, can join. In essence, the network name acts as a shared secret. While being part of a network, stations can keep discovering new networks and may disassociate from the current one in order to associate with a new one (e.g., because it has a stronger signal). On the down side, this behavior makes it easy for the installation of Rogue APs by adversaries.

The primary focus of Wi-Fi security is protecting the confidentiality of the data while it is in the air and providing authentication for the client and the infrastructure, so each knows the other is a trusted entity. With both the client verifying the identity of the AP and the AP verifying the client, attackers have difficulty pretending to be legitimate actors in the network.

The WLAN security is at the link level. Wi-Fi defines two authentication methods: **open system authentication** (OSA) and **shared key authentication** (SKA). OSA is the default authentication protocol for 802.11. As the name implies, OSA authenticates anyone who requests

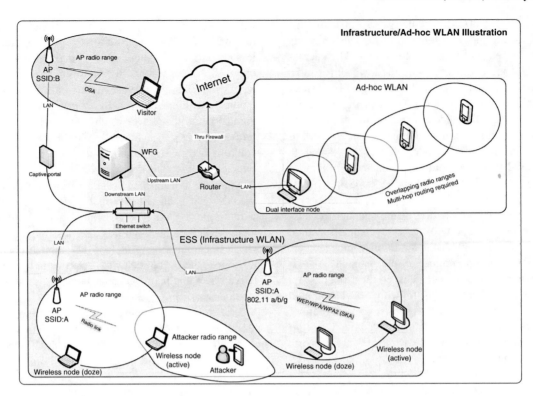

Fig. 1 Infrastructure/ad hoc WLAN network entities and architecture in a typical enterprise setup.

authentication. Essentially, it provides a null authentication process. SKA uses a standard challenge and response along with a shared secret key to provide authentication. A key shared between the AP and many stations is called a *default key*. A key shared between the AP and only one other station is referred to as a *key-mapping key*. The IEEE 802.11 standard does not specify any means for distributing and obtaining the shared secret.

A Wi-Fi device may be in either the awake or doze state. In the doze state, the station can neither transmit nor receive, which reduces power consumption. Consequently, there are two power management modes: **active mode** (AM) and **power save** (PS) mode. The handling of the stations in PS mode differs according to the topology of the Wi-Fi network. In infrastructure mode, a station in AM which wants to switch to PS mode must signal the AP by using the power management bit in the header of its packets. The AP stores all the traffic addressed to stations that are in PS mode; when transmitting the periodic beacon, the AP sends the list of stations in PS mode and whether it has traffic queued for them. At regular and configurable time intervals, the stations in PS mode switch to AM mode in order to receive the beacon. If there is traffic addressed to them, the stations can receive it and then return to PS. In the ad hoc mode, as the central entity is missing, caching of incoming data is not a straight-forward process. The nominal range is 100 m in Wi-Fi. The maximum number of

devices belonging to the network's building block (i.e., the BSS for Wi-Fi) is 2007 for a structured BSS, and unlimited for an IBSS. FCC permitted 1 W of power for unlicensed operators in the 2.4 GHz band.

WIRELESS AD HOC NETWORKS

A wireless ad hoc network (also called peer-to-peer wireless network) is formed temporarily by a group of mobile devices. There is no designated client or server. All members communicate over wireless channels directly without any fixed networking infrastructure or centralized administration. In this structure, all mobile hosts communicate with each other in a wireless multi-hop routing style. Each mobile node maintains all the links within the defined radius (called zone) and acts as a router in the network. If a member is out of its destination member's zone or it is not in the line-of-sight, all messages between them must pass through one or more routers. All members are free to move around and join and leave a network at will. The routing scheme is adjusted dynamically according to the changing network topology.

Wireless ad hoc networks are recognized as an important issue in the military communications arena, for commercial applications such as team collaboration applications, networking intelligent sensors and cooperative robots, etc. It is unlikely that the wireless ad hoc network will be restricted to a small geographical region. Hybrid architecture could be used to expand the range of such networks. For instance, home-networked appliances based on Bluetooth technology can be remotely controlled through the Internet. For military use, a complete networking system, called the AEGIS broadcast network, has been implemented for tactical data systems in the US Navy. It connects, monitors, and controls all military units on both coasts, the Gulf of Mexico, Japan, etc. The interconnection among wireless ad hoc networks through wired relay services creates a complex network topology, where node addressing (aka IP addressing in infrastructure LANs) is an issue in ad hoc network.

WIRED EQUIVALENT PRIVACY

The 802.11 standard for WLAN communications introduced the wired equivalent privacy (WEP) protocol in an attempt to bring the security level of wireless systems closer to that of wired ones. WEP was intended to provide "confidentiality that is subjectively equivalent to the confidentiality of a wired LAN medium that does not employ cryptographic techniques to enhance privacy." IEEE specifications for wired LANs do not include data encryption as a requirement.

WEP relies on a secret key that is shared between a supplicant mobile station (e.g., a laptop with a wireless card) and an AP. The secret key is used to encrypt packets before they are transmitted, and an integrity check is used to ensure that packets are not modified in transit by an intruder. The standard does not discuss how the shared key is established. In practice, most installations use a single key that is shared among all mobile stations and APs. WEP uses the symmetric stream cipher—Rivest cipher 4 (RC4), designed by Ron Rivest,[1] to encrypt all network data traffic. The system is based on a shared secret key k that is distributed out of band. RC4 uses the key to generate a stream of pseudo-random bytes equal in length to the target plaintext. It then combines this stream with the plaintext (P), using the bitwise "exclusive or" function (XOR, or $\oplus$) to produce ciphertext C. The receiver holds a copy of the same key, and uses it to generate an identical key stream. XORing the key stream with the ciphertext yields the original plaintext.

$$C = RC4(k) \oplus P \tag{1}$$

$$P' = RC4(k) \oplus C = RC4(k) \oplus RC4(k) \oplus P = P \tag{2}$$

(Use of the combiner $\oplus$ reveals the original plaintext)

This is an extreme weakness, particularly when the cipher is used to encrypt highly redundant messages,

such as network protocols. To address this weakness, WEP's set of initialization vectors (IVs) attempt to add entropy to the key space by using an IV and produce a different RC4 key for each packet. WEP uses a 3-byte (24-bit) IV and a 40- or 104-bit shared secret to produce the encryption key k. The sender transmits the IV with the encrypted ciphertext so the receiver can produce the full k and decrypt. In addition to its cryptographic protection of the data, WEP uses a 4-byte integrity check value (ICV), computed over the original plaintext of the data portion of the 802.11 frame. The 802.11 transmitter computes the ICV as a cyclic redundancy check, CRC-32 checksum and appends it to the plaintext before encrypting.

While it is recommended that the IV be changed on a per-packet basis, resulting in a different pseudo-random sequence for every packet, many implementations do not follow this recommendation. Since the IV is so short, eavesdropping on a busy network makes it possible to break the cipher in a reasonable length of time. In addition to using WEP for privacy, many vendors utilized access control lists (ACLs) based on the MAC address to prevent unauthorized access to the network. WEP failed to accomplish its security goals, despite deploying the believed-to-be-secure RC4 stream cipher.

WEP VULNERABILITIES

1. Wireless packets with repeated IVs: By XORing two packets that use the same IV, the attacker obtains the XOR of the two plaintext messages. The resulting XOR can be used to infer the contents of the two messages, given that the IP traffic is often very predictable and includes much redundancy. Such redundancy can be used to eliminate many possibilities for the contents of messages. Further, educated guesses about the contents of one or both of the messages can be used to statistically reduce the search space of possible messages, and in some cases it is possible to determine the exact contents, hence cracking the WEP key.

2. The second attack is also a direct consequence of the problems described previously. If an attacker knows the exact plaintext for one encrypted message, such knowledge can be used to forge valid encrypted packets. The procedure involves constructing a new message, calculating the CRC-32, and performing bit flips on the original encrypted message to change the plaintext to the new message. The basic property is that $RC4(X) \oplus X \oplus RC4(Y) = Y$. This fake packet can be sent to the AP or client station, and it will be accepted as a valid packet.

3. In practice, the most serious problem with WEP is that its encryption keys can be recovered through cryptanalysis. WEP uses a common stream cipher, RC4, but

in a nonstandard way: WEP concatenates a base key with a 24-bit per-packet nonce, called the WEP IV, and uses the result as a per-packet RC4 key. Once the first byte of plaintext is known, it is possible to deduce the base RC4 key by exploiting properties of the RC4 key schedule.

4. Decryption Dictionaries: Infrequent re-keying and frames with same IV result in large collection of frames encrypted with same key streams. These are called decryption dictionaries. Therefore, even if the secret key is not known, more information is gathered about the unencrypted frames which may eventually lead to the exposure of the secret key.

5. The WEP implementation of RC4 is flawed in several ways, which allows the algorithm itself to be attacked and the key to be revealed. The first problem with WEP is that the IV is always prepended to the key prior to generation of the keystream by the RC4 algorithm. Second, the IV is relatively small (3 bytes), which produces a lot of repetitions as the scant 16.77 million variations are reused to encrypt millions of packets. Third, some of the IVs are "weak" in the sense they may be used to betray information about the key.

The problems with design of WEP can be summarized as follows:

- 24-bit IVs are too short, and this puts confidentiality at risk.
- The CRC checksum, called the ICV, used by WEP for integrity protection, is insecure, and does not prevent adversarial modification of intercepted packets.
- WEP combines the IV with the key in a way that enables cryptanalytic attacks. As a result, passive eavesdroppers can learn the key after observing a few million encrypted packets.
- Integrity protection for source and destination addresses is not provided.

WEP ATTACKS

- **Passive key recovery:** Scott Fluhrer, Itsik Mantin, and Adi Shamir[2] identified a key scheduling attack, known as FMS attack (named after the above authors) against the RC4 algorithm that, when used with certain keys, renders the cipher vulnerable to key recovery. The entry describes several classes of weak IVs (aka "interesting packets"), one of which is the "B + 3:FF:X" packet. Soon it was suggested that implementations exclude "weak" keys. Although complete filtering prevents the FMS attack, it makes this attack faster than FMS without filtering as the dictionary is smaller because of a smaller IV space—completely negating the counter measure. Utilities

that break WEP encryption by taking advantage of weak IVs are called "FMS utilities" and are freely available on the Internet that can execute this attack on commodity hardware.

- **Small IV space:** WEP's IV space of 24 bits, as Nancy Cam-Winget et al.[3] later identified is too small to provide an adequate number of distinct cryptographic keys, needed for encrypting data. IV reuse, or encrypting multiple messages with the same IV and cryptographic key, results in key stream reuse which, among other things, lets an attacker recover portions of the original, unencrypted message. Under moderate network load, an IV collision occurs every few hours. Given the structure of plaintext (network) messages, it now becomes significantly easier to recover them.

- **Replay attacks:** The WEP protocol has no methods of message authentication, and it easily allows intercepted messages to be replayed or sent again without modification. This lack of authentication can permit attackers to launch potential man-in-the-middle (MITM) attacks or DoS attacks.

- **Message modification attack:** The 802.11 standard uses a 32-bit CRC to provide data integrity. This is a linear function of the plaintext. Since the RC4 encryption is also linear, it becomes possible to modify a message and the CRC without knowing the RC4 pseudo-random sequence.

- **An inductive chosen plaintext attack:** This attack leverages on poor design aspects of WEP—lack of replay protection and the linear nature of the CRC used. It also uses the fact that WEP is a stream cipher rather than a block cipher. The attack involves three steps. The first step, or the base phase, involves recovery of an initial amount of pseudo-random stream from traffic analysis (eavesdropping). The second step, the inductive phase, then uses the initial pseudo-random stream to derive the pseudo-random stream for a full maximum transmission unit (MTU). The final step builds a dictionary of IVs and the corresponding pseudo-random sequences.

- **SKA attack:** The current protocol for SKA is easily exploited through a passive attack by eavesdropping of one leg of an authentication exchange. The attack works because of the fixed structure of the authentication protocol (the only difference between different authentication messages is the random challenge), and the previously reported weaknesses in WEP. The attacker first captures the second and third management messages from an authentication exchange. The second message contains the random challenge in the clear, and the third message contains the challenge encrypted with the shared authentication key. Because the attacker now knows the random challenge (plaintext, P), the encrypted challenge (ciphertext, C) and the public IV, the attacker can derive the pseudo-random stream.

Various publicly available tools have automated these attacks, including Airsnort and WEPCrack. More aggressive attacks of wireless networks include denial-of-service (DoS) attacks, MITM attacks, forced de-authentication of authorized users, WAP MAC address spoofing, to name but a few.

It should be noted that even if the implementation of RC4 was corrected, WEP would still be vulnerable to replay attacks, checksum forging, message integrity check forging, and sundry authentication attacks resulting from the fact that both the plaintext challenge and cipher text response are broadcast.

The question isn not whether WEP can be broken, but how long it takes. Wireless neighborhoods are only as safe as the neighbors.

WI-FI PROTECTED ACCESS

In 2002, The Wi-Fi alliance, a wireless industry organization, created the Wi-Fi protected access (WPA) standard, a subset of the 802.11i draft. WPA was designed to mitigate the existing WEP weaknesses as much as possible, while still being implementable on existing Wi-Fi hardware via firmware updates, without unduly disrupting performance. WPA's quick-fix for WEP insecurities is the temporal key integrity protocol (TKIP). TKIP also utilizes the RC4 stream cipher. The crown jewel of TKIP is the per-packet keying technique which de-correlates the public IVs from weak keys thereby mitigating FMS attacks. It combines the shared key, IV with the packet sequence number and the sender's MAC address to generate a reasonably unique key for each packet.

Next, we list the other improvements in WPA over WEP and document the attacks they help mitigate, within parenthesis. The minimum key length is increased from 40 to 256 bits (dictionary attacks); The IV is doubled to 48 bits (FMS attacks); IV sequencing is enforced (replay attacks); key rotation, providing fresh encryption and integrity keys, is embedded automatically [FMS, pseudo-random number (PRN) generating algorithm (PRGA), and dictionary attacks]; mutual authentication is built in so that both the AP and station have to prove to each other that they are legitimate (spoofing); packet tampering detection is built in with a message integrity code, MIC, called Michael (PRGA injection, forgeries).

For access control, WPA implements the IEEE 802.1x port based access control standard with extensible authentication protocol (EAP) or, alternatively, pre-shared keys (PSKs) for implicit authentication. An early vulnerability arose from the looseness in the way the PSKs were used. The WPA–PSK implementation was meant as a surrogate for authentication servers (aka radius servers), which are uncommon in the Small Office Home Office (SOHO) market. As with WEP, users simply enter the same passphrase on the AP and the client, and the authentication is transparent. Unfortunately, passphrases less than approximately 20 characters give rise to a new Wi-Fi attack vector: hash comparison attacks.

One of the latest WPA–PSK cracking utilities is called coWPAtty. Part of the reason for its superior performance is that only authentication frames are necessary. And if the four handshaking packets prove difficult to come by, other Wi-Fi tools exist that force de-authentication, so the legitimate user has to re-associate and re-authenticate on demand! Now the hackers have all the packets they need. The solution to this problem is long, complex passphrases. According to cryptanalysts after 20 characters the passphrase would begin to be difficult to break. It should be noted that one serious deficiency of WPA is the failure to protect management messages with any sort of authenticity.

IEEE 802.11i

The IEEE 802.11i (WPA2) defines a new type of security architecture called a robust security network (RSN). RSN separates user authentication from message protection and the two are linked using a security context, which is defined by the possession of certain limited-life keys. RSN defines a key hierarchy, which is created as a part of the authentication process. These keys are referred to as temporal or session keys. RSN employs advanced encryption system (AES) using the counter mode-cipher block chaining (CBC) message authentication code mode (CCM) for confidentiality and integrity of 802.11 data messages, replacing the obsolete WEP protocol. IEEE 802.1X/EAP/remote authentication dial-in user service (RADIUS) fits well into the RSN model as an access control mechanism with a strong upper-layer authentication method [such as transport layer security (TLS)]. This upper-layer authentication will occur after a standard OSA at the MAC layer, making SKA obsolete for previously described reasons. RSN requires new hardware to be deployed and is backward-compatible with existing protocols.

IEEE 802.11i defines three data encryption algorithms: WEP, TKIP, and counter mode CBC-MAC (CCMP), where WEP is included for backward compatibility, TKIP is the short-term solution to fix WEP problems, and CCMP is the long-term solution requiring additional hardware capabilities. Table 1 lists the popular wireless encryption algorithms that are in place. The standard defines an RSN association (RSNA) based on IEEE 802.1X authentication. There are two types of authentications provided, one based on an authentication server (AS), such as based on RADIUS or Diameter, and the other between two stations using PSK, e.g., in an ad hoc network or a home/small business network. In using an

Table 1 802.11 wireless encryption algorithms.

	WEP	TKIP	CCMP
Cipher key size(s)	RC4 40- or 104-bit encryption	RC4 128-bit encryption, 64-bit authentication	AES 128-bit
Key lifetime per-packet key	24-bit wrapping IV concatenate IV to base key	48-bit IV TKIP mixing function	48-bit IV not needed
Integrity packet header	None	Source and destination addresses protected by Michael	CCM
Packet data replay detection	CRC-32 none	Michael enforce IV sequencing	CCM enforce IV sequencing
Key management	None	IEEE 802.1x	IEEE 802.1x

AS, a secure network is assumed between the AS and the AP.

The CCMP protocol: the CCMP, like TKIP, addresses all known WEP deficiencies, but without the shackles of already deployed hardware. The AES was selected as its core encryption algorithm. The use of single-key (block cipher) is mainly to provide confidentiality and integrity; reducing key management overhead and minimizing the time spent computing AES key schedules. Another feature provided by CCMP is maintenance of integrity protection for the plaintext packet header, as well as integrity and confidentiality of the packet payload. CCMP employs a 48-bit IV, ensuring that the lifetime of the AES key is longer than any possible association. In this way, key management can be confined to the beginning of an association and ignored for its lifetime. The IV also provides a sequence number to provide replay detection, just like TKIP.

The complete process of an 802.11i authentication consists of handshakes between the supplicant and the authenticator (security capability discovery and 802.1x conversations), between the authenticator and the authentication server (RADIUS de facto), and between the supplicant and the authentication server (EAP-TLS de facto, with the authenticator serving as a relay). After these handshakes, the supplicant and the authentication server have authenticated each other and generate a common secret called the master session key (MSK). The supplicant uses the MSK to derive a pairwise master key (PMK); The authentication, authorization, and accounting (AAA) key material on the server side is securely transferred to the authenticator to derive the same PMK in the authenticator. Alternatively, the supplicant and the authenticator may be configured using a static PSK for the PMK.

IEEE 802.1X

The IEEE 802.1x standard provides an architectural framework to facilitate network access control at the link layer for various link technologies (IEEE 802.11, Fiber distributed data interface (FDDI), Token Ring, IEEE 802.3, Ethernet etc.). The standard abstracts the notion of three entities: the supplicant, the authenticator, and the authentication server. A supplicant is an entity (wireless client) that desires to use a service (link layer connectivity) offered via the notion of a port on the authenticator (AP). In the IEEE 802.11 scenario, a port corresponds to an association between a supplicant (wireless client) and the authenticator (AP). The supplicant authenticates via the authenticator to a central authentication server which directs the authenticator to provide access after successful authentication. Typically, the authentication server and the authenticator (AP) communicate using the RADIUS protocol, or some other AAA protocol. The authentication process between the authentication server and the supplicant (via the authenticator) is carried over the EAP.

The EAP framework may be extended by software and hardware manufactures to facilitate secure authentication and encryption. There are several implementations of EAP, including:

1. **EAP-TLS:** Developed by Microsoft and used in 802.1x clients for Windows XP, EAP-TLS provides strong security, but requires each WLAN user to run a client certificate.
2. **Lightweight EAP (LEAP):** Developed by CISCO and used in their Aironet solution, LEAP supports dynamic WEP key generation, provides for fixed password user authentication, and supports strong mutual authentication between a client and a RADIUS server.
3. **Protected EAP (PEAP):** Co-developed by CISCO, Microsoft and RSA Security, PEAP does not require certificates for authentication. It supports dynamic WEP key generation and provides options for password, token or digital certificate based user authentication.
4. **Tunneled TLS (EAP-TTLS):** Developed by Funk Software and Certicom as a competing standard for PEAP, EAP-TTLS supports password, token or certificate side user authentication. Unlike EAP-TLS, EAP-TTLS requires only the server to be certified.

5. **EAP-message digest 5 (EAP-MD5):** The 802.1x included a baseline version of authentication. The username is passed in clear text; passwords are encoded using an MD5 hash. This scheme may be vulnerable to dictionary attacks.

802.1x authentication occurs when a client first joins a network. Then, periodically, authentication recurs to verify the client has not been subverted or spoofed. 802.1x has the added benefit in a wireless network of not inducing a per-packet overhead. This lightweight implementation is important because it does not adversely affect the relatively low throughput of wireless networks. Keys are generated dynamically, at a per-session basis. In 802.11, the authentication server might be physically integrated into an AP.

IEEE 802.1X SHORTCOMINGS

The shortcomings of IEEE 802.1X, in general, are discussed below.

Absence of mutual authentication. The IEEE 802.1X standard has a fundamentally asymmetrical model of supplicants and access points: It requires a supplicant to authenticate itself to the AP, but not vice versa. This lack of mutual authentication can expose the supplicant to potential MITM attacks with an adversary acting as an AP to the supplicant and as a client to the network AP. EAP-TLS does provide strong mutual authentication, but is not mandatory in the 802.1x state machines and can be overridden in deployments with improperly defined policies. Mutual authentication is usually accomplished by means of an authenticated key establishment protocol.

Session hijacking. In a wireless context, using only the MAC address as identification leaves the supplicant vulnerable to session hijack attacks. With IEEE 802.1X, higher-layer authentication takes place after 802.11 (re) association. Thus, there are two state machines: the 802.11 and the 802.1x state machine. Their combined action dictates the state of authentication. A lack of clear synchronization between these state machines and

message authenticity makes it possible to perform a simple session hijack attack when encryption is not used. The attack proceeds as follows: A legitimate supplicant authenticates itself. An adversary sends an 802.11 MAC disassociate management frame using the AP's MAC address. This causes the supplicant to become disassociated. This message transitions the 802.11 state machine to the unassociated state while the 802.1x state machine of the authenticator (AP) still remains in the authenticated state. The adversary gains network access using the MAC address of the authenticated supplicant because the 802.1x state machine in the authenticator (AP) is still in the authenticated state.

Researchers at the University of Maryland have identified some flaws in 802.1x standard which can be mitigated if the 802.1x authentication is performed within an encrypted channel. The security community is also examining current EAP methods. 802.1x client support is already integrated into Windows XP and Mac OS X. Open1x, an open-source implementation, also is available, and runs on Linux and FreeBSD. RADIUS could provide AAA services, but is still unlikely to resolve all security threats in wireless networks. Therefore, Diameter (IETF RFC 3588) is evolving to improve RADIUS to provide better security. Table 2 lists all the IEEE 802.11 standards that are now in place.

WLAN VULNERABILITIES

Wiretrapping. While wiretapping in a wired network requires physical intrusion, wireless data packets can be received by anyone who is in radio range with an appropriate receiver. The eavesdropping range can be extended with the use of directional antennas. As a result, one must consider WLAN traffic as being delivered to the adversary as well as the intended party, and the adversary with a transmitter has the ability to inject or forge packets onto the network. This is a fundamental difference between wired and wireless network security. In wired networks, the possible attacks are limited to layer 3 and above while they are primarily focused on layer 1 and 2 in wireless networks.

Table 2 Selected IEEE 802.11 standards family.

Standard	Description	Status
IEEE 802.11	WLAN; up to 2 Mbps; 2.4 GHz	Approved in 1997
IEEE 802.11a	WLAN; up to 54 Mbps; 5 GHz	Approved in 1999
IEEE 802.11b	WLAN; up to 11 Mbps; 2.4 GHz	Approved in 1999
IEEE 802.11g	WLAN; up to 54 Mbps; 2.4 GHz	Approved in 2003
IEEE 802.11n	WLAN; up to 540 Mbps; 2.4 or 5 GHz	Draft (2006)
IEEE 802.11i	WLAN security standard aka WPA2	Approved in 2004

MAC address. WLAN standard does not include an access control mechanism. Most vendors have embraced the use of a MAC-address-based ACL. An ACL is essentially a lookup table based on the identity (in this case the MAC address). Recently, device drivers, especially open-source varieties, allow the user to change the MAC address thereby acquiring the identity of victims.

MAC address authentication. Such sort of authentication establishes the identity of the physical machine, not its human user. An adversary can easily acquire the identity of legitimate users by changing the MAC address of his node.

One-way authentication. WEP authentication is client-centered or one-way only. This means that the client has to prove its identity to the AP but not vice versa. Thus a rogue AP may successfully authenticate the client station and then subsequently will be able to capture all the packets sent by that station through it.

Unprotected management frames and susceptibility to carrier-sense attacks. Management frames are sent in clear text even if encryption is turned on for data frames.

Optional security and insecure defaults. The vast majority of 802.11 access points ship without security enabled. Users have the option to turn security on during the initial configuration. However, according to studies, only 30–40% of all access points have security turned on. It appears that users are not taking the time to enable this simple security mechanism.

WLAN ATTACKS

In the following, we will discuss several types of attacks in WLAN. These constitute attacks ranging from PHY layer to application layer.

DoS attacks. Traditionally, DoS is concerned with filling user-domain and kernel-domain buffers. However, in the wireless domain, the adversary is empowered to launch more fundamentally severe types of DoS that block the wireless medium and prevents other wireless devices from even communicating.

Spoofing management frames. Management frames such as de-authentication and disassociation result in the termination of network access for a client. Management frames are always sent in the clear and have no message authentication. An attacker can easily forge a de-authentication or a disassociation frame on behalf of a client or an AP. In response, the client or the AP will exit the authenticated state in the 802.11 state machine and will refuse all additional packets until authentication is re-established, resulting in temporary termination of network services for the client. These attacks can be easily prevented by adding per-packet message authentication mechanisms in new standards. However, for current deployments some system-level measures that mitigate the effect of such attacks are available. In particular, by delaying the effects of the attacks (say queuing the de-authentication/

disassociation messages), the AP can observe subsequent packets from the client. If a data packet arrives after such a request is queued, the request can be discarded because a legitimate client would not generate packets in that order.

Virtual (MAC layer) jamming. The IEEE 802.11 MAC uses a combination of physical carrier-sense and virtual carrier-sense mechanisms to control access to the channel. Each 802.11 frame contains a duration field that indicates the time (in microseconds) the channel is reserved for both transmitting the frame and receiving the ACK. Each node maintains a network allocation vector (NAV) that indicates the virtual time that the node has to wait before it can transmit—the node transmits when the NAV reaches zero. The duration field is used to program the NAV of every node. The NAV concept is used by the request to send (RTS)/clear to send (CTS) method of reserving the channel. An attacker can prevent any node from accessing the channel by setting a large duration field. This can be easily done by just sending an RTS packet with a large duration. The maximum value for the NAV is 32767, i.e., around 32 msec–hence an attacker needs to transmit one packet per 32 msec to bring the network down.

PHY-layer jamming. Wireless networks are susceptible to radio interference attacks and are not addressable through conventional security mechanisms like cryptographic methods. An adversary employing radio interference attacks is capable of preventing users from being able to communicate with, or introducing packet collisions that force repeated backoffs, or even jamming transmissions. An attacker sending a strong jamming signal of duration 1 bit/symbol will make the CRC computation wrong. In a "non-error-correction" encoded data packet a single bit error generates a CRC error, leading to the loss of the entire packet. Resistance to jamming is traditionally achieved by tuning various parameters such as transmission power, spread-spectrum, directional antennas and receiver communication bandwidth. An example could be excessive radio interference caused by 2.4 GHz cordless phones.

Jamming modeling. Based on transmitting patterns, four jammer attack models have been reported in the literature. A constant jammer continually emits a radio signal. The deceptive jammer constantly injects regular packets to the channel without any gap between subsequent packet transmissions. A random jammer alternates between sleeping and jamming. A Reactive jammer stays quiet when the channel is idle, and starts transmitting a radio signal as soon as it senses activity on the channel.

Jamming attacks detection. Wireless nodes can use several metrics to figure out if they are under a jamming attack. It should be noted that these measurements are not immune to false positives and could classify high-levels of network traffic as a DoS attack. Signal Strength and Carrier Sensing Time metrics are first measured. If the values are above a threshold or consistently above a threshold then it is likely that the node is under DoS attack.

Additionally, again each node can generate over time its packet send ratio (PSR) and packet delivery ratio (PDR). Using correlation techniques on above measurements a node can decide if it is under DoS attacks or the network is under heavy load.

Jamming mitigations. Two strategies have been presented in the literature that may be employed by wireless devices to evade a MAC/PHY-layer jamming-style wireless DoS attack. Channel surfing is a form of spectral evasion that involves legitimate wireless devices changing the channel that they are operating on. The second strategy, spatial retreats, is a form of spatial evasion whereby legitimate mobile devices move away from the locality of the DoS emitter. The basic idea of spatial retreat in infrastructure network is that a mobile device will move to a new AP and reconnect to the network under its new AP. In ad hoc networks, it involves physical movement and also change in routes for forwarding packets.

MITM. An intruder can change the route of the traffic, and thus, packets destined for a particular computer can be redirected to the attacking station. In the wireless context, an attacker will convince a victim that it is the AP and at the same time will act as a client to the real AP. Once this setup is done, all traffic from the victim will pass through the attacker.

Impersonation attacks. Most WLAN implementations use MAC and IP addresses to identify the wireless client. Both of these addresses are low layer mechanisms that are capable of being spoofed. Recent device drivers allow a user to change his MAC and IP address at will thus providing no real security. Hackers typically use such techniques to attack a WLAN impersonating as legitimate users. It is also relatively easy to steal MAC addresses or IP addresses by eavesdropping and using them later to gain access to the network.

Rogue access points. A rogue AP is an unauthorized wireless AP within a wireless network. Once a rogue AP without a security feature has been installed, an intruder can get unauthorized access to the entire network. Rogue APs usually use the same SSID as the legitimate network it mimics. A rogue AP can then accept traffic from wireless clients to whom it appears as a valid authenticator. In this way, a rogue AP can seriously harm a network. The WLAN vulnerability that facilitates Rogue Access Points can be attributed to the one-way, client-centered authentication between the client and the AP. Parsing the address resolution protocol (ARP) entries looking for the specific organizationally unique identifier (OUI) of the manufacturer of WLAN cards not installed in the network can help in identifying a rogue AP.

Distributed DoS (DDoS) attack. The DDoS attack is the most advanced form of DoS attacks. As the name suggests, the DDoS attack is distinguished from other DoS attacks by its ability to deploy its weapons in a "distributed" way. The total communication bandwidth can be far less than the total communication capacity of all wireless devices. However, a DDoS attack deliberately coordinates wireless devices to send out synchronized traffic, which can easily consume all spectrum resources or at least significantly reduce the capacity of communication channels for normal traffic. For a more comprehensive list and analysis, see Packet Storm and David Dittrich's articles on the Internet.

WLAN SECURITY SOLUTIONS

From a security architectural point-of-view, the WLAN itself should be placed outside the firewall. In this way, WLAN stations are treated like any other Internet hosts. Using a multi-port firewall, suitable firewall policies can determine which WLAN stations can access the Internet as well as the enterprise network.

We must recognize that the tools for securing communications over wired networks such as Secure Shell (SSH), Secure Sockets Layer (SSL), Pretty Good Privacy (PGP), and Virtual Private Network (VPN) technologies are just as effective over a wireless medium. SSL is an application level protocol that enables secure transactions of data and relies upon public/private keys and digital certificates. SSL has had great success in authenticating Web traffic and is available for many platforms and configurations.

Captive portal. Captive portal techniques are used to redirect clients to a special page, usually an authentication page, until they are authenticated. All traffic irrespective of address and ports is blocked initially. Captive portals are widely deployed in Wi-Fi hot spots and more recently are being deployed to secure wired ports as well. Some drawbacks with captive portal are the following. First, if the captive portal does not use SSL to get the authentication information, passwords or credit card numbers can be stolen by an eavesdropper. Second, after the authentication is completed the MAC address of the authenticated user can be stolen and be spoofed to gain unauthorized entry (MAC addresses on most cards are software updateable). Third, the communication can be overheard by a third party. And finally, along with the above attacks, DoS and hacking can easily be launched.

WIRELESS FIREWALL GATEWAYS

The wireless firewall gateway (WFG) is an all-in-one box that contains a firewall, router, web server, and dynamic host configuration protocol (DHCP) server. All wireless traffic can then be inspected at a centralized location before it is forwarded to other nodes, either wired or wireless. By this way, a wireless network administrator can better visualize the traffic pattern among wireless clients and between the wireless and wired world. Additional software capabilities like intrusion detection and prevention can then easily be added to the setup.

Wireless
LANs—Mesh

VIRTUAL PRIVATE NETWORKS

An additional safeguard that can be used to secure a wireless network is a virtual private network (VPN). A VPN uses a combination of tunneling, encryption, authentication and access control. A VPN establishes a secure, encrypted network tunnel where an authenticated key provides confidentiality and integrity for IP datagrams. VPN works by creating a tunnel, on top of a protocol such as IP. WLAN stations are treated similar to other Internet hosts. VPN solutions are produced using protocols such as layer-2 tunneling protocol (L2TP), IP security (IPSec) etc. IPSec is the most commonly used standard.

On the negative side, VPNs are not a ubiquitous solution; they only support the IP suite. A VPN can also give a false sense of security if it is used only for communicating with a trusted network (such as a remote office). In this case, known as a "split tunnel," the client is still sending data across the network in the clear and is likely still accepting connections from other machines. Unfortunately, VPNs represent layer-3 solutions to layer-2 problems. A layer-3 solution (such as a firewall or VPN) does not necessarily mitigate attacks against layer 2.

PUBLIC KEY CRYPTOGRAPHY

Public key cryptography (PKC) is an ideal candidate to enforce confidentiality on the Internet. Indeed, many security mechanisms in the wired paradigm are based on it. However, PKC is computationally more expensive compared to symmetric-key cryptography. In wireless nodes both the processing power and battery power are at a premium and must be conserved. Therefore, in wireless deployments, PKC is confined to the distribution of encrypted digital certificates as authentication means during a transaction and establishment of symmetric session keys.

Other technologies that can be implemented to secure wireless networks include intrusion detection and prevention systems and vulnerability assessment tools. While cryptography is not the only piece of a secure system, it is definitely an imperative component. The use of encryption assures the protocol stack to protect data and this is seen to be critically important.

HUMAN FACTORS

Technology is only a small portion of the solution. As a security professional, one needs to take a holistic view of wireless security to create a realistic and complete solution to the problem. Even the best security in the system can be subverted by a bad user. If the security features of a wireless product are difficult to configure, a user may very well turn them off rather than deal with the complexity of the process.

WIRELESS SECURITY TOOLS

Sniffers, or protocol analyzers, are the most basic tools available to administrators of any type of network. They provide access to raw protocol packets as user stations (in this case 802.11 wireless cards and APs) transmit and receive them. "Sniffing" is the foundation for most auditors, intrusion detection systems (IDSs), attack tools, and several other applications. Network managers use security auditors to analyze their network's current state with respect to a corporation's security policy. Over the past few years, security community has rolled out honeypots and honeynets. The concept is to allow attackers to gain access to closely watched systems for research purposes and early-warning signs of incoming attack. IDSs are the latest to be included into wireless space to detect abnormal activity. Products like *AirDefense* and *AirMagnet* give users a view of malicious activities at layer 2 and attempt to prevent insecure configurations from being used. In monitor mode, passive sniffers like *AirMagnet* and *Kismet* monitor all wireless transmissions close enough to detect, irrespective of source and destination, without generating any betraying traffic themselves. This is to be distinguished from promiscuous mode, which captures all traffic on the network to which you are associated and is not a default option on all wireless cards. Other wireless sniffers include NetStumbler, Wellenreiter, Mognet, Dstumbler, Ethereal, Wireshark, Airopeek, and Sniffer Wireless. Rest of this section will list details on some of the WEP security tools.

Airsnort is an example of a weak IV encryption cracking utility that targets $B + 3$:FF:X-type packets. Since modern Wi-Fi cards and appliances reduce the percentage of weak IVs that are generated (under the rubric of "WEP+" or "Advanced WEP Encryption"), Airsnort is declining in importance as it takes an unreasonably long time to collect enough packets to break keys. However, recall that there are several classes of weak IVs. Newer utilities like *Aircrack* and *WepLab* use a more sophisticated approach to span a broader spectrum of FMS weak IV classes and also incorporate brute force, dictionary, and an implementation of the "Korek algorithm." Aircrack does not crack live traffic on-the-fly as Airsnort does, but looks for a larger range (five million) of weak IVs. Further, Aircrack is capable of distributing the workload over multiple processors for efficiency.

Other WEP weaknesses that may be exploited include defective key-generation implementations. Replay attacks result when wireless traffic is captured and retransmitted. Aireplay is one such tool for relay attacks. It harvests ARP-like packets from captured ("sniffed") Pcap files, and then re-broadcasts them indefinitely to increase the traffic flow. A variation of this theme is a PRGA injection attack. In WEP, the IV, the key, and the data length value are all used by a PRGA to generate a pseudo-random string exactly the same length as the plaintext data to be encrypted. The ciphertext is the result of XORing this string with the plaintext. If one knew this string, deciphering the text would be trivial because the IV and ciphertext appear in captured packets. Attack vectors that exploit this vulnerability are called PRGA injection attacks. Here's one such strategy.

The hacker sniffs WEP *challenge/response* authentications between the AP and clients. The *challenge* will be in a plaintext form and the response is an encrypted version of the challenge. If we XOR the challenge with the response, the result is the PRGA. Armed with the PRGA and four bytes of control information, one may craft packets at will. Add to that a spoofed, known acceptable IP address for protection, and one can move through the wireless fabric like Code Red through a straw. The only caveat is that the crafted packets must reuse the same IV. So long as IV reuse is accepted by the wireless appliances, this is not a liability. Enter WEPWedgie; WEPWedgie is an automated tool for PRGA injection. It consists of two modules: prgasnarf, which collects WEP challenge/response authentication; and wepwedgie, which crafts the packets (e.g., the SYN packet described two paragraphs earlier). The actual injection is accomplished by a sister product called Airjack. One remaining attack vector of note is the dictionary attack. WEPAttack is designed for just this purpose. You can find more information on tools like Airsnarf, Bluesniff, Kismet, Netstumbler etc., using Google search engine.

INTRUSION DETECTION

When a WLAN is operating in infrastructure mode, intrusion detection methodologies can be usefully applied at a central location. All 802.11 traffic pass through an AP and translates to or is translated from 802.3 frames. Because the layer-3 information remains unchanged during this translation, only the layer-2 frames can provide traces of any intrusion at the AP. One way an intrusion can be detected is by noticing an abnormal signature pattern in a message that has been sent or received and then auditing the system for any anomaly. To detect an abnormal signature pattern, it is important to monitor the wireless traffic at the APs. Research has shown that the following fields in 802.11 frames are relevant to the detection and analysis of intrusion in WLANs.

- **Sequence number:** A 2-byte sequence control field that helps to collect fragments of 802.11 frames.
- **Control type and subtype:** This field distinguishes data, management, and control frames.
- **SSID:** Probe request frames containing an SSID with special value of 0 are used to detect APs in the vicinity.

- **Destination MAC address:** Broadcast packets may be used to determine the available APs in a network.
- **OUI:** A 3-byte field is used to uniquely identify a hardware manufacturer. Knowing the card manufacturer from the publicly available list of OUIs allocated by the IEEE to WLAN card manufacturers, it is possible to identify any messages from the attacker who is likely to change his MAC addresses.
- **Data payload:** The data payload of 802.11 frames can be monitored for information that is out of ordinary.
- **Logical link layer (LLC) protocol-type field:** This field contains the upper-layer protocol message contained in the data payload. It should be monitored to determine what the payload field of the 802.11 frame contains.

WAR DRIVING

War driving is the art of searching for Wi-Fi enabled networks from a moving vehicle. It involves the use of a vehicle such as car/truck, a computer such as laptop/pda. War drivers typically use GPS receivers to mark the location of a wireless network and log it into a central database. The most prominent of these databases is the wigle.net. War driving was named after war-dialing.

An intruder may use either an active scanning or passive scanning method. In the active scanning method, the intruder sends a probe request message on each channel on which it detects wireless activity. APs within the range then respond with a probe response frame that contains the SSID. In the passive method, the intruder sniffs the beacon management frames that are periodically transmitted by the AP. The advantage of this type of attack is that there is, initially, no activity from the intruder side, so the intruder cannot be detected by anyone else on the network.

WIRELESS SECURITY BEST PRACTICES

Specific security methods that can be implemented to secure 802.11 wireless networks include any or all of the following:

1. Turning off the broadcast SSIDs.
2. Enabling encryption.
3. Using MAC address filters: MAC-based access control mechanisms for node admission. This scheme provides an additional security layer.
4. Lowering the power levels of the access points to limit the ability of hackers to connect from outside the specified boundary.
5. Shaping the radio waves by appropriate positioning of the antenna, and thus, preventing the signal from bleeding outside the "designed service area."

CONCLUSIONS—WIRELESS (SECURITY) TRENDS/DIRECTIONS

In this entry we have presented a systematic compilation of all the issues and challenges that are in place for WLANs. We have also identified all the key vulnerabilities and attacks on WLAN. We have also presented a consolidated report on IEEE 802.1x standards and argued on how these standards are in use and their shortcomings on certain protocol aspects. Based on our extensive survey, we may say that general-purpose operating systems running on embedded devices are perfectly amenable for the wireless industry, as applications can be deployed rapidly and on a broad set of devices and can be allowed to interact. However at the same time, attackers have similar advantages! As these operating systems continue to evolve to look like or to mimic their bigger desktop cousins, the problem will only get worse. The bigger the operating system, the more likely is security vulnerability in the core operating system.

The future of Wi-Fi is likely to be multiple input multiple output (MIMO). MIMO systems use multiple transmit and multiple receive antennas. In a scattering-rich environment, each receiving antenna is able to compute a signature of each of the transmitting antennas, and thus distinguish their transmissions. In principle, such a system has an overall capacity proportional to the number of antennas used, at the price of increased complexity. The 802.11n task group is working toward definition of a MIMO PHY layer.

BIBLIOGRAPHIC NOTES

Papers describing several weakness in WEP and attack implementations can be found in Refs. 2–5. Issues and problems pertaining to security risks, threats, problems in wireless networks, in general, with and without security protocols are discussed in Refs. 6–11. The trends and solutions, both in technical and non-technical aspects, in wireless security space are detailed in Refs. 12–18. Work reported in Refs. 19–21 present some common wireless security auditing tools whereas in Ref. 1 the work is more geared toward RC4 stream cipher, the security mechanism behind WEP. Contributions in Refs. 22–25 are extensive in terms of presenting reviews of the flaws, attacks in WEP and 802.1x. Jamming in the PHY and MAC layers of the network stack are presented in Ref. 26 and Ref. 27 discusses several aspects of WLAN. WEP attack material discussed in this entry can be found in Refs. 28–30 and work in Ref. 31 compares two leading wireless protocol namely Wi-Fi and Bluetooth. In Ref. 32 the authors discuss the IEEE 802.11i security standard and the 802.1x port based authentication system. A collection of papers[33–36] present detailed reviews on 802.11i and 802.1x security standards and suggest a new security architecture. A wireless gateway architecture based on IPSec is discussed in Ref. 37. Works in Refs. 38–41 review DoS, DDoS attacks in the wireless space and propose some defenses whereas contributions in Refs. 42–44 discuss some techniques for MAC address spoofing, and intrusion detection.

ACKNOWLEDGMENTS

The authors wish to thank Dr. Konstantinos Anagnostakis (Kostas) at Institute for Infocomm Research, Singapore, for reading through the draft versions of this entry which helped to refine the content. Second, the principal author wishes to thank his colleague at National University of

APPENDIX: GLOSSARY OF WLAN TERMS

Acronym	Description
802.11 a/b/g	A set of IEEE wireless standards for WLAN
802.11i	Wireless security standard aka WPA2
AES	Advanced Encryption Standard
AP	Access Point
ARP	Address Resolution Protocol
CBC	Cipher Block Chaining
CCMP	Counter mode CBC-MAC
EAP	Extensible Authentication Protocol
EAP-MD5	A password-based mechanism for client authentication
EAP-TLS	Certificate based mechanism for client authentication
ICV	WEP integrity check value
IEEE	Institute of Electrical and Electronics Engineers
IEEE 802.1x	Port based access control widely used in multiple protocols
IETF	Internet Engineering Task Force
IV	Initialization vectors
LAN	Local Area Network
LLC	Logical Link Layer
MAC	Media Access Control
MIC	Message Integrity Code
OSA	Open System Authentication
OUI	Organizationally Unique Identifier
PEAP	Protected EAP
PRN	Pseudo-random number

(continued)

Wireless
LANs—Mesh

Appendix
(continued)

Acronym	Description
RADIUS	Remote Authentication Dial-In User Service
RC4	Rivest Cipher #4 (a symmetric encryption algorithm)
SKA	Shared Key Authentication
SSID	Service set identifier, a WLAN ID
Tgi	IEEE-802.11i Task Group
TKIP	Temporal Key Integrity Protocol
TLS	Transport Layer Security
Wardriving	Process of discovering neighborhood wireless networks
WEP	Wired Equivalent Privacy
Wi-Fi Alliance	A wireless industry organization
WPA	Wi-Fi Protected Access
WPA2	Wi-Fi Protected Access, v2 aka 802.11i
XOR	Exclusive OR, a Boolean logic operation that is true if only one of the inputs is true, but not both

Singapore Ms. G. Kavitha for compiling key references used in this entry and Khu Kirk Jon Ong Lopez (Kirk) at Institute for Infocomm Research for reviewing the final draft. Bharadwaj Veeravalli would like to thank Professor Borko Furht for providing him an opportunity to contribute toward this encyclopedia.

REFERENCES

1. The RC4 Encryption Algorithm. Rivest RL, RSA Data Security; Mar 1992.
2. Fluhrer, S.; Martin, I.; Shamir, A. Weakness in the key schedule algorithm of RC4. 8th Annual Workshop on Selected Areas of Cryptography (SAC'01), Ontario, Canada, Aug 16–17, 2001; 1–24.
3. Cam-Winget, N.; Housley, R.; Wagner, D.; Walker, J. Security flaws in 802.11 data link protocols. Commun. ACM. **2003**, *46* (5), 35–39.
4. Stubblefield, A.; Ioannidis, J.; Rubin, A. Using the Fuhrer, Mantin, and Shamir attack to break WEP. 2002 Network and Distributed Systems Security Symposium (NDSS '02), San Diego, CA, Feb 6–8, 2002; 17–22.
5. Walker, J. Unsafe at any key size: an analysis of the WEP encapsulation, IEEE 802.11 doc 00-36; Oct 27, 2000.
6. Berghel, H.; Uecker, J. Wireless infidelity II: airjacking. Commun. ACM. **2004**, *47* (12), 15–20.
7. Berghel, H.; Uecker, J. Wifi attack vectors. Commun. ACM. **2005**, *48* (8), 21–27.
8. Housley, R.; Arbaugh, W. Security problems in 802.11-based networks. Commun. ACM. **2003**, *46* (5), 31–34.
9. Potter, B. Wireless security's future. IEEE Secur. Priv. **2003**, *1* (4), 68–72.
10. Potter, B. Wireless hotspots: Petri dish of wireless security. Commun. ACM. **2006**, *49* (6), 50–56.
11. Hytnen, R.; Garcia, M. An analysis of wireless security. JCSC. **2006**, *21* (4), 210–216.
12. Wireless security—What Is Out There? Frost and Sullivan.
13. Potter, B. Trends in wireless security—the big picture. Netw. Secur. **2002**, *2002* (5), 5–6.
14. Potter, B. Fixing wireless security. Netw. Secur. **2004**, *2004* (6), 4–5.
15. Schmidt, T.; Townsend, A.M. Why Wi-Fi wants to be free. Commun. ACM. **2003**, *46* (5), 47–52.
16. Di Pietro, R.; Mancini, L.V. Security and privacy issues of handheld and wearable wireless devices. Commun. ACM. **2003**, *46* (9), 74–79.
17. Bhagyavati, Summers, W.C.; DeJoie, A. Wireless security techniques: an overview. 2004 Conference for Information Security Curriculum Development Conference (InfoSecCD'04), Kennesaw, GA, Sept 17–18, 2004.
18. Allen, J.; Wilson, J. Securing a wireless network. 30th annual ACM SIGUCCS conference on User services (SIGUCCS '02), Providence, RI, Nov 20–23, 2002.
19. Potter, B. Next generation wireless security tools. Netw. Secur. **2003**, *2003* (9), 4–5.
20. Branch, J.W.; Petroni, N.L., Jr.; Van Dorrn, L.; Safford, D. Autonomic 802.11 wireless LAN security auditing. IEEE Secur. Priv. **2004**, *2* (3), 56–65.
21. Branch, J.W.; Petroni, N.L., Jr.; Van Doorn, L.; Safford, D. Autonomic 802.11 wireless LAN security auditing. IEEE Secur. Priv. **2004**, *2* (3), 56–65.
22. Mishra, A.; Petroni, N.L., Jr.; Arbaugh, W.A.; Fraser, T. Security issues in IEEE 802.11 wireless local area networks: a survey. Wirel. Commun. Mobile Comput. **2004**, *4*, 821–833.
23. Zhaur, Y.; Yang, T.A. Wireless LAN security and laboratory designs. JCSC. **2004**, *19* (3), 44–60.
24. Stubblefield, A.; Ioannidis, J.; Rubin, A.D. A key recovery attack on the 802.11b wired equivalent privacy protocol (WEP). ACM Trans. Inf. Syst. Secur. **2004**, *7* (2), 319–332.
25. Housley, R.; Arbaugh, W.A. Wireless security problems in 802.11-based networks. Commun. ACM. **2003**, *46* (5), 31–34.
26. Lin, G.; Noubir, G. On link layer denial of service in data wireless LANs. Wirel. Commun. Mobile Comput. **2005**, *5*, 273–284.
27. Arbaugh, W.A.; Shankar, N.; Wan, Y.C.J. Your 802.11 wireless network has no clothes. IEEE Wirel. Commun. **2002**, *9* (6), 44–51.
28. Petroni, N.L., Jr.; Arbaugh, W.A. The dangers of mitigating security design flaws: a wireless case study. IEEE Secur. Priv. **2003**, *1* (1), 28–36.
29. Soliman, H.S.; Omari, M. Application of synchronous dynamic encryption system in mobile wireless domains. 1st ACM International Workshop on Quality of Service & Security in Wireless and Mobile Networks (Q2SWinet '05), Montreal, Quebec, Canada, Oct 13, 2005; ACM Press: New York, 2005.
30. Arbaugh, W.A. *An Inductive Chosen Plaintext Attack Against WEP/WEP2*, IEEE doc 802.11-02/230.

Wireless
LANs—Mesh

31. Ferro, E.; Potorti, F. Bluetooth and Wi-Fi wireless protocols: a survey and a comparison. IEEE Wirel. Commun. **2005**, *12* (1), 12–26.

32. Chen, J.-C.; Jiang, M.-C.; Liu, Y.-W. Wireless LAN security and IEEE 802.11i. IEEE Wirel. Commun. **2005**, *12* (1), 27–36.

33. Ahmad, A.; El-Kadi Rizvi, M.; Olariu, S. Common data security network (CDSN). 1st ACM international workshop on Quality of service & security in wireless and mobile networks (Q2SWinet '05), Montreal, Quebec, Canada, Oct 13, 2005.

34. He, C.; Mitchell, J.C. Analysis of the 802.11i 4-way handshake. 2004 ACM workshop on Wireless security (WiSe '04), Philadelphia, PA, Oct 1, 2004.

35. Faria, D.B.; Cheriton, D.R. DoS and authentication in wireless public access networks. 3rd ACM workshop on Wireless security 2002 (WiSe '02), Atlanta, GA, Sept 28, 2002.

36. Arunesh, M.; Arbaugh, W.A. An initial security analysis of the IEEE 802.1x standard. Technical Report CS-TR-4328. University of Maryland, College Park, Feb 2002.

37. Gobher, A.; Dasgupta, P. Secure wireless gateway. 3rd ACM workshop on Wireless security 2002 (WiSe '02), Atlanta, GA, Sept 28, 2002.

38. Xu, W.; Trappe, W.; Zhang, Y.; Wood, T. The feasibility of launching and detecting jamming attacks in wireless networks. 6th ACM International Symposium on Mobile Ad Hoc Networking and Computing (MobiHoc '05), Urbana-Champaign, IL, May 25–27, 2005.

98. Xu, W.; Wood, T.; Trappe, W.; Zhang, Y. Channel surfing and spatial retreats: defenses against wireless denial of service. 2004 ACM workshop on Wireless security (WiSe '04), Philadelphia, PA, Oct 1, 2004; 80–89.

40. Geng, X.; Huang, Y.; Whinston, A.B. Defending wireless infrastructure against the challenge of DDoS attacks. Mobile Netw. Appl. **2002**, *7*, 213–223.

41. Bellardo, J.; Savage, S. 802.11 denial-of-service attacks: real vulnerabilities and practical solutions. 12th USENIX Security Symposium, Washington, DC, Aug 4–8, 2003; 15–28.

42. Sharma, V. Intrusion detection in infrastructure wireless LANs. Bell Labs Tech. J. **2004**, *8* (4), 115–119.

43. Wright, J. *Detecting wireless LAN MAC address spoofing*; White paper: Jan 2003.

44. Wright, J. Layer 2 analyses of WLAN discovery applications for intrusion detection; White paper: Aug 2002.

Wireless Mesh Networks: Architecture, Protocols, and Applications

Vassilios Koukoulidis
Nokia Siemens Networks, Boca Raton, Florida, U.S.A.

Mehul Shah
T-Mobile USA Inc., Bellevue, Washington, U.S.A.

Abstract

Wireless mesh network is a wireless network where any two nodes are connected by at least one path. A wireless mesh network provides access and backhaul capabilities, utilizing exclusively radio links. The mesh network connects to the data backbone using wireless or wired interfaces [the points connecting to an external data network are also called egress (for outgoing traffic) or ingress (for incoming traffic) points.]

INTRODUCTION

Wireless Mesh Networks Overview and Basic Definitions

A mesh network is characterized by a topology where every two nodes have at least one path between them. In a *wireless mesh network*, the paths between any pair of nodes are formed with wireless or radio links. Wireless mesh networks are used to inter-connect user networks such as WLANs, Wi-Fi access points or hot spots, or hot zones [formed through mesh (wireless or wire line) inter-connection of Wi-Fi access points and allowing session continuity]. Such user networks are collectively referred to as *access networks*. Wireless mesh networks are further used to provide transport for interconnected access networks or other mobile/fixed networks (e.g., emulating T1 lines over radio links to inter-connect the elements of a cellular network). Such functionality is also referred to as *backhauling*.

In summary, a wireless mesh network provides access and backhaul capabilities, utilizing exclusively radio links. The mesh network connects to the Internet using a wireless or wired interface. [The points connecting to an external data network are also called egress (for outgoing traffic) or ingress (for incoming traffic) points.]

Fig. 1 illustrates typical use cases where a wireless mesh network is deployed to provide user connectivity to the Internet or other data backbone. To better understand this picture, it is worthwhile to review some commonly used terminology:

- A wireless node with routing capabilities is also referred to as *wireless mesh router*.

- A user connection supported by a wireless network is characterized as:
  - *Nomadic*: The user has to re-establish the connection each time he/she moves to another access point. Typically provided by WLAN.
  - *Portable*: Session continuity (not necessarily in real-time) is provided within well-defined network boundaries. Typically provided by world interoperability for microwave access (WiMAX) (introduced later in this chapter) or WLAN with handover capabilities.
  - *Mobile*: The network provides seamless roaming and handover capabilities. WiMAX has been extended to provide this capability as well.

Presently, commercially available wireless mesh networks provide portability. Mobility is achievable through the use of WiMAX, cellular networks, or convergent networks that combine Wi-Fi and cellular access to provide seamless voice roaming and handover.

Usually, access and backhaul links are created on unlicensed radio frequencies at 2.4 GHz and/or 5.x GHz. However, there is no restriction that only these frequencies must be used. Fig. 2 shows the building blocks and inter-connection in a multipoint-to-multipoint, switched mode, wireless mesh network. Such a network is formed when different frequencies are used for access and backhaul. Further, in these networks backhaul is formed using directional antennae in order to improve capacity and scalability. Various types of wireless connectivity and their relative advantages and disadvantages are examined later in this chapter.

Market Considerations

Mobile technology has quickly become widespread and virtually available to all PCs and notebooks built from

Encyclopedia of Wireless and Mobile Communications DOI: 10.1081/E-EWMC-120043506

Wireless LANs—Mesh

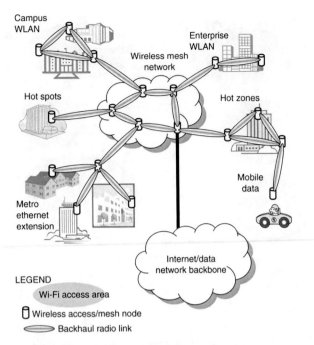

Fig. 1   Data network inter-connect using wireless mesh networks.

providers. The key factors behind this increasing market opportunity for wireless mesh networks are low cost, simplicity, ease of operation, and quick deployment of wireless mesh networks.

At the time of this writing, wireless mesh networks are being deployed by:

- Municipal operations (e.g., to cover own needs and residents' access to municipal services);
- Public utilities (since only power is needed and spectrum is presently unlicensed);
- Traditional operators: wireless Internet service providers (WISPs), cable operators, cellular network operators, fixed network operators.

Served areas can range from campus or city center (up to 25 km^2), cities (25–100 km^2), counties (100–250 km^2), etc.

In this chapter, we introduce the technologies used for wireless mesh networks, present network architectures for various user and operator scenarios, review the relevant standardization work, and finally address how wireless mesh technologies are integrated with other wireless technologies, data network backbones, and security/access control mechanisms. For more detailed information on the subject, refer to the standards referenced throughout this chapter and the April 2006 special issue of *IEEE Wireless Communications*, Refs. 1–6.

2004 onward (e.g., Intel Centrino technology with WLAN capability is used in approximately 90% of the laptop market). The result is an increased demand for wireless Internet access, applications, and services. Intel estimates that approximately 65 million laptops will be Wi-Fi enabled by 2008. An additional 42 million laptops will be WiMAX enabled by the same year. Such wide demand for nomadic or portable data and even mobile access presents an increasing market opportunity for various telecommunications operators and allows introduction of new service

BROADBAND WIRELESS TECHNOLOGIES

Broadband wireless technologies can be divided in two main categories: *fixed wireless* or point-to-point configurations using stationary antennae (e.g., local multipoint

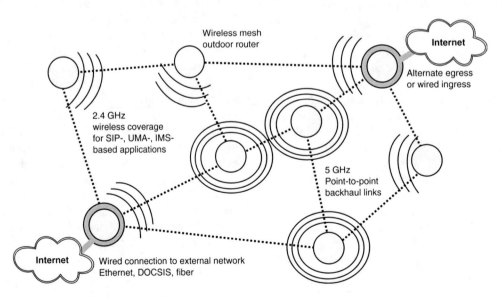

Fig. 2   Building blocks of a wireless mesh network.

distribution services (LMDSs), used as alternative to wired broadband access); *mobile wireless* designed for portable equipment with light-weight transmitters/receivers that operate while the user changes locations.

The main broadband mobile technologies in the market are:

- High-speed packet access (HSPA) which started as high-speed downlink packet access (HSDPA) in 3G/UMTS networks and then enhanced to include high-speed uplink packet access (HSUPA). HSPA technologies support low or high-speed mobility across national or international networks. Throughput depends on the level of mobility and ranges from hundreds of Kbps for high-speed (e.g., vehicular) or remote (e.g., rural) users to 3–4 Mbps for low-speed (e.g., nomadic/pedestrian) users. Deployment of HSPA solutions requires 3G (W-CDMA) infrastructure, which is typically owned by large national or international operators. HSPA requires licensed spectrum that is auctioned by government telecommunications agencies. 3G network technologies are standardized by third generation partnership project (3GPP).
- WLAN, usually following the IEEE 802.11 set of standards. WLAN users are characterized by low mobility (e.g., indoor, access at given urban locations, and roaming through pre-agreed WLAN operators' hot spots). Rates can reach hundreds of Mbps (typically 54 tens for the widely supported IEEE 802.11g standard). WLAN operates in unlicensed spectrum frequencies at 2.4 or 5 GHz.
- WiMAX is standardized by IEEE 802.16 and allows mobility comparable to HSPA but reaches higher throughput (from more than 1 Mbps for vehicular speeds to several tens of Mbps in low-speed scenarios). WiMAX is a true MAN technology, able to cover distances of several kilometers (typical cell diameter is less than 7 km). WiMAX operates in three frequency bands: 5.8 GHz (license-exempt), 3.5 GHz (licensed) and 2.5 GHz (licensed).
- Fast low-latency access with seamless handoff-orthogonal frequency division multiplexing (FLASH-OFDM). The motivation for FLASH-OFDM is to provide mobility and throughput comparable to HSPA but with the much lower latency required by many critical applications (e.g., voice/video communications, key enterprise applications). Standardization for FLASH-OFDM started though IEEE 802.20 but was also included in 802.16 for WiMAX. Following is a comparison between FLASH-OFDM and other mobile technologies:
  - GPRS: throughput 56 Kbps (maximum), latency 600 + ms;
  - 3G [(W-CDMA, evolution-data optimized (EV-DO)]: throughput 384 Kbps (maximum), latency 350 + ms;
  - HSDPA (3GPP Release 5): throughput 3 Mbps (maximum), latency 350 + ms;
  - FLASH-OFDM: throughput 1.5 Mbps, latency less than 50 ms.

Fig. 3 illustrates the throughput and mobility requirements addressed by each technology.

NETWORK ARCHITECTURE AND SCENARIOS

In this section we address mesh network topology and compare various inter-connection techniques. As already mentioned, a wireless mesh network consists of access radio networks interconnected via radio backhaul links.

Access radio networks connect to all the clients in their respective cell range. Depending on the desired coverage area, access networks can use *omnidirectional* or *directional* antennas. An omnidirectional (also called "non-directional") antenna transmits (radiates) or receives in a uniform pattern around the access point. A directional antenna focuses its radio signal energy in a narrow pattern around one or more directions (e.g., bidirectional or figure-8 antennas). For example, one might choose omnidirectional antennas for coverage of a shopping mall or parking lot, and directional antennas in more restricted areas, such as hotels.

Backhaul links are also formed using omnidirectional or directional antennas, depending on the capabilities of the mesh nodes. For example, a mesh node can have an omnidirectional antenna to create an access network at 2.4 GHz frequency, and many directional antennas at 5 GHz frequency to form mesh inter-connections to other nodes.

Therefore, backhaul connections can be of four types:

- *Point-to-point*: Each radio component of a network node connects to one and only one dedicated radio component.
- *Multiple point-to-point*: Multiple radio components, each connecting to one and only one dedicated other radio component.
- *Point-to-multipoint*: A radio component of a network node connects to multiple radio components. This configuration is used for repeater mode operation (e.g., a base station beaming high-speed Internet connections to home or business subscriber stations—a concept used by WiMAX).
- *Multipoint-to-multipoint*: Multiple radio components connecting to multiple other radio components.

Let us further discuss some of these connection configurations in order to get a better understanding. Fig. 4 shows a multipoint-to-multipoint topology, where each node has a single radio, and access and backhaul share this radio on

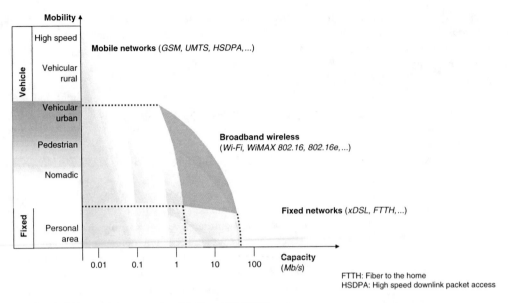

Fig. 3 Addressing bandwidth requirements using Wi-Fi and WiMAX.

the same frequency (2.4 GHz). The mesh inter-connect is formed as each node uses an omnidirectional antenna to send to and receive from all the other nodes within range (i.e., backhaul is a shared network). Clearly, mesh nodes and clients share the available bandwidth thus resulting in relatively low system capacity. Due to the contention from access and backhaul, latency and jitter are not bounded and can be unpredictably high; therefore, this solution is not suitable for real-time applications. Hence, although simple and low-cost, this configuration is not recommended for building high-capacity networks or networks with scalability and real-time requirements.

Fig. 5 represents an improvement over the scenario that we just described. This is also a multipoint-to-multipoint topology but it uses dual radio in order to separate access (2.4 GHz) from backhaul (5 GHz). Backhaul is a shared network operating on a single channel, but access uses different channels to minimize interference. This solution works better than the previous one for small networks, but it is still characterized by high latency and jitter at the backhaul mesh, i.e., it is not suitable for networks with real-time requirements. Due to the shared backhaul, this solution does not scale very well and also cannot be used in high capacity networks.

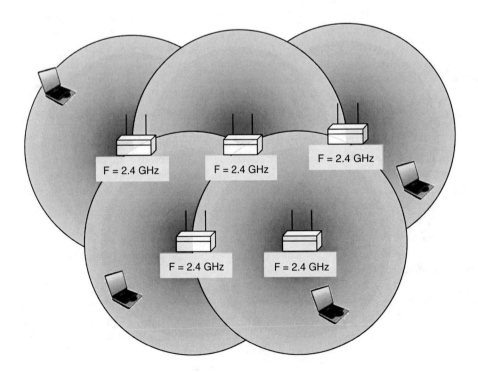

Fig. 4 Multipoint-to-multipoint: single radio with shared mesh backhaul and access.

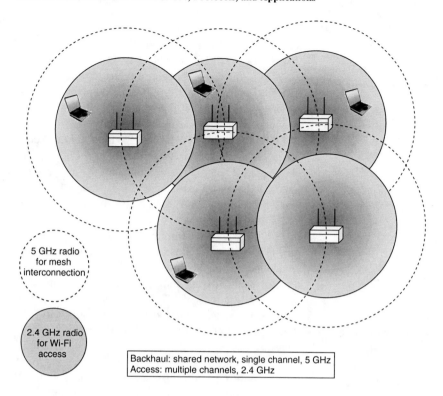

Wireless
LANs—Mesh

Fig. 5 Multipoint-to-multipoint: dual radio with separated mesh backhaul and access.

A significant improvement, functionally resembling wired network inter-connection, is shown in Fig. 6. In this solution, client access and mesh backhaul use completely independent radios. Mesh links are formed using multiple point-to-point links, where each link is created through dedicated radios with directional antennas. The backhaul network links operate on multiple channels to minimize interference, thus allowing scalable, high-capacity configurations (point-to-point links enable high degree of separation of mesh nodes). Such backhaul mesh has the performance of a wired network (typical node throughput is approximately 75 Mbps). This solution is suitable for large operator networks requiring scalability, high capacity, and support for real-time applications (e.g., voice, time division multiplexing (TDM) backhauling, video, and data).

As an example of a special application of the above point-to-point configuration, consider the network scenario in Fig. 7. Here, point-to-point backhauling (e.g., IEEE 802.16 WiMAX) is used to provide connectivity between

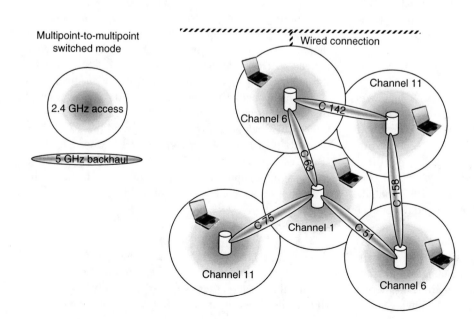

Fig. 6 Point-to-point, switched mode: multi-radio with separated mesh backhaul.

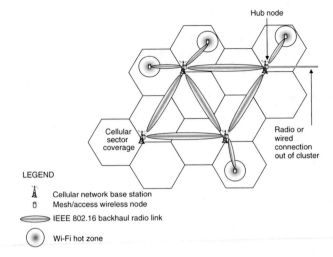

Fig. 7 Wireless mesh network used for cellular backhaul and Wi-Fi hot zones inter-connection.

cellular network base stations by emulating TDM lines. Mesh nodes may or may not be collocated with the cellular network nodes (depending on distance and/or environmental factors). Further, the network services can be expanded over time by deploying Wi-Fi access networks interconnected via the mesh backhaul.

For determining which solution to choose (point-to-point, multipoint-to-multipoint, shared vs. separate radio, etc.) the network designer must consider capacity, throughput, latency/jitter, and scalability—just to name a few—requirements. Currently there are commercial products and standards that cover all the above scenarios.

STANDARDIZATION AND PROTOCOLS

Wireless mesh networks utilize WLAN, WiMAX, and FLASH-OFDM technologies for radio transmission. This chapter presents the standardization related to these technologies only.

Standardization Bodies and Industry Groups

1. Institute of Electrical and Electronics Engineers (IEEE). IEEE has defined the most widely accepted and deployed set of standards for WLAN, which will be presented later in this chapter. Access methods for WLANs were originally defined in 1990 by the IEEE working group 802.11 under the title "Wireless Local Area Network." It was not until 26 June 1997, that the IEEE Standards Activity Board authorized the 802.11 standard.

2. Wireless Ethernet Compatibility Alliance (WECA) or Wi-Fi alliance. The WECA working group was founded in 1999 (see Additional Reading 2) and comprised an international association of wireless Ethernet manufacturers such as Cisco, IBM, Intel, Nokia,

Siemens, etc., with the main objective to develop compatibility criteria for WLAN products based on IEEE 802.11. In October 2002, WECA was renamed to Wi-Fi alliance and began product certification for IEEE 802.11. Presently, more than 200 manufacturers are members of the Wi-Fi alliance and more than 2200 products have been certified.

3. WLAN association (WLANA). WLANA (see Additional Reading 3) consists primarily of service providers, consultants, training companies, and product manufacturers. Its main objective is trade education, public relations, and marketing of IEEE 802.11 standards.

4. The European Telecommunications Standards Institute (ETSI). The ETSI is an international organization headquartered in Sophia Antipolis, France, with the task of standardizing remote data transmission. The main contributions of ETSI is the publication of the high performance radio local area networks (HIPER-LANs) standards and the allocation of radio frequencies in Europe (in the US, frequency allocation is responsibility of the Federal Communications Commission—FCC). In Europe, the frequency band 2.4–2.483 GHz was initially allocated, with bands 5.15–5.35 and 5.47–5.725 added in 2002. In the US, the following bands are allocated: 902–928 MHz, 2.4–2.483 GHz, 5.15–5.36 GHz, and 5.725–5.875 GHz.

Protocol Standards and Working Groups

IEEE 802.11

The WLAN standard 802.11 defines physical (PHY) and partly the data link layers. Therefore, it represents only a substandard of the IEEE 802 standard, which is illustrated in Fig. 8.

The original 1997 version of the 802.11 standard supported three PHY layer specifications: infrared, frequency hopping spread spectrum (FHSS), and direct sequence spread spectrum (DSSS). Both FHSS and DSSS allowed throughputs of 1 and 2 Mbps. In 1999, the original standard was extended by adopting 802.11b, which allowed maximum throughput of 11 Mbps using DSSS. Also in 1999, IEEE adopted 802.11a, which used the 5 GHz frequency band to reach throughput of 54 Mbps. In 2000, IEEE established the 802.11g task group with the objective to reach throughput of 54 Mbps in the 2.4 GHz band. The 802.11g was adopted as standard in 2003.

IEEE 802.11 enhancements

The 802.11 standard is presently undergoing revisions and extensions in order to cover requirements such as compatibility to other standards, quality of service (QoS), improved data rates. Below, we present some of

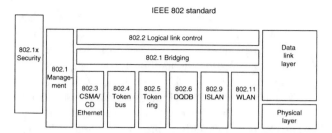

Fig. 8 IEEE 802 standard and some of its secondary standards.

the most important extensions (some may still be at the draft phase).

- IEEE 802.11d: This enhancement aims to guarantee seamless roaming of clients at WLANs of different geographic regions. Country-specific parameters are stored in the access point and communicated to the mobile client device. This enhancement is also referred to as "World Mode."
- IEEE 802.11e: This is an enhancement of the 802.11a and 802.11g in order to implement QoS. QoS is required by real-time applications such as voice over Internet protocol (VoIP), voice over WLAN (VoW-LAN), multimedia applications, streaming. In order to achieve QoS, 802.11e modifies the distributed coordination function defined for carrier sense multiple access/collision avoidance (CSMA/CA—the process used by WLAN). Such modifications result in noticeable improvements to the basic standard and appear very promising for widespread use of real-time applications over WLAN.
- IEEE 802.11f: This standard was aimed at improved inter-operability between access points from different manufacturers but was withdrawn in 2005.
- IEEE 802.11h: This update is largely based on 802.11a (5 GHz band) and contains amendments for spectrum management, namely dynamic frequency selection (DFS) and transmit power control (TPC). These technologies are already used by cellular (GSM) phones and were proposed as part of the ETSI HIPERLAN/2 standard. The adoption of such technologies makes it possible to use 802.11a outdoors as it prevents interference with other devices in the 5 GHz band [e.g., radio detection and ranging (radar) systems].
- IEEE 802.11i: This is a security improvement of the wired equivalent privacy (WEP) process that was used by the 802.11a, b, and g. This improvement uses the temporal key integrity protocol (TKIP) to enhance WEP. The resulting security process is often called Wi-Fi protected access (WPA). For further improvement the advanced encryption standard (AES) is used to create WPA2. Additional optional security enhancements are provided via the use of remote authentication dial-in user server (RADIUS), also defined in 802.11i.

- IEEE 802.11n: This enhancement is not yet completed and contains several improvements on the existing 802.11 standards. The standardization work is carried out by the enhanced wireless consortium (EWC), an industry coalition consisting of 27 Wi-Fi manufacturers. The main contributions from EWC to-date include:
  — Mixed-mode inter-operability between 802.11a/b/g,
  — Data rates up to 600 Mbps,
  — Media access control (MAC) enhancements to provide users with 100 Mbps application-level throughput,
  — Use of 2.4 GHz and/or 5 GHz bands (unlicensed),
  — 20 MHz and/or 40 MHz channels,
  — Spatial multiplexing modes to support simultaneous transmission of up to four antennas,
  — Enhanced range via multiple antennas and advanced coding.
- IEEE 802.11p: This enhancement, also referred to as wireless access for the vehicular environment (WAVE), supports ITSs. ITS include data communication between moving vehicles or vehicles and roadside infrastructure using the licensed ITS band of 5.9 GHz. 802.11p is planned for publication in early 2007 and will be primarily used by the US Department of Transportation (dedicated short range communications (DSRC) project).
- IEEE 802.11r: 802.11r aims at extending the current handover capability of 802.11a/b/g from data only to real-time applications such as voice, video, and gaming. The main application is to allow VoIP handovers in wireless data networks or between WLANs and cellular networks.

Alternative WLAN standards

In addition to the IEEE 802.11, ETSI has defined the HIPERLANs standards:

- HIPERLAN/1: This standard was ratified in 1996 for the 5.15–5.35 GHz band. It supports data rates up to 20 Mbps and defined layers 1 (PHY) and 2 (data link) of the open systems inter-connection basic reference model (OSI model). This standard was hardly adopted by Wi-Fi manufacturers.
- HIPERLAN/2: This standard replaced the unsuccessful HIPERLAN/1. Since most contributions came from the Asynchronous Transfer Mode (ATM) Forum, HIPERLAN/2 is also called "wireless ATM." It supports rates up to 54 Mbps in the 5.15–5.35 GHz band. The PHY layer is identical to IEEE 802.11a. HIPERLAN/2 is connection-oriented and uses a MAC-layer packet data unit (PDU) of 48 bytes (ATM) thus making it possible to introduce bandwidth management and QoS capabilities. It also defines a convergence layer in the data link layer, thus allowing interfaces to Ethernet, ATM, UMTS, IEEE 1394 (multimedia home networking).

Wireless LANs—Mesh

WiMAX and FLASH-OFDM standards

The 802.16 standard was initially defined for the MAN as part of a hierarchy of complementary wireless standards created by IEEE to ensure inter-operability, Fig. 9.

In addition to the wireless PAN, LAN, MAN, and WAN standards, IEEE is working on 802.21 to address inter-network handovers.

- IEEE 802.16 also referred to as broadband wireless access, first became standard in 2001. It only addressed line-of-sight (LOS), point-to-point environments in high frequency bands 10–66 GHz.
- In 2003, IEEE 802.16a adopted major modifications to PHY and MAC layers for non-LOS, point-to-multipoint (i.e., last mile) environments operating in 2–11 GHz bands. 802.16 supported rates up to 100 Mbps in 20 MHz channels.
- IEEE 802.16b was introduced to support unlicensed bands 5–6 GHz.
- IEEE 802.16c defines inter-operability specifications for LOS broadband wireless access. Its peak rate is 70 Mbps with a range of up to 50 km.
- IEEE 802.16e (mobile wireless MAN) was approved in December 2005. It represents a major improvement of the original 802.16 standard as it addresses user mobility. 802.16e covers PHY and MAC layers for combined fixed and mobile operation in licensed bands. It also allows users to connect to a WISP while roaming outside the WLAN covering their home or office (e.g., when they travel to another geographic location). Further, 802.16e supports handovers for users moving at high speeds (up to 120 kmph).
- IEEE 802.16f introduces management information base (MIB) to the original (fixed broadband) standard.
- IEEE 802.16g defines the management plane between the mobile subscriber station (handset) and the base station. It covers procedures such as mobility (roaming and handover), security, accounting, and power management.
- IEEE 802.16h addresses interference and coexistence with other systems in unlicensed bands.

- IEEE 802.20 or mobile broadband wireless access (MBWA) aims to prepare a formal specification for a packet-based air interface designed for IP-based services (also nicknamed as Mobile-Fi). The air interface will operate in licensed bands below 3.5 GHz, with a peak data rate of over 1 Mbps. Further, it aims at full mobility (roaming and handover) up to vehicular speeds of 250 kmph. Another benefit, as already discussed, is low latency. Notice that the goals of 802.20 and 802.16e (also called "mobile WiMAX") are similar and 802.16 is getting a stronger position as the main standard.
- Alternatively to the IEEE, ETSI has also defined the high performance radio metropolitan area network (HIPERMAN) standards. Note: for WAN the ETSI standards cover technologies such as 3GPP (GSM, UMTS, etc.) which are outside the scope of this chapter. HIPERMAN is targeted at small and medium enterprise (SME) and residential users and uses frequencies between 2 and 11 GHz. HIPERMAN has followed closely 802.16. It is also compatible to HIPERLAN and supports ATM, i.e., it offers various service categories and QoS capabilities. HIPERMAN supports point-to-point and point-to-multipoint configurations. IEEE 802.16-2004 and ETSI HIPERMAN share the same PHY and MAC specifications.

INTEGRATION OF BROADBAND WIRELESS ACCESS TECHNOLOGIES

Wireless mesh networks enhanced with functions such as access control, user authentication, and QoS support, can provide an integrated approach to Internet access for data users of cellular second generation (2G) or 3G networks, Wi-Fi, and WiMAX networks, Fig. 10. Integration requirements may include back-end operator's infrastructure and IP multimedia subsystem (IMS). IMS integration with WLAN as well as 3GPP is examined in detail in Ref. 7 and will not be discussed further in this chapter.

In order to integrate a wireless mesh network into an operator's network, we must implement functions that, e.g., control wireless access, authenticate the user, allow

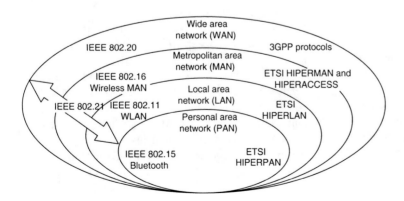

Fig. 9 IEEE and ETSI complementary wireless standards.

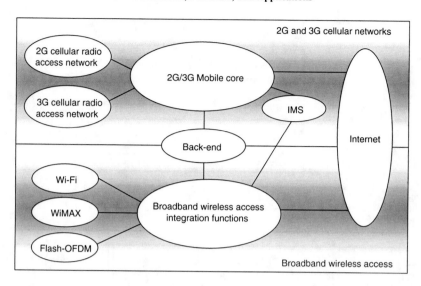

Fig. 10 BWA network integration into the existing 2G/3G and IP infrastructure.

interfacing to the SS7 networks or interface VoIP to core network TDM, perform adaptation of radio access QoS to core network QoS, and provide billing and charging capabilities. Such functions are usually specified by operators (depending on the target market and services), government regulations, or standards. Further, integration functions must allow access to the existing IT infrastructure of the network operator, Fig. 11.

Consider for example the access control function. Typically, a user, trying to access the Internet via a WLAN, is redirected to a web portal where he/she is requested to provide his/her credentials, purchase service using a credit card, or for example redeem a voucher. The web portal may also allow some free web pages or advertising that the user can access without authentication. Fig. 12 illustrates an access control scenario and shows a sample call flow.

SUMMARY

The deployment of wireless mesh networks, although in its early stages, is expanding rapidly by allowing new operators and data service providers to enter the wireless

market. The wireless industry is actively supporting standardization efforts in order to address the user and market needs for bandwidth, real-time performance, reliability and security. Besides user services, wireless mesh networks are applied as a lower-cost alternative to TDM backhauling without sacrificing reliability and performance. As standards mature and QoS is addressed more rigorously, we anticipate widespread use of wireless mesh and access for real-time applications (voice, video, high performance data) as viable alternative to cable/wired networks. In addition to ease and low cost of deployment, mobility standards already in development will provide a better and more efficient alternative to cellular data technologies.

REFERENCES

1. Faccin, S.M.; Wijting, C.; Kenckt, J.; Damle, A. Mesh WLAN networks: concept and system design. IEEE Wirel. Commun. **2006**, *13* (2), 10–17.
2. Lundgren, H.; Ramachandran, K.; Belding-Royer, E.; Almeroth, K.; Benny, M.; Hewatt, A.; Touma, A.; Jardosh, A. Experiences from the design, deployment, and usage of

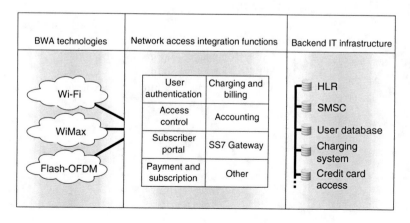

Fig. 11 Concept for integrating BWA into existing network and IT infrastructure.

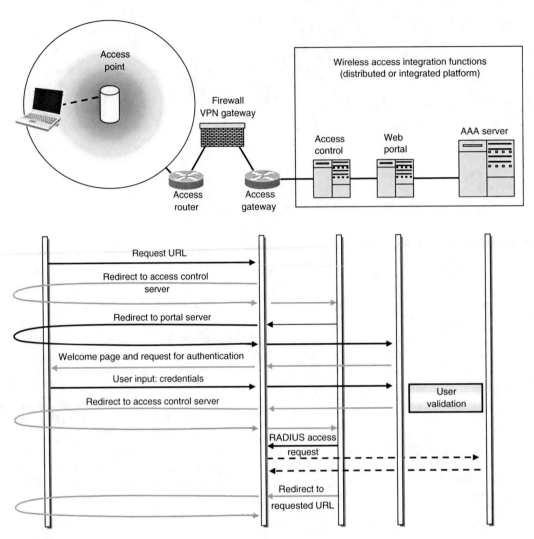

Fig. 12   WLAN access control example.

the UCSB MeshNet testbed. IEEE Wirel. Commun. **2006**, *13* (2), 18–29.

3. Kyasanur, P.; So, J.; Chereddi, C.; Vaidya, N.H. Multichannel mesh networks: challenges and protocols. IEEE Wirel. Commun. **2006**, *13* (2), 30–36.

4. Stine, J.A. Exploiting smart antennas in wireless mesh networks using contention access. IEEE Wirel. Commun. **2006**, *13* (2), 38–49.

5. Salem, N.B.; Hubaux, J.-P. Securing wireless mesh networks. IEEE Wirel. Commun. **2006**, *13* (2), 50–55.

6. Lee, M.J.; Zheng, J.; Ko, Y.-B.; Shrestha, D.M. Emerging standards for wireless mesh technology. IEEE Wirel. Commun. **2006**, *13* (2), 56–63.

7. Koukoulidis, V.; Shah, M. The IP multimedia domain: service architecture for the delivery of voice, data, and next generation multimedia applications. Multimedia Tools Appl. **2006**, *28* (1), 203–220.

BIBLIOGRAPHY

1. The ETSI Homepage, http://www.etsi.org/.
2. The IEEE Homepage, http://www.ieee.org/.
3. The WECA Homepage, http://www.weca.net/.
4. The WLANA Homepage, http://www.wlana.org/.

Wireless Multimedia Systems: Cross Layer Considerations

Yu Chen
Department of Electronic Engineering, Tsinghua University, Beijing, China

Jian Feng
Department of Computer Science, Hong Kong Baptist University, Hong Kong, China

Kwok Tung Lo
Department of Electronic and Information Engineering, Hong Kong Polytechnic University, Hong Kong, China

Xudong Zhang
Department of Electronic Engineering, Tsinghua University, Beijing, China

Abstract

Wireless multimedia applications need networks to smartly allocate resources according to varied wireless environments. Cross layer design (CLD) in a wireless multimedia system is to jointly consider and design each layer of the communication protocol stack so as to improve the system performance by efficient allocation of resources in different layers.

With the rapid development of video coding and wireless technologies, providing multimedia services over wireless link has been one of the most interesting focuses in recent years. Media streaming, video conference and video phone, and wireless-based interactive games are some examples of multimedia services attracting user attention in present and future time. Since existing wireless networks are specially designed based on individual wireless networks, they are not efficiently enough to support each of these attracting services. So it is the main task for beyond third generation (B3G) wireless networks to support those services more efficiently. Meanwhile, for each individual mobile terminal user, more robust and convenient multimedia services through different kinds of wireless networks are required in B3G. Moreover, the network operator in B3G should carefully modulate limited wireless resources to obtain more network capacity in the case that quality of service (QoS) should be reliable firstly, and synthetically allocate total wireless resource to many users in order to make more users get their own desirable QoS levels.

Based on the open systems interconnection (OSI) model, the conventional way to design a wireless network is to establish a protocol stack composed of a number of isolated layers. This strategy has been used widely in existing wireless communication systems. However, it is not smart enough to meet the requirement of next generation mobile systems. The challenges that mobile multimedia transmission meet are from two points: one is the random varied QoS characteristic of wireless channel; another is the dynamic QoS required by different applications and varied media stream content, such as variable bit rate, different priority of protection and transmission for different media units, and different sensitivity for bit error rate (BER), and packet lost rate (PLR). Hence, with respect to these challenges, the conventional solution which optimized mode and parameters for worst wireless situations at design phase cannot always obtain the best performance and make full use of limited resources for varying wireless situations. As an alternative method, the updated configuration information of each layer of the protocol stack is taken into account simultaneously, and they should be adjusted dynamically to adapt to the varying transmission environment.

CROSS LAYER DESIGN

The objective of cross layer design (CLD) for network design is to improve the performance of the wireless multimedia transmission system with joint considering and designing each layer of the protocol stack. CLD is a new solution that the dependency and interactions among different layers of the protocol stack is considered. Its optimization extent is across the layers boundary.[1,2] CLD is not designing network without layers, but a complement to the existing layering approach. The combination of layering and CLD optimization can efficiently adapt the performance of wireless networks. Based on the order in which CLD is performed,[3] the following are different classifications of CLD.

Encyclopedia of Wireless and Mobile Communications DOI: 10.1081/E-EWMC-120043566

1. **Top-down approach:** The protocol of higher layers determines how to find proper parameters in the next lower layer for their most optimized performance. This scheme has been exploited in most existing communication systems. In these systems, the parameters and transmission strategies in medium access control (MAC) are controlled by the application layer, while the MAC layer determines the optimized standards of modulation style in the physical layer.

2. **Bottom-up approach:** Its optimization direction is from lower layer to higher layer. In this approach, with the situation of the limitation of round trip time (RTT), channel error, and fluctuated bandwidth, the lower layers adjust the parameters in higher layers. However, this kind of approach is not suited for quality-prioritized wireless multimedia service, such as high resolution video and image transmission. But it is useful for delay-sensitive low bandwidth multimedia services such as real time video conference and telephone, where the update interactive information among different users is most important and their video quality is not the main concern.

3. **Application-based approach:** The optimization direction in this approach for other layers includes bottom-up (the parameters in physical layer are optimized firstly) and top-down strategy with respect to requirements from the application layer. For longer time interval and coarse multimedia stream data partition resulting in the application layer other than the lower layers, this kind of approach cannot quickly adapt lower layer to achieve optimal performance of the application layer.

4. **MAC-based approach:** In this approach, with the traffic information and certain requirements received from the application layer, the MAC layer determines the transmission order of different parts of the application layer stream, and gives them different QoS levels. In addition, the MAC layer decides the parameters of the physical layer with obtained channel situation. This approach cannot make better source channel coding trade-off for unexpected channel condition and multimedia content.

5. **Integrated optimization approach:** The optimization strategy is determined synthetically for each layer of the protocol stack. This approach can lead to best quality performance by exhaustively finding most proper selection from larger number of combinations of different layer parameters and strategies. Apparently, it is not possible to be used in a practical system for its great complexity. In order to solve this defect, learning and classification are adopted in a simplified scheme to mitigate the complexity of this integrated optimization approach[4] with an assumption that the characteristics of different layers can be easily abstracted by not heavy computation.

The rudiment work on CLD paid more attention to the question of how to achieve best performance of a certain layer. For example, when the performance of the application layer is most important for certain wireless multimedia services, other layers of protocol layers have to modulate their own parameters to meet the requirements of the application layer. For example, the bottom-up approach exploits channel situation to smartly modulate the transmission policy of the application stream.[5] The research on joint CLD consideration for physical layer and data link or MAC layer appeared in Ref. 6.

Recently, many different CLD schemes have been developed. In Ref. 7, referring to a priority function on channel situation, the frame type, queue size, and multiplexing gain, an opportunistic scheduling algorithm for multiple video streams was proposed. In Ref. 8, based on adaptive modulation and channel coding, a cross layer scheduling is derived. In Ref. 9, for each layer of protocol stack, corresponding error control and adaptation strategies are evaluated respectively, such as the re-transmission in data link layer, forward error correction (FEC), scalable video coding, and smartly adaptive packet envelopment in the application layer. A novel algorithm based on competition for wireless communication is proposed in Ref. 3, whose optimization extent is not confined to only one wireless station. In this algorithm, many wireless stations can share their own wireless source situations to modulate their cross layer transmission strategy for desirable multimedia quality and power consumption in these wireless stations.

ARCHITECTURE OF CLD

One CLD solution is to achieve a certain performance of the wireless system by jointly re-configuring multiple protocol layers. The flexible performance of CLD can keep up with randomly varying-channel condition, so the end-to-end multimedia performance can be improved with dynamically allocating wireless resource of CLD.

The design complexity of CLD is much more than that of original separated protocol layer.[3] As we know, the OSI protocol layers are useful for developers to easily design proper optimization scheme for one particular layer. However, CLD does not have this advantage. In order to reduce the implementation complexity, for different wireless video services, each protocol layer can be characterized by some important parameters, which are sent to adjacent layers as guidance to find their optimization configuration for matching the requirement of the current channel, network, and multimedia application stream. In such a design, each layer interacts with other layers to find the optimal point. The essential problem of this approach is how to find a precise principle to define important key information at each layer. For instance, in data link layer, the parameters on channel quality such as signal-to-interference plus noise ratio (SINR), BER or

PLR and the provided transmission date rate, can be used as key information. Also, the shared information between the network and MAC layer include traffic rates and supported link capacities.

We take the low-latency media streaming over an ad hoc network as an example, a block diagram of the CLD architecture is illustrated in Fig. 1,[10] where the shared information among different layers are shown. In the data link layer, under varying-channel situations, a maximum link rate is achieved by adaptive techniques, which can effectively extend the supportable capacity region of the network. For every point of this region, a potential assignment of the different link performance is supported.

For the direction of information from data link layer, the MAC layer assigns time intervals, codecs, or frequency bands to each link in order to find a proper point in the capacity region. Moreover, the MAC layer cooperates with the network layer to determine the set of network flows for control of congestion. Hence, in this ad hoc application, the optimization among these three layers is the main task of the CLD work, whose final objective is to obtain a jointly optimal effect of capacity assignment and network flows. In the transport layer, congestion-distortion based optimization on schedule is carried out to determine the priority of each multimedia packet. Finally, in the top layer of the protocol stack, the highest multimedia encoding efficiency is required in the application layer.

THE CHALLENGES OF CLD

The CLD optimization problem can be considered to select a joint strategy across multiple OSI layers. In this section, for convenient understanding, only one-hop wireless network is discussed, and also only physical, MAC, and application layers are taken into account.[3] In this type of network, the network layer is not as important as the above-mentioned layers in error resilient protection and bandwidth adaptation because the function of the network layer is not so obvious. However, the model presented in the following can easily be extended to include other layers not included in this section.

Let N_P, N_M and N_A represent the number of possible error resilient protection and adaptation strategies in the physical, MAC, and application layer respectively.[3] For the physical layer, the strategies $P_i(1 \leqslant i \leqslant N_p)$ include different existing modulation and channel coding schemes. For MAC layer, the strategies $M_i(1 \leqslant i \leqslant N_M)$ include various envelopment type of packet, automatic repeat request (ARQ), scheduling, admission control, and FEC mechanisms. For application layer, the strategies $A_i(1 \leqslant i \leqslant N_A)$ can be realized by the smart adjustment of multimedia coding parameters, the length of packet, traffic shaping and priority, scheduling, ARQ, and FEC measures. So the optimized strategies aggregation for CLD is

$$\Gamma = \{P_i \ldots P_{N_P}; M_1 \ldots M_{N_M}; A_1 \ldots A_{N_A}\} \tag{1}$$

Apparently, it is seen from Eq. 1 that that there are $N_P \times N_M \times N_A$ combinations for design to select, so the final destination of CLD is to get the most optimal result with the following formula.[3]

$$\Gamma(x^*) = \arg \max Q(\Gamma(x)) \tag{2}$$

where x^* is the most optimized strategy combination for certain performance Q with highest priority in wireless multimedia services. Here, Q can be the best subjective and objective multimedia decoding quality at wireless receiver, the maximum efficient usage ratio of wireless channel bandwidth, and the maximum save for power consumption. Detail requirement of Q in Eq. 2 is determined by the focus of practical wireless multimedia services. In Fig. 2[3] comprehensive wireless station and system constraints for certain Q are given where the optimized Q performance includes video quality, power and RTT.

The selection of the best optimal CLD strategy as Fig. 2 is challengeable for the following reasons.[3]

1. The function for Q is not directly evaluated when only worst and average cases are known, and it is non-linear and there is dependency among the strategies of physical, MAC, and application layers. Therefore, the process to obtain a proper expression for Q as a function of channel conditions is very difficult.

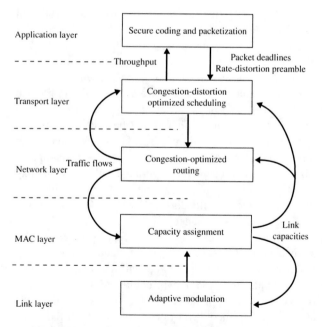

Fig. 1 CLD architecture for low-latency media streaming over an ad hoc wireless network.[10] *Source*: Figure 1, Setton, E.; Yoo, T.; Zhu. X.; Goldsmith, A.; Girod, B. Cross-layer design of ad hoc networks for real-time video streaming. IEEE Wirel. Commun. 2005, 12(4), 59–65. (© 2005 IEEE).

Wireless
Multi—Networks

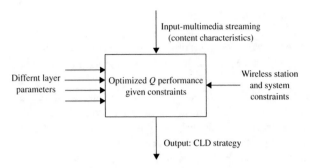

Fig. 2 CLD subject to wireless station and system constraints.[3] *Source*: Figure 1, van der Schaar, M.; Shankar, N. S. Cross-layer wireless multimedia transmission challenges, Principles, and new paradigms. IEEE Wirel. Commun. 2005, 12 (4), 50–58. (© 2005 IEEE).

2. In the existing protocol stack, algorithms and protocols in different layers are originally designed independently for their own objectives. Additionally, the units and types of processing objects each layer process and output are different. As we know, the application layer focuses on high efficiency alignment of semantics, dependency in multimedia stream and fluctuation of multimedia content, while physical layer focuses on symbols heavily dependent on the channel situation.

3. The wireless channel situation and multimedia content are varied randomly and continuously. It is difficult to trace them quickly.

4. For existing and future different kinds of wireless multimedia services, the challenges are—how to choose the extent of stack layers to make CLD, and when the process of CLD is accomplished, how to determine the performing order of selected strategies.

5. Practical environment is changed rapidly, so additional practical consideration is necessary for new wireless standards. For example, in 802.11e QoS MAC standard, unequal error protection (UEP) for different flows or delay sensation is adopted, which does not appear in 802.11a/b/g MAC standards. This new strategy should be given enough attention to perform in CLD.

In fact, the number of feasible error resilient protection and the extent of adaptation strategy in different stack layers are limited.[3] For example, there are only eight modulation modes used in physical layer for 802.11a wireless standard. So the possible solution region is reduced. However, the solving process of the above-mentioned CLD problem is still very complex where iterative optimization and decision-tree approaches are exploited. In such approaches, interesting strategies are optimized with respect to ignorance of all other strategies. In corresponding optimal process, derivative and non-derivative algorithms are used to find final results by linear and non-linear programming. The optimal process can be considered as

complex multivariate optimization with inherent dependency across layers and among strategies, so it is very important to determine the most proper procedure for achieving final $\Gamma(x^*)$ in this process. The considering procedure needs to determine the initialization, the grouping, priority and optimized order of strategies in the protocol stack, and control the rate and value of convergence in solving process. Considering the rapid variation of wireless environments, the speed of convergence is very important.

SIMPLE EXAMPLES FOR CLD

In this section, several simple examples are given to describe the optimization process of CLD in one-hop wireless networks where application, MAC, and physical layers are interacted to find an optimal modulation and power consumption.

Interactive Optimization for Modulation Scheme

As we know, selecting a proper modulation scheme for link adaptation will obtain maximum throughput Φ of the wireless channel as well as the robustness of the MAC frame. This optimization process can be summarized in the following equation.[3]

$$\Gamma^*_{\mathrm{mod}}(x) = \arg \max_{\Gamma} \Phi\big(\Gamma_{\mathrm{mod}}(x)\big) \tag{3}$$

where $\Gamma_{\mathrm{mod}}(x)$ is the collection of different modulation strategies. From Refs. 11 and 3, a higher physical mode can result in more desirable throughput when SINR is high, while the a lower physical mode is suited to lower SINR situation. Moreover, shorter packet length has little efficiency to make full use of wireless bandwidth for much cost of MAC and physical layer overheads used at each transmission attempt. Therefore, the MAC layer can choose proper $\Gamma^*_{\mathrm{mod}}(x)$ in physical layer to get maximum throughput. However, with the above analysis of the MAC-based approach, this optimal result does not pay attention to distortion on multimedia stream. So both the throughput and multimedia quality should be taken into account in CLD simultaneously.

For this requirement, the new CLD problem is described in the following. When the channel condition x_C is determined by the transmitter with the received signal strength indicator (RSSI), by precisely adjusting encoding rates of scalable application layers: the base layer rate R_b and the enhancement layer rate R_e, the length of the MAC layer packet L and the modulation strategy $\Gamma_{\mathrm{mod}}(x)$, the most satisfactory performance Q_m is achieved by

$$\Gamma^*_{\mathrm{mod}}(x) = \arg \max_{\Gamma} \Phi\big(\Gamma_{\mathrm{mod}}(x), R_b, R_e, L\big) \tag{4}$$

This approach is used in the physical-MAC-application optimization for video streaming,[3] MPEG-4 (moving picture experts group-4) fine granular scalability (FGS)[12] and the MAC and physical layer in 802.11b protocol.[13] In them, the requirement of application is dominant. As in Ref. 3, different physical layer modes are used for different SINR, which can make joint MAC-physical layer optimization so as to obtain a better multimedia quality. As a conclusion, the effect for which the application layer is included in CLD for wireless multimedia transmission is obvious. For further extent of above work, in Ref. 14, the joint packetization and re-transmission limit adaptation is developed. And in Ref. 15, the work on how several application (APP) and MAC strategies can be jointly optimized to enhance the multimedia quality is addressed. In Table 1,[15] statistical scores for different transmission strategies are given. It is shown that an obvious performance improvement is obtained for using joint application-MAC optimization when compared with other schemes. More comprehensive results can be found in Ref. 15.

Interactive Optimization for Power Consumption

In this section, we address the issues on interactive optimization for power consumption, which is a very important performance factor for a mobile terminal. We look at the CLD scheme when the modulation scheme is fixed.[3] First, the wireless channel is assumed as a two-state Markov chain where only GOOD and BAD states are available. So the channel capacity derived from the Shannon theorem is

$$C = B \log_2 \left(1 + \frac{P_0 a_0 p_0 + P_1 a_1 p_1}{N_0 B} \right) \quad (5)$$

where 0 and 1 represent the GOOD and BAD states of wireless channel, and a_0 and a_1 are the channel attenuation factor of state 0 and 1, whose occurrence probability is p_0 and P_1 respectively. For the channel state 0 and 1, only two

Table 1 Subjective video quality comparison.[15]

Exploited strategies	Visual score
No optimization at MAC and application layer	1.4
MAC layer optimization (RTRA)	1.9
Application layer optimization	3.8
Joint application-MAC optimization	4.6

Source: Table III, Li, Q.; van der Schaar, M. Providing adaptive QoS to layered video over wireless local area networks through real-time retry limit adaptation, IEEE Trans. Multimed. 2004, 6 (2), 278–290. (© 2005 IEEE)

necessary transmission ways are needed here. When the channel situation is GOOD, there is no additional requirement for normal transmission; while the channel situation is BAD, more power is required to enhance the intensity of the information signal in order to mitigate the lost probability of multimedia packets. Moreover, it is necessary to determine the power constraint

$$P_i p_i a_i \leqslant P_{\max} \quad (i = 0,1) \quad (6)$$

where $P_{\max}$ is the maximum power that the physical layer is tolerable for existing modulation schemes. As in Ref. 3, when Kuhn-Tucker sufficient conditions are used, Eq. 5 is optimized with respect to the limitation of Eq. 6. The final result is obtained as follows

$$P(h_i) = \left[\frac{1}{\lambda} - \frac{BN_0}{a_1 p_1} \right] \quad (i = 0,1) \quad (7)$$

where λ is the Lagrange parameter and can be obtained from Eq. 6. From Eq. 7, the optimal transmission strategy is simple that it is just to transmit data when the channel state is GOOD, while data is not transmitted or transmitted with more power when the channel state is BAD. This strategy is difficult for a practical system for two points: the system should know channel situation before transmission, and the system can smartly adjust the transmission power according to channel states (when channel state is GOOD, less power is needed; when channel state is BAD, more power is necessary).

CLD FOR AD HOC NETWORKS

In the previous section, examples networks have been discussed for one-hop simple wireless networks. As we know, the wireless multimedia transmission through multiple hops is usually required in practical systems. So in this section, with an ad hoc network as example, a complex wireless network environment is considered to make CLD.

An ad hoc wireless network can be considered as a self-configured collection of wireless nodes without the support of any established infrastructure, where some or all of the nodes are not stationary. This kind of network is specially designed for the requirement of smart infrastructure where users can join and leave the wireless network without the limitation of traditional networks. The key characteristics of ad hoc networks include the lack of established infrastructure and the dynamic network and channel situations. Such features make routing of traffic among different nodes change frequently. Hence, unexpected occurrence and disappearance of routing has obvious effect on the QoS of other established data links. Therefore, wireless multimedia service through this kind of network faces more severe challenges.

**Wireless
Multi—Networks**

Adaptive Link Layer Strategy in Ad Hoc Network

In this section, two adaptive link layer strategies for ad hoc wireless networks are disscussed as in Ref. 10. The adaptive packet length strategy is exploited to achieve optimal throughput when current SINR and link layer parameters are fixed. As we know, when the packet length is too long, corresponding packet error rate (PER) turns larger, and the channel throughput is mitigated by frequent re-transmission. On the other hand, shorter packet length cannot make full use of limited wireless channel bandwidth. From Ref. 10, the optimal packet length L^* that can maximize the throughput is

$$L^* = \frac{H}{2} + \frac{1}{2}\sqrt{H^2 - \frac{4bH}{\ln(1 - P_a)}} \qquad (8)$$

where H is the length of packet header that includes multimedia frame header and the overhead cost for necessary protocol, and P_e is the symbol error probability determined by modulation type and link SINR. When the noise level of channel is low, larger symbol rate and little channel coding is feasible. While the noise level of channel is not ignored, add ional redundancy such as heavy channel coding is required.

For certain channel gain condition, when the parameters mentioned above have been determined, the channel link throughput is provided as

$$R_i = W \log_2\left(1 + \frac{SINR}{\Omega}\right) \qquad (9)$$

where W is the channel bandwidth and Ω is a parameter defined by the link layer design. As in Ref. 10, with different channel coding strategies, the difference of performance between this adaptive link layer and Shannon capacity can be constrained.

Optimal Strategy of Network Layer in CLD for Ad Hoc Network

The main objective of this section is to discuss how to properly assign traffic flows to each link whose throughput is optimized as the previous section. If the assigned flow is close to the link capacity, more delay appears in that link. Hence, the trade-off between flow rate and delay should be determined according to the detailed requirement of practical wireless multimedia service. Existing solutions can achieve optimal capacity and flow assignment by minimizing any convex or quasi-convex cost function as the following formula.[16]

$$\Delta(C, f) = \max_{(i,j)} \frac{f_{ij}}{C_{ij}} \qquad (10)$$

where C and f represent the network capacity and flow. Eq. 10 can be solved by a bisection algorithm where a sequence of convex feasibility problem is referred.

When the flow in any pair of transmitter and receiver can be sent through multiple paths, this strategy can obtain broader date width by spatially re-using wireless spectrum. So among these paths, their error patterns are independent, which can be exploited in the application layer such as multiple description coding. However, more number of links results in more contention at the MAC layer, so the cost of keeping multiple routes increases.

As a comparison to advantages of CLD, independent layer design (ILD) is considered here, where channel capacity and practical flow are optimized respectively.[10] In ILD, bi-direction links are adopted between neighboring nodes, and in these links, the corresponding minimum transmission rate is changed to as large as possible. In this situation, optimal flow assignment is achieved by finding the same objective function as in Eq. 10. However, in this design, for some links without support to any traffic, some wireless capacity is still assigned to them. This phenomenon of wasting network resource can be explained by different multi-path routing effect in ILD and CLD. For the former, the channel capacity is fixed and can only be determined by topology. When multi-path routing is used, the data rate in different path can be aggregated into a larger transmission rate. But it can not allocate the entire transmission rate to active links. As modification of this defect, in CLD, the enlarger channel transmission rate can be exploited efficiently but only active links have valid transmission rate.[10]

Congestion-Distortion Scheduling Based CLD for Ad Hoc Network

In this section, optimal focus is put into transport layer where the congestion-distortion scheduling should be optimized. The erasure of the multimedia packet during transmission occurs frequently in existing wireless networks, which not only degrades successive multimedia quality, but also introduces additional delay by requiring repeat transmission for the lost packet. In order to mitigate the effect of a packet lost, a proper assignment for scheduling in transport layer is necessary.

As the fact is that there is no consideration on delay requirement and priority of wireless packet in tradition transport layer protocol, in CLD, this situation can be modified with the application layer constraint taken into account. A more advanced solution wherein the algorithm based on importance classification of different packet in multimedia stream is proposed to find optimal transmission schedule.[17] This algorithm is designed to achieve more satisfactory multimedia decoding quality at receiver in certain transmission rate.

If the congestion situation in ad hoc networks is not severe, the congestion–distortion optimized scheduler in

Ref. 18 can obtain desirable effect as the constraint of end-to-end delay. As the criterion of video distortion reduction, this scheduler can select the most important packets from video stream, and give these packets highest transmission priority where I frames are transmitted with highest priority while B frames may be intentionally dropped according practical channel condition, and re-transmission is only used for I frame and more importantly P frame. Moreover, this schedule can efficiently control increase of queuing delay by avoiding the large burst transmission mode of video packets.

REFERENCES

1. Shakkottai, S.; Rappaport, T.; Karlsson, P. Cross-layer design for wireless networks. IEEE Commun. Mag. **2003**, *41* (10), 74–80.

2. Kawadia, V.; Kumar, P. A cautionary perspective on cross layer design. IEEE Wirel. Commun. **2005**, *12* (1), 3–11.

3. van der Schaar, M.; Shankar, N.S. Cross-layer wireless multimedia transmission challenges, principles, and new paradigms. IEEE Wirel. Commun. **2005**, *12* (4), 50–58.

4. van der Schaar, M.; Tekalp, M. Network and content-adaptive cross-layer optimization for wireless multimedia communication by learning. IEEE International Symposium on Circuits and Systems, Kobe, Japan, IEEE, USA, May 23–26, 2005.

5. Chou, P.; Miao, Z. Rate-distortion optimized streaming of packetized media. IEEE Trans. Multimed. **2006**, *8* (2), 390–404.

6. Girod, B.; Kalman, M.; Liang, N.J.; Zhang, R. Advances in channel-adaptive video streaming. IEEE International Conference on Image Processing, Rochester, IEEE, NY, Sept 22–25, 2002; 9–12.

7. Tupelly, R.S.; Zhang, J.; Chong, E.K.P. Opportunistic scheduling for streaming video in wireless networks. 37th Annual Conference on Information Sciences and Systems, Baltimore, MD, IEEE, USA, Mar 12–14, 2003.

8. Liu, Q.; Zhou, S.; Giannakis, G. Cross-layer scheduling with prescribed QoS guarantees in adaptive wireless networks. IEEE JSAC, **2005**, *23* (5), 1056–66.

9. van der Schaar, M.; Krishnamachari, S.; Choi, S. Adaptive cross-layer protection strategies for robust scalable video transmission over 802.11 WLANs. IEEE JSAC. **2003**, *21* (10), 1752–1763.

10. Setton, E.; Yoo, T.; Zhu, X.; Goldsmith, A.; Girod, B. Cross-layer design of ad hoc networks for real-time video streaming. IEEE Wirel. Commun. **2005**, *12* (4), 59–65.

11. Qiao, D.; Choi, S.; Shin, K.G. Goodput analysis and link adaptation for IEEE 802.11a wireless LAN. IEEE Trans. Mob. Comput. **2002**, *1* (4), 278–292.

12. Radha, H.; van der Schaar, M.; Chen, Y. The MPEG-4 fine-grained scalable video coding method for multimedia streaming over IP. IEEE Trans. Multimed. **2001**, *3* (1), 53–68.

13. Wireless LAN Medium Access Control (MAC) and Physical Layer (PHY) Specifications: Higher-Speed Physical Layer Extension in the 2.4 GHz Band, IEEE Std. 802.11b, Supp. to Part 11; 1999.

14. van der Schaar, M.; Turaga, D. Cross-layer packetization and retransmission strategies for delay-sensitive wireless multimedia transmission. IEEE Trans. Multimed. 2007, 9(1), 185–197.

15. Li, Q.; van der Schaar, M. Providing adaptive QoS to layered video over wireless local area networks through real-time retry limit adaptation. IEEE Trans. Multimed. **2004**, *6* (2), 278–290.

16. Yoo, T.; Setton, E.; Zhu, X.; Goldsmith, A.; Girod, B. Cross-layer design for video streaming over wireless ad hoc networks. IEEE Workshop on Multimedia Signal Processing, Sienna, Italy, IEEE, USA, Oct 2004; 99–102.

17. Chou, P.A.; Miao, Z. Rate-distortion optimized streaming of packetized media, Microsoft Research tech, rep. MSR-TR-2001-35; Feb 2001.

18. Setton, E.; Zhu, X.; Girod, B. Congestion-optimized scheduling of video over wireless ad hoc networks. IEEE International Symposium on Circuits and Systems, Kobe, Japan, IEEE, USA, May 23–26, 2005; 3531–3534.

Wireless
Multi—Networks

Wireless Networks: Cross-layer MAC Design

Amoakoh Gyasi-Agyei
Faculty of Sciences, Engineering and Health, Central Queensland University, Australia

Hong Zhou
Faculty of Engineering and Surveying, University of Southern Queensland, Toowoomba, Queensland, Australia

Abstract

A cross-layer medium access control (MAC) scheme is an ISO/OSI Layer 2 protocol which exploits vertical information from any other layer in its decision making. MAC protocols arbitrate between multiple users sharing a finite communications resource. Single-user systems do not need a MAC scheme. This entry examines the basic philosophy underpinning the design of cross-layer protocols for wireless networks, and how cross-layer protocol differs from traditional non-cross-layer protocols. The article concludes with a case-study that applies neural network concepts to the design of cross-layer MAC protocols for single-hop wireless networks using multicarrier OFDM transmissions.

Wireless networks are communications systems that use the ether to transfer information in the form of electromagnetic signals between two end points. A medium access control (MAC) scheme is a radio resource management (RRM) protocol. RRM is the process of allocating wireless resources such as transmission channels, power, and spectrum to radio nodes in an optimum manner. In traditional wireless communications networks a MAC scheme is part of Layer 2 of the ISO/OSI protocol architecture (Fig. 2). However, other protocol architectures position the MAC at a layer other than Layer 2. For example, MAC operates at Layer 3 in the IEEE 802 standards. MAC protocols are usually tailor-made for specific physical transmission media, which can be wired or wireless. (A transmission medium for a communication signal is usually referred to as a *channel*.) Even MAC protocols for different wired (or wireless) channels differ. However, MAC protocols for wireless networks differ considerably from those of wireline systems as the former use transmission media which are time-varying and have limited capacity. Unlike wired media, the ether cannot be expanded. Hence, different communications systems use different MAC protocols, the reason why a Layer 3 protocol like IP is needed to interconnect networks using different link-layer technologies.

The traditional purpose of MAC is the *efficient coordination* of the sharing of a *finite resource* by multiple *selfish users*. Without MAC the so-called *tragedy of the commons* can easily occur in a finite-resource multi-user system. As resource management schemes in multi-user systems, MAC protocols are essential for quality of service (QoS) provisioning in multi-user networks. Today, MAC protocols provide additional functions needed to enhance efficient resource coordination. A wireless network uses

- transmission media with finite bandwidth and hence finite capacity;
- cordless devices with limited power/energy supply;
- transmission media with time-, spatial-, and frequency-dependent quality due to multi-path fading, shadowing, additive background (thermal) noise, and user mobility.

The above features challenge the provision of QoS to applications in wireless networks. Hence, wireless networks usually provide probabilistic, rather than deterministic, guarantee of QoS. This dynamic and usually random variability in wireless channel behavior is an impetus for cross-layer system design. Also, applications' service requirements may be variable. Hence, cross-layer optimization, allowing one layer to adapt to the "instantaneous" behavior of another layer can improve the overall service experienced by end-user applications.

A MAC protocol can be deterministic or probabilistic (Fig. 1A). Deterministic schemes can be based on scheduling, channelization or hybrids of these two as common with cross-layer MAC (CL-MAC) schemes. While a deterministic MAC protocol uses a fixed algorithm to allocate resources to users, a probabilistic scheme varies its resource allocation pattern pseudo-randomly. Pseudo-random in the sense that a known algorithm is used to generate the randomness in the algorithm. Examples of a class of deterministic MAC schemes are scheduling and channelization schemes, which for the uplink include FDMA, TDMA, CDMA, OFDMA, and the hybrids thereof. The downlink equivalents of these, i.e., multiplexing schemes, are FDM, TDM, CDM, and OFDM. Scheduling usually synergizes with a channelization scheme. Examples of probabilistic MAC schemes are ALOHA, carrier sense multiple access/collision avoidance (CSMA/CA), and their variants. Some

Encyclopedia of Wireless and Mobile Communications DOI: 10.1081/E-EWMC-120043472

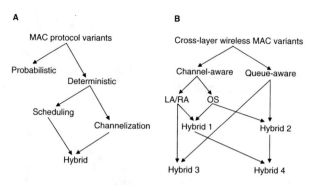

Fig. 1 MAC protocols: (A) variants, (B) design approaches.

wireless networks combine a random with a deterministic MAC. For example, mobile devices attach to the network via random MAC in the GSM network, but information bearers are assigned to devices using a channelization scheme.

This entry provides a brief introduction to cross-layer protocol design at the link layer via a combination of a tutorial and a case study applying neural networks techniques. We start our discussions with a review of traditional communication system design based on layering, abstraction, and modularity in order to contrast it with cross-layer system.

LAYERING: TRADITIONAL COMMUNICATIONS NETWORK DESIGN

Entities, either physical or logical, in communications networks interact via *protocols*. Like a natural language, a protocol comprises a special set of rules including syntax and semantics that must be obeyed for efficient communications. Traditional protocol models (aka architectures) for communications systems have been founded on three basic interrelated concepts: *layering, abstraction,* and *modularity*. Furthermore, there are different protocol models in use today, each of which divides the set of functionalities making up an entire communications system differently. This is illustrated in Fig. 2, which compares some popular protocol architectures in the networking industry, namely the OSI model designed under the auspices of the ISO, the Internet [or TCP/IP] model, the IBM's systems network architecture (SNA) model, and the Cisco's model. The ISO/OSI model is a theoretical construct which is used for benchmarking, as no practical system has so far implemented its exact recommendations.

Layering allows the aggregation of functional components, that interact in some sequential and hierarchical way, to accomplish a specific task or a set of tasks. Each layer usually has an interaction with only its immediate neighboring layers. Layer L receives service from its immediately underlying layer $L - 1$ while offering service to its immediately overlying layer $L + 1$ through standard

Layer	ISO/OSI	TCP/IP	IBM SNA	Cisco	Function
L7	Application		Transaction services		
L6	Presentation	Application	Presentation services	Core layer	Interoperability of network application programs
L5	Session		Data flow		
L4	Transport	Transport			
L3	Network	Internet	Transmission control	Distribution layer	Transmission across different networks (or an Internet)
L2.2	DLC / Data link	Network access	Path control		
L2.1	MAC		Data link control	Access layer	Transmission across a single network
L1	Physical	Physical	Physical		

Fig. 2 Comparison of popular (logical) protocol models for communications systems.

interfaces called service access points (SAPs). Hence, each layer enhances the service offered by its underlying set of layers and offers it to its overlying layer. Protocol layering allows *abstraction* which enables protocol designers to reduce and factor out details and focus on a few concepts at a time. Protocol layering also promotes modularity. *Modularity* allows a complex functionality or task to be broken up into chunks of mini-tasks of similar manageable complexities.

Layering allows easy or (near) transparent updating, design, and management of individual layers in a multilayered communications system. It also allows *peer-to-peer communications,* in which a layer on one node communicates logically with its corresponding layer on another node, possibly attached to a different network. Disadvantages in layering include *overhead* and *redundancy* across the layers of a multi-layered communications system. Overhead results from headers and trailers associated with encapsulation as data (in form of protocol data units or PDUs) move from application layer to the physical layer. Furthermore, layering allows only optimization of independent layers rather than entire system optimization. This reduces protocol design complexity at the expense of overall network performance. This weakness is a motivation toward cross-layer system design. Individual layers in a layered system are designed and optimized independently without considering the specific characteristics of other layers, and above all, the applications they altogether serve. This is referred to as *layered optimization* in Ref. 1. Cross-layer design, on the other hand, focuses on *system-wide optimization* to obtain optimum networked applications' performance. Layered optimization results in duplication of a functionality at multiple layers. This redundancy in itself makes systems *robust* to transmission impairments, albeit at the cost of *efficiency* and *lightweight*.

Layering-based system design is analogous to how this entry is sub-divided into sections. Each section explores a different aspect of the entire theme covered. The sections fit together via binding words or expressions, just as protocol layers do via interfacing. In multi-user communications systems the data link layer (i.e., Layer 2) is broken up into two sub-layers L2.1 (MAC) and L2.2 (data link control or DLC layer), as illustrated in Fig. 2. The DLC sub-layer is designed for link-level (i.e., not end-to-end) error management, while MAC coordinates efficient sharing of system resources by multiple concurrent users. MAC is relevant only in multi-user systems, just as the network layer (i.e., Layer 3) is relevant only in multihop networks. The focus of this entry is on the MAC sub-layer. Details of the functionalities of all the other layers shown in Fig. 2 can be found in, e.g., Ref. 2.

CROSS-LAYERING: RECENT COMMUNICATIONS NETWORK DESIGN

As mentioned above, cross-layer wireless network design is motivated by the desire to optimize system-wide performance and efficiency, contrasting the conventional system design based on abstraction and modularity. It optimizes the interplay between link-layer and networking-layer issues, as illustrated in Fig. 3. Timing includes delay and jitter, while a stable algorithm reduces queuing losses.

Cross-layer system design jointly designs and optimizes the entire system to reduce overheads and redundancy, and provide a better service to meet the QoS requirements of end-user applications. A application's QoS requirements depend on the characteristics of the traffic it generates to the network. A challenge is that both traffic characteristics and wireless channel state can change continually, requiring constant updating in order to achieve an optimum performance. QoS requirements of networked applications can be better met if their characteristics are embedded in the cross-layer design. There are two main cross-layer design approaches: ISO/OSI-friendly and ISO/OSI-hostile design. Unlike the latter approach, the former approach does not disrupt the well-established ISO/OSI protocol model which is founded on protocol abstraction and modularity. Hence, it seems to be the most economical approach. The EU MobileMAN project[3] is an example of

ISO/OSI-friendly cross-layer network design. As cross-layer system design reduces encapsulation overhead and redundancy, it enhances system efficiency and lightweight. These improvements can reduce system response time and energy consumption in hand-held radio devices.

Cross-layer design can be hosted at any protocol layer. Depending on how the protocol layers interact and the location of the algorithm, cross-layer approaches can be categorized into link-layer or MAC-centric, integrated, top-down, and bottom-up approaches. Ref. 1 discusses some pros and cons in each of these design philosophies. MAC-centric cross-layer design, the focus of this entry, has two main perspectives: link/rate adaptation (LA/RA) and opportunistic communications (OC). Schemes that synergize the two philosophies have also been proposed. Cross-layer network design can be limited to an algorithm or entire architecture, such as ECLAIR[4] and MobileMAN.[3] Cross-layer feedback mechanisms can be implemented in the end-user terminals or in the intermediate devices attached to either an infrastructure or ad hoc network.[5]

Consider the cross-layer architecture illustrated in Fig. 4 based on the TCP/IP model and showing possible inter-layer feedback. In this case study the application layer can adapt its encoding and traffic rate to the instantaneous wireless channel state using the feedback signaling $p3$ and the transport layer's packet loss rate from feedback $t1$. The application layer, on the other hand, tells the MAC layer about its current service level and its desired service (e.g., throughput) via the feedback $a2$. The DLC can consider the QoS requirements of the application in the feedback $a1$ and channel conditions in the feedback $p2$ in its error management. The transport layer (here TCP) can exploit the handoff delays and link disruption feedback n from network layer to improve its re-transmission timers, while it sends its end-to-end error management

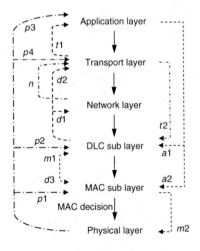

Fig. 4 Potential interactions between layers in cross-layer network design.

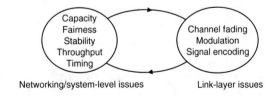

Networking/system-level issues   Link-layer issues

Fig. 3 Interplay between link and networking layers in cross-layer system design.

information $t2$ to the DLC layer to improve the link-level error management. The transport layer also uses the wireless channel state feedback $p4$ to reduce its potential throughput degradation on wireless links due to its congestion management i.e., triggered by long handoff delays, network disconnections, and link losses.[6,7] The signaling feedback $m1$ and $d3$ are exploited by the DLC to optimize its link-level error management operations through the MAC. Through the feedback $m2$ the MAC layer can control the matching of the transmit power at physical layer to the required transmission range. Lastly, the network layer can use Level 2 mobility feedback $d1$ to optimize network-level handoff delays. Upon considering all the feedback information, the MAC layer optimally authorizes the physical layer on service order. Here, the channel feedbacks $p1 - p4$ can be the estimated "instantaneous" bit error or data rates on users' links. The traffic information can be service level feedback to monitor fairness, traffic timing parameters, file size, or goodput. We remark here that using all the interactions shown in Fig. 4 in a single design may be counterproductive[8] (see also the discussions in e.g., Ref. 9). Hence, determining which combinations of inter-layer interactions to be used in a cross-layer design requires optimization skills in itself.

Consider another cross-layer architecture shown in Fig. 5 which depicts a scenario of a multi-service communications in a multi-user system serving M users in a given time frame. This communications paradigm allows the same source to inject traffic of different characteristics into the network. For example, an IP-enabled phone can send e-mail, sms or videophone traffic. At the given time frame shown in Fig. 5 user 1 has three concurrent non-real-time (NRT) applications (i.e., e-mail, HTTP, and FTP) active, user 2 has two concurrently active real-time (RT) applications [i.e., packet voice or voice over IP (VoIP)], while user M has concurrently active a mixture of one RT

(VoIP) and one NRT (e-mail) applications. The figure also shows the service orders of all active users, as well as the individual applications active at each of the users. As a channel-aware MAC scheme, its decisions depend on the "instantaneous" wireless channel qualities of users and other metrics such as traffic QoS requirements. This is a scenario of a multi-service, multi-user wireless communications in which each user can keep multiple applications active concurrently and still expect the network to meet individual service requirements.

CL-MAC DESIGN APPROACHES

This section surveys cross-layer design approaches that are hosted at the MAC layer. In general, CL-MAC design philosophies can be classified into channel-aware, queue-aware or hybrid schemes which exploit both "instantaneous" queue and channel information in their decisions (Fig. 1B). PF, opportunistic round robin and MaxC/I are examples of channel-aware only schemes, while EXP and M-LWDF exploit both queue and channel information. (Detailed information on the algorithms mentioned here can be found in Ref. 10 and references therein.) Channel-ware schemes can be based on LA/RA and opportunistic scheduling (OS). Some non-queue-aware schemes combine both OS and LA/RA (i.e., Hybrid 1).Some schemes embed queue information in OS (i.e., Hybrid 2) or in LA/RA (i.e., Hybrid 3). In the extreme end are Hybrid 4 schemes which combine both OS and LA/RA with queue information. Examples of Hybrid 2 schemes are BLOT,[11] M-LWDF, and OCASD.[12]

LA

LA/RA is applicable to wireless systems that use multiple modulation and signal encoding schemes. Each pair of modulation/coding has a corresponding data transmission rate commensurate with the "instantaneous" channel state. The modulation schemes have different spectral efficiencies, as they map different number of bits onto a symbol. The more the number of bits mapping onto a symbol, the more efficient the scheme but the less resilient it is to transmission medium impairments. The coding schemes also use different rates and hence redundancies. A lower rate coding is more robust to channel impairments but at the cost of spectral efficiency. An example of a cross-layer design based on adaptive modulation and coding (AMC) can be found in Ref. 13. Here, the effects of AMC and MIMO diversity at the physical layer on traffic delay violations and buffer overflow at higher layers are studied using the effective capacity concept.

A more throughput efficient modulation/coding scheme can be used to transmit information at a higher rate at an acceptable transmission quality when the channel condition is good. On the other hand, a lower-order modulation

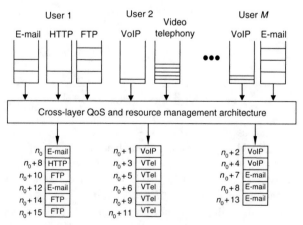

$k + n_0$ = service epoch k in a timeframe starting at time n_0

Fig. 5 A cross-layer architecture serving multiple users with concurrent queues of heterogeneous networked applications in a time-slotted wireless network.

Wireless
Multi—Networks

scheme combined with a more robust error control scheme is used to transmit information at a lower rate at poor channel conditions in order to maintain an acceptable transmission quality. Hence, LA/RA based cross-layer design uses instantaneous wireless channel quality information to adapt data transmission rates to channel state in order to meet the required error rate. Two strategies can be used in LA/RA when wireless channel conditions are poor to maintain a given transmission reliability: 1) use more power to transmit at the same rate or 2) use the same power to transmit at a lower rate. LA/RA is rooted in the *water-filling principle (aka water-pouring principle)* (Ref. 14, p. 610), which dictates that, on any time-varying channel, the average system throughput is maximized by adapting the transmit power and data rate to the instantaneous wireless channel conditions. Techniques used to adapt a wireless user's data transmission rate to its instantaneous channel state include variable rate spreading and adaptive coding/modulation.

OC

The random variability in wireless channel quality is referred to as fading. Fading on wireless channels had been viewed as bad until the advent of OC. OC makes a constructive use of radio fading. In the simplest case without regard to networking issues such as traffic timing constraints and fairness, OC transmits data to/from a user only when its channel conditions are favorable. As channel qualities of multiple users sharing a radio resource vary dynamically independently, at any point in time some of the users are likely to experience good channel conditions while other users experience poor channel conditions. This variability of channel qualities across the user population at a given moment is referred to as *multi-user diversity*, which is the basis of OC.

CASE STUDY OF A CL-MAC DESIGN

We consider a CL-MAC algorithm serving multiple users in a single-hop multimedia wireless system using multi-carrier (OFDM) transmission to support a mixture of delay-sensitive and delay-tolerant applications. OFDM is a technique used to achieve a reliable high-rate communications over otherwise frequency-selective wideband wireless channels. The CL-MAC algorithm should coordinate the efficient use of *finite* system resources while meeting any traffic delay (QoS) constraints. Hence, we formulate the resource allocation task as a constrained optimization problem and solve it using artificial neural networks (ANNs). ANN techniques have been inspired by the operation of biological neurons, and they include a variety of computational models. Application of ANN in communications network design has been motivated by its ability to provide 1) computational power of non-linear techniques, 2) paths for efficient algorithmic implementation in hardware, and

3) learning and RT adaptation capability to solve difficult problems with unknown solution structure.[15] Additionally, ANN uses a parallel technique to quickly search the solution space of a constrained optimization problem. The CL-MAC algorithm described below essentially assigns OFDM sub-channels to active flows, i.e.,

$$f : X(t) \rightarrow Z(t) \qquad (1)$$

where the decision vector $Z(t)$ is defined as $Z(t) = \{J_{im}^1(t), J_{im}^2(t), \ldots, J_{im}^K(t)\}$ and the state vector is $X(t) = \{n_1(t), n_2(t), \ldots, n_1(t)\}$. Here, $n_1(t)$ is the number of non-empty queues of traffic type i in the system at time t and

$$J_{im}^k = \begin{cases} 1 \\ 0 \end{cases} \quad \text{if flow } F_{im} \text{ is} \begin{cases} \text{assigned} \\ \text{not assigned} \end{cases}$$

the k-th OFDM subchannel at time t. The variables used in this algorithm are defined in Table 1. The CL-MAC scheme described here exploits the physical layer information in the form of the data rate and the traffic delay constraints from higher layers in its decisions.

Problem Formulation

We formulate the design as a constrained optimization problem and use ANN techniques to solve it. A similar work has been done in Ref. 16 for CDMA networks, and in Ref. 17 for time-slotted power-constrained networks. The constraint here is the maximum tolerable packet delay for packets of type i, $D_{\max,i}$. We also note the inherent constraint in the design with respect to the finiteness of system resources. The resource allocation problem to be solved at the MAC sub-layer is

$$\text{Minimize } C_{im}^k(t) \qquad (2)$$

$$\text{Subject to } D_{im}(t) \leqslant D_{\max,i} \qquad (3)$$

where

$$C_{im}^k(t) = \frac{B_{i,m}^2(t)}{R_{mk}(t)} \approx \frac{B_{i,m}^2(t)}{\frac{W}{K}\log_2[1 + \frac{2(L-1)}{3}(\frac{0.2}{\text{BER}_{mk}(t)} - 1)]} \qquad (4)$$

is updated as $B_{im}(t+1) = (1 - 1/t_c)B_{im}(t) + 1/t_c J_{im}^k(t) \cdot L_{im}(t)$. Eq. 4 was obtained by a heuristic search for a convex function that achieves the design goals in Eqs. 2 and 3. In essence the CL-MAC policy allocates the kth OFDM subchannel to flow $F_{i,m}$ that fulfils Eqs. 2 and 3. We note from the delay constraint Eq. 3 that the more a flow is delay-insensitive the higher its delay bound. Hence, for NRT traffic class i we have $D_{\max, i} \rightarrow \infty$. In the following we define the vector of optimization variables $x = [x_1, x_2, x_3]^T \triangleq [B_{im}(t), \text{BER}_{m,k}(t), D_{im}(t)]^T$

Table 1 Notation.

Parameter	Meaning
t	Time slot or time index
t_c	Scheduling window (fairness time frame)
m	User index, $m = 1, 2, \ldots, M$
i	Traffic class or queue index, $i = 1, 2, \ldots, I$
$F_{i,m}$	Flow index, i.e., traffic class i at user m
W	Wideband wireless channel bandwidth
K	Number of OFDM subcarriers in the system
$L_{im}(t)$	Size of the head-of-line (hol) packet in flow $F_{i,m}$'s queue
$R_{mk}(t)$	Maximum data rate on user m's kth OFDM subchannel
$B_{im}(t)$	Average amount of traffic scheduled from flow $F_{i,m}$'s queue
$D_{im}(t)$	Queuing delay of hol packet in flow $F_{i,m}$'s queue
$D_{\max,j}$	Maximum tolerable queuing delay of packets of traffic class i
$\text{BER}_{mk}(t)$	Bit error rate on user m's kth OFDM subchannel
L	Signal constellation size

where T denotes the transposition operator. Hence, we set $c(x) \stackrel{\Delta}{=} C_{im}^k(t)$ and $g(x) \stackrel{\Delta}{=} D_{im}(t) - D_{\max,i} \leqslant 0$ in the following.

Comment. At each decision time the vector x is available or estimated for each active user and traffic class combination (i.e., flow) and the CL-MAC policy assigns OFDM sub-tones to active flows that best fulfill Eqs. 2 and 3.

Optimization Using Neural Networks

Optimization neural networks are gradient type networks whose behavior can be modeled by analog electrical circuits.[18] The temporal evolution of these neural networks is a motion in the state space whose trajectory follows the direction of the negative gradient of the system's energy function, which can be made equivalent to the cost function to be minimized. The solution of the optimization problem is equivalent to the point of minimum system energy.

We use neural networks' techniques to compute the vector $x^* = [B_{i,m}^*(t), \text{BER}_{m,k}^*(t), D_{i,m}^*(t)]^T$ in each time slot that exploits the limited radio resource in an optimal way. To do this we first need to convert the constrained optimization problem into equivalent system of differential equations. These differential equations constitute the basic neural network algorithms that must be solved in order to solve the optimization problem online. The Lagrange function $L: \Re^{n+k} \to \Re$ for the problem Eqs. 2 and 3 is

$$L(x, v, \lambda) = c(x) + \lambda[g(x) + v^2] \tag{5}$$

where $\lambda \in \Re^k$, $k = 1$ is the *Lagrangian multiplier*. In accordance with their roles in searching for the optimim solution of Eqs. 2 and 3, v and x are referred to as *variable neurons* or primal variables while λ is the *Lagrangian neuron*. As the neural network dynamically searches for the optimum solution the variable neurons decrease $L(x,v,\lambda)$ while the Lagrangian neuron leads the search trajectory into the feasible region $S = \{x \mid g(x) \leqslant 0, x \geqslant 0\}$. The *Kuhn-Tucker optimality conditions* are

$$\nabla_x L(x^*, v^*, \lambda^*) = 0$$
$$\nabla_v L(x^*, v^*, \lambda^*) = 2\lambda^* v^* = 0$$
$$\nabla_\lambda L(x^*, v^*, \lambda^*) = g(x^*) + v^{*2} = 0$$
$$\lambda^* \geqslant 0 \tag{6}$$

where $\nabla_x L(x, v, \lambda) = \left(\frac{\partial L(\cdot)}{\partial x_1}, \frac{\partial L(\cdot)}{\partial x_2}, \frac{\partial L(\cdot)}{\partial x_3}\right)^T$.

We want to design an ANN with an equilibrium point that fulfils the K-T conditions in Eq. 6. By noting that

$$dx_j/dt = -\partial L(x, v, \lambda)/\partial x_j, \quad j = 1, 2, 3 \tag{7}$$

we can formulate the state equations governing the transient behavior of this ANN by the system of differential equations[19]

$$\frac{dx_1}{dt} = -\frac{2bx_1}{\log_e[1 + a(0.2/x_2 - 1)]} \stackrel{\Delta}{=} f_1(\tilde{x}) \tag{8}$$

$$\frac{dx_2}{dt} = -\frac{abx_1^2}{5x_2(0.2a + x_2 - ax_2)\log_e^2(1 - a + 0.2a/x_2)} \stackrel{\Delta}{=} f_2(\tilde{x}) \tag{9}$$

$$\frac{dx_3}{dt} = -\lambda \stackrel{\Delta}{=} f_3(\tilde{x}) \tag{10}$$

$$\frac{dv}{dt} = -\nabla_v L(x, v, \lambda) = -2\lambda v \stackrel{\Delta}{=} f_4(\tilde{x}) \tag{11}$$

$$\frac{d\lambda}{dt} = \nabla_\lambda L(x, v, \lambda) = g(x) + v^2 \stackrel{\Delta}{=} f_5(\tilde{x}) \tag{12}$$

where $\tilde{x} = (x^T, v, \lambda)^T \in \Re^5$, $b = (K/W)\log_e 2$ and $a = 2(M - 1)/3$. A physically stable neural network has the equilibrium point $\tilde{x}^*$ satisfying $\frac{dx_k}{dt}\big|_{\tilde{x}^*} = \frac{d\lambda}{dt}\big|_{x^*} = \frac{dv}{dt}\big|_{x^*} = 0$ which are equivalent to the Kuhn-Tucker sufficient conditions for optimality, Eq. 6. Hence, the state equations of the neural network, Eqs. 8–12, are referred to as Lagrange Programming Neural Network.[19]

Convexity of the objective function $c(x)$

In order to apply ANN to solve the optimization problem we need to ensure that $c(x)$ is twice continuous differentiable in x. If $c(x)$ is convex then it is twice continuous differentiable in x. A sufficient condition for $c(x)$ to be convex is that its Hessian matrix is positive semi-definite. From Eq. 4 we obtain the Hessian matrix of $c(x)$, $H_c(x) = \nabla^2_{xx}c(x) = [\partial^2 c(x)/\partial x_i \partial x_j]$, as

$$H_c(x) = \begin{pmatrix} D_{11} & D_{12} & D_{13} \\ D_{12} & D_{22} & D_{23} \\ D_{13} & D_{23} & D_{33} \end{pmatrix} \tag{13}$$

where

$$D_{13} = D_{31} = D_{23} = D_{32} = D_{33} = 0 \tag{14}$$

$$D_{11} = \frac{2b}{\log_e[1 + a(0.2/x_2 - 1)]} \tag{15}$$

$$D_{12} = \frac{2abx_1}{5x_2^2[1 + a(0.2/x_2 - 1)]\log_e^2[1 + a(0.2/x_2 - 1)]} \tag{16}$$

and

$$D_{22} = \frac{abx_1^2}{5x_2^3[1 + a(0.2/x_2 - 1)]\log_e^2[1 + a(0.2/x_2 - 1)]}$$
$$\cdot \left(\frac{2a}{5x_2[1 + a(0.2/x_2 - 1)]\log_e[1 + a(0.2/x_2 - 1)]} \right.$$
$$\left. + \frac{a}{5x_2[1 + a(0.2/x_2 - 1)]} - 2 \right) \tag{17}$$

Theorem 1. *We establish conditions under which $H_c(x)$ is a positive semi-definite matrix A sufficient condition for $H_c(x)$ to be a positive semi-definite matrix is that the determinants of all its upper-left sub-matrices are positive. Denote by $det H_c(k)$ the determinant of the square upper-left sub-matrix of $H_c(x)$ of size k. Noting that $H_c(x)$ is a symmetric matrix we require that*

$$\det H_c(1) = D_{11} \geq 0 \tag{18}$$

$$\det H_c(2) = D_{11}D_{22} - D_{12}^2 \geq 0 \tag{19}$$

$$\det H_c(3) = D_{11}D_{22}D_{23} - D_{11}D_{23}^2 - D_{12}^2D_{33} \\ + 2D_{12}D_{13}D_{23} - D_{13}^2D_{22} \geq 0 \tag{20}$$

The validity of Eq. 18 is obvious from Eq. 15. It is obvious from Eq. 14 that, fulfilling Eq. 20. Condition Eq. 19 is fulfilled if

$$\frac{a}{5x_2(1 + a(0.2/x_2 - 1))} \left[1 + \frac{2}{\log_e(1 + a(0.2/x_2 - 1))} \right.$$

$$\left. -2\log_e(1 + a(0.2/x_2 - 1)) \right] \geq 2 \tag{21}$$

Fig. 6 shows that this condition is fulfilled if $x_2(t) = BER_{m,k}(t) \geq 10^{-7}$. The required bit error rate (BER) is realistic for practical wireless systems. Fig. 6 also shows that $H_c(1) > 0$ for all practical BER values. Hence, we have proved that $H_c(x)$ is a positive definite matrix.

Theorem 2. *Positive semi-definiteness of $\nabla^2_{xx}L(x,v,\lambda)$ We prove that $\nabla^2_{xx}L(x,v,\lambda) \geq 0$ from Eq. 5 we obtain $\nabla_x L(x,v,\lambda) = \nabla c(x) + \lambda' \nabla g(x)$. We note that $\nabla g(x) = (0,0,1)^T$. Thus $\nabla^2_{xx}g(x) = (0,0,1)^T$, yielding the Hessian matrix $\nabla^2_{xx}L(x,v,\lambda) = \nabla^2_{xx}c(x) = H_c(x)$. Hence, $\nabla^2_{xx}L(x,v,\lambda)$ if $\rho_m(t) > 3$ dB, as ascertained above.*

The Lyapunov function and stability test

This section establishes the convergence and hence the stability of the set of differential 8-12 that represent our optimization problem via Lyapunov's method. Lyapunov stability theory is based on the idea that if some measure of the energy associated with a dynamic system is decreasing, then the system converges to its equilibrium state. To do this we first need to compute a Lyapunov function, $V(\bullet) = E(\bullet) - E(\bullet)|_{x=x^*} \in C^1$, of the given system, where $E(\bullet) \geq 0$ is the energy function of the system and x^* is the optimum point in the feasible set being searched for. The system's energy function is a function defined on the state space $\{S | g(x) \leq 0\} \subset \Re^3$ which is nonincreasing along the trajectories and it is bounded from below. Let us define the energy function

$$E(x,v,\lambda) = \frac{1}{2}||\nabla_x L(x,v,\lambda)||_2^2 + \frac{1}{2}||g(x) + v^2||_2^2$$

$$= \frac{1}{2}||\nabla c(x) + \lambda \nabla g(x)||_2^2 + \frac{1}{2}||g(x) + v^2||_2^2 \tag{22}$$

whereby $||y||_2 = (\sum_k y_k^2)^{1/2}$ is the L_2-norm of y. It follows that

$$\frac{dE(x,v,\lambda)}{dt} = \frac{\partial E}{\partial x}\frac{dx}{dt} + \frac{\partial E}{\partial v}\frac{dv}{dt} + \frac{\partial E}{\partial \lambda}\frac{d\lambda}{dt} \tag{23}$$

The system is globally stable in the Lyapunov sense, i.e., the trajectories $x_k(t)$, $k = 1,2,3$, $\lambda(t)$ and $v(t)$ converge

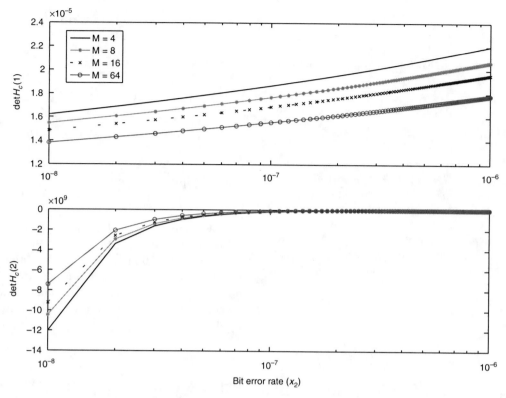

Fig. 6 Plot of det $H_c(k)$ for $K = 1,024$, $W = 5$ MHz and $x_1 = 50$ bytes.

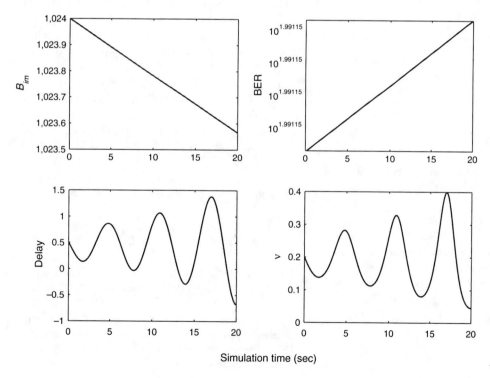

Fig. 7 Transient behaviors of x_1, x_2, x_3 and $x_4 = v$ using the initial conditions $x_1(0) = 1,024$ bits, $x_2(0) = 10^{-3}$, $x_3(0) = 0.5$, $x_4(0) = 0.2$, $x_5(0) = 0.3$.

to stationary points as $t \to \infty$, if and only if $dE/dt = 0 \Rightarrow dx/dt = dv/dt = d\lambda/dt = 0$.

At the stationary point we have $dE/dt|_{x=x^*} = 0$. Proving the stability of the system requires us to establish that

$$E(x, v, \lambda) \geqslant 0 \quad \text{and} \quad dE(x, v, \lambda)/dt \leqslant 0 \tag{24}$$

The sub-sections below prove Eq. 24.

Theorem 3. *Positive semi-definiteness of $E(x,v,\lambda)$. Proving that $E(x,v,\lambda) \geqslant 0$ is an obvious consequence of its definition. Hence, the Lyapunov function $v(x,y,\lambda) = E(x,v,\lambda) = -E(x^*, v^*, \lambda^*)$ is positive definite.*

Theorem 4. *We prove that $dE(x, v, \lambda)/dt \leqslant 0$. We prove that $E(x,v,\lambda)$ is a bounded monotonically decreasing function of time. Using the state conditions 8–12 in Eq. 23 yields*

$$\frac{dE(x, v, \lambda)}{dt}$$

$$= [\nabla_x L(x, v, \lambda)^T \nabla_{xx}^2 L(x, v, \lambda) + v(g(x) + v^2)] \frac{dv}{dt}$$

$$+ (g(x) + v^2)^T \nabla_g(x)] \frac{dx}{dt} + \nabla_x L(x, v, \lambda)^T \nabla g(x)^T \frac{d\lambda}{dt}$$

$$= -\nabla_x L(x, v, \lambda)^T \nabla_{xx}^2 L(x, v, \lambda) \nabla_x L(x, v, \lambda)$$

$$- \lambda v^2 (g(x) + v^2) - (g(x) + v^2) \nabla g(x)^T \nabla_x L(x, v, \lambda)$$

$$+ [(g(x) + v^2) \nabla g(x)^T \nabla_x L(x, v, \lambda)]^T$$

$$= -\nabla_x L(x, v, \lambda)^T \nabla_{xx}^2 L(x, v, \lambda) \nabla_x L(x, v, \lambda)$$

$$- 2\lambda v^2 (g(x) + v^2) \tag{25}$$

As $\lambda \geqslant 0$ and $d^T A d \geqslant 0$ if $A \geqslant 0$ for any nonzero vector d, $dE(x,v,\lambda)/dt < 0$, *as $\nabla_{xx}^2 L(x,v,\lambda) \geqslant 0$ has been proved in Theorem 2.*

SUMMARY

We begin with a brief analysis of the transient behaviors of the parameters used in the discussed CL-MAC over flat Rayleigh fading channels in this section. We apply multi-variate fourth-order Runge-Kutta numerical integration method to evaluate the transient behaviors of the state variables describing the neural network for the dynamic system as given in 8–12. The fourth-order Runge-Kutta method is reputed for its good accuracy and simplicity. For our multi-variate system of differential 8–12 this method can be adapted from (Ref. 20, p. 326) as

$$a_{k,t} = hf_k(\tilde{x}_{i,t}, \tilde{x}_{2,t}, \dots, \tilde{x}_{5,t}), \quad k = 1, 2, \dots, 5 \tag{26}$$

$$xb_{k,t} = hf_k\left(\tilde{x}_{1,t} + \frac{h}{2}a_{1,t}, \tilde{x}_{2,t} + \frac{h}{2}a_{2,t}, \dots, \tilde{x}_{5,t} + \frac{h}{2}a_{5,t}\right) \tag{27}$$

$$c_{k,t} = hf_k\left(\tilde{x}_{1,t} + \frac{h}{2}b_{1,t}, \tilde{x}_{2,t} + \frac{h}{2}b_{2,t}, \dots, \tilde{x}_{5,t} + \frac{h}{2}b_{5,t}\right) \tag{28}$$

$$d_{k,t} = hf_k(\tilde{x}_{1,t} + hc_{1,t}, \tilde{x}_{2,t} + hc_{2,t}, \dots, \tilde{x}_{5,t} + hc_{5,t}) \tag{29}$$

$$\tilde{x}_{k,t+1} = \tilde{x}_{k,t+1} + (a_{k,t} + 2b_{k,t} + 2c_{k,t} + d_{k,t})/6 \tag{30}$$

where $f_k(\tilde{x})$, $k = 1, 2, \dots$ are as given in 8–12, respectively. Fig. 7 shows the effects of inappropriate initial conditions on the transient behaviors of x_1, x_2, x_3 and $x_4 = v$.

This entry presented a brief introduction to the complex issue of CL-MAC design in form of a constrained optimization problem. The solution of the formulated problem is attempted using ANN. It is hoped that this study provokes further studies into the areas touched upon as many issues involved are yet to be resolved.

REFERENCES

1. van der Schaar, M.; Shankar, S. Cross-layer wireless multimedia transmission: challenges, principles, and new paradigms. IEEE Wirel. Commun. **2005**, *12* (4), 50–58.

2. Stallings, W. *Data and Computer Communications*; Prentice Hall: New Jersey, 2007.

3. MobileMAN Project, http://mobileman.projects.supsi.ch/index.html (accessed July 12, 2007).

4. Raisinghani, V.T.; Iyer, S. Cross-layer feedback architecture for mobile device protocol stacks. IEEE Commun. Mag. **2006**, *44* (1), 85–92.

5. Conti, M.; Maselli, G.; Turi, G.; Giordano, S. Cross-layering in mobile ad hoc network design. IEEE Comput. **2004**, *37* (2), 48–51.

6. Gyasi-Agyei, A. Mobile IP-DECT Internet working architecture supporting IMT-2000 applications. IEEE Netw. **2001**, *15* (6), 10–22.

7. Tian, Y.; Xu, K.; Ansari, N. TCP in wireless environments: problems and solutions. IEEE Commun. Mag. **2005**, *43* (3), S27–S32.

8. Kawadia, V.; Kumar, P.R. A cautionary perspective on cross-layer design. IEEE Wirel. Commun. **2005**, *12* (1), 3–11.

9. Srivastava, V.; Motani, M. Cross-layer design: a survey and the road ahead. IEEE Commun. Mag. **2005**, *43* (12), 112–119.

10. Gyasi-Agyei, A.; Kim, S.-L. Comparison of opportunistic scheduling policies in timeslotted AMC wireless networks. IEEE International Symposium on Wireless Pervasive Computing, Phuket, Thailand, Jan 16–18, 2006; 1–6.

11. Gyasi-Agyei, A.; Kim, S.-L. Cross-layer multiservice opportunistic scheduling for wireless networks. IEEE Commun. Mag. **2006**, *44* (6), 50–57.

12. Gyasi-Agyei, A. Multiuser diversity based opportunistic scheduling for wireless data networks. IEEE Commun. Lett. **2005**, *9* (7), 670–672.

13. Zhang, X.; Tang, J.; Chen, H.-H.; Ci, S.; Guizani, M. Cross-layer-based modeling for quality of service guarantees in mobile wireless networks. IEEE Commun. Mag. **2006**, *44* (1), 100–106.

14. Haykin, S. *Communication Systems*; John Wiley & Sons: New York, 2001.

15. Yuhas B., Ansari, N., Eds.; *Neural Networks in Telecommunications*; Kluwer Academic Publishers: Boston, MA, 1994.

16. Fantacci, R.; Forti, M.; Marini, M.; Tarchi, D.; Vannuccini, G. A neural network for constrained optimization with application to CDMA communication systems. IEEE Trans. Circuits Syst. II, Analog Digit. Signal Process. **2003**, *50* (8), 484–487.

17. Gyasi-Agyei, A. Performance analysis of a power-constrained cross-layer scheduling using artificial neural networks. IEEE Wireless Communications and Networking Conference (WCNC 2007), Hong Kong, Mar 11–15, 2007.

18. Sharma, V.; Jha, R.; Naresh, R. An augmented Lagrange programming optimization neural network for short-term hydroelectric generation scheduling. Eng. Optim. **2005**, *37* (5), 479–497.

19. Zhang, S.; Constantinides, A.G. Lagrange programming neural networks. IEEE Trans. Circuits Syst. II, Analog Digit. Signal Process. **1992**, *39*, 441–452.

20. Fröberg, C.-E. *Numerical Mathematics: Theory and Computer Applications*; Benjamin/Cummings Publishing Company, Inc.: Menlo Park, CA, 1985.

Wireless Networks: Error-Resilient Video Transmission

Yu Chen
Department of Electronic Engineering, Tsinghua University, Beijing, China

Jian Feng
Department of Computer Science, Hong Kong Baptist University, Hong Kong, China

Kwok Tung Lo
Department of Electronic and Information Engineering, Hong Kong Polytechnic University, Hong Kong, China

Xudong Zhang
Department of Electronic Engineering, Tsinghua University, Beijing, China

Abstract

Error resilience methods in video coding provide mechanisms for reducing the impact of error propagation in both spatial and temporal domains caused by errors in compressed bit stream. Such methods are necessary for transmission of compressed video over error-prone environments like wireless networks.

WIRELESS VIDEO TRANSMISSION

With the rapid development of wireless communications and video compression techniques in recent years, people can now send and receive video information in different places with more flexible styles. The demand for transmission of various video content over wireless environments has been greatly increasing. Therefore, robust video transmission in wireless environments draws many people's attention from different communities.

It is a very challenging task to make video information robust in wireless transmission. On one hand, the quality of service (QoS) of wireless channel is hardly reliable for its high BER and limited transmission bandwidth. On the other hand, as predictive coding and variable length coding are adopted in most of the existing video codecs to achieve high video compression efficiency, once channel errors occur in compressed video stream, they not only result in spatial error propagation in present frame but also cause temporal error propagation in successive frames. These kinds of error propagation degrade video quality, apparently, for the human visual system (HVS). In Fig. 1 the block diagram of a generic wireless video transmission system,[1,2] shows the transmission rate and distortion. In the figure, an original video stream has first undergone the encoding process, which results in a source encoding distortion $D_s(R_s, \beta)$, where R_s is the source encoding rate, and β is the portion rate of intra-coded macroblock (MB) in present encoding frame. Then, the compressed video stream is processed by channel coding and modulation with additional rate increase R_c. The total transmission rate for the compressed video stream is $R_t = R_s + R_c$. Using packet-switched wireless network as an example, we assume that its packet lost rate is P_p. For correctly received video packets, although many present wireless transmission systems adopt channel coding to constrain channel error, it is hardly possible to ensure that all channel errors in the video stream can be corrected by channel coding. So the occurrence probability of residual error beyond the channel coding ability is assumed as P_r. With P_p, P_r, and R_c, the distortion caused by wireless networks is computed as $D_c(R_c, P_r, P_p)$. With the process of error concealment at video decoder, the distortion caused by wireless networks can be mitigated to $D_{ec}(\beta, D_c)$. As a final result, the reconstructed video distortion at video decoder is $D_r = D_s + D_{ec}$. As such, both the source coding and network will be jointly adapted to achieve an optimum transmission over a noisy wireless channel.

ERROR-RESILIENT VIDEO ENCODING

As mentioned before, acceptable video quality in a wireless environment can be obtained by the adjustment of parameters in video codec and wireless networks. For the former, people have proposed many error-resilient video encoding algorithms to enhance the robust performance of the compressed video stream in wireless networks.[3–5] These algorithms can be divided into three categories: 1) error detection and error concealment algorithms used at video decoder of wireless receiver, 2) error-resilient video encoding algorithms located at video encoder of

Encyclopedia of Wireless and Mobile Communications DOI: 10.1081/E-EWMC-120043562

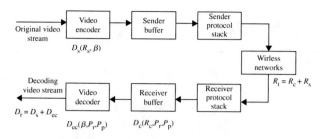

Fig. 1 A generic wireless video transmission system.

wireless transmitter, and 3) robust error control between video encoder and decoder based on (1) and (2). Fig. 2 summarizes different techniques at different parts of a wireless video transmission system.

Since error concealment algorithms are only used at video decoder in wireless receivers, they do not require any modification of video encoder and channel codec. Hence, there is no increase of coding computing complexity and transmission rate. Therefore, error concealment algorithms can be easily realized in the present wireless video transmission systems. However, since error concealment algorithms make full use of spatial and temporal correlation in video stream to estimate the corrupted region of video frames, when the correlation between corrupted region and correctly received frames is weak, error concealment algorithms cannot achieve good effect so there is apparent distortion in repaired reconstructed video frames. In addition, although error concealment algorithms can reduce the intensity of temporal error propagation, it cannot reduce the length of temporal error propagation. As we know, HVS is not very sensitive to short-term obvious error propagation, while even slight long-term error propagation will annoy the observation

of HVS considerably. Therefore, desirable error repaired effect should make the intensity and length of error propagation minimum simultaneously.

In order to compensate the defects of error concealment algorithms, a number of error-resilient video encoding algorithms have been developed in the last decade to make the compressed video stream accustomed to wireless transmission environment. These algorithms can be divided into five categories as shown in Table 1 (Ref. 6).

The first category is concerned about the location of error protection information. Its main purpose is to reduce the length and intensity of spatial and temporal error propagation. In this category, four kinds of representative algorithms are developed based on re-synchronization mode, adaptive intra coding, flexible reference frames, and multiple reference frames.

The second category of error-resilient algorithms utilizes data partition scheme to aggregate same type of syntax information, such as the aggregation of motion vector, and header and texture information. When channel error appears in this type of error-resilient video stream, all information in the same region is not simultaneously wrong, and there is some correct information left in the corrupted region. So, with residual correct information, coarse reconstruction effect is still achieved at the video decoder, which is always more satisfactory than error concealment. The redundant error-resilient video-encoding algorithms can efficiently improve the performance of robust decoding for their inserted redundancy to mitigate the corrupted probability of video stream. Reversible variable length coding (RVLC)[7] can effectively reduce the range of spatial error propagation by reversely decoding from the position of next re-synchronization node at the expense of apparent increase of encoding overheads.

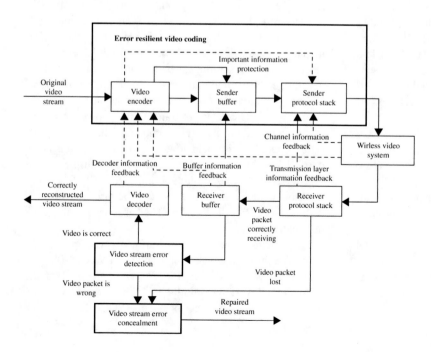

Fig. 2 Error-resilient methods used in packet-switched wireless networks.

Table 1 Different categories of existing error-resilient video encoding algorithms.

Category	Applied extent	Specification algorithms
Location	Reduce the effect of error propagation	—Re-synchronization mode —Adaptive intra coding —Flexible reference frames —Multiple reference frames
Data partition	Unequal protection and transmission	—Frequency coefficients —Motion vector, header and texture information
Redundant encoding	Robust decoding	—Reversible varied length coding —Multiple description coding —Flexible graphic scalable (FGS)
Error concealment based algorithms	Improve repaired effect	—Motion vector compensation —Flexible macroblock (MB) alignment
Channel-based algorithms	The compromise of error resilience and coding rate	—Rate-distortion optimization (RDO) in wireless environment

Multiple descriptions coding (MDC)[8] divides conventional compression video stream into several pieces of sub-video streams, and each of them has the same priority for transmission. When any of them is corrupted or lost in transmission, residual correctly received pieces of video stream are still used to reconstruct coarse picture. Flexible graphic scalable (FGS) coding[9] is another type of error-resilient algorithm, which adopts multiple layers coding for video compression. In FGS coding, there are dependent associations between base layer and enhanced layer, so only if the base layer is correctly decoded can the other enhancement layers be decoded.

The fourth category in Table 1 is developed to compensate the defects of existing error concealment algorithms. For the spatial and temporal correlation in video stream is not always high, and the correct data used as reference by error concealment is not always enough, practical prediction effect of error concealment is not precise, whose final repaired effect is not better than direct replacement and weighted interpolation. In order to avoid this, some essential verification information is necessary to add into original video stream in order to improve the preciseness of error concealment prediction effect.

The last category is the wireless channel-based error-resilient video coding algorithms.[10] With respect to original rate-distortion optimization (RDO) model in conventional video codec, these algorithms are designed to get better video quality and compression efficiency simultaneously. This type of RDO model may not be best suited to the wireless transmission environment. The distortion caused by channel error should be taken into RDO model so that the corresponding optimization parameters in the RDO model can be adjusted according to varied channel parameters, such as, PLR, BER and burst error average length (BEAL).

EXISTING WIRELESS NETWORKS

In the new 3G mobile network, wireless video transmission is feasible as the wireless transmission rate is greatly improved. For the future 4G of mobile networks, it will be operated based on Internet technology with various access technologies such as wireless local area network (WLAN), and much higher wireless transmission rate is obtained in different types of networks.[11]

Table 2 shows different existing wireless networks.[11] Although there are apparent differences in physical (PHY) and transmission layers among different wireless networks, they make full use of adaptive modulation and coding (AMC) to match varied wireless environments. As listed in Table 2, AMC can smartly select modulation techniques, such as binary phase shift keying (BPSK), quadrature PSK (QPSK), 8-PSK, 16-quadrature amplitude modulation (16-QAM) and so on, and adjust the performance of channel coding. When the mobile terminal is close to the base station, for better QoS of wireless channel, higher level modulation is needed while no channel coding is added. If the mobile user is adjacent to the boundary of the cell, weak transmission power results in severe degradation of wireless channel where more powerful channel coding and low level of modulation is required. For instance, in code division multiple access CDMA-2000 1x evolution-data optimized (EV-DO), convolution code and turbo code can be used as channel coding, and QPSK, 8-PSK, 16-QAM are used for downlink modulation.

In order to know the architecture of existing wireless networks, the protocol stack in U-plane of W-CDMA[12] is shown as an example in Fig. 3. The top level is the application layer; the middle includes three sub-layers: packet data convergence protocol (PDCP), radio link

Table 2 Different existing wireless networks.[11]

Radio access	PHY (bps)	Channel bandwidth	Modulation	Channel coding
GPRS	9.06–171.2 K	200 KHz	GMSK	Convolution code for CS1-4 mode None for CS4 mode
EDGE	8.8–473.6 K	200 KHz	GMSK 8-PSK	Convolution code for (CS1-4 MCS1-4)
CDMA-2000 1 × EV-DO	1.25–2 M downlink 144 K uplink	1.25 MHz	QPSK,8-PSK, 16-QAM downlink BPSK uplink	Convolution or turbo code
W-CDMA	64–384 K	5 MHz	QPSK (downlink) BPSK (uplink)	Convolution or turbo code
HSDPA	64 K–14 M downlink	5 MHz	QPSK 16-QAM (downlink)	Rate 1/3-1 Turbo code
802.11	6–54 M (11a) 1–11 M (11b) 1–54 M (11g)	20–22 MHz	OFDM (11a) CCK (11b) OFDM + PBCC (11g)	Convolution code

Source: Table 1, Minoru, E.; Takeshi, Y. Advances in wireless video delivery. Proc. IEEE. 2005, 93, 111–122, IEEE. (© 2005 IEEE).

control (RLC), and media access control (MAC); and the bottom is the PHY channel layer.

The application services can be divided mainly into two categories: circuit switched and packet switched. Voice and video phone, unrestricted digital information (UDI) bearers services belong to circuit-switched services. Voice over Internet protocol (VoIP), video and audio streaming, web browsing, and e-mail are the packet-switched services.

The PDCP sub-layer is only used for packet-switched services, which can hide differences of higher layer protocols from the lower layers. With header compression, it can divide original data stream into certain data units: IP/RTP/UDP packets. These kinds of packets can be fragmented into service data unit RLC-SDU by the RLC sub-layer, where error detection, link-layer re-transmission, and reordering of packet delivery are executed. In addition, there are three types of transmission modes in this sub-layer. They are transparent mode (TM), unacknowledged mode (UM), and acknowledge mode (AM). The TM

is only used for circuit-switched services to process original data stream without header information. The other two modes are used for packet-switched services. In the RLC-UM, error detection and the process of fragmentation or reassembly are made. When there are corrupted bits in corresponding RLC-SDU data unit, the unit will be removed by the receiver. And this mode is only suited to the packet-switched services with restricted delay requirement as it does not give any re-transmission in link-layer. On the contrary, the RLC-AM supports re-transmission in link-layer and orderly packet transmission. This mode can also be used in environments where delay is not very sensitive. The MAC sub-layer assigns a logical channel to each RLC unit and maps logical channels to the bottom layer: PHY channel. The function of this sub-layer is to determine transmission priority based on the situation of RLC buffer and monitor the fluctuation of traffic amount in the wireless channel. In the bottom layer, different PHY channels are given with respect to practical

Wireless Multi—Networks

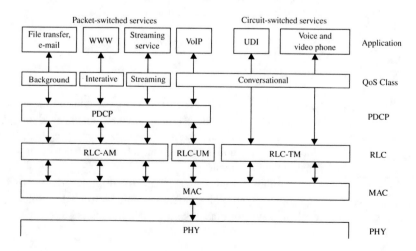

Fig. 3 The protocol stack in U-plane of W-CDMA.

requirement of services and condition of wireless transmission resources.

In the 3G partnership project (3GPP), four QoS classes are defined to support different types of applications. They are conversational, streaming, interactive, and background.[13] These QoS classes are set by the wireless terminal. Generally speaking, the circuit-switched services and VoIP services belong to the conversational class; packet-streaming service is assigned to the streaming class; the best quality and delay-insensitive services such as web browsing and messaging are the interactive and background applications, respectively.

With respect to above-mentioned services, the wireless video transmission services need high channel quality not only in BER or PLR but also for the maximum value of delay and transport feedback. For these different services, the required comprehensive QoS is listed in Table 3 (Ref. 11) and Table 4 (Ref. 2).

In Table 3, UDI and AV bearer are used as examples of circuit-switched services to show the acceptable requirement for BER; conversation and streaming services are used as examples of packet-switched services to show the acceptable requirement for PLR. In Table 4, different types of video services are classified referring to their maximum tolerable end-to-end delay, the availability and usefulness of transport feedback information, and the requirement that the video stream is online or offline encoded.

RELATIONSHIP BETWEEN DISTORTION, COST, AND DELAY IN WIRELESS VIDEO TRANSMISSION

With respect to the previous two sections, it is necessary to select proper error-resilient encoding algorithms S^* and optimized network transmission algorithm N^* from available error-resilient video schemes S and existing network transmission schemes N in order to meet the requirements of different end-users. The selection of S^* and N^* has apparent impact on the end-to-end distortion D_{tot}, the

end-to-end delay T_{tot}, and the total cost C_{tot} for robust transmission of video to the receiver. Here, D_{tot} is identical to D_r in Fig. 1; the total wireless transmission rate R_t and the total transmission power E_{tot} belong to C_{tot}; T_{tot} is the time interval from when one frame is captured at transmitter to when it is displayed at the receiver.

In order to obtain satisfactory video quality at the receiver of wireless video transmission, it is necessary to understand the trade-off between $D_{tot}(S,N)$, $C_{tot}(S,N)$, and $T_{tot}(S,N)$ as shown in Fig. 4 (Ref. 12). In this figure, the minimum distortion achieved by the total cost and delay is shown, where $T_1 < T_2 < T_3$. When the total delay T_{tot} equals certain value, such as T_1, D_{tot} decreases with the increase of C_{tot} for more resources are assigned to video stream to make it more robust in wireless environment. When T_{tot} increases, under fixed total cost C_{tot}, the expected D_{tot} decreases as more delay allows higher probability of re-transmission. So if little T_{tot} is given, fewer bits can be sent to receiver; hence the total distortion D_{tot} is apparent due to source coding. For transmission cost C_{tot} that determines a certain level of distortion D_{tot}, it decreases when total delay T_{tot} is enlarged. As we know, in a wireless system, when additional transmission time is given and transmission power is mitigated, the transmission energy required to keep a fixed BER or PLR will be reduced.

OPTIMIZED FRAMEWORK FOR ERROR-RESILIENT WIRELESS VIDEO TRANSMISSION

In this section, the optimized framework for error-resilient wireless video transmission in different scenarios are given. The objective of this framework is to optimize one system performance, while other performance parameters still meet the basic requirement by efficiently selecting proper source and network algorithms.[12]

When the end-to-end distortion, $D_{tot}(S,N)$, is the key performance as a certain expense of $C_{tot}(S,N)$ and $T_{tot}(S,N)$ in wireless video services, such as download and play, multicast and broadcast, the optimization for this category of wireless video services is drawn as

$$D_{tot}(S^*, N^*) = \min D_{tot}(S,N)\{S^* \in S, N^* \in N\}$$

$$\text{s.t.: } C_{tot}(S^*, N^*) \leqslant C_0 \quad \text{and} \quad T_{tot}(S^*, N^*) \leqslant T_0 \qquad (1)$$

where C_0 is maximum acceptable resource consumption, and T_0 is the longest delay for receiver to display video. Here, the S^* and N^* are the most optimized error-resilient video encoding and transmission schemes in distortion-sensitive environment.

A similar formulation is given to minimize the cost required to send video stream confined to a desirable level of distortion and tolerable delay. This is suited for

Table 3 BER and PLR suggested by third generation partnership project (3GPP).[11]

Video services	BER	PLR
UDI bearer	10^{-6}	N/A
AV bearer	$\leqslant 10^{-4}$	N/A
Packet-switched conversation	N/A	10^{-3}
Packet-switched streaming	N/A	$\leqslant 10^{-4}$

Source: Table 2, Minoru, E.; Takeshi, Y. Advances in wireless video delivery. Proc. IEEE. 2005, 93, 111–122, IEEE. (© 2005 IEEE).

Table 4 Characteristics of available wireless video delivery services.[2]

Video service	Maximum delay	Transport Available?	Feedback Useful?	Encoding
Download and play	N/A	Y	Y	Offline
On-demand streaming	$\geqslant 1$ s	Y	Y	Offline
Live streaming	$\geqslant 200$ ms	Partly	Y	Online
Multicast	$\geqslant 1$ s	Limited	Partly	Both
Broadcast	$\geqslant 2$ s	N	—	Both
Conferencing	$\leqslant 250$ ms	N	—	Online
Telephone	$\leqslant 200$ ms	Limited	Y	Online

Source: Table 1, Stockhammer, T.; Hanuksella, M. M. H. 264/AVC video for wireless communication. IEEE wireless commun. 2005, 12(8), 6–13, IEEE. (© 2005 IEEE).

mobile terminals where transmission power, computation resource, and wireless width are limited. The optimization for this formulation is

$$C_{\text{tot}}(S^*, N^*) = \min C_{\text{tot}}(S, N)\{S^* \in S, N^* \in N\}$$

$$\text{s.t.: } D_{\text{tot}}(S^*, N^*) \leqslant D_0 \quad \text{and} \quad T_{\text{tot}}(S^*, N^*) \leqslant T_0 \quad (2)$$

where D_0 is maximum distortion accepted for observation. Here, the S^* and N^* are most optimized error-resilient video encoding and transmission schemes in cost-sensitive environment. They can be used to obtain a constant quality level with adjustment of resource assignment.[14]

For delay-sensitive services, such as wireless video telephone and video conference, the delay is shortened to the extent possible with the acceptable sacrifice of distortion and cost. Their corresponding optimization formulation is

$$T_{\text{tot}}(S^*, N^*) = \min T_{\text{tot}}(S, N)\{S^* \in S, N^* \in N\}$$

$$\text{s.t.: } D_{\text{tot}}(S^*, N^*) \leqslant D_0 \quad \text{and} \quad C_{\text{tot}}(S^*, N^*) \leqslant C_0 \quad (3)$$

The processes of Eqs. 1–3 need to jointly modulate source and channel parameters. Although the formulas (Eqs. 1–3) give corresponding solution schemes for different application environments, their realization is still difficult for two reasons—One is the huge requirement of computational complexity to access all ways from congregation S and N to find the jointly optimized S^* and N^*. For example, if the separated computation volume for S and N is K_S and K_N, the total computation volume for joint optimization is $K_S \times K_N$, which is obviously larger than that of separated optimization $K_S + K_N$. The other is that the above formulas cannot give a direct solution. It is necessary to make precise adjustments according to requirements of different practical wireless video transmission scenarios, which can be accomplished by deciding how to efficiently allocate limited wireless bandwidth to coding rate of source and channel.

For the general tools used to optimize Eqs. 1–3, based on the traditional RDO model, Lagrange optimization can be exploited to transform the constrained problem in Eq. 1 into an unconstrained problem, as shown in the following fomula:

$$j_{\text{tot}}(S^*, N^*) = \min_{\{S,N\}} j_{\text{tot}}(S, N)$$

$$= D_{\text{tot}}(S, N) + \lambda_{\text{C}} C_{\text{tot}}(S, N) + \lambda_{\text{T}} T_{\text{tot}}(S, N) \quad (4)$$

where $\lambda_{\text{C}} \geqslant 0$ and $\lambda_{\text{T}} \geqslant 0$ are the corresponding Lagrange multipliers for cost and delay, respectively. By carefully selecting λ_{C} and λ_{T}, the solution to Eq. 1 can be computed as a convex-hull approximation problem.[15] For searching

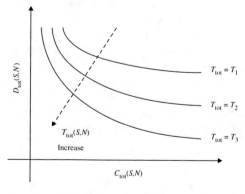

Fig. 4 Relationship between the minimum achievable distortion $D_{\text{tot}}(S,N)$, the total cost $C_{\text{tot}}(S,N)$, and the total delay $T_{\text{tot}}(S,N)$.[13] *Source*: Figure 2, Katsaggelos, A.K.; Eisenberg, Y.; Zhai, F.; Berry, R.; Papps, T.N. Advances in efficient resource allocation for packet-based real-time video transmission. Proc. IEEE. 2005, 93(1), 135–147. (© 2005 IEEE).

proper λ_C and λ_T, cutting plane and sub-gradient methods are effective. When λ_C and λ_T are given fixed value, dynamic programming (DP) is an efficient way to find proper source and channel solution schemes if the dependency of successive video packets is weak.[16] Here the dependency in video stream is determined by inter-prediction encoding and temporal error concealment strategy used at receiver. Even in optimizing situations, the DP solution still needs huge amounts of computations with respect to jointly controlling many parameters in source and channel. For this reason, people proposed iterative decent algorithms to achieve local optimization.[17] In them, firstly, the parameters to be optimized are divided into several categories where little correlation exists. And then, when other categories of parameters equal fixed value, each category of parameters is iteratively configured, respectively, until the minimum value of j_{tot} is achieved. Also, similar processes can be used to solve Eqs. 2 and 3.

In the following, two application scenarios are used as examples to give corresponding optimization processes for wireless video transmissions. They are RDO-based error-resilient video encoding for wireless environment and energy-saved wireless video transmission.

The first scenario can be considered as the algorithm of above-mentioned fifth category of existing error-resilient video coding. Its main objective is to have minimum effect as a result of channel error and error propagation by a proper RDO model. Referring to update wireless channel information such as PLR and packet average burst lost length (PABLL) in packet-switched networks, new RDO model based on original source RDO model is proposed to adapt to wireless environment, where both source and channel parameters S and N are jointly taken into account. Therefore, Eq. 1 changes into

$$D_{tot}(S^*, N^*) = \min D_{tot}(S, N)\{S^* \in S, N^* \in N\}$$

$$\text{s.t.: } T_{tot}(S^*, N^*) \leqslant T_0 \tag{5}$$

For wireless channel with fixed or varied transmission rate, the delay required in Eq. 5 can be converted into a corresponding mitigation to wireless transmission bandwidth. For well-study wireless-based RDO model, its Lagrange multiplier is precisely re-designed to modify original correlation between expected distortion at receiver and source coding rate at transmitter. As a note, the former is computed with the reconstructed effect, which is a result of error concealment and error propagation. Therefore, with this RDO model, each MB in video stream can be encoded as the most wireless adapted reference frame and prediction mode.

Another scenario is concerned with wireless video transmission terminals, where energy supply is limited.

In this case, the total cost, C_{tot}, is energy-sensitive as seen in Eq. 2, which is composed of not only the energy used to transmit and receive video stream but also the energy to accomplish real-time video encoding and decoding. For video stream, its sensitivity to channel errors is varied with the every time interval assigned to each video packet for its time-varying temporal and spatial correlation of video content.

Existing adapting algorithms have adopted joint optimization of source and networks parameters with respect to varied video content and channel situation. In them, based on the analysis results of video content, adapting transmission power is adjusted for different error sensitive video packets. So these video packets are processed by the unequal error protection (UEP) transmission mechanism. Comprehensive UEP transmission power schemes are referred in Ref. 18. In addition, as we know, the transmission power has great effect on practical channel error rate and bandwidth. Although lower transmission power can decrease the cost of C_{tot}, it degrades source encoding quality and makes channel's QoS worse so that the expected distortion D_{tot} is larger than D_0. So trade-off between distortion and transmission power is fundamental.

For energy used to accomplish the real-time video codec, available algorithms consider that the video codec requires more computational cost and more processing energy. For example, if high performance motion compensation algorithm is used, more processing energy will be costly, while the obtained high compression encoding rate needs less transmission energy. So how to make a proper compromise between transmission energy and process energy is still a challenging problem.

CONCLUSIONS

In this chapter, recent advanced techniques on wireless error-resilient video transmission are reviewed. The content of these techniques is focused on how to efficiently allocate limited resources to adapt to different wireless video application scenarios, which is realized by adopting a high-level optimization framework through the modulation of source and networks parameters. Its final effect has an obvious impact on quality and delay of received video, as well as consumption of limited energy at the terminal.

REFERENCES

1. Stuhlmuller, K.; Farber, N.; Link, M.; Girod, B. Analysis of video transmission over lossy channels. IEEE J. Sel. Areas Commun. **2000**, *18* (6), 1012–1032.

2. Stockhammer, T.; Hanuksella, M.M. H.264/AVC video for wireless communication. IEEE Wirel. Commun. **2005**, *12* (8), 6–13.

3. Wang, Y.; Zhu, Q.-F. Error control and concealment for video communication: a review. IEEE. **1998**, *86* (5), 974–997.

4. Wang, Y.; Wenger, S.; Wen, J.; Katsaggelos, A.K. Error resilient video coding techniques. IEEE Signal Process. Mag. **2000**, *17* (4), 61–82.

5. Villasenor, J.; Zhang, Y.-Q.; Wen, J. Robust video coding algorithms and systems. IEEE. **1999**, *87*, 1724–1733.

6. Vetro, A.; Xin, J.; Sun, H.-F. Error resilience video transcoding for wireless communication. IEEE Wirel. Commun. **2005**, *12*, 14–21.

7. Takishima, Y.; Wada, M.; Murakami, H. Reversible variable length codes. IEEE Trans. Commun. **1995**, *43* (2–4), 158–162.

8. Wang, Y.; Reibman, A.R.; Lin, S.-N. Multiple descriptions coding for video delivery. IEEE. **2005**, *93*, 57–70.

9. Jens-rainer, O. Advances in scalable video coding. IEEE. **2005**, *93*, 42–56.

10. Stockhammer, T.; Kontopodis, D.; Wiegand, T. Rate-distortion optimization for JVT/H.26L video coding in packet loss environment. 12th International Packet Video Workshop (PV' 02), Pittsburgh, PA, Apr, 24–26, 2002.

11. Minoru, E.; Takeshi, Y. Advances in wireless video delivery. IEEE. **2005**, *93*, 111–122.

12. Quality of service (QoS) concept and architecture (TS 23.107), Technical Specification Group Services and System Aspects, 3rd Generation Partnership Project (3GPP), Dec 2003.

13. Katsaggelos, A.K.; Eisenberg, Y.; Zhai, F.; Berry, R.; Papps, T.N. Advances in efficient resource allocation for packet-based real-time video transmission. IEEE. **2005**, *93* (1), 135–147.

14. Eisenberg, Y.; Luna, C.E.; Pappas, T.N.; Berry, R.; Katsaggelos, A.K. Joint source coding and transmission power management for energy efficient wireless video communications. IEEE Trans. Circuits Syst. Video Technol. **2002**, *12*, 411–424.

15. Schuster, G.M.; Katsaggelos, A.K. *Rate-Distortion Based Video Compression: Optimal Video Frame Compression and Object Boundary Encoding*; Kluwer: Norwell, MA, 1997.

16. Ortega, A.; Ramchandran, K. Rate-distortion methods for image and video compression. IEEE Signal Process. Mag. **1998**, *15*, 23–50.

17. Zhao, S.; Xiong, Z.; Wang, X. Joint error control and power allocation for video transmission over CDMA networks with multiuser detection. IEEE Trans. Circuits Syst. Video Technol. **2003**, *12*, 425–437.

18. Chan, Y.S.; Modestino, J.W. Transport of scalable video over CDMA wireless networks: a joint source coding and power control approach. IEEE International Conference on Image Processing (ICIP'2001), Thessaloniki, Greece, Oct 7–10, 2001; Vol. 2, 973–976.

Routing in Wireless Networks with Intermittent Connectivity

Ionut Cardei
Cong Liu
Jie Wu
Department of Computer Science and Engineering, Florida Atlantic University, Boca Raton, Florida, U.S.A.

Abstract

Wireless networks with intermittent connectivity [also called delay or disruption tolerant networks (DTN)], are characterized by sporadic availability of end-to-end paths between end hosts. Existing TCP/IP packet routing protocols cannot cope with the lack of reliable end-to-end connectivity. New routing mechanisms are necessary.

The Internet has been exceedingly successful in establishing a global communication network built on the concept of a common set of TCP/IP protocols. Within the last ten years there have been tremendous research efforts spent adapting the TCP/IP stack to various types of wireless and mobile networks. Routing has been recognized as the most challenging problem in networks with a dynamic topology. Protocols, such as ad hoc on demand distance vector (AODV),[1] dynamic source routing (DSR),[2] optimized link state routing protocol (OLSR)[3] and many others have been thoroughly analyzed in multiple scenarios. Their main limitation comes from the fact that, by design, they work only if there is a contemporaneous end-to-end path between endpoints. These protocols are able to find a route only if the destination router can complete the route discovery protocol (for on-demand routing protocols) or successfully disseminate link state advertisements (for table-driven routing). Node mobility or sporadic channel availability increases route instability, causing an increase in routing overhead and a reduction in end-to-end connectivity. Under these circumstances, routing protocols cannot keep up with the topology changes, and the overall network performance is reduced.

In this entry we present an introduction to routing in networks with intermittent connectivity, and we cover several representative routing mechanisms. We begin by describing the main approach for message delivery in case of intermittent connectivity. After a brief overview of the delay-tolerant networking architecture, we describe the main classes of routing protocols for networking with intermittent connectivity and several representative solutions. *Deterministic routing* uses accurate estimates of time intervals when node links, called contacts, are available to schedule transmissions. *Stochastic routing* techniques are either *zero-knowledge*, where nothing is known about node contacts and state, or use *delivery estimation* to approximate a metric for end-to-end message delivery that contributes to more intelligent forwarding decisions. Active stochastic routing techniques rely on controlling the trajectory for some mobile nodes to pickup, carry and deliver messages to improve communication capability in sparse networks. For a comprehensive survey of the state-of-the-art in routing for delay or disruption tolerant networks (DTNs) readers should consult.[4]

A key reason why end-to-end communication is difficult in networks with dynamic topology is that network-layer IP packet delivery works only for as long as the end-to-end path is available. An IP packet will be dropped when it arrives at an intermediary node where, currently, no link to the next hop exists. This protocol design element restricts end-to-end delivery for a range of scenarios where packets can be buffered while in transit to the destination. Fig. 1 illustrates such a scenario.

A route from source node 1 to destination 4 passes through nodes 2 and 3. As there is no time instant when all three links are functioning, the standard IP forwarding approach would cause the packet to be dropped after reception at node 2. With a different approach, the packet could be buffered at node 2 until the 2→3 link becomes available. Similarly, the packet would wait at node 3 before it could be forwarded to the destination 4. Thus,

Encyclopedia of Wireless and Mobile Communications DOI: 10.1081/E-EWMC-120043800

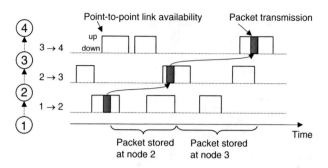

Fig. 1 Packet delivery from source 1 to destination 4 is possible without a contemporaneous end-to-end path, provided nodes buffer packets until a link to the next hop becomes available.

even though no instantaneous source-destination path is ever available for the time considered, a simple *store-and-forward* protocol could deliver packets successfully. If the node is meanwhile moving, this approach is called *store-and-carry*.

Scenarios where network partition is frequent and sporadic connectivity is the norm are very common for a wide range of challenged networks, operating in difficult or "exotic" environments. Initial interest in developing effective communication protocols for networks with intermittent connectivity has been shown by NASA and DARPA in funding the Interplanetary Internet Project (http://www.ipnsig.org/). Their goal is to "define the architecture and protocols necessary to permit interoperation of the Internet resident on Earth with other remotely located internets residing on other planets or spacecraft in transit." An interplanetary Internet encompasses ground fiber and satellite networks, earth–space links, and remote space-based wireless networks. It exhibits a wide range of link delays from the huge propagation distances, as well as sporadic link connectivity from planetary line-of-sight occlusion caused by the orbital motion of space-based communication assets.

Several other challenged networks with similar intermittent connectivity have attracted researchers' attention, such as mobile ad hoc networks (MANETs), wireless sensor networks, acoustic underwater networks and networks for Internet access in undeveloped areas. Wireless ad hoc networks for disaster recovery also suffer from sporadic connectivity, caused mainly by node mobility and by a communication channel with variable quality. A common name for such networks with intermittent connectivity is DTN.

DELAY AND DISRUPTION TOLERANT NETWORK ARCHITECTURE

The DTNRG (http://www.dtnrg.org) has been created as part of the IRTF to address the architectural and protocol design principles needed for interconnecting networks operating in environments where continuous end-to-end connectivity is sporadic. Members of the DTNRG were

instrumental in defining the initial DTN architecture. Kevin Fall was among the first to describe in Ref. 5 the main challenges facing current IP-based networks. He proposed a DTN communication architecture based on a message-oriented overlay implemented above the transport layer. Messages are aggregated in "bundles" that form the protocol data units in a virtual message-switching architecture. Devices that implement this bundle layer, called DTN nodes, employ persistent storage to buffer bundles whenever a proper contact is not available for forwarding. The bundle layer is responsible for implementing reliable delivery and optional end-to-end acknowledgment. In addition, the bundle layer also implements security services and a flexible naming scheme with late binding. For more details on the DTN architecture, the reader should consult Ref. 5. and the Internet draft by Vint Cerf et al.[6]

Since the bundle layer is implemented above several transport layers, it supports interconnecting heterogeneous networks using DTN gateways, similar to how Internet gateways route packets between networks with different data links.

Fall points out in Ref. 5 that routes in a DTN consist of a sequence of time-dependent communication opportunities, called contacts, during which messages are transferred from source towards the destination. Contacts are described by capacity, direction, the two endpoints, and temporal properties such as begin/end time, and latency. Routing in this network with time-varying edges involves finding the optimal contact path in both space and time, meaning that the forwarding decision must schedule transmissions considering temporal link availability in addition to the sequence of hops to the destination. This problem is exacerbated when contact duration and availability are nondeterministic. Contact types are classified in Refs. 5 and 6. Persistent contacts are those always available. On-Demand contacts require some action in order to instantiate. A scheduled contact is an agreement to establish a contact at a particular time, for a particular duration. Opportunistic contacts present themselves unexpectedly. On-Demand contacts require some action in order to instantiate, but then function as persistent contacts until terminated. Predicted contacts are based on a history of previously observed contacts or some other information.

As DTN routing must operate on a time-varying multigraph, message forwarding requires scheduling in addition to next-hop selection. To optimize the network performance, such as delivery rate or latency, DTN routing must select the right contact defined by a next-hop and a transmission time. If a contact is not known when a message is received from the upper layer, the bundle layer will buffer it until a proper contact occurs or until the message is dropped.

In conditions of a DTN with sporadic contact opportunities, the main objective of routing is to maximize the probability of delivery at the destination while minimizing the end-to-end delay. The forwarding decision is more effective when it has better information on the current state of the topology and on its future evolution. At one end of the spectrum is *deterministic DTN routing*, where the current topology is known and future changes can be predicted. With deterministic routing, message forwarding can be scheduled such that network performance is optimal and resource utilization is reduced by using unicast forwarding. At the other end of the spectrum, nodes know very little or nothing about the future evolution of the topology, and node movement is random or unknown. In this case, *stochastic DTN routing* forwards messages randomly hop-by-hop with the expectation of *eventual delivery*, but with no guarantees. In between, there are routing mechanisms that may predict contacts using prior network state information, or that adjust the trajectory of mobile nodes to serve as message ferries. Stochastic routing techniques rely more on replicating messages and controlled flooding for improving delivery rate, trading off resource utilization against improved routing performance in absence of accurate current and future network state.

The next section describes the principles of operation of representative deterministic and stochastic DTN routing mechanisms.

DETERMINISTIC ROUTING TECHNIQUES

Deterministic routing techniques for networks with intermittent connectivity assume that local or global information on how the network topology evolves in time are available to a certain degree. In general, deterministic techniques are based on formulating models for time-dependent graphs and finding a space–time shortest path in DTNs by converting the routing problem to classic graph theory or by using optimization techniques for end-to-end delivery metrics. Deterministic routing protocols use single-copy unicast for messages in transit and provide good performance with less resource usage than stochastic routing techniques. Deterministic routing mechanisms are appropriate only for scenarios where networks exhibit predictable topologies. This is true in applications where node trajectory is coordinated or can be predicted with accuracy, as in interplanetary networking. Major problems facing deterministic routing protocols remain the distribution of network state and mobility information under sporadic connectivity, long delays, and sparse resources.

Jain et al. present in Ref. 7 a deterministic routing framework that takes advantage of increasing levels of information on topology and traffic demand when such information is predictable. A DTN multigraph is defined where vertices represent the DTN nodes and edges describe the time-varying link capacity between nodes. It is

called a multigraph because there may exist multiple directed links between two nodes. Fig. 2A illustrates a DTN scenario with 3 nodes connected by slow dialup links, infrequent and fast satellite links, and a bundle courier riding public transportation, who is capable of delivering a large number of messages with latency measured in hours. The DTN multigraph is shown in Fig. 2B. Fig. 2C illustrates the time-varying edge capacity $c(e_{13}^{sat}, t)$ and $c(e_{13}^{dialup}, t)$ for the four directed edges connecting node 1–3.

The routing objective is to minimize the end-to-end delay. Reducing the message transit times in the network also reduces contention for limited resources, such as buffer space and transmission time.

Four *knowledge oracles* are defined: contacts summary oracle (for aggregate or summary contact statistics), contact oracle (for the time-varying contact multigraph), queuing oracle (for instantaneous queue state) and the traffic demand oracle (for present and future messages injected in the network). The authors adapt the Dijkstra shortest path algorithm to support time-varying edge weights defined by the oracles available and propose six algorithms for finding the optimal contact path.

The first two algorithms from Ref. 7 assume time-invariant edge weights. The first contact (FC) algorithm is a zero-knowledge approach that chooses a random edge to forward a message among the currently available contacts. If no contact is available, the message will be forwarded on the first edge that comes up. The minimum expected delay

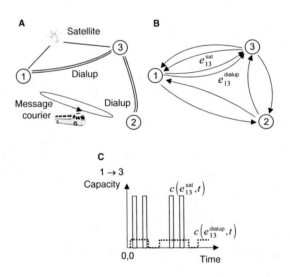

Fig. 2 (A) DTNS topology with time-varying link capacity. Between nodes 1 and 3 the LEO satellite link provides higher throughput several times per day for only $10'$ at a time. The dialup lines are up all during low-fare times. A courier rides the bus and carries message bundles on a high capacity flash memory between nodes 1 and 2, (B) the resulting DTN multigraph, with multiple directional edges between nodes 1 and 3,
(C) comparison of the time-varying capacity for directional edges $1 \rightarrow 3$.

(MED) algorithm applies the Dijkstra algorithm where the edge weight is time-invariant and is determined by the sum of the average waiting time (from the contacts summary oracle), propagation delay and transmission delay. MED ignores congestions and does not re-compute routes for messages in transit.

The next four proposed partial-knowledge algorithms work with a time-varying edge cost, defined as the sum of the waiting, transmission, and propagation delays. The waiting delay includes the time waiting for a contact and the queuing delay. The cost for an edge e at time t is defined for a message of size m and for a root node s: $w'(e,t,m,s) = t'(e,t,m,s) - t + d(e,t)$, where d is the propagation delay, and t' is the earliest time the message for which the route is computed completes transmission. t' is the earliest time the accumulated contact volume $\int_{x=t}^{t''} c(e,t)dt$ exceeds the total queued data. t' includes queuing delay and the time waiting for the corresponding contact:

$$t'(e,t,m,s) = \min\left\{ t'' | \int_{x=t}^{t''} c(e,t)dx \geqslant (m + Q(e,t,s)) \right\}$$

The edge capacity function $c(e,t)$ and the propagation delay d are predicted by the contact oracle. The parameter $Q(e,t,s)$ is the queue size for edge e predicted by node s at time t. The earliest delivery (ED) algorithm uses only the contact oracle and ignores queue occupancy: $Q(e,t,s) \equiv 0$. ED is prone to message loss due to buffer constraints, as the route for a message is not recalculated while the message is in transit.

The ED with local queuing algorithm (EDLQ) defines $Q(e,t,s)$ to be equal to the local queue size at node s, and 0 for all other edges. EDLQ routes around congestion for the first hop and ignores queue occupancy at subsequent hops. Therefore, this algorithm must re-compute the route at every hop. Cycles are avoided by using path vectors. Still, EDLQ is prone to message loss due to lack of available buffer space at reception. The ED with all queues

(EDAQ) algorithm uses the contacts and the queuing oracles. $Q(e,t,s)$ predicts the correct queue space for all edges at all times. In EDAQ, routes are not recomputed for messages in transit since the initial route predicts accurately all delays. EDAQ works only if capacity is reserved for each message along all contact edges. In practice, EDAQ is very difficult to implement in most DTNs with low connectivity, as it requires global and accurate distribution of queuing state. Limited connectivity also severely limits practical implementations of edge capacity reservations.

Simulations results indicate, as expected, that algorithms that use the knowledge oracles (ED, EDLQ, EDAQ) outperform the simpler MED and FC algorithms in terms of latency and delivery ratio. The more constrained the network resources are, the better the performance is for the algorithms that are more informed (i.e., use more oracles). A promising result is that routing with EDLQ (using only local queuing information) performs close to the EDAQ algorithm. This means that similar network performance can be achieved without expensive queue state dissemination and capacity reservations.

In practice, contacts may be deterministically predictable only for a finite time horizon in the future, as trajectories and mission objectives may later change. Merugu et al. propose in Ref. 8 a deterministic routing framework where a space–time graph is built from predicted contact information. It starts with a time-varying link function $L_{ij}(t)$ defined as 1 when the link between two nodes is available and 0 otherwise. This function is defined for time t in a time interval $[0,T]$, where T is the time horizon and the time is discretized in units τ. The space–time graph is built in $\lfloor T/\tau \rfloor$ layers, where the network nodes are replicated at each layer for each time unit τ. Each layer has a copy of each network node. A column of these vertices maps to a single network node. Fig. 3 illustrates the construction of the space–time graph for a simple two-node network where link A→B is available at times 1τ and 3τ for the duration of one time unit. A temporal link in the space–time graph connects graph vertices from the same column at successive time intervals. When it is traversed it

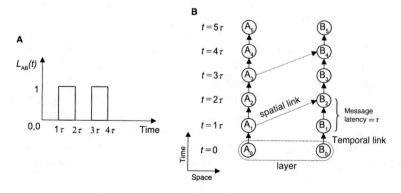

Fig. 3 Construction of the time-space graph. (A) A 2-node topology with time-varying link function $L_{AB}(t)$; (B) the resulting time-space graph organized in node layers (one per time unit), temporal links—connecting the same DTN node across consecutive layers (buffering)—and spatial links, representing message forwarding between two different nodes.

indicates that the message is buffered. A spatial link connects two vertices from different columns, representing message forwarding. Forwarding delay is modeled by the number of layers traversed by a spatial link. In Fig. 3, the message delay is equal to the time unit.

Least cost routing in this DTN becomes equivalent to finding the least cost (shortest) path from the source space–time node (column: layer) associated with the message arrival time to a vertex from the column corresponding to the destination DTN node. The end-to-end latency for a message becomes equal to the length of the path traversed in the space–time graph. The routing problem is solved using the Floyd-Warshall all-pairs shortest paths algorithm, modified to account for the particular characteristics of the space–time graph. Multiple message sizes are supported by a path coloring scheme.

One issue with this approach is that time discretization increases the algorithm complexity by a factor of T, the size of the time horizon. This space–time routing approach is similar to the ED partial-knowledge algorithm from Jain et al.[7] in the way it handles queuing delays with route computation at each hop. Cycles are avoided by verifying the path vector from the message header when computing the next hop.

Gnawali et al. propose in Ref. 9 ASCoT, a dynamic routing mechanism for space networks and the positional link state routing protocol (PLS) to implement position-based routing that enables the prediction of trajectories of satellites and other space assets. Link state updates with predicted contacts and their link performance are disseminated in advance in the network through reliable flooding. Nodes execute a modified Dijkstra algorithm to re-compute routing tables when link state updates are received. To support proximity routing for space assets in close formation, the authors propose a data-centric approach similar to directed diffusion.[10]

Note that in deterministic routing techniques using shortest path algorithms routing tables and forwarding schedules are recomputed whenever the contact graph state has changed, and selection of the next contact is done for a message at each hop along the path, as opposed to source routing. Thus, loops become possible as nodes may use outdated topology information. Cycles are avoided with path vectors. Deterministic DTN routing protocols are effective for a limited range of applications, where the contact schedule can be accurately modeled and predicted. Otherwise it is necessary to frequently disseminate nodes' state throughout the network. In networks with constrained capacity or limited connectivity this becomes very expensive and difficult to implement without an out-of-band broadcast channel. When contacts cannot be accurately predicted, routing must consider stochastic mechanisms that can only hint to predilection for future contacts based on historic information of past experience.

PASSIVE STOCHASTIC ROUTING TECHNIQUES

Stochastic routing techniques can be passive or active, depending on whether node mission is changed in order to support message relay. *Passive routing* techniques do not interfere with node mission and do not change the node trajectory to adapt to traffic demands. Passive routing techniques rely in general on flooding multiple copies of the same message with the objective of *eventual delivery*. In contrast, *active routing* techniques coordinate the mission (trajectory) of some nodes to improve capacity with their store-and-carry capability.

In general, passive routing techniques tradeoff delivery performance against resource utilization. By sending multiple copies of the same message on multiple contact paths, the delivery probability increases and the delay drops at the cost of additional buffer occupancy during message ferrying (MF) and higher link capacity usage during contacts. This approach is appropriate when nothing or very little is known about mobility patterns.

We present first two passive stochastic routing protocols, *Epidemic Routing* and *Spray and Wait*, that do not need any information on the network state. For other routing protocols, nodes can memorize contact history and use it to make more informed forwarding decisions. The section then continues with several passive routing protocols that operate with contact estimation.

Vahdat and Becker propose in Ref. 11 the *Epidemic Routing* protocol for message delivery in a mostly disconnected network with mobile nodes. Epidemic routing implements flooding in a DTN and got its name from a technique for message forwarding that emulates how a disease spreads through direct contact in a population during an epidemic. Even when just one individual of an entire population is initially infected, if the disease is highly contagious and contacts are frequent, over time it will spread exponentially and reach the entire population with a high probability. In epidemic routing, the disease that spreads is a message that must reach one or more destinations.

Each node maintains a *summary vector* with IDs of messages it has already received. When two nodes initiate a contact they first exchange their summary vectors in the *anti-entropy session*. Comparing message IDs, each node decides what messages it has not already received that needs to be pulled from the other node. The second phase of a contact consists of nodes exchanging messages. Messages have a time-to-live (TTL) field that limits the number of hops (contacts) they can pass through. Messages with TTL = 1 are forwarded only to the destination. The main issue with epidemic routing is that messages are flooded in the whole network to reach just one destination. This creates contention for buffer space and transmission time. An approach to mitigate buffer space contention is for nodes to reserve a fraction of their storage for locally originated messages. Even so, older messages in buffers will be dropped when new messages are received, reducing the delivery probability for destination nodes that have a low contact rate. An attempt to reduce resource waste is proposed that uses delivery confirmation (ACK) messages that are flooded starting from the destination and piggybacked with regular messages. Whenever a node receives an ACK it purges the acknowledged message from its buffer, if it is still present.

Epidemic routing uses node movement to spread messages during contacts. With large buffers, long contacts or a low network load, epidemic routing is very effective and provides minimum delay and high success rate, as messages reach the destination on multiple paths. End-to-end delay depends heavily on nodes' contact rate (infection rate), which is in turn affected by the communication range and node speed. Tuning message TTL and buffer allocation allows an epidemic routing implementation to trade off message latency and delivery ratio. In scenarios with a high message load, the increased contention from forwarding mostly redundant messages reduces the protocol performance. Epidemic routing is relatively simple to implement and is used in the DTN research literature as a benchmark for performance evaluation.

An approach to reduce the wasteful flooding of redundant messages in a DTN is presented by Spyropoulos et al. in Ref. 12. A multi-copy, zero-knowledge routing protocol called Spray and Wait is introduced. Similar to epidemic routing, this protocol forwards message copies to nodes met randomly during contacts in a mobile network. The main difference from epidemic routing is that Spray and Wait limits the total number of disseminated copies of the same message to a constant number L. In the spray phase, for every message originated by a source, L copies are forwarded by the source and other nodes receiving the message to a total of L distinct relays. In the wait phase all L nodes storing a copy of the message perform *direct transmission*. Direct transmission[13] is a single-copy routing technique in DTN where the message is forwarded by the current node only directly to the destination node. Direct transmission has been used for wildlife tracking applications and has minimal overhead, but suffers from unbounded delay as there is no guarantee the source will ever have contact with the destination node.

Spray and Wait initially spreads L copies of a message in epidemic fashion in order to increase the probability that at least one relay node would have a direct contact with the destination node. With a simple *Source Spray and Wait* heuristic, the source node forwards all L copies to the first L nodes encountered.

The optimal forwarding policy when nodes move randomly with identical and independent probability distribution (i.i.d.) is called *Binary Spray and Wait*. With this approach, the source node begins with L copies for each message. When a source or relay node A with $n > 1$ copies has contact with another node B that has no copies, A will hand over to $B \lfloor n/2 \rfloor$ copies and will hold on to $\lceil n/2 \rceil$ copies. When $n = 1$ a node will revert to direct transmission, meaning it will wait for a direct contact with the destination node. A message will be physically stored and transmitted just once even when a transfer may virtually involve multiple copies. Each message has a header field indicating the number of copies.

The paths followed by copies of a message can be represented by a binary tree rooted in the source node. Edges in the tree are formed by the transfer contacts. The more that nodes have multiple copies to distribute, the less the expected end-to-end delay will be. With the number of tree nodes fixed to $2^{1+\log L} - 1$, a balanced binary tree has the maximum number of internal nodes and also the maximum number of nodes at every level. Therefore the binary heuristic has the least expected delivery latency in networks with random i.i.d. random mobility. An interesting property of this routing protocol is that as the network node count M increases, the minimum fraction L/M necessary to achieve the same performance relative to the optimal path decreases. This property makes the Spray and Wait approach very scalable. At higher loads it performs much better than epidemic routing, since the limit L of maximum transmissions reduces contention on queue space and transmission time.

Some passive DTN routing protocols use *delivery estimation* to determine a metric for contacts relative to successful delivery, such as delivery probability or delay. Some of these protocols can forgo flooding and deliver single-copy messages by being selective with contact scheduling. The advantage is that considerably less memory, bandwidth, and energy are wasted on end-to-end message delivery. The drawbacks are that nodes must keep track of other nodes' movements and contacts, and that network-wide dissemination of this information imposes additional overhead in a network i.e., already constrained.

A representative routing protocol for DTNs that uses delivery estimation is PROPHET, a Probabilistic ROuting Protocol using History of Encounters and Transitivity, proposed by Lindgren et al. in Ref. 14. PROPHET works on the realistic premise that node mobility is not truly random. The authors assume that nodes in a DTN tend to visit some locations more often than others and that node pairs that have had repeated contacts in the past are more likely to have contacts in the future. A probabilistic metric called delivery predictability, $P(A, B)$, estimates the probability that node A will be able to deliver a message to node B. The delivery predictability vectors are maintained at each node A for every possible destination B.

At the beginning of a contact, the two nodes (A and B) exchange the summary vectors (like in epidemic routing) and also the delivery predictability vectors. Node A then updates its own delivery predictability vector using the new information from B, after which it selects and transfers messages from B for which it has a higher delivery probability than B. The delivery probability is updated during a contact so that node pairs that meet more often have a higher value:

$$P(A,B) = P(A,B)_{\text{old}} + (1 - P(A,B)_{\text{old}})P_{\text{init}}$$

P_{init} is a initialization constant between 0 and 1. For nodes that have not met for a longer time, their delivery probability should be reduced. The delivery probability therefore ages with exponential decay: $P(A,B) = P(A,B)_{\text{old}}\gamma^k$ where $\gamma \in (0,1)$ is the aging constant and k is the length of the time interval since the previous aging. In addition, the delivery predictability has a transitive property that encodes the assumption that if nodes A and B have frequent contacts and nodes B and C have frequent contacts, then node A has a good chance to forward messages intended for node C. After exchanging delivery predictability vectors at the beginning of a contact, nodes A and B update their values for every other node C, using $\beta \in (0,1)$, a scaling constant that controls the impact of transitivity on delivery probability:

$$P(A,C) = P(A,C)_{\text{old}} + \beta(1 - P(A,C)_{\text{old}})P(A,B)P(B,C)$$

When node A begins a contact with node B, it decides to forward a message to B with destination C if $P(B,C) > P(A,C)$. Node A will also keep a copy in its buffer. The buffer has a first in, first out (FIFO) policy for dropping old messages when new messages are received.

Transitive reinforcement of delivery probabilities based on prior contacts make this protocol perform better in simulations than epidemic routing, since it reduces the contention for buffer space and transmission time. Related techniques for delivery probability estimation based on prior contact history are used in meet and visit (MV) routing[15] and Zebranet.[16]

A novel approach for delivery estimation is the use of a virtual Euclidean mobility pattern space, called *Moby-Space*, proposed by Leguay et al.[17] The idea is that messages in a DTN should be forwarded to another node if this next hop has a mobility pattern similar to the destination node. This concept was adapted from the content addressable network peer-to-peer overlay architecture.[18]

Citing studies of user mobility in various scenarios where users tend to follow similar trajectories, the authors suggest a model where the node movement follows a *power law*. This means that the probability that a node is at a location i from a set of N locations is $P(i) = K(1/d)^{n_i}$, where n_i is the preference index for location i, $d > 1$ is the exponent of the power law, and K is a normalization constant. When d is high, nodes tend to visit far more

often in far fewer locations. When $d \to 1$ nodes have similar preference for all locations. The mobility pattern space has a dimension for each possible location, and the coordinate value a node's point in this space (*MobyPoint*) in dimension i is equal to the probability $P(i)$. This model assumes that dwell time at each location is uniformly distributed in a narrow interval.

Two points in MobySpace that have a small distance between them are more likely to have a contact than two nodes situated further apart. With this insight, the forwarding algorithm simply decides to forward a message during a contact to a node that has a shorter distance to the message destination. Thus, the message takes a path through the MobySpace that brings it closer and closer to the destination. Several distance functions have been proposed to measure similarity in nodes' mobility patterns. The Euclidean and the cosine separation distance provide lower delays in simulations.

An example of routing using the mobility pattern space is shown in Fig. 4. The network has three reference locations 1,2,3, visited by at least four nodes A, B, C, D. Each node knows the MobySpace coordinates for all nodes. Node A has a message to send to node D. When it encounters node B it forwards the message as $d(A,D) > d(B,D)$. Through successive contacts, the message eventually arrives at node D. Note that all points in a MobySpace with N dimensions are located on an $(N-1)$ dimensional hyperplane defined by $\sum P(i) = 1$.

The MobySpace approach is effective only if nodes exhibit stable mobility patterns. It also fails if a message reaches a local maximum where the current node has a similar mobility pattern with the destination, but a direct contact with the destination is rare due to trajectory synchronization. Such a case is possible in a DTN where nodes are public transportation buses. While the buses on a line follow the same path and visit the same stations, two buses may get within radio range only at night when they park in the garage. Mobility pattern similarity does not guarantee frequent contacts. A possible solution to

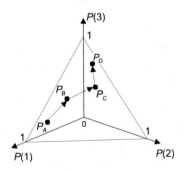

Fig. 4 Routing in MobySpace (3-D for locations 1, 2 and 3) from node A to destination node D. MobyPoints $P_A, \ldots, P_D$ are situated in the plane $P(1) + P(2) + P(3) = 1$. The message follows a contact route with points successively closer to P_D.

this problem is to use the probability (or frequency) of direct contacts with the other nodes as dimensions in the MobySpace. Another approach to deal with the temporal variability of mobility patterns is to supplement MobySpace with conversion of the spatial visit patterns to the frequency domain, representing the dominant visitation frequency and the phase. Other issues with MobySpace are effective dissemination of location probabilities for all nodes in a constrained DTN and high convergence time.

ACTIVE STOCHASTIC ROUTING TECHNIQUES

Active routing techniques rely on controlling the trajectory of some nodes to improve delivery performance with store-and-carry. Mobile nodes pick up messages and ferry them for a distance before another contact brings them closer to the destination. Active routing techniques provide improved flexibility and lower delays with the additional cost of increased protocol and system complexity. Active DTN routing techniques are frequently implemented as optimization problems. In most cases the objective is to maximize network capacity, reduce message latency, and reduce message loss while facing resource constraints. Applications where mobile nodes are controlled to ferry messages can be used in multiple domains. In disaster recovery, mobile nodes [helicopters, unmanned aerial vehicles (UAVs), or personnel] equipped with communication devices capable of storing a large number of messages can be commanded to follow a trajectory that interconnects disconnected user partitions. Similarly in wireless sensor networks, mobile nodes can traverse the sensing area and pickup/deliver measurements, queries and event messages. We review in the remainder of this entry two DTN routing mechanisms that employ active node trajectory control.

Burns et al. introduce in Ref. 15 the MV routing scheme, where node trajectory is adjusted according to traffic demands by autonomous agents. MV aims to improve four performance metrics with a multi-objective control approach. On each controlled mobile node, separate controllers for total bandwidth, unique bandwidth, delivery latency and peer latency, respectively, are combined through multi-objective control techniques such as null space or subsumption. Each controller adjusts the node trajectory such that its own objective is maximized.

- The total bandwidth controller (φ_T) selects the DTN that has the greatest number of unseen messages amortized by the trip time. This prevents making long trips without a matching load of new messages.
- The unique bandwidth controller (φ_U) selects a node that has the largest number of new messages not yet forwarded to any other nodes.

- The delivery latency controller (φ_D) picks the node with the highest average delivery time.
- The peer latency controller (φ_P) selects the node least recently visited by an agent such that. the traveling time to visit this node does not increase the overall peer latency metric.

The four controllers can be composed to optimize agent mission across performance metrics. To do that, controllers are first ordered according to their importance. With the *null space* approach, an agent's *subordinate* controller actions can be optimized without affecting the performance of the dominant controller's actions. To increase the optimal solution space of the dominant controller, a minimum performance threshold method is used. The actions controlled by the subordinate controller are acceptable as long as the dominant controller's performance is above this threshold. A different controller composition approach uses a *subsumption* approach. A controller with a higher priority computes the action space for achieving a specified performance level for its metric. Within this space, the immediate lower priority controller finds its own optimal without changing the performance of any higher priority controllers.

MV implements an epidemic dissemination protocol for the network state necessary for the four controllers. Node information is tagged with a timestamp and flooded during contacts. Simulation results have shown that this approach is sufficient for low bandwidth and latency estimation error, but not enough to estimate correctly "last visit" times and location information. MV routing could be further improved with additional off-line or out-of-band network states. Another limitation of this approach is the key assumption that contact bandwidth is unlimited.

Zhao et al. describe in Ref. 19 a proactive MF routing method with 2-hop forwarding and a single ferry. A message ferry is a special mobile node tasked with improving transmission capacity in a mobile DTN. The authors present two methods for MF in sparse DTNs. In the node-initiated MF (NIMF) scheme the ferry follows a specific trajectory. Nodes that need to send messages adjust their trajectory periodically to meet the ferry for message up/download. The objective of the NIMF node trajectory control mechanism is to minimize message loss due to TTL expiration and buffer limits, while reducing the negative impact of trajectory changes on node mission goals. The first objective can be expressed knowing message generation/drop rates and by estimating contact times. The second objective can be modeled as the work time percentage (WTP). The WTP represents the fraction of time a node performs its main task. It is assumed that during a detour to meet a ferry, a node does not contribute to its main task. The NIMF controller allows node trajectory changes only when the WTP is above a minimum threshold. In the ferry-initiated MF (FIMF) scheme, the ferry responds to requests for contacts broadcast by nodes on a long-range radio channel. The authors show that the ferry trajectory control problem is NP-hard and propose a greedy *nearest neighbor* heuristic and a traffic-aware heuristic that optimizes locally both location and message drop rates. The same authors extend in Ref. 20 their ferry-based DTN routing method for coordinating multiple message ferries such that traffic demands are met and delay is minimized. Approximations are provided for single route and multi route trajectory control. Ferry replacement algorithms for fault-tolerant delivery are further explored in Ref. 21.

CONCLUSIONS

This entry presented an overview of some of the challenges facing routing in networks with intermittent connectivity, and described several routing solutions that used deterministic contact estimation, passive stochastic and active stochastic techniques. DTNs are a new area of wireless ad hoc networking that shows great potential in many important applications. Routing and end-to-end message delivery in DTNs is possibly the most difficult problem in an environment where network resources are very limited and connectivity is scarce. The connectivity limitation affects the ability of distributing network-wide node and link information that could be used to optimize network operations. When contacts cannot be deterministically predicted, routing algorithms must rely on probabilistic methods that estimate future contacts with limited accuracy and on multi-copy forwarding that further strains the reduced network resources. In sparse DTNs, tasking dedicated nodes with ferrying messages improves the overall network capacity and reduces the delay.

Future research in DTN routing may address several remaining problems, such as effective integration of DTN techniques with mostly-connected MANET routing protocols; quality of service (QoS) and policy-based routing; statistical QoS guarantees and routing with probabilistic contact information.

REFERENCES

1. Perkins, C. Ad Hoc on Demand Distance Vector (AODV) Routing, 1997. Internet Draft.

2. Johnson, D.B.; Maltz, D.A. Dynamic source routing in ad hoc wireless networks. In *Mobile Computing,* Imielinski, Korth Eds.; Kluwer Academic Publishers, 1996; Vol. 353.

3. Clausen, T.; Jacquet, P. Optimized Link State Routing Protocol (OLSR), 2003 RFC 3626 IETF Network Working Group.

4. Zhang, Z. Routing in intermittently connected mobile ad hoc networks and delay tolerant networks: overview and challenges. IEEE Commun. Surv. Tutorials **2006**, *8* (1), 24–37.

5. Fall, K. A delay-tolerant network architecture for challenged internets. 2003 Conference on Applications, Technologies, Architectures, and Protocols for Computer Communications (SIGCOMM 2003), Karlsruhe, Germany, Aug 25–29, 2003; ACM Press: New York, NY, 2003; 27–34.

6. Cerf, V.; Burleigh, S.; Hooke, A.; Torgerson, L.; Durst, R.; Scott, K.; Fall, K.; Weiss, H. Delay Tolerant Network Architecture, draft-irtf-dtnrg-arch-05.txt 2004.

7. Jain, S.; Fall, K.; Patra, R. Routing in a delay tolerant network. 2004 Conference on Applications, Technologies, Architectures, and Protocols for Computer Communications (SIGCOMM 2004), Portland, OR, Aug 30–Sept 3, 2004; ACM Press: New York, NY, 2004; Vol. 34, 145–158.

8. Merugu, S.; Ammar, M.; Zegura, E. Routing in space and time in networks with predictable mobility. Technical report, Georgia Institute of Technology, GIT-CC-04-7 2004.

9. Gnawali, O.; Polyakov, M.; Bose, P.; Govindan R. Data centric, position-based routing in space networks. 26th IEEE Aerospace Conference, Big Sky, MT, Mar 5–12, 2005.

10. Intanagonwiwat, C.; Govindan, R.; Estrin, D. Directed diffusion: a scalable and robust communication paradigm for sensor networks. 6th Annual International Conference on Mobile Computing and Networking (MobiCom 2000), Boston, MA, Aug 6–11, 2000; ACM Press: 2000.

11. Vahdat, A.; Becker, D. Epidemic routing for partially connected ad hoc networks. Technical report, Duke University, 2000. CS-200006.

12. Spyropoulos, T.; Psounis, K.; Raghavendra, C.S. Spray and wait: an efficient routing scheme for intermittently connected mobile networks. ACM SIGCOMM Workshop on Delay-tolerant Networking, Philadelphia, PA, Aug 22–26, 2005; ACM Press: New York, NY, 2005; 252–259.

13. Shah, R.C.; Roy, S.; Jain, S.; Brunette, W. Data mules: modeling a three-tier architecture for sparse sensor networks. IEEE International Workshop on Sensor Network Protocols and Applications, Anchorage, AK, May 11, 2003; 30–41.

14. Lindgren, A.; Doria, A.; Schelén, O. Probabilistic routing in intermittently connected networks. SIGMOBILE Mob. Comput. Commun. Rev. **2003**, *7* (3), 19–20.

15. Burns, B.; Brock, O.; Levine, B.N. Mv routing and capacity building in disruption tolerant networks. IEEE INFOCOM 2005, Miami, FL, Mar 13–17, 2005; Vol. 1, 398–408.

16. Juang, P.; Oki, H.; Wang, Y.; Martonosi, M.; Peh, L.S.; Rubenstein, D. Energy-efficient computing for wildlife tracking: design tradeoffs and early experiences with zebranet. 10th International Conference on Architectural Support for Programming Languages and Operating systems (ASPLOS 2002), San Jose, CA, Oct 5–9, 2002; ACM Press: New York, NY, 2002; 96–107.

Wireless
Multi—Networks

17. Leguay, J.; Friedman, T.; Conan, V. Evaluating mobility pattern space routing for DTNs. INFOCOM 2006, Barcelona, Spain, Apr 23–29, 2006.

18. Ratnasamy, S.; Francis, P.; Handley, M.; Karp, R.; Shenker. S. A scalable content addressable network. SIGCOMM 2001, San Diego, CA, Aug 27–31, 2001; ACM Press: New York, NY, 2001.

19. Zhao, W.; Ammar, M.; Zegura, E. A message ferrying approach for data delivery in sparse mobile ad hoc networks. 5th ACM International Symposium on Mobile Ad-Hoc Networking and Computing (MOBIHOC 2004), Tokyo, Japan, May 24–26, 2004; ACM Press: New York, NY, 2004; 187–198.

20. Zhao, W.; Ammar, M.; Zegura, E. Controlling the mobility of multiple data transport ferries in a delay-tolerant network. INFOCOM 2005. 24th Annual Joint Conference of the IEEE Computer and Communications Societies. IEEE, Miami, FL, Mar 13–17, 2005; Vol. 2, 1407–1418.

21. Yang, J.; Chen, Y.; Ammar, M.; Lee, C. Ferry replacement protocols in sparse MANET message ferrying systems. IEEE WCNC, New Orleans, LA, Mar 13–17, 2005.

BIBLIOGRAPHY

1. The Delay-Tolerant Networking Research Group, http://www.dtnrg.org.

2. The Interplanetary Internet Project, http://www.ipnsig.org/.

Wireless Networks: TCP

Sumitha Bhandarkar
Senior Staff Software Engineer, Mobile Devices Technology Office, Motorola Inc., Austin, Texas, U.S.A.

Narasimha Reddy
Department of Electrical and Computer Engineering, Texas A&M University, College Station, Texas, U.S.A.

Abstract

TCP is the predominant transport protocol used on the Internet. This entry inspects the behavior of TCP in wireless networks.

INTRODUCTION

TCP/IP is the primary stack of network protocols on which the Internet operates. Most of the network applications are designed to use the services provided by the TCP/IP family of protocols. The TCP/IP stack has historically offered two main options for use at the transport layer, namely TCP and User Datagram Protocol (UDP), while currently two other options [Stream Control Transmission Protocol (SCTP) and Datagram Congestion Control Protocol (DCCP)] are under investigation by the IETF. TCP[1], is by far the most predominantly used transport protocol on the Internet. Hence, it is desirable to use TCP in wireless networks that connect to the Internet in order to maintain transparency of the medium of the network. However, TCP was designed to work best in wired networks and some issues arise when it is used in wireless networks. In this entry, we provide an overview of TCP and inspect some of the issues and proposed solutions for TCP in wireless networks.

This entry in organized as follows. In the "Background" section, we present a short refresher of TCP congestion control algorithms, followed by an overview of the issues faced by TCP in wireless networks. In subsequent sections, we present an overview of the proposed solutions to each of the issues. In the "Impact of Wireless Channel Errors on TCP" section we discuss the solutions for improving TCP performance in the face of wireless channel errors. The "Impact of Mobility on TCP" section provides an overview of the solutions for addressing TCP performance problems when dealing with mobility in last-hop wireless networks. The "Performance of TCP in Multi-hop Independent Wireless Networks" section is dedicated to the different problems and related solutions for TCP performance in multi-hop wireless networks. We conclude the entry with the "Conclusion" section.

BACKGROUND

Before delving into the issues that TCP faces in wireless networks, we stop for a moment to refresh the basics of the congestion control algorithms of TCP. TCP provides a connection-oriented, reliable, byte-stream service that is both flow and congestion controlled to the upper layer (application layer), without expecting explicit feedback from the lower layers (IP layer and below). While TCP is a complicated protocol with several different algorithms for the different services it provides, the most relevant ones for this entry are those related to congestion control. Hence the focus of this brief TCP primer is on the congestion control algorithms of TCP. Other aspects of TCP relevant to a particular solution discussed in later sections are overviewed in those sections, as and when relevant.

TCP Basics

TCP is a window-based protocol and uses a variable called the *congestion window* to track the available bandwidth along the path of a connection. The congestion control functionality of TCP is provided by four main algorithms namely slowstart, congestion avoidance, fast re-transmit, and fast recovery in conjunction with several different timers. When a flow starts, it uses the slowstart algorithm which exponentially increases the window to quickly bring the flow to speed. In steady state, the flow mostly uses the congestion avoidance algorithm in conjunction with fast re-transmit/recovery. These algorithms implement the additive increase/multiplicative decrease of the congestion window. When no losses are observed, the congestion window is increased by one for the successful ACK of one window of packets. Upon a packet loss, the window is decreased to half its earlier value, to clear out the

Encyclopedia of Wireless and Mobile Communications DOI: 10.1081/E-EWMC-120043604

bottleneck link buffers. Fig. 1 shows how the congestion window of a typical flow changes. The sending rate of the flow is guided by the value of the congestion window and hence follows a similar trend. (The sending rate is determined by the minimum of (*cwnd, rwnd*), where *cwnd* is the congestion window calculated as discussed in this section, and *rwnd* is the window advertised by the receiver based on its buffer availability.)

TCP is a self-sufficient protocol, in the sense that the sender uses only the information provided by the receiver to determine a packet loss. No explicit feedback is expected from the routers. At the sender, every packet that is transmitted is given a unique monotonically increasing sequence number, which is embedded in the packet header. When the receiver receives a packet, it sends an ACK back to the sender. In the header of this packet, the receiver embeds the sequence number of the next packet it expects to receive. This will result in cumulatively acknowledging all the packets that have been received so far. Now suppose a packet is lost in transit. For all subsequent packets that the receiver receives, it repeatedly sends the sequence number of the packet it expects to receive next, which is the packet that has been lost in transit. Since these ACKs repeat the same information, they are called "duplicate acknowledgments" (or dupacks for short). The sender treats the receipt of three consecutive dupacks as an indication that the packet is lost and re-transmits the packet. In addition, it implicitly assumes that the cause for packet loss is network congestion and responds by reducing its sending window by half (as shown in Fig. 1).

In addition to the the above mechanism for determining congestion, to provide robustness against loss of multiple segments, TCP also uses timers for monitoring packets in transit. If no cumulative ACKs indicating successful receipt of the packet or dupacks indicating loss of the packet are received before the timer expires, then the sender assumes *severe* congestion on the network. This timeout, called the re-transmission timeout (RTO, for short), not only results in the re-transmission of the packet presumed to be lost but also re-sets the congestion window to 1, to relieve the *severe* congestion that prevented any form of ACKs from being received. The value of the RTO timer is generally set to max(*RTT* + 4 * *RTTVAR*, 1s) where *RTT* is the round trip delay for the connection and *RTTVAR* is the variance in the RTT. Including the RTT variance in RTO timer calculation reduces the possibility of spurious timeouts. TCP uses several other algorithms to provide many more services to upper layers, in addition to congestion control. An interested reader can find detailed discussion of TCP algorithms in[2,3]. The algorithms briefly discussed above are the core algorithms relevant to the discussion of the issues faced by TCP in wireless networks. Some of the solutions discussed in later entries may use other algorithms in TCP. Details of those algorithms will be discussed in the relevant sections.

Introduction to TCP Woes in Wireless Networks

The core TCP algorithms were designed to work well in wired medium where the channel losses are negligible and it can be safely assumed that the primary cause for packet loss is congestion. When this assumption is not true, as in the case of a wireless network where the transmission medium is highly unreliable, the implicit assumptions made by TCP will result in unnecessary window reductions causing significant degradation in performance. In addition, while the calculation of the RTO timer takes into account some variations in RTT, it is not robust to sudden large increase in delays brought on by host mobility or loss of a physical transmission medium, which are common occurrences in a wireless network. These result in the congestion window being re-set to 1 due to a RTO, and hence a significant performance reduction. In the "Impact of Wireless Channel Errors on TCP" and the "Impact of Mobility on TCP" sections, we will focus on these issues and some of the proposed solutions.

Wireless networks are generally configured in one of the two modes: *infrastructure* mode or *independent* mode. In the *infrastructure* mode, the mobile nodes communicate to the access point which connects to the rest of the network. Hence, there is one wireless hop between the mobile node and the access point. The connection from the access point to the destination may be over wired or wireless medium. The general practice, in most current deployments, is to take advantage of the existing infrastructure by connecting the access point to the wired Internet. Hence, in most general cases, for this mode of operation, only the "last hop" is over a wireless medium.

In the *independent* mode, the mobile nodes form a network by connecting with each other directly, without the need for associating with an access point. Some of the nodes participating in the independent network may be connected to the wired Internet, and may carry out the

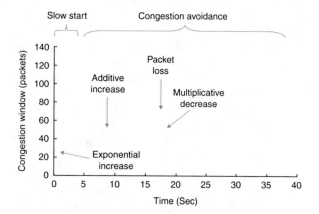

Fig. 1 Congestion window of TCP. Note that the sending rate looks similar to this when the sender is not limited by receiver advertised window.

function of a bridge or a gateway, when necessary. Such networks, in general, are characterized by multiple wireless hops between the source and the destination nodes. Recently, several studies have focused on the throughput of TCP in such multi-hop wireless networks. Several problems have been identified. First, due to the broadcast nature of the wireless medium, in a multi-hop wireless network, channel contention can occur between packets belonging to the *same* connection while traversing different hops within an interference range. Also, TCP data packets flowing in one direction may have to contend with the ACK packets flowing in the reverse direction. Packets of connections that have exclusively different paths can still cause contention to each other, if some of the hops are within each others, interference range. These problems result in increased losses and inefficiency in channel usage. Second, if the nodes within a independent network are mobile, they can result in significantly more complex issues than in an infrastructure wireless network. Node mobility in an independent network is not limited to the mobility of just the end nodes—every node in the path between the source and the destination could potentially be mobile. This can cause frequent changes in routes or cause periods of network partition, where no route exists between the source and the destination. This can result in severely degraded performance for TCP. Finally, it has been shown that there exists severe unfairness among competing TCP flows in a multi-hop wireless network. We dedicate the "Performance of TCP in Multi-hop Independent Wireless Networks" section to discussing TCP issues in multi-hop wireless networks and the proposed solutions for addressing them.

Historically, the issues due to channel errors and end-host mobility in a last-hop wireless network, discussed in the "Impact of Wireless Channel Errors on TCP" and the "Impact of Mobility on TCP" sections, were identified first. Most of these early studies were conducted in the context of infrastructure networks. While some solutions we discuss in these sections were developed specifically for the infrastructure networks, others can easily be applied to independent networks as well. In the "Performance of TCP in Multi-hop Independent Wireless Networks" section, we focus the discussion on those issues that are specific to multi-hop wireless networks, and discuss solutions corresponding to them.

IMPACT OF WIRELESS CHANNEL ERRORS ON TCP

Unlike transmission on a wired medium, transmission on wireless medium is significantly more noisy resulting in higher losses due to channel errors. Since a TCP sender cannot distinguish this loss from congestion, and responds similarly to both, i.e., by reducing the congestion window (and hence, the sending rate), the performance of a TCP

flow in wireless networks is significantly degraded. Several studies[4] have shown this.

Over the past few years, several solutions have been proposed to improve the performance of TCP over wireless networks. Depending on the approach taken, these solutions can be classified into the following broad categories: split-connection approaches, lower layer (IP, link layer or physical layer) schemes, explicit notification approaches, and modifications to TCP that retain end-to-end semantics. We review these solutions in this section.

Split-Connection Approaches

The split-connection approaches are based on the principle that the characteristics of the wired medium and the wireless medium are significantly different. If a connection were to span across both these mediums, then there would be several advantages to split such a connection at the point of intersection, say the wireless gateway, into two separate connections. As pointed out in Ref. 5, this approach would help:

1. use separate flow control and congestion control algorithms on the two connections. Each can be fine tuned based on the characteristics of that particular link.
2. for the wireless link, cross-layer interaction can be added between the link layer and the transport layer, so the transport layer protocol (and the applications above it) can be *link aware* and *location aware.*
3. Splitting the connection allows the usage of an asymmetric transport protocol between the mobile host and the intermediate wireless gateway, so that much of the functionality is shifted to the intermediate gateway, while the mobile node uses a simplified network stack.
4. Splitting the connections at an intermediate node, such as the wireless gateway, helps handle mobility-related issues (as discussed in next section) as well.

Fig. 2 shows an example of using split connections. The connection between the fixed node and the wireless gateway, which is over a wired medium would continue to use existing transport protocols that have been fine tuned for wired medium through years of research and experience. The wireless connection, on the other hand, between the wireless gateway and the wireless node, could use new experimental protocols that are more suitable for the unique nature of the wireless medium. This approach ensures that the network stack on all the fixed hosts will not need modifications to be able to communicate with wireless nodes.

In Indirect TCP (I-TCP)[5], a connection is established between the Mobile Support Router (MSR), which acts *on behalf* of the mobile host, and the fixed host using a standard TCP. A second connection between the MSR and the actual mobile host uses a wireless-aware TCP variant. When data is sent from the fixed host on the first

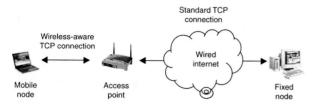

Fig. 2 Split-connection approach.

TCP connection, the MSR caches this data and sends ACK to the sender. It then forwards this data on the wireless connection to the mobile host using an I-TCP connection. If there are any transmission-related errors, they are recovered by re-transmissions only on the second connection between the MSR and the mobile node. The indirection created by splitting the connection, hides these transmission-related errors from the fixed sender, thus preventing window reductions and performance degradation.

In mobile-TCP[6], the authors take this concept of indirection further by proposing an asymmetrical design of transport layer protocol. Based on the observation that the link between the mobile host and the intermediate gateway is essentially a single-hop wireless link, the authors present a highly simplified TCP emulation scheme that moves much of the functionality from the mobile host to the intermediate gateway, while retaining TCP-like behavior. Due to the reduced complication of the protocol on the mobile node, the authors conjecture that the power consumption of the mobile unit can be reduced.

The authors of Mobile-End Transport Protocol (METP)[7], take the concept of simplifying the network stack on the mobile host further by proposing to eliminate the IP and TCP layers from the mobile hosts. The base station handles all the connections on behalf of the mobile node and performs all the IP and TCP operations. It multiplexes/de-multiplexes the data to determine the appropriate port number on the mobile host, and hence the communication between the base station and the mobile host resembles that between an application process and the transport layer protocol. By treating the mobile host as an extension of the base station, the irregularities of the wireless path are shielded from the fixed host. When the communication between the base station and the mobile node slows down due to wireless channel errors and related re-transmissions, the buffer usage at the base station remains high, and the receiver advertised window for the TCP connection between the base station and the stationary host is reduced to control the sender's sending rate.

Another split-connection approach was proposed in Ref. 8, where the authors introduce a session layer protocol at the mobile hosts and the intermediate base stations. This allows the choice of using different underlying transport protocols on the wireless connection. Two alternatives are discussed. In the first alternative called multiple TCP (MTCP), both the wired and wireless connections use

TCP. In the second alternative called Selective Repeat Protocol (SRP), TCP is used for the wired connections, while a selective repeat algorithm over UDP is used for the wireless connection. Through experimental evaluation, the authors show improved performance for both alternatives compared to unmodified TCP, and among the two alternatives, SRP yields slightly better throughput than MTCP.

Lower Layer Schemes

The second category of proposed schemes, namely the lower layer approaches, modify the layers situated below TCP (i.e., IP layer, link layer or the physical layer) in an attempt to hide the transmission errors from TCP. The proposed schemes may or may not be aware of TCP semantics. For instance, AsymmetrIc Reliable Mobile Access In Link-layer (AIRMAIL)[9] is a pure link-layer protocol that focuses on improving the reliability of link-layer transmissions. It uses well-known error recovery techniques such as Forward Error Correction (FEC) in conjunction with Automatic Repeat reQuest (ARQ) for providing improved performance in noisy environments. The authors of Ref. 10 argue that well-designed "pure" link-layer schemes can hide much of the transmission-related errors from higher layer protocols. They propose the use of link-layer re-transmissions that are persistent enough to eliminate non-congestion losses, while providing in-order delivery before the TCP RTO expires to reduce the duplication of efforts between the link-layer re-transmissions and TCP re-transmissions. In order to achieve this, they propose adaptive packet shrinking and adaptive error coding techniques.

SNOOP[11], on the other hand, is a TCP-aware solution that uses a combination of modifications to the network layer and link layer. When using SNOOP, the network layer of the base station caches all the unacknowledged data packets from the fixed host to the mobile host. Simultaneously, it monitors the ACKs from the mobile host to the fixed host. If it detects any packet losses, either by observing dupacks or the expiry of local timers, it looks for the lost packet in its cache. If the packet exists, then the dupacks are held back from further propagating to the fixed host and local link-layer re-transmission is invoked to deliver the packet to the mobile host. When data flows in the opposite direction, SNOOP uses the TCP selective acknowledgment option (SACK)[12] to trigger re-transmission from the mobile sender.

Another approach, Transport Unaware Link Improvement Protocol (TULIP)[13] provides service-aware reliability without being TCP-aware. Before passing the packet to the link layer, the network layer can request the type of service to be either reliable link-layer packet (RLP) or unreliable link-layer packet (ULP). By choosing to send TCP data packets using RLP and the TCP ACKs, UDP and link layer ACKs as ULP, the overhead of a providing link layer reliability is reduced. At the link layer, reliability is provided by using selective-repeat ARQ. When possible, link layer ACKs are piggybacked on TCP ACKs (or data packets) flowing in the reverse direction.

Explicit Notification Approaches

The performance degradation of TCP stems from the fact that the sender cannot distinguish the losses due to congestion from losses due to transmission errors. The schemes in the third category propose to address this problem by using explicit notification mechanisms to allow the sender to distinguish between the different types of losses and respond appropriately. The sender can then de-couple congestion control from packet re-transmission to recover the lost packets. These schemes require the receivers/network routers be able to distinguish the channel errors from congestion losses and be capable of marking the ACKs with appropriate notification. The senders then respond appropriately to the notification.

In the Explicit Loss Notification (ELN)[14] scheme, the base stations use a snooping agent to determine if the packet loss is due to transmission errors on the wireless channel. If so, then the base station marks the ELN bit in the header of the dupack sent by the receiver. When the sender receives dupacks, based on whether the ELN bit is set or not, it can determine whether or not to invoke the congestion control mechanism. Thus, using ELN, congestion control is de-coupled from packet re-transmission.

Alternately, in the Explicit Congestion Notification (ECN)[15] scheme, routers actively monitor their queues and as they start to fill up, they use ECN code points in the TCP header to indicate the onset of congestion. The receivers copy the ECN code point from the data packet to the ACK packet, to explicitly inform the sender of congestion build up. This removes the reliance of the sender on a packet loss as the sole indication of network congestion.

Modifications to TCP that Retain End-to-End Semantics

The pure end-to-end solutions aim to modify the TCP agent at either the sender or the receiver (or both) to improve the performance without sacrificing the end-to-end semantics and without requiring additional support from intermediate nodes.

The delayed dupack scheme[16] employs a mechanism similar to the SNOOP protocol at the TCP receiver, so that the link layer need not be TCP-aware. In this scheme, the third and subsequent dupacks are delayed for a bounded period of time, to allow the link layer time to recover the packet. If the packet is recovered within the delay period, say via link-layer mechanisms, then the dupacks are not sent. If not, then all the dupacks are released. A similar scheme for avoiding the wrongful triggering of congestion

control, but implemented at the TCP sender, is presented in TCP-Delayed Congestion Response (TCP-DCR)[17,18]. TCP-DCR argues that the wait of three dupacks used in TCP is a heuristic, and is not appropriate in todays networks which may contain wireless links or packet re-ordering. In order to allow the maximum time to locally recover any non-congestion-related losses, TCP-DCR proposes that the sender should wait for one round trip time after the receipt of a dupack before concluding that the loss is due to congestion. The wait of one round trip time allows the lower layers sufficient time to recover any transmission losses using local link-layer recovery mechanism, while ensuring that a TCP RTO is not triggered.

In the Wireless TCP (WTCP)[19] scheme, the authors specifically target the unique characteristics of a wireless WAN (WWAN). WWANs can be characterized by asymmetrical links that can result in ACK-bunching, which in turn can result in the transmission of a burst of packets by the sender. So, the authors propose using a rate-based control mechanism, instead of a window-based control mechanism. Instead of using packet loss or timeouts for determining the sending rate, WTCP uses the inter-packet delays for determining the channel rate and predicting the type of loss. The receiver uses the ratio of the inter-packet separation at the receiver and the sender to determine the sending rate. The computation is made at the receiver to protect against ACK losses and to eliminate the impact of delays in the reverse path.

TCP-Westwood[20] uses a slightly different approach for providing an end-to-end solution for improving TCP performance in wireless networks. The argument made in this entry is that the "blind" reduce by half algorithm used by TCP in response to a packet loss is inappropriate. This scheme proposes the use of bandwidth estimation by sampling and exponentially filtering the rate of ACKs, and using this as a guide to determine the appropriate window size after a loss event. In wireless network, loss of packets due to transmission errors does not impact the bandwidth estimation, and hence after a loss event, the sending rate is not drastically reduced, helping maintain better link utilization even in networks with high transmission losses.

IMPACT OF MOBILITY ON TCP

The use of wireless medium for communication offers unique advantages that are not possible with the wired networks—the end-host can now move about since it is not tethered by a wire line. While mobility within a wireless network is generally handled at the link layer or the network layer, it offers some challenges to the transport protocol as well. For instance, while moving from one subnet to another, link-layer associations and possibly, routing needs to be re-established. During this time, there may be periods when communication is not possible. In addition, some of the packets may be lost in transit. While some degradation in performance is to be expected, TCP's response to this situation makes matters worse. As discussed earlier, standard TCP assumes all losses are due to congestion, and its timers are based on the round trip time and observed variations in it. Loss of packets due to the absence of a route or sudden increases in delays, result in triggering the congestion control behavior and reducing the performance. An additional concern is related to the timer management algorithms in TCP. After an RTO, when a packet is re-transmitted, the timer value is doubled to allow for unexpected changes in network conditions. Each unsuccessful re-transmission results in this doubling, causing the timer value to increase exponentially[2]. (A maximum value MAY be placed on RTO provided it is at least 60 seconds[21].) In the case where the route is re-established after a re-transmitted packet is lost, then until the timer (which could have grown to a large value) expires, the connection remains idle resulting in underutilization of network resources, as shown in Fig. 3.

Another issue facing TCP during mobility is maintaining active connections. TCP is an end-to-end protocol that maintains a logical connection between the two end-hosts. Each connection is identified by a four-tuple <source address, source port, destination address, destination port>. When a mobile host moves, if it crosses network boundaries, then its IP address may change. In such a situation, the TCP end points need to be changed, in order to maintain active connections.

In this section we discuss some of the solutions proposed for improving TCP performance during mobility related disconnections and those for maintaining active connections though changes in IP address. It must be noted that many of the solutions presented in the previous section also address mobility-related issues. In this section, instead of going over each solution, we present some of the different approaches that have been proposed. Also, the discussion in this section is limited to the last-hop wireless networks. Mobility issues in multi-hop wireless networks are discussed at length in the "Mobility in Multi-hop Wireless Networks" section.

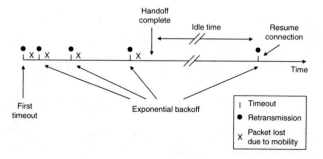

Fig. 3 Impact of exponential increase in timeout value.

Avoiding Performance Degradation during Mobility-Related Disconnections

An early study of the impact of mobility on TCP performance was presented in Ref. 22. In this study the authors show that in the time it takes to complete the handoff after mobility, the TCP sender goes into an RTO. Due to the use of coarse grain timers and exponential backoff, even after the handoff is completed, TCP remains idle. To address this problem, the authors propose triggering fast re-transmission when the switch to the new cell is completed. This is achieved by communication between the IP and TCP layers on the completion of handoff. The two TCP end-hosts are then synchronized by the transmission of three dupacks and the lost packets are re-transmitted. The congestion window is reduced to one and the connection reverts to slow start algorithm to determine the appropriate sending rate in the new cell.

A different approach for dealing with mobility-related disconnections using an existing TCP mechanism called zero window probes was proposed in Freeze TCP.[23] In standard TCP, the rate at which the sender sends data is controlled by the smaller of two windows namely the congestion window and the receiver advertised window. The purpose of the receiver advertised window is to allow a receiver to request the sender to slow down when the process at the receiver is slower than the sender. As the available buffers at the receiver shrink, the receiver advertises smaller and smaller windows, until eventually it advertises a zero window. When this happens the sender enters a persist mode where the window and re-transmit timers are frozen. The sender sends zero window probes at exponentially increasing intervals (limited to 1 min), until the receiver responds. While it is expected that the window advertised by the receiver will gradually decrease, the sender behavior in the unexpected event that the receiver abruptly decreases the window is designed to retain the congestion window and the re-transmission timer values.

Freeze TCP uses this behavior of TCP to avoid slow start after disconnection in a flow. It assumes that the receiver can predict disconnection (e.g., by monitoring signal strength). In such a situation, one RTT before the impending disconnection, the receiver sends a zero window advertisement to the sender, forcing it to freeze its congestion window and timers. When the receiver re-connects, it is possible that it has just missed a zero window probe from the sender. In order to avoid the idle time until the next zero window probe, the receiver wakes up the sender by sending three copies of the ACK to the packet is successfully received prior to sending the zero window advertisement.

Maintaining Active Connections through Mobility-Related Disruptions

In I-TCP[5], the split connection approach is leveraged for providing mobility support as well. The MSR offers an image of the mobile host for the connection with the fixed host. When the mobile host moves away from one MSR to another, the *image* of the mobile host is transferred to the new MSR. The endpoint parameters of the sockets at the new MSR are set to the same value as that in the first MSR. As a result, fixed host continues to communicate with the *image* of the mobile host, unaware that its physical location has changed and hence does not require new connection establishment.

Several solutions have since been proposed to eliminate the need for an intermediate gateway for handling mobility related performance degradation. In Ref. 24, the authors propose a TCP-redirect scheme that does not require the support from intermediate nodes for maintaining TCP connections active if the IP address changes in the middle of a connection. This is done by modifying the address four-tuple that is maintained at both ends of the connection. When the initial connection is setup using TCP's three-way handshake, the two end hosts negotiate the capability for TCP-redirection and exchange authentication keys. Later, when the mobile node detects the change in its IP address, it sends a re-direct message to the fixed host. This is followed by authentication using the authentication keys. Once authentication is completed, both the socket endpoints revise the source/destination address/port four-tuple that identifies the connection, and resume connection using the modified socket endpoints.

A similar approach is also presented in Ref. 25, where the authors present an end-to-end mobility scheme called migrate, and evaluate its benefits compared to network-layer mobility schemes. Similar to the TCP-redirection scheme, mobility support is negotiated between the sender and the receiver during the initial three-way handshake. Also, during the connection establishment, a token that uniquely identifies the connection is exchanged, so that the connection can either be identified by the four-tuple <source address, source port, destination address, destination port> or the triple <source address, source port, token>. Later, during the lifetime of the flow, if the IP address of the mobile node changes, it re-establishes the connection with the fixed sender by sending a special type of syn packet called the *migrate syn* using the triple with the token. One of the main advantages of these schemes is that, mobility is now an option allowed at the discretion of the application whether to use or not, depending on its unique requirements.

In Ref. 26, the authors present a unique split-connection technique for supporting TCP mobility that *does not* break the end-to-end semantics of TCP. Similar to split-connection techniques discussed earlier, the connection between a mobile host and the fixed host are split into two connections, namely mobile-to-proxy and proxy-to-fixed node. The connection splitting is done transparently to the end applications. At the mobile node, the MSOCKS library intercepts the sockets calls and sets up a split connection with the the proxy in between. In case the

connection is originated at the server, the proxy intercepts the connection to set up the split connection with the mobile client. However, unlike in the earlier approaches, once the two connections are established, they are spliced back together to maintain end-to-end semantics. In this approach, the proxy simply acts as a relay. It never generates any ACKs and only forwards the ACKs it receives from the mobile node to the server (or vice versa). This technique, called TCP-splice[27], ensures that even during handoffs, data integrity is maintained, since ACKs are end-to-end.

PERFORMANCE OF TCP IN MULTI-HOP INDEPENDENT WIRELESS NETWORKS

Early research in the area of *independent* wireless networks was motivated by the need for rapid deployment of reliable networks for military operations, emergency operations, disaster relief operations, etc., where no assumptions can be made about the availability of infrastructure. Such networks, commonly referred as mobile ad hoc networks (MANETs), are networks setup on-demand between several wireless nodes without a centralized controlling authority like an access point or a base station. However, in recent years, independent networks have found their way into everyday operations as well. Also referred to as mesh networks, these networks are proposed as a solution for providing wide coverage for spaces such as a large home or offices that have limited infrastructure in place, or as a low-cost alternative to setting up the infrastructure[28].

The chief characteristics of an independent network is that the algorithms for network organization, channel access, and routing are distributed due to the lack of a centralized controller. Also, such a network can have a highly dynamic topology since nodes may be mobile (or temporarily unavailable) resulting in unpredictable route changes. Finally, things can be complicated further by the fact that the different nodes participating in such a network may have different battery spans, transmission power settings or radio capabilities, and processing capacity.

Such a network offers its own unique set of challenges for higher layer protocols such as IP routing and the end-to-end reliable TCP communication. In this section, we focus on the issues faced by TCP in independent multi-hop mobile wireless networks and some proposed solutions. Much of the discussion here assumes that IEEE 802.11 is used at the media access control (MAC) layer.

We first focus on issues related to mobility in an multi-hop independent wireless network. Note that unlike the infrastructure network, where only the source and/or the destination is mobile, in an independent network, the intermediate nodes could also be mobile. This could cause significant disruptions related to mobility as shown in Fig. 4. Also, due to the lack of infrastructure, some of the solutions discussed in the previous section on mobility in wireless networks may not be applicable here. So in the "Mobility in Multi-hop Wireless Networks" section we discuss some of the solutions proposed for avoiding TCP performance degradation specific to mobility in independent networks.

Next, we inspect efficiency of channel use in multi-hop the wireless networks. Unlike the wired medium, wireless medium is a broadcast medium. This causes several issues for TCP performance in a multi-hop wireless network. TCP is an ACK-based protocol. Every data packet (every other data packet, if delayed ACKs are used.) generates an ACK packet in the reverse direction. Due to the broadcast nature of the channel, the data packets and the ACK packets contend with each other for the channel. Also, on a multi-hop path, the packets of the same TCP flow will contend with each other on adjacent hops. Finally, the packets of two flows that do not share the same path, but have intermediate hops that are within each others' interference range can contend for channel access. All these result in increased losses due to contention and consequently, reduced efficiency.

Finally we look at fairness issues. A common problem in a wireless network with multiple hops, is what is generally referred as the "hidden node" or "hidden terminal" problem. Fig. 5A illustrates this. Consider the three nodes A, B, and C. Node C is within the transmission range of node B, but not of node A. Suppose node A starts a transmission to node B. Before this transmission is complete, suppose node C wants to transmit to node B. It first listens to the channel to determine if the channel is busy. Since it cannot listen to transmissions from node A, it determines that the channel is idle and starts its transmission, causing collision at node B. A problem complementary to this, generally referred to as the "exposed node" or "exposed terminal" problem, is illustrated in Fig. 5B. Now suppose that node B is transmitting a packet to node A. If node C wants to send a packet to node D at the same time, this should technically be possible since node D is out of the transmission range of node B. However, when node C listens to the channel before it starts the transmission, it hears the transmission from node B, assumes that the channel is busy and puts off its transmission. This is commonly referred as the "exposed terminal" problem. 802.11 provides several mechanisms (detailed discussion of these mechanisms is beyond the scope of this entry. An interested reader is referred to Ref. 29 for details) to avoid the hidden/exposed node problem. However, even with the use of these mechanisms, it is not always possible to entirely avoid these problems.

Another problem closely associated with the exposed node problem is the "channel capture" problem. In 802.11, upon an unsuccessful attempt to transmit a packet, a node exponentially increases its wait timer before any

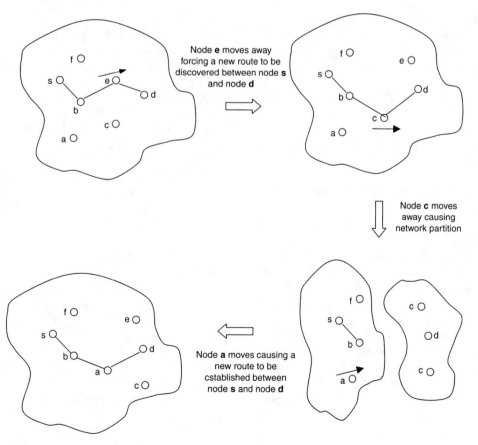

Fig. 4 Problems due to node mobility in multi-hop ad hoc networks.

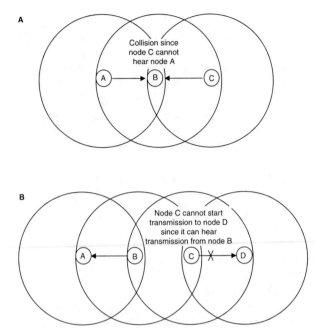

Fig. 5 Node problems. (A) Hidden terminal problem; (B) Exposed terminal problem.

re-attempts. In the example discussed above, suppose that both node B and node C have several packets for transmission. After its first unsuccessful attempt for transmission, node C doubles its wait timer. In the mean time, node B may complete the transmission of its packet. Since the wait time of node C was doubled, it is likely that the wait time of node B expires first, and it gets to send its next packet, effectively "capturing" the channel. Recent studies[30-33] have shown that, these problems can result in severe unfairness among competing TCP flows in a multi-hop wireless networks. In the "Efficiency and Fairness in Multi-hop Wireless Networks" section, we discuss the solutions for improving efficiency and fairness in multi-hop wireless networks.

Mobility in Multi-hop Wireless Networks

Mobility in multi-hop wireless networks presents its own unique challenges.[34] Since independent networks do not have a centralized control, the routing protocols are distributed and use route discovery and caching techniques. When nodes are mobile, it is possible that within the lifetime of a flow, there may be several instances when there exists no route between the source and destination. This could happen because of delays caused by route discovery or alternately, the network may be partitioned and the two end nodes may be in different partitions. In addition, it is possible that if routing updates are not propagated effectively, some of the nodes may contain stale entries in the cache, hampering some packets from reaching their intended destination.

These issues result in degraded TCP throughput due to similar reasons discussed before, i.e., in the absence of explicit information, a TCP sender implicitly assumes that a packet drop is caused by congestion and reduces its congestion window by half. Worse still, in case of an RTO, the congestion window is reduced to 1 and the RTO is backed off exponentially, leading to the possibility that even after a route discovery, a TCP connection may be idle for some time. Explicit Link Failure Notification (ELFN) has been proposed for informing a TCP sender of link failures due to mobility[34]. When a sender receives an ELFN, it immediately goes into a "stand-by" mode where it stops sending any more data packets and disables its timers. It then periodically transmits probe packets to check if a new route is established. The ELFN messages themselves are piggybacked on the routing messages to reduce the overhead. A similar approach is proposed in TCP-Feedback[35] as well. When a node on the path of a connection determines that the route is disrupted, it sends explicit Route Failure Notification (RFN) message to the sender. Intermediate nodes that receive this message check if they have alternate route to the destination. If so, they discard the RFN and switch to the alternate route. Otherwise, they clear their route cache and propagate the RFN back to the sender. When a new route is found, a Route Re-establishment Notification (RRN) is propagated to the sender. Intermediate nodes block duplicate RRN messages.

After a route recovery, it will take some time for a re-transmitted packet to reach the receiver, generate the ACK and for the ACK to arrive back at the sender. If the timers are enabled from their previously frozen values immediately after a route recovery message is received, there is still a possibility that an RTO may occur. Additionally, the above schemes do not ensure the reliable delivery of the route failure/recovery messages which are crucial if any benefits are to be obtained from these schemes. TCP with buffering capability and sequence information (TCP-BuS)[36] has been proposed to address these issues. TCP-BuS assumes that the underlying routing protocol is Associativity-Based Routing (ABR)[37] and proposes modifications to the interface between TCP and routing to improve the performance of explicit RFN schemes. Some of the chief modifications proposed by this scheme are to double the value of the timeout for buffered packets along the source and intermediate nodes, after a route recovery. Also, the reliable delivery of route failure/recovery messages is ensured via the use of timers, re-transmission when necessary, and probe requests by nodes that originate the messages as well as intermediate nodes.

Another solution that uses explicit notification is presented in Ad hoc TCP (ATCP)[38]. ATCP is implemented as a thin layer between IP and TCP at the sender and is proposed to be a complete solution that addresses losses due to channel errors, losses due to mobility and re-ordering that may be caused by some of the ad hoc routing schemes. It defines four different states for TCP—normal, disconnected, congested, and loss. By using multiple types of explicit notifications, the ATCP layer draws inferences and determines what state the sender TCP should be in order to optimize performance. Similar to earlier approaches, this scheme uses explicit RFN based on Internet control message protocol (ICMP) "destination unreachable" messages to determine disconnection. In addition, it uses ECN[15] to distinguish between congestion-related losses and losses due to packet corruption on the channel. To prevent congestion control from being triggered by re-ordering that may be caused by routing protocols such as temporally-ordered routing algorithm (TORA)[39], the ATCP layer hides the multiple dupacks from TCP, but triggers a re-transmission by using the loss state. RTO timers at the TCP layer are disabled when there is no ECN, and the ATCP layer maintains its own timers to determine when a timeout should occur. In addition to the above, this scheme advocates setting the congestion window to 1 after a route re-establishment, so that the appropriate sending rate for the new path may be discovered afresh.

Efficiency and Fairness in Multi-hop Wireless Networks

A problem related to TCP in multi-hop wireless networks is the throughput degradation when the TCP connection traverses several hops. In Ref. 40 the authors propose improving TCP performance by improving spatial re-use. The reduced throughput is observed to be a result of the average window size of TCP flows being higher than the optimal value. Unlike losses due to buffer overflow, contention-related losses increase gracefully and hence may not be sufficient for limiting the congestion window. In order to address this, a combination of two schemes, namely, Link RED and adaptive pacing have been proposed for limiting the window to the optimal value. Link RED applies the concept of RED[41] to the number of transmission re-try attempts to provide explicit early signal to the sender that there is possible network overload due to channel contention. Adaptive pacing modifies the 802.11 backoff algorithm to ensure that larger backoff is used once overload due to channel contention is determined. The resulting scheme is a distributed scheme that is implemented in the MAC layer of the nodes to provide improved TCP throughput.

Some research studies have pointed to severe unfairness among TCP flows competing in a multi-hop wireless network[30,32,33]. When a single-hop flow competes with a multi-hop flow, it was observed that the multi-hop flow could be completely shut down when the single-hop flow is active. The hidden node problem and the exposed node problem along with the exponential backoff scheme in the MAC layer were identified to be the causes of this problem.

It has been observed that flows originating in wired Internet can have an unfair advantage over flows originating in wireless networks. Ref. 31 shows that a flow originating in the wired Internet has an unfair advantage and could starve a competing flow originating in the wireless independent network. First, similar to the previous study, hidden node and the exposed node can result in channel capture behavior. Second, it was observed that the optimal window at which flows share the network in a fair manner is smaller than the optimal window size based on the end-to-end network parameters. As a result, unless the congestion window is not fixed at this value, under dynamic conditions, the flows overshoot this value. When this happens, the flow originating in the wired Internet and the one originating in the wireless independent network behave differently. The flow originating in the wired network gains higher throughput and the flow originating in the wireless independent network loses throughput, eventually leading to very unfair bandwidth sharing.

Neighborhood RED (NRED)[42] has been proposed for reducing the unfairness among competing TCP flows. NRED tries to address the two main causes for unfairness in an wireless independent network, namely, 1) spatial re-use constraint because of which flows that do not even traverse common nodes can interfere with each other causing hidden/exposed node problems, and 2) location dependency constraint because of which nodes in different relative positions in the bottleneck have different perceptions of the level of congestion. In order to counter these causes, NRED responds to congestion in *neighborhood* queue instead of the local queue at any given node. The *neighborhood* of any node is defined as the node itself and all the nodes that can interfere with it, and the neighborhood queue is an estimate of the aggregate queue size of all the nodes within a neighborhood. Using RED-like algorithm, the probability of response is calculated for the changes in the neighborhood queue and propagated to other nodes in the neighborhood. Each node calculates its local probability of response based on the proportion of its channel usage in comparison to its estimate of the neighborhood channel utilization. Results show that NRED can improve the fairness among competing flows at the possible cost of slightly reduced aggregate link utilization.

Another solution, split TCP[43] takes a similar approach to the split-connection approaches discussed earlier. Several of the nodes within the multi-hop wireless network are designated a proxies. A long end-to-end connection is split into several shorter connections for the purposes of congestion control, while retaining end-to-end ACKs for reliability. The congestion window of a standard TCP flow is replaced by two components, namely a congestion window determined by local ACKs from the nearest proxy and an end-to-end window controlled by the cumulative ACK from the receiver. The local ACKs are used for congestion control on local segments, while the end-to-end ACKs are used primarily for providing reliable end-to-end service. By splitting up the long TCP flow into several shorter flows, the scheme reduces the unfairness caused by competing shorter flows within the interference range.

In Fusion[44], three techniques have been employed for improving efficiency and fairness while reducing packet losses due to channel contention. In the first technique, hop-by-hop flow control, each node determines local congestion by monitoring its outgoing queues as well as periodically sampling the channel. If it determines high congestion, it sets a congestion bit in all its outgoing packets. Due to the broadcast nature of the wireless channel all nodes in the vicinity are notified and throttle their sending rate. If congestion is significant, this mechanism propagates in a form of back pressure all the way to the source. This ensures that packets are not sent if there is a high probability that they will be dropped at a downstream node. The second technique imposes rate limiting for traffic originating at a node, if other transit traffic is present, so that priority is given to traffic that has already traversed several hops. This reduces the unfairness against flows with large number of hops. Finally, the third technique uses a prioritized MAC protocol to allow a node with backlogged queues to access the channel with a higher probability.

CONCLUSION

TCP faces several issues when used in wireless networks. This entry provided an overview of these issues and some of the solutions proposed to improve the TCP performance when faced with these issues. We have presented specifically those solutions that improve TCP performance in the presence of channel errors, mobility within both infrastructure networks and independent networks, and issues caused by hidden/exposed terminals in multi-hop wireless networks. A lot of research has been conducted in this area and some topics are still being investigated. It is expected that this entry will provide an introduction to this topic and lead the reader to further exploration in this area.

REFERENCES

1. Allman, M.; Paxson, V.; Stevens, W. TCP congestion control, RFC 2581; Internet Engineering Task Force, 1999: http://www.ietf.org/rfc/rfc2581.txt (accessed June 2006).
2. Stevens, W.R. *TCP/IP Illustrated (Volumes 1, 2 and 3)*; Addison-Wesley Publications: 1994.
3. Duke, M.; Braden, R.; Eddy, W.; Blanton, E. A roadmap for transmission control protocol (TCP) specification documents, RFC 4614; Internet Engineering Task Force, September 2006, http://www.ietf.org/rfc/rfc4614.txt (accessed Nov 2006).
4. Balakrishnan, H.; Padmanabhan, V.; Seshan, S.; Katz, R.H. A comparison of mechanisms for improving TCP performance over wireless links. IEEE/ACM Trans. Netw. **1997**, *5* (6), 756–769.
5. Bakre, A.; Badrinath, B.R. I-TCP: indirect TCP for mobile hosts. 15th International Conference on Distributed Computing Systems (ICDCS 1995), British Columbia, Canada, May 30–June 02, 1995; Computer Society Press: 1995; 126–143.
6. Haas, Z.J.; Agrawal, P. Mobile-TCP: an asymmetric transport protocol design for mobile systems. International Conference on Communications (ICC 1997), Montreat, Canada, June 8–12, 1997; 1054–1058.
7. Wang, K.-Y.; Tripathi, S.K. Mobile-end transport protocol: an alternative to TCP/IP over wireless links. 17th Annual Joint Conference of the IEEE Computer and Communications Societies, San Francisco, CA, Mar 29–Apr 2, 1998.
8. Yavatkar, R.; Bhagawat, N. Improving end-to-end performance of TCP over mobile Internet works. 1st IEEE Workshop on Mobile Computing Systems and Applications (WMCSA 1994), Santa Cruz, CA, Dec 8–9, 1994; 146–152.
9. Ayanoglu, E.; Paul, S.; LaPorta, T.F.; Sabnani, K.K.; Gitlin, R.D. AIRMAIL: a link-layer protocol for wireless networks. ACM Wirel. Netw. **1995**, *1* (1), 47–60.
10. Eckhardt, D.; Steenkiste, P. Improving wireless LAN performance via adaptive local error control. 6th IEEE International Conference on Network Protocols (ICNP 1998), Austin, TX, Oct 13–16, 1998; 327–338.
11. Balakrishnan, H.; Seshan, S.; Amir, E.; Katz, R.H. Improving TCP/IP performance over wireless networks. 1st Annual ACM/IEEE International Conference on Mobile Computing and Networking (MobiCom 1995), Berkeley, CA, Nov 13–15, 1995; 2–15.
12. Mathis, M.; Mahdavi, J.; Floyd, S.; Romanow, A. TCP selective acknowledgment options, RFC 2018; Internet Engineering Task Force, 1996, http://www.ietf.org/rfc/rfc2018.txt (accessed June 2006).
13. Parsa, C.; Garcia-Luna-Aceves, J.J. Improving TCP performance over wireless networks at the link laye. Mobile Netw. Appl. **2000**, *5* (1), 57–71.
14. Balakrishnan, H.; Katz, R.H. Explicit loss notification and wireless web performance. IEEE Global Telecommunications conference (GLOBECOM 1998), Sydney, Australia, Nov 8–12, 1998.
15. Ramakrishnan, K.; Floyd, S.; Black, D. The additon of explicit congestion notification (ECN) to IP, RFC 3168; Internet Engineering Task Force, 2001, http://www.ietf.org/rfc/rfc3168.txt (accessed June 2006).
16. Vaidya, N.H.; Mehta, M.; Perkins, C.; Montenegro, G. Delayed duplicate acknowledgement: a TCP-unaware approach to improve performance of TCP over wireless. J. Wirel. Commun. Mob. Comput. **2002**, *2* (1), 59–70.
17. Bhandarkar, S.; Sadry, N.; Reddy, A.L.N.; Vaidya, N. TCP-DCR: a novel protocol for tolerating wireless channel errors. IEEE Trans. Mob. Comput. **2005**, *4* (5), 517–529.
18. Bhandarkar, S.; Reddy, A.L.N.; Allman, M.; Blanton, E. Improving the robustness of TCP to non-congestion events, RFC 4653; Internet Engineering Task Force, 2006: http://www.ietf.org/rfc/rfc4653.txt (accessed Oct 2006).

19. Sinha, P.; Venkitaraman, N.; Sivakumar, R.; Bhargavan, V. WTCP: a reliable transport protocol for wireless widearea networks. 5th Annual ACM/IEEE International Conference on Mobile Computing and Networking (MobiCom 1999), Seattle, WA, Aug 15–20, 1999; 231–241.

20. Mascolo, S.; Casetti, C.; Gerla, M.; Sanadidi, M.Y.; Wang, R. TCP westwood: bandwidth estimation for enhanced transport over wireless links. 7th Annual ACM/IEEE International Conference on Mobile Computing and Networking (MOBICOM 2001), Rome, Italy, July 16–21, 2001; 287–297.

21. Paxson, V.; Allman, M. Computing TCP's retransmission timer, RFC 2988; Internet Engineering Task Force, 2000, http://www.ietf.org/rfc/rfc2988.txt (accessed June 2006).

22. Caceres, R.; Iftode, L. Improving the performance of reliable transport protocols in mobile computing environments. IEEE J. Sel. Areas Commun. 1995, 13 (5), 850–857.

23. Goff, T.; Moronski, J.; Phatak, D.; Gupta, V.V. Freeze-TCP: a true end-to-end enhancement mechanism for mobile environments. 19th Annual Joint Conference of the IEEE Computer and Communications Societies (INFO-COM, 2000), Tel-Aviv, Israel, Mar 26–30, 2000.

24. Funato, D.; Yasuda, K.; Tokuda, H. TCP-R: TCP mobility support for continuous operation. 5th IEEE International Conference on Network Protocols (ICNP 1997), Atlanta, GA, Oct 28–31, 1997; 229–236.

25. Snoeren, A.C.; Balakrishnan, H. An end-to-end approach to host mobility. 6th Annual ACM/IEEE International Conference on Mobile Computing and Networking (MobiCom 2000), Boston, MA, Aug 6–11, 2000; 155–166.

26. Maltz, D.A.; Bhagwat, P. MSOCKS: an architecture for transport layer mobility. 17th Annual Joint Conference of the IEEE Computer and Communications Societies (INFOCOM 1998), San Francisco, CA, Mar 29–Apr 2, 1998; 1037–1045.

27. Maltz, D.; Bhagwat, P. TCP splicing for application layer proxy performance, Tech. Rep. Research Report RC 21139, IBM, 1998.

28. Eriksson, J.; Agarwal, S.; Bahl, P.; Padhye, J. Feasibility study of mesh networks for all-wireless offices. 4th international conference on Mobile systems, Applications, and services (MobiSys 2006), Uppsala, Sweden, June 19–22, 2006; 69–82.

29. Gast, M.S. 802.11 Wireless Networks: The Definitive Guide, 2nd Ed.; O'Reilly Publications: 2005.

30. Xu, S.; Saadawi, T. Revealing TCP unfairness behavior in 802.11 based wireless multi-hop networks. IEEE PIMRC, San Diego, CA, Sept 30–Oct 3, 2001; E83–E87.

31. Xu, K.; Bae, S.; Lee, S.; Gerla, M. TCP behavior across multihop wireless networks and the wired Internet. International workshop on Wireless Mobile Multimedia (WoW-MoM 2002), Atlanta, GA, Sept 28, 2002; 41–48.

32. Gambiroza, V.; Sadeghi, B.; Knightly, E. End-to-end performance and fairness in multihop wireless backhaul networks. 10th Annual ACM/IEEE International Conference on Mobile Computing and Networking (MobiCom 2004), Philadelphia, PA, Sept 26–Oct 1, 2004; 287–301.

33. Garetto, M.; Salonidis, T.; Knightly, E. Modeling per-flow throughput and capturing starvation in CSMA multi-hop wireless networks. IEEE INFOCOM 2006, Catalunya, Spain, Apr 23–29, 2006.

34. Holland, G.; Vaidya, N. Analysis of TCP performance over mobile ad hoc networks. 5th Annual ACM/IEEE International Conference on Mobile Computing and Networking (Mobi-Com 1999), Seattle, WA, Aug 15–20, 1999; 219–230.

35. Chandran, K.; Raghunathan, S.; Venkatesan, S.; Prakash, R. A feedback-based scheme for improving TCP performance in ad hoc wireless networks. IEEE Pers. Commun. 2001, 8 (1), 34–39.

36. Kim, D.; Toh, C.-K.; Choi, Y. TCP-BuS: improving TCP performance in wireless ad hoc networks. J. Commun. Netw. 2001, 3 (2), 175–186.

37. Toh, C.-K. Associativity-based routing for ad hoc mobile networks. Wirel. Pers. Commun. J.. 1997, 4 (2), 103–139.

38. Liu, J.; Singh, S. ATCP: TCP for mobile ad hoc networks. IEEE J. Sel. Areas Commun. 2001, 19 (7), 1300–1315.

39. Park, V.D.; Corson, M.S. A highly adaptive distributed routing algorithm for mobile wireless networks. 16th Annual Joint Conference of the IEEE Computer and Communications Societies (INFOCOM 1997), Kobe, Japan, Apr 7–11, 1997; 1405–1413.

40. Fu, Z.; Zerfos, P.; Luo, H.; Lu, S.; Zhang, L.; Gerla, M. The impact of multihop wireless channel on TCP throughput and loss. 22nd Annual Joint Conference of the IEEE Computer and Communications Societies (INFOCOM 2003), San Francisco, CA, Mar 30–Apr 3, 2003; 1744–1753.

41. Floyd, S.; Jacobson, V. Random early detection gateways for congestion control. IEEE/ACM Trans. Netw. 1993, 1 (4), 397–412.

42. Xu, K.; Gerla, M.; Qi, L.; Shu, Y. Enhancing TCP fairness in ad hoc wireless networks using neighborhood RED. 9th Annual ACM/IEEE International Conference on Mobile Computing and Networking (MobiCom 2003), San Diego, CA, Sept 14–19, 2003; 16–28.

43. Kopparty, S.; Krishnamurthy, S.; Faloutsos, M.; Tripathi, S. Split-TCP for mobile ad hoc networks. IEEE Global Telecommunications conference (GLOBECOM 2002), Taipei, Taiwan, Nov 17–21, 2002; 138–142.

44. Hull, B.; Jamieson, K.; Balakrishnan, H. Mitigating congestion in wireless sensor networks. 2nd International ACM Conference on Embedded Networked Sensor Systems (ACM SenSys 2004), Baltimore, MD, Nov 3–5, 2004; 134–147.

**Wireless
Multi—Networks**

Wireless Networks: Transmitter Power Control

Nicholas Bambos
Savvas Gitzenis
Stanford University, Stanford, California, U.S.A.

Abstract

Transmitter power control is key in wireless networking for maintaining the quality of each wireless link at a desirable level via adaptation to channel interference and variations, while mitigating the interference the link generates on others and resulting in increased network capacity via higher spatial spectrum reuse.

INTRODUCTION

The advent and rapid expansion of wireless networking in the 1990s have prompted intense investigation of transmitter power control fundamentals, algorithms, and architectures for over a decade. Power control is a key element of wireless networking for various reasons. Adjusting its transmitter power to adapt to interference, mobility variations, and channel impairments, a communication link can maintain a desirable quality of service (QoS). By doing this carefully and systematically, it will generate the minimum possible interference on other links sharing the channel, resulting in higher network capacity via better sharing of the radio bandwidth and denser spectrum reuse. Additionally, the transmitter will minimize the power drain and maximize the battery lifespan between recharges, a key operational limitation in modern mobile devices. QoS support, interference management, battery life, and various other factors make power control a very important element of current and projected wireless network designs.

Early cellular wireless networks circumvented the need for full power control by spatially separating transmissions using the same radio resource [time slot and frequency band in TDMA/FDMA systems, spreading code in CDMA systems, etc.], resulting in low spatial densities of spectrum reuse. Power control was simply used in CDMA systems primarily to mitigate the near–far effect. Due to the scarcity of the radio spectrum and the exponentially increasing user demand for wireless mobile information services, the pressure to pack more radio transmissions into the same fixed radio spectrum is rising fast. To achieve this objective, there are basically two approaches:

1. Use transmitter power control to better manage and mitigate interference in order to increase the spatial spectrum reuse. This increases the network capacity for a given network infrastructure (access point) density.

2. Increase the infrastructure density by deploying more access points (base stations, wireless LANs, etc.) and getting those access (and, perhaps, relay) points closer to the wireless terminals. This makes wireless terminal-to-access links smaller, hence needing less power for a given QoS level. In this way, denser spatial packing of links (and higher spectrum reuse) can be achieved. Besides the higher infrastructure cost, there is also more control and signaling needed in denser infrastructure environments to manage mobility, as more access point handoffs occur per unit of time for a given level of mobility.

Moving forward onto next-generation networks, both approaches are advancing in unison. Therefore, the importance of power control is critical. It is essential to better understand its fundamentals, as well as various important architectural and implementation aspects. In this entry, we mostly focus on the fundamentals of power control, which reveal some important design insights. Particular implementation architectures are not discussed in the limited space of this entry, but the core algorithms presented are specifically chosen to be distributed and autonomous, on-line adaptive (agile), robust and scalable, with an eye toward high-performance design and practical implementation.

It is interesting to consider the logic of power control from an individual link's point of view in order to understand the global nature of power control from the network's point of view. Intuitively speaking, the communication link should choose its transmission power to be large enough to adequately compensate for radio propagation losses and interference/noise effects, yet low enough to generate minimal interference on other network links sharing the same channel and disturb those minimally.

Encyclopedia of Wireless and Mobile Communications DOI: 10.1081/E-EWMC-120043921

According to the above logic, all links are coupled to each other, in the sense that one's benefit (communication) is everyone else's cost (interference). However, interference has a positive aspect too. Indeed, it provides collective feedback information from the channel/network that each link observes in order to assess the stress level of its channel/network environment. This is valuable information that the link can use to control its power, as well as autonomously perform various network control functions like admission, congestion, and handoff control in a distributed scalable manner. It should be mentioned that power control may be considered to operate at the media access control (MAC) layer of wireless networking; however, it touches on and actually couples with many other functions. Those range from signal processing, modulation, transmission, etc., on one side, to admission and handoff control on the other (as seen later).

Early work on power control[1-7] brought attention to the importance of the problem and provided some significant insights and results. A decade of further investigation[8-18] has revealed several interesting key facts and has crystallized the big picture of power control. We aim to sketch out this picture in this entry, from an admittedly particular—yet central and insightful—point of view.

For methodological and presentation purposes, we first consider in the second section the power control problem for continuous streamed delay-sensitive traffic (voice, video) where the main QoS metric is the bit error rate, expressed as a function of the signal-to-interference (plus noise) ratio (SIR). We then consider in the third section the case of packetized delay-tolerant data traffic. In this case, during interference highs the dilemma arises whether data transmission should be halted and delay incurred, expecting that the interference might soon subside and power benefits might be realized by transmitting lots of data at a low power cost. This core *delay vs. power* trade-off is captured below in a simple model and analyzed, yielding some interesting insights about the nature of the power control problem. The overall emphasis is on presenting the fundamental principles of power control, leading to basic core algorithms and protocols, without expanding into various operational and implementation details, some of which can be found in the background references. To achieve this we utilize simple—yet canonical—models of wireless networking, whose analysis provides significant insight as well as several graphs that visually demonstrate important aspects of power control dynamics.

POWER CONTROL FOR STREAMED CONTINUOUS TRAFFIC

In this section, we study the case of power control for streamed continuous delay-sensitive traffic (voice, video). In particular, voice service was the main driver in first-generation cellular wireless networks, while streamed multimedia services are planned for next-generation networks. The SIR of the wireless link is the basic QoS metric managed via control of the transmitter power, mapping to control of the bit error rate on the link.

We consider the wireless network as a collection of radio links, where each link is a single-hop radio transmission from a transmitter to an intended receiver. Chains of consecutive links may correspond to multihop communication paths, but they are also treated as sets of individual links for power control purposes. If there are several communication channels, we assume that the interference between links operating in different ones is negligible. That is, channels are orthogonal, and so only co-channel interference need be considered. We can therefore reduce the network picture to that of a *collection of interfering links* in a *single channel*, rendering the notions of *network admission* and *channel access* equivalent.

In the cellular communication network paradigm, links correspond to upstream or downstream transmissions between mobiles and base stations. In the ad hoc networking paradigm, links may correspond to single-hop transmissions between mobile terminals and fixed access points or other mobile terminals. Each channel is basically a *communication resource* shared by the various interfering links using it. For example, the channels can be nonoverlapping frequency bands in FDMA systems, nonoverlapping time slots in TDMA systems, distinct spreading codes in CDMA systems, or combinations of the above in hybrid systems.

The Target SIR Formulation of the Power Control Problem

A key performance metric of the radio link is its bit error rate, which to first approximation is a decreasing function of the SIR observed at its receiver node. We can therefore consider the QoS of a wireless link as an increasing function of its SIR and use the latter in our problem formulation. Given N interfering links in a channel, we denote the SIR of the ith link by

$$R_i = \frac{G_{ii}P_i}{\eta_i + \sum_{j \neq i} G_{ij}p_j} \quad i, j \in \{1, 2, \ldots, N\} \tag{1}$$

where P_i is the power of the transmitter of link i, η_i is the thermal noise power at the receiver of link i, and G_{ij} is the power gain (actually loss) from the transmitter of the jth link to the receiver of the ith one. The power gain G_{ij} may incorporate free space loss, multipath fading, shadowing, and other radio-wave propagation effects, as well as modulation effects (processing gain, etc.) To keep the presentation simple, in this section we assume that the G_{ij} are deterministic (fluctuations have been averaged out) and stay constant over time (no mobility).

For each link i we consider a target SIR $\gamma_i > 0$ that the link must maintain (or perhaps exceed) throughout the communication in order to operate properly and sustain acceptable QoS. Recalling that the QoS is an increasing function of the SIR, we see that we need to have in order for the links to coexist smoothly in the channel. According to Eq. 1, each SIR is a function of the whole power vector (The symbol $(\cdot)^T$ denotes the transpose column vector.) $\mathbf{P} = (P_1, P_2, \ldots, P_N)^T$. Given a constant power gain matrix $\mathbf{G} = [G_{ij}]$ and noise vector $\eta = (\eta_1, \eta_2, \ldots, \eta_N)$, the issue is to find the optimal power vector $\mathbf{P}$ that satisfies Eq. 2, if any such vector exists.

$$R_i(\mathbf{P}) \geqslant \gamma_1 \quad \text{for all } i \in \{1, 2 \ldots, N\} \tag{2}$$

In particular, the *feasibility* problem is to determine whether Eq. 2 has a solution $\mathbf{P}$ satisfying Eq. 1 and $P_i > 0$ for all $i = \{1, \ldots, N\}$. If that is feasible, then the *optimal power allocation* problem is to select the best power vector among potentially several feasible ones. Before proceeding with the general case, let us first consider a two-link network ($N = 2$), which will provide us with some useful insights.

Example of the simple two-link network

Consider the simplest possible network of Fig. 1, which is comprised of only two links interfering in a channel. For exposition purposes, we analyze it in two steps, addressing the power feasibility and allocation problems. First, let us ignore the thermal noise, setting $\eta_1 = \eta_2 = 0$. The SIR Eqs. 1 and 2 yield $R_1 = \frac{G_{11}}{G_{12}} \frac{P_1}{P_2} \geqslant \gamma_1$ and $R_2 = \frac{G_{22}}{G_{21}} \frac{P_2}{P_1} \geqslant \gamma_2$ which can be put into linear inequality form:

$$G_{11}P_1 - \gamma_1 G_{12}P_2 \geqslant 0 \tag{3}$$

$$G_{22}P_2 - \gamma_2 G_{21}P_1 \geqslant 0 \tag{4}$$

Fig. 2 shows the power feasibility region $\mathbf{D}(\mathbf{G}, \gamma)$ in the (P_1, P_2) power space. Any power vector $\mathbf{P} = (P_1, P_2)$ in $\mathbf{D}$ satisfies both inequalities above. The cone $\mathbf{D}$ is a function of the gain matrix $\mathbf{G} = [G_{ij}]$, the target SIR vector $\gamma = (\gamma_1, \gamma_2)$, and the noise vector η, which is set to zero in the current discussion. Note that ε_1 and ε_2 are the lines

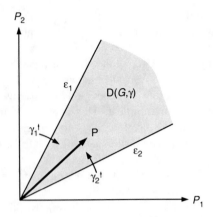

Fig. 2 The power feasibility region $\mathbf{D}(\mathbf{G}, \gamma)$ (shaded region) of the network of Fig. 1 corresponds to the set of power vectors $\mathbf{P} = (P_1, P_2)^T$ satisfying Eqs. 3 and 4. It is a cone with boundary lines ε_1, ε_2 where Eqs. 3 and 4, respectively, hold true with equality. Increasing γ_1 tilts ε_1 toward ε_2 (and conversely for γ_2), resulting in the shrinking of $\mathbf{D}$. The noise vector $(\eta_1, \eta_2) = (0,0)$ is artificially set to zero in this case (it is considered again in Fig. 3).

where Eqs. 3 and 4 become equalities, respectively. The slopes of ε_1 and ε_2 are $\frac{1}{\gamma_1}\frac{G_{11}}{G_{12}}$ and $\frac{\gamma_2 G_{21}}{G_{22}}$, respectively, and according to Fig. 2 we have:

1. When $\frac{1}{\gamma_1}\frac{G_{11}}{G_{12}} > \frac{\gamma_2 G_{21}}{G_{22}}$, the feasibility region $\mathbf{D}$ is a full cone and any power vector in it satisfies Eqs. 1 and 2.
2. When $\frac{G_{11}}{\gamma_1 G_{12}} = \frac{\gamma_2 G_{21}}{G_{22}}$, the two lines ε_1, ε_2 coincide and $\mathbf{D}$ reduces to a line.
3. When $\frac{G_{11}}{\gamma_1 G_{12}} < \frac{\gamma_2 G_{21}}{G_{22}}$, then $D = \emptyset$ is empty; i.e., no power vector $\mathbf{P} = (P_1, P_2)^T$ can satisfy Eq. 2 of both links, so the system is *infeasible*.

The third case is due to the fact that the SIR requirements γ_1 and γ_2 are too high for the system to accommodate, given the power gains $\mathbf{G}$. In general, observe that for any $i = 1, 2$ and $j \neq i$, 1) an increase in any γ_i, 2) an increase in any interlink power gain G_{ij}, or 3) a decrease in any intralink power gain G_{ii} results in line ε_i tilting toward line ε_j (until crossing it) and the feasibility region $\mathbf{D}$ shrinking. The reverse actions make $\mathbf{D}$ expand.

Taking now into consideration the thermal noise $(\eta_1, \eta_2) \neq (0,0)$, the inequalities in Eq. 2 take the following form in the two-link network of Fig. 1:

$$G_{11}P_1 - \gamma_1 G_{12}P_2 \geqslant \gamma_1 \eta_1 \tag{5}$$

$$G_{22}P_2 - \gamma_2 G_{21}P_1 \geqslant \gamma_2 \eta_2 \tag{6}$$

Fig. 3 shows the new feasibility region $\mathbf{D}(\mathbf{G}, \gamma, \mathbf{\eta})$, which is now a function of the gain matrix $\mathbf{G}$, the target SIR vector γ, and the noise vector $\mathbf{\eta}$. The tip of the cone $\mathbf{P}^*$ is the *Pareto-optimal* power vector, in the sense that any other feasible power vector $\mathbf{P}$ requires at least as much

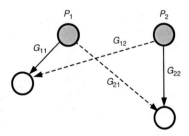

Fig. 1 The simplest possible wireless network consisting of two interfering links in a channel. The power gains G_{ij} from the transmitters (dark circles) to the receivers (light circles) are depicted, as well as the transmitted powers.

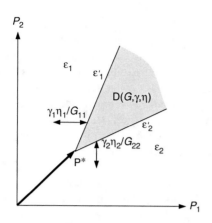

Fig. 3 The feasibility region $\mathbf{D}(\mathbf{G}, \gamma, \mathbf{\eta})$ (shaded area) of the two-link network of Fig. 1 where each power vector satisfies both Eqs. 5 and 6. It is a cone with boundary lines ε'_1, ε'_2 where Eqs. 5 and 6 hold true with equality, respectively. The tip of the cone is the Pareto-optimal power vector $\mathbf{P}^*$ where the target SIRs (γ_1, γ_2) are achieved at minimum power.

power from every transmitter,[17] i.e., $\mathbf{P} \geqslant \mathbf{P}^*$ component-wise. The inclusion of the noise (η_1, η_2) has the effect that cone boundary lines ε'_1 and ε'_2 of $\mathbf{D}$—where Eqs. 5 and 6 respectively, hold true with equality—are shifted right and up by $\gamma_1 \eta_1 / G_{11}$ and $\gamma_2 \eta_2 / G_{22}$ relative to the original ε_1 and ε_2, respectively. Therefore, an increase in the target SIRs γ_i has the effect that not only does ε'_i tilt toward ε'_j, but also the distance of ε'_i from ε_i increases. This results in $\mathbf{P}^*$ moving away from $\mathbf{0}$, and hence all transmitters increasing their powers to satisfy Eq. 2 for the increased γ_i.

The above discussed behavior and the gleaned geometric intuition extend directly to the general case of $N > 2$ links. The feasibility region $\mathbf{D}$ is then an N-dimensional polyhedral cone with boundary hyperplanes where Eq. 2 is satisfied with equality. The tip of the feasibility cone is the Pareto-optimal power vector $\mathbf{P}^*$, at which the links should operate, in order to achieve their target SIRs at minimum possible power. Increasing any γ_i or any interlink gain G_{ij} and decreasing any intralink gain G_{ii} result in $\mathbf{D}$ shrinking and $\mathbf{P}^*$ moving away from $\mathbf{0}$; eventually $\mathbf{D}$ "evaporates" and $\mathbf{P}^*$ disappears at infinity, as the system becomes infeasible. Adding more links in the channel results in introducing more "cutting" hyperplanes that define the feasibility cone, which again shrinks and sees its tip move away from $\mathbf{0}$.

The optimal power vector P*

The inequalities in Eq. 2 can be put into a linear form $G_{ii}P_i - \gamma_i \sum_{i \neq j} G_{ij}P_j \geqslant \gamma_i \eta_i$ and further into a matrix form

$$(\mathbf{I} - \mathbf{F})\mathbf{P} \geqslant \mathbf{u} \quad \text{and} \quad \mathbf{P} > 0 \qquad (7)$$

where

$$\mathbf{u} = \left(\frac{\gamma_1 \eta_1}{G_{11}}, \frac{\gamma_2 \eta_2}{G_{22}}, \ldots, \frac{\gamma_N \eta_N}{G_{NN}} \right)^T \qquad (8)$$

is the column vector of noise powers, normalized by the target SIR requirements over the intralink power gains, and $\mathbf{F}$ is the matrix with entries

$$\mathbf{F}_{ij} = \begin{cases} 0 & \text{if } i = j \\ \frac{\gamma_i G_{ij}}{G_{ii}} & \text{if } i \neq j \end{cases} \qquad (9)$$

where $i, j \in \{1, 2, \ldots, N\}$. Note that the latter is the matrix of interlink power gains only, normalized by target SIRs over intralink gains.

The Perron–Frobenious theory of nonnegative matrices (having positive or zero entries only, like $\mathbf{F}$)[19,20] addresses the existence of feasible power vectors (Eq. 7) as follows. It is reasonable to assume that the nonnegative matrix $\mathbf{F}$ is irreducible; i.e., there are no totally isolated groups of links that do not interact with each other. Then the *maximum modulus eigenvalue* $\rho_{\mathbf{F}}$ of $\mathbf{F}$ is real, positive, and simple, while the corresponding eigenvector is positive component-wise. The following statements are equivalent:

1. There exists a power vector $\mathbf{P} > 0$ such that $(\mathbf{I} - \mathbf{F})\mathbf{P} \geqslant \mathbf{u}$.
2. $\rho\mathbf{F} < 1$.
3. $(\mathbf{I} - \mathbf{F})^{-1} = \sum_{k=0}^{\infty} \mathbf{F}^k$ exists and is positive component-wise.

If Eq. 7 has a solution, by the third fact mentioned above we have that

$$\mathbf{P}^* = (\mathbf{I} - \mathbf{F})^{-1} \mathbf{u} \qquad (10)$$

is the Pareto-optimal solution of Eq. 7, in the sense that any other power vector $\mathbf{P}$ satisfying the equation would require as much power from every transmitter,[17] or $\mathbf{P} \geqslant \mathbf{P}^*$ component-wise.

Therefore, if it is possible to satisfy the SIR requirements for all links simultaneously, the optimal power allocation strategy is to set the transmitter power to $\mathbf{P}^*$ and minimize the overall transmission power throughout the network.

Autonomous Distributed Power Control

Can links achieve the globally optimal power vector $\mathbf{P}^*$ without any interlink communication in a totally distributed and autonomous manner, even without any prior knowledge of the power gains matrix $\mathbf{G}$ and noise vector $\mathbf{\eta}$? Fortunately, the answer is yes, as was explicitly observed in the early 1990s by Foschini and Miljanic[2] and others. To see this, consider first the following iterative algorithm over subsequent time slots $k = \{1, 2, 3, \ldots\}$

$$\mathbf{P}(k + 1) = \mathbf{FP}(k) + \mathbf{u} \qquad (11)$$

and note that by recursively substituting Eq. 11 into itself we get $\mathbf{P}(k) = \mathbf{F}^k\mathbf{P}(0) + \sum_{i=0}^{k-1} \mathbf{F}^i\mathbf{u}$. Therefore, the iteration converges to the the Pareto-optimal power vector

$$\lim_{k\to\infty} \mathbf{P}(k) = \lim_{k\to\infty} \mathbf{F}^k\mathbf{P}(0) + \lim_{k\to\infty} \sum_{l=0}^{k-1} \mathbf{F}^l\mathbf{u} = 0 + \left[\sum_{l=0}^{\infty} \mathbf{F}^l\right]\mathbf{u}$$

$$= (\mathbf{I} - \mathbf{F})^{-1}\mathbf{u} = \mathbf{P}^* \qquad (12)$$

when $\mathbf{P}^*$ exists or, equivalently, $\rho_\mathbf{F} < 1$. This can be easily seen by the basic facts of the Perron–Frobenious theory mentioned in the previous section, and recalling that $\rho_\mathbf{F} < 1$ implies $\lim_{k\to\infty} \mathbf{F}^k\mathbf{P}(0) = 0$.

Observe now that the distributed power control (DPC) algorithm (Eq. 11) can be equivalently written for each link i as the iteration

$$P_i(k+1) = \frac{\gamma_i}{G_{ii}}\left[\sum_{j\neq i} G_{ij}P_j(k)\right] + \frac{\gamma_i}{G_{ii}}\eta_i$$

$$= \frac{\gamma_i}{G_{ii}}\left[\sum_{j\neq i} G_{ij}P_j(k) + \eta_i\right] = \frac{\gamma_i}{R_i(k)}P_i(k) \qquad (13)$$

since from the expression of the SIR (Eq. 1) we have $\sum_{j\neq i} G_{ij}P_j(k) + \eta_i = (G_{ii}P_i(k))/(R_i(k))$ Therefore,

$$P_i(k+1) = \frac{\gamma_i}{R_i(k)}P_i(k) \qquad (14)$$

for each $i = 1, 2, \ldots, N$ and $k = 1, 2, 3, \ldots$. Observe that under Eq. 14 each link independently and autonomously measures its own SIR $R_i(k)$ and adjusts its transmission power to $P_i(k+1)$ according to its own SIR target γ_i, without explicit knowledge of any other transmitter power P_j, any power gain G_{ij}, or any noise level η_j. No interlink communication is required, but only intralink communication of the SIR $R_i(k)$ observed locally at the receiver of each link i back to its transmitter, where the power $P_i(k+1)$ is computed and set. The SIR information communicated is minimal (a single number per power update), so it can be carried on either a low-rate reverse virtual link or a global control channel, which may also carry other control information (acknowledgments, etc.). In "symmetric" networks—like cellular ones—with uplink/downlink communication, it may also be piggy-backed on the reverse link. The lack of any interlink communication justifies calling Eq. 14 distributed, down to the level of the individual link, which is the functional primitive of the wireless network model.

An interesting observation regarding Eq. 14 is that each link i autonomously updates its power trying to hit its SIR target γ_i—as if the interference induced from the other links remained constant because those would not update their powers. If the latter were indeed the case, link i would hit γ_i in a single step. In reality, all other links adjust their

powers too, so $R_i(k)$ will asymptotically converge to γ_i (if feasible) as $k \to \infty$. It is worth noting that the convergence $R_i(k) \to \gamma_i$ is geometrically fast—as easily seen by the equivalent algorithm in Eq. 11—and $\rho_\mathbf{F} < 1$ is the convergence rate. If $\rho_\mathbf{F} > 1$, the DPC algorithm obviously diverges and the link powers explode to infinity. Fig. 4 graphically shows the power vector dynamics of a simple two-link network under DPC, while the evolutions of both power and SIR are plotted for a three-link network in Fig. 5.

DPC with Active Link Protection

The power and SIR evolution dynamics of the raw DPC algorithm (Eqs. 14 and 11, respectively) are problematic because when new links power up in the channel, existing ones may see their SIRs dive significantly below their target SIRs γ for several iterations. This effect may cause established links to actually disconnect, which is unacceptable from a service support point of view. This is clearly shown in Fig. 5, while in Fig. 6 the case of new links powering up and driving even established ones unstable is demonstrated.

The underlying problem is that the raw DPC algorithm (Eq. 14) does not support admission and congestion control in the channel, which are fundamental networking functions.[12,16,17] A novel idea/concept is needed to allow for such support, and this turns out to be the active link protection (ALP), discussed in this section. Indeed, since new links are expected to arrive and power up often in a wireless network, it is essential to protect already active (operational) ones from having their SIRs degrade below their target γ and disconnect. Even more importantly, we need to exercise admission control in the channel, rejecting any new link that makes the network

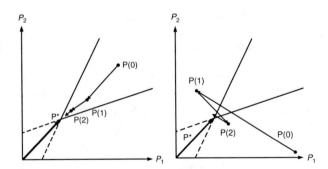

Fig. 4 A graphical view of the convergence of the DPC algorithm (Eq. 14) for the two-link network of Fig. 1. The power vector evolution at consecutive iterations is shown. In the left graph, the initial power vector $\mathbf{P}(0)$ is within the feasibility region/cone, while in the right one it is outside. At each iteration, each transmitter autonomously changes its power to hit the cone boundary, which corresponds to hitting its target SIR if the other transmitters do not change their powers.

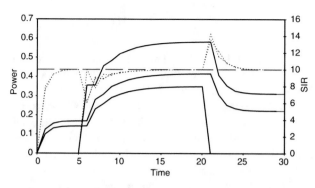

Fig. 5 Power (solid lines, left scale) and SIR (dotted lines, right scale) evolutions under the DPC algorithm (Eqs. 14 and 11, respectively) for a three-link network. All links have the same target SIR $\gamma_i = 10$, which is achievable. In the time interval $\{1, 2, 3, 4, 5\}$ only links 1 and 2 are active. At time 6, link 3 powers up in the channel. At time 20, link 1 terminates transmission and drops out. Note how the SIRs of links 1 and 2 deteriorate temporarily as link 3 powers up in the channel and the powers escalate to accommodate the new link.

infeasible. Can this be done in a distributed, autonomous, scalable fashion, without requiring any interlink communication? Fortunately, the answer is yes,[12,16,17] as discussed below.

Let $\mathcal{L}$ be the set of all interfering links in the channel. At each power update step $k = 1, 2, \ldots$, let us partition $\mathcal{L}$ into the subset $\mathcal{A}_k$ of *active* or *operational* links or the subset $\mathcal{B}_k$ of *inactive* or *new* links according to the criterion of whether the links' SIRs are currently above or below their target γ, that is:

$$\mathcal{A}_k = \{i \in \mathcal{L} : R_i(k) \geqslant \gamma_i\} \tag{15}$$

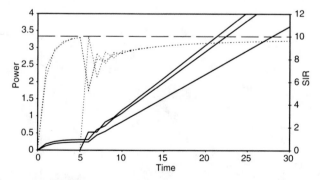

Fig. 6 Power (solid lines, left scale) and SIR (dotted lines, right scale) evolutions under the DPC algorithm (Eqs. 14 and 11, respectively) for a three-link network with a common target SIR $\gamma_i = 10$, which is *infeasible*. In the time interval $\{1, 2, 3, 4, 5\}$ only links 1 and 2 are active and actually admissible at $\gamma = 10$. At time 6, link 3 powers up in the channel. Since all three links are not admissible in the channel at $\gamma = 10$, their powers explode to infinity while their SIRs saturate below the target $\gamma = 10$.

$$\mathcal{B}_k = \{i \in \mathcal{L} : R_i(k) < \gamma_i\} \tag{16}$$

Let also

$$\delta = 1 + \epsilon > 1 \tag{17}$$

be a control parameter ($\epsilon > 0$), whose purpose and significance will become clear below. Then, the **DPC/ALP algorithm** works as follows:

$$P_i(k + 1) = \begin{cases} \frac{\gamma_i}{R_i(k)} P_i(k) & i \in \mathcal{A}_k \\ \delta P_i(k) & i \in \mathcal{B}_k \end{cases} \tag{18}$$

Note that under DPC/ALP, each *active* link updates its power similarly to the standard DPC algorithm (Eq. 14), but shoots for a *boosted SIR target* $\delta\gamma_i$ for reasons explained below; each *new/inactive* link, however, powers up gradually at geometric rate δ. The intuition behind DPC/ALP is the following. Each active link shooting for a slightly higher (than needed) SIR target $\delta\gamma_i$ results in an *SIR protection margin* $\epsilon\gamma_i = (\delta - 1)\gamma_i$ acting as a cushion. It protects the link's SIR from dropping below its true target γ_i, cushioning the jolts induced by the new links powering up. On the other hand, since new links power up gradually, their degrading effect on active ones is bounded and has already been anticipated by raising the SIR target to $\delta\gamma_i$ and introducing the protection/cushioning zone of $(\delta - 1)\gamma_i$ thickness.

Several key properties of the DPC/ALP dynamics can be rigorously established and are leveraged later. First, note that there is a bounded power overshoot property:

$$P_i(k + 1) \leqslant \delta P_i(k) \tag{19}$$

for each link i at each step $k = 1, 2, \ldots$, when the links operate under the DPC/ALP algorithm with any fixed $\delta > 1$. Indeed, this follows immediately from the definition of $\mathcal{A}_k$ and Eq. 18. This property guarantees that active link powers may increase smoothly—up to a factor of δ on each update—while inactive link ones increase by exactly a factor δ in each update.

The second property is key **ALP**. That is, when the links operate under DPC/ALP with any fixed $\delta > 1$, we have

$$R_i(k) \geqslant \gamma_i \Rightarrow R_i(k + 1) \geqslant \gamma_i \tag{20}$$

for each link i at each time $k = 1, 2, \ldots$. That means that if a link i is active ($R_i(k) \geqslant r_i$ and so $i \in \mathcal{A}_k$) in the kth time slot, it will also be active ($R_i(k + 1) \geqslant \gamma_i$ and so $i \in \mathcal{A}_{k+1}$) in the $(k + 1)^{st}$ time slot. Thus, *an active link will remain active forever* (until completing communication), while a new link may become active at some point and remain so forever after; i.e., $i \in \mathcal{A}_k \Rightarrow i \in \mathcal{A}_{k+1}$ and $\mathcal{A}_k \subseteq \mathcal{A}_{k+1}$ $\mathcal{B}_{k+1} \subseteq \mathcal{B}_k$ but at each time k. This property in Eq. 20 can be easily seen as follows. Observe that the interference

$I_i(k+1) = \sum_{j \neq i} G_{ij}P_j(k+1) + \eta_i \leqslant \delta\{\sum_{j \neq i} G_{ij}P_j(k) + \eta_i\} = \delta I_i(k)$ or $I_i(k+1) \leqslant \delta I_i(k)$ at every time k, because of the property in Eq. 19 and $\delta > 1$. Using this fact and Eq. 18, we get for any active link $i \in \mathcal{A}_k$ that $R_i(k+1) = G_{ii}P_i(k+1)/I_i(k+1) = (G_{ii}/I_i(k+1)\delta\gamma_i)/(R_i(k)P_i(k)) = (\delta\gamma_i(I_i(k))/(I_i(k+1)) \geqslant \delta\gamma_i\frac{1}{\delta} = \gamma_i$, hence Eq. 20.

Finally, there is a property of **new link improvement** at every time step. That is, when the links operate under DPC/ALP with any fixed $\delta > 1$, we have

$$R_i(k+1) \geqslant R_i(k) \quad \text{for any new link } i \in \mathcal{B}_k \qquad (21)$$

at each time $k = 1, 2, 3, \ldots$. This is easily seen since $I_i(k+1) \leqslant \delta I_i(k)$ as argued above, and $P_i(k+1) = \delta P_i(k)$ for $i \in \mathcal{B}_k$. Thus, each of the new links increases its SIR at each step, so it may eventually rise above its target SIR γ and become active, staying such forever after. Note that this introduces a natural concept of channel admission, which is exploited later.

Note that the geometric convergence of the plain DPC algorithm, coupled with the geometric power-up of the inactive links, guarantees a geometric convergence of the DPC/ALP. For small δ (close to 1), the ALP component dominates the convergence rate; for large δ, the inherent speed (dominated by ρ_F) of the plain DPC becomes prevalent in the DPC/ALP.

Admission Control under DPC/ALP

Let us now study how the DPC/ALP algorithm (Eq. 18) supports the concept of admission control of new links in the channel. Recall that network admission is equivalent to the link becoming active in the channel by raising its SIR above its target γ. According to the property in Eq. 20, once a link becomes active it is guaranteed to remain so until voluntarily completing transmission. This property is the basis of the network admission concept. A new link arriving and seeking admission to the channel starts powering up from a low initial power and, according to Eq. 21, sees its SIR improve at every step. Hence, it will either rise above its SIR target γ and become active (and stay so forever after) or saturate below its target if it is not admissible in the channel.

Let us examine some key cases of provable[12,16,17] link behavior and make some interesting observations. Suppose that the network starts at time 0 with a set $\mathcal{A} = \mathcal{A}_0$ of active links and a set $\mathcal{B} = \mathcal{B}_0$ of new (inactive) ones, and let us consider their evolution under DPC/ALP.

First consider the case where all links $i \in \mathcal{A} \cup \mathcal{B}$ can be admitted to the channel at SIR levels $\delta\gamma_i$—call them δ-compatible. That means that there is a Pareto-optimal finite power vector $P_\delta^* = (I - \delta F)^{-1}u < \infty$ that the system should operate at. Under DPC/ALP each new link will become active in finite time and be admitted to the channel. The power of each link will eventually converge

$\lim_{k \to \infty} P_i(k) = [P_\delta^*]_i$ to the optimal power value and its SIR to its boosted target value $\lim_{k \to \infty} R_i(k) = \delta\gamma_i$, which is targeted by DPC/ALP. Throughout convergence, no active link will ever drop below its actual target value γ. This is demonstrated in Fig. 7.

What happens if all links $i \in \mathcal{A} \cup \mathcal{B}$ are *not* admissible at SIR levels $\delta\gamma_i$ (hence, $(I - \delta F^{-1})u = \infty$) but are all admissible at SIR levels γ_i (hence, $(I - F^{-1})u < \infty$)? In that case, all new links will actually become active in finite time. However, as active links will keep shooting for $\delta\gamma_i$ under DPC/ALP, all powers will eventually explode to infinity $\lim_{k \to \infty} P_i(k) = \infty$, while all SIRs will converge $\gamma_i < \lim_{k \to \infty} R_i(k) < \delta\gamma_i$ below the target $\delta\gamma_i$. This problematic power explosion situation is demonstrated in Fig. 8.

By DPC/ALP shooting for the boosted SIR target $\delta\gamma_i$ rather than the required γ_i, some network capacity is basically traded for enhanced performance $(\delta - 1)\gamma_i$ per link. However, since δ can be chosen to be arbitrarily small, the capacity loss can be made arbitrarily small as well. In any case, any small loss of capacity is overcompensated by the ALP and the other benefits in network performance discussed before. How should δ be chosen? When the channel is not congested (has low P^*, ρ_F significantly lower than 1, etc.), we can choose a δ significantly higher than 1, so that new links increase their powers fast and gain admission to the channel. On the contrary, as the channel becomes congested (P^* gets high, ρ_F gets close to 1, etc.), the network becomes more sensitive to link admissions and we should choose a δ closer to 1 in order not to lose network capacity and also gracefully admit new links if possible.

In general, during evolution under DPC/ALP each new (inactive) link sees its SIR increase in every step because of the property in Eq. 21; hence, either it will reach its target SIR γ and become active after a few steps or it will

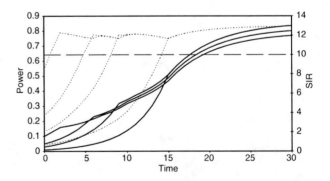

Fig. 7 Power (solid lines, left scale) and SIR (dotted lines, right scale) evolutions under the DPC/ALP algorithm (Eq. 18) of a four-link network with common SIR target $\gamma_i = 10$ and $\delta = 1.3$. All links are admissible at SIR level $\delta\gamma$; i.e., they are δ-compatible. Each link starts inactive and when its SIR rises above γ_i the link turns active and eventually its SIR converges to $\delta\gamma_i$.

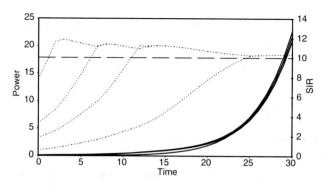

Fig. 8 Power (solid lines, left scale) and SIR (dotted lines, right scale) evolutions under the DPC/ALP algorithm (Eq. 18) of a four-link network with common SIR target $\gamma_i = 10$ and $\delta = 1.3$. All links are admissible at SIR target γ_i but not at $\delta\gamma_i$. Since DPC/ALP shoots for $\delta\gamma_i$, all powers eventually explode to infinity and the SIRs saturate below the DPC/ALP boosted target $\delta\gamma_i$, but still above the absolute target γ_i.

saturate below γ and stick there—this is a clear indication that it is not admissible to the channel. Therefore, some of the new links gain admission, while others are kept out by saturating below their target SIR; it may actually be beneficial to the network if some new links back off (set their power to zero) and try again later for admission. By backing off, they reduce interference on other new links, giving them a better chance of gaining admission. Fig. 9 demonstrates such a case. The network consists of four links, but initially only one of them gains admission. The remaining three links saturate below their SIR target γ_i. However, when a link drops out (the one with the lowest SIR in the plot) by setting its power to zero, the other two links soon get admitted. Had no link dropped out, the power levels of all the links would have exploded to infinity.

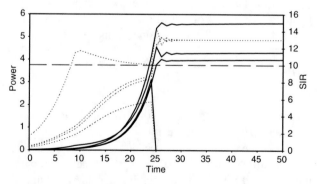

Fig. 9 Power (solid lines, left scale) and SIR (dotted lines, right scale) evolutions under the DPC/ALP algorithm (Eq. 18) of a four-link network with common SIR target $\gamma_i = 10$ and $\delta = 1.3$. One link gains admission soon, but the rest saturate below γ. When the link with the lowest SIR voluntarily drops out at time 25 (sets its power to zero), the other two soon gain admission.

This example shows that we should augment the DPC/ALP algorithm so that new links voluntarily drop out when they sense that the channel is congested; hence, it is unlikely that they will become admitted. These links may either try admission to a different channel immediately (if available) or back off for some time before trying again at the same (or a different) channel. During their back-off, they should remain dormant (set the transmitter power level equal to zero) and then restart their power-up from the beginning. Depending on the criterion used for dropping out, we can design different algorithms for voluntary dropout (VDO) of new links as follows.

VDO

The first algorithm is based on the idea of a *time-out* when a new link tries for a long time to gain admission without success. Specifically, if a link powers up but remains inactive longer than time T_i (time-out), it computes a dropout time horizon D_i as a decreasing function of the distance between its target γ_i and its current SIR $R_i(T_i)$. The link continues to compete for admission for another D_i iterations. If at time $T_i + D_i$ it has not become active, it voluntarily drops out, setting its transmission power to zero. A new admission attempt may be reinitiated later in the same or another channel. The elimination of the link from the channel reduces the interference on other links seeking admission and increases their chances of success. T_i and D_i can be either fixed quantities or chosen at random (e.g., with a geometric distribution). Nonetheless, they are design parameters that should be optimized by testing. Indeed, too short a time out period will cause the link to drop out possibly unnecessarily, whereas too long a time-out will result in the link heavily congesting the channel. Similarly, too short a back-off period will result in frequent readmission trials, which will increase the congestion of the channel and possibly lower the admission chances by having too many inactive links seeking admission; on the other hand, too long a back-off period will result in large delays before a link gets admitted.

Taking a different approach, we present an alternative algorithm for dropping out. The basic idea is that the link should keep trying to get admission as long as it observes some adequate SIR improvement over a time window of M_i steps. If persistently for more than M_i steps no such improvement has occurred, the link decides whether to drop out or to continue by flipping a coin. Coin flips in subsequent steps are independent, and the dropout probability is a decreasing function of the difference between the target and the current SIRs. As we have already seen, the idea behind this algorithm lies in the fact that if a link is inadmissible, then it will saturate below its target. Hence, consecutive power update steps will not bring about any significant SIR improvement, so the link will start the randomized dropout process. If, however, some other link drops out sooner, the link under consideration may experience some significant SIR increase, which will abort the dropout process and give it more time to try for admission. More details can be found in Ref. 17.

Forced dropout

As we have already seen, admission of new links to the channel leads to an increase of the Pareto-optimal power vector $\mathbf{P}^*$, as the Perron–Frobenious eigenvalue ρ_F moves closer to 1. In a real system, however, the transmitters would be limited by some maximum transmission power P_i^{max}. What if the transmitter has to exceed P_i^{max} in order to keep the link active? Clearly, we need a forced dropout (FDO) mechanism[17] that will force the inactive links to dropout when the active links are close to their maximum transmission powers. This mechanism should be as follows: When an active link senses that it is about to exceed its power limit, it should transmit a *distress* signal (which can be a special tone in some control channel/slot) at a certain power level. This signal is received by the links in its vicinity and instructs any inactive link that hears it above a certain power level to drop out immediately in order to de-congest the neighborhood of the active link, permitting it to relax its transmission power level. This distress signal should be transmitted by any active link $i \in A_k$ whenever its transmission power enters the region $(P_i^{max}/\delta, P_i^{max})$ —as it follows from Eq. 19—in order to avoid the for bidden region of (P_i^{max}, ∞). Note that this property is not true for the plain DPC, as the power overshoots are not bounded in the DPC.

FDO is also useful in the case of 1-compatible, but not δ-compatible, links.[17] Although not probable in a dynamic environment where multiple links seek admission, when it occurs, active link powers diverge to infinity, triggering a distress signal from the link that first reaches its power limit. Thus, newly admitted links should monitor for the distress signal for some short time horizon after their admission and drop out if they hear any. After the time horizon has passed, the links should consider themselves permanently admitted.

Initial power level of new links

When the new links initiate their admission process, they start transmitting at some power P_0, which may cause an active link to drop its SIR below its target γ_i.[17] To avoid this link being considered inactive, we should modify the rule of distinguishing between active and inactive links to include a "once active, always active" clause. Thus, even if the active link mildly dips below γ_i for a short time, it will still be considered an active link and rapidly recover due to the fast geometric convergence of the DPC algorithm. However, by choosing P_0 to be small enough, we can limit the disturbance to active links. Indeed, through a

simple calculation[17] we see that if $P_0 \leqslant (\delta - 1)\eta_{min}/G_{max}$ no active link will drop below γ_i upon the appearance of the new one, given that all the active links have attained their enhanced targets $\delta\gamma_i$. G_{max} is the maximum interlink power gain, and η_{min} is the minimum thermal noise among the active links.

Noninvasive Channel Probing and Selection

Another key idea is that of channel probing,[13,21] which can alternatively be employed for admission control, or even in conjunction with the ideas discussed before. Channel probing can be used by a new link to quickly decide whether it is admissible in a channel, by collecting quantitative information about its congestion state in a fully autonomous manner. Hence, when there are several channels available, probing can be used to identify the best channel to seek admission into. The implementation of autonomous channel probing is based on the following provable facts.

Consider a set of initially active links $\mathcal{A} = \mathcal{A}_0$ and a set of initially inactive (new) ones $\mathcal{B} = \mathcal{B}_0$ interfering in a channel. Suppose that:

1. All links follow the DPC/ALP power update algorithm (Eq. 18) with some fixed $\delta > 1$.
2. No new link in $\mathcal{B}$ becomes active before iteration K.
3. All active links $i \in \mathcal{A}$ are such that $26.\gamma_i R_i(0) \leqslant \delta\gamma_i$ (are initially stable).

Then it can be shown[13,21] that the SIR of each new link $i \in \mathcal{B}$ evolves within a certain envelope while the link powers up in the channel, as follows:

$$\frac{1}{\frac{X_i(\delta)}{\delta^k} + Y_i(\delta)} \leqslant R_i(k) \leqslant \frac{1}{\frac{X_i(1)}{\delta^k} + Y_i(1)} \quad \text{for each } i \in \mathcal{B} \quad (22)$$

for all $k \leqslant K$, where $X_i(\delta)$ and $Y_i(\delta)$ are functions of $\mathbf{G}$, γ, η, and $\delta > 1$. Actually, as $\delta \downarrow 1$, we have $X_i(\delta) \downarrow X_i(1) = X_i$ and $Y_i(\delta) \downarrow Y_i(1) = Y_i$; hence, the lower bound converges to the upper bound of the SIR evolution envelope. This means that for $\delta \approx 1$ the envelope shrinks to a trajectory

$$R_i(k) \approx \frac{1}{\frac{X_i}{\delta^k} + Y_i} \quad \text{when } \delta \approx 1 \quad (23)$$

for $k = 0, 1, 2, \ldots, K$. Based on this, each new link $i \in \mathcal{B}$ can *autonomously* perform the following probing functions and make the corresponding decisions. First, by observing its own SIR for a few initial steps, it estimates the parameters X_i and Y_i via curve fitting. Based on these parameter estimates, the link predicts its SIR evolution in the future power-up steps. In principle, two consecutive SIR measurements (at low powers $P_i(0)$ and $P_i(0)\delta$) would be enough to estimate

$$X_i \approx \frac{\delta}{\delta - 1} \left\{ \frac{1}{R_i(0)} - \frac{1}{R_i(1)} \right\} \text{ and}$$

$$Y_i \approx \frac{1}{\delta - 1} \left\{ \frac{\delta}{R_i(1)} - \frac{1}{R_i(0)} \right\} \quad (24)$$

In practice, however, a link may need a few more steps to account for randomness and measurement errors. In any case, the transmitted power during these few initial steps is very low; hence, the probing process is fast and noninvasive, generating minimal interference in the channel.

If the new link i were to continue powering up in the channel without ever becoming active, then from Eq. 23 we see that $R_i(k) \to \frac{1}{Y_i} < \gamma_i$, as $k \to \infty$ (since $\delta > 1$). Indeed, the SIR will keep increasing and eventually saturate below the target γ_i. Having estimated Y_i by probing the channel, the link may immediately decide to drop out when $\frac{1}{Y_i} < \gamma_i$, without stressing the channel any further.

If, however, the estimated parameter Y_l is such that $\frac{1}{Y_i} > \gamma_i$, then the link is admissible in the channel and should proceed to power up in it. Indeed, the link will gain admission and become active at time k_i^* approximately, when its SIR reaches the required threshold $R_i k_i^* \approx \gamma_i$, i.e.,

$$\frac{1}{\frac{X_i}{\delta^{k_i^*}} + Y_i} \approx \gamma_i \quad (25)$$

The link can then autonomously estimate its admission delay time $k_i^* \approx \left(\frac{1}{\log(\delta)} \log(X_i / \frac{1}{\gamma_i} - Y_i) \right)$ and the power level at which it will gain admission $P_i(k_i^*) \approx \delta^{k_i^*} P_i(0) \approx (X_i / (\frac{1}{\gamma_i} - Y_i)) P_i(0)$ using the estimates X_i and Y_i obtained in the probing phase.

To conclude, the link can make a decision to proceed powering up and enter (become active in) the channel, if it finds the admission delay and entry power acceptable. If not, it can drop out from this channel and probe another one. Moreover, if multiple channels are available, the link can probe several of them and decide which one to join according to some criteria, e.g., admission power and delay. We do not elaborate any further on these issues here, but complete details can be found in Ref. 13 and 21.

POWER CONTROL FOR PACKETIZED DATA TRAFFIC

The power control problem formulation employed in the previous section was geared toward wireless links carrying continuous delay-sensitive traffic (primarily voice or video), where power control is used to maintain a desired SIR level. This formulation, however, is not appropriate for packetized data traffic, which may be bursty and delay tolerant. In particular, it does not allow us to capture the fundamental *power vs. delay trade-off* of data traffic,

which entails the following. During high interference periods, when high power is required to reliably transmit data, it may be better to buffer traffic and incur a delay cost, in anticipation that interference will soon subside and power benefits will be realized by reliably transmitting high data volumes at low power. What is needed is an enhanced power control model/formulation,[14,15,18,22,23] capturing the above trade-off. We provide such a model and discuss it in this section, leading to identification of a new class of algorithms for power-controlled multiple access (PCMA).

PCMA: The Basic Model

We take the local perspective of a single link, sharing a wireless channel with many others, which creates a fluctuating interference environment. Time is slotted. Let $I(k)$ be the interference that the link experiences at each receiver throughout time slot $k = 1, 2, 3, \ldots$; the interference remains constant within a slot but may change across different slots. The link's transmitter is equipped with a buffer, where packets may be stored waiting to be transmitted. Let $B(k)$ be the packet backlog in the buffer, i.e., the number of packets in the transmitter buffer awaiting transmission. At the beginning of each time slot k, the link checks the interference $I(k)$ and the transmitter backlog $B(k)$ and decides what power $P(k)$ to use to transmit the head packet (if any) of the transmitter buffer (FIFO). If a packet is transmitted at power P across the channel, when the interference is I, then it is successfully received and decoded with probability $S(P,I)$ and so is removed from the buffer. Successful packet reception events are statistically independent in different time slots. The function $S(P,I)$ should obviously be increasing in P and decreasing in I; two typical functional forms in communications systems are the following:

$$S_1(P, I) = \frac{P}{\alpha P + \beta I} = \frac{\frac{P}{I}}{\alpha \frac{P}{I} + \beta} \quad (26)$$

where $\alpha \geq 1$ and $\beta > 0$, and

$$S_1(P, I) = 1 - e^{-\delta \frac{I}{P}} \quad (27)$$

where $\delta > 0$. Note that $\frac{P}{I}$ is the SIR in the time slot. If the packet is not successfully received at the receiver because it was corrupted during transmission (with probability $\{1 - S(P,I)\}$), then it will have to be transmitted again and again until it is successfully received. We assume that the transmitter is immediately notified at the end of each slot (perhaps over a separate control channel), whether or not the packet transmitted in that slot was successfully received, through some highly reliable ACK/NACK process that takes negligible time. Assume that the transmitter first checks the interference at the beginning of each time slot k, then selects a power $P(k)$ at which to transmit the head packet, and—if successful—removes the packet from the queue at the end of the time slot.

To spotlight and capture the fundamental power vs. delay trade-off, we make the assumption that the interference $I(k)$ induced on the link is *nonresponsive* to the powers $P(k)$ transmitted by it, as if the interference were determined by an extraneous (to the system) agent that is insensitive to the transmitter power of the link under consideration. This is a substantial simplification, since we know that all channel links are entangled to each other through mutual interference. Increasing the power of the link would increase the interference it generates on other channel links; those in turn would increase their powers, resulting in higher interference on the former link. The nonresponsive interference assumption, however, does have high value in the following sense. It allows for obtaining a tractable model whose optimal solution provides substantial insight into the general model of responsive interference. This insight obtained on the reduced model leads to the design of efficient PCMA algorithms, which perform surprisingly well when used in the full system with responsive interference.[14,18,22] This is further discussed later. Based on that, we proceed with the nonresponsive interference assumption, used as a conceptual device and a methodological step in our study. We assume that the interference $I(k)$ evolves according to an irreducible Markov chain with transition probabilities $P[I(k + 1) = J | I(k) = I] = Q_{IJ}$ and steady-state π_I, where I, J are interference states from a finite set of all possible ones.

The power control problem can now be expressed as follows. First, let us define the cost structure of the system. In each time slot k, the link incurs:

1. A power cost $P(k)$, i.e., the power transmitted in the channel in the time slot;
2. A delay cost $D(B(k))$, which is an increasing function of the packet backlog $B(k)$ in the time slot.

Thus the system incurs per time slot a cost equal to the sum of the above two costs. Then, the overall cost is equal to the cost incurred per slot summed over all time slots throughout the evolution of the system. Observing the backlog $B(k)$ and interference $I(k)$ in each slot k, the power control problem is to decide what power $P(k)$ to use for transmitting the head packet of the buffer, so as to minimize the overall average cost throughout the evolution of the system.

Optimally Emptying the Transmitter Buffer

In the simplest form of the problem, we consider the case of starting with B packets in the buffer and interference state I, and optimally controlling the transmitter power $P(1)$, $P(2)$, $P(3)$, ... in order to empty the buffer incurring the minimal overall cost. This problem can be

solved using the framework of dynamic programing,[24] as follows.

Let $V(B, I)$ be the *cost-to-go*, i.e., the minimal expected cost that will be incurred under optimal power control until the buffer empties, given that the system starts with backlog B and interference I. The standard dynamic programming recursion (for $B > 0$) is simply

$$V(B,I) = \inf_{P \geq 0} \left\{ P + D(B) + S(P,I)\left[\sum_J Q_{IJ}V(B-1,J)\right] + [1 - S(p,I)]\sum_J Q_{IJ}V(B,J) \right\} \quad (28)$$

with initial value $V(0,I) = 0$. Q_{IJ} is the probability of the interference state switching from I to J, and the summation above is performed over all possible states J. The first two terms $P + D(B)$ of Eq. 28 correspond to the cost incurred in the current time slot, while the other two terms express the collective future cost conditioned on successful and unsuccessful reception of the packet correspondingly. Rearranging the terms, for $B > 0$, we get

$$V(B,I) = \inf_{p \geq 0} \{P - S(P,I)X(B,I) + Y(B,I)\} \quad (29)$$

where

$$X(B,I) = \sum_J Q_{IJ}[V(B,J) - V(B-1,J)] \quad (30)$$

$$Y(B,I) = D(B) + \sum_J Q_{IJ}V(B,J) \quad (31)$$

Note that $V(B,J) - V(B-1,J) \geq 0$ for every $B > 0, I$, since the cost to empty the buffer increases when there are more packets. Therefore, $X(B,I) \geq 0$. Recursively solving the above equations, we obtain the optimal power control policy $P^* = P^*(B,I)$ as a function of the system state (B,I). Hence, the optimal transmitter power in slot k is $P^*(k) = P^*(B(k), I(k))$.

In order to solve the equations, we need to have a good knowledge (empirical or analytic) of the function $S(P,I)$ and a reliable statistical profile of interference behavior $\{Q_{IJ}\}$. It is then possible to solve the dynamic programming recursion off-line and store the optimal power values $P^*(B,I)$ in a lookup table accessible by the transmitter on-line.

The Case of Per-slot-independent Interference

Dynamic programing recursions can rarely be solved in a closed form. Fortunately, a nice semianalytical solution can be obtained for our model, in the special case when the interference is *independent and identically distributed* (i.i.d.) on different time slots. It is worth studying because it provides significant insight into the structure of optimal power control in the general case.

Under i.i.d. interference, the transition probabilities Q_{IJ} become equal to the stationary distribution probabilities π_J, and thus X and Y in Eqs. 30 and 31, respectively, lose their dependence on I, greatly reducing the complexity of the problem. Indeed, $X = X(B) = \sum_J \pi_J[V(B,J) - V(B - 1, J)]$ and $Y = Y(B) = \sum_J \pi_J[D(B) + V(B,J)]$ under i.i.d. interference with distribution π_J. Revisiting Eq. 29, if $S(P,I)$ is convex in P, then we can see that the solution P^* of the equation

$$\frac{\partial S(P,I)}{\partial P} = \frac{1}{X(B)} \tag{32}$$

if positive, minimizes the right-hand side of Eq. 29. To better study the nature of the solution, let us consider two typical functional forms for $S(P,I)$ as follows. When $S_1(P,I) = \frac{P}{\alpha P + \beta I}$ with $\alpha \leqslant 1$ and $\beta > 0$, from Eq. 32 we get

$$P_1^*(B,I) = \begin{cases} \frac{1}{\alpha}\left[\sqrt{\beta X(B)I} - \beta I\right] & I \leqslant \frac{X(B)}{\beta} \\ 0 & I > \frac{X(B)}{\beta} \end{cases} \tag{33}$$

while when $S_2(P,I) = 1 - e^{-\delta \frac{P}{I}}$ with $\delta > 0$, Eq. 32 results in

$$P_2^*(B,I) = \begin{cases} \frac{1}{\delta}\ln\frac{\delta X(B)}{I} & I \leqslant \delta X(B) \\ 0 & I > \delta X(B) \end{cases} \tag{34}$$

as the optimal power control function. More details can be found in Ref. 18.

Structural Properties: Backlog Pressure and Phased Back-off

Fig. 10 shows plots of the optimal powers P_1^* (Eq. 33) and P_2^* (Eq. 34) against the interference I, for various $X = X(B)$ values. The explicit dependence on B has been suppressed in the graphs and has been incorporated in $X(B)$, which is an increasing function of B reflecting the backlog pressure (observe that the cost-to-go difference $V(B,I) - V(B - 1, I)$ will be increasing in B, as $D(B)$ is increasing in B).

Note that although $S_1(P,I)$ and $S_2(P,I)$ have quite different analytic forms, the plots of the corresponding optimal power levels P_1^* and P_2^* are very similar. Indeed, observe that for a fixed backlog B [and thus a fixed backlog pressure $X(B)$], the optimal power P^* goes through three different phases in response to interference: *aggressive, soft back-off*, and *hard back-off*. In the first phase (low interference zone), the transmitter tries to aggressively match the interference and increases its power so as to maintain the needed success probability $S(P,I)$ and alleviate the delay cost. This effort gradually relaxes and eventually at

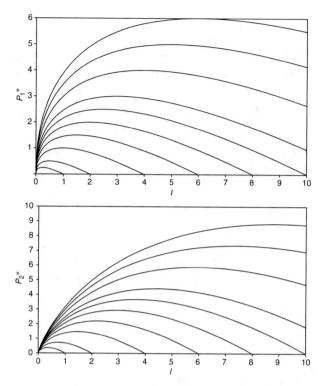

Fig. 10 Plots of optimal power vs. interference function for various fixed back-log pressures $X(B) = 1, 2, 4, 6, 8, 10, 12, 16, 20, 24$. On the left, Eq. 33 is plotted for $\alpha = \beta = 1$, while on the right Eq. 34 is plotted for $\delta = 1$. In both plots, the family of curves are ordered, the lowest being for $X(B) = 1$ and the highest for $X(B) = 24$.

some interference level the power P^* attains a maximum and the second soft back-off phase begins. In this phase, matching the interference is no more cost-efficient, and hence the transmitter starts to withdraw from the channel. The transmitter *softly backs off* of the channel by gradually decreasing its power until it hits zero. In the third phase, the transmitter completely refrains from any transmission (hard back-off) because setting $P^*=0$ as the interference is too high to even make a transmission attempt. Intuitively, the two back-off phases can be explained as being preferable to incur some delay cost instead of spending enormous amounts of power (energy) for successful transmission when the interference is high.

As already mentioned, the effect of the backlog B on P^* is indirectly considered through $X(B)$, which is increasing in B. Observe that the boundaries of the three phases of P^* depend on the value of $X(B)$; an increase in $X(B)$ expands the length of the aggressive phase and the power levels used. In essence, $X(B)$ corresponds to the backlog pressure, i.e., the pressure coming from the delayed packets, which forces the transmitter to be more aggressive at any fixed interference level I.

The above observations regarding the dependence of P^* on I and B, although originating from two particular functional forms of $S(P,I)$, are actually ubiquitous for any regular $S(P,I)$, and thus can be used in designing efficient PCMA algorithms. The properties and behaviors discussed are largely structural in nature and very robust with respect to particular choices of parameters and functions $S(P,I)$.

Incorporating a finite ceiling P_{max} for the transmitter power, that is, $P \in [0, P_{max}]$, the solution changes slightly by having to take into account the new constraint in the minimization. Specifically, $P_1^*(B, I) = \min\{\frac{1}{\alpha}[\sqrt{\beta X(B)I} - \beta I], P_{max}\}\mathbf{1}_{\{I \leq \frac{X(B)}{\beta}\}}$, where $\mathbf{1}_{\{\cdot\}}$ is the standard indicator (The indicator function $\mathbf{1}_{\{x\}}$ is equal to x when $x > 0$ and 0 otherwise) function. Moreover, $P_2^*(B, I) = \min\{\frac{I}{\delta}\ln(\delta X(B)/I), P_{max}\}\mathbf{1}_{\{I \leq \delta X(B)\}}$.

Finally, we can easily extend the model[14,15,18,22–24] to include packet arrivals, a finite buffer and corresponding overflow cost, and even packet deadlines and expiration cost. We can similarly use dynamic programing in all these cases, and the resulting optimal power control solutions have very similar structural properties to those discussed in the simple model analyzed above.

Design of PCMA Algorithms: Responsive Interference

Note that the DPC algorithm (Eq. 14) previously introduced for continuous traffic can be written as $P_i(K + 1) = \frac{\gamma_i}{R_i(k)}P_i(k) = \frac{\gamma_i}{G_{ij}}I_i(k)$ for each link i. At every step, it tries to match the interference and keep the SIR constant. Based on that observation, we can naturally extend it to the case of packetized traffic as

$$P_i(k + 1) = \frac{\gamma_i}{G_{ii}}I_i(k)\mathbf{1}_{\{B_i(k)>0\}} \tag{35}$$

i.e., each link applies DPC in each slot, when its backlog is nonzero.

Now, based on the optimal power control solutions in Eq. 33 and 34, especially their versions including a power ceiling P_{max}, we can introduce the following two PCMA algorithms. Each link i observes its own backlog $B_i(k)$ and interference $I_i(k)$ at its receiver at the current time slot and sets its power in the next time slot as follows:

$$P_i(k + 1) = \min\left\{\frac{1}{\alpha}\left[\sqrt{\beta X(B_i(k))I_i(k)}\right.\right.$$
$$\left.\left. - \beta I_i(k)\right], P_{max}\right\}\mathbf{1}_{\{I_i(k) \leq \frac{X(B_i(k))}{\beta}\}} \tag{36}$$

with $\alpha \geq 1$ and $\beta > 0$. Another PCMA algorithm is to set the power in the next time slot according to

$$P_i(k + 1) = \min\left\{\frac{I_i(k)}{\delta}\ln\frac{I_i(k)}{\delta X(B_i(k))}, P_{max}\right\}\mathbf{1}_{\{I_i(k) \leq \delta X(B_i(k))\}} \tag{37}$$

with $\delta > 0$. The backlog pressure $X(B)$ is taken to be an increasing function of the backlog B, e.g., $X(B) = \Phi B + \Psi$, with $\Phi, \Psi > 0$. These algorithms are designed to operate in the realistic responsive interference environment, where all links are entangled to each other via interference.

It is interesting to note that simulations experiments[14,18] show that both PCMA algorithms perform very well, and certainly outperform by a large margin the DPC algorithm in Eq. 35 in almost all operational scenarios. We do not elaborate further within the limited scope of this entry, but only mention that the topic of optimal power control of wireless data traffic is vast and deserves further investigation.

FINAL REMARKS

We have discussed fundamental aspects and core algorithms for power control of both continuous delay-sensitive traffic and packetized data traffic. The emphasis has been on understanding basic principles and simple—yet robust and scalable—algorithms with high practical potential. Technology-specific implementations of the algorithms have not been discussed—the limited space of this entry would do them injustice.

The topic of transmitter power control is still under intense development and much further research should be expected in the near future, as wireless networking technology advances.

ACKNOWLEDGMENTS

We thank the referees for thoroughly reading this entry and making important suggestions regarding improvements and clarifications.

REFERENCES

1. Zander, J. Distributed cochannel interference control in cellular radio systems. IEEE Trans. Veh. Technol. **1992**, *41* (3), 305–311.
2. Foschini, G.J.; Miljanic, Z. A simple distributed autonomous power control algorithm and its convergence. IEEE Trans. Veh. Technol. **1993**, *42* (4), 641–646.
3. Foschini, G.J.; Miljanic, Z. Distributed autonomous wireless channel assignement with power control. IEEE Trans. Veh. Technol. **1995**, *44* (3), 420–429.
4. Mitra, D. An asynchronous distributed algorithm for power control in cellular radio systems. WINLAB 1993 Workshop, East Brunswick, NJ, Oct 19–20, 1993.
5. Grandhi, S.; Vijayan, R.; Goodman, D. Distributed power control in cellular radio systems. IEEE Trans. Commun. **1994**, *42*, 226–228.

6. Bambos, N.; Pottie, G.J. On power control in high capacity cellular radio networks. GLOBECOM 1992, Orlando, FL, Dec 6–9, 1992; Vol. 2, 863–867.
7. Chen, S.; Bambos, N.; Pottie, G. Admission control schemes for wireless communication networks with adjustable transmitter powers. IEEE INFOCOM 1994, Toronto, Canada, June 14–16, 1994.
8. Andersin, M.; Rosberg, Z.; Zander, J. Soft and safe admission control in cellular networks. IEEE/ACM Trans. Netw. **1997**, *5* (2), 255–265.
9. Grandhi, S.; Yates, R.; Goodman, D. Resource allocation for cellular radio systems. IEEE Trans. Veh. Technol. **1997**, *46*, 581–587.
10. Hanly, S. Capacity and power control in spread spectrum macrodiversity radio networks. IEEE Trans. Commun. **1996**, *44* (8), 247–256.
11. Yates, R. A framework for uplink power control in cellular radio systems. IEEE J. Sel. Areas Commun. **1995**, *13* (7), 1341–1347.
12. Bambos, N.; Chen, S.C.; Pottie, G. Radio link admission algorithms for wireless networks with power control and active link quality protection. IEEE INFOCOM 1995, Boston, MA, Apr 2–6, 1995.
13. Bambos, N.; Chen, S.; Mitra, D. Channel probing for distributed access control in wireless communication networks. GLOBECOM 1995, Singapore, Nov 13–17, 1995; 322–325.
14. Bambos, N.; Kandukuri, S. Power controlled multiple access (PCMA) in wireless communication networks. IEEE INFOCOM 2000, Tel-Aviv, Israel, Mar 26–30, 2000; 386–395.
15. Rulnick, J.; Bambos, N. Mobile power management for wireless communication networks. Wirel. Netw. **1997**, *3*, 314.
16. Bambos, N. Toward power-sensitive network architectures in wireless communications: concepts, issues, and design aspects. IEEE Pers. Commun. **1998**, *5*, 50–59.
17. Bambos, N.; Chen, S.C.; Pottie, G. Channel access algorithms with active link protection for wireless communication networks with power control. IEEE/ACM Trans. Netw. **2000**, *8* (5), 583–597.
18. Bambos, N.; Kandukuri, S. Power-controlled multiple access schemes for next-generation wireless packet networks. IEEE Wirel. Commun. **2002**, *9* (3), 58–64.
19. Seneta, E. *Non-Negative Matrices*. Wiley: New York, 1973.
20. Gantmacher, F.R. *The Theory of Matrices*; Chelsea: New York, 1960.
21. Bambos, N.; Kim, J-W.; Chen, S.; Mitra, D. Non-invasive channel probing for distributed admission control and channel allocation in wireless networks. Technical Report NetLab-2002-01/02. , Engineering Library, Stanford University, Stanford, CA, 2002.
22. Kandukuri, S.; Bambos, N. Multimodal dynamic multiple access (MDMA) in wireless packet networks. IEEE INFOCOM 2001, Anchorage, AK, Apr 22–26, 2001; 199–208.
23. Rulnick, J.; Bambos, N. Power-induced time division on asynchronous channels. Wirel. Netw. **1999**, *5*, 71–80.
24. Bertsekas, D. *Dynamic Programming*; Prentice Hall: New York, 1987.

BIBLIOGRAPHY

1. Alavi, H.; Nettleton, R.W. Downstream power control for a spread spectrum cellular mobile radio system. IEEE GLOBECOM 1982, Miami, FL, Nov 29–Dec 2, 1982.
2. Chuang, J.C-I.; Sollenberger, N.R. Performance of autonomous dynamic channel assignment and power control for TDMA/FDMA wireless access. IEEE J. Sel. Areas Commun. **1994**, *12* (8), 1314–1323.
3. Grandhi, S.; Vijayan, R.; Goodman, D.; Zander, J. Centralized power control in cellular radio systems. IEEE Trans. Veh. Technol. **1993**, *42* (4), 466–468.
4. Kleinrock, L.; Silvester, J. Spatial reuse in multihop packet radio networks. IEEE. **1987**, *75* (1), 156–167.
5. Yates, R.; Huang, C. Integrated power control and base station assignment. IEEE Trans. Veh. Technol. **1995**, *44*, 638–644.
6. Zander, J. Performance of optimum transmitter power control in cellular radio systems. IEEE Trans. Veh. Technol. **1992**, *41* (1), 57–62.
7. Zander, J. Radio resource management in future wireless networks; requirements and limitations. IEEE Commun. Mag. **1997**, *35*, 30–36.

Wireless
Multi—Networks

Wireless Networks: VoIP Service

Sajal K. Das
Kalyan Basu
Center for Research in Wireless Mobility and Networking, Department of Computer Science and Engineering, University of Texas at Arlington, Arlington, Texas, U.S.A.

Abstract

This article presents an overview of voice over IP (VoIP) services, protocols, and architectures.

INTRODUCTION

Currently, voice and data traffic are treated separately in wireless networks such as GSM[1] and GPRS[2] systems. The release 1 definition of third generation (3G) wireless systems also kept this separation. Wireless voice traffic is routed through the circuit-switched infrastructure whereas data traffic is routed through the packet network. To explain this integration, we look to GPRS integration in the existing GSM network. As shown in Fig. 1, two new GPRS dedicated nodes, SGSN (serving GPRS support node) and GGSN (gateway GPRS support node), are added in the existing GSM network. An SGSN is responsible for the delivery of data packets to and from the mobile station within its service area. Its tasks include packet routing and transfer, mobility management (attach/detach and location management), logical link management, and the authentication and charging functions. The location register of the SGSN stores the location information and user profiles of all GPRS users registered within the SGSN. GGSN acts as an interface between the GPRS backbone and the external packet data networks. It converts GPRS packets coming from the SGSN into appropriate PDP (packet data protocol) format and sends them out on the corresponding packet data network. In the other direction, the PDP addresses of incoming data packets are converted to the GSM addresses of the destination. The addressed packets are sent to the responsible SGSN. For this purpose the SGSN stores the current SGSN address of the user and the user's profile in its location register. GGSN also performs the authentication and charging function.

Although wireless and Internet technologies perform quite well within their own domains, the integration of wireless links into the Internet exhibits considerable challenges. Unpredictability in the air-link conditions, such as rapid fading, shadowing, and intermittent disconnection, affect the radio link performance, and the frame error rate (FER) can be as high as 10^{-1}. This causes serious quality degradation to the data users. Packet traffic is very sensitive to the bit error rate (BER). To overcome the higher BER problem, the wireless air-interface protocol has included the definition of a new protocol layer, RLP (radio link protocol), on top of the MAC (medium access control) layer. The RLP layer brings higher reliability on the wireless link by using the ARQ (automatic repeat request) retransmission technique in conjunction with the cyclic redundancy check (CRC).

The signaling for voice over IP (VoIP) should be able to gracefully migrate into the existing infrastructure to converge voice and data networks. Existing GSM voice networks and terminals use SS7 (signaling system 7) and ISDN (integrated signaling digital network) protocols. The ITU has defined H.323 (Ref. 3) as the key protocol that implements this migration to today's PSTN (public switched telephone network) signaling domain. New protocols such as SIP (session initiation protocol),[4] MGCP (media gateway control protocol),[5] and H.248[6] have been proposed recently to implement VoIP services to overcome the limitations of H.323 through gateway decomposition.

SIP, as defined by the IETF, is a signaling protocol for telephone calls over the Internet. Unlike H.323, SIP was specifically defined for the Internet and includes mobility management functions. It exploits the manageability of IP and makes telephony applications development relatively simple. It is basically used for setting up, controlling, and tearing down sessions on the Internet that include telephone calls and multimedia conferences. SIP supports various facets of telecommunications such as user location, user capabilities, user availability, call setup, and call handling.

In our view, multiple VoIP protocols will exist and interact among themselves for the foreseeable future. For our study in this entry, we have considered H.323 and SIP

Encyclopedia of Wireless and Mobile Communications DOI: 10.1081/E-EWMC-120043902

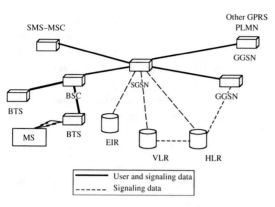

Fig. 1   GPRS network.

as the protocols of choice for VoIP services implementation. The proposed analysis, although based on H.323 implementation, is general in nature and can be easily extended to SIP.

The introduction of VoIP service introduces new impediments because the IP network, as shown in Fig. 2, replaces the network between the two ends. The previous assumption of negligible jitter in the synchronous digital network is no longer valid because of the packet-handling process in the best-effort IP network that will introduce variable amounts of delay between the subsequent packets. This variability must be eliminated at the destination so that the voice packets are offered to the decoder at a constant rate. To achieve this, the receiving end uses a dejitter buffer to delay packets so that all the packets are offered at a constant rate. To compensate for the high degree of network delay variability, the setting of the buffer delay can be high, thus introducing additional delay on the voice path. To reduce this delay jitter, the Internet is introducing new service types, such as DiffServ (differentiated service)[7] and MPLS (multiprotocol label switching),[8] to maintain the performance of VoIP traffic within only a small delay jitter tolerance.

Wireless VoIP introduces additional impediments within the network due to wireless data links in the access. In GSM and CDMA systems today, digitized voice packets are transmitted over wireless links without an ARQ-type protocol. Thus, if there is an error in the voice packets, the system accepts the packet or drops it. Voice quality is not significantly affected by this mechanism. Although the FER in wireless access links can be as high as $10^{-1}-10^{-3}$, voice quality degradation is not significant. But for data communication, no channel error is acceptable unless it is

protected by error correction code. To overcome channel error, the ARQ protocol is used to retransmit the erroneous packets. This retransmission currently happens at the TCP level because the original design of the IP protocol considered a very low error probability on the channel. Application of TCP retransmission for error correction in wireless networks only will significantly slow down the wireless link throughput and increase delay. To overcome this delay and throughput problem, MAC layer retransmission is considered for wireless data links. Wireless data access standards, such as GPRS and cdma2000, use MAC layer retransmission, called RLC/MAC, which causes additional delay jitter for wireless VoIP packets.

BASIS OF VOICE CODING

Voice signals in the telephone device are analog signals within the frequency range of 300 Hz to 3.4 kHz. The landline digital voice network converts these analog signals into digital signals with the help of the PCM (pulse code modulation)[9] scheme. In this scheme, the voice band analog signals are sampled at 8 kHz speed to meet the Nyquist sampling rate, $f_s > 2 \times$ bandwidth. The samples are digitized by the different quantization techniques to 8 bits of data. Thus, the wireline PCM system is a 64-Kbps stream where one voice sample of 8 bits is generated every 125 μs. The quantization of the samples creates error. Successive quantization errors of voice samples can be assumed uncorrelated random noise. Therefore, the quantization error is viewed as noise and expressed as the signal-to-quantization noise ratio (SQR). It is expressed as

$$SQR = \frac{E\{X^2(t)\}}{E\{[Y(t) - X(t)]^2\}}$$

where $E[.]$ denotes the expectation value, $X(t)$ is the analog input signal at time t, and $Y(t)$ is the decoded output signal at time t. The error $[Y(t) - X(t)]$ is limited in amplitude to $q/2$, where q is the height of the quantization interval. If all quantization intervals have equal lengths and the input analog signal is a sinusoidal, the SQR in dB is expressed as

$$SQR(dB) = 10.8 + 20 \log 10[v/q]$$

where v is the RMS value of the amplitude of the input. The SQR values of the signals increase with the sample amplitude and penalize the small sample-size signals. A more-efficient coding can be achieved by not having uniform sample size, but rather by having sample size vary. The process of compounding[9] is used to achieve this non-uniform sampling. The compression algorithm used in North America and Japan for PCM is called μ-law, and a

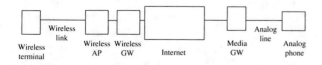

Fig. 2   Schematic diagram for VoIP implementation.

compounding formula recommended by the ITU for Europe and the rest of the world is called A-law. The 64-Kbps voice-coding standard is issued from the ITU as Recommendation G.711.[10] In addition, the ITU developed Recommendations G.726 and G.727 on 40, 32, 24, and 16-Kbps adaptive differential PCM (ADPCM) coding standards and G.722 on 56–64 Kbps, 7-kHz wideband ADPCM standard.

The introduction of digital wireless communication in the early 1980s introduced a new challenge: G.711 and G.721 are not usable for wireless networks due to very limited bandwidth of wireless links. The early digital wireless access standard can support only 108 Kbps over 200-kHz bandwidth in GSM and about 24 Kbps over 30 kHz for North American IS-54 TDMA digital standards. It was felt that lower rate voice coders are needed for wireless voice communications. In addition, the voice coder should be able to provide robust communication under fading channel behavior. The initial answer came from many different solutions such as the code excited linear prediction (CELP) codec at 6.5 Kbps, called the half-rate codec. For example, CDMA introduced the system with a full rate 13-Kbps CELP codec. The ETSI introduced residual excited linear predictive speech coding (RPE-LTP) that is 13 Kbps with a frame size of 20 ms and no look-ahead delay. The ANSI proposed vector-sum excited linear predictive (VSELP) coding at 7.95 Kbps for TDMA IS-54 system that has a frame size of 20 ms and look-ahead delay of 5 ms. This RPE-LTP codec of GSM allows supporting eight voice calls simultaneously within the 200-kHz bandwidths and three voice calls within the 30-kHz bandwidth of the North American IS-54 TDMA system. New coders with improved performance were introduced in the subsequent GSM networks. They include full rate,[11] half rate, enhanced full rate,[12] adaptive multirate,[13] and RECOVC (recognition-compatible voice coding) speech transcoding. Other coders include G.728 16-Kbps speech coding using low delay CELP and G.729 (Ref. 14) 8-Kbps speech coding using conjugate structure algebraic CELP. The Qualcom CELP (QCELP) is a variable bit rate codec of 8.5, 4.0, 2.0, and 0.8 Kbps speeds, a frame size of 20 ms, and a look-ahead delay of 5 ms. Further to this bit rate reduction, a single call also can be provisioned for half-rate coding by taking advantage of the silence period during the conversation. At present a number of lower bit rate coding techniques are under consideration for wireless voice communications.

NETWORK QUALITY REQUIREMENTS

The measurement of voice quality is rather difficult. A subjective rating scale of 1 to 5, called mean opinion score (MOS),[15] is used to state voice quality. Wireline voice quality is normally within 4 to 4.5 MOS, the

current wireless voice quality lies between 3.5 and 4 MOS. To improve the quality of voice signal for wireless networks, G.729 adopted 10-ms frame times. The computation time is 10 ms and look-ahead delay is 5 ms. This results in a total one-way codec delay of 25 ms. In addition to the delay criteria, speech performance depends also on the BER. The objective of performance under random BER $< 10^{-3}$ is recommended not to be worse than that of G.726 under similar conditions.[16] To introduce VoIP, the appropriate selection of coding technology is necessary to meet the criteria of delay and BER of the network.

In network applications of speech coding, coded voice signals are transmitted through multiple nodes and links, as shown in Fig. 2. All these network links and nodes cause impediments to the coded voice signals. The ITU Recommendations G.113 and G.114[17] specify several system requirements, including:

- End-to-end noise accumulation is limited to 14 QDU (quantization distortion unit), where each QDU is equal to the noise of a single 64-Kbps PCM device;
- End-to-end transmission delay budget is 300 ms;
- G.114 limits the processing delay for codec at each end to 10 ms.

Among these network requirements, the most important one in the design of VoIP using wireless is the end-to-end delay budget of 300 ms. In digital networks, because of the synchronous nature of transmission, this delay budget is mostly used for the switching and transmission delay. There is very little variation of this delay, called jitter, within the synchronous digital network for voice signals. In the current wireless voice links, the air interface introduces additional delay because of the air-link multiple access standards. Most of the wireless link designs attempt to meet the delay requirements of 300 ms for single-link voice calls in a national network. So single-link wireless calls without intermediate satellite links perform with a reasonable MOS rating today, however, if two ends of the connection are wireless links, the speech quality deteriorates below MOS 3.0.

There are three important network performance parameters for wireless VoIP service:

1. Performance to set up and tear down the call;
2. Quality of voice payload packets during conversation;
3. Performance of the voice session handoff.

These parameters will depend on the selection of VoIP service protocols to set up the voice session, the voice signal coding and transporting scheme, and the micro/macro mobility protocols used for VoIP services.

The call setup delay depends on the successful transfer of the current Q.931 and ISDN-type messages and the additional message sets for capability check of the terminals and media packet synchronization in the IP network. Both H.323 and SIP signaling protocol implement this function. The average call setup delay of the present wireless voice network is about 3 sec, H.323 and SIP implementations will have to meet this delay requirement.

Network voice quality will depend on the contribution of the different components of a hypothetical connection of VoIP in wireless network. A hypothetical connection, as shown in Fig. 2, includes network components such as a wireless terminal, a wireless link, an access point, a wireless gateway, the Internet, a media gateway, and a landline terminal. The landline phone is assumed connected through an analog line. The wireless terminal includes the codec and the function to map the coded frames into the wireless data channel for communication with the wireless access point. The common wireless channel is shared between multiple users, so to get access to the capacity of the channel is a delay process. The fading and propagation loss of the wireless channel causes the frame error that introduces additional impediments to the transfer of data. Once the packet is received at the wireless access point, it is transferred to the terminating media gateway through the Internet. Internet performance depends on the delay at the different routers and the propagation delay through the transport links. Due to the use of fiber transmission, the transmission related BER or packet loss is almost nonexistent in the links and propagation delay is very small. The delay in the router depends on the long-range dependency of the traffic and link congestion. Using proper engineering techniques, this Internet delay can be maintained within strict limits. The DiffServ and MPLS protocols will be able to support the core Internet with minimum delay and jitter. At the media gateway, the coded voice packets are reconverted to analog voice signals and transferred to the terminating analog voice terminals using copper wire connection. The end-to-end delay of the voice packet for this hypothetical connection can be represented by the following equation:

$$D_{end-to-end} = d_{jitter} + d_{wt} + d_{wc} + d_{w1}$$
$$+ d_{wap} + d_{Internet} + d_{mgw}$$

where

d_{jitter}. Delay introduced by the jitter buffer. To compensate for the fluctuating network conditions, it is necessary to implement a jitter buffer in voice gateways (GW) or terminals. This is a packet buffer that holds incoming packets for a specified amount of time before forwarding them to decoding. This has the effect of smoothing the packet flow, increasing the resiliency of the codec to packet loss, delayed

packets, and other transmission effects. The downside of the jitter buffer is that it can add significant additional delay in the path. It is not uncommon to see jitter buffer settings approaching 80 ms for each direction.

d_{wt}. Delay at the wireless terminal for coding and decoding voice packets and creating voice frames conforming to the Internet frame packet format [TCP or user datagram protocol (UDP)]. Each coding algorithm has certain built-in delay. For example, G.723 adds fixed 30-ms delay. To reduce the IP overhead, multiple voice packets may be mapped to one Internet frame and thus introduce bundling delay.

d_{wc}. Delay at the wireless terminal to get a wireless data channel and map the Internet voice packet to it. This includes delay for buffer allocation such as GSM TBF (temporary buffer flow)2 allocation. In uplink transmission: Before sending the data to the base station, the mobile station must access the common channel in the uplink direction to send the request. Getting permission to send data in the uplink direction takes time, which increases the end-to-end delay.

d_{w1}. Delay to transfer the voice packets to the wireless access point, including the retransmission delay to protect the frame error during propagation.

d_{wap}. Delay at the wireless access point to assemble and reassemble the voice frame from wireless frame formats to the Internet format.

$d_{internet}$. Delay to transfer the packet through the Internet to the media gateway. This is the delay incurred in traversing the VoIP backbone. In general, reducing the number of router hops minimizes this delay. Alternatively, it is possible to negotiate a higher priority to voice traffic than for delay-insensitive data.

d_{mgw}. Delay at the media gateway to convert Internet voice packets to analog voice signals and transfer to analog voice lines.

The TCP retransmission delay impacts the delay parameters of d_{wc} and d_{wap}. The end-to-end delay, $D_{end-to-end}$, of voice packets for conversation should be less than 300 ms, as specified by ITU G.114; one-way delay greater than 300 ms has an adverse impact on conversation, and the conversation seems like half duplex or push-to-talk.

The wireless VoIP session handoff between the two wireless access points is determined by the handoff mechanism supported by the wireless mobility protocol. There are two mobility functions for the wireless IP network: 1) micro-mobility and 2) macro-mobility. The consensus between the different standards bodies is that current mobile IP may be suitable for macro-mobility, but a new technique is necessary for micro-mobility. The potential micro-mobility protocols are GPRS mobility management (GMM), Cellular-IP, Hawaii, terminal independent mobility for IP (TIMIP),[18] intra-domain mobility management protocol (IDMP),[19,20] etc. The challenge of these mobility protocols is to ensure that the voice packets can be routed to the new access point without any packet loss or significant additional delay on the voice path. Most of the protocols in their current form have difficulty meeting the micro-mobility requirements of the voice packets. The current soft handoff of CDMA, the make-before-break mechanism on the GSM, and hard handoff in TDMA maintain the continuity of voice packet flow during handoff. The method of GPRS and 3G packet handoff do not allow similar mechanisms at this time.

OVERVIEW OF THE H.323 PROTOCOL

This section presents a brief introduction to the H.323 and RTP/RTCP.[21] The H.323 standard provides a foundation for audio, video, and data communications across IP-based networks for multimedia communications over local area network (LAN) with no quality of service (QoS) guarantee. The standard is broad in scope and includes both stand-alone devices, embedded personal computer technology, and point-to-point and point-to-multipoint conferences. The standard specifies the interfaces between LANs and other networks, and addresses call control, multimedia management, and bandwidth-management methods. It uses the concept of channels to structure information exchange between the communications entities. A channel is a transport layer connection (unidirectional or bidirectional).

The RTP is used in conjunction with H.323 to provide end-to-end data delivery services with real-time characteristics. RTP can handle interactive audio and video services over a connectionless network. At present, RTP/RTCP along with UDP provides the bare-bones real-time services capability to IP networks with minimum reliability. A typical H.323 network, shown in Fig. 3, consists of a number of zones interconnected by a WAN. The four major components of a zone are terminals, GW, gatekeepers (GKs) and multipoint control units (MCUs). An H.323 terminal is the client and endpoint for real-time, two-way communications with other terminals, GWs, or MCUs. The (optional) GK provides address translation-and-controls access and bandwidth-management functions to the H.323 network. The GW is an endpoint that interconnects the VoIP terminal to the PSTN network. The MCU is the network endpoint for multipoint conferencing. An H.323 communication includes controls, indications, and media packets of audio, moving color video pictures, and data. Thus, H.323 is a protocol suite, as shown in Fig. 4, which includes separate protocol stacks for control and media packet transport. All H.323 terminals must support the H.245 protocol, which is used to negotiate channel usage and capabilities. Three additional components are required in the architecture:

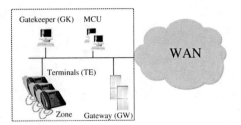

Fig. 3   H.323 system components.

1. H.225 for call signaling and call setup (a variation of Q.931);
2. RAS protocol for communicating with a GK;
3. RTP/RTCP for sequencing audio and video packets.

A typical H.323 call setup is a three-step process that involves call signaling, establishing a communication channel for signaling, and establishing media channels. In the first phase of call signaling, the H.323 client requests permission from the (optional) GK to communicate with the network. Once the call is admitted, the rest of the call signaling will proceed according to one of several call models. Fig. 5 describes the message flows in H.225 and H.245, where Endpoint-1 is the calling endpoint and Endpoint-2 is the called endpoint.

In the direct routing call model, the two endpoints communicate directly instead of registering with a GK. As shown in Fig. 5, Endpoint-1 (the calling endpoint) sends the H.225 setup (1) message to the well-known call signaling channel transport identifier of Endpoint-2 (the called endpoint), which responds with the H.225 connect (3) message. The connect message contains an H.245 control channel transport address for use in H.245 signaling. The H.225 call proceeding (2) message is optional. Once the H.245 control channel (unidirectional) is established, the procedures for capability exchange and opening media channels are used, as shown in Fig. 5. The first H.245 message to be sent in either direction is terminal capability set (5 and 7), which is acknowledged by the

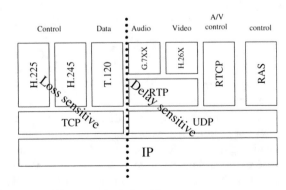

Fig. 4   H.323 protocol relationships.

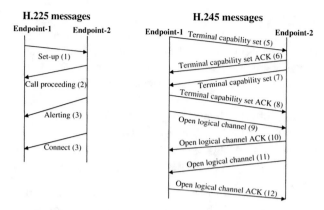

Fig. 5 H.323 messages.

terminal capability set acknowledgment (ACK) (6 and 8) message. There can be an optional master–slave determination procedure invoked at this stage to resolve conflicts between the two endpoints trying to open a bidirectional channel. The procedures of H.245 are used to open logical channels for the various information streams (9 and 11). The open logical channel ACK (10 and 12) message returns the transport address that the receiving end has assigned to the logical channel. Both the H.225 and H.245 messages are transmitted over a reliable transport layer.

After the media channel has been set up and RTCP CNAMEs (canonical name) are exchanged, two TCP sessions are established, one for H.225 and the other for H.245 procedures, so that multiple media streams (e.g., audio and video) to the same user can be synchronized.[21] The setup delay is very large in H.323 for a regular call, which involves multiple messages from its underlying protocols such as H.225 and H.245. This delay is rather prominent on a low-bandwidth, high-loss environment. It is further aggravated by the higher delay margin on wireless access links. The fast call setup method is an option specified in H.323 that reduces the delay involved in establishing a call and initial media streams.

OVERVIEW OF SIP

SIP is an application layer protocol used for setting up and tearing down VoIP sessions. The major difference between SIP and H.323 is SIP is fully based on Internet context and thus does not support the Q.931 or ISUP messages that are currently used for telephony networks. But SIP extends the functionality of telephony signaling and supports mobility, and it is part of the overall IETF multimedia architecture framework that includes protocols such as RTP, RTSP, SDP, SAP, and others. SIP uses a text message format with an encoding scheme very similar to HTTP. Currently, SIP uses SDP[22] to establish the media session and the terminal capabilities, as H.245 in H.323. SDP messages

are carried as the message body of a SIP message. A complete VoIP session includes a number of SDP messages for resource reservation, connection, and ringing in addition to SIP INVITE and BYE messages, as shown in Fig. 6. All SIP messages are transported at the RTP layer, whereas H.323 control messages use TCP. SIP is based on client/server architecture with a SIP user agent and a SIP proxy server. The SIP user agent has two important functions: 1) it listens to the incoming SIP messages and 2) it sends SIP messages on receipt of an incoming SIP message or on user actions. The SIP proxy server relays SIP messages, so that it is possible to use a domain name to find a user. This simplifies the user location determination and allows scalability. The SIP server can be used as a redirect server, in which case it will provide the host location information without relaying the SIP messages, and the SIP user client will set up the session directly with the user.

The SIP mobility architecture components for VoIP are shown in Fig. 7, where we assumed the mobile host and foreign network use DHCP (dynamic host configuration protocol)[23] or one of its variations for subnetwork configuration. An SIP-capable mobile host uses DHCP to register in the network. A mobile host broadcasts a DHCP_DISCOVER message to register to a network. Multiple DHCP servers will respond to this request with the IP address of the server and default gateway in the DHCP_OFFER message. The mobile host selects the DHCP server and sends the DHCP_REQUEST message to register. The registration is confirmed by DHCP_ACK at the DHCP server. As previously mentioned, SIP includes the mobility function, and the mobile host then uses its temporary IP address to register to the visiting register of the foreign network. The registration in a foreign network includes the authentication function, which uses AAA (authentication, accounting, and administration)[21] servers. The foreign AAA server communicates with the home AAA server to get the confirmation from the home register about customer authenticity. Ultimately, the

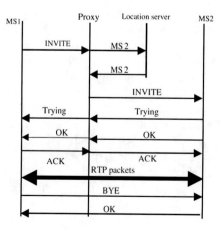

Fig. 6 SIP signaling for VoIP.

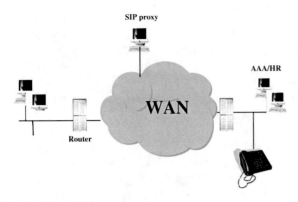

Fig. 7 SIP mobility architecture components.

visiting register receives the authentication response message and, if it is accepted, it sends the 200 OK messages to the mobile host. If authentication fails, it sends 401 messages indicating unauthorized request for registration. After this registration, the mobile host initiates SIP registration for session start by sending an INVITE message to the SIP proxy server. In the case of micro–mobility, authentication with the AAA server is not necessary. The visiting register can authenticate the mobile host (expedited registration). The complete SIP registration sequence is required for macro-mobility. To reduce the macro-mobility registration delay of SIP, a quasiregistration concept is proposed in Schulzrinne.[23] In quasiregistration, whenever the mobile host hands off from an old visiting register to a new visiting register, it informs its home register of its location by sending a REGISTER message. When the home register replies its OK message, it will include the old visiting register's IP address along with the response. Thus, visiting registers will know the adjacent visiting register's address to use for fast registration.

RLP

ETSI and T-1 standards bodies defined the GPRS and 3G standards to carry packet data traffic in addition to voice in wireless access links. A new protocol layer, RLP, is defined on top of the MAC layer to improve BER performance by using an ARQ retransmission technique. In addition to retransmission, GPRS RLC performs block segmentation, reassembly, and buffering. As GPRS provides limited data capability, 3G standards are defined that extend the data capability of radio frequency. The main operation principles of GPRS and 3G RLP standards are very similar, although 3G provides some additional QoS management capability in RLP management at the MAC layer. In our analysis, we will use GPRS and 3G RLP

capabilities interchangeably without significant impact on the results. CDMA2000[24] is one of several 3G air-interface standards under consideration by standards bodies such as the 3GPP/3GPP2 (3G Partnership Project). The MAC layer of CDMA2000 provides two important functions:

1. Best-effort delivery using the retransmission mechanism of RLP that provides reliability;
2. Multiplexing and QoS management by mediating conflicting service requests.

In addition, voice packets are directly given to the multiplex sublayer that bypasses the RLP function. Many transport channels have been defined for CDMA2000 to provide services from physical to higher layers. These channels are unidirectional and either common (shared between multiple users) or fully dedicated to a user for the duration of the service. CDMA2000 has defined many channels for its operation, the following transport channels are of interest:

1. **Forward common control channel (F-CCCH):** Communication from base station to mobile station for layer 3 and MAC messages;
2. **Forward supplemental channel (F-SCH):** Operated in two modes, blind mode for data rate not exceeding 14.4 Kbps, and explicit mode, where data rate information is explicitly provided, individual F-SCH target FERs can be different than other F-SCHs;
3. **Forward fundamental channel (F-FCH):** Transmits at variable data rates, as specified in TIA/EIA-95-B;
4. **Reverse access channel (R-ACH) and reverse-CCCH (R-CCCH):** Used by a mobile station to communicate layer 3 and MAC messages;
5. R-CCCH supports the low latency access procedure required for efficient operation of packet data suspend state;
6. **Reverse supplemental channel (R-SCH):** Operates on two modes, blind and explicit;
7. Reverse fundamental channel (R-FCH) supports 5-and 20-ms frames, the 20-ms frame structures provide rates derived from the TIA/EIA-95-B Rate Set-1 or Rate Set-2

The RLC and MAC layers are responsible for efficient data transfer of both real-time and nonreal-time services. The transfer of nonreal-time data includes the ARQ for low-level data to provide reliable transfers at higher levels. The network layer data PDUs (N-PDUs) are first segmented into smaller packets and transformed into link access control PDUs. The link access control overhead includes a service access point identifier, a sequence number for higher-level ARQ, and other data fields. The link access control PDUs are then transferred to signaling radio burst protocol (SRBP), a connectionless protocol for signaling messages. The data PDUs are segmented into smaller packet RLC PDUs corresponding to the physical layer transport blocks. Each RLC PDU contains a sequence number for lower-level ARQ and CRC fields for error detection. CRC is calculated and appended by the physical layer. When RLP at the receiving end finds a frame in error or missing, it sends back a NAK (negative acknowledgment) request for retransmission of this frame and starts a retransmission timer. When the timer expires for the first attempt, the RLP resets the timer and sends back a NAK request. This NAK triggers a retransmission of the requested frame from the sender. In this way, the number of attempts per retransmission increases with every retransmission trial. As noted in Bao,[25] the number of trials is usually less than four.

The GPRS structures of different radio channels and MAC/RLC are very similar to cdma2000. GPRS uses the same TDMA/FDMA structure as GSM to form the physical channels. For the uplink and downlink direction, many frequency channels with a bandwidth of 200 kHz are defined. These channels are further subdivided into the length of 4.615 ms. Each TDMA frame is further split into eight time slots of equal size. As an extension to GSM, GPRS uses the same frequency bands as GSM and both share the same physical channels. Each time slot can be assigned to either GSM or GPRS. Time slots used by the GPRS are known as packet data channels (PDCHs). The basic transmission unit of a PDCH is called a radio block. To transmit a radio block, four consecutive TDMA frames are utilized. Depending on the message type transmitted in one radio block, a sequence of radio blocks forms a logical channel.

- **PRACH (packet random access channel, uplink):** This common channel is used by the mobile stations to initiate the transfer in the uplink direction;
- **PPCH (packet paging channel, uplink):** The base station controllers (BSC) uses this channel to page the mobile station prior to downlink data transmission.
- **PAGCH (packet access grant channel, downlink):** Resource assignment on the uplink and downlink channel is sent on this channel.
- **PBCCH (packet broadcast control channel):** GPRS specific information is broadcast on this channel.
- **PDTCH (packet data transfer channel):** Data packets are sent on this channel. A mobile station can use one or several PDTCHs at the same time.
- **PACCH (packet associated control channel):** This channel conveys signaling information related

to a given mobile station and the corresponding PDTCHs.

REFERENCES

1. Mouly, M.; Putet, M.B. *GSM System for Mobile Communications*; Mouly and Putet: Palasieu, France, 1992.
2. GSM 03.60, Digital Cellular Telecommunications Systems (Phase 2 +). General Packet Radio Service (GPRS) Service Description Stage-2, ETSI DTS/SMG-030360Q; Global System for Mobile Communications, May 1998.
3. International Telecommunications Union. Packet Based Multimedia Communications Systems, Recommendation H.323; Telecom Standardization Sector, Geneva, Switzerland, Feb 1998.
4. Handley, M.; Schulzrinne, H.; Schooler, E.; Rosenberg, J. SIP: Session Initiation Protocol, Request for Comments (RFC) 2543; Internet Engineering Task Force, Mar 1999.
5. Greene, N.; Ramalho, M.A.; Rosen, B. Media Gateway Control Protocol Architecture and Requirements, RFC 2805; Apr 2000.
6. International Telecommunications Union, Gateway Control Protocol, Recommendation H.248; Telecom Standardization Sector, Geneva, Switzerland, June 2000.
7. Blake, S.; Black, D.; Carlson, M.; Davies, E.; Wang, Z.; Weiss, W. An Architecture of Differentiated Services, RFC 2475; Dec. 1998.
8. Rosen, E.; Viswanathan, A.; Callon, R. Multiprotocol Level Switching Architecture, RFC 3031; Jan. 2001.
9. Bellamy, J.C. *Digital Telephony*; Wiley Interscience: New York, 1991; 98–142.
10. International Telecommunications Union, Pulse Code Modulation of Voice Frequencies, Recommendation G.711; Telecom Standardization Sector, Geneva, Switzerland, 1988.
11. Digital Cellular Telecommunications System (Phase 2 +): Full Rate Speech Transcoding, GSM 06.10; Version 7.0.1; Release 1998.
12. Digital Cellular Telecommunications System (Phase 2 +): Enhanced Full Rate (FER) Speech Transcoding, GSM 06.60, Version 7.0.1, Release 1998.
13. Digital Cellular Telecommunications System (Phase 2 +): Adaptive Multi-Rate (AMR) Speech Transcoding, GSM 06.10, Version 7.1.0, Release 1998.
14. International Telecommunications Union, Coding of Speech at 8 kbits/sec Using Conjugate-Structure Algebraic Code-Excited Linear-Predictive (CS-ACLEP) Coding, Recommendation H.729; Telecom Standardization Sector, Geneva, Switzerland, Mar 1996.
15. Lakaniemi, A.; Parantainen, J. On voice quality of IP voice over GPTS. IEEE International Conference on Multimedia

Wireless Multi—Networks

and Expo, 2000 (ICME 2000), New York, July 30–Aug 2, 2000; 751–754.

16. Cox, R.V. Three new speech coders from the ITU cover a range of applications. IEEE Commun. Mag. **1997**, *39* (9), 40–47.

17. International Telecommunications Union, G.114: Mean One-Way Propagation Time, Recommendation G.114; Telecom Standardization Sector, Geneva, Switzerland Nov. 1988.

18. Grilo, A.; Estrela, P.; Nunes, M. Terminal independent mobility for IP (TIMIP). IEEE Commun. Mag. **2001**, *39* (12), 34–46.

19. Das, S.; McAuley, A.; Dutta, A.; Misra, A.; Chakraborty, K.; Das, S.K. IDMP: An intra-domain mobility management protocol for next generation wireless networks. IEEE Wirel. Commun. (Special issue on Mobile and Wireless Internet: Architectures and Protocols), **2002**, *9* (3),

38–45.

20. Misra, A.; Das, S.; Dutta, A.; McAuley, A.; Das, S.K. IDMP-based fast handoffs and paging in IP-based 4G mobile networks. IEEE Commun. (Special issue on 4G Mobile Technologies), **2002**, *40* (3), 138–145.

21. Schulzrinne, H.; Schulzrinne, H.; Casner, S.; Frederick, R.; Jacobson, V. RTP: A Transport Protocol for Real-Time Applications, RFC 1889; IETF, Jan 1996.

22. Song, J. et al. MIPv6 User Authentication Support through AAA. Internet draft. draft-song-mobileip-mipv6-user-authentication-00.txt, Nov **2001.**

23. Schulzrinne, H. DHCP Option for SIP Servers, Internet draft, draft-ietf-sip-dhep-05.txt, Nov **2001.**

24. http://www.3gpp2.org.

25. Bao, G. Performance evaluation of TCP/RLP protocol stack over CDMA wireless link. Wirel. Netw. **1996**, *3* (2), 229–237.

Wireless Sensor Networks (WSNs)

Gopal Racherla
Advanced Wireless Group, General Atomics, San Diego, California, U.S.A.

Abstract

Wireless sensor networks (WSNs) are a collection of geographically distributed network nodes that perform sensing and computing and use wireless communication to cooperate, share and integrate their individual sensor data.

INTRODUCTION

Wireless sensor networks (WSNs) have been gaining a lot of popularity in the recent past. WSNs have found varied applications ranging from tracking migration of animals, monitoring wildfires, and homeland and defense applications involving surveillance and reconnaissance. A WSN may be ad hoc wireless network or mobile ad hoc networks (MANETs). The typical characteristics of sensor nodes in a WSN are:[1–20]

- Each node performs sensing operation;
- Each node may/may not perform computation on the local sensor data;
- Each node transmits wirelessly the (un)processed/computed sensor data to other nodes typically in order to cooperatively combine and fuse the data for various functional goals;
- Each sensor node is typically small and battery operated and contains low-power system-on-chip (SoC) that may include microprocessors, microsensors, and wireless communication interfaces;
- Each node may or may not be mobile

Fig. 1 shows the block diagram of a typical node in a WSN. The node consists of a wireless transceiver with an antenna. A processor or microcontroller is used to handle the communication protocol and medium access control (MAC) operations. The processor hosts software to process the sensor data and interface with the user and the external world. The processor uses the onboard and external memory (RAM and ROM/flash) to store the data. The sensor performs the sensing operation and uses a transducer to convert the sensor data into electrical signals. A battery and/or an alternate power source like a solar panel provide power to the system.

Fig. 2 shows the configuration of a typical WSN. The WSN usually has some nodes that connect to a gateway which in turn is connected to a computer that is used to collect, display and analyze the integrated and fused sensor information.

The design, implementation, and deployment of WSNs pose a lot of challenges.[1] These include the potential large scale of deployment, limited system resources of a sensor node, and extreme and unpredictable operating environments.

Several technological breakthroughs are responsible for the ease of development, deployment, and adoption of WSNs. These include miniaturization and very large scale integration (VLSI) techniques for developing SoCs, development of very high-density miniature batteries, robust wireless technologies and creation of software engineering ecosystems including real-time embedded OSs, firmware and integrated development and testing tools.

The rest of the entry is organized as follows. Next section reviews related work in the area of design, architecture, and applications of WSNs. Section "Applications" describes various applications of WSNs in detail. Section "Challenges" explores the various challenges in the design of WSNs and the various research efforts in these areas. Section "Supporting Technologies" delves into the various supporting technologies used in the development of WSNs. Section "Summary" summarizes the entry. Relevant abbreviations are presented in Table 1; internet links and references are provided at the end of the entry for completeness.

RELATED WORK

Bulusu et al.'s book[1] is a collection of chapters on WSN including ones on potential field methods for mobile-sensor-network deployment, sensor fault detection and calibration, localization, multi-hop routing, power management, storage and security issues. Callaway's book[3] focuses on architectures and protocols and looks at various issues and solutions for the network. In addition, there is a multitude of books[4–14] covering various aspects of WSNs

Encyclopedia of Wireless and Mobile Communications DOI: 10.1081/E-EWMC-120043612

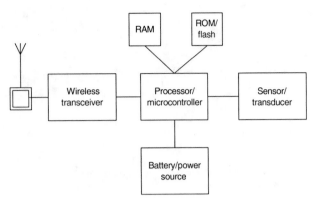

Fig. 1   Block diagram of a typical node in a WSN.

including overview, components of a WSN, applications, technical challenges and various applicable technologies. Akyildiz et al.'s Survey of WSN[16] is a good survey of the state-of-the-art WSNs and the various technical issues involved in the design and development of WSNs. Several technical articles[17–20] provide a very high level view of WSNs and the challenges and opportunities they present. Aetherwire,[21] CrossBow Technologies,[22] Dust Networks[23] and Ember[24] are companies building wireless sensors to be used be in WSNs.

APPLICATIONS

WSNs have widespread applications. All WSN applications have been classified by Culler et al.[15] as:

- Monitoring spaces/environments such as military surveillance/reconnaissance, habitat monitoring, climate control, home automation including indoor climate control etc.;

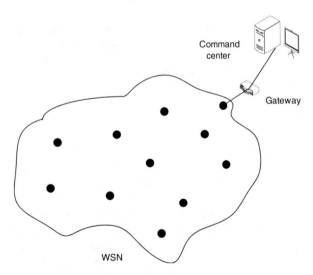

Fig. 2   A typical WSN configuration.

- Monitoring objects such as building monitoring, ecophysiology, terrain mapping, equipment maintenance as needed, medical diagnostics and patient monitoring etc.; and
- Monitoring the interactions amongst objects and between objects and spaces/environments such as wildlife migration, disaster management, emergency response, manufacturing work flow, asset tracking etc.

Real life examples of WSN applications include the UCLA CENS[25] program to monitor treated water for agriculture. Another application is a WSN consisting of vibration sensors in a semiconductor foundry.[26] Aetherwire's[21] sensors have been used in personal safety devices, smart homes, inventory control, IVHS (intelligent vehicle highway system) and next-generation Internet. Crossbow's[22] sensor networks find applications in general aviation and automotive systems, antenna stabilization, unmanned aerial vehicles (UAV), agriculture and construction vibration monitoring and towed sonar arrays. Dust networks'[23] WSN have been target for field equipment maintenance and calibration, machine health monitoring, industrial process control, and monitoring cold storage container or a large distribution facility. Ember's website[24] describes WSN applications like integrated home automation (security systems, entertainment systems, lighting control, HVAC control), building automation (access control, refrigeration control), automatic meter reading (residential demand response, power distribution diagnostics), asset management (cold chain monitoring, container security, asset tracking) and industrial automation (temperature sensing and control, pressure sensing, flow control, level sensing) and homeland/defense (battlefield monitoring, shipboard monitoring).

CHALLENGES

There are numerous technical challenges[1,6,16–18] in the development and the deployment of WSNs. These challenges include:

Scalability. Dense, large scalable deployment of sensor nodes in a WSN is impeded by the capacity limitations of sensor nodes since the WSN is not able to use fully the capability of the individual components and instead relies on partially redundant sensor measurements and their correlation. The spatial and temporal scalability of WSNs is based on the rate of sampling, the coverage of the WSNs, the density of sensor nodes, the activity factor of each sensor node and the granularity of the samples. Bulusu et. al.[1] provide WSN examples that span the extremes of sampling rate, extent of spatial and temporal deployment and the system density of WSNs.

Resource limitations. The sensor nodes are resource constrained. They are usually battery operated and need to conserve power to reduce human intervention. The

Table 1 Abbreviations used in this entry.

Abbreviation	Expansion
ABAM	Associativity-based multicast
ABR	Associativity-based routing
ALARM	Adaptive Location Aided Routing—mines
AMRIS	Ad hoc multicast routing protocol utilizing increasing ID-numbers
AODV	Ad hoc on-demand distance vector routing
AQM	Ad hoc QoS multicast
ARA	Ant-colony-based routing algorithm
ASIC	Application specific integrated circuit
BGR	Blind geographic routing
CAMP	Core-assisted mesh protocol
CBM	Content based multicast
CBRP	Cluster based routing protocol
CEDAR	Core extraction distributed ad hoc routing
CGSR	Clusterhead Gateway Switch Routing
CMOS	Complementary metal–oxide–semiconductor
CSMA	Carrier sense medium access
DART	Dynamic address routing
DBF	Distributed Bellman-Ford routing protocol
DDR	Distributed dynamic routing
DFR	Direction forward routing
DREAM	Distance routing effect algorithm for mobility
DSDV	Dynamic destination-sequenced distance vector routing
DSR	Dynamic source routing
DST	Dynamic spanning tree
ETSI	European Telecommunications Standards Institute
FSR	Fish-eye state routing
FPGA	Field programmable gate array
GLS	Geographic location service
GPS	Global positioning system
GSR	Global state routing
GUI	Graphical user interface
HARP	Hybrid ad hoc routing protocol
HIPERLAN	High performance radio local area network
HomeRF	Home radio frequency
HSLS	Hazy sighted link state routing
HSR	Hierarchical state routing
HVAC	Heating, ventilation, and air conditioning
IARP	Intrazone routing protocol
IDE	Integrated development environment
IEEE	Institute of Electrical and Electronic Engineers
IP	Internet protocol
ISAIAH	Infra-structure AODV for infrastructured ad hoc networks

Table 1

(*continued*)

Abbreviation	Expansion
IVHS	Intelligent vehicle highway systems
LAR	Location-aided routing
LBM	Location based multicast
LCA	Linked cluster architecture
Li-ion	Lithium Ion
LMR	Lightweight mobile routing
MAC	Medium access control
MANET	Mobile ad hoc NETwork
MEMs	Micro electro mechanical systems
MOBICAST	Mobile just-in-time multicasting
MSM	Master state machine
MZR	Multicast zone routing
NiCad	Nickel cadmium
NiMH	Nickel metal hydride
OS	Operating system
PAMAS	Power aware multi-access protocol with signaling
PAN	Personal area network
PARO	Power-aware routing optimization
PHY	Physical layer
RF	Radio frequency
SoC	System-on-chip
SSA	Signal stability based adaptive routing
TORA	Temporally ordered routing algorithm
UAV	Unmanned aerial vehicle
UWB	Ultra-wideband
VLSI	Very large scale integration
WPAN	Wireless personal area network
WSN	Wireless sensor network
ZRP	Zone routing protocol

wireless connectivity of the sensor nodes is required since using wires to connect the sensors may be expensive, difficult, or impractical based on the typical deployment environments for such sensor nodes. Many applications also force the sensors to be small in size, easily transportable, and easily deployable.

Extreme dynamics/conditions. The sensor nodes in WSN may need to deal with extreme conditions—environmental (temperature, humidity, pressure), RF noise and interference, and asynchronous sensing activity.

SUPPORTING TECHNOLOGIES

The popularity of WSNs and their applications has been greatly increased because of the tremendous advances made in many supporting technologies.[1] These technologies include: sensor and actuation devices, high efficiency power systems, wireless communications, embedded devices, embedded software, distributed network architecture, self-organizing ad hoc networks and routing, data-centric systems, and networking standards. The advances in these various technologies are described in more detail below:

Miniaturization and SoC design. The miniaturization of electronic circuits and the advances in VLSI are two of the most important technical advances that have helped in making WSNs economically viable, easy to deploy and transport. In addition, this level of integration reduces the power requirements of the sensor nodes thus reducing the need for bulky batteries and the need to change them often. SoCs can be fabricated by several technologies, including: full-custom application specific integrated circuit (ASIC), standard cell ASIC or field programmable gate array (FPGA). The semiconductor processes have been improving continuously with process nodes improving from $0.25~\mu m$, and $0.18~\mu m$, to 90 nm, 65 nm, and 45 nm. In addition, foundries are constantly improving the power consumption of their semiconductor processes and defining low-power versions of the processes. Also, typically all the digital chips are implemented in complementary metal-oxide-semiconductor (CMOS) while the analog and RF chips are implemented using SiGe and/or GaAs technology. In the recent past because of process improvements, some of the analog and RF chips have been implemented in CMOS thus making it possible to implement the entire SoC using a single CMOS process. The advantage of SoCs is that they consume less power, cost less, and have higher reliability than the multi-chip systems that they replace. Also, the assembly and packaging costs are reduced since there are fewer packages in the system. It should be noted that the total cost is higher for one large chip than for the same functionality distributed over many smaller chips because of lower yields and higher non-recurring engineering (NRE) costs.

Sensor and actuation devices. There has been a lot of progress in the design of sensor/transducer devices (devices that sense and convert signals to data) and actuator devices (devices that act based on the sensor data). Leading the progress has been the design of micro electro mechanical systems (MEMs). MEM is the technology of the very small electrical and/or mechanical machines that generally range in size from a micrometer to a millimeter. MEMs have been used for common applications[27] like inkjet printers (using piezoelectrics), automobile accelerometers, gyroscopes used for dynamic stability control, tire pressure sensors, video displays using micro mirrors, and micro optical switches used for data communications. MEM-based sensors reduce the size and power consumption of the sensor nodes compared to the conventional functional equivalents. MEMs can be used to design small and steerable antenna for communication.

High efficiency power systems. One of the most important advances benefitting WSNs is the design of high-density power/energy systems, and re-usable energy. WSNs typically have strong battery and bandwidth constraints. Power conservation can be achieved on two different fronts—the device and the communication protocols. The power conservation of the device involves reducing the usage of the battery for all the hardware of the device including the CPU, display, and peripherals. The communication protocols can also be designed to be power-aware. Smart batteries have low discharge rates, a long cycle life, a wide operating temperature range and high energy density. These include lithium ion (Li-ion) batteries, fuel cells, NiMH, hydrocarbon fuels on micro-heat engines. In addition, re-usable energy scavenging[1] can be done using solar panels and using vibrations and temperature gradients.

Wireless communications. With the growth of wireless applications in consumer electronics like cell phones, and pagers, sophisticated yet inexpensive CMOS radios have been designed and built. The experience and design from these CMOS radios have been used effectivity to build low-power, small radios for wireless communications in WSNs. Efficient and smart coding and modulation techniques and power-efficient transceivers have contributed to WSNs.

Standardization of networking technologies. WSNs have benefited a lot from several networking technologies that facilitate implementation of WSNs. These include: Bluetooth,[28,29] ultra-wideband (UWB),[30] HIPERLAN/1 and HIPERLAN/2,[31,32] IEEE 802.11 wireless LAN,[33,34] IEEE 802.15.3 wireless personal area network (WPAN),[35] HomeRF,[36] and Zigbee using IEEE 802.15.4 WPAN.[37,38]

Embedded devices and software. A lot of research and development effort has gone in building small embedded devices like the dust motes[23] and Zigbee[37,38] based sensor nodes. Real-life and commercially available sensors built using embedded platforms from Ember, Crossbow, Aetherwire, Dust networks have helped the deployment of WSNs. In addition, the MIT μAMPS node, UCLA mica sensor interface board, motes from Intel are also prime examples of academic/research embedded sensors. Embedded Software including OS, integrated development environments (IDE) and other software tools like compilers, assembler, interactive debugger, modular device drivers, pre-coded libraries, graphical user interface (GUI) toolkit, and graphics image converters have made embedded development easy and reliable. Examples of embedded OS include TinyOS[39,40] and Goeworks[41] and MANTIS OS[42] and IDEs like Keil, VisualLynux. Other relevant applications like TinyDB for embedded database and GNU tool chain have also been popular in the development of WSNs.

Self-organizing ad hoc networks. There has been a strong focus on designing energy-efficient MAC and network routing for WSNs. The data link layer consists of the logical link control (LLC) and the MAC sub-layers. The MAC sub-layer is responsible for channel access and the LLC is responsible for link maintenance, framing data unit, synchronization, error detection and possible recovery, and flow control. The MAC sub-layer tries to gain access to the shared channel such that the frames that it transmits do not collide (and hence get distorted) with frames sent by the MAC sub-layers of other nodes sharing the medium. There have been many MAC sub-layer protocols suggested in the literature for exclusive access to shared channels. Some of these protocols are centralized while others are distributed in nature. With the centralized protocols, there exists a central controller and all other nodes request for channel access to the controller. The controller then allocates time slots (or frequencies) to the requesting nodes. These protocols are reservation–based. The non-reservation based protocols are purely based on contention and are very suitable for MANETs where it is impossible to designate a leader who might be moving all the time. A detailed discussion on wireless MAC schemes can be found in Ref. 43.

The routing algorithms for the MANETs (including WSNs) have received most attention in recent years and a large number of techniques have been proposed to find a feasible path between source and destination node pairs. In the wired environment, the routing protocol[44] can be either *link state* based or *distance vector* based. In link state routing, each router periodically sends a broadcast packet that contains information on the adjacent routers to all the other routers in the network. Upon receiving this broadcast message from all the routers, each router has a complete knowledge of the topology of the network and executes the shortest path algorithm (Dijkstra's algorithm) to determine the routing table for itself. In the case of distance vector routing, which is a modification of the Bellman–Ford algorithm, each router maintains a vector that contains distances that it knows so far (initially infinity for all nodes other than its neighbors) between itself and every other node in the network. Periodically, each node sends this vector to all its neighbors and the nodes that receive the vector update their vectors based on the information contained in the neighbor's vector. In WSNs, routing algorithms based on link state and distance vector face serious issues as outlined below: 1) Executing link state protocol would require each node to send information about its neighborhood as and when it changes; 2) Distributing the distance vector information in a highly dynamic environment is very ineffective; 3) Due to incorrect topological information, both algorithms produce routes containing loops; and 4) Both the algorithms cause severe drain on the batteries due to the excessive amounts of messages needed to construct routing tables at nodes.

Routing in WSNs involves two important problems: 1) finding a route from the source node to the destination node; and 2) maintaining routes when there is at least one session using the route. WSN routing protocols described in the literature can be either *reactive*, *proactive*, *hybrid* (combination of reactive and proactive), or *location based*. In a proactive protocol, the nodes in the system continuously monitor the topology changes and update the routing tables similar to the link state and distance vector algorithms. There is a significant route management overhead in the case or proactive schemes, but a new session can begin as soon as the request arrives. A reactive protocol, on the other hand, discovers a route as a request arrives (on-demand). The route discovery process is performed either on a per packet basis or a per session basis. When routes are discovered on a per packet basis, the routing algorithm has high probability of sending the packet to the destination in the presence of high mobility. A routing algorithm that uses the route discovery process for a session needs to perform local maintenance of severed or broken paths. Location-based routing protocols use the location information about each node to perform intelligent routing. There is a plethora of MANET Routing techniques.[45] The routing algorithms are:

- Proactive (table-driven) like CGSR, DFR, DBF, DSDV, HSR, IARP, LCA;
- Reactive (on-demand) like ARA, ABR, AODV, DSR, LBR, LMR, SSR, TORA;
- Hybrid (proactive/reactive) like HSLS, ZRP;
- Hierarchical like CBRP, CEDAR, DART, DDR, FSR, GSR, HARP;
- Geographical like ALARM, BGR, DREAM, GLS;
- Power aware like PARO, ISAIAH, PAMAS;
- Multicast like ABAM, AMRIS, AQM, CBM, CAMP, MZR;
- Geographical multicast (geocasting) like LBM, MOBICAST.

One of the important decisions that designers and implementors of a WSN need to take is the partition of the WSN's functionality to be done using hardware and software. Below is an example of a partition for network functionality of a sensor node in a WSN.

Fig. 3 shows an example of various functional blocks that make up the PHY, MAC, network (NWK) and application (APP) layers for each of sensor node in a WSN. The functional blocks, logical functionality of each layer and their interfaces is described below. The APP, NWK, and (upper) MAC are typically implemented in software on an embedded processor, and the (lower) MAC and PHY are implemented in hardware such as an FPGA/digital ASIC.

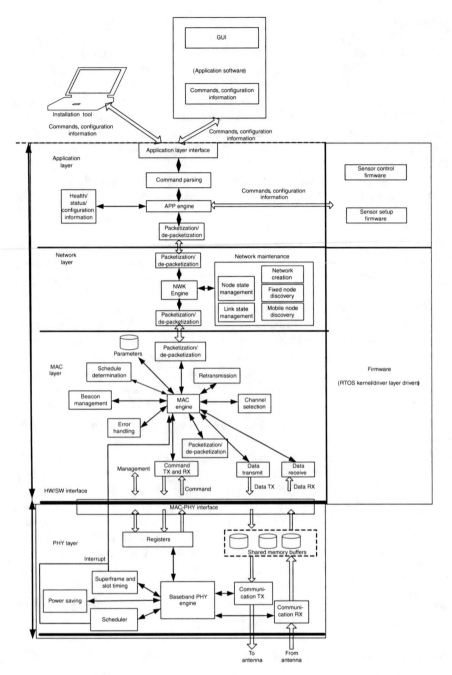

Fig. 3 Example of the network stack of a sensor node.

The system's partitioning (especially the MAC and the PHY) for implementation in hardware and software is driven by three basic design goals:

- Keep the power consumption of the nodes very low;
- Keep the node design easy and flexible;
- Implement critical network timing in hardware.

Fig. 3 also depicts the proposed hardware and software partitioning of the MAC/NWK/APP subsystem.

The PHY layer is responsible for the event scheduler, superframe and slot timing, power consumption, and clock/reset. In addition, the baseband PHY has a master state machine (MSM) which controls the communications transmit and receive chains. The communication chains connect the data being sent or received on the antenna with the MAC layer's data pipes.

The MAC layer is responsible for sending, receiving and re-transmissions of commands, data and management parameters to the PHY using the MAC–PHY interface which doubles as the hardware–software interface. The

MAC Engine is also responsible for packetization/de-packetization, channel selection, determining the schedule for sensor and communication, beacon processing/management, and handling errors.

The NWK layer performs all activities and functions related to the network creation, node discovery, and attachment and detachment. It monitors neighbor nodes and corresponding communication links.

The APP layer receives commands and configuration information from the Command Centre that contains all the sensor information, parses the commands, and appropriately interacts with the various subsystems in the node. It also maintains the health and status of the node.

SUMMARY

In this entry, the basic principles and practices related to WSNs are explored. The characteristics of WSNs, their applications, and the issues related to the design and implementation of WSNs are presented. There has been a significant amount of research done to address the various issues associated with WSNs. The entry presents the technical challenges present in designing and implementing WSNs along with the supporting technical advances that are making it possible to build inexpensive and large scale WSNs. The technical challenges include the constraints on resources in a sensor node along with extreme requirements on performance and the need to achieve scalability. However, significant advances in miniaturization, smart power systems, embedded devices and software, wireless networking, and evolving networking standards are heralding a bright future for WSNs.

REFERENCES

1. Bulusu, N.; Jha, S. *Wireless Sensor Network: a Systems Perspective*; Artech House: Boston, MA, 2005; pp. 321.
2. Wireless Sensor network, http://en.wikipedia.org/wiki/Sensor_network (accessed March 10, 2007).
3. Callaway, E.H., Jr. *Wireless Sensor Networks: Architectures and Protocols*; CRC Press: Boca Raton, FL, 2004; pp. 352.
4. Raghavendra, C.S.; Sivalingam, K.M.; Znati, T. (Eds) *Wireless Sensor Networks*; Kluwer Academic Press: Boston, MA, 2004; pp. 426.
5. Zhao, F.; Guibas, L. *Wireless Sensor Networks: an Information Processing Approach*; Morgan Kaufmann: Amsterdam, San Francisco, 2004; pp. 358.
6. Stojmenovic, I. (Ed.) *Handbook of Sensor Networks; Algorithms and Architectures*; Wiley-Interscience: Hoboken, NJ, 2005; pp. 531.
7. Karl, H.; Willig, A. *Protocols and Architectures for Wireless Sensor Networks*; Wiley-Interscience: Hoboken, NJ, 2005; pp. 526.
8. de Morais Cordeiro, C.; Agrawal, D.P. *Ad hoc and Sensor Networks Theory and Applications*; World Scientific Publishing Company: Singapore, 2006; pp. 664.
9. Krishnamachari, B. *Networking Wireless Sensors*; Cambridge University Press: Cambridge, UK, New York, 2005; pp. 214.
10. Roundy, S.; Wright, P.K.; Rabaey, J.M. *Energy Scavenging for Wireless Sensor Networks: With Special Focus on Vibrations*; Kluwer Academic Publishers: Boston, MA, 2004; pp. 232.
11. Iyengar, S.S.; Brooks, R.R. *Distributed Sensor Networks*; Chapman & Hall/CRC: Boca Raton, FL, 2004; pp. 1144.
12. Ilyas, M.; Mahgoub, I. *Handbook of Sensor Networks: Compact Wireless and Wired Sensing Systems*; CRC Press: Boca Raton, FL, 2004; pp. 672.
13. Nikoletseas, S.; Rolim, J. *Algorithmic Aspects Of Wireless Sensor Networks (Lecture Notes in Computer Science)*, Springer-Verlag: Berlin; Heidelberg, 2004; pp. 201.
14. Shorey, R.; Ananda, A.; Chan, M.C.; Ooi, W.T. *Mobile, Wireless, and Sensor Networks: Technology, Applications, and Future Directions*; Wiley-IEEE Press: Hoboken, NJ, 2006; pp. 422.
15. Culler, D.; Estrin, D.; Srivastava, M. Overview of wireless sensor networks. IEEE Comput. **2004**, *37* (8), 41–49.
16. Akyildiz, I.F.; Su, W.; Sankarasubramanian, Y.; Cayirci, E. A survey on sensor networks. IEEE Commun. Mag. **2002**, *40* (8), 102–114.
17. Estrin, D.; Govindan, R.; Heidemann, J.; Kumar, S. Next century challenges: scalable coordination in sensor networks. ACM Mobile Computing and Networking (Mobicom 1999), Seattle, WA, Aug 15–20, 1999; 263–270.
18. Estrin, D.; Culler, D.; Pister, K.; Sukhatme, G. Connecting the physical world with pervasive networks. IEEE Pervasive Comput. **2002**, *1* (1), 59–69.
19. Pottie, G.J.; Kaiser, W.J. Wireless integrated network sensors. Commun. ACM, **2000**, *43* (5), 51–58.
20. Allan, R. Wireless sensing spawns the connected world. Electron. Des. Mar 30, **2006**.
21. Aetherwire, http://www.aetherwire.com (accessed March 10, 2007).
22. Crossbow, http://www.xbow.com (accessed March 10, 2007).
23. Dust Networks, http://www.dustnetworks.com (accessed March 10, 2007).
24. Ember, http://www.ember.com (accessed March 10, 2007).
25. UCLA Center for Embedded Network Sensing (CENS), http://www.cens.ucla.edu (accessed March 10, 2007).
26. Krishnamurthy, L.; Chhabra, J.; Kushalnagar, N.; Yarvis, M. Wireless Sensor Networks in Intel Fabrication Plants, May 2004, http://www.intel.com.
27. Microelectromechanical systems,, http://en.wikipedia.org/wiki/MEMS (accessed March 10, 2007).
28. Bray, J.; Sturman, C.F. *Bluetooth 1.1—Connect Without Cables,* 2nd Ed.; Prentice Hall PTR: NJ, 2002.

29. Bluetooth Official Website, http://www.bluetooth.com (accessed March 10, 2007).

30. Ultra Wideband Working Group, http://www.uwb.org (accessed March 10, 2007).

31. ETSI, http://www.etsi.org (accessed March 10, 2007).

32. HiperLAN Global Forum 2, http://www.hiperlan2.com (accessed March 10, 2007).

33. Geier, J. *Wireless LANs*; Sams Publishing: Indianapolis, IN, 2001.

34. Fahmy, N.S.; Todd, T.D.; Kezys, V. Ad hoc networks with smart antennas using IEEE 802.11-based protocols. IEEE International Conference on Communications (ICC'02), New York, Apr 28–May 2, 2002.

35. IEEE 802.15 Working Group for WPAN, http://grouper.ieee.org/groups/802/15 (accessed March 10, 2007).

36. HomeRF Official Website, http://www.homerf.org (accessed March 10, 2007).

37. Zigbee Alliance, http://www.zigbee.org (accessed March 10, 2007).

38. Zigbee, http://en.wikipedia.org/wiki/ZigBee (accessed March 10, 2007).

39. TinyOS Community Forum, http://www.tinyos.net (accessed March 10, 2007).

40. Culler, D.E. operating system design for wireless sensor networks. Sensor; May 2006.

41. Geoworks, http://www.geoworks.com (accessed March 10, 2007).

42. Bhatti, S.; Carlson, J.; Dai, H.; Deng, J.; Rose, J.; Sheth, A.; Shucker, B.; Gruenwald, C.; Torgerson, A.; Han, R. MANTIS OS: an embedded multithreaded operating system for wireless micro sensor platforms. ACM Kluwer Mobile Netw. Appl. **2005**, *10* (4), 563–579.

43. Gummalla, A.C.V.; Limb, J.O. Wireless medium access control protocols, IEEE Commun. Surv. **2000**, *3* (2), 2–15 http://www.comsoc.org/surveys.

44. Radhakrishnan, S.; Racherla, G.; Furuno, D. Mobile ad hoc networks: principles and practices. In *Wireless Internet Handbook: Technologies, Standards, and Applications*; Furht, B., Ilyas, M. (Ed.) CRC Press: Boca Raton, FL, 2003; 381–403.

45. List of Ad hoc Routing Protocols, http://en.wikipedia.org/wiki/Ad_hoc_routing_protocol_list, (accessed March 10, 2007).

46. Gutierrez, J.A.; Naeve, M.; Callaway, E.; Bourgeois, M.; Mitter, V.; Heile, B. IEEE 802.15.4: a developing standard for low-power low-cost wireless personal area networks. IEEE Netw. **2001**, *15* (5), 12–19.

BIBLIOGRAPHY

1. Aetherwire, http://www.aetherwire.com.
2. ANRG sensor networks bibliography, http://ceng.usc.edu/~bkrishna/teaching/SensorNetBib.html.
3. Crossbow, http://www.xbow.com.
4. Dust networks, http://www.dustnetworks.com.
5. Ember, http://www.ember.com.
6. Geoworks, http://www.geoworks.com.
7. List of ad hoc routing protocols, http://en.wikipedia.org/wiki/Ad_hoc_routing_protocol_list.
8. Mesh networking, http://en.wikipedia.org/wiki/Mesh_networking.
9. National ecological observatory network, http://www.neoninc.org.
10. References on wireless sensor networks, http://appsrv.cse.cuhk.edu.hk/~yfzhou/sensor.html.
11. TinyOS community forum, http://www.tinyos.net.
12. UCLA CENS, http://cens.ucla.edu.
13. Wireless sensor network, http://en.wikipedia.org/wiki/Sensor_network.
14. Wireless ad hoc networks bibliography, http://w3.antd.nist.gov/wctg/manet/manet_bibliog.html.
15. Wireless sensor networks, http://www.research.rutgers.edu/~mini/sensornetworks.html.
16. Zigbee Alliance, http://www.zigbee.org.
17. Zigbee, http://en.wikipedia.org/wiki/ZigBee.

Wireless Sensor Networks (WSNs): Central Node and Mobile Node Design and Development

Watit Benjapolakul
Ky-Leng
Department of Electrical Engineering, Chulalongkorn University, Bangkok, Thailand

Abstract

A wireless sensor network consists of a large number of low-cost, low-power, multifunctional sensor nodes that are small in size and densely deployed either inside or very close to the phenomenon, use their own processing abilities to locally carry out simple computations, transmit only the required and partially processed data, and communicate between each other using wireless networking techniques.

INTRODUCTION

Recent advances in wireless communications and electronics have enabled the development of low-cost, low-power, multifunctional sensor nodes that are small in size and can communicate between each other using wireless networking techniques.[1] The sensor network represents a significant improvement over traditional sensors, and it can be used in various application areas such as health, military, industry, home, etc. For different application areas, the sensor networks are designed with different technologies.[2–3] It is also a key technology to obtain users' context in the real world because it can enable long-term data collection at scales and resolution that are difficult, if possible, to obtain otherwise. A large number of researchers have been working on sensor network technologies primarily for the realization of large scale environmental monitoring systems for military use and/or scientific use,[2–5] and in recent years, a large number of systems have been proposed for a small network and indoor network such as a wireless platform for deeply embedded networks (MICA)[4] and U3 node that is capable of communicating with other nodes to carry sensing data and queries issued by users. The wireless sensor package developed by Ara N. Knaian can be used only for Intelligent Transportation Systems by counting passing vehicles, measuring the average roadway speed, and detecting ice and water on the road.[6] However, they are not applicable for nonscientific and home use and have no feedback path. A Linux-based robot control system[7] developed by Traig Born and Joel Glidden is very powerful in terms of data collection and remote control; however, it cannot be considered as a sensor network because of its power consumption and its size. The personal sensor network is a private network, easy to update, and allows users themselves to change the configuration. It is an outdoor and indoor network that can be used in a large or small-scale networks such as environmental monitoring and remote security systems. It is very difficult to envision what such future application would be like; however it is desirable to specify some real applications as listed below:

- Measure the water level, gas toxicity, air pollution rate, humidity, and luminosity and control the dam gate;
- Measure the current intensity of the power distribution subcenter, alert the main center with the cause and cut off the dangerous line at the exact place;
- Use as security guard: monitor the gate, doors, windows, hallways, alert the security office and/or police stations with the location where the unpleasant event has occurred;
- Control the electrical appliances in the house.

HARDWARE AND SOFTWARE DESIGN FOR THE MOBILE NODE

Hardware

In order to fulfill the flexible requirements, we divide the system into small modules (modular system) that have specific functionalities, as shown in Fig. 1.

Radio link module. Use a mobile telephone kit as a radio link module to ensure the data transfer between a central node and a mobile node (MN). We use one PIC16F876 (programmable integrated circuits) and one EEPROM (electrically erasable programmable read only memory) to ensure this functionality.

Central processing unit module. Use PIC16F877 to ensure data processing, and sampling and storing of data in the external EEPROM. This module communicates with

Encyclopedia of Wireless and Mobile Communications DOI: 10.1081/E-EWMC-120043442

WSNs–
WSNs: Optimiz

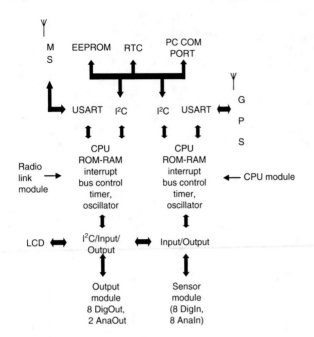

Fig. 1 Mobile node architecture.

the radio link module, some external sensor boards, and real-time clock (RTC) via I²C protocol.

Sensor module. As the CPU module supports digital & analog inputs, some sensors can connect directly to digital input ports and other sensors that give the result as voltage (0–5 V), e.g., humidity sensor, luminosity sensor (photoresistor), pressure sensor, etc., and can connect directly to analog input port. We also reserve one RS-232 port for the GPS receiver.

Output module. We fix I²C-digital to analog converter (DAC) AD5301 or PCF8591 to provide two analog outputs for the user with 8-bit resolution and PCF8574 to provide eight digital outputs. The buffer amplifiers for analog outputs (to protect our DAC) and power switching for digital outputs (for 12 V and/or 220 V load) are implemented but will be added to this MN as and when necessary.

Power module. It can be developed by using solar panel if necessary. Right now, we use adaptor AC-9V to supply the entire system.

Note. All these features are supported by our CPU module. If you want to use it, just connect the corresponding module then enable it from the PC. This implies flexibility and is cost-effective, because we pay for what we want. The modular technique provides not only flexibility but also minimizes power consumption as the CPU module can go to sleep mode ($I < 2$ mA at 5 V, 4 MHz) after processing all its works (it takes about 300 ms per scan, so if we scan the input every 1 min, the CPU can go to sleep mode for 59.7 s, that is, 99.5% of the time). Because the GSM module & radio link module cannot go to sleep

mode, we do not say that our system consumes less power compared to other systems; however, we try to minimize it to the extent possible.

Software

We implement a layer-2 protocol to MN in order to archive the flexibility to change the configuration by the user during or after the network configuration. We also provide feedback control by implementing a script execution function. The program is written by using **PicBasic Pro compiler**.

Radio link module. We use AT-command to send the data and SMS via air interface (mobile telephone); however, as not every mobile telephone supports text modes we implement text-to-protocol data unit (PDU) mode algorithm[8] for we can support the majority of mobile telephone. We also implement a layer-2 protocol on it, in order to ensure data transfer between the central node and MN, as shown in Fig. 2. Sending the collected data, updating the user customization/setting, having an alarm warning ("Critical input detected!" and "The memory is full at mobile node: RobotXXX"), sending input data, having feedback control and sending its report are the functionalities supported by this module. All information about this protocol is shown in Table 1, with other constants used in our work.

Note. All the user customization, warning message, and acquired data are stored in the external EEPROM, which is AT24C512. This EEPROM stores not only the data for radio link module (central mobile telephone number and manager mobile telephone number) but also very useful data for CPU module (options for CPU: scanning period,

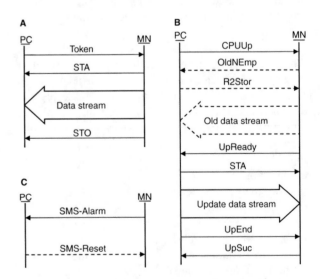

Fig. 2 Time diagram for the exchange of command and response signaling. (A) Send data. (B) Update-data. (C) Alarm by SMS.

Table 1 Protocol constants.

Name	ASCII character	Description
STA	178	Indicate the start of data
STO	187	Indicate the end of data
Token	196	Request for sending data
CPUUp	205	Request for updating data
R2Stor	214	Ready to store the old data
UpEnd	223	Indicate the end of update-data
OldNEmp	232	Data left in EEPROM
UpReady	241	Ready to store update-data
UpSuc	250	Updating is finished successfully
S	167	Synchronizing frame header

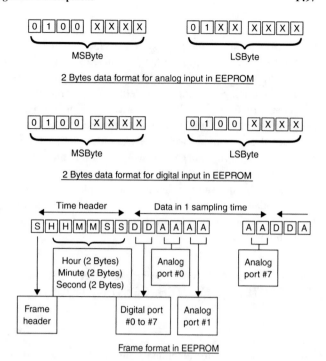

Fig. 3 Data format in EEPROM.

frame size, input masks, input critical values, digital output port's name, analog output port's name, upper bound counter limit, and alarm flag). Here, the sampling period is the same for all input ports because we do not know the exact future application of each port. After the network configuration, the user can only change the options for CPU. By just changing the data at the corresponding memory location, we can allow the user to customize the network very easily during installation as well as after installation.

CPU module. Ensure the data processing function. We sample the input port at every sampling period (integer value from 1 to 60 sec or min) and store the data in the external EEPROM. Note that not all the inputs are sampled; we sample only the valid input ports defined by the input mask (to reduce power consumption and storage). Fig. 3 shows the data format for analog input, digital input, and frame format used in our program. We suppose that the digital input and all the analog inputs are enabled.

The fact that we set bit #6 of each byte to 1 is to avoid the data to be special characters such as character 0, character 10, character 13, and character 26, etc. These characters are considered as a command, not data, for modem and cause modem to disconnect without sending disconnect command because the data themselves have the same meaning as a command. Sending the central node controls data process, and it looks like a "one-to-many" relationship—only the central node can ask and the other nodes can only answer. In case the CPU module stores the data more than FFE0h, the CPU module will inform the radio link module to send the SMS with the message, *"The memory is full at mobile node: **ID**"* where **ID** is the MN itself. At the same time, the CPU continues to store the new data until it reaches the full memory then it restarts. However, this situation rarely occurs because we already thought about this problem and we finally found the solution then program it at the central node See also Fig. 4.

In case any input data reaches the critical value (data is smaller/bigger than threshold value, predefined by the user, for analog signal or data is different from user predefined value for digital signal), we count up the counter by 1. We continue to count up whenever any input is critical, and the counter value is less than the upper bound counter limit. When all inputs change to a normal value, we count down the counter by 1 and continue to count down if all the next sampling inputs are normal and the counter value is bigger than 0. Assume now that the counter reaches the upper bound counter limit; the CPU Module will tell the radio link module to send an alarm message via SMS with the current input value from all ports beginning with the message *"Critical input*

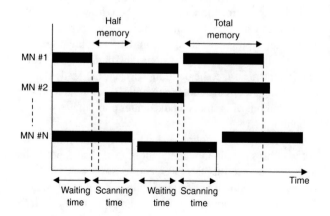

Fig. 4 Scanning process chronogram.

detected!" and then reset the counter. After the central node receives the alarm message, it can send the feedback command via SMS (script format) to the MN to activate some output ports and reset alarm (it's up to the user customizations). Script format is defined by character 60 and character 62 in ASCII mode. Any text starting with ASCII character 60 and ends with character 62 will be considered as command word; for example, <RESET> is used to reset alarm. The reason why we implement this format is to avoid the accidental error when unknown user sends a wrong message to this mobile telephone. In case we do not send the reset alarm, the MN will continue to report the critical port whenever the counter reaches the upper bound.

HARDWARE AND SOFTWARE DESIGN FOR THE CENTER COMPUTER (CENTRAL NODE)

Hardware

We need one mobile telephone connected to the COM port. At present, we use ERICSSON T68. We implement a serial-to-I^2C interface adapter for the central node in order to provide end users the ability to customize their network during installation.

Software

We implement the same layer-2 protocol (as for radio link module at MN) and I^2C protocol in the central node by using the VB programming language (Visual Basic). Moreover, we create graphic user interface and database (user application layer) for storing the MN information such as MN ID, phone number, location (province, district, section, and description), and critical value, etc. and enable the user to customize his own network by changing the input mask, telephone number, etc. In short, the user can change the data needed for radio link module and CPU module during the network installation. We also provide some functionality to the user to save the input data automatically to the user-predefined-folder, plot the record data like in excel, print the record, and show how many MNs this system can support with the current configuration, etc. The latter is implemented to avoid the problem of not having sufficient memory to store new data at MN. After reading the message, the program will automatically delete the message from the center computer's mobile telephone to prevent out of memory in the center computer's mobile telephone.

Note. If we set sampling period to 0, it means that we do not want to save the input data but scan as fast as possible (every 20 ms) so that the GPS feature is disabled. In fact, we can sample the input faster than this; however, because we want to save power and 20 ms sampling period is fast enough, it was decided that it would be the sampling period for our applications.

To determine how many MNs this system can support, and how long we can store the data before sending it to PC, the following formulas are used:

$$N_F = M/[(F_s \times 2) + H], \tag{1}$$

$$T = N_F \times T_s \times F_s, \tag{2}$$

$$N = T/T_0, \tag{3}$$

where
N_F: Number of frames contained in the last half memory;
M: Half memory size constant. It is 32,768 bytes for 64 Kb memory;
F_S: Frame size defined by the user;
H: Time header. It is used to indicate the beginning of new frame. $H = 7$ bytes (Fig. 3);
T: Time needed to fill all that half memory;
T_S: Scanning period defined by the user;
T_0: Scanning time needed for 1 node. It is equal to 4 min for 64 Kb memory;
N: Number of MNs supported by this system.

The longest scanning time must be less than or equal to the waiting time. See Fig. 4 for more details. Now let us assume that the frame size is set at 60, time-scan to 1 min, EEPROM size is 64 Kb, and only one analog input port is activated. Through a simple calculation, we found that our system can finish scanning all MNs within 15,480 min and can support up to 3,870 MNs. However if our system, with the same assumption, just have 15 MNs, we need only 15×4 min $= 60$ min for scanning all the MNsm, but the period of scanning is $15,480 + 60 = 15,540$ min; it means that we scan only 60 min and after that wait for 15,480 min for the new scanning process.

SYSTEM IMPLEMENTATION

Component Study and Hardware Design

PIC 16F87X and modular technique

PIC 16F87X is a product, which has many features as described in the data sheet.[9] For this project, PIC16F876 is chosen as a core component for radio link module and PIC16F877 is chosen as a core component for CPU module. Both of them have 8K of FLASH program memory (14-bit words), 368 bytes of data memory (RAM), and 256 bytes of EEPROM data memory and operate at speeds up to 20 MHz clock. However, a 4 MHz clock is chosen, because the speed is sufficiently fast for our application, and it can reduce power consumption as well, compared to 20 MHz (reducing the number of instruction per second is equivalent to reducing power consumption). The reason why PIC16F876 is chosen as a Microcontroller for radio link module is that it has a big program memory to store the program code, big RAM to store incoming PDU-SMS (SMS in PDU format) and other program variables, big EEPROM to store incoming text-SMS (SMS in text format) and outgoing PDU-SMS, and its input/output pin is enough to communicate with mobile telephone, intelligent LCD, I^2C peripheral (external EEPROM, RTC and output module). With similar reason, PIC16F877 is chosen as a microcontroller for CPU module.

Electrically erasable and programmable read only memory

The AT24C512 is used to store the data of surrounding environment and other options. It provides 524,288 bits of serial EEPROM organized as 65,536 words of 8 bits each. The device's cascadable feature allows up to four devices to share a common two-wire bus. This can be done by selecting the corresponding hardware address (A_1 and A_0). This device is optimized for use in many industrial and commercial applications where low-power and low-voltage operations are essential. It can operate at very low-voltage (down to 1.8 V). It uses two-wire serial interface (I^2C in short) to achieve bi-directional data-transfer protocol and can function up to 1 MHz clock at 5 V power supply. It also has Schmitt Triggers to suppress the inputs noise and other features; however, the most important point for us is the self-timed write cycle, which determines the speed of the scanning process in our project. In general, this self-timed write cycle is about 10 ms but, with this device, we can achieve a very low self-timed write cycle (5 ms maximum). According to our experience, some applications may need high speed of clock for a very short reading process such as a matrix display board, whereas in this application, low self-timed write cycle is needed because we use writing process more often than reading process. Another point to be borne in mind is that we cannot use page write to write the data into the EEPROM because our data length is not fixed, so it will cause a serious problem while writing across the page, therefore, byte write is preferred and self-timed write cycle becomes the most important factor.

RTC

There are so many RTCs available in the market today; however, DS1307 was chosen as our RTC, because it has a programmable square-wave output signal, independence of power supply frequency, and has back-up battery. In fact, this is not the best one but will suffice for our project. The DS1307 is a low-power serial RTC, with full binary-coded decimal clock/calendar plus 56 B of nonvolatile static RAM. Address and data are transferred serially via a two-wire, bi-directional bus. The clock/calendar provides seconds, minutes, hours, day, date, month, and year information. The end of the month date is automatically adjusted

for months with fewer than 31 days, including corrections for leap year. The clock operates in either 24-hr or 12-hr format with A.M./P.M. indicator. The DS1307 has a built-in power-sensing circuit that detects power failures and automatically switches to the battery supply. A lithium battery with 48 mAH or greater will back up the DS1307 for more then 10 yr in the absence of power at 25°C.

DAC

There are many available integrated circuits that assure the digital to analog (D/A) conversion such as AD5301, AD5311, and AD5321. They are single 8-, 10-, and 12-bit buffered voltage-output DACs that operate from a single 2.5 to 5.5 V supply, and consume 120 µA at 3 V. However, PCF8591 is chosen because it is available. The PCF8591 is a single-chip product, single-supply low-power 8-bit CMOS data acquisition device with four analog inputs, one analog output, and a serial I^2C-bus interface. Three address pins A0, A1, and A2 are used for programming the hardware address, allowing the use of up to eight devices connected to the I^2C-bus without additional hardware. Address, control and data to and from the device are transferred serially via the two-line bi-directional I^2C-bus. The functions of the device include analog input multiplexing, on-chip track and hold function, 8-bit analog-to-digital (A/D) conversion and an 8-bit D/A conversion. The maximum conversion rate is given by the maximum speed of the I^2C-bus, but this factor is not important for our application because this time delay is very small compared to the SMS process (read PDU-SMS, convert to text-SMS and run script execution function). Even though PCF8591 supports A/D conversion, we are interested in only D/A conversion. To achieve two analog-output ports, two pieces of PCF8591 are needed. This format is for continuous writing process, which means that after selecting the corresponding I^2C device we can send the digital value to this converter one by one and continue until the stop condition is generated. We send the analog value just once when we get a command via SMS from the central node or manager. The third byte sent to a PCF8591 device is stored in the DAC data register and is converted to the corresponding analog voltage using the on-chip DAC.

PCF8574-remote I/O expander

The PCF8574 is a silicon CMOS circuit. It provides general-purpose remote I/O expansion for most microcontroller families via the two-line bi-directional bus (I^2C). The device consists of an 8-bit quasi-bidirectional port and an I^2C-bus interface. The PCF8574 has a low current consumption and includes latched outputs with high current driving capability for directly driving light emitting diodes (LEDs). It also possesses an interrupt line, which can be connected to the interrupt logic of the microcontroller. By sending an interrupt signal on this line, the remote I/O can inform the microcontroller if there is incoming data on its ports without having to communicate via the I²C-bus. This means that the PCF8574 can remain a simple slave device; however, as our project needs to read and store data every scanning time, this feature seems to be meaningless for us.

GSM module

Besides using ERICSSON T68 as a wireless link for the system, an alternate solution is found. The GM862-GPRS is a GSM GPRS module dual band GSM 900/1800 with easy GPRS embedded (transmission control protocol/internet protocol stack inside). The GM862-GSM wireless data modules are the ready solution for all M2M wireless applications. The GM862-GSM is specifically designed and developed by Telit for original equipment manufacturer (OEM) usage and dedicated to cost-effective voice and telematics applications.

GPS module

In this project, GPS receive engine board, from Starts Navigation Tech. Ltd, is used. This GPS module has many features which can supply this module with 5 Vdc and 3V back-up battery. As we also use one battery back-up for RTC, this battery can be used for both RTC and GPS modules. For National Maritime Electronics Association protocol, we only use global positioning system fixed data format, because this output command can provide us information about latitude, north/south (N/S) indicator, longitude, east/west (E/W) indicator and mean sea level altitude in meters. Other information is not taken into account.

Radio link module hardware design

Fig. 1 shows the block diagram of the whole system and its functionalities are described. We can now start analyzing the schematic of our radio link module (not shown). From this design, we can see that

- The main power supply (7 ~ 10 V dc) is regulated by a regulator L7805. This regulator contains many protection features such as current limitation, thermal protection, etc., which makes it difficult for this regulator to be destroyed. The polarization protection is warranted by a diode, and noise reduction is warranted by three other condensers. However the non-polarized condenser should be connected as near as possible to the PIC to reduce the noise efficiently as it eliminates the high frequency noise.
- Power-on reset is improved by using the NPN transistor C1815 and other auxiliary components such as condenser and resistor. Once the RESET button is pressed, the CPU module, the radio link module and the GPS

module will be reset. We can see that this reset system is separated from one another. The reason is that our system is embedded with in-circuit serial programming (ICSP). To facilitate the design, each module has its own ICSP, implying two reset systems. In the CPU design, the reset system of GPS is connected to the CPU module, without any separation like in this case.

- Have ICSP embedded to facilitate the programmer. With ICPS, we can change the executable code any time we want to update it. This implies the flexibility and mass production support.
- Have LCD to display the inside operations and other information for the user.
- Embedded with debug capability. The connector J-DEBUG provides the programmer sufficient connection to debug/monitor the internal process and to program the EEPROM and PIC. To debug the pin, a 1 kΩ resistor is used to limit the current in case of short circuit at the PC side. Here, the RS-232 level is inverted in software; there is no need for hardware level converters. For I²C device programming, two 4.7 kΩ resistors are used to pull up serial data address (SDA) and serial clock line (SCL) pins to high.
- The connector J-CPU is used to communicate with the CPU module. This connector provides enough connection to communicate with the CPU and other I²C extension module. With this design, we can update/change the radio link or CPU module without mutual interference. This implies the flexibility to update/adapt to the new demand/application. The communication between radio link and CPU is supported by five pins (I²C, Flag, RxTx, SDA, and SCL); other pins are used for ICSP, debug, reset, and LED indicator.
- The internal processes are not only displayed via LCD but also indicated by LEDs. LED-red is used to indicate the PIC's process with mobile telephone such as reading/sending SMS, check new incoming SMS, and check incoming call, whereas LED-green is used to indicate the status of I²C-bus used in radio link module. LED-yellow indicates the CPU operation. They will be turned on every second but with different turn-on delays. This delay depends on the CPU process, if CPU is turned to work, the LED is turned on, and otherwise it is turned off for low-power mode. Turn on/off operations of GSM module are controlled by software. As this GSM module is designed to work with PC (inverted transistor-transistor logic [TTL] level), a level converter is needed. Here, MAX232 is used.

CPU module hardware design

The CPU module, for this design, is characterized as follows:

- The power supply is taken from radio link module via the J-RADIO connector. As described above, this connector allows the CPU module to communicate with radio link, ICSP, I^{2}C programmer, PC, and LED indicator. A 1 kΩ resistor is used to limit the current in case of short circuit at the PC side.

- The analog sensors can be connected to PIC via J-ANAIN, whereas the digital switches can be connected to PIC via J-DIGIN.
- The output board (analog and digital) can be connected to PIC via J-DIGANA. As described, we provide eight digital output pins and two analog output pins, and this is another board not included in this schematic. To communicate with that output board, J-OUTBOARD is used to connect all the digital and analog output pins and I^{2}C-bus (SDA, SCL) from CPU board to output board. We do like this because we want all (input pins and output pins) on the CPU board, so we can see directly that the input/output pins are controlled by CPU module, but the most important reason is that it is easy to update the output board without changing the output connector (modular system).
- As our process is synchronized with time, an RTC is used to generate a 1 Hz square-wave. Normally, this 1 Hz signal can be taken from the GPS module. However, input/output module and GPS module are optional and can be changed easily, and we want the CPU board to work independently, not depend on the GPS module; thus, RTC is needed. This RTC is not only used for generating 1 Hz clock but also for synchronizing the storing data with real-time calendar. This information is also used to display the date and time on the LCD for the user to see/control the module's functioning. Note that, initially, the RTC needs to be set up for it to know what the current time is and which frequency it should generate, so an indicator for this RTC is needed as it plays a very important role in the CPU process. LED-blue will be turned on at the output frequency of the RTC. To keep these data on the RTC, a back-up battery is needed, for which a small 3 V lithium battery is used.
- One EEPROM (24C512) is used to store the data (configuration data and input data from sensor).

Input board

This input board is made for testing purposes. It is used for measuring the temperature and luminosity. In this design (not shown), two types of temperature sensor ICs are used. The first one is LM335 NS and the second is LM35. The LM335[10] is an easily calibrated, precise, IC temperature sensor. Operating as a two-terminal zener, the LM335 has a breakdown voltage directly proportional to absolute temperature at +10 mV/K. With less then 1 Ω dynamic impedance, the device operates over a current range of 400 μA to 5 mA, with virtually no change in performance. When calibrated at 25°C, the LM335 has typically less

then 1°C error over a 100°C temperature range. Unlike other sensors the LM335 has a linear output, which makes interfacing to readout or control circuitry especially easy. The LM335 operates from 40°C to +100°C. The LM35 series are precise integrated-circuit temperature sensors,[10] whose output voltage is linearly proportional to the Celsius (Centigrade) temperature. The LM35 thus has an advantage over linear temperature sensors calibrated in Kelvin, as the user is not required to subtract a large constant voltage from its output to obtain convenient Centigrade scaling. The LM35 does not require any external calibration or trimming to provide typical accuracies of ±1/4°C at room temperature and ±3/4°C over a full −55 to +150°C temperature range. Low cost is assured by trimming and calibration at the wafer level. The LM35's low output impedance, linear output, and precise inherent calibration make interfacing to readout or control circuitry especially easy. It can be used with single power supplies, or with plus and minus supplies. As it draws only 60 μA from its supply, it has very low self-heating, less than 0.1°C in still air. The LM35 is rated to operate over a −55 to +150°C temperature range, while the LM35C is rated for a −40 to +110°C range (−10°C with improved accuracy). For luminosity sensors (photo-resistor), we have no datasheet or information about its characteristics, so we measure its value in a dark place and under sunlight to see the resistance variation. From this measure, a complete testing board is made. This design is to adapt to the analog input characteristics. The other condensers and resistors in this design are used as a filter (low pass filter) to eliminate the noises (electromagnetic noise, high frequency noise from the power supply, and Poisson noise from the light source for photo resistor). For LM35, a series R-C damper from output to ground is used to improve the tolerance of capacitance of the heavy capacitive load (LM35 can drive only 50 pf capacitive load). As this is a simple design for testing the system performance, the result is not so good but sufficient to see the evaluation of the surrounding environment. From our experiments, we can state that LM335 is difficult to use compared to LM35 because LM335 needs calibration, which is difficult to adjust by hand; moreover, if we want it to have a small error (<1% over 100°C range), we have to calibrate it at 25°C. However, LM35 is about thrice as expensive as LM335.

Output board

As described, two PCF8591s are used to assure the two analog outputs with 8-bit resolution and one PCF8574A is used to assure the eight digital outputs. The power switching board with eight outputs can be connected to this eight digital output directly. The design of digital output is easy; however, for analog output, we need a very good amplifier that can operate with 5 V power supply, low power, high input impedance, low output impedance, and rail-to-rail input/output feature such as LMC6001 or MCP6001.[11] The MCP6001 device, from Microchip, has a high phase margin, which makes it ideal for capacitive load applications. The low supply voltage, low quiescent current, and wide bandwidth make the MCP6001 ideal for battery-powered applications. However, for instance, we cannot find this kind of IC in Thailand, but as it does not interfere with our system performance testing, we have not developed this output board.

RS-232 to I²C debug and programming board

This board (not shown) is designed to help the developer easily debug the radio link module and CPU module. This board is not only used to debug but also program the EEPROM and RTC at first use. The EEPROM needs the configuration data to work with the CPU and radio link module, whereas RTC needs the initial time calendar to run its real-time calendar. This board is also used for programming the PIC. The ICSP programmer board is not a part of RS232-I²C board but it is related to our development work, so we just show it, not explain it. For the software (IC-Prog), we can download from http://kudelsko.fr/prog_pic/sommaire.htm free of charge. The PIC serial programmer board can program EEPROM as well as PIC16F8X, PIC16X62X, PIC16X55X, PIC16C6X, and PIC1687X but we still need RS-232 to I²C interface adaptor, because the PIC serial programmer cannot program RTC, and we need to change the cable from serial port (use with mobile telephone) to parallel port (use with EEPROM).

Communication Protocol and Algorithm

I²C-bus

In consumer electronics, telecommunications, and industrial electronics, there are often many similarities between unrelated designs. For example, nearly every system includes

- Some intelligent control, usually a single-chip microcontroller.
- General-purpose circuits like LCD drivers, remote I/O ports, RAM, EEPROM, or data converters.
- Application-oriented circuits such as signal processing circuits for radio and video systems, or DTMF generators for telephones with tone dialing.
- To exploit these similarities to the benefit of both systems designers and equipment manufacturers, as well as to maximize hardware efficiency and circuit simplicity, a simple bi-directional two-wire bus for efficient inter-IC control has been developed. This bus is called the inter-IC or I²C-bus. At present, the IC range includes more then 150 CMOS and bipolar I²C-bus compatible types for performing functions in all

categories (version 1.0—1992, version 2.0—1998, version 2.1—2000). All I²C-bus compatible devices incorporate an on-chip interface, which allows them to communicate directly with one another via I²C-bus. This design concept solves many interfacing problems encountered when designing digital control circuits.

Designer Benefits: I²C-bus compatible ICs allow a system design to rapidly progress directly from a functional block diagram to a prototype. Moreover, since they "clip" directly onto the I²C-bus, without any additional external interfacing, they allow a prototype system to be modified or upgraded simply by "clipping" or "unclipping" ICs to or from the bus. Here are some of the features of I²C-bus compatible ICs, which are particularly attractive to designers:

- Functional blocks on the block diagram correspond to the actual ICs; designs proceed rapidly from block diagram to final schematic.
- No need to design bus interfaces because the I²C-bus interface is already integrated on-chip.
- Integrated addressing and data-transfer protocol allow systems to be completely software-defined.
- The same IC types can often be used in many different applications.
- Design time is reduced as designers quickly become familiar with the frequently used functional blocks represented by I²C-bus compatible ICs.
- ICs can be added to or removed from a system without affecting any other circuits on the bus.
- Fault diagnosis and debugging are simple; malfunctions can be immediately traced.
- Software development time can be reduced by assembling a library of reusable software modules.

In additional to these advantages, the CMOS ICs in the I²C-bus compatible range offer designers some special features that are particularly attractive for portable equipment and battery-backed systems. They all have

- Extremely low current consumption;
- High noise immunity;
- Wide supply voltage range;
- Wide operating temperature range.

RS-232

RS-232 is a "complete" standard. This means that the standard has been set out to ensure compatibility between the host and peripheral systems by specifying

1. Common voltage and signal levels;
2. Common pin wiring configurations;
3. A minimal amount of control information between the host and peripheral systems.

Unlike many standards that simply specify the electrical characteristics of a given interface, RS-232 specifies electrical, functional, and mechanical characteristics in order to meet the above three criteria.

PC to mobile telephone and mobile telephone to radio link module exchange protocol

At this exchange protocol level, AT-command plays a very important role as it is used to establish the connection and to disconnect the communication. Now let us see how we use this command to establish/disconnect the connection. To establish a connection, PC [data terminal equipment (DTE)] sends the command **ATD <phonenumber><CR>** to the mobile telephone [data communications equipment (DCE)] that is connected to the PC via serial port, where **phone number** is phone number to be dialed. After send this command, the PC should wait for response from modem. The possible responses are

- **CONNECT 9600:** Means that the calling modem is now online and data exchange data can be started.
- **Busy:** Means that the called line is busy and the communication cannot be established. If the PC wants to connect to the MN, the PC needs to try again later.
- **No answer:** Means that the receiver did not answer the call and the communication could not be established. So we need to try again later.
- **No carrier:** Means that the modem handshaking has not been successful, so the user needs to check for mobile telephone registration and its signal strength and retry.

So, it is necessary for the PC (central node) to check whether the response **CONNECT 9600** has been received or not, and if not, the connection does not establish. Suppose now that this step has been done successfully, then we go forward to see what happens at MN. When an incoming call is detected, the modem (DCE) at MN will report an unsolicited code to the radio link module (DTE), which may be

- **Ring:** Means the extended format of incoming call indication is disabled and a call (voice or data) is incoming.
- **+CRING: VOICE:** Means the extended format of incoming call indication is enabled and a voice call is incoming.
- **+CRING: ASYNC:** Means the extended format of incoming call indication is enabled and an asynchronous transparent data call is incoming.
- **+CRING: REL ASYNC:** Means the extended format of incoming call indication is enabled and an asynchronous reliable (not transparent) data call is incoming.

- **+CRING: SYNC:** Means the extended format of incoming call indication is enabled and a synchronous reliable (not transparent) data call is incoming.
- **+CRING: FAX:** Means the extended format of incoming call indication is enabled and a fax call is incoming.

By detecting a **RING** code, the radio link can now answer this call by sending the command **ATA<CR>** and then wait for response:

- **CONNECT 9600:** Means the incoming call was a DATA one and called modem is now on line, and then the data exchange can now start.
- **Error:** Means no incoming call is found, so call may have been lost.
- **No carrier:** Means the incoming call was a DATA one and the modem handshaking has not been successful so we need to check for mobile telephone registration and signal strength, and modem settings.
- **Ok:** Means the incoming call was a VOICE call and is now active and the communication can now start.

Note that after the PC calls the MN, even the MN can receive a call with **RING** code, but it does not mean that the communication is established even though the MN answers with **ATA<CR>** command. After the MN answers the call, both sides have to wait for a response **CONNECT 9600** sent by the provider. After both sides detect **CONNECT 9600** responses, the data exchange sent can start. Now suppose that everything is complete and the MN wants to disconnect the connection, **+++ATH<CR>** is used to exit the data mode and enter the command mode then hang up the data call. We note that only the central node can ask for a connection to be established and only the MN decides whether to disconnect, because only the MN knows when to disconnect (finish sending all the data out from EEPROM). However, to prevent an unpleasant case, the central node has the right to disconnect if there is no data received in a period predefined by us (typically 2 s). This predefined delay is chosen because some time the modem at central node receives data and also updates the data into the buffer with practically unlimited size, but the program itself does not know whether the data has been updated because of the multitasking supported by the PC (the PC may not only run our program but also other programs and so if the PC is not fast enough, this problem will occur). To prevent this, 2 s monitoring time is chosen. This 2 s monitoring time is used to monitor whether the incoming data has been changed or not, if yes, the timer will be reset otherwise the timer will be started until it reaches 2 s value, and then the connection will be disabled by the central node.

WSNs–
WSNs: Optimiz

SMS

SMS is a way of sending short messages to mobile telephones and receiving short messages from mobile telephones. "Short" means a maximum of 160 B. According to the GSM Association, "Each short message is up to 160 characters in length when Latin alphabets are used and 70 characters in length when non-Latin alphabets such as Arabic and Chinese are used." The message can consist of text character, in which case the message can be read and written by human beings. SMS text messages have become a staple of wireless communications in Europe and Asia/Pacific and are gradually gaining popularity in North America. The message also can consist of sequences of arbitrary 8-bit bytes, in which case the message is probably created by a computer on one end and intended to be handled by a computer program on the other. The part of our application on the computer/MN creates an SMS message to be sent to the MN/computer. This message is sent to the short message center of the local telephone company to which the destined mobile telephone is subscribed. The telephone company finds the destined mobile telephone and passes the SMS message to it.

ACTUAL WORKS

We complete implementing our layer-2 protocols (Figs. 2A and B) communication link between MN and central node algorithm that we use to avoid the out of memory at MN, send/receive SMS and script execution, so we have tested the following applications:

- **Remote control via SMS:** Dial with receive/send SMS, script execution, and digital output;[12]
- **Read temperature via SMS:** Dial with receive/send SMS, script execution, and temperature sensor.[13]

In both the tested applications above, we are satisfied with the expected results. We also plan to test with the application below:

- **Monitor the current in power transmission line:** Dial with analog input, exchange command and signaling, data format, and scanning process.

CONCLUSION AND FUTURE WORK

These applications are starting points for our future work. It is simple and different from our real target. However, it can let us know and understand more about what we need and what we should do in the future. We intend to design and develop a pilot platform of central node and MNs for wireless sensor networks, which can support the functionalities stated above in the very near future.

ACKNOWLEDGMENT

The authors wish to thank the AUN/SEED-Net's Collaborative Research Support Project (JFY2004) and the Cooperative Project of Research between Department of Electrical Engineering, Chulalongkorn University and Private Sector under Contract Koroau 7/2547 for supporting this work.

REFERENCES

1. Akyildiz, I.F.; Su, W.; Sankarasubramaniam, Y.; Cayirci, E. Wireless sensor networks: a survey. Comput. Netw. **2002**, *38*, 393–422.
2. Akyildiz, I.F.; Su, W.; Sankarasubramaniam, Y.; Cayirci, E. A survey on sensor networks, Georgia Institute of Technology. IEEE Commun. Mag. **2002**, *40* (8), 102–114.
3. Shepherd, D. Networked micro-sensors and the end of the world as we know it. IEEE Technol. Soc. Mag. **2003**, *22* (1), 16–22.
4. Kawahara, Y.; Minami, M.; Morikawa, H.; Aoyama, T. Design and implementation of a sensor network node for ubiquitous computing environment. 58th IEEE International Conference on Vehicular Technology, Orlando, FL, Oct 6–9, 2003; Vol. 5, 3005–3009.
5. Mainwaring, A.; Polastre, J.; Szewczyk, R.; Culler, D.; Anderson, J. Design and Implementation of Wireless Sensor Networks for Habitat Monitoring, Intel Research Laboratory, Berkeley Intel Corporation, EECS Department, University of California at Berkeley, College of the Atlantic Bar Harbor http://www.cs.berkeley.edu/~polastre/papers/masters.pdf, 2003 (accessed June 6, 2007).
6. Knaian, A.N. A Wireless Sensor Network for Smart Roadbeds and Intelligent Transportation Systems, MS Thesis, MIT Department of Electrical Engineering and Computer Science and the MIT Media Laboratory, May 2000, Massachusetts Institute of Technology; http://www.media.mit.edu/resenv/pubs/theses/AraKnaian-Thesis.pdf, 2000 (accessed June 6, 2006).
7. Born, T.; Glidden, J. *Computer Systems Senior Design: A Linux-Based Robot Control System*, University of Arkansas at Little Rock, Donaghey College of Information Science and Systems Engineering, http://theduchy.ualr.edu/classes/syen4386/cs_sd.pdf, 2003(accessed Nov 2006).
8. Guthery, S.B.; Cronin, M.J. *Mobile Application Development with SMS and the SIM Toolkit*; McGraw-Hill: New York, 2002.
9. http://www.microchip.com (accessed June 6, 2007).
10. http://www.national.com (accessed June 6, 2007).
11. http://www.linear.com (accessed June 6, 2007).
12. Rey, D. *Interfaces GSM, Montages Pour Telephones Portables*; Editions Techniques et Scientifiques Françaises: Paris, France, 2004.
13. Benjapolakul, W. *Final Report on Development of Application of Short Message Services in Mobile Telephone System for Telemetering (Year I)*, Cooperative Project of Research between Department of Electrical Engineering; Chulalongkorn University and Private Sector under Contract Koroau 7/2547, Feb 2005 (in Thai).

WSNs—
WSNs: Optimiz

Wireless Sensor Networks (WSNs): Characteristics and Types of Sensors

Dharma P. Agrawal
Ratnabali Biswas
Neha Jain
Anindo Mukherjee
Sandhya Sekhar
Aditya Gupta
OBR Research Center for Distributed and Mobile Computing, ECECS Department,
University of Cincinnati, Cincinnati, Ohio, U.S.A.

Abstract
A wireless sensor network is a collection of tiny disposable devices with sensors embedded in them.

INTRODUCTION

Recent technological advances have enabled distributed information gathering from a given location or region by deploying a large number of networked tiny microsensors, which are low-power devices equipped with programmable computing, multiple sensing, and communication capability. Microsensor systems enable the reliable monitoring and control of a variety of applications. Such sensor nodes networked by wireless radio have revolutionized remote monitoring applications because of their ease of deployment, ad hoc connectivity, and cost effectiveness.

CHARACTERISTICS OF WIRELESS SENSOR NETWORKS

A wireless sensor network (WSN) is typically a collection of tiny disposable devices with sensors embedded in them. These devices, referred to as sensor nodes, are used to collect physical parameters such as light intensity, sound, temperature, etc. from the environment where they are deployed. Each node (Fig. 1) in a sensor network includes a sensing module, a microprocessor to convert the sensor signals into a sensor reading understandable by a user, a wireless interface to exchange sensor readings with other nodes lying within its radio range, a memory to temporarily hold sensor data, and a small battery to run the device. Wireless sensors typically have a low transmission data rate. A small form factor or size of these nodes (of the order of 5 cm^3) limits the size of the battery or the total power available with each sensor node. The low cost of sensor nodes makes it feasible to have a network of hundreds or thousands of these wireless sensors.

A large number of nodes enhance the coverage of the field and the reliability and accuracy of the data retrieved (Fig. 2). When deployed in large numbers, sensor nodes with limited radio communication range form an ad hoc network. An ad hoc network is basically a peer-to-peer multi-hop wireless network where information packets are transmitted in a store-and-forward method from source to destination, via intermediate nodes. This is in contrast to the well-known single-hop cellular network model that

Encyclopedia of Wireless and Mobile Communications DOI: 10.1081/E-EWMC-120043927

WSNs–
WSNs: Optimiz

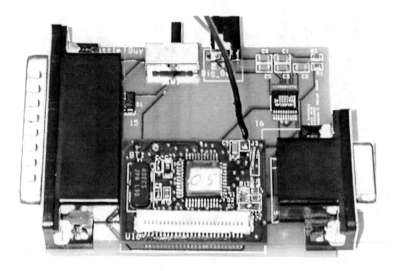

Fig. 1   Sensor mote. *Source*: http://www.sce.umkc. edu/~leeyu/Udic/SCE2.ppt.

supports the needs of wireless communications by installing base stations (BSs) as access points. Sensor nodes are attractive due to the ease of deployment and autonomous ad hoc connectivity that eliminates the need for any human intervention or infrastructure installation. Sensor networks need to fuse data obtained from several sensors sensing a common phenomenon to provide rich, multidimensional pictures of an environment that a single powerful macrosensor, working alone, may not provide.

Multiple sensors can help overcome line-of-sight issues and environmental effects by placing sensors close to an event of interest. This ensures a greater SNR by combining information from sources with different spatial perspectives. This is especially desirable in those applications where sensors may be thrown in an inhospitable terrain with the aid of an unmanned vehicle or a low-flying aircraft. Instead of carefully placing macrosensors in exact positions and connecting them through cables to obtain accurate results, a large number of preprogrammed sensor nodes are randomly dispersed in an environment.

Although communication may be lossy due to the inherent unreliable nature of wireless links, a dense network of nodes ensures enough redundancy in data acquisition to guarantee an acceptable quality of the results provided by the network. These sensor nodes, once deployed, are primarily static; and it is usually not feasible to replace individual sensors on failure or depletion of their battery.

Each node has a finite lifetime determined by its rate of battery consumption. It is a formidable task to build and maintain a robust, energy-efficient multi-hop sensor network in an ad hoc setting without any global control. It is therefore necessary to consider the different types of sensors that are commercially available.

TYPES OF SENSORS

Sensor networks present unprecedented opportunities for a broad spectrum of applications such as industrial automation, situation awareness, tactical surveillance for military applications, environmental monitoring, chemical or biological detection, etc.

Fig. 3 shows a progression of sensors developed by CITRIS investigators (http://www.citris-uc.org). These wireless sensors can be used to sense magnetic or seismic attributes in military security networks; or sense temperature or pressure in industrial sensing networks; strain, fatigue, or corrosion in civil structuring monitoring networks; or temperature and humidity in agricultural maintenance networks. Over a period of 2 yr, the size of these sensors has decreased from a few cubic inches to a few cubic millimeters, and they can be powered using tiny solar cells or piezoelectric generators running on the minute vibrations of walls inside buildings or vehicles.

Microstrain Inc. (http://www.microstrain.com) has also developed a variety of wireless sensors for different commercial applications.

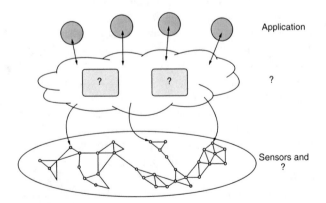

Fig. 2   A generic sensor network. *Source*: http://www.eyes.eu. org/eyes-sa.gif.

February 2001   February 2002   February 2003

Fig. 3 Progression of sensors developed by CITRIS investigators. *Source*: http://www.citris-uc.org/about_citris/annual_report.html.

Fig. 4 SG-link™ wireless strain gauge system. *Source*: http://www.microstrain.com/SG-link.htm.

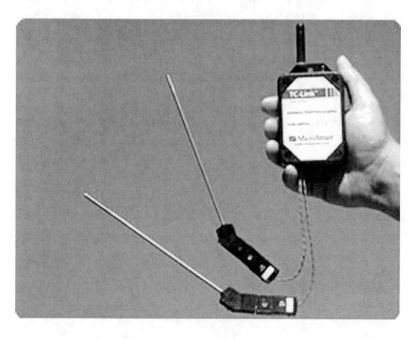

Fig. 5 TC-Link™ wireless thermocouple system. *Source*: http://www.microstrain.com/TCLink.htm.

Fig. 6 EmbedSense™ wireless sensor. *Source*: http://www.microstrain.com/embed_sense.htm.

The SG-link™ wireless strain gauge system (Fig. 4) is a complete wireless strain gauge node, designed for integration with high-speed WSNs. It can be used for high-speed strain, load, torque, and pressure monitoring and finds application in sports performance and sports medicine analysis.

The TC-Link™ wireless thermocouple system (Fig. 5) is a complete, cold junction compensated, multichannel thermocouple node designed to operate as part of an integrated, scalable, ad hoc WSN system. It finds application in civil structures sensing (concrete maturation), industrial sensing networks (machine thermal management), food and transportation systems (refrigeration, freezer monitoring), and advanced manufacturing (plastics processing, composite cure monitoring).

The miniature EmbedSense™ wireless sensor (Fig. 6) is a tiny wireless sensor and data acquisition system that is small enough to be embedded in a product, enabling the creation of smart structures, smart materials, and smart machines. A major advantage is that batteries are completely eliminated, thereby ensuring that the embedded sensors and EmbedSense node can be queried for the entire life of the structure. EmbedSense uses an inductive link to receive power from an external coil and to return digital strain, temperature, and unique ID information. Applications range from monitoring the healing of the spine to testing strains and temperatures on jet turbine engines.

We need to consider how a query can be processed in a sensor network.

Wireless Sensor Networks (WSNs): Key Management Schemes

Jamil Ibriq
Imad Mahgoub
Department of Computer Science and Engineering, Florida Atlantic University, Boca Raton, Florida, U.S.A.

Mohammad Ilyas
College of Engineering and Computer Science, Florida Atlantic University, Boca Raton, Florida, U.S.A.

Mihaela Cardei
Department of Computer Science and Engineering, Florida Atlantic University, Boca Raton, Florida, U.S.A.

Abstract

Advances in sensor technology expand the application potential for wireless sensor networks. However, the widespread use of WSNs is dependent on their security. A wireless sensor network (WSN) is a distributed autonomous network that typically consists of a large number of resource-constrained sensors and one base station in which the base station acts as the network controller and trust authority (TA). Therefore, security in WSNs requires the design and implementation of an efficient key management scheme. A key management scheme or protocol is a security mechanism whereby any group of sensors in the network authenticate each other and jointly establish one or more secret keys over an open communication channel. This entry provides a comprehensive survey of key management schemes designed for WSNs, proposes a classification scheme on the basis of the key management functions, and discusses the future trends in key management design in WSNs

INTRODUCTION

Wireless sensor networks (WSNs) have enormous potential to provide cost-effective solutions to solve a variety of problems encountered in every facet of life. Numerous advantages make these networks very appealing. They are faster and cheaper to deploy than wired networks or other forms of wireless networks. They can be deployed in inaccessible areas and hostile environments and operate unattended throughout their lifetime.

However, despite WSNs' appealing characteristics, they are extremely vulnerable systems;[1] their vulnerabilities stem from two types of constraints: sensor constraints and network constraints. The highly constrained sensor architecture limits the storage and memory capacity, power supply, computing capability, and transmission range of the sensor. The limited storage capacity requires that the code and related storage requirements of the security mechanisms be small. For example, a key management scheme that requires each node to store one secret key for each sensor in the network is not feasible for WSNs. The limited battery power of these disposable

devices makes it difficult for the sensor to use a public-key cryptosystem because of the communication cost of establishing an infrastructure and computation cost of large asymmetric keys. The limited transmission range stipulates that the sensors must use a distributed communication model. The network constraints stem from the sensors' random deployment, their unattended operation, and their mode of communication. The random deployment of sensors and their unattended operations mean that the networks must self-organize using a wireless communication channel, which make them extremely vulnerable to many adversarial attacks. In addition, the harsh environment causes the failure of many sensors that must be replaced. Replacing expired sensors with fresh ones creates a highly dynamic network. The large number of sensors implies that the key management scheme be scalable.

As a result of these constraints, it is impractical to apply existing security solutions that are designed for wired networks or *mobile ad hoc networks* (MANETs) to these networks. Key management protocols such as those in Refs. 2–26 use innovative techniques in solving the key management problem. An important distinction is made

Encyclopedia of Wireless and Mobile Communications DOI: 10.1081/E-EWMC-120043749

here between key distribution and key management schemes. A key distribution scheme defines a mechanism whereby one participant generates one or more secret keys and then transmits them to one or more participants. A key management scheme defines a protocol whereby two or more participants first authenticate each other and then *collaborate* in establishing one or more secret keys by communicating the key or keys over an *open* wireless channel. Open means the communication can be heard by all listeners. The key management schemes reviewed in this work are classified as either 1) *single trust authority* (TA) schemes in which every sensor achieves authentication through the base station; or 2) *pre-distributed* schemes, in which the sensors are pre-loaded with the necessary secret keys to allow them achieve authentication and communication privacy without intervention from the base station; or 3) *hybrid* schemes that combine the single TA and pre-distributed schemes to achieve sensor authentication and key distribution; or 4) *self-enforcing* schemes, in which the authentication is achieved pair-wise using a pre-distributed signed certificate. Self-enforced protocols are asymmetric pair-wise public-key-based protocols.

This entry is organized as follows: section "Terms and Definitions" defines a few important terms used in this entry. Section "Network Model" motivates the network communication model for WSNs. Section "Threat Model" builds the attack model in WSNs. Section "Security Requirements and Evaluation Criteria" discusses the basic security requirements and the criteria used for evaluating a key management scheme in this work. Section "Key Management Schemes in WSNs" formalizes a classification scheme for key management schemes in WSNs and provides an overview of each scheme in accordance with the proposed classification. Section "Future Trends" discusses future direction in key management research in WSNs, and Section "Conclusion" concludes the entry.

TERMS AND DEFINITIONS

The following terms and notations are used throughout this entry:

- **Secret key**: A key that is shared by a group of sensors;
- **Pair-wise key**: A key that is used to secure communication between any pair of sensors in the network;
- **Cryptographic primitives or materials**: A general term that represents the keys, generating functions, and algorithms;
- **Key-chain**: A set of keys derived from a random number;
- **Key-pool**: A set of random keys;
- **Link-key**: A key used to secure a direct wireless communication link established between two sensors;

- **Direct key**: A pair-wise key that is shared between two neighbors;
- **Path-key**: A key that is used to secure multi-hop wireless communication links, through one or more sensors;
- **Indirect key**: A pair-wise key that is established through intermediate sensors;
- **Group key**: A key that is used to secure one-hop or multi-hop communication among a group of sensors;
- **Cluster key**: A key used to secure communication of sensors within a cluster.

NETWORK MODEL

A WSN typically consists of a large number of resource-constrained sensors and one base station in which the base station acts as the network controller and access point for all the sensors. Sensors are usually randomly deployed over the target area but within close proximity of the base station that allow them to communicate with it either directly or indirectly over a public wireless communication channel. The sensors are vulnerable to all types of attacks mounted on the communication channel as well as physical damage and node capture by an adversary.

In every key management protocol presented in this work, the base station is trusted by all sensors in the network, assumed to be completely secure, and has unlimited resources. However, the sensors are highly constrained devices that must self-organize to form an interconnected network. During the self-organization process, they may create one of three network models: 1) *the direct model*; 2) *the multi-hop planar model*; and 3) *the cluster-based hierarchical model*.

In the direct model, every sensor in the network communicates directly with the base station. This model is an infeasible communication model. It has poor scalability because the communication cost grows linearly with the number of sensors in the network, which creates a bottleneck at the base station. Since the sensors have limited transmission range, some sensors cannot reach the base station and therefore cannot become part of the network. Hence, this model has weak connectivity. Since every sensor communicates directly with the base station, this model has low communication efficiency. As a result of these drawbacks, none of the key management schemes that have been investigated use this model. They use either the multi-hop planar model or the cluster-based hierarchical model.

In the multi-hop planar model, a sensor transmits to the base station by forwarding its data to one of its neighbors that is closer to the base station, which in turn sends it to a neighbor that is yet closer to the base station as shown in Fig. 1A. Thus the information travels hop-by-hop from the sensors to base station. In view of the sensor limitations, this is a viable approach. This model has better scalability and connectivity than the direct model. It is also more

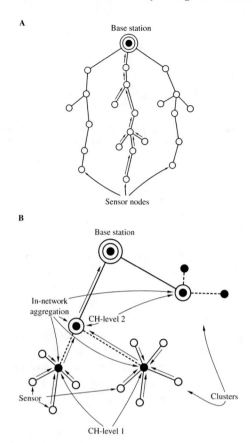

Fig. 1 (A) A multi-hop planar model showing the flow of traffic from the sensors to base station; (B) A cluster-based hierarchical model with two-level clustering. Traffic flows from the sensors to CHs to base station.

efficient. In the cluster-based hierarchical model, the network is divided into clustered layers as shown in Fig. 1B. The sensors are grouped into clusters with a cluster head that has the responsibility of routing messages from the cluster to a higher-level cluster head or to the base station. Sensors' readings are transmitted from the sensor to cluster heads and from cluster heads to the next higher-level cluster head until they reach the base station. This model further improves the scalability of the multi-hop planar model.

To secure communications in both of these models, the scheme must, in addition to establishing trust between the communicating parties, distribute the following keys: 1) pair-wise keys to secure data flow between pairs of sensors; 2) a group key to secure communication among cluster members; 3) a network-wide key to secure broadcasts from the base station to the sensors; and 4) a sensor-to-sink key to secure (unicast) communications between the base station and an individual sensor in the network.

THREAT MODEL

The mode of communication, lack of infrastructure, the hostile environment and constrained architecture of the sensors make WSNs extremely vulnerable to a host of

malicious attacks. Numerous types of attacks in WSNs have been identified and classified in Refs. 27 and 28. Some attacks such as message replay or modification or impersonation can be guarded against while others such as denial-of-service attacks are nearly impossible to guard against. The attacks can be classified either as *outsider attacks* or *insider attacks*.

In the outsider attack model, attackers are neither legitimate users of the network nor authorized users of any of its services. The attacker eavesdrops on the communications between legitimate users and acts in one of two ways. She can act passively by listening on the communication and stealing the information to achieve her own objectives. For example, she can listen to the communications between Bob and Alice in the battlefield and adjust her tactics accordingly. She can also actively attack the communication channel in a variety of ways. She can modify or spoof messages or inject interfering signals to jam the communication channel, thereby disabling the network or portion of it. She can also attack the key management scheme: She can target key trust centers such as cluster heads and compromise them selectively, thereby rendering the whole network useless.

In the insider attack model, attackers are legitimate users of the network and are authorized to use all of the network's services and are referred to as *compromised nodes*. A compromised node may either be a captured node that has been modified by the adversary or a more powerful device that has obtained all the cryptographic keys that enable it to forge legitimate cryptographic associations with other sensors in the network. The node is then employed by the adversary to mount insider attacks.

SECURITY REQUIREMENTS AND EVALUATION CRITERIA

Security Requirements

Sensor constraints and network constraints discussed earlier make WSNs more vulnerable than wired networks and other wireless networks such as MANET. The following are the basic security requirements in WSNs:

- *Entity authentication*: Sensors in a WSN must verify the identities of all network participants they communicate with before giving these participants access to services or information.
- *Source authentication*: Sensors must be assured that the received data are from the source they claim to be from and that the data are not interfered with enroute.
- *Confidentiality*: Sensors must communicate securely and privately over the wireless communication channel, preventing eavesdropping attacks. Data and cryptographic materials must be protected by sending them encrypted with a secret key to the intended receiver only.

- **Availability:** Each member of the network must have unfettered access to all network services at all times.
- **Data integrity:** Sensors must be assured that data received are not modified or altered.
- **Data freshness:** Sensors must be assured that all data are not stale. WSNs are prone to replay attacks. An adversary might re-transmit old data packets that were broadcast by legitimate users of the network to disrupt the normal network operations. A security solution must implement a mechanism that defeats such attacks.
- **Access control:** Sensors seeking to join the network must first be authenticated before given access to information and services.

Evaluation Criteria

The main function of a key management scheme is to provide the security services listed above. In addition to these basic services, a key management scheme will be evaluated in accordance with the following criteria:

- **Scalability:** It is the scheme's ability to support networks with large number of sensors;
- **Flexibility:** It is the scheme's ability to accommodate sensor additions and deletions;
- **Efficiency:** It refers to storage requirement and communication and processing overhead of the key management scheme;
- **Connectivity:** It is the scheme's ability to provide any pair of neighboring sensors a secure communication link;

- **Resiliency:** It is the scheme's ability to prevent and cope with a compromised node.

KEY MANAGEMENT SCHEMES IN WSNs

Communication security in WSNs heavily depends on the key management scheme. Given WSNs' constraints and their security needs, designing an efficient key management scheme for these networks is one of the most challenging problems and one of the most researched.

A Classification of Key Management Schemes

Many key management schemes have been proposed in Refs. 2–7, 9–12, 16–26, and 29. This section proposes a hierarchical classification of these schemes on the basis of the key management functions as shown in Fig. 2. At the higher level, this classification categorizes key management schemes on the basis of their authentication mechanism. If authentication is achieved through the interactive participation of the base station, the schemes are classified as *single TA* schemes. If the sensors are pre-configured to achieve authentication independently without the interactive participation of the base station with a group of sensors in the network, the schemes are classified as *pre-distributed* schemes. If authentication requires both some pre-configuration of the sensors and the involvement of the base station, the schemes are classified as *hybrid* schemes. If the sensors are pre-configured to achieve pair-wise authentication independently without the involvement

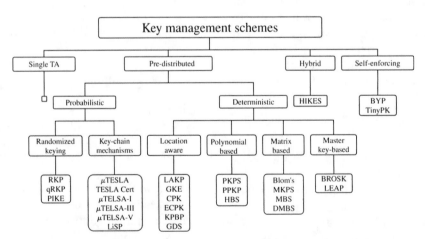

Fig. 2 A classification of key management schemes that are proposed for wireless sensor networks. The following is an alphabetical listing of the schemes and their reference numbers at the end of this entry. Blom's;[3] BROSK,[16] broadcast session key; BYP,[2] Beller-Yacobi protocol; CPK,[19] closest pair-wise key; DMBS,[17] deterministic multiple space Blom's scheme; ECPK,[19] enhanced CPK; GDS,[12] grid-group deployment scheme; GKE,[25] group-based key establishment; HBS,[21] hypercube-based scheme; HIKES,[29] hierarchical key establishment scheme; KPBP,[19] key pre-distribution with bivariate polynomial; LAKP,[9] location-aware keying protocol; LEAP,[26] localized encryption and authentication protocol; LiSP,[22] lightweight security protocol; MBS,[17] multiple space Blom's scheme; MKPS,[10] multiple space key pre-distribution scheme; PIKE,[27] peer intermediaries for key establishment; PKPS,[4] polynomial-based key pre-distribution scheme; PPKP,[21] polynomial pool-based key pre-distribution scheme; qRKP,[6] q-composite random key pre-distribution; RKP,[11] random key-pool; TESLA Cert,[5] timed efficient stream loss-tolerant authentication certificate; TinyPK,[24] Tiny public-key; μTESLA;[23] μTESLA I-V.[18]

of the base station, the schemes are classified as *self-enforcing* schemes.

At lower levels, the classification scheme uses the key distribution attributes as its classification criteria. The pre-distributed schemes are further subdivided into two groups: probabilistic and deterministic. In the probabilistic pre-distributed schemes, the keys are generated randomly, while in the deterministic pre-distributed schemes; the keys are generated using a deterministic mechanism. The probabilistic schemes are also divided on the basis of key generation technique as either *randomized* or *key-chain* schemes. The deterministic schemes are divided on the basis of the salient feature of the key generation technique into four sub-groups: 1) *master key-based* group; 2) *matrix-based* group; 3) *polynomial-based* group; and 4) *location-aware keying* group. The analysis of these schemes is based on this classification.

Using key management functions as the basis for classifying these protocols allows the classification scheme to evolve seamlessly. As new protocols are proposed, they are absorbed in the classification scheme. For example, at present, two protocols reviewed in this work are classified as self-enforcing.[2,24] Both are asymmetric protocols. If a key management scheme is proposed in the group that uses a symmetric key distribution scheme, this group bifurcates into asymmetric and symmetric groups. If a scheme applies a novel deterministic mechanism in its key generation process, then a new category under the deterministic pre-distributed schemes is created. The two examples show the expandability of this classification scheme, which allows the scheme to include new protocols.

Single TA Schemes

In this group, every sensor in the network must be authenticated by the TA, i.e., the base station, before it is given access to services and data. To the best of our knowledge, none of the WSNs' key management schemes that we have investigated falls under this category. However, there are numerous single TA schemes for wired networks and MAN-ETs. Examples of these schemes are: Kerberos,[30] Needham-Schroeder scheme,[31] and Otway–Rees scheme.[32]

Pre-distributed Schemes

In this group, each sensor is heavily pre-loaded with cryptographic materials to use in authenticating other network participants and generating the necessary secret keys with these participants. The main design goal in these schemes is to offset the excessive transmission cost of authentication by intensively pre-loading sensors with cryptographic materials. These schemes are further classified on the basis of their key generation process as either probabilistic or deterministic. In probabilistic schemes, the keys are generated randomly, while in the deterministic schemes the

keys are generated using a deterministic process. The probabilistic schemes are also divided on the basis of the keying technique as either key-chain or randomized schemes. The deterministic schemes are divided on the basis of the salient feature of the keying technique into: 1) master key-based group; 2) matrix-based group; 3) polynomial-based group; and 4) location-aware keying group. The analysis of these schemes is based on this classification.

Probabilistic schemes

These schemes are further classified on the basis of the keying approach as either key-chain schemes or randomized schemes. In the key-chain schemes, the base station probabilistically selects the last key K_n of the chain and hashes it n times to develop an n-key chain. It pre-loads each sensor with the last key of the chain, K_0 called the *commitment*, which the sensor uses for verifying and decrypting messages. In randomized schemes, the base station randomly generates the keys and pre-loads each sensor with a random subset of these keys.

Key-chain schemes. Perrig et al. propose one of the first key-chain schemes for WSNs.[23] It is a scaled-down version of *timed efficient stream loss-tolerant authentication* (TESLA) protocol called μTESLA.[23] μTESLA resolves the key distribution and authentication using the *key-chain approach*. In this approach, the base station generates a random key K_n and successively hashes it to develop a key-chain of n keys as shown in Fig. 3. Then it pre-loads each sensor with the last key, the *commitment* K_0.

The base station reveals the keys in reverse order of their generation, i.e., K_1 then K_2 and so on. The hashing scheme provides perfect forward secrecy, because given K_i, a sensor can generate $K_0, K_1, \ldots, K_{i-1}$ but cannot generate K_{i+1}. The base station broadcasts encrypted messages in interval i, which are stored by the sensors. In interval $i+1$, the base station reveals the encryption key, K_i. For example, in Fig. 3, messages sent in interval 1 that are encrypted with K_1 are first stored by the sensor. In interval 2, K_1 is disclosed. The sensor now verifies the key using successive hashing. If the hashing regenerates K_0, the sensor replaces K_0 with K_1 and decrypts the stored messages. This *delayed key disclosure* provides an efficient mechanism for broadcast authentication.

Although this scheme is flexible and highly resilient, it has scalability and efficiency issues. Consider the situation in which a sensor needs to send data to the base station.

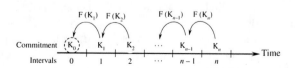

Fig. 3 Key generation and disclosure in μTESLA. F denotes the hashing function and K_i denotes the key for interval i.

The sensor has to generate its own key-chain and pre-load each sensor on the path between it and the base station with its commitment. This process is computationally non-trivial and requires substantial communication overhead. Perrig et al. propose the involvement of the base station in this process. As the network grows in size, the number of commitments a sensor needs to store becomes prohibitively expensive, and communication overhead of this process is substantial. Also, storing messages until the key is disclosed creates storage inefficiency. *Lightweight Security Protocol* (LiSP) uses μTESLA approach in which the encryption key is transmitted before the messages.[22] Hence, it improves its storage efficiency.

Bohge and Trappe use the TESLA Certificate scheme in which the base station assumes the role of main certification authority (CA) in the network.[5] According to this scheme, each node that wants to join the network must have a personal initial certificate (Cert$_j$) that is issued by the CA. The CA generates a certificate Cert$_A$(ID$_A$, t_{n+d}, K, MAC$_{K_n}$()) for sensor A at time t_n. When the certificate expires which occurs at time t_{n+d}, the CA discloses the TESLA K$_n$. The CA becomes a bottleneck creating scalability issues.

In Ref. 18, Liu and Ning address the scalability issues of μTESLA using five extensions, μTESLA-I–μTESLA-V for short. To eliminate communication costs associated with the bootstrapping of commitments, the authors propose long chains or large time intervals. Longer key-chains increase the authentication process for new sensors and since messages must be stored first and then decrypted at least one μTESLA interval later, and larger time intervals require larger message buffering space. Liu and Ning propose a two-level keying scheme that tackles these issues: 1) a higher-level key-chain with time intervals that cover the whole lifetime of the sensor network; and 2) a multiple lower-level key-chain with shorter intervals. The high-level key-chain is used to distribute and authenticate commitment keys of the low-level key-chain. According to this scheme, each sensor is loaded with the commitment key of the high-level key-chain and one-way functions of both levels.

In μTESLA-I, sensor nodes are initialized with the commitment of the high-level chain, time intervals of high-levels and low-level key-chains, and one-way functions of high-level and low-level chains. However, low-level keys are not chained together. As in μTESLA, the loss of a high-level commitment key can be recovered by key disclosure in a later interval, but the loss of a low-level commitment key may result in the loss of all messages in an interval. A fault-tolerant two-level scheme, μTESLA-III, is proposed to address these issues by generating the low-level keys from the high-level keys using a different one-way function as shown in Fig. 4. The paper also proposes a multi-level μTESLA-V scheme that provides smaller time intervals and shorter key-chains. Although these extensions provide some improvement over

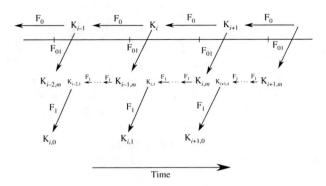

Fig. 4 A fault-tolerant two-level μTESLA-III proposed by Liu and Ning. The downward arrows show the broadcast of the low-level key commitments in each interval. $\mathbf{F_0}$, $\mathbf{F_{01}}$, and $\mathbf{F_1}$ are pseudo random functions. This figure is redrawn from Ref. 18. *Source*: Ref. 18. The two levels of key chains in Scheme III. The downward arrows show the broadcast of the low level key commitments in each interval. $K_{i,m}$ is derived from K_{i+1} using an additional pseudo random function $\mathbf{F_1}$.

μTESLA proposed in Ref. 23, they have the same scalability and efficiency issues of μTESLA.

Randomized keying schemes. Eschenauer and Gligor propose a probabilistic pre-distribution scheme[11] referred to here as *random key-pool* (RKP) scheme. In the pre-distribution phase, a large key-pool consisting of P keys and their identities are generated. In the *key setup phase*, each sensor is loaded with k keys and their identities, which form the sensor's key ring. The probability that two sensors share at least one key is given by: $p_r = 1 - ((P - k)!)^2/(P - 2k)!P!$. In the *shared-key discovery phase*, two neighboring sensors exchange and compare the list of key identities in their key-chains. If a shared key identity is found, the sensors set up a secure communication link between them. In the *path-key establishment phase*, a path-key is assigned to a pair of sensors that do not share a key. These sensors may be two or more hops away from each other.

RKP is an efficient, scalable and resilient key management scheme. However, it has low flexibility and weak connectivity. As the network grows and contracts, the keying mechanism becomes less flexible. Furthermore, not all sensors in the network achieve connectivity.

In **q**-*composite random key pre-distribution* (qRKP) scheme, Chan et al.[6] improve the resiliency of the RKP scheme by stipulating that two sensors can communicate with each other if they share q keys, where $q > 1$. The two sensors create a link-key, K, by hashing all the common keys as follows: K = Hash(K$_1$||K$_2$||K$_3$|| ... K$_q$). This approach improves the network resiliency because the probability of compromising a link decreases from k/P to $\binom{k}{q}/\binom{P}{q}$. But this improvement in resiliency weakens sensors' connectivity. Since q keys are required to establish a link between a pair of sensors, the probability of creating link-keys decreases.

To address scalability in symmetric key pre-distribution schemes, Chan and Perrig[7] developed PIKE, *peer intermediaries for key establishment.* The main idea in PIKE is to use trusted peers as trusted intermediaries to establish shared keys between sensors. Each sensor is pre-loaded with $2(\sqrt{n}-1)$ unique keys, where n is the number of sensors in the network. The sensor ID is represented by (x,y), where $x,y \in \{0, 1, 2, \ldots, \sqrt{n}-1\}$.

In this design shown in Fig. 5, any pair of sensors u and v can find two nodes that will share a pair-wise key with them. So u and v can establish a pair-wise key between them. An example is shown in Fig. 5. In Fig. 5B, sensor 91 can communicate with all the nodes in its row and all the nodes in its column. Suppose 91 wants to establish a key with 13. Since it does not have a shared key with 13, it uses an intermediary to establish the key. It can use either 11 or 93 because both share keys with both 91 and 13. Sensor deployment in PIKE must be batched and in accordance with node ID order. This orderly deployment maximizes the number of intermediates between any two sensors. Simulation results have shown that PIKE has better scalability than RKP and qRKP. However, it has the same flexibility issues of random keying schemes.

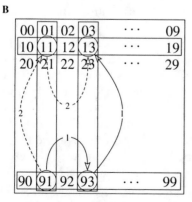

A

00	01	02	03	$\cdots$	09
10	11	12	13	$\cdots$	19
20	21	22	23	$\cdots$	29
90	91	92	93	$\cdots$	99

B

Fig. 5 Sample virtual ID grid for 100 sensors in PIKE. (A) Shows the allocation of the IDs. (B) Shows how node 91 establishes a key with node 13. It uses either 11 or 13 as an intermediary.

Deterministic schemes

Schemes in this group provide the sensor with the deterministic means of achieving authentication and key distribution. The base station in this group pre-loads the sensor with deterministic cryptographic materials that allow the sensor to authenticate other sensors in the network and generate the required keys to secure its communication with those sensors. These schemes are divided into four categories: 1) The master key-based schemes use one global key to solve the key management problem; 2) The matrix-based schemes use Blom's approach to solving the key management problem; 3) The polynomial-based schemes are based on Blundo's scheme in which each sensor has a share of a polynomial that allow it to authenticate other sensors and generate secret keys; and 4) Location-based schemes in which the sensors are pre-loaded with cryptographic materials that allow those that are in the same neighborhood to generate direct keys.

Master-key-based schemes. Lai et al. propose a master-key-based pre-distribution scheme called ***broadcast session key*** (BROSK) *negotiation protocol.*[16] In BROSK, sensors i and j exchange random nonces N_i and N_j and generate a session key $K_{ij} = f(K_m \,||\, N_i \,||\, N_j)$, where f is a pseudo-random function and K_m is the master key. For each link, a sensor develops a session key. Like most master-key-based schemes, BROSK is very efficient. Each sensor uses one master key and a session key for each neighbor; the communication cost is also small involving the exchange of nonces. BROSK is also flexible and scalable; however, it has very low resiliency. The compromise of one sensor compromises the entire network.

In Ref. 26, Zhu et al. propose a more resilient master-key-based key management scheme called **l**ocalized **e**ncryption and **a**uthentication **p**rotocol (LEAP). LEAP requires five keys for its operations: a *global key, pairwise key, cluster key, individual key* and *group key*. Each sensor is pre-loaded with a *global key* K_I, an individual key which is a unique key the sensor shares with the base station, and pseudo-random function, f. Using K_I and f, a sensor can generate its master key and those of other sensors as follows: $K_i = f_{K_I}(i)$, where i represents the sensor's ID. A sensor u generates its pair-wise key with v, K_{uv} using the same mechanism as follows: $K_{uv} = f_{K_v}(u)$ and v can independently compute K_{uv}.

To establish a pair-wise key, a sensor u broadcasts $u \,||\, N_u$ where u is its ID and N_u is a randomly generated nonce. A neighbor v responds with the following message: $v, \mathrm{MAC}(K_v, N_u \,||\, v)$. u generates the K_v and verifies the identity of v by regenerating the MAC value. Both u and v compute K_{uv} using each other ID and K_I. If sensor u needs to establish a secure communication channel with another sensor, c, that is more than one hop away from it, a multi-hop pair-wise key must be established. In this situation, u first finds m intermediate sensors, $v_1, v_2, \cdots, v_m$. Then, it generates a secret key, $S = K_{uc}$, and divides S

into m shares, such that $S = sk_1 \oplus sk_2 \oplus \cdots \oplus sk_m$. It forwards each share to c through the intermediate sensor, v_i. The following illustrates the message exchange: 1) $u \to v_i$: $\{sk_i\}K_{uv_i}, f_{sk_i}(0)$; 2) $v_i \to c$: $\{sk_i\}K_{v_ic}, f_{sk_i}(0)$.

Once the pair-wise keys are established, the cluster head creates the cluster key and sends it to each of its neighbors encrypted with the neighbor pair-wise key. The group key is a global key shared by all sensors in the network. It is designed for maintaining secure communication between the base station and the sensors. LEAP uses μTESLA for broadcast authentication with base station. But for inter-node traffic, a sensor develops a MAC value using the cluster key as the MAC key and transmits the packet with the MAC value. A receiving node first verifies the packet using the same cluster key it has received from the sender during the cluster key establishment and then authenticates the packet with its own cluster key. Thus the message is authenticated hop-by-hop from source to destination, thereby allowing passive participation.

Like BROSK, LEAP is an efficient and scalable scheme but has low resiliency. The entire security is based on a global key, the initial key, K_1. Once a sensor is compromised, an adversary can obtain all pair-wise keys. LEAP also has poor flexibility. Since K_1 is erased immediately after the neighbor discovery phase and the pair-wise key establishment phase, it is not possible to add fresh sensors to the network.

Matrix-based pre-distribution schemes. Several *key matrix-based pre-distribution schemes* based on *Blom's λ-secure key pre-distribution scheme* have been proposed.[3] In Blom's scheme, all links in a network of size N are represented by $N \times N$ key matrix. By storing only a small set of keys in each node, it is possible for the node to calculate the field of the matrix and use it as a link-key. This scheme works as follows: The base station generates a public $(\lambda + 1) \times N$ matrix G and then creates a random $(\lambda + 1) \times (\lambda + 1)$ symmetric matrix D over a finite field Fq, where q is a prime. The base station then computes the $N \times (\lambda + 1)$ private matrix $(D \times G)^T$. It follows that $K = (D \times G)^T \times G = ((D \times G)^T \times G)^T$ is an $N \times N$ symmetric matrix. A node i is pre-loaded with the i-th column of G

matrix and i-th row of $(D \cdot G)^T$ matrix as shown in Fig. 6. Two nodes i and j exchange their public information, the i-th and j-th columns of G. The link-key is generated independently by each node as $K_{ij} = (D \cdot G)^T_{(\text{row } i)} \times G_{(\text{column } j)}$ and $K_{ji} = (D \cdot G)^T_{(\text{row } j)} \times G_{(\text{column } i)}$. In this scheme, each node broadcasts one message and receives one message containing a $\lambda + 1$ vector from each of its neighbor. In a dense network, the size of these vectors may require large storage space. The exchange of vectors entails substantial communication cost particularly in a large network, and the multiplication of the vectors requires substantial computation cost, especially if the vector elements are as large as the cryptographic keys.

To improve the resilience of Blom's scheme, Du et al.[10] propose a ***multiple space key pre-distribution scheme*** (MKPS). MKPS defines two matrices D and G as defined in Blom's scheme. However, G is defined as a set of private matrices, $G_1, G_2, \ldots, G_w$. It also randomly associates a key space, τ, with each sensor from the ω key spaces where $2 \leqslant \tau < \omega$. As in Blom's scheme, for each sensor i, the i-th row of $(D \cdot G^T$ is stored in sensor i. Each sensor has $\tau + 1$ vectors, each consisting of $\lambda + 1$ elements. In the key agreement phase, a pair of sensors first agrees on the common key space by exchanging their IDs, the indices of their respective spaces, and their seeds of the columns in G. If they share a common space, each computes the pair-wise secret key as shown in Blom's scheme. Although MKPS improves resiliency, it degrades network connectivity. A pair of neighboring nodes may not share a common space and as result may not able to establish a secure link. These sensors have to establish a path-key through one of their neighbors. In MKPS, each sensor is required to send an extra message to include its key space (τ) ID. It also has the same scalability issue as that of Blom's scheme.

In Ref. 17, Lee and Stinson propose two matrix-based pre-distribution schemes: ***multiple space Blom's scheme*** (MBS) and the ***deterministic multiple space Blom's scheme*** (DMBS). MBS improves the scalability of Blom's scheme. It differs from Blom's scheme in two ways. First, it divides the network in two sets U and V to form a bipartite key connectivity graph. Therefore, not every pair of sensors shares a secret key. Second, the private matrix D does not have to be symmetric. A pair of sensors μ and v, are assigned public information in the same way as in Blom's scheme. Each sensor $v \in U$ is assigned the secret information $x_u^T D$, and each $v \in V$ is assigned Dx_v. Both u and v are assigned their public column vectors, x_u and x_v. Now, they can exchange their public information and compute their secret keys, $x_u^T Dx_v$ and $x_v^T Dx_u$. DMBS supports larger networks. In DMBS, a network is represented by a regular graph G. Each vertex $u \in V(G)$ is considered as a class of l nodes such that $u = u_1, u_2, \ldots, u_l$. Each edge $e \in E(G)$ is assigned a direction and a random private matrix D_e. For each edge

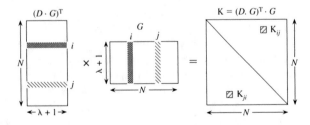

Fig. 6 Blom scheme. Node i stores i-th column of matrix G and i-th row of matrix $(D \cdot G)^T$ and node j stores j-th column of matrix G and j-th row of matrix $(D \cdot G)^T$. i and j exchange their columns and generate $K_{ij} = ((D \cdot G)^T_{(\text{row } i)} \times G_{(\text{column } j)})$ and $K_{ji} = ((D \cdot G)^T_{(\text{row } j)} \times G_{(\text{column } i)})$.

$(u_i, v_j) \in E(G)$, the source node u_i receives $x_{u_i}^T D_e$ and destination node v_j receives $D_e x_{v_j}^T$. Both u_i and v_j can compute the link-key associated with edge e as follows: $K_{u_i v_j} = x_{u_i}^T D_{uv} x_{vj}$. This scheme improves the scalability of Blom's scheme, but decreases the memory efficiency and degrades the resiliency of the network. If each sensor has on average r neighbors, it stores $r \cdot (\lambda + 1)$ vectors. Also, the capture of one sensor compromises $l - 1$ sensors, i.e., the entire class.

Polynomial-based keying schemes. Blundo et al.[4] propose a *polynomial-based key pre-distribution scheme (PKPS) for group keys, which can be used for pair-wise key establishment as follows. The base station randomly generates a bivariate λ-degree polynomial $f(x,y)$ over a finite field F_q, where q is large prime enough to accommodate a cryptographic key. The polynomial also has the property of $f(x,y) = f(y,x)$. During the key setup phase, each sensor receives a polynomial share of $f(x,y)$. A pair of sensors, i and j, obtain a link-key as follows: sensor i evaluates $f_i(i,y)$ at point j to obtain $f(i,j)$ and j evaluates $f_j(j,y)$ at point i to obtain $f(j,i)$. Given that for all x,y, $f(x,y) = f(y,x)$, therefore $f(i,j) = f(j,i)$. Thus both i and j can communicate securely using this key. This scheme requires that each sensor store its share of λ-degree polynomial, which requires $(\lambda + 1) \log q$ memory space. Since each polynomial has $\lambda + 1$ coefficients, it is λ-secure. This means that a coalition of less than $\lambda + 1$ sensors does not have any information about the pair-wise key between two sensors.*

In Ref. 21, Liu et al. combine Blundo's scheme with the randomized key-pool scheme of Eschenauer and Gligor and develop *polynomial pool-based key pre-distribution scheme* (PPKP). The general concept is the same as that of Eschenauer and Gligor's scheme: a pool of symmetric bivariate polynomials is randomly generated and a subset of this polynomial pool is assigned to each sensor. Sensors that have common polynomials in their ring can establish direct keys. This scheme is designed with the aim of improving the resiliency and scalability. PPKP has three phases: *setup phase, direct key establishment phase,* and *path-key establishment phase.*

During the setup phase, the base station randomly generates a set F of bivariate λ-degree polynomials over the finite field F_q and assigns a unique ID to each polynomial. For each sensor node i, the base station assigns a subset F_i of these polynomials ($F_i \subseteq F$). The assignment may be random or deterministic. In the direct key establishment, two methods for establishing a pair-wise key between two sensors are proposed: pre-distribution and real-time discovery. In the pre-distribution method, the base station maps the sensor ID to the IDs of its polynomial share. Thus any pair of sensors can determine whether they can establish a direct key. In the real-time discovery approach, two sensors may exchange challenges as in Ref. 11. When sensor i needs to establish a pair-wise key with sensor j, it

sends sensor j an encryption list, $\alpha, E_{K_v}(\alpha), v = 1, \ldots, |F_i|$, where K_v is computed by evaluating the v-th polynomial share in F_i on point j (a potential pair-wise key sensor j may have). j first computes $\{K'_v\}_{v=1,\ldots,|F_i|}$, where K'_v is computed by evaluating the v-th polynomial share in F_i on point i (a potential pair-wise key node i may have). If a common encryption value is found in both encryption lists, i and j can establish a common key.

If i and j fail to establish a pair-wise key directly, they have to do it indirectly using path-key establishment. Again two methods are proposed: pre-distribution and real-time discovery. In the pre-distribution method, the base station loads each sensor u with certain information so that given the ID of another sensor v, it can determine at least one key path to v directly. The resulting key path is called the pre-determined key path. This method is not resilient. An adversary may take advantage of the pre-distributed information to attack the network. In the real-time discovery method, a sensor that is seeking to discover a path to another sensor picks a set of intermediate sensors with which it has established direct keys. It may send requests to all these intermediate sensors. If one of the intermediate sensors can establish a direct key with the destination node, a key path is discovered. This method is inefficient: It introduces substantial communication overhead. Like the probabilistic schemes in Refs. 6, 10, and 11, it does not guarantee that any pair of sensors can establish a pair-wise key.

To overcome these shortcomings, Liu et al.[21] propose the ***hypercube-based scheme*** (HBS). This scheme guarantees connectivity to any pair of sensors in the network, is more resilient and has lower communication overhead for key establishment. A brief description of the scheme is as follows. Given a network of N sensors, the base station constructs an n-dimensional hypercube with m^{n-1} bivariate polynomials arranged for each dimension, j. The polynomial has the form: $\{f_{i_1,\ldots,i_{n-1}}^j(x,y)\}_{0 \leq i_1,\ldots,i_{n-1} < m}$ where $m = \lceil \sqrt[n]{N} \rceil$. The base station randomly generates nm^{n-1} set of symmetric λ-degree bivariate polynomials $F = \{f_{i_1,\ldots,i_{n-1}}^j(x,y) | 1 \leq j \leq n, 0 \leq i_1, \ldots, i_{n-1} < m\}$ over a finite field, F_q. The base station assigns each sensor a unique n-vector coordinate J on the hypercube of the form $J = (j_1, j_2, \ldots, j_n) \in [0,m)$. It then assigns the sensor its ID and polynomial share $f_{j_2,\ldots,j_n}^1(j_1,y), \ldots, f_{j_1,\ldots,j_{n-1}}^n(j_n,y)$.

To establish a direct key with sensor v, sensor u checks if it has same sub-indices in $n - 1$ dimensions with v, that is the Hamming distance, d_h. If $d_h = 1$, nodes u and v share a common polynomial and can establish a direct key using the PKPS presented earlier. If u and v cannot establish a direct key, they have to go through the *path discovery* phase to establish an indirect key between them. To establish an indirect key with v, u sends a key establishment request to a node that can forward the request to v. The process is secure because the establishment request is encrypted and authenticated by each intermediary using the direct key. Although HBS is more efficient and shows better sensor connectivity than PKPS, it does not guarantee neighbor-to-neighbor connectivity. In general, it shows the same flexibility issues of probabilistic schemes.[6,10,11]

Location-aware pre-distribution schemes. All the pre-distribution schemes presented thus far have assumed that location information prior to sensor deployment is not available. In Ref. 9, Du et al. present location information model referred to here as *location-aware keying protocol* (LAKP). The scheme divides the network into $t \times n$ equal size groups $G_{i,j}$ and deploys them at a *resident point* (x_i, y_j) where $i = 1, 2, \ldots, t$ and $j = 1, 2, \ldots, n$ and where the resident points form a 2-D grid. The resident point of a sensor $v \in G_{i,j}$ follows the pdf $f_v^{i,j}(x, y \mid v \in G_{i,j}) = f(x - x_i, y - y_j)$ where $f(x,y)$ is a Gaussian distribution. In the key setup phase, the key-pool (S) is divided into $t \times n$ key-pools $S_{i,j}$, each of size $|S_g|$ such that $S_{i,j}$ is used for group $G_{i,j}$. Based on a specific deployment distribution, a scheme can be developed such that when the deployment points of two groups $G_{a,b}$ and $G_{c,d}$ are distant from each other, the amount of overlap between $S_{a,b}$ and $S_{c,d}$ becomes smaller or zero.

Given the network construction shown in Fig. 7 with α and β the overlapping factors, two horizontally or vertically neighboring group key-pools share exactly $\alpha|S_c|$ keys, where $0 \leq \alpha \leq 0.25$. Two diagonally neighboring key-pools share exactly $\beta|S_g|$ keys, where $0 \leq \beta \leq 0.25$ and $4\alpha + 4\beta = 1$ while two non-neighboring key-pools share no keys. This scheme has weaker connectivity than the probabilistic schemes proposed in Refs. 6 and 11.

In *group-based key establishment* (GKE) protocol, Zhou et al.[25] remove the deployment assumptions made in Refs. 9 and 12 that the relative location of groups are known prior to sensor deployment. Zhou et al. propose two deployment models: *random region group deployment* (RRGD) and *pure random group deployment* (PRGD). In RRGD, sub-regions are pre-defined on the target area but can receive any sensor group. In PRGD, sub-regions may be deployed anywhere on the target area.

The main idea behind GKE is to pre-load each sensor with a pair-wise key for every sensor in the group. A pair

of sensors that belong to the same group share a common key and are called associated sensors. If the members of the pair belong to different groups and are not associated, they can establish pair-wise key using an intermediate associated pair. This scheme removes the requirement that the deployment points be pre-defined prior to sensor deployment, making GKE structurally less rigid than the schemes proposed in Refs. 9 and 12. Since the communication involved in the key establishment is localized, it is more efficient than both. However, it has weaker connectivity.

Liu et al. propose a scheme similar to GKE, called ***group-based key pre-distribution*** (GKP) *scheme.*[20] GKP has a *pre-distribution phase, direct key establishment phase,* and *path-key establishment phase.* GKP divides the network into n groups $\{G_i\}_{i=1,2,\ldots,n}$ that are evenly and independently deployed on a target area. Sensors in group G_i are deployed at the same time with the same deployment index i. The resident points in G_i follow a pdf $f_i(x,y)$.

In the pre-distribution phase, each G_i is pre-loaded with a key *pre-distribution instance* D_i that consists of G_i sensors, a set of keying materials K, polynomials or matrices, and a function that maps an ID to a subset of K. In addition, *cross-group instances* $\{D_i^1\}_{1,2,\ldots,n}$ are generated. The nodes having the same cross-group instance D_i' form a *cross-group* G_i'. To ensure that a cross-group provides a potential link for deployment groups, GKP stipulates that each cross group G_i' has only one sensor from each deployment group G_i and that for $i \neq j$, $G_i' \cap G_j' = \emptyset$, and that $|G_i' \cap G_j| = 1$. Each group G_i contains the sensors with IDs $\{(i-1)m + j\}_{j=1,2,\ldots,na}$, and each G_i' contains the sensors with IDs $\{i + (i + j)m\}_{j=1,2,\ldots,n}$. This ID assignment allows each sensor node to recognize the deployment group and cross group of any sensor. At the end of this phase, each sensor becomes a member of an in-group instance and a cross group instance.

In the direct key establishment phase, two sensors exchange their IDs. If two sensors belong to the same deployment group, they can establish a key directly. If the sensors do not belong to same group, but belong to the same cross-group, then they can establish a direct key. If they cannot establish a direct key, then they must look for other sensors to help them establish a path-key. When the sensor pair belongs to different groups, the pair needs to find an intermediary to setup an indirect key. The construction presented earlier provides m potential intermediary groups.

To take advantage of location information, Liu and Ning present three location-based key establishment schemes for sensor network.[19] The first scheme, the *closest pair-wise keys* (CPK) *scheme*, is based on the concept that a pair of sensors is pre-loaded with pair-wise keys if they are very likely to be located in a place that is within each other's transmission range. In the *pre-distribution*

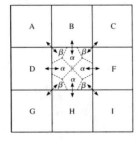

Fig. 7 Key sharing between neighboring key-pools. This figure is redrawn from Ref. 9. *Source*: Ref. 9. Shared keys between neighboring key pools.

phase, the base station first pre-determines the set S of c sensors that are expected to be closest to sensor u. Then for each $v \in S$, it randomly generates a unique key $K_{u,v}$ and pre-loads (v, K_{uv}) in u and (u, K_{vu}) in v. To establish a key with v, u checks if it has a pre-loaded pair-wise key of v. This process is extremely efficient. In adding new sensors, keys are first pre-distributed to the new sensors as discussed in the pre-distribution phase and then the new sensor's keys are sent to the already deployed sensors that are expected to be neighbors to new sensors. The base station achieves this using secure unicasts to each sensor node. If two neighboring sensors u and v do not share a pair-wise key, an intermediate sensor is sought to help establish a session key. One sensor, say u, sends a key request to an intermediate neighboring sensor w which shares keys with both u and v. w generates a session k and encrypts K with K_{uw} and K_{vw} and sends a message to u containing both encryptions: $E_{K_{uw}}(k)$ and $E_{K_{vw}}(k)$. u decrypts $E_{K_{uw}}(k)$ and obtains k and sends $E_{K_{vw}}(k)$ to v, thereby confirming the key setup.

This scheme is more flexible and more resilient than the probabilistic schemes of Refs. 6 and 11. However, it has two drawbacks. First, if the sensors are not uniformly distributed on the target area, some sensors may have large number of neighbors that are not pre-determined neighbors. Such sensors require larger storage capacity. Second, addition of new sensors introduces substantial communication overhead. As a result, Liu and Ning[19] propose an enhanced CPK (ECPK) using a pseudo-random function, f. For each sensor u, the base station generates a master key K_u and determines the set S whose sensors are closest to u. For each $v \in S$, the base station generates a pair-wise key, K_{uv} using f and the master key of $v(K_v)$ as follows: $K_{uv} = fK_v(u)$. ECPK pre-distributes these keys and their IDs to every $u \in S$. This pre-distribution creates a master sensor v, which can compute all pair-wise keys for all $u \in S$ using its master key K_v, and a slave sensor that only stores pair-wise keys.

Direct key establishment in ECPK is the same as in CPK, but sensor addition and revocation are slightly different. To add sensor u, the base station selects the closest c sensors to u's expected location, uses K_v, and computes u's pair-wise key. It pre-loads u with all K_{uv}s and v. To revoke a master sensor, all slaves must remove the pair-wise keys from their memory, but removing a slave requires the master sensor to store the ID of the removed slave. ECPK has lower communication computation overhead than CPK. However, it increases storage requirement as each sensor stores the pair-wise keys and their IDs. This may be substantial if sensor distribution is uneven and network density is high.

To address these issues, *key pre-distribution with bivariate polynomials* (KPBP) is proposed.[19] KPBP combines CPK with the PKPS.[4] In KPBP, the target area

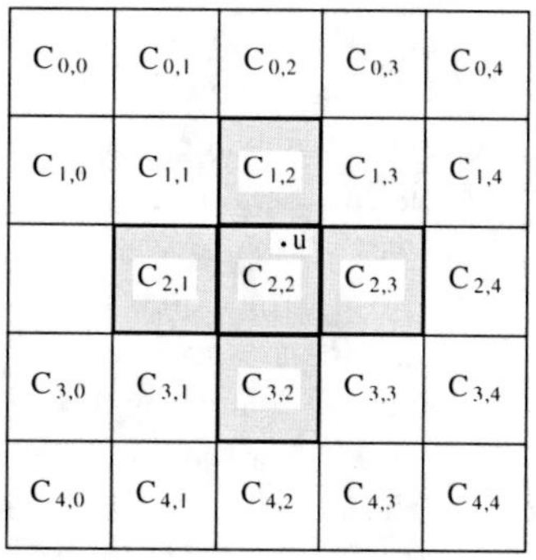

Fig. 8 Partition of a target area. This figure is redrawn from Ref. 19. *Source*: Ref. 19. Partition of a largest field.

is partitioned into equal-sized squared cells $\{C_{i_c,i_v}\}_{i_c=0,1,...,C-1, i_v=0,1,...,R-1}$, where $\rho = R \times C$ is total number of cells. Each cell is denoted by row i_r and column i_c. Fig. 8 shows the partition of the target area. The base station generates ρ symmetric bivariate λ-degree polynomials $\{f_{i_c,i_v}(x,v)\}_{i_c=0,1,...,C-1, i_v=0,1,...,R-1}$ and assigns $f_{i_c,i_v}(x,y)$ to cell C_{i_c,i_v}. The base station determines the expected location cell of each sensor and pre-loads it with the cell coordinates and its polynomial share of the location cell, along with the polynomial shares of $C_{i_c,i_r-1}, C_{i_c,i_{r+1}}, C_{i_c-1,i_r}$, and C_{i_c+1,i_r}. For example, u receives the coordinate $(2,2)$, and the polynomial shares $f_{2,2}(u,y), f_{2,1}(u,y), f_{2,3}(u,y), f_{1,2}(u,y)$ and $f_{3,2}(u,y)$ as shown in the shaded region of Fig. 8.

Two sensors can establish a direct key if they identify a shared polynomial. One way to achieve this is to have a source sensor reveal its home cell coordinate. From the home cell coordinate, the destination sensor can determine the polynomials it shares with a source sensor. This must be done in a secure way perhaps by a challenge as in Ref. 11. Addition and revocation in KPBP are quite simple. To add a new sensor, the base station pre-distributes its polynomial share and home coordinate to the new sensor as stated earlier. To remove a compromised sensor, a sensor simply marks the IDs of that sensor that shares at least one polynomial with it. Huang et al.[12] propose *grid-group deployment scheme* (GDS), a scheme similar to the ones presented by Liu and Ning.[19] Like CPK and KPBP, GDS partitions the deployment area into equal-sized cells and uniformly distributes sensors into these cells.

Hybrid schemes

These schemes combine the attributes of pre-distribution with single TA schemes. In the hybrid protocols, sensor authentication requires involvement of the base station and the pre-distribution of some necessary cryptographic materials. Only one of the protocols that have been reviewed fall under this category.[29]

In Ref. 29, Ibriq and Mahgoub present a hybrid scheme called *hierarchical key establishment scheme* (HIKES). It implements an escrow key management scheme in which one part of the key is made public and the other is kept private. The public part of the key is pre-loaded in the sensor while the private part is escrowed with the base station. The escrowed parts are revealed to specified sensors in the network, thereby delegating those sensors to act on behalf of the base station or the TA.

HIKES has three phases: *key pre-distribution, neighbor discovery* and *key establishment*. During the pre-distribution phase, each node is pre-loaded with the *master key, primary key, cluster key,* and the *precursor of the backup cluster key*. During the node discovery phase, each node establishes *temporary* pair-wise keys with each neighbor as shown in LEAP. Once the cluster head is selected and child-parent associations are formed, each sensor sends a session key request to its parent thus initiating the key establishment phase. The request contains two challenges: one to the cluster head, encrypted with the pre-loaded cluster key and another to the base station, encrypted with the pre-loaded primary key, a unique key each sensor shares with the base station. The parent forwards every key request adding to the request its own challenges to the cluster and base station.

The cluster head keeps the challenges its members have posed to it. However, it transmits an aggregate containing cluster members' challenges posed to the base station along with their IDs to the base station. After verifying the cluster head and the cluster members from the challenges, the base station responds to the cluster head with a group ticket containing the index (I_i) and offset O_i of cluster key of each member i in the cluster. When the cluster head, c, receives the group ticket, for each sensor i, it concatenates the seed S_i corresponding to i's index with its offset, O_i. Then, it hashes the concatenated result to regenerate the cluster key of i, K_{ic}. Using K_{ic} it decrypts the challenge C_i that i has posed to it in its session key request. It returns C_i to i along with the session key, K_{si}, a group key, K_g and life of the key, L, encrypted using K_{ic}. HIKES is a highly resilient, flexible and scalable key management scheme that assures connectivity to all sensors in the network, but it has one main drawback. As the network grows in size and dimension, the communications between the cluster heads and the base station becomes very expensive in spite of the aggregation gain achieved by the cluster heads.

Self-enforcing Schemes

In the self-enforced protocols, authentication is achieved pair-wise using a pre-distributed signed certificate. Self-enforced protocols are asymmetric pair-wise public key based protocols. The goal of these protocols is to minimize the burden of authentication and pre-configuration requirement on the sensors while maintaining strong asymmetric security. The protocols are based on Kohn-felder's scheme and its extension by Denning.[33] This scheme is implemented in *Beller-Yacobi protocol* (BYP).[2] The general scheme is simplified as follows. The central TA, pre-loads each sensor j with a public key and a certificate of the form: $C_j = \{T||\text{ID}_j||K_j^{Pu}\}K_{TA}^{Pr}$ where K_{TA}^{Pr} is the private key of the main TA. To establish a key, j transmits this certificate to any sensor in the network. A recipient sensor decrypts and verifies this certificate using the TA's public key, K_{TA}^{Pu}, as follows: $\{C_j\}_{K_{TA}^{Pu}} = \{T||\text{ID}_j||K_j^{Pu}\}K_{TA}^{Pr} = T||\text{ID}_j||K_j^{Pu}$ where T is a time-stamp. It is designed to prevent an adversary from replaying old certificates and set an expiration time for C_j. The compromise of K_{TA}^{Pr} is equivalent to compromising the master key in a master key-based scheme. Also, BYP is computationally expensive for the constrained sensors.

In Ref. 24, Watro et al. propose another self-enforcing scheme called *TinyPK*, a public-key based key agreement scheme. Its cryptographic operations are based on a fast variant of RSA and are performed by an entity that is external to the sensor network. TinyPK requires a CA with a public and private that is trusted by all the sensors in the network. Any external party (EP) that wishes to communicate with the sensors must have a public and private key and have its own public key signed by the CA to establish its identity to the sensors. Each sensor is also pre-loaded with the necessary software and the CA's public key.

Before the sensors can communicate with each other, they must establish their identities to an EP that is responsible for all cryptographic operations. Sensors communicate with each other via the EP. TinyPK requires the establishment of a public key infrastructure for its operations, which entails substantial communication overhead. Second, since communications between the sensors are done via an EP, it has scalability issues. As the network grows in size, the EPs become bottlenecks.

FUTURE TRENDS

As research in WSNs continues to grow, security will be a central focus of this research. There is a need for a key management scheme that meets all the security requirements and that is also scalable, efficient, flexible, resilient and provides high connectivity to the sensors. The present trend in designing key management schemes is focused on using pre-deployment knowledge to reduce the pre-configuration requirements with the aim of improving the scheme's efficiency, scalability, connectivity, flexibility, and resiliency.

This survey shows that key management schemes in WSNs require more improvements and more research, particularly in relatively unexplored areas. All key management schemes in WSNs share one common goal, namely, to achieve key agreement among the sensors efficiently, i.e., with the smallest energy and storage cost. Given this goal and the substantial communication requirement involved in single TA schemes, none of the WSNs' key management protocols that have been reviewed in this entry fall in this category. Pre-distributed schemes attempt to reduce energy and communication cost of authentication by pre-configuring the sensors with the necessary cryptographic materials for authentication. In the process, pre-distribution negatively affects the scheme's effectiveness as measured by its scalability, resiliency, flexibility, and connectivity. In the randomized probabilistic schemes, efficiency is greatly enhanced, but the flexibility and sensor connectivity are weakened. In key-chain schemes also, efficiency is improved but scalability becomes an issue. In the master key-based schemes, efficiency, scalability and connectivity are improved; however, network resiliency is seriously degraded. The compromise of the master key renders the entire network useless. In the matrix-based schemes, efficiency depends on the size of the pre-distributed matrices; and in the polynomial-based schemes efficiency depends on the number of pre-distributed polynomial coefficients. Communication efficiency, connectivity and resiliency improve with size of pre-loaded cryptographic materials. However, heavily pre-loading the sensor reduces storage efficiency and flexibility. Thus, pre-distribution schemes, while improving the functionality of the key management scheme in one area, degrade it in another.

Since sensors have limited storage capacity, there is a limit to the amount of information that can be pre-loaded in them, which directly affects the network scalability, resiliency, flexibility and efficiency. In the proposed classification in "Introduction," there are two categories of schemes that do not rely on heavily pre-configuring the sensors: the hybrid and the self-enforcing schemes. In view of the sensor limitation, these schemes require more investigation.

The hybrid class presents enormous design challenges. While involving the base station in the authentication process provides strong security, it requires substantial communication cost. Hence, a hybrid scheme that requires the direct interaction between the sensor and base station has low efficiency and poor scalability. Novel, communication-thrifty techniques that achieve the interactive participation of the base station in the authentication process are required. This class has only one protocol, HIKES.[29] In HIKES, the base station empowers specialized sensors to act as local trust authorities using an escrow scheme. Although, this empowerment requires some communication, it provides high resiliency, scalability, flexibility and connectivity. The self-enforcing class has two protocols.[2,24]. Although these protocols have high computation overhead associated with their encryption and high communication overhead incurred in sensor registration with

the CA, they have small storage requirements, high scalability, connectivity, and flexibility. The protocols' effectiveness can be increased by developing innovative techniques that reduce the sensor registration cost and use lightweight encryption. There are many research opportunities in key management for WSNs particularly in these two challenging classes: the hybrid and the self-enforcing.

CONCLUSION

In this entry, we provide a comprehensive survey of key management protocols in WSNs and propose a classification scheme. The classification scheme serves as a framework to analyze and evaluate the present trends in key management design. This entry first introduces the network model and the threat model in WSNs. Then, it states the security requirements and evaluation criteria for key management schemes. It presents the schemes' main features, and evaluate them on the basis of their efficiency, scalability, resiliency, flexibility, and connectivity. In addition, this entry concludes with a discussion of future trends in key management design.

REFERENCES

1. Carman, D.; Kruus, P.; Matt, B. Constraints and approaches for distributed sensor network security. Technical Report TR No. 00-010, NAI Labs, Glenwood, MD, USA, Sept 2000.
2. Beller, M.J.; Yacobi, Y. Fully-fledged two-way public key authentication and key agreement for low-cost terminals. Electron. Lett. **1993**, *29* (11), 999–1001.
3. Blom, R. An optimal class of symmetric key generation systems. In *Advances in Cryptology—Eurocrypt'84, LNCS, Volume 209 of LNCS*; Beth, T., Cot, N., Ingemarsson, I., Eds.; Springer-Verlag: Berlin, Germany, 1984; 335–338.
4. Blundo, C.; De Santis, A.; Herzberg, A.; Kutten, S.; Vaccaro, U.; Yung, M. Perfectly-secure key distribution for dynamic conferences. In *Advances in Cryptology—CRYPTO '92, Volume 740 of LNCS*; Brickell, E.F., Ed.; Springer: Berlin, Germany, 1992; 471–486.
5. Bohge, M.; Trappe, W. An authentication framework for hierarchical ad hoc sensor networks. 4th ACM workshop on Wireless security 2003 (WiSe '03), San Diego, CA, Sep 19, 2003; ACM Press: 2003; 79–87.
6. Chan, H.; Perrig, A.; Song, D. Random key predistribution schemes for sensor networks. IEEE Symposim on Security and Privacy, Berkeley, IEEE, Computer Society, CA, May 11–14, 2003; 197–213.
7. Chan, H.; Perrig, A. PIKE: peer intermediaries for key establishment in sensor networks. 24th Conference of the IEEE Communications Society (INFOCOM '05), Miami, FL, IEEE, Computer Society, Mar 13–17, 2005; Vol. 1, 524–535.
8. Du, R.; Tu, H.; Song, W. An efficient key management scheme for secure sensor networks. 6th International

Conference on Parallel and Distributed Computing, Applications and Technologies (PDCAT '05), Dalian, PR China, The Institute of Electrical and Electronics Engineers, Inc.: Dec 5–8, 2005; 279–283.

9. Du, W.; Deng, J.; Han, Y.; Chen, S.; Varshney, P. A key management scheme for wireless sensor networks using deployment knowledge. 23rd Conference of the IEEE Communications Society (INFOCOM '04), Hong Kong, PR China, The Institute of Electrical and Electronics Engineers, Inc.: Mar 7–11, 2004; Vol. 1, 586–597.

10. Du, W.; Deng, J.; Han, Y.; Varshney, P. A pairwise key pre-distribution scheme for wireless sensor networks. 10th ACM Conference on Computer and Communications Security (CCS '03), Washington, DC, Oct 28–30, 2003; ACM Press, New York, NY, USA; 42–51.

11. Eschenauer, L.; Gligor, V. A key-management scheme for distributed sensor networks. 9th ACM Conference on Computer and Communications Security (CCS '02), Washington, DC, Nov 18–22, ACM Press, New York, NY, USA; 2002; 41–47.

12. Huang, D.; Mehta, M.; Medhi, D.; Harn, L. Location-aware key management scheme for wireless sensor networks. ACM Workshop on Security of Ad Hoc and Sensor Networks (SASN '04), Washington, DC, Oct 25, 2004; ACM Press, New York, NY, USA; 29–42.

13. Ilyas, M., Mahgoub, I., Eds.; Handbook of Sensor Networks: Compact Wireless and Wired Sensing Systems; CRC Press: Boca Raton, FL, 2005.

14. Mahgoub, I., Ilyas, M., Eds.; Sensor Network Protocols; CRC Press: Boca Raton, FL, 2006.

15. Mahgoub, I., Ilyas, M., Eds.; SMART DUST: Sensor Network Applications, Architecture, and Design; CRC Press: Boca Raton, FL, 2006.

16. Lai, B.; Kim, S.; Verbauwhede, I. Scalable session key construction protocol for wireless sensor networks. IEEE Workshop on Large Scale Real-time and Embedded Systems (LARTES '02), Austin, TX, The Institute of Electrical and Electronics Engineers, Inc.: Dec 2, 2002.

17. Lee, J.; Stinson, D.R. Deterministic key predistribution schemes for distributed sensor networks. 11th Annual Workshop on Selected Areas in Cryptography, 2004 (SAC '04), Waterloo, Canada, Aug 9–10, 2004; Springer Berlin/Heidelberg, Germany; 294–307.

18. Liu, D.; Ning, P. Efficient distribution of key chain commitments for broadcast authentication in distributed sensor networks. 10th Annual Network and Distributed System Security Symposium (NDSSS '03), San Diago, CA, Internet Society, Feb 6–7, 2003.

19. Liu, D.; Ning, P. Location-based pairwise key establishments for static sensor networks. ACM Workshop on Security of Ad Hoc and Sensor Networks 2003 (SASN '03), Fairfax, VA, Oct 31, 2003; ACM Press, New York, NY, USA; 72–82.

20. Liu, D.; Ning, P.; Du, W. Group-based key pre-distribution in wireless sensor networks. ACM workshop on Wireless security 2005 (WiSe '05), Cologne, Germany, Sept 2, 2005; ACM Press, New York, NY, USA; 11–20.

21. Liu, D.; Ning, P.; Li., R. Establishing pairwise keys in distributed sensor networks. ACM Trans. Inf. Syst. Secur. 2005, 8 (1), 41–77.

22. Park, T.; Shin, K. LiSP: a lightweight security protocol for wireless sensor networks. Trans. Embedded Comput. Syst. 2004, 3 (3), 634–660.

23. Perrig, A.; Szewczyk, R.; Wen, V.; Culler, D.; Tygar, J. SPINS: security protocols for sensor networks. 7th Annual International Conference on Mobile computing and Networking 2001 (MobiCom '01), Rome, Italy, July 16–21, 2001; 189–199.

24. Watro, R.; Kong, D.; Cuti, S.; Gardiner, C.; Lynn, C.; Kruus, P. TinyPK: securing sensor networks with public key technology. ACM Workshop on Security of Ad Hoc and Sensor Networks (SASN '04), Washington, DC, Oct 25, 2004; ACM Press, New York, NY, USA; 59–64.

25. Zhou, L.; Ni, J.; Ravishankar, C. Efficient key establishment for group-based wireless sensor deployments. ACM workshop on Wireless security 2005 (WiSe '05), Sep 2, 2005; ACM Press, New York, NY, USA; 1–10.

26. Zhu, S.; Setia, S.; Jajodia, S. LEAP: efficient security mechanisms for large-scale distributed sensor networks. 10th ACM Conference on Computer and Communications Security (CCS '03), Washington, DC, Oct 28–30, 2003; ACM Press, New York, NY, USA; 62–72.

27. Karlof, C.; Wagner, D. Secure routing in wireless sensor networks: attacks and countermeasures. 1st IEEE International Workshop on Sensor Network Protocols and Applications (IEEE SNPA '03), Anchorage, Alaska, The Institute of Electrical and Electronics Engineers, Inc.: May 11, 2003; 113–127.

28. Stankovic, J.; Wood, A. Denial of service in sensor networks. Computer. 2002, 35 (10), 54–62.

29. Ibriq, J.; Mahgoub, I. A hierarchical key establishment scheme for wireless sensor networks. IEEE 21st International Conference on Advanced Information Networking and Applications (AINA-07), Niagara Falls, Canada, May 21–23, 2007.

30. Neuman, B.; Ts'o, T. Kerberos: an authentication service for computer networks. IEEE Commun. Mag. 1994, 32 (9), 33–38.

31. Needham, R.; Schroeder, M. Using encryption for authentication in large networks of computers. Comm. ACM. 1978, 21 (12), 993–999.

32. Otway, D.; Rees, O. Efficient and timely mutual authentication. Oper. Syst. Rev. 1987, 21 (1), 8–10.

33. Stallings, W. Cryptography and Network Security, Principles and Practice, 2nd Ed.; Prentice Hall: Upper Saddle River, NJ, 1999.

Wireless Sensor Networks (WSNs): Optimization of the OSI Network Stack for Energy Efficiency

Ahmed Badi
Imad Mahgoub
Department of Computer Science and Engineering, Florida Atlantic University, Boca Raton, Florida, U.S.A.

Abstract

A wireless sensor network (WSN) is a distributed autonomous network that typically consists of one base station and a large number of micro devices that have sensing, processing, and radio communication capabilities.

The Open Systems Interconnection (OSI) network stack is a standard that defines structured layers of functionalities for how data are exchanged between any communicating points in a telecommunication network. This article is a survey of the OSI stack optimizations for energy efficiency in wireless sensor networks.

INTRODUCTION

The OSI (open systems interconnection) model is a standard developed by the ISO for how to transmit messages between any two telecommunicating points in a network.[1,2] The standard defines seven layers of functions that take place at each end of a communication. Each layer is responsible for a number of logical steps that it implements. The advantage of the OSI model is that it serves as a single reference for network communication that furnishes a common ground for equipment manufactures, researchers, and for education.

Wireless sensor networks (WSNs) are one of the fastest developing new technologies.[3–5] They provide a new interface between the computational and physical worlds. This technology is at the frontier of networking research and could be the next multibillion-dollar technology market. Researchers in academia, as well as several hardware and software companies are currently active on the development of these micro sensor devices and associated technologies. The availability of small, cheap low-power embedded processors, radio transceivers and sensors, often integrated on a single chip is leading to the use of sensing, computing and wireless communication for monitoring and interacting with the physical world. These wireless sensor devices, also known as *motes*, are assembled of the hardware components mentioned above, an energy source, in most cases battery together with networking and application firmware and software. Depending on the size of the network and the complexity required of each sensor, the cost of sensor devices can vary from hundreds of dollars to few dollars. The size of a single sensor node can also vary.

Several performance parameters of the OSI stack for wired networks have been optimized. These parameters include latency, fairness and throughput. With the emergence of the new technology of WSNs, a new set of constrains are forcing the optimization of the OSI stack for a different set of parameters.

In Section "The OSI Stack," an overview of the OSI network stack layers is given. Wireless sensors and WSNs are introduced in Section "WSNs." Section "OSI Layers Optimization for Wireless Sensors" is a survey of the different optimizations to the OSI stack classified by layer. And finally Section "Cross-layer Optimizations" introduces new efforts in the OSI stack cross-layer optimizations.

THE OSI STACK

In the OSI model, the communication process between two points in a network is divided into seven layers application, presentation, session, transport, network, medium access control (MAC), and physical layers.[1] An advantage of this view is that the complexity of the communication process is also divided among the different layers making the implementation of such systems manageable. The programming and hardware that furnishes the seven layers, also known as the network stack, is usually found partly in the computer OS, and in several stand alone applications such as web browsers, and in the network firmware and hardware interfaces that are common parts of any computer system.

The OSI layers can be further divided into two groups. The lower three layers (the networking layers) are used

Encyclopedia of Wireless and Mobile Communications DOI: 10.1081/E-EWMC-120043808

whenever a message passes through the host. Messages intended for this particular host will be passed to the upper layers. Other messages destined for other hosts are forwarded by the network layer. On the other hand, the upper layers will be involved in the communication process only when the message is destined to the host device.

The above discussion presented the OSI stack. It describes a fixed, seven layer stack for networking communication protocols. Similarly, there is another layered stack protocol, which is the simpler five layer stack model, also known as the TCP/Internet protocol (IP) stack shown in Fig. 1. There are lots of similarity between the two protocols since they attempt to define the same communication process, but the definition of the different layers are some what different. Wireless sensors network stack has more in common with the TCP/IP stack. These five layers are summarized below.

Application Layer

The application layer sits at the top of the communication stack. It generates the data that will be sent out or it will be the entity that ultimately receive and decodes the data. At this layer the communicating partners are identified, quality of service (QoS) is defined and identified, data encryption and decryption is performed, and user authentication and privacy issues are considered. In the seven layer protocol stack model, this layer is further divided into the presentation and session layers.

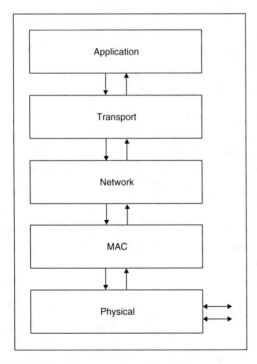

Fig. 1 OSI five layer protocol stack model.

Transport Layer

The transport layer provides transparent data transfer between hosts. It is also responsible for end-to-end error recovery and flow control. It is responsible for providing a reliable, error-free communication over an unreliable communication medium and ensuring complete data transfer. The well-known TCP and user datagram protocol (UDP) are implementations of the transport layer functionality.

Network Layer

The network layer performs error control, source to destination routing by insuring the sending of data messages in the right direction to the right destination on outgoing transmissions, and receiving incoming packet transmissions. This layer is also responsible for flow control, and data segmentation and de-segmentation. IPv4, IPv6, and X.25 are the most commonly used implementations for this layer.

MAC Layer

The MAC layer regulates the usage of the shared communication medium. Before transmitting frames a station must first gain access to the medium. For a LAN this can be the token in a token ring network. In a wireless network scenario the medium is the radio channel. The IEEE 802.11 is a wireless communication MAC standard that is widely adopted. In this standard, as a condition to access the medium, the MAC layer checks the value of its network allocation vector (NAV), which is a counter resident at each station that represents the amount of time that the station accessing the channel needs to send its frames. The NAV must be zero before a station can attempt to send a frame. Prior to transmitting a frame, a station calculates the amount of time necessary to send the frame based on the frame's length and the channel's data rate. The station places a value representing this time in the duration field in the header of the frame. When stations receive the frame, they examine this duration field value and use it as the basis for setting their corresponding NAVs. This process reserves the medium for the sending station.

In general, contention-based medium access is implemented by the distributed coordination function (DCF), which is a random back-off timer that stations use if they detect a busy medium. If the channel is in use, the station must wait a random period of time before attempting to access the medium again. This ensures that multiple stations wanting to send data do not transmit at the same time. The random delay causes stations to wait for different periods of time and prevents them from sensing the medium at exactly the same time, finding the channel idle, transmitting, and colliding with each other. The back-off timer significantly reduces the number of collisions and corresponding retransmissions especially when the number of active users increases.

With radio-based LANs, a transmitting station can't listen for collisions while sending data, mainly because the station can not have its receiver on while transmitting the frame. As a result, the receiving station needs to send an acknowledgement (ACK) if no errors were detected in the received frame. If the sending station does not receive an ACK within a specified period of time, the sending station will assume that there was a collision (or RF interference) and retransmits the same frame again.

Physical Layer

The physical layer is the bottom layer of the OSI stack. It provides the hardware means of sending and receiving data on a carrier and performs services requested by the MAC layer.

The physical layer is the most basic network layer, providing only the means of transmitting raw bits rather than packets over a physical data link. No packet headers or trailers are added to the data by the physical layer. This layer transmits the bit stream through the network as an electrical or electromagnetic signal. It provides bit-by-bit node to node delivery, signal modulation and demodulation, equalization filtering, training sequences, pulse shaping and other signal processing of physical signals. The physical layer determines the bit rate in bits per second (bits/s), also known as channel capacity, digital bandwidth, maximum throughput or connection speed. The physical layer also defines half duplex or full duplex transmission mode.

Since the inception of the ISO OSI layered communication stack model for the wired LAN and wide area network (WAN), the goals have been to achieve compatibility and simplification of functional description of separate units. The optimizations of the stack have also been in the direction of improving the latency, QoS, reliability, and throughput matrices.

With the emergence of ad hoc and wireless networks, the OSI stack is ported as is to this new technology. Research has been active in the study of ways to enhance the stack to optimize it and bring it up to face the new challenges found in the wireless communication field. Token ring protocols have been replaced by a new set of protocols suitable for the wireless communication, e.g., ad hoc on-demand distance vector routing (AODV), dynamic source routing (DSR), and zone routing protocol (ZRP) to mention a few. In the MAC layer, several new editions have been introduced e.g., IEEE 802.11 and IEEE 802.15.4 (ZigBee) protocols. With the emergence of the new technology of WSNs, a new set of constrains is forcing the optimization of the OSI stack for yet new parameters.

WSNs

A WSN is a telecommunication network consisting of spatially distributed sensors to monitor physical or environmental conditions in a cooperative manner. Military applications such as monitoring of troop movement and target tracking originally motivated the development of WSNs. However, currently, WSNs are being considered for use and being used for many civilian applications.

The wireless sensors sensing capabilities cover physical measurements of quantities such as temperature, sound, vibration, pressure, moisture, light intensity, magnetism, motion, radiation, or pollutants among many other physical and environmental quantities. That makes them useful in applications such as security and surveillance applications, smart spaces, monitoring of natural habitats and ecosystems, medical monitoring, battlefield surveillance, healthcare applications, home automation, traffic control, industrial process control, and structural health monitoring. Wireless ad hoc sensor network design requirements include the following:

Low energy: In many applications the sensors are battery powered as shown in Fig. 2. They are usually placed in remote areas where manual service of sensor nodes may not be possible. In this case, the node's lifetime will be dependant on the battery's life thereby requiring the optimization of energy consumption.

Self-configuration: With the large number of nodes and their potential placement in hostile locations, individual node's configuration is not possible. Therefore it is essential that the network be self-configuring. In addition, nodes may fail due to energy exhaustion, malfunction, or destruction and new nodes may be added to the network. For these reasons the network must be able to periodically reconfigure itself so that it can continue to function. Also, depending on the nature of the application the network needs to maintain some degree of connectivity.

Scalability: Wireless senor networks are assumed to have large number of mostly stationary sensors. Networks of 10,000 or even 100,000 nodes are envisioned and network scalability is a major issue.

In-network query processing: The sensor network may collect a large amount of data. This may overwhelm the

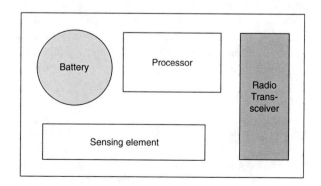

Fig. 2 The subcomponents of a sensor device.

user who may not be able to process all this information. Instead, selected nodes within the network will collect the data from their neighbors and create a representing message.

In-network signal processing: To improve the quality of data collected, it is often useful to fuse data from multiple sources. This requires the transmission of data and control messages to some master node before sending it to the base station. This will impose some requirements on the network's architecture.

The wireless sensor technology still has many limitations that need to be addressed. Their cost and size constraints result in severe limitations on their energy resource, memory, CPU speed and bandwidth. Probably the most important constraint is the limited energy resources. This requires the careful consideration and design of energy-aware, signal and query processing algorithms.

OSI LAYERS OPTIMIZATION FOR WIRELESS SENSORS

A new set of protocols and optimizations have been proposed by researchers to address the energy and other requirements in wireless sensors. Below are the different solutions and ideas classified by OSI layers.

Application Layer

WSNs are known to be application specific. The nature of message exchange between the nodes and the base station is mostly of reporting sensor readings which tend to be short messages. With the vast number of applications envisioned and already in existence for wireless sensors, along with their different traffic patterns, it will be a challenge to achieve general optimization of the application layer for energy performance. Yet, in several publications an indication is given to the need for applications that have small memory footprint due to the limited storage resources.[3] Other application characteristics that can be helpful in performance optimization include the ability to tradeoff between energy and accuracy and dynamic adaptability to node and network resources. The application layer may also assist the rest of the stack layers with hints that will help them optimize their performance in a cross-layer fashion.

Transport Layer

The transport layer provides congestion control and end-to-end reliable data flow. The TCP type transport layer protocols may not be suitable for wireless sensors since they rely on end-to-end acknowledgments and retransmission which wastes valuable energy resources.[4] The requirements for the transport layer reliability can be relaxed to event-to-sink reliability instead of node-to-sink. This is possible due to the fact that the same event will be reported by several nodes.[5] Ref. 6 introduces the sensor TCP (STCP), a generic transport layer protocol for WSNs. STCP is applicable to event driven as well as continuous reporting application communication scenarios. It addresses several requirements of the transport layer and WSNs including scalability and congestion detection and avoidance. In Ref. 7 the PSFQ (pump slowly fetch quickly) transport layer protocol is presented. In PSFQ data recovery and loss detection is done on hop by hop basis instead of the original end-to-end method.

Network/Routing Layer

Energy-efficient protocols for WSNs exploit the fact that these networks are not communication networks in the classical meaning, but rather a distributed system where all the nodes collaborate to perform a given task or set of tasks. This fact can be used to trade per node fairness and other networking qualities for designs that will yield energy-efficient protocols. The networking layer is responsible for the end-to-end routing and delivery of data messages. This makes designing of energy-efficient protocols in the routing layer critical, since this will affect the number of transmissions, the distance covered per transmission and the load placed on nodes participating in the relaying of the message. For these reasons, the network layer attracted more attention than the other layers. Some of the early work on energy-efficient wireless sensor protocols has targeted this layer.[8–11]

Low-energy adaptive clustering hierarchy

Perhaps the first network protocol that is specifically designed for wireless sensors is the low-energy adaptive clustering hierarchy (LEACH) protocol.[8] The main setting that this protocol addresses is that of a large number of homogenous, resource constrained nodes monitoring the environment and periodically sending their readings to a base station located far away from the field as shown in Fig. 3. The protocol achieves its power saving goals by allowing a small percentage of the nodes, called cluster heads, to collect data from their surrounding neighbors, aggregate the data and send a report to the base station representing the combined readings.

The protocol avoids depleting the cluster heads energies by selecting a new set of cluster heads at the beginning of each round. The setup overhead is assumed to be negligible since the setup time is small compared to the round's duration. The protocol uses a randomized routine for each node to elect itself as a cluster head. This routine is run locally at each node to avoid the traffic overhead of a centralized routine. Simulation results show that LEACH can increase the network lifetime by as much as a factor of eight compared with direct transmission. The protocol

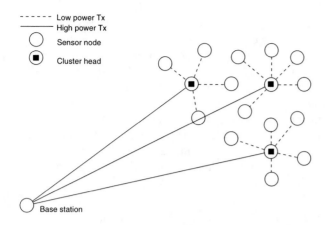

Fig. 3 LEACH protocol configuration.

suffers from few shortcomings including the fact that the energy level and other node resources are not taken into consideration in the election routine. Yet, LEACH is considered the first energy-efficient protocol targeting wireless sensors, and the benchmark against which the performance of other protocols is compared.

Power-efficient gathering in sensor information systems

Power-efficient gathering in sensor information systems (PEGASIS)[9] is an improvement over the LEACH protocol by introducing the following ideas:

1. Nodes transmit only for a short distance to the closest neighbor. Each node defuses its data with the data it receives before transmitting as shown in Fig. 4.
2. Only one node reports the collected data to the base station instead of a group of cluster heads going through the expensive transmission.

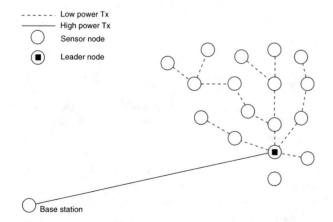

Fig. 4 PEGASIS protocol configuration.

3. The leader node receives at most two messages instead of an average of 20 messages in the case of the LEACH protocol with a 100 nodes network.[9]

PEGASIS achieves 100%–300% energy performance improvement over LEACH. The protocol does not specify how the leader is selected. But since this is an enhancement over LEACH, one can assume that it uses the same random equation by setting the number of cluster heads to one. In which case, issues associated with LEACH cluster heads selection routine can be assumed to be present in PEGASIS.

Threshold-sensitive energy efficient sensor network protocol/adaptive periodic threshold-sensitive energy efficient sensor network protocol

In classifying the routing protocols for WSNs, two classes can be identified, proactive and reactive protocols.[10] The LEACH and PEGASIS protocols discussed above can be considered to be proactive protocols since they periodically send reports to the base station. Reactive protocols, in which reporting is triggered by the occurrence of the event of interest are more suitable for time critical applications where immediate response to changes in the sensed parameter(s) is required.

The threshold-sensitive energy efficient sensor network (TEEN) protocol[10] and the adaptive periodic threshold-sensitive energy efficient sensor network protocol (APTEEN)[11] fall under this reactive category. Similar to LEACH and PEGASIS, TEEN is also a hierarchical protocol. The protocol defines and uses two parameters, a hard threshold and a soft threshold values. The sensors are assumed to monitor the environment continuously. If the value of the sensed parameter reaches or exceeds the hard threshold value, the node will turn on its transmitter and send a report to its cluster head. To prevent the nodes from flooding the network with reports once the hard threshold is reached, the nodes will report a new value only if the value of the sensed parameter exceeds the last reported value by an amount equals to at least the soft threshold value. The TEEN protocol offers the following features:

1. Time critical data is reported immediately to the user.
2. Data transmission occurs only if the threshold value is reached, thus substantial energy conservation is achieved.
3. By varying the values for the hard and the soft threshold parameters, the user has control on the network reporting behavior. The soft threshold can also be adjusted to trade off between accuracy and energy saving.

As already stated in Ref. 10, the main drawback of this protocol is that if the threshold value is never reached, the user will get no repots at all and will not be aware if all the nodes in the network are dead.

The above limitation of the TEEN protocol was removed by introducing a hybrid version of the protocol, the APTEEN protocol.[11] APTEEN defines a new *count time* (CT) parameter that is also under the user's control. The CT is defined as the maximum time between two successive reports. By setting values for this count time APTEEN can act as a pure reactive, a pure proactive, or a hybrid protocol.

Directed diffusion

Directed diffusion[12] is a data-centric protocol where data consists of an attribute-value named pair. It can be considered as a reactive protocol where data is requested by sending an interest in the named data. The protocol relies on local communication between neighbors. To a node, a request arriving from a neighbor will be treated as if it originated from that neighbor and no global routes between source and sink exist. Initially, a node will flood its neighbors with its interest. Later it will enforce the selection of minimum delay routes, or routes that have been constantly delivering timely data. This protocol is applicable to surveillance and target tracking applications.

Geographical and energy aware routing protocol

Geographical and energy aware routing (GEAR) protocol[13] is an energy aware geographical routing protocol for wireless sensors. The GEAR protocol assumes that nodes are aware of their geographical location for its operation. This can be achieved by using global positioning systems (GPS) or some localization algorithms. The GEAR protocol is suitable for applications where the operator is interested in querying a specific geographical region. When there is a neighbor closer to the destination, the protocol forwards the request to that neighbor. When more than one neighbor exist that is closer to the destination, the GEAR picks the one that minimizes some cost function. When all the neighbors are further away from the destination, a hole is said to exist in the path and the GEAR protocol chooses the neighbor that minimizes the cost function to forward the request.

Sensor protocols for information via negotiation

The sensor protocols for information via negotiation (SPIN) protocol[14] is defined as an application-level approach to network communication. It introduces the use of high level data naming (metadata) for message exchange. For the effectiveness of using metadata in this protocol, the metadata messages are assumed to be much shorter that the actual data messages. SPIN uses a simple advertise-request-data handshake to enable a node to send its data to only those nodes that are interested in obtaining it.

A variation of this protocol, named SPIN-2 achieves further energy conservation by requiring nodes to monitor their energy resources and participate in the data exchange phase only if they have adequate amount of residual energy. Simulation of the SPIN-2 protocol shows that %60 more data can be delivered using this setting compared to basic flooding.

Cost-effective maximum lifetime routing

The cost-effective maximum lifetime routing (CMLR) protocol[15] identifies a cost function and a maximum lifetime function and attempt to select a route that minimizes the first function and maximizes the second one. The authors argue that while the path selected will not be the one with the least cost function or the maximum lifetime one, it will be the route that will attempt to optimize both.

Summary of discussed network/routing layer optimization techniques

The energy optimization techniques used by the network protocols discussed in this section are summarized in Table 1.

MAC Layer

As stated in the previous section, the networking layer is responsible for the end-to-end routing and delivery of data messages. Designing energy efficient protocols in the routing layer is important since this will affect the route selected, number of hops per message, the distance covered per transmission, and the load placed on nodes participating in the relaying of the data. At the other end, the MAC layer is responsible for per hop transmission between neighboring nodes. For this reason, and similar to the network layer, the MAC layer attracted more attention.

Sensor MAC

The first protocol that addresses the energy problem at the MAC layer is the sensor MAC (SMAC).[16] SMAC identifies the sources of energy waste at the MAC layer as being due to the following four factors: 1) collision, 2) overhearing, 3) control messages overhead, and 4) idle listening. To reduce the effect of idle listening, SMAC introduces the concept of periodic listen and sleep cycles as shown in Fig. 5. Nodes follow a sleep and listen schedule that synchronizes them together. SMAC also attempts to address the problem of control messages overhead by

Table 1 Energy optimization techniques used by the network/routing layer protocols.

Protocol	Architecture	Proactive/ reactive	Energy optimization techniques
LEACH[8]	Hierarchy	Proactive	Clustering
PEGASIS[9]	Flat	Proactive	Clustering
TEEN[10]	Hierarchy	Reactive	Clustering/reactive. Energy-accuracy tradeoff
APTEEN[11]	Hierarchy	Reactive	Clustering/reactive. Energy-accuracy tradeoff
Directed diffusion[12]	Flat	Reactive	Short distance, local communication between neighbors
GEAR[13]	Hierarchy	Reactive	Short distance communication
SPIN[14]	Flat	Reactive	Subset of nodes is participating in data communication. Short distance
CMLR[15]	Flat	Reactive	Optimizing energy cost function

Abbreviations: APTEEN, adaptive periodic threshold-sensitive energy efficient sensor network protocol; CMLR, cost-effective maximum lifetime routing; GEAR, geographical and energy aware routing; LEACH, low-energy adaptive clustering hierarchy; PEGASIS, Power-efficient gathering in sensor information systems; SPIN, sensor protocols for information via negotiation; TEEN, threshold-sensitive energy efficient sensor network.

reducing the number of control messages needed for data exchange between any sender and receiver pairs. For overhearing and collision issues, SMAC borrows from the IEEE 802.11 medium access standard. The standard defines a pair of control messages, request-to-send and clear-to-send (RTS/CTS) for initiating communication between sender and receiver. SMAC requires all nodes hearing either or both RTS/CTS messages to refrain from accessing the medium to avoid collision. For overhearing, the nodes use the NAV concept introduced in the IEEE 802.11 standard. In this vector, a node will store the duration of time that a communication between its neighbors will take. This time duration can be obtained from the RTS or CTS messages that the node overhears. With that the node can switch off its radio and go to sleep for the duration of time while its neighbors are using the channel. To achieve these energy savings the trade-offs introduced by SMAC are increased delays, and compromised per node fairness.

Delay MAC (DMAC)

The delay problem introduced by SMAC is partially solved in the delay MAC (DMAC)[17] protocol by exploiting the structure of data gathering trees. It solves the message forward interruption by adding an offset to each node's schedule. This offset depends on the node's depth within the forwarding tree as shown in Fig. 6.

Time division multiple access medium access controller protocol

SMAC is further improved by using a variable length active period in the time division multiple access medium access controller (TMAC) protocol.[18] TMAC reduces idle listening by transmitting all queued messages in bursts of variable length and going to sleep directly afterwards. During active time the node will keep listening or transmitting and will go to sleep before the end of the active period if no further activation events are heard within a defined *activation time period* (TA).

Wireless sensor MAC

In the wireless sensor MAC (WiseMAC) protocol,[19] a node will wake up regularly for a very short period to sense the medium. If no activity is detected in the channel, the node will go back to sleep immediately until the next sampling time. The nodes' sampling times are not

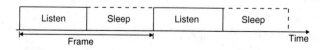

Fig. 5   SMAC periodic listen and sleep schedule.

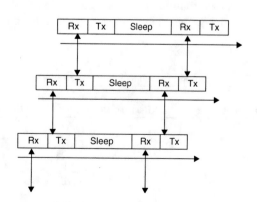

Fig. 6   DMAC staggered listen and sleep schedule.

synchronized together. If a node finds the medium busy, it stays awake to receive the transmitted data. If a node has data, it will precede its transmission with a preamble of length equal to or greater than the network sampling period.

The advantage of using this scheme is that at low traffic levels, nodes will only wake up for very short time at each sampling period. The disadvantages are the high transmission cost of the preamble signal, and that all nodes hearing the preamble must stay awake to hear the data transmission even if it was not meant for them. To minimize this transmission cost, WiseMAC requires the nodes to keep a list of their neighbors and their next wakeup times. Then a transmitting node can start the preamble signal just ahead of the receiver's wakeup schedule keeping the preamble transmission to a minimum.

Group-based MAC

Group-based MAC (GMAC)[20] is a cluster-centric, reservation based MAC protocol. Each frame cycle is divided into a contention period and a contention-free period. A gateway node collects all transmission requests from its members in the form of a future-request-to-send (FRTS) control messages. The gateway then schedules the nodes that submitted requests for transmission during the contention-free period. The gateway node is responsible for storing the transmission request, schedule the transmission time slots, collecting the data messages from its members, and forwarding all the traffic out of the cluster.

To avoid the depletion of the gateway's battery, The GMAC protocol uses the resource adaptive voluntary election (RAVE) scheme to periodically elect a new gateway node. The RAVE is a self election contention back-off algorithm that takes into consideration the node's available levels of energy and other resources. A potential limitation of the GMAC protocol is that a node may be forced to miss its allocated time slot and contend using FRTS message in the next contention round. This will happen if another node belonging to a different gateway and within interference distance from the node uses the channel during that time slot for its own transmission. This scenario can result in excessive delays and high probability of collisions in

densely deployed networks or in applications characterized by high network traffic.

Traffic-adaptive medium access protocol

The traffic-adaptive medium access (TRAMA) protocol[21] is similar to GMAC in that the communication channel is divided into frame cycles. Each frame is divided into a random access (contention) period and a scheduled access (contention-free) period. The scheduled period is divided into time slots. Nodes compete during the random access period to reserve slots for their data transmission during the scheduled access period. To guarantee a collision-free transmission, the TRAMA protocol uses the neighbor protocol (NP), the schedule exchange protocol (SEP), and the adaptive election algorithm (AEA) to obtain and exchange one and two hop information and schedules. Nodes with no data to send will switch off their radios and go to sleep to conserve energy.

Summary of discussed MAC layer optimization techniques

The energy optimization techniques used by the MAC protocols discussed in this section are summarized in Table 2.

Physical and Radio Layer

At the physical and hardware level the focus in complementary metal oxide semiconductor (CMOS) circuit design and optimization is shifting from faster switching circuits to ones that are optimized for power consumption. The work in Ref. 22 summarizes the challenges facing low-power design for WSNs as:

1. Design of low-power low-cost transceiver;
2. Low-power sensing and processing unit design;
3. Energy efficient modulation schemes and strategies to overcome signal propagation effects.

Several projects focused on the radio component and the design of energy efficient transceivers.[23–31] In

Table 2　Energy optimization techniques used by the medium access control (MAC) layer protocols.

Protocol	Architecture	Sleep/active period length	Energy optimization techniques
SMAC[16]	Flat	Fixed	Sleep and active cycles
DMAC[17]	Flat	Fixed	Sleep and active cycles. Improved latency for tree architecture
TMAC[18]	Flat	Variable	Sleep and active cycles. Variable length active period
WiseMAC[19]	Flat	Variable	Sleep and active cycles. Very short active period in low traffic scenarios
GMAC[21]	Hierarchy	Variable	Clustering. Short distance communication
TRAMA[13]	Hierarchy	Variable	Clustering. Short distance, collision-free communication

Table 3 Impact of layer optimization techniques on energy efficiency.

Layer	Optimization techniques used	Impact
Application	Dynamic adaptability to resources. Tradeoff between energy and accuracy	Low
Transport	Loss detection and data recovery on hop by hop instead of end-to-end. Event-to-sink reliability[5,6]	Low
Network	Clustering. Subset of nodes participating in data transmission. LEACH,[8] PEGASIS,[9] TEEN/APTEEN,[10,11] directed diffusion,[12] GEAR,[13] SPIN,[14] CLMR[15]	High
MAC	Low power (sleep) cycles. SMAC,[16] DMAC,[17] TMAC,[18] WiseMAC,[19] GMAC,[20] TRAMA[21]	High
Physical	Energy efficient hardware design. Optimize radio parameters, data package size, and transmission range at design time for energy efficiency[22–41]	Medium

Abbreviations: APTEEN, adaptive periodic threshold-sensitive energy efficient sensor network protocol; CMLR, cost-effective maximum lifetime routing; DMAC, delay MAC; GEAR, geographical and energy aware routing; GMAC, group-based MAC; LEACH, low-energy adaptive clustering hierarchy; MAC, medium access control; PEGASIS, power-efficient gathering in sensor information systems; SMAC, sensor MAC; SPIN, sensor protocols for information via negotiation; TEEN, threshold-sensitive energy efficient sensor network; TMAC, time division multiple access medium access controller; TRAMA, traffic-adaptive medium access; WiseMAC, wireless sensor MAC.

Ref. 32, the power consumption of the sensing and detection technique used is discussed. An example is given where the piezoresistive sensor will draw a large current while a capacitive one will not. Ref. 33 Discusses hardware designs that use several operation states to conserve energy. It also presents available hooks in hardware for power management, and CPU clock-down under OS control. In Ref. 34 power consumption was considered as a design constraint for motion sensors for physiological activity monitoring. Work in Refs. 35 and 36 cover a low-power Analog/Digital converter for wireless sensors and a low-power hardware encryption circuit, respectively. Refs. 37 and 38 proposed design methodology and architectures for low-power sensor design.

Several research projects proposed the redesign of the physical and radio layer parameters for energy optimization. Ref. 39 Discuss low-power algorithms for source coding. In Ref. 40 results are shown for package size optimization for energy efficiency under given communication channel characteristics. Ref. 41 proposed selection of radio parameters at design time with the goal of optimizing the one hop transmission range for energy efficiency.

Summary of OSI Layers Optimization Techniques for WSNs

In Table 3, we summarize the techniques used at the OSI stack layers and the impact of each layer's optimization on energy efficiency.

CROSS-LAYER OPTIMIZATIONS

One of the design philosophies of the OSI layered architecture model is to provide a well defined functionality and interfaces for each layer. This allows changes in the technology and designs in any individual layer to be transparent to the rest of the stack.[42] The OSI model has been adopted for wired networks design. Although arguably not optimal for wireless medium,[43] the OSI model has been migrated unaltered to wireless ad hoc communication systems design.

Cross-layer design is defined as the interaction between the different stack layers and sharing of information with the goal of improving the overall system performance. It has been used in the ad hoc wireless systems to improve throughput, latency, and QoS.[43,44]

In Ref. 40 an optimization agent (OA) is introduced for cross-layer design. As defined, this optimization agent can be thought of as a stack-wide database where each layer can deposit essential parameters that define its current working conditions. Other layers can access these data and use the information to tune and optimize their performance accordingly. In Ref. 45 a similar concept is presented.

A new approach is presented in Ref. 46 in which the communication system is redesigned from the ground up for WSNs. However, it can be argued against these revolutionary and some other cross-layering approaches. The main argument is that the layered approach provides a structured view of the communication network design. Cross-layer designs may produce interdependencies and designs that are hard to debug or upgrade.

REFERENCES

1. Day, J.D.; Zimmermann, H. The OSI reference model. IEEE. **1983**, *71* (12), 1334–1340.
2. Neumann, J. The reality of OSI. 13th Conference on Local Computer Networks, 1988, Minneapolis, MN, IEEE, Oct 10–12, 1988; 157–161.
3. Boulis, A. Models for programmability in sensor networks. In *Smart Dust: Sensor Network Applications, Architecture, and Design*; Ilyas, M., Mahgoub, I., Eds.; CRC Press LLC: Boca Raton, FL, 2006; 4-1-4-12.

4. Su, W.; Cayirci, E.; Akan, O.B. Overview of communication protocols for sensor networks. In *Sensor Network Protocols*; Mahgoub, I., Ilyas, M., Eds.; CRC Press LLC: Boca Raton, FL, 2006; 4-1–4-15.

5. Shah, R.C.; Petrovic, D.; Rabaey, J.M. Energy-aware routing and data funneling in sensor networks. In *Sensor Network Protocols*; Mahgoub, I., Ilyas, M., Eds.; CRC Press LLC: Boca Raton, FL, 2006; 9-1–9-15.

6. Iyer, Y.G.; Gandham, S.; Venkatesan, S. STCP: a generic transport layer protocol for wireless sensor networks. 14th International Conference on Computer Communications and Networks (ICCCN 2005), San Diego, CA, IEEE, Oct 17–19, 2005; 449–454.

7. Wan, C.-Y.; Campbell, A.T.; Krishnamurthy, L. Reliable transport for sensor networks: PSFQ—pump slowly fetch quickly paradigm. In *Wireless Sensor Networks*; Raghavendra, C.S., Sivalingam, K.M., Zanti, T., Eds.; Springer: New York, 2004; 153–182.

8. Heinzelman, W.R.; Chandrakasan, A.; Balakrishnan, H. Energy-efficient communication protocol for wireless microsensor networks. 33rd Hawaii International Conference on System Sciences (HICSS33), Maui, Hawaii, IEEE, Jan 4–7, 2000.

9. Lindsey, S.; Raghavendra, C.S. PEGASIS: power-efficient gathering in sensor information systems. IEEE Aerospace Conference 2002, Big Sky, MT, IEEE, Mar 9–16, 2002; Vol. 3, 3-1125–3-1130.

10. Manjeshwar, A.; Agrawal, D. TEEN: a routing protocol for enhanced efficiency in wireless sensor networks. 15th International Parallel and Distributed Processing Symposium (IPDPS'01), San Francisco, CA, IEEE, Apr 23–27, 2001; 2009–2015.

11. Manjeshwar, A.; Agrawal, D. APTEEN: a hybrid protocol for efficient routing and comprehensive information retrieval in wireless sensor networks. 16th International Parallel and Distributed Processing Symposium (IPDPS'02), Fort Lauderdale, FL, IEEE, Apr 15–19, 2002.

12. Intanagonwiwat, C.; Govindan, R.; Estrin, D.; Heidemann, J.; Silva, F. Directed diffusion for wireless sensor networking. IEEE/ACM Trans. Netw. **2003**, *11* (1), 2–16.

13. Yu, Y.; Govindan, R.; Estrin, D. Geographical and Energy Aware Routing: a recursive data dissemination protocol for wireless sensor networks. UCLA Computer Science Department Technical Report UCLA/CSD-TR-01-0023, May 2001.

14. Heinzelman, W.; Kulik, J.; Balakrishnan, H. Adaptive protocols for information dissemination in wireless sensor networks. 5th ACM/IEEE International Conference on Mobile Computing and Networking (Mobicom '99), Seattle, WA, Aug 15–19, 1999; ACM: 1999; 174–185.

15. Hossain, M.J.; Chae, O.; Mamun-Or-Rashid, M.; Hong, C.S. Cost-effective maximum lifetime routing protocol for wireless sensor networks. Advanced Industrial Conference on Telecommunications/Service Assurance with Partial and Intermittent Resources Conference/ELearning on Telecommunications Workshop, Lisbon, Portugal, July 17–20, 2005; IEEE: 1999; 314–319.

16. Ye, W.; Heidemann, J.; Estrin, D. An energy-efficient MAC protocol for wireless sensor networks. The 21st Annual Joint Conference of the IEEE Computer and Communications Societies (IEEE INFOCOM 2002), New York, IEEE, , June 23–27, 2002; Vol. 3, 1567–1576.

17. Lu, G.; Krishnamachari, B.; Raghavendra, C.S. An adaptive energy efficient and low-latency MAC for data gathering in wireless sensor networks. 18th International Parallel and Distributed Processing Symposium (IPDPS'04), Santa Fe, New Mexico, IEEE, Apr 26–30, 2004; pp. 224.

18. van Dam, T.; Langendoen, K. An adaptive energy-efficient MAC protocol for wireless sensor networks. ACM 1st International Conference on Embedded Networked Sensor Systems (SenSys'03), Los Angeles, CA, Nov 5–7, 2003; 171–180.

19. El-Hoiydi, A.; Decotignie, J. WiseMAC: an ultra low power MAC protocol for multi-hop wireless sensor networks. In *Algorithmic Aspects of Wireless Sensor Networks, Lecture Notes in Computer Science*; Nikoletseas, S., Rolim, J.D.P., Eds.; Springer: Berlin/Heidelberg, 2004; Vol. 3121, 18–31.

20. Brownfield, M.I.; Mehrjoo, K.; Fayez, A.S.; Davis, N.J. Wireless sensor network energy-adaptive MAC protocol. 3rd IEEE Consumer Communications and Networking Conference (CCNC 2006), Las Vegas, NV, IEEE, Jan 8–10, 2006; Vol. 2, 778–782.

21. Rajendran, V.; Obraczka, K.; Garcia-Luna-Aceves, J. Energy-efficient MAC: energy-efficient collision-free medium access control for wireless sensor networks. ACM 1st International Conference on Embedded Networked Sensor Systems (SenSys'03), Los Angeles, CA, IEEE, Nov 5–7, 2003.

22. Akyildiz, I.F.; Su, W.; Sankarasubramaniam, Y.; Cayirci, E. A survey on sensor networks. IEEE Commun. Mag. **2002**, *40* (8), 102–114.

23. Chee, Y.H.; Niknejad, A.M.; Rabaey, J. A 46% efficient 0.8dBm transmitter for wireless sensor networks. 2006 Symposium on VLSI Circuits, Honolulu, HI, IEEE, June 15–17, 2006.

24. Otis, B.P.; Chee, Y.H.; Lu, R.; Pletcher, N.M.; Rabaey, J.M. An ultra-low power MEMS-based two-channel transceiver for wireless sensor networks. 2004 Symposium on VLSI Circuits, Honolulu, HI, IEEE, June 17–19, 2004; 20–23.

25. Joehl, N.; Dehollain, C.; Favre, P.; Deval, P.; Declercq, M. A low-power 1-GHz super-regenerative transceiver with time-shared PLL control. IEEE J. Solid-State Circuits. **2001**, *36* (7), 1025–1031.

26. Molnar, A.; Lu, B.; Lanzisera, S.; Cook, B.W.; Pister, K.S.J. An ultra-low power 900 MHz RF transceiver for wireless sensor networks. Custom Integrated Circuits Conference (CICC'04), Orlando, FL, IEEE, Oct 3–6, 2004.

27. Cook, B.W.; Molnar, A.; Pister, K.S.J. Low power RF design for sensor networks. IEEE Radio Frequency Integrated Circuits (RFIC) Symposium, Long Beach, CA, IEEE, June 12–14, 2005; 357–360.

28. Chien, C.; Elgorriaga, I.; McConaghy, C. Low-power direct-sequence spread-spectrum modem architecture for distributed wireless sensor networks. International Symposium on Low Power Electronics and Design (ISLPED'01), Huntington Beach, CA, IEEE, Aug 6–7, 2001; 251–254.

29. Enz, C.; Scolari, N.; Yodprasit, U. Ultra low-power radio design for wireless sensor networks. 1st IEEE International

Workshop on Radio-Frequency Integration Technology (RFIT'05), Singapore, IEEE, Nov 30–Dec 2, 2005; 1–17.

30. Daly, D.C.; Chandrakasan, A.P. An energy efficient OOK transceiver for wireless sensor networks. IEEE Radio Frequency Integrated Circuits Symposium, Long Beach, CA, June 12–14, 2006; 279–282.

31. Kluge, W.; Poegel, F.; Roller, H.; Lange, M.; Ferchland, T.; Dathe, L.; Eggert, D. A fully integrated 2.4-GHz IEEE 802.15.4-compliant transceiver for ZigBee™ applications. IEEE J. Solid-State Circuits. 2006, 41 (12), 2767–2775.

32. Warneke, B. Miniaturizing sensor networks with MEMS. In Smart Dust: Sensor Network Applications, Architecture, and Design; Mahgoub, I., Ilyas, M., Eds.; CRC Press LLC: Boca Raton, FL, 2006; 5-1–5-19.

33. Sinha, A.; Chandrakasan, A. Dynamic power management in sensor networks. In Smart Dust: Sensor Network Applications, Architecture, and Design; Mahgoub, I., Ilyas, M., Eds.; CRC Press LLC: Boca Raton, FL, 2006; 14-1–14-15.

34. Sadat, A.; Qu, H.; Yu, C.; Yuan, J.S.; Xie, H. Low-power CMOS wireless MEMS motion sensor for physiological activity monitoring. IEEE Trans. Circuits Syst. 2005, 52 (12), 2539–2551.

35. Schroeder, D. Adaptive low-power analog/digital converter for wireless sensor networks. 3rd International Workshop on Intelligent Solutions in Embedded Systems (WISES'05), Hamburg, Germany, IEEE, May 20, 2005; 70–78.

36. Kim, M.; Kim, J.; Choi, Y. Low power circuit architecture of AES crypto module for wireless sensor network. Enformatika Trans. Eng. Comput. Technol. 2005, 8, 146–151.

37. Li, Y.; De Bernardinis, F.; Otis, B.; Rabaey, J.M.; Vincentelli, A.S. A low-power mixed-signal baseband system design for wireless sensor networks. IEEE 2005 Custom Integrated Circuits Conference (CICC'05), San Jose, CA, IEEE, Sept 18–21, 2005; 55–58.

38. Asada, G.; Dong, M.; Lin, T.S.; Newberg, F.; Pottie, G.; Kaiser, W.J.; Marcy, H.O. Wireless integrated network sensors: low power systems on a chip. 24th European Solid-State Circuits Conference (ESSCIRC'98), The Hague, The Netherlands, IEEE, Sept 22–24, 1998; 9–16.

39. Kim, J.; Andrews, J.G. An energy efficient source coding and modulation scheme for wireless sensor networks. IEEE 6th Workshop on Signal Processing Advances in Wireless Communications (SPAWC'05), New York, IEEE, June 5–8, 2005; 710–714.

40. Sankarasubramaniam, Y.; Akyildiz, I.F.; McLaughlin, S.W. Energy efficiency based packet size optimization in wireless sensor networks. 1st IEEE International Workshop on Sensor Network Protocols and Applications (SNPA'03), Anchorage, AK, IEEE, May 11, 2003.

41. Chen, P.; Oapos, D.B.; Callaway, E. Energy efficient system design with optimum transmission range for wireless ad hoc networks. IEEE International Conference on Communications (ICC 2002), New York, IEEE, Apr 26–May 2, 2002; Vol. 2, 945–952.

42. Su, W.; Lim, T.L. Cross-layer design and optimization for wireless sensor networks. 7th ACIS International Conference on Software Engineering, Artificial Intelligence, Networking, and Parallel/Distributed Computing (SNPD'06), Las Vegas, NV, IEEE, June 10–20, 2006; 278–284.

43. Gavrilovska, L. Cross-layering approaches in wireless ad hoc networks. Wirel. Pers. Commun. 2006, 37 (3–4), 271–290.

44. Raisinghani, V.; Iyer, S. Cross layer design optimizations in wireless protocol stacks. Comput. Commun. (ELSEVIER). 2003, 27, 720–724.

45. Safwat, A.M. A novel framework for cross-layer design in wireless ad hoc and sensor networks. IEEE Global Telecommunications Conference Workshops (GLOBECOM 2004), Dallas, TX, IEEE, Nov 29–Dec 3, 2004; 130–135.

46. Akyildiz, F.; Vuran, M.C.; Akan, O.B. A cross-layer protocol for wireless sensor networks. Conference on Information Science and Systems (CISS'06), Princeton, NJ, IEEE, Mar 22–24, 2006.

WSNs–
WSNs: Optimiz

Wireless Sensor Networks (WSNs): Routing

Dharma P. Agrawal
Ratnabali Biswas
Neha Jain
Anindo Mukherjee
Sandhya Sekhar
Aditya Gupta
OBR Research Center for Distributed and Mobile Computing, ECECS Department, University of Cincinnati, Cincinnati, Ohio, U.S.A.

Abstract

This entry presents classification, routing protocols, and architectures of wireless sensor networks.

ROUTING IN WIRELESS SENSOR NETWORKS

There are a few inherent limitations of wireless media, such as low bandwidth, error-prone transmissions, the need for collision-free channel access, etc. These wireless nodes also have only a limited amount of energy available to them, because they derive energy from a personal battery and not from a constant power supply. Furthermore, because these sensor nodes are deployed in places where it is difficult to replace the nodes or their batteries, it is desirable to increase the longevity of the network. Also, preferably all the nodes should die together so that one can replace all the nodes simultaneously in the whole area. Finding individual dead nodes and then replacing those nodes selectively would require preplanned deployment and eliminate some advantages of these networks. Thus, the protocols designed for these networks must strategically distribute the dissipation of energy, which also increases the average life of the overall system.

Query Classification in Sensor Networks

Before discussing routing protocols for sensor networks, let us first categorize the different kinds of queries that can be posed to a sensor network. Based on the temporal property of data (i.e., whether the user is interested in data collected in the past, the current snapshot view of the target regions, or sensor data to be generated in future for a given interval of time), queries can be classified as follows:

Historical queries: This type of query is mainly used for analysis of historical data stored at a remote base station or any designated node in the network in the absence of a base station. For example, "What was the temperature two hours prior in the northwest quadrant?" The source nodes need not be queried to obtain historical data as it is usually stored outside the network or at a node equidistant from all anticipated sinks for that data.

Encyclopedia of Wireless and Mobile Communications DOI: 10.1081/E-EWMC-120043929

One-time query: One-time or snapshot queries provide the instantaneous view of the network. For example, "What is the temperature in the northwest quadrant now?" The query triggers a single query response; hence, data traffic generated by one-time queries is the least. These are usually time critical as the user wants to be notified immediately about the current situation of the network. A warning message that informs the user of some unusual activity in the network is an example of a one-time query response that is time critical.

Persistent: Persistent, or long-running, queries are mainly used to monitor a network over a time interval with respect to some parameters. For example, "the temperature in the northwest quadrant for the next 2 hours." A persistent query generates maximum query responses in the network, depending on its duration. The purpose of the persistent query is to perform periodic background monitoring. Energy efficiency is often traded with delay in response time of persistent queries to maximize utilization of network resources, as they are usually noncritical.

Characteristics of Routing Protocols for Sensor Networks

Traditional routing protocols defined for wireless ad hoc networks[1,2] are not well suited for wireless sensor networks due to the following reasons:[3,4]

Sensor networks are data centric. Traditional networks usually request data from a specific node but sensor networks request data based on certain attributes such as, "Which area has temperature greater than 100°F?"

In traditional wired and wireless networks, each node is given a unique ID, which is used for routing.

This cannot be effectively used in sensor networks because being data centric they do not require routing to and from specific nodes. Also, the large number of nodes in the network implies large IDs,[5] which might be substantially larger than the actual data being transmitted.

Adjacent nodes may have similar data. So instead of sending data separately from each node to the requesting node, it is desirable to aggregate similar data before sending it.

The requirements of the network change with the application and, hence, it is application specific.[6] For example, some applications need the sensor nodes to be fixed and not mobile, while others may need data based only on one selected attribute (i.e., here the attribute is fixed).

Thus, sensor networks need protocols that are application specific, data centric, and capable of aggregating data and minimizing energy consumption.

NETWORK ARCHITECTURE

The two main architecture alternatives used for data communication in a sensor network are the hierarchical and the flat network architectures. The hierarchical network architecture is energy efficient for collecting and aggregating data within a large target region, where each node in the region is a source node. Hence, hierarchical network protocols are used when data will be collected from the entire sensor network. A flat network architecture is more suitable for transferring data between a source destination pair separated by a large number of hops.

Hierarchical Network Architecture

One way of minimizing the data transmissions over long distances is to cluster the network so that signaling and control overheads can be reduced, while critical functions such as media access, routing, and connection setup could be improved. While all nodes typically function as switches or routers, one node in each cluster is designated as the cluster head (CH) and traffic between nodes of different clusters must always be routed through their respective CHs or gateway nodes that are responsible for maintaining connectivity among neighboring CHs. The number of tiers within the network can vary according to the number of nodes, resulting in hierarchical network architecture as shown in Fig. 1.

Fig. 1 shows two tiers of CHs where the double lines represent that CHs of tier-1 are cluster members of the cluster at the next higher level (i.e., tier-2). A proactive clustering algorithm for sensor networks called low-energy adaptive clustering hierarchy (LEACH) is one of the initial data-gathering protocols introduced by MIT researchers Heinzelman et al.[7] Each cluster has a CH that periodically collects data from its cluster members, aggregates it, and sends it to an upper-level CH. Only the CH needs to perform additional data computations such as aggregation, etc., and the rest of the nodes sleep unless they have to communicate with the CH. To evenly

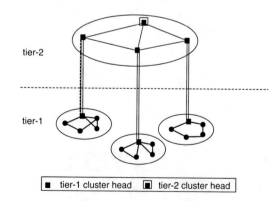

■ tier-1 cluster head   ■ tier-2 cluster head

Fig. 1 Two-tier sensor network architecture. *Source:* Manjeshwar, A. Energy Efficient routing protocols with comprehensive information retrieval for wireless sensor networks; M.S. thesis, 2001.

distribute this energy consumption, all the nodes in a neighborhood take turns to become the CH for a time interval called the cluster period.

Flat Network Architecture

In a flat network architecture as shown in Fig. 2, all nodes are equal and connections are set up between nodes that are in close proximity to establish radio communications, constrained only by connectivity conditions and security limitations. Route discovery can be carried out in sensor networks using flooding that does not require topology maintenance as it is a reactive way of disseminating information. In flooding, each node receiving data packets broadcasts until all nodes or the node at which the packet was originated gets back the packet. But in sensor networks, flooding is minimized or avoided as nodes could receive multiple or duplicate copies of the same data packet due to nodes having common neighbors or sensing similar data. Intanagonwiwat et al.[8] have introduced a data dissemination paradigm called directed diffusion for sensor networks, based on a flat topology. The query is disseminated (flooded) throughout the network with the querying node acting as a source and gradients are set up toward the requesting node to find the data satisfying the query. As one can observe from Fig. 2, the query is propagated toward the requesting node along multiple paths shown by the dashed lines. The arcs show how the query is directed toward the event of interest, similar to a ripple effect. Events (data) start flowing toward the requesting node along multiple paths. To prevent further flooding, a small number of paths can be reinforced (shown by dark lines in the figure) among a large number of paths initially explored to form the multi-hop routing infrastructure so as to prevent further flooding. One advantage of flat networks is the ease of creating multiple paths between communicating nodes, thereby alleviating congestion and providing robustness in the presence of failures.

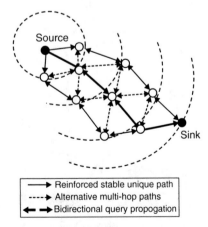

Fig. 2 A flat sensor network that uses directed diffusion for routing. *Source*: Jain, N. energy efficient information retrieval in wireless sensor networks, Ph.D. thesis, 2004.

CLASSIFICATION OF SENSOR NETWORKS

Sensor networks can be classified into two types based on their mode of operation or functionality and the type of target applications:[3,4]

Proactive networks: In this scheme, the nodes periodically switch on their sensors and transmitters, sense the environment, and transmit the data of interest. Thus, they provide a snapshot of the relevant parameters at regular intervals and are well suited for applications that require periodic data monitoring.

Reactive networks: In this scheme, the nodes react immediately to sudden and drastic changes in the value of a sensed attribute and are well suited for time-critical applications.

Once we have a network, we have to come up with protocols that efficiently route data from the nodes to the users, preferably using a suitable medium access control (MAC) sub-layer protocol to avoid collisions.

Having classified sensor networks, we now look at some of the protocols for sensor networks.[3,4]

Proactive Network Protocol

Functioning

At each cluster change time, once the CHs are decided, the CH broadcasts the following parameters (see Fig. 3):

Report time (T_R): The time period between successive reports sent by a node.
Attributes (A): A set of physical parameters about which the user is interested in obtaining data.

At every report time, the cluster members sense the parameters specified in the attributes and send the data to the CH. The CH aggregates this data and sends it to the base station or a higher-level CH, as the case may be. This ensures that the user has a complete picture of the entire area covered by the network.

Important features

Because the nodes switch off their sensors and transmitters at all times except the report times, the energy of the network is conserved.

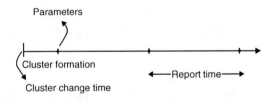

Fig. 3 Timeline for proactive protocol. *Source*: Manjeshwar, A.; Agrawal, D.P. TEEN: a routing protocol for enhanced efficiency in wireless sensor networks. Proc. 15th International Parallel and Distributed Processing Symposium (IPDPS 2001) Workshops, San Francisco, CA, April 23–27, 2001.

At every cluster change time, T_R and A are transmitted afresh and thus can be changed. By changing A and T_R, the user can decide what parameters to sense and how often to sense them. Also, different clusters can sense different attributes for different T_R.

This scheme, however, has an important drawback. Because of the periodicity with which the data is sensed, it is possible that time-critical data may reach the user only after the report time, making this scheme ineffective for time-critical data sensing applications.

LEACH

LEACH is a family of protocols developed by Heinzelman et al. (www-mtl.mit.edu/research/icsystems/uamps/leach).[7] LEACH is a good approximation of a proactive network protocol, with some minor differences.

Once the clusters are formed, the CHs broadcast a TDMA schedule giving the order in which the cluster members can transmit their data. The total time required to complete this schedule is called the frame time T_F. Every node in the cluster has its own slot in the frame, during which it transmits data to the CH. The report time T_R discussed earlier is equivalent to the frame time T_F in LEACH. However T_F is not broadcast by the CH but is derived from the TDMA schedule and hence is not under user control. Also, the attributes are predetermined and not changed after initial installation.

Example applications

This network can be used to monitor machinery for fault detection and diagnosis. It can also be used to collect data about temperature (or pressure, moisture, etc.) change patterns over a particular area.

Reactive Network Protocol: Threshold Sensitive Energy Efficient Sensor Network

A new network protocol called *threshold sensitive energy efficient sensor network* (TEEN) has been developed that targets reactive networks and is the first protocol developed for reactive networks.[3]

Functioning

In this scheme, at every cluster change time, in addition to the attributes, the CH broadcasts the following to its members (see Fig. 4):

Hard threshold (H_T)**:** A threshold value for the sensed attribute. It is the absolute value of the attribute, beyond which the node sensing this value must switch on its transmitter and report to its CH.

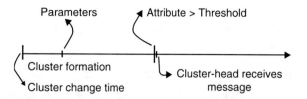

Fig. 4 Timeline for TEEN. *Source*: Manjeshwar, A.; Agrawal, D.P. TEEN: a rouiting protocol for enhanced efficiency in wireless sensor networks. Proc. 15th International Parallel and Distributed Processing Symposium (IPDPS 2001) Workshops, San Francisco, CA, April 23–27, 2001.

Soft threshold (S_T)**:** A small change in the value of the sensed attribute that triggers the node to switch on its transmitter and transmit.

The nodes sense their environment continuously. The first time a parameter from the attribute set reaches its hard threshold value, the node switches on its transmitter and sends the sensed data. The sensed value is also stored in an internal variable in the node, called the *sensed value (SV)*. The nodes will next transmit data in the current cluster period but only when *both* the following conditions are true:

- The current value of the sensed attribute is greater than the hard threshold;
- The current value of the sensed attribute differs from SV by an amount equal to or greater than the soft threshold.

Whenever a node transmits data, *SV* is set equal to the current value of the sensed attribute. Thus, the hard threshold tries to reduce the number of transmissions by allowing the nodes to transmit only when the sensed attribute is in the range of interest. The soft threshold further reduces the number of transmissions by eliminating all the transmissions that might have otherwise occurred when there is little or no change in the sensed attribute once the hard threshold is reached.

Important features

Time-critical data reaches the user almost instantaneously and hence this scheme is well suited for time-critical data sensing applications.

Message transmission consumes much more energy than data sensing. So, although the nodes sense continuously but because they transmit less frequently, the energy consumption in this scheme can be much less than that in the proactive network.

The soft threshold can be varied, depending on the criticality of the sensed attribute and the target application.

A smaller value of the soft threshold gives a more accurate picture of the network, at the expense of increased energy consumption. Thus, the user can control the trade-off between energy efficiency and accuracy.

At every cluster change time, the parameters are broadcast afresh and thus the user can change them as required.

The main drawback of this scheme is that if the thresholds are not reached, the nodes will never communicate. Thus, the user will not get any data from the network and will not even know if all the nodes die. Hence, this scheme is not well suited for applications where the user needs to get data on a regular basis. Another possible problem with this scheme is that a practical implementation would have to ensure that there are no collisions in the cluster. TDMA scheduling of the nodes can be used to avoid this problem. This will, however, introduce a delay in the reporting of time-critical data. CDMA is another possible solution to this problem.

Example applications

This protocol is best suited for time-critical applications such as intrusion detection, explosion detection, etc.

Hybrid Networks

There are applications in which the user might need a network that reacts immediately to time-critical situations and also gives an overall picture of the network at periodic intervals to answer analysis queries. Neither of the above networks can do both jobs satisfactorily and they have their own limitations.

Manjeshwar and Agrawal[3] have combined the best features of the proactive and reactive networks, while minimizing their limitations, to create a new type of network called a *hybrid network*. In this network, the nodes not only send data periodically, but also respond to sudden changes in attribute values. A new routing protocol [adaptive periodic threshold sensitive energy efficient sensor network protocol, (APTEEN)] has been proposed for such a network and uses the same model as the above protocols but with the following changes. APTEEN works as follows.

Functioning

In each cluster period, once the CHs are decided, the CH broadcasts the following parameters (see Fig. 5):

Attributes (A): A set of physical parameters about which the user is interested in obtaining data.

Thresholds: This parameter consists of a H_T and a S_T. H_T is a particular value of an attribute beyond which a node can be triggered to transmit data. S_T is a small change in the value of an attribute that can trigger a node to transmit.

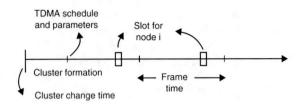

Fig. 5 Timeline for APTEEN. *Source*: Manjeshwar, A.; Agrawal, D.P. APTEEN: a hybrid protocol for efficient routing and comprehensive information retrieval in wireless sensor networks. Proc. Int. Parallel and Distributed Processing Symposium (IPDPS 2002) Workshops, Fort Lauderdale, FL, April 15–19, 2002.

Schedule: A TDMA schedule similar to the one used in Heinzelman et al.,[7] assigning a slot to each node.

Count time (T_C): The maximum time period between two successive reports sent by a node. It can be a multiple of the TDMA schedule length and it introduces the proactive component in the protocol.

The nodes sense their environment continuously. However, only those nodes that sense a data value at or beyond the hard threshold will transmit. Furthermore, once a node senses a value beyond H_T, it next transmits data only when the value of that attribute changes by an amount equal to or greater than the soft threshold S_T. The exception to this rule is that if a node does not send data for a time period equal to the T_C, it is forced to sense and transmit the data, irrespective of the sensed value of the attribute. Because nodes near each other may fall into the same cluster and sense similar data, they may try sending their data simultaneously, leading to collisions between their messages. Hence, a TDMA schedule is used and each node in the cluster is assigned a transmission slot, as shown in Fig. 5.

Important features

It combines both proactive and reactive policies. By sending periodic data, it gives the user a complete picture of the network, like a proactive scheme. It behaves like a reactive network also by sensing data continuously and responding to drastic changes immediately.

It offers a lot of flexibility by allowing the user to set the time interval (T_C) and the threshold values (H_T and S_T) for the attributes.

Energy consumption can be controlled by changing the count time as well as the threshold values. The hybrid network can emulate a proactive network or a reactive network, based on the application, by suitably setting the count time and the threshold values.

The main drawback of this scheme is the additional complexity required to implement the threshold functions and the count time. However, this is a reasonable trade-off and provides additional flexibility and versatility.

A Comparison of the Protocols

To analyze and compare the protocols TEEN and APTEEN with LEACH and LEACH-C, consider the following metrics:[3,4]

Average energy dissipated: shows the average dissipation of energy per node over time in the network as it performs various functions such as transmitting, receiving, sensing, aggregation of data, etc.

Total number of nodes alive: indicates the overall lifetime of the network. More importantly, it gives an idea of the area coverage of the network over time.

Total number of data signals received at BS: explains how TEEN and APTEEN save energy by not transmitting data continuously, which is not required (neither time critical nor satisfying any query). Such data can be buffered and later transmitted at periodic intervals. This also helps in answering historical queries.

Average delay: gives the average response time in answering a query. It is calculated separately for each type of query.

The performance of the different protocols is given in Figs. 6 and 7.

MULTIPLE PATH ROUTING

The above protocols considered a hierarchical network; now consider multiple path routing in a flat sensor network. Multiple path routing aims to exploit the connectivity of the underlying physical networks by providing multiple paths between source destination pairs. The originating node therefore has a choice of more than one potential path to a particular destination at any given time.

The Need for Multiple Path Routing

Classical multiple path routing has been explored for two reasons. The first is *load balancing,* where traffic between the source and destination is split across multiple (partially or fully) disjoint paths to avoid congestion on any one path. The second use of multipath routing is to increase the probability of *reliable data delivery* due to the use of independent paths. To ensure reliable data delivery, duplicate copies of the data can be sent along alternate routes.

Multiple path routing is popularly used to avoid disparity in energy consumption in the network. This suggests that a multiple path scheme would be preferable when there are simultaneous active sources in the network with high traffic intensity. It is typical to have a number of overlapping source sink pairs unevenly distributed in the sensor field. It is a challenging problem to distribute traffic load evenly among a majority of the sensor nodes in the network with such random traffic conditions. Multiple path routing is cost effective in heavy load scenarios, while a single path routing scheme with a lower complexity may be more desirable when the numbers of packets exchanged between the random source sink pairs are few.

Load balancing is especially useful in energy-constrained networks because the relative energy level of the nodes does affect the network lifetime more than their absolute energy levels. With classic shortest path routing schemes, a few nodes that lie on many of these shortest paths are depleted of their energy at a much faster rate than the other nodes. As a result of these few dead nodes, the

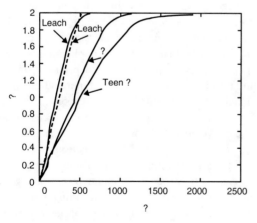

Fig. 6 Comparison of average energy dissipation. *Source*: Manjeshwar, A.; Agrawal, D.P. APTEEN: a hybrid protocol for efficient routing and comprehensive information retrieval in wireless sensor networks. Proc. Int. Parallel and Distributed Processing Symposium (IPDPS 2002) Workshops, Fort Lauderdale, FL, April 15–19, 2002.

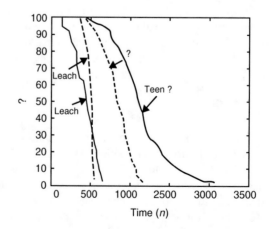

Fig. 7 Comparison of the number of alive nodes. *Source*: Manjeshwar, A.; Agrawal, D.P. APTEEN: a hybrid protocol for efficient routing and comprehensive information retrieval in wireless sensor networks. Proc. Int. Parallel and Distributed Processing Symposium (IPDPS 2002) Workshops, Fort Lauderdale, FL, April 15–19, 2002.

nodes in its neighborhood may become inaccessible, which in turn causes a ripple effect, leading to network partitioning. Chang and Tassiulas[9,10] have proved, assuming each node to have a limited lifetime, that the overall lifetime of the network can be improved if the routing protocol minimizes the disparity in the residual energy of every node, rather than minimizing the total energy consumed in routing. In the multiple path routing protocol proposed by Jain et al.,[11,12] the traffic is spread over the nodes lying on different possible paths between the source and the sink, in proportion to their residual energy. The rationale behind traffic spreading is that for a given total energy consumption in the network, every node should have spent the same amount of energy for data transmission. Their objective is to assign more load to underutilized paths and less load to overcommitted paths so that uniform resource utilization of all available paths can be ensured. They construct multiple paths of variable energy cost, and then design a traffic scheduling algorithm that determines the order of path utilization to enable uniform network resource utilization. They also grade the multiple paths obtained according to their quality of service (QoS), where the QoS metric is delay in response time.[11,12] Thus, they use a reservation-based scheme to provide a good service to time-critical applications, along with dynamic reallocation of network resources to noncritical applications to avoid underutilization of resources.

There is a need to adapt multiple path routing to overcome the design constraints of a sensor network. Important design considerations that drive the design of sensor networks are the energy efficiency and scalability[13] of the routing protocol. Discovery of all possible paths between a source and a sink might be computationally exhaustive for sensor networks because they are power constrained. In addition, updating the source about the availability of these paths at any given time might involve considerable communication overhead. The routing algorithm designed for a sensor network must depend only on local information[14] or the information piggy-backed with data packets, as global exchange of information is not a scalable solution because of the sheer number of nodes.

Service Differentiation

Service differentiation is a basic way to provide QoS by giving one user priority over another. The data traffic is classified based on the QoS demands of the application. QoS parameters for typical Internet applications include bounds for bandwidth, packet delay, packet loss rate, and jitter. Certain additional parameters that deal with problems unique to wireless and mobile networks are power restrictions, mobility of nodes, and unreliable link quality. For a sensor network, node mobility is not high but severe power restrictions may force the packet to be routed through longer paths that have a higher residual energy than the shortest route connecting the source to the

destination. INSIGNIA, a service differentiation methodology developed by Lee and Campbell,[15] employs a field in the packet to indicate the availability of resources and perform admission control. Such a testing is based on the measured channel capacity to the destination or utilization and the requested bandwidth. When a node accepts a request, resources are committed and subsequent packets are scheduled accordingly. If adequate resources are not available, then a flow adaptation is triggered to adjust the available resources on the new requested path.

Service requirements could be diverse in a network infrastructure. Some queries are useful only when they are delivered within a given timeframe. Information provided may have different levels of importance; therefore, the sensor network should be willing to spend more resources in disseminating packets carrying more important information. Service differentiation is popularly used in the Internet[16] to split the traffic into different classes based on the QoS desired by each class. In the multiple path routing protocol proposed by Jain et al.,[12] a priority and preemption mechanism is used to control end-to-end delay for time-critical queries. Some specific examples of applications that could benefit from a sensor network supporting service differentiation are described here.

Battlefield surveillance: Soldiers conduct periodic monitoring for situational awareness of the battlefield. If the network senses some unusual or suspicious activity in the field that requires immediate attention of the military personnel, an alarm is triggered. These warning signals must reach the end user or the soldier immediately to expedite quick decision making.

Disaster relief operations: In case of natural calamities such as floods, wild fires, tornadoes, or earthquakes, or other catastrophes such as terrorist attacks, coordinated operation of sensor nodes could be very useful in conducting efficient rescue operations. Precise information about the location of victims or environmental parameters of risky areas could provide facts that help in the planning of rescue operations. Here, a flexibility to prioritize information retrieval could be beneficial in avoiding communication delays for time-critical responses.

Infrastructure security: A network of sensors could be deployed in a building or a campus that needs to be secured from any intrusion detection. The sensors could be programmed to discriminate among the attacks if they occur simultaneously or associate higher priority for packets confirming intrusion, as compared to other normal data packets containing information related to the usual background monitoring of the building for parameters like light intensity, temperature, or the number of people passing by.

Environmental or biomedical applications: Monitoring presence of certain gases or chemicals in remote areas such as mines, caves, or under water, or in chambers where research experiments are carried out, or radiation levels in a nuclear plant. Alarm signals might be required infrequently to report the presence of attributes in a volume or degree at more than an expected threshold. These warning messages, if received on time, can help in accomplishing the research goals or desired monitoring operations.

Service Differentiation Strategies for Sensor Networks

Bhatnagar et al.[17] discussed the implications of adapting these service differentiation paradigms from wired networks to sensor networks. They suggest the use of adaptive approaches; the sensor nodes learn the network state using eavesdropping or by explicit state dissemination packets. The nodes use this information to aid their forwarding decisions; e.g., low-priority packets could take a longer route to make way for higher-priority packets through shorter routes. The second implication of their analysis is that the applications should be capable of adapting their behavior at runtime based on the current allocation, which must be given as a feedback from the network to the application.

In our work,[3,4] we aim to achieve twofold goals of the QoS-aware routing described by Chen et al.[18] The two goals are 1) selecting network paths that have sufficient resources to meet the QoS requirements of all admitted connections; and 2) achieving global efficiency in resource utilization. In ad hoc networks, static provisioning is not enough because node mobility necessitates dynamic allocation of resources. In sensor networks, although user mobility is practically absent, dynamic changes in the network topology may be present because of node loss due to battery outage. Hence, multi-hop ad hoc routing protocols must be able to adapt to the variation in the route length and its signal quality while still providing the desired QoS. It is difficult to design provisioning algorithms that achieve simultaneously good service quality as well as high resource utilization. Because the network does not know in advance where packets will go, it will have to provision enough resources to all possible destinations to provide high service assurance. This results in severe underutilization of resources.

CONTINUOUS QUERIES IN SENSOR NETWORKS

Multiple path routing is used for uniform load distribution in a sensor network with heavy traffic. Now consider energy optimization techniques for sensor networks serving continuous queries. Queries in a sensor network are spatio-temporal; i.e., the queries are addressed to a region or space for data, varying with time. In a monitoring application, the knowledge of the coordinates of the event in the space, and its time of occurrence,

is as important as the data itself. Queries in a sensor network are usually location based; therefore, each sensor should be aware of its own location.[19] When self-location by GPS is not feasible or too expensive, other means of self-location, such as relative positioning algorithms,[20] can be used. A timestamp associated with the data packet reveals the temporal property of data.

Depending on the nature of the application, the types of queries injected in a sensor network can vary. Queries posed to a sensor network are usually classified as one-time queries or periodic queries. "One-time" queries are injected at random times to obtain a snapshot view of the data attributes, but "periodic" queries retrieve data from the source nodes after regular time intervals. An example of a periodic query is, "Report the observed temperature for the next week at the rate of one sample per minute." We now concentrate on periodic queries that are long running; i.e., they retrieve data from the source nodes for a substantially long duration, possibly the entire lifetime of the network. We now classify queries into three different categories based on the nature of data processing demanded by the application.

Simple queries: These are stand alone queries that expect an answer to a simple question from all or a set of nodes in the network. For example, "Report the value of temperature."

Aggregate queries: These queries require collaboration among sensor nodes in a neighborhood to aggregate sensor data. Queries are addressed to a target region consisting of many nodes in a geographically bounded area instead of individual nodes. For example, "Report the average temp of all nodes in region X."

Approximate queries: These are queries that require data summarization and rely on synopsis data structures to perform holistic data aggregation[14] in the form of histograms, isobars, contour maps, tables, or plots. For example, "Report the contours of the pollutants in the region X." Such sophisticated data representation results in a tremendous reduction in data volume at the cost of additional computation at nodes. Although offline data processing by the user is eliminated, due to a lack of raw sensor data, the user may not be able to analyze it later in ways other than the query results.

Complex queries: If represented in SQL, these queries would consist of several joins nested or condition-based sub-queries. Their computation hierarchy is better represented by a query tree. For example, "Among regions X and Y, report the average pressure of the region that has higher temperature." Sub-queries in a complex query could be simple, aggregate, or approximate queries. In our proposed work, we design a general in-network query processing architecture for evaluating complex queries.

The Design of a Continuous Query Engine

We now present the four basic design characteristics of a continuous query processing architecture:

Data buffering: It is required to perform blocking operations[21] that need to process an entire set of input tuples before an output can be delivered. Examples of blocking operators are GroupBy, OrderBy, or aggregation functions such as maximum, count, etc. A time-based sliding window proposed by Datar et al.[22] is used to move across the stream of tuples and restrict the data buffered at a node at any instant of time for data processing. Time synchronization to process data also requires data buffering.[23] If the data streams arriving from different sources must be combined based on the times each tuple is generated, then synchronization between sensor nodes along the communication path is required. To ensure temporal validity of the results produced while evaluating operators, tuples arriving from the faster stream should be buffered until the tuples in the same time window belonging to the other input stream reach the QP node evaluating the operator.

Data summarization: If data is arriving too fast to be processed by the node, it will not be able to hold the incoming tuples in its limited memory. Therefore, tuples might have to be dropped or a sample of tuples could be selected from the data stream to represent the entire data set.[24]

Sharing: It is important to share processing whenever feasible among multiple queries for maximizing the reuse of resources. This is used to achieve scalability with increasing workload.

Adaptability: In spatio-temporal querying, the number of sources and their data arrival rates may vary during a query's lifetime, thereby rendering static decisions ineffective. Hence, the continuous query evaluation[25] should keep adapting to changes in data properties.

Applications of an Adaptive Continuous Query Processing System

Distributed query processing has numerous applications in remote monitoring tasks over wireless sensor networks. Described here some of the real-world applications that will benefit from the proposed query processing architecture. These examples illustrate the range of applications that can be supported by the proposed query processing architecture. We classify potential commercial applications of wireless sensor networks into two broad classes.

Infrastructure-based monitoring: Sensor nodes are *attached to an existing physical structure* to remotely monitor complex machinery and processes, or the health of civil infrastructures such as highways, bridges, or buildings.

Field-based monitoring: A space in the environment is monitored using a dense network of *randomly scattered* nodes that are not particularly installed on any underlying infrastructure. Environmental monitoring of ecosystems, toxic spills, and fire monitoring in forests are examples of field monitoring.

We now describe infrastructure-based and field-based monitoring in detail.

Infrastructure-based monitoring

Commercial interest in designing solutions for infrastructure-based monitoring applications using sensor networks is growing at a fast pace. EmberNet (Ember Corporation; www.ember.com) has developed an embedded networking software for temperature sensing and heat trace control using wireless sensors. This drastically reduces the installation cost when the number of temperature monitoring points runs into thousands.

We propose that the efficiency of the monitoring task can be enhanced by deploying additional powerful nodes that are able to process the temperature readings and draw useful inferences within the network. As results occupy fewer data bytes than raw data, data traffic between the monitoring points and the external monitoring agency can be reduced. Data is routed faster and processing time at the external monitoring agency is saved, which enables quicker decision making based on the results of data monitoring. Another potential example of infrastructure-based monitoring is *detection of leakage* in a water distribution system. A self-learning, distributed algorithm can enable nodes to switch among various roles of data collection, processing, or forwarding among themselves so that the network continues retrieving data despite the failure of a few nodes. Human intervention would be deemed necessary only when a majority of the nodes in an area malfunction or run out of battery power. Some other infrastructure-based applications that might benefit from the proposed query processing and the resulting real-time decision support designed in this work are as follows:

Civil structure/machine monitoring: Continuous monitoring of civil structures such as bridges or towers yields valuable insight into their behavior under varying seismic activity. By examining moisture content and temperature, it is possible to estimate the maturity of concrete or a corrosive subsurface in structural components before serious damage occurs. Similar principles apply to underground pipes and drainage tiles. Another relevant application is monitoring the health of machines. Thousands of sensor nodes can track vibrations coming from various pieces of equipment to determine if the machines are about to fail.

Traffic monitoring on roadways: Sensors can be deployed on roads and highways to monitor the traffic or road conditions to enable quick notification of drivers about congestion or unfavorable road conditions so that they can select an alternate route.

Intrusion detection: Several sensor nodes can be deployed in buildings at all potential entrances and exits of the building to monitor movement of personnel and any unusual or suspicious activity.

Heat control and conditioning: Precise temperature control is very crucial for many industries, such as oil refineries and food industries. This can be achieved by placing a large number of sensors at specific places in the civil structure (such as pipes).

Intel Research Oregon[26] has developed a heterogeneous network to improve the scalability of wireless sensor networks for various monitoring applications. Intel overlays a 802.11 mesh network on a sensor network analogous to a highway overlaid on a roadway system. Data is collected from local sensor nodes and transmitted across the network through the faster, more reliable 802.11 network. The network lifetime is therefore enhanced by offloading the communication overhead to the high-end 802.11 nodes. This network is capable of self reorganization in case any 802.11 node fails. Similarly, we propose to offload computation-intensive tasks from low-power sensor motes to high-performance nodes.

A few important inferences that can be drawn about infrastructure-based applications are as follows:

- The layout (blueprint) of the infrastructure (like a machine or a freeway) where the sensor nodes are to be placed is known.
- As nodes may have been manually placed or embedded on the structure at monitoring points while manufacturing it, the location of sensors may be known to the user.
- It might be possible to provide a renewable source of power to some of the nodes.
- The communication topology should adapt to the physical limitations (shape or surface area where nodes are placed) of underlying infrastructure.

Field-based monitoring

To perform field-based monitoring, the sensor nodes are randomly dispersed in large numbers to form a dense network to ensure sensor coverage of the environment to be monitored. The purpose is to observe physical phenomena spread over a large region, such as pollution levels, the presence of chemicals, and temperature levels. An example query for field monitoring would be: "Report the direction of movement of a cloud of smoke originating at location (x,y)." The purpose is to monitor the general level

of physical parameters being observed—unlike infrastructure-based monitoring, which is usually applied for high-precision monitoring. The use of a heterogeneous network for field monitoring is not very obvious because node placement in the field usually cannot be controlled. But, at the expense of increasing the cost of deployment, the density of high-performance nodes can be increased to ensure that most of the low-power nodes in the network can access at least one high-performance node. For example, in a greenhouse, the same plant is grown in varying soil or atmospheric conditions, and the growth or health of the plant is monitored to determine the factors that promote its growth. The different soil beds can be considered the different target regions, and their sensor data is compared or combined with each other within the field to derive useful inferences to be sent to a remote server. Some other applications that might benefit from such an automated system of data monitoring are as follows.

Scientific experiments: Sensors can be randomly deployed in closed chambers in laboratories, or natural spaces such as caves or mines, to study the presence of certain gases, elements, or chemicals. Similarly, levels of radioactive materials can be observed to monitor toxic spills.

Examination of contaminant level and flow: The sensor nodes with chemical sensing capabilities can be used to monitor the levels and flow patterns of contaminants in the environment.

Habitat or ecosystem monitoring: Studying the behavior of birds, plants, and animals in their natural habitats using sensor networks has been employed on a small scale.[27]

Wild fire monitoring: This is particularly useful in controlling fires in forests by studying the variation in temperature over the areas affected, and the surrounding habitat.

REFERENCES

1. Broch, J.; Maltz, D.; Johnson, D.; Hu, Y.; Jetcheva, J. A performance comparison of multi-hop wireless ad hoc network routing protocols. 4th Annual ACM/IEEE Int. Conf. on Mobile Computing (MOBICOM), Dallas, Texas, Oct 25–30, 1998.

2. Royer, E.M.; Toh, C.-K. A review of current routing protocols for ad-hoc mobile wireless networks. IEEE Pers. Commun. Mag. **1999**, 6 (2), 46–55.

3. Manjeshwar, A.; Agrawal, D.P. TEEN: a routing protocol for enhanced efficiency in wireless sensor networks. 15th Int. Parallel and Distributed Processing Symposium (IPDPS 2001) Workshops, San Francisco, CA, April 23–27, 2001.

4. Manjeshwar, A.; Agrawal, D.P. APTEEN: a hybrid protocol for efficient routing and comprehensive information

retrieval in wireless sensor networks. Int. Parallel and Distributed Processing Symposium (IPDPS 2002) Workshops, Fort Lauderdale, FL, April 15–19, 2002.

5. Nelson, J.; Estrin, D. An Address-Free Architecture for Dynamic Sensor Networks. Technical Report 00–724, Computer Science Department, University of Southern California, January 2000.

6. Estrin, D.; Govindan, R.; Heidemann, J.; Kumar, S. Next century challenges: scalable coordination in wireless networks. 5th Annual ACM/IEEE Int. Conf. on Mobile Computing and Networking (MOBICOM), Seattle, WA, Aug 15–20, 1999; 263–270.

7. Heinzelman, W.; Chandrakasan, A.; Balakrishnan, H. Energy-efficient communication protocols for wireless micro-sensor networks. Hawaaian Int. Conf. on Systems Science, Wailea Maui, HI, Jan 4–7, 2000.

8. Intanagonwiwat, C.; Govindan, R.; Estrin, D. Directed diffusion: a scalable and robust communication paradigm for sensor networks. 6th Annual ACM/IEEE Int. Conf. on Mobile Computing and Networking (MOBICOM 2000), Boston, MA, Aug 6–11, 2000; 56–67.

9. Chang, J.; Tassiulas, L. Energy conserving routing in wireless ad-hoc networks. IEEE INFOCOM, pp. 22–31, 2000a.

10. Chang, J.; Tassiulas, L. Maximum lifetime routing in wireless sensor networks. Advanced Telecommunications and Information Distribution Research Program, 2000b.

11. Jain, N.; Madathil, D.K.; Agrawal, D.P. Energy aware multi-path routing for uniform resource utilization in sensor networks. IPSN 2003 Int. Workshop on Information Processing in Sensor Networks, Palo Alto, CA, Apr 22–23, 2003a.

12. Jain, N.; Madathil, D.K.; Agrawal, D.P. Exploiting multi-path routing to achieve service differentiation in sensor networks. 11th IEEE Int. Conf. on Networks (ICON 2003), Sydney, Australia, Sept 28–Oct 1, 2003b.

13. Hill, J.; Szewczyk, R.; Woo, A.; Hollar, S.; Culler, D.; Pister, K. System Architecture Directions for Network Sensors. 9th Int. Conf. on Architectural Support for Programming Languages and Operating Systems, Cambridge, MA, Nov 12–15, 2000; 93–104.

14. Ganesan, D.; Estrin, D.; Heidemann, J. DIMENSIONS: why do we need a new data handling architecture for Sensor Networks?. ACM Workshop on Hot Topics in Networks, Princeton, NJ, Oct 28–29, 2002a.

15. Lee, S.; Campbell, A.T. INSIGNIA: in-band signaling support for QOS in mobile ad hoc networks. 5th Int. Workshop on Mobile Multimedia Communications(MoMuC 1998), Berlin, Germany, Oct 13–16, 1998.

16. Vutukury, S. Multipath Routing Mechanisms for Traffic Engineering and Quality of Service in the Internet; Mar 2001 Ph.D. thesis.

17. Bhatnagar, S.; Deb, B.; Nath, B. Service differentiation in sensor networks. 4th Int. Symposium Wireless Personal Multimedia Communications, Aalborg, Denmark, Sept 9–12, 2001.

18. Chen, J.; Druschel, P.; Subramanian, D. An efficient multi-path forwarding method. IEEE INFOCOM, San Francisco, CA, Mar 29–Apr 2, 1998; 1418–1425.

19. HighTower, J.; Borreillo, G. Location systems for ubiquitous computing. IEEE Comput. 2001, 34, 57–66.

20. Doherty, L.; Pister, K.S.J.; Ghaoui, L.E. Convex position estimation in wireless sensor networks. IEEE INFOCOM, Anchorage, AK, Apr 22–26, 2001; 1655–1663.

21. Babcock, B.; Babu, S.; Datar, M.; Motwani, R.; Widom, J. Models and issues in data stream systems. 2002 ACM Symposium on Principles of Database Systems, Madison, WI, June 3–5, 2002; 1–16.

22. Datar, M.; Gionis, A.; Indyk, P.; Motwani, R. Maintaining stream statistics over sliding windows. 13th Annu. ACM-SIAM Symp. on Discrete Algorithms, San Francisco, CA, Jan 6–8, 2002; 635–644.

23. Motwani, R.; Window, J.; Arasu, A.; Babcock, B.; Babu, S.; Data, M.; Olston, C.; Rosenstein, J.; Varma, R. Query processing, approximation and resource management in a data stream management system. First Annu. Conf. on Innovative Database Research (CIDR), Asilomar, CA, Jan 5–8, 2003.

24. Carney, D.; Cetintemel, U.; Cherniack, M.; Convey, C.; Lee, S.; Seidman, G.; Stonebraker, M.; Tatbul, N.; Zdonik, S.B. Monitoring streams—a new class of data management applications. 28th VLDB, Hong Kong, China, Aug 20–23, 2002.

25. Madden, S.; Shah, M.; Hellerstein, J.M.; Raman, V. Continuously adaptive continuous queries over streams. ACM SIGMOD Int. Conf. on Management of Data, Madison, WI, June 3–6, 2002; 49–60.

26. Intel Research Oregon. Heterogeneous Sensor Networks. Technical Report, Intel Corporation, 2003 www.intel.com/research/exploratory/heterogeneous.htm.

27. Mainwaring, A.; Polastre, J.; Szewczyk, R.; Culler, D. Wireless sensor networks for habitat monitoring. ACM Workshop on Sensor Networks and Applications, Atlanta, GA, Sept 28, 2002.

Wireless Sensor Networks (WSNs): Secure Localization

Avinash Srinivasan
Jie Wu
Department of Computer Science and Engineering, Florida Atlantic University, Boca Raton, Florida, U.S.A.

Abstract

Localization is the process by which an object determines its spatial coordinates in a given field.

INTRODUCTION

Wireless sensor networks (WSNs) are shaping many activities in our society, as they have become the epitome of pervasive technology. WSNs have an endless array of potential applications in both military and civilian applications, including robotic land-mine detection, battlefield surveillance, target tracking, environmental monitoring, wildfire detection, and traffic regulation, to name just a few. One common feature shared by all of these critical applications is the vitality of sensor location. The core function of a WSN is to detect and report events which can be meaningfully assimilated and responded to only if the accurate location of the event is known. Also, in any WSN, the location information of nodes plays a vital role in understanding the application context. There are three visible advantages of knowing the location information of sensor nodes. First, location information is needed to identify the location of an event of interest. For instance, the location of an intruder, the location of a fire, or the location of enemy tanks in a battlefield is of critical importance for deploying rescue and relief troops. Second, location awareness facilitates numerous application services, such as location directory services that provide doctors with the information of nearby medical equipment and personnel in a smart hospital, target-tracking applications for locating survivors in debris, or enemy tanks in a battlefield. Third, location information can assist in various system functionalities, such as geographical routing,[1–6] network coverage checking,[7] and location-based information querying.[8] Hence, with these advantages and much more, it is but natural for location-aware sensor devices to become the defacto standard in WSNs in all application domains that provide location-based service.

A straightforward solution is to equip each sensor with a GPS receiver that can accurately provide the sensors with their exact location. This, however, is not a feasible solution from an economic perspective since sensors are often deployed in very large numbers and manual configuration is too cumbersome and hence not feasible. Therefore, localization in sensor networks is very challenging. Over the years, many protocols have been devised to enable the location discovery process in WSNs to be autonomous and able to function independently of GPS and other manual techniques.[9–14] In all these literatures, the focal point of location discovery has been a set of specialty nodes known as *beacon nodes*, which have been referred to by some researchers as anchor, locator, or seed nodes. However, in this entry we shall use the term beacon node without loss of generality. These beacon nodes know their location, either through a GPS receiver or through manual configuration, which they provide to other sensor nodes. Using this location of beacon nodes, sensor nodes compute their location using various techniques discussed in the "Overview of Localization Process" section. It is, therefore, critical that malicious beacon nodes be prevented from providing false location information since sensor nodes completely rely on the information provided to them for computing their location.

There are three important metrics associated with localization: *energy efficiency*, *accuracy*, and *security*. Though the first two metrics have been researched extensively, the security metric has drawn the attention of researchers only recently, and as such has not been addressed adequately. As security is a key metric, we are motivated to survey the existing techniques focusing on secure localization. This entry, in which we review secure localization techniques that have been featured in literature thus far, is intended to be a single point of reference for researchers interested in secure localization.

The rest of the sections in this entry discuss the unique operational challenges in WSNs, the localization process and the security requirements, the classification of localization techniques, the attacker model and attacks that are specific to localization, and the existing secure localization models.

Encyclopedia of Wireless and Mobile Communications DOI: 10.1081/E-EWMC-120043801

OPERATIONAL CHALLENGES IN WSNs

WSNs, unlike their wired counterparts, are often deployed to operate in unattended and hostile environments, rarely encountered by typical computing devices: rain, snow, humidity, and high temperature. When used for military applications like land-mine detection, battlefield surveillance, or target tracking, the conditions further deteriorate. In such unique operational environments, WSNs have to operate autonomously and consequently are faced with unique challenges. An adversary can now capture and compromise one or more sensors physically. Once captured, a node is at the mercy of the adversary. The adversary can now tamper with the sensor node by injecting malicious code, forcing the node to malfunction, extracting the cryptographic information held by the node to bypass security hurdles like authentication and verification, so on and so forth. Now, the adversary can launch attacks from within the system as an insider, and most existing systems would fail in the face of such inside attacks.

For instance, consider a beacon-based localization model. Since sensor nodes are not capable of determining their own location, they have no way of determining which beacon nodes are being truthful in providing accurate location information. There could be malicious beacon nodes that give false location information to sensor nodes compelling them to compute incorrect location. This situation, in which one entity has more information than the other, is referred to as *information asymmetry*. The information asymmetry in beacon-based localization models has been addressed in Ref. 15. Ref. 15 also presents an effective way of resolving insider attacks. The attacker can also launch sybil, worm hole, or replay attacks to disrupt the localization process.

OVERVIEW OF LOCALIZATION PROCESS

Localization is the process by which sensor nodes determine their location. In simple terms, localization is a mechanism for discovering spatial relationships between objects. The various approaches taken in literature to solve this localization problem differ in the assumptions they make about their respective network and sensor capabilities. A detailed, but not exhaustive, list of assumptions made include assumptions about device hardware, signal propagation models, timing and energy requirements, composition of network viz homogeneous vs. heterogeneous, operational environment viz indoor vs. outdoor, beacon density, time synchronization, communication costs, error requirements, and node mobility.[14] In node mobility four different scenarios arise. First, both sensor and beacon nodes ate static. Second, sensor nodes are static while beacon nodes move. Third, sensor nodes move while beacon nodes are static. Fourth, both sensor and beacon nodes move.

In localization models that use GPS as the source, the localization process is straightforward. However, in a localization model that uses beacon nodes to help sensor nodes with location discovery, the beacon nodes are either manually configured with their location or equipped with a GPS receiver which they can use to determine their location. Beacon nodes then provide their location information to sensor nodes and help them in computing their location. The idea of beacon-based localization is presented in Fig. 1. The localization process itself can be classified into two stages. In the first stage, a node merely estimates its distance to other nodes in its vicinity using one or more features of the received signal. In the second stage, a node uses all the distance estimates to compute its actual location. The method employed in stage two to compute the actual location depends on the signal feature used in stage one, and can be classified into three main groups as follows. (Fig. 2).

- Triangulation: A large number of localization algorithms fall into this class. In simple terms, the triangulation method involves gathering angle of arrival (AoA) measurements at the sensor node from at least three sources. Then using the AoA references, simple geometric relationships and properties are applied to compute the location of the sensor node.
- Trilateration: Trilateration is a method of determining the relative positions of objects using the geometry of triangles similar to triangulation. Unlike triangulation, which uses AoA measurements to calculate a subject's location, trilateration involves gathering a number of reference tuples of the form (x, y, d). In this tuple, d represents an estimated distance between the source providing the location reference from (x, y) and the sensor node. To accurately and uniquely determine the relative location of a point on a 2-D plane using trilateration, a minimum of three reference points are needed.
- Multilateration: Multilateration is the process of localization by solving for the mathematical intersection of multiple hyperbolas based on the time difference of

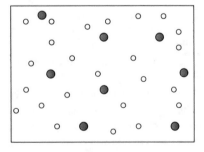

Fig. 1 A network with sensor and beacon nodes. Sensor nodes are represented by hollow circles and beacon nodes are represented by shaded circles.

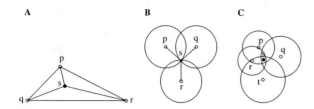

Fig. 2 (A) Triangulation; (B) trilateration; (C) multilateration.

arrival (TDoA). In multilateration, the TDoA of a signal emitted from the object to three or more receivers is computed accurately with tightly synchronized clocks. When N receivers are used, it results in $N-1$ hyperbolas, the intersection of which uniquely positions the object in a 3D space. When a large number of receivers are used, $N > 4$, then the localization problem can be posed as an optimization problem that can be solved using, among others, a least squares method.

Secure localization in sensor networks has become a major focus of research in recent years. Like any other process, localization also has security requirements, which are listed below. The breach of any of these security requirements is a harbinger of compromise in the localization process.

1. Authentication: Information for localization must be provided only by authorized sources. Therefore, before accepting location-related information, the provider has to be authenticated.
2. Integrity: The information provided by the source should be untampered before the sensor nodes can use it to discover their location.
3. Availability: All the information required by a sensor node to compute its location must be available when needed.
4. Non-Repudiation: Neither the source that provides the location information nor the sensor nodes that receive the location information should be able to deny the information exchange at a later time.
5. Privacy: Location privacy is one of the most important security requirements. The source should only help the sensor node in determining its location. Neither the source's location nor the sensor node's location should be disclosed at any point. This constraint helps to prevent malicious nodes from claiming a different legitimate location in the network.

Error in the estimated location of a sensor can be classified into two groups: intrinsic and extrinsic.[16] Intrinsic errors are most often caused by abnormalities in the sensor hardware and software, and can cause many complications when estimating node positions. On the other hand, extrinsic errors are attributed to the physical effects on the measurement channel. This includes

shadowing effects, changes in signal propagation speed, obstacles, etc. Extrinsic errors are more unpredictable and harder to handle. Measurement errors can significantly amplify the error in position estimates. Also, use of lower-precision measurement technology combined with higher uncertainty of beacon locations will augment errors in position estimates.

CLASSIFICATION OF LOCALIZATION TECHNIQUES

In this section, we shall classify localization techniques and discuss their merits and demerits.

Direct approaches. This is also known as *absolute localization.* The direct approach itself can be classified into two types: *Manual configuration* and *GPS-based localization.* The manual configuration method is very cumbersome and expensive. It is neither practical nor scalable for large scale WSNs and in particular, does not adapt well for WSNs with node mobility. On the other hand, in the GPS-based localization method, each sensor is equipped with a GPS receiver. This method adapts well for WSNs with node mobility. However, there is a downside to this method. It is not economically feasible to equip each sensor with a GPS receiver since WSNs are deployed with hundreds of thousands of sensors. This also increases the size of each sensor, rendering them unfit for pervasive environments. Also, the GPS receivers work well only outdoors on earth and have line-of-sight requirement constraints. Such WSNs cannot be used for underwater applications like habitat monitoring, water pollution level monitoring, tsunami monitoring, etc.

Indirect approaches. The indirect approach of localization is also known as *relative localization* since nodes position themselves relative to other nodes in their vicinity. The indirect approaches of localization were introduced to overcome some of the drawbacks of the GPS-based direct localization techniques while retaining some of its advantages, like accuracy of localization. In this approach, a small subset of nodes in the network, called the *beacon nodes*, are either equipped with GPS receivers to compute their location or are manually configured with their location. These beacon nodes then send beams of signals providing their location to all sensor nodes in their vicinity that don't have a GPS receiver. Using the transmitted signal containing the location information, sensor nodes compute their location. This approach effectively reduces the overhead introduced by the GPS-based method. However, since the beacon nodes are also operating in the same hostile environment as the sensor nodes, they too are vulnerable to various threats, including physical capture by adversaries. This introduces new security threats concerning the honesty of the beacon nodes in providing location information since they could have been tampered by the adversary and could misbehave by providing incorrect location information. This particular problem has been addressed well in Ref. 15 where a reputation- and trust-based system is used to monitor such misbehavior.

Within the indirect approach, the localization process can be classified into the following two categories.

1. Range-based: In range-based localization, the location of a node is computed relative to other nodes in its vicinity. Range-based localization depends on the assumption that the absolute distance between a sender and a receiver can be estimated by one or more features of the communication signal from the sender to the receiver. The accuracy of such an estimation, however, is subject to the transmission medium and surrounding environment. Range-based techniques usually rely on complex hardware which is not feasible for WSNs since sensor nodes are highly resource-constrained and have to be produced at throwaway prices as they are deployed in large numbers. Refs. 10–12 and 17–20 are some examples of range-based localization techniques. The features of the communication signal that are frequently used in literature for range-based localization are as follows:

a. AoA: Range information is obtained by estimating and mapping relative angles between neighbors. Refs. 12 and 19 make use of AoA for localization.

b. Received signal strength indicator (RSSI): Use a theoretical or empirical model to translate signal strength into distance. RADAR[21] is one of the first to make use of RSSI. RSSI has also been employed for range estimation in Refs. 18, 22, and 23.

c. Time of arrival (ToA): To obtain range information using ToA, the signal propagation time from souce to destination is measured. A GPS is the most basic example that uses ToA. To use ToA for range estimation, a system needs to be synchronous, which necessitates use of expensive hardware for precise clock synchronization with the satellite. ToA is used in Refs. 24 and 18 for localization.

d. TDoA: To obtain the range information using TDoA, an ultrasound is used to estimate the distance between the node and the source. Like ToA, TDoA necessitates the use of special hardware, rendering it too expensive for WSNs. Refs. 11, 17, 25, and 26 are some localization techniques that make use of TDoA.

Range-free: Range-free localization never tries to estimate the absolute point-to-point distance based on received signal strength or other features of the received communication signal like time, angle, etc. This greatly simplifies the design of hardware, making range-free methods very appealing and a cost-effective alternative for localization in WSNs. Amorphous localization,[27] Centroid localization,[9] APIT,[14] DV-Hop localization,[28]

secure range-independent localization (SeRLoc),[29] and ROCRSSI[30] are some examples of range-free localization techniques. Range-free techniques have also been employed in Ref. 31.

ATTACKER MODEL

Before reviewing the existing secure localization models, we feel it is necessary to analyze the attacker model to understand the attacker's capabilities. The attacker can be either an insider or an outsider. As an insider, the attacker has access to all of the cryptographic keying material held by a node. This is potentially dangerous since the attacker can now claim to be a legitimate part of the network. Authentication or verification via password and other mechanisms give way under this attacker model. On the other hand, in the outsider attack model, the attacker is outside the network and has no information about cryptographic keys and passwords necessary for authentication. The attacker can only capture a node but cannot extract the sensitive information. This model is comparatively less detrimental, but harmful nonetheless. So, for a localization process to be secure it has to be robust in its defense against both outsider and insider attacks. Some attacks that have been discussed for nearly a decade in literature and that are the most common against localization schemes are as follows:

• Replay Attack: A replay attack is the easiest and most commonly used form, by attackers. Specifically, when an attacker's capability is limited, i.e., the attacker cannot compromise more than 1 node, this is the most preferred attack. In a replay attack, the attacker merely jams the transmission between a sender and a receiver and later replays the same message, posing as the sender. The other way to launch a replay attack is, as shown in Fig. 3A: malicious node A re-transmits to node C the message it receives from node B. A replay attack has a two-fold consequence. First, the attacker is replaying the message of another node. Second, the attacker is transmitting stale information. In particular, the chances of the information being stale is higher in networks with higher node mobility. When replay attacks are launched on the localization process, a localizing node will receive an incorrect reference thereby localizing incorrectly. Unlike a wormhole attack, a single node can disrupt the network with a replay attack.

• Sybil Attack: The sybil attack requires a more sophisticated attacker compared to the replay attack. In a sybil attack, a node claims multiple identities in the network. When launched on localization, localizing nodes can receive multiple location references from a single node leading to incorrect location estimation. Like the replay attack, the sybil attack also can be launched by a single

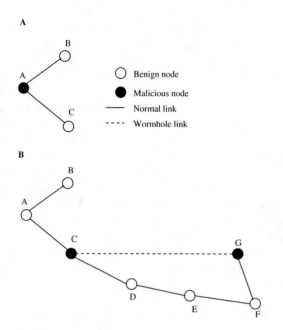

Fig. 3 (A) Replay attack example. Node *A* replays to *C* the message it receives from *B*, (B) wormhole attack example. Nodes *C* and *G* have a wormhole link.

node since there is no need for collusion among nodes to launch this attack.

- Wormhole Attack: A wormhole attack is the most complicated of all the mentioned attacks. To launch a wormhole attack, the attacker has to compromise at least two nodes. In a wormhole attack, the colluding nodes in the network tunnel messages transmitted in one part of the network to their colluding partners in other parts of the network. The effect of a wormhole attack on localization is depicted in Fig. 3B. Here, node *A* is sending its reference to nodes *B* and *C*. However, since there is a wormhole link between *C* and *G*, *G* can locally replay the location reference of *A* in its neighborhood, misleading node *F*. Consequently, *F* will compute its location incorrectly. Intuitively, wormhole attacks pose more serious problems in range-free localization compared to range-based localization.

EXISTING SECURE LOCALIZATION SYSTEMS

In this section we review the existing secure localization techniques, throwing light on their strengths and weaknesses.

SeRLoc

In Ref. 29, Lazos and Poovendran propose a novel scheme for localization of nodes in WSNs in untrusted

environments called SeRLoc. SeRloc is a range-free, distributed, resource-efficient localization technique in which there is no communication requirement between nodes for location discovery. SeRLoc is robust against wormhole attacks, sybil attacks and sensor compromise. SeRLoc considers two sets of nodes: N, which is the set of sensor nodes equipped with omnidirectional antennas, and L, which is the set of locator nodes equipped with directional antennas. The sensors determine their location based on the location information transmitted by these locators. Each locator transmits different beacons at each antenna sector with each beacon containing two pieces of information: the locator coordinates and the angles of the antenna boundary lines with respect to a common global axis. Using directional antennas improves the localization accuracy.

In SeRLoc, an attacker has to impersonate several beacon nodes to compromise the localization process. Also, since sensor nodes compute their own location without any assistance from other sensors, the adversary has no incentive to impersonate sensor nodes. Wormhole attacks are thwarted in SeRLoc due to two unique properties: sector uniqueness property and communication range violation property. In SeRLoc, to improve the localization accuracy, either more locators have to be deployed or more directional antennas have to be used. The authors also make an assumption that no jamming of the wireless medium is feasible. This is a very strong assumption for a real world setting.

Beacon Suite

In Ref. 20, Liu, Ning, and Du present a suite of techniques for detecting malicious beacon nodes that provide incorrect information to sensor nodes providing location services in critical applications. Their suite includes detection of malicious beacon signals, detection of replayed beacon signals, identification of malicious beacon nodes, avoidance of false detection, and finally the revoking of malicious beacon nodes. They use beacon nodes for two purposes: to provide location information to sensor nodes, and to perform detection on the beacon signals it hears from other beacon nodes. A beacon node does not necessarily need to wait passively to hear beacon signals. It can request location information. The beacon node performing the detection is called the *detecting node* and the beacon node being detected is called the *target node*. They suggest that the detecting node should use a non-beacon ID when requesting location information from a target node in order to observe the true behavior of the target node.

Their revocation scheme works on the basis of two counters maintained for each beacon node, namely *alert counter* and *report counter*. The alert counter records the suspiciousness of the corresponding beacon node, and the report counter records the number of alerts this node reported and was accepted by the base station. When a detecting node determines that a target node is misbehaving, it reports to the base station. Alert reports are accepted only from detecting nodes whose report counter is below a threshold and against nodes that are not yet revoked. When this criteria is met, the report counter and the alert counter of the detecting and the target node, respectively, are incremented. These two counters work on a discrete scale and the revocation mechanism is centralized. This has been improved to be more robust in Ref. 15 by employing a continuous scale and a reputation- and trust-based mechanism.

Attack-resistant Location Estimation

In Ref. 32, Liu, Ning, and Du put forward two range-based robust methods to survive malicious attacks against beacon-based location discovery in sensor networks. The first method, attack-resistant Minimum Mean Square Estimation, filters out malicious beacon signals. This is accomplished by examining the inconsistency among location references of different beacon signals, indicated by the mean square error of estimation, and defeats malicious attacks by removing such malicious data. The second method voting-based location estimation quantizes the deployment field into a grid of cells and has each location reference "vote" on the cells in which the node may reside. This method tolerates malicious beacon signals by adopting an iteratively refined voting scheme. Both methods survive malicious attacks even if the attacks bypass authentication.

However, there is a downside to both of these techniques. In the proposed localization technique, an attacker cannot dislodge sensors by compromising a few range estimates. Nonetheless, this localization model fails if the attacker can compromise a simple majority of range estimates. Assume there are k nodes in a neighborhood. Now, if the attacker can compromise $\lfloor k/2 \rfloor + 1$ beacon nodes in that neighborhood, he can generate more malicious location references than benign ones. This will lead to failure of the minimum mean square estimation technique in the neighborhood, the effects of which can propagate throughout the network. Similar attacks are possible for the voting-based location estimation technique.

Robust Statistical Methods

In Ref. 33, Li, Trappe, Zhang, and Nath introduced the idea of being tolerant to attacks rather than trying to eliminate them by exploiting redundancies at various levels within wireless networks. They examine two classes of localization: *triangulation* and *RF-based fingerprinting*. They have presented two statistical methods for securing localization in sensor networks. Both methods are based on the simple idea of filtering out outliers in the range estimates used for location estimation used by sensors.

For the *triangulation*-based localization, they propose to use an adaptive least squares and least median squares estimator. This adaptive estimator switches to the robust mode with least mean squares estimation when attacked, and enjoys the computational advantage of least squares in the absence of attacks. For the *fingerprinting*-based method, the traditional Euclidean distance metric is not secure enough. Hence, they propose a median-based nearest neighbor scheme i.e., robust to location attacks. In this entry, the authors have also discussed attacks that are unique to localization in sensor networks. The statistical methods proposed in Ref. 33 are based on the assumption that benign observations at a sensor always outnumber malicious observations. This is a strong assumption in a real word setting where an attacker can launch sybil attacks or even wormhole attacks to outnumber the benign observations.

Secure Positioning in Sensor Networks

In Ref. 34, Capkun and Hubaux devise secure positioning in sensor networks (SPINE), a range-based positioning system based on verifiable multilateration which enables secure computation and verification of the positions of mobile devices in the presence of attackers. SPINE works by bounding the distance of each sensor to at least three reference points. Verifiable multilateration relies on the property of distance bounding, that neither the attacker nor the claimant can reduce the measured distance of the claimant to the verifier, but only enlarge it. By using timers with nanosecond precision, each sensor can bound its distance to any reference point within range.

If the sensor is within a triangle formed by three reference points, it can compute its position via verifiable multilateration, which provides a robust position estimate. This is based on a strong assumption that no attacker colludes with compromised nodes. Verifiable multilateration effectively prevents location spoofing attacks, wormhole and jamming attacks, and prevents dishonest nodes from lying about their positions. However, SPINE has some drawbacks. In order to perform verifiable multilateration, a high number of reference points is required. SPINE is a centralized approach which creates bottleneck at the central authority or the base station. Also, it is very unlikely that an attacker will not try to collude with other compromised nodes.

Robust Position Estimation

Lazos, Poovendran, and Capkun design robust position estimation (ROPE),[35] a robust positioning system in WSNs. ROPE, a hybrid algorithm, imparts a two-fold benefit to the system. First, it allows sensors to determine their location without any centralized computation. Second, ROPE provides a location verification mechanism by virtue of which the location claims of sensors can be verified prior to data collection. In ROPE, the network consists of two types of nodes: sensors and locators. Each sensor shares a pairwise key with every locator. Since the number of locators is less, it does not impose a large storage overhead on the sensors.

To measure the impact of attacks on ROPE, they introduce a novel metric called maximum spoofing impact. ROPE achieves a significantly lower Maximum Spoofing Impact while requiring the deployment of a significantly smaller number of reference points, compared to Ref. 34. ROPE was second only to Ref. 34 to propose a solution for jamming attacks. ROPE is also resilient to wormhole attacks and node impersonation attacks. The robustness of ROPE has been confirmed via analysis and simulation.

Transmission Range Variation

In Ref. 36, Anjum, Pandey, and Agrawal show a novel transmission-based secure localization technique for sensor networks. They have presented a secure localization algorithm (SLA). Their technique does not demand any special hardware and considers a network with two sets of nodes: the sensor nodes and the beacon nodes. Their scheme works as follows. Beacon nodes associate unique nonce to different power levels at a given time which they transmit securely at the associated power level. As a result, each sensor node receives a set of unique nonce which it will have to transmit back to the sink via the beacon nodes. Then the location of the sensor node can be estimated securely, based on this set of nonce. This is a centralized localization technique where the sink determines the location of the sensor node.

This model has a few drawbacks. The authors have not considered the collaboration of sensor nodes which is very crucial and has to be addressed to suit the real world scenario. They have also assumed that all beacon nodes in the network and the sink are to be trusted and assumed the encryption between beacon nodes and sink to be stronger than that between sensor nodes and sink. They have shown that their model is resilient to replay attacks, spoofing attacks, modification attacks and response delay attacks. Another major drawback arises from the fact that this a centralized model with the base station as the single point of failure. This also causes a significant bottle-neck at the base station.

Distributed Reputation-based Beacon Trust System

Distributed reputation-based beacon trust system (DRBTS)[15] is a distributed reputation and trust-based security protocol aimed at providing a method for secure localization in sensor networks. This work is an extension of Ref. 20. In this model, incorrect location information provided by malicious beacon nodes can be excluded during localization. This is achieved by enabling beacon nodes to monitor each other and provide information so that sensor nodes can choose whom to trust, based on a quorum voting approach. In order to trust a beacon node's information, a sensor must get votes for its trustworthiness from at least half of their common neighbors. Specifically, sensor nodes use a simple majority principle to evaluate the published reputation values of all the beacon nodes in their range.

With this model, it is clearly demonstrated that sensors can accurately guess the misbehaving/non-misbehaving status of any given beacon node, given a certain assumption about the level of corruption in the system. Authors also show that their system grows in robustness as node density increases, and show through simulations the effects of different system parameters on robustness. This distributed model not only alleviates the burden on the base station to a great extent, but also minimizes the damage caused by the malicious nodes by enabling sensor nodes to make a decision on which beacon neighbors to trust, on-the-fly, when computing their location.

High-resolution, Range-independent Localization

Lazos and Poovendran propose another model, a high-resolution, range-independent localization technique called HiRLoc.[37] In HiRLoc, sensors passively determine their location without any interaction amongst themselves. HiRLoc also eliminates the need for increased beacon node density and specialized hardware. It is robust to security threats like wormhole attacks, sybil attacks and compromising of the network entities by virtue of two special properties: antenna orientation variation and communication range variation. In HiRLoc, Lazos and Poovendran have used cryptographic primitives to ensure the security of beacon transmissions. Here, each beacon transmission is encrypted using a global symmetric key, an idea very similar to the one used in Ref. 15.

Unlike SeRLoc, in HiRLoc, sensors receive multiple beacons from the same locator. This relaxation helps in improving the accuracy of location estimation. There are two important observations. First, since no range measurements are required for localization, they are free from attacks aiming at altering the measurements, like jamming to increase hop count. Second, since sensors do not rely on other sensor nodes for computing their location, HiRLoc is robust to sensor compromise attacks.

SUMMARY

Sensor location is vital for many critical applications like battlefield surveillance, target-tracking, environmental monitoring, wildfire detection, and traffic regulation. Localization has three important metrics: *energy efficiency*, *accuracy*, and *security*. Though the first two metrics have drawn the attention of researchers for nearly a decade, the security metric has been addressed only recently. In this entry we have discussed the unique operational challenges faced by WSNs, presented a comprehensive overview of the localization process, and discussed the three localization techniques: triangulation, trilateration, and multilateration. We have also delineated the security requirements of localization, and discussed the merits and demerits of both range-based and range-free localization models that have been proposed as an effective alternative for GPS-based localization. The attacker model and attacks specific to localization have also been discussed in detail. Finally, we concluded the entry with a survey of all secure localization techniques proposed thus far. This entry is intended to serve as a single point of reference to researchers interested in secure localization in WSNs.

ACKNOWLEDGMENTS

We would like to thank the anonymous reviewers for their valuable feedback on the contents and organization of this entry. This work was supported in part by NSF grants, ANI 0073736, EIA 0130806, CCR 0329741, CNS 0422762, CNS 0434533, CNS 0531410, and CNS 0626240.

REFERENCES

1. Wellenhoff, B.H.; Lichtenegger, H.; Collins, J. *Global Positions System: Theory and Practice,* 4th Ed.; Springer Verlag, 1997.
2. Navas, J.C.; Imielinski, T. Geographic addressing and routing. MobiCom 1997, Budapest, Hungary, Sept 26–30, 1997.
3. Ko, Y.-B.; Vaidya, N.H. Location-aided routing (LAR) in mobile ad hoc networks. MobiCom 1998, Dallas, TX, Oct 25–30, 1998.
4. Harter, A.; Hopper, A.; Steggles, P.; Ward, A.; Webster, P. The anatomy of a context-aware application. MobiCom 1999, Seattle, WA, Aug 15–20, 1999.
5. Nagpal, R. Organizing a Global Coordinate System from Local Information on an Amorphous Computer. A.I. Memo1666, MIT A.I. Laboratory. **1999.**
6. Karp, B.; Kung, H.T. Greedy perimeter stateless routing. MobiCom 2000, Boston, MA, Aug 6–11, 2000.
7. Hightower, J.; Boriello, G.; Want, R. SpotON: An Indoor 3D Location Sensing Technology Based on RF Signal Strength. Technical Report 2000-02-02, University of Washington. **2000.**
8. Bulusu, N.; Heidemann, J.; Estrin, D. GPS-less low cost outdoor localization for very small devices. IEEE Pers. Commun. Mag. **2000,** *7* (5), 28–34.
9. Priyantha, N.B.; Chakraborty, A.; Balakrishnan, H. The cricket location-support system. MobiCom 2000, Boston, MA, Aug 6–11, 2000.
10. Bahl, P.; Padmanabhan, V.N. RADAR: an in-building RF-based user location and tracking system. the IEEE INFOCOM 2000, Tel-Aviv, Israel, Mar 26–30, 2000.
11. Mauve, M.; Widmer, J.; Hartenstein, H. A survey on position-based routing in mobile ad hoc networks. IEEE Netw. Mag. **2001,** *15* (6), 30–39.
12. Doherty, L.; Pister, K.S.; Ghaoui, L.E. Convex optimization methods for sensor node position estimation. IEEE INFOCOM 2001, Anchorage, AK, Apr 22–26, 2001.
13. Yu, Y.; Govindan, R.; Estrin, D. Geographical and energy aware routing: a recursive data dissemination protocol for wireless sensor networks. Technical Report UCLA/CSD-TR-01-0023, UCLA, Department of Computer Science. **2001.**
14. Savvides, A.; Han, C.; Srivastava, M. Dynamic fine-grained localization in ad-hoc networks of sensors. ACM MobiCom 2001, Rome, Italy, July 16–21, 2001; 166–179.
15. Xu, Y.; Heidemann, J.; Estrin, D. Geography-informed energy conservation for ad hoc routing. MobiCom 2001, Rome, Italy, July 16–21, 2001.
16. Niculescu, D.; Nath, B. Ad hoc positioning systems (APS). IEEE GLOBECOM 2001, San Antonio, TX, Nov 25–29, 2001.
17. Savarese, C.; Rabay, J.; Langendoen, K. Robust positioning algorithms for distributed ad-hoc wireless sensor networks. USENIX Technical Annual Conference, Monterey, CA, June 10–15, 2002..
18. Savvides, A.; Park, H.; Srivastava, M. The bits and flops of the n-hop multilateration primitive for node localization problems. ACM WSNA 2002, Atlanta, GA, Sept 23–28, 2002.
19. Nasipuri, A.; Li, K. A directionality based location discovery scheme for wireless sensor networks. ACM WSNA 2002, Atlanta, GA, Sept 23–28, 2002.
20. Yan, T.; He, T.; Stankovic, J.A. Differentiated Surveillance Service for Sensor Networks. 1st ACM Conference on Embedded Networked Sensor Systems (SenSys 2003), Los Angeles, CA, Nov 5–7, 2003.
21. Patwari, N.; Hero, A.O.; Perkins, M.; Correal, N.S.; ODea, R. J. Relative location estimation in wireless sensor networks. IEEE Trans. Signal Process. **2003,** *51* (8), 2137–2148.
22. Nagpal, R.; Shrobe, H.; Bachrach, J. Organizing a global coordinate system from local information on an ad hoc sensor network. 2nd International Workshop on Information Processing in Sensor Networks (IPSN 2003), Palo Alto, CA, Apr 22–23, 2003.
23. He, T.; Huang, C.; Blum, B.M.; Stankovic, J.A.; Abdelzaher, T.F. Range-free localization schemes in large scale sensor networks. ACM MobiCom 2003, San Diego, CA, Sept 14–19, 2003.
24. Niculescu, D.; Nath, B. DV based positioning in ad hoc networks. J. Telecommun. Syst. **2003,** *22* (1–4), 267–280.
25. Niculescu, D.; Nath, B. Ad hoc positioning system (APS) using AoA. IEEE INFOCOM 2003, San Francisco, CA, Mar 30–Apr 3, 2003.
26. Gupta, H.; Das, S.R.; Gu, Q. Connected sensor cover: self-organization of sensor networks for efficient query execution. MobiHoc 2003, Annapolis, Maryland, June 1–3, 2003.
27. Lazos, L.; Poovendran, R. SeRLoc: secure range independent localization for wireless sensor networks. ACM

Workshop on Wireless Security (ACM WiSe 2004), Philadelphia, PA, Oct 1, 2004.

28. Liu, C.; Wu, K. Sensor localization with ring overlapping based on comparison of received signal strength indicator. 1st IEEE International Conference on Mobile Ad-hoc and Sensor Systems (MASS 2004), Fort Lauderdale, Florida, Oct 24–27, 2004.

29. Liu, D.; Ning, P.; Du, W. Detecting malicious beacon nodes for secure location discovery in wireless sensor networks. 25th IEEE International Conference on Distributed Computing Systems (ICDCS 2005), Columbus, OH, June 6–10, 2005; 609–619.

30. Liu, D.; Ning, P.; Du, W. Attack-resistant location estimation in sensor networks. 4th International Conference on Information Processing in Sensor Networks (IPSN 2005), Los Angeles, CA, Apr 25–27, 2005; 99–106.

31. Li, Z.; Trappe, W.; Zhang, Y.; Nath, B. Robust statistical methods for securing wireless localization in sensor networks. IPSN 2005, Los Angeles, CA, Apr 25–27, 2005.

32. Capkun, S.; Hubaux, J.-P. Secure positioning of wireless devices with application to sensor networks. IEEE INFOCOM 2005, Miami, FL, Mar 13–17, 2005.

33. Lazos, L.; Poovendran, R.; Capkun, S. ROPE: robust position estimation in wireless sensor networks. the 4th International Symposium on Information Processing in Sensor Networks (IPSN 2005), Los Angeles, CA, Apr 25–27, 2005.

34. Anjum, F.; Pandey, S.; Agrawal, P. Secure localization in sensor networks using transmission range variation. 2nd IEEE International Conference on Mobile Adhoc and Sensor Systems (MASS 2005), Washington, DC, Nov 7–10, 2005; 195–203.

35. Savvides, A.; Garber, W.L.; Moses, R.L.; Srivastava, M.B. Analysis of error inducing parameters in multihop sensor node localization. IEEE Trans. Mob. Comput. 2005, 4 (6), .

36. Srinivasan, A.; Teitelbaum, J.; Wu, J. DRBTS: distributed reputation-based beacon trust system. 2nd IEEE International Symposium on Dependable, Autonomic and Secure Computing (DASC 2006), Indianapolis, IN, Sept 29–Oct 1, 2006; 277–283.

37. Lazos, L.; Poovendran, R. HiRLoc: high-resolution robust localization for wireless sensor networks. IEEE J. Sel. Areas Commun. 2006, 24 (2).

Wireless Sensor Networks (WSNs): Security

Lyes Khelladi
Basic Software Laboratory, CERIST Center of Research, Ben-aknoune, Algiers, Algeria

Nadjib Badache
Computer Systems Laboratory, Department of Computer Science, University of Science and Technology, Algiers, Algeria

Abstract

Security in wireless sensor networks refers to unique features of these emergent networks, including scarce energy resources, small memory, and low computation capability.

INTRODUCTION

Recent advances in micro-electro-mechanical systems (MEMSs) technology, wireless communications, and digital electronics have enabled the development of small, inexpensive, battery-powered micro-sensors (e.g., Berkeley/Crossbow Motes,[1] MIT μAMPS nodes[2]). These tiny devices, which consist of sensing and data processing components, can communicate untethered in short distances, through wireless channel, forming wireless sensor networks (WSNs). WSNs represent a significant improvement over the traditional sensors: the position of sensor nodes need not to be engineered or pre-determined. This allows random deployment over the sensing area, even in hostile or inaccessible terrains. Moreover, while most current sensing networks involve small numbers of sensors supported by centralized processing and analysis hardware, these new networks rely on the collaborative effort of a large number of micro-sensors, offering thus a better sensing granularity.

All the afore-described features ensure a wide range of potential applications for WSNs. Some of the application areas are health, military, and security.[3,4] For example, in a disaster setup, a large number of sensors can be dropped by a helicopter in order to assist rescue operations by locating survivors, identifying risky areas, and making the rescue crew more aware of the overall situation.

Early research efforts in WSNs have focused on the development of new network protocol stack, trying to meet some stringent performance requirements, like energy efficiency, auto-organization, and scalability to high number of nodes. However, most applications of sensor networks are mission-critical and face acute security concerns including forgery of sensor data, eavesdropping, denial of service attacks, and the physical compromise of sensor nodes.[5] Improper use of sensed information or using forged data may cause undesired information leakage and provide inaccurate results. This renders security as

important as other performance issues, so that we can make such networking technology viable.

WSNs are considered as a new class of ad hoc networks, and security in traditional ad hoc networks has already gained a great interest amongst research community. While the knowledge and experiences gained from prior works give us a head start in the quest to secure sensor networks, there exist several features, inherent to these emergent networks that need to be addressed.

First, sensor nodes have a *limited hardware resources* (Table 1), characterized by a scarce memory (static and volatile) and reduced CPU capability. These limitations pose tight constraints on both communication and computing capacity of the employed security mechanisms; they also make WSNs more susceptible to denial of service attacks.[5] Second, the sensor, being a micro-electronic device, can only be equipped with *limited batteries* (<0.5 Ah, 1.2 V). For the most application scenarios, replenishment of power resources might even be impossible.[4] In such a case, the adopted security mechanisms directly affect the system lifetime and must take the energy efficiency as a first design objective. Third, sensor nodes are *stand-alone* and operate in a *non-controlled environment*. As a result, they generally tend to face extreme operating circumstances like vibration, shocks, lightings, power supply fluctuations, user abuse, and so on. This close physical coupling with the operating environment necessitates the use of reliable mechanisms and introduces additional security consideration. And finally, several applications of sensor networks necessitate the dense deployment of thousands, even millions of sensor nodes.[4] The employed security algorithms should support highly localized operations that involve the neighboring nodes only in order to achieve the required scalability.

All the aforementioned unique features impose new challenges for security design in sensor networks; these challenges should gain a more significant attention among research community so that a ripe and effective security

Encyclopedia of Wireless and Mobile Communications DOI: 10.1081/E-EWMC-120043597

Table 1 Hardware limitations of sensor nodes.

	Smart dust[49]	EYES[50]	Mica2Dot[1]
CPU	8 bits, 4 MHz	16 bits, 1 MHz	8 bits, 8 MHz
Flash memory	8 KB	60 KB	128 KB
RAM	512 B	2 KB	4 KB
Bandwidth	10 Kbps	115 Kbps	10 Kbps

solutions can be achieved. In this entry, we show the impact of these new design challenges on the guarantee of the main security services in WSNs, namely cryptography, key management, secure routing, and secure data aggregation. We survey also, through a comprehensive study, the proposed solutions for each area.

The rest of the entry is organized as follows: first, we introduce the communication architecture of a WSN followed by the adopted security threat model. Then, we focus in more detail on the issues of cryptography in WSNs, key management, secure routing, and secure data aggregation. Finally, we summarize the insights gathered in this study and conclude with future research directions.

COMMUNICATION ARCHITECTURE OF WSNs

In a WSN, the sensor nodes are usually scattered in a sensor field as shown in Fig. 1. Each of these sensors has the capabilities to collect data and route back to the sink and the end user. Data are routed back to the end user by a multihop infrastructureless architecture through the sink as shown in Fig. 1. The sink may communicate with the end user via Internet or satellite.

The protocol stack used by the sink and all the sensor nodes are arranged in the following layers:

- *Physical layer*: responsible for carrier frequency generation, frequency selection, signal modulation, and deflection.

- *Data link layer*: responsible for multiplexing of data streams, medium access, data frame detection, as well as transmission error control and correction.
- *Network layer*: takes care of routing the data supplied by sensors to the sink.
- *Transport layer*: helps to maintain a reliable flow of data, if the network application requires it.
- *Application layer*: responsible for specifying how the data are requested and provided for both sensor nodes and end users.

Moreover, the protocol stack of a sensor node can integrate several cross-layer planes like task management, power, or security planes. The task management and power planes help the sensor nodes to coordinate the sensing tasks with an energy-efficient manner, while the security plan should integrate the necessary mechanisms that ensure security in each layer of the protocol stack (Fig. 1). Our study focuses on this plan, the essential services that it should ensure, and the different mechanisms that it can integrate.

THREAT MODEL

In this survey, we essentially distinguish between two types of attackers: mote-class attackers and laptop-class attackers.[6] In the former class, the attacker is of the same nature as sensor nodes, so it has access to some few sensors with the same capability. In contrast, a laptop-class

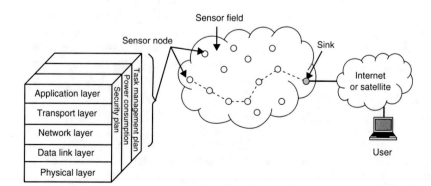

Fig. 1 Sensor network architecture.

attacker may have access to more powerful devices like laptops or their equivalent; it can perform more powerful attacks like jamming or eavesdropping the entire sensor network using its stronger reception power. The conducted attacks on a WSN can also be classified into the following types:

- Outsider vs. insider attacks: outside attacks are defined as an attack from nodes that do not belong to the WSN. In contrast, insider attacks occur when legitimate nodes of a WSN behave in unintended or unauthorized way.
- Passive vs. active attacks: a passive attack is a continuous collection of information that might be used later when launching an active attack. For that, the attacker eavesdrops packets and analyzes them to pick up required information. An active attack requires, however, the attacker to actively interact with the WSN by modifying the data stream for example.

CRYPTOGRAPHIC PRIMITIVES FOR WSNs

Cryptography is the cornerstone of almost all security solutions in sensor networks. Hence, selecting the most appropriate cryptographic methods is vital for the performance of the overall security solution. However, sensor nodes have very limited capabilities in term of available energy, memory capacity, and processing speed. The available range of capability pose tight constraints on cryptographic methods selection in sensor networks. This operation depends essentially on the power consumption factor, which is the first security design objective in WSNs. Indeed, the Moor's law confirms that the processing performance is growing exponentially, while only a linear increase is noticed for batteries' capacity (8% per year).[7] This fact mandates a special care to the energy efficiency factor among other performance concerns in order to reduce the gap between security computing requirements and batteries' supply of sensor nodes.

In this section, we discuss the energy efficiency of both classes of cryptographic primitives, namely symmetric and asymmetric ciphers.

Symmetric Ciphers

Symmetric encryption algorithms can be either stream or bloc ciphers. In the former class, the plaintext is encrypted 1 bit at time. This class is appropriate for applications where buffering is limited, or when characters must be individually processed as received. One of the well-known stream ciphers is rivest cipher 4 (RC4).[8] By contrast to the first class, block ciphers tend to simultaneously encrypt fixed-size blocs of the plaintext message using a fixed encryption transformation. Examples of bloc ciphers include 3DES,[9] AES,[10] and blowfish.[11] For the

encryption/decryption operations, both stream and bloc ciphers use one input secret key, which is expanded, during a key setup phase, in order to derive a distinct and cryptographically strong key for each round. A round denotes the repeated sequence of mathematical computations that form the overall cipher's algorithm.

Concerning the energy efficiency of symmetric ciphers, it is difficult or even impossible to characterize one algorithm as more energy efficient than others. However, based on former studies applied on sensor nodes or other embedded devices,[12–15] we can extract the following pertinent observations:

- Although stream ciphers like RC4 are known to be faster and more efficient compared to block ciphers, they present a more significant encryption/decryption energy cost. This is mainly due to the energy consumed in memory accesses resulting from buffering misses.
- Some secret key encryption algorithms can exhibit a great contrast in dissipated energy between the key setup and the encryption/decryption phases. For example, blowfish is one of the most optimal symmetric ciphers in term of energy consumed during encryption/decryption operations. However, it has a very high energy cost for the key setup phase. This phase encompasses complex operations involving 512 iterations and 4168 bytes of generated sub-keys. Such cryptographic algorithms could be the most appropriate for WSNs' applications that do not require frequent session establishment or those that leverage high volume of exchanged data.
- If we consider the overall cipher's functioning, including both key setup and plaintext encryption/decryption phases, AES appears as a judicious choice because of its competitive energy costs. Moreover, AES cryptanalytic properties have been well studied,[16–18] allowing it to combine energy efficiency with acceptable security level.

Finally, we mention that several benchmarking studies have been conducted to evaluate the performance of symmetric ciphers in WSNs. The work described in Ref. 15 provides a good resource for deciding about the performance of six bloc ciphers on sensor networks including RC5, RC6,[19] AES, MISTY1,[20] KASUMI,[21] and camellia.[22] However, we noticed that authors relied on the computational complexity of the algorithms to evaluate their energy consumption assuming that the energy consumption per CPU cycle is fixed.

Asymmetric Ciphers

Unlike symmetric ciphers, asymmetric cryptosystems use a pair of different, though related, public and private key to encrypt and decrypt plaintext. In these algorithms, guessing the secret key from the public one is equivalent to solve a computationally hard mathematical problem. For example, in RSA, the hardness of this operation is based on that of integer factorization; while in digital signature algorithm (DSA),[23] and Diffie–Hellman (DH),[24] it is based on that of discrete logarithm problem in integer fields. The encryption/decryption algorithms in asymmetric cryptosystems are also based on a more complex and harder to implement mathematical operations than in symmetric cryptography. For example, the basic operation in RSA, DSA, and DH is modular exponentiation.[8] The complexity of such operations has largely limited the utilization of this cipher class in sensor nodes, because they require more important processing capability and higher energy dissipation than symmetric ciphers. In fact, Carman et al.[25] pointed out that on a mid-range processor such as the Motorola MC68328 "DragonBall," the energy consumption for a 1024-bit RSA encryption (respectively signature) operation is much higher than that for a 1024-bit AES encryption operation; i.e., about 42 mJ (respectively 840 mJ) vs. 0.104 mJ. Authors reported also that symmetric-key ciphers and hash functions are between two and four orders of magnitude faster than digital signatures. It has been concluded then that these techniques become the systematic tools of choice for protecting WSNs.

Nevertheless, some works tried to revoke the thesis stated about the inefficiency of public-key cryptography in low-end embedded systems like WSNs. In fact, Gaubatz et al.[26] show, in the context of sensor networks that some low-complexity asymmetric cryptosystems like Rabin's scheme,[27] can be implemented in a low-power co-processor that handles all compute-intensive tasks in an energy-efficient manner. Authors have also pointed out that such low-power design solutions tremendously simplify the implementation of many typical security services using public-key cryptography. They invoke also that the adoption of public-key cryptography permit to reduce the protocol overhead induced by some operations in the symmetric encryption algorithms such as shared key agreement.

Further, certain emergent asymmetric ciphers are continuously gaining attention among the research community because of their low-complexity operations and their potential for offering an acceptable security level at a reduced key size. Elliptic curve cryptography (ECC)[28] is a prominent example of such cryptographic schemes. In fact, a 160-bit ECC key provides the same level of security as a 1024-bit RSA key. As a result, NIST recently approved ECC for use by the United States government.[23] Moreover, several standard organizations such as IEEE, ANSI, OMA, and the IETF have ongoing efforts to include ECC as a recommended security mechanism.

Wander et al.[29] presented a comparison between ECC and RSA cryptography on an Atmel Atmega128L microcontroller, which is used for mica sensors. For this, authors studied the energy dissipation of the two protocols implementation during mutual authentication and key exchange phases. The results showed that the use of ECC allows a significant energy savings compared to RSA. In addition to the computational benefits, authors noticed also that the smaller keys and certificates of ECC lead to considerable minimization in public-key communication costs. In addition, the implementation of RSA and ECC cryptography on mica2 motes proved that a public key-based protocol is very viable for WSNs security.

Overall, and besides the energy efficiency of each cryptographic primitive, we think that a tradeoff should always be considered between the energy consumption and the offered security level of the selected cryptosystem. In fact, certain cryptographic algorithms, especially symmetric ciphers, can provide different security levels, each with its associated energy consumption characteristics. In such cases, the provided security level can be adapted with the current state of the battery in order to achieve the best tradeoff between sensors' lifetime and the guaranteed security level. It has been demonstrated through experimental studies[14] that the tradeoff between security level and power consumption on handheld devices like a PDAs can be affected by several functional parameters, namely the operation mode, the key size, and the number of rounds.

KEY MANAGEMENT ISSUES

Although the "key management" is important for ensuring confidentiality and authentication, it still remains an unsolved problem in WSNs; this is mainly due to the following problems:

- Key pre-deployment: because of the unknown physical topology prior to installation of the sensor network, pre-deployment keying is considered as the only practical option that key distribution phase would have to rely on.[30] However, traditional key pre-deployment schemes are inadequate for WSNs, where the installed keys on each node are either a single mission key or a set of separate $n-1$ keys, each being pair-wise privately shared with another node. In fact, in the former case, the capture of any sensor node may compromise the entire network, because selective key revocation is impossible upon sensor capture detection, whereas the pair-wise key-sharing solution requires storage and loading of $n-1$ keys on each sensor node. This becomes impractical when using >10,000 sensors due to the aforementioned resources-limitations. Moreover, pair-wise private key sharing between any two sensor nodes would be unusable since direct node-to-node communication is achievable only in small node neighborhood.
- Shared key discovery: another challenging issue is that each node needs to discover a neighbor in wireless communication range with which it shares at least one key. Thus, a link exists between two sensor nodes only

if they share a secret key. A good, shared key discovery approach should not permit to an attacker to know shared keys between every two nodes.

- Path-key establishment: for any pair of nodes that do not share a key and are connected by multiple hops, they need to be assigned a path-key to guarantee end-to-end secure communication. It is important that path-key should be different from pair-wise shared keys of intermediary nodes.

In addition to the problems described earlier, key management in sensor networks face other important challenging issues such as building energy-efficient re-keying mechanism when available keys are expired as well as minimizing key establishment delays.

Solutions

To overcome shortcomings of traditional pre-deployed keying approaches, essentially the big size of loaded key ring on each node, several alternatives have been proposed in the literature.

For instance, the probabilistic key-sharing protocol described in Ref. 30 uses a shared key discovery approach that guarantees that every two nodes can, with a chosen probability, share one key while ensuring that only a small number of keys need to be loaded on each sensor node's key ring. This latter is drawn out randomly from a large pool of keys initially generated. However, this protocol suffers from several flaws because of its reliance on trusted controller nodes, where key identifiers of each key ring and corresponding sensor's IDs are saved. Moreover, the resilience of basic probabilistic key-sharing approach to node capture is weak, since only one key is shared between every two nodes, which renders easier for an attacker to break a greater number of secure links by capturing small number of nodes. In an attempt to enhance the resiliency to node capture, Chan et al. [31] have proposed q-composite approach where q keys ($q > 1$) need to be shared between nodes instead of just one. When small number of nodes is compromised, this approach has proved its efficiency compared to the basic scheme.

Another important problem in the described solutions is that a shared key can be located to multiple nodes and thus, is not exclusively known by only two nodes. Hence, this key cannot be used for encrypting any message that is private to only two nodes in the network. That is why Zhu et al. [32] have proposed to harden the basic probabilistic approach, with the use of threshold secret sharing to establish a pair-wise secret key, known exclusively by corresponding nodes.

So far, we can notice that several problems remain challenging. In fact, simulation results in Ref. 33 show that the shared key discovery process in all the discussed approaches cannot resist to the so-called smart attacker that can selectively attack some nodes, which allow him

to compromise the network faster, based on already obtained information (i.e., exchanged key IDs during key discovery steps). This threat model has been considered in Ref. 33, and a new mechanism for key discovery has been proposed. This latter is based on a pseudo-random key pre-deployment strategy that ensures a key discovery phase, which requires no communications, and thus high resilience against the smart attacker model.

SECURE ROUTING

Many routing protocols of sensor networks are quite simple and do not consider security as primary goal. Consequently, these protocols are more susceptible to attacks than in general ad hoc networks. Karlof et al. [34] have shown how attacks against ad hoc, and peer-to-peer networks can be adapted into powerful attacks against sensor networks. They have also introduced sinkholes and Hello floods, two classes of novel attacks against sensor networks:

Sinkholes attack. Depending on the routing algorithm technique, a sinkholes attack tries to lure almost all the traffic toward the compromised node, creating a metaphorical sinkhole with the adversary at the center. For example, the attacker could spoof or replay an advertisement for a high quality route that passes through the compromised node. If the routing protocol employs end-to-end acknowledgment technique to verify routes quality, a powerful laptop-class attacker could then provide a real high quality route by transmitting with enough power to reach the destination (sink node or base station) in a single hop. Since sinkhole attacks imply a great number of nodes (those on or near the high quality route), they can enable many other attacks that need tampering with circulating traffic such as selective forwarding. We should mention that sensor networks are particularly vulnerable to this class of attacks because of their special communication paradigm, where all nodes have to send sensory data to a one particular sink node. Thus, a compromised node has only to provide a single high quality route to the sink node in order to influence a potentially large number of nodes. Sinkholes attacks are difficult to overcome, especially in routing protocols that integrate advertised information such as remaining energy. Besides that, geo-routing protocols are known as one of the routing protocol classes that are resistant to sinkhole attacks. This is because topology is constructed using only localized information and traffic is naturally routed toward the physical location of the sink node. This makes it difficult to lure the traffic elsewhere to create a sinkhole.

Hello floods. This attack exploits Hello packets required in many protocols for announcing nodes to their neighbors. A node receiving such packets may assume that it is in the radio range of the sender. A laptop-class adversary can send this kind of packets to all sensor nodes in the network so that they believe that the compromised node belongs to their neighbors. This causes a large number of nodes sending packets to this imaginary neighbor and thus, into oblivion. Several routing protocols in sensor networks like directed diffusion,[35] LEACH,[36] and TEEN[37] are vulnerable to this class of attacks that may be very critical, especially when hello packets consist of a routing data or a localization information exchange. One simple way to mitigate Hello floods attacks is to verify whether the links are bidirectional. However, if the adversary have a high sensitive receiver, a trusted sink node may opt for limiting the number of verified neighbor for each node in order to prevent Hello attacks.

In addition to Hello floods and sinkholes attacks, sensor routing protocols are also vulnerable to general ad hoc routing threats described in Ref. 34. An example of such threats could be spoofing, wormholes, selective forwarding, etc. These attacks can be achieved more easily and in less time than in other ad hoc networks, especially, if we consider a set of laptop-class adversaries with strong transmitters and a high bandwidth, which is not available to ordinary sensor nodes, allowing such attackers to coordinate their efforts.

Solutions

To overcome the problems discussed, a secure routing protocol needs to ensure integrity, authentication, and confidentiality of the routing messages. However, these services rely heavily on broadcast authentication. This later allows the sink to broadcast authenticated messages to the entire sensor network. The service of authenticated broadcast would be easy to ensure if the sink node could securely distribute its public key, then just sign all its messages with its private key using a public key cryptosystem like RSA. Unfortunately, conventional asymmetric cryptography is too computationally intensive for sensors. In this section, we first discuss μTESLA, a new solution that has been proposed for broadcast authentication in sensor networks. Then, we describe two prominent schemes for secure routing in WSNs, namely systems process improvement networks (SPINS) and INtrusion tolerant routing protocol for wireless SEnsor NetworkS (INSENS).

• μTESLA

μTESLA, the optimized form of TESLA, has been proposed as a block of SPINS[38] in order to provide authenticated broadcast for severely resource-constrained environments. This protocol overcomes overhead and computation problems of asymmetric cryptography by introducing asymmetry through a delayed disclosure of symmetric keys.

In μTESLA, the broadcaster (sink) computes a message authentication code (MAC) on the sent packet using a key that is secret to all receivers. Therefore, no other node can masquerade as the sink during the sending period. At the time of key disclosure, the sink broadcasts the verification key to all receivers. When a node receives the disclosure key, it can easily verify the correctness of the key (using one-way hash functions). If the key is correct, the node can now use it to authenticate the initially broadcasted message.

Nevertheless, μTESLA has several drawbacks: it requires a loose synchronization between broadcasting node and receivers. Moreover, this technique is not suitable for immediate authenticated broadcasting between neighboring nodes because of the μTESLA key disclosure delays. Otherwise, every intermediary receiver would need to wait for the key disclosure time before sending the message to next node. Hence, a message traversing l hops would need l μTESLA delays to arrive at destination.

In an attempt to achieve other major requirements for secure routing in WSN such as confidentiality, some new protocols were proposed in literature under Ref. 6. We briefly describe the following two interesting propositions:

• SPINS

In addition to μTESLA, described earlier, SPINS[38] proposes another security building block called sensor network encryption protocol (SNEP). This security protocol, which was optimized for use in WSNs, provides confidentiality via chaining encryption function (i.e., DES-CBC, cipher-block chaining). Moreover, it employs a shared counter between the sender and receiver to build a one-time encryption key in order to prevent replay attacks and ensure data freshness. SNEP uses also a message authentication code to guarantee two-party authentication and data integrity.

• INSENS

INSENS[39] is a protocol that provides intrusion-tolerant routing in sensor networks, by building multiple redundant paths between sensor nodes and the sink in order to bypass intermediate malicious nodes. In addition, INSENS limits also DOS-style flooding attacks and prevents false advertisement of routing or other control information to overcome sinkhole-style attacks.[40] However, INSENS suffers from several drawbacks: the most important one is that the sink node is supposed to be fault-tolerant and that it cannot be isolated from the rest of the network by an attacker. This assumption is unrealistic in several scenarios.

SECURE DATA AGGREGATION

Data aggregation (or data fusion) is a key emerging theme in the design and deployment of WSNs. In this process, intermediary nodes called "aggregators" collect the raw sensed information from sensor nodes, process it locally and forward the result only to the end user. This important operation permit essentially to reduce the amount of transmitted data on the network and thus prolonging its overall lifetime, the most critical design factor in WSNs. However, this functionality is made even more challenging due to the hostile deployment environment, which makes possible the physical compromise of aggregators and some of the sensor nodes. Indeed, possible threats can vary from denial of service attacks trying to stop completely this service, to stealthy attacks where the attacker's purpose is to make the user accept false aggregation results. This latter is more difficult to detect.

For data aggregation validity assurance, Du et al.[41] have proposed the use of redundant data fusion nodes as witnesses. These nodes conduct the same data fusion operations as aggregators, but send the result as a MACs to the aggregator itself instead of sending it to the base station. In order to prove the validity of the aggregation results, the aggregator has to forward the received proofs from witness nodes along with its calculated result to the base station. If a compromised aggregator wants to send invalid fusion data, it has to forge the proofs on the invalid results. The aggregation result is confirmed when n out of m witness proofs agree with aggregators' results. Otherwise, this latter is discarded and the base station polls one of the witness node to send it the valid aggregation result. We think that this solution is efficient when witnesses are supposed to be trusted enough, otherwise it requires an important additional overhead to get an acceptable aggregation results using the voting scheme. Moreover, authors have not addressed issues about choosing witness nodes.

In literature under Ref. 42, authors have proposed a security framework based on aggregate-commit-prove approach to verify that the answer given by aggregators is a good approximation of the true value, even if aggregators and a fraction of sensor nodes can be corrupted. In this approach, the aggregator commits to the collected data by constructing a Merkle Hash-tree.[43] The commitment ensures that the aggregator uses data provided by sensors and acts as a statement to be verified by the base station about the correctness of aggregation results. Although authors have proposed some concrete protocols for securely computing the median, average, and some other types of specific aggregation operations, we think that the proposed scheme remains somewhat generic and may not be flexible enough to support other kinds of in-network processing, such as in tinyDB.[44]

CONCLUSION

In this entry, we discussed the most important issues related to security in sensor networks. We have shown that addressing these issues proves to be a challenging task and still needs and additional endeavor in order make WSNs secure to the extent required by their applications and without compromising performance, energy consumption, and usability.

Through the survey of existing solutions, we noticed that the domain is still in its infancy, and it opens a large number of contribution areas. In fact, and concerning cryptography, we remarked that few research works have been devoted to the development of analysis models, which permit performance evaluation of encryption schemes in WSNs. We believe that such models[12,45,46] allow designers to project computational limitations and determine the threshold of feasible encryption schemes under a set of constraints for a given sensor network. Moreover, and although that some emergent asymmetric primitives have proven their applicability in WSNs, they still contain expensive operations, especially for private key generation and encryption. These elementary operations need further study and optimization.

Another important research trend in the field of key management consists of investigating new techniques that may permit the use of public key cryptography. Indeed, almost all key management protocols discussed in literature so far are based on symmetric key cryptography. Moreover, several of them suppose the sink node to be trustworthy. This condition is not realistic in many sensor network applications, and a new key management scheme should be developed, considering the sink security.

As for secure routing protocols, all the proposed protocols focus on static sensor networks only, ignoring mobility. However, several applications of sensor networks imply the sink mobility, and some of them extend mobility to all nodes, as in underwater sensor networks.[47] Hence, the design of secure routing protocols, which consider the factor of mobility, constitute an attractive research field.

And finally, we believe that data aggregation is one of the key issues that has to be considered in all layers of the WSN's network protocol stack in order to minimize the energy consumption. However, this operation cannot be efficiently done without being secured. For that, we think that secure data aggregation should consider more energy efficient keying schemes. Moreover, multi-tiers hierarchical aggregation approaches such as in Ref. 48 would be the most efficient scheme when the WSN contains high number of sensor nodes. For that, more research works should be undertaken on how to securely and efficiently construct such schemes and dynamically choose aggregation nodes.

REFERENCES

1. @electronic crossbow, Crossbow Technology Inc., http://www.xbow.com (accessed 2007).
2. Chandrakasan, A.; Min, R.; Bharwaj, M.; Cho, S.; Wang, A. Power aware wireless microsensor systems. ESSCIRC 2002, Florence, Italy, Sept 24–26, 2002.

3. Xu, N. A survey of sensor network applications. University of Southern California, Tech. Rep., 2002.

4. Akyildiz, I.F.; Su, W.; Sankarasubramaniam, Y.; Cayirci, E. Wireless sensor networks: a survey. Comput. Netw. Int. J. Comput. Telecommun. Netw. **2002**, *38* (4), 393–422; http://citeseer.ist.psu.edu/diffie76new.html.

5. Wood, A.D.; Stankovic, J.A. Denial of service in sensor networks. Comput. **2002**, *35* (10), 54–62.

6. Karlof, C.; Wagner, D. Secure routing in wireless sensor networks: attacks and countermeasure. 1st IEEE International Workshop on Sensor Network Protocols and Applications, Anchorage, AK, May 11, 2003.

7. Ravi, P.K.S.; Raghunathan, A.; Hattangady, S. Security in embedded systems: design challenges. Trans. Embedded Comput. Syst. **2004**, *3* (3), 461–491.

8. Schneier, B. *Applied Cryptography*, 2nd Ed.; John Wiley & Sons: 1996.

9. Data Encryption Standard (DES). FIPS PUB Tech. Rep. 46, 1993.

10. Advanced Encryption Standard (AES). FIPS PUB Tech. Rep. 197, 2001.

11. Schneier, B. *Fast Software Encryption* Cambridge Security Workshop; Springer-Verlag: London, UK, 1994; 191–204.

12. Ganesan, P.; Venugopalan, R.; Peddabachagari, P.; Dean, A.; Mueller, F.; Sichitiu, M. Analyzing and modeling encryption overhead for sensor network nodes. 2nd ACM International Conference on Wireless Sensor Networks and Applications (WSNA 2003), San Diego, CA, Sept 19, 2003; ACM Press: New York, NY, 2003; 151–159.

13. Hager, C.T.; Midkiff, S.F.; Park, J.-M.; Martin, T. L. Performance and energy efficiency of block ciphers in personal digital assistants. 3rd IEEE International Conference on Pervasive Computing and Communications (PerCom 2005), Kauai, HI, Mar 8–12, 2005; IEEE Computer Society. Washington, DC, USA, 2005; 127–136.

14. Potlapally, N.R.; Ravi, S.; Raghunathan, A.; Jha, N.K. A study of the energy consumption characteristics of cryptographic algorithms and security protocols. IEEE Trans. Mob. Comput. **2006**, *5* (2), 128–143.

15. Law, Y.W.; Doumen, J.; Hartel, P. Survey and benchmark of block ciphers for wireless sensor networks. ACM Trans. Sen. Netw. **2006**, *2* (1), 65–93.

16. Ferguson, N.; Kelsey, J.; Lucks, S.; Schneier, B.; Stay, M.; Wagner, D.; Whiting, D. Improved cryptanalysis of rijndael. 7th International Workshop on Fast Software Encryption, 2000; Springer-Verlag: London, UK, 2001; 213–230.

17. National policy on the use of the Advanced Encryption Standard (AES) to protect national security systems and national security information. CNSS Policy. **2003**, *15*, no. Fact Sheet No. 1, 2003.

18. Daemen, J.; Rijmen, V. *The Design of Rijndael: AES—The Advanced Encryption standard*; Springer-Verlag, 2002.

19. Ronald, L. Rivest.; Robshaw, M.J.B.; Sidney, R.; Yin, Y.L. The RC6 block cipher, Submitted to NIST as a candidate for the AES, Tech. Rep.

20. Matsui, M. New block encryption algorithm misty. 4th International Workshop on Fast Software Encryption, Haifa, Israel, Jan 20–22, 1997; Vol. 1267.

21. ESTI/SAGE, Specification of the 3gpp confidentiality and integrity algorithms document 2: Kasumi specification, Tech. Rep., 1999.

22. Aoki, K.; Ichikawa, T.; Kanda, M.; Matsui, M.; Moriai, S.; Nakajima, J.; Tokita, T. Specification of camellia: a 128-bit block cipher: specification version 2.0. Nippon telegraphe and telephone corporation and mitsubishi electric corporation, Tech. Rep., 2001.

23. F.I.P.S. Publication, Digital signature standard (DSS). FIPS PUB 186, Tech. Rep., 1994.

24. Diffie, W.; Hellman, M.E. New directions in cryptography. IEEE Trans. Inf. Theory. **1976**, *IT-22* (6), 644–654. http://citeseer.ist.psu.edu/diffie76new.html.

25. Carman, D.W.; Kruus, P.S.; Matt, B.J. Constraints and approaches for distributed sensor network security, NAI Labs, Tech. Rep. 00-010, 2000.

26. Gaubatz, G.; Kaps, J.-P.; Ztrk, E.; Sunar, B. State of the art in ultra-low power public key cryptography for wireless sensor networks. Workshop on Pervasive Computing and Communications Security (PerSec 2005), Kauai Island, HI, Mar 8, 2005; IEEE Computer Society: Washington, DC, USA, 2005; 146–150.

27. Rabin, M. Digital signatures and public key functions as intractable as factorization, MIT Laboratory for computer science, Tech. Rep. 212, 1979.

28. Araki, K.; Satoh, T.; Miura, S. *Overview of Elliptic Curve Cryptography*; Springer-Verlag: 1998; 29–48.

29. Wander, A.; Gura, N.; Eberle, H.; Gupta, V.; Shantz, S. Energy analysis of public key cryptography in wireless sensor networks. PerCom 2005, Kauai, HI, Mar 8–12, 2005; IEEE Computer Society: Washington, DC, USA; 324–328.

30. Eschenauer, L.; Gligor, V.D. A key management scheme for distributed sensor networks. CCS 2002, Washington, DC, Nov 18–22, 2002 of the 9the ACM Conference on Computer and Communications Secutiy, Washington, DC, USA, ACM Press: New York, NY, USA, 2002; 41–47.

31. Chan, H.; Perrig, A.; Song, D. Random key predistribution schemes for sensor networks. IEEE Symposium on Research in Security and Privacy, Berkeley, CA, May 11–14, 2003; IEEE Computer Society: Washington, DC, USA, 2003; 197.

32. Zhu, S.; Xu, S.; Setia, S.; Jajodia, S. Establishing pair-wise keys for secure communication in ad hoc networks: a probabilistic approach. 11th IEEE International Conference on network Protocols, Atlanta, GA, Nov 4–7, 2003; IEEE Computer Society: Washington, DC, USA, 2003; 326.

33. Pietro, R.D.; Mancini, L.; Mei, A. Efficient and resilient key discovery based on pseudorandom key pre-deployment. 18th International Parallel and Distributed Processing Symposium, Santa Fe, NM, Apr 26–30, 2004; Vol. 13; IEEE Computer Society: Los Alamitos, CA, USA, 2004; 217b.

34. Karlof, C.; Wagner, D. Secure routing in wireless sensor networks: attacks and countermeasures. AdHoc Netw. J. **2002**, (Special issue on Sensor Network Applications and Protocols), **2002**, 1 (2–3), 293–315.

35. Intanagonwiwat, C.; Govindan, R.; Estrin, D. Directed difusion: a scalable and robust communication paradigm for sensor networks. ACM MobiCom 2000, Boston, MA,

Aug 6–11, 2001; ACM Press: New York, NY, USA; 2000; 56–67.

36. Heinzelman, W.; Chandrakasan, A.; Balakrishnan, H. 3rd Hawaii International Conference on System Sciences, HICSS, 2000; Vol. 8;. IEEE Computer Society, Washington, DC, USA, 2000; 8020.

37. Manjeshwar, A.; Agrawal, D.P. TEEN: a routing protocol for enhanced efficiency in wireless sensor networks. 15th International Parallel and Distributed Processing Symposium, 2001; IEEE Computer Society, Washington, DC, USA, 2000; 30189a.

38. Perrig, A.; Szewczyk, R.; Wen, V.; Culler, D.; Tygar, J.D. Spins: security protocols for sensor networks. Wirel. Netw.; Kluwer Academic Publishers, Hingham, MA, USA, 2002; Vol. 8; 521–534.

39. Deng, J.; Han, R.; Mishra, S. INSENS: intrusion-tolerant routing in wireless sensor networks. 23rd IEEE International Conference on Distributed Computing Systems (ICDCS 2003), Providence, RI, May 19–22, 2003; IEEE Computer Society: Los Alamitos, CA, USA, 2003.

40. Deng, J.; Han, R.; Mishra, S. A performance evaluation of intrusion-tolerant routing in wireless sensor networks. 2nd IEEE International Workshop on Information Processing in Sensor Networks (IPSN, 2003), Palo Alto, CA, April 22–23, Springer-Verlag: Germany, 2003; 349–364.

41. Du, W.; Deng, J.; Han, Y.S.; Varshney, P.K. A witness-based approach for data fusion assurance in wireless sensor networks. IEEE Global Telecommunications Conference (GLOBECOM 2003), San Fransisco, CA, Dec 1–5, 2003.

42. Przydatek, B.; Song, D.; Perrig, A. SIA: secure information aggregation in sensor networks. 1st ACM Conference on Embedded Networked Sensor Systems (ACM SenSys 2003), Los Angeles, CA, Nov 5–7, 2003.

43. Merkle, R.C. Protocols for public key cryptosystems. IEEE Symposium on Research in Security and Privacy, Oakland, CA, Apr 14–16, 1980; 122–134.

44. Madden, S.; Franklin, M.J.; Hellerstein, J.M.; Hong, W. TAG: a tiny aggregation service for ad-hoc sensor networks. 5th Symposium on Operating Systems Design and Implementation (OSDI 2002), Boston, MA, Dec 9–11, 2002.

45. Venugopalan, R.; Ganesan, P.; Peddabachagari, P.; Dean, A.; Mueller, F.; Sichitiu, M. Encryption overhead in embedded systems and sensor network nodes: modeling and analysis. 2003 International Conference on Compilers, Architecture and Synthesis for Embedded Systems (CASES 2003), San Jose, CA, Oct 29–Nov 1, 2003; 188–197.

46. Xie, G.G.; Irvine, C.E.; Levin, T.E. Quantifying effect of network latency and clock drift on time-driven key sequencing. 22nd International Conference on Distributed Computing Systems Workshop (ICDCSW 2002), Vienna, Austria, July 2–5, 2002; IEEE Computer Society, 2002; 35–42.

47. Akyildiz, I.; Pompili, D.; Melodia, T. Underwater acoustic sensor networks: Research challenges, 2005. http://citeseer.ist.psu.edu/akyildiz05underwater.html.

48. Deng, J.; Han, R.; Mishra, S. Security support for in-network processing in wireless sensor networks. 2003 ACM Workshop on Security of Ad Hoc and Sensor Networks (SASN 2003), Fairfax, VA, Oct 31, 2003.

49. Kahn, J.M.; Katz, R.H.; Pister, K.S.J. Next century challenges: mobile networking for "smart dust." International Conference on Mobile Computing and Networking (MOBICOM 1999), Seattle, WA, Aug 15–20, 1999; 271–278.

50. Eyes Project, http://www.eyes.eu.org (accessed 2005).

Wireless Sensor Networks (WSNs): Self-Organization

Manish Kochhal
Department of Electrical and Computer Engineering, Wayne State University, Detroit, Michigan, U.S.A.

Loren Schwiebert
Department of Computer Science, Wayne State University, Detroit, Michigan, U.S.A.

Sandeep K.S. Gupta
Department of Computer Science and Engineering, Arizona State University, Tempe, Arizona, U.S.A.

Abstract

This entry describes various self-organization schemes for wireless sensor networks. Self-organization involves abstracting the communication nodes into an easily controllable network infrastructure.

INTRODUCTION

The continuing improvements in computing and storage technology as envisioned in *Moore's Law*, along with advances in Micro-Electro-Mechanical Systems (*MEMSs*) and battery technology, have enabled a new revolution of distributed embedded computing where micro-miniaturized low-power versions of processor, memory, sensing, and communication units are all integrated onto a single board. One of the interesting applications of distributed embedded computing is an ad hoc deployed wireless sensor network (WSN)[1] that is envisioned to provide target sensing, data collection, information manipulation, and dissemination within a single integrated paradigm.

WSNs have many possible applications in the scientific, medical, commercial, and military domains. Examples of these applications include environmental monitoring, smart homes and offices, surveillance, intelligent transportation systems, and many others. A WSN could be formed by tens to thousands of randomly deployed sensor nodes, with each sensor node having integrated sensors, processor, and radio. The sensor nodes then self-organize into an ad hoc network so as to monitor (or sense) target events, gather various sensor readings, manipulate this information, coordinate with each other, and then disseminate the processed information to an interested data sink or a remote base station (BS). This dissemination of information typically occurs over wireless links via other nodes using a *multihop* path.[2,3]

The problem of self-organization (or self-configuration) has been a hot topic of research in wireless ad hoc networks, including mobile and sensor networks. Self-organization involves abstracting the communicating entities (or nodes) into an easily controllable network infrastructure. That is, when powered on, it is the ability of the nodes deployed to locally self-configure among themselves to form a global interconnected network. The resulting network organization needs to support efficient networking services while dynamically adapting to random network dynamics. The self-organization problem becomes even more exacting when the devices forming such a collaborative network are crucially constrained in energy, computational, storage, and communication capabilities. The unpredictability of wireless communication media adds a third dimension to this challenge. On top of that, the vision of having unattended and untethered network operation makes it even more complicated to provide even basic network services such as network discovery, routing, channel access, network management, etc. An example of a network that imposes such extreme demands is a wireless ad hoc sensor network.

One of the crucial design challenges in sensor networks is energy efficiency. This is because individual sensor nodes use a small battery as a power source, and recharging or replacing batteries in a remote environment is not feasible. In some cases, sensors may also use solar cells that provide limited power. Thus, to achieve a longer network lifetime, one has to tackle energy efficiency at all levels of the sensor network infrastructure. Because the wireless radio is the major energy consumer in a sensor node, systematic management of network communication becomes critical. Sensor network tasks such as routing, gathering, or forwarding sensing data to a nearby data sink or a remote BS requires network communication. To effectively coordinate these activities, one must address the problems of sensor network organization and the subsequent reorganization and maintenance. However, it

Encyclopedia of Wireless and Mobile Communications DOI: 10.1081/E-EWMC-120043932

should be clear that although energy is currently one of the biggest challenges, the key problems and solutions identified and summarized in this entry will hold equally well for future generations of WSNs.[2,3] This is because in any era of technological progress, there arises a necessity to support new challenging applications that inherently magnify analogous critical technological limitations with similar trade-offs against other objectives.

The focus of this entry is to understand the key problems and their respective solutions toward the various steps involved in self-organization for WSNs. In keeping with this approach, in this entry we first summarize self-organization concepts as applicable to the regime of WSNs. For the sake of simplicity, we consider typical algorithmic aspects of certain elementary network organizations such as a cluster, mesh, tree, and other implicit variations. This provides the necessary platform for exploring existing protocol solutions that consider self-organization as a primary problem. Finally, we provide an elaborate comparative discussion of these self-organization approaches with respect to the metrics used for network formation and the underlying services supported.

The contributions of this entry are thus threefold: 1) we first identify the typical characteristics of WSNs, and discuss their impact on the self organization problem; 2) we present several existing protocols that extend the basic network organization schemes to address the unique requirements of WSNs; and 3) we discuss the pros and cons of these self-organization protocols and provide some insight into their behavior. The remainder of the entry is organized as follows: In the second section, we provide a detailed overview of sensor network self-organization. The "Overview" section includes discussions on the sensor network communication paradigm, the impact of WSN characteristics, and self-organization preliminaries. We provide elementary concepts related to sensing and networking in the section "Self-organization Preliminaries." The third section provides a complete discussion of the currently available approaches and solutions to relevant self-organization issues. The fourth section provides a brief summary of the entire entry. The final section provides a glimpse of the future research directions.

OVERVIEW

The Sensor Network Communication Paradigm

Ad hoc wireless networks, by definition, lack a centralized BS that could otherwise be useful for initial coordination of network start-up and self-organization. In the case of ad hoc WSNs, the remote BS is available only as an application front end rather than as a centralized arbitrator for coordinating basic network activities. Moreover, due to scalability concerns, it becomes difficult, if not impossible,

for a BS to manage in fine detail the entire sensor network consisting of possibly a thousand or more sensors. On the contrary, in most sensing application scenarios, the BS at best expects only application-specific query resolution from a sensor (or a group of sensors) and it leaves the network management responsibility to individual solitary sensors that are isolated from each other by certain geographical distances. For processing on-demand application-specific queries, the sensor network adopts some distributed load-balancing heuristics to select an optimal sink (or a sensor node) to gather and process the sensing information from various sensing sources. This sink can also perform the energy-expending, long-haul query-reply to the BS. Fig. 1 highlights this primary communication paradigm for WSNs. Fig. 1A shows the basic sensing and networking services organized in a traditional layered fashion. It should be noted here that it is the self-organization protocol that engineers the critical preliminary network infrastructure support for these future protocol services that may be required by the sensor network. Hence, a careful design encompassing several critical service requirements at various layers of the protocol stack should be considered in the self-organization protocol. Some of these requirements may be conflicting and, hence, a general trade-off may need to be included in the design. For example, the self-organization protocol may need to control the topology of the network by having more node neighbors, which may be at odds with the energy conservation schemes that turn off idle nodes to increase the lifetime of the network. An integrated approach with some adaptive behavior is therefore warranted in the design of self-organization protocols. In subsequent sections of the entry, we highlight this design aspect by comparing several approaches with respect to basic self-organization requirements. Protocol cross-layering may be required to have reliable context-aware self-organization for certain application-specified tasks. Also, as shown in Fig. 1A, network functions such as energy management and security are inherently cross-layer in nature and must be addressed by all protocol layers in some form or other. However, protocol cross-layering[4,5] is outside the scope of this entry.

Impact of WSN Characteristics

In this section, we discuss some relevant characteristics of WSNs that should be incorporated into the design of self-organization protocols. These characteristics are listed below:

1. Network makeup:

a. *Homogeneous or heterogeneous sensor devices.* The sensor network may consist of specialized nodes having special hardware and software

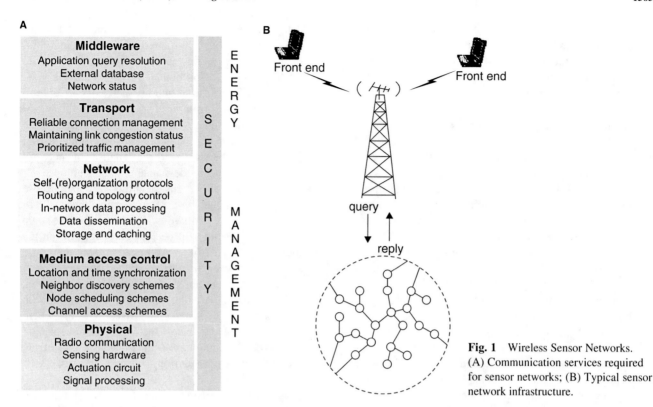

Fig. 1 Wireless Sensor Networks. (A) Communication services required for sensor networks; (B) Typical sensor network infrastructure.

capabilities deployed randomly or deterministically with other low-end sensor devices. This heterogeneous deployment may be required by certain applications, where placement of the sensor nodes is practical. An example of such an application may be monitoring a high-rise building for cracks and other critical hazards or faults. Another example could be some sensor nodes equipped with a GPS that serve as position-estimating beacons for other nodes without GPS. Similarly, nodes with higher processing capability and higher battery power can serve as data sinks for their neighboring underprivileged nodes. A self-organization protocol thus needs to take into account special node capabilities by tasking these nodes with higher responsibilities. As we will see in the "Approaches and Solutions" section, one way of doing this is to organize the network by way of assigning roles to sensor nodes depending upon their performance capabilities.[6,7]

b. *Indoor or outdoor environments.* Sensors deployed for building monitoring fall into the category of indoor environment, whereas sensors deployed to track enemy movements in a tactical military scenario can be categorized as experiencing an outdoor radio environment. In the outdoors, there are few obstructions, so the radio signals do not experience as much attenuation due to reflections and multipath fading. This is not the case for indoor environments, where walls contribute to a drastic reduction in signal strength. Moreover, sensor network applications are driven by

environmental events, such as earthquakes and fires, anywhere anytime following an unpredictable pattern. Sensor node failures are common due to these hostile environments. The radio media shared by these densely deployed wireless sensors may be subject to heavy congestion and jamming. High bit error rate, low bandwidth, and asymmetric channels make the communication highly unpredictable.[8] Frequent network monitoring and feedback should therefore be adopted by the network maintenance part of self-reorganization algorithms to provide a certain degree of fault diagnosis and repair under extreme situations.

c. *Random or controlled node placement.* Biomedical sensor networks[9] are examples of stationary WSNs.[10] In such a network, the placement of sensor nodes is controlled and predetermined. A stationary sensor network normally has little or no mobility. One can also decide in advance the number of neighbors a node may have, depending upon application requirements and the position of the sensor (border or internal node) within the deployment.[11] In contrast, a tactical WSN deployed in a hostile area to track enemy movements in the battlefield is characteristically required to have a random deployment. In this entry, we consider self-organization under the more challenging case of a random node deployment. For stationary sensor networks, interested readers are encouraged to refer to Ref. 10.

d. *Node mobility.* There is an important difference between a WSN and a mobile ad hoc network (MANET).

In general, WSNs have zero to limited mobility compared to the relatively high mobility in MANETs.[12] Network protocols for MANET mostly address system performance for random node mobility rather than for the energy depletion[13] caused by the execution of various network protocols. How ever, for ad hoc sensor networks, energy depletion is the primary factor in connectivity degradation and overall operational lifetime of the network. Therefore, for WSNs, overall performance becomes highly dependent on the energy efficiency of the algorithm. However, mobile sensors make wireless networking solutions extremely challenging. A mobile sensor network essentially becomes a special research challenge in the field of MANETs. In this entry, we discuss only self-organization protocols that *assume fixed or stationary sensors*. Interested readers are encouraged to extend mobility management concepts from MANETs in order to factor mobility into their design of self-organization protocols for mobile WSNs.[14]

e. *Node density (or redundancy)*. Depending on the application scale, tens of thousands of sensors can be deployed in a very large area. Examples of such an application would be deep space probing and habitat monitoring. The highly unpredictable nature of sensor networks necessitates a high level of redundancy. Nodes are normally deployed with a high degree of connectivity. With high redundancy, the failure of a single node has a negligible impact on the overall capacity of the sensor network. Network protocols for self-organization and maintenance need to adapt the topology of the network to its density and redundancy in order to have maximum energy savings. Sparser networks may need special treatment to avoid network partitions due to the existence of several orphan nodes.[15] Higher confidence in data can also be obtained through the aggregation of multiple sensor readings.[16,17] Moreover, a higher density of beacon nodes (i.e., nodes with position information) could be used to reduce localization errors[18] during position estimations by way of localized triangulations or multilaterations. A similar approach could be used to provide fine- or coarse-grained time synchronization.[19–22] A better determination of these network parameters could affect decisions for other primary network operations such as medium access or node scheduling schemes, event localization, topology control, etc.

2. *Data-centric addressing*: Data-centric addressing is an intrinsic characteristic of sensor networks. It is impractical to access sensor data by way of *ID* (or IP, as in the Internet). It is more natural to address the data through content or location. The *IDs* of the sensor nodes may not be of any interest to the application. The naming schemes in sensor networks are thus often data oriented. For example, an environmental monitoring application may request temperature measurements through queries such as "collect temperature readings in the region bounded by the rectangle [(x_1, y_1), (x_2,y_2)]," instead of queries such as "collect temperature readings from a set of nodes with the sensor network addresses *x, y*, and *z*." Thus, in a large-scale sensor deployment, nodes may not have unique *IDs*. However, to pursue localized control of network operations among nodes, unambiguous unique *IDs* become necessary among neighbors. Also, because the scale of a network deployment could be very large, node identification by way of unique hardware *IDs* is not feasible as it becomes an overhead on the media access control (MAC) header. This may be a significant part of the packet payload and hence may represent an important source of energy consumption. Generating unique addresses within the transmission neighborhood of a node and reusing it elsewhere greatly reduces the size of the MAC address size.[23] Auto-address configuration thus becomes one of the preliminary steps during sensor network self-organization.

3. In-network processing: There are essentially two options to communicate event information of interest to the BS application. One is sending individual readings of each sensor to the BS, which is impractical given the constraints discussed earlier. Another more feasible option is to gather, process, and compress neighboring correlated sensing event information within the network and then send it to the BS. This promotes energy efficiency by having a reduced data volume for long-distance transmission to the BS. It also promotes simple multihop reliability as nodes gather and process data in a localized hop-by-hop fashion as opposed to TCP complex end-to-end reliability schemes.[24–26] As discussed in "Self-Organization Preliminaries" section, self-organization algorithms usually achieve in-network processing by organizing nodes according to various aggregation patterns and network layouts or architectures.

4. *Sensing application characteristics*: Some biomedical applications[9] such as a glucose level monitoring require periodic monitoring of a patient's insulin level. In other applications, the sensor network needs to send information to the BS application only when an interesting event has been sensed. There may be other applications that desire both periodic and discrete event-triggered monitoring. A more challenging scenario could be an application requesting an on-demand organization of sensors in a certain region of network deployment for pursuing a fine- or coarse-grained monitoring of current or upcoming sensing events. Highly *resilient* self-organization protocols that (re) organize sensors either around some sensing event traffic or around a geographic region for periodic or

on-demand monitoring by applications are thus warranted.

Self-organization Preliminaries

In this section, we discuss the elementary concepts of network self-organization as applicable to the regime of WSNs. These concepts serve as the necessary foundation for interested readers to pursue future research in the area of sensor network self-organization. For ease of understanding, we have classified these basics into sensing and network organization concepts. We also formalize the necessary steps (or protocols) that fall under the unified umbrella of sensor network self-organization.

The *sensing concepts* (*sensing phenomenon*)[7] are concerned with the characteristics of the sensors, the events to be detected, and their topological manifestations, both in the spatial and temporal domains. For example, it is obvious that sensors in close proximity to each other should have correlated readings. A temporal dual of this observation implies that sensor readings among neighboring sensors also have some correlation within some nearby time intervals. In addition to supporting the properties associated with the sensing phenomenon, it is also necessary to support hierarchical event processing to have an incremental comprehensive global view of an area of deployment at different levels of the self-organized network hierarchy.

As mentioned previously, self-organization involves abstracting the communicating sensor nodes into an easily controlled network infrastructure. Cluster, connected dominating set (CDS), tree, grid, or mesh based organizations are typical. We provide some insight into these organizations for use in WSNs.

Elementary networked-sensing concepts

The sensing phenomena mentioned previously relate to the natural property of sensors sensing events collaboratively as well as individually in a group. Figs. 2 through 5

illustrate these sensing concepts of WSN organization for target detection or tracking. In the following discussion, we use the terms "sensing groups" and "sensing zones" interchangeably.

Fig. 2 illustrates that the sensing capability of sensors sensing events collaboratively or individually in a group depends essentially on the sensitivity of the sensors with respect to the target event. The sensitivity of a sensor diminishes with increasing distance of the sensor from the target. This sensitivity can be characterized theoretically by sensor models that are based on two concepts. One is that the sensing ability (*coverage*) diminishes with increasing distance.* Second is that noise bursts diminish the sensing ability but this effect of noise can be minimized by allowing sensors to sense over longer time periods (more *exposure*). Several algorithms based on the above sensitivity model have been developed that formulate the exposure and coverage properties of sensor networks. These algorithms use traditional computational geometry-based structures such as the Voronoi diagram and the Delaunay triangulation[27,28] to compute sensing coverage and exposure. However, distributed versions of these algorithms are challenging and computationally intensive, and hence are impractical for use during the initial network organization phase.

In general, self-organization protocols usually employ the concept of *redundant sensing* to account for fault-tolerant sensing in the presence of environmental vagaries. By redundant sensing, we mean that an observation of the presence of a nearby target event (i.e., a tank* in Fig. 2) should be supported not only by one sensor, but also by a group of neighboring sensors.[16,17] The sensing range may depend on the dimensions of the observed target; e.g., a seismic sensor can detect a tank at a greater distance than it can detect a soldier on foot. For ease of discussion, we assume the sensing range to be the same for targets of similar dimensions.[29] However, in general, for an application specific sensor deployment, nodes are assumed to be preconfigured for desired targets in terms of their sensing signatures or readings. In the case of on-demand target detection and tracking, the application is free to provide respective target sensing signatures in its queries. This requires selection of a group of neighboring sensors that can take sole responsibility for any event appearing within their region or group. The selection of sensors to form such a group requires quantifying relative proximity distances of each and every neighboring sensor. It also requires an

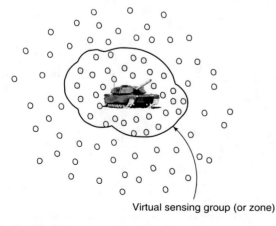

Virtual sensing group (or zone)

Fig. 2 Spatial group sensing concept.

---

* The sensing range may depend on the dimensions of the observed target; e.g., a seismic sensor can detect a tank at a greater distance than it can detect a soldier on foot. For ease of discussion, we assume the sensing range to be the same for targets of similar dimensions.[27] However, in general, for an application specific sensor deployment, nodes are assumed to be preconfigured for desired targets in terms of their sensing signatures or readings. In the case of on-demand target detection and tracking, the application is free to provide respective target sensing signatures in its queries.

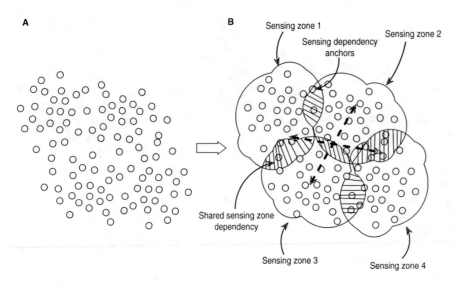

Fig. 3 Sensing group dependency concept. (A) sensor nodes randomly deployed, and (B) preliminary sensing zone based organization.

intelligent discrimination between near and far sensors to avoid grouping sensors from distant locations.

Figs. 3 and 4 illustrate the sensing zone dependency situation during tracking by sensing zones formed around a mobile enemy tank. Specifically, here we are discussing an initial network organization that statically forms sensing zones in anticipation of the occurrence of any future event. In the case of a random sensor deployment scenario, it is not possible to precisely control and place sensors so that they end up in groups having no overlap with neighboring sensor groups. This means that although an attempt was made to form stand-alone sensing groups (or zones) that independently take responsibility for detecting and tracking events, there are some overlapping regions where collaboration among neighboring sensing groups may be needed. However, the boundary nodes in each region can also serve as anchors for tracking events moving from one neighboring region to another. This is essentially a dichotomous scenario because, on the one hand, we need independence between neighboring sensing zones but on the other hand, we also want to efficiently track events moving

across neighboring sensing zone boundaries. An event monitoring and tracking algorithm that runs on top of such a self-organized network would have to analyze this dependency and utilize it to its best advantage. This can be done by either identifying neighboring dependent sensing zones and allowing collaboration among them for events moving around their neighborhood or tracking applications can dynamically specify an on-demand incremental reorganization of a new sensing zone around the moving event as it crosses the old sensing zone boundaries. The *Enviro Track* project[30–32] that is currently being pursued at the University of Virginia is an initial pro-of-of-concept implementation that supports such application-specific

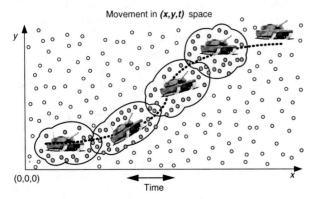

Fig. 4 Tracking a mobile tank around neighboring sensing groups.

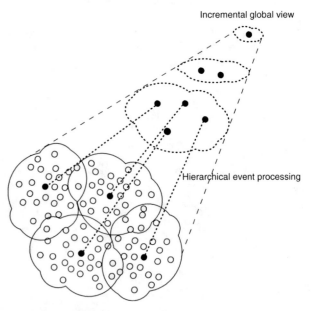

Fig. 5 Hierarchical event processing for incremental global view.

sensing group network (re)organization for tracking in a physical environment.

Fig. 5 illustrates event processing at various levels of a hierarchical sensor organization. It can be seen that as we go higher in the hierarchy, fewer nodes are involved in event processing. However, as we go up the hierarchy, we also lose detail about the event(s). This is because wireless communication is an overhead in terms of draining energy. Also, due to the small form factor of the sensors, memory is also a crucial resource and, hence, a lot of information cannot be maintained by an individual sensor or a small group of sensors. If we assume that the sensors selected for the upper levels of the hierarchy are powerful in terms of both communication energy as well as memory, the problem is still not resolved, due to scalability issues. However, any feedback from the sensing application about the granularity of monitoring would help in reducing overheads in information gathering and processing. In any case, hierarchical processing motivates the concept for distributed gathering, caching, and processing of sensing events where certain nodes in the hierarchy are assigned apt roles[6,7] according to their capability in the current network organization.

Elementary network-organization concepts

The primary objectives of this section are to categorize several elementary network organization architectures and discuss some relevant approximation algorithms that can be extended by self-organization protocols.

Fig. 6 shows a simple classification of various network architectures that can be employed by self-organization protocols. This classification is not complete, as there can be certain combinations of different network architectures. However, it provides the principal categories under which several current implementations can be studied and analyzed. Self-organization protocols can be either proactive or reactive. That is, protocols can organize the sensor network statically in preparation for any future event or they can dynamically configure the network around any current event of interest. Additionally, self-organizing

protocols can pursue either a difficult-to-maintain hierarchical manifestation of the above network architectures or they can simply satisfy requests with their corresponding flat manifestations. Fig. 7 provides a visual blueprint for the above elementary network architectures such as the chain, tree, spine, virtual grid, and role-based virtual regions. We discuss the typical algorithmic aspects of these network formations for general wireless ad hoc networks. This facilitates easier comprehension and analysis of those sensor network organization protocols that extend or modify these algorithms in order to meet various sensing application requirements. However, there are certain concepts that are common across all these network organizations. In all these organizations, nodes adopt certain performance metrics for selecting neighbor(s) in their local network formation heuristics. These performance metrics might be minimum distance, minimum energy, minimum transmission power, maximum/minimum node degree, delay, bandwidth, etc. Some of these metrics can be used collectively in some particular order (depending on priority) to break ties among several eligible competitors. To have an optimally ideal neighbor selection scheme for self-organization, nodes may require complete global state information of the network. However, in an ad hoc network, nodes that execute distributed algorithms for localized self-organization do not have the luxury of gathering, maintaining, and using complete network knowledge. As mentioned previously, this is because there are trade-offs among storage capability, communication costs, computational capability, and time to completion. This effectively results in nodes maintaining network state information for only two- to three-hop neighbors. Using this information, nodes execute local decisions to select neighbors to form a global self-organized network.

The chain-based organization is one of the simplest ways of organizing network communication, where nodes farther from the BS initiate chain formation with their nearest neighbor. The idea is to gather and fuse all the data from every node by forming a chain among them. A leader is then selected from the chain to transmit the fused data to the BS. However, building a chain to minimize its total length is similar to the traveling salesman problem, which is known to be intractable. A greedy chain formation algorithm,[33] when pursued recursively for every node, results in a data-gathering chain oriented toward the BS. As we discuss later, self-organization algorithms for sensor networks usually also include certain sensing metrics in order to form an optimal organization that is efficient from both sensing[34,7] and networking perspectives.[35,36]

The tree type of network formation is similar to the chain and can be considered an extension of the chain-based mechanism. Tree type network organizations utilize the multipoint connectivity nature of the wireless medium, where one source can be heard simultaneously by several nearby receivers that act as its children. If both the sender

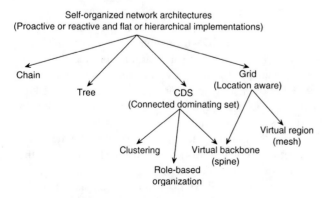

Fig. 6 Self-organized network architectures.

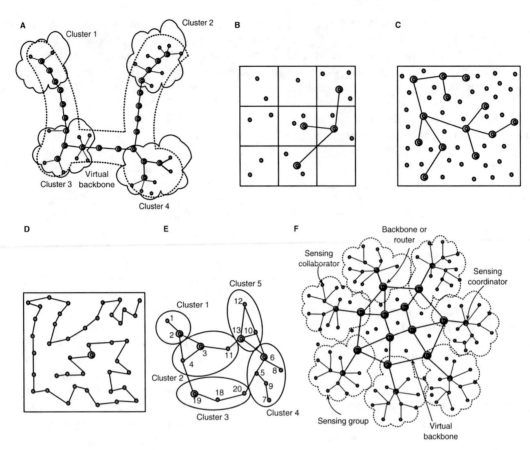

Fig. 7 Example network organizations for (A) spine; (B) virtual grid; (C) tree; (D) chain; (E) clustering; and (F) role-based virtual zones.

and receiver scheduling are made collision-free, then a tree-based network organization can support both *broadcast or multicast* (i.e., dissemination of information from a central node) and *convergecast* (i.e., gathering of information toward a central node) communication paradigms across all application domains.[37] A considerable amount of research work is available for constructing multicast trees[38,39] for dynamic wireless ad hoc networks.[40,41] Algorithms for generating multicast trees typically must balance the goodness of the generated tree, execution time, and the storage requirements. Several cost metrics such as delay, communication costs, etc. are used to generate cost-optimal multicast trees. However, this gives rise to the well-known *Steiner Tree problem*, which is NP-complete.[42] Instead, heuristics are used to generate "good" rather than optimal trees. This is still an active area of research.

Cluster-based organizations (refer to Fig. 6E) partition the entire network into groups called clusters. Each cluster is formed by selecting some nodes based upon some quality metric such as connectivity or distance.[43–45] as cluster members and a group leader, known as the cluster head, is also selected using some metric such as maximum energy to manage that cluster. These cluster heads, when connected, form a virtual backbone or spine as in

Fig. 7B[46,47] or a *CDS* of nodes. Related to clustering is the problem of finding a minimum CDS (*MCDS*) of the nodes, which is NP-complete. An *MCDS* satisfies two properties: 1) each node is either a backbone node or is one hop from a backbone node, and 2) the backbone nodes are connected. There are several approximation algorithms[48] available in the literature that engineer virtual backbone based network configurations satisfying the *MCDS* properties. Of particular importance are Wu's[49,50] distributed and localized algorithms for constructing a hierarchical CDS. This algorithm is inherently distributed and simple in nature. Ideally, it requires only local information and a constant number of iterative rounds of message exchanges among neighboring hosts. The algorithm for CDS formation involves a dominating set reduction process and some elimination rules based on quality metrics that are executed by nodes locally to identify their dominating neighbors. The dominating set reduction process, when executed recursively by an already existing set of dominating nodes, produces a domination hierarchy. Dominating nodes at any level of the hierarchy can serve as local network coordinators for nodes in the next lower level of the hierarchy. Fig. 7F shows a role-based hierarchical CDS organization of nodes where the lower levels of nodes form the cluster, whereas upper levels of dominating nodes can be used for routing or hierarchical information processing. Accordingly, nodes at every level of the domination hierarchy assume respective roles in the network, depending upon the quality metrics used for role selection at that level.[6,7]

Finally, if nodes have location information (e.g., using GPS), then virtual grid based organizations[51,14] typically configure the network by partitioning the area of deployment into uniform grids also known as a mesh. From each grid, a dominating node is chosen using some selection rules. Dominating nodes from each grid, when connected, form a virtual backbone, which can then be used for gathering or forwarding information from one geographic region to another. The grid-based organization allows an implicit and simple naming system by having grids (regions) in the network be addressed by their relative geographic locations. Thus, it avoids the complex and nonscalable address generation mechanisms for individual nodes that are densely deployed in a very large area. However, the efficiency of such an organization depends critically on location accuracy and network partitioning schemes.

Steps to sensor network self-organization

Self-organization or self-configuration is one of the basic and initial steps toward an ad hoc deployment of wireless sensors. The network deployment as mentioned previously can be done deterministically or randomly. In any case, the objective is to have nodes discover their neighbors, establish their positions, and form an easily manageable network architecture. All these self-organization activities must be performed in a localized and distributed manner with high energy efficiency and little or no communication overhead. Moreover, the self-organized network infrastructure should be *adaptive* and *resilient* to being easily reorganized with respect to the ultimate traffic pattern that may run on top of it.

Following the self-organization steps as extended from Refs. 52 and 53 forms the complete basis for any self-organization algorithm for ad hoc WSNs:

- network discovery or initialization phase:
  — each node discovers its set of neighbors,
  — depends on communication transmission radius (Tx_{max}), and
  — random or deterministic initial channel scheduling for neighbor discovery.
- coarse grained estimation phase:
  — location estimation and
  — time synchronization.
- organizational phase:
  — formation of a hierarchical or flat network organization with the help of local group formations,
  — performing group reorganization if necessary,
  — generation of addresses for nodes,
  — generation of routing table at every node,

  — generation of broadcast or multicast trees and graphs within a group,
  — merging of broadcast trees and graphs when groups are aggregated to form hierarchical networks,
  — establish MACschemes both for intra-group and inter-group communication, and
  — establish key setup schemes for secure communication.
- maintenance phase:
  — active or passive monitoring (e.g., by *"I am Alive"* messaging),
  — network quality evaluation schemes such as connectivity and sensing coverage,
  — maintenance of routing tables,
  — maintenance of broadcast infrastructure,
  — fine-grained tuning of network parameters such as location and network time,
  — topology control schemes to maximize network throughput and spectral reuse or network capacity,[46,54,55] and
  — Energy conservation schemes for increased network longevity using dynamic node scheduling.[56,51]
- self-reorganizing phase:
  — redeployment leading to discovery of new node neighbors and
  — fault detection and recovery schemes under node or link failure and group partitions.
  — in general, the steps listed above can also be considered as services provided by self-organization protocols for WSNs. This means that some of these services may be optional whereas some are fundamental to any algorithm that self-organizes the sensor network. Thus, Steps 1, 3, and 5 are necessary. On the other hand, Steps 2 and 4 are optional and can be developed separately. Location estimation protocols are generally referred to as self-configuring localization protocols. Network time synchronization can be considered orthogonal and implemented separately without regard to any specific network design or architecture. Similarly, the maintenance phase can be implemented separately as a suite of network management protocols.

APPROACHES AND SOLUTIONS

In this section we discuss the current approaches and solutions along the same lines as the various self-organization stages mentioned earlier. There is much research in location estimation,[18,57–59] time synchronization techniques,[19–22] MAC,[60–62] and key exchange schemes for secure communication[63–65] for ad hoc sensor networks. Discussion of these are outside the scope of this entry.

Neighbor Discovery or Initialization Protocols

Neighbor discovery is one of the initial steps after node deployment toward sensor network self-organization. However, it can also be periodically used as an intermediate step after self-organization for network maintenance. For network maintenance, a node may have to listen continuously to find dying neighbors or new neighbors. Neighbor discovery at this stage can also be used to evaluate the connectivity and sensing coverage of the network before further network deployment is deemed necessary. Further deployment of nodes may be needed, which again requires additional neighbor discovery to find new nodes as neighbors. Because listening for wireless communication is also a power-consuming process that may exhaust batteries, some randomized or probabilistic node scheduling scheme can be used during the network maintenance stage.

Neighbor discovery messages are essentially broadcast messages that are transmitted at different radio transmission power levels, depending upon the desired number of neighbors for a topology. Due to the wireless medium and dense ad hoc deployment, broadcast radio signals are likely to overlap with others in a geographical area. A straightforward broadcasting by flooding is usually very costly and will result in serious redundancy, contention, and collision problems, which are collectively referred to as the *broadcast storm*[66] problem. Also, initial neighbor discovery by way of sending broadcast messages cannot use sophisticated MAC protocols. As we will see, the presence of a medium access scheme during network maintenance provides important flexibility in terms of avoiding frequent arbitrary channel collisions. This flexibility is not available immediately after initial network deployment, thus affecting the quality of the neighbor discovery process. The performance metrics for the neighbor discovery process include parameters such as the number of neighbors discovered, time to completion of the network discovery process, and energy efficiency. Other relevant network characteristics that affect neighbor discovery include the density of node deployment, radio transmission range, and efficiency of the channel arbitration schemes.

Initial neighbor discovery or network initialization protocols typically employ certain deterministic channel access schemes such as the TDMA-based scheme. Early efforts in using a TDMA-style neighbor discovery can be traced back to the formation of a link clustered architecture (LCA)[67–69] in ad hoc wireless networks. In this case, a BS can initiate network time synchronization and then allow nodes to communicate in TDMA fashion, where a TDMA slot is identified by a node's unique ID. However, there are several limitations to such an approach. One important problem is the convergence speed of such a neighbor discovery algorithm. As the networks become large, the

time required to complete neighbor discovery will increase proportionally. Also, nodes may have to broadcast at their maximum radio transmission power, which will increase energy dissipation.

Neighbor discovery can also be done randomly or probabilistically in order to randomize listening and broadcast times for neighbor discovery messages. McGlynn and Borbash[70] propose an energy-efficient neighbor discovery protocol based on the birthday paradox. In the birthday paradox, we compute the probability that at least two people in the same room will have the same birthday. When there are as few as 23 people, the probability that at least two have the same birthday already exceeds 1/2. This idea is borrowed for a channel access scheme for neighbor discovery. Over a period of n slots, two wireless nodes independently and randomly select k slots. The first node transmits a message on its k slots, and the second node listens on its k slots. For the remaining $n - k$ slots, the nodes are idle or in power-saving mode. The probability that a second node hears the first is nearly 1 when the ratio k/n is relatively small. The authors calculate that even with 93% idle time, there is a high probability that a second node is able to hear the first. For dense networks, the birthday protocol may fail to discover some links. However, by trading energy for minimal discovery loss, the authors show that their protocol is highly effective, even when a node has many neighbors.

Sensor Network Organization Protocols

As discussed previously, sensor network organization protocols usually employ typical organizational designs such as the chain, tree, cluster, or variations of the CDS architecture. Distributed localized algorithms for network self-organization also utilize certain quality metrics that effectively make this problem difficult or even intractable, and the solutions approximate. Although the organizational phase (i.e., Step 3) lists several key requirements for self-organization, these requirements alternatively can be looked upon as services supported by the underlying network infrastructure. Because the main goal of sensor network organization protocols is to provide uniform or balanced conservation of network energy or lifetime, the protocols essentially end up providing some or all of these services at varying degrees of quality-of-service (QoS). All this depends on the energy-saving scheme that is integrated implicitly with the network organization, and the network parameters that are allowed to be in trade-off to gain extra energy savings.

Power-efficient gathering in sensor information systems (PEGASISs)[33] is one of the early representatives of the chain-based design that organizes the sensor network into a chain. The main idea in PEGASIS is for each node to receive from and transmit to close neighbors and take turns being the leader for transmission to the BS. The goal is to distribute the energy load evenly among the sensor nodes in the network for data-gathering services. Because forming a reduced distance chain is similar to the traveling salesman problem, a greedy approach is pursued locally by nodes first selecting its close neighbors in the chain. The chain formation is initiated at the farthest node from the BS. The greedy algorithm assumes that all nodes have global knowledge of the network. Due to the impractical nature of this assumption, the authors also suggest offloading this complex chain formation process to the BS, which then computes and broadcasts the chain organization to all the sensor nodes.

Cluster-based designs are the most favorable approach to sensor network self-organization. One of the preliminary works that uses clusters is the low energy adaptive clustering hierarchy (LEACH) protocol.[71] LEACH is designed to support periodic remote monitoring on ad hoc sensor networks. The LEACH algorithm provides dynamic distributed cluster formation, local processing of sensing data, and randomized rotation of the cluster heads for every round of communication to the BS. All these activities are essentially targeted to support uniform energy usage and to maximize system lifetime. However, there are several limitations in the LEACH design that limit its ability to conserve energy. Because the LEACH protocol selects the cluster heads randomly rather than deterministically, this often results in suboptimal selection of cluster heads. In other words, the distance variation among cluster members in a cluster is large, and the variance becomes more pronounced for sparse network topologies. This effectively translates into nodes communicating information to their clusterhead or BS at larger distances, which results in an appreciable energy loss. Also, there are no checks to avoid large membership overheads in clusters. However, their organization infrastructure does integrate services such as data inter- and intra-cluster gathering, channel access, and network reorganization and maintenance.

Deterministic approaches to cluster-based network organization solve the above problems associated with randomized cluster formations. Krishnan and Starobinski[72] present two algorithms that produce clusters of bounded membership size and low diameter by having nodes allocate local *growth budgets* to neighbors. Unlike the expanding ring approach,[73] their algorithms do not involve the initiator (cluster head) in each round and do not violate the specified upper bound on the cluster size at anytime, thus having a low message overhead compared to the expanding ring approach. Kochhal, Schwiebert, and Gupta[6,7] propose a role-based hierarchical CDS-based self-organization (RBHSO, refer Fig. 7F) that deterministically groups sensors into clusters. To pursue such an organization, they develop fault-tolerant group-sensing metrics to partition the sensor network into several sensing zones (clusters). These sensing zones individually act as an aggregate consisting of sensor nodes collaborating to achieve a common sensing objective with a certain *sensing QoS (sQoS)*. Routing and sensing roles are deterministically assigned to sensor nodes, depending upon their connectivity and sensing quality, respectively. Sensing coordinators are also deterministically elected that act as leaders for their respective sensing zones. The sensor coordinators play the role of systematizing collaboration among members in the sensing zone. To limit the the number of sensing zones as well their membership, the RBHSO protocol uses two specified minimum and maximum sensing zone membership limits. However, in the final stages of their algorithm, orphan nodes are allowed to join any nearest neighboring sensor coordinator or a sensing zone member. This is done to cover the maximum possible number of nodes in the organized hierarchy. Hierarchical cluster-based organization of a network of wireless sensors was also proposed by Chevallay et al.[52] Their proposal essentially builds on the hierarchical self-configuration architecture proposed by Subramanian and Katz.[53] The cluster head election is based primarily on the energy level and processing capability of each sensor node. Within each group, nodes are randomly assigned IDs drawn from numbers of a limited size. However, if this pool of numbers is large enough, then there is a high probability of the identifiers being unique both temporally and spatially. This approach is similar to RETRIs (Random, Ephemeral Transaction Identifiers),[74] which also assigns probabilistically unique identifiers. However, by tagging respective group identifiers to node IDs, Chevallay et al.[52] go one step further to provide a globally unique address for each sensor in the network.

Tree-based organizations are usually used to dynamically group disparate sensing sources with a particular sink node(s). Mirkovic et al.[75] organize a large-scale sensor network by maintaining a dynamic multicast tree-based forwarding hierarchy that allows multiple sinks to obtain data from a sensor. Sohrabi, Pottie et al.[76–78] propose several algorithms for the self-organization of a sensor network, which include the self-organizing MAC, energy-efficient routing, and formation of ad hoc subnetworks for cooperative signal processing functions. Their self-organization technique forms an on-demand minimum-hop spanning tree to a central node or sink elected among neighboring sensors that sense environmental stimuli. They[77,78] propose a self-organization protocol for WSN that trades available network bandwidth by activating only necessary links for random topologies using TDMA-based node scheduling schemes to save energy. Their self-organizing algorithm includes a suite of protocols designed to meet various phases of network self-organization. One protocol, SMACS, forms a joint TDMA-like schedule, similar to LCA,[67] for the initial neighbor-discovery phase and the channel-assignment phase. Other protocols (such as EAR, SAR, SWE, and MWE) take care of mobility management, multihop routing, and the necessary signaling and data-transfer tasks required for local cooperative information processing.

The two tier data dissemination (TTDD) model[14] and geographic adaptive fidelity (GAF)[51] are network organization schemes that use geographic location information to partition the network into a grid. The goal of TTDD is to enable efficient data dissemination services in a large-scale WSN with sink mobility. Instead of waiting passively for queries from sinks, TTDD exploits the property of sensors being stationary and location-aware to let each data source build and maintain a grid structure. Sources then proactively propagate the existence information of sensing data globally over the grid structure, so that each sink's query flooding is confined within a local grid cell. Queries are forwarded upstream to data sources along specified grid branches, pulling sensing data downstream toward each sink. The goal of GAF[51] is to build a predefined static geographical grid only to turn off nodes for energy conservation. Similar algorithms exist that elect or schedule nodes according to certain quality metrics. These algorithms loosely come under the energy-conserving maintenance part of self-organization. Tian and Georganas[56] propose an algorithm that simultaneously preserves original sensing coverage and also autonomously turns off redundant sensor nodes in order to conserve energy. However, they need a specialized multidirectional antenna for measuring the angle of arrival (AoA) of signals to evaluate sensing coverage of neighboring areas that are geometrically modeled as sectors. They also suffer from problems due to the randomization of the backoff timer that is used to avoid having all neighbors turn themselves off, leaving a blind spot. Slijepcevic and Potkonjak[27] propose a heuristic that organizes the sensor network by selecting mutually exclusive sets of sensor nodes that together completely cover the monitored area.

In MANETs, self-organization essentially involves maintaining some form of network organization to support routing infrastructure in the presence of random uncontrollable node mobility. Some relevant research in this area includes the ZRP protocol[79] and the terminodes-based[80] approach. For mobility management, ZRP uses zones that are similar to clusters, whereas the terminodes-based approach uses the concept of self-organized virtual regions. Routing using either of these approaches involves two different schemes, a proactive routing scheme for nodes within a local virtual region or zone and a reactive scheme for nodes located in remote virtual regions or zones. Because in MANETs the availability of the network depends on each user's discretion, an incentive for cooperation by way of virtual money called nuglets is employed in terminodes.

Sensor Network Maintenance Protocols

Frequent network monitoring among neighbors and execution of protocols such as topology control and node scheduling form part of the maintenance phase of the self-organization protocol. The main goal of the maintenance phase is to provide increased network lifetime and also to maintain the infrastructure support for critical networking services. The topology of an ad hoc network plays a key role in the performance of networking services such as scheduling of transmissions, routing, flooding, and broadcasting. In many cases, several network links are not needed for establishing efficient sharing of the channel among neighboring nodes or the routing of data packets. Removing redundant and unnecessary topology information is usually called topology control or topology management.[81]

One way of controlling network topology is to have nodes select their neighbors based on a certain quality metric, say connectivity or available energy, or a combination of both. Nodes that are not selected remain in sleep (or power-saving) mode. In other words, the topology is controlled by having active nodes control their respective node degrees.[82] Another approach is to have nodes adjust their radio transmission power to control their neighboring topology.[83] However, the additional problem of unidirectional links must be solved when nodes are allowed to transmit at different transmission powers.[84] In general, topology control problems are concerned with the assignment of power values to the nodes of an ad hoc network so that the power assignment leads to a graph topology satisfying some specified properties; e.g.,

1. Energy efficiency with an emphasis on increasing network lifetime,
2. Maximize network throughput,
3. Provide strong network connectivity,
4. Maximize wireless spectrum reuse or network capacity.

Xu[85,86] proposes two topology control protocols [(GAF and cluster-based energy conservation (CEC)] that extend the lifetime of dense ad hoc networks while preserving connectivity by turning off redundant nodes. GAF[51] finds redundant nodes by dividing the whole area of deployment into small "virtual grids" and turning ON only those nodes that are essential for communication among neighboring grids. The other protocol, CEC, forms a cluster-based network organization to directly observe radio connectivity to determine redundancy. GAF needs exact location information using GPS and an idealized radio model, which are its limitations. CEC, on the other hand, needs to only monitor network organization (in this case, a cluster) to find redundant nodes. However, this monitoring process employs random sleep and awake timers (i.e., duty cycle) to save energy. Chen et al.[46] propose a topology maintenance algorithm (Span) for energy efficiency in ad hoc wireless networks. Span is a localized power-saving protocol that removes redundant nodes in a dense ad hoc network and adaptively elects only a small number of coordinators that stay awake continuously and perform multihop packet routing. Other nodes remain in power-saving mode and periodically check if they should wake up and become a coordinator. These coordinators form a strongly connected network backbone, also known as a *CDS*. Bao and Garcia-Luna-Aceves[81] propose another topology management algorithm that constructs and maintains a backbone topology based on a minimal dominating set (MDS) of the network. sparse topology and energy management (STEM)[55] accepts delays in path-setup time in exchange for energy savings. It uses a second radio, operating at a lower duty cycle, as a paging channel. When a node needs to send a packet, it pages the next node in the routing path. This node then turns on its main radio so that it can receive the packet.

Using only transmission control techniques for controlling sensor network topology, both centralized[87] and distributed[88] approaches have been proposed. Tseng et al.[87] consider the problem of topology control by tuning the transmission power of hosts to control the structure of the network. The target topology for their protocol is 1-edge, 1-vertex, 2-edge, and 2-vertex-connected graphs. They propose two global centralized algorithms to support minimum spanning tree construction, where host transmission powers are fixed or variable during the lifetime of the network. Kubisch et al.[44] propose two distributed algorithms [local mean algorithm (LMA) and local mean number of neighbors algorithm (LMN)] that dynamically adjust the transmission power level on a per-node basis. The LMA uses a predefined minimum (NodeMinThresh) and maximum (NodeMaxThresh) threshold for the number of neighbors a node can have and continues to increase its transmission power until it finds neighbors within NodeMinThresh and NodeMaxThresh. The LMN goes a step further than LMA by considering the mean number of neighbors each of its neighbors has and increases transmission power accordingly.

Because MAC-level protocols have a very small view of the network, the main approach followed by such energy-efficient protocols has been to turn off radios that are not actively transmitting or receiving packets. Because there is a certain amount of time involved in turning radios back on when they are needed, MAC protocols typically trade off network delay for energy conservation. Energy-efficient MAC and routing protocols can be used together to increase energy conservation. AFECA[89] seeks to maintain a constant density of active nodes by periodically turning radios off for an amount of time proportional to the measured number of neighbors of a node. By following this approach, more energy can be conserved as the density increases. However, AFECA needs to be conservative in its decision to turn off radios as the density measurement is not absolute.

By doing energy conservation with *application-level* information, it is possible to save much more energy, yet the sacrifice is having a network with application-specific characteristics. Adaptive self-configuring sensor networks

topologies (ASCENT)[90] measures local connectivity based on neighbor threshold and packet loss threshold to decide which nodes should join the routing infrastructure based on application requirements.

Self-healing and Recovery Protocols

Self-healing or recovery might also be one of the objectives of network management. The type of network architecture, flat or hierarchical, could also be a factor in deciding the complexity of network management. Mechanisms for detecting or repairing network partitions is necessary for energy-constrained WSNs. Partitions are more prevalent when node density is low. Network partitions can also occur even in dense networks when a number of nodes are destroyed or obstructed. When such network partitions do occur, complementary mechanisms will be needed for detecting and repairing partitions. Prediction-based approaches are essentially an implicit way of recording and maintaining the vital networking stats of a node and its neighboring resources. This history information can then be used to predict network partitions (or holes) due to node failures. Salvage techniques would involve either an explicit redeployment of new nodes around the area of lost coverage or having a subset of nodes move in a controlled manner from a dense coverage area to sparser areas. The latter technique of self-healing is also known as *self-aware actuation*, [91] where the actuation ability of mobile sensors allows the network to adaptively reconfigure and repair itself in response to unpredictable runtime dynamics.

Mini et al.[92,93] present a sensor network state model for each sensor node that is modeled as a Markov chain. This model is then used to predict the energy consumption rate and to create an energy map of a WSN. The energy map is built based on a prediction approach that tries to estimate the amount of energy a node will spend in the near future. Chessa and Santi[94] present an energy-efficient fault diagnosis protocol that can identify faulty or crashed nodes in a WSN. This diagnostic information gathered by the operational sensors can then be used by the external operator for the sake of network reconfiguration or repair, thus extending network lifetime. Zhao et al.[95–97] propose a tree-based monitoring architecture to monitor WSNs with different levels of detail, and focus on the design of computing network digests. Digests represent continuously computed summaries of network properties (packet counts, loss rates, energy levels, etc.) and can serve the need for more detailed, but perhaps energy-intensive, monitoring. Finally, Zhang and Arora[98] propose a virtual hexagonal cell-based organization, similar to the virtual grid-based concept of GAF,[51] for WSNs. In their approach, they also provide self-healing protocols to contain and heal small perturbations, including node joins, leaves, deaths, state corruptions, and node movements, occurring at moderate frequencies. Their approach is similar to the tree monitoring architecture except instead of the root of the tree, the head or cluster head of the cell maintains the state of the cell.

SUMMARY

Self-organization is the ability of the deployed nodes to locally configure among themselves in order to form a global interconnected network. In other words, self-organization involves abstracting the communicating entities (sensor nodes) into an easily manageable network infrastructure. The extreme dynamics of WSNs, along with challenging demands posed by the sensing applications, make self-organization solutions difficult and novel. Energy is one of the crucial design challenges for sensor network self-organization protocols. It is the self-organization protocol that engineers the critical preliminary network infrastructure support for sensor network services at various layers of the network protocol stack. Hence, a careful design encompassing several critical service requirements at various layers of the protocol stack must be considered in the self-organization protocol. Some of these requirements may be conflicting and hence a general trade-off may need to be included in the design. An integrated approach with some adaptive resilient behavior is therefore warranted in the design of self-organization protocols.

The design of sensor network organization protocols includes concepts that are common across all sensing applications. These include the *sensing phenomena* and elementary organizational architectures such as the chain, tree, cluster, or variations of the CDS. Protocols for network self-organization also utilize certain quality metrics that effectively make the self-organization problem difficult or even intractable and the solutions approximate. The energy-conserving capability of a self-organization protocol depends on the energy-saving scheme that is integrated implicitly with the network and the network parameters that are allowed to be traded off with each other.

FUTURE RESEARCH DIRECTIONS

Integrating cross-layer requirements with the self-organization algorithm for supporting diverse concurrent applications is an open research field. Ideally, it requires maintaining and sharing cross-layer network status information across all layers of the protocol stack. It also requires a general formalization of the various QoS requirements and application policies toward network self-(re)organization and management. This also motivates the mapping of these requirements and policies[99,100] into the self-organization algorithm. A role-based self-organization approach,[6,7] where nodes in the network are assigned roles depending upon their performance capabilities, promotes easy integration with the cross-layer design mentioned earlier. It also promotes future evolution of the self-organized infrastructure in terms of new roles that

meet novel services across multiple platforms and applications.

Target event characterization in terms of its dynamics and dimensions is an open field. Tracking multiple targets with different sensing signatures requires dynamic reorganization of the sensing groups or regions. As mentioned previously, this requires an explicit feedback mechanism between the event tracking protocol and the maintenance or reorganization protocol to dynamically control the topological dimension of the overlap among neighboring sensing regions. There is limited knowledge available that provides detailed specifications of the characteristics of the applications, the desired targets, and the required feedback necessary for efficient tracking.

Finally, as discussed previously, mobile sensor networks are a special research challenge that is an outcome of the merger of critical requirements of both the sensor networks and MANETs. However, a considerable amount of research is available in both domains, and interested readers are encouraged to apply mobility management concepts from MANETs to include mobility in their design of self-organization protocols for mobile WSNs.

ACKNOWLEDGMENTS

This material is based upon work supported by the National Science Foundation under Grant ANI-0086020 and a summer dissertation fellowship from Wayne State University.

REFERENCES

1. Pottie, G.J.; Kaiser, W. Wireless sensor networks. Commun. ACM. **2000**, *43* (5), 51–58.
2. Akyildiz, I.F.; Su, W.; Sankarasubramaniam, Y.; Cayirci, E. Wireless sensor networks: a survey. Comput. Netw. **2002**, 38, 393–422.
3. Estrin, D.; Govindan, R.; Heidemann, J.; Kumar, S. Next century challenges: scalable coordination in sensor networks. ACM MOBICOM 1999, Seattle, WA, Aug 15–20, 1999.
4. Conti, M.; Maselli, G.; Turi, G.; Giordano, S. Cross-layering in mobile ad hoc network design. IEEE Comput. **2004**, *37* (2), 48–51.
5. Kawadia, V.; Kumar, P.R. A cautionary perspective on cross layer design. IEEE Wirel. Commun. Mag. **2005**, *12* (1), 3–11.
6. Kochhal, M.; Schwiebert, L.; Gupta, S.K.S. Role-based hierarchical self organization for ad hoc wireless sensor networks. ACM International Workshop on Wireless Sensor Networks and Applications (WSNA 2003), San Diego, CA, Sept 19, 2003; 98–107.
7. Kochhal, M.; Schwiebert, L.; Gupta, S.K.S. Integrating sensing perspectives for better self organization of ad hoc wireless sensor networks. J. Inf. Sci. Eng. **2004**, *20* (3), 449–475.

8. Rappaport, T.S. Mobile radio propagation: large-scale path loss, small-scale fading and multipath (ch.4). Wireless Communications: Principles and Practice, 2nd Ed.; Prentice Hall: 2001; 105–248.

9. Schwiebert, L.; Gupta, S.K.S.; Weinmann, J.; Salhieh, A.; Kochhal, M.; Auner, G. Research challenges in wireless networks of biomedical sensors. 7th Annual International Conference on Mobile Computing and Networking (MOBICOM 2001), Rome, Italy, July 16–21, 2001; 151–165.

10. Salhieh, A. Energy Efficient Communication in Stationary Wireless Sensor Networks; Ph.D. Thesis, Wayne State University: 2004.

11. Salhieh, A.; Weinmann, J.; Kochhal, M.; Schwiebert, L. Power efficient topologies for wireless sensor networks. International Conference on Parallel Processing, Valencia, Spain, Sept 3–7, 2001; 156–163.

12. Mobile Ad hoc Networks (MANET) Charter. MANET, http://www.ietf.org/html.charters/manet-charter.html.

13. Kochhal, M.; Schwiebert, L.; Gupta, S.K.S.; Jiao, C. An efficient core migration protocol for QoS in mobile ad hoc networks. 21st IEEE International Performance Computing and Communications (IPCCC 2002), Phoenix, AZ, Apr 3–5, 2002; 387–391.

14. Ye, F.; Luo, H.; Cheng, J.; Lu, S.; Zhang, L. A two-tier data dissemination model for large-scale wireless sensor networks. International Conference on Mobile Computing and Networking (MOBICOM 2002), Atlanta, GA, Sept 23–28, 2002; 148–159.

15. Schurgers, C.; Tsiatsis, V.; Ganeriwal, S.; Srivastava, M. Topology management for sensor networks: exploiting latency and density. 3rd ACM International Symposium on Mobile Ad Hoc Networking and Computing (MOBIHOC 2002), Lausanne, Switzerland, June 9–11, 2002; 135–145.

16. Van Dyck, R.E. Detection performance in self-organized wireless sensor networks. IEEE International Symposium on Information Theory, Lausanne, Switzerland, June 30–July 5, 2002.

17. Varshney, P.K. Distributed Detection and Data Fusion; Springer-Verlag: New York, 1996.

18. Bulusu, N. Self-Configuring Localization Systems; Ph.D. Thesis, University of California: Los Angeles (UCLA), 2002.

19. Elson, J.; Estrin, D. Time synchronization for wireless sensor networks. International Parallel and Distributed Processing Systems (IPDPS) Workshop on Parallel and Distributed Computing Issues in Wireless Networks and Mobile Computing, San Francisco, CA, Apr 23–27, 2001; 1965–1970.

20. Elson, J.; Römer, K. Wireless sensor networks: a new regime for time synchronization. ACM Comput. Commun. Rev. 2003, 33 (1), 149–154.

21. Ganeriwal, S.; Kumar, R.; Srivastava, M. Timing-sync protocol for sensor networks. ACM Conference on Embedded Networked Sensor Systems (SenSys 2003), Los Angeles, CA, Nov 5–7, 2003; 138–149.

22. Girod, L.; Bychkovskiy, V.; Elson, J.; Estrin, D. Locating tiny sensors in time and space: a case study. IEEE International Conference on Computer Design, San Jose, CA, Nov 10–14, 2002; 214–219.

23. Schurgers, C.; Kulkarni, G.; Srivastava, M.B. Distributed on-demand address assignment in wireless sensor networks. IEEE Trans. Parallel Distrib. Syst. 2002, 13 (10), 1056–1065.

24. Sankarasubramaniam, Y.; Akan, O.B.; Akyildiz, I.F. Esrt: event-to-sink reliable transport in wireless sensor networks. ACM MobiHoc 2003, Annapolis, MD, June 1–3, 2003.

25. Stann, F.; Heidemann, J. RMST: reliable data transport in sensor networks. 1st International Workshop on Sensor Network Protocols and Applications (SNPA 2003), Anchorage, AK, May 11, 2003.

26. Wan, C.; Campbell, A.; Krishnamurthy, L. PSFQ: a reliable transport protocol for wireless sensor networks. 1st ACM International Workshop on Wireless Sensor Networks and Applications (WSNA 2002), Atlanta, GA, Sept 28, 2002.

27. Meguerdichian, S.; Koushanfar, F.; Potkonjak, M.; Srivastava, M.B. Coverage problems in wireless ad-hoc sensor networks. IEEE INFOCOM 2001, Anchorage, AK, Apr 22–26, 2001; Vol. 3, 1380–1387.

28. Meguerdichian, S.; Koushanfar, F.; Qu, G.; Potkonjak, M. Exposure in wireless ad hoc sensor networks. ACM SIGMOBILE (MobiCom 2001), Rome, Italy, July 16–21, 2001; 139–150.

29. Slijepcevic, S.; Potkonjak, M. Power efficient organization of wireless sensor networks. IEEE International Conference on Communications (ICC 2001), Helsinki, Finland, June 11–14, 2001; 472–476.

30. Abdelzaher, T.; Blum, B.; Cao, Q.; Chen, Y.; Evans, D.; George, J.; George, S. Envirotrack: towards an environmental computing paradigm for distributed sensor networks. IEEE International Conference on Distributed Computing Systems, Tokyo, Japan, Mar 23–26, 2004.

31. Blum, B.; Nagaraddi, P.; Wood, A.; Abdelzaher, T.; Son, S.; Stankovic, J. An entity maintenance and connection service for sensor networks. 1st International Conference on Mobile Systems, Applications, and Services (MobiSys 2003), San Francisco, CA, May 5–8, 2003.

32. EnviroTrack: An Enviromental Programming Paradigm for Sensor Networks. EnviroTrack, http://www.cs.virginia.edu/~ll4p/Enviro Track/.

33. Lindsey, S.; Raghavendra, C.S. PEGASIS: power efficient gathering in sensor information systems. IEEE Aerospace Conference, Big Sky, MT, Mar 19–16, 2002.

34. Inanc, M.; Magdon-Ismail, M.; Yener, B. Power Optimal Connectivity and Coverage in Wireless Sensor Networks. Technical Report 03–06, Rensselaer Polytechnic Institute, Dept. of Computer Science, Troy, NY, 2003.

35. Salhieh, A.; Schwiebert, L. Power aware metrics for wireless sensor networks. IASTED International Conference on Parallel and Distributed Computing and Systems, Cambridge, MA, Nov 4–6, 2002; 326–331.

36. Singh, S.; Woo, M.; Raghavendra, C.S. Power aware routing in mobile ad hoc networks. MOBICOM 1998, Dallas, TX, Oct 25–30, 1998; 181–190.

37. Annamalai, V.; Gupta, S.K.S.; Schwiebert, L. On tree-based convergecasting in wireless sensor networks. IEEE Wireless Communications and Networking Conference, New Orleans, LA, Mar 16–20, 2003; Vol. 3, 1942–1947.

38. Ballardie, A.; Francis, P.; Crowcroft, J. Core based trees (CBT). ACM SIGCOMM 1993, Ithaca, NY, Sept 1993; 85–95.

39. Estrin, D. Farinacci, D.; Helmy, A.; Thaler. D.; Deering, S.; Handley, M.; Jacobson, V.; Liu, C.; Sharma, P.; Wei, L.

Protocol Independent Multicast-Sparse Mode (PIM-SM): Protocol Specification. IETF RFC 2362; June 1998.

40. Adelstein, F.; Richard, G.; Schwiebert, L. Building dynamic multicast trees in mobile networks. ICPP Workshop on Group Communication, Aizu-Wakamatsu City, Japan, Sept 21–24, 1999; 17–22.

41. Gupta, S.K.S.; Srimani, P.K. An adaptive protocol for reliable multicast in mobile multi-hop radio networks. IEEE Workshop on Mobile Computing Systems and Applications, New Orleans, LA, Feb 25–26, 1999; 111–122.

42. Winter, P. Steiner problem in networks: a survey. Netw. **1987**, *17* (2), 129–167.

43. Amis, A.D.; Prakash, R.; Vuong, T.H.P.; Huynh, D.T. Max-Min D-cluster formation in wireless ad hoc networks. IEEE INFOCOM 2000, Tel Aviv, Israel, Mar 26–30, 2000.

44. Chen, G.; Garcia, F.; Solano, J.; Stojmenovic, I. Connectivity based k-hop clustering in wireless networks. IEEE Hawaii International Conference on System Sciences, Hilton Waikoloa Village, HI, Jan 7–10, 2002.

45. Steenstrup, M. Cluster-based networks. In *Ad Hoc Networking*; C.E. Perkins, Ed.; Addison-Wesley: 2000; 75–135.

46. Chen, B.; Jamieson, K.; Balakrishnan, H.; Morris, R. Span: an energy-efficient coordination algorithm for topology maintenance in ad hoc wireless networks. MobiCom 2001, Rome, Italy, July 16–21, 2001; 70–84.

47. Sivakumar, R.; Das, B.; Bharghavan, V. An improved spine-based infrastructure for routing in ad hoc networks. IEEE Symposium on Computers and Communications 1998, Athens, Greece, 30, June–July 2, 1998.

48. Guha, S.; Khuller, S. Approximation Algorithms for Connected Dominating Sets. Technical Report 3660, Univ. of Maryland Inst. for Adv. Computer Studies, Dept. of Computer Science, Univ. of Maryland, College Park, 1996.

49. Wu, J. Dominating set based routing in ad hoc wireless networks. In *Handbook of Wireless and Mobile Computing*, L. Stojmenovic, Ed.; John Wiley: 2002; 425–450.

50. Wu, J.; Li, H. On calculating connected dominating set for efficient routing in ad hoc wireless networks. 3rd International Workshop on Discrete Algorithms and Methods for Mobile Computing and Communications, Seattle, WA, Aug 20, 1999; 7–14.

51. Xu, Y.; Heidemann, J.; Estrin, D. Geography informed energy conservation for ad hoc routing. ACM/IEEE International Conference on Mobile Computing and Networking, Rome, Italy, July 16–21, 2001; 70–84.

52. Chevallay, C.; Van Dyck, R.E.; Hall, T.A. Self-organization protocols for wireless sensor networks. 36th Conference on Information Sciences and Systems, Princeton, NJ, Mar 20–22, 2002.

53. Subramanian, L.; Katz, R.H. An architecture for building self-configurable systems. IEEE/ACM Workshop on Mobile Ad Hoc Networking and Computing (MobiHOC 2000), Boston, MA, Aug 11, 2000.

54. Santi, P. Topology control in wireless ad hoc and sensor networks. ACM Comput. Surv. **2005**, *37* (2), 164–194.

55. Schurgers, C.; Tsiatsis, V.; Srivastava, M. STEM: topology management for energy efficient sensor networks. IEEE Aerospace Conference, Big Sky, MT, Mar 9–16, 2002; 78–89.

56. Tian, D.; Georganas, N.D. A coverage-preserving node scheduling scheme for large wireless sensor networks. ACM Workshop on Wireless Sensor Networks and Applications (WSNA 2002), Atlanta, GA, Sept 30–Oct 1, 2002.

57. Doherty, L.; Ghaoui, L.E.; Pister, K.S.J. Convex position estimation in wireless sensor networks. IEEE INFOCOM 2001, Anchorage, AK, Apr 22–26, 2001.

58. Savvides, A.; Han, C.; Srivastava, M. Dynamic fine-grained localization in ad hoc networks of sensors. ACM MobiCom 2001, Rome, Italy, July 16–21, 2001; 166–179.

59. Savvides, A.; Park, H.; Srivastava, M.B. The bits and flops of the n-hop multilateration primitive for node localization problems. 1st ACM International Workshop on Wireless Sensor Networks and Applications, Atlanta, GA, Sept 28, 2002; 112–121.

60. Guo, C.; Zhong, L.C.; Rabaey, J.M. Low power distributed MAC for ad hoc sensor radio networks. IEEE GLOBECOM 2001, San Antonio, TX, Nov 25–29, 2001.

61. Woo, A.; Culler, D. A transmission control scheme for media access in sensor networks. 7th Annual International Conference on Mobile Computing and Networking (MOBICOM 2001), Rome, Italy, July 16–21, 2001.

62. Ye, W.; Heidemann, J.; Estrin, D. An energy efficient MAC protocol for wireless sensor networks. IEEE INFOCOM 2002, New York, NY, June 23–27, 2002; 3–12.

63. Jamshaid, K.; Schwiebert, L. Seken (secure and efficient key exchange for sensor networks). IEEE Performance Computing and Communications Conference (IPCCC 2004), Phoenix, AZ, Apr 15–17, 2004.

64. Perrig, A.; Szewczyk, R.; Wen, V.; Culler, D.; Tygar, J.D. Spins: security protocols for sensor networks. ACM MobiCom 2001, Rome, Italy, July 16–21, 2001; 189–199.

65. Zhou, L.; Haas, Z.J. Securing ad hoc networks. IEEE Netw. **1999**, *13* (6), 24–30.

66. Ni, S.; Tseng, Y.; Chen, Y.; Chen, J. The broadcast storm problem in a mobile ad hoc network. MOBICOM 1999, Seattle, WA, Aug 15–20, 1999; 151–162.

67. Baker, D.J.; Ephremides, A. The architectural organization of a mobile radio network via a distributed algorithm. IEEE Trans. Commun. **1981**, *COM-29* (11), 1694–1701.

68. Gerla, M.; Tsai, J.T. Multicluster, mobile multimedia radio network. ACM Wirel. Netw. **1995**, *1* (3), 255–266.

69. Lin, C.; Gerla, M. Adaptive clustering for mobile wireless networks. IEEE J. Sel. Areas Commun. **1997**, *15* (7), 1265–1275.

70. McGlynn, M.J.; Borbash, S.A. Birthday protocols for low energy deployment and flexible neighbor discovery in ad hoc wireless networks. 2nd ACM International Symposium on Mobile Ad Hoc Networking and Computing, Long Beach, CA, Oct 4–5, 2001; 137–145.

71. Heinzelman, W.; Chandrakasan, A.; Balakrishnan, H. Energy-Efficient Communication Protocol for Wireless Microsensor Networks. International Conference on System Sciences, Maui, HI, Jan 4–7, 2000.

72. Krishnan, R.; Starobinski, D. Message-efficient self-organization of wireless sensor networks. IEEE WCNC 2003, New Orleans, LA, Mar 16–20, 2003.

73. Ramamoorthy, C.V.; Bhide, A.; Srivastava, J. Reliable clustering techniques for large, mobile packet radio networks. IEEE INFOCOM 1987, San Francisco, CA, Apr 11–18, 1987; 218–226.

74. Elson, J.; Estrin, D. Random, ephemeral transaction identifiers in dynamic sensor networks. 21st International Conference on Distributed Computing Systems, Mesa, AZ, Apr 16–19, 2001.

75. Mirkovic, J.; Venkataramani, G.P.; Lu, S.; Zhang, L. A self organizing approach to data forwarding in large scale

sensor networks. IEEE International Conference on Communications (ICC 2001), Helsinki, Finland, June 11–15, 2001.

76. Clare, L.P.; Pottie, G.J.; Agre, J.R. Self-organizing distributed sensor networks. SPIE, Unattended Ground Sensor Technologies and Applications, Orlando, FL, Apr 8–9, 1999; Vol. 3713, 229–237.

77. Sohrabi, K.; Gao, J.; Ailawadhi, V.; Pottie, G. Protocols for self organization of a wireless sensor network. IEEE Pers. Commun. Mag. 2000, 7 (5), 16–27.

78. Sohrabi, K.; Pottie, G. Performance of a novel self-organization protocol for wireless ad hoc sensor networks. 50th IEEE Vehicle Technology Conference, Amsterdam, The Netherlands, Sept 19–22, 1999.

79. Haas, Z.J.; Pearlman, M.R.; Samar, P. The Zone Routing Protocol (ZRP) for Ad Hoc Networks. 2002, Internet Draft draft-ietf-manet-zone-zrp-04.txt, Internet Engineering Task Force.

80. Blazevic, L.; Buttyan, L.; Capkun, S.; Giordano, S.; Hubaux, J.; Le Boudec, J. Self-organization in mobile ad-hoc networks: the approach of terminodes. IEEE Commun. Mag, 39 (6), 166–174.

81. Bao, L.; Garcia-Luna-Aceves, J.J. Topology management in ad hoc networks. 4th ACM International Symposium on Mobile Ad Hoc Networking and Computing (MOBIHOC 2003), Annapolis, MD, June 1–3, 2003.

82. Ramanathan, R.; Rosales-Hain, R. Topology control of multihop wireless networks using transmit power adjustment. IEEE INFOCOM 2000, Tel-Aviv, Israel, Mar 26–30, 2000.

83. Narayanaswamy, S.; Kawadia, V.; Sreenivas, R.S.; Kumar, P.R. Power control in ad-hoc networks: theory, architecture, algorithm and implementation of the COMPOW protocol. European Wireless 2002. Next Generation Wireless Networks: Technologies, Protocols, Services and Applications, Florence, Italy, Feb 25–28, 2002; 156–162.

84. Prakash, R. Unidirectional links prove costly in wireless ad hoc networks. ACM DIALM 1999 Workshop, Seattle, WA, Aug 20, 1999; 15–22.

85. Xu, Y. Adaptive Energy Conservation Protocols for Wireless Ad Hoc Routing; University of Sourthern California (USC): 2002 Ph.D. Thesis.

86. Xu, Y.; Bien, S.; Mori, Y.; Heidemann, J.; Estrin, D. Topology Control Protocols to Conserve Energy in Wireless Ad Hoc Networks. Technical Report 6, University of California, Los Angeles, Center for Embedded Networked Computing, 2003. Submitted for publication.

87. Tseng, Y.C.; Chang, Y.-N.; Tzeng, B.-H. Energy efficient topology control for wireless ad hoc sensor networks. 23rd International Conference on Distributed Computing Systems Workshops (ICDCSW 2003), Providence, RI, May 19–22, 2003.

88. Kubisch, M.; Karl, H.; Wolsz, A.; Zhong, L.; Rabaey, J. Distributed algorithms for transmission power control in wireless sensor networks. IEEE Wireless Communications and Networking Conference (WCNC 2003), New Orleans, LA, Mar 16–20, 2003.

89. Xu, Y.; Heidemann, J.; Estrin, D. Adaptive Energy-Conserving Routing for Multihop Ad Hoc Networks. Technical Report TR-2000-527, University of California, Los Angeles, Center for Embedded Networked Computing, 2000.

90. Cerpa, A.; Estrin, D. ASCENT: adaptive self-configuring sensor networks topologies. 21st International Annual Joint Conference of the IEEE Computer and Communications Societies (INFOCOM 2002), New York, NY, June 23–27, 2002.

91. Ganeriwal, S.; Kansal, A.; Srivastava, M.B. Self-aware actuation for fault repair in sensor networks. IEEE International Conference on Robotics and Automation (ICRA 2004), New Orleans, LA, Apr 26–May 1, 2004.

92. Mini, R.; Nath, B.; Loureiro, A. A probabilistic approach to predict the energy consumption in wireless sensor networks. IV Workshop de Comunica sem Fio e Computao Mvel, Sao Paulo, Brazil, Oct 23–25, 2002.

93. Mini, R.; Nath, B.; Loureiro, A. Prediction-based approaches to construct the energy map for wireless sensor networks. 21 Simpsio Brasileiro de Redes de Computadores, Natal, RN, Brazil, May 2003.

94. Chessa, S.; Santi, P. Crash faults identification in wireless sensor networks. Comput. Commun. 2002, 25 (14), 1273–1282.

95. Zhao, Y.J.; Govindan, R.; Estrin, D. Sensor network tomography: monitoring wireless sensor networks. Student Research Poster, ACM SIGCOMM 2001, San Diego, CA, Aug 27–31, 2001.

96. Zhao, Y.J.; Govindan, R.; Estrin, D. Residual energy scans for monitoring wireless sensor networks. IEEE Wireless Communications and Networking Conference (WCNC 2002), Orlando, FL, Mar 17–21, 2002.

97. Zhao, Y.J.; Govindan, R.; Estrin, D. Computing aggregates for monitoring wireless sensor networks. 1st IEEE International Workshop on Sensor Network Protocols and Applications (SNPA 2003), Anchorage, AK, May 11, 2003.

98. Zhang, H.; Arora, A. GS3: scalable self-configuration and self-healing in wireless networks. 21st ACM Symposium on Principles of Distributed Computing, Monterey, CA, July 21–24, 2002.

99. Perillo, M.; Heinzelman, W. Optimal sensor management under energy and reliability constraints. IEEE Wireless Communications and Networking Conference (WCNC 2003), New Orleans, LA, Mar 16–20, 2003.

100. Perillo, M.; Heinzelman, W. Sensor Management; Kluwer Academic Publishers: 2004.

WSNs: Routing–
Wireless Trans

Wireless Sensor Networks (WSNs): Time Synchronization

Qing Ye
Liang Cheng
Laboratory of Networking Group (LONGLAB), Department of Computer Science and Engineering, Lehigh University, Bethlehem, Pennsylvania, U.S.A.

Abstract

This entry presents various protocols for time synchronization in wireless sensor networks.

INTRODUCTION

Integration of recent advances in sensing, processing, and communication leads to emerging technologies of wireless sensor networks (WSNs), which interconnect a set of ad hoc deployed sensor nodes by their wireless radios. There exist a wide variety of applications such as battlefield surveillance, habitat tracking, inventory management, and disaster assistance.[1] It is envisioned that WSN will be an important platform to perform real-time monitoring tasks.

Time synchronization is essential for WSN applications. For example, data fusion[2] as a basic function in WSN may require synchronized clocks at different sensors. Consider cases where sensors are deployed in a dense fashion. When an event happens, multiple sensors may report observed phenomena at the same time. Based on synchronized timestamps, redundant messages can be recognized and suppressed to reduce the unnecessary traffic across the network. Time synchronization can also be used to realize synchronized sleeping periods for task scheduling. It is desirable to put sensor nodes into sleeping mode and wake them up to exchange information only when necessary to save battery energy. In this case, time synchronization is vital to maintain the accurate schedule among multiple sensors. Last but not least, with time synchronization, some media access control (MAC) layer protocols (e.g., TDMA) can be realized.[3]

By nature, the hardware clocks at different sensors run at different speeds with different time offset to the universal coordinated time (UTC) provided by NIST. Moreover, their clocks drift differently to the environmental conditions. Network time protocol (NTP)[4] is the Internet standard for time synchronization, which synchronizes computer clocks in a hierarchical way by using primary and secondary time servers. Based on multiple data points, clock skew, offset, and drift can be estimated for time synchronization. However, strict resource limitations of sensor nodes make it difficult, if not impossible, to apply the well-studied NTP in WSNs. The resource constraints in WSN require lightweight design of time synchronization mechanisms that work with small storage occupation, low computation complexity, and little energy consumption.

This entry presents multiple proposed protocols for time synchronization in WSN (third section) and our new lightweight approach (fourth section), which work in the sender–receiver way or a similar fashion. The second section presents a theoretical model for time synchronization based on the sender–receiver schemes of time information exchange used by these protocols. Our new approach can efficiently decrease the number of packet exchanges of time information while still maintaining certain synchronization accuracy. This entry also compares the new approach with the conventional time synchronization mechanisms of two-way message exchanging (TWME). The simulation results in the fifth section also show that the new approach can achieve the time synchronization accuracy of 34 μs in one-hop WSN based on a certain delay model, and its maximum time synchronization error increases with the increment of the number of communication hops in multi-hop WSN.

THEORETICAL MODEL OF TIME SYNCHRONIZATION

The principle of the sender–receiver-based time synchronization requires receivers follow the clock of a sender, which can be regarded as a time server. Consider two networked nodes where the sender S sends out a sequence of timestamped reference packets to the receiver A. There are four delay factors along the packet transmission path: 1) processing delay (P part), which includes the time spent at the node S to prepare and process the reference packet and the time for the sensor's microprocessor to transfer the packet to its networking components; 2) accessing delay

Encyclopedia of Wireless and Mobile Communications DOI: 10.1081/E-EWMC-120043933

(A part), the time for the packet staying in the buffer and waiting for wireless channel access; 3) propagation delay (D part), the time spent from sending the packet from the sender's wireless radio to receiving it by the receiver's wireless radio; and 4) receiving delay (R part), which is the time for the node R to retrieve the packet from the buffer, pass it to upper layer applications, and get the reading of its local clock.

We use *PADR delay* to denote these four factors. Although the propagation delay can be calculated based on the physical distance from S to R, the processing, accessing, and receiving delays are uncertain. They are related to the current work load of sensor nodes, implementation of their MAC layer, capabilities of microprocessors and communication radios, and the size of the reference packets. Based on the illustration in Fig. 1, the time reference relationship between the sender and the receiver satisfies Eq. 1, where a_{RS} is the relative clock skew:

$$t_r = a_{RS}t_s + t_0 + P_S A_S D_{SR} R_R + \text{Drift}_R \qquad (1)$$

Therefore, one of the essential tasks of time synchronization algorithms is to estimate the PADR delay accurately because the drift component is much smaller compared to the PADR delay and t_0 is fixed. In general, the PADR delay can be estimated by the conventional method of TWME. Suppose the node S sends out a packet at time t_{s1}, with respect to (w.r.t.) the local clock at S, which is received by the node A at t_{a1}, w.r.t. A's local clock. Node A sends this packet back with the timestamp t_{a1} and S receives the reply at t_{s1}'. Then the sender S can get two time synchronization lines passing through synchronization points (t_{s1}', t_{a1}) to (t_{s2}', t_{a2}), and (t_{s1}, t_{a1}) to (t_{s2}, t_{a2}), respectively, as shown in Fig. 2, where a_{as} is a_{RS} and assume that Drift_R is 0. It has been proved that the difference between these two lines is approximately two times the PADR delay.[5]

EXISTING WSN SYNCHRONIZATION TECHNIQUES

Tiny-sync and Mini-sync

Tiny-sync and mini-sync[6] are proposed to synchronize local clocks in a pair-wise manner where the clock of the node R follows the time of node S using their bidirectional radios. Timestamped packets that contain the information

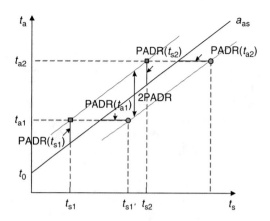

Fig. 2 PADR delay estimation for time synchronization by two-way message exchanging.

of their local clocks are exchanged to estimate the time differences between R and S. Each packet stands for a time point of (t_s, t_r) in the coordinates of the two clocks. After receiving a sufficient number of packets, tiny-sync provides a method to get the upper bound and the lower bound of the time offsets between R and S from a series of time points. In this way, the node R is able to adjust its local clock to follow the node S. Mini-sync extends the idea of tiny-sync to get faster estimation speed with smaller amount of time points. It reduces the heavy traffic overheads of tiny-sync but still requires that at least four packets be exchanged between each pair at every synchronizing cycle.

Neither tiny-sync nor mini-sync keeps the global synchronized time in the WSN. For multi-hop WSNs, they assume that the sensor networks can be organized into a hierarchical structure by a certain level-discovery algorithm and each node only synchronizes with its direct predecessor. The basic idea is shown in Fig. 3. The obvious drawbacks include: 1) heavy traffic of synchronization packets will traverse the WSN because each pair has to perform its own synchronization, and 2) each node has to keep the state information of its ancestors and children. Experiment results show that mini-sync can bound the offset by 945 μs (i.e., time synchronization accuracy) in a little over an hour, corresponding to a drift of 23.3 ms in a day.[6]

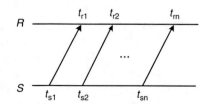

Fig. 1 Referenced time synchronization.

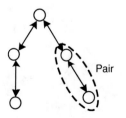

Fig. 3 Time synchronization idea of tiny-sync/mini-sync.

WSNs: Routing—Wireless Trans

Post-facto

Post-facto[7] is a method to synchronize a local neighborhood of sensor nodes in the sender–receiver manner with the help of a "third-party" beacon node. In post-facto, the local clocks inside receivers are normally unsynchronized. When a stimulus from the sender arrives, each receiver records the arrival time using its own local clock. Immediately afterward, a "third-party" node that acts as a beacon broadcasts a synchronization pulse to all the nodes in this area. The receiving nodes then normalize their stimulus timestamps w.r.t. this pulse time reference. Post-facto assumes that the sender's clock is as accurate as that at the "third-party" beacon node. Therefore, each receiver knows the offset between its current local clock and the accurate time. It then adjusts its own time and waits for the next stimulus from the sender. Fig. 4 depicts its basic idea. The solid lines represent the stimulus sent by the sender, and the dotted lines represent the references from the beacon node.

Post-facto does not require any TWME between the sender and receivers. However, it cannot be applied to multi-hop WSN scenarios, and its synchronization range is critically limited by the communication range of the beacon node. In experiments of post-facto presented by Elson and Estrin,[7] the stimulus and sync pulse are sent and received by the standard PC parallel port, which is not wireless, and thus it is reported to achieve an accuracy of 1 μs.

Reference Broadcasting Synchronization

Elson et al.[8] proposed a reference broadcasting synchronization (RBS) that extends the idea of the post-facto. In RBS, a node (the sender) periodically broadcasts reference beacons without an explicit timestamp to its neighbors. The receivers record the arrival timestamps based on their local clocks, and then exchange these observed time values. In this way, all the nodes get acknowledged of the offsets between each other. RBS is different from the conventional sender–receiver synchronization mechanisms because it works in a receiver–receiver manner. The receivers do not synchronize with the sender but try to keep a relative network time among the nodes that receive the same reference from the sender. Fig. 5 illustrates

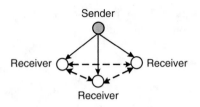

Fig. 5 Time synchronization method of RBS.

the basic idea of the RBS. The solid lines represent the references from the sender while the dotted lines represent the packets exchanged among the receivers.

RBS performs well in synchronizing a neighborhood of receivers with an average accuracy of 29 μs and a worst synchronization case of 93 μs.[9] However, it is not yet applicable for multi-hop WSN. Another

disadvantage is the heavy communication overhead. To synchronize N nodes, the RBS requires at least N packets to be transmitted: one broadcast reference from the sender plus $N-1$ packets from all the receivers to exchange timestamp information. The larger the network scale, the more traffic overhead introduced by the RBS.

Time-sync Protocol for Sensor Networks

A recently proposed sender–receiver time synchronization protocol is time-sync protocol for sensor networks (TPSNs).[9] It is applicable for both single- and multi-hop WSNs. TPSN first organizes a randomly deployed sensor network into a hierarchical structure within a level discovery period. The new nodes that join the network after that period can get their own level values by broadcasting *level request* messages to their neighbors. Time synchronization is performed in a pair-wise fashion along the edges of this tree-like structured network, from the root to all leaf nodes, level by level. TPSN uses the conventional sender–receiver time synchronization mechanism, i.e., to synchronize clocks in a pair by TWME. Its significant contributions include that 1) it provides a good mechanism to accurately estimate the time delays along communication path between two nodes, and 2) it is implemented to be a MAC layer component of TinyOS,[10] which is the operating system for Berkeley Motes.[11] As reported by Ganeriwal et al.,[9] TPSN can achieve the one-hop time accuracy of 44 μs and get almost two times better performance when compared to the RBS.

However, TPSN does not reduce the synchronization traffic overhead, because it is based on two-way handshaking between every synchronized pair. Another disadvantage is that a receiver may have more than one sender at its upper level. In this case, there would be some redundant synchronization operations performed. Moreover, each sender at the upper level must keep all state information

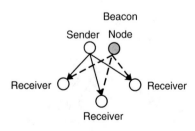

Fig. 4 Time synchronization method of post-facto.

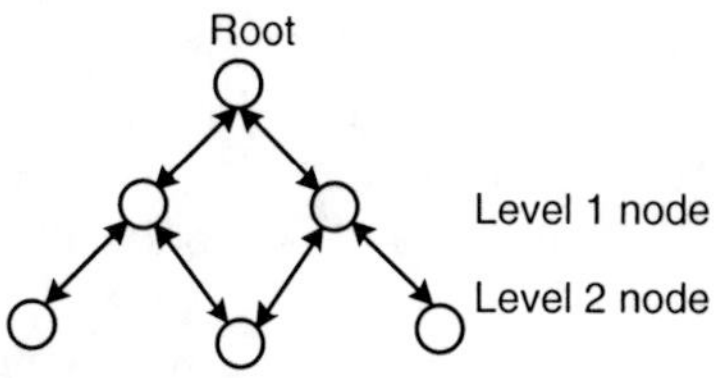

Root

Level 1 node

Level 2 node

Fig. 6 Time synchronization method of TPSN: two-way handshaking is needed between levels.

of its receivers at the lower level when it tries to synchronize them. Fig. 6 shows the basic idea of TPSN.

Delay Measurement Time Synchronization

Ping[12] has proposed a new approach called delay measurement time synchronization (DMTS) for WSNs. DMTS organizes the sensor networks into a hierarchical structure in the same way as TPSN. It is designed to suppress the number of synchronization packets by only performing one-way message transmissionfrom the time leaders at the upper level to the receivers at the lower level, as depicted in Fig. 7. Thus, it synchronizes the global network time with small traffic overhead. For example, only one broadcast packet is needed for a single-hop WSN with N nodes, regardless of the value of N. DMTS estimates the time delays along the communication path of two nodes with an assumption that the distance and bandwidth between these two nodes and the communication packet size are known. However, this assumption may be impractical and system related. Different platforms of WSN may have different communication parameters for different WSN applications. DMTS can achieve an accuracy of 32 μs in one-hop WSNs.[12]

Note that TPSN and DMTS represent two extremes of the time synchronization schemes. TPSN is based on pure two-way message transmissions of time information between any pairs, while DMTS performs one-way message transmissions. In general, time synchronization via a two-way transmission approach has better time synchronization accuracy but may not be lightweight in terms of computation complexity, storage consumption, and traffic overhead, while a one-way transmission approach does the

opposite. There would be trade-offs between decreasing time synchronization traffic overhead and achieving acceptable synchronization accuracy. This motivates the new design approach presented in this entry. Also note that current performance discussions of time synchronization protocols are based on stable WSN scenarios; that is, no sensor node moves around.

A NEW LIGHTWEIGHT APPROACH FOR TIME SYNCHRONIZATION

Time Synchronization in Single-hop WSN

In a single-hop WSN, the sink (the sender) can synchronize a set of receivers by estimating the PADR delay for each receiver. To decrease the number of communication packets, we provide a lightweight solution with the basic idea of selecting an *adjuster* node from the receiver group as depicted in Fig. 8. Only the adjuster node is required to do TWMEs with the sink and estimate the PADR delay for the purpose of time synchronization. All the others will take the estimated PADR delay by the adjuster as their own estimations. In this way, no matter how large the one-hop WSN is, only five packets are needed: two references broadcast by the sender plus two replies from the adjuster, and the fifth packet from the sender synchronizing all the receivers. And the computation and storage tasks in the sink would be simple. Because the adjuster is selected from the receivers' group, it is expected to have properties similar to other receivers, from the type and frequency of internal oscillators to the workload of their jobs. This expectation should be reasonable, considering the WSN applications mentioned in the "Introduction" section. Another estimation error caused by the differences of the receivers' physical distances to the sink is the communication range of the sink divided by the speed of light, which should be a tiny value.

Time Synchronization in Multi-hop WSN

The idea presented in the previous subsection can be extended to multi-hop WSNs. It first organizes the

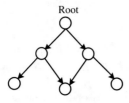

Fig. 7 Time synchronization method of DMTS: only one-way message transmission between levels.

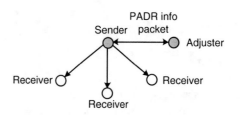

Fig. 8 Basic idea of our approach: two-way handshaking only between the sender and the adjuster.

overall network into a hierarchical structure with different levels. Assuming that each node has a unique ID, then the level discovery can be performed in a flooding way.

The sink as the root is assigned level 0 and broadcasts a *level discovery* packet to its neighbors. The nodes that receive this packet are assigned level 1 and broadcast their own *level discovery* packets containing the new level information to other nodes. A node can receive several such packets but it can only accept the packet with the lowest level number as the one from its predecessor and takes this value plus one as its own level. Then the broadcasting operation continues. Eventually, all the sensor nodes get a unique level number assigned. A node at an upper level (i.e., a node with a smaller level number) means it is closer to the sink in terms of number of hops.

A sender selection algorithm, which is illustrated in "Sender Selection Algorithm" section, is used to select at least one sender per level. The communication ranges of these senders and the sink should guarantee to cover the overall WSN. Our time synchronization approach for multi-hop WSN is performed as follows:

1. The sink is selected as the sender for the nodes at level 1. It periodically broadcasts time references and performs one hop algorithm from level 0 to level 1 by randomly picking up an adjuster from level 1 nodes. Thus, level 1 is synchronized with the sink.
2. At least one sender will then be selected from the level 1 nodes as the senders for the level 2 nodes. The communication ranges of these selected senders must cover all the nodes at level 2. Then, in each sender's neighborhood, a node at the level 2 is randomly picked up to act as the adjuster. Thus, the level 2 is synchronized with the level 1.
3. In general, nodes at level n synchronized with their level $(n-1)$ predecessors may act as the senders for level $(n+1)$ nodes to perform single-hop synchronization. Only randomly picked adjusters conduct TWME with their upper level senders.
4. This process is finished when all the nodes are synchronized. When a time reference packet is received, a node will check the level value of the source. If it is from a sender in the upper level, it accepts the time reference. Otherwise, it discards the packet silently.

Sender Selection Algorithm

Centralized method

A centralized sender selection algorithm can only be performed at the sink node. After the level discovery phase, the sink can gain knowledge of the overall network topology by asking every node to report the information of its neighbors and its level value. To decrease the traffic overhead and keep the entire network synchronized, two goals should be achieved: 1) the minimum number of senders is selected, and 2) the overall network scope should be covered.

This specific network coverage problem can be transferred to a $0-1$ *integer programming* problem. The leveled WSN can be envisioned as a directed graph $G(V, E)$, which consists of a set V of sensors and a set E of communication paths. Consider the node–node adjacency matrix A of G: $a_{ij} = 1$ means that there is an arc (i,j) in E with node i directly connected to node j wirelessly, and node i is exactly located at node j's upper level (i.e., $level_i = level_j - 1$), otherwise $a_{ij} = 0$. a_{ij} also equals 0 if $i = j$. Notice that the node–node adjacency matrix A actually represents the "from–to" relationships of the sensor nodes. Take vector x as the decision variable; then the sender selection problem can be modeled as follows:

$$\{x_1 + x_2 +, \ldots, +x_n\} \tag{2}$$

$$\text{s.t.} \sum_{j=1}^{n} a_{ij}x_j \geqslant 1, \; x_1 = 1, \; x_j \in \{0,1\} \quad \forall j \in V \tag{3}$$

Eq. 2 represents the optimal goal of the sender selection: to select the least number of senders possible. The model is subject to the condition expression (Eq. 3), which means that each node should have at least one arc directed to itself from its upper level, and $x_1 = 1$ because the sink must be selected. This $0-1$ integer programming problem is always solvable because there is at least one feasible solution for selecting all the non-leaf nodes in the graph G. This model can be easily solved by branch and bound algorithm with the commercial optimization solver CPLEX.[13]

Distributed method

Although the centralized method can solve the sender selection problem, it is not efficient due to a large number of communication packets that are needed to gather the overall network topology and a huge computation task has to be done by the sink. We then design a distributed method to locally select senders at each level.

In fact, the above specific network coverage problem can be solved by a greedy select-and-prune method as follows.

1. *Report.* Each node at level n only reports its neighbors at level $(n+1)$ up to its direct predecessor at level $(n-1)$.
2. *Select.* The predecessor then selects the one that has the maximum number of neighbors to be the sender. This node covers its children at level $(n+1)$.

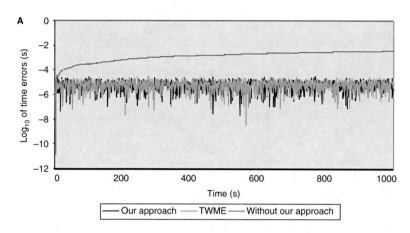

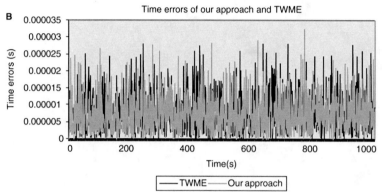

Fig. 9 Time synchronization errors using different approaches in one-hop WSN scenarios.

3. *Prune.* Then the predecessor deletes all the nodes covered by the selected sender from the neighbors of those unselected nodes at level *n,* and goes back to Step 2 until all the nodes at level $(n + 1)$ are covered.

This distributed greedy algorithm is performed in an up down manner from level 0 to the entire network. The distributed sender selection method is able to get the same result as the centralized approach but in a more scalable and practical way.

SIMULATION RESULTS

Network stimulator-2 (NS-2) has been used[14] to simulate our new time synchronization approach in both single- and multi-hop WSN scenarios. The PADR delay associated with each node is taken as an error factor, with the mean value of 50 μs. All simulations run for 1000 sec.

In single-hop scenarios, the average time synchronization errors at the receivers are recorded every second. Under the same network topology, three cases have been studied and compared, including the pure TWME, our approach, and a case without using any time synchronization method at all. Fig. 9 shows the one-hop simulation results. The Fig. 9A results demonstrate that both time synchronization schemes can

effectively synchronize the receiver's local time with the sender.

Fig. 9B illustrates that the maximum time synchronization error of our approach in one-hop is 34 μs, while that of the TWME is 28 μs. This proves that our new approach can achieve a certain acceptable synchronization accuracy but it significantly decreases communication overhead by only requiring adjusters to perform TWME.

In multi-hop WSN scenarios, the performance of our new approach in the worst case is studied, where all the senders are located in a linked list from one to six hops. The senders also act as the adjusters. Fig. 10A depicts that the maximum time synchronization error increases as the number of network hops increases. This phenomenon happens because the approach organizes WSNs into a hierarchical structure. Time errors will be propagated across levels. Fig. 10B shows that the average time synchronization error increases almost linearly with the increase in the number of network hops.

Fig. 11 illustrates an example to solve the sender selection problem by the centralized method. Twenty nodes are randomly deployed in a 10×10 area shown in Fig. 11A with the communication range of four units, and the sender selection results by CPLEX are shown in Fig. 11B. Using the distributed method, the same result has been generated.

The benefit of using the distributed method is that it decreases the communication overhead because every

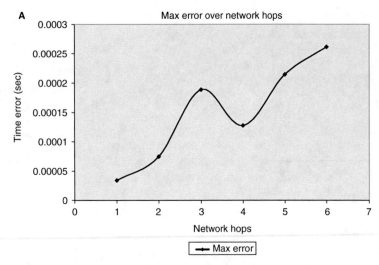

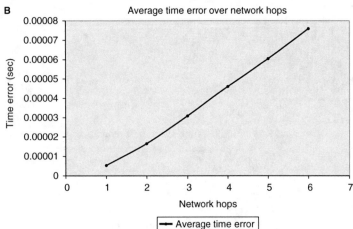

Fig. 10 Maximum and average time synchronization errors from one-hop to six-hop WSN scenarios.

node only needs to care about its children nodes and no global information is required. The overhead comparison between the centralized and distributed methods is shown in Fig. 12. In our simulations, either 100 or 200 sensor nodes are randomly deployed in a 300 × 300 area. The comparison results clearly show that the overheads of both the centralized method and the distributed method increases with increments in the complexity of the network topology in terms of network hops and the number of sensor nodes. However, the overhead of the distributed

sender selection method is much less than that of the centralized one.

CONCLUSION

This entry presented a theoretical model for time synchronization based on the sender–receiver paradigm and summarized the existing time synchronization algorithms proposed for WSN. Then a new lightweight

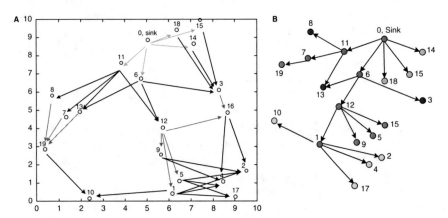

Fig. 11 An example of sender selection by the centralized and distributed methods.

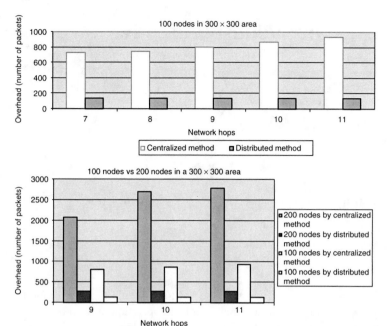

Fig. 12 Overhead comparison between the centralized and distributed sender selection methods.

time synchronization protocol for WSN was presented. The new approach suppresses the communication overhead in terms of the number of time synchronization packets needed by requiring only part of overall nodes in the WSN to perform two-way handshaking while still achieving acceptable time synchronization accuracy. Simulation results have validated the authors' design ideas. Moreover, the simulations reveal that the maximum and average time synchronization errors increase when communication hops increase.

REFERENCES

1. Akyildiz, I.F.; Su, W.; Sankarasubramaniam, Y.; Cayirci, E. A survey on sensor networks. IEEE Commun. Mag. **2002**, *40* (8), 102–116.

2. Heinzelman, W.R.; Chandrakasan, A.; Balakrishnan, H. Energy efficient communication protocol for wireless microsensor network. 33rd Hawaii International Conference on System Sciences (HICSS 2000), Maui, HI, Jan 4–7, 2000; Vol. 8.

3. Claesso, V.; Lönn, H.; Suri, N. Efficient TDMA synchronization for distributed embedded systems. 20th Symposium on Reliable Distributed Systems (SRDS 2001), New Orleans, LA, Oct 28–31, 2001; 198–201.

4. Mills, D.L. Internet time synchronization: the network time protocol. IEEE Trans. Commun. **1991**, *39* (10), 1482–1493.

5. Ye, Q.; Zhang, Y.; Cheng, L. A study on the optimal time synchronization accuracy in wireless sensor networks. Comput. Netw. **2004**, *48*, 549–566.

6. Sichitiu, M.L.; Veerarittiphan, C. Simple accurate time synchronization for wireless sensor networks. IEEE Wireless Communiactions and Networking Conference 2003 (WCNC 2003), New Orleans, LA, Mar 16–20, 2003.

7. Elson, J.; Estrin, D. Time synchronization for wireless sensor networks. IPDPS Workshop on Parallel and Distributed Computing Issues in Wireless Networks and Mobile Computing (IPDPS 2001), San Francisco, CA, Apr 23–27, 2001; 186–191.

8. Elson, J.; Girod, L.; Estrin, D. Fine-grained network time synchronization using reference broadcasts. 5th Symposium on Operating Systems Design and Implementation (OSDI 2002), Boston, MA, Dec 9–11, 2002.

9. Ganeriwal, S.; Kumar, R.; Srivastava, M.B. Timing-sync protocol for sensor networks. 1st ACM International Conference on Embedded Networked Sensor Systems (Sensys 2003), Los Angeles, CA, Nov 5–7, 2003.

10. Levis, P.; Madden, S.; Gay, D.; Polastre, J.; Szewczyk, R.; Woo, A.; Brewer, E.; Culler, D. The emergence of networking abstractions and techniques in TinyOS. 1st USENIX/ACM Symposium on Networked Systems Design and Implementation (NSDI 2004), San Francisco, CA, Mar 29–31, 2004.

11. Warneke, B.A. *Ultra-Low Energy Architectures and Circuits for Cubic Millimeter Distributed Wireless Sensor Networks.* Ph.D. Dissertation, University of California, Berkeley, 2003.

12. Ping, S. Delay Measurement Time Synchronization for Wireless Sensor Networks. Intel Research Berkeley Technical Report 03–013. **2003**.

13. CPLEX, http://www.ilog.com/products/cplex.

14. The Network Simulator—ns-2, http://www.isi.edu/nsnam/ns/.

Wireless Transceivers: Near-Instantaneously Adaptive

Lie-Liang Yang
Lajos Hanzo
University of Southampton, Southampton, U.K.

Abstract

This entry describes several adaptive wireless transceivers based on channel-quality controlled rate adaptation.

INTRODUCTION

There is a range of activities in various parts of the globe concerning the standardization, research, and development of the 3G mobile systems known as the UMTS in Europe, which was termed as the IMT-2000 system by the ITU.[1,2] This is mainly due to the explosive expansion of the Internet and the continued dramatic increase in demand for all types of advanced wireless multimedia services, including video telephony as well as the more conventional voice and data services. However, advanced high-rate services such as high-resolution interactive video and "telepresence" services require data rates in excess of 2 Mbps, which are unlikely to be supported by the 3G systems.[3–7] These challenges remain to be solved by future mobile broadband systems (MBSs).

The most recent version of the IMT-2000 standard is, in fact, constituted by a range of five independent standards. These are the UTRA frequency division duplex (FDD) wideband CDMA (W-CDMA) mode,[8] the UTRA time division duplex (TDD) CDMA mode, the Pan-American multi-carrier CDMA configuration mode known as cdma2000,[8] the Pan-American TDMA mode known as UWT-136, and the Digital European Cordless Telecommunications (DECT)[8] mode. It would be desirable for future systems to become part of this standard framework without having to define new standards. The framework proposed in this contribution is capable of satisfying this requirement.

More specifically, these future wireless systems are expected to cater for a range of requirements. Firstly, MBSs are expected to support extremely high bit rate services while exhibiting different traffic characteristics and satisfying the required quality of service (QoS) guarantees.[3] The objective is that mobile users become capable of accessing the range of broadband services available for fixed users at data rates up to 155 Mbps. Multi-standard operation is also an essential requirement. Furthermore,

these systems are expected to be highly flexible, supporting multimode and multiband operation as well as global roaming, while achieving the highest possible spectral efficiency. These features have to be sustained under adverse operating conditions, i.e., for high-speed users, for dynamically fluctuating traffic loads, and over hostile propagation channels. These requirements can be conveniently satisfied with the aid of broadband mobile wireless systems based on the concept of adaptive software defined radio (SDR) architectures.[9,10]

In the first part of this entry, a broadband multiple access candidate scheme meeting the above requirements is presented, which is constituted by frequency-hopping (FH) based multicarrier direct spectrum (DS)-CDMA (FH/MC DS-CDMA).[11–14] Recent investigations demonstrated that channel-quality controlled rate adaptation is an efficient strategy for attaining the highest possible spectral efficiency in terms of b/sec/Hz,[15–19] while maintaining a certain target integrity. Hence, in the second part of this entry we consider adaptive rate transmission (ART) schemes associated with supporting both time-variant rate and multirate services. These ART techniques are discussed in the context of the proposed FH/MC DS-CDMA system, arguing that SDRs constitue a convenient framework for their implementation. Therefore, in the final part of this contribution, the concept of SDR-assisted broadband FH/MC DS-CDMA is presented and the range of reconfigurable parameters are described with the aim of outlining a set of promising research topics. Let us now commence our detailed discourse concerning the proposed FH/MC DS-CDMA system.

FH/MC DS-CDMA

The transmitter schematic of the proposed FH/MC DS-CDMA arrangement is depicted in Fig. 1. Each subcarrier

Encyclopedia of Wireless and Mobile Communications DOI: 10.1081/E-EWMC-120043955

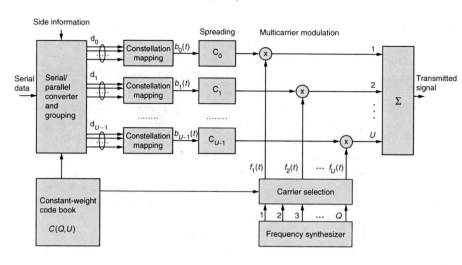

Fig. 1 Transmitter diagram of the frequency-hopping multicarrier DS-CDMA system using adaptive transmission, ©IEEE.[32]
Source: L.L. Yang and L. Hanzo, Software defined radio assisted adaptive broadband frequency hopping multicarrier DS-CDMA, IEEE Commun. Mag., 2002, *40*, 174–183.

of a user is assigned a pseudo-noise (PN) spreading sequence. These PN sequences can be simultaneously assigned to a number of users, provided that only one user activates the same PN sequence on the same subcarrier. These PN sequences produce narrow-band DS-CDMA signals. In Fig. 1, $C(Q,U)$ represents a constant-weight code having U number of 1s and $(Q-U)$ number of 0s. Hence, the weight of $C(Q,U)$ is U. This code is read from a so-called constant-weight code book, which represents the FH patterns. The constant-weight code $C(Q,U)$ plays two different roles. Its first role is that its weight—namely U—determines the number of subcarriers invoked, while its second function is that the positions of the U number of binary 1s determines the selection of a set of U number of subcarrier frequencies from the Q outputs of the frequency synthesizer. Furthermore, in the transmitter, side-information reflecting the channel's instantaneous quality might be employed in order to control its transmission and coding mode so that the required target throughput and transmission integrity requirements are met.

As shown in Fig. 1, the original bit stream having a bit duration of T_b is first serial-to-parallel (S–P) converted. Then, these parallel bit streams are grouped and mapped to the potentially time-variant modulation constellations of the U active subcarriers. Let us assume that the number of bits transmitted by an FH/MC DS-CDMA symbol is M, and let us denote the symbol duration of the FH/MC DS-CDMA signal by T_s. Then, if the system is designed for achieving a high processing gain and for mitigating the inter-symbol-interference (ISI) in a constant-rate transmission scheme, the symbol duration can be extended to a multiple of the bit duration, i.e., $T_s = MT_b$. In contrast, if the design aims to support multiple transmission rates or channel-quality matched variable information rates, then a constant bit duration of $T_0 = T_s$ can be employed. Both multirate and variable rate transmissions can be implemented by employing a different number of subcarriers associated with different modulation constellations as well as different spreading gains. As seen in Fig. 1, after the constellation mapping stage, each branch is DS spread using the assigned PN sequence, and then this spread

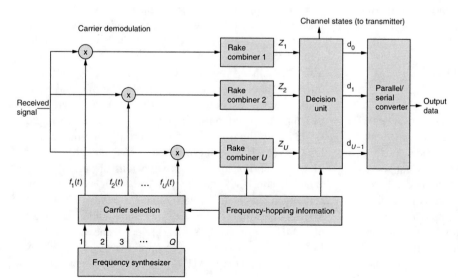

Fig. 2 Receiver block diagram of the frequency-hopping multicarrier DS-CDMA system using conventional RAKE receiver, © IEEE.[32]

signal is carrier modulated using one of the active subcarrier frequencies derived from the constant-weight code C (Q,U). Finally, all U active branch signals are multiplexed in order to form the transmitted signal.

In the FH/MC DS-CDMA receiver of Fig. 2, the received signal associated with each active subcarrier is detected using, e.g., a RAKE combiner. Alternatively, multiuser detection (MUD) can be invoked in order to approach the single-user bound. In contrast to the transmitter side, where only U out of Q subcarriers are transmitted by a user, at the receiver different detector structures might be implemented based on the availability[14] or lack[20] of the FH pattern information. During the FH pattern acquisition stage, which usually happens at the beginning of transmission or during handover, tentatively all Q subcarriers can be demodulated. The transmitted information can be detected and the FH patterns can be acquired simultaneously by using blind joint detection algorithms exploiting the characteristics of the constant-weight codes.[12,13] If, however, the receiver has the explicit knowledge of the FH patterns, then only U subcarriers have to be demodulated. However, if FFT techniques are employed for demodulation, as often is the case in multicarrier CDMA[21] or orthogonal frequency division multiplexing (OFDM)[22] systems, then all Q subcarriers might be demodulated, where the inactive subcarriers only output noise. In the decision unit of Fig. 2, these noise-output-only branches can be eliminated by exploiting the knowledge of the FH patterns.[14,20] Hence, the decision unit only outputs the information transmitted by the active subcarriers. Finally, the decision unit's output information is parallel-to-serial converted to form the output data.

At the receiver, the channel states associated with all the subchannels might be estimated or predicted using pilot signals.[17,19] This channel state information can be utilized for coherent demodulation. It can also be fed back to the transmitter as highly protected side-information in order to invoke a range of adaptive transmission schemes including power control and adaptive-rate transmission.

CHARACTERISTICS OF THE FH/MC DS-CDMA SYSTEMS

In the proposed FH/MC DS-CDMA system, the entire bandwidth of future broadband systems can be divided into a number of subbands, and each subband can be assigned a subcarrier. According to the prevalent service requirements, the set of legitimate subcarriers can be distributed in line with the users' instantaneous information rate requirements. FH techniques are employed for each user in order to evenly occupy the whole system bandwidth available and to efficiently utilize the available frequency resources. In this respect, FH/MC DS-CDMA

systems exhibit compatibility with the existing 2G and 3G CDMA systems and, hence, constitute a highly flexible air interface.

Broadband wireless mobile system—To elaborate a little further, our advocated FH/MC DS-CDMA system is a broadband wireless mobile system constituted by multiple narrow-band DS-CDMA subsystems. Again, FH techniques are employed for each user in order to evenly occupy the whole system bandwidth and to efficiently utilize the available frequency resources. The constant-weight code based FH patterns used in the schematic of Fig. 1 are invoked in order to control the number of subcarriers invoked which is kept constant during the FH procedure. In contrast to single-carrier broadband DS-CDMA systems such as W-CDMA[2] exhibiting a bandwidth in excess of 5 MHz—which inevitably results in extremely high-chip-rate spreading sequences and high-complexity—the proposed FH/MC DS-CDMA system does not have to use high chip-rate DS spreading sequences since each subcarrier conveys a narrow-band DS-CDMA signal. In contrast to broadband OFDM systems[6]—which have to use a high number of subcarriers and usually result in a high peak-to-average power fluctuation—due to the associated DS spreading, the number of subcarriers of the advocated broadband wireless FH/MC DS-CDMA system may be significantly lower. This potentially mitigates the crest-factor problem. Additionally, with the aid of FH, the peak-to-average power fluctuation of the FH/MC DS-CDMA system might be further decreased. In other words, the FH/MC DS-CDMA system is capable of combining the best features of single-carrier DS-CDMA and OFDM systems while avoiding many of their individual shortcomings. Finally, in comparison to the FH/MC DS-CDMA system, both broadband single-carrier DS-CDMA systems and broadband OFDM systems are less amenable to interworking with the existing 2G and 3G wireless communication systems. Let us now characterize some of the features of the proposed system in more depth.

Compatibility—The broadband FH/MC DS-CDMA system can be rolled out over the bands of the 2G and 3G mobile wireless systems and/or in the band licensed for future broadband wireless communication systems. In FH/MC DS-CDMA systems, the subbands associated with different subcarriers are not required to be of equal bandwidth. Hence, existing 2G and 3G CDMA systems can be supported using one or more subcarriers. For example, Fig. 3 shows the spectrum of a FH, orthogonal multicarrier DS-CDMA signal using a subchannel bandwidth of 1.25 MHz, which constitutes the bandwidth of a DS-CDMA signal in the IS-95 standard.[1]. In Fig. 3, we also show that seven subchannels, each having a bandwidth of 1.25 MHz, can be replaced by one subchannel with a bandwidth of 5 MHz ($= 8 \times 1.25/2$ MHz). Hence, the narrow-band IS-95 CDMA system can be supported by a single subcarrier, while the UMTS and IMT-2000 W-CDMA

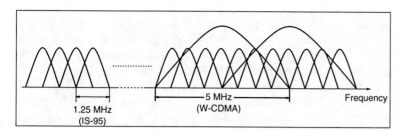

Fig. 3 Spectrum of FH/MC DS-CDMA signal using subchannel bandwidth of 1.25 MHz and/or 5 MHz, ©IEEE.[32] *Source*: L.L.Yang and L. Hanzo, Software defined radio assisted adaptive broadband frequency hopping multicarrier DS-CDMA, IEEE Commun. Mag., 2002, *40*, 174–183. © IEEE. All rights reserved.

systems might be supported using seven subchannels' bandwidth amalgamated into one W-CDMA channel. Moreover, with the aid of SDRs, FH/MC DS-CDMA is also capable of embracing other existing standards, such as the TDMA based global system of mobile communications known as GSM.[8]

FH strategy—In FH/MC DS-CDMA systems, both slow FH and fast FH techniques can be invoked, depending on the system's design and the state-of-the-art. In slow FH, several symbols are transmitted after each FH, while in fast FH, several frequency hops take place in a symbol duration, i.e., each symbol is transmitted using several subcarriers. Moreover, from a networking point of view, random FH, uniform FH, and adaptive FH[14] schemes can be utilized in order to maximize the efficiency of the network. In the context of *random* FH,[14] the subcarriers associated with each transmission of a user are determined by the set of pre-assigned FH patterns constituting a group of constant-weight codewords.[20] The active subcarriers are switched from one group of frequencies to another without the knowledge of the FH patterns of the other users. In contrast, for the FH/MC DS-CDMA system using *uniform* FH,[14] the FH patterns of all users are determined jointly under the control of the base station (BS), so that each subcarrier is activated by a similar number of users. It can be shown that for the downlink (DL), uniform FH can be readily implemented since the BS has the knowledge of the FH patterns of all users. However, for implementing uplink (UL) transmissions, the FH patterns to be used must be signaled by the BS to each mobile station (MS) in order to be able to implement uniform FH. Finally, if the near-instantaneous channel quality information is available at the transmitter, advanced adaptive FH can be invoked, where information is only transmitted over a group of subchannels exhibiting a satisfactory signal-to-interference ratio (SIR).

Implementation of multicarrier modulation—The multicarrier modulation block in Fig. 1 and the multicarrier demodulation block in Fig. 2 can be implemented using FFT techniques, provided that each of the subchannels occupies the same bandwidth. Since not all of the subcarriers are activated at each transmission in the proposed FH/MC DS-CDMA system, the deactivated subcarriers can be set to zero in the FFT or IFFT algorithm. However, if an unequal bandwidth is associated with the subchannels, multicarrier modulation/demodulation can only be implemented using less efficient conventional, rather than FFT based carrier modulation/demodulation, schemes.

Access strategy—When a new user attempts to access the channel and commences his/her transmission, a range of different access strategies might be offered by the FH/MC DS-CDMA system in order to minimize the interference inflicted by the new user to the already active users. Specifically, if there are subchannels which are not occupied by any other users, or if there are subchannels that exhibit a sufficiently high SIR, then the new user can access the network using these passive subchannels or the subchannels exhibiting a high SIR. However, if all the subchannels have been occupied and the SIR of each of the subchannels is insufficiently high, then the new user accesses the network by spreading its transmitted energy evenly across the subchannels. This access scheme imposes the minimum possible impact on the QoS of the users already actively communicating. However, the simplest strategy for a new user to access the network is by randomly selecting one or several subchannels.

Multirate and variable rate services—In FH/MC DS-CDMA systems, multirate and variable rate services can be implemented using a variety of approaches. Specifically, the existing techniques, such as employing a variable spreading gain, multiple spreading codes, a variable constellation size, variable-rate forward error correction (FEC) coding, etc., can be invoked to provide multirate and variable rate services. Furthermore, since the proposed FH/MC DS-CDMA systems use constant-weight code based FH patterns, multirate and variable rate services can also be supported by using constant-weight codes having different weights, i.e., by activating a different number of subcarriers. Note that the above-mentioned techniques can be implemented either separately or jointly in a system.

Diversity—The FH/MC DS-CDMA system includes FH, multicarrier modulation, as well as direct-sequence spreading, hence a variety of diversity schemes and their combinations can be implemented. The possible diversity schemes include the following arrangements.

- If the chip-duration of the spreading sequences is lower than the maximum delay spread of the fading channels, then frequency diversity can be achieved on each of the subcarrier signals.

- Frequency diversity can also be achieved by transmitting the same signal using a number of different subcarriers.
- Time diversity can be achieved by using slow FH in conjunction with error control coding as well as interleaving.
- Time-frequency diversity can be achieved by using fast FH techniques, where the same symbol is transmitted using several time slots assigned to different frequencies.
- Spatial diversity can be achieved by using multiple transmit antennas, multiple receiver antennas, and polarization.

Initial synchronization—In our FH/MC DS-CDMA system, initial synchronization can be implemented by first accomplishing DS code acquisition. The fixed-weight code book index of the FH pattern used can be readily acquired once DS code acquisition is achieved. During DS code acquisition, the training signal supporting the initial synchronization, which is usually the carrier modulated signal without data modulation, can be transmitted using a group of subcarriers. These subcarrier signals can be combined at the receiver using, e.g., equal gain combining (EGC).[23] Hence, frequency diversity can be employed during the DS code acquisition stage of the FH/MC DS-CDMA system's operation, and, consequently, the initial synchronization performance can be significantly enhanced. Following the DS code acquisition phase, data transmission can be activated and the index of the FH pattern used can be signaled to the receiver using a given set of fixed subchannels. Alternatively, the index of the FH pattern can be acquired blindly from the received signal with the aid of a group of constant-weight codes having a given minimum distance.[13]

Interference resistance—The FH/MC DS-CDMA system concerned can mitigate the effects of ISI encountered during high-speed transmissions, and it readily supports partial-band and multitone interference suppression. Moreover, the multiuser interference can be suppressed by using MUD techniques,[15] potentially approaching the single-user performance.

Advanced technologies—The FH/MC DS-CDMA system can efficiently amalgamate the techniques of FH, OFDM, and DS-CDMA. Simultaneously, a variety of performance enhancement techniques, such as MUD,[24] turbo coding,[25] adaptive antennas,[26] space-time coding and transmitter diversity,[27] near-instantaneously adaptive modulation,[16] etc., might be introduced.

Flexibility—The future generation broadband mobile wireless systems will aim to support a wide range of services and bit rates. The transmission rates may vary from voice and low-rate messages to very high-rate multimedia services requiring rates in excess of 100 Mbps.[3] The communications environments vary in terms of their grade of mobility, the cellular infrastructure, the required

symmetric and asymmetric transmission capacity, and whether indoor, outdoor, urban, or rural area propagation scenarios are encountered. Hence, flexible air interfaces are required which are capable of maximizing the area spectrum efficiency expressed in terms of bits/sec/Hz/km^2 in a variety of communication environments. Future systems are also expected to support various types of services based on ATM and IP, which require various levels of QoS. As argued before, FH/MC DS-CDMA systems exhibit a high grade of compatibility with existing systems. These systems also benefit from the employment of FH, MC, and DS spreading-based diversity assisted adaptive modulation.[22] In short, FH/MC DS-CDMA systems constitute a high-flexibility air interface.

ART

Why ART?

There are a range of issues which motivate the application of ARTs in the broadband mobile wireless communication systems of the near future. The explosive growth of the Internet and the continued dramatic increase in demand for all types of wireless services are fueling the demand for increasing the user capacity, data rates, and the variety of services supported. Typical low-data-rate applications include audio conferencing, voice mail, messaging, e-mail, facsimile, and so on. Medium- to high-data-rate applications encompass file transfer, Internet access, high-speed packet- and circuit-based network access, as well as high-quality video conferencing. Furthermore, the broadband wireless systems in the future are also expected to support real-time multimedia services, which provide concurrent video, audio, and data services to support advanced interactive applications. Hence, in the future generation mobile wireless communication systems, a wide range of information rates must be provided in order to support different services which demand different data rates and different QoS. In short, an important motivation for using ART is to support a variety of services, which we refer to as service-motivated ART (S-ART). However, there is a range of other motivating factors which are addressed below.

The performance of wireless systems is affected by a number of propagation phenomena: 1) path-loss variation vs. distance; 2) random slow shadowing; 3) random multipath fading; 4) ISI, co-channel interference, and multiuser interference; and 5) background noise. For example, mobile radio links typically exhibit severe multipath fading, which leads to serious degradation in the link's SNR and, consequently, a higher BER. Fading compensation techniques such as an increased link budget margin or interleaving with channel coding are typically required to improve the link's performance. However, today's cellular systems are designed for the worst-case channel conditions, typically achieving adequate voice quality over 90–95% of the coverage area for voice users, where the signal-to-interference plus noise ratio (SINR) is above the designed target.[16] Consequently, the systems designed for the worst-case channel conditions result in poor exploitation of the available channel capacity a good percentage of the time. Adapting the transmitter's certain parameters to the time-varying channel conditions leads to better exploitation of the channel capacity available. This ultimately increases the area spectral efficiency expressed in terms of bits/sec/Hz/km^2. Hence, the second reason for the application of ART is constituted by the time-varying nature of the channel, which we refer to as channel quality-motivated ART (C-ART).

What is ART?

Broadly speaking, ART in mobile wireless communications implies that the transmission rates at both the BSs and the mobile terminals can be adaptively adjusted according to the instantaneous operational conditions, including the communication environment and service requirements. With the expected employment of SDR-based wireless systems, the concept of ART might also be extended to adaptively controlling the multiple access schemes—including FDMA, TDMA, narrow-band CDMA, W-CDMA, and OFDM—as well as the supporting network structures—such as local area networks and wide area networks. In this contribution, only C-ART and S-ART are concerned in the context of the proposed FH/MC DS-CDMA scheme. Employing ART in response to different service requests indicates that the transmission rate of the BS and the MS can be adapted according to the requirements of the services concerned, as well as to meet their different QoS targets. In contrast, employing ART in response to the time-varying channel quality implies that for a given service supported, the transmission rate of the BS and that of the MS can be adaptively controlled in response to their near-instantaneous channel conditions. The main philosophy behind C-ART is the real-time balancing of the link budget through adaptive variation of the symbol rate, modulation constellation size and format, spreading factor, coding rate/scheme, etc., or, in fact, any combination of these parameters. Thus, by taking advantage of the time-varying nature of the wireless channel and interference conditions, the C-ART schemes can provide a significantly higher average spectral efficiency than their fixed-mode counterparts. This takes place without wasting power, without increasing the co-channel interference, or without increasing the BER. We achieve these desirable properties by transmitting at high speeds under favorable interference/channel conditions and by responding to degrading interference and/or channel conditions through a smooth reduction of the associated data throughput. Procedures that exploit the time-varying nature of the mobile channel are already in place for all the major cellular standards worldwide,[16] including IS-95 CDMA,

cdma2000, and UMTS W-CDMA,[8] IS-136 TDMA, the GPRS of GSM, and the enhanced data rates for GSM evolution (EDGE) schemes. The rudimentary portrayal of a range of existing and future ART schemes is given below. Note that a system may employ a combination of several ART schemes, listed below, in order to achieve the desired data rate, BER, or the highest possible area spectrum efficiency.

- **Multiple spreading codes**—In terms of S-ART, higher rate services can be supported in CDMA based systems by assigning a number of codes. For example, in IS-95B, each high-speed user can be assigned one to eight Walsh codes, each of which supports a data rate of 9.6 Kbps. In contrast, multiple codes cannot be employed in the context of C-ART in order to compensate for channel fading, path-loss, and shadowing unless they convey the same information, and, hence, achieve diversity gain. However, if the co-channel interference is low—which usually implies in CDMA based systems that the number of simultaneously transmitting users is low—then multiple codes can be transmitted by an active user in order to increase the user's transmission rate.

- **Variable spreading factors**—In the context of S-ART, higher rate services are supported by using lower spreading factors without increasing the bandwidth required. For example, in UMTS W-CDMA,[1] the spreading factors of 4/8/16/32/64/128/256 may be invoked to achieve the corresponding data rates of 1024/512/256/128/64/32/16 Kbps. In terms of C-ART, when the SINR experienced is increased, reduced spreading factors are assigned to users for the sake of achieving higher data rates.

- **Variable rate FEC codes**—In a S-ART scenario, higher rate services can be supported by assigning less powerful, higher rate FEC codes associated with reduced redundancy. In a C-ART scenario, when the SINR improves, a higher-rate FEC code associated with reduced redundancy is assigned in an effort to achieve a higher data rate.

- **Different FEC schemes**—The range of coding schemes might entail different classes of FEC codes, code structures, encoding/decoding schemes, puncturing patterns, interleaving depths and patterns, and so on. In the context of S-ART, higher rate services can be supported by coding schemes having a higher coding rate. In the context of C-ART, usually an appropriate coding scheme is selected in order to maximize the spectrum efficiency. The FEC schemes concerned may entail block or convolutional codes, block or convolutional constituent code based turbo codes, trellis codes, turbo trellis codes, etc. The implementational complexity and error correction capability of these codes can be varied as a function of the coding rate, code constraint length, the number of turbo decoding iterations, the puncturing pattern, etc. A rule of thumb is that the coding rate is increased toward unity, as the channel quality improves, in order to increase the system's effective throughput.

- **Variable constellation size**—In S-ART schemes, higher rate services can be supported by transmitting symbols having higher constellation sizes. For example, an adaptive modem may employ binary phase shifting key (BPSK), quadrature phase-shift keying (QPSK), 16-quadrature amplitude modulation (QAM), and 64-QAM constellations,[22] which corresponds to 1, 2, 4, and 6 bits per symbol. The highest data rate provided by the 64-QAM constellation is a factor six higher than that provided by employing the BPSK constellation. In C-ART scenarios, when the SINR increases, a higher number of bits per symbol associated with a higher order constellation is transmitted for increasing the system's throughput.
- **Multiple time slots**—In a S-ART scenario, higher rate services can also be supported by assigning a corresponding number of time slots. A multiple time slot based adaptive rate scheme is used in GPRS-136 (one to three time slots/20 ms) and in enhanced GPRS (EGPRS) (one to eight time slots/4.615GSM frame) in order to achieve increased data rates. In the context of C-ART, multiple time slots associated with interleaving or FH can be implemented for achieving time diversity gain. Hence, C-ART can be supported by assigning a high number of time slots for the compensation of severe channel fading at the cost of tolerating a low data throughput. In contrast, assigning a low number of time slots over benign non-fading channels allows us to achieve a high throughput.
- **Multiple bands**—In the context of S-ART, higher rate services can also be supported by assigning a higher number of frequency bands. For example, in UMTS W-CDMA,[1] two 5 MHz bands may be assigned to a user in order to support the highest data rate of 2 Mbps (=2 × 1024 Kbps), which is obtained by using a spreading factor of 4 on each subband signal. In the context of C-ART associated with multiple bands, FH associated with time-variant redundancy and/or variable rate FEC coding schemes or frequency diversity techniques might be invoked in order to increase the spectrum efficiency of the system. For example, in C-ART schemes associated with double-band assisted frequency diversity, if the channel quality is low, the same signal can be transmitted in two frequency bands for the sake of maintaining a diversity order of two. However, if the channel quality is sufficiently high, two independent streams of information can be transmitted in these bands and, consequently, the throughput can be doubled.
- **Multiple transmit antennas**—Employing multiple transmit antennas based on space-time coding[27] is a novel method of communicating over wireless channels which was also adapted for use in the 3G mobile

wireless systems. ART can also be implemented using multiple transmit antennas associated with different space-time codes. In S-ART schemes, higher rate services can be supported by a higher number of transmit antennas associated with appropriate space-time codes. In terms of C-ART schemes, multiple transmit antennas can be invoked for achieving a high diversity gain. Therefore, when the channel quality expressed in terms of the SINR is sufficiently high, the diversity gain can be decreased. Consequently, two or more symbols can be transmitted in each signaling interval, and each stream is transmitted by only a fraction of the transmit antennas associated with the appropriate space-time codes. Hence, the throughput is increased. However, when the channel quality is poor, all the transmit antennas can be invoked for transmitting one stream of data, hence achieving the highest possible transmit diversity gain of the system while decreasing the throughput.

Above we have summarized the philosophy of a number of ART schemes which can be employed in wireless communication systems. An S-ART scheme requires a certain level of transmitted power in order to achieve the required QoS. Specifically, a relatively high transmitted power is necessitated for supporting high-data rate services, and a relatively low transmitted power is required for offering low-data rate services. Hence, a side effect of an S-ART scheme supporting high data rate services is the concomitant reduction of the number of users supported due to the increased interference or/and increased bandwidth. By contrast, a cardinal requirement of a C-ART scheme is the accurate channel quality estimation or prediction at the receiver as well as the provision of a reliable side-information feedback between the channel quality estimator or predictor of the receiver and the remote transmitter[17,19] where the modulation/coding mode requested by the receiver is activated. The parameters capable of reflecting the channel quality may include BER, SINR, transmission frame error rate, received power, path loss, automatic repeat request (ARQ) status, etc. A C-ART scheme typically operates under the constraint of constant transmit power and constant bandwidth. Hence, without wasting power and bandwidth, or without increasing the co-channel interference compromising the BER performance, C-ART schemes are capable of providing a significantly increased average spectral efficiency by taking advantage of the time-varying nature of the wireless channel when compared to fixed-mode transmissions.

ART in FH/MC DS-CDMA Systems

Above we have discussed a range of ART schemes which were controlled by the prevalent service requirements and the channel quality. Future broadband mobile wireless systems are expected to provide a wide range of services characterized by highly different data rates while achieving the highest possible spectrum efficiency. The proposed FH/MC DS-CDMA based broadband mobile wireless system constitutes an attractive candidate system since it is capable of efficiently utilizing the available frequency resources, as discussed previously, and simultaneously achieving a high grade of flexibility by employing ART techniques. More explicitly, the FH/MC DS-CDMA system can provide a wide range of data rates by combining the various ART schemes discussed above. At the same time, for any given service, the FH/MC DS-CDMA system may also invoke a range of adaptation schemes in order to achieve the highest possible spectrum efficiency in various propagation environments such as indoor, outdoor, urban, rural scenarios at low to high speeds. Again, the system is expected to support different services at a variety of QoS, including voice mail, messaging, e-mail, file transfer, Internet access, high-speed packet- and circuit-based network access, real-time multimedia services, and so on. As an example, a channel-quality motivated burst-by-burst ART assisted FH/MC DS-CDMA system is shown in Fig. 4, where we assume that the number of subcarriers is three, the bandwidth of each subchannel is 5 MHz and the channel quality metric is the SINR. In response to the SINR experienced, the transmitter may transmit a frame of symbols selected from the set of BPSK, QPSK, 16-QAM, or 64-QAM constellations or may simply curtail transmitting information if the SINR is too low.

SDR ASSISTED FH/MC DS-CDMA

The range of existing wireless communication systems is shown in Fig. 5. Different legacy systems will continue to coexist, unless ITU, by some miracle, succeeds in harmonizing all the forthcoming standards under a joint framework while at the same time ensuring compatibility with the existing standards. In the absence of the perfect standard, the only solution is employing multiband,

multimode, multistandard transceivers based on the concept of SDRs.[9,10]

In SDRs, the digitization of the received signal is performed at some stage downstream from the antenna, typically after wideband filtering, low-noise amplification, and down-conversion to a lower frequency. The reverse processes are invoked by the transmitter. In contrast to most wireless communication systems which employ DSP only at baseband, SDRs are expected to implement the DSP functions at an IF band. An SDR defines all aspects of the air interface, including RF channel access and waveform synthesis in software. In SDRs, wide-band analog-to-digital and digital-to-analog converters (ADC and DAC) transform each RF service band from digital and analog forms at IF. The wideband digitized receiver stream of bandwidth W_s accommodates all subscriber channels, each of which has a bandwidth of $W_c (W_c << W_s)$. Because they use programmable DSP chips at both the IF as well as at baseband, SDRs can efficiently support multiband and multistandard communications. An SDR employs one or more reconfigurable processors embedded in a real-time multiprocessing fabric, permitting flexible reprograming and reconfiguration using software downloaded, e.g., with the aid of a signaling channel from the BS. Hence, SDRs offer an elegant solution to accommodating various modulation formats, coding, and radio access schemes. They also have the potential of reducing the cost of introducing new technology superseding legacy systems and are amenable to future software upgrades, potentially supporting sophisticated future signal processing functions such as array processing, MUD, and as yet unknown coding techniques.

FH/MC DS-CDMA systems will be designed in order to provide the maximum grade of compatibility with the existing CDMA based systems, such as IS-95 and W-CDMA based systems. For example, the frequency bands of the IS-95 CDMA system in North America are 824–849 MHz (UL) and 869–894 MHz (DL), respectively. The corresponding frequency bands for the UMTS-FDD wideband CDMA system are 1850–1910 MHz (UL) and

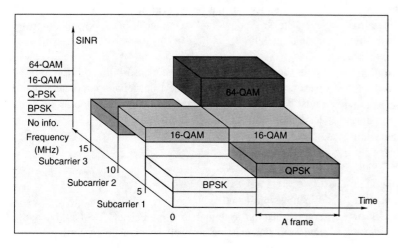

Fig. 4 A frame structure of burst-by-burst adaptive modulation in multicarrier DS-CDMA systems, ©IEEE.[32] *Source*: L.L.Yang and L. Hanzo, Software defined radio assisted adaptive broadband frequency hopping multicarrier DS-CDMA, IEEE Commun. Mag., 2002, *40*, 174–183. © IEEE. All rights reserved.

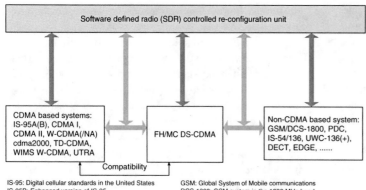

IS-95: Digital cellular standards in the United States
IS-95B: Enhanced version of IS-95
CDMA I: Multiband synchronous DS-CDMA
CDMA II: Asynchronous DS-CDMA
W-CDMA: Wideband CDMA
W-CDMA/NA: North American Wideband CDMA
cdma2000: Multicarrier CDMA system based on IS-95
TD-CDMA: Time-Division synchronous CDMA
WIMS W-CDMA: Wireless Multimedia and Messaging
           Services Wideband CDMA
UTRA: UMTS Terrestrial Radio Access

GSM: Global System of Mobile communications
DCS-1800: GSM system in the 1800 MHz band
PDC: Japanese Personal Digital Cellular system
IS-54: American digital advanced mobile phone system (DAMPS)
IS-136: North American TDMA system
UWC-136: Universal Wireless Communications based on IS-136
UWC-136+: Enhanced version of UWC-136
DECT: Digital Enhanced Cordless Telecommunications
EDGE: Enhanced Data Rates for Global Evolution

Fig. 5 Software defined radio assisted FH/MC DS-CDMA and its reconfiguration modes, ©IEEE.[32]
Source: L.L.Yang and L. Hanzo, Software defined radio assisted adaptive broadband frequency hopping multicarrier DS-CDMA, IEEE Commun. Mag., 2002, *40*, 174–183. © IEEE. All rights reserved.

1930–1990 MHz (DL) in North America, and 1920–1980 MHz (UL) and 2110–2170 MHz (DL) in Europe. In order to ensure compatibility with these systems, the proposed FH/MC DS-CDMA system's spectrum can be assigned according to Fig. 6. Specially, in the frequency band of IS-95, 39 orthogonal subcarriers are assigned, each having a bandwidth of 1.25 MHz, while in the frequency band of UMTS-FDD W-CDMA, 23 orthogonal subcarriers are allocated, each with a bandwidth of 5 MHz. The multicarrier modulation used in the FH/MC DS-CDMA system obeying the above spectrum allocation can be readily implemented using two IFFT subsystems at the transmitter and two FFT subsystems at the receiver, where a pair of IFFT–FFT subsystems carries out modulation/demodulation in the IS-95 band, while another pair of IFFT–FFT subsystems transmits and receives in the UMTS-FDD band. If the chip rate for the 1.25 MHz bandwidth subcarriers is 1.2288 Mcps and for the 5 MHz bandwidth subcarriers is 3.84 Mcps, then the FH/MC DS-CDMA system will be compatible with both the IS-95 and the UMTS-FDD W-CDMA systems.

However, the terminals of future broadband mobile wireless systems are expected not only to support multimode and multistandard communications, but also to possess the highest possible grade of flexibility while achieving a high spectrum efficiency. Hence, these systems are expected to be capable of software reconfiguration both between different standards as well as within a specific standard. In contrast to reconfiguration between different standards invoked, mainly for the sake of compatibility, the objective of reconfiguration within a specific standard is to support a variety of services at the highest possible spectrum efficiency. The SDR assisted broadband FH/MC DS-CDMA system is operated under the control of the software reconfiguration unit shown in Fig. 5. The set of reconfigured parameters of the broadband FH/MC DS-CDMA system may include:

- **Services:** data rate, QoS, real-time or non-real-time transmission, and encryption/decryption schemes and parameters;
- **Error control:** CRC, FEC codes, coding/decoding schemes, coding rate, number of turbo decoding steps, and interleaving depth and pattern;
- **Modulation:** modulation schemes, signal constellation, and partial response filtering;
- **PN sequence:** spreading sequences (codes), chip rate, chip waveform, spreading factor, and PN acquisition and tracking schemes;
- **FH:** FH schemes (slow, fast, random, uniform, and adaptive), FH patterns, and weight of constant-weight codes;
- **Detection:** detection schemes (coherent or non-coherent, etc.) and algorithms [maximum likelihood sequence detection (MLSD) or minimum mean square estimation (MMSE), etc.], parameters associated with space/time as well as frequency diversity, beam-forming, diversity combining schemes, equalization schemes as well as the related parameters (such as the number of turbo equalization iterations, etc.), and channel quality estimation algorithms;.
- **Others:** subchannel bandwidth and power control parameters

In the context of different standards—in addition to the parameters listed above—the transceiver parameters that must be reconfigurable have to include the clock rate, the RF bands, and air interface modes.

FINAL REMARKS

We have presented a flexible broadband mobile wireless communication system based on FH/MC DS-CDMA and reviewed the existing as well as a range of forthcoming

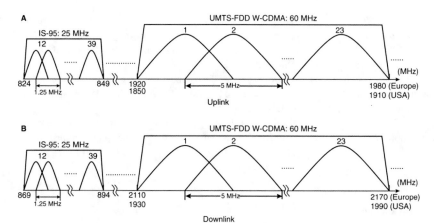

Fig. 6 Exhibition of spectrum compatibility of the broadband FH/MC DS-CDMA system with IS-95 and UMTS-FDD wideband CDMA systems, ©IEEE.[32] *Source*: L.L.Yang and L. Hanzo, Software defined radio assisted adaptive broadband frequency hopping multicarrier DS-CDMA, IEEE Commun. Mag., 2002, *40*, 174–183. © IEEE. All rights reserved.

techniques which might be required to develop broadband mobile wireless systems exhibiting high flexibility and high efficiency. We argued that this broadband FH/MC DS-CDMA system exhibits a high grade of compatibility with the existing CDMA based systems since it is capable of incorporating a wide range of techniques developed for the 2G and 3G mobile systems. At the time of this writing, research is well under way toward the SDR-based implementation of a range of existing systems.[28] It is expected that these efforts will soon encompass a generic scheme not too different from the FH/MC DS-CDMA scheme advocated in this entry. For a range of related topics, the interested reader is referred to Refs. 29–31. This entry is based on L.-L. Yang and L. Hanzo, Software Defined Radio Assisted Adaptive Broadband Frequency Hopping Multicarrier direct sequence code division multiple access (DS-CDMA), ©IEEE.[32]

REFERENCES

1. Zeng, M.; Annamalai, A.; Bhargava, V.K. Recent advances in cellular wireless communications. IEEE Commun. Mag. **1999**, *37*, 128–138.

2. Chaudhury, P.; Mohr, W.; Onoe, S. The 3GPP proposal for imt-2000. IEEE Commun. Mag. **1999**, *37*, 72–81.

3. Progler, M.; Evci, C.; Umehira, M. Air interface access schemes for broadband mobile systems. IEEE Commun. Mag. **1999**, *37*, 106–115.

4. Kleinrock, L. On some principles of nomadic computing and multi-access communications. IEEE Commun. Mag. **2000**, *38*, 46–50.

5. Bender, P.; Black, P.; Grob, M.; Padovani, R.; Sindhushayana, N.; Viterbi, A. CDMA/HDR: a bandwidth-efficient high-speed wireless data service for nomadic users. IEEE Commun. Mag. **2000**, *38*, 70–77.

6. Chuang, J.; Sollenberger, N. Beyond 3G: wideband wireless data access based on OFDM and dynamic packet assignment. IEEE Commun. Mag. **2000**, *38*, 78–87.

7. Dimitriou, N.; Tafazolli, R.; Sfikas, G. Quality of service for multimedia CDMA. IEEE Commun. Mag. **2000**, *38*, 88–94.

8. *Mobile Radio Communications*, 2nd Ed.; Steele, R., Hanzo, L., Eds.; IEEE Press–John Wiley: New York, USA, 1999.

9. Mitola, J., III. The software radio architecture. IEEE Commun. Mag. **1995**, *33*, 26–38.

10. Mitola, J., III. Technical challenges in the globalization of software radio. IEEE Commun. Mag. **1999**, *37*, 84–89.

11. Sourour, E.A.; Nakagawa, M. Performance of orthogonal multicarrier CDMA in a multipath fading channel. IEEE Trans. Commun. **1996**, *44*, 356–367.

12. Yang, L.L.; Hanzo, L. Blind soft-detection assisted frequency-hopping multicarrier DS-CDMA systems. IEEE GLOBECOM 1999, Rio de Janeiro, Brazil, Dec 5–9, 1999; 842–846.

13. Slimane, S.B. MC-CDMA with quadrature spreading for wireless communication systems. Eur. Trans. Telecommun. **1998**, *9*, 371–378.

14. Yang, L.L.; Hanzo, L. Slow frequency-hopping multicarrier ds-cdma over Nakagami multipath fading channels. IEEE J. Sel. Areas Commun. **1999**, *19* (7), 1211–1222.

15. Verdu, S. Wireless bandwidth in the making. IEEE Commun. Mag. **2000**, *38*, 53–58.

16. Nanda, S.; Balachandran, K.; Kumar, S. Adaptation techniques in wireless packet data services. IEEE Commun. Mag. **2000**, *38*, 54–64.

17. Goldsmith, A.J.; Chua, S.G. Variable-rate variable-power MQAM for fading channels. IEEE Trans. Commun. **1997**, *45*, 1218–1230.

18. Alouini, M.S.; Goldsmith, A.J. Capacity of Rayleigh fading channels under different adaptive transmission and diversity-combining techniques. IEEE Trans. Veh. Technol. **1999**, *48*, 1165–1181.

19. Duel-Hallen, A.; Hu, S.; Hallen, H. Long-range prediction of fading signals. IEEE Signal Process. Mag. **2000**, *17*, 62–75.

20. Yang, L.L.; Hanzo, L. Blind joint soft-detection assisted slow frequency-hopping multicarrier DS-CDMA. IEEE Trans. Commun. **2000**, *48*, 1520–1529.

21. Prasad, R.; Hara, S. Overview of multicarrier CDMA. IEEE Commun. Mag. **1997**, *12*, 126–133.

22. Hanzo, L.; Webb, W.T.; Keller, T. *Single- and Multi-carrier Quadrature Amplitude Modulation: Principles and Applications for Personal Communications, WLANs and Broadcasting,* 2nd Ed.; IEEE Press and John Wiley & Sons: London, UK, 1999.

23. Yang, L.L.; Hanzo, L. Parallel code-acquisition for multicarrier DS-CDMA systems communicating over multipath Nakagami fading channels. IEEE GLOBECOM 1999, San Francisco, CA, Nov 27–Dec 1, 2000; 890–894.

24. Verdu, S. *Multiuser Detection*; Cambridge University Press: London, UK, 1998.

25. Berrou, C.; Glavieux, A. Near optimum error correcting coding and decoding: turbo codes. IEEE Trans. Commun. **1996**, *44*, 1261–1271.

26. Li, Y.; Sollenberger, N.R. Adaptive antenna arrays for OFDM systems with cochannel interference. IEEE Trans. Commun. **1999**, *47*, 217–229.

27. Tarokh, V.; Seshadri, N.; Calderbank, A.R. Space-time codes for high data rate wireless communication: performance criterion and code construction. IEEE Trans. Inf. Theory. **1998**, *44*, 744–765.

28. *Software Defined Radio, vols. I and II*; Tuttlebee, W., Ed.; John Wiley & Sons: New York, USA, 2002.

29. Hanzo, L.; Wong, C.; Yee, M. *Adaptive Wireless Transceivers*; IEEE Press and John Wiley, 2002.

30. Hanzo, L.; Cherriman, P.; Streit, J. *Wireless Video Communications: From Second to Third Generation Systems, WLANs and Beyond*; IEEE Press and John Wiley: 2001.

31. Hanzo, L.; Liew, T.; Yeap, B. *Turbo Coding, Turbo Equalisation and Space-Time Coding*; IEEE Press and John Wiley: 2002.

32. Yang, L.L.; Hanzo, L. Software defined radio assisted adaptive broadband frequency hopping multicarrier DS-CDMA. IEEE Commun. Mag. **2002**, *40*, 174–183.

BIBLIOGRAPHY

1. Murotake, D.; Oates, J.; Fuchs, A. Real-time implementation of a reconfiguable imt-2000 base station channel modem. IEEE Commun. Mag. **2000** *37*(9), 148–152.

2. Srikanteswara, S.; Reed, J.H.; Athanas, P.; Boyle, R. A soft radio architecture for reconfigurable platforms. IEEE Commun. Mag. **2000**, *38*, 140–147.

3. Tsurumi, H.; Suzuki, Y. Broadband RF stage architecture for software-defined radio in handheld terminal applications. IEEE Commun. Mag. **1999**, *37*, 90–95.

4. Walden, R.H. Performance trends for analogy-to-digital converters. IEEE Commun. Mag. **1999** *38*(7), 96–101.

Index

1-D RS codes, 159
  MBMS file download, 159
1G wireless systems, 1349
  call processing, 214–215
  narrowband AMPS, European TACS, 215
1R-regeneration, 976
2.5G wireless systems, 1350
2-D RS codes, 159–161
  FEC payload ID determination, 160
  grid dimensions, determination of, 159
  MBMS file download, 159
  packet transmission schedule, 160
  receiver operations, 160
  source and repair packets, generation of, 160
2G cellular systems, 216–219
  CDMA digital cellular system, 217–219
  cordless mobiles, 219
  Pan European global system, 217
    burst format, 218–219
    common channels, 218
    multiple access scheme, 218
  PDC System, 217–218
  United States digital cellular, 216–218
2G wireless systems, 1349
  CDMA technology, 1349
  GSM, 1349
  TDMA technology, 1349
2R-regeneration, 976
3G networks/technology, 1350–1351
  3G packet-switched streaming service, 785
  3G to 4G migration for transmitting multimedia content, 788–794
    mobile wireless networks evolution, 788–789
  3G UMTS networks, 463
  3GPP (3G Partnership Project), 2, 157, 454
  3G-telephony video, See also Video services
    interactive playback control of, 13–25
      See also Interactive playback control
    accounting principles and billing models for, 1–12, See also under Accounting principles
    challenges, 789
    for MMGs, 698
    packet switching vs. circuit switching, 1350
    privacy and authentication in, 138–140
      3GPP based 3G networks, 138
      3GPP2 based 3G networks, 138
      authentication, 139–140
        3GPP based networks, 139
        3GPP2 based 3G networks, 139–140
      privacy, 139
        3GPP networks, 139
        3GPP2 networks, 139
    QoS issues and challenges, 31–32
    QoS negotiations in, 32–33
    streaming multimedia content, 30–31
      challenges faced, 31
      shortcomings, 33
    technology outline, 26

access network domain, 26
core network domain, 26
user equipment, 26
UMTS, 30, 1350–1351
W-CDMA technology, 1350
3GPP specifications information, 1183
3G-telephony control stack, 13–25
  dynamic trans-rating, 21–22
  dynamically controlled frame-rate modification, 19–22
  filter graphs
    for video playback and video recording, 23–24 See also Filter graphs
  timing and synchronization for, 16–19
    extra time, 16
    interactive playback seek, 18–19
    interactive playback speed up and slow down, 16–18
    sample time, 16
    seek points, 16
    transmission time, 16
  video-services demonstration, 22–24
3R-regeneration, 976
4G in mobile communications, 634–642
  from 1G to 4G, 634–635
  4G working groups and forums, 640–641
    mobile IT forum, 641
    NGN, 641
    wireless world research forum, 641
  concept of, 635–639
    key features, 636–637
    user centric' system, 635–639
    user scenarios, 636
  fuzzy logic in RRM of, 43–45
  key 4G technologies, 639–641
    AMC, 639
    multiple access techniques, 639–640, 503–2015, 503–2016, 503–2017
    HSDPA, 640
    MIMO principle, 640
    MIPv6, 640
    OFDM, 639–640
    UWB technology, 640
  research challenges, 637
    mobile station, 637–638
    services, 639
    system, 638
  system architecture and infrastructure, 637
    OWA, 637
    SOA, 637
4G mobile systems architecture, 26–39
  advantages, 37
  future trends, 37–38
  mobile systems, multimedia role in, 26
  multimedia transmission, strategy for, 33–34
  multimedia types supported by, 34
  need for, 27
  QoS negotiations, 35–37
    mobile switching center, 36
    QoS broker, 37
  quality of service, 27–28

audio video applications, 29
  multimedia transmission, importance in, 28
research opportunities, 38
shortcomings, 37
streaming multimedia, for, 33
  challenges in, 31
streaming techniques and processes, 34–35
4G networks
  development, rationale for, 792
  diversity in, 791–792
    application diversity, 792
    network diversity, 791–792
    terminal diversity, 791
    user diversity, 791
  evolution of, 789
  features and advantages, 789–791
  key features of, 792–793
    broadband wireless access, 792–793
    low cost, 793
    wide area coverage, 793
  radio resource management in, 40–49,
    See also RRM
4-QAM (4-quadrature amplitude modulation), 440
8 PSK (8 phase shift keying), 312
136 HS, 176
802.11 technologies, 1201
802.15.4 sensor clusters
  802.15.4 PAN
    defending, 402–403
  802.15.4 standard, 407–408
  MAC layer attacks in, 396–403
  operation of, 397–400 See also CSMA-CA
    network formation and association/disassociation, 398–399
    security services and suites, 399–400
    uplink and downlink communication, 398
  security threats, 400–402
    attacks following MAC protocol, 400–401
    attacks using modified MAC protocol, 401

AAC (advanced audio coding), 742
Ad et al., mechanisms
  in legacy IEEE 802.11 WLANs, 377
    Aad-backoff scheme, 377
    Aad-CWmin scheme, 378
    Aad-DIFS scheme, 378
AASs (attribute authority sets), 83
Absolute localization, 1547
Abstraction, in cross-layer MAC design, 1425
Access methods, 1407, 937
  CDMA, 55–56 See also separate entry
  FDMA, 52–53 See also separate entry
  methods comparison, 56–57
  TDMA, 53–55 See also separate entry
Accounting principles and billing models for 3G systems, 1–12
  3GPP, 2
  accounting for 3G services, 2–6
    accounting services, 5–6

Volume 1: Pages 1–508; *Volume 2*: Pages 509–1094; *Volume 3*: Pages 1095–1598.

I-1

Accounting principles and billing models for 3G
systems (*Cont.*)
accounting using mobile agent technology,
5–6
service-oriented accounting, 5
ARPU average revenue per unit, 7–10
billing and 3G services, 7–10
billing, 2–6
bearer-independent billing, 4
methods, 4–5
standards for, 3–4
call data records, 6
charging in 3G services, 6–7
dynamic pricing models, 6
flat-rate pricing, 6
charging, 2–6
standards for, 3–4
constraints and challenges to, 6–7
flow-based charging, 6–7
IETF (internet engineering task force), 2–3
for mobile games, 8–9
customer profile, 9
detailed billing record, 9
operation flashpoint, 8
price lookup, 9
round-trip delays, 8
mobile payment for 3G services, 9–10
telecommunication standards, 2
Vodafone UK's m-pay bill online payments
solution, 4
ACELP (algebraic code-excited linear
prediction) encoder, 1128
channel coding, 1129
gains quantization, 1129
ACL (asynchronous connectionless link), 145
ACPS (ad-coop positioning system)
EKDF method, 125–126
EKDF vs. HTAP, 128–129
EKDF vs. HTDOA, 129
simulation models, 126–128
statistical models
for RSS estimation, 127–128
fast fading, 128
LOS probability, 127
path loss, 127
shadowing, 127–128
for TOA and AOA estimation errors,
126–127
LOS, 126–127
NLOS, 127
system architecture, 128–129
AOA measurements, 123–124
TDOA measurements, 123–124
TOA measurements, 123–124
using an embedded kalman filter data fusion
method, 120–131
filtering techniques, 122–123
Kalman filter (KF), 122
particles filter (PF), 122
hybrid positioning techniques, 121–122
NLOS error mitigation techniques, 122
ACs (attribute certificates), 83
Active caching, 763
Active constellation extension technique, 852
Active stochastic routing techniques,
1449–1450
Activity based search, of context, 813–814

Ad hoc networks, *See also* ARAN, RIDAN
ad hoc peer communities, 276–278
architecture, 564
applications, WANET, 1222
bluetooth, *See* separate entry
characteristics of, 76
source-initiated on-demand, 77
table-driven, 76
CLD for, 1421–1423
adaptive link layer strategy in, 1422
congestion-distortion scheduling based
CLD for, 1423
optimal strategy of network layer in, 1422
GPS-based routing algorithm, 58–67
AODV-RRS protocol, 62
variations, 63–64
GPS, 60–61
RRS algorithms, 61–64
stable zone and caution zone, 61–62
stable zone radius effect, 62–63
IEEE 802.11, *See* separate entry
intrusion detection in, 84–86
key management in, 81–84
resurrecting duckling, 82–83
self-organized PKI, 82
threshold cryptography, 82
power-conservative designs, 68–75
higher-layer power management, 74
operating system/application layer, 74
transport layer, 74
MAC-layer power management, 70–72
dominating-awake-interval protocol, 71
periodically-fully-awake-interval
protocol, 71, 72
• quorum-based protocol, 72
network-layer power management, 72–74
multicast/broadcast data
communications, 74
unicast data communications, 73–74
transmitter power control mechanisms,
68–70
routing protocols, 59–60
AODV routing protocol, 59–60
DSR protocol, 59
security for, 76–88
ÆTHER, 83–84
fear-based awareness, 85
intrusion detection, 84–86
key management, 81–84 *See also*
individual entry
packet leashes, 81
Pathrater, 84–85
secure ad hoc routing, 78–81
secure AODV routing, 80–81
security solutions, 78
threat model, 77–78
watchdog, 84–85
threat model of, 77–78
routing techniques, in WLANs, 1369–1380,
See also under WLANs
Ad hoc on-demand distance vector routing,
See AODV
Adaptation of quality of service, *See* AqoS
Adaptation techniques, 1211, 925
adaptive allocation, 1100
for multicarrier transceivers, 1102–1103
adaptive array antenna, 497

adaptive bit allocation, 1103
adaptive chaining, 589–590
applications, 110–115
adaptive cell and frequency assignment,
115
adaptive modulation and channel coding,
113
adaptive power control, 113
adaptive receiver algorithms, 110–112
link and transmitter adaptation, 103
for multicarrier transceivers, 1102–1103
parameter measurements, 104–110
channel selectivity estimation, 104–108
channel quality measurements, 108–110
channel decoding, 110
demodulation, 108–109
receiver adaptation, 104
resource allocation, 103–104
Adaptive array antenna, 497
Adaptive bit allocation, 1103
Adaptive chaining, 589–590
Adaptive channel estimation receiver, 111
Adaptive content-driven routing
and data dissemination, 756–757
MP2P system, 756–757
Adaptive continuous query processing system,
1542–1544
field-based monitoring, 1543–1544
contaminant level and flow, 1544
habitat or ecosystem monitoring, 1544
scientific experiments, 1544
wild fire monitoring, 1544
infrastructure-based monitoring, 1542–1543
civil structure/machine monitoring,
1543
heat control and conditioning, 1543
intrusion detection, 1543
traffic monitoring, 1543
Adaptive EXP/PF, 949
Adaptive interference cancellation receivers,
111–112
Adaptive link layer strategy, 1422
Adaptive MIMO systems, 610
Adaptive modulation, 113
Adaptive modulation and coding, *See* AMC
Adaptive periodic threshold-sensitive energy
efficient sensor network,
See APTEEN
Adaptive power control, 113
Adaptive rate transmission, *See* ART
Adaptive receiver algorithms, 110–112
adaptive channel length truncation, 111
adaptive interference cancellation receivers,
111–112
adaptive soft information, 112
channel estimation, 110–111
Adaptive routing, 990
protocols, WANET, 1222
Adaptive soft information, 112
Adaptive thresholds-based SHO, 1090–1091
Ad-coop positioning system, *See* ACPS
Addresses, care-of
types of, 501
ADE (Adaptation decision engine), 687
Ad hoc GSM, *See* A-GSM
Ad hoc on-demand distance vector routing,
See AODV

ADMR (adaptive demand-driven multicast routing), 546
Advance resource reservation, 480–483
  current approaches, 480–483
    preconfigured anchor rerouting, 481
    preconfigured path extensions, 481
    preconfigured tunneling tree, 481
  preallocate network resources, 481–482
    coarse-grained allocation, 482
    history-based allocation, 482
    neighborhood-based allocation, 482
Advanced codec standards, 741
Advanced mobile phone service, *See* AMPS
Advanced radio data information service, *See* ARDIS
ÆTHER instantiations, 83–84
  ÆTHER$_0$, 83
  ÆTHER$_1$, 83
  characteristics of, 83
Affordances of mobile devices, 616
Agent discovery phase, in mobile IP, 726–727
A-GSM (ad hoc GSM)
  cellular system, 326
    call rerouting phases, 329–333
      handover control, 331–333
      handover initiation, 329–331
      measurements, 329
    network entities, 326
      base station subsystem, 326
      integrated dual mode terminals, 327
      internet working unit, 327
      mobile services switching center, 326
      network database, 326–327
    protocol layering, 328–329
      beaconing, 328
      encapsulation protocol, 329
      link layer protocol, 328
      resource manager, 328
    vs GSM, 334–335
AI (artificial intelligence) techniques, 230
AIFS (arbitration inter-frame space), 1382, 1385
Air interface standards, 858
Air-link allocation, 421
AIRMAIL (asymmetric reliable mobile access in link-layer), 1455
AJ (antijam) communications, 1118–1119
  definition, 1121
Alamouti code, 608
ALCAP (access link control application part), 1167
Alert portable telemedical monitor, *See* AMON
Algebraic codebook, 1128–1129
Algebraic code-excited linear prediction) encoder, *See* ACELP
Algorithms, encryption, 133
Allocation
  air-link 421
  channel, 210
    dynamic channel allocation schemes, 210
    fixed channel allocation schemes, 210
    hybrid channel allocation schemes, 210–211
ALP (active link protection), 1468–1470
Amazon.com, 1216
Ambient-audio identification

in coordinated multi-device presentations, 274–285
  audio fingerprinting, 280–281
  personalizing broadcast content, 275–279
  supporting infrastructure, 279–280
  video 'Bookmarks', 278–279
Ambulatory ECG monitoring subsystem, 714
AMC (adaptive modulation and coding)
  4G technologies, 639
AMC (adaptive multi-hop clustering), 523
'Amigo TV' system, 268
AMON (alert portable telemedical monitor), 713–714, 715
AMPS (advanced mobile phone service), 214, 1349
  characteristics, 214
AMR (adaptive multi-rate), 742
Analytical SHO models, 1088 *See also* under Soft handoff
Angle of arrival method, *See* AOA
Angle spread, 107–108
Anomaly detection IDS, 765
Antenna gain, 1312–1313
Antijam communications, *See* AJ (antijam) communications
  and TOA method, combination, 661
AODV (ad hoc on-demand distance vector routing), 59–60, 1370–1371
  performance, 1374
  route reply message, 1370
  route request packet, 1370
  routing algorithm, 1370–1371
  routing protocol, 59–60
    route discovery, 60
  routing table build-up, 1371
AOA (angle of arrival) method, 660
API (application programming interface), 696–697, 1279
Application diversity, in 4G technology, 792
Approximation ratio, 1235
APTEEN (adaptive periodic threshold-sensitive energy efficient sensor network), 1527–1528
AqoS (adaptation of quality of service)
  for mobile devices using MPEG-21, 686–693
    adaptation decision engine (ADE), 687
    adaptation framework, 690
    AqoS tool, 688, 689
    client-side content adaptation, 688
    content adaptation, 687
    experimental results, 691–692
    GRACE 1, 687
    intermediate node content adaptation, 688
    multiple steps adaptation, 688
    related research review, 686–688
    resource adaptation engine (RAE), 687
    server-side content adaptation, 687–688
    UCD tool, 688, 689
    UED tool in, 688, 689
ARAN (authenticated routing for ad hoc networks), 79–80
  route maintenance in, 80
Arbitration schemes, 647
  interpersonal comparisons of utility, 647–648
  Nash arbitration scheme, 647
ARDIS (advanced radio data information service), 1322–1323

network architecture, 1323
ARM 11 processor, 6, 166
ARPU (average revenue per unit), 7–10
Array antenna, adaptive, 497
Array gain, 606
ARSs (ad hoc relay stations), 859
ART (adaptive rate transmission), 1590–1592, 1592–1593, 1593–1595
  in frequency-hopping based multicarrier direct spectrum, 1595
  multiple bands, 1594
  multiple spreading codes, 1593
  multiple time slots, 1594
  multiple transmit antennas, 1594
  variable constellation size, 1594
  variable rate FEC codes, 1593
  variable spreading factors, 1593
Assignment, channel, 210, 940
  CDMA systems, 228–229
  DCA, 227–228
  enhanced FCA, 226–227
  FCA, 225–226
  resource assignment problem, 180–185
Association discovery, 805–806
Asymmetric ciphers, 1557
Asymmetric cryptography, 132, 133
Asynchronous transfer mode, *See* ATM
Asynchronous viewing over a distance, 271
AT commands, 1144
ATM (asynchronous transfer mode), 1251–1263
  ATM PONs, 969
Attack analysis module, in RIDAN, 85
Attacker model, 1548–1549
  replay attack, 1549
  sybil attack, 1549
Attack-resistant location estimation, 1550
Attacks, *See also* individual entries
  attack analysis module, RIDAN, 85
  WEP, 1395–1396 *See also* under WEP (wired equivalent privacy)
  WLANs, 1400–1401 *See also* under WLANs
Audio codecs, 741–742 *See also* AAC; AMR; MP3; MP3PRO
Audio coding, 738
Audio compression, 737
Audio fingerprinting, 280–281
  hashing descriptors, 280
  post-match consistency filtering, 281
  system performance evaluation, 281–283
    empirical evaluation, 281–282
    in-living-room experiments, 282–283
    within-query consistency, 280–281
Audio-based search, content, 812–813
Audio-database server setup, in coordinated multi-device presentations, 279
Audio information, automatic tagging, 808
Authentication, 136–137
  authenticated routing for ad hoc networks, *See* ARAN
  layer, 393
  and privacy in wireless systems, 132–140, *See also* Cryptography
    authentication, 136–137
    challenges and open research issues, 140

Authentication (*Cont.*)
    in global system for mobile
      communications networks, 135–136
      See also separate entry
    in IEEE 802.11 wireless LANs, 134–135
      See also separate entry
    in IEEE 802.16 wireless metropolitan
      area networks, 135 *See also* separate
      entry
    in second generation code division multiple
      access networks, 137–138 *See also*
      separate entry
    in third generation networks, 138–140
      See also 3G networks
    secret keys, 133–134
Automatic monitoring, 713
Automatic protection switching, WDM,
    993–994
Automatic synchronization, in bluetooth
    technology, 142
Automatic tagging, 807–808
    from audio information, 808
    from textual information, 807
    from visual information, 807
AVs (authentication vectors), 139
Awareness, context, 1211
AWGN (additive white Gaussian noise)
    channel, 431, 437–438, 444, 1159,
    1286–1287

Backhaul links, 1409–1410
    types, 1410
      multiple point-to-point, 1410, 1411
      multipoint-to-multipoint, 1410, 1411
      point-to-multipoint, 1410
      point-to-point, 1410
Backhauling, 1407
Backward buffer, 589
Backward handover, 1279–1280
    control flow, 1268–1272
Bad frame masking, 1128
BAN (body area network), 510
    PAN relationship, 511
Bandwidth
    design issue, CDMA, 168
    fluctuation and delay, 254
    requirement, of MMGs, 704
BANs (body area networks), 148–156, *See also*
    IMDs
    applications, 154–155
      healthcare, 155
      lifestyle and sports, 155
      military, 155
      work and emergency services, 155
    challenges, 152–154
      context awareness and multi-sensory data
        fusion, 154
      privacy, 154
      reliability, 154
      safety, 154
      security, 154
      ultra-wideband for ultra-low-power-short-
        range communication, 153–154
    frequency bands used by, 149
    internet connected BANs, 151–152
    invasive bans, 148–150
    network topology, 151 *See also* individual entry

non-invasive BANs, 150–151 *See also*
    separate entry
stand-alone BANs, 151
system architecture employing
    for unobtrusive services, 151–152
    pervasive sensing for, 152
Bargaining problem, 647
Base stations, 204
    base station subsystem, ad hoc GSM, 326
    call from a mobile, 204
    call to a mobile, 204
    registration, 204
Baseband specification, of bluetooth systems,
    144–145
Basic chaining, 589
Battery models, dynamic wireless sensor
    networks, 306–307
Bayesian Nash equilibrium, 766
BCCH (broadcast control channel), 329
BCH (Bose-Chaudhuri-Hocquenghem),
    1260–1261
Beaconing, 328
    Beacon enabled clusters, 407–408
    Beacon suite, 1550
Beamforming, 171
Bearer-independent billing, 4
B-EDCA (B-enhanced distributed channel
    access) mechanism, 381
Bell labs, 259
Bellman–Ford (DBF) algorithm, 566
BER (bit error rate), 492–496, 976
    for coherent binary phase shift keying, 1309
    vs. (dB) for dual-branch, 1320
    initial peak BER threshold calculation, 1106
    peak BER-constrained bit allocation
      algorithm, 1105
    QAM, 1025
Bernoulli/Gilbert-Elliott loss models, 439
Best-effort service, 419, 1137
BI (blocking island), 998–1000
Biconnectivity, 1238–1239
Billing
    bearer-independent, 4
    billing models for 3G systems, 1–12, *See also*
      under Accounting principles
    bearer-independent billing, 4
    methods, 4–5
    standards for, 3–4
Binary phase shift keying, *See* BPSK
Bit allocation method, 431
    adaptive, 1103
    capacity approximation-based, 1103–1104
    with imperfect channel information,
      1107–1108
Bit error rate, *See* BER
Bit rate re-purposing, 255
BL2DF (best link lowest delay first)
    scheduling, 949
Black hole
    attack, 409
    in infrastructureless routing, 77
Blackmail, in infrastructureless routing, 78
Blind power-based techniques, 852
Blog accessing, 779
Blog adapter, 776–777
Blog rendering and accessing instantly,
    See BRAINS

Blokh–Zyablov (BZ) channel coding, 444
Blom scheme, 1516
BlueSnarfing, 1242
Bluetooth 3.0, 95–100, 1153
    data transmission, 97–100
      internet access, 99–100
    network, 95–97
      piconet formation, 96–97
    scatternet, 97
Bluetooth technology/systems, 141–147, 748
    See also Bluetooth 3.0
    applications, 141
      short-range wireless connectivity, 141
      ad hoc networking, 141
      cable replacement, 141
      data and voice access points, 141
    baseband specification, 144–145
    capabilities, 141
    error corrections, mechanism, 145
    L2CAP, 146–147
    link control channels, 145
    link manager channels, 145
      specification, 145–146
    for MMGs, 698
    piconets, 144
    protocol architecture, 142–143 *See also*
      separate entry
    radio specification, 144
    scatternets, 144
    standards documents, 141–142
    standards, 141
    traffic flow specification, parameters, 146
      See also individual entry
    usage models, 143–144 *See also* individual
      entry
    user asynchronous channels, 145
    user scenarios, 142
      automatic synchronization, 142
      briefcase e-mail, 142
      cordless desktop, 142
      delayed messages, 142
      instant digital postcard, 142
      interactive conference, 142
      Internet bridge, 142
      portable PC speakerphone, 142
      three-in-one phone, 142
      ultimate headset, 142
Body area networks, *See* BANs
BoF (buffer overflows), 1244, 1247–1248
Boosting technique, 280
Borrowing, channel, 226
Bose–Chaudhuri–Hocquenghem, *See* BCH
Bottom-up approach, CLD, 1418
BPSK (binary phase shift keying), 1105
BRAINS (blog rendering and accessing
    instantly), 776–780
    analytical comparison, 779–780
    architectural overview, 776
    at user sites, 778–779
    processing flow, 778
    system overview, 776–778
      BC, 777
      BDA, 778
      blog adapter, 776–777
      blog visualizing agent, 777–778
      BRR, 778
      SMS/MMS message receiver, 776

Branch correlation effect, 1313–1314
Branch-point-traversal-based rerouting, 197
Briefcase e-mail, in bluetooth technology, 142
Bristle
    architectural overview, 751–752
    features, 752
    packet forwarding, 752
    routing, 752
Broadband transmission, OFDM for, 934–937
Broadband wireless access, *See* BWA
Broadcast control channel, *See* BCCH
Broadcasting systems, 1240–1241
    OFDM applications in, 931
BROSK (broadcast session key), 1516
BRP (Bordercast Resolution Protocol), 570
BS (base station)
    of MCN, 857–858
BSS (base station subsystem), 326, 1392, 1393
Buffer
    backward, 589
    management mechanism, 874
Burst format, 218–219
Busy, 573
BVA (blog visualizing agent), 777–778
BWA (broadband wireless access), 1409
    4G networks, 792–793
    fixed wireless category, 1409
    HSPA (High-speed packet access), 1409
    integration of, 1417
    mobile wireless category, 1409
    WiMAX, 1409
Byzantine failures, 77

C++ toolkits, 1342
CAB (charging accounting and billing)
        proposal, 2–6
Cable replacement protocol, 143
CAC (call admission control) in mobile
        communications, 643–651, 328
    adaptive bandwidth reservation-based
            CAC, 644
    CAC schemes in time-invariant capacity
            system, 643–645
        shadow cluster concept, 644
    for multi-service cellular networks, 644–645
        dual threshold bandwidth reservation
            (DTBR) scheme, 644
        dynamic partition (DP) scheme, 644
    game theory application to, 645–646 *See also*
            Game theory
    in time-division duplex mode system, 645
    in time-varying capacity system, 645
    system model and, 648–649
Caching operation, 571, 614, 720, 721
    in mobile P2P, 763
        active caching, 763
        passive caching, 763
Call admission control, *See* CAC
Call data records, 6
Call establishment, 54
Call rerouting phases, A-GSM, 329–333
    handover control, 331–333
    handover initiation, 329–331
    measurements, 329
CAMP (core assisted mesh protocol), 546
Capacity approximation-based bit allocation,
        1103–1104

Capacity-related SHO, 1086–1088 *See also* Soft
        handoff in mobile communications
CAR (context-aware information retrieval) for
        nomadic users, 911–917
    Applications
        types, 915
            interactive, 915
            proactive, 915
    CAR paradigm, 913–915
    Context
        context-aware documents matching, 915
        definition of, 913
        in IR, 914
        and mobile IR, 914–915
        stored in documents, 915
    context engine, 911
    context-awareness and ambient
            intelligence, 916
    IR
        performance measurement, 916–917
    document collection, 912
    information retrieval/information filtering
            engine, 911–912
    interface controller, 911
    in mobile environment, 912–913
        context capture, 913
        context definition, 912–913
        infrastructure support, 913
        intermittent wireless connections, 913
        privacy and confidentiality, 913
        reasoning methods, 913
        situation awareness, 913
    mobile users, 911
    mobile web initiatives, 915–916
    practical context-aware retrieval
            scenarios, 912
    query engine, 911
    result engine, 911
    in wireless environment, 912
    wireless network, 911
Care-of addresses
    types of, 501
Carrier sense multiple access mechanism with
        collision avoidance, *See* CSMA-CA
Carrier-to-interference ratio, *See* CIR
Case-based reasoning systems, *See* CBR
Cascade tunneling, 477
Cascaded relaying, 477, 861
CAVE algorithm, 137, 138
CBC (cipher block chaining), 1398
    to reduce PAPR, 926
CBF (content-based filtering), 811
CBR (case-based reasoning) systems, 236
CCDF (complementary cumulative distribution
        function), 925
CDDB (compact disc database),
        806–807
CDMA (code division multiple access)
        technology, 50, 55–56, 228–229,
        905–907, 939, 1349
    2G wireless systems, 1349
    access methods, 55–56
    CDMA2000, 172–174 *See also* separate entry
    design issues, 168–170
        bandwidth, 168
        chip rate, 168
        multi-rate, 168

        spreading and modulation solutions,
            169–170
        coherent detection, 169
        multi-user detection, 170
        pilot channel, 169
        power control, 169
        seamless inter-frequency handover, 170
        transmit diversity, 170
    digital cellular system, 217–219
    microcells, 180–185
    multiple-access interference, 56
    propagation considerations, 55–56
    WCDMA, 170–172 *See also* separate entry
CDMA2000, 172–174
    coherent detection, 174
    forward link channel, 173
    multi-carrier, 172–173
    multirate, 174
    packet data, 174
    reverse link channel, 173
    spreading codes, 173–174
CDP (centralized directory protocol), 753
CDPD (cellular digital packet data),
        1325–1328
    forward link block structure, 1327
    network architecture, 1326
    reverse link block structure, 1327
CDPWFQ (channel dependent parallel
        weighted-fair queuing), 951
CDS algorithm
    Chen's, 516–517
    Lin's, 518–519
    Wu's, 516
Cell forwarding rerouting, 203
Cell sectorization, 212
Cell splitting, 212
Cellular configuration, 209
    macrocell system, 209
    microcell systems, 209
    overlayed system, 209–210
    picocell systems, 209
Cellular data networks
    handoff and rerouting, *See* Rerouting
Cellular digital packet data, *See* CDPD
Cellular message encryption algorithm,
        See CMEA
Cellular network, 1352
    WLAN and
        inter-working between, 454–455
Cellular system
    1G, 214–215
        call processing, 214–215
        narrowband AMPS, European TACS, 215
    2G, 216–219
        CDMA digital cellular system, 217–219
        cordless mobiles, 219
        Pan European global system, 217
            burst format, 218–219
            common channels, 218
            multiple access scheme, 218
        PDC System, 217–218
        United States digital cellular, 216–218
    3G, 220–224
        IMT-2000, *See* separate entry
    fundamentals, 204
        cell splitting and cell sectorization, 212
        cellular configuration, 209

Cellular system (*Cont.*)
    macrocell system, 209
    microcell systems, 209
    overlayed system, 209–210
    picocell systems, 209
    channel characteristics, 204 *See also*
        Channel
    handoff, 211–212
    multiple access schemes, 206
        CDMA scheme, 206–209 SDMA
        FDMA scheme, 206
        scheme, 207–208
        TDMA scheme, 206
    power control, 212
    Using Base Stations, 204
        call from a mobile, 204
        call to a mobile, 204
        registration, 204
Central node for wireless sensor network
    hardware, 567
    software, 567–568
Central processing unit module, in mobile node,
    1496
Centralized and distributed mobile
    communication power controls,
    667–670
Centralized data fusion, 304
Centralized directory protocol, *See* CDP
Centralized P2P approach, 588–592
    adaptive chaining, 589–590
    basic chaining, 589
    batching policy, 591–592
    optimal chaining, 590
    P2Cast transmission, 590–591
Centralized power controls, *See* CPC
Certificate revocation lists, *See* CRLs
CF (collaborative filtering), 811, 1217–1218
    drawbacks, 1217–1218
CF (crest factor), 844–845
CFP (contention-free period), 1384
    packet loss rate bound in, 1386
CFP (cordless fixed part), 292–293
Chaining, adaptive, 589–590
Channel, 204 *See also* individual entries
    allocation and assignment, 210
        dynamic channel allocation schemes, 210
        fixed channel allocation schemes, 210
        hybrid channel allocation schemes,
        210–211
    channel assignment, 225–229
        CDMA systems, 228–229
        DCA, 227–228
        enhanced FCA, 226–227
        FCA, 225–226
        resource assignment problem, 180–185
    channel borrowing, 226
    channel coding, 113
    channel marker, 289
    channel reservation-based SHO, 1089
    channel reuse, 208
    estimation receiver, adaptive, 111
    coding, 1096
    delay spread, 205
    doppler spread, 205
    estimation, in OFDM, 928–929
    fading channels, 205
    link budget and path loss, 205–206

    reservation-based SHO, 1089
Channel characterization
    mobile wireless networks, 817
        large-scale propagation models, 817
            deterministic approach, 817–818
            stochastic approach, 818
        small-scale propagation models, 818–824
            mobile multipath channel parameters,
            818–819
            statistical representation, 820–821
Channel coding, 430
    ACELP encoder, 1129
    adaptive modulation and, 113
Channel dependent parallel weighted-fair
    queuing, *See* CDPWFQ
Channel quality indicator, *See* CQI
Chaperone, 506
Charging in 3G services, 6–7
    dynamic pricing models, 6
    flat-rate pricing, 6
Chen's CDS algorithm, 516–517
Cipher block chaining, *See* CBC
Chip rate design issue, CDMA, 168
Chip time, 1121
Chirp modulation, 1301–1302
CHTML (compact HTML), 697
Ciphering, 135–136
    asymmetric, 1557
CIR (carrier-to-interference ratio), 666
Circuit-switched mobile channels, 891–892
    enhanced circuit-switched data, 892
        bit rates, 893
    high speed circuit-switched data, 891
        network architecture, 891
Civil structure/machine monitoring, 1543
CLD (cross layer design) for wireless
    multimedia systems, 1417–1423
    application-based approach, 1418
    architecture of, 1418–1419
    for ad hoc networks, 1421–1423
        adaptive link layer strategy in, 1422
        congestion-distortion scheduling based
        CLD for, 1423
        optimal strategy of network layer in, 1422
    bottom-up approach, 1418
    challenges of, 1419–1420
    integrated optimization approach, 1418
    MAC-based approach, 1418
    top-down approach, 1418
CLDC (connected limited device configuration),
    700–701
Client/server model, 1214–1215
Client-interface setup, in coordinated multi-
    device presentations, 279
Clipping, 845
CL-MAC design approaches, 1427–1428
Closed-loop power control, *See* CLPC
CLPC (closed-loop power control)
    with inner-loop power controls, 673–674,
    675–676
    with outer-loop power controls, 674–675,
    675–676
Clustered multihop cellular network,
    See CMCN
Clustering/Clustering techniques
    classification, 514–515
        CMC, 515

        DSC, 515
        EEC, 515 *See also* separate entry
        LBC, 515 *See also* separate entry
        LMC, 515 *See also* separate entry
    cluster key, 1510
    cluster merging, 524
    cluster tree network, of BANs, 151
    cluster-based multihop, 303
    clusterguest, 518
    clusterhead, 514
    description, 513
    for MANETs, 513–527
        cost of, 513–514
        need for, 513
    in mobile ad hoc, 872
    passive clustering, 519 *See also* separate entry
    re-clustering, ripple effect of, 514
    and roaming techniques, IEEE 802.11
        WLANs, 365–374, *See also* under
        IEEE 802.11 WLANs
    scheme study, 515–525
        CDS algorithm, *See also* separate entry
    for scalable topology control and routing,
        872–874
    virtual-clusters, 872–873
CMC (combined-metrics-based clustering), 515,
    524–525
ODWCA, 525
CMCN (clustered multihop cellular network),
    863–864
CMEA (cellular message encryption
    algorithm), 137
CMF, 317
    context broker, 317–318
    context source, 317–318
    function of, 318
    internal working of, 318
CMLR (cost-effective maximum lifetime
    routing), 1528
CMOS (complementary metaloxide
    semiconductor), 1490
CN node, 727
CNR (carrier-to-noise ratios), 1307–1308
Coarse alignment, 1121
Coarse-grained allocation, 482
Code division multiple access, *See* CDMA
Codebooks, 1126–1127
Codecs for mobile multimedia transmission,
    736–743
    advanced codec standards, 741
    analog to digital conversion, 738
    audio coding, 738
        audio codecs, 741–742 *See also* AAC;
        AMR; MP3; MP3PRO
    image coding, 738–739
        image codec standards, 741
    mobile multimedia applications, 739–741
    multimedia information, 736–737
    standards, 737
        audio compression, 737
        image compression, 737
        video compression, 737
    text coding, 737–738
    video coding, 739
        video codecs, 742–743
Code-level hardening, 1246
Coding, *See also* individual entries

audio, 738
channel, 113, 1096
code mobility operation, 721
entropy coding, 429
entropy, 429
joint source-channel coding, 429–435,
        See also JSCC
Lossy coding, 429–430
modulation and, 1096–1097
Cognitive radio implementation for wireless
        communication, 230–238
    cognitive radio engines, 230–231
    environmental parameters, 232
    evolutionary algorithms, 235
    expert systems, 234–235
    performance objectives, 233
    radio operating parameters, 231–232
    radio parameter relationships, 232–234
Coherence bandwidth, 819
Coherence time, 819
Coherent binary phase shift keying,
        BER for, 1309
Coherent detection
    CDMA, 169
    CDMA2000, 174
Cold-start problem, 1212
Collaborative filtering, *See* CF
Collaborative handoff-resource reservation, 341
Color code, 1326–1327
Combined-metrics-based clustering, *See* CMC
Commands, AT, 1144
'Common ground' concept, 269
Common public radio interface protocol,
        See CPRI
Common radio resource management,
        See CRRM
Communication, mobile, *See also* Interpersonal
        communication
    content-enriched communication, 267–273
    over AWGN channel, 1286–1287
    technology evolution, 796–797
        digital mobile systems, evolution of,
            796–797
        mobile systems, evolution of, 796
        telecommunications, evolution of, 796
Compact disc database, *See* CDDB
Compact power density spectrum, 835
Complementary cumulative distribution
        function, *See* CCDF
CONFIDANT, 581–582
Congestion window, 1452–1453
Congestion/sending window (swnd), 1333–1334
Congestion-distortion scheduling based
        CLD, 1423
CONNECT 9600, 573
Connected limited device configuration,
        See CLDC
Connection handover, 1273–1274
    in WATM, 1267
        backward handover, 1279–1280
        connection rerouting, 1278–1279
        forward handover, 1280
        handover initiation, 1277
        handover trigger, 1277
        link quality monitoring, 1277
        target RP selection, 1277–1278
Connection rerouting, 1278–1279

Connection-extension rerouting, 197
Connectivity modeling, dynamic wireless sensor
        networks, 307–308
Consulting, 682
Container-based context processing platform,
        684–685
    context information acquisition, 684
    context sharing, 685
Content on mobile devices
    techniques to display, 260–262
Content profiling, 805–806, 806–808
    types, 806
        automatic tagging, 807–808
        manual tagging, 806–807
Content recommendation system
    key elements, 805–806
    typical cases, 810–811
        content-meta-based search, 810, 811
        context-aware search, 810–811
        user-preference-based search, 811
    wireless and mobile communication
        terminals, 811–815
Content re-purposing, 260
    ontology-based approaches to, 264
    with desk tool, 265–266
Content re-structuring, for mobile device web
        content, 262
Content-based filtering, 1218
    requirements, 1218
Content-enriched communication, 267–273
    background, 267–269
        photo and music sharing, 269
        social aspects, 268
        technological aspects, 268–269
    design principles and strategies, 269
    implications, 272
    interpersonal communication, 269–270
    mobile and wireless media content sharing,
        270
    user requirements, 269–271
Content-based filtering, *See* CBF
Contention-free period, *See* CFP
Content-meta-based search, 810, 811–813
    audio-based search, 812–813
    image-based search, 812
    text-based search, 811–812
Context-aware information retrieval, *See* CAR
Context awareness, 1211
    context-aware documents matching, 915
Context capture, 913
Context information, processing
    distributed architecture for, 683–685
        computing context, 684
        context entities, 684
        heterogeneous context sensors, 684
        sensor model, 684
Context learning, 808–809
Context processing platform, container-based,
        684–685
Context profiling, 805–806
Context-aware information retrieval for
        nomadic users, *See* CAR
Context-aware mobile collaboration, 1340–1341
Context-aware mobile computing, 679–685
    collaboration, 1340–1341
    container-based context processing platform,
        684–685

context in distributed teams, 679–680
    human interaction view, 680
    organizational view, 680
    project-centric view, 680
    spatial view, 680
context modeling, 682–683
    distributed composition, 683
    level of formality, 683
    richness and quality of information, 683
interaction context, 680–682 *See also*
        individual entry
processing context information
    distributed architecture for, 683–685
        computing context, 684
        context entities, 684
        heterogeneous context sensors, 684
        sensor model, 684
Context-aware search, 810–811
    location-based search, 813
    personal mood and activity based search,
        813–814
    surrounding-circumstance-based search, 813
    time-based search, 813
Contextual reasoning
    challenges and requirements, 315–316
    context evolution, 314
    context imperfectness, 314
    context interaction features, 314
    context medium, 314
    context persistence, 314
    context relevance, 314
    contextual information, 314
    hybrid reasoners, 321–322
    reasoning techniques, 319–321
        aggregation of context, 319
        inference of context, 319
        prediction of context, 319
        pre-processing of context, 319
    supported in generic context management
        framework, 313–323
        context and context awareness, 313–316
    temporal context, 314
Continuous phase modulation, *See* CPM
Contrasting MANETs and, 759–760
    network characteristics, 761
    node characteristics, 761
    routing characteristics, 762
Controllable mobility in MANETs, 558–560
    relay node bridges with pre-defined
        trajectories, 558–559
    relay nodes
        bridges with general trajectories,
            559–560
        for efficient power consumption, 560
        unified framework for utilizing controlled
            mobility, 560
Controllable network mobility, 557–558
Conventional fixed reserved capacity, 1090
Conventional link-state protocols, 565–566
    OLSR, 565 TBRPF, 565
Conversational multimedia, 480–483
    advance resource reservation, 480–483
        See also separate entry
Coolstreaming, 594
Co-operative caching in MP2P, 754–755
    PReCinC, 755
    super node split operation, 755

Cooperative games, 646–648, 766
Coordinated multi-device presentations
    client-interface setup in, 279
    using ambient-audio identification, 274–285
        audio fingerprinting, 280–281
        personalizing broadcast content, 275–279
            ad hoc peer communities, 276–278
            personalized information layers,
                275–276
            real-time popularity ratings, 278
        supporting infrastructure, 279–280
            audio-database server setup, 279
            client-interface Setup, 279
            social-application server setup, 279–280
        video 'Bookmarks', 278–279
Cordless desktop, 142
Cordless mobiles, 219
    CFP-initiated link setup, 292–293
    CPP-initiated link setup, 292
    CT-2 standard, 286–287 See also separate
        entry
    handshaking, 293–295
    signaling L2, 290–292 See also separate entry
Cordless portable part, See CPP
Cordless telephones-2, See CT-2
CORE, 581
Core-assisted mesh protocol, 540–541
Corporate wireless internet applications,
        1357–1358
Correlation receiver, 1118
Cost/reward function-based SHO, 1089
COUNT register, 137
Counter handoff-resource shortages, 339
Countermeasure module, in RIDAN, 85
Coupling, between WLAN and cellular
        networks
    loose coupling, 454–455
    tight coupling, 454–455, 457–458
    very tight coupling, 454–455, 458–459
Covariance matrix, 1296–1297
CP (contention period), 1384
    average service time, 1385–1386
    guaranteed priority access to voice over data,
        1384–1385
    state transition diagram, 1386
CPC (centralized power controls), 667–670
    basic concepts, 668
CPM (continuous phase modulation), 838–839
CPP (cordless portable part), 292
CPRI (common public radio interface)
        protocol, 193
CPU module hardware design, 1496–1502
CQI (channel quality indicator), 1016–1017
Crest factor, See CF
CRLs (certificate revocation lists), 79
Cross layer design for wireless multimedia
        system, See CLD
Cross-layer MAC design for wireless networks,
        1424–1433
    abstraction, 1425
    CL-MAC design approaches, 1427–1428
    LA, 1427–1428
    layering, 1425–1426
    modularity, 1425
    OC, 1428
    recent communications network design,
        1426–1427

CRRM (common radio resource management),
        47–49
Cryptography, See also Decryption
    asymmetric cryptography 132, 133
    basics, 132–134
    cryptographic primitives, 1510, 1556–1557
        asymmetric ciphers, 1557
        symmetric ciphers, 1556–1557
CSD (circuit switched data), 1143
    high-speed CSD, 1143
CSMA-CA (carrier sense multiple access
        mechanism with collision avoidance),
        397–398, 399
CT-2 (cordless telephones-2), 228
    CT2 standard, 53
    CT2 standard, 53
    main features, 295
    multiplex three burst structure, 290
    standard, 286–287
        radio interface, 286
        signaling layers, 287
        speech coding and transmission, 287
    synchronization patterns, 289
CTS (clear to send), 92–95
Customization framework, for mobile device
        web content, 264
Cyclic redundancy check, 90

Da-BLOT, 949–950
DAC, 1499
    dynamic range of, 847
Dash, 507
Data fusion, centralized, 304
Data link layer, 363–364
Data transmission, in Bluetooth 3.0 97–100
    internet access, 99–100
DCA (dynamic channel assignment), 227–228
    regulated DCA, 227–228
    segregation DCA, 227–228
DCF (distributed coordination function),
        90–101, 376, 1382–1383
    performance, 91–92
        protocol capacity, 91
        synchronization acquisition, 91
        synchronization maintenance, 91
DCMP (dynamic core-based multicast routing
        protocol), 552
DCS (digital cross-connects), 953
DDCA (distributed dynamic clustering
        algorithm), 520–521
Decision phase, of handovers, 452–453
Decryption, 132
DECT (digital enhanced cordless
        telecommunication), 228, 325
Dedicated inquiry access code, See DIAC
Dedicated-link protection, 972
Dedicated-path protection, 972
Dedicated physical control channel,
        See DPCCH
Dedicated physical data channel, See DPDCH
Defuzzification methods, 44
Dehopping, 1121
Delay spread, 105–107, 205, 819
    channel, 205
Delayed dupack, 1456
Delayed events, 17
Delayed messages, 142

Delayed synchronization, 17–18
Delay-locked loop, 1121
Delegation support, in mobile internet
        technologies, 720
Demodulation, 108–109, 109–110
    of DSSSMA UWB system, 1159–1160
Deng et al., mechanism
    in legacy IEEE 802.11 WLANs, 377
Denial of service, in infrastructureless
        routing, 78
Deployment coverage models, dynamic wireless
        sensor networks, 308–309
Deployment operation, 721
Design-level hardening, 1246
Desktop, cordless, 142
Despreading, 1121
Destination mobile host, 196
Destination-based rerouting, 197
Destination sequenced distance-vector routing,
        See DSDV
De-synchronization, 337
Deterministic approach, in PAPR reduction
        technique, 850–851
Deterministic coverage, 309
Deterministic multiple space Blom's scheme, 1517
Deterministic routing techniques, 1444–1446
Deterministic schemes, 1515–1520
    location-aware pre-distribution schemes,
        1518
    matrix-based pre-distribution schemes, 1516
    polynomial-based keying schemes, 1517
Device-specific authoring, for mobile device
        web content, 262
DHT (distributed hash table), 1004, 1005
DIAC (dedicated inquiry access code), 96
DICOM (Digital Imaging and Communication
        in Medicine), 716–717
Differential pulse coded modulation, See DPCM
Diffie–Hellman key exchange, 133
DiffServ model—prioritization categorizes, 798
DIFS (distributed interframe space), 90
Digital communication
    basics, 1284–1286
    functional blocks of, 1285
Digital cross-connects, See DCS
Digital enhanced cordless telecommunication,
        See DECT
Digital entertainment in home networking
        scenario, 1332
Digital Imaging and Communication in
        Medicine, See DICOM
Digital mobile systems, evolution, 796–797
Digital signal processing, See DSP
Digital traffic channel, See DTC
Direct key, 1510
Direct sequence CDMA system, See DS-CDMA
Direct sequence modulation, See DS
Direct sequence SS, See DSSS
Directed retry, 226
Disconnected operations mobile internet
        techniques, 721
    caching operation, 720, 721
    code mobility operation, 721
    deployment operation, 721
    discovery operation, 721
    hoarding operation, 720, 721
    queuing operation, 720, 721

replication operation, 721
Disconnection mitigation support, in mobile
    internet technologies, 720–721
Disconnection prevention support, in mobile
    internet technologies, 719–720
Discovery operation, 721
Disruption tolerant network architecture,
    See DTN
Distance routing effect algorithm for mobility,
    See DREAM
Distance-vector protocols, 566–567
    DSDV, 566
    WRPs, 566
Distortion/Distortion models
    estimation methods
        in image and video communication,
            443–445
    models
        for image and video communication,
            445–450
        for JPEG compressed images, 445–446
    for JPEG, 445–446
        JPEG encoder, 446
        model parameters, 448
Distributed admission control, 339–340
Distributed coordination function, See DCF
Distributed DoS attack, WLANs, 1401
Distributed dynamic clustering algorithm,
    See DDCA
Distributed hash table, See DHT
Distributed P2P, 592
    adaptive layered streaming, 593–594
    Coolstreaming, 594
    multiple sender distributed video streaming,
        592–593
Distributed power control, See DPC
Distributed reputation-based beacon trust
    system, 1551–1552
Distributed teams, context in, 679–680
    human interaction view, 680
    organizational view, 680
    project-centric view, 680
    spatial view, 680
Diversities
    in 4G networks, 791–792
        application diversity, 792
        network diversity, 791–792
        terminal diversity, 791
        user diversity, 791
    diversity gain, 1312
    exploitable, 942–944
        multi-carrier diversity, 944
        multiuser diversity, 942–944
        spatial diversity, 944
    in wireless communications
        diversity gain, in MIMO, 607
        frequency diversity, 607
        space diversity, 607
        time diversity, 607
        types of, 607
DLBC (degree-LBC), 524
DLC (data linkcontrol), 1276
DOA (direction of arrival), 496–497
Document collection, 912
Document routing protocol, See DRP
Domain-independent algorithms, location,
    827–830

LZ-based predictors, 828–830
Dominating-awake-interval protocol, 71
Dominating-set based clustering, See DSC
Doppler shift, 818
Doppler spread, 104–105, 205, 819
DoS attacks, WLANs, 1400
Download services
    FEC design for, 159–161
        1-D RS codes and 2-D RS codes usage,
            159–161
        simulation results, 166
    and real-time flows on wireless hec,
        1332–1334
    on a wireless hec, 1332–1334
DPC (distributed power control),
    669–670, 1468
    with active link protection, 1468–1470
    admission control, 1470–1472
        VDO, 1471–1472
DPCCH (dedicated physical control channel),
    172
DPCM (differential pulse coded modulation)
    JSCC technique for, 433
DPDCH (dedicated physical data channel),
    172
DREAM (distance routing effect algorithm for
    mobility), 572
DroPS (dynamically reconfigurable protocol
    stacks), 798
DRP (document routing protocol), 753
DS (direct sequence) modulation,
    1115–1116
DSA (dynamic spectrum access), 1100
DSC (dominating-set based clustering),
    515–517
DS-CDMA (direct sequence CDMA system),
    668–669
    operation concept, 906
DSDV (destination sequenced distance-vector
    routing), 64–65, 1374–1375
    performance, 1375
DSP (digital signal processing), 567–568,
    1062–1063, 1381
DSR (dynamic source routing), 59,
    1377–1379
    performance, 1378–1379
    query classification, 1534–1535
        historical queries, 1534
        one-time query, 1535
        persistent queries, 1535
DSSS (direct sequence SS), 905
DTBR (dual threshold bandwidth reservation)
    scheme, of CAC, 644
DTC (digital traffic channel), 216
DTN (disruption tolerant network architecture)
    architecture, 1443–1444
    topology, 1444
DTV (digital TV), 267
Dual-band direct-conversion CMOS transceiver
    frequency synthesizer design,
        248–250
    for IEEE 802.11a/b/g WLAN, 239–252
    radio architecture, 239 See also separate entry
    transmitter design, 244
        LO leakage calibration circuit, 244,
            246–247
        RF front-end, 245–246

Dual threshold bandwidth reservation scheme,
    See DTBR
Duplexing, 901–902
    multiplexing, 901–902
Dynamic channel assignment, See DCA
Dynamic constellation shaping
    techniques, 852
Dynamic core-based multicast routing protocol,
    See DCMP
Dynamic index in MP2P network, 755–756
    analysis, 756
    design and implementation, 756
Dynamic partition (DP) scheme, of CAC, 644
Dynamic per-host routing, 478, 486
Dynamic plug-ins, 1056–1057
Dynamic source routing, See DSR
Dynamic spectrum access, See DSA
Dynamic trans-frame-rating, 19–21, 21–22
Dynamic wireless sensor networks, 300–311
    connectivity modeling and topology
        optimization, 307–308
    deployment and sensing coverage models,
        308–309
    energy and battery models, 306–307
    modeling of, 304
    performance metrics, 304–306
        flexibility, 305
        lifetime/energy efficiency, 305
        quality, 305
        robustness, 305
        scalability, 305
        throughput, 305
    traffic models, 306
Dynamic trans-rating, 3G-telephony control
    stack, 21–22
Dynamically controlled frame-rate modification
    3G-telephony control stack, 19–22
Dynamically reconfigurable protocol stacks,
    See DroPS

EAP (extensible authentication protocol), 1397
    EAP-message digest 5, 1399
    EAP-TLS, 1399
    lightweight EAP, 1399
    protected EAP, 1399
Early blocking schemes, 338–340
Early deadline first, See EDF
Eavesdropping, 132
Ecosystem monitoring, 1544
ECS (efficient clustering scheme), 517–518
ECSD (enhanced circuit-switched data), 312
EDCA (enhanced distributed channel access),
    379, 1382–1383
EDF (early deadline first), 875
EDFA (erbium-doped fiber amplifiers), 967
EDGE (enhanced data rates for GSM evolution),
    463, 1143, 960
    key features of, 567
EDS system, 808
EEC (energy-efficient clustering),
    515, 521–523
    cost comparison of, 523
    IDLBC (ID load balancing clustering),
        521–522
    Wu's algorithm, 522
Efficient clustering scheme, See ECS
Efficient double-cycle, 97–98

EIR (equipment identity register), 327
EIRP (equivalent isotropically radiated power), 1151
EKDF (Kalman filter data fusion) method, 120–131, 125–126 *See also* Embedded Kalman filter data fusion method
in ACPS, 125–126
EKDF vs. HTAP, 128–129
EKDF vs. HTDOA, 129
filtering techniques, 122–123
Kalman filter (KF), 122
particles filter (PF), 122
vs. HTAP, 128–129
vs. HTDOA, 129
hybrid positioning techniques, 121–122
NLOS error mitigation techniques, 122
ELFN (explicit link failure notification), 1460
Embedded Kalman filter data fusion method, for ACPS, 120–131
filtering techniques, 122–123
Kalman filter (KF), 122
particles filter (PF), 122
hybrid positioning techniques, 121–122
NLOS error mitigation techniques, 122
Embedded network build-up, 1056–1057
Emergency, m-health for, 714–715
Encapsulation
algorithm, RC4, 388–389
IP, 485
protocol, ad hoc GSM, 329
Encoding
purpose of, 1097
in wireless video streaming, 1202–1204
Encryption, 132
algorithms, 133
symmetric encryption, 135
Energy-efficient clustering, *See* EEC
Energy models, dynamic wireless sensor networks, 306–307
Enhanced circuit-switched data, *See* ECSD
Enhanced data rates for GSM evolution, *See* EDGE
Enhanced distributed channel access, *See* EDCA
Enhanced GPRS, 893
bit rates, 895
Enhanced radio message system, *See* ERMS
Entities, context, processing, 684
Entropy coding, 429
Envelope, 835
Environmental parameters, wireless communication, 232
EoD (education-on-demand), 795
Eolutionary algorithms, in cognitive radio, 235
E-OTD (enhanced observed time difference), 503
Equal gain combining, 1312
Equivalent isotropically radiated power, *See* EIRP
Erbium-doped fiber amplifiers, *See* EDFA
Ergodic capacity, 1289
Erlang distribution, 1310
ERMS (enhanced radio message system), 1324
Error concealment multimedia streaming, 886
Error corrections mechanism, in bluetooth technology, 145

Error resilient coding multimedia streaming, 886
Error, *See also* individual entries
error concealment, 886
error resilient coding, 886
error-resilient re-purposing, 255
Error-resilient re-purposing, 255
Error-resilient video transmission in wireless networks, 1434–1441
distortion, cost, and delay relationship, 913–915
existing wireless networks, 1436–1438
in packet-switched wireless networks, 1435–1442
optimized framework for, 915
video encoding, 1434–1436
categories, 1435
channel-based algorithms, 1436
error concealment based algorithms, 1436
location, 1435
redundant encoding, 1435–1436
syntax information, aggregation, 1435–1436
ESS (extended service set), 1393
ETDM (electric time-domain multiplexing), 981
E-Textile garments for health monitoring, 716
Ethernet PONs, 969
Event generation module, in RIDAN, 85
EVM, 1031
Evolutionary algorithms, wireless communication, 235
Execution phase, of handovers, 452–453
Expert systems, wireless communication, 234–235
Explicit link failure notification, *See* ELFN
Extended service set, *See* ESS
Extensible authentication protocol, *See* EAP
Extension, 197
Extra time, 3G-telephony control stack, 16

Fade margin, SHO and, 1088
Fading, 818
channels, 205
fast fading, 820
flat fading, 819
frequency-selective fading, 819–820
slow fading, 820
Fairness attack, 409
Fairness in communication systems, 945
Jain's fairness index (JFI), 945
temporal fairness, 945
Family Locator, 506
Fast beacon packets (FastBP), 347–348
Fast cell selection, *See* FCS
Fast collision resolution mechanism, *See* FCR
Fast fading, 820
Fast join mechanism, 348
Fast-DOA-tracking algorithm, 497–498
Fast location updates, in mobile IP, 477
Fast-paced games, 695
FCA (fixed channel assignment), 225–226
enhanced FCA, 226–227
FCC (Federal Communications Commission), 1150
EIRP limits, 1151
emission restrictions, 1151–1157
FCR (fast collision resolution) mechanism, 380

FCS (fast cell selection), 1017
FDD (frequency division duplex), 206
FDMA (frequency division multiple access), 50, 52–53, 668, 904
access methods, 52–53
antenna height influence, 53
channel considerations, 52
FDPC (fully distributed power control), 669–670
Fear-based awareness, in ad hoc network security, 85
Feature phones, 1066
FEC (forward error correction) model, 157, 882, 997–998
decoding, 157
FEC payload ID, 160
FEC overhead, 157
MBMS download services, 159–161
1-D RS codes and 2-D RS codes usage, 159–161
complexity–performance tradeoff, 161
for MBMS streaming services, 161–165
complexity–performance tradeoff, 165
fixed size RTP packets, 161
high streaming bit rates and longer protection periods, 164–165
variable size RTP packets, 161–164
payload ID determination, 160
2-D RS codes, 159–161
Feedback power control model, 671
Federal Communications Commission, *See* FCC
Feed-forward neural network, 662
Ferrying scheme, 558–559
FGMP (forwarding group multicast protocol), 539–540
FH (frequency hopping) modulation, 1116–1117
FHSS (frequency-hopping SS), 905
Field-based monitoring
adaptive continuous query processing system, 1543–1544
contaminant level and flow, 1544
habitat or ecosystem monitoring, 1544
scientific experiments, 1544
wild fire monitoring, 1544
Filter graphs for video playback and video recording, 23–24
3GPP-file reader (ap1/vp1) and writer (u2), 23
3G-telephony control stack, , 23–24
AMR decoder (ap2) and AMR or G.711 encoder (ap4), 23
Audio TSM filter (ap3), 23
RTP receiver (au1/vu1) and sender (ap5/vp3), 23
Video TSM filter (vp2), 23
Filtering techniques, ACPS, 122–123
Fingerprinting, audio, *See* Audio fingerprinting
Finite-state Markov chain model, 821, 822
Firepad software, 1357
First generation (1G) mobile devices, 259
Fisheye routing, 1371–1372
performance, 1371–1372
routing algorithm, 1371
Fixed channel assignment, *See* FCA
Fixed wireless category, 1409
vs. mobile communications, 797

FLASH-OFDM standards, 1414–1417
Flat ad hoc multi-hop network, 303
Flat fading, 819
Flat network architecture, 1536
Flooded request protocol, 753
Flooding model, 1214
Fluctuation and delay, bandwidth, 254
Folk computing, 1208
Forced dropout, 1472
Forced handover, 453
Forward buffer, 589–590
Forward common control channel, 1484
Forward error correction model, See FEC
Forward fundamental channel, RLP, 1485
Forward handover, 1272–1273, 1280
Forward link block structure, CDPD, 1327
Forward link channel, CDMA2000, 173
Forward supplemental channel, RLP, 1484
Forwarding group multicast protocol,
    See FGMP
Forwarding group multicast protocol,
    539–540
Forward-link aspects
  in capacity-related SHO, 1087–1088
    log-linear path loss, 1087
    log-normal shadowing, 1087
    non-uniform and uniform traffic
        distributions, 1087
    power control, 1087
    rake receiver, 1087
    SHO area/SHO threshold, 1087
    short-term fading, 1087
  of soft handoff, 1087–1088 See also under
        Soft handoff
FPLA (future probable location area), 833
Free riders problem, 748
Freescale i.Smart, 1068–1072
Free-space propagation model, 817
Freeze TCP, 1457–1458
Frequency bands used by BANs, 149
Frequency division duplex, See FDD
Frequency division multiple access, See FDMA
Frequency hopping, 363, 1302
  in Bluetooth, 144–145
  frequency-hopping based multicarrier direct
        spectrum
    ART in, 1595
Frequency synthesizer design, direct-conversion
        WLAN, 248–250
  synthesizer loop architecture, 247, 250
  wide band VCO, 249–250
Frequency-domain-based processing, in PAPR
        reduction technique, 852–853
Frequency-hopped spread spectrum, 1350
Frequency-hopping based multicarrier direct
        spectrum, Wireless transceivers,
        1588–1590, 1589–1599 See also
        under Wireless transceivers
Frequency-selective fading, 819–820
Frequent handover, in mobile IP, 477
FRP (flooded request protocol), 753
FRTS (future-request-to-send), 1530
FSLS (Fuzzy Sighted Link State) routing, 567
FSMCs (finite-state Markov channels), 433
FSR (Fisheye State routing), 567
Full rerouting, 199–200
Fully distributed power control, See FDPC

Fuzzy logic, in RRM of 4G mobile systems,
        43–45
Fuzzy-neural methodology, 45

G.modem, 964
Gabriel graph, 529–530
Gains quantization, ACELP, 1129
Game theory, 585
  application to CAC, 645–646
    game theoretic framework
        for fair and efficient CAC problem,
            649–650
    non-cooperative and cooperative games,
            646–648
    two-person cooperative games, 646–648
        Arbitration schemes, 647 See also
            separate entry
    two-person non-cooperative games, 149
        equilibrium point, 646
        security level, 646
  models for network intrusions detection,
        764–770, See also IDS
    co-operative game, 766
    game definition, 767
    game formulation, 768
    game objectives and constraints, 767
    network assumptions and constraints, 767
    non-cooperative games, 766
        Bayesian Nash equilibrium, 766
        dominant strategy for, 766
        Nash equilibrium for, 766
    solution of the game, 768–769
    strategies of players, 767–768
  packet forwarding based on, 585
Games, See also Game theory
  fast-paced games, 695
  slow-paced games, 695
GAs (genetic algorithms), 231
  in cognitive radio implementation,
        231, 236
Gateway GPRS support node, See GGSN
Gaussian minimum shiftkeying, See GMSK
Gaussian sub-carrier SNR error model, 1107
Gauss-Markov and video source models, 439
GEAR (geographical and energy aware
        routing), 1528
General inquiry access code, See GIAC
General packet radio service, See GPRS
Generic connection framework, See GFC
Genetic algorithms, See GAs
Geocaching, 507, 614
Geocast, 532–533
Geodashing, 507
Geographic distance routing, 530
Geographical and energy aware routing,
        See GEAR
Geographical packet leashes, 81
Geographical routing algorithm, 530
Geonotes, 615
Geostationary earth orbit, 224
Geotagging, 507
GERAN (GSM EDGE radio access network)
  handover to, 1179–1180
GFC (generic connection framework),
        700–701
GGSN (gateway GPRS support node), 324
GIAC (general inquiry access code), 96

Global system for mobile communication,
        See GSM
GLS (grid location service), 572
GMAC (group based MAC), 1530
GMSK(Gaussian minimum shiftkeying),
        839–841
  frequency shaping pulse, 840
GPRS (general packet radio service), 892, 893,
        961, 1143, 1151
  bit rates, 894
  delay classes, 894
  GPRS/UMTS networks and WLANs
    handovers in, 453–454
  quality of service, 894
  reliability classes, 895
  residual error probabilities, 895
  short message service and other procedures,
        1183
  signaling and data transfer, 1180–1181
GPS (global positioning system), 502–504, 528,
        659–660, 1545
  assisted-GPS (A-GPS), 502–503
  GPS-based routing algorithm for ad hoc
        networks, 58–67
    AODV-RRS protocol, 62
    variations, 63–64
    GPS, 60–61
    RRS algorithms, 61–64
        stable zone and caution zone, 61–62
        stable zone radius effect, 62–63
  GPS-based localization, 1547
  for position location of mobile station,
        659–660
GPS-based routing algorithm
  for ad hoc networks, 58–67
    AODV-RRS protocol, 62
    variations, 63–64
    GPS, 60–61
    RRS algorithms, 61–64
        stable zone and caution zone,
            61–62
        stable zone radius effect, 62–63
GPSR (Greedy Perimeter Stateless Routing),
        529–530, 572
  greedy forwarding, 529
  perimeter forwarding, 529–530
GRACE 1, 687
Granular errors, 1107
Graph-based clustering, 369–372
Grayhole attack, 409
Greedy and on-demand proxy discovery
        algorithms, 863
Greedy dual least distance, See GD-LD
Greedy forwarding, 529
Greedy Perimeter Stateless Routing, See GPSR
Grid framework
  in mobile commerce, 627–632 See under
        Mobile commerce
Grid dimensions
  2-D RS codes, determination, 159
Grid location service, See GLS
GRID protocol, 531–532
  packet relay, 531
  route discovery, 531
  route maintenance, 531
Group based MAC, See GMAC
Group key, 1510

Grouping-error, 669
GSM (Global system for mobile
    communication), 55, 892, 1349
  2G wireless systems, 1349
  A-GSM, 325–326, 334–335 See also separate
      entry
  GSM/EDGE radio access network, 899
    user plane protocol stack, 898
  GSM–MANET dual mode terminal, 327
  privacy and authentication in, 135–136
    privacy, 135–136
      ciphering, 135–136
      temporary identify information, 136
  radio resource assignment, 226
  for telemetering technology, 1141–1149,
      See also Telemetering
  2G wireless systems, 1349
GSM-to-MANET internal handover signaling, 332
Guard period, 287–288

H.323 protocol, 1482–1483
  messages, 1483
Habitat or ecosystem monitoring, 1544
Handoff, 211–212 See also Rerouting
  hard handoff and soft handoff, 212
  mobile-assisted handoff, 211–212
  in mobile communications, 1086–1094,
      See also Hard handoff; Soft
      handoff
  mobile-controlled handoff, 211–212
  network-controlled handoff, 211
Handoff prioritization using early blocking,
    336–344
  counter handoff-resource shortages
      characteristics, 339
  handoff resource shortages, 337–338
    application-based approaches, 337
    handoff prioritization approaches, 338
    network-based approaches, 338
  resource reservation
    approaches for, 340–343
      collaborative handoff-resource
          reservation, 341
      and distributed admission control,
          339–340
      local handoff-resource reservation,
          340–341
      per-mobile vs. aggregated
          reservations, 343
      static handoff-resource
          reservation, 340
Handovers
  backward, 1279–1280
    control flow, 1268–1272
  break-before-make handover, 453
  connection, See Connection handover
  control in A-GSM, 331–333
  detection, in mobile IP, 487
  execution phase of, 452–453
  control, 331–333
  decision phase, 452–453
  design algorithms, 331
  detection, 487
  execution phase, 452–453
  forced handover, 453
  in GPRS/UMTS networks and WLANs,
      453–454

GSM-to-MANET internal handover
    signaling, 332
  hard handover, 453
  initiation phase, 452–453
  initiation, A-GSM, 329–331
  inter-technology handover, 453
  intra-technology handover, 453
  location-based vertical handover, 455, 456
  make-before-break handover, 453
  management
    in integrated WLAN/cellular networks,
        452–461
    in mobile networks, 452–453
    solutions, 455–459
      similarities and differences between,
          459–460
  mobile-assisted handover, 453
  mobile-controlled handover, 453
  network-assisted handover, 453
  network-controlled handover, 453
  policy-based handover, 455
  in RRM of 4G mobile systems, 41, 45–46
  seamless handover, 453
  soft hard handover, 453
Handshake, 293–295, 394
Hard handoff, 212, 453
Hardening, See Security hardening
Hardware design
  of central node
    for wireless sensor network, 567
  for mobile node wireless sensor network,
      564–567
    central processing unit module, 1496
    hardware, 1495–1496
    output module, 564
    power module, 564
    radio link module, 1495–1496
    sensor module, 1496
  and software design for the mobile node,
      564–567 See also Central node;
      Mobile node
Hartley architecture, 1040
Hash chains, 78–79
  one-way hash function, 78–79
Hash function, 80
Hash-based message authentication code
      message digest 5, See HMAC-MD5
Hashing descriptors, audio fingerprinting, 280
Hazy Sighted Link State, See HSLS
HCCA, 379
HCF (hybrid coordination function), 1382
Headset, 142
Healthcare applications of BANs, 155
Healthcare wireless sensor networks, 404–411
  intrusion detection, 406–407
  possible attacks, 409
  security, 405–406
    architecture, 407
    secure blocks, 406
  SKKE protocol, 408–409
Heat control and conditioning, 1543
HECs (home entertainment centers), 1331
Heidelberg QoS model, 798
Hello floods, 1559
Heterogeneity
  challenge of multimedia streaming, 885
  in personal mobility, 475

Heterogeneous WSNs, 345–353, 636,
    1203–1204
  benefits, 346–347
  number and placement of supernodes,
      349–350
  routing and data gathering, 347–349
  target coverage application, 351–352
  topology design and control, 350–351
Hierarchical location prediction, 832, 838
Hierarchical MCN, See HMCN
Hierarchical state routing, 1372
  drawbacks, 1373
  performance, 1373
  routing algorithm, 1372–1373
Higher-layer power management
  in ad hoc networks, 74
    operating system/application layer, 74
    transport layer, 74
High-level security hardening, 1247
High speed circuit-switched data, 891
High speed downlink packet access,
      See HSDPA
High-speed packet access, See HSPA
High-speed wireless internet access, 354–361
  mobile data access, 354–355
    capacity and bandwidth efficiency, 354
    space, 354–355
  multiantenna technology, 355 See also
      separate entry
Highway microcells, 597–601
  spectral efficiency, 597–600
  teletraffic issues, 600–601
HIPERLAN/2 standard, 1133–1135
  MAC Protocol, 1133–1135
    broadcast phase, 1134
    contention phase, 1134
    downlink phase, 1134
    uplink phase, 1134
  physical layer, 1133
HLR (home location register), 324, 327
HMAC-MD5 (hash-based message
      authentication code message
      digest 5), 728
HMCN (hierarchical MCN), 864–866
  cell and multihop cell in, 865
Hoarding operation, 720, 721
Holistic QoS management, 798–799
Home entertainment center, wireless
  digital entertainment in home networking
      scenario, 1332
    iTunes music download, 1332
    video streaming, 1332
    video-chat and massive multiplayer on-line
        games, 1332
  downloading and real-time flows on a
      wireless hec, 1332–1334
  protocol communications and architecture,
      1331–1338
  queuing delays, 1334–1335
  simulation assessment, 1335–1336
Home location register, See HLR
Home networking scenario, digital
      entertainment in, 1332
Home subscriber server, See HSS
HomeZone, 572
Hop time, 1121
Host-based IDS, 765

HSDPA (high speed downlink packet access), 463, 640, 708, 957
  4G technologies, 640
H-Shirt, 715
HSLS (Hazy Sighted Link State), 567
HSPA (high-speed packet access), 1409
HSS (home subscriber server), 139
HTML (hypertext markup language), 262, 697
HTTP (Hypertext transfer protocol), 505, 696
Hybrid coordination function, *See* HCF
Hybrid multicast, 1012–1014
Hybrid networks, 1538–1539
  functioning, 1538
  important features, 1539
Hybrid positioning techniques, ACPS, 121–122
Hybrid reasoners, 321–322
Hybrid relay
  MCN with, 864–866
    HMCN (hierarchical MCN), 864–866
Hybrid routing protocols, 570
  LANMAR, 570
  ZRP, 570
Hypertext markup language, *See* HTML
Hypertext transfer protocol, *See* HTTP

I^2C-bus, 1502
IARP (IntrAzone Routing Protocol), 570
IBSS (independent basic service set), 1393
IBT (in-band-terminator), 997
ICAR (integrated cellular and ad hoc relaying), 859–861
IDEs (Integrated development environments), 505
IDLBC (ID load balancing clustering), 521–522
IDS (intrusion detection systems), 406–407, 1543
  in ad hoc networks, 84–86
  categorized, 764–765
  game theoretic models for, 764–770, *See also* under Game theory
    anomaly detection, 765
    host-based, 765
    misuse detection, 765
    network-based, 765
IEEE 802.11
  architecture and protocols, 89–95, 362
    distributed coordination function, 90–101, 90–92
      performance, 91–92
      request to send/clear to send, 92–95
      RTS/CTS effectiveness, 93–95
  bluetooth, *See* separate entry
  data link layer, 363–364
  enhancements, 1413–1414
  history, 1392–1393
  IEEE 802.11i, 1397–1398
  physical layer, 363
  standard, 362
  wireless encryption algorithms, 1398
IEEE 802.11 WLANs, 362–364, 365–374, *See also* IEEE 802.11
  clustering and roaming techniques, 365–374
    graph-based clustering, 369–372
    interaccess point protocol, 366
    location-based clustering, 366–369
    quasihierarchical routing, 372–373
    strict hierarchical routing, 373–374

limitations, 1382–1383
  overhead reduction, 1383
  voice multiplexing, 1383
privacy and authentication in, 134–135
  authentication, 134, 135
  PMK (pairwise master key), 135
  privacy, 134–135
QoS Support in, 375–386, *See also* QoS support
security standards, 387–388
  WEP, *See* separate entry
IEEE 802.11 WLAN, QoS in, 375–386
  comparative performance evaluation, 382–386
  IEEE 802.11e standard, 378–382
    EDCA, 379
    HCCA, 379
    QoS enhancements to, 379–382
  legacy IEEE 802.11
    WLANs, 375–378 *See also* separate entry
IEEE 802.11a/b/g WLAN
  dual-band direct-conversion CMOS transceiver for, 239–252, *See also* Dual-band direct-conversion CMOS transceiver
IEEE 802.11e standard, 378–382
IEEE 802.11i
  security architecture, 392–395
    authentication and access control, 393
    key generation process, 393–394
    TKIP security protocol, 394–395
      data integrity protection, 394–395
      data confidentiality protection, 395
IEEE 802.15.4, 150–151
IEEE 802.15.4a, 1153
IEEE 802.16
  QoS mechanisms in, 1135–1139
    bandwidth request mechanisms, 1138
      base on contract, 1138
      piggyback, 1138
      polling, 1138
    downlink subframe, 1136
    flow types, 1137
      best-effort service, 1137
      nrtPS, 1137
      rtPS, 1137
      UGS, 1137
    grants, 1138–1139
      GPC (grant per connection), 1138
      GPSS (grant per SS), 1139
    MAC layer, 1135–1137
    physical layer, 1135
    uplink sub-frame, 1136
  wireless metropolitan area networks
    privacy and authentication in, 135
      privacy, 135
  wireless metropolitan area networks, 417–427
    QoS provisioning, 425
IEEE 802.16 MAC, 417–421
  mesh mode, 420–421
  PMP mode, 418
  resource management in, 421–425
    cross-layer design, 423–425
    QoS provisioning, 425
    scheduling schemes, 421–423
IEEE 802.1X, 1398–1399
  EAP-message digest 5, 1399

EAP-TLS, 1399
  lightweight EAP, 1399
  protected EAP, 1399
  shortcomings, 1399–1400
    mutual authentication absence, 1399
    session hijacking, 1399
  tunneled TLS, 1399
IEEE802.15.2, 1160
IEEE802.15.3, 1162
IEEE802.15.4, 1162
IERP (IntErzone Routing Protocol), 570, 1376
IETF (internet engineering task force), 2–3
IF (intermediate frequency), 1038–1041
  advantages, 1040
  disadvantages, 1040
  low-IF transceiver architecture, 1039
IFFT/FFT
  OFDM implementation by, 921, 922, 930
IIP2 (input second-order intercept point), 1028
IIP3 (input third-order intercept point), 1030
Image and video communication
  joint source-channel coding for, 429–435, *See also* JSCC
  distortion estimation methods in, 443–445
  distortion models for, 445–450
  JSCC for, 429–435
    bit allocation method for, 431
    channel coding, 430
    methods for, 431–434
    over wireless channels, 434
    source coding, 429–430
    source-channel rate allocation method, 432
    energy/power optimization, 437–438
    multi-resolution JSCC, 439
'Image-based communication practices', 269
Image coding, 738–739
  image codec standards, 741
Image compression, 737
Image-based search, 812
Image-rejection filter, 1043
  Hartley architecture, 1040
  polyphase filtering, 1040
  Weaver architecture, 1040
Imaging systems, 1151
  GPR systems, 1151
  medical systems, 1151
  surveillance systems, 1151
  through-wall imaging systems, 1151
  typical parameters, 1152
  wall-imaging systems, 1151
IMDs (implanted medical devices), 148
  RF transceivers operation in
    low-frequency inductive coupling, 149
    MICS band, 149–150
IMEI (international mobile equipment identification), 746
I-Mode Java, 701–702
Imperfect channel information, BIT allocation with, 1107–1108
Imperfectness, context, 314
Impersonation attacks, WLANs, 1401
Impersonation risk, 132
Implanted medical devices, *See* IMDs
Imprecise network state model, 543
Imprinting procedure, 82–83
Impulse radio, 1150

IMSI (international mobile subscriber identity), 746
IMS—IP multimedia subsystem, 462–473
  as enhancement to mobile domain for 3G UMTS networks, 463
  IMS architecture, 463–467
    application server, 465–466
    call session control function, 464
    gateway control functionalities and gateway functionalities, 466–467
    HSS, 466
    I-CSCF, 465
    IM-SSF AS, 466
    media server, 467
    OSA SCS AS, 466
    P-CSCF, 465
    S-CSCF, 464
    SIP AS, 466
  interfaces and protocols in, 467
    basic communication, 468
    charging, 469–470
    registration, 468
    security, 469
    service control, 468
    user and service identities, 467–468
  mobile domain characteristics, 470
    GPRS-specific concepts of, 470
IMT-2000, 1588
  features, 220–221
  planning considerations, 222–223
    intelligent networks, 205
    radio access, 222
    regulatory environments, 205
    security, 204
    spectrum requirements, 204
  satellite operation, 223–224
  services, 221–223
    network services, 222
    supplementary services, 222
In-band-terminator, See IBT
Incremental bit allocation, 1104–1105
Independent basic service set, See IBSS
Independent mode, 1453–1454
Indirect key, 1510
Indoor location modeling, 680
Indoor microcells, 601
Inductive chosen plaintext attack, WEP, 1396
Information retrieval for nomadic users, context-aware, See CAR
Information, multimedia, 736–737
Infrastructure mode, 1453
Infrastructure-based monitoring
  adaptive continuous query processing system, 1542–1543
  civil structure/machine monitoring, 1543
  heat control and conditioning, 1543
  intrusion detection, 1543
  traffic monitoring, 1543
Infrastructureless routing, 76–77
  attacks that target, 77
    black hole, 77
    blackmail, 78
    denial of service, 78
    location disclosure, 77
    replay, 77
    routing table poisoning, 78
    wormhole, 78

black hole in, 77
  location disclosure in, 77
  service denial in, 78
Inherent networking, MMGs in, 694
Initialization protocol, 1572–1573
Initiation phase, of handovers, 452–453
Inner-loop power controls, CLPC with, 673–674, 675–676
Input sequence envelope scaling technique, 852
INSENS, 1560
Instant digital postcard, 142
Integer overflow, 1244
Integrated cellular and ad hoc relaying, See ICAR
Integrated development environments, See IDEs
Integrated dual mode terminals, of ad hoc GSM, 327
Integrated optimization approach, CLD, 1418
Integrated personal mobility architecture, See IPMoA
Integrated presence, in personal mobility, 475
Integrated WLAN/cellular networks
  handovers management in, 452–461, See also Handovers
Interaccess point protocol, 366
Interaction context, in context-aware mobile computing, 680–682
  team awareness, 681–682
Interactive conference, 142
Interactive mobile applications, 1340
Interactive playback control of video, See also Video services
  under the 3G-telephony control stack, 13–25
    dynamic trans-rating, 21–22
    dynamically controlled frame-rate modification, 19–22
    filter graphs
      for video playback and video recording, 23–24 See also Filter graphs
    timing and synchronization for, 16–19
      extra time, 16
      interactive playback seek, 18–19
      interactive playback speed up and slow down, 16–18
      sample time, 16
      seek points, 16
      transmission time, 16
    video-services demonstration, 22–24
Interference reduction, in MIMO, 606–607
Interference-based channel assignment, for SHO, 1089–1090
Interleaving techniques, 853, 1127–1128
Intermediate time-variability channels, link adaptation for, 1292
Internet applications, wireless, 1339–1345
  context-aware mobile collaboration, 1340–1341
  interactive mobile applications, 1340
  mobile WS performance limits, 1343–1344
  networked devices and appliances at home, 1340
  real-time multimedia, 1340
  service interactions, 1341–1342
    mobile client, 1341–1342
    mobile server, 1342
    mobile proxy, 1342

WS for .NET compact framework, 1343
WS on mobile devices, 1342–1343
  C++ toolkits, 1342
  J2ME
    toolkits for, 1342–1343
Internet bridge, 142
Internet connected BANs, 151–152
Internet engineering task force, See IETF
Internet protocol (IP)-based platforms, See IPTV
Internet protocol (IP) multimedia subsystem (IMS), See IMS–IP multimedia system
Internet technologies, mobile, 718–725, See also SOA
  delegation support, 720
  disconnected operations techniques, 721
    caching operation, 720, 721
    code mobility operation, 721
    deployment operation, 721
    discovery operation, 721
    hoarding operation, 720, 721
    queuing operation, 720, 721
    replication operation, 721
  disconnection mitigation support, 720–721
  disconnection prevention support, 719–720
  mobile web service support framework, 721–724 See also separate entry
  selection support, 719
Internet working unit, ad hoc GSM, 327
Inter-operability formats, 749
Interpersonal communication, 269–270
  asynchronous, 269–270
  interaction and visual design, 270–271
  mobile and wireless media content sharing, 270
  presence, awareness, and social bonding, 270
  synchronous, 269–270
  user-generated content, 270
Intersymbol interference, 205
Inter-technology handover, 453
IntErzone Routing Protocol, See IERP
Intra-technology handover, 453
IntrAzone Routing Protocol, See IARP
Intrusion detection systems, See IDS
Invasive BANs, 148–150
IP (internet protocol)
  encapsulation, 485
  IP multicast vs. overlay multicast, 1003
  personal mobility challenges, See under Personal mobility
  terminal mobility challenges, See Terminal mobility
  user mobility, See separate entry
IP1dB (input 1-dB compression point)
IPMoA (integrated personal mobility architecture), 475
IPTV (Internet protocol (IP)-based platforms), 268, 1331
IR
  performance measurement, 916–917
    fall-out, 916–917
    precision, 916
    recall, 916
IRUWB (impulse radio ultrawideband) systems, 1160

IS-95B/CDMA2000/W-CDMA SHO, 1091–1092
IS-95B/CDMA2000/W-CDMA SHO, 1091–1092
ISMA (internet streaming media alliance), 889
Isolated game, 694–695
ITunes, 269, 1332
ITV (interactive TV)
  definition, 267
  social uses of, 267–273, *See also*
    Content-enriched communication

J2ME (Java 2 Micro Edition), 700–701
  toolkits for, 1342–1343
Jammer-to-signal power ratio, 1115
Jamming attacks detection, WLANs, 1401
Jamming mitigations, WLANs, 1401
Jamming modeling, WLANs, 1401
Java-enabled wireless devices, 1356
JFI (Jain's fairness index), 945
Jitter, 1480
Joint photographic experts group, *See* JPEG
Joint source-channel coding, *See* JSCC
Joint source-channel distortion modeling
  for MPEG-4 coded video sequences, 449–450
    model performance, 450
JPEG (joint photographic experts group), 429–430, 741
  distortion models for, 445–446
    JPEG encoder, 446
    model parameters, 448
JSCC (joint source-channel coding)
  for DPCM coded images, 433
  for image and video communication, 429–435
    bit allocation method for, 431
    channel coding, 430
    methods for, 431–434
    over wireless channels, 434
    source coding, 429–430
    source-channel rate allocation method, 432
  in image and video communication
    energy/power optimization, 437–438
    multi-resolution JSCC, 439
  operational rate-distortion behavior of, 432
JVM (java virtual machine), 700

Kalman filter, *See* EKDF (Kalman filter data fusion)
K-Edge-connectivity, 1239
Key management, in ad hoc networks, 81–84
  resurrecting duckling, 82–83
  self-organized PKI, 82
  threshold cryptography, 82
Key, cluster, 1510
Key-chain schemes, 1510, 1514
Key-pool, 1510
Knowledge oracles, 1444
Kuhn-Tucker optimality conditions, 1429
KVM (kilobyte virtual machine), 700

L1, 60
L2, 60
L2CAP (logical link control and adaptation protocol), 143, 146–147
Lagrangian multiplier, 432, 1429
  Lagrangian-based optimization problem, 437

LAN Connect, 958
Land earth stations, 224
Land mobile communication systems, 491–500
  phased-array antennas for, 491–492
    beam pattern, 492
  macrocell control, array antenna for, 492–496
    See also Macrocell system
  microcell control, array antenna for, 496–500
    See also Microcell system
Laptops, games for, 696
LAR (location-assisted routing) protocol, 529
  request and expected zones, 529
Latency requirement, of MMGs, 703–704
Layering, in cross-layer MAC design for wireless networks, 1425–1426
LBC (load-balancing clustering), 515
  AMC, 523
  cost comparison of, 525
  DLBC, 524
LBSs (location-based services), 501–508
  conceptual notions of, 502
  current commercial trends, 506–507
    opportunistic M-Commerce, 506–507
    people-tracking and personalized services, 506
    recreational and grassroots, 507
  emerging support
    in middleware and, programming tools, 505–506
    in standards bodies, 505
  LBS ecosystem, 504–505
  positioning technologies, survey of, 502–504
  types of, 504
LEACH (Low-energy adaptive clustering hierarchy), 1526–1527, 1537
LEAP (localized encryption and authentication protocol), 1516
Learning, context, 808–809
Legacy IEEE 802.11 WLANs, 375–378
  DCF, 376
  PCF, 376
  QoS-aware schemes for, 376–378
    Aad et al., mechanisms, 377
    Aad-backoff scheme, 377
    Aad-CWmin scheme, 378
    Aad-DIFS scheme, 378
    Deng et al., mechanism, 377
    TCMA mechanism, 378
LeZi-update, 829
Lifestyle and sports applications of BANs, 155
Lifetime-refining energy efficient of multicast trees, *See* L-REMiT
Lightpath, 970–971, 975–976
  lightpath restoration, WDM, 994
Light-tree, 973
Lin's CDS algorithm, 518–519
Line powered supernodes, 349
Linear model, 1295
Link adaptation techniques, 103
  for channels with intermediate time-variability, 1292
  for rapidly time-varying channels, 1290–1292
  for wireless channels communication, 1284–1293

AWGN (additive white Gaussian noise) channel, 1286–1287
  digital communication basics, 1284–1286
  link-adaptive communication system, 1288–1290
  non-adaptive communication system, 1287–1288
  Rayleigh fading channels, 1290
  time-varying channels
    modeling of, 1287
Link budget, 205–206
Link capacity, 1089–1090
Link control channels, in bluetooth, 145
Link heterogeneity, 349
Link layer assisted handover detection, in mobile IP, 479–480
Link layer protocol, 328
Link manager channels, in bluetooth, 145
Link manager protocol (LMP), 142
Link protection, 972
Link restoration, 972
Link-key, 1510
Link layer assisted handover detection in mobile IP, 479–480
Links, backhaul, 1409–1410
  types, 1410
    multiple point-to-point, 1410, 1411
    multipoint-to-multipoint, 1410, 1411
    point-to-multipoint, 1410
    point-to-point, 1410
Linux, 1086–1087
  Mobilinux 4.0, 1086
  PalmSource, 1087
  qtopia phone edition components, 1084
Live multimedia streaming, 884
LLC (logical link control), 362, 363
LMC (low-maintenance clustering), 515, 517–520, 519–520
  cost comparison of, 520
  ECS, 517–518
LMP (link manager protocol), 142
LO (local oscillator) leakage calibration circuit, 244, 246–247
Load-balancing clustering, *See* LBC
Local handoff-resource reservation, 340–341
Local oscillator, 1035–1036, 1043
Localization, absolute, 1547
Localized encryption and authentication protocol, *See* LEAP
Location-based services, *See* LBSs
Location disclosure, in infrastructureless routing, 77
Location management, 1273
  in WATM, 1264–1265
    and control flow, 1265–1273
      connection forwarding, 1266–1267
      connection redirect, 1266–1267
      location query, 1266–1267
      network entities, 1265
      authentication server, 1275
      EMAS, 1275
      location server, 1275
      mobile terminal, 1275
Location prediction algorithms, 825–834
  domain-independent algorithms, 827–830
    LZ-based predictors, 828–830
  domain-specific heuristics, 830–833

Location prediction algorithms (*Cont.*)
    hierarchical location prediction, 832
    mobile motion prediction, 831
    segment matching, 831–832
  preliminaries, 826–827
    approach, 827
    movement history, 826–827
Location registration, 486
Location service protocols, 572
Location-assisted routing protocols, 528–530
  geographic distance routing, 530
  geographical routing algorithm, 530
  greedy perimeter stateless routing, 529–530
  LAR protocol, 529
Location-aware routing in MANETs, 528–536
  Geocast, 532–533
  location services, 533–534
  location-assisted broadcasting, 534
  location-assisted tour guide, 534–535
Location-based clustering, 366–369
Location-based search, of context, 813
Location-based vertical handover, 455, 456
Location-limited channels concept, 83
Location registration, in mobile IP, 486
Log-distance path loss model, 817–818
Logical link control, *See* LLC
Logical link control and adaptation protocol,
    See L2CAP
Log-linear path loss and log-normal
    shadowing, 1087
Long-haul networks, 970–978
  call admission control, 976
  fault management, 971–973
    protection, 972
    restoration, 972–973
  IP over WDM, 975–976
  multicasting, 973–974
    multicast-capable OXC architectures,
      973–974
    multicast RWA, 974
  network control and signaling, 976–977
  OPS, 977–978
    optical burst switching, 978
  RWA, 970–971
  traffic grooming, 974–975
Longley–Rice model, 1320
Loop removal, extension with, 197
Loose coupling architectures, in WLAN and
    cellular networks, 455–457
Loose source routing, 486
LOS
  for RSS estimation, 127
  for TOA and AOA estimation errors, 126–127
Lossy coding, 429–430
Low noise amplifier, 1041
Low-energy adaptive clustering hierarchy,
    See LEACH
Low-frequency inductive coupling, 149
Low-latency handoff, 730–731
  in mobile IPv4, 730–731
Low-level security hardening, 1247–1248
  BoF vulnerabilities, 1247–1248
  file management vulnerabilities, 1248
  integer vulnerabilities, 1248
  memory management vulnerabilities, 1248
Low-maintenance clustering, *See* LMC
Low probability of intercept, *See* LPI

LPI (low probability of intercept), 1119
L-REMiT (lifetime-refining energy efficient of
    multicast trees), 547, 549–550
  refinement operation in, 550
LSR (label-switching router), 997–998
LT (Luby transform) code, 1194–1195
  decoding, 1194–1195
  encoding, 1194
LTV channels, 1301
Luby transform, *See* LT
Lyapunov function and stability test, 1430–1432
LZ-based predictors, 828–830
  parsing algorithm, 828
  prediction, applying to, 828–830

MAC (medium access control), 1382
  guaranteed priority access to voice over data,
    1384–1385
  power management, 70–72
  service interval structure, 1384
  voice overhead reduction, 1385
  voice traffic multiplexing, 1384
MAC (mobility-aware clustering), 515, 520–521
    See also under HIPERLAN/2 standard
  802.15.4 sensor clusters, attacks in, 396–403
  cost comparison of, 522
  DDCA, 520–521
  MAC-based approach, CLD, 1418
  MAC layer attacks
    in 802.15.4 sensor clusters, 396–403,
      See also 802.15.4 sensor clusters
  MOBIC, 520
MAC-layer power management
  in ad hoc networks, 70–72
    dominating-awake-interval protocol, 71
    periodically-fully-awake-interval protocol,
      71, 72
    quorum-based protocol, 72
Macrocell system, 209, 492–496
  channel capacity improvement, 494–496
    computer simulation model, 495
    numerical results, 495–496
  conventional and ideal zone configurations,
    493
  conventional cell arrangements, 493
  prototypes developed, 493
Magic WAND system, 1276
MAHO (mobile-assisted handover), 332
Make-before-break handover, 453
Managed fiber services, *See* MFS
Managed wavelength service, 957–959
MANETs (mobile ad hoc networks), 1487, 1491
    See also Contrasting MANETs
  BAN, 510
  clustering techniques for, 513–527, *See also*
    Clustering techniques
  controllable mobility in, 558–560
    relay node bridges with pre-defined
      trajectories, 558–559
    relay nodes
      bridges with general trajectories, 559–560
      for efficient power consumption, 560
      unified framework for utilizing
        controlled mobility, 560
  location-assisted routing protocols, 528–530
    See also separate entry
  location-aware routing, 528–536

  Geocast, 532–533
  location services, 533–534
  location-assisted broadcasting, 534
  location-assisted tour guide, 534–535
  multicast protocols, 537–545, *See also*
    separate entry
  multimedia applications
    in QoS for, 870–879, *See also* QoS
  packet forwarding, selfish behavior on,
    See Packet forwarding
  PAN, 510–511
  routing protocols in, 564
  topological changes in, 563
  WLAN, 511
  zone-based routing protocols, 530–532
    See also separate entry
MANSI (multicast for ad hoc networks with
    swarm intelligence), 546, 553–554
Manual configuration, 1547
Manual tagging, 806–807
MAODV (multicast ad hoc on-demand distance
    vector), 538–539, 546, 547
  non-duplicate JOIN_DATA propagation, 548
  RouteREPly (RREP) propagation, 547
  RouteREQuest (RREQ) propagation, 547
Marker, channel, 289
Marking process, in clustering, 516
Markov conception, 1088
Markov predictor, 830
M-ary binary-orthogonal keying, *See* M-BOK
Massively multiplayer mobile game,
    See MMMG
Mass personalization, 274
Master key, 393–394
Master state machine, *See* MSM
Matching, 805–806, 810
  NB classifier, 809–810
  other approaches, 810
  VSM, 809
Matrix channel model, 823
Max hop count, 80
MaxC/I (maximum carrier-to-noise ratio), 947
Maximum ratio combining, 1310
Maximum ratio combining-3, *See* MRC-3
MBMS (multimedia broadcast multicast
    service), 880–883, 157
  1-D RS codes, 159
  2-D RS codes, 159
  architecture and concepts, 880
  download services, FEC design for, 159–161
    1-D RS codes and 2-D RS codes usage,
      159–161
    simulation results, 166
  FEC model, 882
  media codecs and formats, 882
  protocols and codecs, 881
  QoE metrics, 882–883
  RTP, 882
  SDP, 881–882
  streaming services, 161–165
    complexity–performance tradeoff, 165
    fixed size RTP packets, 161
    high streaming bit rates and longer
      protection periods, 164–165
    simulation results, 165–166
    variable size RTP packets, 161–164
  video streaming, 1197–1199

delivery, 1197
post-delivery, 1197
user service discovery, 1197
MBOA (multiband OFDM alliance), 1159
M-BOK (M-ary binary-orthogonal keying),
1159–1160
MBPS (measurement-based prioritization
scheme), 1089
M-CAN (mobile content addressable network),
752–754
centralized Directory Protocol, 753
document routing protocol, 753
flooded request protocol, 753
hierarchy and peer community,
753–754
lookup process, 754
performance issues and analysis, 754
MC-CDMA, 908–909
MCHO (mobile-controlled handover), 332
MCM (multicarrier modulation), 1100
disadvantages, 1100–1101
orthogonal FDM, 1101–1102
MCN (multihop cellular network) architectures,
856–869
air interface standards, 858
architectures, comparison, 866
BS, 857–858
database, 858
gateway to wireline system, 858
MCN architectures, proposed, 859
MCN with fixed relay, 859–861
cascaded relaying, 861
iCAR, 859–861
primary relaying, 860
secondary relaying, 860–861
MS, 857
need for, 858–859
routes establishment in, 862
security mechanism, 858
single-hop cellular network and, 858
with hybrid relay, 864–866 See also Hybrid
relay
with mobile relay, 861–864
cMCN, 863–864
MCN-b, 861–862
MCN-p, 861
UCAN, 862–863
with relays, comparison, 867
fixed relay, 867
hybrid relay, 867
mobile relay, 867
M-Commerce, See Mobile commerce
MDBS (mobile data base station), 1326
MDU (max-delay-utility) scheduling, 950
Measurement-based prioritization scheme,
See MBPS
Media content sharing, mobile and wireless, 270
Media streaming, peer-to-peer networks,
See P2P
Medical implant communication service,
See MICS
Medical systems, 1151
Medium access control, See MAC
MEMs (micro electromechanical systems), 708,
1490
Merging, cluster, 524
Mesh mode, 420–421

IEEE 802.16 MAC, 420–421
scheduling schemes, 422–423
Mesh networks, wireless
architecture, protocols, and applications,
1407–1416
of BANs, 151
basic definitions, 1407–1408
access networks, 1407
backhauling, 1407
wireless mesh router, 1407
broadband wireless technologies, 1409
See also BWA
data network inter-connect using, 1408
market considerations, 1408–1409
mobile network, 1407
network architecture and scenarios,
1409–1412
backhaul links, 1409–1410 See also
separate entry
nomadic network, 1407
portable network, 1407
standardization and protocols, 1412–1417
standardization bodies and industry groups,
1412–1413
ETSI, 1412
IEEE, 1412
protocol standards and working groups,
1413–1417
FLASH-OFDM standards, 1414–1417
IEEE 802.11 enhancements,
1413–1414
IEEE 802.11, 1413
WiMAX, 1414–1417
WECA, 1412
WLANA, 1412
Mesh protocol, core-assisted, 540–541
Message modification attack, WEP, 1396
Messaging wireless internet applications, 1356
Metro optical services, 955–959
enterprise applications, 955–957
service evolution, 955
Metropolitan area networks, wireless
privacy and authentication in, 135
QoS provisioning, 425
Metropolitan networks, 969–970
MFS (managed fiber services), 957
MFSK (multiple-frequencyshift keying), 1116
M-Health (mobile health)
ambulatory ECG monitoring
subsystem, 714
benefits, 708, 709
configurations of, 709
e-textile garments for health monitoring, 716
for emergency, 714–715
historical overview of, 707–708
h-Shirt, 715
motivations behind, 708–709
prospect and challenges for, 715–717
PTT-based BP monitoring, 714
WBSN for, 710–711 See also WBSN
wearable sensors for, 709–710
wireless body sensor network
and, 707–717
Micro mobility, 477–478
cascade tunneling, 477
dynamic per-host routing, 478
overlay routing, 478

Microcell systems, 209, 496–500
CDMA networks, 180–185
fast-DOA-tracking algorithm, 497–498
highway microcells, 597–601
spectral efficiency, 597–600
teletraffic issues, 600–601
indoor microcells, 601
infrastructure, 601–603
microcellular infrastructure, 601–603
miniaturized microcellular BSs,
602–603
radio over fiber, 602
prototypes developed, 496–497
Microelectromechanical systems, See MEMs
Micro-mobile IP, 733
MICS (Medical implant communication service)
band, 149–150
Middleware security
related works, 1243
security hardening, 1246
code-level hardening, 1246
design-level hardening, 1246
high-level security hardening, 1247
low-level security hardening, 1247–1248
BoF vulnerabilities, 1247–1248
file management vulnerabilities, 1248
integer vulnerabilities, 1248
memory management vulnerabilities,
1248
operating environment hardening,
1246–1247
software process hardening, 1246
security requirements and vulnerabilities,
1243–1244
authentication, 1243
authorization, 1243
BoF, 1244
confidentiality, 1243
dynamic analysis, 1245–1246
file management, 1244
integer overflow, 1244
integrity, 1243
memory management, 1244
non-repudiation, 1243
security code inspection, 1245
security evaluations, 1244–1246
static analysis, 1245
in wireless applications, 1242–1250
Military applications of BANs, 155
MIMO (multiple input multiple output)
systems, 823
4G technologies, 640
adaptive MIMO systems, 610
adaptive 610
alamouti code, 608
applications of, 610
array gain, 606
capacity of, 604–606
diversity gain, 607
interference reduction, 606–607
matrix channel model, 823
ML decoding, 609
multi-antenna systems
benefits of, 606
multiplexing gain, 607–608
physical scattering model, 823–824
principle, 640

MIMO (multiple input multiple output)
    systems (*Cont.*)
  SIC decoding, 609
  space-time codes, 608
  spatial multiplexing, 608–609
  sphere decoding, 610
  STBCs (space time block codes), 608
  STTCs (space time trellis codes), 608
  in wireless communications, 604–612
  WIMAX—IEEE 802.16, 610–611
  WLANs—IEEE 802.11n, 611
Minimum mean-squared error, *See* MMSE
Minimum shiftkeying, *See* MSK
Minimum spanning tree, 1235–1236
Minkowski distance measure, 808
MIPv4 (mobile IPv4), 454
  4G technologies, 640
MIPv6, 640
Misuse detection IDS, 765
ML decoding, 609
MMGs (multiplayer mobile games), 694–706,
    See also MMMG
  devices and networks, 695–699
    3G, 698
    Bluetooth, 698
    BREW, 702
    laptops, 696
    mobile phones, 696–698
    PDAs, 696
    portable gaming consoles, 698–699
    windows mobile, 702
  fast-paced games, 695
  future directions, 705–706
  in inherent networking, 694
  in mobile devices ubiquity, 694
  life cycle, 702–703
  network configuration for, 697
  requirements, 703–705
    technical requirements, 703
      high bandwidth, 704
      latency, 703–704
      reliability, 704
  slow-paced games, 695
  social aspects, 694
  technical characteristics for, 699–700
    game development platforms, 700–702
  Tibia ME, 695
  types, 694–695
  in a Wi-Fi network, 699
MMS (multimedia message service), 1143
MMSE (minimum mean-squared error), 440
MOBIC (mobility of blind and elderly people
    interacting with computers), 520
Mobile ad hoc networks
  categorization of, 564
  clustering in, 872
  mobile ad hoc multicasting, *See* Multicasting
  overlay multicast in, 546–555, *See also*
    MAODV
  reactive routing protocols in, 567–570
    See also Reactive routing protocols
  routing protocols in, 563–575, *See also*
    Routing protocols
Mobile ad hoc networks, *See* MANETs
Mobile advertisements, 631
Mobile agents
  accounting using, 5–6

in personal mobility, 475
Mobile and wireless media content sharing, 270
Mobile auctions, 631
Mobile channels, circuit-switched, *See* Circuit-
    switched mobile channels
Mobile client, 1341–1342
Mobile commerce, 912, 1356–1357
  applications, wireless internet, 1356–1357
  evolution and main features, 623–627
    data portability, 625
    localization, 624
    personalization, 625
    reachability, 625
    ubiquity, 625
    value chain, 625–627
      content and transaction services, 627
      network operators, 626
      technology providers, 626
  strategic grid framework, 627–632
    high-potential applications, 628, 632
      mobile games and entertainment, 632
      mobile distance education, 632
    key operational applications, 627–628,
      630–631
      inventory management, 630–631
    strategic and customer-focused
      applications, 628, 631–632
      mobile advertisements, 631
      mobile auctions, 631
      mobile financial applications, 631
      reengineered applications, 631–632
    support applications, 627, 628–630
      mobile access, 630
      mobile office, 628–630
      personal life management, 628
Mobile communications, call admission control
    in, 643–651, *See also* CAC
Mobile computing, context-aware, 679–685,
    See also Context-aware mobile
    computing
Mobile content addressable network,
    See M-CAN
Mobile devices
  4G technology, 260
  evolution of, 259–260
  first generation (1G) mobile devices, 259
  second generation (2G) digital
    technology, 259
  web content re-purposing for, ontology-based
    approach to, 259–266
Mobile devices ubiquity, 694
Mobile distance education, 632
Mobile environment, context-aware retrieval in,
    912–913
  context capture, 913
  context definition, 912–913
  infrastructure support, 913
  intermittent wireless connections, 913
  privacy and confidentiality, 913
  reasoning methods, 913
  situation awareness, 913
Mobile financial applications, 631
Mobile games, 632
  accounting principles and billing
    models for, 8–9
    customer profile, 9
    detailed billing record, 9

    operation flashpoint, 8
    price lookup, 9
    round-trip delays, 8
Mobile health, *See* m-Health
Mobile host, destination, 196
Mobile internet protocol, *See* Mobile IP
Mobile internet technologies, *See* Internet
    technologies, mobile
Mobile IP (mobile internet protocol), 726–735,
    486–487 *See also* Mobile IPv4
  enhancements, 476
    fast location updates, 477
    frequent handover, 477
    link layer assisted handover detection,
      479–480
    route optimization, 476–477
  entities and relationship, 727
  handover detection, 487
  location registration, 486
  mobile IP route optimization scheme,
    731–732
  packet forwarding, 486–487
  paging extensions for, 504
  phases in, 729
    agent discovery, 726–727
    registration, 501, 726–727
      Qthrough foreign agent, 727
    tunneling, 726–727
  routing operation in, 728
  security aspect in, 728
Mobile IPv4
    low-latency handoff in, 730–731
    regional registration, 730
Mobile IT forum
  4G in mobile communications, 641
Mobile motion prediction, 831
Mobile multimedia
  applications, 739–741
  Codecs for, 736–743
    advanced codec standards, 741
    analog to digital conversion, 738
    audio coding, 738
      audio codecs, 741–742 *See also* AAC;
        AMR; MP3; MP3PRO
    image coding, 738–739
      image codec standards, 741
    mobile multimedia applications, 739–741
    multimedia information, 736–737
    standards, 737
      audio compression, 737
      image compression, 737
      video compression, 737
    text coding, 737–738
    video coding, 739
      video codecs, 742–743
  future of, 803
  transmission
    codecs for, 736–743, *See also* Codecs
Mobile multipath channel, 818–819
  statistical models, 821–824
    finite-state Markov chain model, 822
    MIMO channel models, 823
    motif model, 821–822
    satellite channel model, 822
    two-ray fading channel model, 821
Mobile network, 1407, 899
  multimedia streaming, *See* separate entry

Mobile node for wireless sensor network
  design and development of, 1495–1504
    hardware design, 564–567
      central processing unit module, 1496
      hardware, 1495–1496
      output module, 564
      power module, 564
      radio link module, 1495–1496
      sensor module, 1496
    software design, 566–567
      radio link module, 564–565
Mobile P2P system, See MP2P
Mobile payment for 3G services, 9–10
Mobile phones, games for, 696–698
Mobile proxy, 1342
Mobile relay, MCN with, 861–864
Mobile server, 1342
Mobile services in wireless communications
    environment, 774–781
  blog rendering and accessing instantly,
    776–780 See also BRAINS
  complete mobile service model, 780–781
  concerns in, 774–776
    networks, 775
    services, 774–775
    user devices, 775–776
Mobile services switching center, for ad hoc
    GSM, 326
Mobile station
  in 4G mobile communications, 637–638
  position location of, 659–665, See also
    Position location
  soft handover, 184
  transmit power, 184
Mobile streaming
  standards, 784–787
    release 4 PSS, 784
      control and scene description elements,
        784–785
      media elements, 785–786
      protocol stack, 785
    release 5 PSS, 786–787
Mobile systems, evolution of, 796
Mobile technology evolution, 28
Mobile telephones, 1142
  MT0 (mobile terminal 0), 1142
  MT1, 1142
  MT2, 1142
Mobile terminals, recommendation services for,
    805–816
  technological overview, 805–811
    content recommendation system, 805–806
      See also separate entry
Mobile web service support framework,
    721–724
  context, 722
  deployment time, 722
  design time, 722
  discovery time, 722
  immediate vs. deferred, 724
  mobile web service management, 722
  mobility strategies, 721
  provider/client impact, 724
  real-time vs. non-real-time, 724
  run time, 722
  service clients and providers, 721
  stateless service, 724

unique, multiple, 724
Mobile web services, 1358
  applications, wireless internet, 1358
Mobile wireless networks
  channel characterization, 817
    large-scale propagation models, 817
      deterministic approach, 817–818
      stochastic approach, 818
    small-scale propagation models, 818–824
      mobile multipath channel parameters,
        818–819
      statistical representation, 820–821
  evolution of, 788–789
  location prediction algorithms, See separate
    entry
Mobile WS performance, limits, 1343–1344
Mobile-assisted handoff, 211–212
Mobile-assisted handover, 453
Mobile-controlled handover, See MCHO
Mobile services switching center, See MSC
Mobiles, cordless, See Cordless mobiles
Mobile stations, See MS
Mobile Ubiquitous LAN Extensions,
    See MULEs
Mobilinux 4.0, 1086
Mobility controllable nodes, 487
  as data relays, 561
  as data sink, 561
  controllable network mobility, 557–558
  frameworks that utilize, 559
  in MANETs, 558–560
  in sensor networks, 560–562
Mobility impact, TCP, 1457–1458
  mobility-related disruptions, 1458
  performance degradation, 1457–1458
Mobility of blind and elderly people interacting
    with computers, See MOBIC
MOBITEX, 1323–1324
MobySpace, 1448
Modeling, context, 682–683
  distributed composition, 683
  level of formality, 683
  richness and quality of information, 683
Modeling of dynamic wireless sensor networks,
    304
Modem architectures
  ABL, 940
  channel estimation, 940
  FFT implementation, 938–940
  for OFDM-based systems, 934–941, See also
    under OFDM
Modem services, 1321–1322
Modulation methods, 835–842
  adaptive, 113
  analog FM, 836
  and coding, 1096–1097
  CPM, 838–839
  GMSK, 839–841
  modulated signals, 835–836
  MSK, 838–839
  orthogonal frequency division multiplexing,
    841–842
  PSK, 836–838
  Π/4-QPSK, 836–838
Monitoring, automatic, 713
Monitoring-based reactive solutions
  for packet forwarding, 577–580

ABO, 578
  end-to-end ACKs, 577
  probing, 579–580
  two-hop ACK, 578–579
  watchdog, 577–578
MOSFET (metal-oxide semiconductor
    field-effect transistor), 1158–1159
Motif model, 821–822
Moving picture experts group, See MPEG
MP2P (mobile P2P) system, 751
  adaptive content-driven routing and data
    dissemination, 756–757
  computing challenges, 757–758
    decentralized and transparent resource
      management, 758
    middleware support for MP2P computing,
      757
      metrics for evaluating, 758
    real-time support, 758
    resource discovery and data dissemination,
      757
  co-operative caching in, 754–755
    PReCinC, 755
    super node split operation, 755
  dynamic index in, 755–756 See also Dynamic
    index
  M-CAN, 752–754 See also separate entry
  MP2P architecture, 751–752 See also Bristle
MP3, 742
MP3PRO, 742
MPEG (moving picture experts group), 1067
  mobile devices using, AqoS for, 686–693
    adaptation decision engine (ADE), 687
    adaptation framework, 690
    AqoS tool, 688, 689
    client-side content adaptation, 688
    content adaptation, 687
    experimental results, 691–692
    GRACE 1, 687
    intermediate node content adaptation, 688
    multiple steps adaptation, 688
    related research review, 686–688
    resource adaptation engine (RAE), 687
    server-side content adaptation,
      687–688
    UCD tool, 688, 689
    UED tool in, 688, 689
MPEG-21, 686–693, See also AQoS
MPEG-4 AAC (moving picture experts group-4
    advanced audio coding), 785
MPEG-4 coded video sequences
  joint source-channel distortion modeling for,
    449–450
MPLS (multi-protocol label switching),
    797–798, 997–998
MPSK–OFDM, 848
MQAM–OFDM, 848–849
MRC-3 (maximum ratio combining-3), 1087
MS (mobile stations), of MCN, 857
MS position estimation methods, of mobile
    station, 659–664
  AOA method, 660
  GPS method, 659–660
  neural network method, 662–664
  signal strength method, 661
  TDOA method, 661
  TOA method, 660–661

MSC (mobile services switching center), 214, 326
MSK(minimum shiftkeying), 838–839
  premodulation filtered MSK, 840
MSM (master state machine), 1493
M-teams, 682
MTMR (multipletransmit multiple-receive), 355, 358
MULEs (Mobile Ubiquitous LAN Extensions), 560–561
Multi-antenna systems, 606, 911
  benefits of, 606
  models, 355
  MTMR potential realisation, 359–360
  single-user throughput, 357–358
    MTMR architectures, 358
    receive diversity, 357–358
    single-user bandwidth efficiency, 357
    transmit diversity, 357
  system throughput, 358–359
Multiband OFDM alliance, See MBOA
Multi-carrier, CDMA2000, 172–173
Multi-carrier diversity, 944
Multicarrier modulation, See MCM
Multicarrier transceivers, adaptive allocation for, 1102–1103
Multicast ad hoc on-demand distance vector, See MAODV
Multicast for ad hoc networks with swarm intelligence, See MANSI
Multicast protocols, 537–545
  core-assisted mesh protocol, 540–541
  forwarding group multicast protocol, 539–540
  MAODV routing protocol, 538–539
  on-demand multicast routing protocol, 537–538
  QoS multicast, 542–543
  reliable multicast, 543–544
Multicast, 973–974, 1240–1241
  to neighbors, 198
Multicast/broadcast data communications, 74
Multicasting, 1002–1015, See also P2P (peer-to-peer) overlay multicast
  in mobile ad hoc networks, 546–555, See also MAODV
    design goals, 549
    mobile ad hoc multicasting, 547–551
    new trend, 551–554
      DCMP, 552
      MANSI, 553–554
      NSMP, 552
      overlay multicast, 552
      PHAM, 553
      STMP, 554
    ODMRP, 548–549 See also separate entry
    PBM routing, 549 See also separate entry
    PUMAs, 550–551 See also separate entry
Multicast-join-based rerouting, 197
Multi-device presentations, coordinated
  ambient-audio identification, 274–285
    audio fingerprinting, 280–281
    personalizing broadcast content, 275–279
    supporting infrastructure, 279–280
    video 'Bookmarks', 278–279
Multidimensional birth-death process, 648
Multihop, cluster-based, 303
Multihop cellular network, See MCN

Multihop packet communication, WANET, 1221–1222
Multi-hop planar model, 1511
Multilateration, 1547
Multimedia
  in 3G and 4G, 28
    applications in 3G and 4G, 28
    mobile technology evolution, 28
    multimedia broadcast multicast service, See MBMS
    multimedia coding standards re-purposing, 255
    multimedia information, 736–737
    multimedia message service, See MMS
    multimedia rich product, 747–748
    role in 4G mobile systems architecture, 26
Multimedia, conversational, 480–483
  advance resource reservation, 480–483
    See also separate entry
Multimedia streaming, 884–889
  challenges of, 885–886
    delay, 886
    heterogeneity, 885
    packet loss/error control, 885–886 See also separate entry
    rate control/bandwidth fluctuation, 886
  coding standard for, 887–888
    H.263, 888
    H.264, 888
    MPEG-4, 888
  heterogeneity challenge of, 885
  live multimedia streaming, 884
  mobile networks, 890–899
    challenges, 890–899
      circuit-switched mobile channels, 891–892
      packet-switched mobile channels, 891–892
    end-to-end system architecture, 890
      one-way data distribution, 890
      offline media encoding, 890
  multimedia streaming client or player, 884–885
  multimedia streaming server, 884
  on-demand multimedia streaming, 884
  protocols for, 886–887
    RTCP, 887
    RTP, 887
    RTSP, 887
    SAP, 887
    SCTP, 887
    SDP, 887
    SIP, 887
    TCP, 887
    UDP, 887
  streaming standardizations, 888–889
    3GGP, 888–889
    ISMA, 889
Multimedia systems, wireless, cross layer considerations for, 1417–1423, See also CLD
Multimedia transmission strategy
  for 4G mobile systems architecture, 33–34
Multipath LAR MLAR, 571
Multiplayer mobile games, See MMGs
Multiple access techniques, 900–910

4G in mobile communications, 639–640, 503–2015, 503–2016, 503–2017
  HSDPA, 640
  MIMO principle, 640
  MIPv6, 640
  OFDM, 639–640
  UWB technology, 640
  access schemes, 206
    CDMA scheme, 206–209 SDMA
    FDMA scheme, 206
    scheme, 207–208
    TDMA scheme, 206
  CDMA, 56, 905–907
  code set properties, 907–908
  duplexing, 901–902
  FDMA, 904
  MC-CDMA, 908–909
  multiple access division, 901–902
  multiplexing, 901–902
  radio channel coverage area, 902–903
  resource assignment, 900–901
  resource division, 900
  SDMA and spatial diversity, 903–904
  TDMA, 904–905
  WANET, 1221
Multiple bands, ART, 1594
Multiple-frequencyshift keying, See MFSK
Multiple input multiple output systems, See MIMO
Multiple layer protection, WDM, 994–995
Multiple path routing, 1539–1541
  need, 1539–1540
  service differentiation, 1540–1541
    battlefield surveillance, 1540
    disaster relief operations, 1540
    environmental or biomedical applications, 1541
    infrastructure security, 1541
    strategies, 1541
Multiple point-to-point, in backhaul links, 1410, 1411
Multiple space Blom's scheme, 1517
Multiple spreading codes, ART, 1593
Multiple time slots, ART, 1594
Multiple transmit antennas, 1594
Multiplexing gain, 607–608
Multipoint-to-multipoint, in backhaul links, 1410, 1411
Multi-rate design issue, CDMA, 168, 171–172
Multi-sensory data fusion challenge, of BANs, 154
Multi-service cellular networks
  CACs for, 644–645
    dual threshold bandwidth reservation (DTBR) scheme, 644
    dynamic partition (DP) scheme, 644
Multi-tier cellular networks, 338
Multiple transmit multiple-receive, See MTMR
Multiuser diversity, 942–944
Multi-user detection, CDMA, 170
Multiuser water-filling principle, 950–951
Music scenario, mobile
  current scenario, 745–746
    communication infrastructure, 745
    copyright protection, 745–746
    pricing strategy, 746
  enrichment approaches, 747–748

all-in-one file, 747
  independence of media objects, 748
  multi-channel distribution, 748–749
future, challenges and directions, 746–749
  free riders problem, 748
  inter-operability formats, 749
  payment-based mechanism, 748
  reputation system, 748
problems and challenges, 744–750
Musical fingerprinting, 808
Mutual coupling, 1314–1317
  T-equivalent, 1321
MyBest TV, 1216
MyBestBets for music, 1217

NA (network adaptation), 962
Nakagami fading channel, 820–821
Napster-like phenomenon, 748
Narrowband AMPS, European TACS, 215
Nash arbitration scheme, 647
Nash equilibrium, 766
NAV (network allocation vector), 1524
NB classifier, 809–810
Neighbor discovery protocol, 1572–1573
Neighbor supporting ad hoc multicast routing
    protocol, See NSMP
.NET compact framework (CF), 1343
Network, See also individual entries
  Bluetooth 3.0, 95–97
    piconet formation, 96–97
  network allocation vector, See NAV
Network build-up, embedded, 1056–1057
Network database, 326–327
Network database, ad hoc GSM, 326–327
Network diversity, in 4G technology, 791–792
Network entities, of ad hoc GSM, 326
  base station subsystem, 326
  integrated dual mode terminals, 327
  internet working unit, 327
  mobile services switching center, 326
  network database, 326–327
Network intrusions detection, game theoretic
    models for, 764–770, See also
    Game theory
Network layer mobility, 485
  in mobile ad hoc and sensor networks,
    See Mobility controllable nodes
Network topology, of BANs, 151
  cluster tree network, 151
  mesh network, 151
  point-to-point network, 151
  star network, 151
  star-mesh hybrid network, 151
Network-assisted handover, 453
Network-based IDS, 765
Network-controlled handoff, 211
Networked devices and appliances at home, 1340
Networked multimedia systems, 795–796
Network-layer power management
  in ad hoc networks, 72–74
    multicast/broadcast data
      communications, 74
    unicast data communications, 73–74
Neural networks, 236, 321, 662–664
  in RRM of 4G mobile systems, 43–45
Nexperia system solution 9100, 1069–1070
Next generation network, See NGN

Next-generation wholesale service, in optical
    broadband services networks,
    959–960
NF (noise figure), 1027–1028
NGN (next generation network), 462, 641
  4G, 641
  fundamentals of, 463
NLOS error mitigation techniques, ACPS, 122
No answer, 1540
No carrier, 1540
Node mobility, in wireless video streaming,
    1204–1205
Node, mobile, See also Central node; Mobile
    node
  architecture, 1496
  communication protocol and algorithm,
    569–570
  system implementation, 570–573
    component study and hardware design,
      568–569
      CPU module hardware design,
        1496–1502
      DAC, 1039–1041, 1142, 1146, 1499
      electrically erasable and programmable
        read only memory, 1499
      GPS module, 1500–1501
      GSM module, 1496, 1500
      input board, 1501
      output board, 1501–1502
      PCF8574-remote I/O expander, 1499
      PIC 16F87X and modular technique,
        1498
      radio link module hardware design,
        1500
      RS-232 to I2C debug and programming
        board, 1499
      RTC, 1407
Nomadic network, 1407
  nomadic users, information retrieval for,
    See CAR
Non-adaptive communication system,
    1287–1288
Non-cooperative games, 646–648, 766
Non-duplicate JOIN_DATA propagation, 548
Non-invasive BANs, 150–151
  wireless medical telemetry service (WMTS)
    bands, 150
  WLAN/WPAN technologies, 150–151
Non-propagating request packet, 569–570
Non-real-time polling service, See NnrtPS
Non-uniform and uniform traffic distributions,
    1087
NnrtPS (non-real-time polling service), 419,
    1137
NSMP (neighbor supporting ad hoc multicast
    routing protocol), 552
Nuglets, 583–584
Numerical cost functions, in RRM of 4G mobile
    systems, 42–43
Nyquist principle, 738

OADMs (optical add drop multiplexers), 968
OBS (optical burst switching), 996–998
  switching techniques, 997
  vs. circuit and packet switching, 996–997
OBSAI (open base station architecture
    initiative), 186–192

conformance test specifications, 192
functional blocks, 187
mechanical specifications, 192
modules, 188
  BBM, 188
  CCM, 188
  GPM, 188
  RFM, 188
  TM, 188
OAM&P specifications, 192
OBSAI RP3-01 interface, 194
reference point interfaces, 188–192 See also
    RPI
RRHs using CPRI and OBSAI RP3-01,
    192–194
  CPRI overview, 192–193
  CPRI interface, 193–194
ODMRP (on-demand multicast routing
    protocol), 548–549
ODWCA (on-demand weighted clustering
    algorithm), 525
OFDM (orthogonal frequency division
    multiplexing) technology, 106, 235,
    489, 639–640, 841–842, 1158–1159
  4G technologies, 639–640
  applications, 931–932
  basic principles of, 918–921
  demodulation, 920–921, 922
  design, 929–931
    modulation and coding schemes,
      930–931
    number of sub-carriers, 930
    symbol duration, 930
  evolution of, 918–919
  guard interval and cyclic prefix, 921–924
  history, 918, 919
  modem architectures for, 937–938
    for broadband transmission, 934–937
  modulation, 919–920, 921
  OFDM framework, 843–844
    receiver, 844
    transmitter, 844
  PAPR reduction for, 843–855, See also PAPR
  realization, 921–931
    implementation by IFFT/FFT, 921, 922
    peak-to-average power ratio (PAPR), 924
      See also PAPR
  synchronization, 926–929
  windowing, 929
  sub-bands in, separation, 919
    filters for, 919
    using DFT and IDFT, 919
    using QAM, 919
  transmitter, 841
Ok, 441
Okumura–Hata model, 1320
OLAM (on-demand location aware multicast
    protocol), 541
OLPCs (outer-loop power controls), 673
  CLPC with, 674–675
OLSR (optimized link-state routing),
    564–565
OM systems
  categorization, 1004–1006
    implicit, 1005
    mesh-first, 1005
    tree-first, 1005

OM systems (*Cont.*)
    performance improvement via routing level
        proximity node selection,
        1006–1009
    proactive proximity-aware tree
        reformation, 1008–1009
On-demand location aware multicast protocol,
    See OLAM
On-demand multicast routing protocol,
    See ODMRP
On-demand multimedia streaming,
    537–538, 884
On-demand weighted clustering algorithm,
    See ODWCA
Ontology-based approach to web content
    re-purposing for mobile devices,
    259–266
    content re-purposing, 264
    content re-structuring, 262
    customization framework, 264
    device-specific authoring, 262
    different domain objects, 261
    different spatial relations among sub-tasks,
        261–262
    different task de-composition, 261
    domain ontology, 264
    OntoWeaver tool suite, 264
    presentation ontology, 264
    site-view ontology, 264
    thumbnailing, 262–263
Ontoweaver, 264
    articulation layer, 265
    components, 264
    composition layer, 265
    generation layer, 265
    integration layer, 265
    OntoWeaver tool suite
        for mobile device web content, 264
Opaque optical crossconnect, 974
Opaque switches, 973–974
Open base station architecture initiative,
    See OBSAI
Open standards for cellular base stations,
    186–195, *See also* OBSAI
Open system authentication, *See* OSA
Open systems interconnection stack, *See* OSI
Operating environment hardening, 1246–1247
Operating systems for smartphones,
    See Smartphone operating systems
Operation of 802.15.4 sensor clusters, 397–400
    See also CSMA-CA
    network formation and association/
        disassociation, 398–399
    security services and suites, 399–400
    uplink and downlink communication, 398
Operation flashpoint, in mobile games billing, 8
Opportunistic M-Commerce, 506–507
Opportunistic scheduling, *See* OS
OPS (optical packet switching), 977–978
Optical add drop multiplexers, *See* OADMs
Optical broadband services networks, 953–965
    current optical transport networks, 953–955
        as an optical services network, 955
        traditional assumptions, 954–955
        transport network evolution, 953–954
    emerging applications and network
        requirements, 959–960

    next-generation wholesale service,
        959–960
        point-to-point circuit, 960
        service granularity and flexibility,
        959–960
        service growth and provisioning
        time, 960
        service transparency, 960
        private line services, 959
        wholesale wavelength services, 959
    metro optical services, 955–959
        enterprise applications, 955–957
        service evolution, 955
    new service opportunities, 964–967
        bandwidth on demand, 965
        connection granularity, 965
        OVPN, 965–966
        packet/optical inter-working, 966–967
        rapid provisioning, 965
            auto discovery, 965
            mesh flexibility, 965
            planning tools, 965
    service and network adaptation, 962–964
        NA requirements, 963
        SA requirements, 963
        SA/NA technologies, 964
            OTN wrapping, 964
            SONET wrapping, 964
    services architectural framework, 960–962
        high-level architecture, 960
        optical switching and intelligent control
        plane, 961–962
        POP/CO consolidation, 961
        service adaptation and management, 962
    vertical markets and applications, 956
        MFS, 957
        managed wavelength service, 957–959
            LAN Connect, 958
            storage connect, 958
            Video Connect, 958–959
Optical burst switching, 978
Optical circulators, 983
Optical communication networks, 966–980
    access networks, 968–969
        optical wireless technology, 969
        point-to-point topologies, 969
        PONs, 969
    all-wave fiber vs. conventional fiber, 968
    long-haul networks, 970–978
        call admission control, 976
        fault management, 971–973
            protection, 972
            restoration, 972–973
        IP over WDM, 975–976
        multicasting, 973–974
            multicast-capable OXC architectures,
            973–974
            multicast RWA, 974
        network control and signaling, 976–977
        OPS, 977–978
            optical burst switching, 978
        RWA, 970–971
        traffic grooming, 974–975
    metropolitan networks, 969–970
    optical crossconnect, 968
    point-to-point link, 968
    WDM technologies, enabling, 967–968

Optical crossconnects, 968
Optical networks
    architectures, 981–988
        optical access networks, 986–987
        regional/metro optical networks, 984–986
    resource management and allocation,
        989–1001
        BI paradigm, 998–1000
        enabling technology, 989–990
        reconfiguration and QoS, 995–1000
            IP-over-OBS WDM, 998
            load balancing, 995–996
            MPLS, 995–996
Optical wireless technology, 969
Optimal band-limited modulation, 1097–1099
Optimal chaining, 590
Optimized link-state routing, *See* OLSR
Orthogonal FDM, 1101–1102
Orthogonal frequency division multiplexing,
    See OFDM
OS (opportunistic scheduling), 942–951
    categorization of, 946
    design issues, 944–945
        delay and jitter, 945
        effective throughput, 944–945
        energy consumption, 945
        fairness, 945
        goodput, 944–945
        throughput, 944–945
        wireless channel state estimation, 945
    diversities exploitable in wireless systems,
        942–944 *See also* under Diversities
    generalized framework for, 945–951
        Type F OS schemes, 950–951
        Type I opportunistic schedulers, 947–949
            MaxC/I scheduling, 947
            multi-service fair OS schemes, 947
            OSASP-QC family of OS, 947–948
            PF family of OS, 947
            score-based OS, 948
            TAOS family of OS, 948
            wireless credit-based fair queuing,
            948–949
        Type II OS schemes, 949–950
            OCASD family of OS, 949–950
        Type III OS schemes, 950
        Type IV OS schemes, 950
            MDU (max-delay-utility)
            scheduling, 950
            MC-EXP/OCASD, 950
        Type V OS schemes, 950
        Type VI OS schemes, 950
    single-hop multiuser wireless system serving
        multi-flow users, 943–953
OSA (open system authentication), 1393–1394
OSI (open systems interconnection) stack,
    1523–1533
    application layer, 1524
    five layer protocol stack model, 1524
    MAC layer, 1524–1525
    network layer, 1524
    physical layer, 1525
    transport layer, 1524
    wireless sensors, layers optimization for,
        1526–1531
OTDoA (observed time difference
    of arrival), 503

Outage capacity, 1289
Outage probability, 1088
Outer-loop power controls, *See* OLPCs
Output module, in mobile node, 564
Over wireless channels, 434
Overhead reduction, 1383
Overlay routing, 478
Overload errors, 1107
OVPN (optical virtual private network),
    965–966
OWA (open wireless architecture), 637

P2Cast transmission, 590–591
P2P (peer-to-peer), *See also* individual entries
    centralized P2P approach, 588–592
        adaptive chaining, 589–590
        basic chaining, 589
        batching policy, 591–592
        optimal chaining, 590
        P2Cast transmission, 590–591
    distributed P2P, 592
        adaptive layered streaming, 593–594
        Coolstreaming, 594
        multiple sender distributed video
            streaming, 592–593
    media streaming, 588–595
    P2P (peer-to-peer) computing, mobile,
        See MP2P (mobile P2P) system
P2P (peer-to-peer) overlay multicast, 1002–1015
    applications, 1003
    design requirements, 1002–1004
    dynamic adaptation with split streaming,
        1011–1012
    hybrid multicast, 1012–1014
    IP multicast vs. overlay multicast, 1003
    network heterogeneity with overlay routing
        adaptation, 1009–1011
    OM systems, 1004–1006 *See also* separate
        entry
    performance metrics, 1006
    using distributed hash table (DHT), 1004
P2P networks, 748
    of BANs, 151
    Wi-Fi walkman, 1208
P2P, mobile
    caching, 763
    contrasting MANETs and, 759–760
        network characteristics, 761
        node characteristics, 761
        routing characteristics, 762
    data retrieval and caching, 759–763
        challenges, 760–763
            complex queries, 762–763
            epidemic data dissemination, 762
            lookup services and key-based
                routing, 762
            parameters, 761–763
            structured overlay network, 762
            unstructured overlay network, 762
    mobility, 759
    reliable data delivery, 763
PACCH (packet associated control channel),
    1485
Packet associated control channel, *See* PACCH
Packet broadcast control channel, *See* PBCCH
Packet data, CDMA2000, 174
Packet data protocol, *See* PDP

Packet data transfer channel, *See* PDTCH
Packet drops, 875
Packet forwarding, 486–487, 576–587
    in mobile IP, 486–487
    preventive techniques, 583–585
        data dispersal, 584–585
        economic-based, 583–584
            nuglets, 583–584
            SPRITE, 584
        game theory based, 585
    reactive solutions, 577
        monitoring-based, 577–580
            ABO, 578
            end-to-end ACKs, 577
            probing, 579–580
            two-hop ACK, 578–579
            watchdog, 577–578
        reputation-based, 580–582
            CONFIDANT, 581–582
            CORE, 581
            friends and foes, 582
            signed token, 580–581
Packet leashes, 81
    temporal, 81
    geographical, 81
Packet losses, 1201–1202
    packet loss/error control, challenge of
        multimedia streaming, 885–886
        error concealment, 886
        error resilient coding, 886
        forward error correction, 885–886
        retransmission, 885
    rate bound in CFP, 1386
Packet paging channel, uplink, *See* PPCH
Packet partition algorithm, *See* PPA
Packet pursue model, *See* PPM
Packet random access channel, uplink,
    See PRACH
Packet relay, 531
Packet-based access network connectivity, 470
Packet-switched mobile channels, 891–892
    in 3G networks/technology, 785
    vs. circuit switching, 1350
    enhanced GPRS, 893
    GPRS, 892
    GSM/EDGE radio access network, 899
Packet-switched streaming service, 3G, 785
Packet-switched wireless networks, 1435–1442
Packet trade model, *See* PTM
Packet transmission schedule, 2-D RS
    codes, 160
PADR delay estimation, 1580–1581
PAGCH (packet access grant channel,
    downlink), 1485
Page Rank, 807
Paging extensions for mobile IP, 504
Paging mobile IP, 732–733
Pair-wise key, 1510
    pairwise master key, *See* PMK
Palm OS Cobalt 6.1, 1079–1083
    architecture, 1083
    technical specifications, 1080–1083
        applications, 1081
        communication infrastructure, 1081–1082
        desktop and synchronization, 1082
        multimedia, 1081
        network services, 1081

personal area networking, 1082
software development, 1082–1083
security, 1082
WiFi, 1082
PalmSource, 1087
PAMAS (power-aware multi-access protocol
    with signaling), 70
PAN (personal area network), 407, 510–511
    802.15.4 defending, 402–403
    BAN relationship with, 511
Pan European global system
    of 2G cellular systems, 217
        burst format, 218–219
        common channels, 218
        multiple access scheme, 218
PAPR (peak-to-average power ratio),
        in OFDM, 924
    baseband PAPR, 845–846
        continuous-time PAPR, 845–846
        discrete-time PAPR, 846
    CBC technique to reduce, 926
    definition of, 844–847, 924–925
    deterministic approach in, 850–851
    distribution of, 947
    frequency-domain-based processing in,
        852–853
    PAPR reduction, motivation for, 847
        power savings, 847
    passband PAPR, 846–847
    peak-power-limited OFDM
        achievable information rate with, 853–854
    reduction for OFDM, 843–855
        efficient PAPR, definition, 853
        reduction techniques, 850–853
    reduction, 926
    selection criteria, 853–854
        deterministic approach, 850–851
        MPSK–OFDM, 848
        MQAM–OFDM, 848–849 *See also*
            separate entry
        probabilistic approach, 850–851
            active constellation extension
                technique, 852
            blind power-based techniques, 852
            dynamic constellation shaping
                techniques, 852
            frequency-domain-based processing,
                852–853
            input sequence envelope scaling
                technique, 852
            interleaving techniques, 853
            selective scrambling technique, 853
            signal set expansion technique, 852
            time-domain-based processing, 852
    theoretical bounds on, 847–850
Parameter measurements, in adaptive
        techniques, 104–110
    channel selectivity estimation, 104–108
    channel quality measurements, 108–110
        channel decoding, 110
        demodulation, 108–109
Pareto-optimal power, 1467
Parsing algorithm, 828
Partial rebuild, 198
Partial rerouting, 200–202
Particles filter (PF), ACPS, 122
Pas, dynamic range of, 847

Passive caching, 763
Passive clustering, 519
  mobile node, state in
    clusterhead, 519
    gateway, 519
    initial, 519
    ordinary node, 519
Passive key recovery, WEP, 1396
Passive optical networks, See PONs
Passive stochastic routing techniques,
    1446–1449
Patching, 590
Path extension scheme, 198
Path loss, 181, 205–206
  RSS estimation, 127
Path protection, 972
Path rerouting, 198
Path restoration, 972–973
Path-key, 1510
Pathrater, 84–85
PBCCH (packet broadcast control channel),
    1485
PBM (position-based multicast), 546
  PBM routing, 549
    L-REMiT, 549–550
PC to mobile telephone and mobile telephone to
    radio, link module exchange protocol,
    1503
PCF (point coordination function), 376, 1382
PCF8574-remote I/O expander, 1499
PCM (pulse-coded modulation), 433
PCMA (power-controlled multiple access),
    1473–1474
  responsive interference, 1476
PCS (personal communications service), 1254
PDA (personal digital assistants), 696, 1065
  games for, 696
PDC (personal digital cellular), 217–218
PDP (packet data protocol), 32–33
PDTCH (packet data transfer channel), 1485
Peak BER-constrained bit allocation algorithm,
    1105
Peak-to-average power ratio, See PAPR
Peer-to-peer, See P2P
Peer communities, ad hoc, 276–278
PEGASIS (Power-efficient gathering in sensor
    information systems), 265, 1527
People-tracking, wireless, 506
Performance metrics, of dynamic wireless
    sensor networks, 304–306 See also
    under Dynamic wireless sensor
    networks
Per-group assignment technique, 669
Perimeter forwarding, 529–530
Periodically-fully-awake-interval
    protocol, 71, 72
Per-mobile vs. aggregated reservations, 343
Personal alarm system, 713
Personal area network, See PAN
Personal communications service, See PCS
Personal digital assistants, See PDA
Personal digital cellular, See PDC
Personal handy phone system, 492–501
Personalizing broadcast content, 275–279
Personal mobility, 487–490, 906
  challenges and recent developments, 474
    heterogeneity, 475

integrated presence, 475
  integrated personal mobility
    architecture, 475
  mobile agents, 474
computing paradigm, 788
support personalization, 489–490
support user location, 489
Personal mood based search
  and activity based search, 813–814
  of context, 813–814
Personalization endeavor
  factors imperative to, 284
    accessibility of personalized content, 284
    guaranteed privacy, 284
    integrity of mass-media content, 284
    minimized installation barriers, 284
    relevance of personalized content, 284–285
Personalized services, wireless, 506
PHAM (physical hierarchic ad hoc
    multicast), 553
Phases in mobile IP, 729
  agent discovery, 726–727
  registration, 501, 726–727
    Qthrough foreign agent, 727
  tunneling, 726–727
Phase shift keying, See PSK
Philips Nexperia system solution 9100,
    1069–1070
Physical hierarchic ad hoc multicast, See PHAM
Photo and music sharing, 269
PHY-layer jamming, WLANs, 1401
Physical layer, 363
Physical scattering model, 823–824
PIC 16F87X and modular technique, 1499
PicBasic Pro compiler, 566–567
Picocell systems, 209
Piconet, 141, 144, 1352
  Bluetooth 3.0, 96–97
  transmissions in, 97
Piece, 516
  black piece, 516
  white piece, 516
Pilot channel, CDMA, 169
PINV (matrix pseudo-inversion decoding), 609
PKI-based key management approach, for
    mobile ad hoc networks, 82
Playback control of video, See Interactive
    playback control of video
Plazes, 615
PMK (pairwise master key), 135
PMP (point to multipoint) mode, 418
  IEEE 802.16 MAC, 418
  scheduling schemes, 421–422
Point-to-multipoint, backhaul links, 1410
Point-to-point, backhaul links, 1410
Point-to-point topologies, 969
Policy-based decisions, in RRM of 4G mobile
    systems, 45–47
Policy-based handover, 455
Polyphase filtering, 1040
PONs (passive optical networks), 969
Portable device usage-based pervasive
    accounting, See PUPA
Portable gaming consoles, 698–699
Portable network, 1407
Portable PC speakerphone, 142
Position-based multicast, See PBM

Position location of mobile station, 659–665
  evaluation criteria, 664
    accuracy, 664
    capacity of location service, 664
    capacity reduction of other services, 664
    cost, 664
    positioning coverage, 664
    power consumption, 664
    response time, 664
    user privacy, 664
  MS position estimation methods, 659–664
    AOA method, 660
    GPS method, 659–660
    neural network method, 662–664
    signal strength method, 661
    TDOA method, 661
    TOA method, 660–661
Position-based routing protocols, 570–573
  DREAM (distance routing effect algorithm
      for mobility), 572
  GPSR (Greedy Perimeter Stateless
      Routing), 572
  LAR, 571
  location service protocols, 572
  multipath LAR MLAR, 571
Positioning technologies, survey of,
    502–504
  E-OTD, 503
  GPS, 502–504
  OTDoA, 503
  U-TDoA, 503
  UWB, 503
  Wi-Fi positioning, 503–504
Post-facto, 1582
Post-match consistency filtering, 281
Post-registration handoff method, 730–731
Power allocation-based SHO, 1091
Power assignment, 1234–1241
  biconnectivity, 1238–1239
  broadcast and multicast, 1240–1241
  k-edge-connectivity, 1239
  strong connectivity, 1235–1236
  symmetric connectivity, 1236–1238
  symmetric unicast, 1239
Power control in mobile communications,
    666–678, See also CPC; DPC;
    Feedback power control model; SIR
  adaptive, 113
  categories of, 666, See also CLPC; OLPCs
  centralized and distributed power controls,
      667–670
  uplink and downlink power controls, 667
Power delay profile, 818–819
Power module, in mobile node, 564
Power-aware multi-access protocol with
    signaling, See PAMAS
Power-controlled multiple access, See PCMA
Power-conservative designs
  for ad hoc networks, 68–75
    higher-layer power management, 74
      operating system/application layer, 74
      transport layer, 74
    MAC-layer power management,
        70–72
      dominating-awake-interval protocol, 71
      periodically-fully-awake-interval
          protocol, 71, 72

quorum-based protocol, 72
network-layer power management, 72–74
multicast/broadcast data
communications, 74
unicast data communications, 73–74
transmitter power control mechanisms,
68–70
Power-optimized image and video
communication over wireless systems,
436–442
methods, 436–440
AWGN channel, 437–438
JSCC and transmission power
management, 437–438
Lagrangian-based optimization
problem, 437
multimedia communication over multiple
antenna systems, 439
multi-resolution JSCC, 439
power–rate–distortion analysis, 438
target bit error rate minimization,
438–439
unequal power allocation, 439
MIMO systems
unequal power allocation method, 440–441
Power–rate–distortion analysis, in image and
video communication, 438
PPA (packet partition algorithm), 592–593
PPCH (packet paging channel, uplink), 1485
PPM (packet pursue model), 583
PPM (pulse-position modulation), 1159
PPS (precise positioning service), 61
PRACH (packet random access channel,
uplink), 1485
Preallocate network resources, 481–482
coarse-grained allocation, 482
history-based allocation, 482
neighborhood-based allocation, 482
PreCinCt (proximity region for caching in
co-operative), 755
content discovering and caching, 755
handling mobility, 755
using Hash function, 755
Precise positioning service, See PPS
Preconfigured anchor rerouting
resource reservation, 481
Preconfigured path extensions
resource reservation, 481
Preconfigured tunneling tree
resource reservation, 481
Preliminary certification process, 79
Premodulation filtered MSK, 840
PRF (pulse repetition frequency), 1152–1153
Price lookup, in mobile games billing, 9
Primary relaying, 860
Primitives, cryptographic, 1510, 1556–1557
asymmetric ciphers, 1557
symmetric ciphers, 1556–1557
Priority assignment–based SHO, 1089
Privacy and authentication
in wireless systems, 132–140, See also
Cryptography
authentication, 136–137
challenges and open research issues, 140
in global system for mobile
communications networks, 135–136
See also separate entry

in IEEE 802.11 wireless LANs, 134–135
See also separate entry
in IEEE 802.16 wireless metropolitan area
networks, 135 See also separate entry
in second generation code division multiple
access networks, 137–138 See also
separate entry
in third generation networks, 138–140
See also 3G networks
secret keys, 133–134
Privacy and authentication, in 3G networks,
138–140
3GPP based 3G networks, 138
3GPP2 based 3G networks, 138
authentication, 139–140
3GPP based networks, 139
3GPP2 based 3G networks, 139–140
privacy, 139
3GPP networks, 139
3GPP2 networks, 139
Privacy challenge of BANs, 154
Private line services, in optical broadband
services networks, 959
Proactive networks, 1536–1537
functioning, 1536–1537
important features, 1537
Proactive protocol, 76
Proactive proximity-aware tree reformation,
1008–1009
Proactive resource caching, 1211
Proactive routing protocols, of mobile ad hoc
networks, 564–567
conventional link-state protocols,
565–566
OLSR, 565
TBRPF, 565
distance-vector protocols, 566–567
DSDV, 566
WRPs, 566
issues in, 567
Probabilistic approach, in PAPR reduction
technique, 850–851
Probabilistic schemes, 1513–1515
Processing gain, 1121
Processor-independent transport, 1056
Profiling, context, 805–806
Programmable read only memory, 1499
Propagation considerations, in
CDMA, 55–56
Protocol architecture, of bluetooth systems,
142–143
adopted protocols, 143
baseband, 142
cable replacement protocol, 143
link manager protocol (LMP), 142
logical link control and adaptation protocol
(L2CAP), 143
radio, 142
service discovery protocol (SDP), 143
Protocol for unified multicasting through
announcement, See PUMAs
Protocol layering
ad hoc GSM, 328–329
beaconing, 328
encapsulation protocol, 329
link layer protocol, 328
resource manager, 328

Proximity node selection, 1006–1009
Proxy-based re-purposing approach, 255
re-purposing proxies, 257
system architecture, 256–257
server, 256–257
content/server profile, 256
proxy profile, 256–257
re-purposing service profile, 257
terminal and user's profile, 256
Pseudo-inversion decoding, See PINV
PSK (phase shift keying), 836–838
PSTN (public switched telephone
networks), 1381
PTM (packet trade model), 583
PTT-based BP monitoring, 714
PTV, 1217
Public key cryptography, 1402
Public switched telephone networks, See PSTN
Pulse-coded modulation, See PCM
Pulse-position modulation, See PPM
PUMAs (protocol for unified multicasting
through announcement), 550–551
mesh and data forwarding in, 551
PUPA (portable device usage-based pervasive
accounting), 5–6

QAM (quadrature amplitude modulation),
1025, 1098
QBIC, 807
q-Composite random key pre-distribution,
See QRKP
QoS (quality of service) delivery/support
in 3G networks/technology, 32–33
issues and challenges, 31–32
negotiations in, 32–33
in 4G networks, 35–37
audio video applications, 29
multimedia transmission, importance in, 28
concept, 796
networked multimedia systems,
795–796
fixed wireless vs. mobile communications,
797
for multimedia applications
in MANETs, 870–879
buffer management mechanism, 874
necessary components, 871–880
new scheduling, 874
QoS routing, 876–878
QoS-aware MAC protocol, 870–872
rate-based scheduler, 874
negotiations, in 4G mobile systems
architecture, 35–37
mobile switching center, 36
QoS broker, 37
holistic QoS management, 798–799
in IEEE 802.11 WLAN, 375–386
comparative performance evaluation,
382–386
IEEE 802.11e standard, 378–382
EDCA, 379
HCCA, 379
QoS enhancements to, 379–382
legacy IEEE 802.11
WLANs, 375–378 See also separate entry
in IEEE 802.16, 1135–1139 See also under
IEEE 802.16 issues and challenges

QoS (quality of service) delivery/support
(*Cont.*)
in 3G networks/technology, 31–32
mobile communication technology evolution,
796–797
over mobile systems, 795–804
QoS architectures and frameworks, 798
QoS management technologies, 797–800
for mobile networks, 799–800
management user interfaces for mobile
devices, 800–803
QCTT model, 800
QoS communication protocols, 797–798
TRAQS Model, 800
user centred framework, 800
in TDMA/TDD wireless networks, 1131–1140
adapted request, 1139
HIPERLAN/2 standard, 1133–1135
See also separate entry
IEEE 802.16, 1135–1139 *See also*
separate entry
non-adapted request, 1139
QPSK (quadrature phase shift keying), 836–838,
919–920
signal-space constellations, 838
QRKP (q-composite random key
pre-distribution), 1515
Qtopia phone edition components, 1084
Quadrant number generation, 1007
Quadrature amplitude modulation, *See* QAM
Qualitative analysis, in RRM of 4G mobile
systems, 47
Quality of service adaptation for mobile devices
using MPEG-21, *See* AQoS
Quantization reproduction level placement
technique, 1108
Quasihierarchical routing, 372–373
Queuing delays, 1334–1335
Queuing operation, 720, 721
Quorum, 72

RAA (rate allocation algorithm), 592–593
Radio architecture
direct-conversion wireless local area network
(WLAN) radio architecture, 239
DC offset voltage issue, 240
low pass filter and variable amplifier,
243–244
receiver design, 241–244
successive DC offset cancellation, 244
wideband RF front-end, 241–243
Radio channel, 102–120
coverage area, 902–903
Radio frequency communication,
See RFCOMM
Radio frequency identification, *See* RFID
Radio interface, 287–288
burst formats, 288
guard period, 288
multiple access and burst structure, 287–288
power ramping, 288
propagation delay, 288
transmission, 287
Radio link module, 1495–1496
Radio link protocol, *See* RLP
Radio operating parameters, wireless
communication, 231–232

Radio resource management, *See* RRM
Radio specification, of bluetooth systems, 144
Radiometer, 1121
RAE (Resource adaptation engine), 687
Rake receiver, 1087
Random key-pool, *See* RKP
Rapidly time-varying channels, link adaptation
techniques for, 1290–1292
Raptor codes, 1195–1197
systematic Raptor codes, 1196–1197
decoding, 1197
encoding, 1196
Raster Scan, 739
Rate allocation algorithm, *See* RAA
Rate compatible punctured convolutional codes,
See RCPC
RAVE (resource adaptive voluntary
election), 1530
Rayleigh fading channel, 1289–1290
with muliple paths and receive antennas, 1290
Rayleigh fading channel, 820
Rayleigh fading FM signals detection, 1302
RC4(Rivest Cipher 4)
encryption algorithm, 388–389
keys, 389–390
RCPC (rate compatible punctured
convolutional) codes, 430, 1260–1261
RDP (route discovery packet), 79
Reactive protocol, 77, 1537–1538
Reactive routing protocols, in mobile ad hoc
networks, 567–570
AODV, 568–569
route discovery, 568–569
route maintenance, 569
DSR, 567–568
issues in, 569–570
route discovery, 568
route maintenance, 568
Real time-fast collision resolution, *See* RT-FCR
Real-time flows on wireless hec, 1332–1334
Real-time intrusion detection for ad hoc
networks, *See* RIDAN
Real-time multimedia, 1340
Real-time polling service, *See* RtPS
Real-time services over WLANs with QoS
support, 1381–1391
background, 1382–1383
numerical results, 1387–1390
average voice collision number, 1388
packet loss rate, 1389, 1390
service time in CP, 1388–1390
simulation parameters, 1387
voice traffic capacity, 1390
performance analysis, 1385–1387
average service time in CP, 1385–1386
Session Admission Control, 1386–1387
proposed MAC scheme, 1383–1385
guaranteed priority access to voice over
data, 1384–1385
service interval structure, 1384
voice overhead reduction, 1385
voice traffic multiplexing, 1384
related work, 1383
Real-time streaming protocol, *See* RTSP
Real-time transport control protocol, *See* RTCP
Real-time transport protocol, *See* RTP
Reasoning techniques, context, 319–321

aggregation of context, 319
inference of context, 319
prediction of context, 319
pre-processing of context, 319
Receive array theory, 1305–1306
Receiver adaptation, 104
Receiver algorithms, adaptive, 110–112
adaptive channel length truncation, 111
adaptive interference cancellation receivers,
111–112
adaptive soft information, 112
channel estimation, 110–111
Receiver operations, 2-D RS codes, 160
Re-clustering, ripple effect of, 514
Recommender system, 1210–1216
basic system, 1212–1216
client/server model, 1214–1215
flooding model, 1214
implementation, 1215–1216
peers and play-lists, 1213
comparison and analysis, 1219
advantages and drawbacks, 1218
definition, 1210
research issues, 1211–1212
adaptability, 1211
cold-start problem, 1212
context awareness, 1211
proactive resource caching, 1211
scalability, 1211–1212
sparsity, 1211
state-of-the-art, 1216–1219
basic solutions, 1217–1219
commercially available systems,
1216–1217
user and product information, 1210–1211
Redundancy, 1285
Reed–Solomon code, 1260–1261, 1326–1327
Reference point interfaces, *See* RPI
Refinement operation in L-REMiT, 550
Reflection coefficients, 1126
Registration phase, in mobile IP, 501,
726–727
Regular cell search, 366
Relative neighborhood graph, *See* RNG
Relays, *See also* under MCN
cascaded relaying, 861
iCAR, 859–861
primary relaying, 860
secondary relaying, 860–861
hybrid relay, 864–866 *See also* Hybrid relay
mobile relay, 861–864
comparison, 867
cascaded, 477, 861
Release 4 PSS, 784
Release 5 PSS, 786–787
Reliability challenges of BANs, 154
Reliability requirement, of MMGs, 704
Reliable multimedia transport
for broadcasting over 3G and future mobile
networks, 157–166
application layer FEC, 157–158
performance evaluation, 166
RS code kernel, 158–159, 164
simulation results, 165–166
Reliable route selection, *See* RRS
Reliable wireless video streaming
with digital fountain codes, 1193–1199

LT-code, 1194–1195
 decoding, 1194–1195
 encoding, 1194
 Raptor codes, 1195–1197 *See also*
  separate entry
Replay attack, 1549
 WEP, 1396
Replay, in infrastructureless routing, 77
Replication operation, 721
Re-purposing techniques, 255
 bit rate, 255
 content, 260
  ontology-based approaches to, 264
  with desk tool, 265–266
Re-purposing, multimedia content, 253–258
 bit rate re-purposing, 255
 error-resilient re-purposing, 255
 multimedia coding standards
  re-purposing, 255
 proxy-based re-purposing approach, 255
  See also separate entry
 re-purposing techniques, 255
 temporal and spatial resolution re-purposing,
  255
Rerouting, 196–203
 analysis, 198
  cell forwarding rerouting, 203
  common handshaking signals, 198–199
  full rerouting, 199–200
  partial rerouting, 200–202
  tree rerouting, 202–203
 classification, 196
  branch-point-traversal-based
   rerouting, 197
  connection-extension rerouting, 197
  destination-based rerouting, 197
  extension, 197
  loop removal, extension with, 197
  multicast to neighbors, 198
  multicast-join-based rerouting, 197
  partial rebuild, 198
  path extension scheme, 198
  path rerouting, 198
  total rebuild, 198
  virtual connection tree, 198
  virtual-tree-based rerouting, 197
 performance evaluation, 1018–1023
  comparison, 1018
   bandwidth allocation, 1019
   buffering requirement, 1020
   connection length, 1018
   handoff, messages exchanged
    during, 1018
   rerouting completion time, 1020
   rerouting, messages exchanged
    during, 1022
   service disruption time, 1020
   special metrics, 1022
   user connections, 1019
Reputation-based reactive solutions
 for packet forwarding, 580–582
  CONFIDANT, 581–582
  CORE, 581
  friends and foes, 582
  signed token, 580–581
Rerouting, destination-based, 197
Reservation-based SHO, channel, 1089

Resource adaptation engine, *See* RAE
Resource adaptive voluntary election,
  See RAVE
Resource reservation, advance, 480–483
 current approaches, 480–483
  preconfigured anchor rerouting, 481
  preconfigured path extensions, 481
  preconfigured tunneling tree, 481
  preallocate network resources, 481–482
   coarse-grained allocation, 482
   history-based allocation, 482
   neighborhood-based allocation, 482
Resource reservation protocol, *See* RSVP
Re-structuring, content, 262
Resurrecting duckling, in ad hoc networks,
  82–83
Retransmission multimedia streaming, 885
Reuse, channel, 208
Reverse access channel, 1485
Reverse supplemental channel, RLP, 1485
Reverse-link aspects
 block structure, CDPD, 1327
 in capacity-related SHO, 1086–1087
  log-linear path loss and log-normal
   shadowing, 1087
  non-uniform and uniform traffic
   distributions, 1087
  power control, 1086–1087
  sectorization, 1086
 channel, CDMA2000, 173
 of soft handoff, 1086–1087 *See also* under
  Soft handoff
RF systems design
 for wireless communications, 1023–1034
 system requirements, 1023, 1024–1035
  RF receiver, 1027–1028
   input second-order intercept point,
    1030
   input third-order intercept point, 1030
   IP1dB, 1030–1031
   noise figure, 1027–1028
  RF transmitter, 1031
   synthesizer, 1031–1033
    IQ imbalance, 1033
 RF transceiver, 1035–1045
  direct-conversion architecture, 1041–1043
   advantages, 1042
  low-IF architecture, 1038–1041
  super-heterodyne, 1035–1038
RFCOMM (radio frequency
  communication), 143
RFD (reserve-a-fixed-duration), 997
RFID (radio frequency identification), 615
Ricean fading channel, 820
RIDAN (real-time intrusion detection for ad hoc
  networks), 85–86
 architecture, 85
 attack analysis module, 85
 countermeasure module, 85
 event generation module, 85
 traffic interception module, 85
Ring, 1503
Ripple effect of re-clustering, 514
RKP (random key-pool), 1515
RLP (radio link protocol), 1484–1485
 forward common control channel, 1484
 forward fundamental channel, 1485

forward supplemental channel, 1484
 reverse access channel, 1485
 reverse supplemental channel, 1485
RNG (relative neighborhood graph), 572
ROADMs (reconfigurable OADMs), 968
Roaming, 1323–1324
Robust position estimation, 1551
Robust security network, *See* RSN
Robust statistical methods, 1550
ROF (radio over fiber), 602
Rogue access points, WLANs, 1401
Round-trip delays, in mobile games billing, 8
Route discovery process, 79
Route discovery packet, *See* RDP
Route maintenance in ARAN, 80
Route optimization, in mobile IP, 476–477
Route reply message, AODV, 348
RouteREPly (RREP) propagation, 547
Route request packet, AODV, 1370
RouteREQuest (RREQ) propagation, 547
Routing, 1534–1544, *See also* individual entries
 adaptive, 990
 flat network architecture, 1536
 hierarchical network architecture,
  1535–1536
 multiple path routing, *See* separate entry
 query classification, 1534–1535
  historical queries, 1534
  one-time query, 1535
  persistent queries, 1535
 routing protocols characteristics, 1535
 in wireless networks with intermittent
  connectivity, 1442–1451
  active stochastic routing techniques,
   1449–1450
  deterministic routing techniques,
   1444–1446
  DTN architecture, 1443–1444
  passive stochastic routing techniques,
   1446–1449
Routing algorithm, of hierarchical state,
  1372–1373
Routing and data gathering, of heterogeneous
  WSNs, 347–349
Routing operation in mobile IP, 728
Routing protocols in mobile ad hoc networks,
  59–60, 563–575
 AODV routing protocol, 59–60
 DSR protocol, 59
 hybrid routing protocols, 570 *See also*
  separate entry
 position-based routing protocols, 570–573
  See also separate entry
 proactive routing protocols (table-driven
  protocols), 564–567 *See also*
  individual entry
 reactive routing protocols, 567–570 *See also*
  separate entry
Routing table build-up, AODV, 1371
Routing table overflow, 78
Routing table poisoning, 78
Routing techniques, deterministic,
  1444–1446
Routing/proxy maintenance protocol, 863
RPI (reference point interfaces), 188–192
 reference timing, standards for delivering,
  194–196

RPI (reference point interfaces) (*Cont.*)
  RP1, 188–189
  RP2, 189
  RP3, 189–191
  RP4, 191–192
RRM (radio resource management) in 4G
    mobile systems, 40–49
  fuzzy logic and neural networks, 43–45
  mechanisms, 41–42
  numerical cost functions, 42–43
  policy-based decisions, 45–47
    context parameters, 46
  qualitative analysis, 47
  signaling between RRM entities, 47–49
RRS (reliable route selection), 61–64
  stable zone and caution zone, 61–62
RS code kernel, 158–159
  decoding, 158–159
    complexity, 165–166
  encoding, 158
RS-232 to I²C debug and programming
    board, 1499
RSN (robust security network), 1398
RSVP (resource reservation protocol), 35,
    797–798
RTC, 1407
RTCP (real-time transport control protocol),
    782–783
RT-FCR (real time-fast collision resolution)
    mechanism, 381
RTP (real-time transport protocol) packets, 882
  fixed size, 161
    source block construction, 161
  variable size, 161–164
    hybrid-padding, 162–163
    receiver operation, 164
    simple-padding, 161–162
    source block construction, 162
    transmitter operation, 163–164
RtPS (real-time polling service), 1137
RTS (request to send), 92–95
RTSP (real-time streaming protocol), 783
Rule-based systems, 321
RVLC (reversible variable length coding),
    1435–1436
RWA (routing and wavelength assignment),
    970–971, 990

SA (service adaptation), 962
SACCH (slow associated control channnel), 329
Sace time block codes, *See* STBCs
Safety challenges of BANs, 154
Salvaging, 569–570
Sample time, 3G-telephony control stack, 16
SAN Connect, 958
SAODV (secure AODV routing), 80–81
SAP-LAW (Smart Access Point with Limited
    Advertised Window), 1336,
    1337–1338
SAT (supervisory audio tone), 214
Satellite ATM, 1255
Satellite channel model, 822
Satial time–frequency distributions, *See* STFD
Saturation throughput, 1227
SBM (subnet bandwidth management), 798
SB-OS (score-based OS), 948
Scalability, 1211–1212

issue, in networks, 874
Scaling law, 1224–1225
Scatternets, 144
  bluetooth, 97
SCFQ (self-clocked fair queueing), 380
Schemes, deterministic, 1515–1520
  location-aware pre-distribution schemes, 1518
  matrix-based pre-distribution schemes, 1516
  polynomial-based keying schemes, 1517
SCO (synchronous connection-oriented)
    link, 145
SCTP (stream control transmission
    protocol), 887
SD (sphere decoding), 610
SDCCH (stand alone dedicated control
    channel), 329
SDH (synchronous digital hierarchy), 954
SDKs (Software development kits), 505–506
SDMA and spatial diversity, 903–904
SDP (service discovery protocol), 143, 881–882
SDR (software defined radio) technology, 1100,
    1595–1597
  assisted FH/MC DS-CDMA, 1595–1597
  RF transceiver, 1043–1044
Seamless handover, 453
Seamless inter-frequency handover, 170
Second generation code division multiple access
    networks
  privacy and authentication in, 137–138
    authentication, 137–138
    privacy, 137–138
    signaling privacy, 137
    voice privacy, 137
Secondary relaying, 860–861
Secret key, 1510
  privacy in wireless systems, 133–134
Secure and open mobile agent, *See* SOMA
Secure localization systems, 1549–1552
  attack-resistant location estimation, 1550
  beacon suite, 1550
  distributed reputation-based beacon trust
    system, 1551–1552
  high-resolution, range-independent
    localization, 1552
  robust position estimation, 1550
  robust statistical methods, 1550
  secure positioning, 1551
  SeRLoc, 1549–1550
  transmission range variation, 1551
Security, *See also* individual entries
  for ad hoc networks, 76–88
    ÆTHER, 83–84
    fear-based awareness, 85
    infrastructureless routing, 76–77 *See also*
      separate entry
    intrusion detection, 84–86
    key management, 81–84 *See also*
      individual entry
    packet leashes, 81
    Pathrater, 84–85
    secure ad hoc routing, 78–81
    secure AODV routing, 80–81
    security solutions, 78
    threat model, 77–78
    watchdog, 84–85
  challenges of BANs, 154
  management, 1273

module, 583
Security architecture, IEEE 802.11i, 392–395
Security hardening, 1246
  code-level hardening, 1246
  design-level hardening, 1246
  high-level security hardening, 1247
  low-level security hardening, 1247–1248
    BoF vulnerabilities, 1247–1248
    file management vulnerabilities, 1248
    integer vulnerabilities, 1248
    memory management vulnerabilities, 1248
  operating environment hardening, 1246–1247
  software process hardening, 1246
Security threats
  802.15.4 sensor clusters, 400–402
    attacks following MAC protocol, 400–401
    attacks using modified MAC protocol, 401
Seek points, 3G-telephony control stack, 16
Segment matching, 831–832
Selection combining, 1308–1310
Selection support, in mobile internet
    technologies, 719
Selective scrambling technique, 853
Self-clocked fair queueing, *See* SCFQ
Self-healing ring, WDM, 994
Self-organized PKI, 82
Semantic web stack, 263
Sender selection algorithm, 1584–1585
  centralized method, 1584
  distributed method, 1584–1585
Sensing coverage models, dynamic wireless
    sensor networks, 308–309
Sensing nodes, 351–352
Sensor module, 1496
Sensor networks
  controllable mobility in, 560–562
    for energy efficient data delivery, 560–561
    for mission-specific sensor deployment,
      561–562
Sensor protocols for information via
    negotiation, *See* SPIN
Sequence and topology encoding for multicast
    protocol, *See* STMP
Series 60 platform, on symbian OS, 1079
SeRLoc (secure range-independent
    localization), 1549–1550
Service discovery protocol, *See* SDP
Service management, 1274
Service-oriented architecture, *See* SOA
Services, in 4G mobile communications, 639
Session initiation protocol, *See* SIP
Set partitioning into hierarchical trees,
    See SPIHT
Set-top box, *See* STB
SGSN (serving GPRS support node), 324
Shadow cluster concept, 644
Shadowing
  RSS estimation, 127–128
Share refreshing techniques, 82
Shared key authentication, *See* SKA
Shared-link protection, 972
Shared-path protection, 972
Short message service, *See* SMS
Short-range wireless connectivity applications
  of blue tooth technology, 141
    ad hoc networking, 141
    cable replacement, 141

data and voice access points, 141
SIC (successive interference cancellation)
 decoding, 609
Signal processing
 STFD, *See* separate entry
 TFSP applications, 1301–1303
 time–frequency array, 1294–1301,
  1295–1305
Signal regeneration, 976
Signal set expansion technique, 852
Signal strength method, 661
Signal to interference and noise ratio, *See* SINR
Signal-to-interference ratio, *See* SIR
Signaling L2, 290–292
 fixed format packet, 291–292
 general message format, 290–291
SIM card, 135
Simple object access protocol, *See* SOAP
Simulation assessment, of wireless home
  entertainment, 1335–1336
Simulation models
 ACPS, 126–128
Single-hop, 1230
Single-user throughput
 in multi-antenna systems, 357–358
  MTMR architectures, 358
  receive diversity, 357–358
  single-user bandwidth efficiency, 357
  transmit diversity, 357
Sinkhole attack, 409
Sinkholes, 1558–1559
SINR (signal to interference and noise
  ratio), 491
SIP(session initiation protocol), 488–489,
  1483–1484
 call establishment, 488
 mobility architecture, 1484
SIR (signal-to-interference ratio),
  109, 668, 1465
 SIR based power controls, strength-based
  and, 671–672, 672–673
Situated interactions, 613–621
 affordances of mobile devices, 616
 mobile and wireless technologies, 614–617
 situations, characteristics of, 618–620
  bounded-ness, 619
  linkage, 619
  presence, 619
  separation, 619
  temporality, 619
 spatial settings, 617–618
  bounded-ness, 617
  linkage, 617
  physical separation, 617
  presence, 617
  temporality, 617
 spatial tagging, 615
 wireless networks and nodes, 616–617
SKA (shared key authentication), 1393–1394
 WEP, 1397
SKKE (symmetric-key key establishment),
  408–409
Sleep attack, 409
Sleep deprivation torture, 78
Slow-fading channel, 820
 link-adaptive communication system for,
  1288–1290

Slow-paced games, 695
Smart inter-processor communicator protocol
  stack, 1055–1064
 architecture, 1057
 current systems, co-existence with, 1057
 device and router layer, 1059–1060
 dynamic plug-ins and embedded network
  build-up, 1056–1057
 experimental data, 1060–1062
 network topology, 1058
 network-wide service advertisement, 1057
 OS-independence and ease of migration, 1056
 performance, 1057
 processor-independent transport, 1056
 session layer, 1057–1059
 supported socket types, 1059
 use case example, 1062–1063
 vs different inter-processor communication
  schemes, 1061, 1062
 vs transmission control protocol, 1060
Smart radio, 1217
Smartphone operating systems,
  1073–1085, 1067
 design requirements, 1073–1074
  access to source code, 1074
  conventional and mobile computing
   paradigms, supporting, 1074
  mobile device reliability, 1073–1074
  mobile devices, supporting, 1073
  product differentiation and innovation,
   supporting, 1074
  real-time kernel, implementing, 1074
  small devices, supporting, 1073
  third-party application, service, and
   technology development,
   encouraging, 1074
 palm OS Cobalt 6.1, 1079–1083 *See also*
  separate entry
 series 60 platform, 1079
 Symbian OS, 1074–1078 *See also* separate
  entry
 UIQ (user interface IQ), 1078–1079
 windows mobile 2003 second-edition
  software, 1083–1086
  technical specifications, 1084–1086
   applications, 1085
   communication infrastructure, 1085
   multimedia, 1084
   network services, 1084
   personal area networking, 1085–1086
   security, 1086
   software development, 1086
   telephony or telephony API, 1085
  windows mobile smartphone, 1086 *See also*
   separate entry
Smartphones, 1065–1072, *See also* Smartphone
  operating systems
 applications, 1070–1071
  digital data collection, 1067–1072
  internet applications, 1067–1072
  standardized platform, 1068–1072
  user notification applications,
   1068–1072
 operating systems, *See* Smartphone operating
  systems
 reference designs, 1067
  Freescale i.Smart, 1068–1072

Philips Nexperia system solution 9100,
 1069–1070
 TI TCS2500, 1067–1072
 TI TCS2600, 1067–1072
 TI TCS3500, 1068–1072
SMIL (synchronized multimedia integration
  language), 30
SMIRA (stepwise maximum-interference
  removal algorithm), 668
SMS (short message service), 615, 1142,
  1142–1143
 SMS data ransmission
  elemetering system with, 1144
 SMS/MMS message receiver, 776
SNOOP, 1455–1456
SNR quantization error model, 1107–1108
SOA (service-oriented architecture),
  718–725, 637
 challenges, 718–719
  mobility-related, 719
  technology-related, 718
 concept, 719
SOAP (simple object access protocol),
  1342–1343, 1344
Social aspects of TV, 267–273, *See also*
  Content-enriched communication
Social-application server setup, in coordinated
  multi-device presentations,
  279–280
'Social TV' system, 268
SOCUR (stabilized online constraint-based
  routing), 876–877, 878
Soft handoff in mobile communications,
  1086–1094, 212
 adaptive thresholds-based SHO, 1090–1091
 analytical SHO models, 1088
  Markov conception, 1088
  outage probability, 1088
 capacity-related SHO, 1086–1088
  forward-link aspects, 1087–1088
   log-linear path loss, 1087
   log-normal shadowing, 1087
   non-uniform and uniform traffic
    distributions, 1087
   power control, 1087
   rake receiver, 1087
   SHO area/SHO threshold, 1087
   short-term fading, 1087
  reverse-link aspects, 1086–1087
   log-linear path loss and log-normal
    shadowing, 1087
   non-uniform and uniform traffic
    distributions, 1087
   power control, 1086–1087
   sectorization, 1086
  channel reservation-based SHO, 1089
  cost/reward function-based SHO, 1089
  and fade margin, 1088
  interference-based channel assignment for,
   1089–1090
  power allocation-based SHO, 1091 *See also*
   separate entry
  priority assignment–based SHO, 1089
Soft handover, 181
 for mobile stations, 184
Soft hard handover, 453
Soft information, adaptive, 112

Soft-QoS model, 1279
  control in the WATMnet system, 1280
Software defined radio technology, See SDR
Software design
  of mobile node
    of central node, 567–568
    for wireless sensor network, 566–567
Software development kits, See SDKs
Software process hardening, 1246
Software-defined radio technology, 230
Sojourn time, probability distribution function
    of, 1088
SOMA (Secure and open mobile agent), 489
SONET (synchronous optical network), 954,
    964, 969–970
Source coding, 429–430
Source-initiated on-demand, in ad hoc
    networks, 77
Source mobile host, 196
Source-channel distortion modeling
  for image and video communication, 443–451
    distortion estimation methods, 443–445
      See also Distortion models
    adaptive mode selection, 444–445
    optimal mode-selection method, 444
  joint source-channel distortion modeling
    for MPEG-4 coded video sequences,
      449–450
    model performance, 450
  limitations, 445
Source-channel rate allocation method, 432
Source-initiated on-demand characteristics of ad
    hoc networks, 77
Space-time codes, 608
Space time trellis codes, See STTCs
Sparsity, 1211
Spatial diversity, 944
  wireless communications, 1305–1318,
    See also under Wireless
      communications
Spatial group sensing concept, 1567
Spatial multiplexing, 608–609
Spatial settings in mobile and wireless
    technologies, 613–621, See also
      Situated interactions
Spatial time–frequency distributions, See STFD
Specialized mobile radio, 1320
Spectral efficiency, 596, 1095–1099
  coding and modulation, 1096–1097
  optimal band-limited modulation, 1097–1099
  theoretical bounds, 1095–1096
Spectrally-efficient wireless multicarrier
    communications, 1100–1113
  adaptive allocation for multicarrier
    transceivers, 1102–1103
  BIT allocation with imperfect channel
    information, 1107–1108
  capacity approximation-based bit allocation,
    1103–1104
  incremental bit allocation, 1104–1105
  peak BER-constrained bit allocation
    algorithm, 1105
  simulation results, 1108–1112
    imperfect sub-carrier SNR information,
      1110–1112
    perfect sub-carrier SNR information,
      1109–1110

Speech coder, 1127
Speech service
  blocking, 183
  transmit power, 183
Sphere decoding, See SD
SPIHT (set partitioning into hierarchical
    trees), 433
SPIN (sensor protocols for information via
    negotiation), 1528, 1559
Split streaming, dynamic adaptation with,
    1011–1012
Split-connection, 1454–1455
Spoofing management frames, WLANs, 1400
Spray and wait protocol, 1447
Spread spectrum communications, 1114–1121
  applications, 1118–1121
    commercial, 1119–1121
      interference rejection, 1120–1121
      multiple access communications,
        1119–1120
    military, 1118–1119
      AJ communications, 1118–1119
      LPI, 1119
  basic concepts and terminology,
    1114–1115
  techniques, 1115–1118
    DS modulation, 1115–1116
    FH modulation, 1116–1117
    hybrid modulations, 1118
    TH modulation, 1117–1118
Spreading codes, CDMA2000, 173–174
Spreading codes, multiple, 1593
Sprint Nextel, 506
SPRITE, 584
SPS (standard positioning service), 61
SS signal transmission
  DSSS, 905
  FHSS, 905
  THSS, 905
SSID (service set Id), 1393
Stable zone radius effect, of ad hoc networks,
    62–63
Stand alone dedicated control channel,
    See SDCCH
Stand-alone BANs, 151
Standard positioning service, See SPS
Star network, of BANs, 151
Star topology, 1353
Star-mesh hybrid network, of BANs, 151
Static handoff-resource reservation, 340
Statistical models, of ACPS
  for RSS estimation, 127–128
    fast fading, 128
    LOS probability, 127
    path loss, 127
    shadowing, 127–128
  for TOA and AOA estimation errors, 126–127
    LOS, 126–127
    NLOS, 127
STB (set-top box), 268
STBCs (space time block codes), 608
Stepwise maximum-interference removal
    algorithm, See SMIRA
STFD (spatial time–frequency distributions),
    1294–1296
  autoterms and cross-terms, 1297
  chirp modulation, 1301–1302

direction-of-arrival estimation, 1298
LTV channels, precoding for, 1301
mobile velocity/Doppler estimation,
    1302–1303
source separation, 1298–1300
  convolutive mixtures, 1300
  instantaneous mixture, 1299
  sources than sensors, separating more,
    1299–1300
  STFD-based separation, 1300–1301
  vs. covariance matrix, 1296–1297
  wireless communications, structure in, 1296
STMP (sequence and topology encoding for
    multicast protocol), 554
Stochastic coverage, 309
Stop cell search, 366
Store-and-carry, 1442–1443
Stream control transmission protocol, See SCTP
Streamed continuous traffic, for transmitter
    power control, 1465–1473
Streaming multimedia
  for 4G mobile systems architecture, 33
    challenges in, 33
  content, in 3G networks, 30–31
    challenges faced, 31
    shortcomings, 33
Streaming services, in MBMS, 161–165
    See also Video streaming
  complexity–performance tradeoff, 165
  fixed size RTP packets, 161
  high streaming bit rates and longer protection
    periods, 164–165
  simulation results, 165–166
  variable size RTP packets, 161–164
Street microcells, 596–597
Strength-based power control, 670–671
Strict hierarchical routing, 373–374
Strong connectivity, 1235–1236
Structured overlay network, contrasting
    MANETs, 762
STTCs (space time trellis codes), 608
Subcarrier clustering, 950–951
Successive interference cancellation, See SIC
Super-heterodyne, 1035–1038
  advantages, 1038
  disadvantages, 1038
Supervised learning, 662
Supervisory audio tone, See SAT
Surrounding-circumstance-based search, 813
Surveillance systems, 1151
SVCs (switched virtual connections), 797
Switched combining, 1308
Switching techniques, 996–998
Sybil attack, 409, 1549
Symbian OS, 1074–1078, 1354
  architecture, 1078
  Symbian OS v9.1, 1075
    technical specifications, 1074–1078
      application framework, 1076
      communication infrastructure, 1076
      Java, 1075
      multimedia, 1076
      network services, 1075
      personal area networking, 1077
      security, 1078
      software development, 1078
      Telephony/Telephony API, 1076

Symmetric ciphers, 1556–1557
Symmetric connectivity, 1236–1238
Symmetric cryptography, 132
Symmetric encryption, 135
Symmetric-key key establishment, *See* SKKE
Symmetric unicast, 1239
Synchronization
  automatic, 142
  in bluetooth systems, 144
  OFDM and, 926–929
    channel estimation, 928–929
    frequency synchronization, 927
    time synchronization, 927–928
Synchronized multimedia integration
    languagem, *See* SMIL
Synchronous connection-oriented link, *See* SCO
Synchronous digital hierarchy, *See* SDH
Synchronous optical network, *See* SONET
Synchronous viewing over a distance, 271
System architecture
  4G in mobile communications, 637
  ACPS, 128–129
    AOA measurements, 123–124
    TDOA measurements, 123–124
    TOA measurements, 123–124
  of BANs for unobtrusive services, 151–152
System model and CAC, 648–649
System, in 4G mobile communications, 638
Systematic Raptor codes, 1196–1197

T_ADD, 1090–1091
T_DROP, 1090–1091
Table-driven characteristics of ad hoc
    networks, 76
TACS (total access communication system), 214
TAG (tell-and-go), 997
Tagging mechanisms, 747
Tagging, automatic, 807–808
  from audio information, 808
  from textual information, 807
  from visual information, 807
TAOS (traffic-aided OS) family of OS, 948
Tau–dither loop, 1121
TBRPF, 565
TCM (Trellis-coded modulation), 712
TCMA (Tiered contention multiple access)
    mechanism, 378
TCP (transmission control protocol),
    604, 1333
  basics, 1452–1453
  mobility impact, 1457–1458
    mobility-related disruptions, 1458
    performance degradation, 1457–1458
  multi-hop independent wireless networks,
    1458–1462
    efficiency and fairness, 1461–1462
    mobility, 1460–1461
  wireless channel errors, 1454
    explicit notification approach, 1456
    lower layer scheme, 1455–1456
    modifications, 1456
    split-connection approach, 1454–1455
  woes, 1453–1454
TCP/IP (transmission control protocol/ Internet
    protocol), 505, 797
TD (trend discovery), 805–806
TDD (time division duplex), 178, 206, 419

TDMA (time division multiple access), 50,
    53–55, 378, 668, 904–905, 945
  access methods, 53–55
  2G wireless systems, 1349
  algebraic codebook, 1128–1129
  bad frame masking, 1128
  bandwidth expansion, 1126
  based schemes, 174–178
  carrier spacing, 174–175
  channel coding and interleaving, 1127–1128
  channel coding, 1124
  codebooks, orthogonalization of,
    1126–1127
    excitation and signal gains, 1127
  digital voice and data signals modulation,
    1122–1123
    generic modulation circuit, 1123
  frame structures, 176–177
  initial channel assignment, 54–55
  linear prediction analysis and quantization,
    1124
  logical frame format, 1275
  long-term filter search, 1126
  magic WAND TDMA, 1276
  modulation, 175–176
  multi-rate scheme, 177
  propagation considerations, 54
  radio resource management, 177–178
  reflection coefficients, 1126
    VSELP codebook search, 1124
  speech coding fundamentals, 1123–1124
  symbol rate, 174–175
  technology, 1349
  time division duplex, 178
  VSELP encoder, 1124
TDMA/TDD wireless networks, QoS in,
    1131–1140
  adapted request, 1139
  HIPERLAN/2 standard, 1133–1135 *See also*
    separate entry
  IEEE 802.16, 1135–1139 *See also* separate
    entry
  non-adapted request, 1139
TDOA (time difference of arrival) method, 661,
    1155
TDOA method, 661
Teaching and learning, wireless, 1358
Technology evolution, in mobile
    communication, 796–797 *See also*
    under Communication, mobile
Technology outline, 3G networks, 26
  access network domain, 26
  core network domain, 26
  user equipment, 26
TEEN (threshold sensitive energy efficient
    sensor network), 1527–1528,
    1537–1538
  functioning, 1537
  important features, 1538
Telecommunications
  evolution of, 796
Telemetering
  GSM-based distributed telemetering systems,
    1144
  GSM systems and technologies for,
    1141–1149
    data transmission, 1142–1144

    with SMS data transmission, 1144
    wireless transmission of, 1141
      AT commands, 1144
      CSD, 1143
      EDGE, 1143
      GPRS, 1143
      MMS (multimedia message service), 1143
      SMS, 1142–1143
Telephony-compatible control stack, 14
Telephony-oriented control,
    disadvantages of, 15
Temporal and spatial resolution
    re-purposing, 255
Temporal context, 314
Temporal fairness, 945
Temporal key integrity protocol, *See* TKIP
Temporal packet leashes, 81
Temporary international mobile identification,
    See TMSI
Tenet architecture, 348
Terminal diversity, in 4G technology, 791
Terminal mobility, 485, 905
  challenges and recent developments, 476
    conversational multimedia, 480–483
      advance resource reservation, 480–483
      See also separate entry
    higher layer mobility management, 479
    mobile IP enhancements, 476 *See also*
      separate entry
Terminals, wireless and mobile communication,
    811–815
TESLA (timed efficient stream loss-tolerant
    authentication), 1514
  TESLA, 1559
  fault-tolerant two-level, 1514
  key generation and disclosure, 1513
  TESLA, 1559
TETRA (terrestrial trunked radio), 1329
Text coding, 737–738
Text-based search, 811–812
Textual information, automatic
    tagging in, 807
TF/IDF (term frequency/inverse document
    frequency), 807
TFSMs (timed finite state machines), 85
TH(time hopping) modulation, 1117–1118
Third generation (3G) networks, *See* 3G
    networks
Threat model, 1511, 1556
  ad hoc networks, 77–78
Three-in-one phone, 142
Threshold cryptography, 82
Threshold sensitive energy efficient sensor
    network, *See* TEEN
Through-wall imaging systems, 1151
THSS (time-hopping SS), 905
Thumbnailing, in mobile device web content,
    262–263
TI TCS2500 (Texas instruments TCS2500),
    1067–1072
TI TCS2600 (Texas instruments TCS2600),
    1067–1072
TI TCS3500 (Texas instruments TCS3500),
    1068–1072
Tibia ME, 695
Tiered contention multiple access mechanism,
    See TCMA

Tight coupling architectures, in WLAN and
  cellular networks, 457–458
Time division multiple access, See TDMA
Time slots, multiple, 1594
Time division multiple access medium access
  controller, See TMAC
Time division multiple access, See TDMA
Time hopping, See TH
Time of arrival method, See TOA
Time-based search, of context, 813
Timed finite state machines, See TFSMs
Time-division duplex mode system, CAC
  schemes in, 645
Time-domain-based processing, in PAPR
  reduction technique, 852
Time-invariant capacity system
  CAC schemes in, 643–645
    shadow cluster concept, 644
Time-varying capacity system, CAC schemes
  in, 645
Time-varying channels, modeling of, 1287
Timing and synchronization
  for 3G-telephony control stack, 16–19
    extra time, 16
    interactive playback seek, 18–19
    interactive playback speed up and slow
      down, 16–18
    sample time, 16
    seek points, 16
    transmission time, 16
Tiny-sync/mini-sync, 1581
TiVo, 1217
TKIP (temporal key integrity protocol),
  394–395
TMAC (time division multiple access medium
  access controller), 1529–1530
TMSI (temporary international mobile
  identification), 136
TNCP (transport network control plane), 1167
TOA (time of arrival) method, 660–661, 1155
Top hash, 80
Top-down approach, CLD, 1418
Topology optimization, dynamic wireless sensor
  networks, 307–308
Tork maintenance protocols, 1011
Total rebuild, 198
Traffic flow specification, bluetooth
  parameters, 146
    delay variation, 147
    latency, 147
    peak bandwidth, 147
    service type parameter, 146
    token bucket size parameters, 146
    token rate parameter, 146
Traffic grooming, 969–970, 974–975, 992–993
  node architecture, 975
Traffic interception module, in RIDAN, 85
Traffic models, 306
Traffic monitoring, 1543
TRAMA (traffic-adaptive medium access), 1530
Transcoding devices, 261
  functions of, 262
Transmission control protocol, See TCP
Transmission range variation, 1551
Transmission technologies, UWB, 1159–1160
Transmission time, 3G-telephony control stack, 16
Transmit antennas, multiple, 1594

Transmitter adaptation, 103
Transmitter power control, 1464–1477
  in ad hoc networks
    mechanisms, 68–70
    for streamed continuous traffic, 1465–1473
      autonomous distributed power
        control, 1468
      backlog pressure, 1475–1476
      DPC, See separate entry
      noninvasive channel probing and selection,
        1472–1473
      PCMA, 1473–1474
      per-slot-independent interference, 1475
      phased back-off, 1475–1476
      target SIR formulation, 1465
        optimal power vector, 1467
        two-link network, 1466–1467
      transmitter buffer emptying, 1474–1475
Transparent optical crossconnect, 974
Transparent switches, 974
  link aliveness, 783
  RTSP signaling issues, 783
  transport control protocol, 782–783
Transport layer, power management, 74
Transport network control plane,
  See TNCP
Transport networks, optical, 953–955 See also
  under Optical broadband services
  networks
Transport unaware link improvement protocol,
  See TULIP
TRAQS Model, 800
Trellis-coded modulation, See TCM
Tree network, cluster, of BANs, 151
Tree rerouting, 201
  tree-group rerouting, 202
  tree-virtual rerouting, 202
Tree-group rerouting, 201, 202
Tree-virtual rerouting, 202
Triangulation, 1546
Trilateration, 1546
Trusted third party, See TTP
TTP (trusted third party), 79
TULIP (transport unaware link improvement
  protocol), 1456
Tunnel header, 485
Tunneling
  across QoS domains, 478–479
  cascade tunneling, 477
  phase, in mobile IP, 726–727
  tunneling phase, in mobile IP, 726–727
TV, mobile TV, See also ITV
  'Amigo TV' system, 268
  'Social TV' system, 268
  social aspects of TV, 267–273, See also
    Content-enriched communication
TV kingdom, 807
  overview, 814
Twelve-tone analysis, 808
Two-link network, 1466–1467
Two-person cooperative games, 646–648
Two-person non-cooperative games, 149
Two-ray fading channel model, 821
TXOP (transmission opportunity), 1384

UCAN (unified cellular and ad hoc network),
  862–863

UDP (user datagram protocol), 1333
UEP (unequal error protection), 444
UGS (unsolicited grant service), 1137
UIQ (user interface IQ), on symbian OS,
  1078–1079
Ultimate headset, 142
Ultra wide band technology, See UWB
Ultra-low-power-short-range communication
  challenges, of BANs, 153–154
UMA (unlicensed mobile access) architecture
  and protocols, 1168–1181
  3GPP specifications information, 1183
  discovery and registration, 1175–1177
  emergency services, support for, 1182–1183
  GPRS short message service and other
    procedures, 1183
  GPRS signaling and data transfer, 1180–1181
  IPSec tunnel establishment procedure,
    1174–1175
  network architecture, 1169–1170
  procedures, basic, 1174
    EAP-SIM authentication procedure, 1173
    UMAN discovery and registration, 1175
  protocols, 1172
    standard 3GPP protocols, 1172
    standard IP protocols, 1172
    UMA specific protocols, 1172
      UMA-RR, 1172
      UMA-RLC, 1172
    unlicensed spectrum protocols, 1173
  signaling and user plane architecture, 1170
  working of, 1168
UMAN (unlicensed mobile access network)
  discovery and registration, 1175
  handover to, 1177–1179
UMTS (universal mobile telecommunications
  service), 713–714, 799, 1332, 959
  3G networks, 30, 463, 1350–1351
  architecture, 27–40
  network reference architecture, 1166
  quality of service, 897
  services in, 32
  traffic classes, 896
  UCN architecture, 1165
  UMTS terrestrial radio access, See UTRA
  UMTS terrestrial radio access network,
    See UTRAN
  user plane protocol stack, 898
  UTRAN, See separate entry
Unequal error protection, See UEP
Unequal power allocation, 439
Unicast data communications, 73–74
Unified cellular and ad hoc network, See UCAN
Unitary model, 1295–1296
United States digital cellular, 216–218
Universal mobile telecommunications service,
  See UMTS
Universal personal telecommunication, See UPT
Unlicensed mobile access, See UMA
Unobtrusive services
  of BANs, system architecture, 151–152
  pervasive sensing for, 152
Unsolicited grant service, 419
Unstructured overlay network, contrasting
  MANETs, 762
Unsupervised learning, 662
Uplink and downlink communication

in 802.15.4 sensor clusters, 398
power controls, 667
UPT (universal personal telecommunication), 488
architecture, 488
Usage models, of bluetooth systems, 143–144
file transfer, 143
headset, 144
internet bridge, 143
LAN access, 144
synchronization, 144
three-in-one phone, 144
User asynchronous channels, in bluetooth, 145
'User centric' system
4G in mobile communications, 635–639
User datagram protocol, See UDP
User diversity, in 4G technology, 791
User interface IQ, See UIQ
User mobility, 484–490
personal mobility, 487–490
terminal mobility, 485
mobile IP, 486–487
network layer mobility, 485
User preference learning, 805–806, 809
User scenarios
4G in mobile communications, 636
User-generated content, in interpersonal communication, 270
User-preference-based search, 811, 812, 814
U-TDoA (uplink time difference of arrival), 503
UTRA (UMTS terrestrial radio access), 222
UTRAN (UMTS terrestrial radio access network), 896, 1166–1167
logical interfaces, 1167
UWB (ultra wide band) technology, 503, 440, 606, 640
4G technologies, 640
advantages, 1150
devices types, 1151
communications and measurement systems, 1151
imaging systems, 1151 See also separate entry
applications, 1162–1163
characteristics, 1157–1159
for ultra-low-power-short-range communication, 153–154
principles, standardization and applications, 1150–1156
asset/inventory tracking, 1155
commercial applications, 1154
communication applications, 1154–1155
defense/military applications, 1154
locating cars in a parking lot, 1155
location/positioning, 1155
non-US regulatory effort, 1154
real-time call forwarding, 1155
regulatory status, 1153–1154
security, 1155
sports, 1155
system design considerations, 1152–1153
tracking people, 1155
wireless body area networking, 1155
pulse waveform, 1152
standards, 1160–1162

IEEE802.15.2, 1160
IEEE802.15.3, 1162
IEEE802.15.4, 1162
transmission technologies, 1159–1160
vehicular radar systems, 1151

Variable rate FEC codes, 1593
Variable-hop overlay, 1011
VCC (virtual-cluster centre), 872–873
VCRs (video cassette recorders), 1331
VDO (voluntary dropout), 1471–1472
Vector Graphics, 739
Vector-sum excited linear prediction, See VSELP
Vehicular radar systems, 1151
Verizon Wireless, 506
Very tight coupling architectures, in WLAN and cellular networks, 458–459
types, 459
VGSN (virtual GPRS support node), 457
advantage, 458
important feature of, 457–458
Video 'Bookmarks', 278–279
Video cassette recorders, See VCRs
Video coding, 739
video codecs, 742–743
Video compression, 737
Video conference, 740–741
Video Connect, 958–959
Video encoding, 1434–1436
categories, 1435
channel-based algorithms, 1436
error concealment based algorithms, 1436
location, 1435
redundant encoding, 1435–1436
syntax information, aggregation, 1435–1436
Video playback and video recording
filter graphs for, 23–24 See also Filter graphs for video playback and video recording
Video services
demonstration, 22–24
3G-telephony control stack, 22–24
video-enabled SIP clients, 22
video-enabled SIP services, 22–23
in 3G-networks, 14–16
direct video-streaming, 14
streaming-video services, 14
Video streaming, 1332
in MBMS, 1197–1199
delivery, 1197
post-delivery, 1197
user service discovery, 1197
Video streaming, wireless, 1193–1199, 1200–1206, 1332
available bandwidth, 1200–1201
encoding, 1202–1204
with MBMS, 1197–1199
delivery, 1197
post-delivery, 1197
user service discovery, 1197
node mobility, 1204–1205
packet losses, 1201–1202
Video transmission, error-resilient
in wireless networks, 1434 See also Error-resilient video transmission

Video-chat and massive multiplayer on-line games, 1332
Virtual (MAC layer) jamming, WLANs, 1400–1401
Virtual connection tree, 198
Virtual GPRS support node, See VGSN
Virtual private networks, See VPN
Virtual-clusters, See VCC
Virtual-tree-based rerouting, 197
Visiting location register, See VLR
Visual information, automatic tagging in, 807
VisualSEEK, 807
VLR (visiting location register), 327
VMAC (voice mobile attenuation code), 214
Vodafone UK's m-pay bill online payments solution, 4
Voice mobile attenuation code, See VMAC
Voice multiplexing, 1383
Voice overhead reduction, MAC, 1385
Voice over internet protocol, See VoIP
Voice privacy, in CDMA networks, 137
Voice traffic multiplexing, MAC, 1384
VoiceXML+, 14, 15
VoIP (voice over internet protocol), 1381
network quality requirements, 1480–1482
RLP, See separate entry
voice coding basis, 1479–1480
wireless networks, 1478–1479
Voluntary dropout, See VDO
VoWLAN (voice over WLAN), 1381
challenges for, 1382
VPN (virtual private networks), 1402
VSELP (vector-sum excited linear prediction), 1124
speech encoder in, 1124
VSM (vector space model), 809

WADMs (wavelength add-drop multiplexers), 968
WAL (wireless markup language), 262
Wall-imaging systems, 1151
WANETs (wireless ad hoc networks)
key features
ad hoc network applications, 1222
adaptive routing protocols, 1222
multihop packet communication, 1221–1222
multiple access mechanism, 1221
wireless physical communication, 1221
performance analysis, 1221–1233
bluetooth performance, 1228–1229
capacity of, 1224
stochastic capacity, 1225–1227
scaling law, 1224–1225
spatial reuse and connectivity, 1223–1224
TCP controlled transfers, performance of, 1229–1232
single-hop performance, 1230
topology, 1223–1224
WANET graph, 1223
WAP (wireless application protocol), 1354–1356
WAP stack, 1355
WAP topology, 1355–1356
War driving, 1404
WASP (wireless application service providers), 1358

Watchdog, of ad hoc networks security, 84–85
WATM (wireless asynchronous transfer mode), 1251–1263
  connection handover in, 1267 *See also* under Connection handover
  mobility management, 1264–1272
    connection handover, 1267 *See also* separate entry
    location management, 1264–1265 *See also* separate entry
    security management, 1273
    service management, 1274
  location management in, 1264–1265 *See also* under WATM
  PCS-to-ATM interworking, 1256–1260
    architecture and reference model, 1256–1258
    signaling link evolution, 1258–1260
  QoS support, 1260–1262 quality of service, 1273–1283
    ATM QoS model, 1273–1276
    MAC layer functions, 1274–1279
      resource reservation and allocation mechanisms, 1214
      scheduling, 1276
    network and application layer functions, 1277–1281
    QoS mechanisms, 1275
    soft-QoS model, 1279
  wireless interworking, 1252–1256
    ad hoc networks, 1256
    digital cellular, 1255
    integrated wireless–wireline ATM network architecture, 1253–1254
    PCS, 1254
    satellite systems, 1255–1256
    wireless access technology options, 1254
    wireless LAN, 1254–1255
Wavelength-division multiplexing, *See* WDM
WBSN (wireless body sensor network)
  for m-health, 710–711
  network technologies of, 711–712
  technical challenges facing, 712–713
    automatic monitoring, 713
    network security and reliability, 712–713
    personal alarm system, 713
  topology of, 711
W-CDMA (wide code division multiple access), 170–172, 207, 1350
  beamforming, 171
  coherent detection, 171
  multirate, 171–172
  packet data, 172
  spreading codes, 171
W-CDMA (wideband CDMA) system, 670, 708
  3G networks/technology, 1350
WCDS (weakly connected dominating set), 516
WCFQ (wireless credit-based fair queuing), 948–949
WDM (wavelength-division multiplexing), 966
  basic elements, 990–995
    opaque vs. transparency, 995
    optical layer survivability, 993–995
      automatic protection switching, 993–994
      lightpath restoration, 994

      multiple layer protection, 994–995
      self-healing ring, 994
    RWA, 990
      route computation, 990
      wavelength selection, 991
  in optical communication networks, 967–968
Weakly connected dominating set, *See* WCDS
Wearable intelligent sensors and systems for m-health, *See* WISSH
Wearable sensors for m-HEALTH, 709–710
Weaver architecture, 1040
Web content re-purposing for mobile devices
  ontology-based approach to, 259–266, *See also* Ontology-based approach
WEP (wired equivalent privacy), 1395
  attacks, 1395–1396
    inductive chosen plaintext attack, 1396
    message modification attack, 1396
    passive key recovery, 1396
    replay attacks, 1396
    SKA attack, 1397
    small IV space, 1396
  authentication, 390–391
    authentication security problems, 390–391
  characteristics, 388–392 RC4
  confidentiality protection, 391
  encryption algorithm, 388–389
  integrity protection, 391–392
  RC4 keys, 389–390
  vulnerabilities, 1395–1396
Wideband channels, 819
Wi-Fi network, 503–504
  multiplayer game in, 699
  security, 1082
Wi-Fi protected access, *See* WPA
Wi-Fi walkman, 1207–1220
  folk computing, 1208
  in client/server model, 1212
  music recommendation, 1209
  Peer-to-Peer Networks, 1208
    advantages, 1208
  personalized services, 1207–1208
  problem definition and formalization, 1209
    context, 1209
    interest, 1209
    services, 1209
  recommender system, 1210–1216 *See also* separate entry
  scenarios, 1209
  system diagram, 1215
  user preferences, 1210
    demographic data, 1210
    music-to-music correlation information, 1210
    people-to-people correlation information, 1210
Wild fire monitoring, 1544
WiMAX (worldwide interoperability for microwave access), 857, 1409, 1414–1417
  WiMAX—IEEE 802.16, 610–611
WiMedia/IEEE 802.15.3a, 1153
Windowing, in OFDM, 929
Windows mobile, 702
  2003 second-edition software

    for smartphones, 1083–1086 *See also* under Smartphone operating systems
  smartphone
    description, 1086
    LINUX, 1086–1087
Windows mobile 2003 second-edition software, in smartphone, 1083–1086
  technical specifications, 1084–1086
    applications, 1085
    communication infrastructure, 1085
    multimedia, 1084
    network services, 1084
    personal area networking, 1085–1086
    security, 1086
    software development, 1086
    telephony or telephony API, 1085
Wired equivalent privacy, *See* WEP
Wireless ad hoc networks, 1394–1395
  power assignment, *See* separate entry
Wireless ad hoc networks, *See* WANETs
Wireless application service providers, *See* WASP
Wireless asynchronous transfer mode, *See* WATM
Wireless body sensor network, *See also* WBSN
  mobile health integrated with, 707–717, *See also* m-Health
Wireless channel errors, TCP, 1454
  explicit notification approach, 1456
  lower layer scheme, 1455–1456
  modifications, 1456
  split-connection approach, 1454–1455
Wireless channels communication
  link adaptation techniques for, 1284–1293
    AWGN (additive white Gaussian noise) channel, 1286–1287
    digital communication basics, 1284–1286
    link-adaptive communication system, 1288–1290
    non-adaptive communication system, 1287–1288
    Rayleigh fading channels, 1290
    time-varying channels
      modeling of, 1287
Wireless communications, *See also* individual entries
  cognitive radio implementation for, 230–238, *See also* Cognitive radio implementation
  diversities in, *See* Diversities
  modulation techniques, 1347–1348
  performance elements, 1348
    actual data rate, 1348
    bandwidth, 1348
    mobility, 1348
    range, 1348
  RF systems design, *See* separate entry
  RF transceiver, 1035–1045, *See also* separate entry
  signal processing tools, 1294–1304, *See* Signal processing
  spatial diversity, 1305–1318
    antenna gain, 1312–1313
    branch correlation effect, 1313–1314
    combining techniques, 1306–1308
      equal gain combining, 1312
      maximum ratio combining, 1310

selection combining, 1308–1310
  diversity gain, 1312
  general receive array theory, 1305–1306
  mutual coupling, 1314–1317
  wireless system topologies, 1348
  wireless technologies, 1347
Wireless credit-based fair queuing,
  See WCFQ
Wireless data, 1319–1330
  cellular digital packet data, 1325–1328
    See also separate entry
  characteristics, 1319–1321
    radio propagation characteristics,
      1320–1321
  digital cellular data services, 1328–1329
  market issues, 1321
  modem services, 1321–1322
  packet data, 1322–1325
    advanced radio data information service,
      1322–1323
    MOBITEX, 1323–1324
  paging and messaging networks, 1324–1325
  terrestrial trunked radio, 1329
Wireless devices, 1353–1354
  communication appliances, 1354
  two-way pager, 1354
  voice portals, 1354
  Web PCs, 1354
  web phone, 1354
  wireless handheld device, 1354
Wireless firewall gateways, 1402
Wireless home entertainment center, *See* Home
  entertainment center, wireless
Wireless internet fundamentals, 1346–1359,
  1352
  applications, *See* Internet applications,
    wireless
  wireless communications, principles of, 1346
    See also separate entry
  wireless devices and standards, 1351–1353
    Java-enabled wireless devices, 1356
    WAP, *See* separate entry
  wireless internet applications, 1351–1353
    corporate applications, 1357–1358
    messaging applications, 1356
    mobile commerce, 1356–1357
    mobile web services, 1358
    wireless application service providers,
      1358
    wireless teaching and learning, 1358
  wireless internet architectures, 1351–1353
    wireless internet networks, 1352 *See also*
      separate entry
    wireless internet topologies, 1353
  wireless systems, generations
    of, 1349–1351
    1G wireless systems, 1349
    2G wireless systems, 1349 *See also*
      separate entry
    3G wireless systems, 1350–1351 *See also*
      separate entry
    basic characteristics, 1351
  wireless technology future, 1358–1360
  wireless WAN, 1352
  WLAN, 1352
  W-PANs, 1352
Wireless local area networks, *See* WLANs

Wireless medical telemetry service, *See* WMTS
Wireless mesh networks, *See* Mesh networks,
  wireless
Wireless metropolitan area networks,
  See WMANs
Wireless multimedia
  challenges of, 254–255
    heterogeneity, 254
      bandwidth fluctuation and delay, 254
      diversity, 254
      limited bandwidth, 254
      limited capability and resources, 254
    of wireless terminal, 5
    user preferences and mobility,
      254–255
  cross layer design for, 1417–1423,
    See also CLD
Wireless networks
  1G systems, 1349
  2G systems, 1349
  2.5G systems, 1350
  with intermittent connectivity, routing in,
    1442–1451
    active stochastic routing techniques,
      1449–1450
    deterministic routing techniques,
      1444–1446
    DTN architecture, 1443–1444
    passive stochastic routing techniques,
      1446–1449
  cross-layer MAC design for, *See* Cross-layer
    MAC design
  and nodes, 616–617
    routing, *See* separate entry
    TCP, *See* separate entry
    transmitter power control, *See*
      separate entry
    VoIP service, 1478–1486, *See also*
      separate entry
Wireless personal area networks, *See* WPANs
Wireless physical communication,
  WANET, 1221
Wireless sensor networks, *See* WSN
Wireless systems
  authentication and privacy in, 132–140,
    See also Authentication and privacy
Wireless transceivers, 1588–1598
  frequency-hopping based multicarrier direct
    spectrum, 1588–1590, 1589–1599
  characteristics, 1590–1592
    access strategy, 1591
    broadband wireless mobile
      system, 1591
    compatibility, 1591
    diversity, 1591
    frequency-hopping strategy, 1591
    multicarrier modulation, 1591
    multirate and variable rate services,
      1591
Wireless transmission, of telemetering
  data, 1141
Wireless USB, 1153–1154
Wireless video streaming, *See* Video streaming,
  wireless
Wireless wide area networks, *See* WWAN
Wireless world research forum
  4G in mobile communications, 641

Wiretrapping, 1400
WISSH (wearable intelligent sensors and
  systems for m-health), 715
WLAN and cellular networks
  inter-working between, 454–455
    types, 454–455
      loose coupling, 454–455
      tight coupling, 454–455, 457–458
      very tight coupling, 454–455, 458–459
WLANs (wireless local area networks), 1352
  ad hoc routing techniques, 1369–1380
    ad hoc on-demand distance vector routing,
      1370–1371 *See also* AODV
    destination-sequenced DSDV, 1374–1375
      performance, 1375
    DSR, 1377–1379
      performance, 1378–1379
    fisheye routing, 1371–1372 *See also*
      separate entry
    hierarchical state routing, 1372 *See also*
      separate entry
    zone routing protocol, 1375–1377
      performance, 1376
  attacks, 1400–1401
    Distributed DoS attack, 1401
    DoS attacks, 1400
    impersonation attacks, 1401
    jamming attacks detection, 1401
    jamming mitigations, 1401
    jamming modeling, 1401
    MITM, 1401
    PHY-layer jamming, 1401
    rogue access points, 1401
    spoofing management frames, 1400
    Virtual (MAC layer) jamming,
      1400–1401
  handovers in, 453–454
  implementation, 511
  infrastructure/ad hoc WLAN network, 1394
  OFDM applications in, 931
  security and privacy, 1392–1406
    human factors, 1402
    intrusion detection, 1403–1404
    security best practices, 1404
    security solutions, 1401–1402
    security tools, 1402–1403
    Wi-Fi protected access, 1397
    wired equivalent privacy, 1395 *See also*
      WEP
    wireless ad hoc networks, 1394–1395
  supporting voice and data services
    architecture for, 1382
  vulnerabilities, 1397
    address, 1400
    MAC address authentication, 1400
    one-way authentication, 1400
    optional security, 1400
    unprotected management frames, 1400
    wiretrapping, 1400 MAC
WLANs—IEEE 802.11n, 611
WMANs (wireless metropolitan area networks),
  OFDM applications in, 931–932
WMTS (wireless medical telemetry service)
  bands, 150
World Wide Wireless Web, *See* WWWW
Work and emergency services, applications of
  BANs, 155

Wormhole
in infrastructureless routing, 78
wormhole attack, 409, 1549
WPA (Wi-Fi protected access), 1397
WPANs (wireless personal area networks),
511, 1352
WSN (wireless sensor networks)
applications, 1488
architecture, 301–304
communication mode-based classification,
303–304
data fusion, 304
functional layers, 302
homogeneous vs. heterogeneous, 302
challenges, 1488–1490
extreme dynamics/conditions, 1489–1490
resource limitations, 1489
characteristics, 301
extended coverage and easier
deployment, 301
mobility, 301
monitoring capabilities and information
quality, 301
reliability and flexibility, 301
communication architecture, 1555
constraints, 1509
continuous queries, 1541
adaptability, 1542
continuous query engine design, 1542
data buffering, 1542
data summarization, 1542
sharing, 1542
design requirements, 1525
low energy, 1525
in-network query processing, 1525–1526
in-network signal processing, 1526
scalability, 1525
self-configuration, 1525
future trends, 1521
key management schemes in, 1509–1522
classification, 1512, 1513
deterministic schemes, 1515–1520
hybrid schemes, 1520
pre-distributed schemes, 1513–1520
probabilistic schemes, 1513–1515
self-enforcing schemes, 1510
single TA schemes, 1513
network model, 1510–1511
OSI layers optimization, 1526–1531
application layer, 1526
cross-layer optimizations, 1531–1532
MAC layer, 1528–1530
delay MAC, 1529
group-based MAC, 1530

sensor MAC, 1528–1529
TMAC protocol, 1529–1530
TRAMA protocol, 1530
wireless sensor MAC, 1530
network/routing layer, 1526–1528
CMLR, 1528
directed diffusion, 1528
GEAR, 1528
LEACH protocol, 1526–1527
PEGASIS, 1527
SPIN protocol, 1528
TEEN/APTEEN protocol, 1527–1528
physical and radio layer, 1530–1531
transport layer, 1526
routing, See separate entry
secure localization, survey on, 1545–1553
attacker model, 1548–1549
classification, 1547–1548
localization process, 1546–1547
operational challenges, 1546
security in, 1554–1562
cryptographic primitives, 1556–1557
key management, 1557–1558
solutions, 1558
secure data aggregation, 1560
secure routing, 1558–1560
threat model, 1556
security requirements and evaluation criteria,
1511–1513
access control, 1512
availability, 1512
confidentiality, 1512
connectivity, 1512
data freshness, 1512
data integrity, 1512
efficiency, 1512
entity authentication, 1512
flexibility, 1512
resiliency, 1513
scalability, 1512
source authentication, 1512
self-organization, 1563–1579
approaches and solutions, 1572–1576
maintenance protocols, 1575–1576
neighbor discovery, 1572–1573
organization protocols, 1573–1575
self-healing and recovery protocols,
1576
characteristics, 1564–1567
data-centric addressing, 1566
in-network processing, 1566
network makeup, 1564
sensing application characteristics,
1567

preliminaries, 1567–1572
elementary networked-sensing concepts,
1567–1569
elementary network-organization
concepts, 1569–1571
sensor network self-organization,
1571–1572
sensor networks classification, 1536–1539
hybrid networks, 1538–1539
proactive networks, 1536–1537 See also
separate entry
reactive network protocol, 1537–1538
sensor node, network stack of, 1492
supporting technologies, 1490–1493
embedded devices and software,
1490–1491
high efficiency power systems, 1490
miniaturization and SoC design, 1490
networking technologies, standardization
of, 1490
self-organizing ad hoc networks, 1491
sensor and actuation devices, 1490
wireless communications, 1490
terms and definitions, 1510
threat model, 1511
time synchronization, 1580–1587
existing synchronization techniques,
1581–1583
delay measurement time
synchronization, 1583
post-facto, 1582
reference broadcasting synchronization,
1582
time-sync protocol, 1582–1583
tiny-sync and mini-sync, 1581
lightweight approach, 1583–1585
in multi-hop WSN, 1583–1584
in single-hop WSN, 1583
sender selection algorithm,
1584–1585
theoretical model, 1580–1581
Wu's CDS algorithm, 516, 522
WWAN (wireless wide area networks), 1352
WWRF (wireless world research
forum), 641
WWWW(World Wide Wireless Web), 260

Zerotree wavelet packetization method,
433–434
Zone-based routing protocol, 530–531,
1375–1377
performance, 1376
ZRP (zone routing protocol), 570